HANDBOOK OF
AIR POLLUTION TECHNOLOGY

HANDBOOK OF
AIR POLLUTION TECHNOLOGY

Edited by
SEYMOUR CALVERT
Consultant
San Diego, California

HAROLD M. ENGLUND
Air Pollution Control Association
Pittsburgh, Pennsylvania

A Wiley-Interscience Publication

JOHN WILEY & SONS

New York · Chichester · Brisbane · Toronto · Singapore

Copyright © 1984 by John Wiley & Sons, Inc.

All rights reserved. Published simultaneously in Canada.

Reproduction or translation of any part of this work
beyond that permitted by Section 107 or 108 of the
1976 United States Copyright Act without the permission
of the copyright owner is unlawful. Requests for
permission or further information should be addressed to
the Permissions Department, John Wiley & Sons, Inc.

Library of Congress Cataloging in Publication Data
Main entry under title:
Handbook of air pollution technology.

    "A Wiley-Interscience publication."
    Includes index.
    1. Air—Pollution—Handbooks, manuals, etc.
2. Pollution control equipment—Handbooks, manuals,
etc. I. Calvert, Seymour, 1924–   II. Englund,
H. M.
TD883.H356 1984   628.5′3   83–19797
ISBN 0-471-08263-5

Printed in the United States of America

10 9 8 7 6 5 4 3 2 1

CONTRIBUTORS

Patrick R. Atkins
Aluminum Company of America
Pittsburgh, Pennsylvania

Lee Beck
U.S. Environmental Protection Agency
Research Triangle Park, North Carolina

T. R. Blackwood
Monsanto Chemical Company
St. Louis, Missouri

Norman E. Bowne
TRC Environmental Consultants, Inc.
East Hartford, Connecticut

David V. Bubenick
GCA Corporation
Bedford, Massachusetts

Seymour Calvert
Consultant
San Diego, California

Thomas K. Corwin
PEDCo Environmental, Inc.
Cincinnati, Ohio

B. B. Crocker
Monsanto Chemical Company
St. Louis, Missouri

Edward DeKiep
University of Michigan
Ann Arbor, Michigan

Timothy W. Devitt
PEDCo Environmental, Inc.
Cincinnati, Ohio

James A. Eddinger
U.S. Environmental Protection Agency
Research Triangle Park, North Carolina

Harold M. Englund
Air Pollution Control Association
Pittsburgh, Pennsylvania

Richard W. Gerstle
PEDCo Environmental, Inc.
Cincinnati, Ohio

Murray S. Greenfield
Dofasco, Inc.
Hamilton, Ontario, Canada

Donald J. Henz
PEDCo Environmental, Inc.
Cincinnati, Ohio

John H. Hirt
Hirt Combustion Engineers
Montebello, California

William F. Kemner
PEDCo Environmental, Inc.
Cincinnati, Ohio

James D. Kilgroe
U.S. Environmental Protection Agency
Research Triangle Park, North Carolina

Albert J. Klee
U.S. Environmental Protection Agency
Cincinnati, Ohio

Kenneth T. Knapp
U.S. Environmental Protection Agency
Research Triangle Park, North Carolina

Brian Lamb
Washington State University
Pullman, Washington

Gregory Leonardos
Environmental Odor Consultant
Arlington, Massachusetts

William Licht
University of Cincinnati
Cincinnati, Ohio

K. J. Lim
Acurex Corporation
Mountain View, California

Alan Lloyd
Environmental Research & Technology,
   Inc.
Westlake Village, California

John D. McKenna
ETS, Inc.
Roanoke, Virginia

J. D. Mobley
U.S. Environmental Protection Agency
Research Triangle Park, North Carolina

Donald J. Patterson
University of Michigan
Ann Arbor, Michigan

Ronald G. Patterson
Calvert Environmental Equipment Co.
San Diego, California

Michael J. Pilat
University of Washington
Seattle, Washington

John J. Roberts
Argonne National Laboratory
Argonne, Illinois

Elmer Robinson
Washington State University
Pullman, Washington

Karl B. Schnelle, Jr.
Vanderbilt University
Nashville, Tennessee

Leigh Short
Environmental Research & Technology,
   Inc.
Houston, Texas

Leslie E. Sparks
U.S. Environmental Protection Agency
Research Triangle Park, North Carolina

A. R. Stankunas
TRC Environmental Consultants, Inc.
East Hartford, Connecticut

Richard F. Toro
Recon Systems, Inc.
Three Bridges, New Jersey

Michael Treshow
University of Utah
Salt Lake City, Utah

James H. Turner
Research Triangle Institute
Research Triangle Park, North Carolina

William M. Vatavuk
U.S. Environmental Protection Agency
Research Triangle Park, North Carolina

Norman J. Weinstein
Recon Systems, Inc.
Three Bridges, New Jersey

H. H. Westberg
Washington State University
Pullman, Washington

Harry J. White
Consultant
Carmel, California

John E. Yocom
TRC Environmental Consultants, Inc.
East Hartford, Connecticut

PREFACE

This handbook is intended to present the best available practical information on air pollution and its control. Engineers and other professionals, in a self-study situation, should be able to use this book to learn how to do the work required to define, analyze, and control air pollution.

The objective is to provide a basis for doing something about the problem. Given a completely interdisciplinary topic as air pollution, the desirability of understanding all aspects of the problem must be balanced against the need for reasonable book size. The approach is to provide enough information to enable a useful design or survey, and to present a guide to specific detailed sources of information for studies and designs in depth.

Each section of the handbook presents a description of the concepts in its area, but the emphasis is on providing the essential information that will be useful for the neophyte and practicing professional alike.

The authors have directed their chapters toward giving the reader a clear concept of what can be done to solve his or her problem and what limitations he or she faces. We assume that the reader is pressed to resolve the problem quickly and has no time (and perhaps no inclination) to read an equivocal, scholarly review of all the facts and viewpoints. Our approach is to use judgment and opinion in order to prepare forthright statements of the best approaches, what is known, and what is not known.

Engineers in all disciplines must now include interactions with the environment as design and operating criteria. They have to define the undesirable effects of the contaminated environment, the relationships between these and their activity, and means for achieving the desired end at optimum cost.

The engineer faced with an air pollution problem does not have the "simplicity" of design criteria that are part of the usual process or mechanical situation. For one thing, his or her objective may be very unclear. Is it to prevent a threat to human health, or plant life, or visibility, or deterioration? Is it to meet a legal limitation, or to live at peace with the local community? If one can set his or her objective there may be many routes to it, ranging from using a different process, or adding controls, to using a high smoke stack. Some of these problems are inside the plant, some outside. Obviously, one has to make sure that he or she is solving the right problem and in the best way.

Work on many aspects of the air pollution problem has proceeded considerably beyond the exploratory phase and has resulted in methods for routine application. Much of air pollution control engineering design can be presented in the form of simple charts, tables, and formuli which can be used for purposes of approximation.

Such things as estimates of atmospheric dispersion can be made by people without special education in micrometeorology.

The handbook will help the practicing engineer to:

1. Understand what is happening or may happen.
2. Take action to define a specific problem.
3. Develop a course of corrective action (i.e., do the design or specification and cost estimation).
4. Prepare for discussions.
5. Do further reading.

The overall approach is to present the essential principles, design, methods, examples, useful data, and guides to more information. We have emphasized the complete coverage of general concepts and methods, such as the chapters on basic principles for gaseous and particulate pollutants, control methods, sampling and analysis, atmospheric dispersion and chemistry, air quality standards, and air quality management.

Not all of the significant sources are covered, although an effort was made to include the major sources. In some cases where an outstanding author could make a contribution in a specific area we took that option rather than attempting to cover a broader range and possibly compromising on quality.

The individual authors deserve and have our gratitude for the work they have done and for their patience during the long process of assembling the many chapters. In some cases the authors have made recent revisions on their chapters to incorporate new developments.

Several of our authors are employed by the United States Environmental Protection Agency and it should be pointed out that the views expressed in their chapters do not necessarily reflect official positions of the Agency.

Dr. George M. Hidy, a member of the Handbook Editorial Advisory Board, made valuable suggestions concerning content and prospective authors. His contributions during the formative period of this handbook are appreciated.

Finally, the index was prepared by Williamina T. Beery and the editors gratefully acknowledge her skill in this enhancement of the value of this handbook.

SEYMOUR CALVERT
HAROLD M. ENGLAND

San Diego, California
Pittsburgh, Pennsylvania
January 1984

CONTENTS

HANDBOOK OF
AIR POLLUTION TECHNOLOGY

CHAPTER 1
INTRODUCTION

SEYMOUR CALVERT

Consultant
San Diego, California

The handbook is primarily concerned with the control of air pollution so we will begin by specifying what that involves. Historically, the definition of air pollution has stemmed from the observation of adverse effects. That is, if some concentration and duration of air contamination causes observable effects on human health, animals, plants, materials, atmospheric properties, or aesthetic factors, it is air pollution. The mere act of emitting some substance into the ambient air does not, in itself, constitute the act of air pollution.

This approach, which is based to a large extent on the concept of common law nuisance, can be rationalized in terms of our experience. The air has some capacity for holding contamination without causing observable effects, because there is some natural purification process, or because the rate of pollutant emission is low and would cause a negligible increase in its ambient concentration. We also assume for the time that the final form or fate of the pollutant is harmless, whether it be on the land or in the water.

Air pollution control efforts based on this approach are responsive to the perception of effects. In the most primitive form, one would have to wait until the effect is perceived and proved before being able to cause the abatement of the air pollution.

1 NEED FOR RAPID RESPONSE

In the face of increasing amounts of pollution, it is logical that some rapid and effective means of controlling pollution be developed, and this requires the ability to anticipate the effects of air contamination. One fundamental reason for this is that industrial systems are not built so the emissions can be controlled continuously in response to a feedback signal describing undesirable effects.

Systems capable of meeting the most severe requirement must be designed in advance, so that they can provide the necessary performance. For another thing, control agency efforts can be more effective if they are applied to the prevention of anticipated pollution rather than to the cure of existing problems perceived only from their effects. Therefore, there has been a great effort to specify the conditions that cause undesirable effects.

The first step in relating effects to air composition is to develop *air quality criteria* (AQC), which more accurately should be called "air quality *corollaries.*" These are the cause-and-effect relationships between various concentration–duration exposures to air pollution and effects of all types on receptors of all types. The next step of the control agency is to establish *air quality standards* (AQS) based upon the AQC, along with any other factors that may be weighed in the balance to determine the best course of action for society to take at a given time.

The federal law, PL91–640 (1970), marked the first time that the federal government assumed the responsibility for setting AQS. It defines *primary and secondary standards* as follows:

National primary ambient air quality standards define levels of air quality which the administrator judges are necessary, with an adequate margin of safety, to protect the public health. National secondary ambient air quality standards define levels of air quality which the administrator judges necessary to protect the public welfare from any known or anticipated adverse effects of a pollutant. Such standards are subject to revision, and additional primary and secondary standards may be promulgated as the administrator deems necessary to protect the public health and welfare.

1

The law also states that these standards are not intended to allow significant deterioration of the existing air quality, and that the states may adopt more stringent standards. Since its enactment, this law has evolved into a complex system which continues to be interpreted and modified.

Given the AQS, we might control on the basis of a feedback signal from an analysis of the ambient air composition, or we might go a step further and relate the air composition to the emission rate for any pollutant in question. Computed from knowledge of the effect of dilution due to dispersion in the atmosphere, this type of information permits us to establish the allowable rates of emission in order to meet an AQS and, consequently, to prevent primary or secondary air pollution effects. The design engineer needs this knowledge of what can be sent out of the smoke stack in order to design the process or device and any necessary air pollution controls.

2 BEST AVAILABLE CONTROL TECHNOLOGY

A different approach to air pollution control is the requirement of the application of the best available control technology. This may either be very permissive or very restrictive, depending on how it is enforced. If one is permitted to exceed standards on the basis that technology is not available to do better, it is permissive. If, on the other hand, one is required to do a better job than would be required based on standards, because the technology is available, it is restrictive. The federal requirement is that primary air quality standards must be met, while the secondary standards may be negotiated, depending on economics and other factors.

Control legislation has been written to utilize all of these approaches or criteria for establishing that a state of pollution exists or may be presumed possible. It is necessary for the engineer to be able to understand and work with any of these ways of limiting air pollution.

3 THE ENTIRE PICTURE

A framework for understanding the total air pollution problem is given in Figure 1. Arranged in a column on the left-hand side of this figure are the links in the air pollution chain. Along the right-hand side of the figure are brief descriptions or examples of the types of things included in each link.

The beginning is always some *source* of pollution that may be in the form of gases or particles. In very general terms some major sources are: combustion, materials handling and conveying, process and storage vents and evaporation, ventilation exhausts, and processes that reduce particle size. Generally, there are some *alternatives* available, although they may not be used, which is why this block is surrounded by a dashed rather than a solid line. Conceivably one can change the process, the raw materials, the rate of production, or the type of energy utilized.

It is always possible to use some *control process or device* at some cost to prevent or reduce the amount of emission. Potential pollutants might be recycled, sold, or discharged to the land or water. Control processes might also destroy the pollutant or might change the chemical or physical properties of the pollutant so that it is innocuous, or at least less hazardous.

The *emission method* can sometimes provide a significant means of control of atmospheric dispersion. Smoke stacks high enough to give good dilution and to avoid the effects of air flow around buildings can be very helpful for the close-range problems. In some instances, the emissions may be regulated to occur at times when atmospheric dispersion is rapid, and shut off or reduced during periods of poor dispersion. The temperature and other factors influencing the buoyancy of the pollutant plume may also be an important means for emission control. Plant or stack location are additional major means for the control of emissions.

Fugitive emissions can be difficult to contain for either pollutant collection or controlled dispersion. These occur when sources are spread over a large area, such as wind-blown dust from fields and streets, quarrying operations, stock piles, and coke oven batteries.

Atmospheric dispersion is an inescapable part of air pollution, and it is influenced by a variety of factors. Weather patterns, terrain features, time of day, the place of emission, the location of receptors, and many other factors must be considered. To a useful extent the quantitative nature of atmospheric dispersion can be defined and predicted.

Chemical and physical reactions may occur to a very significant extent for pollutants as they move with the wind. Gases may undergo photochemical (light-catalyzed), dark (not requiring light), or catalyzed reactions with other pollutants. Or, they may combine with naturally occurring constituents of the atmosphere, such as oxygen, carbon dioxide, and water vapor, and may form other gases or particulate matter. Particulate matter may undergo chemical reactions with gases, or by themselves due to radiation or temperature, or they may undergo physical changes such as agglomeration with other particles to form larger particle clusters.

Pollutants are always lost from the air by some means and to some extent. Particulate matter may be deposited due to gravitational settling. Both particles and gases may be included in raindrops or snowflakes by the process of nucleation and growth of precipitates. Pollutants may also be collected

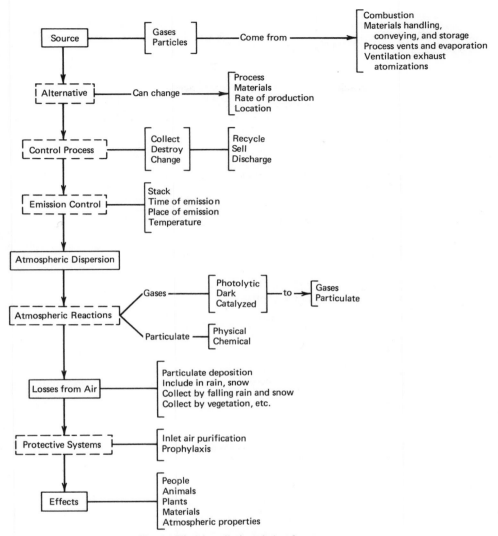

Fig. 1 The air pollution chain of events.

by the falling raindrops or snowflakes by impaction of other mechanisms. Objects in the path of air flow, such as vegetation and buildings, may also collect pollutants.

Protective systems may be inserted between the ambient air and the receptors. One frequent means for human protection is the purification of inlet air. Another approach is prophylaxis, such as treating plants with some substance, making them less susceptible to pollution, or painting materials to prevent corrosion.

Finally, there are the *effects* of air pollution on people, animals, plants, materials, and atmospheric properties. Because there is generally a wide range in individual susceptibility to pollution, it is also possible in some situations to control the receptors so that the resistant species are present and the susceptible ones absent.

4 AGENCIES FOR CONTROL

Another facet of the air pollution picture is the social or political side, from which stem the agencies for air pollution control. Figure 2 presents the framework of a chain similar to Figure 1. As mentioned before, the beginning of air pollution is the *perception* of some undesirable effect. In our more sophisticated (or polluted) state, it may also begin with the apprehension of air pollution effects by citizens.

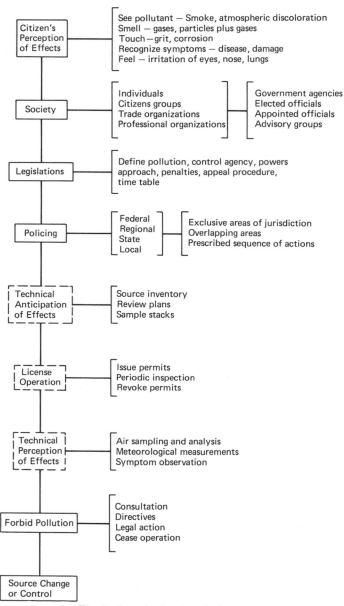

Fig. 2 Agencies for air pollution control.

These perceptions or apprehensions will be based on visual observation, smell, damage symptoms, health symptoms, and the tactile sense if particles are large enough to be felt.

If the observation of the individual citizen is to be effective, it must be transmitted through the *mechanisms* of organized society, or politics. The components of society that might be involved are citizen's groups, professional groups, business and trade organizations, communications with elected and appointed public officials, and legal action.

Society may initiate action by passage of *legislation*. This legislation may define pollution, the course of action and timetable to abate pollution, the agencies for enforcement, the powers of these agencies, appeals procedure and board, and the penalties.

Policing or *enforcement* of air pollution legislation is generally done by an air pollution control agency or may be allocated among several agencies. Federal, state, regional, interstate, county, and local air pollution control agencies exist. In some instances, their areas of jurisdiction are exclusive, in some they overlap, and in others they take turns in a prescribed sequence of actions.

Agency control may begin by the *anticipation of effects* on a technical basis. This may involve the review of plans, stack testing, and inventorying the sources of air pollution. Following the evaluation of the importance of a given source of pollution, its operation may or may not be allowed through some permit or licensing procedure.

Control agencies generally operate *surveillance* programs for the perception of air pollution effects through technical methods. These may include ambient air sampling and analysis, measurement of meteorological parameters, and observation of air pollution symptoms on well-defined and/or controlled receptors.

Once some air pollution effects are observed, the control agency must act in some prescribed way to forbid air pollution. This *enforcement* may be in the form of orders or directives, or consultation, or closing of an operation. The final consequence of this chain of events will be that the source of air pollution is either changed or controlled.

5 QUESTIONS FOR THE ENGINEER

Practicing engineers who have the responsibility for dealing with air pollution must be able to answer a number of questions. They may be concerned either with an existing process, or proposed change, or a new process. If with an existing process, they need to know whether it is causing pollution, either from the standpoint of known effects or because it violates some statutory requirement. If with a proposed change, they need to know if it will cause pollution.

To start, engineers must be able to define the source of pollution; what is the pollutant, what are its characteristics, at what rate is it emitted, and how does it vary with raw materials, process cycle, production rate, and other factors? They have to know how it is dispersed in the atmosphere, so as to predict where it goes and what concentrations may exist at various places and times.

The potential for process change must be explored—different raw materials to be used, different machinery, different production rates, different fuel, perhaps even a different product. What methods exist for pollution control? What may be done in the line of emission control, such as higher stack or higher gas temperature?

Above all is the question of what are the costs for all of the possible alternatives. It is also important to know what are the costs of the air pollution effects.

This book provides some answers to these questions, as shown in Table of Contents. Additional sources of information can and should be consulted when the need arises. In each chapter of the book, sources of more specific information pertaining to the subject matter of that chapter are identified. The last chapter describes general sources of information and means for searching and retrieval.

CHAPTER 2

EFFECTS OF AIR POLLUTANTS ON PLANTS

MICHAEL TRESHOW

Department of Biology
The University of Utah
Salt Lake City, Utah

1 INTRODUCTION AND BACKGROUND

Plants have evolved in harmony with their total environment. The temperatures in a particular area, the amount of precipitation, the soil characteristics, the biotic parameters, and even the atmospheric conditions, all interacted to shape the character of the landscape and the plants that are part of it. Should the environment change, the plant community changes. Even the variation in rainfall from one year to the next is reflected in commensurate changes in the plant community. Should the changes be extreme, the plants, those most sensitive to the change, become stressed, less able to compete, and may ultimately die. Sufficient changes in even a single parameter may disrupt the plant system (Mansfield, 1976; Treshow, 1980).

Normal air contains a myriad of gaseous and particulate components. In addition to the principal components of nitrogen and oxygen, and such important but less abundant components as carbon dioxide, the air contains an array of chemicals that can be literally considered as air pollutants. These include certain hydrocarbons emitted by the plants themselves, and sulfur compounds arising largely from the activity of bacteria. Such biogenic sources of sulfur are estimated to contribute about 11% of all the sulfur dioxide in the atmosphere (National Academy of Sciences, 1978). The remainder comes from anthropogenic, or human, activities.

Oxides of nitrogen also are commonly formed in the natural atmosphere. They originate from lightning and biological oxidation, principally by bacteria. By comparison, artificial sources contribute only about 10% of the total oxides. Nevertheless, by being concentrated near urban centers these sources are significant (Taylor et al., 1975). The anthropogenic oxides are formed during any combustion when air is oxidized to NO. The higher the temperature, the more oxides are produced. During daylight, the NO is further oxidized to NO_2 by photochemical reactions. The NO_2 may be partly consumed in the production of ozone, peroxyacyl nitrates (PANs), and other pollutants.

Thus we see that the precursors for many of the major air pollutants already exist in the natural atmosphere. But since plants have evolved in the presence of such concentrations, they are rarely damaged at background concentrations. Only when the concentrations exceed the boundaries of adaptation and tolerance is plant life adversely affected.

There are many instances where this has occurred. The most striking examples have been near smelters where the high concentrations of sulfur oxides and heavy metals in the air have altered the environment to where sensitive plants could no longer survive. The incident near Ducktown, Tennessee was among the best documented. The rolling hills there were covered with a diverse population of trees and shrubs together with a rich understory of herbaceous forbs and grasses. When the smelter operations began in 1864, much of the surrounding large old oak and hickory forest was cut for fuel; the sulfur released from the roasting high-sulfide ores drifted over the countryside to kill many more. At least 7000 acres of forest were killed and another 17,000 acres devastated to where only the more tolerant grasses survived. The unprotected soils were soon eroded, leaving an infertile mantel of red clay barely able to support a population of mixed grasses (Hursh, 1948).

This story was repeated in Montana, and later in British Columbia, following the discovery and smelting of copper ores in the West. The dominant trees were conifers which were particularly susceptible to sulfur dioxide. Trees reportedly were damaged nearly 20 miles from one smelter and as far as 65

7

miles from the other (Treshow, 1970). These and many similar incidents gradually drew attention to and concern about air pollution.

In the subarctic pine forests of Ontario, there is an area rich in nickel, copper, and iron. The smelting of these ores has led to even greater plant damage since the atmosphere was altered beyond the tolerance of many plant species. White pine trees died up to 30 miles away (Linzon, 1978) and the sensitive lichen populations were affected well beyond this.

Any population of plants consists of numbers of individuals, each one different from the other. Just as one species of plant may be more or less sensitive to a pollutant than another, a similar variation in tolerance exists within the population of a single species. Thus when a pollutant is present, the less tolerant species and individuals may be killed or weakened, while the more tolerant persist to produce the next generation of plants. A similar variation in tolerance will occur in this generation, and so the selection process continues, the plant population responding to one more environmental parameter.

Unfortunately, not all plant populations possess the genetic makeup to tolerate existing concentrations of all pollutants. Too often our capacity to add pollutants to the atmosphere has exceeded the ability of the plant population to evolve genetically with sufficient speed to keep pace and adapt to the changing environment. Such species have disappeared in polluted environments.

This chapter is intended to provide a general idea of how air pollutants affect plants. Major pollutants, and combinations of pollutants, those that probably cause over 95% of the problems, will be discussed. Lesser pollutants are far too numerous to be included and tend to be a problem more in limited areas. I shall begin by discussing the basic mechanisms of how pollutants act, first at the molecular level of organization. As basic metabolic processes are altered, effects appear in the cells, the plant tissues, and such affected organs as the leaves and fruits. Where this happens, symptoms appear that can be diagnosed, and the methods of diagnosis will be treated next. Ultimately, the plant population and community are impaired and damage is reflected in changing ecosystems and agricultural production.

The impact of air pollution damage, and the need for controls, has led to the development of air quality standards. Basic considerations of these standards will be reviewed.

2 BIOCHEMICAL AND CELLULAR EFFECTS

The impact on the ecological system, whether principally a desert, grassland, or forest, is first reflected not on the total system or organism; rather, any imbalance or stress is felt first at the molecular level of the individual plant and plant system. The decline originates when processes within the plant cell are stressed, altering the metabolic state, and affecting the cell itself. Every pollutant acts in a somewhat different way; but certain fundamental processes, most notably water balance, seem to be universally affected (Heath, 1980). Primarily affected are the mechanisms regulating the entry of the pollutant and the chemical reactions responsible for photosynthesis, respiration, and energy production.

2.1 Sulfur Dioxide

A pollutant first enters the plant through the stomates, or openings in the leaf, where normal gas exchange takes place. Sulfur dioxide first affects the cells that regulate the aperture of these openings. The extent to which the stomates are open, and the factors influencing this, are initially most critical to the severity of pollution effects. Even in very low concentrations, sulfur dioxide can stimulate the stomates to remain open when the relative humidity is high. The open stomates can facilitate entry of more of the pollutant. The stomates close, however, if the carbon dioxide level is high. The reaction is even more complex since SO_2 concentrations that can cause the stomates to open at high humidities cause them to close when the relative humidity is low. Furthermore, different plant species respond oppositely when exposed to the same SO_2 concentrations (Biggs and Davis, 1980).

Once within the intercellular spaces of the leaf, the pollutant comes into contact with the membrane, the plasmalemma, that surrounds the cell. This intricate layer of fats and protein regulates the flow of substances into and out of the cell's interior. When the integrity of this semipermeable membrane is disrupted, nutrient balance and ion flow are also disrupted.

After passing into the cell, SO_2 interacts with such organelles as mitochondria and chloroplasts, including their membranes; the consequences may be even more serious (Ziegler, 1975). Here it interferes with the permeability and structure of the intracellular membranes, particularly the chloroplast membrane (Malhotra and Hocking, 1976). Within the chloroplasts, where photosynthesis, the process of converting light energy to carbohydrates, takes place, the first observable changes of SO_2 toxicity occur. This is the granulation of the stroma and swelling of the associated membranes. The chloroplasts then swell and ultimately degenerate (Wellburn et al., 1972).

Within the leaf SO_2 is converted into bisulfite, sulfite, and sulfate, or stays as aqueous SO_2. All of these, but especially the SO_2, inhibit photosynthesis (Silvius et al., 1975).

Photosynthetic pigments and many enzymes are associated with the chloroplasts. Malhotra and Hocking (1976) suggest that even low SO_2 concentrations will affect such vital enzymes as chlorophyllase and gradually cause the senescence, or breakdown of the related enzyme systems. Sulfur dioxide may cause a conversion of chlorophyll to phaeophytin (Rao and LeBlanc, 1961). This conversion may be caused simply by a change in the acidity since a lower pH will cause the loss of magnesium from the chlorophyll molecule. The effect is greatest at high humidities, suggesting that SO_2 is acting as sulfurous acid.

Further, SO_2 disrupts enzyme configuration and thereby affects many biochemical processes in the cell. Direct interference with metabolism by inhibiting mitochondrial ATP production and by splitting disulfide bonds in proteins and enzymes is among the serious effects (Horsman and Wellburn, 1976). Such splitting destroys the structural integrity of proteins. However, the precise mechanism of action remains unknown and a continuing subject for study in several laboratories.

Sulfur is essential to normal plant growth, and the action of SO_2 may relate to this role of sulfur. Sulfur is required in the reduced state. Sulfite is presumably an intermediate in the reduction of sulfate. Normally sulfide, cysteine, and methionine are also intermediates. But when plants are exposed to SO_2, the intermediates include sulfate, cysteine, glutathione, and at least one uncharacteristic compound (reported in Mudd, 1975). Sulfate is the major product. This imbalance in the reduced and oxidized forms of sulfur may prove harmful.

Enzyme inactivation has also been considered. Sulfur dioxide inhibits various biochemical reactions. Sulfite (SO_3) especially can act as a weak acid and inhibit certain enzymes by binding certain key sites on the enzyme, making them unavailable to the required chemical, a process known as competitive inhibition. Sulfur dioxide acts as a competitive inhibitor of ribulose diphosphate carboxylase and thus interferes with photosynthetic CO_2 fixation (Malhotra and Hocking, 1976).

While no definite conclusions can be drawn to positively establish the way SO_2 acts at the molecular level, the mechanism does seem to involve the excess of oxidized forms of sulfur, their imbalance with reduced sulfur, and the actions on critical enzymes.

2.2 Fluoride

Fluoride affects cellular metabolism in much the same general way, although the fundamental mechanisms naturally differ. All plants contain some fluoride, but too much can be toxic. If fluoride accumulates rapidly, toxic amounts may be as little as 10–20 ppm (parts per million) above the background concentrations of 1–10 ppm in the leaf. This, of course, applies only to the most sensitive plants. Most plants can accumulate 100–200 ppm fluoride or more in the leaves with no ill effects. A few, including tea and camellia, have an affinity to accumulate fluoride and normally concentrate several hundred parts per million in their leaves.

For most plants though, accumulation of over 50–100 ppm fluoride can exceed a toxic threshold and alter cell metabolism and structure. Granulation, plasmolysis, and collapse of the chloroplasts are the earliest microscopic expressions. In the case of pine needles, hypertrophy of the food-conducting phloem cells and associated transfusion tissue occurs, but this response is also evoked by such other stresses as senescence and drought (Stewart et al., 1973).

Before this happens though, fluoride affects a number of enzymes and metabolic systems. Plants fumigated with HF can show changes in organic acids, amino acids, free sugars, DNA and RNA phosphorus, and starch and nonstarch polysaccharides before any visible symptoms appear. McCune and Weinstein (1971) reported that fluoride alters the pathways of glucose breakdown and suggested that this could modify normal leaf development.

The inhibition of an enzyme results in the inhibition of the reaction mediated by that enzyme. Although only one step along a pathway of reactions may be affected directly, the entire process is impaired. Such is the case with photosynthesis which has long been known to be inhibited by fluorides (Thomas, 1958). One mechanism of this reduction is in the inhibition of chlorophyll. Fluoride may tie up the magnesium ion, central to the molecule. If larger amounts of magnesium are provided such inhibition *in vitro* can be offset (Ballantyne, 1972). Fluoride also may act on photosynthesis by interfering with the energy processes that involve adenosine phosphates and nucleotides (Chang, 1975).

2.3 Ozone

Ozone, the third pollutant of major consequence, also first affects plants at the molecular level. Again, the earliest impact seems to be on the stomates and membranes. Ozone tends to cause stomatal closure, but the degree of this effect is greatly influenced by ozone concentrations prior to the main exposure. Stomates of plants raised in filtered air close most rapidly in response to high ozone (Runeckles and Rosen, 1977).

The earliest visible histological effect is on the chloroplasts, which become disrupted, granular, and lighter green in color soon after exposure. Specifically, the stroma are first affected and granulation may result from ionic alterations within the chloroplast or from the ozone-induced permeability changes. The chloroplast membrane breaks down, chlorophyll is dispersed in the cytoplasm, the nuclear membrane is damaged, and the cell becomes plasmolyzed (Figure 1) (Treshow, 1970).

Ozone is extremely reactive and might theoretically be expected to react completely with the first molecules contacted. This would be in the cell wall and underlying cell membrane. However, such metabolic processes as photosynthesis also are seriously affected (Heath, 1975).

Disruption of the cell wall and membrane alters the normal metabolism tremendously, causing increased water loss and ionic imbalance. In turn, numerous other processes become affected (Heath, 1975). Ozone has been found to modify amino acids, alter protein metabolism, and change the unsatu-

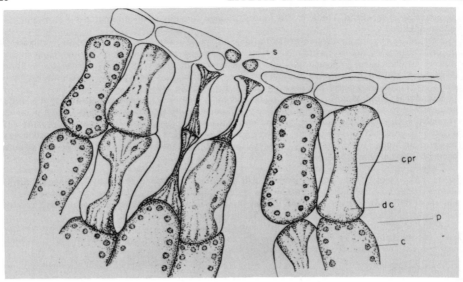

Fig. 1 Early impact of ozone toxicity showing the principal area of entry, the stoma (s), the collapse of the protoplast (cpr), and the breakdown of the chloroplasts (dc) as contrasted with the normal chloroplast (c). The leaf tissue involved is the palisade layer (p).

rated fatty acid composition and sulfhydryl residues. Furthermore, there seems to be a clear connection between the level of oxidant pollution and reduction of chlorophyll and some soluble proteins (Rabe and Kreb, 1979). The ATP levels drop almost immediately following exposure. This may be a primary response or may be caused by ionic imbalance.

Ozone also severely inhibits CO_2 fixation, although at lower concentrations the ozone would probably react entirely with the outer membranes and not reach the chloroplasts.

Many workers believe that the unsaturated fatty acid residues of the membrane lipids provide the primary site of ozone injury (Heath, 1975). However, lipid oxidation products are formed mainly after the death of the cell.

It is also possible that all the lipid changes might arise via sulfhydryl oxidations. Sulfhydryls may well be a primary site of ozone attack (Mudd, 1973). Ozone can oxidize the sulfhydryls to disulfides (—S—S) and sulfenic acid groups (—SO_2H) and then irreversibly oxidize those to sulfonic acid groups (—SO_3H). Since sulfhydryls can protect lipids from oxidation, their oxidation would facilitate lipid and membrane damage. Other work, however, found no change in cytoplasmic sulfhydryl after ozone exposure (Tingey et al., 1973).

3 EFFECTS ON THE WHOLE PLANT

3.1 Diagnosis

Once a sufficient number of plant cells have been injured, the symptoms become visible to the naked eye. At first insidious and scarcely discernible, symptom expressions characteristic for each pollutant gradually appear (Jacobson and Hill, 1970). However, in many cases, symptoms caused by different pollutants may be sufficiently similar to require some expertise to distinguish between them. Equally serious, symptoms caused by certain other environmental stresses also can be similar and mistaken for pollutant injury. Stresses imposed by high or low temperatures, moisture deficiency, and chemical treatments are most commonly responsible for mimicking symptoms. Each can cause a general chlorotic yellowing or necrosis of affected leaves. The distinctions in the specific injury pattern between causal agents are not always obvious. Diagnosis must consist of evaluating the total syndrome. The possible presence of an organism or virus, the part of the plant affected, the distribution of affected plants, the species of plants affected and their susceptibility to a suspected pollutant, characteristics of the terrain or locale, and the crop history or ecosystem background all must be considered (Treshow, 1970; Lacasse and Treshow, 1976).

3.1.1 Susceptibility

One of the most helpful criteria in diagnosis is the relative susceptibility of different plant species or varieties to a given pollutant. Susceptibility is a relative value, but under the same conditions, certain

Table 1. Plant Species Most Sensitive to Major Air Pollutants

Sulfur Dioxide

Alfalfa (*Medicago sativa*)	Ponderosa pine[a] (*Pinus ponderosa*)
Barley (*Hordeum vulgare*)	Soybeans (*Glycine max*)
Cotton (*Gossypium hirsutum*)	Wheat (*Triticum* sp.)
Douglas fir[a]	
(*Pseudotsuga menziesii*)	

Fluoride

Apricot (Chinese variety)	Oregon grape (*Mahonia repens*)
(*Prunus armeniaca*)	Peach (fruit) (*Prunus persica*)
	Ponderosa pine[a] (*Pinus ponderosa*)
Gladiolus (some varieties)	St. John's wort (*Hypericum*
Grape (some European varieties)	*perforatum*)
(*Vitis vinifera*)	

Ozone

Alfalfa (*Medicago sativa*)	Quaking aspen (*Populus tremuloides*)
Barley (*Hordeum vulgare*)	Spinach (*Spinachia oleraceae*)
Bean (*Phaseolus vulgaris*)	Tobacco (some varieties)
	(*Nicotiana tobacum*)
Green ash (*Fraxinus pennsylvanica*)	Wheat (*Triticum* sp.)
Oats (*Avena sativa*)	White pine (*Pinus strobus*)

[a] While these species are traditionally regarded as "sensitive," they are far more tolerant than the herbaceous species listed in this category.

species typically are injured by the lowest concentrations of a pollutant. The most sensitive plants differ for each pollutant.

Ideally it would be convenient if the species most sensitive to a given pollutant occurred universally, but such is not the case. The most sensitive plants may occur in only a few areas. Thus, it is desirable to know the relative sensitivity of a number of species (Table 1).

Coniferous species, but particularly pine trees, tend to be among the more sensitive plant groups to many pollutants. But there are generally other species that are still more sensitive. Where pollutants are present in toxic concentrations, some of these species may be injured. More tolerant species are rarely injured except where pollutant control equipment is lacking at a source, where there has been a breakdown in pollutant control equipment, or, in the case of urban pollution, when conditions are exceptionally favorable for ozone formation.

Lichens, those symbiotic associations of fungi and algae, are among the most sensitive of all organisms to air pollution (Ferry et al., 1973). The more sensitive species have long since disappeared from urban and industrial areas, and continue to be inpacted wherever new pollutant sources arise.

3.1.2 Thresholds of Visible Injury

Much work has been directed toward learning the effect of air pollutants on whole plants. The injury threshold, or dose, that is, the concentration of a pollutant times the duration of exposure that can cause injury, has been studied extensively.

Plants grown in chambers or greenhouses, or more recently in open-top chambers, have been exposed to known concentrations of pollutants for various periods of time, or plants grown in the field have been exposed to pollutants released beside or over them through pipes. In other field studies, where injury has been observed, atmospheric pollutant concentrations have been monitored and related to the incidence and severity of injury. The basic objectives have been the same: to learn how much of a given pollutant it takes to cause injury. At the same time the data would help to establish a concentration threshold below which injury was absent and so help provide a basis for air quality standards.

Such a threshold is difficult to establish. Not only is every plant species unique in its sensitivity to every pollutant, each variety, cultivar, or even individual plant has a different genetic tolerance. To add to the complexity, susceptibility varies with the stage of growth, or plant maturity, and the time of year. A threshold value is also complicated in that every environmental parameter affects susceptibility. Moisture regime, humidity, temperature, soil conditions including nutrition, light conditions, and everything else modifies the plant response. Therefore a threshold can never be just one value; it is best considered as a concentration range, the exact threshold being dependent on the interacting environmental parameters.

The degree of poor health, or damage, is generally regarded to be a function of both the concentration of the pollutant and the duration of exposure. Multiplied together, these values are the dose. This dose threshold would seem to provide an ideal value to estimate when plants might first be threatened. Unfortunately, it does not. It is the peak concentrations to which the plant is exposed that seem most critical. Exposures of long duration are also important, especially if the exposure is continuous, but their significance seems to be outweighed by the peaks, even if they are only for an hour or so, and how frequently the exposures occur.

3.1.3   *Symptoms*

3.1.3.1   Sulfur Dioxide Symptoms.   Wherever large quantities of coal are burned or ores high in sulfur are smelted, SO_2 concentrations may exceed the toxic thresholds of sensitive plant species.

In broad-leaved plants, the spongy mesophyll cells closest to the lower epidermis are the first to be injured. As the affected areas become more extensive, the classic symptoms appear. These consist of the gradual development of pale green or yellow areas between the large veins of the leaves, or sometimes along the margin. This is often called chronic injury.

Should concentrations be higher for even a brief period, the chlorophyll disappears rapidly, the cells break down, and the tissues become necrotic and turn brown. The affected area dries out, leaving straw-colored to reddish-brown lesions. This is called acute injury. There is often no clear demarcation between chronic and acute injury; they may occur together or separately.

We have seen that numerous variables influence the susceptibility of a plant, but in a general way, under ideal environmental conditions for injury, the most sensitive species such as alfalfa, soybeans, and barley, may show some injury when SO_2 concentrations in a range above 0.3–0.5 ppm persist for a period of at least 2 or 3 hr (Figure 2). Exposures to even lower concentrations, if they are continuous for several weeks, may cause adverse metabolic effects including growth suppression in the absence of visible symptoms (Keller, 1977).

Needle-leaved species, the pines, spruce, and fir trees react somewhat differently from broad-leaved species since their anatomy differs. The most pronounced symptoms of SO_2 injury are the reddish-brown discoloration of the needles generally beginning at the tip and progressing toward the base. Occasionally only a band of necrosis appears on the affected needles. A severe exposure to SO_2 may cause almost complete necrosis of the younger leaves (Figure 3). Damaged needles may be shed over a period of days or weeks when injury is this severe, thus giving the injured tree a sparse-foliaged appearance.

Exposures to lower but continuous SO_2 fumigations may cause a more general chlorosis or chlorotic stippling, most characteristically involving the older needles. The younger, more metabolically active needles are most susceptible to intermittent exposures.

Fig. 2   Sulfur dioxide injury on alfalfa plants illustrating the characteristic bleaching and interveinal yellowing.

Fig. 3   Severe SO₂ injury of younger short-leaf pine (*Pinus echirata*) needles.

Continuous exposure of conifers to low concentrations of SO₂ under rigidly controlled greenhouse conditions has shown that growth may be impaired, and the summer wood produced under these conditions made more dense, in the absence of needle necrosis (Keller, 1977).

3.1.3.2 Fluoride Symptoms.   Fluoride air pollution can be a problem near ceramic or glass manufacturing plants or phosphate reduction plants, but primarily it is a problem near aluminum reduction plants. Since large amounts of energy are required for aluminum production, these plants often are situated where inexpensive electric power, frequently hydroelectric, is available. This is significant in that, typically, hydroelectric power is available near areas of high precipitation which favors the growth of forests, or is available near agricultural areas. Since coniferous forests are among the most sensitive to fluoride, damage has sometimes been extensive (National Academy of Sciences, 1971).

Fluoride injury symptoms on Ponderosa pine and Douglas fir are quite similar to those caused by SO₂. Necrosis begins at the tips of the current year's needles and progresses toward the base as toxic amounts of fluoride accumulate. Where fumigations are successive, a darker reddish-brown band often delimits the paler necrosis caused by successive exposures. The affected parts of the needles ultimately have a reddish-brown appearance. The severity may range from a few discolored needle tips to involvement of the entire needle. Even when necrosis is this severe on current year's needles, the older needles may remain resistant and not be injured. Needles are by far the most sensitive when they first emerge and elongate in the spring. After this period, lasting but a few weeks, they become relatively tolerant.

The sensitive broad-leaved plants can be injured by far lower fluoride concentrations than even the most sensitive of the conifers. The earliest sign of injury to such sensitive plants as gladiolus or apricot is a dull, gray-green, water-soaked discoloration of tissues at the leaf tip or margin, but the affected tissues die rapidly, and within a day or two of exposure, the leaf turns light to dark brown (Figure 4).

Fig. 4 Characteristic fluoride injury of apricot leaf with necrotic area breaking away and interveinal yellowing and necrosis.

The size of the necrotic area may range from a mere fleck 1–3 mm across along the leaf margin to involvement of the entire leaf. Successive exposures are typically delimited by wavy, dark, reddish-brown bands. Often the dead tissue will drop out along this band. The leaf itself tends to hang on the tree even when entirely necrotic.

In some plants, such as citrus, corn, poplar, and sometimes sweet cherry trees, chlorosis or leaf yellowing is the common symptom. Sometimes necrosis accompanies the yellowing. In these cases the yellow area tends to extend from the leaf margin toward the center of the leaf between the larger veins, somewhat similarly to the pattern caused by SO_2.

Fluoride also can affect fruits, although apparently indirectly. The classic example is the peach, and the disease caused is known as soft suture or suture red spot. An area of the tissue along the lower third of the suture ripens and reddens prematurely so that by the time the fruit ripens, the affected suture area is rotten. A remarkable feature of this is the extreme sensitivity of some peach varieties that can be damaged at fluoride concentrations scarcely above background. The disease is initiated early in fruit development when the calcium needs in the fruit are greatest. Calcium is apparently tied up by the fluoride and rendered unavailable. Symptoms also have been described on apricot, cherry, and pear fruits but only when fruits were exposed to far higher fluoride concentrations.

Flowers themselves, even those of sensitive gladiolus varieties, are quite resistant to direct fluoride injury. But there is an exception. The pollination process may be adversely affected, so that the blossoms may drop rather than set fruit (Facteau et al., 1973). Again, calcium deficiency, this time in the pollen tube, is involved. This can be at least a theoretical problem in some tree fruit crops, but the fluoride concentrations found to cause this effect are higher than those usually found in commercial orchards.

3.1.3.3 Ozone Symptoms. The most significant impact of ozone, either alone or in combination with nitrogen or sulfur oxides, on crops may well be reduced yields. This is difficult to establish though, and visible symptoms are more commonly sought. Presumably if yields are suppressed, at least some of the affected plants will express some symptoms visible to the naked eye.

Several such symptoms may appear, depending on the plant species and the conditions of exposure. Since minute areas of cells in the upper part of the leaf, the palisade layer, are the first to be affected, small flecks on the upper surface are the most common visible expression. These lesions appear between the smallest veins and may range in color from pale to dark brown or even purplish or reddish in plants where anthocyanin pigment formation is stimulated by ozone (Figure 5).

As the lesions enlarge and coalesce, much of the upper surface of the leaf often takes on a pale brownish or straw-colored cast, or sometimes a bronzed appearance (Figure 6). As the injury becomes more severe, cells in the lower part of the leaf, the spongy parenchyma, become affected and visible symptoms appear on the lower surface as well.

Fig. 5 Soybean leaflet with reddish-brown flecking of the type caused by ozone, but possibly in combination with SO₂.

In cases where the ozone concentrations are scarcely above the threshold of visible injury, the chloroplasts may be damaged before the cell dies. Symptoms then are characterized by a pale yellowing or chlorosis. Chlorotic mottling or flecking is especially characteristic of ozone injury on pine. But chlorosis has also been observed on grasses and occasionally alfalfa. Chlorosis tends to be accompanied by premature senescence and defoliation.

It is significant for diagnosis that symptoms appear first on the older leaves of a plant. The injury

Fig. 6 Bronzing and chlorosis of oak leaves (*Quercus gambeli*) exposed to 10 pphm ozone for 4 hr.

first develops covertly when the leaves are in an earlier, more sensitive stage of development. Leaves have been reported to be most sensitive when they are 65–95% of their full size (Hill et al., 1970).

The concentrations of ozone that can harm plants is remarkably low. In fact, they can be in the same concentration range that exists in remote areas of the country. That is, at least under environmental conditions ideal for causing injury, growth suppression and early senescence may occur when ozone concentrations are only in the range of 0.10 ppm. Visible symptoms of stippling and glazing can also occur at such concentrations (Treshow, 1970).

3.1.3.4 Nitrogen Oxides.

Nitrogen oxides rarely occur in concentrations that are directly harmful to plants. There may be an infrequent spill associated with certain industrial processes such as nitric acid manufacture, but urban concentrations rarely exceed a toxic threshold. Rather, oxides of nitrogen are important mostly as precursors of ozone or in combinations with ozone or SO_2.

When symptoms do appear, they first consist of deep, dull-green, water-soaked lesions that gradually become straw colored or bronzed. They appear mostly between the larger veins but can also develop along the margins. Both upper and lower leaf surfaces are affected.

Under the most predisposing conditions, NO_2 at concentrations as low as 0.25 ppm for an hour, might cause some chronic physiological injury to the most sensitive plants, but only if the exposures were repeated and even then no leaf symptoms would be likely. More typically, over 1 ppm would be required to produce injury.

3.1.3.5 Peroxyacyl Nitrates.

The symptoms of peroxyacyl nitrate (PAN) toxicity were the first caused by photochemical pollutants to be described, yet the most elusive to detect. Many years passed before the cause of the symptoms was known. The bronzed or silvery glaze that caused such serious damage to the leaves of sensitive vegetable crops in the Los Angeles basin beginning in the 1940s was thought to have been associated with automobiles. But it remained until infrared analytic methods were refined over the next decade before PAN and its homologs were identified.

Formed by photochemical reactions between nitrogen oxides and hydrocarbons, PAN is widespread wherever these substrates occur in the presence of sunlight. PAN is thought to be toxic because it may react with the sulfhydryl (SH) groups and in turn affect certain SH-containing enzymes that are vital to photosynthesis. The enzymes concerned are located in the stroma of the chloroplasts and these are the first part of the cell to show damage (Thomson et al., 1965).

The more sensitive plant species, including pinto bean (*Phaseolus vulgaris,* Pinto), lettuce (*Lactuca sativa*), petunia (*Petunia hybrida*), and oats (*Avena sativa*), may be injured by as little as 20 ppb (parts per billion) for 2–4 hr. Symptoms usually involve a glazing or bronzing of the lower leaf surface, but when injury is more severe, the upper surface also may be affected.

3.1.3.6 Pollutant Combinations.

Air is a mixture of gases. The gases may typically exist in varied combinations. Fluoride and SO_2 from point sources, so often considered to occur alone, occasionally exist together such as when a coal-fired power plant is constructed to provide electricity for a neighboring aluminum plant. More commonly though, it is sulfur and nitrogen oxides that occur together. The sulfur oxides come from burning coal and the ozone most typically comes from urban areas where automobile pollution is serious.

The concentration of each gas in the combination is quite apparently of paramount importance in determining the effect on plants; but the ratio of the concentrations of each gas to the other is also significant. A third variable is the sequence in which the pollutants occur, that is, if they occur simultaneously or intermittently (Reinert et al., 1975). The possibility that one pollutant may predispose a plant to injury by another also must be considered.

The effects of these combinations can be additive, as would seem most reasonable; antagonistic, actually less than the combined additive effects of the pollutants; or synergistic, that is, greater than the additive effects.

3.1.3.6a Sulfur Oxides and Ozone.

Ozone and sulfur dioxide are the most universal air pollutants, and their combined effects have been studied most intensively. The synergistic effects of these two gases occurring together can visibly injure plants that would not be injured by the same concentrations of either alone. The symptoms produced are characteristic of those produced by ozone alone. Sulfur dioxide symptoms appear only when concentrations exceed the accepted injury threshold.

Soybeans are acutely sensitive to both pollutants, and injury has been observed in the field when ozone concentrations were only in the 0.05–0.1 ppm range and SO_2 below 0.3 ppm (personal observation). Yield reductions from SO_2—O_3 concentrations of 0.1 ppm for 6 hr/day for 19 weeks were demonstrated in field exposure chambers (Heagle et al., 1974).

Eastern white pine is another species of note that is sensitive to these combinations. Costonis (1973) has observed that the greatest injury to needles occurred following exposure to 0.05 ppm ozone for 2 hr and then 0.05 ppm SO_2 for 2 hr, followed 24 hr later by a 2-hr exposure to a mixture of the gases at 5 pphm. Such concentrations are common in many forests of the northeastern United States. Indeed, it is very likely that O_3—SO_2 combinations drifting far from their sources of origin may cause some of the most serious crop losses and ecosystem stresses of any pollutants.

3.1.3.6b Sulfur Oxides and NO₂. Since SO_2 and NO_2 are both components of coal combustion, they are often found together. In urban areas where automobiles are also prevalent, NO_2 concentrations may be further elevated.

Bennett et al. (1975) have fumigated representative plant species with realistically low concentrations of these gases in the range of 0.1–1.0 ppm. The more sensitive plants were injured at concentrations between 0.75 and 1.0 ppm of each pollutant. The gases alone did not produce injury at these concentrations.

Symptoms resembled those caused by ozone which raises the question of whether or not the symptoms so commonly observed in the field really are caused by ozone alone, as commonly assumed, or rather by combinations with NO_2.

3.1.3.6c Sulfur Oxides and HF. The interactions of fluoride with SO_2 are generally additive, at least on citrus (Matsushima and Brewer, 1972; Weinstein, 1977). However, others have reported some synergistic effects using sweet corn and barley on which foliar lesions were intensified (Mandl et al., 1975).

3.1.4 Chemical Analysis

The amount of a polluting chemical found in the tissues of a presumably affected plant can sometimes serve as a diagnostic aid. Absence of significant amounts of the chemical does not necessarily mean noninvolvement, but neither does its presence establish a cause-and-effect relation. In other words, the presence of a pollutant does not necessarily prove that it is injurious in and of itself.

Chemical analysis has been most helpful in diagnosing fluoride injury. Background concentrations of fluoride are normally quite low, typically less than 10 ppm in the leaves. Thus a buildup of 50–100 ppm or more is, with some exceptions, indicative of some fluoride pollution. In severe cases, fluoride concentrations may actually reach several hundred parts per million, but even then, it is not necessarily toxic.

Low fluoride concentrations, roughly below 10 ppm, suggest that any injury was caused by other agents. However, this is not necessarily the case. It is possible that high concentrations of atmospheric fluorides caused injury rather rapidly before much fluoride accumulated. Occasionally mild injury may appear even though fluoride concentrations are barely above background. Soft suture of peach, for instance, can occur at very low fluoride concentrations. And conceivably blossom drop could occur prior to leaf maturity with little fluoride accumulation in the leaves at the time.

Chemical analysis as a diagnostic aid is still more tenuous with SO_2. Sulfur is a major, normal plant constituent and comprises roughly 0.1–0.3% of the dry weight of leaves depending on the sulfur content of the soil and other environmental parameters. Thus a substantial amount of atmospheric sulfur must be incorporated by the leaf before the background concentrations are altered significantly. Should the foliar sulfur concentrations reach two or three times the probable background values, or over 0.3% as total sulfate sulfur, then SO_2 pollution might be suspected. However, this does not prove that it is causing any harm since plants can be highly tolerant of sulfur if it is assimilated gradually.

4 ACID "RAIN"

In recent years, another dimension in air pollution biology has been recognized—acid "rain," or more correctly acid deposition, a subject that has caught the fancy of the media and public (Likens et al., 1979). Precipitation is normally acid, theoretically around 5.5 to 5.6, due to the carbon dioxide in the air as well as the nitrogen and sulfur oxides produced in nature. Rain, snow, or dusts can be made more acid by excessive anthropogenic sources of oxides. Deposition can also be alkaline in areas where basic components, especially calcium ions, occur. There is still considerable question as to the relative contribution of anthropogenic and biotic sources to acid deposition, but it is generally, although by no means unanimously (Liljestrand and Morgan, 1979), agreed that precipitation has become more acid in recent decades, at least in northern Europe and the northeastern United States. Much has yet to be learned.

The major biologic question though is, what is the impact of this acidification? The impact on aquatic systems appears to be well documented, especially where the pH of unbuffered young lakes is below 5.0 (Overrein, 1980).

But what of terrestrial ecosystems and agriculture? It is obvious that rain can injure plants if it is sufficiently acid. Studies have shown that simulated rain with a pH range of 2.7–3.4, far more acid than found naturally, produced lesions on poplar leaves (Evans et al., 1977). About 1% of the leaf area was injured at a pH of 3.4. Bean and sunflower were among the plants found to be most sensitive. Much concern is expressed about the impact on conifers, and yet no lesions were produced on pine needles exposed to simulated rain at a pH of 2.5. According to Evans, pine trees were most tolerant, woody and broad-leaved plants next, and herbaceous species the most sensitive to acidified rain.

Simulated acid rain at a pH of 3.1 applied to a soybean field throughout the growing season failed to cause visible leaf injury or yield reductions (Patricia Irving, personal communication). Any effects even in this extremely low pH range seem to be more advantageous than deleterious. In an extensive survey of the effects of exposing a wide variety of agricultural crops to simulated acid precipitation down to pH 3.0 (Lee et al., 1980), some crops were found to be adversely affected, but the growth of legumes and fruits was stimulated and grains remained the same.

The most acid fallout occurs immediately after a rain or snowfall has begun. Initially the reaction may be well below normal, but as the rain continues, the air is cleansed and the reaction becomes normal. Throughout much of the eastern United States and Canada, and southern Scandinavia, pH values down to 4.2 have been measured. During individual storms the pH may drop to between 3.0 and 4.0 for briefer periods.

It is generally agreed (Overrein, 1980) that even in the most polluted areas anthropogenic additions of acid rainfall have had no impact on agriculture. In natural terrestrial systems, normal biological cycling produces far more acidity than that contributed by precipitation. The potential for long-range impact exists, but it has not been demonstrated.

5   AGRICULTURE

The possible impact of air pollutants on crop production and agriculture is a major concern and fundamental impetus for improving air quality. Agricultural crop losses have been sustained for many years, but the extent of this loss is not well known.

Along the Eastern Seaboard and in the Great Lakes region of the United States, ozone probably has been the major air pollutant. Concentrations have increased gradually over the years and the impact was gradual. The effect on some of the most sensitive plants, such as tobacco, grapes, pines, and potatoes was obvious. The flecking symptoms of tiny dead areas on the upper surface of tobacco leaves and grapes and the general yellowing and early senescence of sensitive potato varieties all were striking. The response of many other species was more insidious. These plants simply did not look healthy. They were unthrifty and yielded poorly. Any plant breeder raising commercial seed or other nursery stock in a polluted area would simply discard the weak, intolerant individuals. Thus, it is very likely that some selection against the more pollutant-sensitive species and for the tolerant plants has been going on for some time. What we grow today may well be only the more tolerant varieties, or cultivars.

It is possible that some leaf injury and even defoliation may occur without a significant reduction in yield or fruit quality but this is unusual (Heggestad et al., 1978). More often such injury, especially if over 5% of the leaf area is affected, is accompanied by decreased production, depending on the crop. Even more critical, air pollution can possibly reduce yields in the absence of visible markings on the leaves or fruits of plants; such losses especially are difficult to estimate quantitatively. Yield reductions can be discerned only by utilizing special methods of investigation.

In order to determine the effects of air pollutants on plants under ambient field conditions, ideally the crop is grown following standard commercial practices. The pollutants are then removed over parts of the field. This method, known as "reverse fumigations," or "pollutant exclusion," can best be accomplished by placing large chambers over replicated areas and filtering the air. However, the environment, other than for the absence of the pollutant, should be altered as little as possible, so open-top chambers are preferred. Even better, filtered air can be piped between rows of crops to the exclusion of the background atmosphere (Figure 7) (Jones et al., 1977; V. Runeckles, personal communication). Anti-oxidant sprays have also been tried in the case of ozone studies.

Relatively few studies have met these requisites. One of the earlier studies that did was conducted in a citrus grove in which plastic-covered greenhouses were installed over 16-yr-old lemon and orange trees. Although no distinctive leaf markings developed, yields of both species were reduced as much as 50% by the oxidants in the Los Angeles basin atmosphere (Thompson and Taylor, 1969).

Reductions in grape yields were demonstrated under ambient conditions in southern California. Cotton yields were reduced 20–30% in the San Joaquin Valley (Thompson et al., 1969).

When soybeans were grown in open-top chambers at Queenstown, Maryland, yields were reduced an average of 20% even when the more tolerant varieties were included (Howell et al., 1979). Visible symptoms in all of these studies were not especially serious or in some instances, even present.

Although soybeans are among the species most sensitive to SO_2, some of the varieties have a fair degree of tolerance. In Maryland, the state average production for soybeans of all varieties was 28 bushels per acre, but the more SO_2-resistant York variety yielded 60 bushels per acre (Howell, 1975; Howell et al., 1979). The yield difference was attributed to differences in SO_2 tolerance. It is apparent that SO_2 will, and already is, serving as a selective force for growing certain soybean varieties in the East. The principle applies to any crop, and over a period of time, sensitive varieties will simply no longer be grown.

Snap beans and tomatoes were grown in open-top chambers for 43 and 99 days, respectively, in Yonkers, New York (MacLean and Schneider, 1976). Yields were reduced 33 and 26%, respectively, in the unfiltered chambers that were exposed to the ambient air containing an array of pollutants presumably containing most significantly ozone and SO_2.

Howard Heggestad and his colleagues with the U.S. Department of Agriculture in Beltsville, Mary-

Fig. 7 Layout of aluminum pipes for distribution of SO_2 over a study plot at the University of British Columbia. Part of the research program of Dr. V. C. Runeckles.

land, also used filtered, open-top chambers. Snap bean varieties that were sensitive, or intermediate in sensitivity, to ozone toxicity, were grown for 5 yr. During this time yields of the sensitive varieties were reduced about 15% due to oxidant pollution. Yields of the intermediate variety were unaffected. The significance of this study was that the ozone concentrations were not particularly high. They were above 10 pphm only 18 hr each year in all during the 5 yr of the study. Three of the years, the ozone concentration exceeded 10 pphm only 10 hr in all. No visible injury symptoms developed.

An air exclusion system for field studies of SO_2 effects on crop production has been described by Jones et al. (1977). The system would seem to be the most realistic since it eliminates the need for chambers or continuous air circulation. Filtered air is circulated through the plant canopy only when the pollutant reaches a predetermined concentration. Results of preliminary tests conducted near a coal-fired power plant showed that the yield of soybeans of the variety Forrest was 28.6% lower in the plot without the air exclusion system.

Studies were conducted the next year at a site remote from any SO_2 sources and the exclusion system was operated continuously. Over a 2-yr period, soybean yields in the filtered air averaged 25% higher than plants exposed to ambient air (J. C. Noggle, personal communication). The inference was that ozone was the air pollutant most likely responsible for the yield loss.

A different approach was used by Oshima and his associates at the University of California at Riverside (1976). Utilizing the ozone concentrations that existed along a gradient with distance in the Los Angeles basin, alfalfa plants were grown in uniform containers with the same soil mixture. A high negative correlation was found between ozone dose and yield.

6   ECOSYSTEM RESPONSES

The ultimate success of any population depends on genetic diversity. The variation among individuals in the population gives a species the capacity to adapt to changes in the environment, and thereby helps assure its continued survival. Over the course of time, the best adapted individuals and species will come to dominate and provide the stable components of the ecosystem.

Genetic diversity in a population assures that changes in the environment will favor some individuals over others. Under the stress of the most severe air pollution conditions, all the plants may be killed. Such instances are extremely rare today. Should the pollutant concentrations be only slightly less, the more sensitive species may be eliminated. These conditions are also infrequent. What we are more likely to find today are much lower concentrations of any pollutants. Under these conditions a more subtle response could be expected, largely among individuals of a sensitive species. Under such conditions, genetic diversity continues to operate. Species or individuals lacking pollutant tolerance cannot compete as well and may not persist to perpetuate any genetic traits. The more tolerant individuals are relatively unaffected and survive to reproduce their genetic traits to the next generation (Treshow, 1980).

Almost any pollutant can provide such a selective force. The most natural are the heavy metals. Such elements as nickel, cadmium, copper, cobalt, and molybdenum occur naturally in high concentrations. Where they do, populations of certain species have evolved that are tolerant to these chemicals. In fact, some plants thrive at concentrations that are toxic to others.

This is equally true where the deposition of metals or other pollutants is of recent origin. Mine tailings or deposition of metals from smelters provide an apt illustration where contamination has provided a selective force that precluded recolonization by all but the most tolerant individuals. Seed-sowing experiments have provided some excellent examples of how rapidly evolution can occur under such conditions (Bradshaw, 1976). Fully tolerant populations of *Agrostis tenuis* developed on metal-contaminated soils can evolve in only two or three generations from the approximately 2% of the original population that had some degree of tolerance.

The same principles apply to other pollutants. This is well illustrated with SO_2. Populations of sensitive species near smelters, most conspicuously conifers such as pine and fir, have been strongly impacted. But except under the most extreme conditions, a few tolerant individuals have survived to reproduce. More distant from the pollutant sources, changes are more subtle; the plant population looks the same, but the component plants are more tolerant of the pollutant.

This is well illustrated by the studies of Bell and Mudd (1976) in England who showed that over the course of time, rye grass (*Lolium perenne*) indigenous to the chronically SO_2-polluted Helmshore region of Lancashire evolved resistance to coal smoke. No symptoms or growth suppression occurred on plants in this population, while commercial strains of rye planted in the same area exhibited characteristic SO_2 injury.

Where a seed-producing population has evolved some tolerance to a pollutant, the new generation will arise from these seeds. But the development of aerial, sexually reproducing parts may be impaired by continued high SO_2 concentrations, giving an advantage to the plants that can reproduce asexually, particularly by underground stolons, runners, or sucker shoots. In this way clones of tolerant individuals may spread and populate fairly heavily polluted areas.

This is striking with grasses. Near one smelter, two perennial grasses, muhly grass (*Muhlenbergia asperifolia*) and salt grass (*Distichlis striata*), comprised nearly the entire plant cover. Neither were prevalent in nearby undisturbed areas where competition from other species, better able to compete in the absence of pollution, restricted their development. Close to a second smelter, clones of quaking aspen (*Populus tremuloides*) appeared to have spread from SO_2-resistant individuals (Figure 8).

Although the most striking examples of forest ecosystems altered by SO_2 are historic, there are notable current examples. One of these has been near gas plants in central Alberta, Canada. Here, SO_2 is released during the removal of hydrogen sulfide from the natural gas. White spruce (*Picea glauca*) stands are dominant in this area. During the course of studies between 1968 and 1975, the

Fig. 8 Patches of quaking aspen relatively tolerant to SO_2 which has been common over the area for about 90 yr.

percentage cover for all understory plants, that is, those growing beneath the tree canopy, was found to decrease markedly toward the source (Winner and Bewley, 1978). The number of white spruce seedlings was also reduced, but the most conspicuous effect was on mosses which were entirely absent nearest the SO_2 source; further out mosses showed a reduced canopy cover, depth, dry weight, and reproductive structures. Unfortunately the SO_2 concentrations to which the effects were attributed were not known.

Studies in Pennsylvania (Rosenberg et al., 1979) also have demonstrated that plant species' richness and diversity decrease with distance from an SO_2 source, and provide a more sensitive indicator of pollution stress than growth assessment.

Fluorides also can thin out or eliminate sensitive plants, thereby encouraging survival of more tolerant species and individuals. This has been demonstrated in many areas including especially forest communities surrounding aluminum reduction plants. In the Pacific Northwest, sensitive ponderosa pine stands were eliminated in the early 1950s at one site, and pines and Douglas fir near another only a few years later. Recovery has been slow in both areas, and the dominant vegetation consists of more tolerant shrubs and grasses (personal observation). A similar alteration of a Douglas fir community surrounding a phosphate reduction plant in Idaho was especially striking, and recovery also extremely slow (Treshow, 1980).

Photochemical pollutants have also taken their toll of forest ecosystems. The most vivid example is in the San Bernadino Mountains north and east of the Los Angeles basin in southern California. Here, the apparently most sensitive species, ponderosa pine, is also the dominant plant. Hence the impact is especially critical. The most sensitive individuals began to decline in the early 1950s. Needles became chlorotic and dropped prematurely, leaving the trees too little food-producing tissue to survive (Figure 9). Within the next decade these trees had died, and increasingly, more tolerant individuals began to decline and succumb. As the population of ponderosa and Jeffery pines thinned out, the

Fig. 9   Thin, sparse-foliaged ponderosa pine trees characteristic of ozone toxicity in southern California.

more tolerant conifers including coulter pine (*P. coulteri*) and sugar pine (*P. lambertiana*), and incense cedar (*Libocedrus decurrens*) became more conspicuous. But these species cannot adequately fill the more xeric habitat once filled by the more heat- and drought-tolerant ponderosa pines. Hence the forest has become more open, and even more dry.

7 AIR QUALITY STANDARDS

Air quality standards, emission standards, and fines, to say nothing of law suits, have all encouraged the installation of pollution-control equipment that has come a long way toward reducing air pollution damage in many industrial areas of the world. The high concentrations of SO_2 and fluoride near point sources are largely gone. The potential SO_2 and soot pollution from large new coal-fired power generating plants was never fully realized. Photochemical pollution has been held partially in check with the requirements of controls on automobile emissions. However, many controls are not entirely adequate, and pollution sources are more numerous than ever before. Thus the need persists for continued vigilance. This comes only at considerable economic costs, and it is helpful to understand some of the criteria and values that must be considered in establishing the air quality standards that are necessary to protect vegetation.

The threshold of pollution concentrations at which plants may first be injured is one of the many significant criteria. It is a difficult threshold to establish and, as discussed earlier, many factors must be considered.

The concentration at which any pollutant begins to affect plant health is most significant. The duration and frequency of exposure are also important, but development and severity of injury depends on far more. To be harmful, a number of factors, acting in concert, are necessary. The more predisposing the environmental parameters are, the more likely that a plant will be injured.

In laboratory fumigation studies, the most predisposing conditions have generally been provided— the most sensitive plants, an ideal nutrient and moisture regime, optimal light and temperature, and so on. In the field, conditions are not likely to be so conducive or favorable for injury. In other words, plants are less predisposed and less susceptible to pollutant injury when growing under field conditions.

The thresholds of injury then, depend on the plant species in question, the conditions under which they grew and were exposed, and the concentration and duration of exposure as well as whether the exposure was continuous or intermittent. All these variables make the establishment of air quality standards difficult and necessarily imprecise. To protect the most sensitive individuals under the most predisposing conditions would require a virtually pristine atmosphere. The average population of even sensitive species growing under normal conditions could tolerate modest amounts of pollution before being measurably harmed. Thus the amount of acceptable injury must be considered in setting any air quality standard. Should it be the faintest measurable physiological response, some visible leaf or fruit injury, or should the amount of injury reflect an economic loss? Also, should the standards protect only plants growing in a given region, or the most sensitive crops whether grown in a particular area or not?

In other words, "What plants are we trying to protect, and to what extent?" In any population of plants there tends to be a normal distribution curve for injury. A few individuals are most sensitive, a few most tolerant, and the majority intermediate in sensitivity to any given stress including air pollution. When a particular species is sensitive to a pollutant, the more sensitive individuals tend to be hypersensitive, and pollutant concentrations scarcely above background might have some deleterious effect. Should this segment of the population be protected?

Air pollutants have adversely and severely damaged plants and plant communities for many decades. They are likely to continue to do so well into the future. Sulfur and nitrogen oxides, ozone, fluoride, various combinations of these, and many more pollutants all threaten plant health. As populations, affluence, and industry increase, the risks of pollution also grow. Only by maintaining a constant vigil, strict controls, and the will to provide a clean atmosphere will it be possible to protect the health and welfare of our natural ecosystems and agriculture on which all life depends.

ACKNOWLEDGMENTS

I should like to thank Dr. Donald D. Davis of the Pennsylvania State University and Dr. Andrew Schoenberg at the University of Utah for their valuable suggestions and comments in reviewing the manuscript.

REFERENCES

Ballantyne, D. J. (1972), Fluoride Inhibition of the Hill Reaction in Bean Chloroplasts, *Atmos. Environ.* **6**, 267–273.

Bell, J. N. B., and Mudd, C. H. (1976), Sulphur Dioxide Resistance in Plants: A Case Study of *Lolium perenne,* pp. 87–105, in *Effects of Air Pollutants on Plants* (T. A. Mansfield, ed.), Cambridge University Press, Cambridge, 209 pages.

Bennett, J. H., Hill, A. C., Soleimani, A., and Edwards, W. H. (1975), Acute Effects of Combinations of Sulfur Dioxide and Nitrogen Dioxide on Plants, *Environ. Pollut.* **9**, 127–132.

Biggs, A. R., and Davis, D. D. (1980), Stomatal Response of Three Birch Species Exposed to Varying Acute Doses of SO_2, *J. Amer. Soc. Hort. Sci.* **105**(4), 514–516.

Bradshaw, A. D. (1976), Pollution and Evolution, pp. 135–138, in *Effects of Air Pollutants on Plants* (T. A. Mansfield, ed.), Cambridge University Press, Cambridge, 209 pages.

Chang, C. W. (1975), Fluorides, Chapter 4, pp. 57–95, in *Responses of Plants to Air Pollutants* (J. B. Mudd and T. T. Kozlowski, eds.), Academic Press, New York, 383 pages.

Costonis, A. C. (1973), Injury of Eastern White Pine by Sulfur Dioxide and Ozone Alone and in Mixtures, *Eur. J. Forest Pathol.* **3**, 50–55.

Evans, L. S., Gmur, N. F., and Dacosta, F. (1977), Leaf Surface and Histological Perturbations of Leaves of *Phaseolus vulgaris* and *Helianthus annuus* After Exposure to Simulated Acid Rain, *Amer. J. Bot.* **64**, 903–913.

Facteau, T. J., Wang, S. Y., and Rowe, K. E. (1973), The Effect of Hydrogen Fluoride on Pollen Germination and Pollen Tube Growth in *Prunus avium* L.C.V. "Royal Ann," *J. Amer. Soc. Hort. Sci.* **98**(3), 234–236.

Ferry, B. W., Baddeley, M. S., and Hawksworth, D. L. (eds.) (1973), *Air Pollution of Lichens,* University of Toronto Press, Toronto, Canada.

Heagle, A. S., Body, D. E., and Neely, G. E. (1974), Injury and Yield Responses of Soybeans to Chronic Doses of Ozone and Sulfur Dioxide in the Field, *Phytopathol.* **64**, 132–136.

Heath, R. L. (1975), Ozone, Chapter 3, pp. 23–56, in *Responses of Plants to Air Pollutants* (J. B. Mudd and T. T. Kozlowski, eds.), Academic Press, New York, 383 pages.

Heath, R. L. (1980), Initial Events in Injury to Plants by Air Pollutants, *Ann. Rev. Plant Physiol.* **31**, 395–431.

Heggestad, H. E., Heagle, A. S., Bennett, S. H., and Koch, E. J. (1978), *The Effects of Photochemical Oxidants on the Yield and Quality of Crops,* 71st Am. Mtg. Air Poll. Cont. Assoc., Houston, Texas, June 25–30, 1978.

Hill, A. C., Heggestad, H. E., and Linzon, S. N. (1970), Ozone, in *Recognition of Air Pollution Injury to Vegetation: A Pictorial Atlas* (J. S. Jacobson and A. C. Hill, eds.), Air Pollution Control Association, Pittsburgh, Pennsylvania.

Horsman, D. C., and Wellburn, A. R. (1976), Appendix II. Guide to Metabolic and Biochemical Effects of Air Pollutants on Higher Plants, pp. 185–199, in *Effects of Air Pollutants on Plants* (T. A. Mansfield, ed.), Cambridge University Press, Cambridge.

Howell, R. K. (1975), *Air Resources and Land Use Planning. Effects of Air Pollution on Food Production,* 30th Ann. Meg. Soil Cons. Soc. Amer., Aug. 10–13, 1975, San Antonio, Texas.

Howell, R. K., Koch, E. J., and Rose, L. P. (1979), Field Assessment of Air Pollution-Induced Soybean Yield Losses, *Agron. J.* **71**, 285–288.

Hursh, C. R. (1948), *Local Climate in the Copper Basin of Tennessee as Modified by Removal of Vegetation.* U.S. Dept. of Agriculture Circ. 774, 38 pages, illus.

Jacobson, J. S., and Hill, A. C. (1970), *Recognition of Air Pollution Injury to Vegetation: A Pictorial Atlas,* Air Pollution Control Association, Pittsburgh, Pennsylvania.

Jones, H. C., Lacasse, N. L., Liggett, W. S., and Weatherford, F. (1977), *Experimental Air Exclusion System for Field Studies of SO_2 Effects on Crop Productivity,* TVA E-L-P-77-5, EPA-600/7-77-122, 67 pages.

Keller, T. (1977), The Effect of Long Duration Low SO_2 Concentrations upon Photosynthesis of Conifers, pp. 81–83, in *Proc. 4th Int. Clean Air Congress.*

Lacasse, N. L., and Treshow, M. (1976), *Diagnosing Vegetation Injury Caused by Air Pollution,* Environmental Protection Agency, 139 pages plus illus. and appendix.

Lee, J. J., Neely, G. E., and Perrigan, S. C. (1980), *Sulfuric Acid Rain Effects on Crop Yield and Foliar Injury,* EPA-60013-80-016.

Likens, G. E., Wright, R. F., Galloway, J. N., and Butler, T. J. (1979), Acid Rain, *Sci. Amer.* **241**, 43–51.

Liljestrand, H. M., and Morgan, J. J. (1979), Error Analysis Applied to Indirect Methods for Precipitation Acidity, *Tellus* **31**, 421–431.

Linzon, S. N. (1978), Effects of Airborne Sulfur Pollutants on Plants, pp. 109–162, in *Sulfur in the Environment. Part II. Ecological Impacts* (J. O. Nriagu, ed.), Wiley, New York.

MacLean, D. C., and Schneider, R. E. (1976), Photochemical Oxidants in Yonkers, New York; Effect on Yield of Bean and Tomato, *J. Environ. Quality* **5**, 75–78.

Malhotra, S. S., and Hocking, D. (1976), Biochemical and Cytological Effects of Sulphur Dioxide on Plant Metabolism, *New Phytol.* **76**, 227–237.

Mandl, R. H., Weinstein, L. H., and Keveny, M. (1975), Effects of Hydrogen Fluoride on Sulfur Dioxide Alone and in Combination on Several Species of Plants, *Environ. Pollut.* **9,** 133–143.

Mansfield, T. A. (ed.) (1976), *Effects of Air Pollutants on Plants,* Cambridge University Press, Cambridge, 209 pages, illus.

Matsushima, J., and Brewer, R. F. (1972), Influence of Sulfur Dioxide and Hydrogen Fluoride as a Mix or Reciprocal Exposure on Citrus Growth and Development, *J. Air Pollut. Control Assoc.* **22,** 710–713.

McCune, D. C., and Weinstein, L. H. (1971), Metabolic Effects of Atmospheric Fluorides on Plants, *Environ. Pollut.* **1,** 169–174.

Mudd, J. B. (1973), Biochemical Effects of Some Air Pollutants on Plants, *Advan. Chem. Ser.* **122,** 31–47.

Mudd, J. B. (1975), Sulfur Dioxide, pp. 9–22, in *Responses of Plants to Air Pollutants* (J. B. Mudd and T. T. Kozlowski, eds.), Academic Press, New York, 383 pages.

National Academy of Sciences (1971), *Fluorides,* Committee on Biological Effects of Air Pollutants, National Research Council, Washington, D.C., 195 pages.

National Academy of Sciences (1978), *Sulfur Oxides,* Committee on Sulfur Oxides, National Research Council, Washington, D.C., 209 pages.

Oshima, R. J., Poe, M. P., Braegelmann, P. K., Baldwin, D. W., and Vanway, V. (1976), Ozone Dosage—Crop Loss Function for Alfalfa; A Standardized Method for Assessing Crop Losses from Air Pollutants, *J. Air Poll. Control Assoc.* **26,** 862–865.

Overrein, L. (1980), Results of the Norwegian Study of the Effects of Acidic Precipitation on Forest and Fish Resources. Effects of Air Pollutants on Mediterranean and Temperate Forest Ecosystems: An International Symposium, Riverside, California, June 22–27, 1980.

Rabe, R., and Kreb, K. H. (1979), Enzyme Activities and Chlorophyll and Protein Content in Plants as an Indicator of Air Pollution, *Environ. Pollut.* **19,** 119–137.

Rao, D. N., and Leblanc, F. (1961), Effects of Sulfur Dioxide on Lichen Algae with Special References to Chlorophyll, *Bryologist,* 69–75.

Reinert, R. A., Heagle, A. S., and Heck, W. W. (1975), Plant Responses to Pollutant Combinations, pp. 159–178, in *Responses of Plants to Air Pollutants* (J. B. Mudd and T. T. Kozlowski, eds.), Academic Press, New York, 383 pages.

Rosenberg, C. R., Hutnik, R. J., and Davis, D. D. (1979), Forest Composition at Varying Distances from a Coal-Burning Power Plant, *Environ. Pollut.* **19,** 307–317.

Runeckles, V. C., and Rosen, P. M. (1977), Effect of Ambient Ozone Pretreatment on Transpiration and Susceptibility to Ozone Injury, *Can. J. Bot.* **55,** 193–197.

Silvius, J. E., Ingle, M., and Baer, C. H. (1975), Sulfur Dioxide Inhibition of Photosynthesis in Isolated Spinach Chloroplasts, *Plant Physiol.* **56,** 434–437.

Stewart, D., Treshow, M., and Harner, F. M. (1973), Pathological Anatomy of Conifer Needle Necrosis, *Can. J. Bot.* **51,** 983–988.

Taylor, O. C., Thompson, C. R., Tingey, D. T., and Reinert, R. A. (1975), Oxides of Nitrogen, pp. 122–140, in *Responses of Plants to Air Pollutants* (J. B. Mudd and T. T. Kozlowski, eds.), Academic Press, New York, 383 pages.

Thomas, M. D. (1958), Air Pollution with Relation to Agronomic Crops, I. General Status of Research on the Effects of Air Pollution on Plants, *Agron. J.* **50,** 545–550.

Thompson, C. R., Hensel, E. G., and Kats, G. (1969), Effects of Photochemical Oxidants on Zinfandel Grapes, *Hort. Sci.* **4,** 222–224.

Thompson, C. R., and Taylor, O. C. (1969), Effects of Air Pollutants on Growth, Leaf Drop, Fruit Drop and Yield of Citrus Trees, *Environ. Sci. & Tech.* **3,** 934–940.

Thomson, W. W., Duggar, W. M., and Palmer, R. L. (1965), Effects of Peroxyacyl Nitrate on Ultrastructured Chloroplasts, *Bot. Gaz.* (Chicago) **126,** 66–72.

Tingey, D. T., Fites, R. C., and Wickliff, C. (1973), Ozone Alteration of Nitrate Reduction on Soybean, *Physiol. Plant* **29,** 33–38.

Treshow, M. (1970), *Environment and Plant Response,* McGraw-Hill, New York, 422 pages.

Treshow, M. (1980), Pollution Effects on Plant Distribution, *Environ. Conserv.* **7,** 279–286.

Weinstein, L. H. (1977), Fluoride and Plant Life, *J. Occup. Med.* **19,** 49–78.

Wellburn, A. R., Majernick, O., and Wellburn, F. A. M. (1972), Effects of SO_2 and NO_2, Polluted Air on the Ultrastructure of Chloroplasts, *Environ. Pollut.* **3,** 37–49.

Winner, W. E., and Bewley, J. D. (1978), Contrasts Between Bryophyte and Vascular Plant Responses in an SO_2-Stressed White Pine Association in Central Alberta, *Oecologia* **33,** 311–325.

Ziegler, I. (1975), The Effect of SO_2 Pollution on Plant Metabolism, *Residue Rev.* **56,** 79–105.

CHAPTER 3

POLLUTANT EFFECTS ON MATERIALS

JOHN E. YOCOM

ALEXANDER R. STANKUNAS

TRC-Environmental Consultants, Inc.
East Hartford, Connecticut

1 INTRODUCTION

Material damage related to air pollution has been recognized for well over 300 yr. In 1661 John Evelyn, a public-spirited citizen of London, described the air pollution situation in London.

> *This is the pernicious Smoake which sullyes all her Glory, superinducing a sooty Crust or Fur upon all that it lights, spoyling the moveables, tarnishing the Plate, Gildings and Furniture, and corroding the very Iron-bars and hardest Stones.* [1]

The author's perception appears prophetic, yet probably had little to do with a knowledge of the chemistry related to air pollution effects. Today scientists are studying the causative agents of material damage, recognizing that there is still great uncertainty in determining the quantitative relationships between pollutants and damage. Table 1 lists types of air pollution damage, emphasizing the variety of materials damaged by common air pollutants; sulfur oxides as a predominant pollutant type; and air pollution as only one environmental factor in material damage. [2]

2 SULFUR OXIDES

Moisture and oxidation to sulfates are usually involved in the mechanisms by which SO_2 and other sulfur oxides damage materials. One of these mechanisms is the "acid rain syndrome" in which acidic rainwater falls on sensitive surfaces in otherwise pristine areas, resulting in erosion, corrosion, or other chemical changes.

A second mechanism is the "acid gas syndrome." In this case the direct solution and reaction of SO_2 in the moisture film on the surface of exposed materials lead to acid formation and damage. Finally there is the "acid particle syndrome" in which the settling or impaction of acid sulfates or nitrates in the absence of wet precipitation leads to damage. Currently it is difficult, if not impossible, to distinguish which mechanism is most important. However, since high SO_2 levels often coincide with areas of abundant synthetic materials (i.e., cities), the acid gas mechanism is most likely to dominate total damage costs.

2.1 Ferrous Metals

Corrosion of iron and steel in polluted atmospheres has been the subject of a great deal of research over the years. Documents prepared by the U.S. Environmental Protection Agency (EPA) to support

Table 1   Air Pollution Damage to Materials

Materials	Type of Damage	Principal Air Pollutants	Other Environmental Factors
Metals	Corrosion tarnishing	Sulfur oxides, other acid gases	Moisture, air, salt
Building stone	Surface erosion and discoloration	Sulfur oxides and acid gases, particulate matter	Moisture, temperature fluctuations, salt, vibration, microorganisms, CO_2
Paint	Surface erosion, discoloration	Sulfur oxides, hydrogen sulfide, ozone, particulate matter	Moisture, sunlight, microorganisms
Textiles	Reduced tensile strength, soiling	Sulfur oxides, nitrogen oxides, particulate matter	Moisture, sunlight, physical wear
Textile dyes	Fading, color change	Nitrogen oxides, ozone	Sunlight
Paper	Embrittlement	Sulfur oxides	Moisture, physical wear
Leather	Weakening, powdered surface	Sulfur oxides	Physical wear
Ceramics	Changed surface appearance	Acid gases, HF	Moisture

air quality standards for particulate matter and sulfur oxides (PM/SO_x) contain an extensive review of the possible mechanisms for metal corrosion in the presence of sulfur oxides.[3]

Some of the earliest work on the nature of iron corrosion in atmospheres containing sulfur oxides was that of Vernon.[4] He showed the relative roles of SO_2 and humidity and proposed that there are critical humidities for the corrosion of metals above which SO_2-induced corrosion proceeds rapidly. He also showed that the corrosion of iron proceeds from randomly distributed "centers" which are associated with the deposition of particulate matter. However, it may also be likely that these centers represent differences in the grain structure of the sample. These differences produce electrochemical cells which, with a moist, conducting surface, allow current to flow and corrosion to advance.

According to Nriagu, once corrosion has been initiated, the progress of the reaction is controlled largely by sulfate ions produced from the oxidation of absorbed or adsorbed SO_2.[5] However, the actual mechanism of SO_2 oxidation on the surface is poorly understood. From the work of Johnson it appears that sulfur or sulfates are only a minor constituent of the products of steel corrosion.[6] Mild steel samples were exposed to two urban areas near Manchester, England. One area was heavily polluted, and the other lightly polluted. The relative amount of corrosion produced was strongly dependent on whether or not the sample was wet at the beginning of exposure.

Rust can protect iron and steel from further corrosion under some circumstances. Nriagu and Sydberger showed that the steel samples initially exposed to low concentrations of sulfur oxides were more resistant than samples continuously exposed to high concentrations.[5,7]

2.1.1   Laboratory Studies

Exposing iron and steel samples to SO_2 and humidity under controlled laboratory conditions permits isolation of the specific effect of the pollutant in relation to other environmental conditions. While many of the early experiments clearly show a correlation between corrosion rates and both SO_2 and humidity, exposure conditions usually consisted of SO_2 concentrations many times higher than those found in the ambient atmosphere. In addition, other test conditions such as the temperature and time of wetness were not representative of actual exposure.

The most realistic set of laboratory experiments was conducted by Haynie and others.[8] Although synthetically polluted atmospheres were used, the concentrations of pollutants to which materials were exposed were within the range found in the ambient air.

Steel corrosion was determined in terms of the weight change of steel panels. The results show that there is a strong, statistically significant relationship between steel corrosion and SO_2 concentration, together with high humidity.

2.1.2 Field Studies

Field studies are by definition representative of real exposures. However, they do not offer the degree of control over variables that is possible in the laboratory.

Upham exposed mild steel samples in a number of sites in and around St. Louis and Chicago.[9] He showed that corrosion correlates well with sulfur oxide levels and increases with the duration of exposure. In 1963 Haynie and Upham carried out a 5-yr program in which three different types of steel were exposed in eight major metropolitan areas in the United States.[10] Multiple regression analyses showed significant correlation between average SO_2 concentrations and corrosion to all three types of steel.

In 1964 Haynie and Upham exposed enameling steel samples for 1 and 2 yr at 57 stations of the National Air Sampling Network.[11] Pollutants of interest were SO_2, total suspended particulate matter (TSP), and the sulfate and nitrate content of particulate matter. Multiple linear regression and nonlinear, curve-fitting techniques were used to analyze the relationship between the corrosion behavior of steel and pollutant concentrations. A "best-fit" empirical damage function (see Section 7) was developed relating SO_2 and humidity to corrosion as follows:

$$cor = 325 \, te^{[0.00275 \, SO_2 - (163.2/RH)]}$$

where

$$cor = \text{depth of corrosion, } \mu m$$
$$t = \text{time, yr}$$
$$SO_2 = SO_2 \text{ concentration, } \mu g/m^3$$
$$RH = \text{average relative humidity, } \%$$

This damage function demonstrates that the sensitivity of corrosion to humidity is far greater than that for SO_2, especially at the levels of SO_2 normally experienced in urban areas. (See also Chandler and Kilcullen,[12] and Guttman and Sereda.[13])

A more recent study of material damage in the St. Louis area in 1974–1975 by Mansfeld included the use of special atmospheric corrosion monitors which measure the length of time that a corrosion panel is wet enough for electrochemical corrosion to take place.[14] Mansfeld's sample exposure array included weathering steel, galvanized steel, house paint, and marble. The concentrations of SO_2 measured in this study were an order of magnitude lower than those measured by Upham's 1967 study.[9] Low concentrations of SO_2 may be one reason a significant correlation between corrosivity and pollutant levels was not shown.

Gerhard and Haynie have given limited consideration to the possible role of air pollutants (notably SO_2) in the catastrophic failure of metals—largely as a result of stress corrosion.[15]

2.2 Nonferrous Metals

The corrosion rates of commercially important nonferrous metals in polluted atmospheres cover a wide range, although generally lower than for steel. The ASTM Committee on Atmospheric Corrosion compared corrosion rates for different metal formulations and under different exposure conditions over the period 1930–1954.[16]

Samples of copper, aluminum, brass, nickel, lead, and zinc were exposed at six different locations representing different environmental conditions. Locations presumably representing the highest air pollution levels show the highest corrosion rates with nickel and zinc being the most susceptible. Of the two, zinc is by far the more important because of its extensive use in galvanizing for the protection of steel in the ambient atmosphere.

In the work of Sydberger and Vannenberg it was shown that the adsorption of SO_2 by iron and three nonferrous metals over a period of time at 90% relative humidity correlates with the relative susceptibility to SO_2 attack of each of the metals.[17]

2.2.1 Aluminum

Aluminum is considered quite resistant to SO_2-induced corrosion because of the highly inert oxide layer formed on its surface. However, the presence of high concentrations of particulate matter can produce a mottled or pitted surface in the presence of SO_2. In view of efforts to control emissions of SO_2 and particulate matter in the United States, especially larger particles or agglomerates that could act as centers for corrosion initiation, the SO_2-induced surface corrosion of aluminum is not a significant problem.

The possible stress corrosion of aluminum in the presence of SO_2 is potentially a more serious problem. During extensive chamber studies Haynie found that stressed aluminum samples exposed to 79 and 1310 $\mu g/m^3$ of SO_2 lost, respectively, approximately 8.6 and 27.6% of their bending strength.[8]

2.2.2 Copper

Copper and copper alloys in most atmospheres develop thin, stable surface films which inhibit further corrosion. Initial atmospheric corrosion is a brown tarnish of mostly copper oxides and sulfides which can thicken to a black film; in a few years the familiar green patina forms. This film is either basic copper sulfate or, in marine atmospheres, basic copper chloride, both of which are extremely resistant to further atmospheric attack.

2.2.3 Zinc

The primary use of zinc other than as an alloying metal with copper to produce brass, is in galvanizing steel. Zinc is anodic with respect to steel so that when zinc and steel are in contact with an electrolyte, current flow protects the steel from corrosion at the expense of some oxidation of zinc.

Because of its economic importance, the behavior of zinc in polluted atmospheres has been studied intensively by a number of workers. Guttman developed an empirical damage function for zinc corrosion in relation to SO_2 concentrations and "time of wetness."[18] Time of wetness was measured by means of a dew detector. SO_2 was measured by lead peroxide sulfation candles and conductiometric SO_2 measurements.

Haynie and Upham carried out an extensive zinc corrosion study in eight cities where zinc panels were exposed, while collecting data on SO_2, temperature, and humidity.[19] They developed the following empirical damage function relating zinc corrosion to SO_2 levels and relative humidity.

$$y = 0.001028(48.8/RH)SO_2$$

where

$$y = \text{corrosion rate, } \mu m/yr$$
$$RH = \text{average annual relative humidity}$$
$$SO_2 = \text{average } SO_2 \text{ concentration, } \mu g/m^3$$

The damage coefficients for these two functions and functions developed from four other studies were compiled by EPA[3] (see Table 2). Note that the SO_2 coefficients vary over a factor of over 4, and that the chamber study (Haynie[8]) has by far the lowest SO_2 coefficient. The chamber study isolated the specific influence of SO_2. The studies carried out in the ambient atmosphere include the joint effect of SO_2 and particulate matter containing sulfates, chlorides, nitrates, and other anions. A difference in the SO_2 coefficient (and its direction) between the chamber study and those for outdoor exposures thus appears reasonable.

2.3 Paint

Paint consists of pigment and vehicle. Pigments, such as titanium dioxide and zinc oxides, serve to conceal and increase the durability of a material. The vehicle provides the film-forming properties of

Table 2 Experimental Regression Coefficients with Estimated Standard Deviations from Small Zinc and Galvanized Steel Specimens Obtained from Six Exposure Studies

Study	Time-of-Wetness Coefficient, $\mu m/yr$	SO_2 Coefficient, $(\mu m/yr) / (\mu g/m^3)$	Number of Data Sets
CAMP (Haynie and Upham)[19]	1.15 ± 0.60	0.081 ± 0.005	37
ISP (Cavender)[20]	10.5 ± 0.96	0.073 ± 0.007	173
Guttman[18]	1.79	0.024	>400
Guttman and Sereda[13]	2.47 ± 0.86	0.027 ± 0.008	136
Chamber study (Haynie)[8]	1.53 ± 0.39	0.018 ± 0.002	96
St. Louis (Mansfeld)[14]	2.36 ± 0.13	0.022 ± 0.004	153

Table 3 Paint Erosion Rates and *t*-Test Probability Data for Controlled Environmental Laboratory Exposures

Type of Paint	Mean Erosion Rate (mil Loss $\times 10^{-5}$/hr with 95% Confidence Limits) for Unshaded Panels and Probability that Differences Exist		
	Clean Air Control	SO_2 (1.0 ppm)	O_3 (1.0 ppm)
House paint			
Oil	20.1 ± 7.1	141.0 ± 19.0 99%	44.7 ± 10.5 99%
Latex	3.5 ± 1.5	11.1 ± 1.0 99%	8.5 ± 5.9 93%
Coil coating	11.9 ± 2.3	34.1 ± 4.7 99%	14.9 ± 2.5 94%
Automotive refinish	1.8 ± 0.8	3.1 ± 2.6 75%	5.1 ± 1.3 99%
Industrial maintenance	18.6 ± 5.1	22.4 ± 7.0 66%	28.1 ± 14.0 85%

the paint and contains resin binders, solvents, and additives. The most important potential effects of SO_2 on paints are the interference with the drying process and acceleration of the normal erosion process.

Holbrow has reported on a number of experiments to determine the effects of relatively high concentrations of SO_2 (1–2 ppm) on newly applied paints.[21] The drying time for oil-based paints increases and dried paint films have softer or more brittle finishes. The bleaching of Brunswick green paints was also noted. No experiments have been reported on the effect of SO_2 on the drying time of water-based latex paints.

Campbell carried out an extensive study of paint erosion for various paint types and exposure conditions (including SO_2 and O_3).[22] Both chamber and field experiments were conducted. The researchers evaluated four important types of paint: (1) house paints (acrylic latex and oil-based); (2) a urea–alkyd coil coating; (3) a nitrocellulose–acrylic automotive refinishing paint; and (4) an alkyd industrial maintenance coating. In the chamber studies paints were exposed to five separate controlled environmental conditions: (1) "clean" air (unpolluted); (2) SO_2 at 262 μg/m^3 (0.1 ppm); (3) sulfur dioxide at 2620 μg/m^3 (1 ppm); (4) ozone at 196 μg/m^3 (0.1 ppm); and (5) ozone at 1960 μg/m^3 (1 ppm). The exposure chamber operated on a 2-hr, dew/light cycle. Erosion measurements were made after exposure periods of 400, 700, and 1000 hr, and rates were calculated.

Table 3 presents erosion rate data.[22] Exposures to 0.1 ppm pollutants did not produce significant erosion rate increases over clean air exposures.

Field exposures were conducted at four locations with different environments: (1) rural—clean air, (2) suburban, (3) urban—sulfur-dioxide-dominant (annual mean level 60 μg/m^3), and (4) urban—oxidant-dominant (annual mean ozone level, 40 μg/m^3). Panels were exposed facing north and south, and were evaluated after 3-, 7-, and 14-month exposures. Table 4 shows the erosion rate results for the various coatings exposed facing south.[22] In most cases, southern exposures produced somewhat larger erosion rates. It is noteworthy that the oil-based house paint and coil coating experienced the largest erosion rate in both the field and laboratory SO_2 exposures. These coatings were the only ones that contained a calcium carbonate extender, a substance that is sensitive to attack by acids.

Spence summarized the results of paint exposure to several gaseous pollutants for the full-scale chamber studies reported by Haynie.[23] Four classes of painted surfaces were evaluated: oil-based house paint, vinyl–acrylic latex house paint, vinyl coil coating, and acrylic coil coating. A strong correlation was found between paint erosion for the oil-based house paint and SO_2 and humidity. Vinyl and acrylic coil coating were unaffected, but blistering was noted on the latex house paint. It was not certain if the blistering was the result primarily of SO_2 or moisture.

A multiple linear regression relationship was developed for the joint influence of SO_2 and relative humidity on the oil-based house paint:

$$E = 14.3 + 0.0151SO_2 + 0.388RH$$

where

$$E = \text{erosion rate, } \mu\text{m/yr}$$
$$SO_2 = \text{concentration of } SO_2, \ \mu\text{g/m}^3$$
$$RH = \text{relative humidity, } \%$$

Table 4   Paint Erosion Rates and t-Test Probability Data for Field Exposures

	Mean Erosion Rate (mil Loss $\times 10^{-5}$/hr with 95% Confidence Limits) for Unshaded Panels and Probability that Differences Exist			
Type of Paint	Rural (Clean Air)	Suburban	Urban (SO$_2$-Dominant)	Urban (Oxidant-Dominant)
House paint				
Oil	4.3 ± 7.5	14.8 ± 2.6 99.3%	14.2 ± 4.9 98.1%	21.0 ± 6.2 99.2%
Latex	1.8 ± 0.5	3.0 ± 0.7 99.2%	3.8 ± 0.3 97.8%	6.5 ± 5.6 94.3%
Coil coating	2.1 ± 0.8	10.0 ± 1.9 99.9%	9.5 ± 0.8 99.9%	8.8 ± 1.7 99.9%
Automotive refinish	0.9 ± 1.1	2.3 ± 0.7 97.6%	1.6 ± 0.4 86.2%	1.7 ± 0.4 91.6%
Industrial maintenance	3.6 ± 1.6	8.2 ± 4.2 97.3%	6.6 ± 3.9 91.2%	7.8 ± 2.4 99.7%

The relationship indicates that paint erosion is significantly more sensitive to changes in humidity than SO$_2$.

2.4   Textiles

Sulfur oxides are capable of weakening natural and some synthetic fibers. Cotton and its relative, viscose rayon, are cellulose fibers and can be weakened by SO$_2$ by the breakage of the cellulose chain at the glucoside linkage. Polyamides such as Nylons 6 and 66 can be weakened by SO$_2$ and acid sulfate aerosols, especially when fibers are small in diameter and under tension. Polyester, acrylic, and polypropylene fibers are resistant to SO$_2$ damage, but some may be susceptible to damage from acidic particles containing sulfates.

Brysson exposed cotton fabrics to the outdoor atmosphere for 1 yr at several sites in St. Louis, Missouri and Chicago.[24] Sulfur oxide concentrations were estimated by means of lead peroxide sulfation plates. The data showed that a loss in breaking strength correlated well with relative sulfur oxide concentration.

Zeronian carried out laboratory exposures in which he exposed cotton and rayon fabrics for 7 days to clean air with and without 250 μg/m^3 (0.1 ppm) SO$_2$.[25] Both controlled environments included a continuous exposure to artificial light (xenon arc) and a water spray turned on for 18 min every 2 hr. A loss in strength for all fabrics exposed to clean air averaged 13%, while the fabrics exposed to air containing SO$_2$ averaged 21%. Zeronian also exposed fabrics made from artificial fibers (nylon, polyester, and modacrylic) to controlled environmental conditions similar to the cotton exposures, except that the SO$_2$ level was 486 μg/m^3 (0.2 ppm).[26] Only the nylon fabrics were affected, losing 80% of their strength when exposed to SO$_2$, and only 40% when exposed to clean air conditions.

2.5   Building Materials and Works of Art

Sulfur oxides are involved in the deterioration of building materials, which may ultimately result in a loss in structural integrity, but more important, in the disfiguration or destruction of structures of great artistic and historical value. In the United States such damage is not as crucial as it is in Europe and Asia where buildings date from antiquity. Americans tend to be utilitarian and demolish rather than save buildings. Recently, however, more old buildings are being retained to avoid high construction costs and to preserve our architectural and historical heritage.

Air pollution damage to the building and monuments on the Acropolis is one of the most serious and widely publicized air pollution damage situations in the world and has been the subject of several international conferences.[27,28,29]

The Coliseum and Arch of Titus in Rome, and the San Marco Basilica in Venice show accelerated decay from air pollution. The situation in Florence, Italy, has been described as disastrous. In France, conservation specialists have replaced exterior statues with copies. A team of experts has been fighting the decay and corrosion that is destroying the massive, twin-spired Cologne Cathedral, the most magnificent church building of the German High Gothic era (c. 1200 A.D.). Polluted atmospheres threaten century-old shrines and temples in the industrial areas of Japan. Cleopatra's Needle, the large, stone obelisk moved from Alexandria to London, has suffered more deterioration in London's air in 85 yr than in 3000 yr in Egypt. An ancient Egyptian obelisk that stands in Central Park in New York City has been similarly affected.

The protection and repair of England's buildings and cathedrals has been discussed by Schaffer and Feilden.[30,31] Besides air pollution, vibrations from traffic cause severe damage, accelerated by pollutants and weather conditions such as freezing and thawing.

In Germany, Luckat has investigated the damage to various limestone types and has tested systems that protect exposed stone.[32] He found a silicic acid ester together with a water repellent more effective than several other systems tested.

Longinelli and Bartelloni have done research on the damage to building stone in Venice.[33] They used isotopic analyses of oxygen, sulfur, and carbon in rainwater and stone to show that the damage to buildings, which has been most severe since 1900, is caused primarily by sulfur-bearing air pollutants. Despite Venice's proximity to the sea, sea water spray and aerosols play only a minor role in deterioration.

Recently it was found that sulfur-consuming bacteria may be directly or indirectly responsible for a significant amount of stone damage (Riederer).[34] Hansen points out that the microbe, *Thiobacillus thioparus*, one of several sulfur-converting organisms, converts atmospheric SO_2 to sulfuric acid which it uses as a digestive fluid in attacking carbonate stone.[35] The CO_2 gas produced is the food of the organism. A method proposed to stop this damage is to treat the surface with an antibiotic or bactericide.

Glass or ceramic surfaces are thought to be impervious to damage from all air pollutants except perhaps fluoride. However, the glass for certain Medieval stained glass windows was made from potash obtained from wood ash[36] since these artisans were isolated from sources of soda ash. The potassium salts in such glass are more soluble than sodium salts, and the glass surface simply leaches away.

2.6   Other Effects of Sulfur Oxides

While sulfur oxides can damage a number of other materials or material systems, the effects can generally be considered less significant than those discussed here. Damage to leather, paper, or electrical contacts is physically present, but does not usually result in a significant economic impact because the pollution-induced damage is minor in relation to other factors.

3   PARTICULATE MATTER

The primary effect of particulate matter on materials is the soiling of surfaces. However, particulate matter under proper conditions can also increase permanent surface damage insofar as it serves as a center for the condensation and adsorption of gaseous pollutants. While soiling itself does not necessarily create damage, removal of the particulate matter by cleaning can increase wear and shorten the life of the article.

It is extremely difficult to isolate the specific influence of particulate matter from the damage caused by other pollutants and environmental factors since:

1.   High levels of particulate matter and other pollutants such as sulfur oxides tend to coexist.

2.   Particles can interact synergistically with other pollutants to increase their damage capability but the damage caused may be indistinguishable from that caused by other pollutants acting alone.

3. SO_2 converts to sulfates in the atmosphere and on surfaces. Such sulfates are in particulate form, but are related more to atmospheric emissions of SO_2 than to particulate matter.

3.1   Corrosion of Metals

Some of the experiments of Vernon show that moist air polluted with SO_2 and particles of charcoal produce corrosion much more rapidly than air containing SO_2 and moisture alone.[4] Vernon reasoned that the effect of the particles was primarily physical in that they increased the concentration of SO_2 by adsorption and created "hot spots" of SO_2 concentration. Sanyl and Singhania stated that particulate matter had a profound effect on corrosion rates.[37] They believed that the influence of particulate matter on corrosion was related to its electrolytic, hygroscopic and/or acidic properties, and its ability to absorb corrosive pollutant gases.

While these laboratory studies show the strong influence of particulate matter on corrosion, field studies have not confirmed the effect.

3.2   Building Materials

The primary effect of particulate matter on building materials is soiling. The ultimate impact is aesthetic rather than a loss in physical strength or integrity. Dark particulate matter, largely from combustion, has blackened buildings in many urban areas of the world. Rainwater tends to remove such deposits in areas exposed to weather, giving buildings a streaked or mottled appearance, depending on the surface characteristics of the building. In spite of qualitative observations, the physical effects of particulate matter on building materials are not well understood.

3.3 Paints

As with building materials, the primary effect of particulate matter on paint is soiling. Soluble salts such as the iron sulfate contained in deposited particles can also produce staining.

Large, chemically active particles such as acid smut from oil-fired boilers, mortar dust from building demolition, or iron particles from grinding operations can severely damage automotive paint. The effects range from discoloration of the paint film to the ultimate penetration of the paint film and corrosion of the underlying metal near individual particles. Large particles that become imbedded in a freshly painted surface can act as wicks to transfer moisture and corrosive pollutants such as SO_2 to the underlying material.

3.4 Fabrics

Large particles imbedded in fabrics are capable of causing physical damage to fabrics that receive considerable flexing. Soiling from particulate matter does not normally cause direct damage to fabrics, but increased cleaning reduces the service life of a fabric. However, stress from the natural environment, such as sunlight and water vapor, and gaseous pollutants such as NO_2, ozone, and SO_2, are much more important than the effects of particulate matter. Furthermore, most articles made of fabric are used indoors where levels of particulate matter and other pollutants are lower than outdoors. Therefore the extent of physical damage to fabrics from outdoor levels of particulate matter is difficult to assess. Furthermore, new, more resistant fabrics and fabric treatments or coating processes have minimized the deleterious effects of particulate matter on fabrics used outdoors.

4 NITROGEN OXIDES

Nitrogen oxides (principally NO_2) are capable of damaging several types of materials. The most significant effect is on certain types of fabric dyes. NO_2 is also involved in the weakening of some fabrics, causing the deterioration of certain types of plastic materials, and in the corrosion of metals.

The ability of nitrogen oxides to be converted to nitrates in the atmosphere is presumed to be part of the acidic deposition or acid rain problem. While elevated levels of nitrates are often found in acidic rainwater, the relative importance of anthropogenic nitrogen oxides in any material damage caused by acidic rain (e.g., metal corrosion or stone damage) is not known but is believed to be small in comparison with the effects related to sulfur oxides.

4.1 Dyed Fabrics

The fading of textile dyes by air pollutants, primarily nitrogen dioxide and ozone, has been a particularly vexing problem for the textile industry for most of the twentieth century. Just prior to World War I, a German dye manufacturer investigated the unusual fading of stored woolen goods traceable to nitrogen oxides in the air from open electric-arc lamps and incandescent gas mantles.[38]

In the mid-1920s a newly developed synthetic fiber, cellulose acetate rayon, required the development of disperse dyes. Many of these derive from anthraquinone and therefore contain amino groups. Fading began to show on blue, green, and violet shades of dyed acetate goods. This mysterious fading was called "gas fading" because it was frequently observed in rooms heated by unvented gas heaters.

In the 1930s acetate fading became a serious problem. Dye and fiber chemists (apparently unaware of the earlier German work) expended considerable effort to find a solution. In 1937 Rowe and Chamberlain systematically investigated the fundamental chemistry of dye degradation; independently they reached the same conclusions as the German team.[39]

Upham and Salvin summarized much of the information on this subject.[40] They showed that the ability of the fiber to absorb NO_2 plays an important role in the dye-fading mechanism. Salvin showed that cellulose acetate is an excellent absorber of NO_2.[41] Polyester and polyacrylic fibers have low absorption rates for NO_2. Nylon, cotton, viscose rayon, and wool have intermediate rates.

The American Association of Textile Chemists and Colorists (AATCC) exposed a wide range of fibers and dyes to the atmosphere of four U.S. cities representing different air quality and meteorological regimes.[42] Exposures were made in cabinets to isolate the samples from sunlight. Air quality was monitored for SO_2, NO_x, and O_3. Fading was demonstrated on a range of fabrics including cotton and rayon, and several types of dyes (vat, sulfur, and fiber-reactive) applicable to cellulosics exhibited color change. However, the results did not show the degree of responsibility of each pollutant in the effects noted.

In an expansion of the AATCC field program, Beloin exposed 67 dye–fabric combinations, using 56 dyes, at 11 urban and rural sites in the United States.[43] Fabric samples were exposed in covered, louvered cabinets to exclude light, and ambient levels of SO_2, NO_2, and O_3 were monitored. Urban sites produced more fading than rural sites. Fading also increased with both temperature and humidity. The data did not show the effects of specific pollutants, but a statistical analysis identified NO_2 concentrations as a significant variable.

Beloin performed a laboratory study in which 20 dyed fabrics were exposed to individual pollutants (SO_2, NO, NO_2, and O_3) at two levels of concentration, relative humidity, and temperature.[44] Severe fading occurred with high humidity and temperature (90% RH and 32°C) and a high NO_2 level (940 μg/m³) in eight of nine samples of cotton or viscose. Significant fading also occurred with high humidity and a moderate NO_2 level (94 μg/m³). These laboratory results were confirmed in chamber tests conducted by Upham.[45]

Dyed nylon fabrics are also susceptible to fading in polluted atmospheres but are not as sensitive in this respect as cotton, rayon, and cellulose acetate. Dyed polyester fabrics are more resistant to NO_2 damage than nylon. Field exposure of a range of dyes on nylon to urban atmospheres containing NO_2, SO_2, and O_3 produced significant fading. Dyed polyester fabrics were unaffected.

The fading of permanent press and double-knit polyester in chamber tests does not occur in the fiber matrix. Dye that migrates into the surface resins and surfactants applied to these fabrics during manufacture is affected by NO_2.

In addition to dye fading, NO_2 can be responsible for the yellowing of white fabrics and is often exacerbated by fabric whiteners, softeners, static compounds, and other surface treatment materials.

4.2 Loss of Fiber Strength

Nitrogen oxides (especially NO_2) just as sulfur oxides can weaken cellulose and nylon fibers by acid hydrolysis. However, the question must be asked as to how important this effect is in view of the predominant use of fabrics indoors and the trend to use more resistant fabrics outdoors.

Field and laboratory studies of the effects of nitrogen oxides on fiber strength are somewhat inconclusive. One chamber study by Morris exposed cotton samples in two chambers to sunlight, but the air for one chamber was filtered through activated carbon which presumably removed NO_2, O_3, SO_2, and other pollutants.[46] Since the study was carried out in Berkeley, California, it was reasoned that little SO_2 was present and that the principal gaseous pollutant in the atmosphere was NO_2. Loss of fiber strength was greatest in the presence of unfiltered air and was attributed to the presence of NO_2.

Zeronian exposed various fibers to xenon arc radiation and air containing 376 μg/m³ of NO_2 at several combinations of temperature and relative humidity.[26] While loss in strength occurred, the differential effect of NO_2 and radiation could not be isolated.

4.3 Effects on Plastics and Elastomers

Synthetic organic materials such as polyolefins, polyvinyl chloride, polyacrylonitrile, and polyamides are increasingly used in a variety of consumer and manufactured products. Plastic and elastomeric materials are formulated to result in maximum aging properties. The aging of these materials is usually dominated by the effect of heat and sunlight, the importance of which is intensified by the increased use of plastics in the interior and exterior of automobiles.

Polymers representing the structures in plastics, as well as textiles, were exposed by Jellinek to the action of SO_2, NO_2, and ozone.[47] The polymers tested included polyethylene, polypropylene, polystyrene, polyvinyl chloride, polyacrylonitrile, butyl rubber, and nylon. All polymers suffered deterioration in strength. Butyl rubber was more susceptible to SO_2 and NO_2 than other polymers. However, the effect of O_3 on rubber was more pronounced than that of SO_2 or NO_2.

Jellinek examined the reaction of linear polymers, including nylon and polypropylene, to NO_2 at concentrations of 1880 to 9400 μg/m³.[48] Nylon 66 suffered chain scission. Polypropylene tended to cross-link. Chain scission of polymers caused by small concentrations of SO_2 and NO_2 took place in the presence of air and ultraviolet radiation.

The action of NO_2 and O_3 on polyurethane was also investigated by Jellinek.[49] The tensile strength of linear polyurethane was reduced by NO_2 alone and also by NO_2 plus O_3. Chain scission resulting in lower molecular weights and formation of nitro and nitroso groups along the polymer backbone occurred upon exposure to NO_2.

4.4 Corrosion of Metals

Nitrogen oxides, as potentially acidic pollutants, are implicated in several types of corrosion: pitting, selective leaching, and stress corrosion. However, their role in the outdoor corrosion of metals is much less important than that of sulfur oxides. The U.S. EPA has summarized principal research results on the effects of nitrogen oxides and nitrates on metals.[50]

Probably the most significant form of nitrogen oxide damage to metals is the stress corrosion of nonferrous springs where deposits containing nitrate appear to be implicated. Hermance and his associates at the Bell Telephone Laboratories researched this problem which has caused failures in telephone-switching equipment.[51,52,53] This group showed that nitrate salts are more hygroscopic than chlorides or sulfate salts. Thus nitrate deposits may lower threshold humidity requirements for the production of electrolytic corrosion effects.

5   OZONE

Ozone is an extremely reactive gas with the ability to react with many organic materials, both natural and artificial. Materials especially susceptible to ozone attack are elastomers (rubber), textile fibers and dyes, and some types of paints. While there are other strong oxidizing materials in polluted atmospheres, ozone is the most abundant of such active oxidants. Depending on the structure of susceptible organic polymers, the effect of ozone is through chain-scissioning or cross-linking, or both. The first of these effects produces a reduction in molecular weight and a decrease in tensile strength, while the second tends to increase the rigidity of some polymers, making them more brittle and less resilient. In the air quality criteria document for ozone, EPA has summarized the principal effects of ozone on materials.[54]

5.1   Elastomers

The aging or cracking of natural rubber has long been associated with exposure to the atmosphere, and was originally thought to be solely the result of sunlight. Williams[55] and later, van Rossem[56] showed that ozone rather than sunlight was the major cause of rubber cracking from atmospheric exposure. In the 1940s and 1950s Norton, Newton, and Crabtree recognized that in addition to ozone, other reactive species in atmospheres polluted with photochemical smog could produce rubber cracking.[57,58,59,60] In work performed in the early days of the Los Angeles smog problem, Bradley and Haagen-Smit developed a method for measuring atmospheric ozone concentrations using stressed rubber samples and measuring crack depth at the end of specified exposure periods.[61] For several years this method was considered the most reliable and specific method for atmospheric ozone measurement.

The cracking of natural rubber is initiated by ozone attack at the double bond. Therefore, synthetic elastomers such as styrene–butadiene, polybutadiene, and polyisoprene are also susceptible to ozone damage. Elastomers with saturated chemical structures or which contain halogen atoms tend to be ozone resistant.

Mueller and Stickney reviewed the physical and economic effects of air pollution on elastomers, and the development of rubber additives to combat ozone attack.[62] These are generally aromatic amines and phenols.

Considerable research has been conducted to develop dose-response relationships for the exposure of rubber to ozone, but most of the work has been done at elevated ozone levels with susceptible rubber formulations (without antiozonants). One exception is the work of Edwards and Storey, who determined the effects of ozone at approximately 490 μg/m^3 on two styrene–butadiene rubber formulations at several levels of antiozonant.[63] In the course of this work a damage function was developed relating the dose needed to produce visible cracks to the level of antiozonant in the rubber. In the chamber studies of Haynie, rubber samples of the type used in the white sidewalls of top quality steel-belted tires were exposed to ozone at 160 and 1000 μg/m^3 and two levels of strain (10 and 20%) for 1000 hr.[8] Table 5 shows clearly that ozone concentrations, and to some degree strain, influence the cracking rate.

Table 5   Cracking Rates of White Sidewall Tire Specimens

Ozone Concentration, μg/m^3 (ppm)	Strain, %	Mean Cracking Rate ±S.D.,[a] μm/yr
160 (0.08)	10	10.36% ± 7.76
	20	11.70% ± 7.22
1000 (0.5)	10	19.80% ± 9.64
	20	24.09% ± 6.24

[a] S.D. = Estimated standard deviation of the mean.

5.2   Fabrics

While the effects of nitrogen oxides on fabric dyes were first noted before World War I, it was not until 1955 that Salvin and Walker saw similar effects from atmospheric ozone on specially developed blue disperse dyes designed to resist the effects of NO$_2$.[64] However, the ozone fading of dyes was manifest in a bleached or washed-out appearance rather than in the reddening which is characteristic of the effects of NO$_2$ on the most sensitive dyes. The term "0-fading" described this newly discovered phenomenon.

Salvin noted another type of fading on polyester–cotton/permanent-press fabrics stored in warehouses in California, Texas, and Tennessee. The combination of fibers and dyes was thought to be resistant to the effects of air pollution. It was found that the fading had occurred in dye which had migrated into the permanent-press materials and not in the fiber itself; Salvin also showed that humidity is an important factor in the ozone fading of textile fibers.

Beloin carried out both field and laboratory exposures of several fabric-dye combinations.[43,44] Although field studies did not help to distinguish the relative effects of NO_2, O_3, or SO_2, the controlled environment laboratory studies at 100 and 1000 $\mu g/m^3$ ozone did show fading. The higher concentration produced the greater fading. This study also confirmed that humidity, and to a lesser extent temperature, are factors in the ozone fading of dyes. The chamber studies of Haynie (including the exposure of drapery materials) also confirm these results.

Cellulose fibers are susceptible to oxidation. Therefore one would expect that cotton and other cellulose-based fibers would be vulnerable to ozone attack. Bogarty exposed samples of cotton duck and print cloth to air containing ozone in the range between 40 and 120 $\mu g/m^3$ in the absence of light.[65] Samples were exposed both wet and dry. Damage was measured in terms of fluidity (reciprocal of viscosity) of dissolved samples of the cloth. An increase in fluidity indicates depolymerization from chemical attack (e.g., O_3, SO_2, or NO_2). Biological attack, while affecting fiber strength, does not change the average size of the fiber molecules and fluidity.

After 50 days exposure, the cloth samples exposed to both ozone and high moisture levels showed increased fluidity. In addition, these samples exhibited a 20% reduction in breaking strength. The samples exposed dry to ozone showed no significant change. It was believed that ozone's relatively high solubility in water promotes the action of ozone on cotton. In spite of these results, ambient concentrations of ozone were considered a minor influence when compared with the effects of sunlight, heat, alternate wetting and drying, and biological attack. The work of Morris conducted at concentrations 10 times higher than those of Bogarty, but at ambient humidity levels, showed no appreciable changes in fluidity or breaking strengths.[46]

Kerr studied the joint effect of ozone and simulated wash/dry cycles on dyed print cloth over a 60-day period.[66] Ozone concentrations averaged 1500 $\mu g/m^3$. Breaking strength decreased 18% for samples exposed to both ozone and the wash/dry cycles, compared with 9% for samples exposed only to the wash/dry cycles.

The work of Zeronian showed that ozone at 400 $\mu g/m^3$ over a 7-day exposure period had no effect on modacrylic (Dynel) and polyester (Dacron) fibers, and a slight effect on acrylic (Orlon) and Nylon 66.[26]

5.3 Paints

The primary deleterious effect of ozone on paint is the accelerated erosion of the paint film. Weight loss over a specified exposure period is the usual method of measuring this effect on test panels.

As pointed out earlier in the discussion of paint and SO_2, there are two research efforts that represent the bulk of knowledge on the effects of ozone on paint. Campbell, working at the Sherwin-Williams Paint Company, conducted both laboratory and field studies on five different types of paints: oil and latex house paints, urea–alkyd coil coating, nitrocellulose–acrylic automotive paint, and alkyd industrial maintenance coating.[22] Spence conducted laboratory chamber studies on four classes of painted surfaces: oil-based house paint; vinyl–acrylic latex house paint; vinyl coil coating; and acrylic coil coating.[23]

Campbell's study showed that exposures of both shaded and unshaded samples, at 2000 $\mu g/m^3$, produced measurable erosion effects. Oil-based house paint showed the greatest erosion rate; the industrial maintenance coating showed a moderate erosion rate; and the latex, coil, and automotive coatings showed the least effect. The unshaded samples showed the greatest erosion rates. Field exposures at several U.S. locations did not produce results as obvious as the laboratory tests produced. However, the higher erosion rates of paints exposed in Los Angeles when compared with other locations, were attributed to higher ozone levels.

The Spence study was designed to show the separate and joint influence of several gaseous pollutants (SO_2, O_3, and NO_2) and other environmental factors. Statistically significant effects of ozone were measured for the vinyl and acrylic coil coating. The oil-based house paint contained a calcium carbonate filler which was strongly affected by moisture and SO_2. This effect masked any possible effect by ozone on this paint.

A damage function for acrylic coil coating at 90% relative humidity was developed as follows: erosion rate = $0.159 + 0.000714 O_3$ where erosion rate is in $\mu m/yr$ and O_3 is in $\mu g/m^3$.

While the effect of ozone on this type of coating is statistically significant, at an average O_3 level of 100 $\mu g/m^3$, this function predicts that a 20-μm-thick paint film would last over 80 yr.

6 OTHER POLLUTANT EFFECTS

The foregoing sections describe the effects of those criteria pollutants which are capable of producing effects on materials. Volatile organic materials and carbon monoxide are believed to have no significant effects on materials. However, several other gaseous pollutants that have from time to time been emitted from industrial and natural sources are briefly noted.

6.1 Hydrogen Sulfide

Hydrogen sulfide may be emitted from a number of anthropogenic sources: polluted salt water, coke-making, oil refineries, pulp and paper manufacture, chemical manufacture, and other sources. Natural sources include decomposing vegetable and animal matter, volcanoes, and geothermal operations.

Yocom and Upham[2] have summarized some of the effects on materials which include the tarnishing of nonferrous metals, discoloration of lead-based paint, and damage to electrical contacts. While such problems have been noted in the past in many urban areas, in recent years the control of H_2S sources and changes in exposed materials (e.g., the use of titanium oxide rather than lead oxide in paints) have significantly reduced their severity.

6.2 Fluorides

Hydrogen fluoride and other fluorine-containing materials are emitted from a number of industrial processes including glass or frit manufacture, fertilizer production, and aluminum manufacture. Under extremely high concentration levels, HF can etch glass and other high-silica materials.[67] One would also expect other types of material damage, especially metal corrosion, at these high levels. However, efforts to control fluoride emissions because of the extreme sensitivity of vegetation to fluoride damage have essentially eliminated any problem of material damage from ambient concentrations.

6.3 Ammonia

Ammonia occurs in the atmosphere from natural sources, especially anaerobic decomposition of protein-aceous material and from a number of industrial processes, for example, by-product coke manufacture and a variety of chemical processes. However, its extreme solubility in water and ability to react with acidic components of the atmosphere assure that it normally exists in low atmospheric concentrations except near strong sources.

At high concentrations it is able to corrode a number of nonferrous metals such as copper, tin, zinc, and their alloys. Ferrous alloys are generally resistant, but stress corrosion can be produced in carbon steels.

Ammonia, again at high concentrations, can soften wood through interaction with cellulose fibers and can soften natural rubber. Some plastics also can be adversely affected by ammonia, notably epoxy fiberglass, nylons, and polyvinyl chloride under certain conditions of temperature and concentration.

In general, ammonia is not a significant air pollutant with respect to material damage.

7 ECONOMIC ESTIMATES

The change in the cost to society of repair or replacement of damaged materials, or the cost of substitution or protection of sensitive materials at different levels of air pollution, can be quantified. The quantitative expression for the relationship between pollution and its effects, expressed in monitoring terms, is called an economic damage function. The aesthetic costs of materials damage are much more difficult to quantify, but are at least partially reflected in the costs paid to avoid them.

A number of economic damage function studies have been carried out. In general, the results are highly variable and uncertain. This uncertainty is due largely to the unavailability of key information regarding such factors as the quantities of materials exposed to pollutants, the important mechanisms of damage, and the actual practices followed in responding to damage at critical damage levels.

So many materials in use are potentially exposed to a variety of pollutants and natural damage influences that most studies have attempted to limit their scope to those materials for which pollution-induced damage has serious economic implications. That is, studies are limited to those materials for which pollution-induced damage significantly affects the useful life of the system of which the material is a part. Studies of paint, structural metals, electrical components, fabrics, elastomers, nonmetallic building materials, and works of art and historical monuments have been reported.

7.1 Paint

Spence and Haynie conducted a survey and economic assessment of the deterioration of exterior paints ("trade" paints) caused by air pollution.[68] The total annual economic damage to exterior household paints was estimated at $540 million (1972), including paint loss, and a labor factor three times the cost of the paint.

Salmon estimated that the annual cost of soiling of household paint would be $35 billion if surfaces were maintained on a schedule which would keep them as clean as they would be in a clean environment.[69] The annual direct cost of deterioration damage to paints was estimated to be $1.2 billion.

A 1974 study by Midwest Research Institute recalculated the annual cost of damage to be $22 billion for soiling and $753 million for deterioration. These figures appear to be based on a hypothetical definition of cleanliness similar to Salmon's. However, Census Bureau information for this period indicates a total value of $2.5 billion for the annual production of household paints—a value far less than the indicated level of damage, even with a labor factor of 6 to 8.

Michelson and Tourin investigated the frequency of house repainting as a function of suspended particulate matter concentration using a comparative approach.[70] Questionnaires were sent to the residents of three suburbs of Washington, D.C. (Suitland, Rockville, and Fairfax) and two cities in the upper Ohio Valley (Steubenville and Uniontown) regarding the maintenance intervals for exterior repainting in each of the five communities.

In Fairfax, where the mean annual particulate matter concentration was 60 μg/m^3, repainting was reported to occur every 4 yr. In Steubenville, where the mean annual particulate matter concentration was 235 μg/m^3, the estimated repainting frequency was greater than once a year. A linear function relating particulate matter concentration and painting frequency was derived. However, the principal result of the investigation is the suggestion that although a significant relationship may exist between the frequency of repainting and particulate matter concentration, it is difficult to accurately quantify it by survey techniques.

Booz, Allen, and Hamilton studied painting maintenance frequencies in several zones of the Philadelphia metropolitan area having various population characteristics, climates, and industry.[71] Socioeconomic factors were delineated by pollution zone; however, paint types were not reported. No statistically significant correlation between painting frequency and particulate matter level was found. However, it was found that there is a socioeconomic correlation between the pollution level and the number of households with an annual income less than $6000.

7.2 Structural Metals

A recent study published by the U.S. Department of Commerce estimates that the cost of metallic corrosion in the United States was $70 billion in 1975.[72] However, the cost of corrosion specifically associated with ambient air pollution was not separated from other types of corrosion. Uncertainty in the total corrosion cost figure was estimated at ± 30%. Analysis suggests that the avoidable cost of metallic corrosion represents about 15% of the total, but could range from 10 to 45%.

Fink estimates that corrosion caused by air pollution to external metal structures costs $1.45 billion annually.[73]

7.3 Electrical Components

Robbins estimated that 15% of the gold and platinum used in the United States for electrical contacts in 1970 was for the specific purpose of combating SO_2 corrosion, with the remainder for protection against other environmental pollutants.[74]

In areas where electrical instrumentation and computers are used, and air is dehumidified and purified to help protect against corrosion, as well as to provide proper operating conditions, Robbins suggested that the use of activated carbon filters and high-efficiency fine particle filters represents a cost attributable to SO_2 and particulate contaminants.

7.4 Fabrics

The chief problem with prior economic estimates of the damage to fabrics is that a large proportion of fading and physical wear were often attributed arbitrarily to pollution rather than to other environmental factors. Surveys of use and impact have found that for most textiles the lifetime of the material is determined primarily by the amount of exposure to wear, sunlight, humidity levels, and fashion changes rather than to air pollution. Most indoor and outdoor textile soiling results from factors other than outdoor levels of particulate matter.

However, according to Brysson high pollutant levels (mean sulfation 5 mg SO_3/100 cm^2/day, and/or SO_2 concentrations of 0.2 ppm or 520 μg/m^3) can reduce the effective life of fabrics by 84% when compared to low pollution sites (0.5 mg SO_3/100 cm^2/day and/or 0.02 ppm or 60 μg/m^3 SO_2 concentrations).[24]

Of the 1257 trillion lb of fiber used for industrial purposes in 1965, the textile industry reported that 583 million lb were cotton and that 300 million lb had the outdoor uses shown in Table 6.

In a report of a telephone survey of consumer awareness of damage to textiles due to air pollution, Upham and Salvin noted that Philadelphia respondents did not perceive the soiling of fabrics as a damage effect.[75]

7.5 Plastics and Elastomers

No studies exist that quantitatively show any cause-and-effect relationship between ambient pollution and the damage to many of the major plastic formulations on the market today. Thus, no estimate

Table 6   Amounts of Cotton Fiber Used for Various
Outdoor Purposes

Use	Amount, 10^6 lb
Automotive upholstery and seat covers	56
Fire hose	20
Cordage	56
Tarpaulins, tents, awnings, etc.	70
Bags and bagging	63
Miscellaneous (agricultural cloth, flags)	35
Total	300

of the economic impact of pollution damage on plastics is possible. Studies based on the assumed pollutant damage are purely speculative and have not been considered here. Damage to elastomers, such as rubber, have a firmer basis but are still relatively speculative.

7.6   Nonmetallic Building Materials

With the possible exception of marble, nonmetallic building materials do not have well-defined physical damage functions for the irreversible damage related to ambient air pollution. All may suffer from soiling and discoloration, and the costs incurred for cleaning can be significant. However, excepting glass, cleaning is done infrequently and is generally not part of a routine maintenance program. Cleaning costs are highly variable and depend on aesthetics. Since cleaning a building usually represents a major undertaking, an interesting inversion of normal practice is sometimes observed. As pollution from particulate matter decreases, expenditures for cleaning increase. At high pollution levels, the appearance degrades so quickly that the cost of cleaning is not justified. As pollution levels decline, the foregone maintenance is perceived as worthwhile and is implemented.

The fact that some materials (such as light-colored or carbonate-rich stone) are not considered suitable for building is a potential aesthetic loss.

The inconvenience of soiling, especially on windows where cleanliness is especially desired, may be considered an aesthetic cost. However, in most cases the costs incurred for cleaning are sufficient to lower uncompensated aesthetic costs.

7.7   Works of Art and Historical Monuments

The type of damage suffered by an object is a function of the material of which the object is made and the way in which the importance of the object is affected by damage.

Where replication is possible and benefits warrant reconstruction or rehabilitation, the costs of rebuilding provide an estimate of the damages caused by pollutants. Where structures cannot be duplicated exactly, because of a lack of skill or materials, rebuilding costs may far exceed actual damage estimates. However, where tastes have changed from intricate baroque designs in marble toward steel or glass or other materials, rebuilding costs may greatly exceed the aesthetic value.

Finally, there are cases where a large portion of the value of a particular object is its originality. A replica of an object of historical significance is simply not the same in value as the original. If the object is lost due to damage, the cost is virtually impossible to quantify in monitoring terms.

Recent articles by Yocom[29] on damage to buildings on the Acropolis and other irreplaceable works of art address these issues in qualitative terms, but no quantitative work has been done to date.

In closing, it appears that much more development of information and techniques is necessary before reasonably quantitative estimates of the economic damage due to air pollution may be made.

REFERENCES

1. Evelyn, J., *Fumifugium: Or the Smoake of London Dissipated 1661,* National Smoke Abatement Society Reprint, 1933.
2. Yocom, J. E., and Upham, J. B., Effects on Economic Materials and Structures, *Air Pollution,* Vol. 2, 3rd edition (A. C. Stern, ed.), Academic Press, New York, 1977, pp. 65–116.
3. "Air Quality Criteria for Particulate Matter and Sulfur Oxides," Vol. 3, *Welfare Effects, External Review Draft No. 1,* U.S. Environmental Protection Agency, Research Triangle Park, NC, April 1980.
4. Vernon, W. H. J., *Trans. Faraday Soc.* **31,** 1668 (1935).
5. Nriagu, J. O., *Sulfur in the Environment, Pt. II: Ecological Impacts,* Wiley, Toronto, 1978.

6. Johnson, J. B., Elliott, P., Winterbottom, M. A., and Wood, G. C., Short-Term Atmospheric Corrosion of Mild Steel at Two Weather and Pollution Monitored Sites, pp. 691–700, *Corrosion Science*, Vol. 17, Pergamon Press, Elmsford, N.Y., 1977.

7. Sydberger, T., "Influence of Sulphur Pollution on the Atmospheric Corrosion of Steel," University of Gothenburg, Department of Inorganic Chemistry, Sweden, 1976.

8. Haynie, F. H., Spence, J. W., and Upham, J. B., "Effects of Gaseous Pollutants on Materials— A Chamber Study," EPA-600/3-76-015, U.S. Environmental Protection Agency, Research Triangle Park, NC, February 1976.

9. Upham, J. B., Atmospheric Corrosion Studies in Two Metropolitan Areas, *J. Air Poll. Control Assoc.* **17**, 398–402 (1967).

10. Haynie, F. H., and Upham, J. B., Effects of Atmospheric Pollutants on Corrosion Behavior of Steels, *Mater. Prot. Perform.* **10**, 18–21 (1971).

11. Haynie, F. H., and Upham, J. B., Correlation Between Corrosion Behavior of Steel and Atmospheric Pollution Data, *Corrosion in Natural Environments*, ASTM STP 558, American Society for Testing and Materials, 1974, pp. 33–51.

12. Chandler, K. A., and Kilcullen, M. B., Survey of Corrosion and Atmospheric Pollution in and Around Sheffield, *Br. Corros. J.* **3**, 80–84 (1968).

13. Guttman, H., and Sereda, P. J., Measurement of Atmospheric Factors Affecting the Corrosion of Metals, *Metal Corrosion in the Atmosphere*, ASTM STP 435, American Society for Testing and Materials, Philadelphia, 1968, pp. 325–359.

14. Mansfeld, F. B., "Regional Air Pollution Study. Effects of Airborne Sulfur Pollutants on Materials," EPA-600/4-80-007, U.S. Environmental Protection Agency, Research Triangle Park, NC, January 1980.

15. Gerhard, J., and Haynie, F. H., "Air Pollution Effects on Catastrophic Failure of Metals," EPA-650/3-74-009, U.S. Environmental Protection Agency, Research Triangle Park, NC, November 1974.

16. American Society for Testing and Materials, *Symposium on Atmospheric Corrosion of Non-Ferrous Metals*, held in Atlantic City, NJ, June 29, 1955. ASTM STP 175, American Society for Testing and Materials, Philadelphia, 1955.

17. Sydberger, T., and Vannenberg, N. G., The Influence of the Relative Humidity and Corrosion Products on the Adsorption of Sulfur Dioxide on Metal Surfaces, *Corros. Sci.* **12**, 775–784 (1972).

18. Guttman, H., Effects of Atmospheric Factors on the Corrosion of Rolled Zinc, *Metal Corrosion in the Atmosphere*, ASTM STP 435, American Society for Testing and Materials, Philadelphia, 1968, pp. 223–239.

19. Haynie, F. H., and Upham, J. B., Effects of Atmospheric Sulfur Dioxide on the Corrosion of Zinc, *Mater. Prot. Perform.* **9**(8), 35–40 (1970).

20. Cavendar, J. H., Cox, W. M., Georgevich, M., Huey, N., Jutze, G. A., and Zimmer, C. E., "Interstate Surveillance Project: Measurement of Air Pollution Using Static Monitors," APTD-0666, U.S. Environmental Protection Agency, Air Pollution Control Office, Research Triangle Park, NC, May 1971.

21. Holbrow, G. L., Atmospheric Pollution: Its Measurement and Some Effects on Paint, *J. Oil Color Chem. Assoc.* **45**, 701–718 (1962).

22. Campbell, G. G., Shurr, G. G., Slawikowski, D. E., and Spence, J. W., Assessing Air Pollution Damage to Coatings, *J. Paint Tech.* **46**, 59–71 (1974).

23. Spence, J. W., Haynie, F. H., and Upham, J. B., Effects of Gaseous Pollutants on Paints: A Chamber Study, *J. Paint Tech.* **47**, 57–63 (1975).

24. Brysson, R. S., Trask, B. J., Upham, J. B., and Booras, S. G., The Effects of Air Pollution on Exposed Cotton Fabrics, *J. Air Poll. Control Assoc.* **17**, 294–298 (1967).

25. Zeronian, S. H., *Text. Res. J.* **40**, 695–698 (1970).

26. Zeronian, S. H., Alger, K. W., and Omaye, S. T., *Proceedings of the Second International Clean Air Congress* (H. M. Englund and W. T. Beery, eds.), Academic Press, New York, 1971, pp. 468–476.

27. *Proceedings of the Second International Symposium on the Deterioration of Building Stone*, Athens, Greece, September 27–October 1, 1976.

28. *International Meeting on the Restoration of the Erechthion*, Athens, Greece, December 8–9, 1977.

29. Yocom, J. E., Air Pollution Damage to Buildings on the Acropolis, *J. Air Poll. Control Assoc.* **29**, 333–338 (1979).

30. Schaffer, R. J., The Effects of Air Pollution on Buildings and Metalwork, *Air Pollution* (M. W. Thring, ed.), Butterworths Scientific Publications, London, 1957.

31. Feilden, B., The Care of Cathedrals and Churches, *J. Royal Soc. Arts* (*London*) **123** (5224), 196–215 (1975).

32. Luckat, S., Investigations concerning the Protection against Air Pollutants of Objects of Natural Stone, *Staub-Reinhaltung der Luft* **32**(5), 30–33 (1972).

33. Longinelli, A., and Bartelloni, M., Atmospheric Pollution in Venice, Italy, as Indicated by Isotopic Analysis, *Water, Air and Soil Pollution* **10**, 335–341 (1978).

34. Riederer, J., The Lack of Effect of Air Pollutants on Stone Decomposition, *Staub-Reinhaltung der Luft* **23**, 15–19 (1973).

35. Hansen, J., Ailing Treasures, *Science* **80** (September/October), 59–61 (1980).

36. Specialties Help Restore Glory to the Glass, *Chemical Week* **125**, 82–84 (December 19, 1979).

37. Sanyal, B., and Singhania, G. K., Atmospheric Corrosion of Metals: Part I, *J. Sci. Indus. Res. (India)* **15B**, 448–455 (1956).

38. Giles, C. H., The Fading of Colouring Matter, *J. Appl. Chem.* **15**, 541–550 (1965).

39. Rowe, F. M., and Chamberlain, K. A. J., The "Fading" of Dyeings on Cellulose Acetate Rayon—The Action of "Burnt Gas Fumes" (Oxides of Nitrogen, and so Forth, in the Air) on Cellulose Acetone Rayon Dyes, *J. Soc. of Dyers and Colourists* **53**, 268–278 (1937).

40. Upham, J. B., and Salvin, V. S., Effects of Air Pollutants on Textile Fibers and Dyes, *Ecological Research Series*, EPA-650/3-74-008, U.S. Environmental Protection Agency, Washington, D.C., 1975.

41. Salvin, V. S., Paist, W. D., and Myles, W. J., Advances in Theoretical Practical Studies of Gas Fading, *Am. Dyestuff Reporter* **41**, 297–302 (1952).

42. Salvin, V. S., Relation of Atmospheric Contaminants and Ozone to Light Fastness, *Am. Dyestuff Reporter* **53**, 33–41 (1964).

43. Beloin, N. J., A Field Study—Fading of Dyed Fabrics by Air Pollution, *Text. Chem. Color* **4**, 43–48 (1972).

44. Beloin, N. J., A Chamber Study—Fading of Dyed Fabrics Exposed to Air Pollutants, *Text. Chem. Color* **5**(7), 29–34 (1973).

45. Upham, J. B., Haynie, F. H., and Spence, J. W., Fading of Selected Drapery Fabrics by Air Pollution, *J. Air Poll. Control Assoc.* **26**(8), 790 (1976).

46. Morris, M. A., "Effect of Weathering on Cotton Fabrics," Bulletin No. 823, California Agricultural Experiment Station, Davis, CA, 1966.

47. Jellinek, H. H. G., Flajeman, F., and Kryman, F. J., Reaction of SO_2 and NO_2 with Polymers, *J. App. Polym. Sci.* **13**, 107–116 (1969).

48. Jellinek, H. H. G., Chain Scission of Polymers by Small Concentrations (1–5 ppm) of Sulfur Dioxide and Nitrogen Dioxide, Respectively, in the Presence of Air and Near UV Radiation, *J. Air Poll. Control Assoc.* **20**, 672–674 (1970).

49. Jellinek, H. H. G., "Degradation of Polymers at Low Temperatures by NO_2, O_3, and Near-Ultraviolet Light Radiation," USN TIS AD 782950/OGA, Cold Regions Research Engineering Laboratories, Hanover, NH, 1974.

50. "Air Quality Criteria for Oxides of Nitrogen," External review draft, U.S. Environmental Protection Agency, Research Triangle Park, NC, June 1979.

51. Hermance, H. W., Russell, C. A., Bauer, E. J., Egan, T. F., and Wadlow, H. V., Relation of Air-Borne Nitrate to Telephone Equipment Damage, *Environ. Sci. Tech.* **5**, 781–789 (1971).

52. McKinney, N., and Hermance, H. W., Stress Corrosion Cracking Rates of a Nickel-Brass Alloy Under Applied Potential, *Stress Corrosion Testing*, ASTM STP 452, American Society for Testing and Materials, Philadelphia, 1967, pp. 274–291.

53. Hermance, H. W., Combatting the Effects of Smog on Wire-Spring Relays, *Bell Lab. Rec.* **44**, 48–52 (1966).

54. "Air Quality Criteria for Ozone and Other Photochemical Oxidants," EPA-600/8-78-004, U.S. Environmental Protection Agency, Research Triangle Park, NC, April 1978.

55. Williams, I., Oxidation of Rubber Exposed to Light, *Ind. Eng. Chem.* **18**, 367–369 (1926).

56. Rossem, A. van, and Talen, H. W., The Appearance of Atmospheric Cracks in Stretched Rubber, *Kautschuk* **74**, 79–86, 115–117 (1931).

57. Norton, F. J., Action of Ozone on Rubber, *Rubber Age* **47**, 87–90, New York (1970).

58. Newton, R. G., Mechanism of Exposure-Cracking of Rubbers, with a Review of the Influence of Ozones, *J. Rubber Res.* **14**, 27–39, 41–62 (1945).

59. Crabtree, J., and Biggs, B. S., Cracking of Stressed Rubber by Free Radicals, *J. Polymer Sci.* **11**, 280–281 (1953).

60. Crabtree, J., and Kemp, A. R., Accelerated Ozone Weathering Test for Rubber, *Ind. Eng. Chem., Anal. Ed.* **18**, 769–774 (1946).

61. Bradley, C. E., and Haagen-Smit, A. J., The Application of Rubber in the Quantitative Determination of Ozone, *Rubber Chem. Technol.* **24**, 750–755 (1951).

62. Mueller, W. J., and Stickney, P. B., "A Survey and Economic Assessment of the Effects of Air Pollution on Elastomers," NAPCA Contract No. CPA-22-69-146. Work performed by Battelle Memorial Institute, Columbus, OH, U.S. EPA, Research Triangle Park, NC, June 1970.

63. Edwards, D. C., and Storey, E. B., A Quantitative Ozone Test for Small Specimens, *Chem. in Canada* 11(11), 34–38 (1959).

64. Salvin, V. S., Ozone Fading of Dyes, *Text. Chem. Color* 1, 245–251 (1969).

65. Bogarty, H. S., Campbell, K. S., and Appel, W. D., The Oxidation of Cellulose by Ozone in Small Concentrations, *Text. Res. J.* 22, 81–83 (1952).

66. Kerr, N., Morris, M. A., and Zeronian, S. H., The Effect of Ozone and Laundering on a Vat-dyed Cotton Fabric, *Am. Dyestuff Reporter* 58, 34–36 (1969).

67. Robinson, J. M., and others, "Engineering and Cost Effectiveness Study of Fluoride Emissions Control," Resources Research, Inc., and TRW Systems Group. Prepared for Office of Air Programs, U.S. Environmental Protection Agency, Research Triangle Park, NC, Contract No. EHSD 71–14, January 1972.

68. Spence, J. W., and Haynie, H. F., "Paint Technology and Air Pollution: A Survey of Economic Assessment," Pub. No. AP-103, U.S. Environmental Protection Agency, Research Triangle Park, NC, February 1972.

69. Salmon, R. L., *Systems Analysis of the Effects of Air Pollution on Materials,* Midwest Research Institute (APTD-0943), Kansas City, MO, January 1970.

70. Michelson, I., and Tourin, B., "Report on Study of Validity of Extension of Economic Effects of Air Pollution Damage from Upper Ohio River Valley to Washington, D.C. Area," Environmental Health and Safety Research Association, August 1967.

71. Booz, Allen and Hamilton, Inc., "Study to Determine Residential Soiling Costs of Particulate Air Pollution," APTD-0715, NAPCA, Washington, D.C., October 1970.

72. Bennett, L. H., Kruger, J., Parker, R. L., Passaglia, E., Reimann, C., Ruff, A. W., and Yakowitz, H., Economic Effects of Metallic Corrosion in the United States, *Part I: A Report to the Congress by the National Bureau of Standards,* NBS Special Publication 511-1, U.S. Department of Commerce, National Bureau of Standards, Washington, D.C., May 1978.

73. Fink, F. W., Buttner, F. H., and Boyd, W. K., "Technical Economic Evaluation of Air Pollution Corrosion Costs on Metals in the United States," Final Report, prepared for the Air Pollution Control Office, Environmental Protection Agency (APTD-0654), Battelle Memorial Institute, Columbus, OH, 1971.

74. Robbins, R. C., "Inquiry into the Economic Effect of Air Pollution on Electrical Contacts," Stanford Research Inst. Contract PH-22-68-35, April 1970.

75. Upham, J. B., and Salvin, V. S., "Effects of Air Pollutants on Textile Fibers and Dyes," Environmental Protection Agency, Office of Research and Development, Ecological Research Series, EPA-650/3-74-008, U.S. Government Printing Office, Washington, D.C., 1975.

CHAPTER 4
EFFECTS OF AIR POLLUTANTS ON THE ATMOSPHERE

ELMER ROBINSON

Air Pollution Research Section
College of Engineering
Washington State University
Pullman, Washington

1 INTRODUCTION

The environmental impacts of air pollutants include a number of effects on the earth's atmosphere and on atmospheric processes. These effects are the subject of this chapter so that the air pollution control engineer can appreciate some of the reasons for control operations and can recognize some of the problems that may be associated with specific pollutants. The topics covered in this chapter relate to atmospheric impacts and possible long-term changes in the atmosphere. The discussion includes the visibility problem and urban haze, clouds and precipitation mechanisms, possible climatic impacts, and potential effects on the stratosphere.

The importance of these air pollutant impacts can be seen in a number of aspects of present-day life. For example, the degradation of visibility has an important aesthetic impact as well as being an inconvenience or even dangerous when visibility reductions are severe. These impacts from visibility restrictions can translate into economic losses in terms of tourism, transportation schedules, and real estate values. Changes in precipitation patterns may affect agricultural production as well as urban activities and both of these could be related to economic losses. In the area of climatic change, the possible impacts are not yet clearly defined; but there is certainly no indication that a beneficial result could be produced. Thus there are very strong arguments for minimizing air pollutant impacts on the atmosphere through appropriate control programs.

2 VISIBILITY IMPACTS

2.1 Introductory Comments

Air pollutant impacts on visibility cause a loss in the visual clarity of the atmosphere. In fact, the apparent clarity of the atmosphere has been used for years by the scientist and the general public as an indicator of air pollution. The meteorologist quantifies atmospheric clarity by the visual observation identified as "visibility." By definition, daytime visibility in U.S. weather observing practice is the greatest distance in a given direction at which it is just possible to see and identify with the unaided eye a prominent dark object against the horizon sky.[1]

Although visible pollutants and urban haze are a long-standing feature of polluted areas, probably coinciding in time with the beginning of the Industrial Revolution, a relatively new dimension to the visibility situation has been emphasized in the United States with the adoption of the 1977 Clean Air Act Amendments. This act specifically addresses the problem of visibility impairment due to air pollution and sets as a national goal the prevention of any visibility degradation by air pollutants in presently clean, scenic areas, such as the National Parks. Many of these areas are identified specifically because the clarity of the atmosphere and the distant scenic vistas are important to the enjoyment of these areas and thus are valuable to the people of the United States. A brief discussion of the 1977 Clean Air Act Amendments with regard to pollutant impacts on visibility has been presented by Lewis.[2] Some of the regulatory processes under development by the U.S. EPA in the field of visibility have been reviewed by Horne.[3]

Not all air pollutants pose a threat to atmospheric visibility. In general, to affect visibility or the clarity of the atmosphere a pollutant should be an aerosol particle, a gas capable of forming aerosol particles through reactions in the atmosphere, or a colored gas. Thus the common pollutants with potential visibility impact are: (1) particulate emissions of dusts, fumes, fly ash, and other solid particles generally classed as total suspended particles (TSP); (2) SO_2 and other gaseous sulfur compounds because these gases react at a significant rate in the atmosphere to form sulfate and sulfuric acid aerosol particles; (3) NO and NO_2 because these gases react to form nitrate and HNO_3 aerosol particles and also because the red-brown color of NO_2 can cause a visible coloration in plumes and urban haze clouds under certain conditions; and (4) photochemical air pollution (e.g., "Los Angeles smog") because of the submicrometer-sized aerosol particles formed by these reactions. Although these materials are the most important factors for air pollution visibility a wide variety of contributors are possible. The general characteristics of the materials that affect the visibility will become clear in the following discussion.

2.2 Visibility Theory

As mentioned previously, visibility has the meteorological definition as the greatest distance in a given direction at which it is just possible to see and identify with the unaided eye, in the daytime, a prominent dark object against the horizon sky. This means that visibility measurements in both meteorology and air pollution problems are visual, or human observer measurements, and are not instrumental. This situation is probably unique in the air pollution measurement category in that it is a noninstrumental measurement. This does not mean that instruments have not been developed to approximate a visibility measurement; but the official measurement, as reported by the National Weather Service and accumulated in climatological statistics of visibility, comes from an actual human observer going outside, looking at the horizon, and determining how far he can see in terms of a set of known landmarks. The landmarks are supposed to be large and relatively dark; but obviously, for any given area the visibility targets are targets of opportunity.

The restriction, or the specification, for the target to be dark is to give maximum contrast between the target and the horizon and then to use the ability to distinguish that contrast visually as the guide to the visibility distance. It is fortunate that the objects that are available in most situations are relatively dark, and thus they are a reasonable approximation of the specification of a dark target. Obvious departures from this specification would be a snow-capped mountain or a light colored building which are quite often brighter than the horizon sky; but these would still be used as visibility targets.

The visibility observation is a subjective evaluation by the observer of the attenuation of light by an evaluation of the contrast of a target in a particular direction. When an observer looks at a target on the horizon, he evaluates the contrast because the ability to see an object is determined by the contrast between that object and its background. The old story about "you can't find a black cat in a coal bin" is an example of such a situation, that is, you need some contrast between the object and its background in order to see it. In the case of visibility targets, the background is the horizon sky and the object is the target. Visibility, as mentioned above, is that distance at which the observer can just identify the target; thus it is the distance at which the contrast has been decreased to a value where it can no longer be detected visually. This contrast limit, if the observer has perfect eyesight, as described by Middleton[4] has been estimated to be 2%. However, the ordinary observer is often believed to require closer to a 5% contrast in order to identify an object against the background. Thus in practical terms, visibility is then that distance at which the contrast between the dark object and the horizon sky falls to 5% or less—that point at which it can no longer be distinguished. Middleton[4] discusses this topic in detail and shows that there is no real "standard" contrast limit that can be validated and that under various conditions and with different observers the observed contrast threshold value may range between 1 and 10%.

The loss, or decrease of contrast, and thus the loss of clarity of the atmosphere is caused mainly by scattering and absorption due to fine particles in the atmosphere. There is also some scattering caused by the air molecules which is the process of Rayleigh scattering. Extremely clean regions— high-altitude atmospheres or remote areas—may approach a Rayleigh scattering situation. However, for urban air pollution conditions, the visual problems are due to scattering and absorption primarily by fine particles and not Rayleigh scattering. At this point we will also mention that the colored pollutant gas, nitrogen dioxide (NO_2), can also play a role in visibility because of its strong absorptivity, but it is the only gaseous compound that may be present in high enough concentrations to contribute to air pollution visibility problems.

From a mathematical point of view, contrast can be expressed in terms of a ratio involving the brightness of the target B and the brightness of the horizon sky B' where contrast C is the expression

$$C = \frac{B - B'}{B'}$$

Now if the target is black, or nearly so, its brightness B is zero and thus the contrast with the horizon of a black visibility target is equal to -1.

Let us look at this contrast relationship again and what happens as the distance to the target approaches the visibility. Since contrast approaches zero at the visibility distance, this ratio must also approach zero, and that means the apparent brightness of our target object has to approach the brightness of the horizon. This also means that in appearance, the black target seems to get brighter. Most people can probably recall looking at a series of mountain ridges, for example, or similar distant objects where the more distant ridges in the mountainous area increase in brightness until they essentially equal the brightness of the sky and disappear. This comes about because of the influence of the light from the sky and the sun being scattered into the sight path of the observer; this scattered radiation is known as *air-light*. Thus, because of the air-light the sight path between a distant object and an observer picks up light due to sky radiation; when this light is scattered toward the observer, it changes the apparent brightness of the object in proportion to the distance. Finally, at the visibility limit the air-light contributes sufficiently to the brightness of the sight path that it is no longer possible to distinguish apparent brightness of the object from the brightness of the horizon.

The horizon is bright beside a visible object because it is, one might say, completely saturated with scattered or air-light. The sight path that terminates at a nearby target is not saturated with horizon light. As an observer moves away from the target, the sight path picks up more and more air-light from scattering until the amount of air-light in this target sight path is also approximately equal to the saturated value at the horizon. The visibility in this case is determined by the intensity of the scattering from the sight path between the target and the observer.

The mathematical description of this particular physical situation is generally attributed to Koschmieder,[5] who published his first paper along these lines in 1924. This is the basic theory which has been used to develop visibility relationships.

The apparent contrast of a visibility target compared to the adjacent horizon and at a distance x from an observer has been shown by Middleton[4] and a number of other authors to follow a Beer's law attenuation relationship, namely,

$$C_x = C_0\, e^{-\sigma x}$$

where C_x is the apparent contrast between the horizon and the target at distance x, C_0 is the actual or inherent contrast, and σ is the attenuation coefficient along the sight path. For a dark or ideal visibility target, $C_0 = -1$ and at the visibility limit, $x = V$; thus the threshold or limiting contrast can be expressed as

$$C_v = -e^{-\sigma v}$$

At the visibility limit the threshold contrast C_v is frequently assumed to be 0.02 and thus

$$0.02 = -e^{-\sigma v}$$

or, solving for what is frequently called the Koschmieder equation,

$$V = \frac{3.9}{\sigma}$$

where V is now the distance at which the contrast between a black target and the horizon is 2% and σ is the atmospheric extinction or attenuation coefficient. Both V and σ must be expressed in compatible units, for example, m and m^{-1}. Thus when there is a way of measuring or of judging the attenuation coefficient, it is possible to calculate a visibility, and vice versa.

A complete expression for the attenuation coefficient is a combination of scattering and absorption of both gases and particles, namely,

$$\sigma = b_r + b_{sp} + b_{ap} + b_{ag}$$

Here b_r is the Rayleigh molecular scattering, b_{sp} is the scattering by particles, b_{ap} is the absorption by particles, and b_{ag} is the absorption due to gases, for example, NO_2. In this relationship most of the emphasis is put on the particle scattering component, b_{sp}, because it seems to be the largest of the four components in urban areas. As we will see this is a fortunate situation because the scattering coefficient can be measured relatively easily with a commercially available instrument, the nephelometer.

In terms of the relative magnitude of the factors contributing to an urban air pollution situation, particle scattering or b_{sp} is generally most important and commonly will range in value from 1 to 5

$\times\ 10^{-4}$ m^{-1}. Data on the particle absorption attenuation coefficient are not commonly available for atmospheric situations. In England in the 1940s when coal, smoke, soot, and fly ash were characteristic of the atmospheric particulate concentrations, Waldram[6] determined that attenuation by absorption was approximately equal to that due to scattering. At present our urban and rural areas differ significantly from the era of heavy, uncontrolled coal usage and b_{ap}, the particle absorption attenuation coefficient, is probably 10–20% of b_{sp}, and thus may be in the range of 5 to 10×10^{-5} m^{-1}. For absorption by NO$_2$ the gaseous absorption attenuation coefficient for a 10-mile visibility and 0.1 ppm NO$_2$ concentration, Robinson[7] shows that there is about a 10–20% increase in attenuation. Thus the value of b_{ag} would typically be in the range from 5 to 10×10^{-5} m^{-1}. The Rayleigh scattering coefficient b_r has a value of 1.8×10^{-5} m^{-1} at a wavelength of 0.5 μm.[8] If this value is put into the Koschmieder visibility equation, the result is

$$V = \frac{3.9}{1.5 \times 10^{-5}} = 260 \text{ km}$$

By contrast, in a typical urban situation, with a visibility of 30 km or 20 miles—not a particularly low visibility—the attenuation coefficient using the Koschmieder expression is 1.3×10^{-4} m^{-1}. Thus in any sort of realistic urban visibility situation Rayleigh scattering is 10% or less of the total extinction. For this reason, Rayleigh scattering, while it is always present, drops out as far as an important factor is concerned. Likewise, as mentioned above, the two absorption terms are relatively low or small in most of our current air pollution situations. Here, absorption would increase with air pollution, but research has shown that even in a situation such as Los Angeles, in only 20–30% of the observed situations was there an indication that NO$_2$ absorption played an important role in the atmospheric extinction coefficient, as shown by Charlson et al.[9]

2.3 Visibility in Clean Regions

Although Rayleigh scattering has a minor influence on most air pollution visibility situations, there are conditions in the United States and probably other areas at the present time where both Rayleigh scattering and particulate scattering are important and the combined effects need to be considered. These situations are the present and prospective problems in the United States resulting from increased development and pollutant emissions in some of the remote areas of the western and southwestern United States, for example, Wyoming, Arizona, Utah, and the Four-Corners Area. In these areas, where, for example, a 130-km or 80-mile visibility is frequently recorded, the Rayleigh scattering part of the attenuation coefficient becomes a major part of the total atmospheric attenuation. Winslow, Arizona is one station with a high frequency of reported visibility equal to or greater than 130 km. Under these conditions, and using the Koschmieder expression in the form

$$V = \frac{3.9}{b_r + b_{sp}}$$

with $b_r = 1.8 \times 10^{-5}$ m^{-1} and $V = 130$ km, results in a value for b_{sp} of 1.2×10^{-5} m^{-1}. Thus half or less of the attenuation is due to particle scattering (or other particle effects) and the remainder is due to the Rayleigh component. Research by Ursenbach et al.[10] provides another example of the potential impact of pollutant emissions in a very clean area. Ursenbach's data on extinction coefficient and particle mass ratios shows that a non-Rayleigh extinction of 1.2×10^{-5} m^{-1}, as estimated above for a clean area, could result from about 30 μg/m^3 of soil dust, from about 35 μg/m^3 of fly ash, from 1.2 μg/m^3 of sulfate particles, or from 1.6 μg/m^3 of soot. As will be shown subsequently, the greater attenuation effectiveness of the sulfate and soot particles is due to their submicrometer size range. The very small concentrations that are capable of these changes is a clear indication of the sensitivity of clean areas to the intrusions of air pollutants.

2.4 Particle Scattering

This now leads into a more complete discussion of the particle scattering portion of the extinction coefficient. This is generally covered as Mie scattering, identified with Gustav Mie,[11] who in 1908 developed the mathematics to explore scattering where the size of the particle is about the same as the wavelength of light. There are a number of things that result from this particular relationship. Mie's analysis showed that particle scattering b_{sp} is a function of the cross-sectional area of the particle

$$b_{sp} = K\pi r^2$$

where K is the scattering efficiency ratio dependent on the nature of the particle, and is a multiplier for the area to give the scattering intensity. K has a value as low as 2 for quite large particles, and for some materials, as high as 6 for some optimum-sized submicrometer particles. Thus the scattering intensity is larger than might be expected on the basis of just the cross-sectional area of the particle. This occurs because the scattering of the particle is caused not only by the light that penetrates the particle and is then scattered and dispersed, but also by light that passes close to the particle and is diffracted around its edges. With the proper combination of wavelength, particle size, and refractive index of the particle, a maximum area effect of the particle occurs that is several times larger than the physical area of the particle. This happens in the submicrometer size range, from about 0.1–1 μm, depending on the refractive index, the nature of the particle, and the wavelength of the light.

If $b_{sp} = K\pi r^2$ is the extinction for a single particle, obviously this can be multiplied by N, the number of particles, if all of them are the same, and the scattering for a volume containing N similar particles is

$$b_{sp} = NK\pi r^2$$

If there is a heterogeneous population of particles with different sizes and different values of K, the scattering function is still the sum over all of the particles

$$b_{sp} = \sum_{i=1}^{n} N_i K_i \pi r_i^2$$

When the scattering function of the atmosphere is measured, we are actually measuring an integrated scattering, or a summation of the particle scattering, because the atmosphere is made up of a heterogeneous aerosol which has a large number of different types of particles, different refractive indices, and different sizes that contribute to the scattering potential. K is a complex function of the size of the particle, the refractive index of the particle, and the wavelength of light. Calculated values are available for a number of compounds in books by Middleton,[4] McCartney,[12] and van de Hulst[13] to name several.

The total scattering intensity affecting a particular observation is also a function of the angle between the incident light and the observer's sight path. McCartney[12] describes the nature of this angular dependence of scattering intensity which forms a characteristic three-dimensional pattern in space around the particle and is strongly dependent on the ratio of particle size to the wavelength of the incident light. With relatively small particles, diameter less than 10% of the wavelength, the Rayleigh region, the particle tends to scatter equally forward and backward along the direction of the incident light. For somewhat larger particles, diameter about one-fourth the light wavelength, the total scattering is increased and the pattern is biased in the forward direction. With very much larger particles, diameter greater than the light wavelength, the total scattering increases further and the fraction scattered in the forward direction also increases. Scattering normal to the incident beam is generally less than either the forward or back scatter for all particle sizes and wavelength ratios according to McCartney.[12] Thus, depending on the angle between the light and the sight path, a different scattering intensity is observed.

For air pollution applications this becomes important primarily for forward-scattering situations. In other words, when an observer looks in the general direction of the sun—the usual illumination source—scattered light from the atmospheric aerosol is maximized. With the maximum scattering potential along that sight path there is minimum visibility. It also follows that as an observer looks away from the sun, there can be much better visibility, without necessarily any change in the total mass or character of particulate material or air pollution concentrations in the atmosphere. So visual estimates of areas of pollutant concentration around a given observation point might be inclined to estimate maximum air pollutant concentrations in the direction of the sun and minimum air pollution concentrations in the direction opposite of the sun, merely because of the change in scattering intensity on an angular basis.

This angular scattering factor becomes important in the estimation of plume opacity. There is still a degree of importance given to visual observations of apparent plume concentration or plume opacity.[7] The Ringelmann scale of visual plume opacity is still used, although there is strong pressure to have it removed. Here, if an observer tries to evaluate a plume where the plume is subject to forward scattering, in other words, with the sun behind the plume, he will tend to maximize the estimate of plume opacity because of the illumination. Therefore, in air pollution regulations, where the Ringelmann scale of plume opacity is used, there should be a specification that during a plume density observation, the sun should be behind the observer. This will prevent the observation from being overly sensitive to the intensity of the forward scatter.

There is another point that should be made here with regard to Mie scattering. Namely that since small particles, for example, in the submicrometer-size range, tend to maximize the scattering considerable significance is often made of statements such as "small particles are more important than large particles in terms of visibility." That is only true if a unit mass of small particles is compared to a similar mass of much larger particles. A unit mass of material, divided up into submicrometer particles, rather than 10-, 20-, or 30-μm particles, will provide much more visibility restriction, because you will have a greater cross-sectional area from the small particles. Thus the actual effective scattering area is increased for smaller particles, not only because of an increase in K for smaller particles but

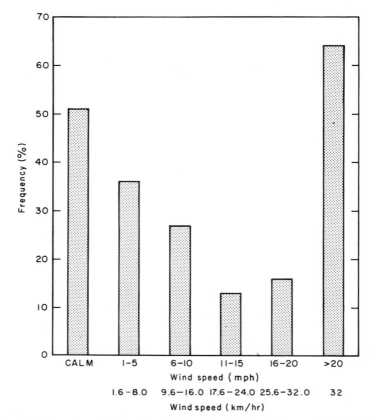

Fig. 1 Percent frequencies of 0–10 mile visibilities by wind speed classes, Bakersfield, California, 7:30 A.M.–5:30 P.M.; May, July, September, November, 1948–1957. Adapted from Holzworth and Maga, 1960.[14]

also on the basis of a volume/area ratio. If this aerosol mass versus extinction ratio concept is carried on further, it is possible to maximize the visibility restriction for a given mass and this occurs with submicrometer particles. For water, the most effective size is in the 0.6–0.8-μm range. For some solid materials, the most effective size is down to 0.3 μm.[7] The refractive index is also a significant factor in the final determination of the size for maximum visibility restriction.

Thus, the statement that the maximum visibility effect is caused by small particles rather than large particles must also specify that equal *masses* of material are involved. In terms of equal *numbers* of particles, it is easy to show that a dust storm or a fog, where the particles are relatively large, is obviously a serious visibility hindrance.

2.5 Meteorological Impacts on Visibility

In the determination of the extinction coefficient and thus the visibility there are a number of meteorological factors that are important because they may affect both the number of particles that are present in the atmosphere and their size. The three most important factors are wind, mixing height, and relative humidity.

Wind is important in two ways. At relatively low wind speeds, an increase in wind tends to increase the dilution of haze or air pollutants, and therefore increasing wind produces an increase in the visibility. In practice, there is a limit to this, because if winds become too strong, they begin to pick up and entrain dust and large particles from the ground surface. This obviously can cause low visibility. There have been a number of verifications of wind effects. Figure 1, from Robinson[7] and based on the studies of Holzworth and Maga[14] shows one such set of data. As would be expected, this plot of wind speed versus visibility shows increasing visibility with increasing wind speed up to a value of about 20 miles/hr, and then a significant decrease in visibility for higher wind speeds, the region of entrained dust.

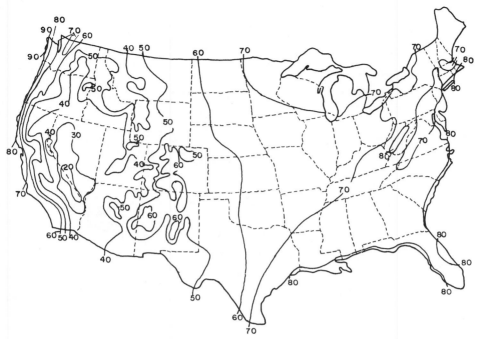

Fig. 2 Mean relative humidity, percent, for July. From U.S. Department of Commerce, 1968.[16]

The mixing height, the depth through which the surface-generated convective or mechanical turbulence elements are active, is also a dilution factor for ground-generated pollutants. The stronger and deeper the mixing, all the other factors being equal, the more dilute the particulate materials become and the better the visibility.

Relative humidity is a more complex factor because its effect depends on the chemistry of the particulate materials in the atmosphere. Many pollutant materials, as well as natural aerosol particles are hygroscopic—they absorb water and thus they increase in size. Many of these begin to deliquesce or to show a hygroscopic reaction in a relative humidity range of 70–80%.[15] Sea salt, for example, under laboratory conditions begins to increase in size rapidly at about 72%. Sea salt is one of the components of the natural aerosol that is quite sensitive to relative humidity, with the radius of the particle and thus the cross-sectional area of the particle increasing with an increase in relative humidity above 72%. These particles can rapidly evolve into a dense haze situation before they actually reach saturation or a true fog or water-droplet condition. In air pollution regulations that deal with visibility criteria for judging air pollution visibility situations, it is common to specify that the visibility should only be considered as an air pollution indicator at relative humidities below 70%.

The hygroscopic limit works for most materials; there is, however, one important air pollutant that does not fit this pattern. This is sulfuric acid which is hygroscopic and does not have a sharp deliquescent point like sea salt. Sulfuric acid aerosol is in continual equilibrium in terms of ion concentration, and thus size, with the atmospheric humidity. Sulfuric acid aerosol as an air pollutant is usually formed by the oxidation in the atmosphere of sulfur dioxide. It is considered to be one of our more important pollutants at the present time, both in urban and nonurban atmospheres, because of the SO_2 emissions from fossil fuel combustion, particularly. When SO_2 is converted to sulfuric acid the result is an aerosol, and these particles have a continuous growth pattern with relative humidity. Thus, the atmospheric sulfuric acid aerosol system is hygroscopic, but it does not have a break point in the hygroscopic curve. However, it shows a continuous response reaching an equilibrium size determined by the relative humidity.[15] Therefore, in dealing with the SO_2 oxidation products from the combustion of fossil fuel containing sulfur, it is necessary to be concerned with the complete spectrum of relative humidity, and not merely with a question of whether the relative humidity is above 70%.

Synoptic meteorology factors that tend to increase pollutant concentrations and decrease visibility on a regional scale are associated with the slow-moving high-pressure systems or anticyclones. These weather systems have relatively weak winds because of an absence of strong pressure gradients. Mixing depths are generally reduced because of the presence of inversions in the surface boundary layer. In addition, clear skies will favor photochemical pollutant reactions, and if the anticyclonic system includes

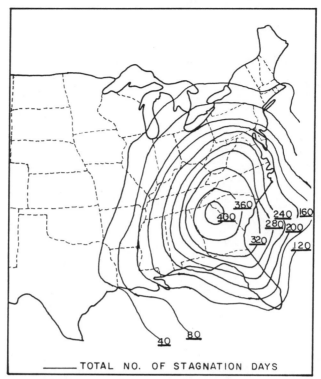

—— TOTAL NO. OF STAGNATION DAYS

Fig. 3 Total number of air stagnation days in episodes of 4 or more days, 1936–1965. Adapted from Korshover, 1967.[17]

air with a southerly or Gulf of Mexico history, relative humidities will be high. For these several reasons visibility will generally be reduced in slow-moving anticyclones. Some examples of these are described in a following section.

In different sections of the United States the general climate contributes to different potential visibility effects. The hot, high-humidity conditions in the South would produce a different visibility situation with a given natural aerosol mixture than would occur with the same aerosol in Arizona, with its low humidity. In a similar fashion the climatology of slow-moving anticyclonic systems shows that they are more frequent in the southeastern states than in other regions.

These two points are illustrated by Figures 2 and 3. In Figure 2 the nationwide pattern of July or midsummer mean relative humidity is shown.[16] Note that there are large areas in the East and South where the mean relative humidity is between 70 and 80%. By way of contrast, west of the Rockies the mean relative humidity is generally less than 50% except for the immediate coastal areas. Figure 3 shows the distribution of air mass stagnation days for the 30-yr period from 1936 through 1965 as determined by Korshover.[17] These stagnation days were based on periods of 4 or more days of adverse anticyclonic stagnation in the area east of the Rockies. It is clear that the southeastern states are the most strongly affected by these high air pollution potential weather situations.

2.6   Air Pollution Color Effects

Air pollutants may also bring about a change in the color of the atmosphere. This, as has been stated, is primarily due to absorption by nitrogen dioxide. The color occurs because this gas absorbs in the blue-green wavelengths and therefore transmits light that is overbalanced in the yellow and red. Thus, an illuminated volume of NO_2 is a reddish-brown, yellow-red-brown colored gas. There has been considerable discussion by Charlson et al.[9] and others with regard to how much of the typical brown haze that is seen over urban areas is due to NO_2. The confusion comes about from the fact that this color is not related just to NO_2, but it is also possible to produce a brownish colored haze through submicrometer or fine-particle Mie scattering. Thus there may be a debate as to the cause of a particular "brown haze" situation unless there are actual pollutant measurements. As mentioned earlier, NO_2 measurements in Los Angeles by Charlson et al.[9] have indicated that on perhaps 20–30% of the total days examined there was significant NO_2 absorption relative to the aerosol particle scattering. On these occasions NO_2 could contribute some to the coloration of the pollutant cloud. As shown by Robinson[7] if computations are carried out on the color effect of NO_2, it is found that at a rather high concentration of about 0.25 ppm, it is possible to expect to see some brown coloration. Initially, in California air quality regulations, which were the first to have an NO_2 criterion, a 0.25-ppm concentration of NO_2 was specified just on the basis of the probable visible color effect. Thus, visible color is the basis of the 0.25-ppm limit that appears in some discussions of NO_2 air quality and visibility.

2.7 Sources of Visibility Data

Almost all of the available visibility data on a global basis come from the climatological records of the various national weather services. Since global weather observations are closely linked with aviation operations, these visibility data apply mainly to airport situations, and thus these records have obvious limitations for urban air pollution analyses because of the typical distance between airports and urban centers.

Since the visibility data are part of the official weather observation record access to the data is available through official channels. In the United States all climatic data, including visibility, is archived by the National Climatic Center, National Oceanic and Atmospheric Administration, Asheville, North Carolina. Copies of the basic data record and various statistical summaries can be obtained from this source. Canadian data may be obtained through Environment Canada, Ottawa. Offices of the national weather services can usually be found at major airports, and if there is doubt about how to obtain data one of these offices can probably be of assistance.

2.8 Analyses of Visibility Data

Although this discussion is not the place for a detailed presentation of the numerous studies of visibility data available in the meteorological and air pollution literature, it will be useful to illustrate the several applications of visibility data through a brief presentation of the results of several investigators. Additional research results are being published at increasingly frequent intervals and the reader is urged to review recent copies of the technical journals in the field of air pollution and meteorology. Prominent English-language journals that frequently publish air pollution visibility studies are: *Atmospheric Environment, Journal of the Air Pollution Control Association,* and *Journal of Applied Meteorology.*

Visibility climatological data have been analyzed for air pollution impacts in two general types of analyses. First, the data from individual reporting stations for an extended period have been used in a statistical analysis for trends that might be correlated to air pollutant emissions. Patterns for individual cities and regional areas have been established. Second, visibility has been assessed on a regional basis from daily synoptic weather maps, and individual case studies have been made of the buildup, transport, and decay of regional air mass air pollution situations.

2.8.1 Climatological Trend Studies

Assessments of the visibility and air pollution trends at individual stations and for regional groups have been common for many years. Holzworth and Maga[14] analyzed the data from three California cities for a time period extending from the mid-1930s to the late 1950s. For Sacramento and Bakersfield a trend toward decreasing visibility was identified with wind from the urban area sector. For Los Angeles there was declining visibility from 1932 to 1947 and then essentially no trend in the period 1948 to 1959, the years of considerable photochemical smog development.

This paper by Holzworth and Maga[14] is especially useful because of the method presented for analyzing visibility data. The method provides for the fitting of trend lines to annual frequency plots of specified ranges of visibility, for example, 0–10 miles, 11–19 miles, and so on. Net changes in range classes and shifts in frequency from one class to another could be evaluated from linear trend lines. The resulting "conservation of frequencies" was considered to be generally indicative of the trend in visibility at the given station.

After the 1960 paper, Holzworth[18] carried out a similar study for 28 weather bureau airport stations with an emphasis on visibilities less than 7 miles during two periods of record separated by about 15 to 20 yr. The stations were selected to be those stations where significant changes in location had not occurred, and the earlier and later periods were probably comparable. Monthly frequencies in two periods, typically 1930–1938 and 1955–1961, were compared, giving a total of 336 station-months of comparative data. Of these only 89 or 26% showed poorer visibility in the 1955–1961 period, and of these only 13 showed an increase of more than 10% in the frequency of visibility less than 7 miles.

Although these studies of pre- and post-World War II data tended to show more improvement than deterioration of visibility, studies of more recent data seem to indicate another deteriorating cycle. The study by Miller et al.[19] dealing with visibility at Akron, Ohio, Lexington, Kentucky, and Memphis, Tennessee for the period 1962–1969 is one such example. This study indicates that these three stations all showed an increase in the frequency of less-than-7 miles visibility in the period 1966–1969 compared to the period 1962–1965.

Recent research on visibility trends in the United States has been much more extensive, as the importance of visibility as an air pollution impact has increased. An extensive study of long-term (i.e., 25 years—from about 1950 to 1975) trends in the northeastern United States has been carried out by Trijonis and Yuan.[20] This study, in line with Holzworth's earlier study, showed improving visibility from the late 1940s to the mid-1950s, an era when coal was declining as a fuel in home heating and transportation. Since the mid-1950s metropolitan areas showed only a small, about 5%, decrease in visibility but nonurban areas showed much more marked decreases of 10–40%. Trijonis and Yuan[20] also show that seasonal visibility patterns have changed between the 1950s and 1970s. In the earlier years summer and fall were seasons of relatively good visibility while in the 1970s these same seasons tend to show the greatest deterioration of visibility. This is especially evident in the southern states. These authors show a strong case for attributing this visibility pattern to SO_2 emissions with the subsequent formation of sulfate aerosol particles.

2.8.2   Regional Synoptic Visibility Studies

At least since the Donora, Pennsylvania air pollution incident in October 1948, and the London, England smog episode in December 1952, there has been continued and increasing interest in regional air pollution episodes that can result from the impact of stagnant or slow-moving, high-pressure weather systems.

Because of the large amount of observational data and the high pollutant source density, the Midwest and in particular the Ohio Valley has been an area where regional visibility degradation has been noted. One of the first descriptions of such an incident was given by Hall et al.[21] with a case study of an incident that occurred in August 1970. Figures 4, 5, and 6 are taken from this study by Hall. Figure 4 shows the relative locations of estimated source areas in the Ohio Valley and the receptor or impact areas in the Great Plains. Some of the wind trajectories that were calculated for this study are shown in Figure 5. These are based on the wind at about 1.5 km (850 mb) above ground and occur over a 4-day travel period. The fact that these trajectories indicated a general cyclonic or counter-clockwise rather than anticyclonic curvature is due to the fact that a small, weak low-pressure area

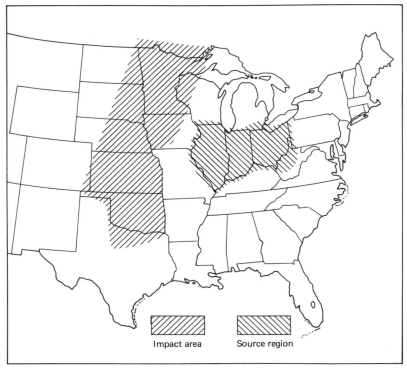

Impact area   Source region

Fig. 4 Relative locations of estimated source areas and pollutant impact areas, August 8–11 1970. Adapted from Hall, 1973.[21]

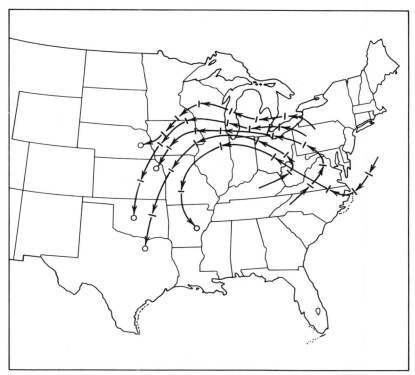

Fig. 5 Individual 850-mb wind trajectories for the period 0000 GMT August 8–1200 GMT August 11 1970. Tick marks show 12-hr positions. Adapted from Hall, 1973.[21]

drifted across the southern states—northern Texas and Oklahoma—while the Ohio Valley and northern Great Plains areas were under a high-pressure, anticyclonic system. Figure 6, also from Hall et al.,[21] shows the advection of the "visible pollution front" during this episode. The pollution front location was based on the occurrence of restricted visibility due to smoke and heavy haze at National Weather Service stations. The movement of the front from a location in Illinois on August 5 to the Texas–Oklahoma border on August 10 is consistent with the wind trajectories shown in Figure 5.

A number of subsequent regional or air mass visibility episodes have been reported by other investigators. Husar et al.[22] describe a visibility impact and pollutant buildup episode covering the period from June 25 to July 5, 1975 which was characterized by noon visibilities of less than 3 miles over large areas. In this time period the visible pollutant accumulation was over the Great Lakes on July 27, covered much of the Ohio Valley on July 3, and finally moved southeast across Georgia and Florida on July 5, 1975. This late June and early July 1975 episode was intense enough to be detected by a weather satellite as reported by Lyons and Husar.[23]

2.9   Methods of Visibility Measurement

2.9.1   *Visual Observations*

As mentioned previously the standard weather service visibility observation is an appraisal of "prevailing visibility" obtained from a visual survey of the horizon and available target objects around the particular station. These observational data are available in the national climatic records. The observation of "prevailing" visibility is defined as the greatest visibility that is equaled or exceeded over at least half of the horizon circle. Thus the standard observational data are not an average for the horizon and may not necessarily be affected by a sector of low visibility due to smoke unless the sector covers a significant fraction of the horizon. For individual studies it is of course possible to make special observations using standardized visual techniques. Details are available in a number of publications, for example, Robinson[7] and Hewson,[24] or in official government manuals, for example, U.S. National Weather Service, Circular N.

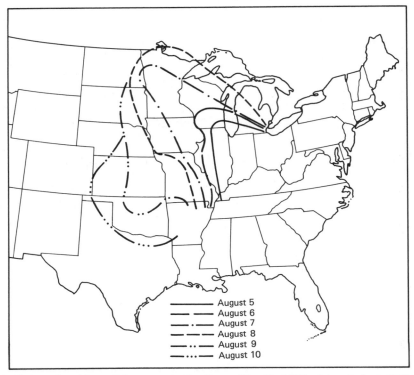

——————— August 5
— — — — August 6
—— · —— August 7
—— —— —— August 8
—— ·· —— August 9
—— ··· —— August 10

Fig. 6 Estimated positions of the transported pollutant "front" at 24-hr intervals from 1800 GMT, August 5–1800 GMT, August 10 1970. Adapted from Hall, 1973.[21]

2.9.2 *Photometric Instruments*

Since the estimation of the visibility can be related to the apparent contrast between a target and the horizon sky, it is possible to make this measurement instrumentally and to calculate an estimate of the visibility. A number of commercial instruments are available for this purpose.

A visibility photometer is a device for assessing, at a distance, the brightness of a field of view. Using proper lens systems the field can be restricted so that it can be completely filled by a reasonably sized target and then a photometric sensor can measure the apparent brightness of the image. Readings in succession can record the brightness of the target and the adjacent horizon. Since the inherent contrast C_0 between the target and the horizon would be assumed to be −1 for most dark targets and the distance x to the target is known, the attenuation coefficient b for the atmosphere between the sensor and the target can be calculated from the formula

$$C = -C_0 e^{-bx}$$

From the calculated value of the attenuation coefficient, the apparent visibility can be calculated using an assumed value of the contrast threshold. If 2% is assumed, the expression, as given previously, is

$$V = \frac{3.9}{b}$$

In actual practice the use of more than one target along a given line of sight can prove useful because of possible inhomogeneity of the atmosphere and instrument threshold problems.

Comparisons between photometry data and visual observations of prevailing visibility have been reported by Tombach and Allard[25] for midwestern situations of low to moderate visibility ranges. Their results showed a good correlation between simultaneous visual and photometer data. The method, of course, is limited to daytime conditions.

2.9.3 Photographic Techniques

Photographic methods are also available for measuring apparent target contrast and thus serve as a visibility measurement system. One of the early systems was described by Steffens.[26] The photographic methods are essentially the same as the telephotometer measurement except for the substitution of photographic film to record the image brightness. The camera can, of course, record several target images and the horizon on the same film frame. In the photographic method, precision and quality control is very important at all stages of the film processing and the image density measurements. Tombach and Allard[25] describe the techniques used in their midwestern visibility study. When they compared photographic results with visual observations of prevailing visibility, the correlations were acceptable but showed more scatter than simultaneous photometer data. They attributed the greater scatter to the increased probability of an accumulation of random errors in the complex film processing and evaluation procedure.

2.9.4 Transmissometers

A transmissometer uses a photometric sensor to measure the apparent intensity of a light source located at a known distance. The visibility is then calculated from this value using calibration data from a clear day. Most transmissometers are designed and operated for fog detection at airports and are not suited for the greater visibilities characteristic of air pollution situations. Research transmissometer systems using laser light sources have been constructed and used in research studies; however, commercial units are not available.

2.9.5 Scattering Measurements

The previous discussions of the attenuation coefficient and visibility emphasized the importance of the part of the attenuation due to scattering, that is, b_{SCAT}. The scattering of a sample volume of air can be measured relatively easily using techniques known as nephelometry. From a measured value of b_{SCAT} an estimated visibility value can be calculated from

$$V = \frac{3.9}{b}$$

if it is assumed that there is no significant absorption, and thus $b_{SCAT} = b$ is a reasonable approximation.

A commercial nephelometer is available for air pollution studies and is illustrated schematically in Figure 7. The commercial unit follows the design of Ahlquist and Charlson.[27] In operation the nephelometer integrates the light scattered over an angle from about 8° in the forward direction to 170° backscatter. The integrating nephelometer can be calibrated against the known Rayleigh scattering of a pure gas, such as CO_2, by substituting the gas for the sample air.

Many studies, such as those reported by Charlson[28] and Samuels et al.,[29] showed reasonable correlations between the urban scattering coefficient measurement of the nephelometer and local visibility. More recently, however, the rural midwestern study of visibility measurement methods by Tombach and Allard[25] showed relatively poor correlations between prevailing visibility and the nephelometer measurement. Here the nephelometer-derived visibility was consistently higher than the observed visibility. This was attributed to the effect of the ambient humidity and either the loss of large, moist particles or particle desiccation in the instrument inlet.[25] It would not be unexpected for the atmospheric aerosol in a humid region to have a significant absorption component in the attenuation coefficient

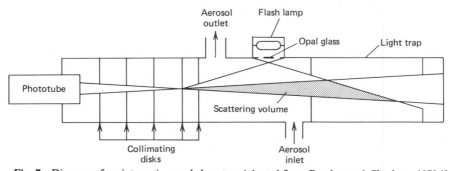

Fig. 7 Diagram of an integrating nephelometer. Adapted from Butcher and Charlson, 1972.[15]

because of the influence of hygroscopic particles. A nephelometer reading which includes only the scattering coefficient would be larger than the ambient visibility in such a situation.

The nephelometer is convenient to use as a continuous recorder of the scattering component of the attenuation coefficient. In many urban and rural areas this can give a good indication of the prevailing visibility. Thus it has proved to be very useful in many situations. However, where large particles and/or high humidity (over 60%) conditions are prevalent the nephelometer signal may underestimate the attenuation coefficient of the ambient air and thus overestimate the visibility.

3 IMPACTS ON PRECIPITATION

3.1 Introductory Comments

The previous sections of this chapter have described the relationships between air pollutants and visibility. Although one cannot argue the importance of visibility as an aesthetic factor, there is relatively little long-term impact from air pollution visibility degradation. However, this is not the case with regard to some of the other impacts of air pollutants on atmospheric processes. One example is the possible impact of air pollutants on precipitation, because the climatic precipitation cycle is a vital one for human existence on the earth. If there were to be large-scale impacts on precipitation, the results would certainly be extremely serious. One precipitation impact that has shown relatively widespread detectability is lowered rainfall pH, or as it is more commonly called, "acid rain." Although many serious problems may occur in the biosphere because of altered precipitation pH, as of this writing (1980) the quantification of biospheric problems is still an active research topic.

This section describes the ways in which air pollutants can effect changes in precipitation, including the mechanisms of precipitation formation. Some of the classic research studies are cited as examples, but, as in the case of the visibility section, this is not designed as a detailed review of the very extensive current research programs.

3.2 Condensation and Precipitation Formation

In the lower atmosphere, water vapor is omnipresent. When the air becomes saturated with water vapor, condensation occurs. Condensation is in the form of small liquid water droplets at warmer temperatures while at sufficiently low temperatures ice crystals are formed and in the atmospheric sciences the formation process is referred to as sublimation. At any reasonable conditions of saturation in the atmosphere this condensation or sublimation takes place on the surfaces of nuclei present as fine particles. The particle formation process is different depending on temperature and whether liquid or ice particles are formed. Condensation processes and the subsequent formation of precipitation along with the possible impacts of air pollutants on the mechanisms can be separated into warm cloud processes and cold cloud processes. This is the organization of the following discussion.

3.2.1 Warm Cloud Condensation and Precipitation

When an air parcel reaches saturation at temperatures above freezing, liquid water begins to condense on selected or preferential nuclei surfaces. These are "cloud condensation nuclei" or CCN to designate their special role in the cloud formation process. As described by Wallace and Hobbs[30] the CCN are typically soluble, hygroscopic salts and are usually present in the atmosphere at concentrations ranging from one to several hundred per cm^3. A value of 100 CCN/cm^3 is considered to be a typical ambient atmospheric concentration. It should be noted that the total aerosol concentration is usually many times greater than the number of CCN and that typical dust and haze particles are not ordinarily active in the atmospheric condensation process. There is frequently some confusion on this point because the total aerosol concentration is often referred to as the "condensation nuclei" concentration because extreme super saturation conditions are used to count them using an Aitken particle counter. More properly, the fine particle fraction of the atmospheric aerosol should be identified as "Aitken nuclei," and only a size range of less than 0.1-μm diameter is inferred, without any chemical influence. Thus the CCN would be included in the Aitken particle concentration but in most cases the CCN are only a small fraction of the total atmospheric aerosol concentration.

When saturation is reached, cloud droplets form on the CCN and a cloud becomes visible; as more water vapor becomes available the cloud expands as the number of droplets increases. The typical cloud droplet is in the size range of tens of micrometers and is not large enough to fall as precipitation, which typically requires millimeter-sized droplets. Continued growth by condensation is not fast enough to produce precipitation particles. Precipitation, however, is produced in liquid droplet clouds when there is a broad range of droplet sizes present in the cloud. This range of droplet sizes can be produced by a broad range in the CCN size distribution. Under these conditions the larger particles move relative to the smaller droplets, collide, and thus grow by collision and agglomeration at the expense of the smaller droplets. This can be a self-propagating chain reaction as droplets grow and then break up due to aerodynamic forces with the fragments still being large enough to serve as collision nuclei for further precipitation drop formation. Turbulence in the cloud can enhance the potential contact between the large and small droplets. This is called a "warm" process because the particles are in liquid form rather than ice.

Air pollutants can impact on the warm cloud process in at least two ways.[30] First, an addition of hygroscopic fine particles to the cloud can increase the CCN concentration and thus the concentration of cloud droplets in the cloud. An increase in the number of droplets will, however, be compensated by a decrease in droplet size since the same amount of available water vapor is divided among a larger number of particles. Studies of similar sized cumulus clouds over the ocean, where CCN numbers are relatively low, and over land where the CCN are more numerous have shown a smaller droplet size range in the clouds with the higher CCN concentration.[30] An increased number of smaller cloud droplets would under most circumstances lead to a reduced efficiency of precipitation formation. A second, and more or less opposite impact of air pollutants may occur if the pollutants cause an increase in the number of large CCN. An increased number of large CCN could provide an increase in the efficiency of the collision precipitation mechanism. Thus air pollutants can potentially cause either decreased or increased precipitation. Research on specific situations have not been consistent, some studies have shown decreases and others increases in precipitation. Robinson[7] and Wallace and Hobbs[30] review some of these instances of precipitation changes with an apparent correlation to air pollutant emissions.

3.2.2   Cold Cloud Processes

A cold cloud is one consisting of ice particles or a mixture of ice and liquid particles. Precipitation formation in cold cloud systems, which include a large fraction of midlatitude clouds, are basically different from the warm cloud process described above. The cold cloud process is frequently called the Bergeron–Findeisen process after the two European meteorologists who described the mechanism in the 1930s.[30] The mechanism rests on several features of the physics of water particles in the atmosphere. First, water droplets can be supercooled to temperatures as low as −40°C in the atmosphere before freezing into an ice particle. Second, there is a vapor pressure difference and thus a moisture gradient between adjacent, equal-temperature surfaces of supercooled water and ice. Thus an ice surface can grow by gradient transfer of vapor from the supercooled liquid. Third, to form an ice particle by direct sublimation of water vapor or by freezing of a supercooled liquid droplet (above −40°C) requires the presence of nuclei with special properties. These are called freezing or ice nuclei, or IN. Fourth, the concentrations of IN in the atmosphere are typically about $10^{-3}/cm^3$ and thus supercooled cloud droplets formed on CCN can outnumber the ice particles formed by the IN by a factor of 10^5. However, since water vapor will migrate from the supercooled droplets to the much fewer ice particles, the ice particles can accumulate mass rapidly and reach precipitation size as an ice crystal. Snowflakes can result from the agglomeration of several crystals, and if the snow passes into a warm layer the particle will melt and fall as rain.

Potentially, air pollutants can influence the cold cloud precipitation process in ways analogous to the warm cloud, namely, in increasing the concentration of IN, and, in fact, this is the goal of cloud-seeding operations for weather modification purposes. For small increases in IN, such as are caused intentionally by cloud seeding, the expected effect is to increase precipitation by increasing the rate of formation of precipitation-sized particles. However, a vast increase in IN has the potential effect of "overseeding," the creation of an excess number of ice particles, and reducing the probability of forming precipitation-sized particles. Although overseeding is possible in a cold cloud system, there is no apparent research that shows that these conditions have resulted from urban pollutants. The logical reason is that the IN apparently require a very special chemistry or at least crystal structure and this is not a common property. Studies by Ogden[31] on steel mill effluents in Australia did not indicate any increase in IN in the downwind plume. Hobbs et al.[32] also failed to find any enhancement in smelter and other industrial plumes in the U.S. Pacific Northwest. Hidy et al.[33] failed to find elevated IN concentrations in the Los Angeles, California air pollution area and there may have been a decrease of the IN concentration by a very small amount.

3.3   Cloud and Precipitation Chemistry

The impact of air pollutants on the chemistry of precipitation through the pickup of pollutants by cloud droplets and precipitation particles is probably much more important than any of the impacts on the precipitation formation mechanisms described above. The major effect is probably a lowering of the pH of precipitation as a result of the entrainment of acidic compounds.

Pollutant entrainment into precipitation can occur within the cloud by diffusion of the pollutants to cloud droplets. This process is called rainout and because of the long contact times that can occur in clouds this is probably the more effective accumulation mechanism for pollutant absorption, especially when the pollutants are relatively uniformly mixed through the atmosphere as would be the case far downwind from a source. When precipitation falls through a polluted layer, the entrainment process is called "washout." Washout may be important where a heavily contaminated layer is present close

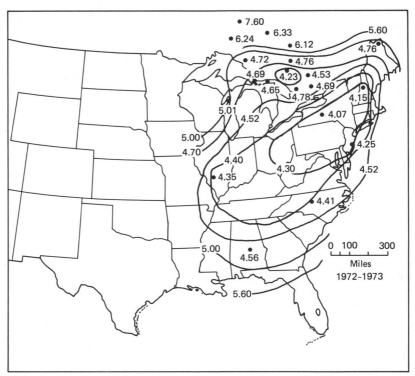

Fig. 8   Distribution of observed precipitation pH during 1972–1973. Adapted from Cogbill, 1976.[35]

to the ground, but the process is not too effective for a well-mixed atmosphere. Junge[34] provides a detailed discussion of the precipitation-scavenging processes of rainout and washout.

Concern about possible changes in precipitation chemistry is due to the potential damage to the biosphere resulting from changes in aquatic and terrestrial ecosystem conditions. Figure 8 from Cogbill[35] shows the observed pattern of average precipitation pH over the United States during 1972–1973. The area of relatively low pH in the Northeast including southeastern Canada is frequently attributed to the long-range influence of combustion emissions, especially from the Ohio Valley states of Ohio, Indiana, Illinois, and Kentucky, on precipitation chemistry. This has apparently become an increasingly severe problem in the past few decades, according to Galloway et al.[36] The change in precipitation pH can be due to a number of materials but the most important are the sulfur oxide and nitrogen oxide pollutants. Sulfur gas emissions may be picked up by precipitation either in gaseous SO_2 form or as a sulfate or H_2SO_4 aerosol. The result is an increase in the acidity of the precipitation as well as an increase in surface concentration. Husar et al.[37] have estimated that roughly half the sulfur gas emission from a typical midwestern power plant will be removed from the atmosphere by precipitation and thus it can play a major role in the chemistry and pH problem. Roughly two-thirds of the precipitation pH problem in the northeastern United States is attributed to sulfur pollutant emissions, and the remainder is attributed to nitrogen oxide emissions. Nitrogen oxides, such as NO and NO_2, are oxidized in the atmosphere to nitrates and HNO_3 and when these compounds are picked up by cloud and precipitation particles they also contribute to low pH values.

This problem of excessively low pH in precipitation was originally identified in Scandinavia in the late 1960s. It was quickly dubbed as "acid rain" for obvious reasons. The problem in Scandinavia seems strongly related to the increased sulfur pollutant emissions in the rest of Europe although at least some of the impact is due to local emissions. More research on precipitation pH probably has been done in Europe than in North America where full identification of pH as a serious problem did not seem to occur until the late 1970s. In many areas excessive acidity in precipitation seems to cause little impact on the natural soils and water bodies because there is sufficient natural buffering in the soil to counter the pH change in the precipitation. Deep organic soils seem to have this buffering capacity. However, thin glacial soils, such as are found in New England and in Scandinavia, do not have sufficient buffering capacity, and in these areas low-precipitation pH can cause changes in soils and this can in turn induce both pH and chemical changes in water bodies. Changes in both soil and water can bring about changes in the biosphere. Although widespread effects have apparently not been fully documented, eventual identification can be expected.

4 IMPACTS ON GLOBAL METEOROLOGY

4.1 Introductory Comments

The air pollutant effects on the atmosphere discussed in the previous sections have been essentially small-scale effects. Visibility is primarily a local problem with some regional situations; precipitation chemistry changes are essentially regional in scope. There are, however, some air pollutant situations that can have an impact on much larger scales and up to a potential global scale. These large-scale problems include the impact of increasing CO_2 concentrations in the atmosphere, possible climate change from widespread haze layers, and the accumulation in the atmosphere of essentially nonreactive organic chlorine compounds such as the fluorocarbons. The potential atmospheric impacts of these materials will be described briefly in this section.

4.2 Impact of CO_2 in the Atmosphere

Since the beginning of the Industrial Revolution humans have been gradually increasing the CO_2 content of the atmosphere through the combustion of fossil fuels. This change has been monitored by determinations of CO_2 concentration since late in the nineteenth century when the CO_2 concentration was about 290 ppm, a value that was probably quite close to preindustrial levels. Since then, however, the CO_2 concentration in the atmosphere has been steadily increasing, as shown in Figure 9.[38] Much more detailed data gathered since the mid-1950s show that the present annual rate of increase may be as much as 1 ppm and that global concentrations are around 340 ppm in the late 1970s. This increase indicates that about half of the CO_2 emitted by fuel combustion remains in the atmosphere. It is usually concluded that the remainder is either incorporated into the earth's biomass or the oceans.

The concern over an increase in atmospheric CO_2 is due to the fact that CO_2 plays a vital role in the thermal radiation regime of the earth. The impact is that an increase in atmospheric CO_2 content is generally expected to cause an increase in the earth's average temperature, although the relationships are complex. A variety of increasingly sophisticated modeling studies are being undertaken to examine this problem. In general the results, as summarized by Schneider,[39] indicate that an approximate doubling of CO_2 may occur by about 2040 and that this could raise the average global temperature by 2 to 3°C. In the polar regions the increase would be several times larger. By the year 2000, only about 20 yr hence at this writing, CO_2 levels of about 400 ppm are considered likely and this could produce an increase of about 1°C in the global average temperature, with larger increases in the polar areas.

The possible impacts of anthropogenic-induced temperature changes are still a subject of much debate. Certainly the impact is more than just "turning up the thermostat" because the motions of

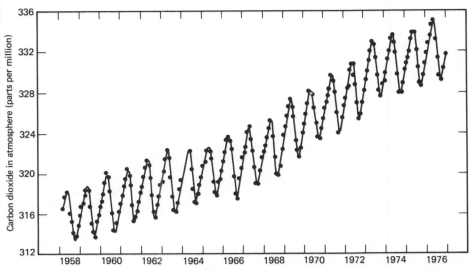

Fig. 9 Trend in atmospheric CO_2 as measured over the time period 1958–1976 by C. D. Keeling at Mauna Loa Observatory, Hawaii. Adapted from Woodwell, 1978.[38]

the atmosphere, for example, the winter storms, are responsive to the global temperature distribution, and this will be altered by the radiation balance response to CO_2. Water vapor contents can also change with temperature. Thus a set of complex temperature and precipitation climate changes could be the end result to the expected changes in atmospheric CO_2 concentrations.

4.3   Impacts of Atmospheric Aerosols

Some investigators, in particular Bolin and Charlson,[40] suggested that the extensive haze layers that can accompany large-scale air mass stagnation may also have some climatic implications. In the preceding discussion of visibility it was pointed out that atmospheric aerosol particles can cause the intensity of solar radiation to decrease due to both scattering and absorption losses. Depending on the balance between scattering and absorption and on the albedo of the earth's surface, as pointed out by Mitchell,[41] the impact of the typical air pollution aerosol may be either cooling or heating.

The investigations of Bolin and Charlson[40] have emphasized the radiation-scattering properties of the atmospheric aerosol because of their conclusion that submicrometer sulfate particles dominate the regional air mass aerosols. Their calculations indicate that radiation-scattering losses could cause temperature to decrease as much as 1°C in the air mass regions affected. They argue further that while this is a regional effect it is acting over a sufficiently large area so as to have some global impact. Thus, as also pointed out by Schneider and Dennett,[42] although there is little evidence of significant global increases of anthropogenic aerosols, it is not clear that the potential regional effects are too small to have an impact on the global radiation balance.

4.4   Impacts of Fluorocarbons and Other "Nonreactive" Materials

For many years it was a basic tenet of atmospheric chemistry that there were two general classes of chemical compounds—"reactive" and "nonreactive"—and that nonreactive compounds such as the fluorocarbons could be ignored as far as potential anthropogenic impacts on the atmosphere were concerned. The modeling studies of Molina and Roland[43] on the potential impacts of the fluorocarbons F-11 and F-12 on the ozone layer in the stratosphere changed that picture by showing that these compounds broke down under stratospheric UV radiation and the resulting chlorine release could increase ozone-scavenging processes. The end result could be a net reduction of the ozone layer by 5–10% as summarized in the 1979 report by the U.S. National Academy of Sciences.[44] The NAS report also summarizes the impact of a diminished ozone layer and indicates that an increased probability of skin cancer seems to be one of the serious results. In addition, changes in the biosphere and also global climate cannot be ruled out.

Fluorocarbons reach the atmosphere as a result of their end product uses which terminate with release to the atmosphere. This has seldom been controlled or regulated because the compounds are nontoxic and nonflammable. Although the most popular application, as a spray can propellant, is apparently being limited by regulation, the fluorocarbons are still essential in refrigeration systems and several industrial processes; thus their release to the atmosphere is still widespread, essentially global.

The rate of change of atmospheric fluorocarbon concentrations has been around 10%/yr for the past 5 or more years, the time limit of available, reliable measurement data. Figure 10 shows a time history of F-11 and F-12 concentrations in the Northern and Southern Hemispheres for the period from 1975–1980 as developed by Robinson.[45] A continuation of the regression line for the period after 1980 is expected to show a continued decrease in the rate of growth of the concentrations as a result of the legislation restricting the uses of fluorocarbon propellants. More data will be necessary before this is confirmed. The differences between the Northern Hemisphere, Pullman, Washington (47°N) and the Antarctic concentrations result from the 1–2 yr lag in mixing between the hemispheres and the fact that 90% or more of the fluorocarbons are probably released in the Northern Hemisphere, probably north of 30°N.

The initial atmospheric data on F-11 and F-12 was gathered by Lovelock et al.[46] Since that time a number of investigators have made concerted efforts to identify and monitor for similar halocarbon compounds in the atmosphere. A relatively large number have been identified, see Singh et al.;[47] but the most important seem to be some of the chlorocarbon solvents, in particular carbon tetrachloride, CCl_4, and methylchloroform, CH_3CCl_3. The concentrations of both seem to be due to anthropogenic uses. Ambient concentrations of CCl_4 are in the range of 125–130 ppt (parts per 10^{12}) with little evidence of an annual trend. Concentrations of CH_3CCl_3, however, have been increasing rapidly, perhaps at a rate of 20–30% annually, because of rapidly expanding usage. Usage has grown because this solvent had little rapid photochemical activity and thus was "safe" for use and release in urban photochemical smog situations. The growth in usage came as CH_3CCl_3 was substituted for more reactive solvents. However, CH_3CCl_3 may pose a long-term threat to the stratospheric ozone layer because of the reactions similar to those for F-11 and F-12.

From a meteorological standpoint, pollutants diffuse slowly from the troposphere to the stratospheric ozone layer, which is mainly between 20 and 50 km. The transport time is measured in years because

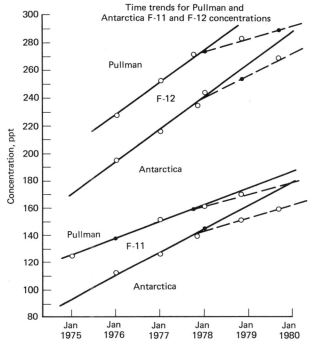

Fig. 10   Time trends for fluorocarbons 11 and 12 over the time period 1975–1980 in Antarctica and at Pullman, Washington. Adapted from Robinson, 1980.[45]

of the general resistance in the stratosphere to vertical motions. Thus the Academy report[44] indicates that the major impacts on the ozone layer may not occur until the 1990s, even from past emissions. However, since current arguments are mainly on the severity of the ozone layer impact rather than on whether one might occur, it would certainly seem prudent to include increased controls on halocarbon compounds at an early date.

One other material, nitrous oxide (N_2O), has also been of concern with regard to an impact on the ozone layer. N_2O has been known for a considerable time to be a constant constituent of the atmosphere that was produced by the biosphere and destroyed in the ozone layer accompanied by ozone destruction.[34,48] Thus the ozone layer has adapted to this situation over geologic time. It has recently been proposed, however, that increased use of nitrogen fertilizers in agriculture may accelerate the release of N_2O, increase the N_2O concentration in the atmosphere, and thus change the ozone layer balance. At present, small-scale research studies seem to show some increase in N_2O release from the soil with increased fertilizer applications. However, there are no detectable time or geographic trends in the N_2O monitoring data that would indicate that a significant change has yet occurred in atmospheric N_2O concentrations.

5 CONCLUDING REMARKS

The atmosphere is the transport medium for air pollutants linking the pollutant source with receptors. The atmosphere also plays the role of a receptor in that its characteristics may exhibit an adverse response to air pollutants, just as do human populations, vegetation, and animals.

Visibility effects have long served as a qualitative indication of air pollution severity for the average citizen; now with the greater recognition of long-range transport of pollutants concern about visibility impacts have moved well beyond urban areas and include the effects of pollutants on distant scenic areas. This doubtless indicates an important trend in air pollution programs, not only in the United States but in other industrial nations, toward a more comprehensive appraisal of air pollution impacts. Visibility impacts are mainly aesthetic, although considerable economic losses have sometimes been related to reduced visibility.

It is important for the air pollution engineer to understand the nature of visibility impairment by air pollutants so that the most effective and efficient control programs can be installed and operated. The fact that visibility problems may only become evident at some distance from pollutant sources is an additional reason why there should be an understanding of the air pollution visibility mechanisms.

The other impacts of air pollutants on the atmosphere—pH, precipitation processes, and trace chemistry—are generally not apparent to the general citizen or to anyone except an experienced investigator. This does not make the effects any less real however, and as research provides more substantiation of these effects engineers will be required to include considerations of them in their control options. This could be so extreme as a need to control CO_2 emissions from combustion processes as a step to slow the growth of CO_2 in the atmosphere. Currently, even relatively small releases to the atmosphere of materials previously considered to be harmless are now suspect after the ozone layer impacts of the fluorocarbons have been recognized.

Thus, in conclusion, engineers must include the recognition of potential hazards to the atmosphere and atmospheric processes in their evaluation of air pollution control options and control strategies. Since the atmospheric impacts may not be readily visible nor have the emphasis of possible health impacts, it is all the more important for engineers to understand why these impacts occur.

REFERENCES

1. Huschke, R. E. (ed.), *Glossary of Meteorology,* American Meteorological Society, Boston, Massachusetts, 1959.

2. Lewis, W. H., Jr., Protection Against Visibility Impairment Under the Clean Air Act, *J. Air Pollut. Control Assoc.* **30**(2), 118–120 (1980).

3. Horne, J., Visibility Regulations, *J. Air Pollut. Control Assoc.* **30**(2), 120–122 (1980).

4. Middleton, W. E. K., *Vision through the Atmosphere,* University of Toronto Press, Ottawa, Canada, 1952.

5. Koschmieder, H., Theorie der Horizontalen Sichtweite, *Beitr. Phys. frei. Atmos.* **12,** 33–35, 171–181 (1924).

6. Waldram, J. M., Measurement of the Photometric Properties of the Upper Atmosphere, *Quart. J. R. Meteorol. Soc.* **71**, 319–336 (1945).

7. Robinson, E., Effect (of air pollution) on the Physical Properties of the Atmosphere, Chapter 1, in *Air Pollution,* Vol. 2, 3rd edition (A. C. Stern, ed.), Academic Press, New York, 1977.

8. Friedlander, S. K., *Smoke, Dust, and Haze—Fundamentals of Aerosol Behavior,* Wiley, New York, 1977.

9. Charlson, R. J., Covert, D. S., Tokiwa, Y., and Mueller, P. K., Multiwave Nephelometer Measurements in Los Angeles Smog Aerosol, III. Comparison to Light Extinction by NO_2, in *Aerosols and Atmospheric Chemistry* (G. M. Hidy, ed.), Academic Press, New York, 1972.

10. Ursenbach, W. O., Hill, A. C., Edwards, W. H., and Kunen, S. M., "Atmospheric Particulate Sulfate in the Western United States," Paper No. 76–7, Air Pollution Control Association, 1976 Annual Meeting, Portland, Oregon, June 1976.

11. Mie, G., Beitrage zur Optak Truber Medien, speziell kolloidaler Metallosugen, *Ann. der Phys.* **25**, 377–445 (1908).

12. McCartney, E. J., *Optics of the Atmosphere,* Wiley, New York, 1976.

13. van de Hulst, H. C., *Light Scattering by Small Particles,* Wiley, New York, 1957.

14. Holzworth, G. C., and Maga, J. A., A Method for Analyzing the Trend in Visibility, *J. Air Pollut. Control Assoc.* **10**, 430–435 (1960).

15. Butcher, S. S., and Charlson, R. J., *An Introduction to Air Chemistry,* Academic Press, New York, 1972, pp. 195–206.

16. U.S. Department of Commerce, *Climatic Atlas of the United States,* U.S. Government Printing Office, Washington, D.C., 1968.

17. Korshover, J., "Climatology of Stagnating Anticyclones East of the Rocky Mountains," 1936–1965, U.S. Public Health Service, Pub. No. 999-AP-34, Cincinnati, Ohio, 1967.

18. Holzworth, G. C., Some Effects on Visibility in and Near Cities, in *Air Over Cities—A Symposium,* Robert A. Taft Sanitary Engineering Center, SEC Tech. Report A62-5, Cincinnati, Ohio, 1962, pp. 69–88.

19. Miller, M. E., Canfield, N. L., Ritter, T. A., and Weaver, C. R., Visibility Changes in Ohio, Kentucky, and Tennessee from 1962 to 1969, *Mon. Weather Rev.* **100**, 67–71 (1962).

20. Trijonis, J., and Yuan, K., "Visibility in the Northeast," EPA-600/3-78-075, Environmental Sciences Research Lab., Environmental Protection Agency, Research Triangle Park, NC, August 1978.

21. Hall, F. P., Jr., Durhon, C. E., Lee, L. G., and Hagan, R. R., Long-Range Transport of Air Pollution: A Case Study, August 1970, *Mon. Weather Rev.* **101**, 404–411 (1973).

22. Husar, R. B., Gillani, N. V., Husar, J. D., Paley, C. C., and Turcu, P. N., Long-Range Transport of Pollutants Observed Through Visibility Contour Maps, Weather Maps and Trajectory Analysis,

American Meteorological Society Preprints, Third Symposium on Atmospheric Turbulence, Diffusion, and Air Quality, Raleigh, NC, October 1976, pp. 344–347.

23. Lyons, W. A., and Husar, R. B., SMS/GOES Visible Images Detect a Synoptic-Scale Air Pollution Episode, *Mon. Weather Rev.* **104,** 1623–1626 (1976).

24. Hewson, E. W., Meteorological Measurements, Chapter 11, in *Air Pollution,* Vol 1, 3rd edition (A. C. Stern, ed.), Academic Press, New York, 1977.

25. Tombach, I., and Allard, D., Intercomparison of Visibility Measurement Methods, *J. Air Pollut. Control Assoc.* **30**(2), 134–142 (1980).

26. Steffens, C., Measurement of Visibility by Photographic Photometry, *Ind. Eng. Chem.* **41,** 2396–2399 (1949).

27. Ahlquist, N. C., and Charlson, R. J., A New Instrument for Evaluating the Visual Quality of Air, *J. Air Pollut. Control Assoc.* **17,** 467–479 (1967).

28. Charlson, R. J., Atmospheric Visibility Related to Aerosol Mass Concentration, a Review, *Environ. Sci. Technol.* **3,** 913–918 (1969).

29. Samuels, H. J., Twiss, S., and Wong, E. W., "Visibility, Light Scattering and Mass Concentration of Particulate Matter," Air Resources Board, State of California, Sacramento, California, 1973.

30. Wallace, J. M., and Hobbs, P. V., *Atmospheric Science,* Academic Press, New York, 1977, 467 pages.

31. Ogden, T. L., The Effect on Rainfall of a Large Steelworks, *J. Appl. Meteorol.* **8,** 585–591 (1969).

32. Hobbs, P. V., Radke, L. F., and Shumway, S. E., Cloud Condensation Nuclei from Industrial Sources and Their Apparent Influence on Precipitation in Washington State, *J. Atmos. Sci.* **27,** 81–89 (1970).

33. Hidy, G. M., Green, W., and Alkegweeny, A., Inadvertent Weather Modification and Los Angeles Smog, in *Aerosols and Atmospheric Chemistry* (G. M. Hidy, ed.), Academic Press, New York, 1972, pp. 339–344.

34. Junge, C. E., *Air Chemistry and Radioactivity,* Academic Press, New York, 1963.

35. Cogbill, C. V., The History and Character of Acid Precipitation in Eastern North America, *Water, Air, and Soil Pollut.* **6,** 407–413 (1976).

36. Galloway, J. N., Likens, G. E., and Edgerton, E. S., Acid Precipitation in the Northeastern United States: pH and Acidity, *Science* **194,** 722–724 (1976).

37. Husar, R. B., Patterson, D. E., Husar, J. D., Gillani, N. V., and Wilson, W. E., Sulfur Budget of a Power Plant Plume, *Atmos. Environ.* **12,** 549–568 (1978).

38. Woodwell, G. M., The Carbon Dioxide Question, *Sci. Amer.* **238,** 34–43 (1978).

39. Schneider, S. H., On the Carbon Dioxide-Climate Confusion, *J. Atmos. Sci.* **32,** 2060–2066 (1975).

40. Bolin, B., and Charlson, R. J., On the Role of the Tropospheric Sulfur Cycle in the Shortwave Radiative Climate of the Earth, *Ambio* **5,** 47–54 (1975).

41. Mitchell, J. M., Jr., The Effect of Atmospheric Aerosols on Climate with Special Reference to Temperature Near the Earth's Surface, *J. Appl. Meteorol.* **10,** 702–714 (1971).

42. Schneider, S. H., and Dennett, R. D., Climatic Barriers to Long-Term Energy Growth, *Ambio* **4**(2), 65–74 (1975).

43. Molina, M. J., and Rowland, F. S., Stratospheric Sink for Chlorofluoromethanes: Chlorine Atom Catalysed Destruction of Ozone, *Nature* **249,** 810 (1974).

44. National Academy of Science, *Stratospheric Ozone Depletion by Halocarbons,* Chemistry and Transport, Washington, D.C., 1979, 25 pages.

45. Robinson, E., Atmospheric Trace Gases of Antarctic Ocean Areas, *Antarctic Journal of the United States* **15,** 176–177 (1980).

46. Lovelock, J. E., Maggs, P. J., and Wade, R. J., Halogenated Hydrocarbons in and over the Atlantic, *Nature* **241,** 194 (1973).

47. Singh, H. B., Salas, L. J., Shigeishi, H., and Scribner, E., Atmospheric Halocarbons, Hydrocarbons, and Sulfur Hexafluoride: Global Distribution, Sources, and Sinks, *Science* **203,** 899 (1979).

48. Bates, D. R., and Witherspoon, A. E., The Photo-Chemistry of Some Minor Constituents of the Earth's Atmosphere, *Mon. Notices Roy. Astron. Soc.* **112,** 101–124 (1952).

CHAPTER 5
GASEOUS POLLUTANT CHARACTERISTICS

BRIAN LAMB

Laboratory for Atmospheric Research
College of Engineering
Washington State University
Pullman, Washington

1 INTRODUCTION

In air pollution control technology, the physical and chemical properties of gaseous pollutants provide the basis for understanding formation mechanisms, describing atmospheric transformations, preparing sensitive and accurate analytical procedures, and, as a final result, developing effective, economical control techniques and strategies. This chapter outlines the physical properties of gaseous pollutants and describes chemistry pertinent to their formation, control, and analysis. Atmospheric transformations are considered in detail in Chapter 35. Because the abundance of an air pollutant in the atmosphere has a direct impact on control efforts, atmospheric chemistry, and also measurement procedures, the occurrence of the various pollutants will also be considered.

Primary air pollutants for which air quality standards have been formulated include sulfur dioxide (SO_2), the nitrogen oxides (NO and NO_2), carbon monoxide (CO), and gaseous hydrocarbons (HC).[1-4] Other pollutants emitted directly to the atmosphere include sulfur trioxide (SO_3), reduced sulfur compounds such as hydrogen sulfide (H_2S) and carbon disulfide (CS_2), ammonia (NH_3), specific hazardous hydrocarbons such as benzene (C_6H_6), and a variety of halogenated gases including the chlorofluorocarbons, hydrogen fluoride (HF), hydrogen chloride (HCl), and vinyl chloride (CH_2CHCl).

Generally, the formation of these gaseous pollutants can be considered in three categories: combustion sources, industrial manufacturing processes, and natural emission mechanisms. In the case of sulfur dioxide, emissions from fuel combustion sources account for 78% of the SO_2 emission rate in the United States.[5] Natural sources of sulfur in the northeastern United States are estimated to contribute only 0.6% of the anthropogenic flux.[6] Hydrocarbons, controlled because of their role as precursors in ozone formation, are formed during the combustion of fossil fuels and also during the processing of petroleum products. However, a wide variety of reactive hydrocarbon compounds are also released from vegetation during growth and decay.[7] On a global basis, natural sources of hydrocarbons are estimated to release 117×10^6 tons/yr while anthropogenic sources are estimated to contribute 100×10^6 tons annually.[8] However, because of the concentrated nature of anthropogenic sources, urban hydrocarbon concentrations are dominated by combustion products.[9]

2 SULFUR COMPOUNDS

Sulfur occurs primarily as metal sulfides and pyrites in the earth's surface, as sulfate salts in the oceans, and as sulfur dioxide, hydrogen sulfide, and sulfate aerosol in the atmosphere. Reduced sulfur organics constitute the greater portion of sulfur in the biosphere and in soils. The chemistry associated with the global sulfur cycle involves sulfide oxidation by weathering processes to sulfates, microbiological reduction of sulfates to organic sulfides, and atmospheric oxidation of organic sulfides to sulfur dioxide and then to sulfate aerosols.

Table 1 Physical Properties of Sulfur Pollutants

	SO_2	SO_3	H_2S	CH_3SH
Molecular weight (g)	64.06	80.10	34.08	48.10
Melting point (°C)	−75.5	16.77 (α) 62.2 (γ)	−85.6	−121.0
Boiling point (°C)	−10.2	44.8	−60.75	5.96
Critical temperature (°C)	−157.12	218.3	100.4	196.8
Critical pressure (atm)	77.7	83.8	88.9	71.4
Density (g/ml) (liq)	1.46	1.992	0.993	0.896 (0°C)
(g/l) (g)	2.93 (0°C)	—	1.53	2.14 (15°C)
Solubility (H_2O)	10.5 g/100 g (20°C)	H_2SO_4 formed	0.34 g/100 g (25°C)	2.40 g/100 g (15°C)
Bond length (Å)	1.43	1.43	1.35	
Bond energy (kcal/mole)	131	—	90	88 (S—H)
Vapor pressure relation (liq)	2.34 atm (21°C)^a	—	17.14 atm	0.748 atm (21°C)
ΔH_f° (kcal/mole)	−70.96	—	4.80	—
Specific heat C_v (cal/deg-mole)	9.51	—	6.53	21.13 (C_p)
Odor threshold (ppm)	1000–5000	—	0.9–8.5	0.6–40.0
Viscosity (cP)	0.01242 (18°C)	—	0.01166 (0°C)	—

$^a \log P = a + b/T - cT + dT^2$

$a = 12.07540 \quad c = 1.5865 \times 10^{-2}$

$b = 1867.52 \quad d = 1.5574 \times 10^{-5}$

2.1 Physical Properties and Occurrence of Sulfur Compounds

2.1.1 Sulfur Dioxide

Gaseous sulfur oxides occur as sulfur dioxide (SO_2) and sulfur trioxide (SO_3) under ambient conditions. Although other oxides of sulfur [SO, S_2O, $(SO)_2$, SOO] and polymeric oxides [S_mO_n, $(SO_3)_n$] exist or are postulated, only the very reactive monoxide SO is assumed to play a role in the air pollution chemistry of sulfur.

Sulfur dioxide is a colorless gas with an irritating, pungent odor. It is nonexplosive and nonflammable. The threshold odor concentration (TOC) for recognition by 50% of the population equals 1000 ppm.[10] Sulfur dioxide is detectable by taste at levels of 0.3–1 ppm. Sulfur dioxide is very soluble in water (10.5 g/100 ml at 20°C) which is the basis for flue gas scrubbing processes and for the ready formation of sulfuric acid in water droplets. Physical properties of SO_2 are presented in Table 1.

Sulfur dioxide has an S—O bond length of 1.432 Å and a bond angle O—S—O equal to 119.5°. The SO_2 molecule is polar with a dipole moment of 1.62 D. For commercial purposes, SO_2 is formed by combustion of elemental sulfur, combustion of H_2S from natural gas, and roasting of metal sulfides with oxygen or metal oxides. Iron pyrite (FeS_2), marcasite (FeS_2), magnetic pyrites (FeS), copper pyrites ($CuFeS_2$), and zinc blende (ZnS) are raw materials. Sulfur dioxide is used primarily as feedstock in the production of sulfuric acid. Sulfur dioxide is also used for the preparation of sulfites, dithionites, and sulfuochlorinated hydrocarbons, as a bleach, disinfectant, and preservative. There is a continuing effort to employ SO_2 in new products. Chief among these are sulfur asphalts, sulfur foams and polymers, and sulfur plastics.

As a global air pollutant, concentrations of SO_2 are estimated to range from 0.04 to 6 ppb.[11] In contrast, hourly maximum concentrations in urban areas range from 0.26 ppm in San Francisco to 1.69 ppm in Chicago.[12] Altshuller has reported a decreasing trend in SO_2 concentrations in the eastern United States which corresponds to a decrease in SO_2 emissions.[13]

A variety of estimates have been prepared which attempt to detail the various atmospheric sources of sulfur dioxide. As Table 2 illustrates, stationary combustion sources account for 76% of all U.S. SO_x (90% SO_2 and 10% SO_3) emissions, while metal and petroleum refining contribute another 19%.[14] The remainder of SO_x emissions are distributed among mobile sources, sulfuric acid production, coking, and forest fires. On a worldwide basis, volcanoes and geothermal activity constitute the major natural source of SO_2 with an estimated annual emission rate of 3×10^6 tons/yr.[11] The violent eruption of Mt. St. Helens in Washington during May 1980 emitted SO_2 into the atmosphere at a rate of 9 kg/sec over a period of a few hours.[15]

Table 2   Estimated SO_x Emissions in the United States (1970)[14]

Source Category	SO_x Emission, 10^6 tons/yr	Percentage of Total
Transportation	1.0	2.89
Fuel combustion in stationary sources	26.5	76.52
Industrial process losses		
Pulp and paper	0.077	0.22
Calcium carbide	0.002	0.01
Sulfuric acid plants	0.474	1.37
Claus sulfur plants	0.875	2.53
Coking	0.474	1.37
Petroleum refining		
Fluid catalytic cracking	0.354	1.02
Thermal catalytic cracking	0.005	0.01
Nonferrous metals		
Copper	3.57	10.31
Zinc and lead	0.9	2.60
Solid wastes disposal	0.1	0.29
Agricultural burning	0.1	0.29
Miscellaneous		
Coal refuse burning	0.2	0.58
Total	34.63	100.00

2.1.2 Sulfur Trioxide

Sulfur trioxide is produced along with SO_2 in the combustion of sulfur at approximately 1–10% of the SO_2 concentration. In air, SO_3 has a very short lifetime (10^{-6} sec) because of its rapid reaction with water vapor to form sulfuric acid. Physical properties of SO_3 are tabulated in Table 1. The SO_3 molecule in the gas phase is monomeric with an SO bond length of 1.43 Å and an O—S—O bond angle of 120°. Because of the symmetry of its planar structure, the molecule has a zero dipole moment.

The colorless gas is produced commercially by catalytic oxidation of SO_2. It is stored as a liquid or as oleum (25% SO_3 in H_2SO_4). Because of its short lifetime, the atmospheric importance of sulfur trioxide lies in its rapid conversion to H_2SO_4. Analysis of SO_3 in air is difficult and ambient data are not routinely reported.

2.1.3 Hydrogen Sulfide

Hydrogen sulfide is a colorless, flammable gas that is highly toxic and has a characteristic rotten egglike odor. Physical properties of H_2S are shown in Table 1. In all states, H_2S is a bent molecule with a bond angle of 92.3° and a H—S bond length of 1.35 Å.

Hydrogen sulfide is emitted in large quantities in nature from biological decay processes and from volcanoes and geothermal activities. It is the major form of natural atmospheric sulfur emissions. Because of its relatively rapid oxidation to sulfur dioxide, H_2S indirectly is a large natural source of SO_2.

Anthropogenic sources of H_2S include kraft pulp mills, natural gas and petroleum refining facilities, rayon and nylon manufacturing plants, and coke ovens. As a result of the strong odor of H_2S, these industries generally employ procedures to convert H_2S to either sulfuric acid or elemental sulfur.

In the atmosphere H_2S has an estimated lifetime of 1 day. Average ambient concentrations of H_2S are estimated to equal 6 $\mu g/m^3$. Urban concentrations have been observed to range from average values of 3.0 $\mu g/m^3$ for New York City to a maximum level of 390 $\mu g/m^3$ detected in Houston, Texas.[12]

2.1.4 Reduced Sulfur Compounds

In addition to hydrogen sulfide, sulfur in a reduced state is emitted from natural and anthropogenic sources as carbon sulfide (CS), carbonyl sulfide (COS), carbon disulfide (CS_2), and a variety of organosulfides and disulfides (RSH, RSR, RS_2R; R=C_1—C_5 paraffins). In general, the methyl mercaptans and sulfides are odorous gases (COS is an exception) which can undergo a series of oxidation reactions in the atmosphere. Physical properties for methyl mercaptan are presented in Table 1. Threshold odor concentrations for reduced sulfur gases are as low as 0.1 ppb for dimethyl sulfide (CH_3SCH_3).[10]

Reduced sulfur gases are emitted naturally as a result of biological decay and geothermal activity. Anthropogenic sources include those indicated for H_2S previously. The atmospheric lifetimes of various reduced sulfur gases range from less than 3 hr for methyl mercaptan to almost 2 yr for carbonyl sulfide.[12] Atmospheric concentrations of these gases are difficult to measure and have not been widely reported.[16]

2.2 Formation Mechanisms

2.2.1 Combustion

Sulfur emissions in the form of SO_2 and SO_3 arise from combustion sources because of trace amounts of inorganic and organic sulfur contained in the fuel. Coal combustion accounts for approximately 70% of global SO_2 emissions. Sulfur content in coal varies widely from deposit to deposit. Typically the total sulfur content in coal ranges from 1.0 to 4.0 wt%.[17] The ratio of inorganic to organic sulfur in coal varies from 4:1 to 1:3 with an average value of approximately 2:1. Inorganic sulfur consists primarily of disulfides and sulfates. Following coal pulverization, these heavy mineral compounds can be removed mechanically to some extent. Organic sulfur is bound to the hydrocarbon matrix of coal and exists as sulfides and thiphenic material. The exact structure of sulfur in coal is poorly understood. Organic sulfur can only be removed from coal through coal gasification techniques.

Petroleum oils contain sulfur primarily as mercaptans, organic sulfides, cyclic compounds, thiophene, and polysulfides. Sulfur content in oils varies widely; typical sulfur content ranges between 0.1 and 3%.[18]

Heating coal volatilizes the organics and, to a lesser extent inorganics, forming H_2S, some CS_2, COS, and small amounts of thiophenes, thiols, and organic sulfides. Combustion mechanisms for H_2S have been outlined in detail and also in some detail for CS_2 and COS. Very little is known about the combustion of the remaining sulfur gases evolved in coal combustion.

The key steps in the combustion of H_2S involve the consumption of H_2S and SO:

$$H_2S + O \rightarrow SO + H_2 \tag{1}$$
$$SO + O_2 \rightarrow SO_2 + O \tag{2}$$

As H_2S is consumed, O atom concentrations increase and hydroxyl radicals are formed

$$O + H_2 \rightarrow OH + H \tag{3}$$

which serve to further oxidize sulfur

$$SO + OH \rightarrow SO_2 + H \tag{4}$$

A more detailed mechanism is shown in Table 3.[19] Combustion of COS has also been investigated and is expected to involve conversion of COS to CO, SO_2, and CO_2 as indicated in Table 3.[20] Similarly, CS_2 combustion yields CO via a pathway involving the production of COS and CS as shown in Table 3.

Table 3   Combustion of Reduced Sulfur Compounds[19,20]

1. H₂S Combustion Reactions

$H_2S + O_2 \rightarrow HO_2 + SH$	(a)
$H_2S + M \rightarrow SH + H + M$	(b)
$SH + O_2 \rightarrow OH + SO$	(c)
$H_2S + OH \rightarrow SH + H_2O$	(d)
$SO + O_2 \rightarrow SO_2 + O$	(e)
$H_2S + O \rightarrow H_2 + SO$	(f)
$H_2S + O \rightarrow SH + OH$	(g)

2. COS Combustion Reactions

Initiation $\rightarrow$ O	(a)
$COS + O \rightarrow CO + SO$	(b)
$COS + O \rightarrow CO_2 + S$	(c)
$S + O_2 \rightarrow SO + O$	(d)
$SO + O_2 \rightarrow SO_2 + O$	(e)
O $\rightarrow$ termination	(f)

3. CS Combustion Reactions

Initiation $\rightarrow$ O	(a)
$CS_2 + O \rightarrow CS + SO$	(b)
$CS_2 + O \rightarrow COS + S$	(c)
$CS + O \rightarrow CO + S$	(d)
$COS + O \rightarrow CO + SO$	(e)
$O_2 + SO \rightarrow SO_2 + O$	(f)
$O_2 + S \rightarrow SO + O$	(g)
$CS + O_2 \rightarrow CO + SO$	(h)
$CS_2 + S \rightarrow CS + S_2$	(i)
$S_2 + O \rightarrow SO + S$	(j)
O $\rightarrow$ termination	(k)

Sulfur trioxide, as indicated previously, is an important intermediate species in the sulfur oxidation chain leading to sulfuric acid and sulfates. Low concentrations of SO_3 (less than 10% of total emissions) are produced during combustion via the reaction

$$SO_2 + O + M \rightarrow SO_3 + M \tag{5}$$

At high temperatures, SO_3 may then be consumed by the reactions

$$SO_3 + O \rightarrow SO_2 + O_2 \tag{6}$$
$$SO_3 + H \rightarrow SO_2 + OH \tag{7}$$

The balance between the production and consumption steps quantitatively account for the low SO_3 levels present downstream of the flame.

2.2.2   Industrial Processes

Sulfur oxides are emitted in significant quantities from the smelting of sulfide ores to produce copper, lead, and zinc and also from petroleum refinery processes. A large portion of SO_2 emissions from the refinery sector is associated with combustion in boilers or process heaters; strictly process-related emissions arise from Claus sulfur plant and regenerators of catalytic cracking units. Sulfuric acid plants produce significant SO_x emissions through combustion of sulfur-containing raw materials and subsequent losses during conversion to SO_3 and then to H_2SO_4. Key reactions in the smelting of sulfide ores include

$$Cu_2S + O_2 \rightarrow 2Cu + SO_2 \tag{8}$$
$$FeS + \tfrac{3}{2} O_2 \rightarrow FeO + SO_2 \tag{9}$$

in the production of copper and iron, and

$$2PbS + 3O_2 \rightarrow 2PbO + 2SO_2 \tag{10}$$
$$ZnS + \tfrac{3}{2}O_2 \rightarrow ZnO + SO_2 \tag{11}$$

in the production of lead and zinc.[21] The resulting SO_2 gas streams are typically used as feedstock to sulfuric acid plants and elemental sulfur processes. Sulfur oxides are also produced during pulp and paper manufacturing as a result of combustion of sulfur-containing materials.

Reduced sulfur compounds, including hydrogen sulfides and organosulfur gases, are oxidized to SO_2 in most combustion processes. However, there are several major industrial processes that produce significant amounts of these odorous gases. In the production of fuel gases, coal carbonization yields varying amounts of carbon disulfide, thiophene (C_4H_4S), and carbonyl sulfide. Thermal cracking of hydrocarbon streams yields hydrogen sulfide through reactions such as

$$C_4H_9SH \rightarrow C_4H_8 + H_2S \tag{12}$$

Hydrogen sulfide is also the product in a variety of control procedures for organosulfur gases.[10] Catalytic oxidation of carbon disulfide under low O_2 conditions occurs according to

$$CS_2 + 3O_2 \rightarrow 3SO_2 + CO_2 \tag{13}$$
$$SO_2 + 3H_2 \rightarrow H_2S + 2H_2O \tag{14}$$

Hydrogenation of reduced sulfur compounds can be accomplished catalytically to yield hydrogen sulfide:

$$RSH + H_2 \leftrightarrows RH \tag{15}$$
$$C_4H_4S + 4H_2 \leftrightarrows C_4H_{10} + H_2S \tag{16}$$
$$CS_2 + 2H_2 \leftrightarrows C + 2H_2S \tag{17}$$
$$COS + H_2 \leftrightarrows CO + H_2S \tag{18}$$

Carbonyl sulfide and carbon disulfide can also be hydrolyzed catalytically with H_2S as the product

$$CS_2 + 2H_2O \leftrightarrows CO_2 + 2H_2S \tag{19}$$
$$COS + H_2O \leftrightarrows CO_2 + H_2S \tag{20}$$

Significant amounts of organosulfur gases are produced in kraft pulp production. Cellulose in wood chips is dissolved in a digester containing sodium sulfide and sodium hydroxide. A portion of the cellulose can be demethylated to form methyl mercaptan, dimethyl sulfide, and dimethyl disulfide. Subsequent evaporation of the spent cooking liquor through direct contact with hot furnace gases produces hydrogen sulfide:

$$Na_2S + CO_2 + H_2O \leftrightarrows H_2S + Na_2CO_3 \tag{21}$$

Hydrogen sulfide is also produced in a special liquor recovery furnace.[10]

Carbon disulfide is the primary feedstock in the production of viscose rayon. Cellulose is reacted with sodium hydroxide and then carbon disulfide to yield cellulose xanthate

$$- - - - - CH_2OH + NaOH \rightarrow - - - - - CH_2ONa + H_2O \tag{22}$$
$$- - - - - CH_2ONa + CS_2 \rightarrow - - - - - CH_2OCSSNa \tag{23}$$

which is then filtered, aged, and extruded into an acid bath to produce rayon.[10] Sodium sulfate and carbon disulfide are by-products of the extrusion process.

2.2.3 Natural Formation Mechanisms

In the natural environment, hydrogen sulfide and the mercaptans are produced from the decomposition by enzymes of amino acids from proteins.[22] For example, cysteine ($HS \cdot CH_2 \cdot CHNH_2 \cdot CO_2H$) is produced from bacterial decomposition of protein. In turn, cysteine can be decomposed by an enzyme classified as desulfhydrases with ammonia and hydrogen sulfide as by-products:

$$\underset{\text{cysteine}}{HS \cdot CH_2CHNH_2 \cdot CO_2H} \xrightarrow{-NH_3} \underset{\substack{\beta\text{-mercaptopyruvic} \\ \text{acid}}}{HS \cdot CH_2 \cdot COCO_2H} \xrightarrow{-H_2S} \underset{\text{pyruvic acid}}{CH_3 \cdot CO \cdot CO_2H} \tag{24}$$

Since organic matter is comprised of large amounts of proteins, the decomposition of proteins by bacteria in soils and water account for a significant amount of the natural emission of reduced sulfur compounds.[6,22,23,24]

2.3 Chemistry of Sulfur Compounds

Sulfur dioxide oxidizes hydrogen sulfide to produce elemental sulfur via the Claus reaction:

$$2H_2S + SO_2 \rightarrow \tfrac{3}{8}S_8 + 2H_2O \tag{25}$$

As a reducing agent, SO_2 reacts only slowly to form SO_3:

$$SO_2 + \tfrac{1}{2}O_2 \leftrightharpoons SO_3 \tag{26}$$

The equilibrium which is pushed to the left with increasing temperatures is moved far to the right in the presence of a catalyst such as V_2O_5.[25]

Decomposition of SO_2 can occur thermally (>1000°C) or photochemically (1950 Å) as given by

$$SO_2 + energy \rightarrow SO_2{}^* \quad \begin{array}{l} \rightarrow SO + O \\[4pt] \rightarrow SO + SO_3 \end{array} \tag{27}$$

The monoxide rapidly disproportionates to sulfur and SO_2. Reaction of SO_2 with metal oxides yields sulfates and sulfides. For example,

$$4MgO + 4SO_2 \rightarrow 3MgSO_4 + MgS \tag{28}$$
$$ZnO + SO_2 \rightarrow ZnS \tag{29}$$
$$PbO_2 + SO_2 \rightarrow PbSO_4 \tag{30}$$

At high temperatures, some metals react with SO_2 to form sulfides:

$$SO_2 + 3M \rightarrow MS + 2MO \tag{31}$$

Sulfur dioxide combines with fluorine spontaneously to form SO_2F_2 and with chlorine catalytically to yield SO_2Cl_2. With elevated temperatures (350°C), SO_2 reacts with carbon to yield sulfur and CO_2. Sulfur is also obtained by catalytically reacting roasting gases with SO_2:

$$SO_2 + 2CO \rightarrow S + 2CO_2 \tag{32}$$
$$SO_2 + 2COS \rightarrow S + 2CO_2 \tag{33}$$
$$SO_2 + CS_2 \rightarrow 3S + CO_2 \tag{34}$$

In the chamber sulfuric acid process, oxidation of SO_2 with NO_x yields SO_3. Hydrogen peroxide also oxidizes SO_2 to form sulfuric acid. This is the basis for the hydrogen peroxide method of measuring SO_2. Bubbling SO_2 through a dilute solution of sodium tetrachloromercurate yields a stable sulfite ion which in turn reacts with pararosaline to produce red methysulfonic acid. This sequence forms the basis for the West–Gaeke colorimetric method for measuring SO_2 concentrations in air. Further descriptions of sulfur chemistry are available from a number of sources.[26,27]

3 NITROGEN COMPOUNDS

Just as sulfur pollution occurs against the backdrop of a natural sulfur cycle, the air pollution chemistry of nitrogen is a perturbation of an ongoing natural nitrogen cycle. Diatomic nitrogen (N_2) constitutes

Table 4 Valence States of Nitrogen[28]

Valence State	Typical Compound(s)
−3	Ammonia (NH_3)
−2	Hydrazine (NH_2NH_2)
−1	Hydroxylamine (H_2NOH)
0	Nitrogen (N_2)
+1	Nitrous oxide (N_2O)
+2	Nitric oxide (NO)
+3	Nitrogen trioxide (N_2O_3); nitrous acid (HNO_2); nitrites (NO_2^-)
+4	Nitrogen dioxide (NO_2)
+5	Dinitrogen pentoxide (N_2O_5); nitric acid (HNO_3); nitrates (NO_3^-)
+6	Nitrogen trioxide (NO_3)

approximately 78% of our atmosphere and, as an element of the amino acid chain, nitrogen plays a dominant role in the chemistry of all living organisms. In the atmosphere, nitrogen exists at trace levels as ammonia (NH_3), nitrous oxide (N_2O), nitric oxide (NO), and nitrogen dioxide (NO_2). In the environment, nitrogen occurs in the reduced state as amides, amines, amino acids, and nitriles and in the oxidized state as a variety of oxides, nitro, nitroso, nitrite, and nitrate compounds. Table 4 lists the valence states of nitrogen along with representative compounds.

3.1 Physical Properties and Occurrence of Nitrogen Compounds

Nitrogen forms seven different oxides including nitrous oxide (N_2O), nitric oxide (NO), nitrogen dioxide (NO_2), nitrogen trioxide (NO_3), nitrogen sesquioxide (N_2O_3), nitrogen tetraoxide (N_2O_4), and nitrogen pentaoxide (N_2O_5). However, only NO and NO_2 exist as important air pollutants. Typically, the atmospheric concentrations of NO and NO_2 are lumped and reported as NO_x. Nitrous oxide is a stable, naturally emitted nitrogen compound which may be important in stratospheric chemistry. The remaining oxides are considered to be important only as intermediates in photochemical smog mechanisms. Formation of nitric acid (HNO_3) is an important sink for nitrogen in the atmosphere and nitric acid is one of the main components of acid precipitation.

3.1.1 Nitrous Oxide

Although N_2O is the primary nitrogen oxide emitted naturally, it is not generally considered an air contaminant because of its harmless nature. Nitrous oxide is a colorless gas with a slightly sweet taste and odor. Breathing low concentrations of N_2O in air produces a kind of hysteria and N_2O is familiarly known as "laughing gas." Physical properties of N_2O are listed in Table 5. The molecule has a linear structure and is a resonance hybrid:

$$N-N= O \leftrightharpoons N=N-O \tag{35}$$

Nitrous oxide is produced during decomposition of nitrogen compounds in the soil by anaerobic bacteria. The estimated global emission rate is 54×10^7 tons/yr.[29]

At wavelengths less than 2200 Å, N_2O is dissociated to nitrogen

$$N_2O + h\nu \rightarrow N_2 + O \tag{36}$$

and may also be decomposed to nitric oxide by reaction with oxygen atom

$$N_2O + O \rightarrow 2NO \tag{37}$$

These reactions occur in the stratosphere with the result that N_2O levels are well mixed in the troposphere and decrease above the tropopause. Typical background concentrations are shown in Figure 1. The participation of N_2O in stratospheric nitrogen and oxygen chemistry has raised questions, still unanswered, concerning N_2O loading of the atmosphere through the increased use of nitrogen fertilizers.

Table 5 Physical Properties of Nitrogen Gases

	N_2O	NO	NO_2	NH_3
Molecular weight (g)	44.02	30.01	46.01	17.03
Melting point (°C)	−90.84	−163.6	−11.3	−77.7
Boiling point (°C)	−89.5	−157.1	21.15	−33.37
Critical temperature (°C)	36.5	−93	158.0	132.45
Critical pressure (atm)	71.7	64	100	112.3
Density (g/ml) (liq)	1.23 (b.p.)	1.27 (b.p.)	1.45 (20°C)	0.674 (b.p.)
(g/l) (g)	3.10 (b.p.)	1.34	3.40 (21°C)	0.89 (b.p.)
Solubility (H_2O) g/100 g	0.256 (0°C)	0.0098 (0°C)	—	47.3
Bond length (Å)	1.19	1.14	1.19	1.008
Bond energy (kcal/mole)	40 (O—N)	150	73	103
Vapor pressure (atm)	50.68 (21°C)	1 (−151.7°C)	1 (21°C)	a
ΔH_f° (kcal/mole)	19.61	21.58	7.91	−11.04
Specific heat C_v (cal/deg-mole)	6.77 (15°C)	4.99 (15°C)	74.5 (C_p)	6.80
Viscosity (cP)	0.01362 (0°C)	0.0178	—	0.00918

$^a \log_{10} P = 9.95028 - 1473.17/T - 3.863 \times 10^{-3}T$

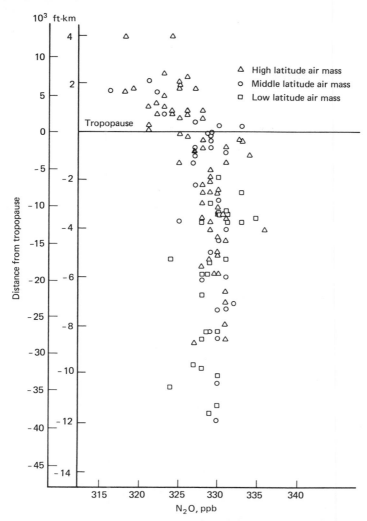

Fig. 1 N₂O concentration distribution relative to tropopause height, North America flights.[29]

3.1.2 Nitric Oxide

Nitric oxide is a colorless, odorless gas which is nonflammable and slightly soluble in water. In air, NO is oxidized to NO_2; concentrations and emission rates are generally lumped under the designation NO_x. Both NO and NO_2 are extremely toxic gases. Physical properties of NO are listed in Table 5. Nitric oxide absorbs light at wavelengths less than 2300 Å. The N—O double bond has a bond length of 1.14 Å with a bond strength of 149.7 kcal/mole. The NO molecule can be ionized to NO^+ with a relatively low ionization potential of 9.5 eV.[30]

In the environment NO is emitted as a product of bacterial action and as a result of naturally occurring combustion processes. The global natural emission rate of NO is estimated at 45×10^7 tons/yr.[31] In comparison, the global anthropogenic emission rate is estimated to equal 4.8×10^7 tons/yr which is reported as NO_x.[31] Background NO concentrations measured in the equatorial Pacific averaged approximately 4 ppt during noontime conditions.[32]

3.1.3 Nitrogen Dioxide

Nitrogen dioxide is a reddish-orange-brown gas with a sharp, pungent odor. NO_2 is a toxic gas and is highly corrosive. Physical properties of NO_2 are presented in Table 5. The NO_2 molecule absorbs

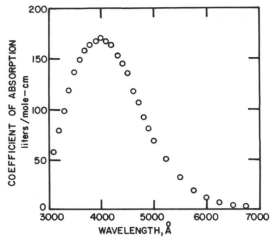

Fig 2   NO₂ absorption spectrum.[2]

light over much of the visible spectrum as indicated in Figure 2.[2] As a result, NO₂ in the atmosphere can produce a yellowish or orange haze. Curves of light transmittance as a function of wavelength for different products of NO₂ concentration times distance are shown in Figure 3.[2] Background concentrations of NO$_x$ have been reported to range from 2.7 to 4.1 ppb in the continental United States and from 0.9 to 3.6 ppb in Hawaii and Panama.[31] Nitrogen dioxide concentrations in maritime air in northern Europe have been observed in the range 0.5–2.8 ppb.[33] A clean air NO$_x$ concentration of 0.015 ppb has been used in a modeling study which suggested that significant peroxyacetyl nitrate (PAN) formation may occur in remote areas as an atmospheric reservoir for NO$_x$.[34] In comparison, average urban concentrations range from 29 ppb in St. Louis to 54 ppb in Chicago during 1967–1971.[31] The average of the second highest annual value ranged from 142 to 265 ppb for these same cities and time periods. Similar values are shown in Table 6 for NO and Table 7 for NO₂.

3.1.4   *Ammonia*

Like nitrous oxide, ammonia at typical ambient levels is not generally considered an air pollutant. However, NH₃ is produced in large amounts by bacterial decomposition of amino acids in organic waste, and NH₃ plays a key role through the formation of ammonium salts in the fate of many gaseous air pollutants. Ammonium salts are major components of atmospheric particulates.

Ammonia is a colorless gas with a pungent odor detectable at concentrations above 50 ppm. Physical properties of the gas are given in Table 5. The NH₃ molecule has a pyramidal structure with a H—N—H bond angle of 106°47′. Light is absorbed only at wavelengths below 2168 Å. Ammonia has an ionization potential of 10.5 eV corresponding to a wavelength of 1222 Å.[35]

3.2   Formation Mechanisms

3.2.1   *Combustion*

Nitrogen oxides formed during combustion occur as a result of oxidation of atmospheric N₂ or, to a lesser degree, oxidation of organic nitrogen in the fuel. In the first case, the primary product is nitric oxide

$$N_2 + O_2 \leftrightarrows 2NO \tag{38}$$

The predominant steps in this equilibrium were first proposed by Zeldovich[36] to be

$$O_2 + M \leftrightarrows 2O + M \tag{39}$$
$$O + N_2 \leftrightarrows NO + N \tag{40}$$
$$N + O_2 \leftrightarrows NO + O \tag{41}$$

This formation mechanism has a strong temperature dependence with increased NO production favored at higher temperatures. Rapid cooling following combustion freezes the equilibrium and NO persists in the exhaust gases. As a result of the temperature dependence, NO and NO₂ production is also dependent on the air-to-fuel ratio as indicated in Figure 4.[37] Under rich (high-fuel) conditions, O₂

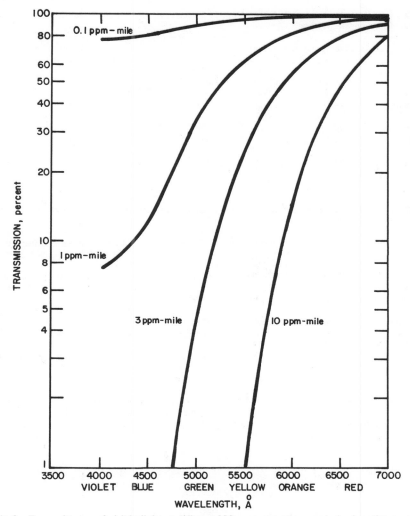

Fig 3 Transmittance of visible light at different NO_2 concentrations and viewing distances.[2]

levels and flame temperatures are low; NO production is also low. Under lean (low-fuel) conditions, flame temperatures are reduced through dilution with excess air and NO levels are decreased.

Downstream of the flame where NO is mixed with air, significant amounts of NO_2 can be formed:

$$2NO + O_2 \rightarrow 2NO_2 \tag{42}$$

As a result, between 0.5 and 10% of the total NO_x emissions exist as NO_2. The mechanism for NO conversion is considered to occur in two steps:

$$NO + O_2 \rightarrow NO_3 \tag{43}$$
$$NO_3 + NO \rightarrow 2NO_2 \tag{44}$$

Because nitrogen trioxide is very unstable, little NO_2 is formed at high temperatures. As discussed in detail in Chapter 35, NO is further converted to NO_2 at moderate rates in clean air and at high rates in dirty air.

With some fuels, oxidation of organically bound nitrogen to produce NO can account for a significant portion of NO_x emissions. Recent studies have suggested that because of differences in bond strength NO is formed more readily from combined nitrogen (C—N and N—H) than from atmospheric N_2 (N≡N).[36,38] During thermal decomposition of coal, the evolution of ammonia, hydrogen cyanide,

Table 6   Five-Year Averages of Nitrogen Dioxide Concentrations, Measured by the Continuous Saltzman Colorimetric Method[35]

Station	Average Concentration, $\mu g/m^3$, 25°C			Average of Annual Second Highest Value, $\mu g/m^3$, 25°C		
	1962–1966	1967–1971	Change, %	1962–1966	1967–1971	Change, %
Chicago	86.1	101.2	+18	444	499	+12
Cincinnati	62.0	60.0	−3	391	367	−6
Denver	66.0	67.9	+3	498	493	−1
Philadelphia	67.7	77.6	+15	361	414	+15
St. Louis	58.5	54.2	−7	320	267	−16
Average	68.1	72.2	+6	403	408	+1

Table 7   Five-Year Averages of Nitric Oxide Concentrations, Measured by the Continuous Saltzman Colorimetric Method[33]

Station	Average Concentration, $\mu g/m^3$, 25°C			Average of Annual Second Highest Value, $\mu g/m^3$, 25°C		
	1962–1966	1967–1971	Change, %	1962–1966	1967–1971	Change, %
Chicago	122.6	125.4	+2	731	969	+32
Cincinnati	43.8	53.6	+22	782	1067	+36
Denver	44.9	54.4	+21	633	620	−2
Philadelphia	55.2	65.4	+18	1331	1395	+5
St. Louis	39.8	47.6	+19	541	578	+7
Average	61.2	69.3	+13	804	926	+15

and nitrogen occurs. It might be assumed that in combustion higher order nitrogen compounds are pyrolyzed to these low molecular weight gases. In turn these species are further oxidized to NO. Unfortunately, the mechanisms for converting organically bound nitrogen to NO have yet to be clearly elucidated.

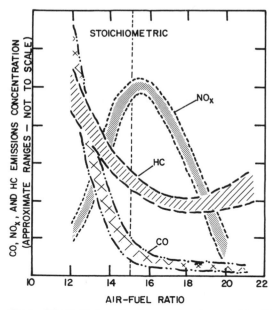

Fig 4   Effects of air–fuel ratio on exhaust composition.[37]

3.2.2 Industrial Processes

Nitric acid production is the major noncombustion source of NO_x emissions. Even so, the NO_x emissions from nitric acid plants account for only approximately 1% of the national NO_x emission total.

Nitric acid is produced by the catalytic oxidation of ammonia in air.[39] Typically, the reactants are heated, mixed, and passed over a platinum–rhodium catalyst where nitric oxide is produced:

$$4NH_3 + 5O_2 \rightarrow 4NO + 6H_2O \tag{45}$$

The gas stream is cooled, and NO reacts with oxygen to form NO_2 which dimerizes at high concentrations to yield nitrogen tetroxide

$$2NO + O_2 \rightarrow 2NO_2 \leftrightarrows N_2O_4 \tag{46}$$

Addition of water in an absorber tower yields nitric acid

$$3NO_2 + H_2O \rightarrow 2HNO_3 + NO \tag{47}$$

The excess nitric oxide is reoxidized for further processing. Emissions from the process occur as absorber tail gases and contain NO and NO_2.

Ammonia is produced industrially via the Haber process which involves the direct reaction of nitrogen with hydrogen:

$$\tfrac{1}{2}N_2 + \tfrac{3}{2}H_2 \rightarrow NH_3 \tag{48}$$

where iron–potassium aluminate mixtures are used as the catalyst.

3.2.3 Natural Formation of NH_3 and N_2O

Both ammonia and nitrous oxide are formed in nature in significant quantities and both are formed as part of the organic nitrogen cycle. Key steps in the cycle are presented in Table 8 and in Figure 5.[35] Nitrogen exists in the biosphere primarily in proteins and nucleic acids. Decomposition of organic matter leads to the enzyme catalyzed release of ammonia. For example, an amino acid may undergo the following reaction

$$\underset{\underset{NH_2}{|}}{R-CH-COOH} + \tfrac{1}{2}O_2 \overset{enzyme}{\rightarrow} \underset{\underset{O}{\|}}{R-C-COOH} + NH_3 \tag{49}$$

Table 8   Processes of the Nitrogen Cycle[35]

Mineralization				
RNH_2 + O_2 $\rightarrow$ CO_2 + H_2O + NH_4+				
organic nitrogen oxygen carbon dioxide water ammonium				
Nitrification				
NH_4^+ + O_2 $\rightarrow H_2O + NO_2$				
ammonium oxygen water nitrite				
Nitrite oxidation				
NO_2 + O_2 $\rightarrow NO_3^-$				
nitrite oxygen nitrate				
Denitrification				
$[HCHO]$ + NO_3^- $\rightarrow$ CO_2 + H_2O + N_2				
organic matter nitrate carbon dioxide water nitrogen gas				
Nitrate reduction				
NO_3^- + $[HCHO]$ $\rightarrow$ NH_4^+ + CO_2				
nitrate organic matter ammonium carbon dioxide				
(or amino or amide nitrogen)				
Nitrogen fixation				
N_2 + $[HCHO]$ $\rightarrow$ NH_4^+ + CO_2				
nitrogen gas organic matter ammonium carbon dioxide				
(or amino or amide nitrogen)				

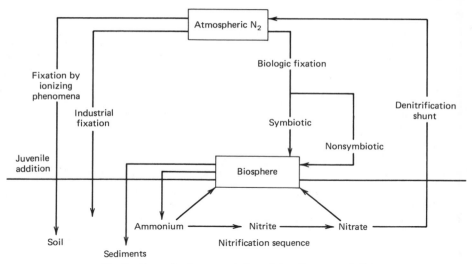

Fig. 5　Generalized representation of the nitrogen cycle.[35]

Similarly urea, which is found in high concentrations in sewage, can be broken down catalytically to release ammonia:

$$\overset{\text{urease}}{CO(NH_2)_2 + H_2O \;\rightarrow\; 2NH_3 + CO_2} \tag{50}$$

The global emission rate of ammonia from soils and from plants and animals is estimated at 75×10^6 tons/yr.[35] Pollution sources primarily from ammonia production plants and from fertilizer application account for 0.32×10^6 tons/yr in the United States.

Denitrification processes carried out by anaerobic organisms release nitrogen and nitrous oxide. The following reaction is a generalized scheme of carbohydrate oxidation with N_2O formation:

$$NO_3^- + HCHO \rightarrow \tfrac{1}{2}N_2O + \tfrac{1}{2}H_2O + CO_2 + OH^- \tag{51}$$

This type of oxidation and release can also occur with fats, fatty acids, amino acids, and methane. Nitrous oxide formation typically amounts to approximately 10% of the nitrogen returned to the atmosphere. At low pH, nitric oxide is also produced. Under conditions of heavy fertilization and periodic flooding, nitrous oxide formation may be considerable.

3.3 Chemistry of Nitrogen Compounds

3.3.1 *Nitrogen Oxides*

Nitrous oxide is an inert gas that is unreactive to halogens, alkali metals, and ozone at room temperature. It does decompose to oxygen and nitrogen at high temperatures. Its role in combustion processes has been shown to be unimportant. Because of its inertness, it has been used as an aerosol propellant in some consumer products. Nitrous oxide has an appreciable electron affinity as evidenced by its ready detection by electron capture gas chromatography.

Nitric oxide is moderately reactive and will form nitrosyl halides (XNO) when reacted with F_2, Cl_2, or Br_2. It can be oxidized to nitric oxide by a number of oxidizing agents; NO is reduced to N_2O by sulfur dioxide. In the atmosphere, NO participates in the photolytic cycle involving NO_2 and O_3:

$$NO_2 + h\nu \rightarrow NO + O \tag{52}$$
$$O + O_2 + M \rightarrow O_3 + M \tag{53}$$
$$O_3 + NO \rightarrow NO_2 + O_2 \tag{54}$$

Because of the extreme reactivity of the oxygen atom, an equilibrium between NO, NO_2, and O_3 is reached and the steady-state ozone concentration is given by

$$O_3 = \frac{k_1[NO_2]}{k_3[NO]} \tag{55}$$

In the presence of reactive hydrocarbons, usually emitted along with NO and NO_2, the equilibrium favors NO_2 and, in turn, O_3 production via radical reactions such as

$$NO + HO_2^{\cdot} \rightarrow NO_2 + HO^{\cdot} \tag{56}$$
$$NO + RO_2^{\cdot} \rightarrow NO_2 + RO^{\cdot} \tag{57}$$
$$NO + RCOO_2^{\cdot} \rightarrow NO_2 + RCO_2^{\cdot} \tag{58}$$

where R represents a variety of alkyl groups. As a result of this chemistry, NO and NO_2 concentrations exhibit a diurnal pattern in urban areas. In the morning, NO is emitted from early morning traffic and NO levels are high. Later in the day as the sunlight intensity increases, NO is converted to NO_2 and O_3 is produced as shown in Figure 6.

Nitrogen dioxide exists at high concentrations at temperatures below 140°C in equilibrium with its dimer nitrogen tetroxide N_2O_4. In the atmosphere, the dimer exists at negligible levels. NO_2 is rather reactive and attacks metals rapidly. In water NO_2 forms nitric and nitrous acid

$$2NO_2 + H_2O \rightarrow HNO_3 + HNO_2 \tag{59}$$

which can convert with warming to nitric oxide

$$3HNO_2 \rightarrow HNO_3 + 2NO + H_2O \tag{60}$$

The surface reactivity of both NO and NO_2 has been measured experimentally.[40] Results in Table 9 indicate the effectiveness of different surfaces in removing the oxides. Generally, the rates increase at higher humidity, and NO_2 is removed more rapidly than NO. Surface adsorption of NO_x by molecular sieves has been suggested as a control technique for removing and concentrating low concentrations of NO_x in gas streams.[41]

The oxides can be controlled through catalytic reduction via reactions such as

$$CH_4 + 4NO_2 \rightarrow 4NO + CO_2 + 2H_2O \tag{61}$$
$$CH_4 + 4NO \rightarrow 2N_2 + CO_2 + 2H_2O \tag{62}$$
$$8NH_3 + 6NO_2 \rightarrow 7N_2 + 12H_2O \tag{63}$$
$$4NH_3 + 6NO \rightarrow 5N_2 + 6H_2O \tag{64}$$

The latter two reactions occur in the limited temperature range 210–271°C.[39]

The analytical chemistry of NO is generally based on its conversion to NO_2 and subsequent analysis of the NO_2.

The Griess–Saltzman method for determining NO_2 depends on the reaction of NO with sulfamic acid to form a diazonium salt which in turn is combined with a diamine salt to form a colored azo dye. More recently, the chemiluminescent method has taken advantage of the light-emitting reaction of NO with O_3.[42] In this case, NO_2 can be measured by conversion in a thermal catalytic reactor to NO.

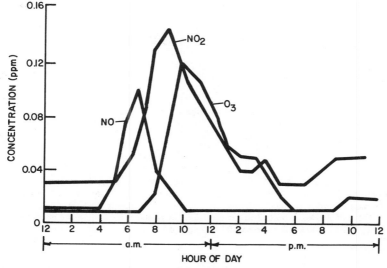

Fig. 6 Example of photochemical smog buildup in Los Angeles, California: the diurnal variation of nitric oxide, nitrogen dioxide, and ozone concentrations on July 19, 1965.[31]

Table 9   Experimental Reactivities for NO_2[40]

Materials	$\phi \times 10^5$ for Relative Humidities of[a]	
	0%	40–50%
Charcoal	160	—
MnO_2	100	—
Al_2O_3	30	—
PbO	≥ 10	≥ 10
Fly ash	≥ 10	≥ 10
CaO	≥ 10	≥ 10
Al_2O_3	≥ 10	≥ 10
ZnO	≤ 0.03	≥ 10
Adobe clay soil	8.4	8.4
Sandy loam soil	6.4	6.4
Cement	3.4	3.4
Cu_2O	≤ 0.03	3
Fe_2O_3	≤ 0.03	3
Sand	≤ 0.03	1
V_2O_5	≤ 0.03	0.1
$(NH_4)_2SO_4$	≤ 0.03	0.1
H_2SO_4	≤ 0.03	0.1

[a] ϕ is normalized to the removal rate of NO_2 passed over a bed of sand.

Table 10   Reactions for the Photochemistry of Nitrogen Oxides[43]

1. $NO_2 + h\nu \rightarrow NO + O$
2. $O + O_2 + M \rightarrow O_3 + M$
3. $O_3 + NO \rightarrow NO_2 + O_2$
4. $O + NO_2 \rightarrow NO + O_2$
5. $O + NO_2 + M \rightarrow NO_3 + M$
6. $NO_3 + NO \rightarrow 2NO_2$
7. $O + NO + M \rightarrow NO_2 + M$
8. $2NO + O_2 \rightarrow 2NO_2$
9. $NO_3 + NO_2 \rightarrow N_2O_5$
10. $N_2O_5 \rightarrow NO_3 + NO_2$
11. $NO_2 + O_3 \rightarrow NO_3 + O_2$
12. $N_2O_5 + H_2O \rightarrow 2HNO_3$
13. $HNO_3 + NO \rightarrow HNO_2 + NO_2$
14. $HNO_3 + HNO_2 \rightarrow 2NO_2 + H_2O$
15. $NO + NO_2 + H_2O \rightarrow 2HNO_2$
16. $HNO_2 + h\nu \rightarrow NO + OH^\bullet$
17. $OH^\bullet + NO_2 \rightarrow HNO_3$
18. $OH^\bullet + NO \rightarrow HNO_2$
19. $OH^\bullet + CO \rightarrow CO_2 + H^\bullet$
20. $H^\bullet + O_2 + M \rightarrow HO_2^\bullet + M$
21. $HO_2^\bullet + NO \rightarrow NO_2 + OH^\bullet$
22. $HO_2^\bullet + HO_2^\bullet \rightarrow H_2O_2 + O_2$
23. $H_2O_2 + h\nu \rightarrow 2OH^\bullet$

As indicated previously, the higher nitrogen oxides exist at low levels as transient species in combustion and atmospheric reactions. For example, the reactions listed in Table 10 are considered to be important in photochemical smog formation and in the global nitrogen cycle.[43] In particular, the reaction of NO_2 with hydroxyl radical to form nitric acid is considered the most important nitrogen sink in the atmosphere. Nitric acid is removed through wet or dry deposition at the surface.

3.3.2 Ammonia

Ammonia dissolves readily in water to form the ammonium ion

$$NH_3 + H_2O \rightarrow NH_4^+ + OH^- \tag{65}$$

The solubility of ammonia in water is available for a wide range of conditions.[44,45,46] Ammonia undergoes a variety of addition and substitution reactions including

$$Cu^{2+} + 4NH_3 \leftrightarrows Cu \cdot (NH_3)_4^{2+} \tag{66}$$
$$HgCl_2 + 2NH_3 \leftrightarrows Cl^- + NH_4^+ + ClHgNH_2 \tag{67}$$

and

$$\begin{array}{c} Cl \\ \\ NO_2 \end{array} + 2NH_3 \rightarrow \begin{array}{c} NH_2 \\ \\ NO_2 \end{array} + NH_4Cl \tag{68}$$

Ammonia can also reduce metal oxides to the free metal:

$$3CuO + 2NH_3 \leftrightarrows 3Cu + 3H_2O + N_2 \tag{69}$$

Although ammonia does not absorb visible radiation, it can be dissociated by ultraviolet radiation.[47,48]

$$NH_3 + h\nu \rightarrow NH_2 + H \tag{70}$$
$$NH_3 + h\nu \rightarrow NH + 2H \tag{71}$$

Because these reactions occur only in the far ultraviolet, photochemical decomposition of NH_3 is not of importance in the troposphere.

4 CARBON COMPOUNDS

The carbon species important in the global carbon cycle include carbon dioxide (CO_2), carbon monoxide (CO), methane (CH_4), and the immense variety of volatile hydrocarbons produced by vegetation. Carbon compounds that are important in local or regional air pollution problems are carbon monoxide and the number of hydrocarbons emitted as a result of fossil fuel combustion.

On a planetary basis, the carbon cycle revolves around the photosynthetic conversion of atmospheric carbon dioxide to oxygen and carbohydrates with subsequent decomposition of the organic material back to carbon dioxide. The decomposition occurs via oxidation pathways which convert methane, higher molecular weight hydrocarbons, and carbon monoxide to carbon dioxide. Because CO_2 effectively absorbs reflected long-wave radiation from the earth's surface, the presence of CO_2 in the atmosphere acts as a blanket to decrease the heat loss of the planet. The "greenhouse" effect of CO_2 is thus an important temperature regulator for the globe. Since industrial activities may be disrupting the existing CO_2 cycle in the atmosphere, anthropogenic CO_2 is considered a global pollutant.

On a local scale, carbon monoxide is emitted in large quantities from motor vehicles and can be a serious pollutant problem. Higher molecular weight hydrocarbons are also emitted from motor vehicles as a result of inefficient combustion. Hydrocarbons in an atmosphere containing nitrogen oxides significantly increase the efficiency of the photochemical production of ozone and thus are controlled as oxidant precursors.

4.1 Physical Properties and Occurrence of Carbon Compounds

4.1.1 Carbon Dioxide

The physical properties of carbon dioxide are given in Table 11.[49] It is a nontoxic, colorless, odorless gas which is soluble in water. The molecule has a linear structure (O=C=O) with a C=O bond length equal to 1.163 Å. Carbon dioxide absorbs radiation in the infrared region from 12 to 18 μm. Since the earth re-radiates solar energy over the range 5–100 μm with a maximum intensity near 10 μm, CO_2 along with water vapor effectively traps a significant portion of the outgoing radiation.

The global distribution of carbon, shown in Table 12,[50,51] includes 7×10^{17} g C as CO_2 in the atmosphere, 1.8×10^{18} g C in the terrestrial biosphere, and 3.9×10^{19} g C in the ocean waters. An additional 1.2×10^{19} g C exists as fossil fuels; approximately 60% of this amount is considered recoverable. The amount of carbon dioxide consumed by photosynthetic processes is estimated to be 3×10^{16} g C/yr, and the emission rate of CO_2 from fossil fuel combustion is currently 4.1×10^{15} g C/yr. The exchange of CO_2 between the atmosphere and the ocean occurs at an estimated rate equal to 9×10^{16} g C/yr. Although these numbers are subject to considerable uncertainty, the carbon cycle clearly involves the annual exchange of significant amounts of carbon between the atmosphere, terrestrial biosphere, and the oceans.

Since 1900 the ambient CO_2 concentration has increased from an estimated 290 ppm to approximately 330 ppm in 1977.[51] From 1968 to 1977, this increase occurred at a rate equal to 1 ppm/yr (0.3%/yr). The elevation of CO_2 concentrations mirrors an increase in the annual anthropogenic CO_2 emission rate. In 1900 the estimated CO_2 release rate from fossil fuel combustion equaled 1×10^{14} g C/yr

Table 11   Physical Properties of Carbon Monoxide and Carbon Dioxide[49]

	CO	CO_2
Molecular weight (g)	28.01	44.01
Melting point (°C)	−205.06 (1 atm)	56.6 (5.2 atm)
Boiling point (°C)	−191.50 (1 atm)	−78.5 (sublimes)
Critical temperature (°C)	−140.21	31.00
Critical pressure (atm)	34.529	75.282
Density (g/l)	1.25 (0°C, −1 atm)	1.977
	—	1.101 g/cm³ (liq, −37°C)
	—	1.56 g/cm³ (solid, −70°C)
Solubility (H_2O) g/100 cm³	0.4425 (0°C)	0.385 (0°C)
	0.2675 (25°C)	0.097 (40°C)
Bond length (Å)	1.1282	1.1632
Bond energy (kcal/mole)	255.8	127
Dipole moment	0.12×10^{-19} (c.s.u.)	—
Ionization potential (eV)	14.01	13.79
Vapor pressure relation (liq)	$\log_{10} p = a - b \log_{10} T - C/T$	$p = a(T/100 - 6)^n$
	$a = 13.7179$	$a = 7.856$
	$b = 2.893$	$b = 1.261$
	$c = 432.8$	$n = 3.917$
	p in kg/cm²	p in kg/cm²
$\Delta H_f°$ (kcal/mole)	− 26.4157	94.0518
Specific heat C_v (cal/deg-mole)	4.98 (formation)	$C_v = 0.691 + 0.889\rho + 1.42\rho^2$

Table 12   Distribution of Planetary Carbon[50,51]

	g C
Atmosphere (as CO_2)	7.00×10^{17}
Terrestrial biosphere	1.76×10^{18}
Oceans	3.90×10^{19}
Fossil fuels and shale	1.20×10^{19}
(recoverable)	0.73×10^{19}
Ocean life and sediments	2.05×10^{18}

	Fluxes of Carbon, g C/yr
CO_2 consumed in photosynthesis	3.1×10^{16}
CO_2 released from fuel combustion	4.1×10^{15}
CO_2 equilibrium flux with oceans	9.0×10^{16}
CO_2 increase in atmosphere	1.4×10^{15} (1.0 ppm/yr)

and in 1977 the estimated rate exceeded 5×10^{15} g C/yr. The accumulation of CO_2 in the atmosphere has been approximately half the amount released by industrial activities which suggests that carbon storage pools in the land and oceans have absorbed half the CO_2 released.[51]

The primary effect of increasing CO_2 concentrations in the atmosphere on the global climate is predicted to be a warming of the average global temperature. Both simple and complex models of global circulation predict that doubling the CO_2 concentration will cause an increase in the global temperature of approximately 2°C.[52] However, these model predictions are strongly tempered by uncertainties in the interrelationships among CO_2 levels, temperature, cloud processes, the surface albedo, and aerosol formation.

4.1.2   Carbon Monoxide

Carbon monoxide is a colorless, odorless, flammable, and toxic gas. Physical properties of CO are given in Table 11.[49] At 25°C, CO is slightly soluble in water (2.14 ml/100 ml H_2O). The toxicity of the gas arises from its ability to react with hemoglobin in the blood some 200 times more effectively than oxygen.[2]

The CO molecule is a resonance hybrid of three forms—$\overset{+}{C}$—$\overset{-}{O}$, C=O, $\overset{-}{C}$≡$\overset{+}{O}$—with a low dipole moment, a short bond distance (1.13 Å), and a strong molecular bond.[53] Carbon monoxide absorbs infrared radiation with a maximum absorption band at 4.7 μm, but it does not absorb in the visible or near-ultraviolet regions. In the far-ultraviolet, CO weakly absorbs radiation from 0.155 to 0.125 μm.[54]

Carbon monoxide is the most abundant and most widely distributed of all air pollutants. It is formed during combustion when carbon is burned with deficient oxygen. Anthropogenic emissions in the United States were estimated to equal 86.9×10^6 metric tons/yr in 1975;[53] motor vehicles accounted for 66% of the total as shown in Table 13. On a global basis, the situation is similar with motor vehicles contributing approximately 55% of the total anthropogenic emissions. It is estimated that the total global emission rate caused by human activities equals 359×10^6 metric tons/yr.[55] Robinson and Moser[56] estimated that approximately 95% of the global emissions originates in the Northern Hemisphere.

The importance of natural sources of CO was not recognized until after 1970 when Stevens et al.[57] showed that the ambient distribution of different isotropic forms of CO did not match the estimated natural and anthropogenic source distribution. Weinstock and Niki[58] proposed that these differences could be explained by the oxidation of naturally emitted methane by hydroxyl radical in the troposphere. The estimated tropospheric production rate from methane oxidation is 5000×10^6 metric tons/yr. Oxidation of formaldehyde and biogenic hydrocarbons, and release of CO directly from vegetation during the decay of chlorophyll, and from the surface waters of the oceans are also considered significant natural sources of carbon monoxide.[59]

Depending on the existing meteorological conditions, ambient concentrations of CO in urban areas can vary from 1 to 140 ppm. Commuters in heavy traffic can be exposed to concentrations exceeding 50 ppm.[60] For comparison the 1-hr CO air quality standard is 35 ppm and the 8-hr standard is 9

Table 13   Summary of Carbon Monoxide Emissions in the United States, 1975[53]

Source Category	Estimated Carbon Monoxide Emissions, 10^6 metric tons/yr
Transportation	
Gasoline vehicles	57.23
Diesel vehicles	0.55
Total road vehicles	57.78
Railroads	0.27
Vessels	0.97
Aircraft	0.78
Other nonhighway use	6.73
Total transportation	66.53
Solid-waste disposal	
Municipal incineration	0.21
On-site incineration	1.89
Open burning	1.31
Total solid waste	3.41
Industrial process losses	13.28
Agricultural burning	0.77
Total controllable	85.24
Fuel combustion in stationary sources	
Steam-electric	0.24
Industrial	0.45
Commercial and institutional	0.07
Residential	0.46
Total fuel	1.25
Miscellaneous	
Forest fires	1.64
Structural fires	0.30
Total miscellaneous	1.67
Total	86.91

ppm. In urban areas, CO concentrations follow a diurnal pattern which depends on traffic volume and traffic speed. Generally, CO concentrations reach a maximum between 7 and 9 A.M. due to the early morning traffic and then fall to an elevated steady level during the day. A second peak is usually observed corresponding to the late afternoon traffic period, and CO concentrations decrease to low levels during the night. This pattern, without the early morning maximum, also occurs on the weekends and on holidays.

Remote background concentrations of CO range from 0.01 to 1 ppm.[61,62] Concentrations in the Northern Hemisphere average almost twice the levels observed in the Southern Hemisphere. Observations of CO in the upper troposphere are similar to those measured near the surface, which indicates that CO is relatively well mixed throughout the troposphere. Above the tropopause, the CO mixing ratio decreases rapidly presumably because of recombination with OH radical.

In spite of the increased consumption of fossil fuels ambient CO concentrations have not increased during the past century. As a result, the discovery of the high natural CO production rate necessarily implies an equally effective natural sink for CO. Among a variety of possible removable mechanisms— including loss to the stratosphere, biological uptake, surface adsorption, and oceanic sinks—oxidation by hydroxyl radical is assumed to be a major pathway for CO removal. Hydroxyl radical is naturally present in the troposphere as a result of the photochemically driven reaction of water and oxygen atoms:

$$O_3 + h\nu \rightarrow O('D) + O_2 \tag{72}$$
$$O('D) + H_2O \rightarrow 2OH \tag{73}$$

The hydroxyl radicals produced are present in sufficient concentrations ($\sim 1.5 \times 10^6$ molecules/cm^3) to act as an efficient removal mechanism for carbon monoxide

$$OH + CO \rightarrow CO_2 + H \tag{74}$$

Thus, hydroxyl radical acts both as a natural source of carbon monoxide through the oxidation of methane and as a sink through oxidation of CO to produce carbon dioxide.[63]

4.1.3 Hydrocarbons

The diverse chemistry of carbon gives rise to an almost infinite variety of hydrocarbons ranging from methane to long-chain polymers. In the atmosphere, volatile hydrocarbons typically range from C_1 to C_{10} compounds as illustrated in Table 14. The ambient hydrocarbon composition includes the un-burned hydrocarbons from fuels such as gasoline, species formed during combustion, and natural hydrocarbons emitted by vegetation. Types of hydrocarbons are alkanes (paraffins), alkenes (olefins), aromatics, and the oxygenated species: aldehydes, ketones, and alcohols. The list in Table 14 is not exhaustive and many hydrocarbons formed in the atmosphere through photochemical reactions are not included.[64]

The hydrocarbons emitted from motor vehicles, stationary combustion devices, and industrial processes are not controlled as criteria pollutants because of their inherent toxicity. At typical ambient concentrations, even hydrocarbons of known toxicity may not be important. Rather, hydrocarbons are controlled because they are precursors to the toxic photochemical oxidants such as ozone and peroxyacetyl nitrate (PAN). As a result, hydrocarbons are typically classified in terms of their atmospheric reactivity with respect to specific reactive gases as shown in Table 15. The U.S. EPA has grouped hydrocarbons in three classes: (1) low reactivity, (2) moderate reactivity, and (3) high reactivity. Considering hydrocarbons in terms of reactivity not only provides a guideline for control purposes, but it also yields methods for modeling complex photochemical mechanisms using a limited set of reaction steps and species.[43]

Anthropogenic and biogenic emission inventories conducted for hydrocarbons involve a large degree of uncertainty. Emission rates from mobile sources are difficult to measure directly and typically involve extrapolating rates observed from individual vehicles to the total traffic mix. Similarly, the specification of hydrocarbon emission rates from vegetation usually involves extrapolating measurements from a small area to a large region. Present estimates of anthropogenic hydrocarbon emissions for the United States are shown in Table 16. The total anthropogenic hydrocarbon emission rate is estimated to be approximately 30×10^6 metric tons/yr.[65] Biogenic hydrocarbon emission rates for the United States have been estimated to equal 65×10^6 metric tons/yr; this does not include methane emissions estimated at 50×10^6 metric tons/yr.[66]

Methane is considered an unreactive hydrocarbon and it occurs at high concentrations relative to individual hydrocarbons. For these reasons, hydrocarbon concentrations are reported as total nonmethane hydrocarbons (NMHC). In urban areas, nonmethane hydrocarbon concentrations are in the range 0.1 to more than 10 ppm. Methane concentrations can be half to two times these levels. Concentrations of individual hydrocarbons are typically determined via flame ionization gas chromatography. Concentrations and distributions of hydrocarbons for seven urban areas are shown in Table 17.

Table 14 Hydrocarbons Identified in Ambient Air[65]

Carbon Number	Compound	Carbon Number	Compound
1	Methane		2,2-Dimethylbutane
2	Ethane		2,3-Dimethylbutane
	Ethylene		cis-2-Hexane
	Acetylene		trans-2-Hexane
3	Propane		cis-3-Hexane
	Propylene		trans-3-Hexane
	Propadiene		2-Methyl-1-pentene
	Methylacetylene		4-Methyl-1-pentene
4	Butane		4-Methyl-2-pentene
	Isobutane		Benzene
	1-Butene		Cyclohexane
	cis-2-Butene		Methylcyclopentane
	trans-2-Butene	7	2-Methylhexane
	Isobutene		3-Methylhexane
	1,3-Butadiene		2,3-Dimethylpentane
5	Pentane		2,4-Dimethylpentane
	Isopentane		Toluene
	1-Pentene	8	2,2,4-Trimethylpentane
	cis-2-Pentene		o-Xylene
	trans-2-Pentene		m-Xylene
	2-Methyl-1-butene		p-Xylene
	3-Methyl-1,3-butene	9	m-Ethyltoluene
	2-Methyl-1,3-butadiene		p-Ethyltoluene
	Cyclopentane		1,2,4-Trimethylbenzene
	Cyclopentene		1,3,5-Trimethylbenzene
	Isoprene	10	sec-Butylbenzene
6	Hexane		α-Pinene
	2-Methylpentane		β-Pinene
	3-Methylpentane		3-Carene
			Limonene

Table 15 Classification of Organics with Respect to Their Oxidant-Related Reactivity in Urban Atmospheres[65]

Class I (Low Reactivity)	Class II (Moderate Reactivity)	Class III (High Reactivity)
C_1–C_3 paraffins	Tert-monoalkyl benzenes	Primary, secondary monoalkyl benzenes
Acetylene	Cyclic ketones	Dialkyl benzenes
Benzene	Tolualdehydes	Styrene
Benzaldehyde	Tert-alkyl acetates	N-Methyl pyrrolidone
Acetone	2-Nitropropane	Partially halogenated olefins
Methanol	C_4, paraffins, cycloparaffins	Aliphatic olefins
Isopropanol	Ethanol	Tri-, tetra-alkyl benzene
Tert-alkyl alcohols	Primary, secondary C_2, alkyl	Methyl styrene
Methyl acetate	N, N-dimethyl acetamide	Branched alkyl ketones
Methyl benzoate	n-alkyl C_5,-ketones	Unsaturated ketones
Ethyl amines		Aliphatic aldehydes
N, N-dimethyl formamide		Diacetone alcohol
Perhalogenated hydrocarbons		Ethers
Partially halogenated paraffins		2-Ethoxy-ethanol
Mono, dichlorobenzenes		
Methyl-ethyl-ketone		

Table 16 1974 Nationwide Estimates of Total Hydrocarbon Sources and Emissions[65]

Source Category	Emissions, 10^6 metric tons/yr	
	1974	1975 (Preliminary)
Transportation (total)	(11.3)	(10.6)
Highway	9.8	9.1
Nonhighway	1.5	1.5
Stationary fuel combustion (total)	(1.6)	(1.3)
Electric utilities	0.1	0.1
Other	1.5	1.2
Industrial processes (total)	(3.3)	(3.2)
Chemicals	1.6	1.5
Petroleum refining	0.8	0.8
Metals	0.2	0.2
Others	0.7	0.7
Solid waste (total)	(0.9)	(0.8)
Miscellaneous (total)	(12.7)	(12.2)
Forest wildfires	0.5	0.5
Forest-managed burning	0.2	0.2
Agricultural burning	0.1	0.1
Coal refuse burning	0.1	0.1
Structural fires	0.0	<0.1
Organic solvents	8.1	7.5
Oil and gas production and marketing	3.7	3.8
Total	29.8	28.0

Table 17a Average Nonmethane Hydrocarbon Concentrations (ppb C) Measured Between 6–9 A.M. in Seven Urban Areas[67]

Site	Average NMHC	Range of Concentrations	Number of Samples
Houston	1414	356–6284	36
Newark	732	89–6946	132
Washington, D.C.	715	210–2953	64
Philadelphia	669	205–1710	82
Baltimore	598	51–2798	140
Boston	569	83–4750	160
Milwaukee	324	24–3116	150

Table 17b Percentage Paraffins, Olefins, and Aromatics in Samples Collected Between 6–9 A.M. in Seven Urban Areas[67]

City	Paraffins	Olefins	Aromatics	Number of Samples
Houston	64	11	23	36
Newark	61	11	26	132
Washington, D.C.	59	6[a]	31	64
Philadelphia	67	6[a]	25	82
Baltimore	58	10	29	140
Boston	66	8	23	160
Milwaukee	64	10	24	7

[a] Ethylene not included in olefin total.

Somewhat like carbon monoxide, hydrocarbons exhibit a diurnal pattern in urban areas. During the early morning traffic period, concentrations are elevated. As the sun rises, the hydrocarbons begin to be consumed in the production of photochemical oxidants and the concentrations decrease. In the later afternoon, a second, smaller maximum may be observed and at night concentrations are relatively low.

4.2 Formation Mechanisms

4.2.1 Combustion

Complete combustion of fossil fuels yields CO_2 and H_2O as the only exhaust products. However, in the internal combustion engine and in most stationary combustion burners, the dynamics of fuel–air mixing and engine operation prevents the complete oxidation of the fuel hydrocarbons. As a result, the exhaust stream is comprised of a mixture of unburned fuel, lower order hydrocarbons produced in the flame, hydrocarbons oxidized in the exhaust stream, carbon monoxide, carbon dioxide, and water. The mix of these different gases depends markedly on the air–fuel ratio and to lesser degrees on the ignition timing, combustion chamber design, and fuel type.

Gasoline, the prevalent fuel for the internal combustion engine, is the 40–200°C fraction from petroleum oil and contains approximately 2000 compounds. Generally, these include C_4 to C_9 paraffins, olefins, and aromatics; the composition of gasoline varies from approximately 4% olefins and 48% aromatics to 22% olefins and 20% aromatics.[68] The distribution of reactive hydrocarbons in gasoline is important because the overall photochemical reactivity of the exhaust gas mixture depends partly on the initial mix of olefins and aromatics in the fuel. Unleaded fuel has a higher aromatic content than leaded fuel and the exhaust stream from unleaded gasoline has a higher photochemical reactivity than that from leaded fuel.[69]

The primary source of unburned hydrocarbons in the exhaust stream is quenching of combustion reactions at the wall of the combustion chamber. In the internal combustion engine, these unreacted hydrocarbons are swept out of the cylinder with each engine cycle. In the hot exhaust stream, some of these species will be oxidized to lower order hydrocarbons. For the 10 most prevalent hydrocarbons in the exhaust stream, six compounds are volatile species not originally in the gasoline, and these hydrocarbons account for approximately 64% of the exhaust mixture.[43] These include ethylene, methane, propylene, acetylene, 1-butene, and ethane. Fuel hydrocarbons observed in the exhaust stream include toluene, p-, m-, and o-xylene, i-pentene, and n-butane. In terms of the photoreactivity of the exhaust stream, approximately two-thirds of the reactive hydrocarbons are formed in the engine and these combustion products account for 75% of the reactivity.

The complexity of the starting fuel mixture has prevented a clear understanding of the chemical mechanism of hydrocarbon combustion. At high temperatures (>1000°C) for the simplest hydrocarbon, methane, combustion involves radical attack on methane

$$CH_4 + OH \rightarrow CH_3 + H_2O \tag{75}$$

with subsequent reaction of the methyl radical to form formaldehyde and CO:[36]

$$CH_3 + O \rightarrow H_2CO + H \tag{76}$$
$$H_2CO + OH \rightarrow HCO + H_2O \tag{77}$$
$$HCO + OH \rightarrow CO + H_2O \tag{78}$$

The CO formed can further react to yield CO_2

$$CO + OH \rightarrow CO_2 + H \tag{79}$$

These steps are representative of the combustion mechanism which, in fact, is much more detailed and involves many other possible reaction sequences. At the relatively low temperatures found in the internal combustion engine, hydrocarbon combustion involves radical production of the type

$$R + O_2 \rightarrow RO_2 \tag{80}$$
$$RO_2 + R'H \rightarrow R' + RO_2H \tag{81}$$
$$R + O_2 \rightarrow olefin + HO_2 \tag{82}$$
$$HO_2 + R'H \rightarrow H_2O_2 + R' \tag{83}$$

Again, these steps represent only a small portion of the mechanism that produces a variety of olefins, ketones, aldehydes, ethers, and acids.[36]

4.2.2 Industrial Processes

As indicated in Table 16, hydrocarbons emitted from industrial processes account for approximately 10% of the total U.S. anthropogenic hydrocarbon emission rate. These sources include chemical manu-

Table 18   CO and Hydrocarbon Emissions from Refinery
Processes[71]

Process[a]	CO, kg/day	Hydrocarbons, kg/day
Crude distillation	242	242
Heavy naphtha reformer	130	131
Catalytic cracking	48	82
CO boiler	116	73
Hydrocracker	50	489
Heavy hydrocracker reformer	129	130
Hydrogen plant	156	153
Storage	—	5,759
Miscellaneous	159	27,777
	1,030	34,836

[a] Basis: 200,000 bbl/day crude feed.

facturing facilities, petrochemical refineries, and metallurgical operations. Hydrocarbons are emitted from the chemical production of elastomers, dyes, flavors, perfumes, plastics, resins, plasticizers, pigments, pesticides, and rubber processing. The kinds and amounts of hydrocarbons emitted from these operations obviously depend strongly on the specific manufacturing process, the age and maintenance of the plant, and existing control equipment. In 1978, the total production rate of the top 50 chemicals produced was 525 billion lb; organics accounted for 172 billion lb of the total.[70]

Petroleum refining and petrochemical production also emit a wide variety of hydrocarbons dependent on the refining process and the plant facilities. Types of refineries are the topping refinery, a simple distillation to yield naphtha, middle distillate, and fuel oil; the fuel oil refinery, a process optimized to produce 40–60% fuel oil with gasoline as remainder; the gasoline refinery, which includes extensive conversion (cracking) facilities to give motor-quality gasoline; the lube oil refinery, which incorporates lube-oil processing; and the petrochemical refinery, which is any basic refinery configuration with the addition of olefin and aromatic plants.[71] The hydrocarbon composition from these refineries will include volatile raw materials, products, and by-products of the refinery operation. Petroleum refineries perform four basic operations on crude oil. These include separation via distillation of the hydrocarbon fractions, conversion of hydrocarbons, treatment to remove impurities, and blending of hydrocarbon stocks to yield specific grades of fuel products.[72] Specific conversion processes are alkylation, amination, hydrogenation, dehydrogenation, halogenation, and polymerization, among others.

The majority of hydrocarbon emissions within a refinery occur as general fugitive emissions from the individual processes and storage tanks. Typical emissions for different processes are listed in Table 18. For refineries located in Southern California, the typical composition observed in these emissions included 21% methane, 37% ethane-propane-butane, 32% C_4 to C_8 paraffins, and 9% toluene-benzene.[73]

4.3   Chemical Properties of Carbon Compounds

4.3.1   *Carbon Monoxide*

Carbon monoxide can be prepared in the laboratory by dropping formic acid in warm concentrated sulfuric acid or by the decomposition of nickel tetracarbonyl at 200°C:

$$Ni(CO)_{4(g)} \rightarrow Ni_{(s)} + 4CO_{(g)} \tag{84}$$

Carbon monoxide is manufactured in the form of producer gas (25% CO, 70% N_2, 4% CO_2, trace H_2, CH_4, O_2) by blowing air through incandescent coke:

$$C + \tfrac{1}{2}O_2 \rightarrow CO + H_2 \qquad \Delta H = +32 \text{ kcal/mole} \tag{85}$$

Water gas (40% CO, 50% H_2, 5% CO_2, 5% N_2 and CH_4) is made by passing steam over incandescent coke:

$$C + H_2O \rightarrow CO + H_2 \qquad \Delta H = +32 \text{ kcal/mole} \tag{86}$$

Other methods for producing CO include the reduction of CO_2 with zinc dust,

$$CO_2 + Zn \rightarrow CO + ZnO \tag{87}$$

heating charcoal with metal oxides

$$C + ZnO \rightarrow CO + Zn \tag{88}$$

and heating carbon with barium carbonate

$$C + BaCO_3 \rightarrow BaO + 2CO \qquad \text{(Refs. 49, 53)} \tag{89}$$

Carbon monoxide is an important feedstock in a variety of commercial operations. Sodium formate is produced when CO is passed over hot caustic soda:

$$CO + NaOH \rightarrow HCOONa \tag{90}$$

and carbon monoxide is an effective reducing agent for many metal oxides:

$$CO + PbO \rightarrow Pb + CO_2 \tag{91}$$
$$CO + HgO \rightarrow Hg + CO_2 \tag{92}$$

Carbonyl halides are formed in the presence of light as in the formation of highly toxic phosgene:

$$CO + Cl_2 \rightarrow COCl_2 \tag{93}$$

Metallic carbonyls are also produced commercially with carbon monoxide:

$$Ni + 4CO \rightarrow Ni(CO)_4 \tag{94}$$

Carbon monoxide is also important commercially in the synthesis of a number of hydrocarbons. Hydrogenation of CO (Fischer–Tropsch) yields methane, benzene, paraffins, olefins, methanol, ethylene, and a number of other species. The efficiency of different catalysts in promoting these reactions has been studied extensively.[74] Carbon monoxide reacts with alcohols to give formate esters or carboxylic acids, and CO combines with olefins and hydrogen to yield aldehydes and alcohols.

Carbon monoxide can be detected continuously in the atmosphere using nondispersive infrared spectrometry. In the 0–50 ppm range, measurements are reproducible to within ±3.5 ppm with a minimum detection limit of 0.3 ppm.[53] An alternate method involves separation of CO from other species in a gas chromatography column, subsequent catalytic oxidation to methane, and detection via a flame ionization detector.[53]

4.3.2 Hydrocarbons

The chemical properties of hydrocarbons are as diverse as the thousands of hydrocarbon molecules which occur in nature or which have been synthesized by humans. Detailed discussions of organic chemistry are available in any number of textbooks,[75] and descriptions of atmospheric organic chemistry are provided in Chapter 35.

Hydrocarbon chemistry can be presented in a general way by classifying the types of compounds and the kinds of chemical reactions they can undergo. Alkanes are saturated hydrocarbons which can be prepared by hydrogenation of alkenes:

$$C_n H_{2n} \rightarrow C_n H_{2n+2} \tag{95}$$

and by reduction of halogenated hydrocarbons:

$$RX + Mg \rightarrow RMgX \rightarrow RH \tag{96}$$

Alkanes can be halogenated:

$$RH + X_2 \rightarrow RX + HX \tag{97}$$

and they can undergo pyrolysis reactions to form hydrogen and smaller alkanes and alkenes.

Double-bonded hydrocarbons are called alkenes or unsaturated hydrocarbons. Alkenes are prepared via dehydrohalogenation:

$$RX-R' + KOH \rightarrow R{=}R' + KX + H_2O \tag{98}$$

and also by dehydration:

$$ROH-R' + acid \rightarrow R{=}R' + H_2O \tag{99}$$

Alkenes are also produced by reduction of alkynes and through dehalogenation of halocarbons:

$$RX—R'X + Zn \rightarrow R{=}R' + ZnX_2 \tag{100}$$

Generally, alkenes are more reactive than alkanes. Double-bonded species undergo a variety of addition reactions: addition of H

$$R{=}R' + H_2 \rightarrow RH—R'H \tag{101}$$

addition of halogens

$$R{=}R' + X_2 \rightarrow RX—R'X \tag{102}$$

addition of sulfuric acids

$$R{=}R' + H_2SO_4 \rightarrow RH—R'OSO_3H \tag{103}$$

addition of water

$$R{=}R' + H_2O \rightarrow RH—R'OH \tag{104}$$

and addition of hydrogen halides.

$$R{=}R' + HX \rightarrow RH—R'X \tag{105}$$

Glycols can be formed from alkenes by hydroxylation:

$$R{=}R' + KMnO_4 \rightarrow ROH—R'OH \tag{106}$$

and aldehydes and ketones can be produced via oxonolysis:

$$R{=}R' + O_3 \rightarrow R{=}O + R'{=}O \tag{107}$$

Alkynes are triple-bonded molecules ($C{\equiv}C$) which are formed by dehydrohalogenation of alkenes. Alkynes can react to form alkanes, halogenated alkenes and alkanes, and ketones.

Aromatics are compounds containing one or more benzene rings. Aromatics can undergo a number of addition reactions including nitration, sulfonation, halogenation, and alkylation where the added species replaces a hydrogen on the aromatic ring (C_6H_5Cl, for example). Cyclic alkanes are prepared by hydrogenation of aromatic rings, and aromatic carboxylic acids are produced by oxidation of substituted aromatic rings:

$$C_6H_5— R \rightarrow C_6H_5—COOH \tag{108}$$

Aromatic rings can also be modified via substitution reactions of the type

$$C_6H_5—CH_3 + Cl_2 \rightarrow C_6H_5—CH_2—Cl \tag{109}$$

Oxygenated hydrocarbons include alcohols (ROH), aldehydes (RH$=$O), and ketones (RR'$=$O). Preparation of these compounds has been mentioned above. Alcohols react to form alkenes, esters (ROOR'), ketones, and carboxylic acids. Aldehydes and ketones can be oxidized to form carboxylic acids and reduced to form alcohols or alkanes.

5 HAZARDOUS AIR POLLUTANTS

Among the vast number of chemicals released into the air are a large number of known or suspected toxic gases. These hazardous air pollutants are not controlled as criteria pollutants. In some cases where the hazardous nature of a gas has been amply demonstrated, specific chemicals are controlled as hazardous air pollutants.[76] Currently, designated hazardous species include mercury vapor, vinyl chloride, and benzene among the gaseous chemicals. However, there is a much longer list of suspected mutagens and carcinogens which are found in urban air. As shown in Table 19, this list includes halomethanes, ethanes, and propanes, chloroalkenes, chloroaromatics, oxygenated gases, and nitrogenated compounds.[77] There is also a growing recognition of hazardous gases which are of particular concern as indoor pollutants. Radioactive radon gas and formaldehyde vapor are examples.

Results from recent ambient monitoring surveys are given in Table 19. These data indicate that hazardous gases are present in urban atmospheres at concentrations ranging from parts per trillion

Table 19 Potentially Hazardous Air Pollutants[77]

Chemical Name	Chemical Formula	Toxicity	Average Concentration,[a] ppt
Halomethanes			
Methyl chloride	CH_3Cl	BM[b]	788
Methyl bromide	CH_3Br	BM	141
Methyl iodide	CH_3I	SC, BM	2.7
Methylene chloride	CH_2Cl_2	BM	978
Chloroform	$CHCl_3$	SC, BM	346
Carbon tetrachloride	CCl_4	SC, NBM	221
Haloethanes and halopropanes			
Ethyl chloride	C_2H_3Cl	—	100
1,2-Dichloroethane	CH_2ClCH_2Cl	SC, BM	558
1,2-Dibromoethane	CH_2BrCH_2Br	SC	32
1,1,1-Trichloroethane	CH_3CCl_3	Weak BM	512
1,1,2-Trichloroethane	$CH_2ClCHCl_2$	SC, NBM	29
1,1,2,2-Tetrachloroethane	$CHCl_2CHCl_2$	SC, BM	10
1,2-Dichloropropane	$CH_2ClCHClCH_3$	BM	60
Chloroalkenes			
Vinylidene chloride	$CH_2{=}CCl_2$	SC, BM	19
Trichloroethylene	$CHCl{-}CCl_2$	SC, BM	143
Tetrachloroethylene	$CCl_2{=}CCl_2$	SC	401
Allyl chloride	$ClCH_2CH{=}CH_2$	SC	<5
Hexachloro-1,3-butadiene	$Cl_2C{=}CCl{-}CCl{=}CCl_2$	BM	5
Chloroaromatics			
Monochlorobenzene	C_6H_5Cl	—	280
a-Chlorotoluene	$C_6H_5CH_2Cl$	BM	<5
o-Dichlorobenzene	$o\text{-}C_6H_4Cl_2$	—	12
m-Dichlorobenzene	$m\text{-}C_6H_4Cl_2$	—	6
1,2,4-Trichlorobenzene	$1,2,4\text{-}C_6H_3Cl_3$	—	5
Aromatic hydrocarbons			
Benzene	C_6H_6	SC	3883
Oxygenated and nitrogenated species			
Formaldehyde	HCHO	SC, BM	14,200
Phosgene	$COCl_2$	—	<20
Peroxyacetyl nitrate (PAN)	$CH_3COOONO_2$	Phytotoxic	589
Peroxypropionyl nitrate (PPN)	$CH_3CH_2COOONO_2$	Phytotoxic	103
Acrylonitrile	$CH{\equiv}CN$	SC	—

[a] Average from 2 weeks of measurements in Houston, St. Louis, Denver, and Riverside.
[b] BM: Positive mutagenic activity based on Ames salmonella mutagenicity test (Bacterial Mutagens).
NBM: Not found to be mutagens in the Ames salmonella test (Not Bacterial Mutagens).
SC: Suspected Carcinogens.

to parts per billion. In general, however, information is lacking concerning the sources, emission rates, ambient concentration patterns, and the health effects of these air pollutants. In particular, knowledge concerning the synergistic effects of exposure to the variety of chemicals listed in Table 19 is not currently available. In the remainder of this section, the characteristics of several hazardous pollutants are discussed. Because of the wide range of potentially hazardous gases, this presentation is necessarily limited.

5.1 Benzene

Benzene is a clear, colorless, volatile, flammable liquid with a characteristic odor somewhat like that of gasoline. The benzene molecule is a planar hexagon with the formula C_6H_6 which exists as a series of resonance hybrids. The resonant double-bonded nature of benzene makes it a stable, relatively unreactive hydrocarbon. The C—C bond length is 1.39 Å and the C—H bond length is 1.08 Å. Physical properties of benzene are listed in Table 20.[78]

Table 20   Properties of Benzene[78]

Constant	Value
Freezing point (°C)	5.553
Boiling point (°C)	80.100
Density at 25°C (g/ml)	0.8737
Vapor pressure at 26.075°C (mm Hg)	100
Refractive index, n_D^{25}	1.49792
Viscosity (absolute) at 20°C (cP)	0.6468
Surface tension at 25°C (dyn/cm)	28.18
Critical temperature (°C)	289.45
Critical pressure (atm)	48.6
Critical density (g/ml)	0.300
Flash point (closed cup) (°C)	−11.1
Ignition temperature in air (°C)	538
Flammability limits in air (vol%)	1.5–8.0
Heat of fusion (kcal/mole)	2.351
Heat of vaporization at 80.100°C (kcal/mole)	8.090
Heat of combustion at constant pressure and 25°C (liquid C_6H_6 to liquid H_2O and gaseous CO_2) (kcal/g)	9.999
Solubility in water at 25°C (g/100 g water)	0.180
Solubility of water in benzene at 25°C (g/100 g benzene)	0.05

Although benzene is unreactive under ambient conditions, it is a major feedstock for the chemical manufacturing industry. Products using benzene as starting material include ethylbenzene, phenol, cyclohexane, and maleic anhydride among others. Benzene is produced in the distillation of crude petroleum. It is a significant component of gasoline.

Approximately 1 billion lb of benzene were produced in the United States in 1973.[79] The estimated emission rate of benzene from commercial activities was 760 million lb in 1971. However, emissions from motor vehicle operation were estimated to exceed 1000 million lb in 1971. Benzene comprises 4% (wt) of automobile exhaust emissions and approximately 1% (wt) of fuel evaporative emissions.

Benzene can be detected in air using flame ionization gas chromatography at concentrations as low as 0.1 ppb. Typical concentrations in urban areas range from 0.1 to 37 ppb.[77] Levels observed immediately downwind of benzene commercial facilities range from 0.5 to 58 ppb. Mara and Lee[78] conducted estimates of population exposure to benzene. The results shown in Table 21 indicate that exposure in urban areas far exceeds that caused by commercial activities.

In the atmosphere, benzene is considered a low reactive hydrocarbon. Benzene is not photolyzed at wavelengths above 2800 Å. Because benzene is soluble in water, significant concentrations have been observed in natural waters. Microbial activity has been cited as a means of benzene degradation in the environment.

Table 21   Summary of Estimated Total Exposures[a] of People Residing in the Vicinity of Atmospheric Benzene Sources[78]

Vicinity of Residence	Number of People Exposed Annual Average Benzene Concentrations (ppb)				Total	Comparison among Sources, 10^6 ppb-person-yr
	0.1–1.0	1.1–4.0	4.1–10.0	>10.0		
Chemical manufacturing	3,900,000	3,100,000	200,000	80,000	7,300,000	10.0
Coke ovens	200,000	100,000			300,000	0.2
Petroleum refineries	3,250,000	1,750,000			5,000,000	4.5
Urban areas		110,000,000			110,000,000	250.0

[a] The term "total exposures" is used to mean the sum of an individual's exposure to atmospheric benzene from a variety of activities during a year.

5.2 Vinyl Chloride

Vinyl chloride (CH_2CHCl) is a colorless, flammable gas with a faintly sweet odor. It boils at $-13.4°C$ and has a vapor pressure of 2600 mm Hg at 25°C. Vinyl chloride decomposes to phosgene upon heating. Vinyl chloride is only slightly soluble in water.[80]

The recommended standard for occupational atmospheres is 1 ppm on an 8-hr average and 5 ppm for 15-min exposures. Vinyl chloride is classified as a hazardous air pollutant by the EPA. Workers exposed to concentrations above 1 ppm have developed angiosarcoma of the liver, and studies have suggested that vinyl chloride may lead to other forms of cancer, liver dysfunction, and other disorders. As a hazardous air pollutant, vinyl chloride is controlled via an emissions standard of 10 ppm in vent gas streams.

Vinyl chloride is produced through dechlorination of 1,2-dichloroethane (ethylene dichloride) which, in turn, is manufactured by either chlorination of ethylene or oxychlorination of ethylene:

$$2CH_2{=}CH_2 + 2Cl_2 \rightarrow 2CH_2ClCH_2Cl \tag{110}$$
$$2CH_2{=}CH_2 + O_2 + 4HCl \rightarrow 2CH_2ClCHCl + 2H_2O \tag{111}$$
$$CH_2ClCH_2Cl \rightarrow CH_2{=}CHCl + HCl \tag{112}$$

The production rate of vinyl chloride increased from 45 million kg in 1943 to 2.4 billion kg in 1973 because of its role as a feedstock for polyvinyl chloride (PVC), a plastic which appears in a multitude of products.[81] Vinyl chloride was used as an aerosol propellant until 1974 when its carcinogenicity was recognized. Emissions from the manufacture of vinyl chloride, polyvinyl chloride, and associated products are estimated to be 100 million kg/yr. Approximately 90% of the emissions are from PVS facilities.

Ambient concentrations of vinyl chloride range from less than 5 ppt[82] in urban air to approximately 1 ppm in areas near VC and PVC manufacturing plants. As a hydrocarbon, vinyl chloride can participate in photochemical smog reactions. Its reactivity is moderate compared to similar hydrocarbon molecules. Products of vinyl chloride oxidation in the atmosphere include formaldehyde, formic acid, and hydrogen chloride.

5.3 Mercury

Mercury is a silvery liquid with an atomic weight of 200.6 g which melts at $-38.9°C$ and boils at 356.9°C. The liquid density of mercury is 13.546 g/ml. To a large degree, the environmental impact of mercury arises from its relatively high vapor pressure of 0.002 mm at ambient temperatures. In a room, the saturation concentration of the vapor is approximately 15 mg Hg/m^3. Mercury is not soluble in water (2×10^{-5} g/1).[80]

Mercury vapor and the volatile organic mercury compounds are extremely toxic. The primary effect of exposure to mercury is damage to the central nervous system. Exposure can be via inhalation, adsorption through the skin, or ingestion of foods. Accumulation of mercury in aquatic food chains has been clearly documented.[83] The threshold limit value for occupational air is 0.1 mg/m^3.

Mercury occurs in nature primarily as HgS. Anthropogenic sources of mercury are associated with the smelting of mercury and other metal ores, with the utilization of mercury in the chlorine–alkali and electrical products industries, and with the combustion of fossil fuels. Total mercury production exceeds 10,000 tons/yr. The loss of mercury from the combustion of coal is estimated to be 3000 tons annually. Mercury content in coals range from 0.07 to as much as 300 ppm for coals mined in areas of mercury deposits. Average values are of order 3 ppm.

In the atmosphere, mercury concentrations over the ocean are less than 2 ng/m^3 and up to 9 ng/m^3 over nonmineralized land areas. Above mineral deposits, mercury concentrations ranging up to 53 ng/m^3 have been observed. Mercury undergoes an ecological cycle involving weathering of minerals, evaporation, biological transformation of inorganic and organic mercury compounds, and atmospheric removal via precipitation and dry deposition. As noted above this cycle is perturbed by human activities in the form of industrial consumption and fossil fuel combustion.

REFERENCES

1. *Air Quality Criteria for Sulfur Oxides,* U.S. Department of Health, Education and Welfare, AP-50, Washington, D.C., 1970.

2. *Air Quality Criteria for Nitrogen Oxides,* U.S. Department of Health, Education and Welfare, AP-84, Washington, D.C., 1971.

3. *Air Quality Criteria for Carbon Monoxide,* U.S. Department of Health, Education and Welfare, AP-62, Washington, D.C., 1970.

4. *Air Quality Criteria for Hydrocarbons,* U.S. Department of Health, Education and Welfare, AP-64, Washington, D.C., 1970.

5. Committee on Sulfur Oxides, *Sulfur Oxides,* National Academy of Sciences, Washington, D.C., 1978.

6. Adams, D. F., Farwell, S. O., Robinson, E., and Pack, M. R., *Biogenic Sulfur Emissions in the SURE Region,* EPRI, EA-1516, Palo Alto, CA, 1980.

7. Rasmussen, R. A., What Do the Hydrocarbons from Trees Contribute to Air Pollution? *J. Air Pollut. Contr. Assoc.* **22,** 537–543 (1972).

8. Robinson, E., Hydrocarbons in the Atmosphere, *Pageoph 116,* 372–384 (1978).

9. Dimitriades, B., The Role of Natural Organics in Photochemical Air Pollution, *J. Air Pollut. Contr. Assoc.* **31,** 229–235 (1981).

10. Bhatia, S. P., Organosulfur Emissions from Industrial Sources, in *Sulfur in the Environment* (J. O. Nriagu, ed.), Wiley, New York, 1978.

11. Granat, L., Rodhe, H., and Hallberg, R. O., The Global Sulfur Cycle, in *Nitrogen, Phosphorus, and Sulfur—Global Cycles* (B. Svensson and R. Soderlund, eds.), SCOPE Report 7, Ecol. Bullet. No. 22, Stockholm, 1975.

12. Moss, M. R., Sources of Sulfur in the Environment: The Global Sulfur Cycle, in *Sulfur in the Environment* (J. O. Nriagu, ed.), Wiley, New York, 1978.

13. Altshuller, A. P., Regional Transport and Transformation of Sulfur Dioxide to Sulfate in the U.S., *J. Air Pollut. Contr. Assoc.* **26,** 318–324 (1976).

14. Semrau, K., Controlling the Industrial Process Sources of Sulfur Oxides, in *Sulfur Removal and Recovery from Industrial Processes* (J. B. Pfeiffer, ed.), American Chemical Society, Washington, D.C., 1975.

15. Hobbs, P. V., Radke, L. F., Eltgroth, M. W., and Hegg, D. A., Airborne Studies of the Emissions from the Volcanic Eruptions of Mount St. Helens, *Science* **211,** 816–818 (1980).

16. Farwell, S. O., Gluck, S. J., Bamesberger, W. L., Schutte, T. M., and Adams, D. F., Determination of Sulfur-Containing Gases by a Deactivated Cryogenic Enrichment and Capillary Gas Chromatographic System, *Anal. Chem.* **51,** 609–615 (1979).

17. Attar, A., and Corcoran, W. H., Sulfur Compounds in Coal, *Ind. Eng. Chem.* **16,** 168 (1979).

18. Meyer, B., *Sulfur, Energy, and Environment,* Elsevier, New York, 1977.

19. Cullis, C. F., and Mulcahy, M. F. R., The Kinetics of Combustion of Gaseous Sulfur Compounds, *Comb and Flame* **18,** 225 (1972).

20. Kramlich, J. C., "The Fate and Behavior of Fuel-Sulfur in Combustion Systems," Ph.D. Thesis, Washington State University, Pullman, WA, 1980.

21. *Control Techniques for Sulfur Oxide Air Pollutants,* U.S. Department of Health, Education and Welfare, AP-52, Washington, D.C., 1969.

22. Galloway, J. N., and Whelpdale, D. M., An Atmospheric Sulfur budget for Eastern North America, *Atmos. Environ.* **14,** 409–418 (1980).

23. Cullis, C. F., and Hirschler, M. M., Atmospheric Sulfur: Natural and Man-made Sources, *Atmos. Environ.* **14,** 1263–1278 (1980).

24. Adams, D. F., Farwell, S. O., Pack, M. R., and Robinson, E., Biogenic Sulfur Gas Emissions from Soils in Eastern and Southeastern United States, *J. Air Pollut. Contr. Assoc.* **31,** 1083–1089 (1981).

25. Postgate, J. R., The Sulphur Cycle, in *Inorganic Sulphur Chemistry* (G. Nickless, ed.), Elsevier, New York, 1968.

26. Senning A. (ed.), *Sulfur in Organic and Inorganic Chemistry,* Vols. 1–3, Dekker, New York, 1971.

27. Oae, S., *Organic Chemistry of Sulfur,* Plenum Press, New York, 1977.

28. Drake, R. L., Hales, J. M., Mishima, J., and Dreurs, D. R., *Mathematical Models for Atmospheric Pollutants, Appendix B: Chemical and Physical Properties of Gases and Aerosols,* EPRI, EA-1131, Palo Alto, CA, 1980.

29. Saunders, W., "Concentrations of Fluorotrichloromethane and Nitrous Oxide in the North American Troposphere and Lower Stratosphere," M.S. Thesis, Washington State University, Pullman, WA, 1978.

30. Cotton, F. A., and Wilkinson, G., *Advanced Inorganic Chemistry,* Interscience Publishers, New York, 1966.

31. Committee on Medical and Biologic Effects of Environmental Pollutants, *Nitrogen Oxides,* National Academy of Sciences, Washington, D.C., 1977.

32. McFarland, M., Kley, D., Drummond, J. W., Schmeltekopf, A. L., and Winkler, R. H., Nitric Oxide Measurements in the Equatorial Pacific Region, *Geophys. Res. Let.* **6,** 605–607 (1979).

33. Platt, U., Perner, D., and Pätz, H. W., Simultaneous Measurement of Atmospheric CH_2O, O_3, and NO_2 by Differential Optical Absorption, *J. Geophys. Res.* **84**, 6329–6335 (1979).

34. Singh, H. B., and Hanst, P. L., Peroxyacetylnitrate (PAN) in the Unpolluted Atmosphere: An Important Reservoir for Nitrogen Oxides, *Geophys. Res. Let.* **8**, 941–944 (1981).

35. Subcommittee on Ammonia, *Ammonia,* National Research Council, University Park Press, Baltimore, 1979.

36. Palmer, H. B., and Seery, D. J., Chemistry of Pollutant Formation in Flames, in *Ann. Rev. Phys. Chem.* 24 (H. Egring, C. J. Christensen, and H. S. Johnston, eds.), Annual Reviews, Inc., Palo Alto, CA, 1973.

37. *Control Techniques for CO, NO_x, and HC—Mobile Sources,* U.S. Department of Health, Education, and Welfare, AP-66, Washington, D.C., 1970.

38. Vogt, R. A., and Laurendeau, N. M., Preliminary Measurements of Fuel Nitric Oxide Formation in a Pulverized Coal Transport Reactor, *J. Air Pollut. Contr. Assoc.* **28**, 60–62 (1978).

39. *Control Techniques for Nitrogen Oxide Emissions from Stationary Sources,* U.S. Department of Health, Education and Welfare, AP-67, Washington, D.C., 1970.

40. Judeikis, H. S., Siegel, S., Stewart, T. B., Hedgpeth, H. R., and Wren, A. G., Laboratory Studies of Heterogeneous Reactions of Oxides of Nitrogen, in *Nitrogenous Air Pollutants* (D. Grosjean, ed.), Ann Arbor Science, Ann Arbor, 1979.

41. Lewis, W. H., *Nitrogen Oxides Removal,* Noyes Data Corporation, Park Ridge, NJ, 1975.

42. Environmental Protection Agency, *Federal Register* **43**, 121, 26962–68. See also Fontijn, A., Subadell, A. J., and Ronco, R. J., Homogeneous Chemiluminescence Measurement of Nitric Oxide with Ozone, *Anal. Chem.* **42**, 6 (1976).

43. Seinfeld, J. H., *Air Pollution Physical and Chemical Fundamentals,* McGraw-Hill, New York, 1975.

44. Jones, M. E., Ammonia Equilibrium Between Vapor and Liquid Aqueous Phases at Elevated Temperatures, *J. Phys. Chem.* **67**, 1113–1115 (1963).

45. Morgan, O. M., and Maass, O., An Investigation of the Equilibria Existing in Gas-Water Systems Forming Electrolytes, *Can. J. Res.* **5**, 162–199 (1931).

46. Polak, J., and Lu, B. C. Y., Vapor-Liquid Equilibria in System Ammonia-Water at 14.69 and 65 psia, *J. Chem. Eng. Data* **20**, 182–183 (1975).

47. Jones, K., Ammonia, in *Comprehensive Inorganic Chemistry* (J. C. Bailar, H. J. Emeleus, R. Nyholm, and A. F. Trotman-Dickenson, eds.), Pergamon Press, New York, 1973.

48. McConnell, J. C., Atmospheric Ammonia, *J. Geophys. Res.* **78**, 7812–7821 (1973).

49. Holliday, A. K., Hughes, G., Walker, S. M., Green, M. L. H., and Powell, P., *The Chemistry of Carbon: Organometalic Chemistry,* Pergamon Press, Oxford, 1975.

50. Singer, S. F. (ed.), *The Changing Global Environment,* D. Reidel Publishing Co., Boston, 1975.

51. Baes, C. F., Jr., Goeller, H. E., Olson, J. S., and Rotly, R. M., Carbon Dioxide and Climate: The Uncontrolled Experiment, *Am. Sci.* **65**, 310–320 (1977).

52. Climate Modeling, *Environ. Sci. Tech.* **14**, 501–507 (1980).

53. Committee on Medical and Biologic Effects of Environmental Pollutants, *Carbon Monoxide,* National Academy of Sciences, Washington, D.C., 1977.

54. Herzberg, G., *Molecular Spectra and Molecular Structure, 1. Spectra of Diatomic Molecules,* Van Nostrand, New York, 1950.

55. Jaffe, L. S., Carbon Monoxide in the Biosphere: Sources, Distributions, and Concentrations, *J. Geophys. Res.* **78**, 5293–5305 (1973).

56. Robinson, E., and Moser, C., Global Gaseous Pollutant Emissions and Removal Mechanisms, in *Proceedings of the Second International Clean Air Congress* (H. Englund and W. Beery, eds.), Academic Press, New York, 1971.

57. Stevens, C., Krout, L., Walling, D., Venters, A., Engelkemeir, A., and Ross, L. E., The Isotopic Composition of Atmospheric Carbon Monoxide, *Earth Planet. Sci. Lett.* **16**, 147–165 (1972).

58. Weinstock, B., and Niki, H., Carbon Monoxide Balance in Nature, *Science* **176**, 290–292 (1972).

59. Jaffe, L. S., The Global Balance of Carbon Monoxide, in *The Changing Global Environment* (S. F. Singer, ed.), D. Reidel Publishing Company, Boston, 1975.

60. Lynn, P. A., Tabor, E., Ott, W., and Smith, R., Present and Future Commuter Exposures to Carbon Monoxide, Paper no. 67-5 presented at the 60th Annual Meeting of the Air Pollut. Contr. Assoc., Cleveland, 1967.

61. Seiler, W., and Junge, C., Carbon Monoxide in the Atmosphere, *J. Geophys. Res.* **75**, 2217–2226 (1970).

62. Junge, C., Seiler, W., and Warneck, P., The Atmospheric ^{12}CO and ^{14}CO Budget, *J. Geophys. Res.* **76**, 2866–2879 (1971).

63. Cadle, R. D., Atmospheric Carbon Monoxide, in *Carbon and the Biosphere* (G. M. Woodwell and E. V. Pecan, eds.), U.S. Atomic Energy Commission, NTIS CONF-720510, 1973.

64. Lamb, S. I., Petrowski, C., Kaplan, I. R., and Simoneit, B. R. T., Organic Compounds in Urban Atmospheres: A Review of Distribution, Collection, and Analysis, *J. Air Pollut. Contr. Assoc.* **30**, 1098–1115 (1980).

65. Office of Research and Development, *Air Quality Criteria for Ozone and Other Photochemical Oxidants,* U.S. Environmental Protection Agency, EPA-600/8-78-004, Research Triangle Park, NC, 1978.

66. Zimmerman, P., *Testing of Hydrocarbon Emissions from Vegetation, Leaf Litter, and Aquatic Surfaces, and Development of a Methodology for Compiling Biogenic Emission Inventories,* U.S. Environmental Protection Agency, EPA-450/4-79-004, Research Triangle Park, NC, 1979.

67. Westberg, H., personal communication, 1982.

68. Faust, W. J., and Sterba, M. J., Minimizing Exhaust Emissions—A Realistic Approach, in *Effect of Automotive Emission Requirements on Gasoline Characteristics,* American Society for Testing and Material, ASTM Pub. No. 487, Philadelphia, PA, 1970.

69. Doelling, R. P., Gerber, A. F., and Walsh, M. P., Effect of Gasoline Characteristics upon Automotive Exhaust Emissions, in *Effect of Automotive Emission Requirements on Gasoline Characteristics,* American Society for Testing and Material, ASTM Pub. No. 487, Philadelphia, PA, 1970.

70. *Chemical and Engineering News,* June 11, 1979, p. 35.

71. Sittig, M., *Petroleum Refining Industry Energy Saving and Environmental Control,* Noyes Data Corporation, Park Ridge, NJ, 1978.

72. *Control Techniques for Hydrocarbon and Organic Solvent Emissions from Stationary Sources,* U.S. Department of Health, Education and Welfare, AP-68, 1970.

73. Sonnichsen, T. W., Taback, H. J., and Brunetz, N., Hydrocarbon Emissions from Petroleum Production Operations in California's South Coast Air Basin, Paper 78-36.3 presented at 71st Annual Meeting of the Air Pollut. Contr. Assoc., Houston, 1978.

74. Kugler, E. L., and Steffgens, F. W. (eds.), *Hydrocarbon Synthesis from Carbon Monoxide and Hydrogen,* American Chemical Society, *Advances in Chemistry* **178**, Washington, D.C., 1979.

75. Morrison, R. T., and Boyd, R. N., *Organic Chemistry,* 2nd edition, Allyn and Bacon, Boston, 1966.

76. Pedco Environmental, Inc., *National Emission Standards for Hazardous Air Pollutants,* EPA-340/1-78-008, Washington, D.C., 1978.

77. Singh, H. B., Salas, L. J., Smith, A., Stiles, R., and Shigeishi, H., *Atmospheric Measurements of Selected Hazardous Organic Chemicals,* Project Summary, EPA-600/53-81-032, Research Triangle Park, NC, 1981.

78. Mara, S. J., and Lee, S. S., *Assessment of Human Exposures to Benzene,* EPA Contract 68-02-2835, Stanford Research Institute Report 30R, Menlo Park, CA, 1978.

79. Howard, P. H., and Durkin, P. R., *Benzene, Environmental Sources of Contamination, Ambient Levels, and Fate,* EPA 560/5-75-005, NTIS PB 244 139, Springfield, VA, 1974.

80. Sax, N. I., *Dangerous Properties of Industrial Materials,* Van Nostrand Reinhold, New York, 1968.

81. Sittig, M., *Vinyl Chloride and PVC Manufacture, Process and Environmental Aspects,* Noyes Data Corporation, NJ, 1978.

82. Graedel, T. W., *Chemical Compounds in the Atmosphere,* Academic Press, New York, 1978.

83. Friberg, L., and Vostal, J. (eds.), *Mercury in the Environment,* CRC Press, Cleveland, Ohio, 1972.

CHAPTER 6

PARTICULATE POLLUTANT CHARACTERISTICS

RONALD G. PATTERSON

Calvert Environmental Equipment Co.
San Diego, California

1 INTRODUCTION

Small dust particles are usually kept suspended in air for long periods of time by the viscous force or drag of the air which resists any forces tending to precipitate them. Sometimes forces such as electrostatic charge also act to keep the particles suspended.

Dusts suspended in air are not in a stable state and will tend to separate from the air in time. Particle separation processes and equipment are designed to make this time as short as possible. Primary particles that reach the atmosphere and secondary particles formed in the atmosphere can affect human health, visibility, and climate. Particle size, concentration, and chemical composition are usually the most important factors in determining these effects.

2 PARTICLE CHARACTERISTICS

2.1 Dispersion Aerosols

Dispersion aerosols are formed by the grinding or atomization of solids and liquids and by the transfer of powders into a state of suspension through the action of air currents or vibration. Condensation aerosols are formed when supersaturated vapors are condensed or when gases react chemically to form a nonvolatile product. Condensation aerosols are on the order of 0.5 μm in diameter or smaller. Dispersion aerosols are, in most cases, considerably coarser and contain a wider range of particle sizes than condensation aerosols. When the dispersed phase is solid, the dispersion aerosol usually consists of individual or slightly aggregated particles of irregular form. In condensation aerosols, solid particles are often loose aggregates of a very large number of primary particles of a regular crystalline or spherical form.

Dispersion aerosols with solid particles are called dusts. Condensation aerosols with a solid disperse phase or a solid and liquid disperse phase are called smokes and fumes. Condensation and dispersion aerosols with a liquid disperse phase are called mists. This terminology applies regardless of particle size. Distinction between smokes and condensation mists is sometimes difficult, but is preferable to calling both types of aerosols smoke. In many cases a combination of dispersion and condensation aerosols are encountered.

2.2 Particle Size

Each of the particle types mentioned above may involve particles in a wide range of sizes, between 10^{-3} and 10^3 μm. This range represents a variation of 10^6 in size and 10^{18} in mass. Other physical properties are greatly affected over this wide size range. It is convenient to subdivide the range of aerosol particle size into three groups: 10^{-3} μm to λ_g, λ_g to 1 μm, and 1 to 10^3 μm, where λ_g is the mean free path of gas molecules (0.065 μm at 20°C and 760 mm Hg). Particles in the different size groups behave differently with respect to the different physical properties, like resistance of the medium to particle movement, light scattering, rate of evaporation and cooling, and dominant mechanism of particle removal from the medium.

Examination of particles with the aid of an optical or electron microscope involves the measurement of a linear dimension of the particle. The measured particle size is related to the particle perimeter or to the particle projected area diameter. Particle size measurement in this manner does not account for variation in particle density or shape. All laws describing the properties of aerosols can be expressed most simply for particles of spherical shape. To accommodate nonspherical particles it is customary to define a coefficient of sphericity, which is the ratio of the surface area of a sphere with the same volume as the given particle to the surface area of the particle.

An estimate of the particle volume can be obtained from microscopic sizing, and by assuming a density, an estimate of particle weight may be obtained. Because of the large variation in particle density and the aggregate nature of atmospheric particles, it is useful to define particle size based on aerodynamic behavior of the particle. The Stokes diameter d_{St} is defined as the diameter of a sphere having the same settling velocity as the particle and a density equal to that of the bulk material from which the particle was formed. Since dispersion and condensation aerosols are usually formed from many materials of different densities, it is more useful to define particle size based on the aerodynamic diameter d_{pa}. The aerodynamic diameter is the diameter of a sphere having the same settling velocity as the particle and a density equal to 1 g/cm^3. The aerodynamic diameter will be discussed in more detail in Section 2.7.

In calculating the average size of measured particles, attention should always be given to the purpose of these calculations. It is desirable to use averages relating to definite physical properties such as surface or volume. For example, if the surface of a certain dust is to be determined, more weight should be given to the finer particles. Hatch and Choate (1929) defined an average diameter as the diameter of a hypothetical particle which in some way represents the total number of particles in the sample. Their definitions are as follows:

$$\text{mean diameter} = d_n = \frac{\Sigma n d_p}{\Sigma n} \tag{1}$$

$$\text{length mean diameter} = d_{pln} = \frac{\Sigma n d_p^2}{\Sigma n d_p} \tag{2}$$

$$\text{surface mean diameter} = d_{psn} = \left(\frac{\Sigma n d_p^2}{\Sigma n}\right)^{1/2} \tag{3}$$

$$\text{volume mean diameter} = d_{pvn} = \left(\frac{\Sigma n d_p^3}{\Sigma n}\right)^{1/3} \tag{4}$$

$$\begin{aligned}\text{volume–surface mean diameter} = d_{vs} = \frac{\Sigma n d_p^3}{\Sigma n d_p^2}\\ \text{(Sauter mean diameter)}\end{aligned} \tag{5}$$

For example, Eq. (3) gives the surface area of that particle which when squared and multiplied by the total number of particles will give the total particle surface area.

2.3 Particle Size Distribution

The number or mass distribution of sizes may be plotted against small equal intervals of size to derive a histogram as shown in Figure 1. The choice between representing the particle size distribution by number or mass depends on the method of measurement, that is, whether the data were collected by counting or weighing. The area of the histogram represents the total number of particles measured. The histogram approaches the classical frequency distribution curve as the particle size increments are decreased to infinitesimal size.

It is often more informative and more convenient to express particle size frequencies as percentages of the total number of measurements. For this presentation, the total area under the curve represents 100%. Integration of these data results in the cumulative percentage curve.

The distribution of sizes in a particulate cloud arises from a number of chance causes which make it likely that the distribution curve would follow the normal or Gaussian error curve. Actually it is rare for the sizes to be distributed equally about the mean. In general there is more skewness of the distribution and the probability curve follows a geometric or logarithmic form. By plotting the log of the particle size against frequency, the skew curve is converted into one of the symmetrical types. The log probability distribution is expressed as

$$f(d_p)_n = \frac{\Sigma n}{\log \sigma_g \sqrt{2\pi}} \exp\left[\frac{-(\log d_p - \log d_g)^2}{2 \log^2 \sigma_g}\right] \tag{6}$$

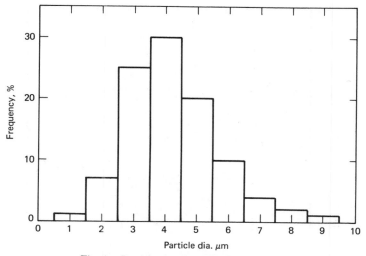

Fig. 1. Particle size distribution histogram.

where

$$f(d_p)_n = \text{frequency of occurrence of diameter } d$$
$$n = \text{total number of particles}$$
$$\sigma_g = \text{geometric standard deviation obtained from}$$

$$\log \sigma_g = \left(\frac{\Sigma[n(\log d_p - \log d_g)^2]}{n}\right)^{0.5} \tag{7}$$

where d_g is the geometric number mean diameter and is defined as

$$d_g = \sqrt[n]{d_1, d_2, \ldots, d_n} \tag{8}$$

From the log–probability curve, the number median or 50% diameter, which by definition corresponds to the geometric mean d_g, can be derived. Another useful property, the geometric standard deviation σ_g, is calculated from

$$\sigma_g = \frac{84.13\% \text{ size}}{50\% \text{ size}} = \frac{50\% \text{ size}}{15.87\% \text{ size}} \tag{9}$$

The geometric mean diameter d_g and the standard deviation σ_g are determined from the particle size distribution as measured with a particle sizing instrument, such as a cascade impactor. The particle size data shown in Table 1 were determined from cascade impactor measurements of an emission source.

Table 1 Particle Size Distribution Data

Aerodynamic Particle Diameter, μmA	Cumulative Percentage	Cumulative Mass Concentration, g/Nm³
	100	2.29
4.5	92	2.10
2.5	70	1.60
1.4	40	0.92
0.8	15	0.34
0.5	4	0.09

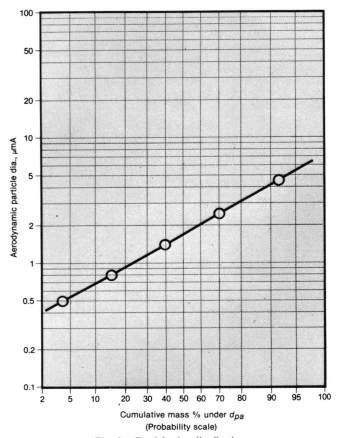

Fig. 2. Particle size distribution.

The particle diameters given in this table are in aerodynamic particle size (μmA), since this is the particle diameter measured by a cascade impactor. The particle size distribution is then plotted on log–probability paper as shown in Figure 2.

The aerodynamic diameter is plotted versus the cumulative mass percent of particles smaller than that size. Fifty percent of the particle mass is smaller than the mass median diameter d_{gm}.

From Figure 2, the mass median diameter is $d_{gm} = 1.7$ μmA and the geometric standard deviation σ_g is

$$\sigma_g = \frac{d_{84.13}}{d_{50}} = \frac{3.4\ \mu\text{mA}}{1.7\ \mu\text{mA}} = 2.0 \tag{9a}$$

An important property of the log normal distribution is that if the number distribution is log normal, the weight distribution is also log normal and with the same standard deviation. This also holds true for other properties of the particle size distribution, such as the distribution with respect to surface area or particle volume.

The relationship between the various means are

$$\ln d_l = \ln d_{gm} - 2.5 \ln^2 \sigma_g \tag{10}$$
$$\ln d_s = \ln d_{gm} - 2\ \ \ln^2 \sigma_g \tag{11}$$
$$\ln d_m = \ln d_{gm} - 1.5 \ln^2 \sigma_g \tag{12}$$
$$\ln d_{vs} = \ln d_{gm} - 0.5 \ln^2 \sigma_g \tag{13}$$

or, in general

$$\ln d_{qp} = \ln d_{gm} + 0.5(q + p - 6) \ln^2 \sigma_g \tag{14}$$

Table 2 Mean Diameters

Symbol	Name of Mean Diameter	p	q	Order	Field of Application
d_l	Linear (arithmetic)	0	1	1	Comparison, evaporation
d_s	Surface	0	2	2	Absorption
d_v	Volume	0	3	3	Comparison, atomizing
d_m	Mass	0	3	3	Comparison, atomizing
d_{sd}	Surface–diameter	1	2	3	Adsorption
d_{vd}	Volume–diameter	1	3	4	Evaporation, molecular diffusion
d_{vs}	Volume–surface	2	3	5	Mass transfer, reactions
d_{ms}	Mass–surface	3	4	7	Combustion equilibrium

The definitions for the various mean diameters as given by Mugele and Evans (1951) are shown in Table 2.

The relationship between the means given in Eq. (14) proves very useful in practical work to estimate other properties of the particles in a given size distribution. Several of the means are shown in Figure 3.

The various means can be used to determine different properties of the aerosol or other dispersed phase, such as the spray drops in a wet scrubber. For example, it may be necessary to determine the total surface area of the spray drops in a scrubber for a mass transfer calculation. The volume–surface mean diameter d_{vs}, also known as the Sauter mean diameter, is the drop diameter which relates the surface area of the spray to the total spray volume. For a spray with $d_{vs} = 380$ μm being produced at the rate of 3800 cm³/min (approximately 1 gpm), the total surface area of the spray will be 3800 cm³/min/380 × 10⁻⁴ cm = 10⁵ cm²/min.

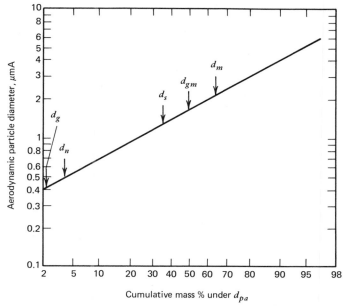

Fig. 3. Several characteristic particle diameters.

2.4   Particle Concentration

Particle concentrations are expressed as either a number or mass concentration and are defined in a manner similar to gas density. The number concentration of particles is the ratio of the number particles in a given volume to the volume. Particle concentrations in a polluted atmosphere may be of the order of $10^6/cm^3$, with source concentrations several orders of magnitude greater. In polluted air, most of the particles by number fall in the size range between 0.01 and 0.1 μm. Concentrations of particles larger than 1 μm typically fall to less than $10/cm^3$. Particle number concentrations are typically measured with light scattering, electrical mobility analyzers, and condensation nuclei counters.

Particle mass concentrations are defined as the ratio of the mass of particles in a given volume to the volume. The particle mass concentration may be determined by filtering a known volume of gas and weighing the collected particles. Particle mass concentrations in polluted atmospheres are usually well below 1000 μg/m^3. Particle mass concentrations in process gases are normally much greater than 1 g/m^3 (2.29 g/m^3 = 1 g/ft^3).

2.5   Chemical Composition

A single aerosol particle may be composed of many chemical compounds, but chemical analysis is difficult because the amount of material to be analyzed is so small. Some success in measuring the elemental composition of individual particles has been achieved with the electron microprobe and the quadrapole mass spectrometer, but identification of individual particles is beyond the capabilities of the normal laboratory.

Two sampling methods routinely available for determining chemical composition are total filter and cascade impactor analysis. Filtration is the most common method of sampling particles for the determination of chemical composition. If the filter is perfectly efficient, it provides information on the chemical composition averaged over all particle sizes and over the time interval of sampling.

Information on the chemical composition as a function of particle size can be obtained from cascade impactors by analyzing each of the impactor substrates separately. This information can be used to develop a particle control strategy to remove the most hazardous particles from the stack gas.

2.6   Nondimensional Parameters

Several nondimensional parameters that characterize aerosol systems will be given here, since they will be used throughout the remainder of this chapter. Nondimensional parameters that characterize particle collection efficiency in air pollution control systems are given in Section 6, which describes particle collection mechanisms.

2.6.1   *Flow Reynolds Number* (Re_g)

The flow Reynolds number is the ratio of inertial-to-viscous forces of a flowing fluid. When the Re_g < 2100, viscous forces dominate and the flow is laminar. For Re_g > 4000, the flow becomes turbulent. The flow Reynolds number is

$$Re_g = \frac{Du_g\rho_g}{\mu_g} \qquad (15)$$

where

D = diameter or other characteristic dimension of the gas-containing device
u_g = gas velocity
ρ_g = gas density
μ_g = absolute viscosity of the gas

2.6.2   *Particle Reynolds Number* (Re_p)

Particle motion in the gas stream is characterized by the particle Reynolds number which is defined by

$$Re_p = \frac{d_p(u_p - u_g)\rho_g}{\mu_g} \qquad (16)$$

where

$$d_p = \text{particle diameter}$$
$$u_p = \text{particle velocity}$$

Note that the particle Reynolds number is dependent on the velocity of the particle relative to the gas stream and the fluid properties, even if the particle is a liquid. Typically the particle Reynolds number will be on the order of 10^{-4} to 10^2.

2.6.3 Knudsen Number (Kn)

Another nondimensional parameter that characterizes particle motion is the Knudsen number. This number is the ratio of the mean free path of gas molecules to particle diameter and is

$$\text{Kn} = \frac{2\lambda_g}{d_p} \tag{17}$$

where

$$\lambda_g = \text{mean free path of gas molecules, cm}$$

From the kinetic theory of gases the mean free path, λ_g, is

$$\lambda_g = \frac{kT}{2\pi P d_{mo}^2} \tag{18}$$

where

$$k = \text{Boltzmann constant} = 1.38062 \times 10^{-23} \text{ J K}^{-1}$$
$$d_{mo} = \text{molecular diameter}$$
$$P = \text{gas pressure, kPa}$$
$$T = \text{absolute gas temperature, °C}$$

The value of the mean free path of air molecules at 20°C and 760 mm Hg is approximately 6.53×10^{-2} µm.

For very small particles the gas appears discontinuous and the particles tend to slip between the gas molecules. This occurs when $\text{Kn} > 0.1$.

Four size regimes are typically used to characterize the effect of gas molecules on the motion of particles. The size regimes in order from large particles to small particles are: continuum, slip flow, transition, and free molecule regimes. The approximate diameters for particles in air at standard conditions are shown in Table 3.

2.6.4 Cunningham Slip Correction Factor (C')

When the Knudsen number, Kn, is greater than about 0.1, particles slip between gas molecules and the resistance of the air is thought of as discontinuous. In the Cunningham slip flow regime a correction factor is applied to account for the slippage. The correction factor, which is based on the Millikan oil drop studies, includes thermal and momentum accommodation factors and is empirically fit to a wide range of Kn values.

The Cunningham slip correction factor can be calculated from

$$C' = 1 + A\frac{2\lambda_g}{d_p}$$
$$= 1 + [1.257 + 0.4 \exp(-1.1\frac{d_p}{2\lambda_g})]\frac{2\lambda_g}{d_p} \tag{19}$$

Table 3 Particle Size Regimes for Air at 20°C and 760 mm Hg (Hesketh, 1979)

Size Regime	Kn	d_p, µm
Continuum (Stokes)	<0.1	>1.3
Slip flow (Cunningham)	≤0.3	>0.4
Transition	10–0.3	0.01–0.4
Free molecule	>10	<0.01

A simplified equation given by Calvert et al. (1972) for use in air at normal pressure is

$$C' = 1 + \frac{6.21 \times 10^{-4}}{d_p} T \tag{20}$$

The Cunningham slip correction factor becomes negligible for particles larger than approximately 1 μm under normal conditions.

2.6.5 Stokes Number (St)—Impaction Parameter (K_p)

The nondimensional ratio of the particle stopping distance to the characteristic dimension of the system is the Stokes number, St. This parameter represents the ratio of drag-to-viscous forces on the particle. In this chapter the Stokes number will be defined as equal to the impaction parameter, which will be developed in Section 6.2.

$$St = K_p = \frac{2X_s}{d_c} \tag{21}$$

where

$$X_s = \text{particle stopping distance}$$
$$d_c = \text{collector diameter}$$

Several other definitions of the Stokes number and impaction parameter are used in the literature, so care should be exercised in comparing values of these numbers between references.

2.7 Aerodynamic Diameter

Particle size, shape, and density can vary among the particles which make up a specific aerosol. Variations in these particle properties strongly affect the collection efficiency in particle collection systems which rely primarily on the inertia of the particle for capture, such as wet scrubbers and cyclones.

The aerodynamic diameter is a measure of how the particle will react to inertial forces. It can be obtained with cascade impactor measurements and does not require specific knowledge of the physical size, shape, or density of the particle. Particles of different size, shape, and density can have the same aerodynamic size. If two particles have the same aerodynamic size, they will be collected at the same efficiency by the impaction mechanism as discussed in Section 6.2.

Several aerodynamic diameters have been used as defined below:

1. *Stokes Diameter.* The diameter of a sphere having the same terminal settling velocity and density as the particle and corrected for the Cunningham slip correction factor based on this diameter.

2. *Classical Aerodynamic Diameter.* The diameter of a unit density sphere ($\rho_p = 1$ g/cm^3) having the same terminal settling velocity as the particle, at low particle Reynolds number which has been corrected for the Cunningham slip correction factor based on this diameter.

3. *Aerodynamic Impaction Diameter.* The diameter of a unit density sphere having the same settling velocity as the particle, at low Reynolds number. The diameter in this case is not corrected for the Cunningham slip correction factor. This makes it the easiest to obtain from cascade impactor measurements and the most representative aerodynamic diameter for inertial separation of particles.

The relationship between the various aerodynamic diameters is given in Table 4 and Eq. (24). A comparison between the Stokes or physical diameter and the aerodynamic impaction diameter is shown in Figure 4 for several particle densities. This figure also shows the effect of the Cunningham slip correction factor on the aerodynamic particle size for $d_p < 1.0$.

2.8 Cut Diameter

Cut diameter d_{pc} is defined as the diameter for which collection efficiency (and penetration) equals 50%. In the collection of particles, this becomes a convenient method of indicating the particle removal efficiency of a collector for a dust with a specific particle size distribution. Cut diameter is most applicable to systems such as scrubbers or cyclones that collect particles mainly by a single mechanism. For these two devices, the particle collection efficiency is a continuous function of the particle size. A filter, which operates with one mechanism for larger particles and another for smaller may not be categorized as accurately by the cut diameter–efficiency relationship.

Table 4. Equations Used for Particle Size Conversions—Classical Aerodynamic, Stokes Diameter, and Aerodynamic Impaction Diameter [a]

Diameter Definition, Given	Conversion Equation		
	Stokes Diameter, d_p	Classical Aerodynamic Equivalent Diameter, d_{pac}	Aerodynamic Impaction, d_{pa}
Stokes diameter, d_p (Note 1)	1.0	$d_{pac} = d_p \left[\dfrac{\rho C(d_p)}{C(d_{pac})} \right]^{1/2}$	$d_{pa} = d_p [C(d_p)\rho]^{1/2}$
Classical aerodynamic diameter, d_{pac}	$d_p = d_{pac} \left[\dfrac{C(d_{pac})}{\rho C(d_p)} \right]^{1/2}$	1.0	$d_{pa} = d_{pac} [C(d_{pac})]^{1/2}$
Aerodynamic impaction diameter, d_{pa} (Note 2)	$d_p = d_{pa} \left[\dfrac{1}{C(d_p)\rho} \right]^{1/2}$	$d_{pac} = d_{pa} \left[\dfrac{1}{C(d_{pac})} \right]^{1/2}$	1.0

[a] Notation: d_p = Stokes diameter, μm
d_{pac} = classical aerodynamic equivalent diameter, μm
d_{pa} = aerodynamic impaction diameter, μmg$^{1/2}$cm$^{-3/2}$
ρ = Particle density, g/cm^3
$C(d_p)$, $C(d_{pac})$, $C(d_{pa})$ = Cunningham slip correction factor

Note 1: The Stokes diameter is normally taken as approximation to the physical diameter of the particle.

Note 2: The aerodynamic impaction diameter is easily determined by cascade impactor measurements and is most useful in design of air pollution control equipment in which the impaction parameter is dominant.

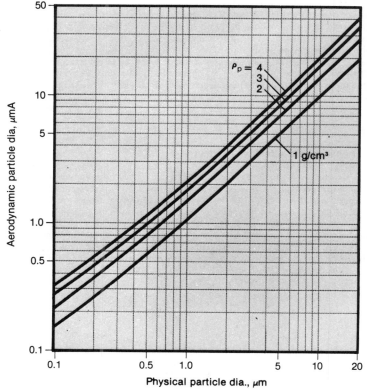

Fig. 4. Relation between physical and aerodynamic diameter.

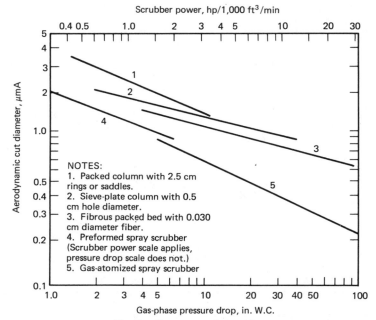

Fig. 5. Cut power relationship.

The 50% cut diameter expressed as a 50% aerodynamic cut diameter d_{50} is used to compare the relative effectiveness of wet scrubbers as shown in Figure 5 (Calvert, 1977). The Calvert cut/power method, described more fully in Chapter 10, may be used to select the wet scrubber with the lowest energy requirement for a given application.

The 50% cut diameter is also used to characterize the particles collected on a cascade impactor stage as discussed in Chapter 31. The stage cut diameter may be plotted as a function of the fractional cumulative amount collected on each of the cascade impactor stages to obtain the particle size distribution as shown in the example in Section 2.3.

3 TRANSPORT PROPERTIES

3.1 General Motion of Individual Particles

Three types of forces act on an aerosol particle: external forces (gravitational, electrical, magnetic, etc.), resistance of the medium, and interaction of the particles. The third is, in most cases, considerably less than the others and the particles can be considered independent of one another.

Particles in air extend over a wide range of sizes from those much larger in size, to those much smaller, than the mean free path of the gas molecules, which is 6.53×10^{-2} μm at 20°C and 760 mm Hg. The resistance of a gas to the motion of particles much larger than the mean free path is caused by hydrodynamic forces. For rigid spherical particles where there are no inertial effects arising from fluid displaced by the moving particle and where the medium is infinite in size, this resistance is expressed by Stokes law:

$$F = -3\pi\mu_g d_p u_r \tag{22}$$

where

$$\begin{aligned}
F &= \text{fluid resistance force} \\
\mu_g &= \text{gas absolute viscosity} \\
d_p &= \text{particle diameter} \\
u_r &= \text{relative velocity between particle and gas}
\end{aligned}$$

3.2 Terminal Settling Velocity

The terminal settling velocity of a spherical particle is derived by equating the gravitational and resistance forces.

$$u_t = \frac{(\rho_p - \rho_g)g d_p^2}{18\mu_g} \tag{23}$$

where

$$\rho_p = \text{particle density}$$
$$\rho_g = \text{gas density}$$

When the particle size approaches the mean free path of the gas molecules, particles slip between air molecules and the resistance of the air is thought of as discontinuous. A correction factor called the Cunningham correction factor is applied to account for the increased rate of particle fall. The particle terminal settling velocity then becomes

$$u_t = \frac{d_p^2(\rho_p - \rho_g)g C'}{18\mu_g} = \frac{d_{pac}(\rho_p - \rho_g)g C'_{ac}}{18\mu_g} = \frac{d_{pa}^2}{18\mu_g} \tag{24}$$

where

$$C' = \text{the Cunningham slip correction factor}$$
$$d_p = \text{Stokes' diameter}$$
$$d_{pac} = \text{classical aerodynamic diameter}$$
$$d_{pa} = \text{aerodynamic impaction parameter}$$

The Cunningham correction factor can be calculated from Eq. (19).

Stokes law is valid so long as there are no inertia effects arising from displacement of the air by the moving sphere, $Re_p < 0.05$. ($Re_p < 0.05$ corresponds to a 77-μm sphere of unit density falling through air at room temperature and pressure.) At higher Reynolds numbers, the deviation becomes greater and Stokes law tends to overestimate the velocity. From consideration of dynamical similarity, the drag coefficient C_D, which is defined as the ratio of the resistance to the product of cross-sectional area and dynamic pressure, can be applied for larger Reynolds numbers. For the simple case of Stokes law,

$$C_D = \frac{24}{Re_p} \tag{25}$$

Since C_D and Re_p both contain a velocity term, it is not convenient to use a graph of Re_p versus C_D to calculate the velocity. To eliminate the velocity term, the expression ($C_D\ Re_p^2$) is used to obtain a direct solution for the velocity. Eq. (25) becomes

$$Re_p = \frac{C_D\ Re_p^2}{24} \tag{26}$$

For the inverse problem of finding the size of particles from their rate of settling, a graph that correlates Re_p and C_D is used. Figure 6 gives the terminal settling velocity of particles and drops in air. The settling velocity for drops larger than approximately 2000 μm in diameter deviates from that of rigid spheres because the drag forces distort the shape of the drop, which changes the drag coefficient.

3.3 Brownian Motion and Diffusion

The terminal settling velocities of particles much smaller than the mean free path of the gas molecules are very low, the random bombardment of air molecules deflects these particles and a random motion is superimposed on the settling motion. This random motion is called Brownian motion. Removal of these small particles is affected mainly by diffusion.

The diffusion coefficient D_p is related to particle diameter by Fuchs (1964):

$$D_p = \frac{RTC'}{3\pi\mu_g d_p N} \tag{27}$$

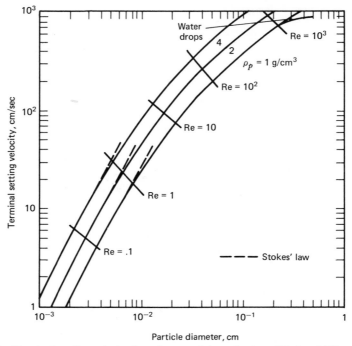

Fig. 6. Terminal settling velocity for spheres and drops in air at 20°C and 760 mm Hg.

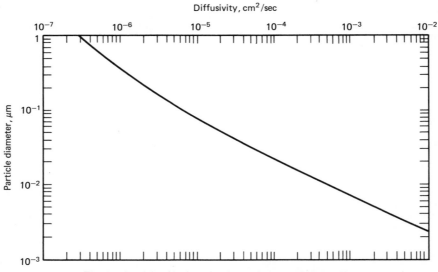

Fig. 7. Particle diffusivity in air at 20°C and 760 mm Hg.

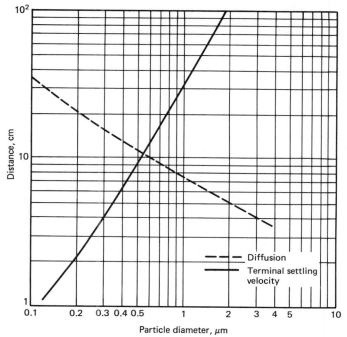

Fig. 8. Comparison between Brownian displacement and settling distance of unit density spherical particles in air.

where

$$N = \text{Avogadro's number}$$

Figure 7 shows the diffusivity as a function of particle size for several operating conditions.

The mean square displacement of a particle, $\Delta^2 x$, in a given interval of time t along the x axis is

$$\Delta^2 x = 2 D_p t \tag{28}$$

Figure 8 compares the root mean square displacement of a particle in 1 sec in a given direction with the distance a spherical particle of unit density falls in 1 sec in air at 760 mm Hg pressure and 20°C under gravity, calculated from Stokes equation. It can be seen that particles of unit density smaller than approximately 0.5 μm in diameter have a larger Brownian displacement than displacement due to gravitational settling.

3.4 Electrical Migration

Aerosol particles are charged in one of three ways:

1. During the process of generation.
2. By free ion diffusion.
3. By corona discharge.

Different aerosols have different charge distributions according to their generation methods. Dust particles become charged, either by friction or separation of the particles, with the result that two groups of oppositely charged particles are formed. The dispersing medium does not become appreciably charged and any electrification detected would arise from fine particles suspended in it. The total electrification of the two sets of charged particles and the fine particles in the air accompanying them is zero.

In general, only small levels of charge would be found on the small particles. The average charge per particle increases somewhat more slowly than the square of the particle diameter. In homogeneous systems, when the dust and the container from which it is dispersed are of the same material, the charges of opposite sign are approximately equal and there is no net charge. In an inhomogeneous system, asymmetry of charge distribution is not usually large and depends on the proportion of particles striking the surface of the containing vessel relative to those just separated.

When mists are produced by atomization, drops from polar liquids are charged considerably more than those of nonpolar liquids. The drop charge depends on the number of positive and negative ions that happen to be in that volume of the liquid when it is released. This gives a random distribution of charges in the drops. In nonconducting liquids the charges produced follow a normal distribution with the average charge of zero and the mean square charge proportional to drop volume.

Condensation aerosol particles generated at moderate temperatures are uncharged, but they become gradually charged by gaseous ions diffusing to them, until a steady state is reached. The process is a random one and the proportion of particles becoming charged will depend on particle size, but the number of positive and negative charges will be equal in the absence of extraneous effects.

Aerosol charging during generation and through free ion diffusion usually produces charges of both signs in equal quantities. Corona discharge, on the other hand, will lead to charged particles of a single sign. Corona discharge affects the aerosol through two mechanisms: field charging and diffusion charging. Field charging predominates for particles larger than about 0.5 μm in diameter while diffusion charging predominates for smaller particles. In the intermediate range, charging is due to both mechanisms and the rate of particle charging shows a minimum at a particle diameter of about 0.3 μm.

When a conducting particle is suspended in an electric field, its charge will increase until the field caused by its own charge equals the imposed field. At that point the particle will have reached its saturation charge and will not be further charged. Oglesby and Nichols (1970) give the following equation for the saturation charge q_s on a conducting sphere:

$$q_s = 3\pi d_p^2 \epsilon_0 E \tag{29}$$

where

$$\epsilon_0 = \text{permittivity of free space} = 8.954 \times 10^{-12} \text{ F/m}$$
$$E = \text{electric field, V/m}$$

For a nonconductive particle,

$$q_s = 3\frac{\epsilon_p}{\epsilon_p + 2}\pi\epsilon_0 d_p^2 E \tag{30}$$

where ϵ_p is the relative dielectric constant of the particle. The charge as a function of time for either a conducting or nonconducting sphere is

$$q(t) = q_s \frac{1}{1 + (4\epsilon_0/N_0 eBt)} \tag{31}$$

where

$$N_0 = \text{number of ions per unit volume, cm}^{-3}$$
$$e = \text{electronic charge} = 1.6 \times 10^{-19} \text{ C}$$
$$B = \text{particle mobility, sec/g}$$
$$t = \text{time, sec}$$

For diffusion charging the same authors give the following equation:

$$q(t) = \frac{d_p kT}{2e} \ln\left(1 + \frac{\pi d_p V_i N_0 e^2 t}{2kT}\right) \tag{32}$$

where

$$V_i = \text{root mean square thermal velocity, cm/sec}$$

The root mean square thermal velocity of the ions is given by

$$V_i = \left(\frac{8kT}{\pi m}\right)^{1/2} \tag{33}$$

where

$$m = \text{ion mass, g}$$

In the intermediate range of particle sizes both charging mechanisms contribute significant charges. In this case, it is more nearly correct to sum the rates of charging from the two mechanisms and then solve for the charge. Oglesby and Nichols (1970) present several solutions to the combined nonlinear differential equation.

The electrostatic force acting on a charged particle in an electric field E is given by

$$F = qE \tag{34}$$

For a quasistationary motion, neglecting all forces except electrostatic and drag, the velocity of a particle obeying Stokes law in an electrostatic field will be

$$u_{pE} = \frac{qEC'}{3\pi d_p \mu_g} \tag{35}$$

An equation for particle velocity for particles charged by diffusion, field, or both can be obtained by substituting the appropriate equation for q in Eq. (35).

An aerosol consisting of particles all bearing a charge of the same sign will uniformly disperse under the action of its own bulk charge. The rate of scattering of a charged aerosol cloud is described by

$$\frac{1}{n} - \frac{1}{n_i} = \frac{4q^2 t C'}{3 d_p \mu_g} \tag{36}$$

where

$$n_i = \text{initial particle concentration, cm}^{-3}$$
$$n = \text{particle concentration at time } t, \text{ cm}^{-3}$$

Since the concentration of the aerosol is the same in all places at any instant during electrostatic scattering, motion of the medium has no effect on the scattering, so that the equation given above is valid for an aerosol experiencing either laminar or turbulent flow.

When the aerosol contains both positive and negative particles, not in equal concentrations, the particles in the minority are driven to the center of the aerosol cloud where their concentration consequently increases. Then the charge density in this central nucleus decays asymptotically, until it becomes practically neutral. At the same time scattering of the purely unipolar part of the aerosol outside the nucleus continues.

3.5 Magnetic Fields

When a particle with no intrinsic magnetic properties moving with the gas at a velocity u_g is charged to an electric charge q and then introduced into a magnetic field of strength H, it will be acted on by a force at right angles to both the direction of the field and its direction of motion. The particle will thus be given a terminal drift velocity which may be calculated from Stokes' law:

$$u_{pM} = \frac{Hqu_g C'}{3\pi \mu_g d_p} \tag{37}$$

Here the drift velocity is a function of the gas velocity. The higher the gas velocity, the higher the particle drift velocity will be.

If small magnetic particles were introduced into a magnetic field they would align themselves in the magnetic field, their opposite ends being attracted to the two poles of the magnet. The net magnetic force acting on the particle can be calculated by the algebraic sum of the attractive and repulsive forces.

3.6 Acoustic Fields

A particle suspended in a vibrating medium is itself set in oscillation by the drag on the particle. Attempts have been made to use acoustic agglomeration as a method of growing particles into larger sizes so they may be readily removed in a cyclone or other particulate removal device. These attempts have been successful, but a detailed discussion of the transport of particles in an acoustic field will not be developed here. Detailed accounts can be found in Fuchs (1964) and Chou et al. (1981).

3.7 Thermophoresis

For $d \ll \lambda$ thermophoresis is the result of gas molecules impinging on the particle from opposite sides with different mean velocities. The net force caused by the difference in momentum received by the particle from each side as calculated by Waldmann (1959) is

$$F_T = -\frac{d_p^2 P\lambda}{T} \nabla T \tag{38}$$

The thermophoretic velocity in this regime is given by

$$u_{pT} = \frac{6\mu_g}{(8 + \pi a)T\rho_g} \nabla T \tag{39}$$

where

$$\alpha = \text{fraction of gas molecules reflected}$$
$$\text{diffusively by the particle}$$

For a freshly formed and very smooth surface of amorphous particles and for liquid particles, $\alpha \sim 0.9$. For particles with rough surfaces obtained by mechanical working, or with aged particles, $\alpha \sim 1.0$.

When the size of a particle becomes large compared to the mean free path ($\lambda_g \ll d$), the molecule–particle interaction changes, since the molecule no longer sees the particle as another molecule but as a solid surface. In the layer of gas adjacent to the nonuniformly heated wall, a similar tangential temperature gradient will be established.

If the temperature is higher on the left, the gas molecules impinging on the element of surface from the left have on the average a higher velocity than those coming from the right. As a result the wall receives an impulse directed to the right, against the temperature gradient, and an equal impulse directed to the left is imparted to the gas, making it slip along the surface toward the higher temperature. From these considerations, Epstein (1929) derived the following equation for the thermophoretic force:

$$F_T = -\frac{9\pi d_p \mu_g^2}{2\rho_g T} \left[\frac{k_g}{2(k_g + k_p)} \right] \nabla T \tag{40}$$

where

$$k_g = \text{gas thermal conductivity}$$
$$k_p = \text{particle thermal conductivity}$$

The steady-state velocity of a large particle can be determined by equating this force to the Stokes' drag:

$$u_{pT} = -\frac{3\mu_g}{2\rho_g T} \left[\frac{k_g}{2k_g + k_p} \right] \nabla T \tag{41}$$

Epstein's equation agrees with experimental measurements for particles of low thermal conductivity ($k_p/k_g < 10$). However, for particles of high thermal conductivities, experimentally measured thermophoretic velocities were 30–100 times those predicted by the Epstein equation. Brock (1962) extended the theory to account for high thermal conductivities and obtained the following equation for the thermophoretic velocity:

$$u_{pT} = \frac{3\mu_g C'}{2\rho_g T} \left[\frac{1}{1 + (6C_m \lambda/d_p)} \right] \left[\frac{(k_g/k_p) + C_t(2\pi/d_p)}{1 + 2(k_g/k_p) + 4C_t(\lambda/d_p)} \right] \nabla T \tag{42}$$

Brock indicates that C_t ranges from 1.875 to 2.48 and C_m from 1.00 to 1.27. He showed that this equation satisfactorily accounts for experimental measurements for particles having both high and low conductivity to within 25%. Derjaguin and Bakanov (1962), using a different approach, developed the following equation for the thermophoretic velocity of large particles:

$$u_{pT} = -\frac{(4k_g + 0.5k_p)\mu_g}{(2k_g + k_p)\rho_g T} \nabla T \tag{43}$$

Goldsmith and May (1966) compared these three equations with experimental results for an aerosol composed mainly of chromium particles and found the Derjaguin and Bakanov equation agrees best with the velocities measured in air and helium. The Epstein equation modified by the addition of the Cunningham slip correction factor gave good agreement in helium and poorer agreement with the velocities measured in air. Brock's equation gave a better approximation in air and somewhat worse than the modified Epstein equation in helium.

3.8 Photophoresis

The motion of aerosol particles illuminated from one side, photophoresis, is a special case of thermophoresis. The phenomena observed depend on the temperature distribution in the illuminated particle, which can be very different depending on particle shape, size, transparency, and refractive index. In transparent particles the rear side may be heated more than the side facing the light source with the result that the particle will move toward the light. When the side facing the light is hotter than the back side, the particle will move away from the light source.

3.9 Diffusiophoresis and Stephan Flow

At the surface of particles wetted with a volatile liquid, a dust-free space is formed even when a temperature difference is lacking. This is due to the diffusiophoretic force on the particle, which is the sum of the force due to the Stephan flow and a force due to the gas momentum transfer processes. Stephan flow is the hydrodynamic force necessary to maintain a uniform total pressure in a diffusing gaseous system and is directed toward the liquid surface when vapor is condensing. In a binary system of a vapor and a carrier gas the velocity of the Stephan flow is

$$u_s = \frac{D_v}{p_g} \frac{dp}{dx} \tag{44}$$

where

$$p_g = \text{gas partial pressure, kPa}$$
$$D_v = \text{vapor diffusivity in gas, cm}^2/\text{sec}$$

$$\frac{dp}{dx} = \frac{p_0 - p_v}{\Delta x} \tag{45}$$

where

$$p_0 = \text{saturation vapor pressure of the drop}$$
$$\text{liquid at the drop temperature, kPa}$$
$$p_v = \text{vapor partial pressure, kPa}$$
$$\Delta x = \text{distance from the drop surface to the gas bulk}$$
$$\text{phase where the vapor pressure is } p_v$$

The gradient is positive for an evaporating drop, causing the Stephan flow velocity to be negative and the particles to move away from the drop. In condensing systems, the flow of vapor and particles is positive and toward the drop.

If the molecular mass of the diffusing vapor molecules is different from the molecular mass of the carrier gas molecules, the motion of small particles is affected by gas momentum transfer processes. Waldmann (1959) has treated this phenomenon for small particles $(d < \lambda)$. When the two mechanisms, namely, Stephan flow and the difference in molecular impact, are combined, the diffusiophoretic velocity for a dilute vapor system is

$$u_{pD} = - \frac{\sqrt{M_v}\,P}{p_v \sqrt{M_v} + p_g \sqrt{M_g}} \frac{D_v}{p_g} \nabla P_v \tag{46}$$

where

$$M_v = \text{molecular weight of the diffusing vapor}$$
$$M_g = \text{molecular weight of the stationary gas}$$

The Waldmann equation has been developed for $d < \lambda$, but, according to Waldmann and Schmitt (1960), it agrees with experimental results where (d/λ) ranged from 2 to 12 within 9%. A similar equation was also derived by Bakanov and Derjaguin (1960).

4   NONSTEADY-STATE PARTICLE MOTION

4.1   Relaxation Time

The velocity of a particle will change whenever a force is applied or removed from a particle. The time required for the velocity change to take place is dependent on the particle characteristics. The relaxation time τ is the time during which most of the motion change occurs. A particle accelerated from rest will reach two-thirds of its terminal velocity when $t = \tau$. The relaxation time τ is defined by

$$\tau = \frac{d_p^2 \rho_p C}{18\mu_g} \tag{47}$$

4.2   Accelerating Particles

The velocity of an accelerating particle u_p can be determined from a force balance:

$$u_p = \tau a[1 - \exp(-t/\tau)] \tag{48}$$

where

$$u_p = \text{particle velocity, cm/sec}$$
$$a = \text{particle acceleration, cm/sec}^2$$

When the particle is accelerating under the force of gravity, Eq. (48) gives the terminal settling velocity of the particle as

$$u_p = u_g = \frac{d_p^2 \rho_p C_g}{18\mu_g} \tag{49}$$

which is the same as Eq. (23).

For large particles and high Reynolds numbers, the velocity of an accelerating particle according to Fuchs (1964) is

$$\int_{\mathrm{Re}_{pi}}^{\mathrm{Re}_{pf}} \frac{d\mathrm{Re}_p}{A - C_D \mathrm{Re}_p^2} = \frac{3\mu_g t}{4d^2 \rho_p} \tag{50}$$

where

$$\mathrm{Re}_p = \text{particle Reynolds number}$$
$$A = 4a\rho_p \rho_g d^3 / 3\mu_g^2$$

This expression can be solved graphically by integrating the area under the curve $1/(A - C_D \mathrm{Re}_p^2)$ versus Re_p such that it equals the right-hand side of the equation. This is done by trial-and-error selection of Re_p. The final Re_p gives the value of u_p for the time t chosen.

4.3   Decelerating Particles

When a force is removed from a particle, the particle will decelerate. A force balance on the decelerating particle gives the following particle velocity, u_p:

$$u_p = u_i \exp(-t/\tau) \tag{51}$$

For larger particles, Fuchs (1964) gives the following equation:

$$\int_{\mathrm{Re}_{pi}}^{\mathrm{Re}_{pf}} \frac{d\mathrm{Re}_p}{C_D \mathrm{Re}_p^2} = -\frac{3\mu_g t}{4d^2 \rho_p} \tag{52}$$

This equation may be solved for u_p by trial-and-error integration, similar to Eq. (50).

4.4   Distance Traveled

The distance traveled by a particle accelerating from rest in still air is given by

$$X = \tau(at - u_p) \tag{53}$$

For larger particles, the distance traveled according to Fuchs (1964) is

$$\int_{\text{Re}_{pi}}^{\text{Re}_{pf}} \frac{\text{Re}_p d\text{Re}_p}{A - C_D \text{Re}_p^2} = \frac{3\rho_g X}{4 d \rho_p} \tag{54}$$

The distance traveled by a decelerating particle is given by

$$X = u_i \tau [1 - \exp(-t/\tau)] \tag{55}$$

From this equation, the stopping distance X_s is

$$X_s = u_i \tau \tag{56}$$

The distance traveled by decelerating large particles is given by

$$\int_{\text{Re}_{pi}}^{\text{Re}_{pf}} \frac{d\text{Re}_p}{C_D \text{Re}_p} = -\frac{3\rho_g X_s}{4 d \rho_p} \tag{57}$$

5 COAGULATION, CONDENSATION, AND CHEMICAL REACTION

Coagulation occurs when aerosol particles come into contact and coalesce or adhere to one another. The most common cause for coagulation is thermal motion, which is also known as Brownian coagulation. Superimposed on the Brownian motion, there may be motion produced by hydrodynamic, electrical, gravitational, or other forces. The velocity of approach imposed by these forces may set a coagulation rate which is much greater than Brownian motion. Thermal (Brownian) coagulation and coagulation of charged aerosols are spontaneous processes and go on continuously so that the aerosol becomes larger.

5.1 Brownian Coagulation

Experimental data show that coagulation of many aerosols follows a simple rate model. If n is the number of particles per cm³ at any time t and n_0 is the number present immediately after the aerosol is formed, then

$$\frac{1}{n} - \frac{1}{n_0} = kt \tag{58}$$

where

$$k = \text{coagulation constant}$$

From Eq. (58) it is easy to see that the rate at which particles disappear by coagulation is

$$\frac{dn}{dt} = -kn^2 \tag{59}$$

This equation was originally developed by Smoluchowski (1916, 1917), who made the following assumptions:

1. Particles adhere at every collision.
2. Particles collide due to Brownian motion.
3. Aerosol particles are spherical.
4. All particles are of the same size (monodispersed aerosol).
5. The shape and size of the particles are preserved though they coagulate with other particles.
6. The radius of the sphere of influence of the particles is $2r$.

Making all these assumptions Smoluchowski derived the equation

$$\frac{dn}{dt} = -kn^2 = -\frac{4RTC'}{3\mu_g N} n^2 \tag{60}$$

where N is Avogadro's number. Table 5 lists experimentally determined coagulation constants k for monodispersed aerosols in air at 760 mm Hg pressure and 25°C.

Table 5   Experimental Coagulation Constants of Various Aerosols (Air at Ambient Conditions—760 mm Hg, 25°C) (Based on Green and Lane, 1964)

Substance	$k \times 10^9$ cm³/sec
Ferric oxide	0.66
Magnesium oxide	0.83
Cadmium oxide	0.80
Stearic acid	0.51
Oleic acid	0.51
Resin	0.49
Paraffin oil	0.50
p-Xylene-azo-* naphthol	0.63
Ammonium chloride (ambient)	0.51
Ammonium chloride (46% relative humidity)	0.43
Zinc oxide formed in arc	1.9
Silica powders in electric field	2.8–3.7
Theoretical value	0.51

The rate of aerosol coagulation can better be appreciated by calculating the time taken to halve the number of particles as shown in Table 6 for a stearic acid smoke with $k = 0.51 \times 10^9$ cm³/sec. Figure 9 shows the change in stearic acid aerosol concentration with time for various initial concentrations while Figure 10 shows the half-life $t_{1/2}$ as a function of concentration. Figure 10 shows that regardless of initial particle concentration, within about 1 min all clouds of high initial concentration will have almost the same particle concentration.

Schumann (1940) gave an asymptotic solution for the mass distribution of a coagulating polydispersed system with an arbitrary initial particle mass distribution $n(m_p t = 0)$. By assuming k to be constant and $kn_0 t \gg 1$,

$$m(m_p) = \frac{c}{\overline{m}_p^2} e^{(-m_p / \overline{m}_p)} \tag{61}$$

where

$$c = \text{aerosol weight concentration, g}$$
$$\overline{m}_p = \text{particle mean mass at time } t$$

$$\overline{m}_p = \frac{c}{n} = \frac{c(1 + 0.5kn_0t)}{n_0} \sim \frac{kct}{2} \tag{62}$$

5.2   Coagulation of Particles in Laminar Flow

Smoluchowski (1916) made the first attempt to account for the influence of fluid nonuniformities on coagulation rate. In laminar flow, particles may collide and agglomerate because of the various speeds with which they move relative to one another. Neglecting Brownian diffusion and assuming no aerodynamic interaction between particles, Smoluchowski showed that the number of particles touching an absorbing sphere per second is

$$\phi = \frac{4}{3} n(r_i + r_j)^3 \frac{du_x}{dZ} \tag{63}$$

Table 6   Particle Coagulation Rate

N_0, particles/cm³	$t_{1/2}$
10^{11}	0.02 sec
10^6	33 min
10^5	5.5 hr

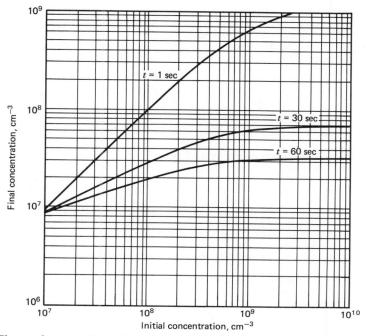

Fig. 9. Change of stearic acid particle concentration with time for various initial concentrations.

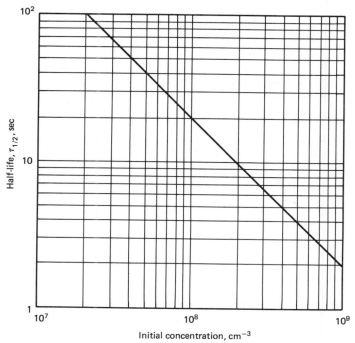

Fig. 10. Coagulation half-life versus initial particle concentration for stearic acid.

where r_i and r_j are the radii of the absorbing sphere and the collected particle and du_x/dZ is the gradient of the x-directed velocity in the Z direction.

If curvature of streamlines near the particle and the relative velocity are allowed for, the numerical coefficient of Eq. (63) is reduced.

The velocity gradient coagulation rate exceeds the Brownian diffusion rate when the following condition is met:

$$\frac{du_x}{dZ} > \frac{\pi D_p}{r_p^2} = \frac{\pi kTC'}{3\pi \mu_g r_p^3} \tag{64}$$

Thus a small change in particle size has a large effect on the required velocity gradient.

5.3 Coagulation of Aerosols in Turbulent Flow

Turbulence affects coagulation by two mechanisms. The first mechanism is the difference in fluctuation velocity of the medium at two points a distance $r_0 + r_j$ apart, which is similar to the mechanism described earlier for laminar flow. The second mechanism causes coagulation through the difference in the fluctuation velocity of the particles caused by unequal inertia. Both mechanisms occur only when the characteristic particle size is smaller than the characteristic length scale of the eddies in the turbulent medium. Levich (1962) and Saffman and Turner (1956) both have made estimates of the coagulation rate caused by these two mechanisms. Their results apply to cases where the particle radii are smaller by an order of magnitude than the size of the smallest eddies (Hidy and Brock, 1970).

Since particle inertia depends on size, the first mechanism can give rise to collision between both equal and unequal sized particles while the second mechanism results only in encounters between particles of different sizes.

Saffman and Turner (1956) showed that the number of particles of radius r_i colliding with a sphere of radius r_j per second is given by

$$\phi = 1.67(r_i + r_j)^3 \, n \left(\frac{\epsilon}{u_g}\right)^{1/2} \tag{65}$$

where

$$\epsilon = \text{rate of dissipation of turbulent energy, cm}^2/\text{sec}^3$$

For coagulation by the second mechanism, Saffman and Turner (1956) showed that the number of particles colliding with one particle per second is

$$\phi = 5.7n(r_i + r_j)^2 \, (\tau_Q - \tau_W)\epsilon^{3/4} u_g^{-1/4} \tag{66}$$

From the limitation regarding the scale of turbulence it follows that Eqs. (65) and (66) hold for particles up to 20 μm in radius even for strong turbulence in which $\epsilon \sim 1000$ cm^2/sec^3.

5.4 Coagulation of Aerosols Containing Electrically Charged Particles

Fuchs (1964) derived an equation predicting the rate of coagulation for monodisperse spherical aerosols where the particles have charges q_i and q_j. The electrostatic forces, superimposed on the Brownian motion, enhance or slow the coagulation, depending on the sign of the charge on the particles. β is the ratio of (ϕ/ϕ_0) where ϕ_0 is the number of particles contacting one particle per second under Brownian coagulation and ϕ is the number of particles contacting one particle per second due to both forces.

For uncharged particles,

$$\phi_0 = 8\pi Drn \tag{67}$$

For a charged aerosol,

$$\phi = \beta\phi_0 \tag{68}$$

For a charged aerosol Fuchs neglects induction forces and shows that in the case of bipolar charging for repulsion between particles due to equal sign charge β is given by

$$\beta = \frac{\gamma}{\exp(\gamma) - 1} \tag{69}$$

For charge attraction,

$$\beta = \frac{|\gamma|}{1 - \exp(-|\gamma|)} \tag{70}$$

where

$$\gamma = \frac{q_i q_j}{d_p kT} \tag{71}$$

The effect of electrical charge can be readily evaluated by considering an example where $\gamma = 0.5$, then $\beta = 1.271$ for unlike charges and $\beta = 0.77$ for like charges. This shows that particles having unlike charges coagulate faster than neutral ones while those with like charges coagulate more slowly. The arithmetic mean of the two values is 1.02 so that in a case of a charged aerosol with no net charge the overall effect is extremely small.

5.5 Polarization Coagulation of Aerosols

When an aerosol is under the influence of an electric field the particles are inductively charged and the rate of coagulation is affected. The force field between polarized particles causes attraction between particles when they are arranged along the field direction and repulsion when they are at right angles.

The mechanism of coagulation in an electric field consists of diffusion of particles toward one another until they are close enough for electric forces to act.

Fuchs (1964) showed that β for this case is a function of α where α is defined as

$$\alpha = \left[\frac{E^2 d_p^3}{32 kT}\right] \left(\cos^2 \theta - \frac{1}{2} \sin^2 \theta\right)^{1/3} = \alpha_1 \left(\cos^2 \theta - \frac{1}{2} \sin^2 \theta\right)^{1/3} \tag{72}$$

where θ is the angle between the field direction and the line of centers of the particles. Table 7 gives the effect of electric field on the coagulation rate of mists. This table shows that the coagulation of mists is accelerated appreciably only in the presence of very strong electric fields.

The coagulation of solid particles in an electric field is unique. A doublet formed by the coagulation of two primary particles orientates with its long axis parallel to the field, and its dipole moment in this position is much greater, increasing the rate of coagulation.

5.6 Acoustic Coagulation of Aerosols

It has been known for many years that small particles suspended in a gas are readily agglomerated by ultrasonic waves, particularly if standing waves are set up in a resonant enclosure. If the frequency and intensity of the vibrations are suitable, the aerosol coagulates in a few minutes or even seconds, forming annular deposits on the walls of the tube at the antinodes. A few of the finest particles usually remain suspended and coagulate extremely slowly under continued action of the ultrasonic waves.

It was shown experimentally by Brandt and Hiedemann (1936) and St. Clair et al. (1968) that the increase in size of the particles depends on the sound intensity. However, there are indications that coarse aggregates develop only when the vibrations are not too intense since the aggregates are not strong. The addition of water or oil mists to the smoke or dust being coagulated causes the aggregates to stick together.

Two main stages can be distinguished in the coagulation process. In the first stage, the particles oscillate under the influence of the sound waves and take part in the general circulation between node and antinode increasing in size due to collision with other particles. In the second stage, the particles are so much enlarged that they can no longer follow the oscillation. During this stage coagulation continues by collisions between enlarged particles and also between enlarged and small particles.

A review of the theory of acoustic coagulation can be found in Chou et al. (1981). The behavior of suspended particles in a vibrating gas may be considered a combination of the following factors:

1. Covibration of particles in the vibrating gas.
2. Hydrodynamic forces of attraction and repulsion between neighboring particles.
3. Sonic radiation pressure.

Table 7 Effect of Electrical Field on the Coagulation Rate of Mists (Based on Fuchs, 1964)

α_1	1	2	3	5	10	20	>20
β	1.0	0.95	1.07	1.7	3.4	6.8	$\beta \sim \frac{1}{3} \alpha_1$

5.7   Kinematic Coagulation of Aerosols

Relative motion between particles of unequal size, owing to different speeds acquired under the influence of external forces, lead to collision and coagulation. This process is called kinematic or orthokinetic coagulation. Kinematic coagulation takes place in gravitational electric and sound fields, under the action of centrifugal forces, when velocity gradients occur in the gas stream, and so forth.

The theory of kinematic coagulation is different for aerosols with large and fine particles although, in the case of coarse particles, diffusion can be neglected. A coarse spherical particle falling freely with a velocity u_s through an aerosol consisting of finer particles can be considered equivalent to the case of an aerosol particle moving with a velocity u_p past a stationary sphere, when the rate of fall of the fine particles is negligible. The number of finer particles captured in 1 sec by the coarse sphere of diameter d_p is

$$\phi = \frac{\pi}{4} d_p^2 u_p \eta \tag{73}$$

where

$$\eta = \text{target efficiency}$$

If the sizes of the coarse and fine particles approach one another and the rate of settling of the fine particles cannot be neglected, then the relative velocity and a different velocity distribution in the gas flowing around the coarse particle should be used. The gravitational coagulation of aerosols with a particle radius not exceeding 15 μm involves only a small inertial displacement. In this case the target efficiency can be calculated from Eq. (94), or a similar relationship for viscous flow, neglecting the inertia of the particle and assuming that its center passes around the sphere along a streamline. The capture coefficients actually measured are usually higher than those calculated using Eq. (94) because of inertial effects and changes in particle flow lines in the wake of the falling larger particle.

5.8   Condensation and Evaporation

Droplets of smoke and mist may increase or decrease in size due to condensation or evaporation. Aerosol particles when in saturated or supersaturated atmosphere may act as nuclei, enhance condensation, and grow this way. These changes in particle size and particle size distribution may have a pronounced effect on aerosol behavior.

Vapors may self-nucleate and condense into drops in the absence of condensation nuclei if the degree of supersaturation S is great enough. The equilibrium saturation ratio over a curved surface of a pure substance is given by the Kelvin equation

$$S = \exp \frac{2\sigma M}{RT\rho_L r} \tag{74}$$

where

$$S = \text{supersaturation ratio}$$
$$r = \text{particle radius of curvature, cm}$$

Since S is a decreasing function of r for a given concentration, water drops grow by condensation if they are larger than a critical size, which is the condensation nuclei. Vapors reach a reasonable rate of self-nucleation only when saturations exceed 200–400%.

If an insoluble surface is present, as in the case of insoluble aerosol particles, drops of the critical radius will form on this surface at a higher rate than they will self-nucleate since there is less condensate volume in the critical drop and fewer molecular collisions are needed. This explanation is not completely adequate since it has been shown by Banghali and Saweris (1938) that a supersaturated vapor will not condense on a plane surface which has absorbed a liquid film. The Volmer (1939) theory of condensation on a flat surface predicts that the critical supersaturation for a given nucleation rate should rise with increasing contact angle between the liquid and the solid.

Fletcher (1958) showed that the increase in nucleation rate in the presence of aerosol particles is a function of the surface wettability of the aerosol particle by the condensing liquid and the ratio of drop radius to particle radius. Since for the same rate of nucleation lower saturations are required in the presence of wetted particles than in their absence, this process will dominate over homogeneous nucleation.

When the aerosol particles are of irregular shape they may have convex and concave surfaces. From Kelvin's equation the vapor pressure on a concave surface is less than for a film on a plane surface. Hence vapor will condense more readily within a concave surface than upon a convex one.

But condensation upon a concave surface will not make the particle a nuclei; it will just consume a certain portion of the vapor.

When soluble aerosol particles are present, nucleation will occur even more readily, since the equilibrium vapor pressure over a solution is less than that over pure water. Howell (1949) showed that soluble particles can act as nucleation sites without having to depend on chance agglomeration of water molecules to reach a critical size. If the ambient supersaturation for a given particle is below the peak of the equilibrium saturation curve, two equilibrium drop sizes are possible. The point to the left of the peak represents stable equilibrium. Growth by condensation will occur only if the ambient saturation increases or if statistical fluctuations in size cause the drop to go over the peak. If, however, the ambient saturation is above the peak of the curve, spontaneous nucleation, condensation, and growth will occur. Thus, soluble particles will serve as nucleation sites at very low supersaturation ratios and will form drops much more readily than insoluble particles.

After a drop of critical size has been nucleated it grows at a rate determined by the vapor pressure in the gas phase and the conditions at its surface. Assuming stationary condensation for a spherical motionless droplet (constant drop temperature and vapor pressure) and ideal behavior of the vapors, drop diameter can be calculated from Maxwell's equation:

$$d_d^2 - d_{do}^2 = \frac{8}{\rho_L} \frac{DMt}{RT} (p_v - p_0) \tag{75}$$

where

$$
\begin{aligned}
d_d &= \text{drop diameter, cm} \\
d_{do} &= \text{initial drop diameter, cm} \\
D &= \text{vapor diffusivity, cm}^2/\text{sec} \\
p_0 &= \text{vapor partial pressure at drop surface, kPa} \\
P_v &= \text{vapor partial pressure in gas, kPa}
\end{aligned}
$$

This equation applies equally well to evaporation and condensation of vapor on drops larger than 1 μm radius and pressure greater about 1 cm Hg. The equation holds best for 1 mm drops at 1 atm pressure.

For drops condensing while moving relative to the gas stream and assuming all of the above conditions, Frossling's equation best describes the rate of growth:

$$d_d^2 - d_{do}^2 = \frac{8}{\rho_L} \frac{DMt}{RT} (p_v - p_0)(1 + 0.276\text{Re}_p^{1/2}\text{Sc}^{1/3}) \tag{76}$$

where

$$
\begin{aligned}
\text{Re}_p &= \text{particle Reynolds number} \\
\text{Sc} &= \text{Schmidt number}
\end{aligned}
$$

Theoretical considerations show that the vapor pressure gradient reaches a value considerably greater than the normal value of the gradient at a distance on the order of the mean free path of the gas molecules from the drop surface. The effect of the concentration change on the rate of evaporation and condensation can therefore be calculated to a first approximation by assuming that Maxwell's equation and a reciprocal concentration gradient $(c - c/c_0 - c) = r/\rho$ are valid only at a distance greater than Δ from the surface of the drop. Δ was found experimentally to be several times the mean free path. In the layer of thickness Δ adjacent to the surface, the evaporation or condensation proceeds unhindered as in a vacuum. Following this line of derivation, the evaporation of drops 2–20 μm radius is

$$r\frac{dr}{d\theta} = \frac{MD}{\rho_L RT} (p_0 - p_v)\left[\frac{1}{(D/rA\alpha) + 1}\right] \tag{77}$$

where

$$
\begin{aligned}
A &= (RT/2M)^{1/2} \\
\alpha &= \text{condensation or evaporation coefficient expressing} \\
&\quad \text{the fraction of molecules striking a surface that} \\
&\quad \text{condenses}
\end{aligned}
$$

For water, $\alpha = -0.034$ sec/cm² (Fuchs, 1959). Equation (77) and experimental results show (Birks and Bradley, 1949) that an aerosol grows at a rate roughly proportional to the inverse of its radius. Growth continues until the ambient vapor pressure is lowered to equilibrium due to condensation of the available water vapor or due to particle movement to different regions. All of the above equations can also be used to predict evaporation rates, after the sign has been changed.

5.9 Chemical Reaction

Aerosol buildup as well as changes in particle size and particle size distribution may result from a chemical reaction between two gaseous reacting components, between the particles and a reacting component, or between two particles reacting with one another to form a third particle.

Gillespie and Johnstone (1955) studied particle size distributions in hygroscopic aerosols and reached the following conclusions. In their work, the particle size distribution in all the aerosols studied gave straight lines on log probability paper. The particle size of the aerosol depends on the method of formation. If the hygroscopic vapors react with water to form a new compound with the release of considerable energy, the particle size is small (< 2 μm in diameter). If the water vapor simply condenses on the hygroscopic aerosol and the liquid is diluted without the release of much energy, the particles are large (2–6 μm in diameter). Within these two classifications the particle size is directly related to the difference between the vapor pressure and the equilibrium vapor pressure of the final product. The larger this difference, the smaller the particles and the larger their number. The reaction between the hygroscopic vapors and the water vapor is very fast (less than 3×10^{-3} min as measured in this experiment).

For a given mass concentration, the particle size decreases as the particle number concentration increases. The larger the number concentration, the faster the agglomeration rate. Thus if enough time is allowed there is a tendency toward equalization of the particle size distribution. The addition of foreign nuclei to the hygroscopic aerosol prior to the introduction of water vapor usually results in a smaller particle size and larger particle number.

When particle growth takes place as a result of the diffusion of a reacting gaseous component to the particle and the growth is diffusion controlled, the equations describing condensation growth are applicable. Where reaction follows particle collision, the agglomeration equations derived earlier in this section will describe the changes in particle size adequately. Additional information on particle coagulation, condensation, and chemical reaction can be found in Friedlander (1977), Fuchs (1959, 1964), and Hidy and Brock (1970, 1971, and 1972).

6 PARTICLE COLLECTION MECHANISMS

6.1 General Equations for Particle Deposition

All types of air pollution control equipment for particle collection can be described in terms of the same equations for the two cases of:

1. No mixing of particles normal to the flow axis in the gas stream.
2. Complete mixing.

The equations are derived below in terms of variables defined in Figure 11, where

u_g = gas velocity, cm/sec
L = collector length, cm
H_f = flow stream height, cm
W = flow stream width, cm
A_d = deposition area = WL
u_d = deposition velocity, cm/sec
Q = volumetric flow rate = $u_g H_f W$
n = particle concentration averaged across a plane normal to the flow, cm^{-3}
n_d = particle concentration at deposition plane, cm^{-3}

A material (particle) balance over a differential control volume gives

$$-u_g H_f W \, dn = u_d n_d H_f \, dL \tag{78}$$

or

$$-Q \, dn = u_d n_d \, dA_d \tag{79}$$

If there is no mixing, n_d remains constant at n (inlet conc.) until it abruptly becomes zero. Thus,

$$-dn = \frac{u_d n_d}{Q} \, dA_d \tag{80}$$

$$\frac{n_1 - n_2}{n_d} = \frac{n_1 - n_2}{n_1} = \eta = \frac{u_d A_d}{Q} \tag{81}$$

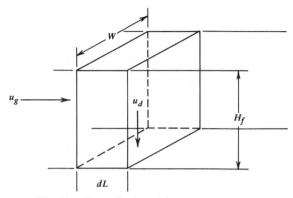

Fig. 11.   Generalized particle deposition model.

If there is complete mixing on every cross-sectional plane normal to the flow axis, $n_d = n$ and

$$-\frac{dn}{n} = \frac{u_d dA_d}{Q} \tag{82}$$

$$\frac{n_2}{n_1} = \exp\left(-\frac{u_d A_d}{Q}\right) = \text{Pt} \tag{83}$$

where

$$\eta = \text{particle collection efficiency}$$
$$\text{Pt} = \text{particle penetration} = 1 - \eta$$

Equations (81) and (83) hold for any device in which u_d is constant with L. Note that the general deposition parameter $(u_d A_d/Q)$ is the same, except for sign, in both equations. Consequently, an equation defining collection efficiency or penetration for one case can easily be rewritten for the other case.

The actual deposition velocity for a given device is highly dependent on the particle size, collector geometry, gas properties, and flow characteristics. The following sections of this chapter develop the individual paticle collection mechanisms which operate in air pollution control systems to remove particles from dirty gas streams.

6.2   Inertial Deposition

When an obstacle such as a water drop or filter fiber is introduced into a flowing aerosol, the large particles collide with the obstacle because of their inertia, whereas the smaller particles are able to follow the gas streamlines around the obstacle. This mechanism of particle collection by inertia is called impaction. Many examples of it can be found in nature as well as in the air pollution control field, for example, collection of snowflakes on the windshield of a moving car, or particle collection on the stage of a cascade impactor.

When the moving aerosol stream approaches the obstacle, the fluid streamlines spread around it. At the same time, inertia forces carry particles across the streamlines, so that the particles hit, and in many cases stick, to the obstacle. The collection efficiency due to inertia η is called the target efficiency and is defined as the ratio of the number of particles striking the obstacle to the number which would strike it if the streamlines were not diverted by the obstacle. The assumption is always made that all the particles adhere on striking. If the particles are uniformly distributed throughout the approaching gas stream, then the collection efficiency equals the ratio of the area swept clean to the cross-sectional area of the obstacle

$$\eta = \left(\frac{y_0}{r_c}\right)^2 \quad \text{for spheres}$$

$$= \frac{y_0}{r_c} \quad \text{for cylinders} \tag{84}$$

where

$$y_0 = \text{distance from the axis of the limiting streamline for impaction}$$
$$r_c = \text{collector radius}$$

The particle collection efficiency can be greater than 100% when other forces are present that force particles to the surface of the collector. A collection efficiency greater than 100% means that the area swept clean by the collector is larger than the cross-sectional area of the collector.

Three factors determine the inertial collection efficiency. The first is the velocity distribution of the gas flowing by the collector, which varies with the Reynolds number of the gas with respect to the collector. The second factor is the trajectory of the particle. This depends on the mass of the particle, its air resistance, the size and shape of the collector, and the flow rate of the gas stream. The third factor is the adhesion of the particles to the collector—usually assumed to be 100%.

The Reynolds number of the air with respect to the collector is

$$\mathrm{Re}_c = \frac{u_0 \rho_g d_c}{\mu_g} \tag{85}$$

where

u_0 = undisturbed upstream air velocity relative to the collector, cm/sec
ρ_g = gas density, g/cm³
μ_g = absolute gas viscosity, poise
d_c = collector diameter, cm

At high values of Re_c (potential flow), the parting of the gas streamlines occurs close to the collector. Except near the collector surface, the flow pattern corresponds to that of an ideal gas. When the Reynolds number is low, flow is governed by viscosity (viscous flow) and the effect of disturbance created by the collector is noticed at relatively large distances upstream. The sudden spreading of the streamlines at high Reynolds number enhances the influence of particle inertia and therefore causes a higher collection efficiency.

Ignoring all field forces and assuming that the particles obey Stokes law, the equation of motion for the particles is derived by equating the particle inertia force to the air drag on the particle. The equations in dimensionless form are

$$K_P \frac{du'_p}{dt'} = (u'_g - u'_p) \tag{86}$$

$$K_P \frac{dv'_p}{dt'} = (v'_g - v'_p) \tag{87}$$

where

$$K_P = \frac{C' \rho_p d_p^2 u_0}{9 \mu_g d_c} \tag{88}$$

$$u'_p = \frac{u_p}{u_0} \tag{89}$$

$$v'_p = \frac{v_p}{u_0} \tag{90}$$

$$u'_g = \frac{u_g}{u_0} \tag{91}$$

$$v'_g = \frac{v_g}{u_0} \tag{92}$$

$$t' = \frac{2u_0 t}{d_c} \tag{93}$$

where

$d_c/2$ = a characteristic dimension of the collector. For a sphere it is the radius.
u_p and v_p = particle velocity components in the X and Y directions, respectively
u_g and v_g = gas velocity components in the X and Y directions, respectively, relative to the collector
C' = Cunningham slip correction factor

The dimensionless inertia parameter K_P, which characterizes the motion of the particle, has a physical meaning too. It is the ratio of the particle stopping distance to the radius of the collector.

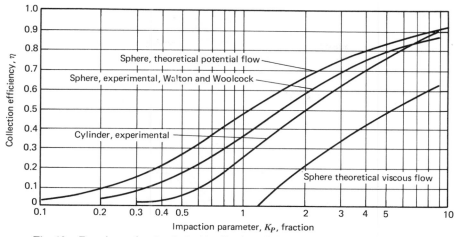

Fig. 12. Experimental and calculated collection efficiencies for spheres and cylinders.

The particle stopping distance is the distance it would travel before coming to rest if injected into a still gas at velocity u_0, when all forces on the particle except the drag force are zero.

The solutions to the equations of motion will depend on the velocity field assumed. These equations have been solved for several collectors of different shapes under given boundary conditions, mostly by numerical methods. Figure 12 shows theoretical and experimental collection efficiency for a spherical collector for potential and a viscous flow.

For potential flow and for values of K_P greater than 0.2, the experimental values of inertial collection efficiency for spheres can be approximated by the correlation

$$\eta_P = \left(\frac{K_P}{K_P + 0.7}\right)^2 \tag{94}$$

Most of the theoretical solutions to Eqs. (86) and (87) yield a critical K_P below which no inertial deposition takes place. For cylinders, $K_{crit} = 0.125$ and for spheres $K_{crit} = 0.083$. This was shown by many investigators to be a theoretical limit. In reality, in cases of turbulent flow, particles are collected on the back of the collector and collection efficiency for $K_P \leq K_{crit}$ is not zero.

As in the case of spheres, inertial collection by cylinders is a function of the impaction parameter and the flow regime as defined by the Reynolds number. When $Re_c < 1$ the flow field around the cylinder is considered to be viscous, while for $Re_c > 100$ the flow around the cylinder can be approximated by potential flow. Values between 1 and 100 represent transition conditions.

A number of theories have been developed to predict the particle collection efficiency of single fibers due to impaction: Sell (1931), Albrecht (1931), Langmuir and Blodgett (1946), Bosanquet (1950), Davies (1952), Glanert (1957), Pearcey and Hill (1957), Das (1950), Hocking (1959), Pemberton (1960), Picknett (1960), and Walton and Woolcock (1960). However, no theories produce results completely compatible with experimental results.

Whitby (1965) has analyzed the data of Wong and Johnstone (1953) and has developed a plot of single fiber impaction efficiency as a function of the impaction parameter, with Reynolds number as parameter, as shown in Figure 13. This figure covers the transition from laminar to turbulent flow and can be used for $0.2 < Re_c < 150$.

6.3 Interception

A particle is collected by interception when its surface (not center) touches the surface of the collector, which can happen without crossing any gas streamlines. This occurs when the streamline the particle is traveling on passes within a distance of one particle radius from the collector surface.

In the model for particle collection by inertial impaction, Section 6.2, it was assumed that a particle is a massive point, and is captured when that point touches the collector's surface. However, a particle is actually captured when its surface, rather than its center, hits the collector. For this reason there is not a unique curve of collection efficiency versus the inertia parameter, K_P. Rather, a family of curves exists, depending on the ratio of particle diameter to collector diameter.

Ranz and Wong (1952) developed a relation for interception efficiency for the inertia parameter

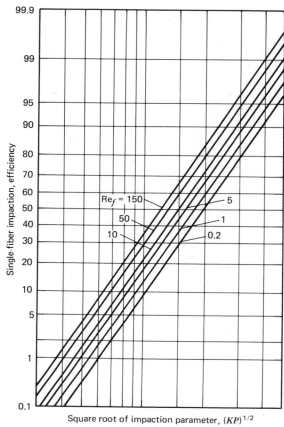

Fig. 13. Single-fiber impaction efficiency as a function of impaction parameter and Reynolds number.

$K_P = 0$, that is, a particle follows a streamline in potential flow. Using this model they showed the interception efficiency for spherical collectors to be

$$\eta_I = (1 + K_I)^2 - \left(\frac{1}{1 + K_I}\right) \tag{95}$$

where the interception parameter K_I is defined by

$$K_I = \frac{d_p}{d_c} \tag{96}$$

For particle collection on cylinders in the laminar region, Torgenson gives

$$\eta_I = 0.0518 \left(\frac{C_D \, \mathrm{Re}_c}{2}\right) K_I^{1.5} \tag{97}$$

An alternate approach is given by Ranz (1953) who used a modification of Langmuir's (1942) viscous flow equation. Ranz and Wong (1952) give the collection efficiency due to interception in the turbulent region as

$$\eta_I = 1 + K_I - \left(\frac{1}{1 + K_I}\right) \tag{98}$$

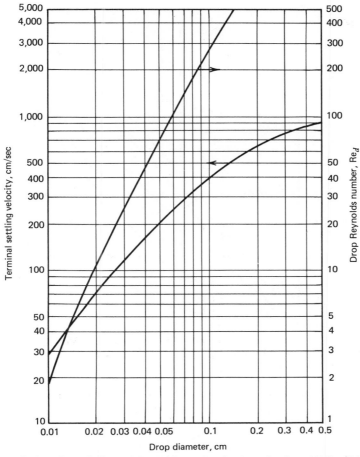

Fig. 14. Terminal settling velocity and Reynolds number for drops in air at 20°C and 760 mm Hg.

6.4 Gravitational Deposition

For the collection of particles by free falling drops, collection efficiencies may be calculated for drops of radius r_d falling at their terminal velocities. It is assumed that drops will rapidly approach terminal velocity. The settling velocities of the particles are assumed negligible compared to those of the collecting drops, that is, the drops are much larger than the particles. The air is assumed to be saturated so that no drop evaporation or growth takes place and diffusion and interception are neglected. Drop collection efficiencies are read from Figure 12 for potential and viscous flow (η_{pot} and η_{vis}). Drop velocities are read from Figure 14. The actual efficiency is then calculated from Langmuir's (1948) interpolation formula:

$$\eta_g = \frac{\eta_{vis} + \eta_{pot}\,(Re_c/60)}{1 + (Re_c/60)} \tag{99}$$

Figure 15 shows efficiency as a function of particle diameter with drop diameter as a parameter, and may be used alternatively to Eq. (98). It can be seen that the highest collection efficiency per drop is for drops of 0.1 to 0.2 cm in diameter.

The efficiency of a single cylindrical fiber due to the gravitational settling of particles on it, η_G,

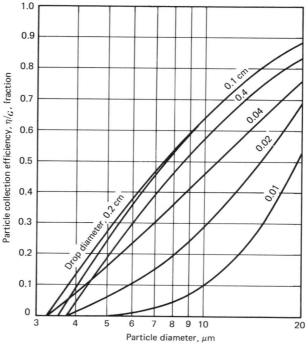

Fig. 15. Collection efficiency of particles ($\rho_p = 2 \text{g/cm}^3$) by free falling water drops.

may be significant for large particles and low gas velocities. Ranz and Wong (1952) give the following expression:

$$\eta_G = \mathrm{K}_G = \frac{C' d_p^2 \rho_p g}{18 \mu_g u_0}$$

(100)

Here K_G is a gravity parameter, equal to the settling velocity of the particle divided by the velocity of the gas stream flowing past the fiber. This equation applies when the flow past the fiber is laminar.

6.5 Diffusion

Small particles, in the submicrometer size range, are rarely collected by inertial impaction or interception because they follow the gas streamlines surrounding the collecting body, and their ratio of d_p/d_c is small. In a still gas, small particles move freely across gas streamlines due to random bombardment from gas molecules. Some of the particles hit the collector and are removed from the gas.

The aerosol concentration near the collector surface takes the form (Hidy and Brock, 1970)

$$c = c_i \left(1 - \frac{r_c}{\delta}\right)$$

(101)

where

r_c = collector radius, cm
δ = distance from collector center, cm
c_i = concentration of particles in main body of the gas stream, cm^{-3}

The concentration gradient at the collector surface is

$$\left.\frac{\partial c}{\partial \delta}\right|_{\delta = r_c} = \frac{c_i}{r_c}$$

(102)

The rate of deposition ϕ (g/sec if c_i is g/cm³) on a spherical collector for a gas at rest is equal to

$$\phi = 4\pi D_p r_c c_i \tag{103}$$

Particle diffusivities are several orders of magnitude smaller than diffusivities of gases, and are closer to the values obtained for diffusivities of solutes in liquids. Particle diffusivity may be calculated using the Einstein (1905) equation [see Eq. (24)]. For an aerosol flowing past a spherical collector the rate of diffusion increases and becomes a function of the Reynolds and the Schmidt numbers. The Schmidt number Sc is a dimensionless group indicating the ratio of convective and diffusive transfer rates at constant Re_c. The Schmidt number is defined as

$$Sc = \frac{\mu_g}{\rho_g D_p} \tag{104}$$

For low diffusivities, leading to high Schmidt number (Sc = 10⁶) and low Reynolds numbers ($Re_c <$ 3), Levich (1952) showed that

$$\phi = 2\pi D_p r_d c_i \, Re_c^{1/3} Sc^{1/3} \tag{105}$$

For large Reynolds numbers (600–2600) and the same value of Schmidt number, Akselrud (1953) showed that

$$\phi = 1.6\pi D_p r_d c_i \, Re_c^{1/2} Sc^{1/3} \tag{106}$$

For Reynolds number = 100–700 and Schmidt number = 10³, Garner et al. (1958a and 1958b) showed that

$$\phi = 1.9\pi D_p r_d c_i \, Re_c^{1/2} Sc^{1/3} \tag{107}$$

Equations for the prediction of single-fiber efficiency due to Brownian diffusion are usually a function of the Peclet number, which is a ratio of bulk transport to diffusive transport,

$$Pe = \frac{u_0 d_f}{D_p} = \frac{u_0 d_f \, 3\pi \mu_g d_p}{C' kT} \tag{108}$$

where

$$d_f = \text{fiber diameter}$$

Equations predicting the single-fiber efficiency due to diffusion in the laminar region are given by several authors. The Torgenson expression is

$$\eta_D = 0.75 \left(\frac{C_D Re_f}{2} \right)^{0.4} (Pe)^{-0.6} \tag{109}$$

where C_D is the drag coefficient of a cylinder which is a function of the Reynolds number Re_f available in Perry's *Handbook* (1973) and elsewhere. Alternatively, η_D can be calculated using Lamb's equation for cylinders in the laminar region:

$$C_D = \frac{8\pi}{Re_f [2 - \ln(Re_f)]} \tag{110}$$

Other equations for calculating the efficiency of single fibers due fo particle diffusion in the laminar region are given by Bosanquet (1950), Davies (1952), Landt (1956), Langmuir (1942), Natanson (1957), and Ranz (1953).

6.6 Electrostatic Deposition

The electrostatic properties of small particles affect their motion and collection in an electric field. Kraemer and Johnstone (1955) solved the equations of motion for particles in an electric field neglecting all other forces. They calculated the collection efficiency of conducting and nonconducting aerosol particles by a conducting sphere, due to electrostatic forces, for potential and viscous flows and for several interception parameters. Gas streamline flow and low-amplitude Brownian motion relative to the collector diameter were assumed.

There are five aspects of electrical forces acting in a system of particles approaching a collector. For each case a collection parameter, the ratio of the electrostatic force to the Stokes–Cunningham drag force, has been defined by Kraemer and Johnstone.

For the coulombic force between a charged collector and a charged particle,

$$K_{E1} = \frac{C' q_p q_c}{3\pi^2 \mu_g d_p u_0 \epsilon_0 d_c^2} \tag{111}$$

For the induction force between a charged spherical collector and an uncharged particle,

$$K_{E2} = \left(\frac{\epsilon - 1}{\epsilon + 2}\right) \frac{2C' d_p^2 q_c^2}{3\mu_g u_0 d_c \epsilon_0 \pi^2 d_c^4} \tag{112}$$

For the induction force between charged particles and uncharged spherical collector,

$$K_{E3} = \frac{C' q_p^2}{3\pi^2 \mu_g d_p u_0 \epsilon_0 d_c^2} \tag{113}$$

For the case of the repulsion force exerted by unipolar charged particles on the aerosol particle being deposited,

$$K_{E4} = \frac{C' q_p^2 d_c n}{18\pi \mu_g d_p u_0 \epsilon_0} \tag{114}$$

For the attraction between a charged aerosol particle and a grounded collector that has a charge induced by the surrounding unipolar aerosol particles,

$$K_{E5} = \frac{C' q_p^2 n b^2}{3\pi \mu_g d_p u_0 d_c \epsilon_0} \tag{115}$$

where

q_p = charge on the particle, C
q_c = charge on the collector per unit area, C/cm²
ϵ_0 = permittivity of free space = 8.85×10^{-21} F/m
ϵ = dielectric constant
u_0 = relative velocity between the particle and the gas stream, cm/sec
n = particle concentration, cm⁻³
b = radius of the spherical aerosol cloud which influences the collector, cm

In most cases only one of the five mechanisms is important. When the collector is maintained at a uniform charge and the aerosol particles are oppositely charged, the coulomb attraction forces represented by K_{E1} are dominant. On the other hand, when only the aerosol is charged, K_{E2} becomes dominant. The force represented by K_{E3} is small and in most cases can be neglected. The forces corresponding to K_{E4} and K_{E5} depend on aerosol concentration and become important only at high aerosol concentrations ($n > 10^7$).

Lundgren (1962) and Whitby and Lundgren (1964) have shown that the only mechanism of importance in nonelectrified fiber beds is the attractive force between a charged aerosol particle and its image in a collecting fiber. Lundgren has shown experimentally that the collection efficiency due to this force can be expressed as

$$\eta_E = 1.5(K_E)^{1/2} \tag{116}$$

where K_E is the cylindrical analog of Eq. (106) and is given by

$$K_E = \frac{[(\epsilon - 1)/\epsilon + 1)] q_p^2 C'}{12\pi^2 \epsilon_0 d_f^2 \mu_g d_p u_0} \tag{117}$$

6.7 Magnetic Collection

Magnetic effects can be included in the equations of particle motion in the same manner electrostatic forces were treated. Magnetic forces result when a magnetic particle is introduced into a magnetic field or close to a ferromagnetic metal, or when a particle is charged and then introduced into a magnetic field. The equation of motion, when solved, will give collection efficiency for different velocity distributions as a function of the magnetic dimensionless number.

NOMENCLATURE

a acceleration, cm/sec^2

A_d deposition area, cm^2

b radius of spherical aerosol cloud influencing a collector, cm

B particle mobility, sec/g

C particle weight concentration, g/cm^3

C' Cunningham slip correction factor, dimensionless

C_D drag coefficient, dimensionless

C_t thermal coefficient

d_c collector diameter, cm

d_d drop diameter, cm

d_{do} initial drop diameter, cm

d_f fiber diameter, cm

d_g geometric number mean diameter, μm

d_{gm} mass median diameter, μm

d_l length mean diameter, μm

d_m mass mean diameter, μm

d_{mo} molecular diameter, cm

d_{ms} mass–surface mean diameter

d_n mean diameter, μm

d_p particle diameter, μm

d_{pa} aerodynamic diameter, μmA

d_s surface mean diameter, μm

d_{sd} surface–diameter mean diameter, μm

d_{st} Stokes diameter, μm

d_v volume mean diameter, μm

d_{vd} volume–diameter mean diameter, μm

d_{vs} Sauter (volume–surface) mean diameter, μm

D characteristic dimension, dimensionless

D_p particle diffusion coefficient, cm^2/sec

D_v vapor diffusivity in gas, cm^2/sec

e electronic charge $= 1.6 \times 10^{-19}$ C

E electric field, V/m

$f(d_p)_n$ frequency of occurrence of diameter d, dimensionless

F fluid resistance force, N

F_T thermophoretic force, N

g gravitational constant

H magnetic field strength

H_f flow stream height, cm

L collector length, cm

k Boltzmann constant $= 1.38062 \times 10^{-23}$ J/K^{-1}

k_g gas thermal conductivity, cal/g K

k_p particle thermal conductivity, cal/g K

Kn   Knudsen number, dimensionless

K_E electrostatic parameter, dimensionless

K_G gravitational parameter, dimensionless

K_I interception parameter, dimensionless

K_P impaction parameter, dimensionless

m ion mass, g

m_p particle mass, g

$\bar{m}_p$ particle mean mass, g

M_g gas molecular weight, g/g-mole

M_v vapor molecular weight, g/g-mole

n number of particles, dimensionless

n particle concentration, cm^{-3}

n_i initial particle concentration, cm^{-3}

N Avogadro's number $= 6.022 \times 10^{23}$/mole

N_d particle number concentration at deposition plane, cm/sec

N_0 number of ions per unit volume, cm^{-3}

p_g gas partial pressure, kPa

p_0	saturation vapor pressure, kPa
p_v	vapor partial pressure, kPa
p_{vd}	vapor partial pressure at drop surface, kPa
P	absolute gas pressure, kPa
Pe	Peclet number, dimensionless
Pt	particle penetration $= 1 - \eta$, fraction
q_c	collector charge per unit area, C/cm²
q_p	particle charge, C
q_s	saturation charge, C
Q	volumetric flow rate, m³/sec
r	particle radius of curvature, cm
r_c	collector radius, cm
r_d	drop radius, cm
R	gas constant $= 8.314$ J/mole-°C
Re_c	collector Reynolds number, dimensionless
Re_g	flow Reynolds number, dimensionless
Re_p	particle Reynolds number, dimensionless
S	supersaturation ratio, dimensionless
Sc	Schmidt number, dimensionless
St	Stokes number, dimensionless
t	time, sec
T	absolute gas temperature, °C
Tu_d	deposition velocity, cm/sec
u_g	gas velocity, cm/sec
u_i	initial velocity, cm/sec
u_0	undisturbed upstream air velocity relative to collector, cm/sec
u_p	particle velocity, cm/sec
u_{pD}	diffusiophoretic velocity, cm/sec
u_{pE}	particle velocity in an electric field, cm/sec
u_{pT}	thermophoretic velocity, cm/sec
u_r	relative velocity between particle and gas, cm/sec
u_S	particle Stephan flow velocity, cm/sec
u_t	particle terminal settling velocity, cm/sec
v_g	gas velocity in the y direction, cm/sec
v_p	particle velocity in the y direction, cm/sec
V_i	ion root mean square thermal velocity, cm/sec
W	flow stream width, cm
X_s	particle stopping distance, cm
y_0	distance from axis of the limiting streamline for impaction, cm
α	fraction of gas molecules reflected diffusively by the particle, fraction
Δx	distance, cm
$\Delta^2 x$	particle mean square displacement, cm²
ϵ	rate of dissipation of turbulent energy, cm²/sec³
ϵ_0	permittivity of free space $= 8.854 \times 10^{-12}$ F/m
ϵ_p	particle relative dielectric constant, dimensionless
η	target efficiency, fraction
η_E	electrostatic efficiency, fraction
η_G	gravitational efficiency, fraction
η_I	interception efficiency, fraction
λ_g	mean free path of gas molecules, cm
μg	gas absolute viscosity, poise
ρ_g	gas density, g/cm³
ρ_L	liquid density, g/cm³
ρ_p	particle density, g/cm³
σ	surface tension, dyn/cm²
σ_g	geometric standard deviation, dimensionless
τ	relaxation time, sec
ϕ	particle deposition rate, sec⁻¹
ΔT	temperature gradient, K/cm

REFERENCES

Akselrud, G. (1953), *Zh. Fiz. Khim.* **27**, 1445.

Albrecht, F. (1931), *Z. Physik* **32**, 48.

Bakanov, S. P., and Derjaguin, B. V. (1960), *Discuss. Faraday Soc.* **30**, 130.

Banghali, D. H., and Saweris, Z. (1938), *Trans. Faraday Soc.* **34**, 554.

Birks, J., and Bradley, R. S. (1949), *Proc. Roy. Soc.* **198A**, 226.

Bosanquet, C. H. (1950), *Trans. Inst. Chem. Engry. (London)* **28**, 130. Appendix to paper by C. J. Stairmand.

Brandt, O., and Hiedemann, E. (1936), *Trans. Faraday Soc.* **32**, 1101.

Brock, J. R. (1962), *J. Colloid Sci.* **17**, 768.

Calvert, S. et al. (1972), *Scrubber Handbook,* NTIS PB 213–016.

Calvert, S. (1977), *Chemical Engineering,* August 29.

Chou, K. H., Lee, P. S., and Shaw, D. T. (1981), *J. Colloid. Interface Sci.* **83–2**, 335–353.

Das, P. K. (1950), *Indian J. Met. Geophys.* **1**, 137.

Davies, C. N. (1952), *Proc. Inst. Mech. Engrg.* **1B**, 185.

Derjaguin, B. V., and Bakanov, S. P. (1962), *Dokl. Akad. Nauk. USSR (Phys. Chem.)* **147**, 139.

Einstein, A. (1905), *Ann. Physik.* **17**, 549.

Epstein, P. (1929), *Z. Phys.* **54**, 537.

Fletcher, N. H. (1958), *J. Chem. Phys.* **29**, 572.

Friedlander, S. K. (1977), *Smoke, Dust and Haze,* Wiley-Interscience, New York.

Fuchs, N. A. (1959), *Evaporation and Droplet Growth in Gaseous Media,* Pergamon Press, New York.

Fuchs, N. A. (1964), *The Mechanics of Aerosols,* Pergamon Press, New York.

Garner, F., and Suckling, R. (1958a), *A.I.Ch.E. Journal* **4**, 114.

Garner, F., and Keey, R. (1958b), *Chem. Eng. Sci.* **9**, 119.

Gillespie, G. R., and Johnstone, H. F. (1955), *Chem. Eng. Prog.* **51**, 74F.

Glanert, M., Aeronautical Research Comm. Report No. 2025 (London), H.M.S.O.

Goldsmith, P., and May, F. G. (1966), *Aerosol Science* (C. N. Davies, ed.), Academic Press, New York.

Green, H. L., and Lane, W. R. (1964), *Particulate Clouds: Dusts, Smokes, and Mists,* 2nd edition, Van Nostrand, Princeton, NJ.

Hatch, T., and Choate, S. (1929), *J. Frankl. Inst.* **207**, 369.

Hesketh, H. (1979), *Air Pollution Control,* Ann Arbor Science, Ann Arbor, MI.

Hidy, G. M., and Brock, J. R. (1970), *The Dynamics of Aerocolloidal Systems,* Pergamon Press, New York.

Hidy, G. M., and Brock, J. R. (1971), *Topics in Current Aerosol Research,* Pergamon Press, New York.

Hidy, G. M., and Brock, J. R. (1972), *Topics in Current Aerosol Research,* Part 2, Pergamon Press, New York.

Hocking, L. M. (1959), *Quart. J. Royal Met. Soc.* **85**, 44.

Howell, J. (1949), *Meteorology* **6**, 134.

Kraemer, H. F., and Johnstone, H. F. (1955), *Ind. Eng. Chem.* **47**, 2426.

Landt, E. (1956), *Gasundheits Ing.* **77**, 139.

Langmuir, I. (1942), O.S.R.P. Report No. 865.

Langmuir, I. (1948), *J. Met.* **5**, 175.

Langmuir, I., and Blodgett, K. (1946), Army Air Force, Tech. Rep. No. 5418.

Levich, V. (1952), Physico-Chem. Hydrodyn. Acad. Sci. USSR, Moscow.

Levich, V. (1962), *Physiochemical Hydrodynamics,* Prentice-Hall, Englewood Cliffs, NJ.

Lundgren, D. A. (1962), Masters Thesis, University of Minnesota, MN.

Mugele, R. A., and Evans, H. D. (1951), Droplet Size Distribution in Sprays, *Ind. Eng. Chem.* **43**, 1318.

Natanson, G. L. (1957), *Dokl. Akad. Nauk. USSR (Phys. Chem.)* **112**, 100, English edition **112**, 21.

Oglesby, S., Jr., and Nichols, G. B. (1970), *A Manual of Electrostatic Precipitator Technology,* Part 1, Southern Research Institute, Birmingham, AL, p. 75.

Pearcy, T., and Hill, G. W. (1957), *Quart. J. Royal Met. Soc.* **83**, 77.

Pemberton, C. S. (1960), *Int. J. Air Pollution* **3**, 168.

Perry, R. H. (ed.) (1973), *Chemical Engineer's Handbook,* 5th edition, McGraw-Hill, New York.

Picknett, R. G. (1960), *Int. J. Air Pollution* **3**, 160.

Ranz, W. E. (1953), Technical Report No. 8, January 1, University of Illinois, Eng. Exptl. Sta.

Ranz, W., and Wong, J. (1952), *Ind. Eng. Chem.* **44**, 1371.

Saffman, P., and Turner, J. (1956), *J. Fluid Mech.* **1**, 16.

Schumann, T. (1940), *Quart. J. Roy. Met. Soc.* **66**, 195.

Sell, W. (1931), *Ver. Deut. Ing. Forschungsheft 347.*

Smoluchowski, M. (1916), *Z. Phys.* **17**, 557, 585.

Smoluchowski, M. (1917), *Z. Phys. Chem.* **92**, 129.

St. Clair, H. W., Spendlove, M. J., and Potter, E. V. (1968), U.S. Bureau of Mines R. I. 4218.

Torgeson, W. L., Paper J-1057, Applied Science Division, Litton Systems, Inc., St. Paul, MN.

Volmer, M. (1939), *Kinetic der Phasenbildung,* Steinkopff, Leipzig.

Waldmann, L. (1959), *Z. Naturforsch* **14A**, 589.

Waldmann, L., and Schmitt, K. (1960), *Z. Naturforsch* **15A**, 843.

Walton, H. W., and Woolcock, A. (1960), *Int. J. Air Pollution* **3**, 129.

Whitby, K. T. (1965), *ASHRAE Journal,* September.

Whitby, K. T., and Lundgren, D. A. (1964), Presented 1964 Ann. Meeting Amer. Soc. of Agricultural Engrs., Ft. Collins, CO (June 21–24).

Wong, J. B., and Johnstone, J. F. (1953), University of Illinois, Eng. Exptl. Sta. Tech. Report No. 11 (October).

CHAPTER 7

CONTROL OF GASES BY ABSORPTION, ADSORPTION, AND CONDENSATION

B. B. CROCKER

Monsanto Company
St. Louis, Missouri

K. B. SCHNELLE, JR.

Vanderbilt University
Nashville, Tennessee

1 APPLICABILITY OF METHOD

There are five basic ways of controlling gaseous pollutants where applicable: absorption, adsorption, condensation, chemical reaction (Chapter 9), and incineration where combustible (Chapter 8). Absorption requires the use of a liquid solvent for the gas to be removed. It is particularly attractive for control of pollutant gases present in appreciable concentration (such as several percent by volume or more), but is applicable to gases at quite dilute concentrations when the gas is highly soluble in the absorbent. While water is the most commonly used absorbent, nonaqueous liquids of low vapor pressure may be used for gases with low water solubility such as hydrocarbons or H_2S. Water used for absorption may frequently contain other chemicals, such as acids, alkalies, oxidants, or reducing agents, to react with the gas being absorbed and reduce its equilibrium vapor pressure. This could be considered a combination of absorption and chemical reaction, but will be treated as absorption in this chapter.

Adsorption is quite adaptable to removal of many contaminant gases, especially organics, down to extremely low levels (less than ppmv). Its greatest applicability is (1) handling large volumes of gases with very dilute pollutant levels, and (2) removing contaminants down to only trace quantities. Polar adsorbents such as activated aluminas, silica gel, and molecular sieves have high selectivity for polar gases. However, the presence of common polar materials such as water vapor can reduce their capacity for other substances or even render them ineffective through competition. Activated carbon, a nonspecific adsorbent, receives the greatest utilization, and is one of the few adsorbents that can be used on wet gases. Since it is nonspecific, it tends to absorb all trace gases roughly in proportion to their concentration and adsorption capacity must be provided for those not specifically desired to be adsorbed. Where the adsorbent is to be regenerated and reused, desorption should receive equal consideration in process selection. Some materials adsorb so strongly that they can be removed only by removal of some of the adsorbent molecules at the same time (chemisorption). Other materials may polymerize in the pores of the adsorbent, become unremovable, and reduce the adsorbent surface area, rendering it ineffective.

Condensation may be a suitable control method for materials with low vapor pressures at moderately high temperatures, or for materials with high vapor pressures if recovery down to ppmv levels is not required. The need for refrigeration as a final condensation step greatly reduces the economic attractiveness of this route as a control method.

The selection of the control method may be dictated in part by the size of the gas stream and the concentration of the pollutant. Costs may be minimized by keeping the gas stream volume as small as possible and the contaminant concentration as high as possible (minimizing the quantity of inert diluents). For streams of appreciable concentration, consideration can be given to the merits of pretreatment systems and the combined economics of pretreatment and control system as contrasted with a larger capacity control system alone. Unless there are other considerations, such as recovery of a valuable material, or a process need to precool a hot gas stream, one system will generally be cheaper in initial cost than two or more. However, the economics of separate treatment systems should be considered before arbitrarily combining small concentrated streams and large-volume dilute streams. If kept separate, sometimes the more dilute streams can be introduced partway through a treatment system, thus preserving the concentration driving force.

1.1 Complicating Situations

The presence of particulate matter in the gas stream can be a problem in operation of gaseous control equipment, especially if the loading is heavy. Particulates can result in plugging of absorber packing and granular adsorption beds, and fouling of condenser heat transfer surfaces. One approach is to provide for its prior removal with dry particulate collectors such as cyclones, bag filters, and electrostatic precipitators. The other alternative is to design for the handling of gases and particulate matter together with suitable equipment. Packed absorption equipment can handle particulates that are readily soluble in the absorbing liquid if all impacting surfaces are well flushed and the solids loading is not excessive. An upper limit of solids concentration is generally around $11.4 \ g/m^3$ ($5 \ g/ft^3$). Cross-flow packed towers can be designed to handle insoluble (but nonsticky) particulates at light loading. For higher solids loading, equipment capable of simultaneous particulate collection and gas absorption may be indicated such as open spray towers, impingement tray towers, venturi scrubbers, impaction sphere fluid-bed collectors, and wet electrostatic precipitators. Where inadequate absorption efficiency is obtained in such devices, the design is set to give efficient particulate removal before the gases enter subsequent efficient gas contactors. For condensation, direct-contact jet condensers are more satisfactory when solids are present.

The presence of a variety of pollutant gases can also complicate design and selection of a control system. Generally, a method capable of handling all of the contaminant gases is chosen. The control system must be designed with adequate capability for each individual component as well as all in combination. The need for separation of the collected substances before reuse in the process (or disposal) can further complicate the design.

All of the control methods to be discussed capture and recover the pollutant in some form, often as a concentrated material. Thus, they are most useful and attractive where the captured pollutant has some value either as a recovered material for reuse or as a raw material for another marketable product. Disposal of the recovered material can be a real problem when there is no practical use for it. In such situations, other control methods may be more attractive, such as incineration when the pollutant is combustible.

2 ABSORPTION

Absorption is a chemical engineering unit operation involving mass transfer between a soluble gas component and a solvent liquid in a gas–liquid contacting device. The driving force for absorption is the difference between the partial pressure of the soluble gas in the gas mixture and the vapor pressure of the solute gas in the liquid film in contact with the gas. If the driving force is not positive, no absorption will occur. If it is negative, desorption or stripping will occur and pollution of the gas being treated can actually be enhanced.

2.1 Absorption Systems

Absorption systems can be divided into (1) those that use water as the primary absorbing liquid and (2) those that use a low volatility organic liquid. Further, the system can be a simple absorption in which the liquid is used single pass and then disposed of while still containing the absorbed pollutant. Alternatively, the pollutant can be separated from absorbing liquid and recovered in a pure, concentrated form by stripping or desorption. The absorbing liquid is then used closed circuit and is continuously regenerated and recycled. Other regeneration alternatives to stripping are removal of the absorbed pollutant through precipitation and settling; chemical destruction through neutralization, oxidation, reduction, or hydrolysis; solvent extraction, liquid adsorption, and so on.

2.1.1 Aqueous Systems

In water-based absorption, the absorbing gas must have adequate solubility in water at the resulting temperature of the gas–liquid system. For pollutant gases with limited water solubility such as SO_2 or benzene vapors, very large quantities of water are required. Such schemes are usually impractical but may occasionally be employed in unusual circumstances. Scrubbing SO_2 from flue gas with normally alkaline river water at the Battersea and Bankside electric power stations in England[1] are two early examples of removal of a limited solubility gas with extremely large quantities of water on a single-pass basis.

Water by itself is quite efficient for removing acidic soluble gases such as HCl, HF, and SiF_4, especially if the last contact is made with water having a slightly alkaline pH. Conversely, NH_3 can be readily recovered in water scrubbing, particularly when the pH is acidic in the last contact stage. Gases of more limited solubility such as SO_2, Cl_2, and H_2S can be absorbed more readily in an alkaline solution such as dilute NaOH or lime–water than in pH 7 water alone. Addition of reactive chemicals, which change the molecular species of the absorbing pollutant by reaction in the water film, is often desirable to increase the driving force for mass transfer.

Water-based systems are generally suitable for absorption of organics only when they have appreciable water solubility.

2.1.2 Nonaqueous Systems

While water is the most commonly used absorbing liquid for acidic gases, organic liquids such as dimethylanaline and amines (mono-, di-, and triethanolamine and methyldiethanolamine) can be used. Such absorbents are generally limited to solid particulate-free systems since the presence of solids would contaminate the organic liquids and possibly produce sludges difficult to handle. Because of scrubbing liquid cost, absorbent separation and scrubbing liquid regeneration will be required in most cases.

Absorption with an organic liquid appears at first glance to be the preferred means for removal of organic pollutant vapors because of improved solubility and miscibility. The lower heat of vaporization of organic liquids is an additional plus for energy conservation when regeneration must occur by stripping. Many heavy oils, hexadecane (no. 2 fuel oil) or heavier, and low vapor pressure solvents can do extremely well in reducing organic vapor concentrations to very low levels. However, the possible loss of absorbing liquid vapor to the atmosphere must be considered. Scrubbing liquids with appreciable vapor pressure could saturate the effluent gas and violate state or EPA limitations on emission of nonmethane hydrocarbons. As gas quantity to be scrubbed increases, lower vapor pressure liquids will be needed, but, in general, scrubbing liquid vapor pressure at the gas outlet will need to be below 13.3 Pa (0.1 mm Hg).

Other aspects of organic solvent scrubbing requiring consideration are stability of the solvent in the gas–solvent system (such as its resistance to oxidation) and possible fire and explosion hazards.

2.1.3 Simple Absorption Versus Regeneration

In simple absorption, fresh absorbing liquid on a single-pass basis is continually supplied to the absorption column or chamber. The effluent liquid, containing the absorbed or chemically combined pollutant gases, must be used in a process, sold, or disposed of in an environmentally acceptable manner. Possible disposal methods are mentioned in Section 2.5. The alternative is to regenerate and reuse the absorbing liquid.

Stripping and recovery of the pollutant gas in concentrated form is one such regeneration method. Stripping is almost always practiced with organic absorbing liquids since their cost requires reducing losses to negligible amounts. Figure 1 is a flow sheet of an absorber–stripper used for the removal and recovery of benzene vapors from vent gases.

Alternatives to stripping for regeneration are complexing, precipitation, and settling of the absorbed material; destruction by chemical means, or separation in other ways such as crystallization, solvent extraction, and so on. Such alternatives are mentioned further in Section 2.5.

2.2 Types and Arrangements of Absorption Equipment

Absorption requires intimate contact between a gas and a liquid. It is usual to provide a means of breaking the liquid up into small droplets or thin films (which should be constantly renewed through turbulence) to provide a high liquid surface area for mass transfer and a fresh, unsaturated film for high driving force. The most commonly used devices are packed and plate columns, open spray chambers and towers, cyclonic spray towers, and combinations of sprayed and packed chambers. Baffle towers are sometimes used where scaling and plugging problems could be severe and mechanical contacting devices have been used extensively for absorption in Europe.

2.2.1 Packed Absorbers

Packed towers give excellent gas–liquid contact and efficient mass transfer. For this reason, they can generally be smaller in size than open spray chambers. Figure 2 illustrates some of the more common and more frequently used types of packing. Saddles generally give more efficient contact and lower pressure drop but are more costly in initial purchase cost than ring packing. Pall rings also give

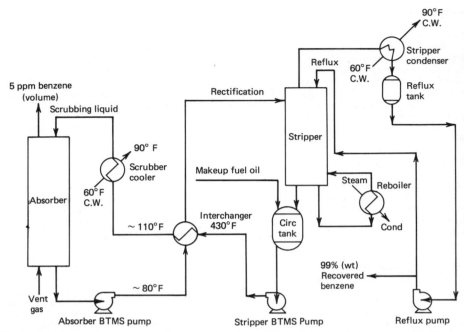

Fig 1 Absorber–stripper system using recirculated no. 2 fuel oil to absorb benzene vapors from a vent air stream down to concentrations of 7 ppmv.[2]

higher mass transfer rates than Raschig and Lessig rings but are generally more expensive and must be fabricated of a corrosion-resistant metal. The ceramic and carbon rings are generally quite resistant to many chemical environments. Larger extruded ceramic shapes are available for very large packed chambers and a variety of shapes and meshes are available where metals can be employed. The Tellerette packing and other more intricate molded plastic packings can be used where temperatures will not exceed 355–360 K to give high mass transfer rates and low pressure drop. The extended surface of such packing is designed to provide constant liquid film renewal and the high void space generally assures minimum pressure drop. Figure 3 shows two types of packed absorbers. A countercurrent packed tower (Figure 3a) maximizes driving force since it brings the least concentrated outlet gas into contact with the fresh absorbing solution. The compact size, high mass transfer, and high driving force make this type of tower the best choice when the inlet gas is essentially particulate free. However, where some insoluble particulate will be present, countercurrent flow cannot be recommended since rapid plugging of the packing frequently ensues. Parallel downflow of gas and liquid will reduce plugging

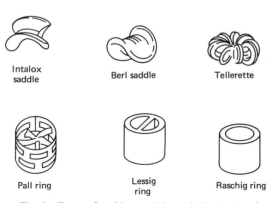

Fig. 2   Types of packing used in packed absorbers.[3]

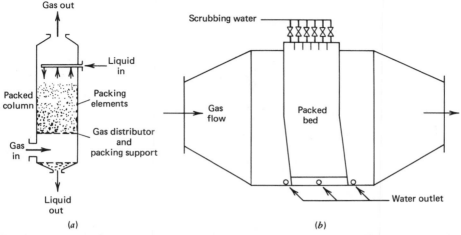

Fig. 3 Two types of gas absorbers: (a) countercurrent packed tower,[3] and (b) cross-flow packed absorber.[4]

tendencies. The tower of Figure 3a could be converted to parallel flow by revising the direction of the gas flow. In a water-based system absorbing SiF_4, Grant[5] reports six months plug-free operation of a parallel-flow bed before the packing must be removed and cleaned. In this system, SiF_4 reacts with water to produce gelatinous silicic acid which can easily adhere to the packing.

The cross-flow packed scrubber, Figure 3b, described by Teller,[6] has been found to be even more resistant to solids plugging. In this absorber, gas passes through horizontally, while liquid flows down vertically over the packing. Tellerette packing is frequently used to give low pressure drop, high mass transfer, and because this packing does not redistribute liquid flow. The latter feature permits varying liquid rate with bed position. When used purely as a gas absorber, typical design parameters are gas mass flow rate, $G = 2.44$ kg/sec-m² (1800 lb/hr-ft²), and liquid flow rate, $L = 2.03$ kg/sec-m² (1500 lb/hr-ft²). When particulates are present, L is increased to 2.71 kg/sec-m² throughout (2000 lb/hr-ft²) and to 13.56 kg/sec-m² (10,000 lb/hr-ft²) over the first 300 mm of packing in the direction of gas flow. Solids loadings up to 11 g/m³ have been handled successfully. (Areas for mass flow rates in cross-flow scrubbers are taken normal to the direction of each flowing fluid.)

Scale-up is a major problem with cross-flow scrubbers. Each cubic unit of the bed has a different mass transfer driving force gradient and no simple mathematical expression can be derived for predicting an integrated overall value. Cabibbo and Teller[7] have described a computer program for carrying out the calculations.

References 8 and 9 give general information on packed tower internals and design. References 10 through 14 give more general information on absorber design including plate and spray towers as well as packed columns.

2.2.2 Plate Columns

Figure 4 illustrates a plate column used for gas absorption. Contact between gas and liquid is obtained by forcing the gas to pass upward through small orifices bubbling through a liquid layer flowing across a plate. The bubble cap plate shown is the classical contacting device. A variation is the valve tray which permits larger variations in gas flow rate without dumping of the liquid through the gas passages. Sieve plates are simple plates perforated with small holes. The advantages are lower cost and higher plate efficiency, but they have narrower gas flow operating ranges. A number of plates are used in series to obtain the required overall absorption efficiency.

A plate column, by its nature, requires countercurrent flow of liquid and gas. Generally, the liquid flows across the plate and its level is maintained with a weir on the discharge side. A downcomer conducts the liquid to the next lower plate and forms a seal to prevent gas bypassing from plate to plate.

Duo-flow plates have larger holes than sieve plates such that two-way flow of liquid and gas occurs through the same orifice. Such plates do not need downcomers. Other variations of the sieve plate sometimes employ a small baffle or liquid-submerged impaction target above the gas orifice as in the Peabody scrubber, or an expanded mesh grid instead of a perforated plate. These latter modifications will generally be found in plate scrubbers designed for particulate collection and probably are of little benefit in gas absorption alone.

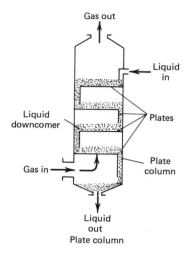

Plate column

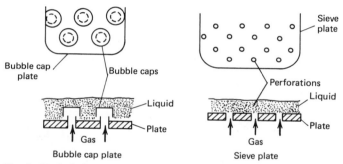

Fig. 4   Typical plate column absorber and two types of plate internals.[3]

The gas space above the plate liquid is filled with foam and entrained liquid spray. This adds to the contacting efficiency and is beneficial as long as liquid reentrainment is not excessive. In extreme cases, mesh pad mist eliminators have been utilized in the gas space above the plate to minimize entrainment.

References 10 through 14 give design and performance information on plate tower absorbers with Refs. 12 and 14 more comprehensive.

2.2.3   Spray Towers and Chambers

Figure 5 illustrates various types of spray chambers. They are considerably more resistant to plugging when particulates are present. However, difficulties with spray nozzle plugging may be encountered if the liquid is recycled. Thorough settling followed by fine strainers or even coarse filters is beneficial. Horizontal spray chambers (Figure 5a) have been discussed by Grant.[5] Both these and vertical spray towers (Figure 5b) have been used extensively to control gaseous emissions with particulates present. Cyclonic spray towers (Figures 5c and 5d) may provide slightly better scrubbing of the gas (higher mass transfer coefficients and more transfer units per tower).

While there is theoretically no limit to the number of transfer units which can be built into a countercurrent vertical packed or plate column if it is made tall enough, there are limits to the number of transfer units which can be achieved in a single-spray tower. Above certain tower heights and at higher gas velocities, spray particles are entrained upward resulting in a loss of true countercurrency. While achievable limits have not been clearly defined in the literature, some experimental achievable values have been reported. Lunde[16] has reported 5.8 transfer units in a vertical spray tower of Figure 5b design and 3.5 transfer units in a horizontal spray chamber (Figure 5a). Jewell and Crocker[17] have attained 7 transfer units in a single commercial cyclonic spray tower (Figure 5d). When more transfer units are required, it is customary to use spray towers in series.

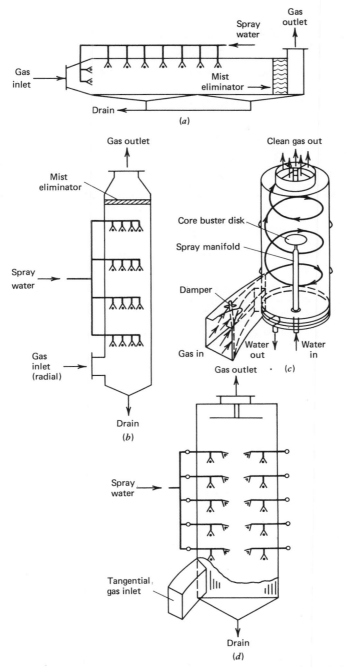

Fig. 5 Various types of spray towers: (*a*) horizontal spray chamber; (*b*) simple vertical spray tower; (*c*) cyclonic spray tower, Pease–Anthony type; (*d*) cyclonic spray tower, external sprays.[15]

In addition to the references cited above, Ref. 18 gives some theoretical discussion of spray tower performance and a design equation for cyclonic spray towers of the Pease–Anthony type. The addition of nozzles and a high spray concentration in the cyclonic inlet of a spray tower can quite effectively increase the amount of absorption.

2.2.4 Miscellaneous Absorption Contactors

When heavy particulate loads are present or submicron particulates must be recovered along with the gaseous pollutants, it is common to use wet collection devices having high particulate collection efficiencies which also have some capability for gas absorption. The venturi scrubber (Chapter 10) is one of the more versatile of such devices, but it has limitations for absorption since the gas and liquid have parallel flow. Lunde[16] indicates venturi scrubbers may be limited to 3 transfer units for gas absorption. The liquid-sprayed wet electrostatic precipitator (Chapter 12) is another device suitable for gas absorption and efficient collection of fine particles. Research tests have indicated that the presence of the corona discharge enhances the mass transfer absorption rates, but the mechanism through which this is brought about has not been established.

Another tower contacting device is the baffle tower (Figure 6) which has been employed occasionally when scaling and plugging problems could be severe. Gas passing up the tower has to pass through sheets of cascading liquid, providing some degree of contact. At higher velocities, there will be some atomizing of the liquid sheets into spray droplets which enhance contact surface. Baffle tower design may use alternating segmental baffles as illustrated, or disk and doughnut plates where the gas alternately flows upward through central orifices and annuli traversing through water curtains between each change in direction. Generally, mass transfer in baffle towers is poor and information on design parameters extremely scarce.

In Europe, considerable gas–liquid absorption has occurred in mechanically agitated contactors which produce spray *in situ*. Frequently, the gas passes through the top of a chamber half-filled with liquid. High-speed horizontal shafts equipped with disks partially dip into the liquid and spray a shower of droplets into the gas space. The spinning disks may have plain surfaces or radial grooves to increase the quantity of water entrained. Other units[19] have been developed in which the absorbing liquid is supplied to the center of a horizontal rotating disk. Concentric rotating vanes on the disk extend between stationary vanes causing the gas to pass through a succession of centrifugally atomized sprays as illustrated in Figure 7. Published methods for predicting the capability of such devices are very limited.

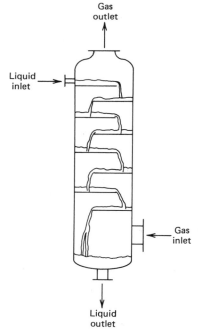

Fig. 6 Baffle tray tower used for gas absorption.

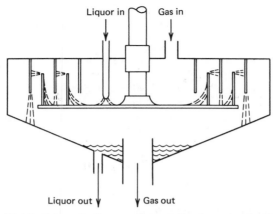

Fig. 7 Schematic diagram of a centrifugal gas absorber.[19]

2.3 Theory and Design Equations

Air contaminants are present in off-gases most usually in low concentrations. Thus, the absorption design problem in air pollution focuses on low-concentration gas streams which produce liquid effluents of low concentration. Design equations are simple for this dilute gas case. Furthermore, heat of solution effects are negligible and the column operates nearly isothermally, making the design process even more simple. The difficulty in air pollution control usually stems from the great quantities of gas that must be treated, especially from coal-fired electric utility boilers. In this case absorption columns can become quite large and expensive to build and operate.

Design equation development will be presented for isothermal countercurrent packed towers only. Design factors in this case where input gas flow and concentration are known include the size and type of packing, the absorption liquid flow rate, the height of packing required, and the tower diameter. In addition to mass transfer considerations, packing type and size are determined by the need to protect against corrosion and the pressure loss through the column that can be tolerated.

The height of packing is determined by the rate of mass transfer which is dependent on the phase equilibrium relationships. The amount of mass transferred is determined by the product of rate of mass transfer and the contact time. Contact time is dependent on column size and flow rates.

Tower diameter is determined by the quantity of gas passing up the tower. The upper limit of flow occurs when the tower begins to flood. Most efficient mass transfer occurs at flow rates just short of flooding. Higher flow rates result in higher pressure losses. Thus, the economic optimum column size can be determined by balancing pressure loss and pumping cost versus column and packing cost.

Nomenclature for equations used in Section 2.3 is defined in Section 2.3.7.

2.3.1 Equilibrium Relationships

2.3.1.1 Vapor–Liquid Phase Equilibria. Smith and Van Ness[20] is a widely used text describing phase equilibria thermodynamics with material similar to the following:

Vapor–liquid equilibria for miscible systems are determined from the equality of fugacity in both phases at the same temperature and pressure;

$$\hat{f}_i^v = \hat{f}_i^L \tag{1}$$

For component i in the liquid phase,

$$\hat{f}_i^L = x_i \gamma_i f_i^0 \tag{2}$$

and in the vapor phase,

$$\hat{f}_i^v = y_i \hat{\phi}_i P \tag{3}$$

According to Eq. (1) therefore,

$$y_i \hat{\phi}_i P = x_i \gamma_i f_i^0 \tag{4}$$

This system is structured so that only the fugacity coefficient $\hat{\phi}_i$ and the activity coefficient γ_i depend on the compositions at any given temperature and pressure. The standard-state fugacity f_i^0 is a property of the pure component only.

The Gibbs–Duhem equation which involves partial molar solution properties is extremely useful in dealing with phase equilibria. Smith and Van Ness give this equation in general form as follows:

$$\left(\frac{\partial M}{\partial T}\right)_P dT + \left(\frac{\partial M}{\partial P}\right)_T dP - \sum_i x_i \, d \ln \overline{M}_i = 0 \tag{5}$$

Here M is a solution property and $\overline{M}_i$ is the corresponding partial molar property. By using excess Gibbs free energy changes, this equation can be written in terms of the enthalpy and volume changes on mixing ΔH and ΔV, respectively, and the activity coefficient,

$$-\left(\frac{\Delta H}{RT^2}\right) dT + \left(\frac{\Delta V}{RT}\right) dP = \sum_i x_i \, d (\ln \gamma_i) \tag{6}$$

In the case of solutions of nonpolar molecules or solutions at low concentration, the enthalpy and volume changes on mixing are small and Eq. (6) for a binary mixture becomes

$$x_1 \, d \ln \gamma_i + x_2 \, d \ln \gamma_2 = 0 \tag{7}$$

For a solution at low pressure where the vapor is an ideal gas, and presuming f_i^0 to be equivalent to the pure component fugacity at the temperature and pressure of the solution, Eq. (7) can be rewritten at constant temperature:

$$x_1 \, d \ln (y_1 P) + x_2 \, d \ln (y_2 P) = 0 \tag{8}$$

Note that $(y_1 P) = \overline{P}_1$ and $(y_2 P) = \overline{P}_2$, the partial pressures, and $x_2 = 1 - x_1$; thus

$$\int_{P_2^0}^{\overline{P}_2} d \ln \overline{P}_2 = -\int_0^{\overline{P}_1} \frac{x_1}{(1 - x_1)} \, d \ln \overline{P}_1 \tag{9}$$

$$\ln\left(\frac{\overline{P}_2}{P_2^0}\right) = -\int_0^{\overline{P}_1} \frac{x_1}{(1 - x_1)} \frac{d\overline{P}_1}{\overline{P}_1} \tag{10}$$

Therefore, from partial pressure data of one component of a binary solution, the partial pressure of the second component may be calculated. Use of this technique is recommended when limited experimental data are available. Smith and Van Ness (p. 348) present an example problem using this technique.

2.3.1.2   Ideal Solutions, Henry's Law.   A general relationship for fugacity of a component in a liquid mixture in terms of the pure component fugacity is given by the following equation,

$$\ln\left(\frac{\hat{f}_i^L}{x_i f_i^L}\right) = \frac{1}{RT} \int_0^P (\overline{V}_i - V_i) \, dP \tag{11}$$

According to Amagat's law, an ideal solution is one in which $(\overline{V}_i - V_i) = 0$, and thus the value of the integral in Eq. (11) is also zero. This makes

$$\hat{f}_i^L = \hat{f}_i^{id} = x_i f_i^L \tag{12}$$

for an ideal solution. This relationship is known as the Lewis and Randall fugacity rule. More generally to be consistent with Eq. (2), the ideal fugacity would be defined as

$$\hat{f}_i^{id} = x_i f_i^0 \tag{13}$$

where f_i^0 is the fugacity of component i in the standard state at the same temperature and pressure of the solution. If the standard states adopted were that of the pure substance designated by i, then $f_i^0 = f_i$.

Equation (13) is plotted in Figure 8 in such a manner that both broken lines are valid representations of the equation. Each broken line represents one model of the ideal solution while the solid line represents the value of f_i^L for a real solution as given by Eq. (11). From the figure, it can be seen that while $f_i^0(B)$ and $f_i^0(A)$ are both fugacities of pure component i, only $f_i^0(A)$ represents the fugacity of a pure component as it actually exists. Thus $f_i^0(B)$ represents an imaginary state of the real solution.

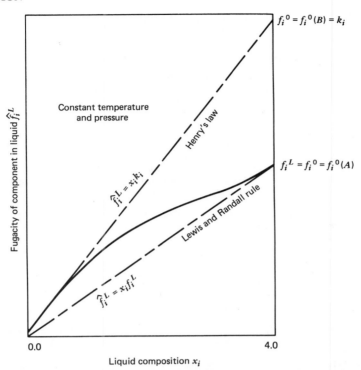

Fig. 8 Defining Henry's law and the Lewis and Randall fugacity rule in relation to the true fugacity.

In either case the broken line is drawn tangent to the solid curve for the real solution where it meets the curve. Real solution behavior is represented by $f_i^0(A)$ as $x_i \to 1.0$ and by $f_i^0(B)$ as $x_i \to 0$.

$$\lim_{x_i \to 1.0} \left(\frac{\hat{f}_i^L}{x_i} \right) = f_i^0(A) = f_i \tag{14}$$

$$\lim_{x_i \to 0} \left(\frac{\hat{f}_i^L}{x_i} \right) = f_i^0(B) = k_i \tag{15}$$

Equation (15) is a statement of Henry's law which is valid for a dilute solution and may be written as

$$\hat{f}_i^L = x_i k_i \tag{16}$$

where k_i is Henry's law constant. Equation (4) now becomes

$$y_i \hat{\phi}_i P = x_i k_i \tag{17}$$

For solutions at low pressure $\hat{\phi}_i = 1.0$ and $y_i P = \bar{P}_i$, thus

$$\bar{P}_i = x_i k_i \tag{18}$$

which is a more common form of Henry's law. The composition x_i could be expressed as a weight fraction as well as a mole fraction and k_i must have units to make the equation consistent. Putting the equation into mole fraction terms on both sides of the equation results in the following equation;

$$y_i = \left(\frac{k_i}{P} \right) x_i \tag{19}$$

which reduces to the following equation with $m_i = (k_i/P)$;

$$y_i = m_i x_i \tag{20}$$

an equation that has application to the case of absorption of dilute solutions.

Henry's law is an adequate method of representing vapor–liquid phase equilibria for many dilute solutions encountered in absorption work. However, in the case of weak electrolytes dissociation of the electrolyte may take place, upsetting the normal equilibrium relationship. In this case, Henry's law must be altered. The discussion that follows describes some up-to-date work which results in a reasonable, though much more complex, method to handle weak electrolytes.

2.3.1.3 Thermodynamics of Aqueous Solutions Containing Volatile Weak Electrolytes.

More recent thermodynamic analysis of experimental data on volatile weak electrolytes has been conducted by Edwards et al.[21-23] These papers lay out the framework for a rigorous molecular thermodynamic treatment following the methods of Prausnitz.[24] The electrolytes examined include ammonia, carbon dioxide, hydrogen sulfide, sulfur dioxide, and hydrogen cyanide, all of which have significance for air pollution control. The temperature range covered was from 273 to 443 K with concentration ranging up to 20 M. The third paper[23] deals with the multicomponent problem. Calculated equilibria are in agreement with the limited experimental data for the ternary systems ammonia–carbon dioxide–water and ammonia–hydrogen sulfide–water.

The objective of the work by Edwards et al. was aimed at industrial interests of stripping weak electrolytes from aqueous streams before discharge into receiving waters. However, the thermodynamic analysis is equally valid for absorption of weak electrolytes. In the theory, a weak electrolyte in the liquid phase is assumed to exist in molecular and ionic forms. A chemical equilibrium between these two forms is described by a dissociation constant. Vapor phase dissociation of the electrolyte is appreciable only at high temperature and is thus neglected. Vapor–liquid equilibrium is assumed to be described by Henry's law, and the thermodynamic analysis is based on a macroscopic and microscopic description.

The macroscopic basis neglects dissociation and describes the system in terms of the bulk properties. The bulk property that characterizes the liquid phase is the stoichiometric or total electrolyte concentration. This concentration is measured by standard quantitative analysis techniques and is the sum of the molecular and ionic concentrations. Deviations from nonideality are described by a stoichiometric activity coefficient.

The microscopic basis deals with the solution on a molecular basis. Concentrations of the molecular solutes and ions are used to characterize the liquid phase. Dissociation in the liquid phase is accounted for. The success of the theory is exhibited in its ability to accurately predict the measured data.

2.3.2 Defining Mass Transfer

2.3.2.1 Steady-State Molecular Diffusion.

Molecular diffusion occurs as the result of the random motion of molecules. A net movement of molecules takes place from a region of high concentration to a region of low concentration. The driving force for this diffusion can be considered to be due to this difference in concentration. This driving force continues to act until the concentrations are the same throughout the vessel holding the molecules.

A quantitative description of molecular diffusion is found in many texts on the subject. Faust et al.[25] detail the derivation of the following equations.

Relative to the interface between two phases, it is assumed that only one component is in motion across the interface. The molecular diffusivity is interpreted in terms of a relative velocity of the diffusing component with respect to an average velocity of the entire stream in a direction normal to the interface. Mass transfer of a typical component A in a mixture with component B is depicted as a movement to or from a fixed surface that is parallel to the interface and stationary in relation to the interface. The molecular motion is accounted for by Fick's law and the total flux is the sum of this molecular diffusion and the convective bulk flow.

Defining $\overline{N}_A$ as the flux* of component A and $\overline{N}_B$ as the flux of component B, the bulk flux is given by

$$\overline{N}_{\text{bulk}} = \overline{N}_A + \overline{N}_B \tag{21}$$

For component A with mole fraction y_A, $\overline{N}_A$ is given by

$$\overline{N}_A = y_A (N_A + N_B) - D_{AB} \rho_M \frac{dy_A}{dz} \tag{22}$$

---

* Note, flux has the units of mass per unit time per area through which the material is diffusing. An overbar will designate a flux. A variable without the overbar is a rate of mass transfer in mass per unit time.

where the second term on the the right-hand side of the equation represents molecular diffusion as described by Fick's law.

In absorption of a single component, for example, A, it is reasonable to assume that component B will remain stagnant. Thus $N_B = 0$, and

$$\bar{N}_A = y_A \bar{N}_A - D_{AB}\rho_M \frac{dy_A}{dz} \tag{23}$$

This equation can be rearranged and integrated over a distance L, with the following boundary conditions:

$$y_A = y_{AO} \quad \text{when } z = 0$$
$$y_A = y_{AL} \quad \text{when } z = L$$

The result is Eq. (24);

$$\bar{N}_A = \left(\frac{D_{AB}\rho_M}{L}\right) \ln\left(\frac{1 - y_{AO}}{1 - y_{AL}}\right) \tag{24}$$

Note that

$$y_{AO} + y_{BO} = 1.0$$
$$y_{AL} + y_{BL} = 1.0$$

and

$$(1 - y_{AO}) = y_{BO} \qquad (1 - y_{AL}) = y_{BL}$$

Thus Eq. (24) can be rewritten by defining a logarithmic mean concentration y_{BM},

$$y_{BM} = \frac{(y_{BO} - y_{BL})}{\ln(y_{BO}/y_{BL})} = \frac{(y_{AO} - y_{AL})}{\ln(y_{AO}/y_{AL})} \tag{25}$$

and a mass transfer coefficient k_{my}

$$k_{my} = \left(\frac{D_{AB}\rho_M}{L y_{BM}}\right) \tag{26}$$

$$\bar{N}_A = k_{my}(y_{AO} - y_{AL}) \tag{27}$$

2.3.2.2 Whitman Two-Film Theory. In the past, several sophisticated theories of interphase mass transfer have been developed by Higbe[26] and Dankwerts.[27] These theories serve as a realistic description of the dynamic action at the interface between the two fluids. However, the Whitman[28] "two-film" concept remains an excellent basis for defining mass transfer coefficients and as a picture of interfacial action.

It is presumed that the gas and liquid in the adsorption apparatus are in motion relative to each other with an interface between them. The Whitman theory assumes two film forms, one in each phase as in Figure 9. Material is transferred in the bulk of each phase by turbulent convection. Concentration differences are negligible except in the vicinity of the interface. On either side of the interface the convection currents die out and a thin film of fluid forms. Transfer of material takes place through these films by a mechanism similar to molecular diffusion.

The interface offers no resistance to mass transfer and the liquid and gas are considered to be in equilibrium, which can be described by Henry's law in the dilute solution cases.

In a packed absorption tower the fluid is in tumultuous and turbulent conditions. Mass transfer is much greater than in the molecular diffusion case. However, this turbulent mass transfer may be expressed by an equation similar to Eq. (27) for molecular diffusion.

$$\bar{N}_A = k_y(y_A - y_{Ai}) = k_x(x_{Ai} - x_A) \tag{28}$$

In this case k_y and k_x are mass transfer coefficients defined in relation to the interfacial concentrations y_{Ai} and x_{Ai}, respectively, as shown in Figure 9. Thus mass transfer through the gas film is controlled by the value of k_y and similarly through the liquid film by the value of k_x.

2.3.2.3 Overall Mass Transfer Coefficients. Mass transfer coefficients based on interfacial concentrations have little practical value because of the difficulty in determining interfacial concentration. It

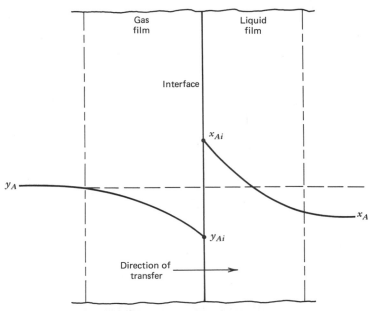

Fig. 9 Whitman two-film theory of interphase mass transport. y_A = mole fraction of A in bulk gas stream, x_A = mole fraction of A in bulk liquid stream, y_{a_i} = mole fraction of A in the gas film at the interface, x_{a_i} = mole fraction of A in the liquid film at the interface.

is customary to define an overall mass transfer coefficient that can be found more readily from experimental data. Two new fictitious quantities y_A^* and x_A^* are created to define these overall mass transfer coefficients.

   y_A^* = equilibrium mole fraction of the solute in the vapor corresponding to the mole fraction of the solute in the liquid
   x_A^* = equilibrium mole fraction of the solute in the liquid corresponding to the mole fraction of the solute in the vapor

The relationship between these quantities is presented in Figure 10. Note that according to Henry's law,

$$y_A^* = m x_A \tag{29}$$
$$x_A^* = y_A/m \tag{30}$$

The overall mass transfer equations become

$$\overline{N}_A = K_y\,(y - y_A^*) = K_x\,(x_A^* - x) \tag{31}$$

Equation (31) can be rearranged and expanded;

$$\frac{1}{K_y} = \frac{(y_A - y_A^*)}{N_A} = \frac{(y_A - y_{Ai}) + (y_{Ai} - y_A^*)}{N_A} \tag{32}$$

Since equilibrium is presumed to exist at the interface,

$$y_{Ai} = x_{Ai} \tag{33}$$

Therefore,

$$\frac{1}{K_y} = \frac{(y_A - y_{Ai})}{N_A} + \frac{m(x_{Ai} - x_A^*)}{N_A} \tag{34}$$

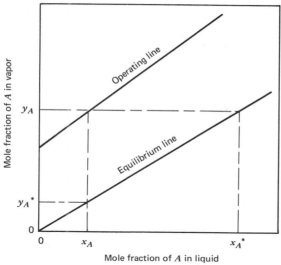

Fig. 10 Equilibrium relationships.

which can be rewritten by combining with Eq. (28).

$$\frac{1}{K_y} = \frac{1}{k_y} + \frac{m}{k_x} \tag{35}$$

Equation (32) can be treated similarly for the liquid film to produce

$$\frac{1}{K_x} = \frac{1}{k_x} + \frac{1}{mk_y} \tag{36}$$

Equations (35) and (36) are relationships between the film coefficients and the overall relationship.

Note that the units of these mass transfer coefficients must be equivalent to mole/hr-m²-mole fraction where the area represented is the interfacial area of mass transfer.

When discussing absorption in terms of the Whitman two-film theory, it has become customary to refer to either the gas or liquid film as the controlling resistance to mass transfer. In the case where the gas is very soluble, the interface picture is like that of Figure 9. In this case, the equilibrium mole fraction in the liquid phase is much greater than the equilibrium mole fraction in the vapor phase. The slope of the equilibrium curve m is small and the gas film controls. When the slope of the equilibrium curve is large, the liquid film controls. Table 1 details several common systems and the resistance to mass transfer exhibited by these systems.

2.3.3 Countercurrent Packed Tower Equations

2.3.3.1 Mass Balance—Operating Line.
A mass balance is made around a cross section of a packed absorption tower in which it is assumed that the liquid and gas flow rates are constant as indicated in Figure 11.

$$y_A G + x_A L = (y_A - dy_A)G + (x_A + dx_A)L \tag{37}$$
$$G\, dy_A = L\, dx_A \tag{38}$$

If it is supposed that G and L will vary throughout the column, then it is reasonable to rewrite Eq. (38):

$$d(Gy_A) = d(Lx_A) \tag{39}$$

Equation (39) is integrated from the top of the tower down to the tower cross section AA as shown in Figure 11.

$$y_A G - y_{A1} G_1 = x_A L - x_{A1} L_1 \tag{40}$$

Table 1   Sources of Major Resistance to Mass Transfer for Some Common Systems

Gas film controls	Gas very soluble, mole fraction in liquid phase is much greater than mole fraction in gas phase ($x > y$, m is small).
Liquid film controls	Gas only slightly soluble, mole fraction in gas phase is much greater than mole fraction in liquid phase ($y > x$, m is large).

Gas Film Controlling	Both Films Controlling	Liquid Film Controlling
1. Absorption of NH_3 in H_2O or aqueous NH_3.	1. Absorption of SO_2 by water.	1. Absorption of CO_2, O_2, or H_2 in water.
2. Stripping of NH_3 from aqueous NH_3.	2. Absorption of acetone by water.	2. Absorption of CO_2 in aqueous alkali solution.
3. Absorption of dilute (5% by volume) NH_3 in acids.	3. Absorption of nitrogen oxide by strong sulfuric acid.	3. Absorption of chlorine by water.
4. Absorption of water vapor in strong acids.		
5. Absorption of SO_3 in strong H_2SO_4.		
6. Absorption of SO_2 in alkali and NH_3 solutions.		
7. Solution of H_2S in aqueous caustic.		
8. Evaporation and condensation of liquids.		

Solute free flow rates can be defined:

$$G_B = (1 - y_A)G \quad \text{or} \quad G_B = (1 - y_{A1})G_1$$
$$L_B = (1 - x_A)L \quad \text{or} \quad L_B = (1 - x_{A1})L_1 \tag{41}$$

Rewriting the material balance based on these flow rates,

$$\left(\frac{y_A}{1 - y_A}\right) G_B + \left(\frac{x_{A1}}{1 - x_{A1}}\right) L_B = \left(\frac{x_A}{1 - x_A}\right) L_B + \left(\frac{y_{A1}}{1 - y_{A1}}\right) G_B \tag{42}$$

In the case of dilute solutions, $1 - y_A \approx 1.0$, and so on, and Eq. (42) reduces to

$$y_A G + x_{A1}L = x_A L + y_{A1}G \tag{43}$$

Equation (42) is an operating line that can be used in the general case, while Eq. (43) applies to the dilute solution case.

2.3.3.2   Volume Bases Mass Transfer Coefficients.   A quantity a can be defined for tower packing where

$$a = \frac{\text{interfacial area}}{\text{packing volume}}$$

Thus a is a function of the packing characteristics and the liquid retained in the void spaces in the packing. The flow rate of both phases, and the viscosity, density, surface tension, and the size and shape of packing affect the value of a. These same variables determine the value of the mass transfer coefficients k_y, k_x, K_y, and K_x. Therefore the values $k_y a$, $k_x a$, $K_y a$, and $K_x a$ can be determined as a combined quantity.

The tower packing volume is Az and the total interfacial area A_i,

$$A_i = aAz \tag{44}$$

Equation (44) can be written in differential form

$$dA_i = aA \, dz$$

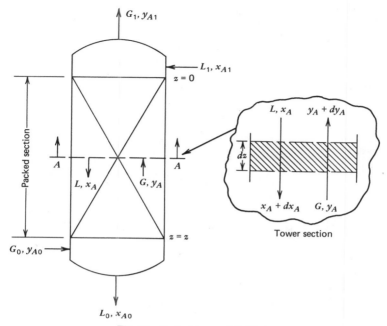

Fig. 11 Packed tower definitions.

Equation (31) can now be written in terms of a as a differential rate of mass transfer,

$$dN_A = \bar{N}_A \, dA_i = K_y a(y - y^*)A \, dz \tag{45}$$

The units of this equation are now mole/hr, and the combined mass transfer coefficient $K_y a$ has the units mole/hr-m³-mole fraction. Here the volume represented is the volume occupied by the tower packing.

2.3.3.3 Determining Tower Height. The rate of interfacial mass transfer is equated to the rate of change of mass within a phase for the gas phase,

$$dN_A = \bar{N}_A \, dA_i = d(GY_A) \tag{46}$$

Note that from Eq. (41),

$$d(Gy_A) = d\left[\frac{G_B y_A}{(1 - y_A)}\right] = \frac{G \, dy_A}{(1 - y_A)} \tag{47}$$

Therefore,

$$G\left[\frac{dy_A}{(1 - y_A)}\right] = K_y a(y_A - y_A^*)A \, dz \tag{48}$$

becomes the basic design equation. A similar equation based on liquid flow rates and mole fractions and $K_x a$ could be written which would be useful in stripping calculations. Note that a mass transfer coefficient, $K_y^0 a$ could be defined by analogy to Eq. (27) that was independent of concentration. Thus,

$$K_y a = \frac{K_y^0 a}{y_{BM}} = \frac{K_y^0 a}{(1 - y_A)_{LM}} \tag{49}$$

where the logarithmic mean (LM) mole fraction is given by

$$(1 - y_A)_{LM} = \frac{(1 - y_A) - (1 - y_A^*)}{\ln\left[(1 - y_A)/(1 - y_A^*)\right]} \tag{50}$$

It would be expected that $K_y^0 a$ would not vary with composition and thus would be essentially constant throughout the column. In addition it has become customary to define liquid and gas flow rates on the basis of tower cross section. Therefore,

$$\bar{G} = G/A \quad \text{and} \quad \bar{L} = L/A \tag{51}$$

Basic Eq. (48) now can be rewritten

$$\bar{G}\left[\frac{dy_A}{(1 - y_A)}\right] = \frac{K_y^0 a}{(1 - y_A)_{LM}}(y_A - y_A^*)\,dz \tag{52}$$

and then integrated (see Figure 11).

$$z = \int_0^z dz = \left(\frac{\bar{G}}{K_y^0 a}\right)\int_{y_{A1}}^{y_{A0}}\frac{(1 - y_A)_{LM}}{(1 - y_A)(y_A - y_A^*)}\,dy_A \tag{53}$$

The ratio of flow rate to mass transfer coefficient has been designated as the height of a transfer unit (HTU) or H_{OG} in this case for the gas flow rate and overall mass transfer coefficient.

$$H_{OG} = \left(\frac{\bar{G}}{K_y^0 a}\right) \tag{54}$$

Note that the way in which H_{OG} is defined, it should be constant throughout a column; hence the ratio was removed from under the integral during the integration. There remains the integral of Eq. (53) which is designated the number of transfer units N_{OG} in this case of overall mass transfer in the gas phase.

$$N_{OG} = \int_{y_{A1}}^{y_{A0}}\frac{(1 - y_A)_{LM}}{(1 - y_A)(y_A - y_A^*)}\,dy_A \tag{55}$$

The integral can be found by graphical or numerical techniques. In the case of a dilute solution, the integral can be simplified as described in the following section.

Tower height is then found from H_{OG} and N_{OG} as follows:

$$z = H_{OG}N_{OG} \tag{56}$$

2.3.3.4 Dilute Solution Case.

In most cases where absorption is a practical means of removal of a pollutant from a gas stream, the concentrations are low enough to be considered dilute. Under dilute conditions, the design equations are considerably simplified, flow rate remains constant since only a small quantity of mass is exchanging, and Henry's law of equilibrium will apply. Hence,

$$\begin{aligned} y^* &= mx^* \\ (1 - y_A) &\approx (1 - y_A)_{LM} \approx 1.0 \end{aligned} \tag{57}$$

Therefore,

$$N_{OG} = \int_{y_{A1}}^{y_{A0}}\frac{dy_A}{(y_A - y_A^*)} \tag{58}$$

$$y_A = (\bar{L}/\bar{G})(x_A - x_{A1}) + y_{A1} \tag{59}$$

where the ratio (L/G) is constant and equivalent to $(\bar{L}/\bar{G})$. Refer to Figure 10 where it can be seen that

$$y_A^* = mx_A \tag{60}$$

Rewriting the operating line,

$$y_A = \left(\frac{\bar{L}}{m\bar{G}}\right)(mx_a - mx_{A1}) + y_{A1} \tag{61}$$

Define an absorption factor A:

$$A = \left(\frac{\bar{L}}{m\bar{G}}\right) = \left(\frac{L}{mG}\right) \tag{62}$$

where A will be constant in the dilute solution case. Then

$$y_A = A(y_A^* - mx_{A1}) + y_{A1} \tag{63}$$

$$y_A^* = \left(\frac{y_A - y_{A1}}{A}\right) + mx_{A1} \tag{64}$$

and

$$N_{OG} = \int_{y_{A0}}^{y_{A1}} \frac{dy_A}{y_A(1 - 1/A) + (y_{A1}/A - mx_{A1})} \tag{65}$$

can now be analytically integrated to give

$$N_{OG} = \frac{\ln\{[(y_{A0} - mx_{A1})/(y_{A1} - mx_{A1})](1 - 1/A) + (1/A)\}}{[1 - (1/A)]} \tag{66}$$

A number of texts, such as Ref. 29, present a graphical solution of this equation.

The height of the transfer unit H_{OG} is found from Eq. (54). For this dilute solution case note that

$$K_y a \approx K_y^0 a \tag{67}$$
$$\bar{G} \approx G \tag{68}$$

and

$$H_{OG} = G/k_y a \tag{69}$$

2.3.4 Countercurrent Packed Tower Design

2.3.4.1 Operations of Packed Towers. Most gas absorption columns are operated as countercurrent contactors with gas entering at the bottom. This gas flow causes some liquid to be held up in the column. Liquid holdup in a typical column is illustrated by Figure 12 where the holdup is shown as a function of liquid flow rate and column pressure drop.[30]

Examination of current data shows a continually increasing pressure drop with gas flow until the slope of the curve becomes infinite. This point is the upper limit to the gas rate. The velocity corresponding to this upper limit is known as the flooding velocity.

Flooding can be observed as a crowning of the packing which begins at the bottom and progresses to the top of the packing where a layer of liquid is visible above the packing. At gas rates greater than the flooding velocity, the columns act as a gas bubbler. Flooding occurs at lower gas velocities as liquid rate is increased. Larger, more open packings flood at higher velocities than smaller, more dense packings.

Packed towers may be operated above 90% of the flooding velocity when pressure controls are provided to maintain a fixed maximum pressure drop not to exceed the 90% flooding velocity figure. Typical operating velocities as a percentage of flooding used in commercial operations are listed in Table 2.

In a packed tower operating in countercurrent flow at constant liquid rate, the pressure drop varies with the gas mass velocity at a rate proportional to the 1.7 to 1.9 power of the gas rate. With liquid holdup in the packing, a higher pressure drop exists than with dry packing, due to the lower void space in the column. Absorbers and absorption liquid regenerators are designed to operate at pressure drops of 2–4 cm H_2O per meter of packed depth (0.25–0.50 in. water per foot of packing); vacuum distillation columns from 0.08 to 4.0 cm of H_2O pressure drop per meter of packing; and pressure stills at 6.2 to 9.3 cm of H_2O pressure drop per meter of packing. If packed beds are operated over 8.3 cm of H_2O pressure drop per meter of packed depth, there is danger of the tower going into a flooded condition, which is discussed in the following paragraphs. A typical set of constant liquid flow rate pressure drop versus gas rate curves is given in Figure 13. The publication by the Norton Company[30] and Bhatia[9] give general information on packed tower internals as well as the design of packed towers. More general information on packed absorption tower design can be found from Treybal,[10] Sherwood et al.,[11] Sherwood and Pigford,[12] and Wilke and Von Strockbar[13] and the *Chemical Engineers' Handbook.*[14]

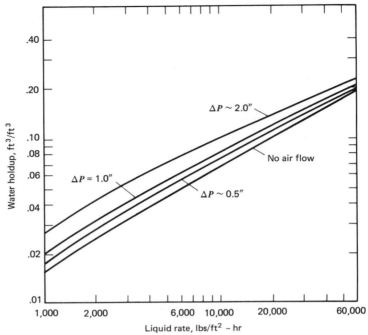

Fig. 12   Liquid holdup in a 30-in. (0.76-m) diameter column packed with 2-in. (0.0507-m) ceramic Intalox saddles [courtesy of the Norton Company, Akron, Ohio].[30] (1 in. = 25.4 mm; 1 lb/ft²-hr = 0.0013574 kg/m²-sec).

2.3.4.2   Choosing a Liquid–Gas Flow Ratio.   When designing an absorption tower for the cleanup of an off-gas, most commonly the following variables will be known:

G = actual gas flow rate
y_{A0} = mole fraction of A, gas at inlet coming from process
y_{A1} = mole fraction of A, gas at outlet (maximum value specified by control-agency regulation if discharged to the atmosphere)
x_{A1} = mole fraction of A, liquid into tower, a very low value, quite frequently zero if solvent is used only once (no regeneration and recycle)

The material balance for the dilute solution case around the entire column is given by Eq. (59) rearranged:

$$x_{A0} = \frac{G}{L}(y_{A0} - y_{A1}) + x_{A1} \tag{70}$$

Thus, if L or (G/L) can be found, the entire material balance will be set since the equation may be solved for x_{A0}.

In absorption, on a plot of y_A versus x_A, two situations can develop in regard to the equilibrium curve. The curve can be concave up or nearly a straight line, or the curve can be s-shaped, having a

Table 2   Packed Tower Operation

Type of Packing	Typical Operating Velocity as a Percentage of Flooding
Raschig rings	60–80%
Berl saddles	65–80%
Intalox saddles	65–85%
Tellerettes	75–100%

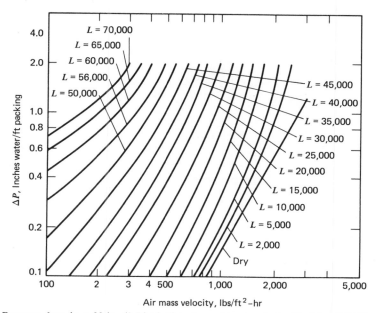

Fig. 13 Pressure drop in a 30-in. (0.76-m) diameter column packed with 2-in. (0.0507-m) ceramic Intalox saddles [courtesy of the Norton Company, Akron, Ohio].[30] (1 in./ft = 83.33 mm/m; 1 lb/ft²-hr = 0.0013574 kg/m²-sec).

concave downward portion. These situations are illustrated in the two curves (Figure 14) on which the operating line as specified by the material balance is plotted and assumed to be straight.

In the case of Figure 14a with a concave equilibrium line, three possible operating lines are shown. Note on these plots (L/G) is the slope of the operating line. As can be seen, the line furthest to the right with the smallest slope represents the limit to which the slope of the operating line can be drawn. Thus, this slope represents a minimum (L/G) ratio in which x_{A0} and y_{A0} are actually in equilibrium. Since true thermodynamic equilibrium can never be reached practically, this point represents a theoretical minimum (L/G) which never can be attained in practice.

Real (L/G) ratios are set at 1.1 to 1.7 times this minimum rate. Thus, if G is known, (L/G) is set; and L and x_{A0} can be found. In the case of Figure 14b with s-shaped equilibrium curve, the minimum value of (L/G) is reached at some point within the column where the operating and equilibrium lines are tangent and thermodynamic equilibrium would be "exceeded" if the (L/G) ratio were any smaller.

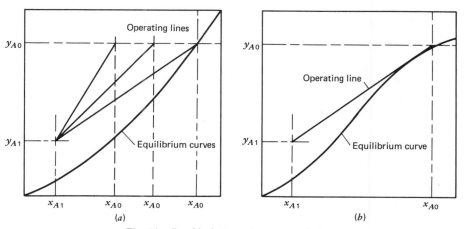

Fig. 14 Graphical absorption tower relations.

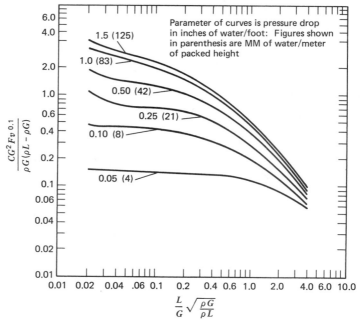

NOMENCLATURE FOR CORRELATION AXIS

Property	Symbol	British units	Metric units
Gas rate	G	lbs/ft^2 sec	kg/m^2 sec
Liquid rate	L	lbs/ft^2 sec	kg/m^2 sec
Gas density	ρ_G	lbs/ft^3	kg/m^3
Liquid density	ρ_L	lbs/ft^3	kg/m^3
Liquid viscosity	v	centistokes	centistokes
Conversion factor	C	1,000	10.764
Packing factor	F	—	—

Fig. 15 Pressure drop correlation. [Courtesy of the Norton Company, Akron, Ohio.][30]

2.3.4.3 Determining Tower Diameter. Knowing the (L/G) ratio and the overall material balance, and having selected the type of packing and the pressure drop per foot of packing, the tower diameter can now be determined. Current accepted practice is to use a modified Sherwood flooding correlation to determine tower diameter. Such a correlation is presented by the Norton Company,[30] and shown in Figure 15. See also Table 3.

Since (L/G) is known, we can calculate the value of the abscissa if ρ_G and ρ_L are known. Although ρ_G and ρ_L may vary through the column, the variations will be small, especially in the dilute solution case, and an average of the top and bottom values from the column should be sufficiently accurate.

From the abscissa, the packing factor, and the required pressure drop, the ordinate of the curve can be read, and the value of $\overline{G}$ calculated. The column cross-sectional area is then

$$A = \frac{G}{\overline{G}} \tag{71}$$

and the column diameter

$$D = \left(\frac{4G}{\pi \overline{G}}\right)^{0.5} \tag{72}$$

2.3.4.3 Mass Transfer Coefficients. The measurement of mass transfer coefficients has occupied many experimenters for the past 40 yr. The surprising thing is that there is not much data available in the open literature. The problems of making satisfactory and consistent measurements are threefold:

Table 3 Packing Factors for Use with Generalized Pressure Drop Correlation[a] (Courtesy of the Norton Company)[30]

Packing Type	Material	Nominal Packing Size, in.										
		1/4	3/8	1/2	5/8	3/4	1 or #1	1 1/4	1 1/2	2 or #2	3	3 1/2 or #3
Hy-Pak™	Metal						43		26	18		15
Super Intalox® saddles	Ceramic						60			30		
Super Intalox® saddles	Plastic				97		33			21		16
Pall rings	Plastic				70		52		40	24		16
Pall rings	Metal						48		33	20		16
Intalox® saddles	Ceramic	725	330	200		145	92		52	40	22	
Raschig rings	Ceramic	1600	1000	580	380	255	155	125	95	65	37	
Raschig rings	$\tfrac{1}{32}$-in. metal	700	390	300	170	155	115					
Raschig rings	$\tfrac{1}{16}$-in. metal			410	290	220	137	110	83	57	32	
Berl saddles	Ceramic	900		240		170	110		65	45		

[a] Packing factors determined with an air–water system in 30-in. I. D. tower.

(1) difficulty in determining the interfacial area, (2) difficulty in determining individual coefficients from measurements of overall coefficients, and (3) entrance effects.

The rate of mass transfer depends on the available contact area between the phases, that is, on the degree of subdivision of the phases. Tower packing provides the skeleton over which the liquid flows in thin turbulent films. The packing area becomes effective only if it is wetted. Thus, $K_y a$ must include a hidden factor that reflects the proportion of the packing area that is effective. This effective area is proportional to: (1) liquid rate, (2) type of packing, (3) method of packing the tower, (4) nature of the liquid distribution, and (5) height and diameter of the tower.

Data are available in the literature for some systems, such as that summarized in the *Chemical Engineers' Handbook*.[14] Data are given in graphical form or analytical correlations as either HTU's (H_{OG}, etc.) or mass transfer coefficients ($K_g a$, etc.). The HTU's and mass transfer coefficients are related as follows.

Both sides of Eq. (35) are multiplied by $\bar{G}/[a(1 - y_A)_{LM}]$, and the last term on the right-hand side by the identity $\bar{L}(1 - x_A)_{LM}/[\bar{L}(1 - x_A)_{LM}]$, producing

$$\frac{\bar{G}}{K_y a(1 - y_A)_{LM}} = \frac{\bar{G}}{k_y a(1 - y_A)_{LM}} + \frac{m\bar{G}}{k_x a(1 - y_A)_{LM}}\left(\frac{\bar{L}(1 - x_A)_{LM}}{\bar{L}(1 - x_A)_{LM}}\right) \tag{73}$$

This equation can be rewritten using Eqs. (49) and (54).

$$H_{OG} = H_G + \left(\frac{m\bar{G}}{\bar{L}}\right) H_L \left[\frac{(1 - x_A)_{LM}}{(1 - y_A)_{LM}}\right] \tag{74}$$

where H_G and H_L are individual film coefficients based on $k_y^0 a$ and $k_x^0 a$ similar to H_{OG}. Equation (74) reduces to the following, for dilute solutions:

$$H_{OG} = H_G + \left(\frac{m\bar{G}}{\bar{L}}\right)^{H_L} \tag{75}$$

2.3.5   *Cross-Flow Packed Scrubber*

Although cross-flow packed scrubbers are more commonly used when a gas to be cleaned contains particulate matter, they can also be used for gas absorption. The design problem with cross-flow scrubbers has been discussed in Section 2.2.1, where a computer program for design is referenced.[7]

2.3.6   *Spray Towers and Plate Columns*

General design conditions for spray towers and plate columns have been discussed in Sections 2.2.2 and 2.2.3. Design of plate towers depends on the equilibrium stage concept. This concept is discussed by Treybal,[16] Sherwood and Pigford,[12] and in the *Chemical Engineers' Handbook*.[14] Spray towers can be designed using either an equilibrium stage or HTU approach. Once the number of equilibrium stages are determined, a stage efficiency must be found. Very little data exist for stage efficiencies in the open literature. Thus, manufacturers' literature must be relied on to determine efficiencies.

Spray towers operate with a high degree of back mixing due to the lack of a stabilizing pressure drop. Because of this back mixing, some designers assume a spray tower will operate as a single equilibrium stage regardless of height.

However, when using the transfer unit concept, the literature indicates 5–7 transfer units are possible with careful design.

2.3.7   *Nomenclature*

a interfacial surface per active volume, m²/m³
A cross-sectional area, m²
        (also absorption factor = L/mG, unitless)
A_i interfacial area of mass transfer, m²
D_{AB} diffusivity of component A in component B, m²/sec
$\hat{f}_i$ fugacity of component i in solution, Pa
$\hat{f}_i^{id}$ fugacity of component i in an ideal solution, Pa
$\hat{f}_i^0$ fugacity of component i in the standard state, Pa
G total gas flow rate, mole/sec
G_B solute-free gas flow rate, mole/sec
$\bar{G}$ total gas flow rate, mole/sec-m²
H_G height of a local gas phase transfer unit, m
H_L height of a local liquid phase transfer unit, m

H_{OG} height of an overall gas phase transfer unit, m
k_i Henry's law constant, Pa/mole fraction
k_{my} mass transfer coefficient in molecular diffusion, mole/sec-m²-mole fraction
k_x individual liquid film mass transfer coefficient, mole/sec-m²-mole fraction
k_y individual gas film mass transfer coefficient, mole/sec-m²-mole fraction
K_y overall gas phase mass transfer coefficient, mole/sec-m²-mole fraction
K_x overall liquid phase mass transfer coefficient, mole/sec-m²-mole fraction
K_{ya} overall volumetric gas phase mass transfer coefficient, mole/sec-m³-mole fraction
L total liquid flow rate, mole/sec
L_B solute-free liquid flow rate, mole/sec
$\overline{L}$ total liquid flow rate, mole/sec-m²
m Henry's law constant, or slope of the equilibrium curve, mole fraction/mole fraction
M general solution property
$\overline{M}_i$ partial molar solution property
$\overline{N}_A$ flux of component A, mole/sec-m²
$\overline{N}_B$ flux of component B, mole/sec-m²
N_{OG} number of overall transfer units
P pressure, Pa
P_i^o vapor pressure of component i, Pa
$\overline{P}_i$ partial pressure of component i, Pa
R gas constant, (various units)
T temperature, K
V volume, m³
$\overline{V}_i$ partial molar volume of component i, m³
x_i mole fraction of i in the liquid
y_i mole fraction of i in the vapor
z molecular diffusion distance, or packed tower height, m

2.3.7.1 Greek
Δ difference between final and initial thermodynamic state
γ_i activity coefficient of component i in solution, unitless
ρ_G density of gas, kg/m³
ρ_L density of liquid, kg/m³
ρ_M density of solution, mole/m³
$\hat{\phi}_i$ fugacity coefficient of component i in solution, unitless

2.3.7.2 Superscripts
L liquid
0 ideal solution property, standard state
v vapor

2.3.7.3 Subscripts
A component
B component B in solution
i component i in solution
LM   logarithmic mean
O conditions at bottom of absorption tower, or conditions at top of absorption tower
1 component 1 in solution
2 component 2 in solution

2.4 Absorption—Specific Applications

One major application for absorption is removal of water-soluble pollutant gases from process gaseous waste streams—pollutants such as HCl, SO_2, NO_2, HF, SiF_4, NH_3, and H_2S. Scrubbing of flue gases to remove SO_2 from combustion of sulfur-bearing fossil fuels has received extensive study and some commercial testing.[31-45] Because of limited solubility of SO_2 in water, the absorbing liquid is usually made alkaline to neutralize the SO_2 in the absorbing liquid film, reducing the SO_2 back-vapor pressure and increasing driving force. However, since combustion gases contain CO_2, if the absorbing liquid is made highly alkaline (pH > 9), large quantities of CO_2 will be absorbed, resulting in inordinate caustic usage and solids for waste disposal. Therefore, the pH of the absorbing liquid is usually maintained in the range of 8.0–8.5 at the point of final contact with the gas. An adequate ratio of liquid to gas will be employed to maintain the spent absorbing liquid at pH 7 or above at the outlet. Countercurrent contact of liquid and gas is employed to maximize driving force.

Packed and plate columns give the most efficient contact between gas and liquid and should be the preferable choice in the absence of solids (either from particulates in the gas stream or precipitated reaction products). Lime (and also limestone) is a plentiful and inexpensive alkali which is frequently first choice for absorbing liquid pH adjustment. However, many calcium salts such as the sulfates, sulfites, and fluorides have limited solubility. Plugging problems can result from these solutions plating out deposits in pipelines, spray nozzles, on packing, and so on. This has been a severe problem in flue gas desulfurization systems. Open spray chambers are often helpful as in HF absorption[5,16,17] but difficulties have even been encountered here with flue gas desulfurization. Successful contacting has been carried out in venturi scrubbers[46-49] and in turbulent contactors[50] in which deposits on the hollow, fluidized packing are constantly ground off.

Another approach to the plugging problem is to operate the absorber with an alkali having highly soluble reaction products such as NaOH. The liquid effluent from the sodium system absorber is limed and clarified externally in reaction and settling tanks, which regenerates the NaOH for recycle to the absorber. The chemical reactions for SO_2 absorption are shown.

In the absorber:

$$SO_2 + 2NaOH \rightarrow Na_2SO_3 + H_2O$$

In the neutralizer–clarifier:

$$Na_2SO_3 + Ca(OH)_2 \rightarrow CaSO_3 \downarrow + 2NaOH$$

Sometimes the liquid will be aerated to oxidize the sulfite to sulfate.

Where there is a suitable and valuable use for the product produced, a completely soluble absorption process may be employed throughout. Absorption of SO_2 with caustic soda or sodium carbonate solution could be very attractive where the sodium bisulfite or sulfite produced could be used in a nearby sulfite pulp and paper process.

The Cominco process[51] uses an alkaline ammonium salt solution to absorb SO_2 and produces ammonium bisulfite, sulfite, bisulfate, or sulfate where there is suitable market for such salts. In this process, the gases to be scrubbed are contacted in a first-stage scrubber with a more alkaline solution to effect complete removal of the SO_2, and finally in a second-stage scrubber which is neutral or slightly acidic to prevent loss of NH_3 to the atmosphere.

Other regenerable absorption systems have been developed and tested for SO_2 such as a Chemico process utilizing MgO[52] and others scrubbing with citric acid, phosphate, or carbonate solutions. Murthy et al.[38] have reviewed problems with these processes. Absorption systems for removal of fluorides have been discussed in Refs. 5, 6, 16, 17, and 53.

Care must be exercised in design of the initial contact stages of an absorber that utilizes water containing a volatile neutralizing agent which can react readily with the gaseous pollutant in the vapor phase to produce a solid particulate reaction product. Examples are absorption of NH_3 into water acidified with HCl or HNO_3 or HCl being absorbed into water containing NH_3. Vapor phase reactions can occur at the initial contact point, which produces a submicrometer smoke, difficult to wet, that can carry through the absorber without appreciable collection. Such problems can be prevented by control of the neutralizing agent concentrations (vapor pressure) at the initial contact points. Utilization of water alone, or water with exceedingly dilute neutralizing agent is helpful, with further addition of neutralizing agent to the absorbing liquid as the partial pressure of the pollutant in the gas stream is reduced. Problems of this type are aggravated when the gas to be scrubbed is quite hot, resulting in initial vaporization of neutralizing agent. Precooling of the gas is beneficial.

While pH adjustment of water is the most common method employed in aqueous absorption for pollutant gases, other chemical reagents can be and are employed to aid the process where indicated by the specific chemical situation. Trace organic compounds, including odors, may be removed or modified with the use of oxidizing agents in a water solution. Typical scrubbing solutions deserving consideration are $KMnO_4$, $NaClO_4$, $NaClO$, HNO_3, and H_2O_2. Where reducing agents are more effective, Na_2SO_3 may be considered. Concentrated inorganic liquids are very effective in specific applications such as removal of olefins with H_2SO_4.

Absorption of NO_2 is especially troublesome in that 3 mole, when absorbed into water, produce nitric acid and free 1 mole of NO. The NO must be reoxidized to NO_2 (a slow reaction) for reabsorption. The entire process slows down as nitrogen oxide concentration becomes more dilute. While this operation is practiced extensively in both packed and plate columns, removal is never complete and traces of brown plumes are frequently visible in effluent gases. Final scrubbing with caustic solutions are helpful but disposal of reaction products can produce extreme difficulties. Sulfuric acid will absorb NO_x readily, but its utilization or disposal again becomes a problem. Maynard and Heinze[54] have proposed NO_2 absorption in stripped nitric acid utilizing a catalytic packing that reoxidizes the NO.

Water-insoluble organic vapors are usually absorbed in organic fluids of low volatility. Absorption of benzene vapors and other cancer-suspect organics has been discussed by Crocker.[2] Processes are in use for removing H_2S and acidic compounds by absorption in organic amines in purifying hydrocarbon gases.

2.5   Handling, Utilization, and Disposal of Recovered Materials

Removal of gaseous pollutants by absorption poses the question of disposal of the recovered contaminants, now in liquid form and usually quite dilute. Concentration and recovery or conversion of the pollutants to a useful form is preferred since further disposal problems are avoided. If the pollutant is purely absorbed (not reacted to a different chemical species), it can practically always be obtained in concentrated form by stripping of the absorbing solution. It may be possible to reutilize the recovered material in the process or some other process. Sometimes conversion to another form can convert it to a useful raw material either for the producer or for other manufacturers or consumers in the area. Neutralized pollutants may be precipitated or concentrated in solution or dried to produce a product with some marketable use, even if its value is only enough to defray transportation costs. Recovered sodium sulfate and sulfite compounds may be usable by a pulp or paper mill either in liquid or solid form. Reaction of sulfate and sulfite compounds with ammonia can produce disposable fertilizer materials.

The pollutant may be precipitated from the absorbing solution as an insoluble sludge through the addition of lime or other reagents. The sludge may be thickened by settling and dewatered further by further filtration or centrifugation. However, it is not unusual for the dewatered sludge to contain no more than 30–50% solids which can create real disposal problems. Disposal to streams is usually not feasible or permitted. Impounding in land fill or tailing ponds may be possible where land is available, but the harm done to the land for future use is frequently of concern. The sludge may not support adequate bearing loads for future utilization of the land for construction.

Recovery of chlorides, nitrates, and nitrites present a greater problem since no inexpensive precipitates can be formed. Deep-well injection has been practiced in some locations but may not be permitted in the future. Barging to sea at coastal locations has been practiced for such solutions as well as sludges in the past, but is being looked at with greater disfavor. For chlorides, purification and utilization as a raw material for chlorine manufacture should be considered. Nitrites could be oxidized to nitrates, concentrated, and utilized as a fertilizer ingredient.

With organic pollutants, stripping and recovery in concentrated form may be the best solution, even if the recovered material is valuable only as a fuel. Where the organic is water soluble and biodegradable, absorption in water on a once-through basis followed by bio-oxidation may be more economical than stripping when the organic concentration is very dilute. Breakeven points between bio-oxidation and stripping are generally at pollutant concentration levels of 100–300 ppmv in the vapor being treated.

Other separation techniques are possible to regenerate the absorbing liquid instead of stripping. Among them are chemical oxidation utilizing ozone and chemical oxidants (H_2O_2, HClO, $KClO_4$, $KMnO_4$, HNO_3, etc.), or hydrolysis (possibly at high temperature and pressure), or reaction with other reagents to produce: (a) nontoxic waste materials, (b) nonmiscible materials separable by settling, or (c) insoluble precipitates. Other alternatives to distillation stripping are adsorption of organics on solids such as clays, filter aid, sawdust, carbon char, and resin materials; or separation of the pollutant by means such as solvent extraction, contact with solid or liquid-ion exchange materials, and cooling with crystallization of the pollutant or absorbing liquid.

While water can sometimes be employed on a single-pass basis as an absorbent, very special situations are generally needed for single-pass utilization of an organic absorbing liquid. However, such possibilities should not be overlooked, such as using an organic raw material or fuel oil as an absorbent prior to its intended use in processing or combustion.

2.6   Absorption on Solids

Pollutant gases have been removed with beds of solid absorbents such as iron oxide (Fe_2O_3) granules for the removal of H_2S. The iron oxide becomes sulfided and later must be regenerated by treatment with air. Zinc oxide has also been used for sulfur removal. Similarly, organic sulfur compounds have been hydrogenated with solid catalysts of cobalt, nickel, and molybdenum sulfides to effect freeing the sulfur from the organic molecule. Most of these processes are old and date from manufactured gas technology, but have been described by Nonhebel.[55] Static beds were generally used although Nonhebel has described one iron oxide fluid bed. Reaction efficiencies were generally low because solids with high surface areas had not been developed. Because of the age and inadequacies of the technology, such methods have generally been in poor repute among engineers. However, with present national needs to develop processes for coal gasification and liquefaction, there may be a resurgence of development in this area. The development of high surface area, solid absorbents suitable for use in fluid and traveling beds would appear logical.

More recently, lime and limestone in solid form have been tried in flue gas desulfurization, but the completeness of the reaction (extent of SO_2 removal) seldom exceeded 50–60%. The much higher SO_2 removal efficiency achieved with liquid slurries has discouraged further work with dry solids in this area.

3   ADSORPTION

The attractive forces between atoms, molecules, and ions which hold a solid together are unsatisfied at the surface of a solid and thus available for holding other materials such as gases and liquids. This phenomenon is called adsorption. If the solid is produced in a highly porous form with extensive pores and microstructure, its adsorptive capacity is greatly enhanced.

In being adsorbed, a gas molecule travels to the surface of the solid where it loses much of its molecular motion, releasing heat that is often close in magnitude to its heat of condensation. Thus, adsorption is always exothermic. Desorption is a reversal of the adsorption process and heat must be supplied to cause it to proceed. A heat sink for the energy released aids the adsorption process and increases the adsorptive capacity of the solid. Thus, cooling for the bed or precooling of the gases is desirable.

Some adsorption processes occur so strongly that they are irreversible, that is, the adsorbed material can only be desorbed by removal of some of the solid substrate. Such a process is referred to as *chemisorption*. For example, oxygen can be adsorbed so strongly on activated carbon that it can only be removed in the form of CO or CO_2. The adsorbing solid is called the adsorbent or sorbent, the adsorbed material, the adsorbate or sorbate.

3.1   Types of Adsorbents

Commercially important sorbents are activated carbon, other simple or complex oxides, and impregnated sorbents. *Activated carbon* is composed largely of neutral atoms of a single species with no electrical gradients between molecules. Thus, there are no significant potential gradients to attract and orient polar molecules in preference to nonpolar molecules. For this reason, carbon is effective in adsorbing organic compounds from a humid gas stream. Carbon has less selectivity than other sorbents and is one of the few that will work in a humid gas. Since the polar water molecules attract each other as strongly as the neutral carbon, the latter tends to be slightly selective for organic molecules. However, some water is adsorbed, especially if its partial pressure is greater than that of the organic molecules and its effect on capacity of the carbon must be taken into account in design. Typical sources of activated carbon are coconut and other nut shells, fruit pits, bituminous coal, hardwoods, and petroleum coke and residues.

Other adsorbents are simple or complex oxides that contain an inhomogeneous, molecular-scale charge distribution and are polar. These adsorbents have much greater selectivity than carbon with a great preference for polar molecules. Thus, they can be very useful for removal of a particular species from a gas stream but suffer from their inability to be effective when moisture is present. Most of these adsorbents are excellent desiccants.

The siliceous adsorbents comprise one class of oxide adsorbent such as silica gels, fuller's and diatomaceous earth, synthetic zeolites, or molecular sieves. They are available in a wide range of capacities with the best equaling that of the best activated carbons. Synthetic zeolites can be prepared with specific and uniform pore sizes, giving adsorptive specificity based on size and shape of sorbate molecules. However, even this property does not overcome the chemical preference for polar molecules so that they will not absorb organic molecules of a size matching their pore structure in the presence of water vapor. Metallic oxide adsorbents such as activated alumina are even more polar than the silicas and are seldom used directly for pollution control adsorption.

Impregnated sorbents fall into three general classes: (1) those where the impregnant is a chemical reagent or reactant, (2) those where the impregnant acts as a continuous catalyst for pollutant oxidation or decomposition, and (3) those where the impregnant acts only intermittently as a catalyst. Reagent impregnants chemically convert the pollutant to a harmless or adsorbable product. Carbon may be impregnated with 10–20% of its weight of bromine, which reacts with olefins. This is especially useful for removing ethylene as a pollutant. Ethylene is poorly absorbed from an airstream because of its low molecular weight. However, in contact with the brominated surface, it is converted to 1,2-dibromo-ethane which is readily adsorbed. Once all the free bromine is consumed, the system loses its effectiveness and must be discarded or restored. Other reagent impregnations include iodine to collect mercury vapor, lead acetate for hydrogen sulfide, and sodium silicate for hydrogen fluoride.

With a catalytic impregnant acting continuously, the only reactions that are commonly catalyzed are those that can occur between the pollutant and its carrier gas. If the absorbent is carbon, limitations must be imposed that would prevent the oxidation of the carbon itself. Highly oxidative catalysts can make the carbon pyrophoric. Other more polar adsorbent substrates are subject to limitations caused by their polar nature. Continuous catalyst impregnations have been made on activated carbon using chromium, copper, silver, palladium, and platinum. The impregnation is affected by in situ decomposition of impregnated complex salts. Such preparations have been used for ambient temperature destruction of readily oxidizable pollutants and specific war gases (chloropicrin and lewisite). The development of oxygen-carrying catalysts (oxcar catalysts) raises the potential for adsorbent impregnation with a system that would oxidize pollutants in an oxygen-free carrier gas. The oxidation-catalyst adsorbent would have to be regenerated at intervals by exposure to air.

With intermittently acting catalyst impregnants, pollutants are first collected by ordinary physical adsorption. In a second operation, the temperature is raised to activate the catalyst and oxidation of the sorbate occurs. If the adsorbent is carbon, care must be taken not to render it pyrophoric at the catalytic operating conditions. Turk[56] has described impregnant oxides of chromium, molybdenum, and tungsten used in this manner.

Adsorbents used for gas treatment in fixed beds are generally granular and sized to produce little resistance to gas flow. Figure 16 shows pressure drops as a function of superficial velocity for typical grades of granular carbon. In moving and fluid bed adsorption, the hardness of the sorbent particles is also important to prevent attrition and minimize sorbent replacement costs. Typical physical properties of adsorbents are given in Table 4.

3.2 Adsorption System Principles

Table 5 lists factors that affect the capacity of an adsorbent bed. The rate of adsorption is affected by the partial pressure of the sorbate, its diffusion into the pores of the sorbent, and the degree of saturation of the bed. General principles that must be considered in the design of the system are: (1) ample contact time (low superficial velocities and adequate bed depth) between the gas stream and the bed to provide the required removal efficiency; (2) adequate adsorbent capacity to provide for a reasonable adsorption on-stream time or service life; (3) reasonably low resistance to gas flow to conserve energy; (4) uniform air flow distribution to ensure complete utilization of the sorbent; (5) possible pretreatment to remove particulates, interfering or competing vapors, or materials that could not be desorbed, or to precool the gases; and (6) provision for renewing or replacing the spent saturated sorbent.

Removal of the pollutant from the carrier gas in a sorption bed can be illustrated in the form of an "adsorption wave" shown in Figure 17. Curve (1) shows the pollutant concentration as a function of bed position with a freshly regenerated bed. The dotted horizontal line C_b is the maximum permissible discharge concentration of the pollutant in the treated gas. With a properly designed bed, the outlet concentration is well below this limit. As time passes, the sorbent at the inlet of the bed becomes saturated and the pollutants penetrate more deeply into the bed before appreciable adsorption occurs (curve 2). Finally, the permissible outlet concentration is reached (curve 3) and the bed must be regenerated.

As indicated, adsorption is exothermic and adsorption capacity is lost with increase in temperature. Therefore, means to cool the bed is desirable. Cooling pipes imbedded in stationary beds have sometimes been used, but heat transfer is generally poor except over short distances. Precooling of the gas is often employed. Removal of competing vapors, such as water vapor, which would reduce bed capacity may also be practiced.

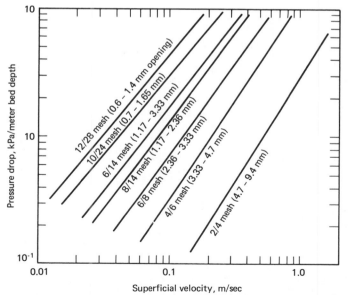

Fig. 16 Pressure drop through dry packed granular carbon beds. Air flowing downward, 100 kPa pressure, 294 K.[57]

Table 4 Physical Properties of Adsorbents[58]

Material and Typical Use	Shape of Particles[h]	Size Range, U.S. Standard Mesh	Internal Porosity χ, %	Bulk Dry Density, lb/ft³	Average Pore Diameter, Å	Surface Area, m²/g	Adsorptive Capacity, g/g Dry Solid	Trade Designation
Aluminas								
Active alumina (transition alumina) Uses: drying gases and liquids; catalyst; catalyst support; defluoridation of alkylates; neutralization[c] of lube oils	G	Various	25-30	50	35-45	235	0.15[a]	Alcoa F-1; Reynolds R-2101 = RA-1, R2101 = RA-3
	S	3-8, etc.	50-60	47-50	40-50	400	0.22[a]	Alcoa H-151; Kaiser KA-201; Pechiney Alumina A
	T	1-⅛ in.[c]	30, 47	~50	136, 99	90, 190	—	Harshaw Al-0104T, -1404T
$CoCl_2$-impregnated	G	8-14	30	54	45	200	0.14	Alcoa F-6 (indicator grade)
Desiccant (single-use), $CaCl_2$-impregnated	G	3-8, etc.	30	57	45	200	0.22[a]	Alcoa F-5
Catalytic alumina, low soda	S	Various	62	47	45	300	—	Pechiney CR
Dry column grade	G	—	55	55	—	—	—	Woelm Dry Column Al_2O_3
Activated bauxite	G	8-20, etc.	35	~53	~50	—	0.04-0.2[d]	Florite
Chromatographic alumina[j]	G	80-200	30	58	45	225	0.14	Alcoa F-20
	S, P	30-140	—	~50	—	—	—	Pechiney CBT, CBL
		70-270	—	54	—	—	—	Woelm Al_2O_3-W200, Al_2O_3-TLC
Siliceous adsorbents								
Aluminosilicates Use: selective adsorption based on molecular size and shape	C, S, P	4-8, 8-12; 1/16, 1/8	40-55	40-55	—	770	—	Molecular sieves: Davison 3A, 4A, 5A, 13X, 700 (acid-resistant); Linde 3A, 4A, 5A, 10X, 13X, AW-300, AW-500
	C, S	1/16, 1/8 in.	30	44	3, 4, 5	600-700	0.22[f]	Siliporite NK30, NK10, NK20
	P	<400	—	31	3, 4, 5	700	—	Siliporite NK10AP, NK20AP
Acid-treated clay	P	—	—	30-45	—	225-300	—	Clarsil
	S	4-8	—	53	—	—	—	Filtrol 120
Magnesia-silica gel	G	Various	33	~30	—	300	—	Florisil
Fuller's earth Uses: same as for clay	G	Various	~54	~40	—	130-250	—	Cecacite; Clarsil PSC-G, Clarsol ATC
	P	<200	—	50	—	—	—	Clarsil PCS; Florex
Diatomaceous earth	P	—	—	9	—	—	—	Celite; Dicalite (powder and granules); Sorbo-cel (for emulsified oil)
Silica gel Uses: drying of gases, separation of hydrocarbons, catalyst base	G	10-140[c]	—	25-27	—	4	—	Chromosorb P (straight calcined)
	G	30-140	—	11-12	—	1.0	—	Chromosorb W (flux calcined)
	G	10-140	—	30-22	—	0.5	—	Chromosorb G (flux calcined)
	G	10-80	—	25-27	—	2.7	—	Chromosorb A (flux calcined)
	P	<200	—	6-25	—	—	—	Clarcel
	G	Various	Various	~27-45	Various	300-800	Various	Davison Silica Gel
	G	Various	35-50	40-48	20-40	650-900	0.4-0.6[f]	Cecagel, Sorbsil
	S, P	1/8 in., etc.	34-51	41-52	21-28	650-700	0.4-0.6[f]	Cecagel; Mobil Sorbead R, H
	G	70-140	—	~45	—	—	—	Woelm (adsorption)
	P	<270	—	—	—	—	—	Woelm TLC, F-TLC, G-TLC, GF-TLC
	S	4-8	45	46-51	72	250	0.27[f]	Mobil Sorbead W
Other inorganic materials								
Anhydrous $CaSO_4$	G	Various	38	60	—	—	0.12[d]	Drierite
Calcium silicate (fatty-acid removal)	P	—	—	12	—	100-110	—	Micro-Cel T-49, T-13
Magnesium silicate (decolorizing)	P, G	—	—	13	—	180	—	Celkate T-21; Woelm TLC

Carbons

Uses (for all carbons: water treatment, gas purification, solvent recovery and purification, decolorizing)

Source	Form	Size						Trade names
Shell-based	G	Various	50–60	27–34	20	800–1100	0.45[g]	Cochranex FCB; Pittsburgh PCB
	G	Various	50–80	25–35	20–30	1200–1600	0.5–0.95[b]	Picactif T.A., T.E.
	G, (P)	Various	60–65	27–36	18–19	1000–1500	~0.4	Acticarbone NC, WNC
	G	Various	~50	27–32	20	800–1100	0.45[g]	Barnebey-Cheney AC, KE, VG, PC, PL
	G	4–6, 8–30	~50	33–35	~30	800–900	—	Girdler 32E (Fe-impregnated), 32W (Cu-Cr-impregnated)
	P	¼, ⅛ in.	60–80	20–22	22–24	1200	—	Barnebey-Cheney YF, JF, JU (high capacity)
	C, P	Various	70–75	12–28	5–10	750–1450	—	Acticarbone AC, etc.; Anticromos
Wood-based	P	Various	30–50	9–35	8–30	600–1200	—	Darco KB, G60; Nuchar Aqua, Nuchar WA, B-100, C-115, CEE, C-190, C, C-1000
	—	5–7	—	24	3–10	1400	—	Supersorbon W
	P	Various	—	21–27	20–40	1000–1500	—	Carboraffin (various grades)
	P	Various	—	27–29	20–100	750–900	—	Brilonit (various grades)
	G	10 × 30, etc.	60	~20	—	600–1000	—	Cochranex FCN-1, FCN-2, FCA, FCC
	G	100–300	40–50	15–20	30–40	800–1200	40–70%	Picactif C. O.
	P	Various	40–60	27–33	5–20	600–1200	80–130%(I_2)	Picactif CM
Peat-based	C, G, P	5–7	~55	15–32	—	500–1600	—	Norit (various grades)
	G	<200	—	20–24	—	1300–1400	—	Supersorbon (various grades)
Coal-based	P	Various	65–75	28–31	20–38	700–900	—	Acticarbone AM, AH
	G	Various	60	20–30	60–65	500–1200	0.4[e]	Darco Granular; Permutit Carbo-Dur
	G	12 × 40, etc.	56–67	30	20	800–1000	—	Cochranex FCP-1, FCP-2, FCW-V
	G	Various	80	25–30	22	1000–1400	50[g]	Pittsburgh BPL, CAL, SGL, etc. (RB, RC, BL)
	G, (P)	—	—	28	5–15	110	30–60%	Barnebey-Cheney MN
	G	Various	—	27–37	—	850–1350	—	Acticarbone LM; Nuchar WV-W, WV-L, WV-G, WV-H
	G	5–7	—	20–24	18–22	1300–1500	—	Contarbon (various grades)
	G	Various	—	25–30	—	600–700	—	Darco BG, DC, S-51; Hydrodarco B
Petroleum-based	C, P	Various	65–85	~30	—	800–1100	—	Columbia (various grades)

Organic materials

Source	Form	Size						Trade names
Porous resin (decolorizing)	S, G	16–50	—	20–45	—	3	—	Asmit 224, Duolite S-35; Permutit S-360; Wofatit E
	S, (P)	16–50	—	30–40	—	—	0.1, (0.2)	Ionex RV (Micro-Ionex RV)
Cross-linked polystyrene	G	50–140[c]	40–45	19–22	500	15–35	—	Chromosorb 101, 103
	S	20–50	50–55	40–44	90	330	—	Amberlite XAD-2
	S	20–50	50–55	39	50	750	—	Amberlite XAD-4
Cross-linked polystyrene	G	60–200[c]	—	18–20	85	300–400	—	Chromosorb 102
Phenolic	S	16–50	50–55	22–25	80	—	—	Duolite S-30, ES-33, S-37 (general adsorbent)
Acrylic ester	S	20–50	50–54	41	250	450	—	Amberlite XAD-7
	S	20–60	—	43	—	140	—	Amberlite XAD-8
Aromatic-amine resin	G	10–50	—	40–50	—	—	—	Asmit 173N
Quaternary amine chloride resin	S	16–50	~65	40–50	—	—	—	Asmit 259N, 261; Duolite ES-111, A-140
Copper-amine resin (O_2 removal)	S	10–50	—	30	—	—	0.12[i]	Duolite S-10
Cellulose	G, (P)	100–200	—	—	—	3×10^4	—	Whatman CC31 (CF1, 2, 11, 12)

[a] Water at 60% humidity.
[b] Carbon tetrachloride; test conditions not specified.
[c] Various sizes available within stated range.
[d] Water; test conditions not specified.
[e] Benzene, at 20°C and 7.5 mm partial pressure.
[f] Water, at 100% relative humidity.
[g] Accelerated chloropicrin test.
[h] C, cylindrical pellets; G, granular; P, powder; S, spherical beads; T, tablets.
[i] Oxygen.
[j] Separate grades specified for use at pH 4, 7.5, or 10.

Table 5   Factors Affecting the Capacity of an Adsorbent[2]

1. Concentration of the adsorbate in the bed passages.
2. Total surface area of the adsorbent.
3. Relative concentration of competing adsorbable molecules.
4. Temperature of the gas and bed (the lower the better).
5. Adsorbate molecule characteristics (molecular weight, electrical polarity, chemical activity, size and shape).
6. Sizes and shapes of adsorbent pore microstructure.
7. Chemcial activity of adsorbent (polarity and chemical activity).

3.2.1   Desorption or Disposal

After pollutants have been adsorbed, disposal may consist of discarding the saturated sorbent. This is attractive when the quantity of adsorbate is small or occurs infrequently. The cost of fresh adsorbent is insignificant compared to the cost or inconvenience of regeneration. The adsorbent may be contained in a paper carrier or disposable cartridge. Disposal might be to a landfill but if the pollutant is toxic or carcinogenic and leachable, and if the sorbent is carbon, then disposal by incineration is preferable.

Generally, economics will dictate regeneration of the sorbent and recovery of the pollutant. Desorption may occur by (1) heating the bed, (2) evacuating the bed, (3) stripping with an inert gas, (4) displacing the sorbate with a more adsorbable material, and (5) a combination of two or more of the above methods.

Heating is the method most commonly practiced and it allows recovery of the sorbate in undiluted form and perhaps in a form ready for process reuse. Raising the bed temperature to the atmospheric boiling point of the sorbate results in its boiling off as a vapor that can be condensed. Where temperatures required for atmospheric boiling could result in decomposition, a combination of temperature and partial vacuum may be indicated. Bed evacuation alone is apt to result in regeneration which is less thorough than if heat is applied. In addition, a refrigerated condenser or a combination of compression and condensation may be needed to recondense the stripped adsorbate with vacuum alone.

Stripping with a gas, inert from the standpoint of the adsorption process (such as air or nitrogen), has the disadvantage that the pollutant is again dispersed in a gas stream. However, there are situations where such a method is desirable. Absorption can be used as a concentrating technique for a very dilute pollutant which is to be incinerated or the pollutant may be removed from a process gas stream which is now to be stripped from the absorbent with air for combustion.

Steaming a carbon bed is often cited as a displacement method of regeneration. Water vapor is adsorbed, displacing the adsorbate which is present at a lower equilibrium partial pressure. The desorbed

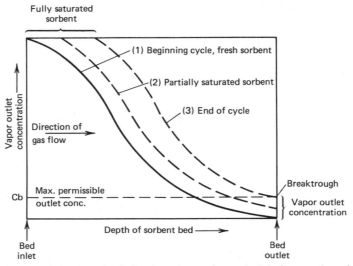

Fig. 17   Passage of an adsorption wave through a stationary bed during an adsorption cycle.

pollutants are condensed along with excess steam. Following bed saturation with steam, a second regeneration step is required to remove the water by drying with hot air. In reality, steaming is a combination method. Displacement may be the major driving force, but the steam raises the bed temperature and the excess steam serves as a stripping gas as well.

During regeneration, no sorbate will desorb completely. Therefore, excess adsorptive capacity must be allowed to provide for capacity loss in reuse following the initial cycle. Other regeneration problems can occur which should be investigated early in the process of control method selection. Some organics, especially monomers like styrene vapor, may polymerize in the pores of the adsorbent during attempted regeneration and render it inactive for further use. At higher regeneration temperatures, some organics can crack, coating the sorbent with tars and soot, rendering it inactive. Aside from using other treatment methods, possible solutions to such desorption problems may lie in utilizing moderate temperatures, reduced pressures, and steaming, or by removal of the bed material for regeneration in an oxidation furnace. Here, under controlled conditions, the organics can be burned from the surface and pores of the sorbent. Often the sorbent manufacturer will provide such reworking.

Chemisorption provides such strong bonding that desorption can occur only with loss of some of the adsorbent structure. Special regeneration conditions are often required. The Reinluft process[60] for adsorption of SO_2 on activated carbon is an example. A portion of the SO_2 becomes permanently attached to carbon as sulfuric acid. This is desorbed by heating to 370°C in an inert atmosphere where the acid and carbon react, producing CO_2 and SO_2.

3.3 Types of Adsorption Equipment

Five distinct categories of gaseous adsorption devices are available: (1) disposable and rechargeable canisters, (2) fixed regenerable beds, (3) traveling bed adsorbers, (4) fluid bed adsorbers, and (5) chromatographic baghouses.

3.3.1 Disposable and Rechargeable Canisters

For very small vent flows, flows of an intermittent or infrequent nature, and effluents with very low sorbate concentration, utilization of disposable canisters of carbon or other sorbent may be an economical and desirable control approach. Disposable paper and fiber cartridges containing the sorbent as a fine powder dispersed in and on an inert carrier of paper, organic or inorganic textiles, and plastic filaments have been developed. Paper can contain 50–75% carbon by weight and cellulose filters with up to 80% activated carbon are available. However, such elements have seldom received significant industrial usage. A more frequently used element is a replaceable and disposable canister filled with granular sorbent held in a permanent container in the vent line and operated until the sorbent approaches saturation. At such times, the canister is manually removed much like a cartridge filter and disposed of as solid waste (landfill, solid waste incinerator, etc.). Consideration should be given to the disposal method and the possible release of the sorbate as a further environmental pollutant.

For emission control in remote locations such as breathing losses from a storage tank containing volatile solvents or hydrocarbons, 55-gal steel drum containers of granular carbon are available from several sources. These drums are specially fitted with inlet and outlet nozzles. Such a drum installed on the vent of a benzene storage tank continuously venting 0.1 ft³/min of vapor containing 13.5% benzene by volume would have a useful life of 2 weeks. When saturated, the spent carbon is replaced with fresh carbon. The spent carbon can be handled as combustible solid waste, but often arrangements can be made to return it to the carbon manufacturer for regeneration.

3.3.2 Fixed Regenerable Beds

Regenerable beds of sorbent are utilized when the volume of gas to be treated or the quantity (concentration) of sorbate is high enough to make recovery attractive, or cost of fresh sorbent would be expensive compared to regeneration. Two types of stationary fixed beds are generally considered in design: thin beds and thick beds. In either case, regeneration generally follows the methods described in Section 3.2.1.

3.3.2.1 Thin Beds.
In thin beds, the adsorbent (only a few inches thick) is usually retained between vertical screens or preferably perforated sheet metal for rigidity, with the gas passing through horizontally as in an air filter. Thin beds have very low pressure drop, but extreme care must be taken to maintain uniform bed thickness to prevent gas channeling. Such bed elements have been designed using flat, cylindrical, or pleated shapes. Figure 18 shows examples. Commercial cylindrical elements are available for about 0.012 m³/sec (25 ft³/min) while larger, pleated cells can handle 0.35–0.50 m³/sec (750–1000 ft³/min). Thin beds have low adsorptive capacity and are best suited to large gas volumes of low pollutant concentration. Higher concentrations can be handled under automatic control with the use of very short time cycles. The amount of adsorbate that can be held in a unit area of a thin bed is limited so removal of the heat of adsorption may be of little importance. Frequently, precooling of the inlet gas is adequate for this aspect.

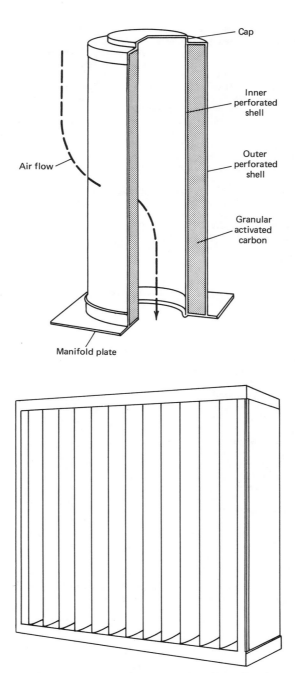

Fig. 18 Typical thin bed adsorber elements.[64] (*a*) Cylindrical thin bed canister adsorber. (Courtesy of Connor, Inc., Danbury, Conn.) (*b*) Pleated cell thin bed adsorber. (Courtesy of Barneby-Cheney Co., Columbus, Ohio.)

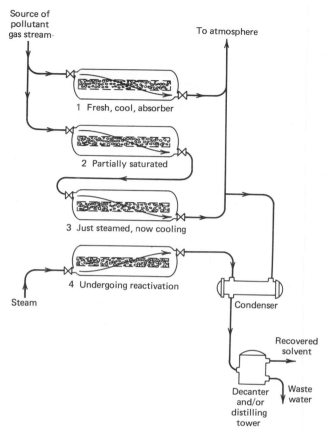

Fig. 19 Four-unit deep bed adsorption train.[64]

3.3.2.2 Thick Beds.
Thick adsorption beds are generally attractive when pollutant concentrations exceed 100 ppm or the flow quantity exceeds 4.7 m³/sec (10,000 ft³/min). They are typically 0.3–1 m (1–3 ft) thick with granular material supported horizontally in long horizontal tanks. Gas flow is generally downward through the bed. The deeper bed permits longer operation before breakthrough and the greater pressure drop reduces the difficulty with channeling, but bed uniformity is still important. Problems with adsorption heat release within the bed become greater than in thin beds and horizontal cooling coils have sometimes been buried in the bed. However, cooling effectiveness is seldom very good in a static bed.

Unless the flow of gas to be treated is cyclical, a minimum of two identical deep beds is required: one adsorbing the pollutant and the other undergoing cyclical regeneration and cooling. More elaborate arrangements are used and Figure 19 illustrates a four-unit adsorption train in which the first, freshly regenerated unit is adsorbing pollutant. A second, partially saturated unit is performing similarly, but its effluent is passing through a regenerated bed to provide cooling. The fourth unit is actively undergoing regeneration by steaming. A condenser and separator handle the effluent from this bed. Obviously, suitable piping and valving (not shown) is required for proper sequencing of each bed.

3.3.3 Traveling Bed Adsorbers

Continuous countercurrent column adsorbers have been designed and used with limited operation. Both downward traveling packed column units and tray units have been utilized. In both, freshly regenerated adsorbent is elevated and added at the top of the column at a rate to maintain a constant height of solids. A mechanism is provided at the bottom for steady removal of saturated sorbent which is regenerated in another vessel before return to the top. Gas to be treated enters at the bottom and passes up the column being purified and leaves at the top. In a traveling packed bed unit, all the bed rests on a bottom support grill or grate and the gas passes upward through the voids in the granular bed of solids. With the tray design, solids are fluidized on each tray by the gas rising through sieve-plate holes in the tray. The solids slowly travel across the tray to a downcomer which conducts the solids to the tray below so that the design much resembles that of a plate-type absorption column. Another embodiment of traveling bed adsorption would be utilization of equipment designed as traveling bed granular filters (see Chapter 11) with downflow of solids and cross-flow of gas.

Such devices are most suitable for high concentrations of adsorbable materials requiring high sorbent to gas ratio and are seldom needed for pollution control applications. However, pilot processes for adsorbing SO_2 from flue gases have used similar approaches such as the Reinluft Process[60] which involved both adsorption and catalytic oxidation of SO_2 on a traveling bed of activated carbon, and the U.S. Bureau of Mines alkalized alumina process[60,61] which used sorbent granules of 56% Al_2O_3 and 37% Na_2O. Attrition and loss of the sorbent can be a sizable operating problem and expense. Generally very hard and abrasive-resistant granules are required. Westvaco Pulp and Paper has had considerable experience in design of such equipment and manufacture of suitable granular carbons.

3.3.4 *Fluid Bed Adsorbers*

Unless bed staging is practiced, a single-stage fluid bed adsorber would not appear to be a desirable adsorption contactor at first glance. All sorbent particles in such a bed are well back-mixed and a typical "adsorption wave" does not occur in the bed. Since all particles are in equilibrium with the outlet gas, low outlet pollutant concentrations can be attained only if all the bed particles are kept relatively unsaturated. This makes the adsorptive capacity of the bed low and the concept would be unattractive were it not for the ease with which adsorbent particles can be removed, externally regenerated, and returned to the bed on a continuous basis. An advantage of the fluid bed is the ability to obtain high heat transfer rates with cooling tubes submerged in the bed to remove the heat of adsorption. It also has merit in a situation where frequent sorbent regeneration is needed. It might have application for adsorption of organics from a very moist gas stream where frequent carbon regeneration is needed to remove adsorbed water.

Kunii and Levenspiel[62] have described a two-fluid bed transport reactor using silica gel adsorbent for drying a gas stream. Moisture adsorption occurs in one bed and regeneration in the other with continuous transport of the solids between the two beds. Such a principle can be applied to the adsorption of other pollutants, including the possible use of an oxcar catalyst to adsorb and destroy toxic organics as mentioned in Section 3.1. A novel adaptation of transported fluid bed adsorption–desorption all in one vessel is the Linde Purasiv HR System[63] shown in Figure 20. A granular carbon is continuously recirculated within a single partitioned vessel having two separate fluid beds, one for adsorption and one for regeneration.

3.3.5 *Chromatographic Baghouses*

In this approach, granular adsorbent is introduced continuously at a controlled rate into the gas stream to be treated and the gas stream conveys the suspended adsorbent through a line of adequate length and residence time to provide appreciable contact and adsorption before entering a conventional baghouse. The sorbent is filtered from the gas stream on the surface of the bags and further adsorption may occur as additional gas passes through the collected cake of sorbent. Periodically, the sorbent is removed from the bags in the conventional manner. Since the flow of sorbent and gas is cocurrent, the exit concentration of pollutant is controlled by the ratio of sorbent to gas used, and to a lesser extent by the contact time provided. Such a system cannot completely saturate the sorbent and reach extremely low outlet pollutant concentrations. However, with suitable sorbents, the spent solids removed from the baghouse can be regenerated and recirculated.

3.4 Theory and Design

When a gas is brought into contact with an unsaturated solid surface, a part of the gas is taken up by the solid. Some of the gas molecules enter into the solid and are said to be absorbed as in the case of liquid absorbing molecules. The molecules that remain in the surface of the solid are said to be adsorbed. If the process occurs at constant volume, the gas pressure decreases, and the volume decreases.

The techniques used in adsorption processes include stagewise and continuous contacting applied to both continuous and semicontinuous operations. These operations are analogous to absorption when only one component of the gas is strongly absorbed. When more than one component is strongly adsorbed, the operation is more analogous to fractionation and, in particular, extraction.

In developing design equations, we will deal with continuous contactors. Assuming dilute solutions and negligible heat effects, the result will be a design for isothermal operation. Equation development will be limited to steady-state traveling bed adsorbers, and to unsteady-state fixed-bed adsorbers, both with only one component adsorbed.

In the steady-state traveling bed continuous contactor, the solid adsorbent acts like the liquid

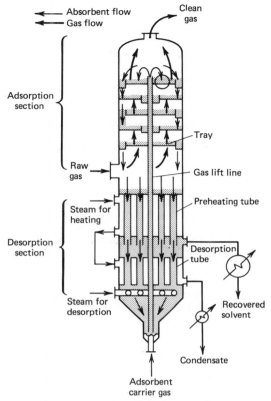

Fig. 20 Linde Purasiv HR adsorption system.[63]

solvent in absorption. The adsorbate is picked up by the descending solid which is removed from the adsorbing section of the column and is regenerated for reuse in the column. Design proceeds in the same manner as in absorption. The height of adsorbent required is determined from mass transfer calculations.

Unsteady-state fixed-bed adsorber design depends on determining the breakthrough point for a stationary bed of adsorbent. The breakthrough point occurs when the constantly changing concentration of the effluent from the absorber reaches a predetermined value, equal to an emission standard, for example. The time to reach the breakthrough point is found from mass transfer and equilibrium considerations which determine the amount of adsorbent required.

Bed diameters are determined from pressure loss considerations as in the case with absorption. Nomenclature for equations in Section 3.4 is defined in Section 3.4.5.

3.4.1 Adsorption Mechanisms

When the gas on the surface is held by a weak interaction with the solid, the process is termed physical adsorption or van der Waals' adsorption since the force of attraction is like the van der Waals forces. Some gases are bound to the surface by forces similar to those in chemical bonding. This type of adsorption is known as chemical adsorption or chemisorption and has the synonym activated adsorption. The energy in the case of physical adsorption is similar to the energy required for condensation of a gas, while in the case of chemisorption the energy is similar to the energy of activation in a chemical reaction.

Physical adsorption may be called surface condensation, and chemisorption may be called surface reaction. Since the processes are so different, the fundamental laws that deal with the mechanisms are different. At lower temperature, physical adsorption takes place between any surface and any gas, and the rate of adsorption is rapid. Chemisorption demands that the energy of activation must be supplied before the adsorbent–adsorbate complex can form. The adsorption isotherm in chemisorption always indicates unimolecular layers, while in physical adsorption the process may be multimolecular layered. An isobar of gases that can be adsorbed by the two processes shows two regions in which the adsorption decreases with temperature reflecting the two types of adsorption. The lower temperature region corresponds to the physical adsorption process.

Another factor that distinguishes the two types of adsorption is the ability to readily reverse the physical adsorption process by simple evacuation with heating. Heating will usually remove physically adsorbed molecules, leaving the chemisorbed molecules behind. Much more severe conditions are required to remove chemisorbed molecules. The laws that deal with thermodynamic equilibrium only, such as the Clausius–Claperyon equation, may be used to calculate the heat release for both types of adsorption. Similarly, models such as the Freundlich equation, which merely describes the shape of an isotherm without implying any mechanism, may be applied to both types of adsorption.

An additional mechanism that absorbed gases may undergo is capillary condensation. Most adsorbents are full of capillaries and the gases make their way into these pores adsorbing on the sides of the pore. If a liquid wets the walls of a capillary, the vapor pressure will be lower than the bulk vapor pressure. Thus, it can be assumed that adsorption in capillaries takes place at a pressure considerably lower than the vapor pressure. The capillaries with the smallest diameters fill first at the lowest pressure. As the pressure is increased, larger capillaries fill until, at saturation pressure, all pores are filled with liquid.

Most experimenters agree that capillary condensation plays a role in physical adsorption. Multimolecular adsorption and capillary condensation are necessarily preceded by unimolecular adsorption. A complete theory must be applicable to all the ranges of adsorption from capillary condensation to multimolecular adsorption. The theory due to Brunauer, Emmett, and Teller, known as the BET theory, describes this entire range of adsorption. Reference 65, although an older text, gives an excellent description of the theory.

3.4.2 Equilibrium Data of Adsorption

When a gas or vapor is admitted to a thoroughly evacuated adsorbent, its molecules are distributed between the gas phase and the adsorbed phase. The rate of adsorption is so fast in some cases that it is difficult to measure. In other instances, the rate is moderate and can be readily observed. After a time the process stops, and a state of stable equilibrium is reached. The amount of gas adsorbed per gram of adsorbent at equilibrium is a function of temperature and pressure and the nature of the adsorbent and the adsorbate. For a given amount of adsorbent and with the temperature held constant, the data are referred to as the adsorption isotherm. The amount adsorbed is usually expressed as the volume taken up per gram of adsorbent at 0°C and 760 mm Hg pressure or as the mass of gas adsorbed per unit mass of adsorbent. The latter method is preferred for engineering design techniques. At constant temperature, the adsorption of a gas or vapor increases with increasing pressure. Since the adsorption process is exothermic, the amount absorbed at equilibrium must always decrease with increasing temperature. Figure 21 exhibits these points for the adsorption of ethyl chloride on charcoal.

Another method of plotting is shown in Figure 22. In this technique the pure material being

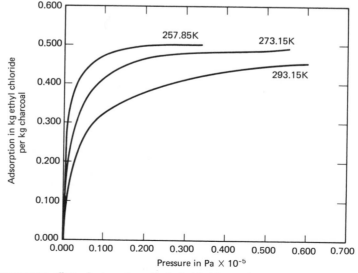

Fig. 21 Temperature effect of adsorption of ethyl chloride on charcoal. [Data of F. Goldman and M. Polyani, *Z. Phys. Chem.* **132**, 321–370 (1928).]

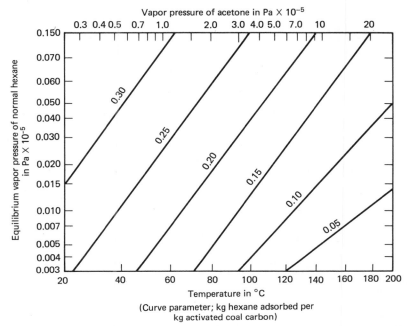

Fig. 22 Reference substance plot equilibrium adsorption of hexane on activated carbon. [Data of S. Josefowitz and D. F. Othmer, *Ind. Eng. Chem.* **40**, 739 (1948.]

adsorbed is used as a reference substance. The plot shows the adsorption isosteres for hexane adsorbed on activated carbon using acetone as the reference substance.

As a result of the shape of the openings in capillaries and pores of the solid or of other complex phenomenon, such as wetting the adsorbent, different equilibria result on desorption than on adsorption. Thus the adsorption process exhibits hysteresis with desorption pressure always lower than that obtained by adsorption. Figure 23 illustrates this effect.

In some cases it has been found that hysteresis disappears upon a thorough evacuation of the absorbate. Therefore hysteresis must be attributed to impurities that are removed by the evacuation.

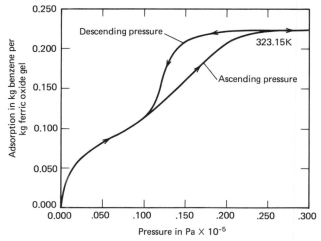

Fig. 23 Hysteresis in adsorption of benzene on ferric oxide gel. [Data of B. Lambert and A. M. Clark, *Proc. Roy. Soc.* **A122**, 497–512 (1929.]

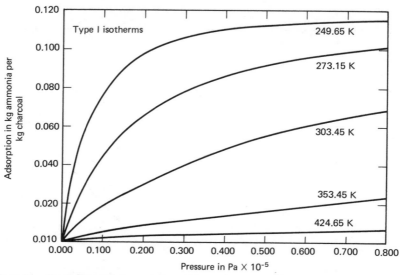

Fig. 24   Adsorption isotherms of ammonia on charcoal at several temperatures. [Data of A. Titoff, *Z. Phys. Chem.* **74,** 641–678 (1910).]

3.4.2.1 Types of Adsorption Isotherms. Five types of adsorption isotherms have been described by Brunauer[65] for physical or van der Waals adsorption. Examples are presented here with brief comments on their applicability. The Freundlich, Langmuir, and BET equations mentioned will be presented in the next section.

Type I   (see Figure 24)

1. Assumes unimolecular layer.
2. Freundlich equation and Langmuir equation good only for this type of isotherm.
3. BET equation can be applied.

Type II   (see Figure 25)

1. Typical s-shaped curve approaching a saturation pressure with an infinite amount adsorbed.
2. Indicates multilayer adsorption.
3. Only BET equation can be applied.

Type III   (see Figure 26)

1. Multilayer adsorption indicated.
2. Only BET equation can be applied.

Type IV   (see Figure 27)

1. Multilayer adsorption with capillary condensations indicated.
2. Requires BET equation.

Type V   (see Figure 28)

1. Multilayer adsorption with capillary condensation indicated.
2. Requires BET equation.

In chemisorption the situation is much simpler. Only Type I isotherms are found. Multilayers and capillary condensation do not occur. The Langmuir or Freundlich equations can be used to describe the equilibrium.

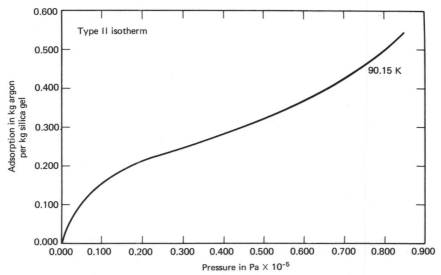

Fig. 25 Adsorption of argon on silicia gel. [Data of S. Brunauer and P. H. Emett, *J. ACS* **59**, 2682–2689 (1937).]

3.4.3   *Adsorption Isotherm Equations*

A number of theories have been devised to produce mathematical models describing the adsorption isotherm. Brunauer[65] discusses these theories and the derivation of the equations for the models. A brief summary of more commonly used equations is presented here. These equations are useful for chemisorption, monomolecular adsorption, and multilayer adsorption without capillary condensation. The BET model indicating capillary condensation is discussed in Brunauer.[65]

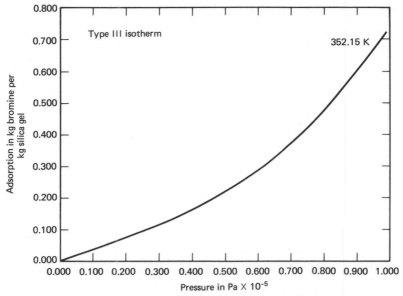

Fig. 26 Adsorption of bromine on silica gel. [Data of L. H. Reyerson and A. E. Cameron, *J. Phys. Chem.* **39**, 181–190 (1935).]

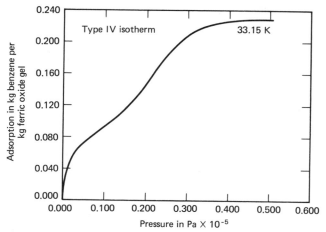

Fig. 27   Adsorption of benzene on ferric oxide gel. [Data of B. Lambert and A. M. Clark, *Proc. Roy. Soc.* **A122**, 497–512 (1929.]

3.4.3.1   The Freundlich Equation.   Valid for monomolecular physical and chemical adsorption, the Freundlich equation is an empirical relationship. When it is valid, it gives a concise analytical equation for experimental facts, rather than an accurate description of the mechanism of adsorption. It is the oldest isotherm equation and is widely employed in industrial design. In this equation, the isotherm is expressed as a power law where v is the volume adsorbed and P is the pressure:

$$v = kP^{1/n} \tag{76}$$

Taking the logarithm of both sides of the expression produces a linear equation:

$$\ln v = \ln k + \frac{1}{n} \ln P \tag{77}$$

Thus on a logarithmic plot of v_a versus P, experimental data must produce a straight line if the model is valid. The slope of the straight line is $1/n$ and the intercept is $\ln k$.

3.4.3.2   The Langmuir Equation.   Langmuir's equation was one of the first to be derived from theory. Derived for monomolecular physical adsorption, it can describe chemisorption as well. In

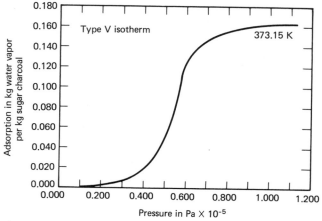

Fig. 28   Adsorption of water vapor on sugar charcoal. [Data of A. S. Coolidge, *J. ACS* **49**, 708–721 (1927).]

this equation v_m is the volume adsorbed when the surface is covered with a monomolecular layer. Then,

$$v = \frac{bPv_m}{1 + bP} \tag{78}$$

There are two rearrangements of the equation that can be used to evaluate data

$$\frac{P}{v} = \frac{1}{bv_m} + \frac{P}{v_m} \tag{79}$$

Here P/v is plotted versus P, where the slope is $1/v_m$ and the intercept is $1/bv_m$,

$$\frac{1}{v} = \frac{1}{bv_m P} + \frac{1}{v_m} \tag{80}$$

When $1/v$ is plotted versus $1/P$ in this equation, the slope is $1/bv_m$ and the intercept is $1/v_m$.

3.4.3.3 The BET Equation. The BET theory produces equations valid for all five types of isotherms previously discussed. This includes the region of monomolecular adsorption, multimolecular adsorption, and capillary condensation. It was derived by reference to physical adsorption. Since it describes the Type I isotherm, it could mathematically describe chemisorption. If adsorption takes place in a limited space, a finite number of layers builds up at saturation. In the following equation, n represents the number of layers adsorbed and c is a constant related to the heat of adsorption.

$$\frac{v}{v_m} = \left(\frac{cx}{1-x}\right)\left[\frac{1 + nx^{n+1} - (n-1)x^n}{1 + (c-1)x - cx^{n+1}}\right] \tag{81}$$

where x is a function of the pressure. For the case where $n = 1$, Eq. (81) reduces to

$$\frac{v}{v_m} = \frac{cx}{1 + cx} \tag{82}$$

Because cx is a function of adsorption pressure, this equation is essentially the Langmuir isotherm. If the adsorption takes place on a free surface, an infinite number of layers can build up at saturation. As $n \to \infty$ and the surface becomes saturated, $x \to P/P_0$ where P_0 is the saturation pressure

$$\frac{v}{v_m} = \frac{cP}{(P_0 - P)[1 + (c-1)P/P_0]} \tag{83}$$

For the purposes of plotting, Eq. (83) can be put in the form

$$\frac{P}{v(P_0 - P)} = \frac{1}{v_m c} + \frac{(c-1)}{v_m c}\left(\frac{P}{P_0}\right) \tag{84}$$

Here $P/v(P_0 - P)$ is plotted versus (P/P_0) and the slope is $(c-1)/v_m c$ and the intercept is $1/v_m c$. The technique of fitting data is to use Eq. (83) up to a point where $P/P_0 = 0.35$. At this point significant deviation from data can be expected. The values of v_m and c are first evaluated by Eq. (84). Then, using these values in Eq. (81), n is evaluated by trial and error.

3.4.4 Traveling Bed Adsorbers at Steady State

In presenting design techniques for traveling bed adsorbers at steady state, only isothermal conditions where one component is adsorbed will be considered. For a single component the operation is analogous to gas absorption with the solid adsorbent taking the place of the liquid solvent. In the following development, flow rates and concentrations will be defined on a solute-free basis, and weight ratio will be used instead of weight fractions. Thus,

$$Y = \text{weight of solute/weight of solvent}$$
$$G_s = \text{weight of solvent gas/(unit time) (unit area)}$$
$$X = \text{weight of adsorbate/weight of adsorbent}$$
$$L_s = \text{weight of adsorbent/(unit time) (unit area)}$$

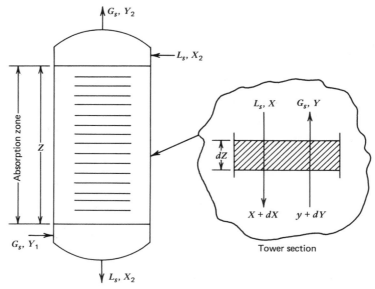

Fig. 29 Variables in a continuous countercurrent traveling bed adsorber.

Figure 29 is a schematic of a continuous countercurrent traveling bed adsorber removing one component. It illustrates the relationship among the variables.

A solute balance around the entire column produces

$$G_s(Y_1 - Y_2) = L_s(X_1 - X_2) \tag{85}$$

and around the upper part of the tower, a similar equation results:

$$G_s(Y - Y_2) = L_s(X - X_2) \tag{86}$$

These two equations set the operating line for the system. Equation (86) can be solved for Y:

$$Y = (L_s/G_s) X + [Y_2 - (L_s G_s)X_2] \tag{87}$$

and then plotted on a Y versus X curve as a straight line where the slope is (L_s/G_s) and the intercept is $[Y_2 - (L_s/G_s)X_2]$.

Heat effects can become significant for a large amount of material adsorbed. Calculation of these heat effects is very complex. Dilute gas adsorption is assumed in this development. Similar to absorption, the mass transfer is characterized by an overall gas mass transfer coefficient based on the external surface area of solid particle per packing volume, a_p. Then the rate of solute transfer over the differential height of the adsorbent, dZ, may be written

$$L_s\, dX = G_s\, dY = K_y a_p\, (Y - Y^*)\, dZ \tag{88}$$

Then, in the usual manner,

$$Z = \int_{Y_2}^{Y_1} \left(\frac{G_s}{K_y a_p}\right) \frac{dY}{(Y - Y^*)} \tag{89}$$

The height of a transfer unit is defined as

$$H_{OG} = \frac{G_s}{K_y a_p} \tag{90}$$

and the number of transfer units is given by the integral:

$$N_{OG} = \int_{Y_2}^{Y_1} \frac{dY}{(Y - Y^*)} \tag{91}$$

When the equilibrium line is straight as well as the operating line, the logarithmic mean technique for evaluating N_{OG} can be used.

$$N_{OG} = \frac{Y_1 - Y_2}{\Delta Y_{LM}} \tag{92}$$

$$\Delta Y_{LM} = \frac{(Y_1 - Y_1^*) - (Y_2 - Y_2^*)}{\ln\left[\dfrac{Y_1 - Y_1^*}{Y_2 - Y_2^*}\right]} \tag{93}$$

3.4.5 Unsteady-State Fixed Regenerable Beds

In fixed-bed adsorbers, the fluid is passed continuously over the adsorbent which is initially free of adsorbate. At first the adsorbent contacts a strong solution entering the bed. Initially, the adsorbate is removed by the first portion of the bed and nearly all the solute is removed from the solution before it passes over the remaining part of the bed. Figure 30 shows the situation that occurs in a downflow situation. In Figure 30a the effluent is nearly solute free.

The uppermost part of the bed becomes saturated and the bulk of the adsorption takes place over a relatively narrow portion of the bed in which concentration changes rapidly. This narrow adsorption zone moves down the bed as a concentration wave at a rate much slower than the linear velocity of the fluid through the bed. As time progresses the concentration of the solute in the effluent increases. When the effluent solute concentration reaches a predetermined value, set by emission standards, for example, the breakthrough point is reached. The solute concentration in the effluent now rises rapidly as the adsorption zone passes out the end of the bed and solute concentration in the effluent essentially reaches the initial concentration. The concentration–volume of effluent curve in this portion is known as the breakthrough curve.

If a vapor is being adsorbed adiabatically from a gas mixture, the evolution of the heat of adsorption causes a temperature curve to flow through the bed similar to the adsorption curve. The rise in temperature of the effluent stream may be used to predict the breakpoint.

The shape and time of appearance of the breakthrough curve greatly influence the method of operating a fixed-bed adsorber. The actual rate and mechanism of the adsorption process, the nature of the adsorption equilibrium, the fluid velocity, the concentration of the solute in the feed, and the bed depth contribute to the shape of the curve produced for any system. The breakpoint is very sharply defined in some cases and in others poorly defined.

Generally the breakpoint time decreases with decreased bed height, increased particle size of adsorbent, increased rate of flow through the bed, and increased solute concentration in the feed. Design

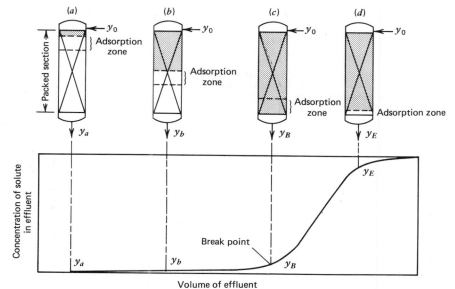

Fig. 30 Unsteady-state concentration in a fixed adsorption bed.

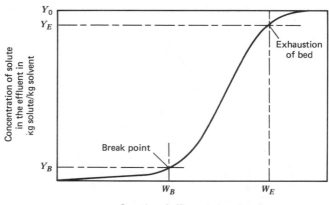

Fig. 31 Breakthrough curve.

consists in part of determining the breakpoint time which means that a breakpoint curve must be evaluated.

A design method due to Michaels[66].with the following assumptions is discussed by Treybal;[10]

1. Isothermal operation.
2. Dilute solutions.
3. The adsorption isotherm is concave to the solution–concentration axis, Type III isotherm.
4. The adsorption zone is constant in thickness as it travels through the adsorption column.
5. The height of the adsorption bed is large relative to the height of the adsorption zone.

3.4.5.1 Determining the Breakthrough Curve. Consider the idealized breakthrough curve shown in Figure 31. The time for the adsorption zone to move down the column is θ_A and the time for the adsorption formation zone to form and move out of the bed is θ_E.

$$\theta_A = \frac{W_A}{G_s} \tag{94}$$

$$\theta_E = \frac{W_E}{G_s} \tag{95}$$

We now define θ_F as the time required for the formation of the adsorption zone of thickness Z_A.

A quantity f, the fractional ability of the adsorbent to adsorb solute, is defined. If Q_B is the quantity of solute received from the gas as the concentration changes from Y_B to Y_E, then

$$Q_B = \int_{W_B}^{W_E} (Y_0 - Y)\, dW \tag{96}$$

and

$$f = \frac{Q_B}{Y_0 W_A} \tag{97}$$

where $Y_0 W_A$ is the total quantity of solute that can be removed. With $f = 0$, the time required for the formation of the adsorption zone θ_E should be equal to θ_A, the time required for the zone to travel a distance equal to its own thickness Z_A. On the other hand, if $f = 1.0$, meaning that there is no solute adsorbed in the adsorption zone, θ_F should be zero. Thus,

$$\theta_F = (1 - f)\,\theta_A \tag{98}$$

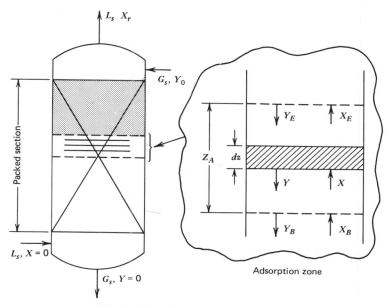

Fig. 32 Variables in a fixed-bed adsorber.

where $(1 - f)$ represents the fraction of nonsaturated adsorbent through which the adsorbent zone has passed, and we assume

$$\frac{Z_A}{Z} = \frac{\theta_A}{\theta_E - \theta_F} = \frac{\theta_A}{\theta_E - (1-f)\,\theta_A} \qquad (99)$$

$$\frac{Z_A}{Z} = \frac{W_A}{W_E - (1-f)\,W_A} \qquad (100)$$

For the situation where the adsorption has taken place so that the effluent concentration is Y and the total effluent is W,

$$\frac{Z_A}{Z} = \frac{W_A}{W - W_B} \qquad (101)$$

In the fixed bed of adsorbent, the adsorption zone moves down the column. We can envision instead the solid moving upward through the column countercurrent to the fluid at a rate so that the adsorption zone remains stationary in the column as illustrated in Figure 32.

Figure 33 is a plot of the equilibrium and operating lines. In this diagram the solid leaving at the top is shown to be in equilibrium with the fluid entering ($Y_0 = X_T$) and all solute has been removed from the gas before it leaves the column (column not saturated). This is an ideal situation which would require an infinitely tall column. However, we are concerned primarily with the adsorption zone only. Over the entire adsorption tower the operating line is

$$G_s\,(Y_0 - 0) = L_s\,(X_r - 0)$$
$$G_s Y_0 = X_T L_s \qquad (102)$$

and furthermore,

$$G_s Y = X L_s \qquad (103)$$

We can now write a different mass balance and introduce the HTU concept:

$$G_s\,dY = K_y a_p (Y - Y^*)\,dZ \qquad (104)$$

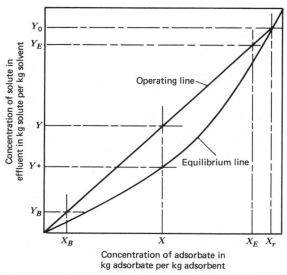

Fig. 33 Equilibrium and operating line relationship in a fixed-bed adsorber.

For the adsorption zone then

$$Z_A = \int_0^{Z_A} dZ = \int_{Y_B}^{Y_E} \left(\frac{G_s}{K_y a_p}\right) \frac{dY}{(Y - Y^*)} \tag{105}$$

The number of transfer units is given by

$$N_{OG} = \int_{Y_B}^{Y_E} \frac{dY}{(Y - Y^*)} \tag{106}$$

with the height of a transfer unit defined as

$$H_{OG} = \frac{G_s}{K_y a_p} \tag{107}$$

Therefore,

$$Z_A = H_{OG} N_{OG} \tag{108}$$

The breakthrough curve can be calculated from the following relationship assuming H_{OG} remains constant throughout the bed:

$$\frac{Z}{Z_A} = \frac{W - W_B}{W_A} = \frac{\displaystyle\int_{Y_B}^{Y} \frac{dY}{(Y - Y^*)}}{\displaystyle\int_{Y_B}^{Y_E} \frac{dY}{(Y - Y^*)}} \tag{109}$$

3.4.5.2 Determining the Breakpoint Time. We first define a degree of bed saturation (DBS). Note that the volume of adsorbate V is given by

$$V = AZ \tag{110}$$

and therefore, $Z = V/A$ and represents the volume of adsorbate per cross-sectional area of the bed. Then, knowing the bed density,

$$\text{weight of adsorbent} = ZA\rho_s$$
$$\text{weight of solute adsorbed at equilibrium} = ZA\rho_s X_T$$
$$\text{weight of solute adsorbed in part of column which is saturated} = (Z - Z_A) A\rho_s X_T$$
$$\text{weight of solute adsorbed in part of column not saturated} = Z_A A\rho_s X_T (1 - f)$$

The degree of bed saturation can now be defined:

$$\text{DBS} = \frac{\text{solute adsorbed}}{\substack{\text{total possible solute adsorbed} \\ \text{or equilibrium concentration}}}$$

$$\text{DBS} = \frac{(Z - Z_A)\, A\rho_s X_T + Z_A A \rho_s X_T\, (1 - f)}{Z A \rho_s X_T}$$

$$\text{DBS} = \frac{Z - Z_A f}{Z} \tag{111}$$

The time to breakthrough T_B is then given by

$$T_B = \frac{\substack{\text{weight of solute accumulated} \\ \text{in bed per unit area}}}{\substack{\text{weight of solute in incoming} \\ \text{gas per unit time per unit area}}}$$

$$T_B = \frac{(\text{DBS})Z\rho_s X_T}{G_s Y_0} \tag{112}$$

Hougan and Marshall[67] have discussed a design method where the equilibrium line is straight. The method was reviewed by Fair[68] who also presents a correlation for mass transfer coefficients in adsorption. More recently the technique has been extended by Ufrecht, Somerfeld, and Lewis.[69]

3.4.6   Nomenclature

A_p external surface area of solid particles per unit volume, m^2/m^3
A cross-sectional area, m^2
b constant in Langmuir's equation, Pa^{-1}
c constant in BET equation, dimensionless
DBS   degree of bed saturation, dimensionless
f fractional ability of adsorbent zone to adsorb solute, dimensionless
G_s rate of solvent gas flow, kg/sec-m^2
H_{OG} overall gas transfer unit, m
k constant in Freundlich equation, $m^3/(Pa)^{1/n}$
$K_g a_p$ overall volumetric gas phase mass transfer coefficient, mole/sec-m^2-mole fraction
L_s rate of adsorbate-free adsorbent flow, kg/sec-m^2
n power in Freundlich equation, or number of layers adsorbed in BET equation, dimensionless
N_{OG} number of overall gas phase transfer units, dimensionless
P pressure, Pa
P_0 saturation pressure, Pa
Q_B quantity of solute transferred from gas phase, kg/m^2
T_B time to breakthrough, sec
v volume adsorbed, m^3, or volume adsorbed per unit weight of adsorbent, m^3/kg
v_m volume adsorbed in a monomolecular layer, m^3, or volume adsorbed in a monomolecular layer per unit of weight of adsorbent, m^3/kg
V volume of adsorbate, m^3
W total quantity of effluent from fixed-bed adsorber, kg/m^2
W_A quantity of effluent from fixed-bed adsorber between breakpoint and bed exhaustion, kg/m^2
W_B quantity of effluent up to breakpoint, kg/m^2
W_E quantity of effluent up to bed exhaustion, kg/m^2
x fraction of saturation pressure in BET equation, dimensionless
X weight fraction of adsorbate per weight of adsorbent, kg/kg
X_T weight fraction of adsorbate per weight of adsorbent in a fixed-bed adsorber when in equilibrium with entering fluid, kg/kg
Y weight fraction of solute per weight of solvent, kg/kg
Z height of adsorbate in adsorber, m
Z_A height of adsorption zone, m

3.4.6.1   Greek
θ_A time required for adsorption zone to move down the column a distance Z_A, sec
θ_E time for the adsorption zone to form and move out of the bed, sec
θ_F time of formation of adsorption zone, sec
ρ_s density of adsorbent bed, kg/m^3

3.4.6.2 Superscripts
* equilibrium condition

3.4.6.3. Subscripts
B   breakthrough point
LM   logarithmic mean
S   solute-free flow rates
1   conditions at bottom of traveling bed adsorber
2   conditions at top of traveling bed adsorber

3.5   Specific Applications

Adsorption finds its greatest usefulness for controlling pollutants which must be removed down to very low trace concentration ranges (ppb to a few ppm). Many odorants are still detectable and objectionable at concentrations as low as 100 ppb so that complete odor control is beyond the capability of many other control methods. Odor control applications for adsorption have included food processing (odor removal from canning, coffee roasting, fish processing, rendering, fermentation, frying, and baking), chemical and process manufacturing (especially glue manufacture and operations involving natural materials such as blood and glands, tanning, and pulp and paper manufacture), and other operations such as foundries, animal laboratories, and painting and coating operations.

Adsorption is less attractive for removing high concentrations of pollutants because of the large amount of adsorptive capacity required. However, when the pollutant concentrations are low and the air column to be treated is large, it is frequently indicated as excellent control means for volatile hydrocarbons and solvents. It is especially useful for control of toxic and cancer-suspect vapors where effluents must be purified to a few ppm or less. Typical applications are for the removal of solvent vapors from exhaust air in automobile body painting operations, removal of organic resin and solvent vapors from ventilation air in bonded fiberglass manufacture, and in control of ether, acetone, and other solvent vapors in nitrocellulose and smokeless powder manufacture. It is used to control evaporative emissions from automobiles and for control of toxic emissions such as H_2 from laboratory hoods. Adsorption is used to control radioactive gases from nuclear activities such as removal of radon and radioiodine.

Adsorption is useful where some selectivity in removal of specific gaseous components is of importance. In addition to gas drying, impregnated adsorbents as discussed in Section 3.1 have specific applications in being quite selective such as removal of ethylene from a vent stream. Molecular sieves have been used to control mercury vapor concentrations in mercury cell chlor–alkali plants and Kovsky et al.[70] have experimented with molecular sieves for SO_2 removal. Adsorption processes have been mentioned (Section 3.3.3) for removing inorganic pollutants from flue gases. These have been developed and pilot planted but have not won commercial acceptance.

3.6   Adsorption Concentration for Incineration

Occasionally, incineration of combustible vapors is the desired end control method to destroy toxic organics, but their concentration is extremely dilute and they are contained in very large volumes of air. The utility cost of heating such large volumes of air to the required ignition temperature can be very high even with a gas incinerator designed for heat interchange. Incineration operating costs can be reduced appreciably by concentration of the pollutants by adsorption prior to incineration. Cost benefits are most pronounced when pollutant concentration is in the range of 20–100 ppmv although concentrations up to 300 ppmv can still be good candidates for this approach. The gases to be treated are passed through an adsorption bed in the usual manner for removal of the pollutant, but the bed is regenerated by stripping with air which is then incinerated. A fortyfold increase in pollutant concentration can be achieved. Grandjacques[71] has presented comparative economics of incineration alone and adsorptive concentration for C_6 hydrocarbons in streams of 3.8, 7.6, and 28.3 m³/sec (8000, 16,800, and 60,000 ft³/min). In all cases, adsorption concentration was more attractive at concentrations up to 300 ppmv but the attractiveness increases rapidly with increase in gas volume.

3.7   Handling, Utilization, and Disposal of Recovered Materials

The problem of utilization or disposal of recovered material from adsorption will depend greatly on the process utility of the material but can be influenced to some extent by the desorption method chosen for sorbent regeneration. If the recovered pollutant has process value, desorption by heating alone or in combination with partial evacuation may be best, since the pollutant will boil off as a concentrated vapor and can be condensed as a concentrated liquid for process reuse. When more than one pollutant species is being recovered, such as a mixture of solvent vapors, further separation processing, such as by distillation, may be required if the mixture cannot be reutilized directly. Since adsorption will frequently be used for recovery of organic pollutants, the recovered liquid can frequently be burned as a fuel or admitted with a fuel for combustion if it has no other useful benefit. Combustion of substituted hydrocarbons must be carefully considered as to whether they will produce corrosive combustion products or ones that would be environmentally objectionable.

Desorption by line steaming is frequently practiced. This can be troublesome when the organic material is highly water soluble since considerably further treatment may be required to separate the condensed steam and organics. Even when the organic is but slightly soluble and separation can occur by decantation of the nonmiscible phases, thought should still be given to the problems of disposal of the aqueous phase. If the aqueous quantity is small and the dissolved organics are biodegradable, conventional sewerage treatment may be economical and acceptable. However, where the organics are highly refractory or the waste water quantity is large, other methods of desorption may be more desirable.

Desorption by inert gas stripping, except in the case of adsorption concentration for gas incineration (Section 3.6), is likely to create more problems than it solves since the pollutant is again dispersed in a gas stream.

Where the pollutants being adsorbed are inorganic or noncombustible in nature and have no useful purpose, disposal can present many of the same problems such materials would present if recovered by absorption and the discussion in Section 2.5 should be noted.

4 CONDENSATION

Condensation can be an applicable control method for gaseous vapors fairly close to their dewpoints. The method is most suitable for hydrocarbons and organic compounds having reasonably high boiling points compared to ambient conditions and present in the vapor phase in appreciable concentrations. Pollutants having reasonably low vapor pressures at ambient temperatures may be controlled satisfactorily in water- or air-cooled condensers. For somewhat more volatile solvents, two-stage condensation may be required, using cooling water in the first stage and refrigeration in the second stage. Refrigeration to extremely low temperature levels is seldom attractive for pollution control alone (unless needed for other process reasons) and alternate control methods will usually be more attractive. Minimizing the presence of inert or noncondensable gases in the effluent will increase the practicability and economic attractiveness of condensation, because it will become less necessary to cool to very low dewpoints.

Condensation is not a practical method of total control for reasonably volatile hazardous or toxic organics in appreciable concentrations in streams of noncondensables if control levels of a few ppmv are required. However, condensation may be useful as a preliminary treatment method to recover valuable solvents or reduce the capacity required in the final treatment method. Partial condensation can be useful in those cases where the stripped gas can be recycled to the process rather than vented, or where it can be used as primary combustion air. (The remaining pollutants become incinerated.) Condensation as a preliminary pretreatment is also attractive when the gas must be precooled for the final control method such as might be required for adsorption.

4.1 Types of Equipment

Condensation may be carried out either by direct contact or by indirect cooling. In direct contact, the vapor to be cooled is brought into intimate contact with a cooled or refrigerated liquid. With indirect cooling, a surface condenser containing metal tubes is employed. The tubes are cooled with another fluid on the other side of the wall. When noncondensable gases are present, compression of the stream to be treated prior to cooling can be beneficial in that less low temperatures need be employed to reach equivalent pollutant partial pressures. However, gas compression for pollution control alone is seldom economical unless higher pressures are needed for other process reasons.

When low temperatures must be used to achieve adequate low dewpoints for a particular pollutant, consideration must be given to the possible presence of other materials that could solidify at those temperatures. An example might be the presence of water vapor in a gas stream contaminated with a low-boiling organic solvent requiring condensation at temperatures below 273 K. Icing on a surface condenser could shortly foul the heat transfer surface. In such a situation, a direct contact condenser using a suitable antifreeze liquid would be preferable.

4.1.1 Direct Contact Condensers

In direct contact condensers, a cooled liquid is frequently recirculated through an external heat exchanger to remove the heat of condensation and any sensible heat picked up from the gas by the liquid. Within the condenser itself, intimate mixing and contact is brought about between the liquid and gas to effect as close an approach to thermal and mass transfer equilibrium as possible. The cold liquid may be sprayed into the gas in a spray tower or jet eductor or a tower with gas–liquid contacting internals can be used. The contacting tower can be a packed tower, sieve plate tower, disk and doughnut or segmental baffle plate tower, or even a slat-packed chamber. Since the liquid becomes heated by both the gas and the condensing vapor, lower dewpoints can be reached with counterflow of liquid and vapor, but many parallel flow devices are used.

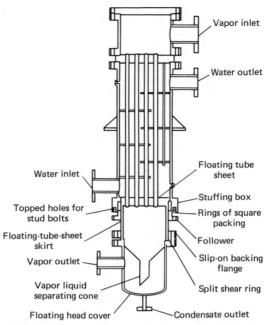

Fig. 34 Outside-packed-head condenser with vapor–liquid separator. (Downingtown Iron Works, Inc., Division of Pressed Steel Tank Co.).[72]

The recirculated cold liquid is often water if temperatures near its freezing point are not required. However, it can be any appropriate liquid of low vapor pressure and may even be the substance being condensed. The latter offers the advantage that no further steps are required to separate the condensing material from the cooling liquid. When a different liquid is used, it must be remembered that the treated gas will be close to being in vapor–liquid equilibrium with the cold liquid. Therefore, if it has appreciable vapor pressure, it could in turn pollute the gas being purified. Further, the economic cost of loss of the cooling liquid through vaporization must be considered.

When using a different cooling liquid, the method of separating the condensed material from the cooling liquid must also be considered in its selection. Utilization of a liquid having low solubility or miscibility with the condensed material is often helpful since a simple phase separator can be used. When refrigeration must be used externally to cool the recirculating liquid, consideration must be given to the possibility of freezing out solids on the heat transfer surfaces. If water is the cold liquid used, obviously an antifreeze must be added. However, when water is not used in the system, condensing moisture from the gas or other vapors present with relatively high freezing points can condense and cause problems.

4.1.2 Surface Condensers

Surface condensers are most frequently used when the vapors to be condensed constitute the major portion of the gas stream and only a small amount of noncondensables are to be vented. In such cases, ordinary tubular condenser-type heat exchangers may be used. Figure 34 illustrates a typical vertical tubular condenser. When noncondensables predominate, finned tubes on the gas side will give better heat transfer unless the condensing material tends to scale the surface. In such cases, tubular exchangers designed for easy cleaning of the gas-side heat transfer surface should be considered. Condensers can be either vertical or horizontal. Horizontal condensers are frequently pitched with a slight slope to provide for tube drainage. Where dewpoints 10 K or more above maximum ambient air temperatures are satisfactory for control, air-cooled heat exchangers as illustrated in Figure 35 can be used.

4.2 Theory and Design

In the process of cooling a multicomponent complex gas mixture containing common noncondensable gases, the gases will first be cooled by convection with the heat transfer surface (either tube wall in a surface condenser or cold liquid drop or film in a direct contact condenser), giving up its sensible

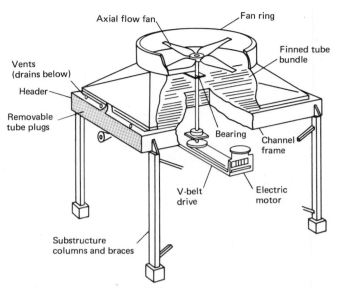

Fig. 35 Air-cooled heat exchanger. Process fluid inlet nozzle is on top of the header, with process fluid outlet nozzle below. (*Chem. Eng.*, Oct. 31, 1960, p. 91.)[73]

heat until the gas becomes saturated with one or more of its condensable components. As additional cooling proceeds, condensable gases must diffuse to the heat transfer surface where they condense, giving up latent heat. The initial dewpoint or saturation temperature for each component can be predicted from the temperature–vapor pressure curve for the component and its mole fraction in the vapor: $y_A \pi = (p_A)_g$ where y_A is the mole fraction of component A in the vapor, π is the total absolute gas pressure, and $(p_A)_g$ is the partial pressure of component A in the vapor, all in consistent units. Component A will start to condense, once the gas temperature has been lowered to the temperature at which component A has a vapor pressure, $p_A = (p_A)_g$.

Once condensation starts, the gas temperature will drop only as both an adequate amount of sensible and latent heat has been removed to permit the gas to remain saturated with component A as the temperature is dropped. Since vapor A must diffuse to the heat transfer surface, the process is both heat and mass transfer controlled. In a system of additional condensable components, B, C, and so on, each additional component will begin to condense as the gas becomes saturated with these components, fulfilling equivalent partial pressure relationships to those above for each of these components.

To determine the temperature to which the gas must be cooled to reach an allowable emission level for component A, one uses the following equations, remembering that concentration expressed as a volumetric fraction is the same as the mole fraction:

$$v_{Aa} = y_{Aa} \quad \text{and} \quad y_{Aa} \pi = p_{Aa} \tag{113}$$

where v_{Aa} is the allowable volumetric fraction of component A in the emission, y_{Aa} is the allowable mole fraction of A in the emission, π is the absolute total pressure of the gas, and p_{Aa} is the allowable vapor pressure of component A, all in consistent units. The required gas temperature is the temperature at which component A has a vapor pressure equal to p_{Aa} on its vapor pressure curve. In the case of multiple condensable pollutants, the component requiring the lowest temperature would be controlling.

If condensation is being practiced to avoid producing a visible plume in the atmosphere, the condenser must cool the gas to a low enough dewpoint that the path of the mixing operation as the gas disperses in a cold atmosphere does not cross the saturation curve for any of the condensable components.[74] One method is to cool to the atmospheric temperature before release. Another is to provide adequate superheat[74] after condensation and before release.

Design of both direct contact and surface condensers follows advanced chemical engineering design principles for gas dehumidification, which are beyond the scope of an air pollution handbook. However, the following sections give short summaries and appropriate references.

4.2.1 Design of Direct Contact Condensers

References 75 and 76 describe precise methods for sizing direct contact condensers involving calculation of both heat and mass transfer. While Ref. 75 discusses several methods, the simplest is the approximate method as supplemented by Ref. 77. In this method, the interfacial temperature between the gas and liquid films at the condensing surface is assumed to be equal to the bulk temperature of the liquid in the film or droplet. The following equations result:

$$Z = \frac{G_G}{K'_{OG}a} \int \frac{di_G}{i_L - i_G} \qquad (114)$$

and

$$Z_{OG} = \frac{G_G}{K'_{OG}a} \qquad (115)$$

where Z is the height of the transfer area in a vertical tower, in m (ft); G_G is the mass gas velocity, in kg/sec-m² (lb/hr-ft²); $K'_{OG}a$ is the overall mass transfer coefficient for the gas film, in kg of vapor condensed/sec-m² (unit humidity potential difference) [lb of vapor condensed/hr-ft² (unit humidity potential difference)]; i_L is the enthalpy of the gas–vapor mixture at the bulk temperature of the liquid, in kJ/kg (Btu/lb); i_G is the enthalpy of the gas–vapor mixture at the gas temperature, in kJ/kg (Btu/lb); and Z_{OG} is the height of a single overall mass transfer unit, m (ft).

To use the above equations, one must have values for $K'_{OG}a$ or Z_{OG}. These values are a function of the type and design of the gas–liquid contacting internals, the superficial mass velocity for the liquid G_L, in kg/sec-m² (lb/hr-ft²), and the value of G_G. These values must either be determined or estimated from the literature for the system involved or determined experimentally from pilot-plant or full-scale operating units. Reference 78 plots values of Z_{OG} for Raschig ring packed water-cooling towers for values of G_G and parameters of G_L. Other data are given in Ref. 76. The integral in Eq. (114), which is the number of transfer units required, is determined by graphical or digital integration of values taken from a plot of the saturation curve and the operating line for the condenser.

Fair[79,80] has proposed an alternative calculational approach substituting heat transfer analogies for mass transfer rates and has proposed equivalent empirical overall heat transfer rate coefficients for baffle plate packed columns, packed columns, and spray towers.

4.2.2 Design of Surface Condensers

Surface condensers for concentrated, saturated, single-component vapors are designed in accordance with standard engineering approaches[81,82] to heat transfer where the heat transfer flux $q/A = U_0 \Delta t_m$. The heat transfer flux q/A is in W/m² (Btu/hr-ft²), and the mean temperature driving force between the condensing vapor and the coolant fluid, Δt_m, is in K (°F). The overall heat transfer coefficient, U_0, in W/m²-K (Btu/hr-ft²-°F), is the composite coefficient made up of the condensing-side film coefficient, the coolant-side film coefficient, the separating wall conductivity, and any fouling resistance allowances desired.

However, these design methods are not applicable when even relatively small quantities of noncondensable gas are present. In situations involving surface condensation of vapors from noncondensable gases, special methods considering both mass and heat transfer rates are required. Reference 82 discusses a trial-and-error design procedure recommended by Colburn and Hougen in which two heat transfer coefficients, U' and h_g, and a mass transfer coefficient, K_G, are used.

U' is a partial overall coefficient including the resistances of the condensate film, the tube wall, fouling resistances, and the coolant film. The coefficient h_g is the mean coefficient for sensible heat transfer through the gas film and K_G is the mass transfer coefficient for vapor across the gas film. A heat balance over a differential element of surface area dA is employed:

$$WC_L \, dt_L/da = U'(t_i - t_L) = h_G(t_v - t_i) + \lambda K_G(p_v - p_i) \qquad (116)$$

where W is the mass rate of coolant flow, C_L is the mean coolant heat capacity, t_L is the bulk temperature of the coolant, t_i is the temperature of the condensate–vapor film interface, t_v is the bulk temperature of the vapor, λ is the latent heat of condensation of the vapor at saturation temperature, p_v is the partial pressure of the vapor in the body of the gas stream, and p_i is the partial pressure of the vapor at t_i. Values of t_i are assumed, as are corresponding values of the corresponding vapor pressure p_i, until the equation balances. The total heat transfer area A is then obtained by the graphical or numerical integration below:

$$A = \int_0^q \frac{dq}{U'(t_i - t_L)} \qquad (117)$$

The mass transfer coefficient K_G is obtained from heat transfer analogy by substituting the dimensionless terms

$$\frac{K_G p_{nm}}{G} \left(\frac{\mu_v}{\rho_v D_v} \right)^{2/3} \quad \text{for} \quad \frac{h_G}{C_{pv}G} \left(\frac{C_p \mu}{k_v} \right)^{2/3}$$

where p_{nm} is the logarithmic mean of the vapor pressure of the noncondensables at the gas–condensate interface and vapor pressure in the main body of the gas stream, G is the superficial mass velocity of the gas, $(\mu_v / \rho_v D_v)$ is the Schmidt number for the gas–vapor stream, C_{pv} is the heat capacity at constant pressure for the vapor, and $(C_p \mu / k_v)$ is the Prandtl number for the vapor.

Reference 84 describes a stepwise modification of the above method by Volta which provides for a tabular method of calculation that eliminates trial and error. This reference also describes several other alternate calculation approaches. References 85 through 87 also discuss design methods for partial condensation of vapors.

4.2.3   Problem Area—Fog Formation

In the preceding sections (4.2.1 and 4.2.2) suitable design methods were discussed for direct contact and surface cooling condensers. In these methods, both the rate of heat and mass transfer are considered. When the gas must be cooled more than 40–50 K below its initial dewpoint to achieve adequate pollutant condensation, it is possible for condensate fog to form in the bulk of the gas stream. Fog so produced will generally be 1.0 μm in particle size or smaller, resulting in difficult collection. Fog occurs when the rate of heat transfer appreciably exceeds the rate of mass transfer and the bulk of the gas subcools appreciably below the dewpoint of the condensable vapor. The vapor then nucleates and condenses in droplets in the bulk gas stream rather than having time to migrate to a cold surface and condense. Fog seldom occurs in direct contact condensers because of the close proximity of the bulk of the gas to cold heat transfer surfaces (the liquid droplets or films). Fog in surface condensers can be predicted by calculating the rates of heat and mass transfer as the gas proceeds through the condenser. When the bulk of the gas stream becomes supersaturated for the temperature, fog is likely. Fog may be avoided by switching to direct condensation. Another alternative is to produce the fog and then remove it in a suitable fine particle mist eliminator (Brownian diffusion particle filter, electrostatic collector, or venturi scrubber—see Chapters 10–12).

REFERENCES

1. Frankenberg, T. T., Sulfur Removal for Air Pollution Control, *Mech. Eng.* **87**, 36–41 (Aug. 1965).

2. Crocker, B. B., Removal of Hazardous Organic Vapors from Vent Gases, in *Proceedings: Control of Specific Toxic Pollutants*, APCA Specialty Conference, Gainesville, Fla., Feb. 1979, Air Poll. Control Assn., Pittsburgh, 1979, pp. 360–376.

3. Stern, A. C. (ed.), *Air Pollution*, 3rd edition, Vol. IV, Academic Press, New York, 1977, p. 270.

4. Crocker, B. B., Novak, D. A., and Scholle, W. A., Air Pollution Control Methods, in *Encyclopedia of Chemical Technology*, 3rd edition, Vol. 1, Wiley, New York, 1978, p. 654.

5. Grant, H. O., Pollution Control in a Phosphoric Acid Plant, *Chem. Eng. Prog.* **60**, 53–55 (Jan. 1964).

6. Teller, A. J., Control of Gaseous Fluoride Emissions, *Chem. Eng. Prog.* **63**, 75–79 (Mar. 1967).

7. Cabibbo, S. V., and Teller, A. J., The Crossflow Scrubber—A Digital Model for Absorption, Paper No. 69–186, APCA 62 Annual Meeting, New York, N.Y. (June 22–26, 1969).

8. Leva, M., *Tower Packings and Packed Tower Design*, 2nd edition, U.S. Stoneware Co., Akron, OH., 1953.

9. Bhatia, M., Packed Tower and Absorption Design, in *Air Pollution Control and Design Handbook*, Part 2, Dekker, New York, 1978, Chapter 24.

10. Treybal, R. E., *Mass Transfer Operations*, 3rd edition, McGraw-Hill, New York, 1980.

11. Sherwood, T. K., Pigford, R. L., and Wilke, C. R., *Mass Transfer*, McGraw-Hill, New York, 1975.

12. Sherwood, T. K., and Pigford, R. L., *Absorption and Extraction*, 2nd edition, McGraw-Hill, New York, 1952.

13. Wilke, C. R., and Von Stockar, V., Absorption, in *Encyclopedia of Chemical Technology*, 3rd edition, Vol. 1, Wiley, New York, 1978, pp. 53–96.

14. Perry, R. H., and Chilton, C. H. (eds.), *Chemical Engineers' Handbook*, 5th edition, McGraw-Hill, New York, 1973, pp. 18–3 through 18–58.

15. See Ref. 4, p. 656.

16. Lunde, K. E., Performance of Equipment for Control of Fluoride Emissions, *Ind. Eng. Chem.* **50**, 293–298 (1958).

17. Jewell, J. P., and Crocker, B. B., Control of Fluoride Emissions, *Proc. 8th Annual Sanitary and Water Resources Engineering Conf.*, Vanderbilt University, Nashville, Tenn., 1969, pp. 211–228.

18. See Ref. 13, pp. 268–277.

19. Nonhebal, G., *Gas Purification Processes*, George Newness Ltd., London, 1964, pp. 219–228.

20. Smith, J. M., and Van Ness, H. C., *Introduction to Chemical Engineering Thermodynamics*, 3rd edition, McGraw-Hill, New York, 1975.

21. Edwards, T. J., Newman, J., and Prausnitz, J. M., Thermodynamics of Aqueous Solutions Containing Volatile Weak Electrolytes, *AIChE J.* **21**, 248–259 (1975).

22. Edwards, T. J., Newman, J., and Prausnitz, J. M., Thermodynamics of Vapor–Liquid Equilibria for the Ammonia–Water System, *Ind. Eng. Chem. Fundam.* **17**, 264–269 (1978).

23. Edwards, T. J., Maurer, G., Newman, J., and Prausnitz, J. M., Vapor–Liquid Equilibria in Multicomponent Aqueous Solutions of Volatile Weak Electrolytes, *AIChE J.* **24**, 966–976 (1978).

24. Prausnitz, J. M., *Molecular Thermodynamics of Fluid-Phase Equilibria*, Prentice-Hall, New York, 1969.

25. Foust, A. S., Wenzel, L. A., Clump, C. W., Maus, L., and Andersen, L. B., *Principles of Unit Operations*, 2nd edition, Wiley, New York, 1980.

26. Higbie, R., The Rate of Absorption of Pure Gas into a Still Liquid During Short Periods of Exposure, *Trans. AIChE* **31**, 365–389 (1935).

27. Danckwerts, P. V., Significance of Liquid Film Coefficients in Gas Absorption, *Ind. Eng. Chem.* **43**, 1460–1467 (1951).

28. Whitman, W. G., The Two-Film Theory of Gas Absorption, *Chem. & Met. Eng.* **29**, 146–148 (1923).

29. See Ref. 10, p. 117.

30. *Design Information for Packed Towers*, Bulletin DC-11, Norton Co., Akron, OH, 1977.

31. EPA Flue Gas Desulfurization Pilot Study Reports NTIS Pub PB-295001/2BE, PB-295002/OBE, PB-295003/8BE, PB-295004/6BE, PB-295006/1BE, PB-295007/9BE, PB-295008/7BE, PB-295010/3BE, PB-295011/1BE, 1979.

32. Rosenberg, H. S., How Good Is Flue Gas Desulfurization?, *Hydrocarbon Processing* **57**(5), 132–135 (May 1978).

33. Campbell, I. E., State of the Art of Flue Gas Desulfurization Technology, Am. Min. Congr., Session Paper, set n-5, Environ. Controls 1 & 2, Las Vegas, Nev., Oct. 9–12, 1978, Am. Min. Cong., Washington, D.C., 1978.

34. *Definitive SO₂ Control Process Evaluation: Limestone Double Alkali, and Citrate FGD Process*, EPA Pub. EPA-600/7-79-177, Aug. 1979.

35. Kerlson, H. T., and Rosenberg, H. S., Technical Aspects of Lime/Limestone Scrubbers for Coal Fired Power Plants, *JAPCA* **30**, 710–714 (1980).

36. Bakke, E., Application of the Lime/Limestone Flue Gas Desulfurization Process to Smelter Gases, *JAPCA* **30**, 1157–1160 (1980).

37. Getler, J. L., Shelton, H. L., and Furlong, D. A., Modeling the Spray Absorption Process for SO₂ Removal, *JAPCA* **29**, 1270–1274 (1979).

38. Murthy, K. S., Rosenberg, H. S., and Engdahl, R. B., Status and Problems of Regenerable Flue Gas Desulfurization Processes, *JAPCA* **26**, 851–855 (1976).

39. Slack, A. V., Flue Gas Desulfurization: An Overview, *CEP* **72**, 94–97 (Aug. 1976).

40. *Proceedings, EPA Flue Gas Desulfurization Symposium*, New Orleans, March 1976, U.S. EPA, Research Triangle Park, NC, 1976.

41. *Proceedings, Symposium on Flue Gas Desulfurization*, Atlanta, GA, U.S. EPA, Research Triangle Park, NC, 1974.

42. Pyler, E. L., and Maxwell, M. A. (eds.), *Proceedings, Flue Gas Desulfurization Symposium—1973*, New Orleans, Pub. EPA-650/2-73-038, U.S. EPA, Research Triangle Park, NC, 1973.

43. Rai, C., and Siegel, R. D. (eds.), Air II, Control of NO_x and SO_x Emissions, *AIChE Symp. Ser.* **71**(148), AIChE, New York, 1975.

44. *Sulfur Dioxide Processing*, AIChE, New York, 1975.

45. *Control Technology: Gases and Odors*, APCA, Pittsburgh, 1973.

46. Nannen, L. W., West, R. E., and Kreith, F., Removal of SO₂ from Low Sulfur Coal Combustion Gases by Limestone Scrubbing, *JAPCA* **24**, 29–39 (1974).

47. Slack, A. V., Lime-Limestone Scrubbing: Design Considerations, *CEP* **74**, 71–75 (Feb. 1978).

48. Grimm, C., et al., The Colstrip Flue Gas Cleaning System, *CEP* **74**, 51–57 (Feb. 1978).

49. Murthy, B. N., Harris, D. B., and Philips, J. L., SO₂ Absorption Studies with EPA In-House Pilot-Scale Venturi Scrubber, in *Proceedings, 2nd Intern. Lime/Limestone Wet Scrubbing Symp.*, New Orleans, LA, Nov. 1971, Vol. 2, U.S. EPA, Research Triangle Park, NC, 1972. Also: Bondor, F. S., Jones, G., and Saleem, A., Flue Gas Desulfurization in Venturi Scrubbers and Spray Towers, in 2nd Pap. Intern. Conference **1**, III, Univ. Salford, Salford, England, 1976.

50. Epstein, M., Leivo, C. C., and Rowland, C. H., Mathematical Models for Pressure Drop, Particulate Removal, and SO₂ Removal in Venturi, TCA, and Hydrofilter Scrubbers, in *Proceedings, 2nd Intern. Lime/Limestone Wet Scrubbing Symp.*, New Orleans, LA, Nov. 1971, Vol. 2, U.S. EPA, Research Triangle Park, NC, 1972. Also: Wen, C. Y., and Chang, C. S., Absorption of SO₂ in Lime and Limestone Slurry: Pressure Drop Effect on Turbulent Contact Absorber Performance, *Environ. Sci. Technol.* **12**(6), 1978. Also: Wen, C. Y., and Fong, F. K., Analysis and Simulations of Recycle SO₂–Lime Slurry in TCA, U.S. NTIS Rpt. PB-266104, 1977.

51. Lehle, W. W., in *The Manufacture of Sulfuric Acid* (W. W. Duecker and J. R. West, eds.), Reinhold, New York, 1959, pp. 348–352.

52. Koehler, G. R., Alkaline Scrubbing Removes SO₂, *CEP* **70**(6), 63–65 (June 1974).

53. Djololian, C., and Billand, D., Absorbing Fluorine Compounds from Waste Gases, *CEP* **74**, 46–51 (Nov. 1978).

54. Mayland, B. J., and Heinze, R. C., Continuous Catalytic Absorption for NOₓ Emission Control, *CEP* **69**, 75–76 (May 1973).

55. See Ref. 19, Chapter 8, Solid Chemical Absorbents for Gases, pp. 275–327.

56. Turk, A., Catalytic Reactivation of Activated Carbon in Air Purification Systems, *Ind. Eng. Chem.* **47**, 966–971 (1955).

57. Strauss, W., *Industrial Gas Cleaning*, 2nd edition, Pergamon Press, Oxford, 1975, p. 138, from Union Carbide published data.

58. See Ref. 14, pp. 16–5 and 16–6.

59. See Ref. 4, p. 658.

60. Bienstock, D., Field, J. H., Katell, S., and Plants, K. D., *JAPCA* **15**, 459–464 (1965).

61. Bienstock, D., Field, J. H., and Myers, J. G., *J. Eng. Power* **86**(3), 353 (1964).

62. Kunii, D., and Levenspiel, O., *Fluidization Engineering*, Wiley, New York, 1969, pp. 31–33.

63. Anon., Beaded Carbon Ups Solvent Recovery, *Chem. Eng.* **84**, 39–40 (Aug. 29, 1977).

64. See Ref. 3, pp. 341, 342, and 345.

65. Brunauer, S., *The Adsorption of Gases and Vapors*, Princeton University Press, Princeton, NJ, 1943.

66. Michaels, A. S., Simplified Method of Interpreting Kinetic Data in Fixed-Bed Ion Exchange, *Ind. Eng. Chem.* **44**, 1922–1930 (1952).

67. Hougan, O. A., and Marshall, W. R., Jr., Adsorption from a Fluid Stream Flowing Through a Stationary Granular Bed, *Chem. Eng. Prog.* **43**, 197–208 (1947).

68. Fair, J. R., Sorption Processes for Gas Separation, *Chem. Eng.* **76**, 90–110 (July 14, 1969).

69. Ufrecht, R. H., Sommerfeld, J. T., and Lewis, H. C., Design of Adsorption Equipment in Air Pollution Control, *JAPCA* **30**, 1348–1352 (1980).

70. Kiovsky, J. R., Koradia, P. B., and Hook, D. S., Sulfur Compound Cleanup: Molecular Sieves for SO₂ Removal, *Chem. Eng. Prog.* **72**, 98–103 (Aug. 1976).

71. Grandjacques, B., Carbon Adsorption Can Provide Air Pollution Control with Savings, *Poll. Eng.* **9**, 28–31 (Aug. 1977).

72. Perry, R. H., Chilton, C. H., and Kirkpatrick, S. D., *Chemical Engineers' Handbook*, 4th edition, McGraw-Hill, New York, 1963, p. 11–8.

73. See Ref. 14, p. 11–23.

74. Crocker, B. B., Water Vapor in Effluent Gases: What to Do About Opacity Problems, *Chem. Eng.* **75**, 109–116 (July 15, 1968).

75. McAdams, H. M., *Heat Transmission*, 3rd edition, McGraw-Hill, New York, 1954, pp. 356–365.

76. Jakob, M., *Heat Transfer*, Vol. II, Wiley, New York, 1957, pp. 342–354.

77. Sherwood, T. K., and Reed, C. E., *Applied Mathematics in Chemical Engineering*, McGraw-Hill, New York, 1939, p. 68.

78. McAdams, H. M., *Heat Transmission*, 2nd edition, McGraw-Hill, New York, 1942, p. 289.

79. Fair, J. R., Designing Direct-Contact Coolers/Condensers, *Chem. Eng.* **79**, 91–100 (June 12, 1972).

80. Fair, J. R., Process Heat Transfer by Direct Fluid-Phase Contact, *AIChE Symp. Ser.* **68**(118), 1–11 (1972).

81. See Ref. 14, Chapter 10.

82. Rohsenow, W. M., and Hartnett, J. P., *Handbook of Heat Transfer*, McGraw-Hill, New York, 1973.

83. See Ref. 75, pp. 355–356.

84. See Ref. 82, pp. 12–29 through 12–33.

85. Mizushina, T., in *Heat Exchangers: Design and Theory Sourcebook* (N. Afgan and E. J. Schlunder, eds.), McGraw-Hill, New York, 1974.

86. Bell, K. J., and Ghaly, M. A., An Approximate Design Method for Multicomponent/Partial Condensers, *AIChE Symp. Ser.* **69**(131), 72–79 (1973).

87. Ward, D. J., How to Design a Multiple Component Partial Condenser, *Petrol./Chem. Eng.* **32**, C-42 (1960).

CHAPTER 8
CONTROL OF GASES BY COMBUSTION

JOHN HIRT

Hirt Combustion Engineers
Montebello, California

Combustion—actually a thermal oxidation ("incineration") process—provides a control method to effectively destroy hydrocarbon components of waste gas (or liquid) streams by oxidation to CO_2 and H_2O. Other stream components (e.g., halogen–or sulfur–containing organic compounds) are chemically altered by thermal oxidation to enable removal (or recovery) from the exhaust gas stream.

The net result of proper application of combustion technology to waste gas streams is an environmentally acceptable exhaust gas to atmosphere, with minimal discharge of unconverted hydrocarbons, NO_x, SO_x, halogens, and other original contaminants.

This chapter deals with the practical extension of combustion (thermal oxidation) technology to control of waste gas streams. Key factors influencing combustion equipment design are emphasized, rather than theories of combustion, that is, the emphasis is on extension of combustion theory to practical control equipment design.

1 BACKGROUND

With the coming of the "hydrocarbon age," discharge by the process industries of a wide variety of hydrocarbon-containing fumes to atmosphere or hydrocarbon-containing liquid effluent streams is commonplace. Control techniques include a wide variety of chemical (thermal oxidation, absorption) and physical (e.g., adsorption, electrostatic precipitation) processes, which are discussed elsewhere in this book.

Alert technical and management personnel will recognize that, when hydrocarbon-containing fumes (and/or liquids) are handled on-site via thermal oxidation coupled with energy recovery from the hot exhaust gas stream, the hydrocarbon-containing fumes actually represent a "plant asset" rather than an operating/disposal cost or liability. Advanced thermal oxidation systems can handle both fume and liquid hydrocarbon-containing process streams, so that hydrocarbon discharge from the entire plant is minimized and energy is recovered.

Thus, rather than hydrocarbons representing simply a control problem, the heat value of the organics is effectively used to reduce plant energy consumption, and costs for outside disposal of any liquid wastes (and its associated liability) are eliminated.

2 THERMAL OXIDATION

Thermal oxidation (or "incineration") systems must be properly designed to fit the particular process conditions and requirements. If not correctly applied, the system can constitute an explosion hazard and a process capacity limitation and can require substantial maintenance. If properly applied, and coupled with appropriate type of energy recovery, a safe, reliable, and cost-effective system results.

2.1 Combustion Chemistry

Theoretical aspects of combustion chemistry are discussed elsewhere in this book.

2.2   System Design Parameters

2.2.1   *Process/Thermal Oxidizer Interface*

The thermal oxidizer system must be designed to handle all process conditions, including startup, shutdown, and any possible upset condition in addition to normal operating conditions. For both gas and liquid waste streams, key necessary design parameters for *each* of these process conditions include (on a maximum, minimum, and normal basis):

> Origin of waste streams
> Volatile organic compound ("VOC") mass flow rates
> Liquid waste mass flow rates
> Volumes
> Temperatures
> Pressures
> Composition
> Heat content (expressed in a variety of ways, e.g., percent lower explosive limit (%LEL), Btu/scf, Btu/lb)
> O_2 content (of fume stream)
> Condensable material (or aerosol material)
> Inorganic material
> Particulate matter (particulate size and composition)

The design requirement that the thermal oxidizer system be able to handle all possible process conditions, rather than simply a "normal" process condition, substantially complicates design.

In some cases, field testing of process situations may be required to develop the above process information. It is mandatory that the designer have a thorough understanding of the upstream process, with all ramifications, to design a safe, effective, and reliable system. Note that it is not acceptable to rely on, for example, "average" oven or coating tower hydrocarbon loadings calculated from normal process throughput data. In many cases, excursions of hydrocarbon levels into or close to the lower explosive limit (LEL) can occur with changing process conditions, rendering hazardous a system designed to only handle normal process situations. Care should be taken to consider all sources of organic emissions.

2.2.2   *Exhaust Flow Rates*

Many processes have been designed without control equipment in mind. These are typically characterized by high-exhaust flow rates to ensure "safe" operation, that is, to ensure organic concentrations well below the LEL.

High-exhaust flow rates with associated low percentage of the LEL are generally not practical. What is needed, again for the most stringent design conditions, is the lowest possible exhaust flow rate to provide the highest percent LEL consistent with insurance requirements. Generally, most insurance underwriters will permit process operation up to 25% LEL. Systems can be designed to meet insurance requirements at higher percent LELs, with LEL monitoring equipment generally required.

2.2.3   *LEL Levels*

Extreme caution should be used in analyzing oven and coating tower designs to make certain that all exhaust systems are picking up the maximum hydrocarbon loading (or maximum percentage of LEL) to the thermal oxidizer. There have been many instances where hydrocarbon "pockets" exist in ovens/towers, that constitute a potentially dangerous situation. It is also not generally recognized that most organic solvents have vapor densities substantially heavier than air, such that higher hydrocarbon concentrations (and higher percent LELs) can exist in the lower portions of the oven.

Generally, the process equipment should be, if at all possible, designed for operation under a slight negative pressure in order to prevent any fume leakage to the atmosphere.

The following table shows the interrelationship of exhaust flow and percent LEL levels, and impact on maximum auxiliary fuel consumption for the thermal oxidizer (at 1400°F) fumes at 100°F, ignoring any minimum burner requirements:

Percent LEL	Exhaust Flow Rate, scfm	Auxiliary Fuel Consumption, MM Btu-hr
2.5	10,000	14.9 MM
25	1,000	0.8 MM
40	625	0.3 MM

Since both capital and operating costs of thermal oxidizer equipment are significantly reduced, reducing exhaust flow rates to give the highest possible percent LEL consistent with insurance requirements is mandatory.

2.2.4 Other Fume Stream Characteristics

In some cases, with differing fume or waste stream compositions, it is desirable to have separate thermal oxidizer units to achieve the most effective and economical control. A typical example would be mixed streams containing halogens and inorganics. Higher thermal oxidation temperatures provide more favorable equilibrium shift to the halogen acid. However, inorganics are best handled at lowest destruction temperature so that they will carry through as solid particulate (versus molten salts which cause refractory attack, deposition on downstream equipment, etc.).

3 COMBUSTION SYSTEMS

3.1 General Principles

The combustion system is the heart of an effective thermal oxidizer system. With proper and complete process information, design parameters can be developed.

3.2 Design Parameters

3.2.1 Burner

Burner turndown requirement (i.e., maximum and minimum output) is determined from extremes in flow rates, hydrocarbon loadings, and temperature changes.

Many continuous processes have fixed flow rates, hydrocarbon loadings, and off-gas temperatures. Design is straightforward.

At the other extreme are noncontinuous processes with many sources of fumes of varying compositions, flow rates, and temperatures. A much more sophisticated burner system is required in these cases.

Whenever possible, the O_2 content of the fumes should be used for combustion (i.e., "inside air") since typical fuel savings are above 30% versus use of "outside" combustion air. When insufficient O_2 is present in the fumes, then outside air must be introduced for complete oxidation. Combustion products such as CO_2 and H_2O dilute the O_2 content—this dilution effect must be considered in the thermal oxidizer design.

By preheating the fumes, the burner capacity requirements are reduced. The burner design must be capable of handling these preheat temperatures without destroying itself or causing preignition.

Preheating fumes widens the explosive limits; care must be exercised with high percentages of LEL and/or high-temperature fumes to ensure that no flashback occurs to the process fume source. When there is the remotest possibility of such flashback, then the system must be capable of automatic isolation from the process.

Some process fumes can be within the explosive range or above the upper explosive limit (UEL). These can be handled in one of three ways:

1. By supplying them to the burner through a pipe train and manifold and treating them as a "fuel gas."
2. By nozzle mixing with air so air velocity plus mix velocity is always above the flashback velocity.
3. By dilution with air to a safe concentration below the LEL.

3.2.2 Waste Liquids

Waste organic liquids, if fired in a thermal oxidizer system, generally reduce the burner's auxiliary fuel requirements. In many cases, elimination of outside disposal costs (and possible future liability) and the energy credit for the heat content of the wastes themselves combine to result in very short payback for the thermal oxidizer system.

3.2.3 Multiple Waste Streams

If multiple gas and liquid streams exist which are not compatible, the burner must be designed to accommodate this situation. In such cases, the thermal oxidizer becomes an "energy integrator," calling for heat from the auxiliary fuel or waste streams to meet its requirements.

3.2.4 Waste Compositions

Conversion of halogenated materials to the halogen acids is favored by increasing temperatures. However, thermal oxidizer design is the key to achieving required destruction. Burner ignition chamber temperatures of higher than 2000°F can initiate oxidation such that much lower thermal oxidizer combustion chamber temperatures can be used to effect destruction. Also, halogen acids raise the dewpoint temperature. All steel in contact with acidic fumes must be above the dewpoint (i.e., above approximately 350°F). Acidic gases generally go through an absorption step for neutralization or recovery of halogen acid. Salts (inorganics) must be carefully considered on a case by case basis. Minimizing refractory attack and deposition of molten salts on downstream equipment are keys. Metallic oxides, carbonates, and silicons can often be "frozen" to solid particulate prior to hitting the refractory—lower operating temperatures obviously favor this.

Sulfur-containing wastes are converted primarily to SO_2, which is removed in downstream exhaust gas cleaning equipment.

Equipment downstream of the thermal oxidizer and any heat recovery system is designed to handle the specific exhaust gas components (e.g., particulates, halogen acids), prior to final discharge to atmosphere.

4 COMBUSTION EQUIPMENT

4.1 Engineering

Design of effective thermal oxidizer systems requires many engineering disciplines. In many cases, these specialty disciplines are only developed over a long period of time. To the extent that the system is more complex, the extent of the burner and control design determines the success of the system.

4.2 Burner Types

4.2.1 Gas Burners

If gaseous fuels are used, generally the gas main pressure (or gas booster) will be sufficient to get the gas through the metering devices and into the burner.

4.2.2 Package Burners

Conventional package burners use a fan to boost combustion air through metering devices and into the burner. If the process fumes are "dirty" (i.e., contain condensables, particulate, etc.) and "inside air" is used, cleaning and maintenance problems can occur.

4.2.3 Wing Burners

A wing burner has a high heat resisting "V" shape, with fuel gas discharged at the base of the "V." This burner relies on a combustion air flow toward the point of the V with velocities generally of 2000 to 3000 ft/min to force sufficient air into the V to ensure combustion. Any change in fume flow to the thermal oxidizer requires a change in the fuel gas supply which is readily accomplished. However, a change in silhouette area is also required to retain the necessary velocity—this should generally be accomplished automatically to handle varying process conditions.

4.2.4 Multijet Burners

A multijet burner system consists of a number of venturi "tiles." Behind each venturi is a gas jet. Combustion air (either fresh or supplied by the fume stream) and fuel gas stabilize the combustion process in the expanding venturi tube. The burning gases are discharged into a refractory-lined, high-temperature ignition tube. When the O_2 of the fumes is high enough for "inside air" to be used, the fume stream is carefully split or metered into a segment going to the burner with the remainder directed around the burner to dilute or quench the burning gases to the correct incineration temperature in the main combustion chamber. Once this "metering" is set, changes in fume flow rates change the percentage of flow to the burner as well as the Btu requirements—hence the system automatically accommodates changes in total fume flow rate.

Liquid wastes and/or fuel oils can effectively be fired into the burner ignition tube. The high operating temperature in the ignition tube (generally well in excess of 2000°F) provides complete atomization, vaporization, and ignition of the liquids before they enter in gas form the main combustion chamber for completion of the thermal oxidation process.

Fumes of high Btu content (i.e., close to the LEL, or within or above the explosive range) can be taken through a separate burner pipe train and manifold and fired into the multijet venturis. Another manifold supplies natural gas to satisfy system heating requirements. In essence, the rich fume stream can be treated as a fuel gas. The major requirements with this approach are that the fumes do not contain condensables that would deposit out in the fume gas pipe train and/or manifold, that the fumes can be supplied at sufficient pressure to the system (e.g., approximately 2 psig), and that proper precautions are taken to prevent preignition and flashback to the process source.

4.2.5 Premix Burners

Premix burners provide for mixing of the fume, fuel, or liquid waste streams. A big advantage is that combustion stoichiometry can be precisely controlled (e.g., this type of burner is used to generate "inert gas" from stoichiometric combustion of fuels). Disadvantages of this type burner are that they may have an internal cleaning problem, and require an additional power source to move the gases through the system.

4.2.6 Liquid Waste/Oil Burners

In many cases, it is desirable to include firing capabilities for fuel oil, liquid wastes, and/or other hydrocarbon fuel sources, in addition to fuel gas, in the system. An effective method is to fire these through burner guns into a high-temperature ignition tube, which facilitates faster combustion of these slower burning liquids.

High-pressure (80–100 psig) air or steam-atomized burner guns are effective in achieving complete atomization even with high viscosity liquids, and provide greater turndown. Fuel oils and/or liquid wastes of different compositions (i.e., nonmiscible wastes) can be handled in the same burner/ignition tube by using a special gun configuration, with both liquids having their individual atomizing air or steam. The gun tip must be carefully designed to integrate the liquid and atomizing passages to obtain sufficient atomization. Also, it must be designed to be readily removable from the burner system for maintenance and service.

Low pressure air atomizing burners (approximately 1 psig) require a much larger volume of atomizing air for the same energy release and hence do not get as much intimate contact between the liuqid and the atomization medium, and thus require a lower liquid viscosity to ensure reasonable atomization.

4.3 Controls

Control systems include three component control systems—safety controls, temperature controls, and necessary process safety controls interlocked from the thermal oxidizer to the upstream and downstream process equipment. These must be fully integrated into a unified and safe control system.

Severity of an explosion is exponentially related to the volume that accommodates the explosion. Since total system volume (i.e., thermal oxidizer, ducting to thermal oxidizer from process, and the process equipment itself) can be significant, careful attention to system safety is mandatory. A typical thermal oxidizer control system is as follows.

4.3.1 Safety Controls

Insurance requirements for the specific installation generally dictate the type of burner control systems. The optimum system complies with the prevailing regulations and is also practical.

Most thermal oxidizer systems are operated unattended after startup, so that controls must be fail-safe. Thus, despite the desire to keep the control systems simple, protection against each failure mode results in quite complicated safety control systems with increased system cost (offset by increased safety, reliability, and reduced operating costs). The basic safety control system theory is as follows.

Prior to system startup, confirmation of complete fuel safety valve shutoff is necessary. Approved automatic valves (which are extremely reliable) are required. These valves have their own assured power to close, and are unable to be opened until the system has been purged and the pilot properly lit.

The safety system is based on a draft to move a safe amount of air. A fresh air purge prior to ignition, an automatic system to get maximum flow when excessive thermal oxidizer temperatures are reached, and automatic disconnect of the thermal oxidizer from the process whenever hazardous upset conditions occur are requirements. Pressure switch and flow indicators must be applied carefully to ensure accurate and reliable operation.

Since a major concern is that proper draft be established at all times, power to the air/fume mover as well as the draft switch are both interlocked to the safety shutdown system. Controls must be provided to ensure that the draft switch is always functioning.

High- and low-fuel gas pressure, low-pressure liquid (e.g., fuel oil) and atomizing medium, and low-combustion air pressure are monitored and limit controlled to provide safe startup, operation, and shutdown. Combustion safeguard systems that provide automatic instantaneous relights should be locked out so that a new ignition trial cannot occur until a new purge cycle has been completed.

Limits must be installed in the correct sequence in the control circuit for proper operation and to facilitate troubleshooting. The cause of failure of any limit or component that indicates abnormal operation must be capable of being identified, isolated, and repaired or corrected.

4.3.2 Temperature Controls

Overheating of the thermal oxidizer is indicative of excessive heat content in the fume or waste streams. When this occurs, a fail-safe design must be incorporated into the system (e.g., the incoming stream must be diverted from the thermal oxidizer and, if necessary, fresh air used to purge the system prior to attempting relight).

4.3.3 Thermal Oxidizer/Process Interface

The design of the thermal oxidizer/process interface and extent of necessary interlocks is determined by the characteristics of the fume or waste stream (Section 2.2.1). For this purpose, wastes may be categorized as follows:

1. Low Btu fumes: fumes where the maximum instantaneous hydrocarbon loading is below 25% LEL under any conceivable process condition.
2. High Btu fumes: fumes that exceed 25% LEL, but do not exceed approximately 40% LEL.
3. Explosive fumes: fumes above approximately 40% LEL up to 100% LEL, fumes within the explosive range, and fumes above the UEL.
4. Fumes well above the UEL, which can be considered as organic fuels.
5. Liquid organic-containing wastes.
6. Any/all combinations of the above.

4.3.3.1 Low Btu Fumes. Generally in these systems the thermal oxidizer is connected directly to the process fume source. Provision must be made for correct ducting design from the process to the thermal oxidizer, especially if any condensable or particulate material is present. Ducting should be properly drained and cleaned so that there is no accumulation that could reach the thermal oxidizer in a slug or support a flashback.

4.3.3.2 High Btu Fumes. Whenever there is the remotest possibility that the fume hydrocarbon levels could concentrate to a sufficient level to cause a flashback, it is necessary to make provision for ensuring a safe system shutdown. This will include keeping air flow maximized, all dampers driven wide open, process interlocks that provide for process (e.g., oven) shutdown, process vents open, and process heat and vapor input shutdown.

In essence, the thermal oxidizer must be immediately and reliably removed from the process source, with assurance that the source is adequately vented or shut down until the two systems can be safely started and reconnected.

4.3.3.3 Explosive Fumes. These fumes can be diluted with air to lower the organics loadings to safe levels. A safer, more efficient and lower cost approach is to take this type fume through a separate pipe train and burner manifold or nozzle mixing burner design with adequate provision to prevent flashback to the source, and essentially treat these high Btu fumes as a fuel gas.

4.3.3.4 Organic Fuels. These fumes can be handled as other gaseous fuels but with the addition of backflash prevention when there is the possibility of upstream process upsets causing a hazardous condition (including the presence of oxygen in the fumes).

4.3.3.5 Liquid Organic-Containing Wastes. The interface for liquid wastes is more straightforward, but with combinations of fumes and liquid wastes the earlier considerations apply.

In many cases, available process data only reflect "average" conditions, rather than the maximum instantaneous conditions (e.g., fume flow rate, hydrocarbon loadings) necessary for design of a safe, effective, and reliable system. Consequently, to accurately define the above process information, testing is mandatory.

4.4 Combustion Chamber Design

Combustion chamber design includes provisions for proper entry of all process streams, an ignition source, a means of mixing fuels with proper amounts of oxygen, a means of mixing hot combustion gases with dilution air or low Btu fumes, proper geometry to enable holding the mixed wet gases at design temperature until oxidation is essentially complete, and proper safety and temperature controls.

4.4.1 Flame Contact

It is a common misconception that flame contact with the fume stream is necessary for destruction of hydrocarbon fumes. Actually, destruction is a thermal oxidation process, with kinetics governed by temperature (i.e., a time/temperature relationship). Temperature is the controlling factor, since the visible flame has been well extinguished before the combusted gases are mixed into the portion of the fume stream going around the burner and into the main combustion chamber. This is why the term "thermal oxidizer" more accurately describes this type of equipment. Direct flame contact is not necessary.

4.4.2 Temperature/Residence Time/Gas Phase Mixing

A 1200–1400°F range is about the lowest temperature range that will destruct hydrocarbons within a practical residence time in the combustion chamber zone. Since the time constants rapidly increase from fractions of a second to minutes as the temperature is reduced, it is extremely important that the gas phase be intimately mixed to obtain maximum destruction with the minimum combustion chamber size. Some additional considerations are:

1. Increasing temperatures decrease residence time, but increase auxiliary fuel consumption, may affect heat exchanger reliability, and affect formation of side-reaction products such as NO_x.
2. Inadequate mixing increases residence time requirements since some organic molecules may be O_2 starved or not reach a sufficiently high temperature for oxidation to occur.
3. Particulates and aerosols in the fume stream require longer residence time.
4. Each organic compound has its own autoignition temperature and reaction rate time constant, which in turn affects residence time.

4.4.3 Destruction Efficiency

Destruction efficiency is a calculated term generally misunderstood with reference to fume incineration. Fumes with high organic loadings lead to high destruction efficiencies, whereas very low organic loadings in fumes lead to lower destruction efficiencies.

For example, assume a thermal oxidizer system capable of destroying hydrocarbon provides a 10-ppm hydrocarbon content in the exhaust stream. With a hydrocarbon content of 10,000 ppm in the inlet fumes, destruction efficiency is 99.9%. However, with only 100-ppm hydrocarbon in the incoming fumes, destruction efficiency is 90.0%. Thus, the important factor from a pollution control standpoint is the absolute mass flow of undestructed organic materials in the gases discharged to atmosphere from the thermal oxidizer, not the so-called destruction efficiency.

4.4.4 Combustion Chamber Controls

When mixing is complete and temperatures are uniform, a temperature controller should measure the temperature. This instrument should be connected to a control system with proper response to modulate the Btu input to keep a precise minimum temperature that will optimally destroy the organic materials being supplied to the thermal oxidizer.

4.5 Fan Selection

Thermal oxidizer systems require a controlled movement of gases. This is usually accomplished upstream or downstream of the thermal oxidizer by a fan (but sometimes by pressure release from upstream equipment).

4.5.1 Forced Versus Induced Draft

The thermal oxidizer exhaust gases are "clean." However, since they are usually at higher temperatures than incoming fumes (even with heat recovery equipment where it is economically desirable to exhaust to atmosphere at a temperature not greater than 350°F), a larger fan and driver are required with an induced draft system. With induced draft systems, however, any system leakage is inward. This reduces potential atmospheric contamination when incinerating toxic fume streams (e.g., HCN), but will increase heating load should leaks develop. Forced draft fans are available to handle "dirty" fume streams containing particulates and/or condensables, but they may require additional cleaning and maintenance.

4.5.2 Fan System Interlocks

The thermal oxidizer system safety depends on a properly and continually working fan—thus the fan is the first piece of equipment to be started and the last to shut down. The fan should be interlocked electrically, as well as mechanically, with a flow proving switch.

4.6 Refractories

There are generally four areas of the thermal oxidizer system where separate refractory design considerations apply.

4.6.1 Inlet to Thermal Oxidizer

Some systems handle fumes at high temperatures, thus requiring insulation. If the inlet is internally insulated, the need for expansion joints is minimized. However, where fumes containing condensables are handled, external insulation with expansion joints may more than offset the maintenance of internal insulation.

4.6.2 Fume Combustion Zone

The refractory lined ignition tube or combustion zone requires high-temperature refractories. This is especially true when firing organic liquid waste and/or fuel oil, since high temperatures promote droplet vaporization prior to destruction. Improperly firing oil/liquid waste burners impinging on refractories in this zone can greatly reduce refractory life.

4.6.3 Transition to Combustion Chamber

The area where the hot burner gases mix into the balance of fumes taken around the burner is subject to thermal shock. Refractories should have high insulating qualities, low expansion factors, and low porosity.

4.6.4 Combustion Chamber

Combustion chamber refractories require high insulating qualities to ensure the combustion gases maintain a proper and uniform temperature (i.e., to minimize heat losses). The insulation must also withstand the abrasive effect of high-velocity hot gases, thermal shock, and should not form lower melting eutectics with any inorganic components in the combustion gases.

5 HEAT RECOVERY

Hot thermal oxidizer exhaust discharge gases are an extremely valuable resource. For example, 1000 scfm of 1400°F exhaust contains heat equivalent to approximately 1000–1100 lb/hr of 100–psig steam (which at $6.00/1000 lb steam and 8000 operating hr/yr is equivalent to $50,000/yr or more). Most companies emitting hydrocarbons, when faced with air pollution control compliance, do not recognize that these hydrocarbon emissions, properly controlled, can significantly reduce overall plant energy costs. The heat recovery technique depends on the specific process and plant operating requirements. Several major options are as follows.

5.1 Heat Exchangers

Preheating fumes with the hot thermal oxidizer exhaust gases provides a system always in balance at any given flow rate. Heat exchanger effectiveness, and consequently heat exchanger selection, is controlled by the percent LEL and temperatures of the incoming fume stream. As percent LEL and/or temperatures are increased, the maximum permissible heat exchanger effectiveness (i.e., without overheating the system) is lowered—which also lowers capital and operating costs. This is shown in Table 1. Thus, even though fume preheat is limited with high-LEL/high-temperature fumes, other techniques should be used for energy recovery.

The practical requirement for a minimum burner fire rate to ensure satisfactory and stable system operation will further reduce the above maximum permissible effectiveness figures. When percent LEL levels vary widely in a process, a heat exchanger bypass system may be required to provide maximum heat recovery at low percent LEL levels, and avoid overtemperaturing the system at high percent

Table 1   Maximum Permissible Heat Exchanger Effectiveness (%)[a]

	% LEL		
Inlet Fume Temperature	2	25	40
100°F	91	54	26
400°F	88	40	4

[a] At 1400°F combustion temperature ignoring any minimum burner requirement.

LELs. Multiple pass shell and tube (recuperative) heat exchangers can achieve up to approximately
80% effective heat recovery. Rotary heat wheel (regenerative) heat exchangers can achieve approximately
85% effectiveness, generally at lower pressure drops. On rotary wheels, the system should be designed
such that any leakage is from the clean gas side to the incoming fume side to avoid discharge of any
contaminants to atmosphere.

As in any engineering design, the system should be optimized to enable maximum heat recovery
at minimum cost (minimizing horsepower requirements).

Secondary heat exchanger systems can also be used (i.e., for fresh air preheat) when sufficient
temperatures exist in the exhaust gas discharge from the primary heat exchanger system. From a
practical standpoint, final discharge temperature from the entire system should be as low as possible,
consistent with the consideration of minimizing dewpoint corrosion. Normally, this is approximately
350°F unless provision for exhaust dewpoint corrosion is made in design.

5.2   Waste Heat Boilers

When process steam can be used, a waste heat boiler provides economical low maintenance heat
recovery. Duty on existing direct-fired boilers then can be reduced.

When boiler discharge gases are not close to their dewpoint, then boiler feed water can be preheated
with an economizer. Cogeneration of electricity and process steam can also be employed with waste
heat boilers or condensing turbine generators (which are less economically effective).

5.3   Organic Fluid Heaters

Organic fluid heaters can be used when process requirements call for high temperatures. These systems
provide high temperatures at low system pressures, but require more meticulous attention to design
and operating details.

It is obvious that many factors must be considered to ensure a system that will operate safely,
reliably, economically, and effectively in handling destruction of fumes and/or liquid wastes by combustion. However, it is most important to recognize that the hydrocarbons that have to be controlled
constitute a valuable resource. Properly designed equipment for the combustion of gases and/or liquid
wastes can actually reduce overall plant energy consumption—as many case histories will attest.

CHAPTER 9

CONTROL OF GASES BY CHEMICAL REACTION

J. D. MOBLEY

Industrial Environmental Research Laboratory
U.S. Environmental Protection Agency
Research Triangle Park, North Carolina

K. J. LIM

Energy and Environmental Division
Acurex Corporation
Mountain View, California

1 INTRODUCTION

As pollution control requirements are becoming more stringent, the control of gaseous pollutants by chemical reaction is receiving increasing attention as an alternative or supplementary means of increasing pollution control, process reliability, energy efficiency, or cost effectiveness. A number of chemical reaction control techniques have been under development. This paper reviews some of the developing chemical techniques potentially available for the control of NO_x and SO_x emissions from stationary combustion sources.

Chemical reaction control is defined here as a process in which a chemical reaction is the dominant unit operation rather than absorption, adsorption, condensation, or combustion (incineration). Since most processes involve more than one unit operation, it is difficult to categorize a control mechanism by only one of these classifications. For example, SO_2 control by lime/limestone scrubbing, even though the process includes a chemical reaction step(s), is not included here because of the dominance of the absorption unit operation in the scrubbing process. The definition of chemical reaction control used here is somewhat arbitrary, and introduced solely for convenience in discussion.

The purpose of this chapter is to identify and review promising chemical reaction systems for the control of gaseous pollutants from stationary combustion sources. Applicability of these techniques is potentially widespread, so it is hoped that by highlighting some of the more important ones, potential users can then consider the technology transfer possibilities for their respective processes or industries. The following discussion, which is organized by pollutant, reviews the various process options, control effectiveness, applicability, status of development, factors affecting performance, and some of the advantages and disadvantages.

2 NO_x CONTROL

NO_x control processes are perhaps the best examples of pollution control by chemical reaction. Thus, the emphasis of the chapter is on these NO_x control techniques, which include noncatalytic homogeneous reduction by NH_3 and selective catalytic (heterogeneous) reduction by NH_3.

2.1 Ammonia Injection (Noncatalytic)

The ammonia injection process controls NO_x in combustion flue gas by reducing NO to N_2 and H_2O in the presence of oxygen with injection of NH_3 as the reducing agent. The relevant overall reactions are as follows:[1,2]

$$NO + NH_3 + \tfrac{1}{4}O_2 \rightarrow N_2 + \tfrac{3}{2}H_2O \tag{1}$$

$$NH_3 + \tfrac{5}{4}O_2 \rightarrow NO + \tfrac{3}{2}H_2O \tag{2}$$

The first reaction dominates at flue gas temperatures ranging from 1070 to 1270 K (1470 to 1830°F). A likely mechanism which results in the overall reaction (1) is given elsewhere.[3] Above 1370 K (2000°F), however, reaction (2) becomes significant and the undesirable formation of NO occurs.[4] Thus the ammonia injection process is very temperature sensitive with maximum NO_x reductions occurring in a very narrow temperature window around 1240 ±50 K (1770° ±90°F). The data in Figure 1, taken from tests on flue gas from a gas-fired pilot-scale combustor, dramatically display this temperature sensitivity.[5] The addition of hydrogen can lower and thus extend the effective temperature window. With a $H_2:NH_3$ ratio of about 2:1, NO_x reduction can be forced to proceed rapidly at about 970 K (1290°F).[4]

The ammonia injection NO_x control technique is a commercially offered system,[4] potentially applicable to a broad spectrum of stationary combustion systems, including utility and industrial boilers and process furnaces. The process has been demonstrated on several gas- and oil-fired boilers and process heaters in Japan and on an enhanced oil recovery steam generator in the United States.[6] Figure 2 shows the 40–60% NO_x reductions typical for these installations.[4,7] Figure 3 is a schematic of a typical application of NH_3 injection to a utility boiler, with the NH_3 injection nozzles located in the convective section.[8] To meet stringent local regulatory needs, several refinery heaters, industrial boilers, and oil recovery steam generators in California are being fitted with ammonia injection systems. No coal-fired boilers have employed the technology to date.[6]

The major factors affecting performance include:

1. Combustion system/fuel characteristics.
2. Residence time at optimal temperatures (flue gas).
3. Temperature profile across boiler.
4. $NH_3:NO_x$ ratio and NO_x concentration.
5. Mixing.

In practical applications, the limiting phenomenon is the injection of NH_3 in the proper boiler location so as to achieve maximum NH_3/NO_x (and possibly H_2) mixing within the "desired" temperature window, consistent with normal boiler operation. This will require boiler temperature profile mapping as a function of load. Residence times of 0.2 to 0.3 sec are adequate. A $NH_3:NO_x$ ratio of 1.5 is commonly used for initial NO_x levels of 200 ppm or less. This ratio is reduced toward 1.0 as the initial NO_x concentration increases.[4] Minimizing excess air tends to favor the effectiveness of ammonia injection as a NO_x control technique (although the reaction chemistry would indicate a minimum

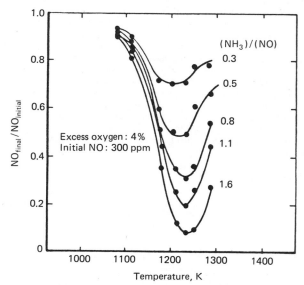

Fig. 1   Effect of temperature on NO reduction with ammonia injection.[5]

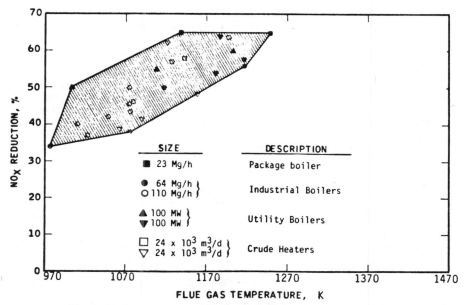

Fig. 2   Performance of ammonia injection in commercial applications.[4]

O$_2$ concentration is necessary). This is compatible with the combustion modification NO$_x$ control technique of low excess air boiler firing.

Since ammonia injection is still a developing technology for NO$_x$ control, there are several potential implementations and operational problems that can occur:

1. Optimal effectiveness for noncatalytic reduction of NO by NH$_3$ occurs over a very narrow temperature range; hence the precise location of NH$_3$ injection ports is crucial.

2. Since the temperature in a boiler changes with load and/or fuel fired, NO$_x$ control with NH$_3$ may dictate load restrictions.

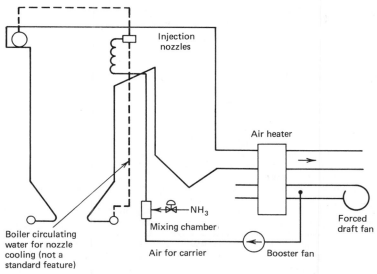

Fig. 3   Schematic diagram of the NH$_3$ injection system.[8]

3. Emissions of NH_3 (generally below 50 ppm) and by-products.
4. Possible boiler equipment (e.g., air preheater) fouling by ammonium bisulfate when firing high-sulfur oil or coal.
5. Cost of the process can be much higher than for combustion modification techniques.

However, these potential problems can be, and have been in specific installations, minimized with such modifications as multiple ammonia injection grids for load following capability, as well as water washing and soot blowing for fouling control. Moreover, the major strengths of the technique are its potential for high NO_x removal (40–60%), and its applicability as an additional control that can be combined with conventional combustion techniques for large NO_x reductions. If even greater reductions are necessary, postcombustion flue gas treatment techniques, with their associated higher equipment and operating costs, may need to be considered.

2.2 Selective Catalytic Reduction (SCR)

A large number of postcombustion flue gas treatment techniques have been investigated for reduction of NO_x emissions. Dry (nonscrubbing) NO_x processes have emerged as the most promising because the wet processes are inherently more complex, and have associated water pollution problems and significantly higher costs.[9,10] Some of the more advanced commercial systems were discussed at a recent EPA/EPRI symposium.[11-19]

Selective catalytic reduction refers to a process that chemically reduces NO_x with NH_3 over a heterogeneous catalyst in the presence of O_2. The process is termed selective because the reducing agent NH_3 preferentially attacks NO_x rather than O_2. However, the O_2 enhances the reaction, and is indeed a necessary part of the reaction scheme. Thus SCR is potentially applicable to flue gas from fuel-lean combustion systems; that is, the flue gas is under oxidizing conditions (e.g., greater than 1% O_2). Nonselective catalytic reduction (NCR) processes are applicable to fuel-rich firing combustion conditions (i.e., reducing conditions in the flue gas). The overall SCR reactions can be expressed as[9]

$$2NH_3 + 2NO + \tfrac{1}{2}O_2 \rightarrow 2N_2 + 3H_2O \qquad (3)$$

$$2NH_3 + NO_2 + \tfrac{1}{2}O_2 \rightarrow \tfrac{3}{2}N_2 + 3H_2O \qquad (4)$$

Equation (3) represents the predominate reaction since approximately 95% of the NO_x in combustion flue gas is in the form of nitric oxide (NO). Therefore, in theory, a stoichiometric amount of NH_3 is sufficient to reduce NO_x to harmless molecular nitrogen (N_2) and water vapor (H_2O). In practice, an $NH_3:NO$ molar ratio of about 1:1 has typically reduced NO_x emissions by 80–90% with a residual NH_3 concentration of less than 20 ppm. SCR has been extensively employed in Japan on gas- and oil-fired industrial and utility boilers.[9,20] A form of SCR has been in use commercially for NO_x removal from the tail gas from nitric acid plants and other chemical processes in the United States and Canada.[21] Moreover, in the United States, a number of demonstration projects on coal-, oil-, and gas-fired utility boilers are under way.[9,11,13]

The major factors affecting SCR performance include:

1. Combustion system/fuel characteristics.
2. Catalyst composition.
3. Catalyst activity, selectivity, and longevity.
4. Catalyst/reactor configuration.
5. $NH_3:NO_x$ ratio and NO_x concentration.
6. SCR reactor temperature.
7. Space velocity.

The SCR processes require a reactor, a catalyst, and an ammonia storage and injection system. Due to increased pressure drop across the SCR reactor, some increase in boiler fan capacity, or possibly an additional fan, may be necessary.

The optimum temperature for the catalytic reaction is in the temperature range 570–720 K (570–845°F). To obtain flue gas temperatures in this range, the reactor is usually located between the boiler economizer and the air preheater. A typical flow diagram is shown in Figure 4. Obviously, the reactor and catalyst are the critical elements of the process and warrant further discussion.

Vanadium compounds have been found to promote the reduction of NO_x with NH_3 and to be unaffected by the presence of SO_x. Titanium dioxide (TiO_2) has been found to be an acceptable carrier, since it is resistant to attack from SO_3. Thus, many SO_x-resistant catalysts are based on TiO_2 and V_2O_5; however, constituents and concentrations of most catalysts are proprietary.[9,21]

Reactor and catalyst configurations also vary with the application, primarily to accommodate the

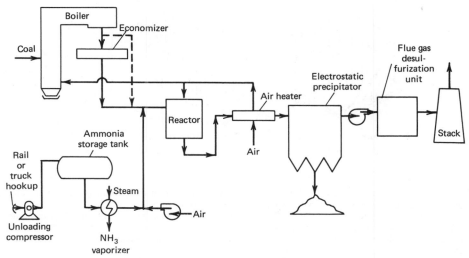

Fig. 4 Typical flow sheet for NO_x selective catalytic reduction processes.[22]

different particulate concentrations. Natural-gas-fired boilers employ SCR catalysts as spherical pellets, cylinders, or rings, and reactor vessels as fixed packed beds. However, designs for use with oil- and coal-fired boilers have to be capable of tolerating particulates (fly ash) in the flue gas stream. For these applications, a parallel-flow catalyst is preferred. Parallel flow means that the gas flows straight through the open channels parallel to the catalyst surface. The particulates in the gas remain entrained while NO_x reaches the catalyst surface by turbulent convection and diffusion. An alternative to the parallel-flow catalysts is the parallel passage reactor. In this design, the catalyst material is arranged in channels and held in place by a metal screen. The operating principle is similar to that of the parallel-flow catalyst.

The various parallel flow catalyst shapes are shown in Figure 5. The catalyst may be a homogeneous material or may be composed of an active material coated on the surfaces of a metallic or ceramic carrier. The parallel-flow catalysts are normally manufactured in a unit cell configuration about 1 m³ as shown in Figure 6. The cells are stacked in banks in the reactor as shown in Figure 7.

Even though much progress has been made in catalyst and reactor design, some problems still remain. The catalysts may not be resistant to all contaminants in flue gas or be able to tolerate high particulate loadings. In addition, fine particulates, smaller than about 1 μm, may blind the catalyst surface. Long-term operation without catalyst plugging or catalyst erosion needs to be demonstrated

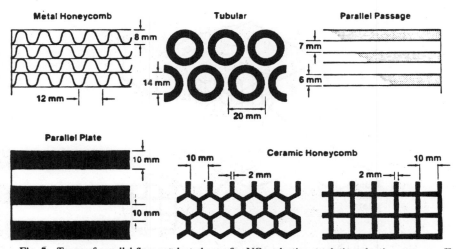

Fig. 5 Types of parallel-flow catalyst shapes for NO_x selective catalytic reduction processes.[20]

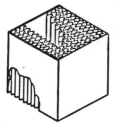

Tubular configuration   Metallic honeycomb configuration

Fig. 6   Unit cells of parallel-flow catalysts for NO_x selective catalytic reduction processes.[20]

for coal-fired applications. Catalyst life also needs to be extended from the current guarantees of 1 to 2 yr of applications with SO_x and particulates in the gas stream.

The $NH_3:NO_x$ molar ratio and the flue gas space velocity are the major operating variables affecting the level of NO_x control achieved for a given boiler condition and SCR system. An $NH_3:NO$ molar ratio of 1:1 can achieve about 90% NO_x reduction, with higher NH_3 rates resulting in higher undesirable NH_3 emissions.

One of the major concerns with SCR processes is the formation of solid ammonium sulfate $[(NH_4)_2SO_4]$ and liquid ammonium bisulfate (NH_4HSO_4) downstream of the reactor. The formation conditions are difficult to completely avoid since some unreacted NH_3 from an SCR system and some SO_3 from combustion of sulfur-containing fuels are expected. The biggest problem seems to be deposition of $(NH_4)_2SO_4$ and NH_4HSO_4 on the air preheater. These compounds are corrosive and can form deposits that plug the air preheater. Air preheater problems have been found to be most severe on high-sulfur oil-fired applications and coal-fired units that employ a hot electrostatic precipitator (ESP) ahead of the SCR reactor and the air preheater.[9] Systems that accept full particulate charge of a coal-fired boiler through the air preheater should have less difficulty with pluggage of the air preheater. In these systems, the fly ash apparently scours the surface of the air preheater to remove any deposits, or the ammonium compounds deposit on the fly ash and are carried through the air preheater by the fly ash. However, increased soot blowing, from both the hot and cold sides, and water washing of the air preheaters appear necessary for most applications. The wash water from cleaning air preheaters as well as purge streams from a flue gas desulfurization (FGD) unit may require treatment to remove NH_3 before being discharged. Modifications to the conventional air preheater design are also being developed to address the ammonium sulfate problem; however, these units have not yet been applied on full-scale systems.

Other concerns and potential problems include: emission of NH_3 and NH_3 compounds; causing or increasing the emission of undesirable compounds such as SO_3; affecting the performance of down-

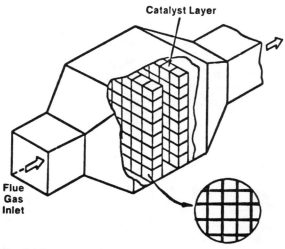

Fig. 7   Parallel-flow reactor for NO_x selective catalytic reduction processes.[20]

stream pollution control equipment such as FGD processes, ESPs, and baghouses; lack of proven NH$_3$ analytical control systems; sensitivity of the process to temperature changes due to boiler load swings; disposal or reclamation of spent catalysts in an environmentally acceptable manner; and reliability of the process and its effect on the boiler system's availability.

Despite these potential problem areas and uncertainties, the processes have been successfully installed and operated in Japan on gas- and oil-fired boilers, and coal-fired applications are being constructed. If 80% or more NO$_x$ reduction is required, SCR is the only near-term option under development. It has good potential as a supplement to conventional combustion modification control techniques.

2.3 Nonselective Catalytic Reduction (NCR)

In nonselective catalytic reduction processes, a reducing agent other than NH$_3$, such as H$_2$, CO, or hydrocarbons, is used. However, these other reducing agents act nonselectively, reacting with O$_2$ and SO$_x$ present in the flue gas, in addition to NO$_x$, and thus a large amount of reductant must be added.[21] One practical application of NCR is to operate the combustor near stoichiometric (so-called "fuel lean") so that exhaust gas is reducing.[23] Then the CO and unburned hydrocarbon present in the exhaust can be used to reduce the NO$_x$ using an NCR system, without the need for additional reductant. However, such operation of a combustion system will require careful control of the air/fuel ratio near stoichiometry; for example, use of a computer-controlled oxygen trim system. This in fact is being used in a number of current production automobile engine systems.[23,24]

Technology transfer to stationary combustion systems such as fuel rich-running internal combustion engines is still in an early development stage,[25,26] and therefore will not be discussed further here.

2.4 Electron Beam Irradiation

Another potential means of NO$_x$ control is the electron beam irradiation process. The radiation process simultaneously removes NO$_x$ and SO$_x$ with NH$_3$, activated by electron beam irradiation, without the need for catalysts.[27,28] An artist's rendition of the process applied to a large boiler is shown in Figure 8. The process first requires removal of fly ash from the flue gas. Then ammonia is added, and the gas enters the electron beam reactors. There, ammonia in the presence of electrons converts NO$_x$ and SO$_x$ into a dry powder of ammonium sulfate [(NH$_4$)$_2$SO$_4$] and ammonium nitrate sulfate [(NH$_4$)$_2$SO$_4$

2NH$_4$NO$_3$]. The exact chemistry is not well established. After the dry powder is removed by a second particulate collection device, the treated flue gas leaves the system hot enough to be exhausted through the stack without reheat.

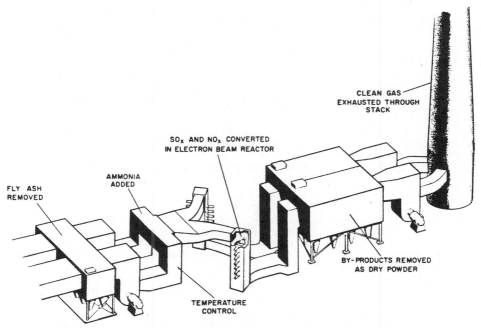

CLEAN GAS
EXHAUSTED THROUGH
STACK

SO$_x$ AND NO$_x$ CONVERTED
IN ELECTRON BEAM REACTOR

AMMONIA
ADDED

FLY ASH
REMOVED

BY-PRODUCTS REMOVED
AS DRY POWDER

TEMPERATURE
CONTROL

Fig. 8 Potential application of electron irradiation process to boiler flue gas.[23]

A pilot plant operating on a heavy oil-fired flue gas has achieved simultaneous removal of over 85% of NO_x and over 95% of SO_x.[27] Although this technology is in early development, it is worthy of attention because of the potential for high simultaneous NO_x and SO_x removal and the production of a dry, easily disposed of waste (e.g., as potential fertilizer feedstock). Large-scale application and the economics of the radiation process need to be demonstrated, however.

3 SO_x CONTROL

There are several processes for control of SO_x emissions that employ chemical reaction as the dominant unit operation. However, these processes also generally require adsorption of SO_2 prior to the occurrence of the chemical reaction. Nevertheless, these processes are included in this discussion since the chemical reaction seems to dominate the control mechanism.

3.1 CuO/CuSO₄ Process

This simultaneous NO_x and SO_x process uses copper oxide (CuO) supported on stabilized alumina placed in two or more parallel-passage reactors. The reactions, which characterize process operation, can be expressed as[15]

Acceptance

$$CuO + \tfrac{1}{2}O_2 + SO_2 \rightarrow CuSO_4 \tag{5}$$

$$2NO + 2NH_3 + \tfrac{1}{2}O_2 \xrightarrow{\text{CuSO}_4} 2N_2 + 3H_2O \tag{6}$$

Regeneration

$$CuSO_4 + 2H_2 \rightarrow Cu + SO_2 + 2H_2O \tag{7}$$

$$Cu + \tfrac{1}{2}O_2 \rightarrow CuO \tag{8}$$

Flue gas is introduced at 660 K (725°F) into one of the reactors where the SO_2 reacts with CuO to form copper sulfate (CuSO₄). The CuSO₄ and, to a lesser extent, the CuO act as catalysts in the reduction of NO_x with NH₃. When the reactor is saturated with CuSO₄, flue gas is switched to a fresh reactor for acceptance of the flue gas, and the spent reactor is regenerated. In the regeneration cycle, hydrogen (H₂) is used to reduce the CuSO₄ to copper (Cu), yielding an SO_2 stream of sufficient concentration for conversion to sulfur or sulfuric acid. The Cu in the reactor is oxidized, preparing the reactor for acceptance of the flue gas again. Between acceptance and regeneration, steam is injected into the reactor to purge the remaining flue gas or H₂ to eliminate any possibility of combustion. The process can be operated in the NO_x-only mode by eliminating the regeneration cycle, or in the SO_x-only mode by eliminating the NH₃ injection. A utility boiler demonstration project is under way in the United States, with a target of 90% simultaneous removal of NO_x and SO_x emissions.[9]

3.2 Coal/Limestone Fuel Mixtures

With major increases in coal consumption projected for utility and industrial boilers,[29] there is a strong need to find environmentally acceptable means of burning high-sulfur coal in existing as well as new boilers. Two active areas of research in utilizing limestone to capture SO_2 emissions from coal combustion are coal/limestone pellets for stoker coal boilers and pulverized coal/limestone feed for pulverized coal boilers. The concept is especially promising because of the current interest in controlling NO_x as well as SO_x emissions. Indeed, laboratory results have indicated that the combustion conditions that minimize NO_x formation (longer residence times in fuel rich zones and lower peak flame temperatures) are also favorable for high-sulfur capture via calcium addition to the fuel.[30] The net reaction of interest may be

$$CaCO_3 + \tfrac{1}{2}O_2 + SO_2 \rightarrow CaSO_4 + CO_2 \tag{9}$$

However, the exact mechanism and sequence of reactions is subject to speculation.[31]

Coal/limestone pellets with a Ca:S ratio of 3:5 have achieved sulfur captures up to 70% for a laboratory combustor and up to 50% on a small industrial stoker.[32] Although these values are lower than the 80–90% SO_x reduction usually cited with established wet scrubbing systems, the results are significant in that the coal/limestone pellets could enable a large portion of the industrial boiler population (coal-fired stokers) to be amenable to SO_2 control. Current research efforts on the coal/limestone pellet system include improving the physical handling characteristics of the pellets, and elucidating the details of the SO_2 removal reaction in a stoker bed.

Sulfur capture research results have also been promising for firing coal/limestone mixture in a low NO_x pulverized coal burner.[33] Preliminary results on a pilot-scale low NO_x burner under development by EPA (limestone injection multistage burner) have shown over 80% SO_2 removal with a 3:1

Ca:S mole ratio.[34] The high SO$_2$ removal has been attributed to the fuel-rich conditions and low peak flame temperatures exhibited with the low NO$_x$ burner.

For both limestone techniques, key research areas include: sulfur capture as a function of coal composition and sorbent type and concentration; the effects of boiler design and operating variables; potential problems such as slagging, fouling, and convective section plugging; and increased particulate collection and waste disposal requirements. A major driving force for the development of these techniques is the opportunity for major reduction in capital and operating costs for SO$_2$ control, as compared to wet scrubbing systems. Furthermore, the dry systems have a greater potential for retrofit applications and also may become attractive SO$_2$ control options for small industrial boilers.

3.3 Dry Sorbent Injection

Another promising technique for SO$_2$ control is dry sorbent injection, which can reduce emissions from boilers by about 50%. This technique, in which a dry alkaline sorbent is pneumatically injected into the flue gas duct upstream of a particulate collection device, is closer to commercial application than the sorbent/fuel mixtures discussed earlier. Figure 9 shows a schematic flow diagram. A fabric filter is the perferred particulate collection device since much of the SO$_x$ capture appears to occur on the bags, which have unspent sorbent on their surface. Sodium-based sorbents have proved to be more effective than calcium-based sorbents.[34] Overall reactions are

$$2NaHCO_3 + SO_2 \longrightarrow Na_2SO_3 + 2CO_2 + H_2O \tag{10}$$

$$Na_2CO_3 + SO_2 \longrightarrow Na_2SO_3 + CO_2 \tag{11}$$

Either nahcolite, a naturally occurring form of bicarbonate, or trona, a naturally occurring mineral (Na$_2$CO$_3$ · NaHCO$_3$ · 2H$_2$O), is a likely sorbent candidate based on reactivity, cost, and availability.[34] Bag temperatures of 380–430 K (230–320°F) should be adequate, although higher temperatures result in better sorbent utilization and greater SO$_2$ removal.[34]

Although dry sorbent injection is nearing commercial application, tests to date have been limited to small industrial boilers. The principal advantages of dry injection are:

1. Near-term commercial availability.
2. Lower capital costs than current FGD systems.
3. Better retrofit capability than current FGD systems.
4. Easier disposal of dry waste.

Major drawbacks include:

1. Lower SO$_2$ reduction capability than current FGD systems.
2. Not demonstrated on large boilers, as are current FGD systems.
3. Sorbent collected with fly ash as a soluble salt with potential for groundwater pollution.

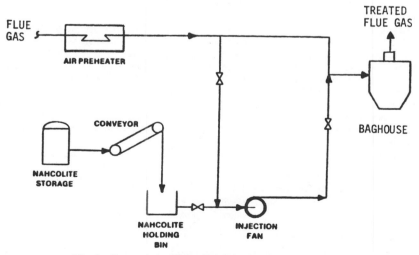

Fig. 9   Dry sorbent (Nahcolite) injection flow schematic.[34]

Spray dryer SO_x control is another technique based on the reaction of SO_2 with lime/sodium carbonate; however, it is not discussed in this paper since, like wet lime/limestone scrubbing, it is predominately an absorption process rather than a chemical reaction process.

4 CONCLUDING REMARKS

This chapter has reviewed some of the emerging chemical reaction techniques potentially available for the control of NO_x and SO_x emissions. In certain situations, chemical reaction control may be the preferred, indeed only, means of meeting stringent control requirements. The emphasis here has been to identify developing techniques, and to discuss their advantages and disadvantages, rather than to rank their utility. As these are primarily developing technologies, it is not possible to make definitive comparisons. It is hoped that practicing environmental engineers can be alerted to new processes and encouraged to keep abreast of developments, so that they can make their own site-specific evaluations.

REFERENCES

1. Bartok, W., Non-Catalytic Reduction of NO_x with NH_3, in *Proceedings of the Second Stationary Source Combustion Symposium: Volume II*, EPA-600/7-77-073b (NTIS-PB 271 756), U.S. EPA, Industrial Environmental Research Laboratory, Research Triangle Park, NC, July 1977, pp. 145–161.

2. Hunter, S. C., and Robinson, J. M., NO_x Controls in Industrial Boilers and Furnaces, presented to the American Flame Research Committee International Symposium on NO_x Reduction, Houston, TX, October 22–23, 1979.

3. Lyon, R. K., Communication to the Editor: The NH_3-NO-O_2 Reaction, *International Journal of Chemical Kinetics* **8**, 315–318 (1976).

4. Hurst, B. E., and Schleckser, C. E., Jr., Applicability of Thermal $DeNO_x$ to Large Industrial Boilers, in *Proceedings of the Joint Symposium on Stationary Combustion NO_x Control: Volume V*, EPA/IERL-RTP-1087, U.S. EPA, Industrial Environmental Research Laboratory, Research Triangle Park, NC, October 1980, pp. 441–470.

5. Muzio, L. J., et al., *Homogeneous Gas Phase Decomposition of Oxides of Nitrogen*, EPRI Report FP-253 (NTIS-PB 257 555), Electric Power Research Institute, Palo Alto, CA, August 1976.

6. Mason, H. B., NO_x Control in Utility Boilers, presented to the ASME Joint Power Generation Conference, Phoenix, AZ, September 30, 1980.

7. Castaldini, C., et al., *Technical Assessment of Thermal $DeNO_x$ Process*, EPA-600/7-79-117 (NTIS-PB 297 947), U.S. EPA, Industrial Environmental Research Laboratory, Research Triangle Park, NC, May 1979.

8. *Non-Catalytic NO_x Reduction Process Applied to Large Utility Boilers*, Mitsubishi Heavy Industries, Ltd., Tokyo, Japan, November 1977.

9. Mobley, J. D., Assessment of NO_x Flue Gas Treatment Technology, in *Proceedings of the Joint Symposium on Stationary Combustion NO_x Control: Volume II*, EPA/IERL-RTP-1084, U.S. EPA, Industrial Environmental Research Laboratory, Research Triangle Park, NC, October 1980, pp. 1–23.

10. Faucett, H. L., et al., *Technical Assessment of NO_x Removal Processes for Utility Application*, EPRI Report AF-568, Electric Power Research Institute, Palo Alto, CA, March 1978.

11. Cichanowicz, J. E., and Giovanni, D. V., Empirical Evaluation of Selective Catalytic Reduction as an NO_x Control Technique, in *Proceedings of the Joint Symposium on Stationary Combustion NO_x Control: Volume V*, EPA/IERL-RTP-1087, U.S. EPA, Industrial Environmental Research Laboratory, Research Triangle Park, NC, October 1980, pp. 322–349.

12. Nakabayaski, Y., et al., Development of Flue Gas Treatment in Japan, in *Proceedings of the Joint Symposium on Stationary Combustion NO_x Control: Volume V*, EPA/IERL-RTP-1087, U.S. EPA, Industrial Environmental Research Laboratory, Research Triangle Park, NC, October 1980, pp. 350–399.

13. Johnson, L. W., et al., Status of SCR Retrofit at Southern California Edison Huntington Beach Generating Station Unit 2, in *Proceedings of the Joint Symposium on Stationary Combustion NO_x Control: Volume II*, EPA/IERL-RTP-1084, U.S. EPA, Industrial Environmental Research Laboratory, Research Triangle Park, NC, October 1980, pp. 24–46.

14. Itoh, H., and Kajibata, Y., Countermeasures for Problems in NO_x Removal Process for Coal-Fired Boilers, in *Proceedings of the Joint Symposium on Stationary Combustion NO_x Control: Volume II*, EPA/IERL-RTP-1084, U.S. EPA, Industrial Environmental Research Laboratory, Research Triangle Park, NC, October 1980, pp. 47–69.

15. Pohlenz, J. B., et al., Treating Flue Gas from Coal-Fired Boilers for NO_x Reduction with the Shell Flue Gas Treating Process, in *Proceedings of the Joint Symposium on Stationary Combustion*

NO_x Control: Volume V, EPA/IERL-RTP-1087, U.S. EPA, Industrial Environmental Research Laboratory, Research Triangle Park, NC, October 1980, pp. 400–411.

16. Wiener, R. S., et al., The Hitachi Zosen NO_x Removal Process Applied to Coal-Fired Boilers, in *Proceedings of the Joint Symposium on Stationary Combustion No_x Control: Volume II*, EPA/IERL-RTP-1084, U.S. EPA, Industrial Environmental Research Laboratory, Research Triangle Park, NC, October 1980, pp. 70–105.

17. Narita, T., et al., Babcock-Hitachi NO_x Removal Process for Flue Gases from Coal-Fired Boilers, in *Proceedings of the Joint Symposium on Stationary Combustion No_x Control: Volume II*, EPA/IERL-RTP-1084, U.S. EPA, Industrial Environmental Research Laboratory, Research Triangle Park, NC, October 1980, pp. 106–128.

18. Aoki, N., and Cvicker, J. S., Test Summary of an Integrated Flue Gas Treatment System Utilizing the Selective Catalytic Reduction Process for a Coal-Fired Boiler, in *Proceedings of the Joint Symposium on Stationary Combustion No_x Control: Volume II*, EPA/IERL-RTP-1084, U.S. EPA, Industrial Environmental Research Laboratory, Research Triangle Park, NC, October 1980, pp. 129–153.

19. Sengoku, T., et al., The Development of a Catalytic NO_x Reduction System for Coal-Fired Steam Generators, in *Proceedings of the Joint Symposium on Stationary Combustion No_x Control: Volume V*, EPA/IERL-RTP-1087, U.S. EPA, Industrial Environmental Research Laboratory, Research Triangle Park, NC, October 1980, pp. 412–440.

20. Ando, J., *NO_x Abatement for Stationary Sources in Japan*, EPA-600/7-79-205 (NTIS-PB 80-113 673), U.S. EPA, Industrial Environmental Research Laboratory, Research Triangle Park, NC, August 1979.

21. Kiovsky, J. R., et al., Evaluation of a New Zeolitic Catalyst for NO_x Reduction with NH_3, *Industrial Engineering Chemistry R&D* 19, 218–225 (1980).

22. Maxwell, J. D., et al., *Preliminary Economic Analysis of NO_x Flue Gas Treatment Processes*, EPA-600/7-80-021 (NTIS-PB 80-176 456), U.S. EPA, Industrial Environmental Research Laboratory, Research Triangle Park, NC, February 1980.

23. Hightower, J. W., Catalysts for Automobile Emission Control, in *Preparation of Catalysts* (B. Delmon et al., eds.), Elsevier, Amsterdam, the Netherlands, 1976.

24. *Buick Chassis Service Manual*, General Motors Corporation, Warren, MI, January 1980.

25. Grandy, D. M., and Reese, R. E., *Proposed Strategy for the Control of Oxides of Nitrogen Emissions from Stationary Internal Combustion Engines*, California Air Resources Board, Sacramento, CA, October 1979.

26. Cherry, S. S., and Hunter, S. C., *Cost and Cost Effectiveness of NO_x Control in Petroleum Industry Operations*, API Publication No. 4331, American Petroleum Institute, Washington, D.C., October 1980.

27. *Avco/Ebara E-Beam Dry Scrubber*, Avco Everett Research Laboratory, Inc., Everett, MA, 1980.

28. Kawamura, K., et al., Pilot Plant Experiment on the Treatment of Exhaust Gas from Sintering Machine by Electron Beam Irradiation, *Environmental Science and Technology* 14, 288–293 (1980).

29. *Annual Report to Congress 1979, Volume Three: Projections*, DOE/EIA-0173(79)/3, U.S. Department of Energy, Energy Information Administration, Washington, D.C., 1979.

30. Maloney, K. L., et al., Sulfur Capture in Coal Flames, presented to the AIChE 87th National Meeting, Boston, MA, August 19–22, 1979.

31. Giammar, R. D., et al., Evaluation of Emissions and Control Technology for Industrial Stoker Boilers, in *Proceedings of the Third Stationary Source Combustion Symposium: Volume I*, EPA-600/7-79-050a (NTIS-PB 292 539), U.S. EPA, Industrial Environmental Research Laboratory, Research Triangle Park, NC, February 1979, pp. 3–34.

32. Giammar, R. D., et al., Evaluation of Emissions and Control Technology for Industrial Stoker Boilers, in *Proceedings of the Joint Symposium on Stationary Combustion No_x Control: Volume III*, EPA/IERL-RTP-1085, U.S. EPA, Industrial Environmental Research Laboratory, Research Triangle Park, NC, October 1980, pp. 1–38.

33. NO_x Control Review, Volume 5, U.S. EPA, Industrial Environmental Research Laboratory, Research Triangle Park, NC, Fall 1980, pp. 1–2.

34. Blythe, G. M., et al., *Survey of Dry SO_2 Control Systems*, EPA-600/7-80-030 (NTIS-PB 80 166853), U.S. EPA, Industrial Environmental Research Laboratory, Research Triangle Park, NC, February 1980.

CHAPTER 10

PARTICLE CONTROL BY SCRUBBING

SEYMOUR CALVERT

Consultant
San Diego, California

1 INTRODUCTION

The *Scrubber Handbook*[1] defined a wet scrubber as any device that uses a liquid in the separation of particulate or gaseous contaminants from a gas. In view of this very general definition, there are as many types of scrubbers as there are ways of contacting a liquid and a gas.

Scrubber manufacturers offer a variety of products over a range of designs, sizes, advertised performance capabilities, and capital and operating costs. Choosing the optimum scrubber for a particular job requires an understanding of the fundamental principles underlying the various designs.

1.1 Wet Scrubbers

Wet scrubbers are defined as devices that use a liquid to clean a gas. This chapter is restricted to the collection of particles; the scrubbing of gaseous contaminants is covered in Chapter 7.

Some scrubbers use water as the primary collection surface for particles and others use it to wash solid surfaces on which particles have collected. In the latter case the primary means for particle collection is some type of dry collector, such as a filter or cyclone separator, which is discussed in another chapter as well as this one. The reader should consult the appropriate chapter(s) when this is applicable.

1.2 Overall System

The wet scrubber is one component of an overall scrubber system that picks up the gas stream at the source and emits it into the atmosphere. As shown in Figure 1, a schematic diagram of a generalized scrubber system, some components are common to all air pollution control systems and others are specific to scrubbers. Because particle properties are susceptible to change (see Chapter 6), what happens to the gas before it reaches the particle collection zone can be extremely important.

Intimate, agitated contact of gas and liquid occurs in wet scrubbers, with the consequent generation of liquid drops. Entrainment of drops in the gas can result in liquid being carried out of the scrubber into any reheater, duct, fan, and stack and then into the atmosphere. Corrosion, erosion, plugging, fan damage, and pollutant emission can result unless the entrainment is separated from the gas after particle collection.

The contaminant that accumulates in the scrubbing liquid must be discharged from the system. Various methods can be used to remove solids from the scrubber system, depending on the specific situation, and these are not discussed in this chapter. Changes in the liquid (which also includes suspended solids) can cause scaling, corrosion, erosion, and other serious problems in the scrubber. Because of this and the environmental consequences, liquid treatment and waste disposal deserve careful attention.

1.3 Particle Behavior

Whatever the type of scrubber, particles are separated by means of one or more of several basic mechanisms. Those that are important to scrubbing are described briefly here and discussed in detail in Chapter 6.

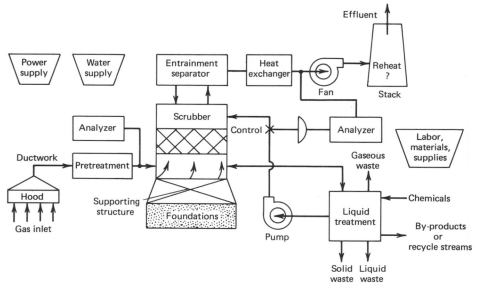

Fig. 1 Scrubber system.

1. *Gravitational Sedimentation.* This mechanism is usually of little consequence for any particles small enough to require consideration of a scrubber.

2. *Centrifugal Deposition.* Particles may be "spun out" of a gas stream by centrifugal force induced by a change in gas flow direction. Large-scale changes in flow direction, as would be encountered in a cyclone separator, are not very effective on particles smaller than about 5.0 μm in diameter.

3. *Inertial Impaction and Interception.* When a gas stream flows around a small object, the inertia of the particles causes them to continue to move toward the object, and some of them will be collected. (This is really the same thing as centrifugal deposition, and the distinction is made here on the basis of common usage.) Inertial impaction customarily describes the effects of small-scale changes in flow direction.

Because inertial impaction is effective on particles as small as a few tenths of a micrometer in diameter, it is the most important collection mechanism for the usual particle scrubber. All important particle properties may be lumped into one parameter, the *aerodynamic diameter,* defined as

$$d_{pa} = d_p(\rho_p C')^{1/2} \tag{1}$$

where

d_{pa} = particle aerodynamic diameter, μmA
d_p = particle physical diameter, μm
C' = the Cunningham correction factor, dimensionless

Most methods for measuring particle size (such as the cascade impactor) determine the aerodynamic diameter. Since this is the most important parameter where inertial impaction is at work, one need not know the actual physical diameter or particle density (ρ_p).

4. *Brownian Diffusion.* When particles are small enough (i.e., less than about 0.1 μm in diameter), they are buffeted by gas molecules, and they begin to move like gas molecules. That is, they diffuse randomly through the gas because of their Brownian motion. In general, inertial impaction and Brownian diffusion are the two principal mechanisms operating in particulate scrubbers. As a consequence, there is generally a minimum point when collection efficiency is plotted against particle diameter. Above about 0.3 μm in diameter, inertial impaction becomes important and efficiency rises with particle diameter. Below 0.3 μm in diameter, diffusion begins to prevail and efficiency rises as particle diameter falls below that size.

5. *Thermophoresis.* If there is heat transfer from the gas to the liquid, there will be a corresponding temperature gradient, and fine particles will be given toward the cold surface by differential molecular bombardment due to the gradient. This effect will rarely be of much significance in a scrubber.

6. *Diffusiophoresis.* Mass transfer within the scrubber, as might be caused by condensation of water vapor from the gas onto a cold liquid surface, will exert a force on particles that causes them to deposit on the surface. Diffusiophoretic deposition can be significant; the fraction of particles removed will roughly equal the fraction of the gas stream condensed.

7. *Electrostatic Precipitation.* If an electrostatic charge is induced on the particles, they can be precipitated from the gas stream by the influence of a charge gradient. This mechanism can be effective on all particle diameters and can provide high collection efficiency.

8. *Condensation on Particles.* While it is not a collection mechanism in itself, the enlarging of particle mass by such means as having water condense in a film around it makes the particles more susceptible to collection by inertial impaction. This phenomenon, in combination with diffusiophoresis and thermophoresis, can take place in scrubbers where condensation occurs. The combination of mechanisms is referred to as *flux force/condensation* (F/C) scrubbing.

9. *Coagulation.* Particles can stick together if they collide under the influence of Brownian motion or turbulence. This coagulation (or agglomeration) results in an increased particle size and, therefore, greater susceptibility to separation by mechanisms other than diffusion.

2 CLASSIFICATION

The classification system used here[1] separates scrubbers into generic groups, more or less according to their basic mechanism(s) of particle collection. To determine which group a specific scrubber belongs in it may be necessary for the reader to use his or her own judgment after studying its design and operating mechanism.

2.1 Plate Scrubbers

A plate scrubber consists of a vertical tower with one or more horizontal plates (trays) inside. Gas comes in at the bottom of the tower and must pass through perforations, valves, slots, or other openings in each plate before leaving through the top. Usually, liquid is introduced to the top plate, and flows successively across each plate as it moves downward to the liquid exit at the bottom. Gas passing through the openings in each plate mixes with the liquid flowing over it. Gas–liquid contacting causes the mass transfer or particle removal for which the scrubber was designed.

Figure 2 shows several types of plates and a tower. Plate scrubbers are generally named for the type of plates they contain. For example, a tower containing sieve plates is called a sieve plate tower.

Impingement baffles can be placed a short distance above each perforation on a sieve plate, forming an impingement plate. The impingement baffles are below the level of liquid on the perforated plates, so they are continuously washed clean of collected particles. Particle collection is mainly by inertial impaction from gas jets impinging on the liquid or on solid members. It possibly may be aided by atomization of liquid in contact with the gas jets flowing through the plate.

Collection efficiency increases as the perforation diameter decreases and can enable a cut diameter of about 1.0 μmA for 3.2-mm (0.125-in.) diameter holes in a sieve plate. The cut diameter is that of a particle that is collected at 50% efficiency.

A plate does not have the same efficiency for all particle sizes, but rather shows a sharp efficiency change around the cut diameter. Once particles larger than this size are removed from the gas, additional plates can do little good. This kind of behavior is characteristic of most types of scrubbers and should be kept in mind whenever one is tempted to try two scrubbers in series.

Plate scrubber capacity, entrainment, pressure drop, and stability properties are given in Chapter 7, chemical engineering literature,[2] and manufacturer's literature. Care must be taken in selecting plates for systems that have a tendency toward scaling or adherence of solids to the plates, which can result in plugging of the perforations.

2.2 Massive Packing

Packed bed (or "packed column") scrubbers are familiar as gas absorbers or fractionators (see Chapter 7) and can also be used as particle scrubbers. They may be packed with a range of manufactured elements, such as various ring- and saddle-shaped packings. The gas–liquid contacting may be cocurrent, countercurrent (Figure 3), or cross flow. Mist collection in packed beds with subsequent drainage can be accomplished without additional liquid flow.

Collection in packing works mainly by centrifugal deposition due to curved gas flow through the pore spaces and around packing edges, and by inertial impaction due to gas jet impingement within the bed. The good mass transfer characteristics of packings can also make for efficient collection of particles by diffusion if the particles are small enough.

Collection efficiency for particles in the inertial size range (larger than 0.3 μmA) rises as packing size falls. A cut diameter around 1.5 μmA can be attained by columns packed with 2.5-cm (1-in.) Berl saddles or Raschig rings. Smaller packing such as the 1.25-cm (0.5-in.) size can achieve 0.7-μmA cut diameter at 9.2 m/sec (30 ft/sec) gas velocity, which implies cocurrent flow.

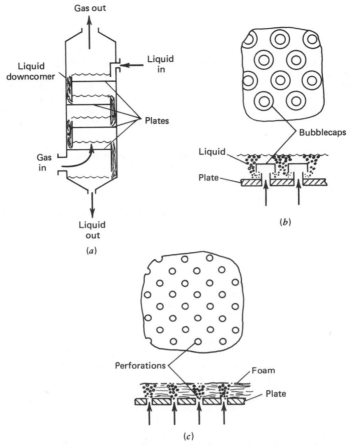

Fig. 2 Plates. (*a*) Plate column. (*b*) Bubblecap plate. (*c*) Sieve plate.

Packings are subject to plugging, but can be removed for cleaning. Temperature limitations are of special importance when plastics are used and likewise, corrosion can have a severe effect on metallic packings. As for plate columns, data on capacity, pressure drop, and other characteristics can be found in Chapter 7, chemical engineering literature,[2] and manufacturer's literature.

2.3 Fibrous Packing

Beds of fiber have been employed in various static or moving configurations with continuous or intermittent washing (Figure 4). Fibers can be made from materials such as plastic, glass, and steel. Fibrous packing usually has a very large void fraction ranging from 97 to 99%. Fibers should be small in diameter for efficient operation, but strong enough to support collected particles or drops without matting together. Mixtures of large- and small-diameter fibers have been used to combine these properties. Liquid flow flushes away collected material from the fibers in cocurrent, countercurrent, or cross-flow arrangements similar to those for massive packings.

When collection is by inertial impaction accompanying the gas flow around the fibers, efficiency rises as fiber diameter decreases and as the gas velocity increases. Diffusional collection can be very efficient for very small particles, and the efficiency of this mechanism will improve as gas velocity diminishes. Inertial *cut diameters* can run as low as 1.0 or 2.0 μmA for knitted wire mesh, 0.28-mm (0.011-in.) diameter wire, and to around 0.5 μmA for very fine wires and/or higher gas velocities.

Fibrous beds are very susceptible to plugging and can be impractical where scaling exists or where conditions favor deposition of suspended solids. Obviously, they will also be especially sensitive to chemical, mechanical, and thermal attack. They are successfully used in many applications. Consult the same references as for plates and packings for capacity and related information.

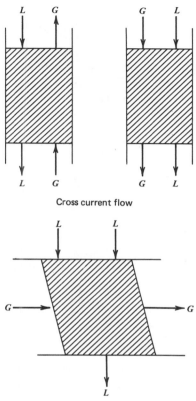

Fig. 3 Contacting in packing.

2.4 Preformed Spray

A preformed spray scrubber collects particles on liquid drops that have been atomized by spray nozzles. For single-fluid atomizers, properties of the drops are determined by the configuration of the nozzle, the liquid to be atomized, and the pressure to the nozzle. With two-fluid atomizers, the flow rates and pressures of both phases (air and water or steam and water) are important. Sprays leaving the nozzle are directed into a chamber that has been shaped so as to conduct the gas past the atomized droplets. Horizontal and vertical gas flow paths have been used, as well as spray-introduced cocurrent, countercurrent, or cross flow to the gas (Figure 5).

Ejector venturis use high-pressure spray both to collect particles and to move the gas. Preformed sprays have also been installed in venturi scrubbers which use a fan to provide high gas-phase pressure drop.

Particle collection in these units results from inertial impaction on the droplets. *Efficiency* is a complex function of droplet size, gas velocity, liquid–gas ratio, and droplet trajectories. There is often an optimum droplet diameter that varies with fluid flow parameters. For droplets falling at their terminal settling velocity, the optimum droplet diameter for fine particle collection is around 100–500 μm; for droplets moving at high velocity within a few feet of the spray nozzle, the optimum is smaller.

Spray scrubbers that take advantage of gravitational settling can achieve *cut diameters* around 2.0 μm at moderate liquid–gas ratios. High-velocity sprays can give cut diameters down to about 0.7 μmA. Efficiency improves with higher spray-nozzle pressures and liquid–gas ratios.

Spray scrubbers are practically immune to plugging on the gas flow side but are subject to severe problems on the liquid side. The liquid–gas ratio required is high, usually running 4 to 13 l/m³ (30–100 gal/Mcf) of gas treated, depending on efficiency.

The recirculating scrubber liquor can erode and corrode nozzles, pumps, and piping. Nozzles can plug with pieces of scale or agglomerates of particles. By their nature, sprays generate a heavy loading

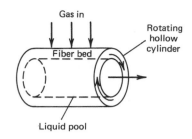

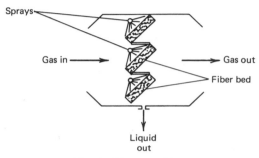

Fig. 4 Fibrous packing.

of liquid entrainment, which must be collected. Gas-phase pressure drop (ΔP) is generally low or may even be positive, and thus enhances the flow of gas.

2.5 Gas-Atomized Spray

Gas-atomized spray devices use a moving gas stream to first atomize liquid into droplets, and then accelerate the droplets. Typical of these devices are the *venturi* scrubber and the various *orifice*-type scrubbers. High gas velocities of 60–120 m/sec (200–400 ft/sec) raise (Figure 6) the relative velocity between the gas and the liquid drops, and promote particle collection. The Calvert Collision Scrubber (Figure 7) increases relative velocity and liquid atomization by causing two opposing gas-atomized streams to collide.

Liquid may be introduced in various places and in different ways without having much effect on collection efficiency as long as it results in a uniform spray distribution. Usually it is introduced at the entrance to the throat through several straight pipe nozzles directed radially inward. Other gas-atomized spray designs distribute a liquid film over the scrubber walls upstream from the throat.

Gas-atomized scrubbers have about the simplest and smallest configurations of all the scrubbers. While fairly difficult to plug up, they are susceptible to erosion because of their high throat velocity. They can be built with adjustable throat openings to permit variation of pressure drop and collection efficiency.

2.6 Centrifugal Scrubbers

Centrifugal scrubbers impart a spinning motion to the gas passing through them. The spin may come from tangential introduction of gases into the scrubber, or from direction of the gas stream against stationary swirl vanes. In a dry centrifugal collector (cyclone), the walls can be wetted to decrease reentrainment of particles and to wash off deposits. Drops can be projected through the rotating gas stream to catch particles on drops by impaction. The spray can be directed outward from a central manifold, or inward from the collector wall. Nozzles directed inward from the wall are more easily serviced, since they can be made accessible from the outside of the scrubber.

One particle collection mechanism is *centrifugal deposition* caused by the rotating gas stream. In the absence of spray, the efficiency will be the same as for a dry collector. A particle *cut diameter* of

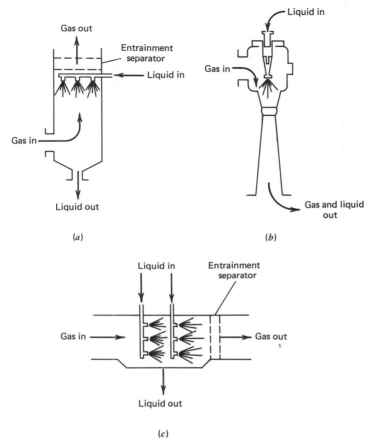

Fig. 5 Preformed spray. (*a*) Countercurrent spray. (*b*) Ejector venturi. (*c*) Cocurrent spray.

4.0 or 5.0 μmA can be obtained with a centrifugal scrubber without spray. As spray is introduced or generated inside, *impaction on drops* occurs and the efficiency approaches that of a preformed spray scrubber.

Centrifugal scrubbers are fairly simple in form and have no small passages. They are not very susceptible to plugging, although solids can deposit on sections of the wall that are not adequately washed. If designed properly, centrifugals have the advantage of a built-in potential for entrainment separation. Tangential gas velocity should not exceed about 30 m/sec (100 ft/sec), and the internal configuration must prevent flow of the liquid along the wall into the gas exit.[3]

2.7 Baffle and Secondary Flow Scrubbers

Baffled scrubbers cause changes in gas flow direction and velocity by means of solid surfaces. Either the major direction of flow may be altered, or secondary flow patterns may be set up, as shown in Figure 8. Louvers, zigzag baffles, and disk and donut baffles are examples of surfaces that produce changes in main flow direction. If the material to be collected is liquid, it runs down the baffle into a collection sump. Solid collected particles may be washed intermittently from the baffle plates.

Particle collection is by the centrifugal deposition caused by change in the main flow direction, or by rotating secondary flows. The *cut diameter* can go as low as 5.0 to 10.0 μmA for continuous and discontinuous zigzags and similar arrangements.

Baffles are used as precleaners and as entrainment separators. Heavy-particle or slurry loadings can cause solids deposition, which can lead to plugging and corrosion.

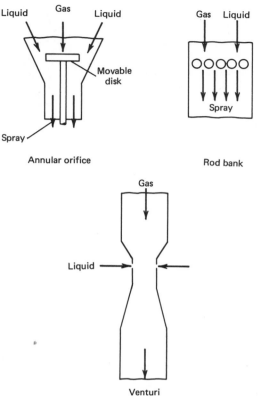

Fig. 6   Gas-atomized spray.

2.8   Impingement–Entrainment Scrubbers

Impingement–entrainment (self-induced spray) scrubbers force the gas to impinge on a liquid surface to reach a gas exit (Figure 9). Some of the liquid atomizes into drops that are entrained by the gas and collecting particles. The gas exit is usually designed so as to change the direction of the gas–liquid mixture flowing through it and reduce drop entrainment.

Particle collection is generally by inertial impaction caused by impingement on the liquid surface and the atomized drops. Drop size and the liquid–gas flow ratio inside the scrubber depend on scrubber geometry and gas flow rate, but are not controllable or measurable.

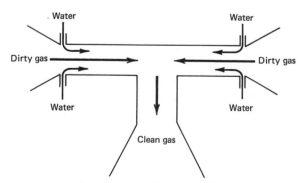

Fig. 7   Calvert collision scrubber.

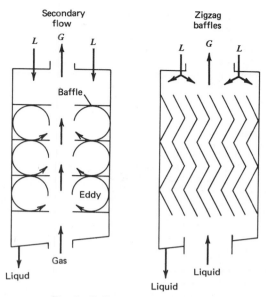

Fig. 8 Baffles and secondary flow.

Generally, the performance of an impingement–entrainment scrubber seems to be comparable to a gas-atomized scrubber operating at the same gas-phase pressure drop. *Cut diameter* ranges from several micrometers for low-velocity impingement to a few tenths of a μmA for high-velocity impingement, as in the Doyle scrubber.

Liquid flow in this type of scrubber is induced by the gas, so that liquid pumping requirements are mostly for makeup and purge streams. Solids deposition can pose a problem on the bottom and on portions of the wall that are not well washed. Good entrainment separation is required because of the amount of spray generated.

2.9 Mechanically Aided Scrubbers

Mechanically aided scrubbers incorporate a motor-driven device such as fan blades which cause gas flow and on which particles collect by impaction. When liquid is introduced at the hub of the rotating fan blades, some atomizes upon impact with the fan and some runs over, and washes, the blades. The liquid is recaptured by the fan housing, which drains into a sump, from which it may be recirculated.

Disintegrator scrubbers draw on a submerged, motor-driven impeller to atomize liquid into small drops. The drops spin off the impeller across the gas stream, collecting particles on the way.

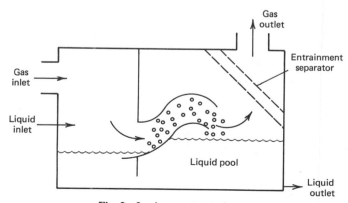

Fig. 9 Impingement–entrainment.

Particle collection mechanisms, in probable order of importance, are: inertial impaction on the atomized liquid, inertial impaction on the rotor elements, and centrifugal deposition on the housing. *Cut diameters* down to about 2.0 μmA have been achieved with devices having fine sprays and low-speed fans. Diameters as low as 1.0 μmA can be reached with a disintegrator-type scrubber.

Generally, there seems to be no power advantage for mechanically aided units over other types. Disintegrators require more power than a gas-atomized scrubber with comparable efficiency. Further, high-speed impaction of liquid and slurry on the scrubber parts promotes severe abrasion and corrosion conditions. Rotating parts are also subject to vibration-induced fatigue caused by solids deposition, or wear leading to unbalancing.

2.10 Moving Bed Scrubbers

Moving bed scrubbers provide a zone of mobile packing, usually plastic or glass spheres, where gas and liquid can mix intimately. The vessel shell holds a support grid on which the movable packing is placed. As shown in Figure 10, gas passes upward through the packing, while liquid is sprayed up from the bottom and/or flows down over the top of the moving bed. Gas velocities are sufficient to move the packing material around when the scrubber operates. This movement aids in making the bed turbulent and keeps the packing elements clean. When hollow or low-density spheres are used, the bed fluidizes and bed depth becomes about double that when quiescent.

Particle collection is mainly by inertial impaction on atomized liquid and on the packing elements. *Cut diameters* down to about 1.0 μmA are attained in the fluidized "ping-pong ball" type of bed having three stages in the scrubber column. Performance of the less violently agitated "marble" type bed resembles that of massive packing beds unless the gas velocity rises so high as to cause significant liquid atomization and entrainment from the bed.

Moving bed scrubbers prove beneficial where good mass transfer characteristics are needed, as well as particle collection. The agitation cleans the packing and reduces problems with solids deposition. Ball wear can be severe, and the hydrodynamic stability of the scrubber is limited by fluidization and surging characteristics.

2.11 Combination and Enhanced Scrubbers

Numerous combinations of the scrubber types already discussed have been contrived and used. No attempt will be made here to describe these essentially geometric variants.

Condensation of water vapor caused by steam addition to a saturated gas, or by scrubbing of a saturated gas with cold liquid, gives rise to F/C effects. Particles may grow to several micrometers in diameter with steam addition, enhancing scrubbing efficiency, although the cost of purchased steam will generally be prohibitive. Scrubbers utilizing sonic or supersonic *steam jets* can induce sonic agglomeration as well as other collection mechanisms.

Scrubbing a hot, humid gas with relatively cold liquid can be economically attractive in cases where the gas is hot enough. Such a scrubber operating at moderate pressure drop can provide the same efficiency as a high-energy venturi scrubber, depending on how much condensation occurs. The scrubber design and economics have to be worked out for each case.

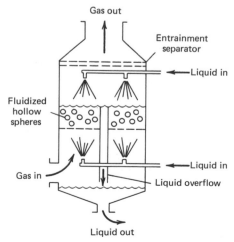

Fig. 10   Mobile bed.

Electrostatically augmented scrubbers can be very efficient, depending on design and operating parameters. Most of the benefit from charging is on the collection of submicrometer particles. The variations available include: wet electrostatic precipitators, charged-dust/grounded-liquid scrubbers, charged-drop scrubbers, and charged-dust/charged-liquid types. Performance prediction is not very reliable, so pilot tests are generally needed. Corrosion and voltage isolation problems can be severe with this class of scrubbers.

3 PERFORMANCE ANALYSIS

The particle scrubber performance characteristics of primary importance are collection efficiency as a function of particle size (often called: "grade efficiency") and power requirement. For a given scrubber, performance will depend on gas and liquid flow rates and other factors. The sources of this kind of information have been:

1. Scrubber manufacturer's proprietary test results as presented in their brochures and published literature.
2. Original experimental research; usually by academic personnel and others doing government-supported research which is presented in journal articles, books, and government reports.
3. Analytical studies that apply basic principles and whatever reliable data are available for the development of mathematical models of performance.
4. Pilot plant tests of a scrubber with a portion of the actual gas stream or a simulated gas.

3.1 Approach

For most scrubbers the amount of reliable, experimentally determined performance information is inadequate and it must be extended by engineering analysis. Manufacturer's test data are generally not well defined and too limited. Academic research is often done with small-scale apparatus which may not be representative of large equipment.

A number of performance tests on full-scale scrubbers in various industries have been done under government contracts but these were largely "one of a kind" tests. In a realistic situation there are many factors that can change the effective size of the particles in the collection zone, so one must attempt to account for them and extract a true definition of the scrubber's capability. The reverse situation also applies; one must be able to account for changes in effective particle size when predicting performance for a specific application.

The performance information that is presented is the result of a combination of engineering analysis and the best available experimental data. The knowledge of basic principles and their application can be important for the satisfactory design and/or operation of a scrubber, so the outlines of the performance modeling concepts are given in the following sections.

Rapid design or selection methods are also presented so that the reader can make a quick estimate, which may be adequate for preliminary evaluations or useful for comparison with other performance predictions.

3.2 Unit Mechanisms

Much of the engineering analysis that led to the development of the performance models has been along the lines of the unit mechanisms approach[1] or comparable logical paths. Unit mechanisms are the basic gas–liquid contacting situations in which particles are removed from the gas. They are the smallest elements that are representative of the way the scrubber works; to understand them is to understand the scrubber.

By recognizing the unit mechanisms and their functioning, we have a rational method for analyzing and predicting scrubber performance. One can examine any scrubber and reduce it to an equivalent set of one or more unit mechanisms that will account for its particle collection ability. The unit mechanisms are:

1. Collection by drops (of liquid) moving through gas.
2. Collection by cylinders (usually solids, like wires).
3. Collection by sheets of liquid (usually flowing over solid surfaces).
4. Collection from bubbles of gas (usually rising through liquid).
5. Collection from gas jet impingement (either on liquid or solid surfaces).

For each unit mechanism, the particles are separated from the gas by one or more of the following particle collection mechanisms: gravitational sedimentation, centrifugal deposition, inertial impaction, interception, Brownian diffusion, thermophoresis, diffusiophoresis, and electrostatic precipitation. Particle collection also may be enhanced by increasing the particle size through agglomeration, condensation, or other particle growth mechanisms.

Examples of mathematical models that were developed by means of a unit mechanisms approach are given in a subsequent section of this chapter. Generally, the models relate particle collection efficiency to particle diameter and other properties.

A specific performance relationship between collection efficiency and particle size (often called a "grade-efficiency curve") can be integrated over the particle size distribution to yield the overall collection efficiency [see Eq. (2)]. Alternatively, a generalized and somewhat idealized method, which is described below, can be used for rapid prediction of scrubber performance.

3.3   Cut Diameter Method

The cut diameter method for scrubber performance prediction[1] is based on the idea that the most significant single parameter to define both the difficulty of separating particles from gas and the performance of a scrubber is the particle diameter for which collection efficiency is 50%; that is, the cut diameter (d_{pc}). When a range of sizes is involved, the "overall collection efficiency" of a control device will depend on the amount of each size present and on the efficiency of collection for that size. In mathematical expression, the overall (integrated) penetration, Pt, of any device on a dust of any type of size distribution is

$$\overline{Pt} = \int_0^w \frac{Pt_d}{w} dw \qquad (2)$$

where

Pt = overall penetration, fraction
Pt_d = penetration for particles with diameter d_{pa}, fraction
w = total dust loading, g

The right-hand side of the above equation is the integral of the product of each weight fraction of dust times the penetration of that fraction.

Penetration for many types of inertial collection equipment can be expressed as

$$Pt_d = \exp(-A_e d_{pa}{}^{B_e}) = 1 - E \qquad (3)$$

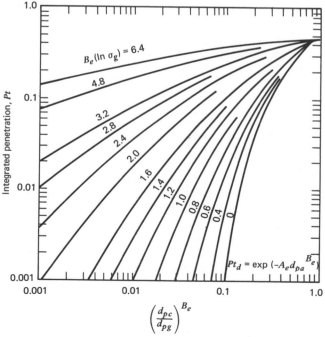

Fig. 11   Overall penetration generalization.

where

$$A_e = \text{constant}$$
$$B_e = \text{constant}$$
$$E = \text{efficiency, fraction}$$

In some cases, where one is concerned with particles larger than 1 μm in diameter, or where the particle size distribution is log-normal in terms of physical rather than aerodynamic diameter, it may be convenient to use the simplifying assumption that penetration is related to physical diameter by

$$Pt_d = \exp(-A_p d_p{}^{B_e}) \qquad (4)$$

where

$$A_p = \text{constant}$$

Packed towers, centrifugal scrubbers, and sieve plate columns follow the first relationship. For the packed tower and sieve plate column, B_e has a value of 2. For centrifugal scrubbers, B_e is about 0.67. Venturi scrubbers also follow the above relationship and B_e is approximately equal to 2 when the throat impaction parameter is between 1 and 10.

Equations have been solved[1] for a variety of log-normal size distributions and the results presented in graphical form in Figures 11 and 12. Figure 11 is a plot of Pt versus $(d_{pc}/d_{pg})^{B_e}$ with $B_e \ln(\sigma_g)$ as a parameter. Figure 12 is presented as a plot of Pt versus (d_{pc}/d_{pg}) with σ_g as the parameter when $B_e = 2$.

To illustrate the use of these graphs, assume the scrubber cut diameter determined from the cut/power plot (see next section) is 0.63 μmA and the particles from an emission source have a size distribution of $d_{pg} = 10$ μmA and $\sigma_g = 3.0$. Then, $d_{pc}/d_{pg} = 0.063$. From Figure 12, the overall penetration corresponding to $d_{pc}/d_{pg} = 0.063$ and $\sigma_g = 3.0$ is 0.01. Since penetration is 100% minus the percent efficiency, the overall collection efficiency of the control device is 99%.

3.4 Cut/Power Relationship

Analysis of the scrubber power law,[4,5] which states that scrubber penetration depends only on power input,[6,7] showed that it does not account for the effect of particle size. However, it has been shown[1] that performance cut diameter could be related to gas-phase pressure drop, or power input to the scrubber. This is called the "cut/power" relationship.

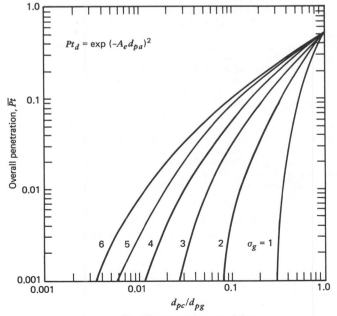

Fig. 12 Overall penetration, special case.

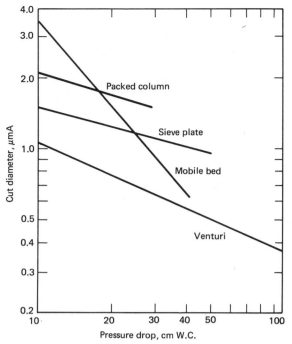

Fig. 13   Cut/power plot.

Subsequent performance tests on a variety of scrubbers in industrial installations, combined with mathematical modeling[8,9,10] has led to the refinement of the cut/power relationship, shown in Figure 13. The curves give the cut diameter as a function of gas-phase pressure drop [cm water column (W.C.)] for a number of typical installations; sieve plate column with 0.32-cm holes and foam density = 0.4, packed column with 2.5-cm diameter rings or saddles, venturi, and mobile bed scrubbers.

The cut/power relationship appears to be an accurate and reliable tool for scrubber selection and performance prediction.

The cut/power plot can be used to predict scrubber performance if its operating pressure drop is known. For example, consider a venturi scrubber with a pressure drop of 33 cm W.C. From Figure 13, the performance cut diameter of the venturi scrubber is 0.63 μmA. Suppose the particle size distribution has $d_{pg} = 10$ μmA and $\sigma_g = 3$, then the overall penetration is 0.01 (from Figure 12).

3.5   Deposition Velocity Generalization

Particle collection by all mechanisms and in all kinds of equipment can be described by either of two general relationships if a few conditions are met. The basic idea is that the velocity at which a particle moves toward a collection surface (i.e., its deposition velocity, u_{pd} can be computed if all the forces acting on the particle are known. From this one can compute the rate of deposition for a given particle concentration (c_p) deposition area (A_d), and gas flow rate (Q_G).

It is assumed that u_{pd} is constant over the deposition area and the c_p at the deposition surface is due to either of two simplified cases:

1.   There is no mixing of particles on planes normal to the gas flow.
2.   There is complete mixing, as for highly turbulent flow, so there is uniform concentration on any plane normal to the gas flow.

For *no mixing*,

$$E = 1 - \frac{c_{pe}}{c_{pi}} = \frac{u_{pd}A_d}{Q_G} \qquad (5)$$

Table 1 Particle Deposition Velocity

Collection Phenomenon	Particle Deposition Velocity
Gravitational sedimentation	$u_S = \dfrac{C' d_p^2 (\rho_p - \rho_G) g}{18 \mu_G}$
Centrifugal deposition	$u_C = \dfrac{C' d_p^2 (\rho_p - \rho_G) u_i^2}{18 \mu_G R}$
Brownian diffusion	$u_{BD} = 1.13 \left(\dfrac{D_p}{\theta} \right)^{0.5}$
Thermophoresis	$u_T = - \left(\dfrac{3 C' \mu_G}{2 \rho_G T} \right) \left(\dfrac{k_G}{2 k_G + k_p} \right) \nabla T$
Diffusiophoresis	$u_D = \left(- \dfrac{M_v^{0.5}}{p_v M_v^{0.5} + p_G M_G^{0.5}} \right) \dfrac{P D_{VG}}{p_G} \nabla p_v$
Electrical migration	$u_E = \left(\dfrac{E_p}{E_p + 2} \right) \left(\dfrac{C' E_o E_c E_p d_p}{4 \pi \mu_G} \right)$
Magnetic precipitation	$u_M = \dfrac{C' \mu_o H q_p \mu_G}{3 \pi \mu_G d_p}$

For *complete mixing,*

$$Pt = \frac{c_{pe}}{c_{pi}} = \exp \left(- \frac{u_{pd} A_d}{Q_G} \right) \tag{6}$$

Deposition velocities for several phenomena can be computed by means of the equations given in Table 1. So long as the particles are small enough that Stokes law applies (i.e., frictional resistance is a linear function of particle velocity), one can add deposition velocities for the phenomena that are active.

The deposition velocity relationships are useful for making rapid estimates of the efficiency possibilities for proposed systems or of the effect of scale changes. Note that the use of Eq. (6) can give fractional efficiencies larger than 1.0, which means that the device is 100% efficient and has reduced c_{pe} to zero, but the equation does not know it.

4 MATHEMATICAL MODELS

The efficiency estimated by the cut diameter method is only an approximation because the inlet particle size distribution might not follow the log-normal distribution closely and Eq. (3) is correct only for packed beds and similar devices and is an approximation for others. To accurately predict the scrubber overall collection efficiency, one should perform the integration in Eq. (2) with the actual size distribution and the grade penetration for the scrubber. In the following sections, design equations for predicting the grade penetration will be presented for several scrubber types.

The power requirement for particle scrubbing is mainly due to the gas pressure drop. Preformed sprays and mechanically aided scrubbers have significant power inputs to pumps and other devices. Equations for predicting the gas-phase pressure drop for some scrubbers are presented in the following sections. Additional information can be found in sources previously cited.

4.1 Plates

Particle separation in sieve (perforated) and impingement plates can be defined mathematically by starting from the unit mechanisms of particle collection in bubbles, on drops, and from impacting jets.

Experimental data on the collection of hydrophilic particles by water on sieve plates in laboratory equipment are correlated by

$$Pt = \exp(-40FK_p) \tag{7}$$

$$K_p = \frac{u_h d_{pa}^2}{9\mu_G d_h} \tag{8}$$

where

K_p = inertial impaction parameter, dimensionless
F = foam density, g/cm³
u_h = gas velocity through sieve plate hole, cm/sec
μ_G = gas viscosity, g/cm-sec
d_h = diameter of sieve plate hole, cm

Foam density generally ranges about 0.38 and 0.65 and can be estimated[1] by the equation below:

$$-\ln F = 0.184 u_{Gs}(\rho_G)^{1/2} + 0.45 \tag{9}$$

where

u_{Gs} = superficial gas velocity, based on plate cross-sectional area, cm/sec

By setting $Pt = 0.5$ in Eq. (7), the following relationship for cut diameter in a sieve plate is obtained:

$$d_{pc} = 0.4 \left(\frac{\mu_G d_h}{u_h F}\right)^{1/2} \tag{10}$$

Pressure drop in a sieve plate column under usual operating conditions runs about 2.5 to 10 cm W.C. per plate.

For *impingement plates* there are no reliable experimental data available so the efficiency is predicted, based on the impingement of round jets on plane surfaces. This approach gives a lower efficiency than claimed by the manufacturer. The cut diameter is given by

$$d_{pc} = \left(\frac{1.37 \mu_G n_h d_h^3}{Q_G}\right)^{1/2} \tag{11}$$

The total *pressure drop* per plate can be divided into three main components: dry plate pressure drop, wet plate pressure drop, and frictional losses in the scrubber. The wet plate pressure drop can be estimated from liquid depth above the plate. Frictional losses must be evaluated from experimental work.

The dry plate pressure drop that is mainly due to jet exit can be approximated by the following equation:

$$\Delta p_j = 588 \left(\frac{d_h}{d_{pc}^2}\right)^2 \tag{12}$$

where:

Δp_j = dry plate pressure drop, cm W.C.

4.2 Massive Packings

Particle collection in packed columns can be described in terms of gas flow through curved passages, and laboratory performance data for a variety of packing shapes, such as saddles, rings, and spheres, can be correlated simply by the packing diameter, as shown in Eq. (13).

$$Pt_d = \exp\left(-7.0\frac{ZK_p}{d_c}\right) \tag{13}$$

where

$$Z = \text{height of packing, cm}$$
$$K_p = \text{inertial impaction parameter, dimensionless}$$
$$f_v = \text{void volume, fraction}$$
$$d_c = \text{nominal packing diameter, cm}$$

K_p is defined as in Eq. (8), but with gas velocity equal to the superficial velocity through the total bed area, u_{Gs}, and collector dimension, d_c, taken as the packing diameter. Aerodynamic cut diameter is given by

$$d_{pc} = \left(\frac{f_v d_c^2 \mu_G}{u_{Gs}} \right)^{1/2} \tag{14}$$

The *pressure drop* in a packed column can be predicted by methods given in standard chemical engineering texts.[2] Gas flow capacity is usually limited to about 50% of the flooding velocity. Superficial velocities for gases at about atmospheric pressure range around 12 m/sec (4 ft/sec) for packed column operation. The pressure drop at flooding is in the range of 12 to 35 cm W.C./m of packing height (1.5–4 in. W.C./ft) for most packings, with 15 cm W.C./m (2 in. W.C./ft) being an average value for rough estimation purposes.

A substantial amount of pressure drop information is available in packing manufacturers' literature. These data can be useful if they are for the operating conditions of the scrubber.

4.3 Fibrous Packing

The particle collection efficiency of a bed of clean fibrous packing for a given particle size is given by

$$Pt = \exp(-\eta_s S) \tag{15}$$

$$S = \frac{4 z f_s}{\pi d_f} \tag{16}$$

where

$$\eta_s = \text{collection efficiency of a single fiber, fraction}$$
$$z = \text{length of fiber bed in gas flow direction, cm}$$
$$f_s = \text{volume fraction solids (fiber) in bed}$$
$$d_f = \text{fiber diameter, cm}$$
$$S = \text{solidarity factor, fraction}$$

Approximate values of η_s for inertial impaction and for diffusion plus interception can be computed by means of Eqs. (17) and (18). The combined effects of inertial impaction, diffusion, and interception can be estimated by Eq. (19).

$$\eta_p = \left(\frac{K_p}{K_p + 0.85} \right)^{2.2} \tag{17}$$

$$\eta_p = [6(N_{Sc})^{-2/3}(N_{Ref})^{-1/2}] + [3(K_I)(N_{Ref})^{1/2}] \tag{18}$$

$$(1 - N_{pDI}) = (1 - \eta_p)(1 - \eta_{DI}) \tag{19}$$

where

$$\eta_p = \text{collection efficiency for inertial impaction}$$
$$\eta_{DI} = \text{collection efficiency for diffusion and interception, fraction}$$
$$\eta_{pDI} = \text{collection efficiency for inertial impaction, diffusion, and interception, fraction}$$
$$N_{Sc} = \text{Schmidt number} = \mu_G/\rho_G D_p, \text{ dimensionless}$$
$$D_p = \text{particle diffusivity, cm}^2/\text{sec}$$
$$N_{Ref} = \text{Reynolds number based on fiber diameter} = d_f u_G \rho_G/\mu_G, \text{ dimensionless}$$
$$K_I = \text{interception parameter, } d_p/d_f, \text{ dimensionless}$$

Fibrous packing performance can be predicted with the equations given above for most scrubber applications. Accumulation of solids and/or liquid in the bed can influence both the efficiency and the pressure drop. Prediction of pressure drop for wetted beds is not simple and it will generally be best to use experimental data.

4.4 Preformed Spray

Preformed sprays can be used in many combinations of spray and gas flow directions and there is no general relationship for predicting particle collection efficiency in all of them. Three important simplified cases will be discussed here; vertical counterflow, cross flow, and cocurrent flow.

In each case it will be assumed that the drops are of uniform size, reach their terminal settling velocity at once, are evenly distributed over the flow cross section, and do not coalesce. Modifications to account for departures from these assumptions, wall losses, and other factors must be made for each case where warranted.

For *vertical counterflow* (spray down, gas up) the particle penetration for collection by inertial impaction can be predicted with Eq. (20).

$$Pt_d = \exp\left[-\frac{3Q_L u_t Z \eta_d}{4 Q_G r_d (u_t - u_G)} \right] = \exp\left[-0.25\left(\frac{A_d u_t \eta}{Q_G}\right)\right] \tag{20}$$

where

$A_d = \dfrac{3 Q_L Z}{r_d (u_t - u_G)}$ = total surface area of all drops in the scrubber, assuming no liquid reaches

the scrubber wall, cm²

Q_L = liquid volumetric flow rate, cm³/sec

u_G = gas velocity relative to the duct, cm/sec

u_t = drop terminal settling velocity, cm/sec

η_d = collection efficiency of a single drop, fraction

r_d = drop radius, cm

Single-drop collection efficiency is greatly influenced by the drop Reynolds number and is calculated from an interpolation between the target efficiencies for viscous and potential flows around the drop. It can be approximated by means of Eq. (21)

$$\eta = \left(\frac{K_p}{K_p + 0.7}\right)^2 \tag{21}$$

Only a small fraction of the drops remain in suspension, and as little as 20% of Q_L may be effective, depending on scrubber size. The pressure drop through a counterflow spray scrubber is very low, around 1–2 cm W.C./m of column height (0.4–0.8 in. W.C./ft of column) for gas velocities of 15–45 cm/sec (0.5–1.5 ft/sec), with the liquid flow rate having negligible effect for $L = 20,000–80,000$ kg/hr-m² (4100–16,000 lb/hr-ft²).

In the *cross-flow case*, the water is sprayed at the top of the spray chamber while the gas flows horizontally. For collection by inertial impaction Eq. (22) predicts the penetration

$$Pt_d = \exp\left(-\frac{3Q_L Z \eta_d}{4 Q_G r_d}\right) = \exp\left(-0.25 \frac{A_d u_r \eta_d}{Q_G}\right) \tag{22}$$

where

$$u_r = \text{drop velocity relative to gas, cm/sec}$$

For *crosscurrent* and *cocurrent flow*, one can estimate the *pressure drop* on the basis of the change in momentum of the liquid when it is accelerated from zero velocity to the gas velocity.

$$\Delta p = 1.0 \times 10^{-3}\left(\frac{Q_L}{Q_G}\right) u_G^2 \tag{23}$$

Cocurrent spray scrubbing is often done with the drops being projected into the gas at a velocity appreciably greater than u_T. In the derivation of Eq. (24) for this case,[11] it was assumed that particle collection occurs only along a drop trajectory of length (range) R_d.

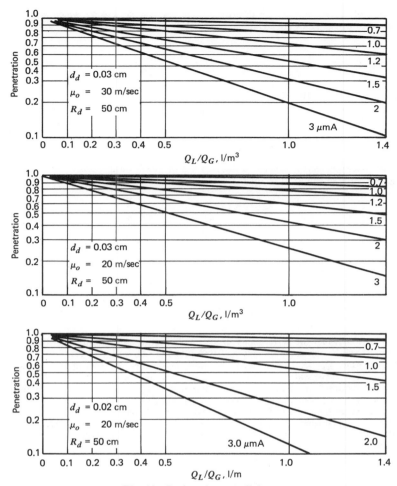

Fig. 14 Projected spray efficiency.

The average collection efficiency E_a for inertial impaction over the drop range depends on the initial drop velocity, drop diameter, particle aerodynamic diameter, and gas properties. To compute E_a one must account for drop dynamics and the relationship between collection efficiency and parameters such as drop velocity.

$$Pt = \exp\left(\frac{-3R_d Q_L E_a}{2d_d Q_G}\right) \qquad (24)$$

Figure 14 presents plots of particle penetration as a function of liquid–gas ratio with particle aerodynamic diameter as a parameter and for combinations of R_d, d_d, and initial gas velocity u_{Go}. Experimentally determined single-drop efficiencies[12] were used for computing these relationships. The use of larger R_d or u_{Go} would give lower penetrations, but these depend on apparatus configuration, spray nozzle characteristics, and atomization pressure.

4.5 Gas-Atomized Spray

Particle collection in gas-atomized spray scrubbers is predominantly by inertial impaction on drops. Diffusion may have some significance for particles smaller than 0.1 μm in diameter, depending on the drop size and the extent of gas–liquid contact.

Gas-atomized spray scrubbers are inherently cocurrent flow devices for which particle collection can be described by the following equation:

$$-\frac{dc}{c} = \frac{1.5|u_p - u_d|}{u_d d_d} \frac{Q_L}{Q_G} \eta_d \, dz \tag{25}$$

where

$$u_p = \text{velocity of particle, cm/sec}$$
$$u_d = \text{velocity of drop, cm/sec}$$

4.5.1   Simple Solution

A useful and relatively *simple solution* to Eq. (25) was obtained[1] on the basis of several assumptions, including the use of an empirical factor f

$$Pt = \exp\left[\frac{Q_L u_G \rho_L d_d}{55 Q_G \mu_G} F(K_{po}, f)\right] \tag{26}$$

where

K_{po} = inertial impaction parameter evaluated for the gas velocity at the throat entrance

and

$$F(K_{po}, f) = \left[-0.7 - K_{pa}f + 1.4 \ln\left(\frac{K_{pa}f + 0.7}{0.7}\right) + \frac{0.49}{0.7 + K_{pa}f}\right]\frac{1}{K_{po}} \tag{27}$$

It has been found that the performance of a variety of large-scale venturi and other gas-atomized spray scrubbers can be correlated by means of Eqs. (26) and (27) with $f = 0.5$.

Liquid–gas ratio, Q_L/Q_G, has an important influence on the collection of large particles. If too little liquid is used, there will not be enough to contact the gas stream and even if each drop were 100% efficient, the scrubber efficiency would be less than 100%. To illustrate, assume that standard air and water are used in a venturi scrubber with $u_G = 61$ m/sec (200 ft/sec) and we want to see the effect of varying Q_L/Q_G on Pt for particles larger than 4 μmA.

The impaction parameter will be about 50 or larger for these conditions and $F(K_{po}, f) = -0.413$, for $f = 0.5$. Drop diameter is evaluated by means of Eq. (32) for several liquid–gas ratios. The results are as follows:

Q_L/Q_G (l/m³)		7.5	10	15
Pt_d for d_{pa}	$\geq$ 4 μmA	5.64%	1.46%	0.026%

As shown by the illustrative example, there is an asymptotic limit to Pt_d for larger particles at a given Q_L/Q_G. It can also be shown that collection efficiency for submicrometer particles increases somewhat as Q_L/Q_G decreases. Thus, there is a conflict between large- and small-particle efficiency requirements. It must generally be resolved by using more liquid to get the large particles and more pressure drop to get the small ones.

4.5.2   Alternate Solution

Another solution[13] of Eq. (25) utilized the standard drag coefficient relationship rather than the simplified one which was used to get Eq. (26). The result for the straight throat section is given in Eq. (28), which does not include any empirical constant.

$$\ln Pt_d = \frac{B_v}{K_{po}(1 - u_{dl}^*) + 0.7} \times \left[4K_{po}(1 - u_{dl}^*)^{1.5} + 4.2(1 - u_{dl}^*)^{0.5} - 5.02 K_{po}^{0.5}\right.$$

$$\times \left(1 - u_{dl}^* + \frac{0.7}{K_{po}}\right) \tan^{-1}\left[\frac{(1 - u_{dl}^*)K_{po}}{0.7}\right]^{0.5} - \frac{B_v}{K_{po} + 0.7}$$

$$\left. \times \left[4K_{po} + 4.2 - 5.02 K_{po}^{0.5}\left(1 + \frac{0.7}{K_{po}}\right) \tan^{-1}\left(\frac{K_{po}}{0.7}\right)^{0.5}\right]\right] \tag{28}$$

where

$$B_v = \frac{Q_L \rho_L}{Q_G \rho_G C_{Do}}$$

C_{Do} = drag coefficient at throat entrance, dimensionless
u_{dl}^* = drop velocity at throat exit, dimensionless

$$K_{po} = \frac{d_{pa}^2 (u_{po} - u_{do})}{9 \mu_G d_d} \tag{29}$$

where

u_{po} = particle velocity at throat entrance, cm/sec
u_{do} = drop velocity at throat entrance, cm/sec

The particle velocity is assumed equal to the gas velocity. Drag coefficient at the throat entrance C_{Do} is determined from the standard curve using a Reynolds number calculated on the basis of the relative velocity applying at the throat entrance. Drop velocity at the throat exit is calculated from the following equations:

$$u_{dl}^* = 2 [1 - x^2 + (x^4 - x^2)^{0.5}] \tag{30}$$

$$x = \left(\frac{3 l_t C_{Do} \rho_G}{16 d_d \rho_L} \right) + 1 \tag{31}$$

where

l_t = venturi throat length, cm

Particles are collected only by atomized liquid drops. The drop diameter is the Sauter mean diameter predicted by the empirical correlation of Nukiyama and Tanasawa:[15]

$$d_d = \frac{0.0585}{u_G} \left(\frac{\sigma}{\rho_L} \right)^{0.5} + 1884 \left(\frac{\mu_L}{(\sigma \rho_L)^{0.5}} \right)^{0.45} \left(\frac{Q_L}{Q_G} \right)^{1.5} \tag{32}$$

where

d_d = Sauter mean drop diameter, cm
σ = surface tension, dyn/cm

Although in actuality a distribution of drop sizes will exist, the use of a single representative size simplifies calculations and gives reasonable results.

4.5.3 Comparison with Experiment

Comparison of Eqs. (26) and (28) *with experimental results* shows that while there is some difference between the predictions by the two equations, the experimental data scatter too much to warrant the selection of one over the other. Equation (26) is simpler to work with and, with $f = 0.5$, it gives predictions close to those of the complete model of Yung et al.,[13] which includes both the straight throat and the diffuser.

Equation (29) is useful for showing the effect of throat length on performance. As l_t increases, u_{dl} increases and Pt_d decreases. However, as will be shown later, the pressure drop also increases with u_{dl}^*. The net effect is that while Pt_d and ΔP vary with throat length, the relationship between Pt_d and ΔP stays nearly the same.

Drop size is not correctly defined by Eq. (32) for all operating conditions[16] but alternative relationships do not do as well when used with Eqs. (26) or (28) to predict particle collection. Again, it appears that Eq. (26) with Eq. (28), despite all of the assumptions and approximations, is the best model for prediction of performance. Equation (29) is better for designing gas-atomized spray scrubber proportions.

4.5.4 Pressure Drop

The *pressure drop* for gas flowing through a gas-atomized spray scrubber is due to the frictional loss along the wall of the scrubber and the acceleration of liquid drops. Frictional loss depends largely on the geometry of the scrubber. Acceleration loss, which is frequently predominant in the venturi scrubber pressure drop, is fairly insensitive to scrubber geometry and in most cases can be predicted theoretically.

There are several correlations available, both theoretical and experimental, for the prediction of pressure drop in a venturi scrubber.[16,17] Calvert's equation[1] neglects the pressure loss due to wall friction and pressure recovery by the gas in the divergent section. This simplification is acceptable since wall friction is compensated to some extent by the pressure recovery. The pressure loss in a venturi scrubber is equal to the momentum expended to accelerate the liquid in the venturi throat and is given[14] by

$$\Delta P = -\left(\frac{2\rho_L u_G^2 Q_L}{g_c Q_G}\right)[1 - x^2 + (x^4 - x^2)^{0.5}] \tag{33}$$

For *illustration,* one can evaluate u_{dl}^* for typical venturi scrubber conditions of $u_G = 59$ m/sec (190 ft/sec) and $Q_L/Q_G = 1.33$ l/m³ (10 gal/Mcf). Assuming standard air and water properties, one can compute $d_d = 0.013$ cm, $N_{Reo} = 500$, and $C_{Do} = 0.55$. Note that N_{Reo} generally ranges between 500 and 600 and $C_{Do} = 0.55$ for usual scrubber conditions.

If $l_t = 30$ cm (1 ft), $u_{dl}^* = 0.77$ and if $l_t = 60$ cm (2 ft), $u_{dl}^* = 0.87$. This means that the drops will attain 77% of the gas velocity at the end of a 30-cm throat and 87% at the end of a 60-cm throat. As an approximation for standard air and water properties one can reduce Eq. (33) to

$$\Delta P = 0.8 \times 10^{-3} \frac{Q_L}{Q_G} u_G^2 \tag{34}$$

4.5.4.1 Optimization. Because of the wide use of venturi and similar scrubbers, there has been considerable attention given to defining their characteristics and optimizing their design. Optimization studies have led to three general approaches:

1. Select the best combination of u_{Go} and (Q_L/Q_G) for collecting a given particle size.
2. Generate optimum-sized drops and introduce these into the throat.
3. Use a series of two or more stages of scrubbing with different throat conditions.

An analysis[7] based on Eq. (26) concluded that the optimum condition is when $(K_{pa}f) = 1.1$. If one uses a value of $f = 0.5$, the optimum would occur when $K_{po} = 2.2$. Since the inertial parameter is a function of particle size, one still has the problem of how to define the optimum when there is a range of particle sizes to collect.

For practical purposes, it is simplest to use the cut/power correlation and the cut diameter design approach to find the approximate pressure drop required and then to find the best combination of u_{Go} and (Q_L/Q_G) for the case at hand. The best prediction can be made by integrating the grade penetration values over the given particle size distribution.

Introduction of *preformed spray* appears to have no advantage, especially when the extra power and complexity is considered. Multiple stages can be advantageous when a range of particle sizes is to be collected. The reason is that small drops are more effective for collecting submicrometer-sized particles, while large drops are more effective for large particles. As Eq. (32) shows, smaller drops are formed when u_G increases and (Q_L/Q_G) decreases. Thus, one can set the throat sizes and/or liquid flow rates for a series of scrubbers so that large drops are formed in one and small drops in another.

The Calvert Collision Scrubber is, in effect, a two-stage scrubber in which first large drops and then small drops are employed. Small drops will generally provide more surface area than large ones (after accounting for the difference in liquid flow rates and holdup) so that diffusional collection of particles (and gas absorption) will be improved.

5 ENHANCEMENT

5.1 Flux Force/Condensation

When a hot and saturated gas is in contact with cold water or a cold solid surface, condensation of water vapor occurs. Part of the vapor will be condensed on the particles which serve as condensation nuclei. Thus, the particles will have grown in mass due to the layer of water they carry and will be more susceptible to collection by inertial impaction. While condensation occurs, there will be diffusiophoretic and thermophoretic deposition on the cold surfaces as well as some inertial impaction. The particle growth by condensation in combination with diffusiophoresis and thermophoresis is referred to as flux force/condensation (F/C) scrubbing.

The gas leaving the source is hot and has a water vapor content that depends on the source process. The first step is to saturate the gas by quenching it with water. This will cause no condensation if the particles are insoluble, but will if they are soluble. There will be a diffusiophoretic force directed away from the liquid surface.

Condensation is required in order to have diffusiophoretic deposition, any growth on insoluble particles, and extensive growth on soluble particles. Contacting with cold water or a cold surface is employed to cause condensation.

Subsequent scrubbing of the gas will result in more particle collection by inertial impaction. This will be more efficient than impaction before particle growth because of the greater inertia of the particles. There may be additional condensation, depending on water and gas temperatures, and its effects can be accounted for as discussed above.

One can apply this general outline of F/C scrubbing to a variety of scrubber types. The condenser may be a separate unit or can be part of the scrubber.

Several phenomena are simultaneously involved in a F/C scrubber and the mathematical model is complex and cumbersome. A simplified performance prediction and design method, based on the conclusion that the flux force effects and condensation effect can be treated separately is summarized in the following paragraphs.[18]

5.1.1 Diffusiophoretic Deposition

Particle deposition by diffusiophoresis can be described by the following equation.

$$u_D = \frac{(M_1)^{1/2} D_G}{[y(M_1)^{1/2} + (1-y)(M_2)](1-y)} \frac{dy}{dr} \tag{35}$$

or

$$u_D = C_1 D_G \left(\frac{1}{1-y}\right) \frac{dy}{dr} \tag{36}$$

where

D_G = diffusivity of water vapor in carrier gas, cm^2/sec
M_1 = molecular weight of water, g/mole
M_2 = molecular weight of nontransferring gas, g/mole
y = mole fraction water vapor, dimensionless
r = distance in the direction of diffusion, cm
u_D = diffusiophoretic deposition velocity, cm/sec

The molecular weight and composition function represented by C_1 describes the effect of molecular weight gradient on the deposition velocity corresponding to the net motion of the gas due to diffusion (the sweep velocity). For water mole fraction in air ranging from 0.1 to 0.5, C_1 varies from 0.8 to 0.88, and one can use a rough average of 0.85 for computing u_D and consequent particle collection efficiency by integrating over the period of condensation.

Whitmore[19] concluded that the *fraction of particles removed* from the gas by diffusiophoresis is *equal to* either the mass fraction or the mole *fraction condensing,* depending on what theory is used for deposition velocity. In other words, it is not necessary to follow the detailed course of the condensation process, computing instantaneous values of deposition velocity, and integrating over the entire time to compute the fraction of particles collected. One can simply observe that if some fraction of the gas is transferred to the liquid phase it will carry along its load of suspended particles.

5.1.2 Particle Growth

Particle growth is dependent on how well the particles can compete with the cold surface for the condensing water. There are several transport processes at work simultaneously in the condenser section of an F/C scrubber:

1. Heat transfer
   a. from the gas to the cold surface
   b. from the particles to the gas
2. Mass transfer
   a. from the gas to the cold surface
   b. from the gas to the particles

A mathematical model that accounts for these transport processes in addition to particle deposition has been solved for sieve plates under various situations to predict the fraction of the total condensate which goes to the particles (this fraction defined as f_p). It was found that f_p depends heavily on n_p, the particle number concentration, liquid temperature, and liquid-phase heat transfer coefficient.[18] It decreases significantly with n_p below about 10^6 particles/cm^3 and does not change much for particle number concentration greater than 10^7/cm^3.

The predicted f_p varied between 0.1 and 0.4 and an average of 0.25 fit experimental data for a sieve plate scrubber. It should be noted that there is considerable uncertainty about this factor and that it should be evaluated for the scrubber and operating conditions in use.

5.1.3 *Prediction of Performance*

The procedure for predicting the collection efficiency of an F/C scrubber system depends on whether the condensation and particle growth occurred within the scrubber or before the scrubber. The sequence of steps to be followed for *before the scrubber* is outlined in the following:

1. Determine the initial particle size distribution at the condenser inlet.
2. Calculate the condensation ratio corresponding to the condenser operating conditions.
3. Calculate the penetration due to diffusiophoresis according to the following equation. Collection by other mechanisms may be neglected.

$$Pt_D = 1 - 0.85 f_v = 1 - \left(\frac{0.85q'}{H_1 + 18/29} \right) \tag{37}$$

where

$$
\begin{aligned}
Pt_D &= \text{penetration due to diffusiophoresis, fraction} \\
f_v &= \text{volume fraction of gas condensing, fraction} \\
q' &= \text{condensation ratio, g condensed/g dry gas} \\
H_1 &= \text{original humidity ratio, g } H_2O/\text{g dry gas}
\end{aligned}
$$

4. Calculate the grown particle size distribution at the condenser outlet, assuming that an equal amount of vapor condensed on each particle.
5. Compute the grade penetration for the scrubber and calculate the overall penetration for the grown particle size distribution leaving the condenser.
6. Calculate the total overall fractional penetration for the F/C scrubber system. Overall penetration is equal to the product of steps 3 and 5.

5.2 Electrostatic Charge

There have been a number of electrostatically augmented scrubbers introduced for application in recent years. Generally, the manufacturer's data have been the only performance information available and the prediction of performance has been rather difficult.

The *theoretical basis* for computing particle collection by electrostatically augmented scrubbers is outlined in the following. While the theory is useful for interpreting performance and for estimating the magnitude of enhancement one may expect, there are often discrepancies between the claims for performance and what one can compute.

When electrostatic charges are present, the deposition of particles by inertial impaction and other mechanisms can be enhanced by one of the following:

1. The coulombic force between a charged particle and a charged collector.
2. The electrical image force between a charged particle and a neutral collector.
3. The electrical image force between a neutral particle and a charged collector.
4. The force on a charged particle in the presence of a neutral collector by a uniform external electric field directed parallel to the flow field.
5. The electric dipole interaction force between a neutral particle and a neutral collector, both polarized by a uniform external electric field directed parallel to the flow field.

For particle *collection by drops,* the dimensionless electrical force parameters for the above five conditions are

$$K_c = \frac{Q_d Q_p C'}{3\pi^2 E_G \mu_G u_o d_d^2 d_p} \tag{38}$$

$$K_{ic} = \left(\frac{E_d - E_G}{E_p + 2E_G} \right) \frac{Q_p^2 C'}{3\pi^2 \rho_o \mu_G u_o d_d^3 d_p} \tag{39}$$

$$K_{ip} = \left(\frac{E_d - E_G}{E_p + 2E_G} \right) \frac{2Q_d^2 d_p^2 C'}{3\pi^2 E_G d_d^5 \mu_G u_o} \tag{40}$$

$$K_{ex} = \frac{Q_p E_o C'}{3\pi \mu_g u_o d_p} \tag{41}$$

$$K_{icp} = \left(\frac{E_p - E_G}{E_p + 2E_G} \right) \left(\frac{E_d - E_G}{E_d + 2E_G} \right) \left(\frac{E_G d_p^2 E_o^2 C'}{d_d \mu_G u_o} \right) \tag{42}$$

where

K_c = coulombic force parameter, dimensionless
K_{ic} = charged particle image force parameter, dimensionless
K_{ip} = charged collector image force parameter, dimensionless
K_{ex} = external electric field force parameter, dimensionless
K_{icp} = electric dipole interaction force parameter, dimensionless
Q_d = drop collector charge, C
Q_p = particle charge, C
C' = Cunningham slip correction factor, dimensionless
d_d = drop diameter, cm
d_p = particle diameter, cm
u_o = gas velocity relative to collection body, cm/sec
E_o = uniform external electric field strength, V/cm
E_d = dielectric constant of the drop, F/cm
E_p = dielectric constant of the particle, F/cm
E_G = dielectric constant of the gas, F/cm
μ_G = gas viscosity, g/cm-sec

The single-drop collection efficiency, in the presence of electrostatic force, can be predicted by performing a force balance and solving the resulting equation. Kraemer and Johnstone[20] numerically solved the equations of motion and obtained approximate collection efficiencies for potential flow and Stokes flow around a spherical collector in the absence of particle inertia. For the collection of a charged aerosol by a charged collector, considering only the coulombic force, the collection efficiency is

$$\eta = -4K_c \tag{43}$$

For the collection of uncharged aerosol particles by a charged spherical collector considering only the induced charge on the particles, the collection efficiency is

$$\sigma = \left(\frac{15}{8} \pi K_{ip} \right)^{0.4} \tag{44}$$

The equations of motion for the collection of inertialess particles on spheres with electrical force have been solved numerically.[21] The collection efficiency under the influence of coulombic force is the same for potential flow and viscous flow. The collection efficiency for external electric field force is

$$\eta = \left(1 + 2 \frac{E_d - E_o}{E_d + 2E_o} \right) \left(\frac{K_{ex}}{1 + K_{ex}} \right) \tag{45}$$

The flow field does affect the collection efficiencies for the image force cases. The results are plotted in Figure 15.

5.2.1 Model for Charged Spray

Nielsen[21] presented the results of his computations of predicted particle collection efficiency for single drops for the cases of

1. Inertial impaction (NP/ND), neutral particle neutral drop.
2. Coulombic attraction (CP/CD) plus impaction.
3. Charged particle image force (CP/ND) plus impaction.
4. Charged collector image force (ND/CD) plus impaction.

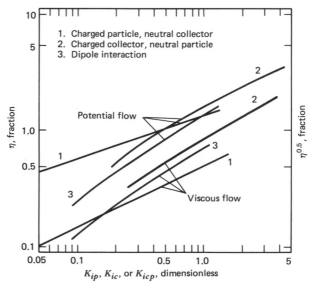

Fig. 15   Electrostatic collection efficiencies.

Nielsen's plots of collection efficiency against inertial impaction parameter and his electrostatic deposition parameters[21] can be used to predict the efficiency of a spray drop at various points along its trajectory. They show that *coulombic attraction* is the only mechanism that would cause a significant increase in the collection efficiency of an electrostatically augmented spray scrubber.

The dependence of particle collection efficiency on K_c and K_p is illustrated in Figure 16. Note that the values of K_c are negative, signifying that the particles and drops are oppositely charged. It can be seen that the greatest effect of coulombic attraction occurs when K_p is small, which corresponds to small values of relative velocity, u_o. Since K_c is inversely proportional to u_o, it increases as K_p decreases, thus intensifying the predicted influence of coulombic attraction.

5.2.2   *Average Drop Efficiency*

The efficiency shown in Figure 16 is an instantaneous value, while a spray scrubbing model requires accounting for particle collection over the total path the drop travels. Drop trajectories can be computed from the initial velocity, drop size, and gas properties in terms of drop velocity versus drop range (i.e., the distance traveled by the drop). A 100-μm diameter drop sprayed into air at a velocity of 30 m/sec would travel 30 cm relative to the air before stopping, while a 200-μm diameter drop would go about 90 cm.

Given the drop velocity at all positions along its range and the instantaneous efficiency correlation of Figure 16, one can compute collection efficiency for all points on the range. Figure 17 is a plot of predicted particle collection efficiency (for $d_{pa} = 0.6\ \mu$mA) versus drop range for charged particles with charged drops (CP/CD) and for neutral particles and drops (NP/ND). The two sets of curves shown are for 100- and 250-μm diameter drops.

The charge level on the drops was computed from experimental data which indicated a charge of approximately 5×10^{-7} C/g for drops of 200–500 μm diameter, with induction charging. For 250-μm diameter drops this is about 15% of the Rayleigh limit, while for 100-μm diameter drops it is only 3.7% of the limit. The particle charge levels were computed for a corona charger with a field strength of 4×10^5 V/m.

Figure 17 shows that the *greatest effect of charging* occurs after the drops have slowed so much that inertial impaction becomes unimportant. Since the average efficiency is obtained by integrating over the drop range, it is strongly influenced by how much of the drop range is considered. In many spray scrubber configurations the drop range R_d may be about 50 cm. This is the distance the drop would go from the point of atomization to the scrubber wall or to a collision with other drops.

A 50-cm range limit presents no problem for a 100-μm diameter drop, whose range would only be 30 cm if its initial velocity were 30 m/sec. For a charged 250-μm diameter drop the 50-cm range limit eliminates the most effective part of the drop trajectory. This fact has obvious importance in the designing of a charged spray scrubber.

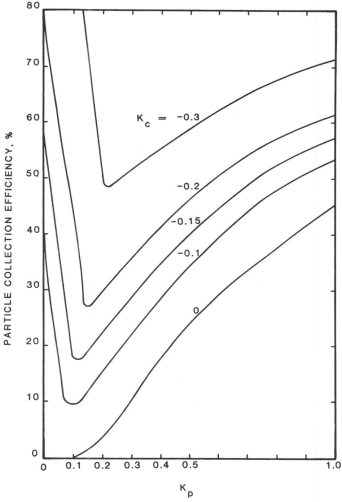

Fig. 16 Electrostatic and inertial efficiencies.

Average efficiencies for 100- and 250-μm diameter drops were computed by integration over drop trajectories and are given in Table 2. Note that the range R_d used is the smaller of the drop range and the range limit. Thus, R_d = 30 cm for d_d = 0.01 cm and 50 cm (limit) for d_d = 0.025 cm.

5.2.3 Discussion

While the theoretical model predicts no significant effect of charging for 250-μm diameter drops, experiments[11] have shown a large increase in efficiency for the smaller particles. The discrepancy may be due in part to the effects of drop range and drop diameter. If the drops travel farther than 50 cm before striking a wall and if the effective average drop diameter is smaller than 250 μm, the predicted efficiency would be higher. The influence of drop diameter is clearly shown. Despite their smaller range, 100-μm diameter drops would be much more effective than 250-μm diameter drops.

A higher charge on the drops would have a large effect. For instance, if the charge level on 100-μm diameter drops were 10% of the Rayleigh limit, the average efficiencies for 0.6-μmA particles would be 30 and 47%, respectively. Comparison with the values given in Table 2 for a lower drop charge level shows the magnitude of improvement better charging could cause.

Spray scrubbing efficiency for fine particles is significantly improved by electrostatic charging of the spray, and even more by charging the particles as well. The use of small drops and higher charge

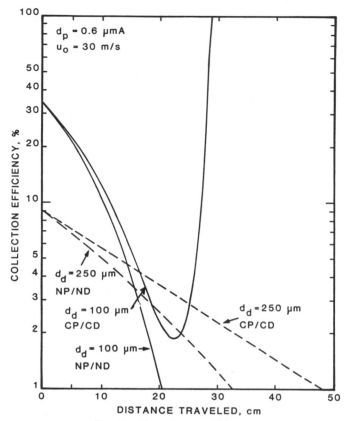

Fig. 17 Charged spray efficiency.

levels on the drops is desirable. Charging causes little improvement for particles larger than a few μm diameter.

Mathematical modeling of charged spray scrubbing has not yet yielded a method for predicting the magnitude of improvement caused by spray charging. Better accounting for drop size, trajectory, and charge level appear to be required in order to improve the model.

5.3 Surfactant

Surface active agents have been used in many attempts to enhance the efficiency of scrubbers, with varying success. While some experimental studies have shown that efficiency increases as particle wettability increases, other research and large-scale tests have shown little effect of interfacial tension.

The reasons for the apparently contradictory results become clear when the experimental findings are examined. A summary of the main observations is as follows:

1. When an effect has been found, it was seen[22] that nonwetted particles formed a layer on the surface of the liquid. It was hypothesized that particles impacting on the solid layer would bounce rather than being retained by the liquid. Wettable particles tended to move into the liquid body, with the result that a liquid surface is presented to depositing particles.

2. Mixing the liquid or replacing the liquid surface rapidly has the same effect as increasing wettability. The surface deposit is either mixed into the body of the liquid or replaced with fresh liquid surface.

3. Studies of behavior on clean liquid surface have led to the conclusion that interfacial tension cannot be a significant factor. It is obvious from the first point that such studies do not address the proper problem.

4. Surfactants can influence surface properties if there is sufficient time for a high enough concentration to build up at the surface. In many scrubbers fresh liquid surface is created very rapidly and

Table 2 Average Collection Efficiencies of Drops and Spray Penetrations

d_{pa}, μmA	d_d, μm	R_d, cm	u_o, m/sec	η, %, NP/ND	η, %, CP/CD
1.0	250	50	20	12.4	15.0
0.6	250	50	20	1.4	1.8
1.0	250	50	30	22.2	24.0
0.6	250	50	30	2.8	3.4
1.0	100	30	30	28.0	33.0
0.6	100	30	30	8.6	11.5

the surfactant cannot diffuse to the surface in time to do any good. Attempts to improve atomization by adding surfactant to the liquid have failed for this reason, despite the fact that a pure liquid with a lower surface tension would yield smaller drops.

5. The role of wettability in a scrubber is most obvious in stable foam-type scrubbers in which an extensive surface is produced and particle collection is by diffusion within the foam cells to the relatively quiescent liquid films.

6 ENTRAINMENT SEPARATION

An unfortunate consequence of thorough and vigorous liquid–gas contacting in the scrubber is that some liquid is atomized and carried out of the scrubber by the gas that has been cleaned. The liquid entrainment, or mist, will generally contain both suspended and dissolved solids.

In many cases, excessive entrainment imposes a limitation upon scrubber capacity. That is, while the scrubber itself might be capable of handling a larger gas flow rate, the rate at which entrainment becomes excessive will dictate a limit on capacity.

6.1 Separation Principles

Drops entrained from the scrubber contacting zone may be separated from the gas by the same mechanisms described earlier for particles. Since drops are usually larger than particles, dominant collection mechanisms are gravitational sedimentation, centrifugal deposition, and inertial impaction.

6.2 Equipment Types

Apparatus used for entrainment separation can be grouped into categories according to the mechanism of operation:

1. *Gravitational Sedimentation.* Within the scrubber and its outlet ducting, sedimentation is always active and important. However, entrainment separators using sedimentation rarely follow a scrubber.

2. *Centrifugal Deposition.* Cyclone separators of various designs are commonplace. Radial baffles and other types of guide vanes can be installed to induce a rotary motion of the gas stream within the scrubber shell. Zigzag baffles, chevrons, corrugated sheets, and similar devices force one or more abrupt changes in the gas flow direction. Drops deposit on the baffles, and the collected liquid film either runs down the baffles (if the major axis is vertical) or drips off as large drops (if the axis is horizontal). A directional change in the gas duct can likewise cause considerable deposition of large drops.

3. *Inertial Impaction.* Beds of massive packing such as saddles, rings, and other elements are used in either vertical or horizontal gas flow configurations. Beds of fibrous packings such as knitted wire mesh, screens, and glass fiber are also used for this purpose. Other impaction devices employ banks of round tubes, stream-lined struts, and other shapes in vertical and horizontal orientations. Trays (perforated, valve, impingement, and others) have been used for special purposes. Because these are scrubbing devices, they generate entrainment. Their use after another type of scrubber may be redundant and should be carefully assessed.

6.3 Primary Efficiency

Primary collection is defined as the collection of the drops present in the original entrainment by various mechanisms and is reported in terms of the mass fraction collected. The reentrainment of these collected drops or the subsequent collection of these reentrained drops does not affect the primary collection efficiency even though it affects the net entrainment collection efficiency.

Methods for predicting primary collection efficiency for various types of separators have been

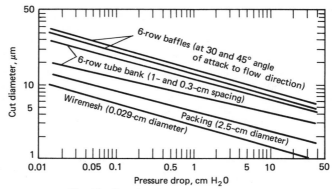

Fig. 18   Entrainment separator cut/power.

developed.[23] A concise representation of the primary efficiencies of several types of separators can be shown with the same cut/power relationship used for characterizing particle collection efficiencies in the scrubber itself. This relationship is given by a plot of the cut diameter against the gas-phase pressure drop, or power input for the separator (Figure 18).

Plots shown in Figure 18 are based on design equations and experimental correlations. Curves are given for baffles at two angles of attack to the flow direction, tube banks with two different spacings between tubes within a row, packing of one particular size, and knitted mesh with a certain wire diameter.

6.4   Reentrainment

The overall collection efficiency of an entrainment separator is equal to the mass ratio of liquid collected in the entrainment separator to the liquid present in the inlet. It can also be expressed as the difference between the primary collection and reentrainment.

Reentrainment from an entrainment separator may take place by any one or more of the following mechanisms:

1. Shattering of liquid drops upon impaction.
2. Creeping of liquid along the solid surface and movement into the gas exit in the entrainment separator.
3. Rupture of bubbles at the gas–liquid interface and subsequent drop formation.
4. Transition from separated flow to entrained flow caused by high gas velocity.

The first three mechanisms of reentrainment depend on the design of entrainment separators. The last mechanism represents the upper limit of the operation of entrainment separators.

The rate of reentrainment depends on separator geometry, gas velocity, liquid flow rate, and separator orientation. Increasing the gas or liquid flow rate will increase the rate of reentrainment. Liquid drainage is best when gas flow is horizontal and collection surface is near-vertical, and with this configuration, reentrainment occurs at higher flow rates than for horizontal elements.

Some approximate values of superficial gas velocity at the onset of entrainment are given here to illustrate the range possible for well-designed equipment at moderate liquid loadings.[23]

Separator	Superficial Gas Velocity, m/sec
Zigzag with upward gas flow and horizontal baffles	3.7–4.6
Zigzag with horizontal gas flow and vertical baffles	4.6–6.1
Cyclone (inlet gas velocity)	30.0–40.0
Knitted mesh with vertical gas flow	3.0–4.6
Knitted mesh with horizontal gas flow	4.6–7.0
Tube bank with vertical gas flow	3.7–4.9
Tube bank with horizontal gas flow	5.5–7.0

6.5 Solids Deposition

Industrial experience with entrainment separator fouling and plugging, and experimentation on sus-pended solids deposition, have yielded quantitative guidelines for design.[24] Vertical collection surfaces stay cleaner than horizontal ones due to better liquid drainage. Intermittent washing with sprays is beneficial, but the details of the washing system and procedure depend on the specific case. Precipitation scaling must be controlled through the system chemistry.

7 SOURCES OF ADDITIONAL INFORMATION

The reader who requires additional information can find a considerable amount within a few specialized sources and can identify what is available through printed or computerized information systems. Re-search supported by federal agencies continues to produce useful information on subjects that are of current importance to the agencies.

Table 3 presents a listing of subjects and reference numbers for some good sources of information. While some chapters of this handbook are listed, the reader should check others that may be pertinent, but omitted from the table.

Table 3 Information Sources

Subject	References	Chapters
1. Performance	1, 3, 8, 9, 25, 26 27, 28, 32	
2. Mathematical models	1, 6, 13, 25, 26	6
3. Test methods	8, 9, 32	6, 21
4. Design information	1, 2, 3, 23, 24 29, 30	
5. Operation and maintenance	29, 31	
6. Proprietary scrubbers	29, 32	
7. Applications	1, 2, 3, 29, 32	15–28
8. Auxiliaries	1, 2, 29, 32	
9. Costs		14
10. Literature searches		38

8 NOMENCLATURE

A_d deposition area, cm^2

A_p constant in Eq. (4)

B_v liquid/gas ratio parameter, dimensionless

c concentration, g/cm^3

c_p particle concentration, g/cm^3 (c_{pi} = at inlet, c_{pe} = at exit)

C_{Do} drag coefficient at venturi throat entrance, dimensionless

C' Cunningham slip correction factor, dimensionless

CD/CP   charged drop and charged particle

CD/UP   charged drop and uncharged particle

d_c packing diameter (nominal), cm

d_f fiber diameter, cm

d_h diameter of sieve plate hole, cm

d_j diameter of jet, cm

d_p physical particle diameter, μm

d_{pa} aerodynamic diameter of particle, μmA

d_{pg} geometric mean particle diameter, μmA

d_s Sauter (surface) mean diameter, cm

d_{pc} performance cut diameter, μmA

d_{RC} required cut diameter, μmA

D_p particle diffusivity, cm^2/sec

E collection efficiency, fraction

E_a average efficiency over drop range, fraction

E_d dielectric constant of drop, F/cm

E_G dielectric constant of drop, F/cm

E_o permittivity of free space = 8.854×10^{-12} F/m

E_p dielectric constant of particle, F/cm

f empirical factor

f_s volume fraction solids

f_v volume fraction voids

F foam density, g/cm³

F/C   flux force/condensation scrubbing

k_G thermal conductivity of gas, cal/g-K

k_p thermal conductivity of particle, cal/g-K

K particle relative dielectric constant, dimensionless

K_c coulombic force parameter, dimensionless

K_{ex} external electric field force parameter, dimensionless

K_I interception parameter, dimensionless

K_{ic} charged particle image force parameter, dimensionless

K_{icp} electric dipole interaction parameter, dimensionless

K_{ip} charged collector image force parameter, dimensionless

K_p inertial impaction parameter, dimensionless

K_{po} inertial impaction parameter at venturi throat entrance, dimensionless

l_t venturi throat length, cm

L liquid mass velocity, kg/hr-m² or lb/hr-ft²

n_h number of holes

N_{Re} Reynolds number, dimensionless

N_{Ref} Reynolds number based on fiber diameter, dimensionless

ND/CP   neutral drop and charged particle

ND/UP   neutral drop and uncharged particle

ΔP pressure loss, N/m², or cm W.C.

ΔP_j dry plate pressure drop, cm W.C.

Pt penetration, fraction

Pt_d penetration for a given particle diameter, fraction

$\overline{Pt}$ overall penetration, fraction

Q_d drop charge, C

Q_p charge on particle, C

Q_G gas volumetric flow rate, m³/sec

Q_L liquid volumetric flow rate, m³/sec

R_d range of drop travel, cm

S solidarity factor, dimensionless

t time, sec

T absolute temperature, K

u_d drop velocity, cm/sec

u_{do} drop velocity at venturi throat entrance, cm/sec

u_G gas velocity relative to duct, cm/sec

u_{Gs} superficial gas velocity, cm/sec

u_h gas velocity through sieve plate hole, cm/sec

u_o gas velocity relative to collector, cm/sec

u_p particle velocity, cm/sec

u_{pd} particle deposition velocity, cm/sec

u_{po} particle velocity at venturi throat entrance, cm/sec

u_j jet velocity, cm/sec

u_r drop velocity relative to gas, cm/sec

u_{dt}^* drop velocity at throat exit, dimensionless

u_{BD} particle deposition velocity for Brownian diffusion, m/sec or cm/sec

u_C centrifugal deposition velocity, m/sec or cm/sec

u_D diffusiophoretic deposition velocity, m/sec or cm/sec

u_m magnetic deposition velocity, m/sec or cm/sec

u_s gravitational settling velocity, m/sec or cm/sec

z length of fiber bed, cm

Z height of packing, cm

η collection efficiency, fraction

η_d efficiency for single drop

η_p efficiency for inertial impaction

η_s efficiency for single fiber

η_{DI} efficiency for diffusion and interception

η_{pDI} efficiency for impaction, diffusion, and interception

μ_o magnetic permeability, $4\pi \times 10^{-7}$ V-sec/A-m

μ_m micrometer

μmA $\mu m\sqrt{g/cm^3}$ = aerodynamic micrometer

μ_G gas viscosity, P

μ_L liquid viscosity, P

ρ_L liquid density, g/cm³

ρ_p particle density, g/cm³

σ surface tension, dyn/cm

σ_g geometric standard deviation of particle size

8.1 Dimensionless Numbers

N_{Sc} $\mu_G/\rho_G D_G$, Schmidt number

N_{Re} $\rho\, du/\mu$, Reynolds number

K_p $\dfrac{C'\rho_p d_p^2 u_p}{9\mu_G d_c}$, impaction parameter for shaped collectors

K_p $\dfrac{C'\rho_p d_p^2 u_j}{9\mu_G d_j}$, impaction parameters for jets

REFERENCES

1. Calvert, S., Goldshmid, J., Leith, D., and Mehta, D., *Wet Scrubber System Study, Vol. I: Scrubber Handbook*, NTIS #PB 213-016, U.S. Environmental Protection Agency, Washington, D.C., 1972.

2. Perry, R. H. (ed.), *Chemical Engineers' Handbook*, 4th edition, McGraw-Hill, New York, 1963.

3. Nonhebel, G., *Process for Air Pollution Control*, Butterworth and Co., London, 1972.

4. Lapple, C. E., and Kamack, H. J., Performance of Wet Dust-Scrubbers, *Chem. Eng. Prog.* **51**, 110–121 (1955).

5. Semrau, K. T., Correlation of Dust Scrubber Efficiency, *J. Air Pollut. Control Assoc.* **10**, 200–207 (1960).

6. Calvert, S., Source Control by Liquid Scrubbing, in *Air Pollution*, Vol. III, 2nd edition (A. C. Stern, ed.), Academic Press, New York, 1968, pp. 457–496.

7. Leith, D., and Cooper, D. W., Venturi Scrubber Optimization, *Atm. Envir.* **14**, 657 (1980).

8. Calvert, S., Yung, S.-C., Barbarika, H., Monahan, G., Sparks, L., and Harmon, D., Field Evaluations of Fine Particle Scrubbers, in *Second EPA Fine Particle Scrubber Symposium*, NTIS PB 273-828, 1977.

9. Calvert, S., Yung, S.-C., Barbarika, H., and Patterson, R. G., *Evaluation of Four Novel Fine Particulate Collection Devices*, EPA-600/2-78-062, U.S. Environmental Protection Agency, Washington, D.C., 1978, 106 pages.

10. Yung, S.-C., Chmielewski, R., and Calvert, S., *Mobile Bed Flux Force/Condensation Scrubbers*, EPA-600/7-79-071, U.S. Environmental Protection Agency, Washington, D.C., 1979, 241 pages.

11. Yung, S.-C., Curran, J., and Calvert, S., *Spray, Charging, and Trapping Scrubber for Fugitive Particle Emission Control,* EPA-600/7-81-125, U.S. Environmental Protection Agency, Washington, D.C., 1981.

12. Walton, W. H., and Woolcock, A., The Suppression of Airborne Dust by Water Spray, *International Journal of Air Pollution* 3, 129–153 (1960).

13. Yung, S.-C., Calvert, S., and Barbarika, H. F., Venturi Scrubber Performance Model, *Environmental Science and Technology* 12, 456–459 (1978).

14. Yung, S.-C., Barbarika, H. F., and Calvert, S., Pressure Loss in Venturi Scrubbers, *J. Air Pollut. Control Assoc.* 27, 348–351 (1977).

15. Nukiyama, S., and Tanasawa, Y., *Transactions of the Society of Mechanical Engineers, Tokyo* 4, 86 (1938).

16. Boll, R. H., Particle Collection and Pressure Drop in Venturi Scrubbers, *Industrial and Engineering Chemistry Fundamentals* 12, 40 (1973).

17. Behie, S. W., and Beekmans, J. W., On the Efficiency of a Venturi Scrubber, *Canadian Journal of Chemical Engineering* 51, 430 (1973).

18. Calvert, S., and Gandhi, S., *Fine Particle Collection by a Flux-Force/Condensation Scrubber: Pilot Demonstration,* EPA-600/2-77-238, U.S. Environmental Protection Agency, Washington, D.C., 1977, 777 pages.

19. Whitmore, P. J., Diffusiophoretic Particle Collection under Turbulent Conditions, Ph.D. Thesis, University of British Columbia, 1976.

20. Kraemer, H. F., and Johnstone, H. F., Collection of Aerosol Particles in the Presence of Electric Fields, *Industrial and Engineering Chemistry* 47, 2426 (1955).

21. Nielsen, K. A., *Effect of Electrical Forces on Target Efficiencies for Spheres,* Engineering Research Institute, Technical Report 74127, Iowa State University, 1979.

22. Goldshmid, Y., and Calvert, S., Small Particle Collection by Supported Liquid Drops, *AIChE J.* 9, 352 (1963).

23. Calvert, S., Yung, S., and Leung, J., *Entrainment Separators for Scrubbers—Final Report,* EPA-650/2-74-119b, U.S. Environmental Protection Agency, Research Triangle Park, NC, 1975.

24. Conkle, H. N., Rosenberg, H. S., and Di Novo, S. T., *Guidelines for the Design of Mist Eliminators for Lime/Limestone Scrubbing Systems,* EPRI FP-327, December 1976.

25. First EPA Fine Particle Scrubber Symposium, *JAPCA* 24(10) (1974).

26. *Second EPA Fine Particle Scrubber Symposium,* EPA-600/2-77-193, NTIS PB 273-828, U.S. Environmental Protection Agency, Washington, D.C., 1977.

27. *Symposium on the Transfer and Utilization of Particulate Control Technology,* EPA-600/7-79-044c, U.S. Environmental Protection Agency, Washington D.C., 1978.

28. *Second Symposium on the Transfer and Utilization of Particulate Control Technology,* EPA-600/9-80-093c, U.S. Environmental Protection Agency, Washington D.C., 1980.

29. McIlvaine, R., *The McIlvaine Scrubber Manual,* The McIlvaine Co., Northbrook, IL, 1974.

30. Rosenberg, H. S., et. al., *Construction Materials for Wet Scrubbers,* EPRI CS-1736, 1981.

31. Rimberg, D. B., *Management and Technical Procedures for Operation and Maintenance of Air Pollution Control Equipment,* EPA-905/2-79-002, U.S. Environmental Protection Agency, Washington, D.C., 1979.

32. Sittig, M., *Particulates and Fine Dust Removal—Processes and Equipment,* Noyes Data Co., Park Ridge, NJ, 1977.

CHAPTER 11

CONTROL OF PARTICLES BY FILTERS

JAMES H. TURNER

Research Triangle Institute
Research Triangle Park, North Carolina

JOHN D. McKENNA

ETS, Inc.
Roanoke, Virginia

1   PRINCIPLES OF OPERATION

1.1   Introduction

Filtration is one of the oldest and most widely accepted methods used for particle removal from dusty gas streams. In its modern forms, filtration removes a huge variety of particles from the near-visible sizes down to the Angstrom range. It is unrivaled for applications requiring extremely high efficiency capture of very small particles at moderate cost.

Basic principles of particle behavior and control are given in Chapter 6. The present chapter will deal with uses, design, operation, and costs of fabric filters, and to a lesser extent, granular bed filters and fiber mat filters.

Fabric filters consist of semipermeable woven or felted materials that constitute the support (or substrate) for the approaching dust. In most cases, it is the deposited dust layer that enables the high-efficiency capture of the particles once a uniform surface layer has been established. The fabric is periodically partially cleaned of accumulated particles.

Gas cleaning can be accomplished by granular bed filters by passing the gas through one or more layers of granules that are large in comparison to the particles to be separated from the gas stream. Granular beds are also periodically cleaned of accumulated particles.

Although fiber mat filters are similar in some respects to fabric filters, they do not depend on accumulated particles for high efficiency. Depending on the density and/or depth of the fiber packing, relatively high capture efficiencies, ~90%, are attainable without an accumulated surface dust layer. They are generally discarded rather than cleaned except for heavy duty wire screen designs.

1.1.1   Appropriate Uses for Filters

Fabric filters are normally used where high efficiency is required and where process conditions enable the filter to operate without harm to itself or to the process. Limitations on the filter are imposed by temperatures greater than 500–600°F that can destroy the fabric or shorten its life to an uneconomical degree; by gas or particle constituents that attack the fabric; and by gas or particle characteristics that prevent proper cleaning of the fabric, for example, sticky particles not dislodged because of failure to precoat fabric with absorbing nonsticky material. Although liquids in the system have often been cited as detrimental to filter operation, they can be tolerated with proper design. Small particles can be collected, even in the Angstrom range[1,2] because of high diffusibilities. Many field problems can be circumvented by altering operating procedures, for example, cooling hot gases to a temperature suitable for the installed filter medium.

Gravel bed filters are used primarily at temperatures above the 500–600°F range. They tend to be bulky and heavy, as well as requiring careful design and operation in order to maintain high efficiencies for small particles.

Fiber mat filters are generally not cleaned; they are ordinarily used where particle concentrations are low (fractions of a grain per cubic foot) so that reasonable service life can be attained before discarding the filter.

Alternate cleaning devices are electrostatic precipitators (ESPs) and wet scrubbers where efficient collection is desired. Mechanical collectors play secondary roles usually as precleaning devices. ESPs are competitive with fabric filters in those cases where particle resistivity ranges from about 10^{11} to 10^{12} ohm-cm, the required collection efficiency is less than 99.5%, the total gas handling capacity is greater than a few hundred thousand acfm, and the micrometer size particle population is relatively small. Scrubbers are competitive when small equipment size for large gas volumes is required, a liquid and sludge effluent can be tolerated, and where there is not a requirement for high capture efficiency for small particles. Mechanical collectors are used for collecting large particles at relatively low cost, where there is no requirement for high efficiency for the particle size fraction less than 5–10 μm.

1.2 Theoretical Aspects

1.2.1 Collection Mechanisms and Their Importance

When a particle approaches a fabric, there are several mechanisms by which it may be captured:

1. Interception
2. Impaction (inertial impaction)
3. Diffusion
4. Electrical precipitation
5. Thermal precipitation
6. Gravitational settling
7. Sieving

Figures 1 through 3 illustrate the more important of these mechanisms. In each case, the particle is assumed to be captured if it strikes yarn, fiber or a previously collected particle.

Interception. the particle is carried by a gas streamline directly toward a yarn or fiber (target). As the gas follows a path around the target, and within a particle radius of it, the particle contacts the target and is collected.

Impaction. The particle is on a gas streamline that would carry the particle around the yarn or fiber (target) without contact, but the particle's inertia causes it to leave its parent streamline. The net result is a collision course followed by impact on the target. The larger (more massive) the particle, the more its inertia, and the greater the likelihood of capture. This mechanism is not effective for particles smaller than about a micrometer at ordinary filter velocities.

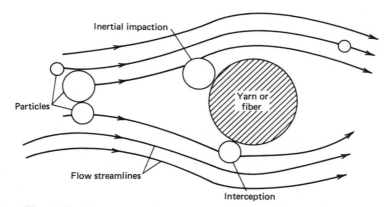

Fig. 1 Particle capture mechanisms—interceptions and inertial impaction.

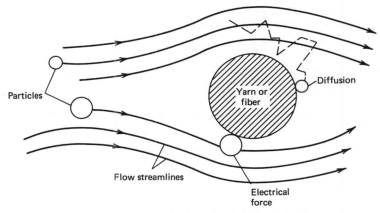

Fig. 2 Particle capture mechanisms—diffusion and electrical force.

Diffusion.   The particle is so small that its path, due to Brownian motion, is highly erratic. Collection may occur when a random excursion carries the particle to a target fiber. This mechanism becomes more important as particle size falls below about 0.1 μm.

Electrical Precipitation.   The particle and target have electrical charges of opposite sign on adjacent surfaces and, therefore, the particle is attracted to the target.

Thermal Precipitation.   The particle drifts toward a target because of a flow established by temperature gradients.

Gravitational Settling.   The particle mass and target mass cause sufficient attractive force to move the particle from a streamline near the target to the target itself; or the particle trajectory is influenced by the earth's gravitational force with subsequent capture by the target. This effect is very minor.

Sieving.   The particle is trapped because it is too large to pass through the specific pore or channel in which it finds itself.

In single-particle/single-fiber collection, interception, impaction, and diffusion are probably the most important mechanisms. Gravitational and thermal forces are usually minor, electrical forces can range from minor to major, and sieving does not apply. For fabric filters, a large portion of the collection process takes place in the particle mass (cake) that accumulates on and in the fabric. Dennis and his coworkers[3] state that the conventional mechanisms (interception, impaction, and diffusion)

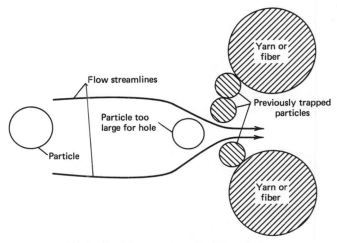

Fig. 3 Particle capture mechanisms—sieving.

are effective for only a brief time during each filtration cycle, and that sieving becomes the dominant mechanism as soon as the cake reforms after the filter is cleaned.

1.2.2 Drag

Energy is required to force a gas stream through a fabric. A common measure of this energy is the pressure drop or pressure differential across the fabric. For operating systems, pressure drop is often measured by manometer from inlet to outlet of the baghouse, or inlet duct to outlet duct for a single compartment in a baghouse.

Pressure drop correlates strongly with gas velocity. The two terms are combined to give a term called drag:

$$S = P/V \tag{1}$$

where

$S =$ filter drag
$P =$ pressure drop across filter (media only)
$V =$ velocity at filter surface (superficial face velocity)

Williams, Hatch, and Greenberg[4] used the concept of drag to develop a relationship for pressure drop through the combination of fabric and dust collected on the fabric. In modified form,

$$\Delta P = S_E V + K_2 C_i V^2 t \tag{2}$$

where

$S_E =$ effective residual drag of the fabric
$K_2 =$ specific resistance coefficient of the dust
$C_i =$ concentration of dust in the gas stream
$t =$ filtration time.

S_E is a measure of drag for a cleaned filter fabric whose residual dust loading is firmly imbedded in the fabric and very difficult to dislodge. Figure 4, a plot of drag versus weight of dust accumulated on the fabric, shows S_E. The specific resistance coefficient, K_2, is primarily a characteristic of the dust and should be measured. Table 1 gives values of K_2 reported by various investigators. These values, which have been taken from many sources, are of varying accuracy. If no experimental values of S_E are available, a default value of 350 N-min/m³ at 25°C for a residual fabric loading of 50 g/m² has been used.

1.2.3 Penetration

Dennis and Klemm[5] present a series of equations for predicting outlet concentration and particle penetration through the filter:

$$C_o = [Pn_s + (0.1 - Pn_s)e^{-aW}]C_i + C_R \tag{3}$$

$$Pn_s = 1.5 \times 10^{-7} \exp[12.7(1 - e^{1.03\,V})] \tag{4}$$

$$a = 3.6 \times 10^{-3}\ V^{-4} + 0.094 \tag{5}$$

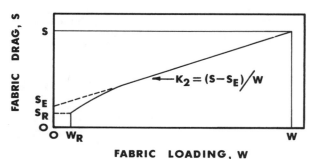

FABRIC LOADING, W

Fig. 4   Drag versus fabric loading.

Table 1 Values of K_2 Specific Cake Resistance

Application	Fabric	Cleaning	G/C, ft/min	G/C, m/min	K_2, in. H_2O ft-min/lb	K_2, N-min/g-m
Abrasives	Polyester	S	3	0.92		0.033
Alumina	Polyester	S	1.9	0.58		
			2.25	0.67	0.2	0.033
Aluminum	Cotton, nylon	S				
Aluminum hydrate	Polyester	P	1	0.30	0.2	0.033
Aluminum oxide	Cotton	S	2	0.61		
Asphalt plant drier	Glass, Nomex, cotton, wool	P, R, S	2.5–7.5	0.76–2.23	1.7	0.284
Asbestos	Cotton, nylon	S	2.75	0.84	2.18	0.364
Baking powder			2.5	0.76		
Bauxite	Cotton		2.5	0.76		
Beryllium sinter	Polypropylene, acrylic	P, S	6–20	1.83–6.1		
Bismuth and cadmium	Cotton		6–8	1.83–2.44	2.7	0.451
B.O.F.	Cotton, polyester	R, S	1.8–4.5	0.55–1.37	120–233	20.04–38.91
Bronze powder			2.0	0.61		
Buffing operation			3.0–3.25		0.92–0.99	
Calcimine			2.6	0.79		
Calcium sulfate	Polyester		7.5	2.28	0.4	0.067
Carbon	Acrylic		2.0	0.61		
Carbon black	Glass, Nomex Teflon, acrylic	R, S	1.1–1.6	0.34–0.49	22–56	3.67–9.35
Cement	Cotton, glass, acrylic, poly-ester	R, S	1.5–2.1	0.46–0.64	12–70	2.00–11.69
Cement	Cotton	R	2.5	0.76	350	58.34
Ceramics		S	2.5	0.76		
Charcoal		S	2.25	0.69		
Chocolate		S	2.25	0.69		
Chrome ore		S	2.5	0.76		
Chrome salts	Wool		2.4	0.73	15	2.51
Clay	Cotton	S	2.25	0.69		
Cleanser		S	2.25	0.69		
Cocoa		S	2.25	0.69		
Coke		S	2.25	0.69		
Copper	Glass, acrylic	S	0.6–2.7	0.18–0.82	15–65	2.51–10.86
Corn					0.62–8.8	0.10–1.47
Cork		S	3.0	0.92		
Cosmetics		S	3.0	0.92		
Cotton		S	3.5	1.07		
Dolomite	Polyester	R	3.3	1.00	670	112
Electric furnace	Glass, acrylic	R, S	1.5–4.0	0.46–1.22	45–715	7.5–119
Feldspar		S	2.5	0.76	6.30–27.3	1.05–4.56
Fertilizer		S	2.0–2.4	0.61–0.73		
Flint		S	2.5	0.76		
Flour	Cotton		2.5	0.76	4.3	0.717
Fly ash (coal)	Glass, Teflon	P, S, R	1.9–6.0	0.58–1.8	7–15	1.17–2.51
Fly ash (oil)	Glass	R	6.5–7.7	1.98–2.35	4.7	0.79
Fly ash (oil)	Glass	R	6	1.82	127	21.12
Fly ash (incinerator)	Glass, Teflon	P, S, R	2–6	0.61–1.83		

Table 1 Cont.

Application	Fabric	Cleaning	G/C, ft/min	G/C, m/min	K_2, in. H_2O ft-min/lb	K_2, N-min/g-m
Fly ash (incinerator)	Glass	R	2.5	0.76	180	30.00
Foundry	Glass, polyester, polypropylene, nylon	S	2.1	0.64	0.62–120	0.10–20
Glass	Polyester	S	2.5	0.76		
Grain	Cotton	S	3.25	0.99		
Granite		S	2.0	0.61		
Graphite		S	2.0	0.61		
Grinding dust		S	2.25	0.69		
Gypsum	Cotton, acrylic	S	2.5	0.76	6.3–18.9	1.05–3.16
Hypochlorite mfg.	Acrylic	P	3.3	1.0	15	2.51
Iron ore		S	2.0	0.61		
Iron oxide	Nomex	P	2.1	0.64	121	20.17
Iron oxide		S	2.0	0.61		
Iron oxide	Acrylic, polyester	R, S	1.4–3.3	0.43–1.00	3–715	0.50–119
Iron oxide, Zinc oxide	Glass	R	1.9	0.58	66	11.00
Kish	Polyester	S	2.5	0.76	230	38.34
Lampblack		S	2.0	0.61	47.2	7.88
Lead blast furnace	Polyester	R, S	1.0	0.31	57	9.52
Lead dust	Acrylic		2.4–35	0.73–10.7		
Lead oxide	Acrylic	S	2.25	0.69		
Lead oxide	Polyester	R, S	1.0	0.30	57	9.50
Leather dust		S	3.5	1.07		
Lime kiln	Glass	R	2.3	0.70	9	1.50
Limestone		S	2.75	0.84		
Magnesium trisilicate			0.5	0.15		
Manganese		S	2.25	0.69		
Marble		S	3.0	0.92		
Mica	Cotton	S	2.25	0.69		
Milk powder					4.5	0.75
Molybdenum	Wool					
Oats					1.58–11.0	0.26–1.84
Oyster shell		S	3.0	0.92		
Paper		S	3.5	1.07		
Perlite	Polyester, glass	S	3	0.92		
Pigments	Cotton	S	2.0	0.61	2.3–2.9	0.38–0.48
PVA	Wool	R	10	3.05	25	4.18
PVC	Wool, polyester					
Plastics		S	2.5	0.76		
Quartz		S	2.75	0.84		
Resin	Cotton	S	2.7	0.82	0.62–25.2	0.10–4.21
Rock dust		S	3.25	0.99		
Sand, scale	Cotton		5.0	1.52	3	0.50
Sanding machine		S	3.25	0.99		
Silica	Nomex	S	2.75	0.84		
Sinter dust	Glass	R	2.3	0.70	12.5	2.08
Soap	Polyester, acrylic	S	2.25	0.69	1.6–3.1	0.27–0.52
Soapstone		S	2.25	0.69		
Starch	Cotton, wool	S	2.25	0.69		
Stucco	Cotton, polyester	R, S	3.4	1.04	9	1.50
Sugar		S	2.25	0.69		

Table 1 Cont.

Application	Fabric	Cleaning	G/C, ft/min	G/C, m/min	K_2, in. H_2O ft-min/lb	K_2, N-min/g-m
Talc		S	2.25	0.69		
Titanium dioxide	Cotton, acrylic				94–206	15.7–34.4
Tobacco	Cotton, polyester	S	3.5	1.07	36	6.01
Wood	Cotton	S	3.5	1.07	2.8–6.3	0.47–1.05
Zinc	Acrylic, cotton, Nomex	R, S	1.8–3.0	0.55–0.92	7–50	1.17–8.35
Zinc oxide					15.7	2.62
Zinc oxide	Glass	R, S	0.6	0.18	40	6.67
Zinc oxide and lead chloride	Glass	R	1.2	0.36	18.5	3.08

Note: Table compiled from various sources. For a given application other fabrics, cleaning methods, and gas-to-cloth (G /C) ratio may be used. Many other sources use baghouses. Many cotton applications have been replaced by polyester.
Key: S = shake; P = pulse-jet; R = reverse air.

where

$$C_o = \text{outlet dust concentration, g/m}^3$$
$$Pn_s = \text{dimensionless constant}$$
$$V = \text{local face velocity, m/min}$$
$$C_i = \text{inlet dust concentration, g/m}^3$$
$$C_R = \text{slough-off concentration (constant), g/m}^3$$
$$W = \text{dust loading on fabric, g/m}^2$$

These equations were developed for use in an iterative computer model and for woven glass fabrics and fly ash. Extrapolation to other systems gives results of unknown accuracy. The value for C_R used by Dennis and Klemm was 0.5 mg/m^3.

Dennis and Klemm also reported that penetration for woven glass fabrics was largely due to leakage through unblocked pores. Size distribution of the leaking particles was essentially unchanged from that of the dust approaching the fabric. Apparent decreases in median particle size through a fabric filter system such as they studied were attributed to drop out of larger particles between the upstream sampling point and the filter.

1.2.4 Limitations

As with many technologies, filtration is limited by imperfect knowledge and lack of understanding of all the complex interactions that take place in an industrial model of a filter used in the field.

Design of industrial filters has traditionally been based on past experience without the aid of design equations. Many proposed equations are empirical in nature; there is no dependable, strictly theoretical approach for the design of fabric filters or the estimation of key variables. Measured values of K_2 and S_E must be used for calculation of pressure drop, with accuracy of Eq. (2) suffering if theoretically derived values are used.

1.3 Mechanical Aspects

The major differences in fabric filtration equipment lie in the areas of filtration media, cleaning methods, and equipment geometry. These differences normally result in differences in the operational mode.

1.3.1 Filtration Media

The heart of the fabric filter is a porous flexible web of textile material. This element is employed in the separation of the solid particulate matter from the gas stream. There are a variety of filter media to choose from and the selection is made on the basis of the least costly material that is compatible with the temperature, pH, and other physical and chemical characteristics of both the gas stream and the particulate matter. In addition, the fabric design must be adaptable to the fabric filter geometry and cleaning method, as well as the efficiency requirements. Both natural and synthetic fibers are employed in fabric filters. The use of synthetics now dominates because of their greater range of temperature and pH adaptability. Table 2 is a chart that can be employed in the selection of the filter medium. It includes both natural and synthetic fibers, as well as indicating the recommended maximum operating temperature and the fibers' suitability with acids and alkalies. On the low end of the temperature scale is cotton and polypropylene and on the high end Teflon and glass. Both the average operating temperature and the excursion limits must be considered in fiber selection. Exposure of the filter medium to either temperatures or pH conditions outside its operable range can result in the loss of bag function in a matter of hours or less. Upset conditions must also be accommodated in the system design to keep the baghouse operating. For example, if the gas stream to be cleaned is above 550°F (today's upper limit of commercially and economically available bags), then cooling of the gas stream must be employed before applying the baghouse. Although this adds another system cost component (the cooling hardware), the gas volume and thus the baghouse cost may be reduced.

Once the fiber has been selected, the next step is to determine what fabric design and weight (density and thickness) should be employed, for example, should a felt or woven filter medium be employed. Here felt refers to a random array of fibers interlocked by a needling process, whereas a woven filter medium is a symmetrically woven series of yarns (twisted fine fiber strands). The fabric decision is very much a function of the house–hardware design and especially the cleaning system to be employed. These relationships will be made clearer in the cleaning method section that follows.

1.3.2 Cleaning Methods and Physical Configurations

There are three common cleaning methods of which the oldest is the shaking of the filter media to dislodge the dust. The other two consist of blowing the dust away from the filter media, that is, backflushing with low pressure air or impacting the dust with a jolt of compressed air. In addition to the above cleaning methods, other techniques have been employed, which either for economic or technical reasons, are not widely used today. Combining two cleaning methods is sometimes used, for example, the combination of reverse air and shake in difficult cleaning applications.

1.3.3 Shake Cleaning

Shake cleaning, as one might guess, is the dislodging of dust from the filter media by imparting a rapid motion to the filter media via rocking or shaking the framework supporting the filtration media. This normally produces an undulating motion in the bags. The amplitude, shaking frequency, and bag tensioning are the parameters that can be varied in order to achieve effective cleaning.

Shaking should be only as vigorous and lengthy as necessary to dislodge the dust, since the gentler the shake, the longer the bag life.

As shown in Figure 5, shaking is usually carried out with bags that are open at the bottom and closed at the top. The dust-laden gas enters the bag interior through the tube sheet inlet at the bottom. As the gas passes through the filter medium, the dust is deposited on the inside of the bag while the cleaned gas passes through the bag to the clean air manifold and, hence, to fan and stack. The inside to outside gas flow distends and maintains the bag in a tubular form while filtration is in progress. During cleaning, the gas flow is stopped, thus causing the bag to collapse to an extent determined by the applied tension. No cage support or anticollapse rings are employed since these would inhibit the shaking motion.

Woven fabrics are used almost exclusively with shakers in the United States. The gas-to-cloth ratio is relatively low, normally below 4 to 1, with the shake clean-woven fabrics. In Europe, felts have been used with shakers at somewhat higher gas-to-cloth ratios.

Often shakers are employed where the dust loading and plant operations allow the collector to be on-stream for a number of hours, perhaps an entire shift, before being taken off-stream for cleaning. If one wishes to apply a shaker to a process requiring continuous gas cleaning, it is then necessary to design the baghouse with many individually isolatable compartments. This will allow one compartment to be taken off-line for cleaning while the balance remain on-stream. The design gas-to-cloth ratio must accommodate the off-stream cloth requirement.

While most shakers employ a motor and an eccentric to impart the shaking motion, other methods are sometimes employed. In the case of very small units needing only intermittent cleaning, manual shaking may be practiced. "Fluttering" of bags by use of an air shake represents another approach while high-frequency agitation can be achieved by means of sonic and ultrasonic devices. The motor and eccentric, however, appear to be the most economic automated shaker alternative to date.

1.3.4 Reverse Flow Cleaning

Reverse flow cleaning is the dislodging of the dust from the filter media by stopping the flow in the normal filtration direction and then reversing the flow so that, in effect the gas flow backflushes the

Table 2 Fiber Selection Chart

Common Name or Trademark	Generic Name	Fiber Properties					Recommended Operating Temperatures, °F		Elongation % at Break
		Tensile Strength	Abrasion Resistance	Chemical Acids	Resistance Alkalies	Supports Combustion	Continuous	Surges	
Cotton	Natural fiber cellulose	Good	Average	Poor	Excellent	Yes	+180	+225	
Polypropylene	Polyolefin	Excellent	Good	Excellent	Excellent	Yes	+190	+190	3–4
Glass	Glass	Excellent	Poor	Good	Poor	No	+500	+550	
Nylon	Polyamide	Excellent	Excellent	Poor	Excellent	Yes	+200	+250	
Dacron	Polyester	Excellent	Excellent	Good	Fair	Yes	+275	+325	
Orlon	Acrylic	Average	Average	Very good	Fair	Yes	+240	+260	
Microtain	Acrylic	Average	Average	Very good	Fair	Yes	+260	+280	
Wool	Natural fiber protein	Poor	Average	Fair	Poor	No	+200	+250	
Nomex	Aromatic polyamide	Very good	Very good	Fair	Very good	No	+400	+425	16–18
Teflon	Fluorocarbon	Average	Below average	Excellent	Excellent	No	+450	+500	15–32
PBI	Polybenzimidazole	Excellent	Excellent	Good	Fair	No	+1000	+1200	
Ryton (PPS)	Polyphenylene sulfide	Average	Good	Good	Good	No	+340		25–35
Dralon-T	Homopolymer acrylic	Good	Good	Good	Excellent		+284		
Huyglas	Fiberglass (felted)	Excellent	Average	Good	Poor	No	+450	+500	
Goretex	Polytetrafluoroethylene (film on felt or woven)	Average	Below average	Excellent	Excellent	No	+450	+500	
Bekipor	Stainless steel 316L felt and woven	Excellent	Excellent	Good	Good	No	+850	+1000	

SIDE   VIEW   FRONT   VIEW

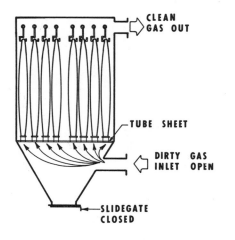
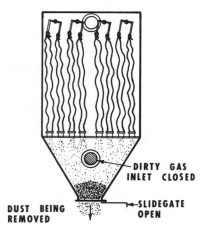

A— DUST COLLECTING MODE   B — BAG   CLEANING   MODE

Fig. 5 Shaker.

filter media. The duration and frequency are usual cleaning parameters that can be varied, although the deflation rate and reverse gas pressure can also be varied to a limited extent.

As shown in Figure 6, reverse flow cleaning is normally employed in conjunction with bags that are open at the bottom and closed at the top. Unlike the shaking system, bags cleaned by reverse flow typically have metal caps attached at the top end where tension adjustments are made. As with the shaking method, the dust-laden gas enters the bottom of the bag through the tube sheet thimble and then proceeds up the bag and through the filter media to the clean air side. As the gas passes from inside to outside the bag, the dust is collected on the inside of the bag. During cleaning, the normal flow to the bag is stopped and the reverse air flow is initiated. The bag is now flexed as the deflation and reverse flow begins. Complete collapse is prevented by a series of rings sewn into the bag as shown in Figure 6. The dislodged dust descends through the bag interior and exits via the tube sheet into the hopper.

Woven fabrics are normally employed in reverse air flow. The few inches of water reverse pressure gradient normally employed would not be sufficient to backflush felted bags. As with shaken bags, gas-to-cloth ratios of less than 4 to 1 are most common.

Large capacity reverse flow systems are applied to continuous processes by the use of the modular system design as shown in Figure 7. By proper damper and duct arrangement, the total gas flow is accommodated continuously by allowing the gas-to-cloth ratio to rise during cleaning. The figure applies to a hot gas where a portion of the just cleaned gas is used to clean the off-line module. This same gas stream, along with the dust loading acquired during backflushing, is cycled through the on-stream modules for filtration before being discharged to the atmosphere.

Most reverse flow systems employ a separate fan for providing the reverse flow. Often each module will have its own reverse flow fan, even though there may be a single main system fan.

1.3.5 *Pulse Jet Cleaning*

Pulse cleaning also involves a backflushing of the filter medium. In this cleaning method, a vigorous cleaning action is obtained since the backflushing employs high-pressure compressed air, normally in the 60–100 psi range. The brief pulse of compressed air creates a shock wave that runs down the bag, thus dislodging the dust from the felted fabric. With this cleaning method, it is possible to overclean the bags, thereby producing emissions above the allowable limit. Normally, this can be overcome by reducing the pulse pressure or the pulse duration. The cleaning parameters that can be varied are the pulse pressure, frequency, and sequence of cleaning and to a limited extent, the duration of cleaning.

As shown in Figure 8, pulse cleaning is usually employed in conjunction with bags open at the top and closed at the bottom. In some instances, a venturi (as seen in Figure 9) is employed in an attempt to provide entrained air for backflushing immediately following the shock due to the burst of compressed air. It appears that the initial fabric flexure provides the major dust dislodgement. Contrary to the common shake and reverse air system, the bag in this instance collects the dust on the outside with the cleaned gas passing inward and upward through the bag, exiting the top of the

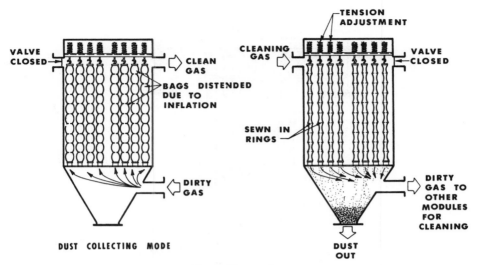

Fig. 6 Reverse flow.

bag through the tube sheet hole in the venturi region. Cages within the bag are necessary to keep the bag from collapsing. It is possible to clean the bags with this cleaning method simply by initiating the compressed air pulse while the bags are still on-stream. This is so because the compressed air will momentarily halt the flow in the normal filtration direction. Thus, pulse jets do not require the usual compartmentalization and one or more rows of bags may be cleaned while adjacent bags are collecting dust. In difficult cleaning applications, off-stream cleaning via house subdivision is employed in order to eliminate reentrainment (see Figure 10).

Felted fabrics are normally employed in pulse jets with gas-to-cloth ratios of 6 to 1 and higher, and are permissible with larger and easier to clean particles.

Although this cleaning method is the newest of the three methods described here, it has been applied for more than 30 yr. It is widely employed in the lower and medium capacity ranges (≤ 100,000 acfm) while recently it has seen use in some fairly high capacity installations.

2 PRELIMINARY DESIGN

2.1 Problem Definition

In order to design and/or select a baghouse, a thorough definition of the emission problem is required. The information needed falls into two categories. The first category is that of absolutely essential information, that is, information without which it is impossible to design or select the baghouse. The second category is information that should be obtained if at all possible, because without it the risk of operational difficulties is high.

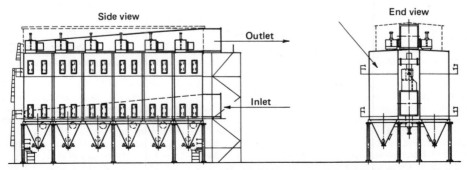

Fig. 7 Reverse flow—modular design. Copyright © 1980 Enviro-Systems & Research, Inc. All rights reserved.

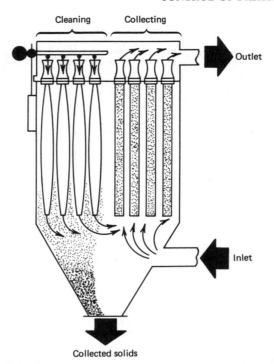

Fig. 8 Pulse jet—on-stream cleaning and collecting dust simultaneously.

The first category, absolutely essential information, includes volume flow, temperature, and specific application. With these data, one can size the unit, select the filter medium, and probably decide on the cleaning method. When determining the gas volume to be treated, one needs to establish the actual gas volume at fan and filtration temperatures. One must be careful to specify whether the volume is constant or variable and if variable, the minimum and maximum values. Similarly, in determining temperature, it is necessary to determine not only the average temperature and the normal temperature operating range, but also what excursions and peaks and durations might be anticipated. The volume variations influence the gas-to-cloth ratio selection, while temperature variations impact on both filter media selection and system design parameters.

General information about the application sometimes suffices to allow one to select the cleaning method and baghouse design. This is true if it can be determined that the same application has already been successfully treated numerous times at other plants. In such instances, knowledge of such empirical data as shown in Table 3 will allow selection of the baghouse and gas-to-cloth ratio to proceed quickly. This approach is often employed in the choice of small capacity units where the capital expenditure does not justify the substantial cost associated with extensive problem definition. In straightforward applications, this approach is often successful, although the risk of misapplication is always present. The cost of correcting subsequent operational problems often outweighs the corresponding problem definition costs.

Other helpful but less critical problem definition factors include the gas stream dust loading, particle size, the chemical and physical characteristics of the dust, the gaseous constituents, pressure drop and back pressure limitations, space, available utilities, and the production process cycles.

The baghouse emissions are by comparison to other particulate control devices relatively insensitive to inlet gas stream dust loadings. In spite of this fact, a knowledge of the inlet dust loading is very desirable since it greatly influences the collected dust conveying and discharge systems. In addition, the inlet dust loading comes into play in determining the filter's operating pressure drop. The particle size and density of the dust are, in a sense, of greater importance than the dust loading. High-density, submicrometer particles are, in general, very difficult to capture. Therefore, they are normally collected by fabric filter systems operating at low gas-to-cloth ratios as is the case with certain metallic oxides. Selection of too high a ratio in this instance can result in either very high pressure drop or very high emission levels, depending on the filter media type and the method of cleaning. Conversely, a relatively large but very low density particle can give presssure drop problems if too high a ratio is

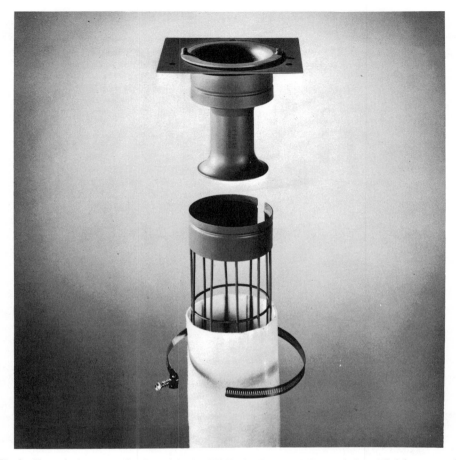

Fig. 9   Venturi—bag assembly. Copyright © 1980 Enviro-Systems & Research, Inc. All rights reserved.

chosen or if improper gas flow patterns occur. Reentrainment and bridging of the dust within the collector and excessive dust discharge can occur as a result.

The chemical and physical properties of the dust can be crucial to the successful operation of a baghouse. For example, if the dust is sticky or hygroscopic, it can lead to inadequate cleaning that eventually causes blinding of the bags. Particularly, abrasive dust can result in reduced bag life or extreme wear if the incorrect collected dust discharge system is chosen.

Gaseous constituents can be crucial in the selection of fabric media and baghouse materials of construction while determining whether or not the system requires insulation. The presence of acid gases or condensable hydrocarbons is a very important consideration and the nature and concentration of these constituents should be determined in the problem definition stage.

Similarly, any restriction regarding the pressure drop across the baghouse system should be defined since it comes into play in the selection of the gas-to-cloth ratio. A lower gas-to-cloth ratio, for a given case, results in a lower ΔP. Likewise, any backpressure restrictions need to be defined early on since they can influence the baghouse design and cleaning method chosen. For example, on-line pulse jet cleaning causes a momentary increase in house pressure drop when rows of bags are off-line for cleaning; thus, it may be desirable to choose an off-line cleaning design in order to avoid backpressure.

Available space can also influence the baghouse design. In particular, it can greatly influence the use of top access and, in some instances, pushes one toward high gas-to-cloth designs.

A comprehensive questionnaire that one should fill out in as complete a manner as possible in the problem definition phase is shown in Table 4. In general, and particularly where large capital expenditures are involved, it is recommended that sufficient hard data be obtained on the specific source to be controlled to minimize the risk of misapplication. Even in cases of relatively small capital

Fig. 10 Pulse jet—modular design.

expenditure, a significant expenditure on problem definition is often justifiable if a shutdown of the fabric filter can result in a shutdown of the production process for any length of time.

2.2 Gas-to-Cloth Ratio Estimates

Equation (6) can be used in trial-and-error fashion if pressure drop is limiting, and if S_E and K_2 are known. Time for a filtration cycle must be chosen, and inlet concentration is assumed to be known. K_2 and S_E are not usually known, and values must be estimated or else G/C must be found in some other manner. Table 1 may be used to select G/C or K_2, but the tabulated values are suspect because of lack of knowledge about system details.

Table 3 Empirical Application Information

Application	Gas-to-Cloth Ratio	Dominant Cleaning Method[a]	Cloth[a]	Dust Size	Dust Density
Grain	12–14	RA	F	Large	Low
Limestone (quarry)	6–8	PJ	F	Large	Medium
Lead oxide	1.5–2	S	W	Fine	High
Coal fly ash (utility boiler)	2–3	RA	W	Fine	Medium
Coal fly ash (industrial boiler)	4–5	PJ	W/F	Medium	Medium
Cement (kiln)	2–3	RA	W	Medium	Medium

[a] Key: RA = reverse air; PJ = pulse jet; S = shaker; F = felt; W = woven.

Table 4   Customer Inquiry Data Sheet

Firm Name _____	Telephone _____
Street _____	Tel. Ext. _____
City _____	Per _____
State _____ Zip _____	Title _____

Essential Information Required

1. Description of Application _____

2. Gas Inlet Volume _____ acfm   3. Gas Inlet Temp. _____ °F
4. Description of Dust _____

Other Information Desired

1. Space Available for Equipment (Sketch or Dwgs.) _____
2. Equipment Auxiliaries Wanted _____
   _____
3. Particulate Pollutants: Type of Dust _____ Particulate Size _____
   Abrasive? _____ Explosive? _____ Inflammable? _____
   Sticky? _____ Moisture Content (% by wt.) _____
   Bulk Density _____ Specific Gravity _____
   Inlet Loading _____ gr/scf or lb/min
   Allowable Outlet Loading _____ gr/scf or lb/min
4. Gaseous Pollutants: Type(s) and Concentration(s) _____
   _____
5. Additional Comments (i.e., Location, etc.) _____
   _____
6. Type of Duty: Intermittent _____ Continuous _____
7. Operating Pressure _____ in. water gauge
8. Existing Electrical and Compressed Air Capabilities _____
   _____

Once the system pressure drop is known, fan energy requirements can be estimated from

$$P = 0.000218(Q)(\Delta P)(t) \tag{6}$$

where

P = fan power, kWhr/yr
Q = system flow rate, acfm
ΔP = system pressure drop, in. H_2O
t = operating time, hr/yr

System pressure drop includes pressure loss through the bags, ductwork, breechings, and manifold sections. Ductwork pressure loss can be estimated by methods described in the *Air Pollution Engineering Manual*.[6] Pressure loss through the baghouse structure, which varies with design and size, is roughly ½ to 2 in. of water. System pressure drop may be designed from 2 to 3 in. of water and up to about 20, in unusual cases. Five or six inches would be typical. Cleaning energy for reverse air systems can be based on number of compartments to be cleaned at one time, and reverse gas-to-cloth ratio. This reverse ratio normally ranges from about the same value as the forward gas-to-cloth ratio to 2 times higher. Pulse jet cleaning requires compressed air at about 60 to 100 psig, delivered at a rate of about 2 scfm/1000 cfm of gas filtered. Mechanical shake cleaning usually is performed by one shaker motor per compartment or pair of compartments with 1 hp serving about 2000 ft² of fabric. Typical energy consumption is

$$E = 0.053A \tag{7}$$

where

E = energy consumption, kWhr/yr
A = total fabric area, ft²

2.3   Media Selection

Historically, natural fibers, wool and cotton, were the first employed in fabric filter baghouses. The advent of synthetic fibers in the 1940s broadened the applicability of the baghouse to higher temperatures and a wider range of chemically reactive aerosols. Expanded development and application of synthetics continues today in efforts to increase the temperature capabilities of filter media by the use of new and different fibers.

As mentioned in Section 2.1, selection of the filter media is possible once the working temperatures and application are known. This is often true since a knowledge of the application frequently includes information on what bag materials have previously performed successfully. A detailed definition of the gas stream components is highly desirable. The chemical nature of suitable filter media is primarily determined by the gas temperature and the alkalinity or acidity of the gas stream. Table 2 provides a listing of filter media types indicating both their temperature and chemical limitations.

Commercially available filter media are currently limited to about 550°F, although a number of firms are developing higher temperature fabrics. Both metal and ceramic fiber-based media are being evaluated at this time. Currently, however, the upper limit of the commercial temperature applications are being met with either glass (550°F) or Teflon (500°F). Of the two, Teflon is the more resistant to chemical attack, but at the same time more expensive. Teflon is employed both in felted and woven form. Glass, until recently, has been available only in the woven form. The development of a felted glass has provided industry with a high-temperature filter media that is anticipated to be more durable than woven glass (see Figure 11). Another high-temperature product development of recent vintage is the microporous PTFE film on both woven and felted substrates. Good dust cake release properties have been reported for this material. The use of the felt backing seems to be dominating in attempts to obtain greater bag life.

If the gas stream temperature exceeds 550°F, and a baghouse is determined to be the preferred particulate control device, cooling of the gas stream is necessary. An economic evaluation is often advisable to determine whether radiant cooling, a heat exchanger, dilution air, or spray cooling (each a feasible choice) should be employed. Spray cooling is potentially risky because of the possibility of wetting the bags and/or condensing out an acid on the baghouse structure and/or on the bags. *

In general, the higher the fabric temperature resistance, the higher will be its initial cost. Therefore, one normally selects a bag material meeting, but not exceeding, the temperature requirement. Table 5 shows the relative cost and temperature limitations for a number of filter media.

The physical nature and/or fabric construction of the filter media is chosen in conjunction with the cleaning method. Normally, woven fabrics are employed with shake and/or reverse air cleaning. On the other hand, bags made of felted fabrics are used with pulse jet cleaning. The latter choice is

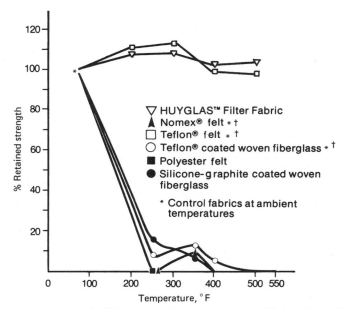

Fig. 11   Percent retained strength of fabrics after repeated exposure to sulfuric acid at indicated temperatures. Courtesy of Huyck Felt. (†) Registered trademark of duPont.

Table 5   Filter Media Comparison

Fabric	Relative Cost, $^{a}	Temperature, °F
Polyester	6	275
Nomex	14	400
Teflon	45	450
Woven glass	25	500
Huyglas	30	500

^{a} Basis: all materials assumed to be at $G/C = 5/1$ except woven glass at $G/C = 2/1$.

necessary because most woven fabrics are less durable under repeated high-energy pulsing and more importantly, their pore structures allow excessive particle penetration immediately after pulsing (see Figure 12).

2.4   Penetration

2.4.1   Codes

Most existing emission regulations deal with total outlet loading and opacity and a few with particle size. In the future, there may be increasing use of codes based on particle size.

Federal air pollution control emission regulations for particulate matter are based on measured capabilities of control equipment. For fabric filters, measured penetration varies depending on the industry surveyed. Emission codes developed from these measurements are usually based on conservative design; low gas-to-cloth ratios, moderately high-pressure drops, and good quality fabrics. Representative penetration values, or outlet loadings, for several industries are given in Table 6.

2.4.2   Gas-to-Cloth Ratio

Penetration generally increases with increasing gas-to-cloth ratio. Figures 13 and 14 are examples taken from combustion sources. Under unusual circumstances, however, penetration may also decrease

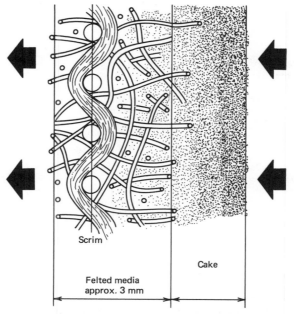

Fig. 12   Felted media showing distribution of dust.

Table 6 Performance of Fabric Filters in Several Industries [a]

Industry Application	Inlet Loading, g/dscm (gr/dscf)	Outlet Loading, g/dscm (gr/dscf)	Type of Filter/GC[b]	% of Dust Less Than 5 μm
Aluminum ore plant			P/8	
Fine ore storage		0.007 (0.003)		
Copper ore plant				
Truck loading	1.597 (0.698)	0.030 (0.013)		30
Truck dump	0.304 (0.133)	0.040 (0.018)	S/2.9	67
Crusher	5.42 (2.37)	0.013 (0.006)	S/2.9	2
Electric arc furnace[c]	0.137–0.343 (0.06–0.15)	0.005 (0.002)	S/3.22	40–65
Feldspar				
Pebble mill	13.8 (6.05)	0.011 (0.005)	R/3.0	
Fuller's earth				
Fluid energy mill	2.38 (1.04)	0.007 (0.003)	R/5.2	92
Raymond mill	11.99 (5.24)	0.005 (0.002)	R/6.0	18
Glass melting furnace[d]				
Pressed and blown				
soda–lime	0.09 (0.039)	0.025 (0.011)	R/0.65	
soda–lead–borosilicate	0.40 (0.173)	0.021 (0.009)	R or S/0.6	
Wool fiberglass				
borosilicate		0.034 (0.015)	R or S/0.85	
soda–lime borosilicate		0.135 (0.059)	R or S/0.5	
Gold ore plant				
storage reclaim	0.39 (0.17)	0.015 (0.007)	P/9.1	12–21
Industrial boiler	1.24 (0.54)	0.085 (0.037)	R/~5	~40
Iron ore plant				
Ore car dump		0.007 (0.003)		
Secondary/tertiary crusher		0.007 (0.003)		
Fine crusher conveyor	2.99 (1.31)	0.009 (0.004)		
Kaolin				
Raymond mill	10.36 (4.53)	0.037 (0.016)		70
Roller mill	4.03 (1.76)	0.016 (0.007)		70
Limestone				
Primary crusher		0.013 (0.006)	P/5.3	
Primary crusher screen		0.005 (0.002)	P/7	
Primary crusher transfer		0.004 (0.002)	P/7	
Primary crusher scalping				
screen and hammermill	2.84 (1.24)	0.005 (0.002)	S/2.3	
Secondary crusher		0.008 (0.004)	S/2.1	
Secondary crusher screen		0.002 (0.001)	P/5.2	
Sizing screen transfer		0.005 (0.002)	S/2.0	
Municipal refuse boiler	1.14 (0.5)	~0.0016 (~0.0007)	R/3	20–65
Traprock				
Secondary/tertiary				
crushers/screens		0.019 (0.009)	S/2.8	
Finishing screens		0.029 (0.013)	P/7.5	
Transfer points		0.007 (0.003)	S/2.8	
Utility boiler[e]				
Pulverized fired	~5.7 (~2.5)	0.005 (0.002)	R/1.9	23–65
Stoker fired	4.6 (2.00)	0.009 (0.004)	R-S/2.8	5–60

[a] Data generally from reports of EPA Method 5 tests.
[b] P = pulse jet; R = reverse air; S = shaker; GC = gas-to-cloth ratio.
[c] New source performance standard limit = 12 mg/dscm (0.0052 gr/dscf).
[d] New source performance standard limit = 0.13–0.65 g/kg of glass produced (0.26–1.30 lb/ton).
[e] New source performance standard limit = 13 mg/J (0.03 lb/10⁶ Btu) input.

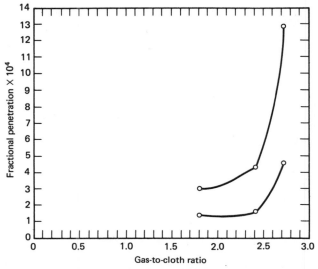

Fig. 13   Penetration versus gas-to-cloth ratio for an industrial boiler pilot baghouse.

as gas-to-cloth ratio increases as shown in Figure 15. Equations (3) through (5) indicate an adverse velocity effect. (See Section 2.4.)

2.4.3   *Media*

Intuitively, one would predict higher penetration through more open fabrics. Such is not necessarily the case. Cake builds up on the filter fabric as a consequence of individual particles being trapped on yarns or fibers, thus enabling bridging across pores bounded by the fibers. Tomaides[7] shows cases of

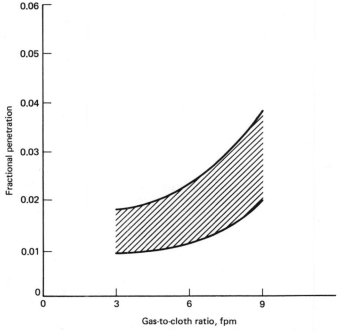

Fig. 14   Penetration versus gas-to-cloth ratio for a small utility boiler baghouse.

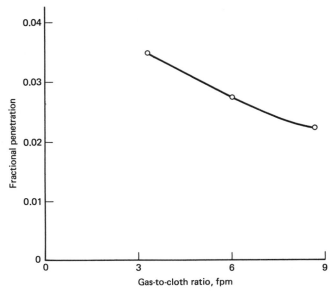

Fig. 15   Penetration versus gas-to-cloth ratio for homopolymer acrylic on an industrial boiler pilot plant.

particles being able to bridge across pores up to about 10 particle diameters wide. As pores are bridged, cake builds up as a permeable mass on the fabric. Because the pore spectrum for the cake is usually of smaller average diameter than that for the underlying fabric, penetration through the cake is much less than that through the fabric before the cake is formed. Figure 16 shows penetration versus time for a single filtration cycle. Choosing a fabric so as to minimize peak and total penetration is an art. Some general rules are: monofilament fabrics tend to have higher penetration than staple fabrics, napped fabrics tend to have lower penetration than unnapped fabrics, high-count woven fabrics tend to have lower penetration than low count fabrics, electrostatic interaction between particles and fabrics may overwhelm all other contributing factors, and experience is the best teacher. Filter fabrics are often purchased with a specified permeability (ASTM test D-737, or Frazier permeability) that usually confirms fabric identity but rarely affords a good index of final resistance to air flow and/or penetration.

2.4.4   Penetration Versus Bag Life

Penetration may be directly or indirectly involved with bag life. As gas-to-cloth ratio increases, penetration increases while service life tends to decrease. It is suspected that reduced bag life may result from the concurrent action of several factors, for example, greater stress on the fabric at higher pressure drops, more frequent cleaning of the fabric, or increased abrasion of the fabric due to greater numbers of particles transiting the bag and contacting fibers.

2.5   Ancillaries

While the baghouse proper may be the major component of the pollution control system, there are normally many other components whose smooth operation is essential to the system. Perhaps the most important of these ancillary components is the fan and motor system that one might consider to be the heart of the system, since it is in effect the pump which keeps the gases flowing. A major design consideration is the location of the fan with respect to the dust collector. While it can be located either before or after the baghouse, the most common practice is to locate it after the baghouse, to minimize damage by dust abrasion or wheel imbalance. The second major design question is what volume and pressure drop should the fan be designed to handle. Since there is usually some uncertainty as to the exact baghouse operating pressure drop as well as the need to operate during some limited range of upset conditions, it usually makes economic sense to select a fan with excess capacity. Of course, the fan being one of a few continuously moving parts in the system, it should be given a high priority on the maintenance schedule.

   The system hoods and exhaust locations should be designed in accordance with the engineering practices spelled out in *Industrial Ventilation.*[8] Other than obtaining the proper face velocity for each

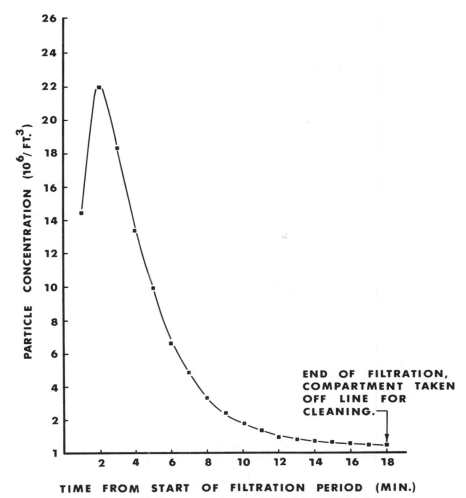

Fig. 16 Outlet concentration versus time for filtration cycle.

hood, the major design considerations relate to the interplay of each pickup point on the system as a whole. Balancing is of primary importance, but of equal magnitude are considerations of temperature and moisture differences at the various hoods on the resultant dew points when these streams join and mix.

The ductwork size is determined by noting the lower velocity limit set by the minimum velocity required to transport the dust and the upper velocity limit set by the abrasive nature of the dust or by fan operating costs. Special care needs to be exercised in the duct design at the entry to the baghouse. Gas velocity and location of the point of entry to the baghouse to avoid localized structure or fabric abrasion are important factors.

Any dampers that are integral to the baghouse cleaning system are usually specified as to type and size by the baghouse manufacturer; however, the system dampers within the ductwork need to be selected in accordance with their usage; for example, butterfly dampers may be suitable for balancing flows but guillotine dampers may be required for isolating the baghouse modules for off-line maintenance.

Dust discharge valves and conveying systems for removing the collected dust from the baghouse are usually selected and sized based on the type of dust and the anticipated maximum quantity to be removed. Thus, while rotary air locks and screw conveyors are commonly employed in rock quarries, double dump valves and pneumatic conveyors are often the practice in fly ash handling. A design consideration related to dust discharge is whether or not to store dust in the baghouse hopper. Although there are a few applications in which this practice is tolerable, it is generally recommended that dust not be stored in the baghouse hopper. Reasons for this recommendation include fire hazards, pluggage, and difficult dust removal after "setting up" and reentrainment of the dust, resulting in high operating pressure drops if storage levels are too high.

Spray nozzles, explosion venting, and electrically grounded bags, while not generally employed, are common practice in the woodworking and grain industries.

Cooling and heating equipment also find use in certain high-temperature applications. As stated in Section 2.3, if the gas stream temperature is greater than 550°F and it is desired to employ a baghouse, the gas stream must be cooled prior to entering the baghouse. Three alternative techniques are evaporative cooling, indrafting ambient air, and use of some form of heat exchanger. The choice of cooling method is usually an economic decision based on the gas volume and temperature. It is emphasized that evaporative cooling is probably the most difficult of these to operate successfully, especially if process variations are common. An auxiliary heater may be employed during startup and shutdown in the case of high-temperature applications in order to preheat the baghouse before starting it up and to purge the baghouse at shutdown. This procedure, which minimizes condensation within the baghouse, is especially desirable when acid gases are present.

Continuous monitoring instruments are becoming more common baghouse components, especially for the higher temperature applications. Monitoring and recording of the key parameters such as temperature, pressure, volume, and so on, as well as alarms and automated shutdowns can protect against loss of expensive bags or avoidable man-hours spent in cleaning or repairing bags and baghouse. Identification and continuous monitoring of only the key parameters is a good first approach.

2.6 System Considerations

Neither the design and selection of a baghouse nor its operation should be undertaken without full regard to the normal dust-producing operations. Dust source characteristics will impact not only on the baghouse design proper, but also on the entire turnkey baghouse system. The baghouse system might be very simple; for example, a single hood picking up the dust, a limited amount of ductwork between the hood and baghouse, and following the baghouse a fan and stack. The baghouse system might also be very complex as with fly ash collection from a coal-fired boiler where the system also includes a system bypass, auxiliary heaters, numerous dampers, insulation, temperature, pressure, and operational alarms, automatic shutdown and bypass modes, and extra modules for maintenance and off-line cleaning. Thus, it is obvious that the production process must be taken into account during the problem definition stage. When parameters, such as temperature, dust load, and gas volume, are specified, not only should the average levels be noted, but also the maximum operating ranges and possible upset conditions.

Considerations at this stage should include the following:

1. *System Objectives.* Efficiency, operating and maximum pressure drop, maximum and future production rates, present and future process effluents.
2. *System Constraints.* Ordinances limiting bypass, height, and so on, plant height and space limitations, backpressure limitations.
3. *System Operation.* Continuous versus intermittent; if continuous, is on-stream maintenance and, therefore, a modular approach desired.

At startup, it is necessary to determine whether the problem definition used in the design phase is compatible with the actual operating conditions. Then all system hardware must be checked for proper operation and positioning. One needs to avoid any startup conditions that fall outside the original design specification; for example, extremely high velocities causing initial blinding or temperature fluctuations causing dew point excursions. Likewise, a major decision at this point is whether or not to initiate the process with the baghouse on- or off-stream.

After a baghouse has been on-stream and a change in the production process or source operation is proposed, a review of the capabilities of the baghouse system, flange-to-flange baghouse, and filter medium is a must. Key parameters to be reexamined include the following:

1. *Gas Volume.* Too high a G/C can blind the bags while too low values can cause dust dropout in the ducts.
2. *Temperature.* Too high can destroy the bags and/or gasketing, too low might cause dewpoint excursions.
3. *Dust Load.* Too high may exceed the unit's capacity to convey the dust from the baghouse, too low may cause extended emissions after each cleaning cycle.
4. *Particle Size.* Too fine can cause blinding of the bags or excessive emissions at the design G/C.

When a previously trouble-free baghouse begins to require heavy maintenance or becomes inoperable, a first step in troubleshooting should be a review of the dust-producing operation in order to identify any process changes that may cause baghouse problems. If this step does not establish the cause, then one should make a systematic check of all baghouse system hardware to be sure it is being operated as designed. The production system and the baghouse are interdependent such that baghouse problems can cause a production shutdown and production changes or problems will often cause the baghouse to malfunction.

3   LIMITATIONS OF PRELIMINARY ESTIMATION METHODS

3.1   Pressure Drop

Equation (2) was used in modified form by Dennis and Klemm[5] in developing their improved computer model for predicting baghouse performance, and in the precurser version by Dennis et al.[9] The earlier version was validated against data from measurements at two full-scale field installations. For minimum, maximum, and average pressure drops for the two installations, predicted values ranged from −7.46 to +15.3% of the measured values. The predicted values were not completely free of bias, since data from the field installations were used indirectly in development of certain model input parameters, but the bias should be of small magnitude. Of more importance was the need for accurate knowledge of S_E and K_2 for the field installations. As estimates of S_E and K_2 become less accurate, so does prediction of pressure drop. Field installations designed by reputable manufacturers have had pressure drops that may be two or three times the design value. Other systems based on a conservative design approach show pressure drops half of the design value. It is seldom clear whether the discrepancies are due to poor estimating by the designer, or to changes in operating conditions after the original design has been fixed.

3.2   Penetration

Penetration is harder to predict than pressure drop. Dennis' validation for penetration at two field installations showed predicted values ranging from +233 to −9.51% of the measured values.[9] The overprediction may have been caused by service life. Bags that had been in service for 2 yr showed only 40% as much penetration as bags that had been in service for 1.5 days. Compared to the measured value at 1.5 days, the predicted value was −33%. This behavior is consistent with results found by Donovan, Daniel, and Turner[10] showing that penetration can decrease with bag life for an extended period of time (thousands of cycles) before bag wear allows penetration to increase.

It is not unusual to find field installations with penetration as little as half of the value guaranteed by the manufacturer. This fact indicates the present state of ignorance about fabric filter behavior.

4   IMPROVING THE DESIGN ESTIMATES

4.1   Computer Models

Dennis and his coworkers[5,9,11,12] have prepared a mathematical/computer model suitable for simulation of a multicompartment baghouse. The model is based primarily on data from field and laboratory work with fly ash and glass fabrics, but is probably suitable for most nearly spherical dusts and for woven fabrics. Output from the program includes plots and/or printed results showing pressure drop and penetration for the baghouse. Average and maximum values for one filtration cycle may be obtained, or more detailed information may be printed such as changes (with time) of drag, flow, and loading throughout the baghouse.

In order to use the model, one must know or estimate the following information:

1. Number of compartments.
2. Compartment cleaning time (minutes to clean one compartment).
3. Cleaning cycle time (minutes to clean all compartments in sequence).
4. Time between cleaning cycles.
5. Limiting pressure drop.
6. Reverse flow velocity.
7. Shaking frequency.
8. Shaking amplitude.
9. Average gas-to-cloth ratio.
10. Gas temperature.
11. Inlet dust concentration.
12. Temperature at which 11 was measured.
13. Specific resistance coefficient (K_2 in Figure 4)

14. Temperature at which 13 was measured.
15. Velocity (gas-to-cloth ratio) at which 13 was measured.
16. Effective residual drag (S_E in Figure 4)
17. Temperature at which 16 was measured.
18. Residual fabric loading (W_R in Figure 4).

The model will supply a default value if none is given for items 2, 6, 12, 14, 15, 16, 17, and 18.

For very large systems, a single compartment may process on the order of 60,000 acfm. Smaller field-erected baghouses may handle about 15,000 acfm per compartment, while small packaged units may handle about 2000 acfm.

Compartment cleaning times are on the order of 2–4 min for most erected baghouses.

Time between cleaning cycles is the time during which all compartments are filtering before sequential cleaning of the compartments is initiated.

Limiting pressure drop may be determined by system constraints, or may be the designer's choice. Values from 5 to 20 in. H_2O are common.

Reverse flow velocity usually ranges from one to two times the filtration velocity. Shaking frequency and amplitude (half-stroke) typically average 4 cps and 2 in., respectively, but may be less than half to more than double in some applications.

Reference 5 contains a program listing in FORTRAN suitable for use on an IBM 370 computer with Calcomp plotter.

4.2 Estimation of K_2 Values

In using the above mentioned computer model, Dennis and Klemm[5] caution the user that predictive accuracy suffers as estimates of K_2 become less accurate. Properly measured values provide the best estimate, but less accurate estimates can be made if dust particle size distribution, bulk density, and discrete particle density are known. Dennis and Klemm[5] use the following equation for estimating K_2:

$$K_2 = \frac{\mu S_0^2}{6\rho_P C_c} \frac{3 + 2(\beta)^{5/3}}{3 - 4.5(\beta)^{1/3} + 4.5(\beta)^{5/3} - 3(\beta)^2} \tag{8}$$

where

μ = gas viscosity, P

S_0 = Specific surface parameter

$\quad = 6 \left(\dfrac{10^{1.151 \, \log^2 \sigma_g}}{\mathrm{MMD}} \right)$, cm^{-1}

MMD = particle mass median diameter, cm (in this equation)

σ_g = particle geometric standard deviation

ρ_p = discrete particle density, g/cm^3

C_c = Cunningham correction factor (≈ 1 for MMD > 5)

$\beta = \bar{\rho}/\rho_p$

ρ = cake bulk density, g/cm^3

This equation provides values for K_2 that are three times larger than they should be when predicted by theory alone.

To obtain K_2 at one set of conditions when its magnitude is known at another set, Dennis and Klemm[5] give

$$(K_2)_2 = (K_2)_1 \frac{(S_{0_2})^2}{(S_{0_1})^2} \tag{9}$$

A further estimate of K_2 as it changes with velocity was given by

$$K_2 = 1.8 V^{1/2} \tag{10}$$

Equation (10), which was developed specifically for the case of fly ash with MMD = 9 μm, $\sigma_g = 3$, and temperature = 25°C, illustrates that K_2 varies with filtering velocity. The velocity exponent may vary with the type and properties of a dust.

5 DESIGN OUTLINE

The purpose of this section is to provide in outline form a design approach and a designer's checklist. The steps to be taken in the design or selection of a baghouse system are listed in the author's recommended order. This order should normally be followed to ensure proper design.

Problem Definition:
Determine
  Dust properties
    Type   Density (bulk and particulate)
    Size distribution K_2
    Concentration range   Stickiness or other
                                characteristics

  Gas properties
    Temperature range   Flow rate range
    Humidity range   Chemical constituents
  Exhaust requirements
    Concentration   Allowable pressure drop
    Opacity
  Design
    Choose filter medium
      Compatible with temperature and other gas and dust properties—use fabric selection chart in
      Section 2.3
    Choose cleaning method
      Compatible with fabric—as discussed in Section 1.3.1, et. seq.
    Calculate gas-to-cloth ratio
      Equation (2), Section 2.2, Section 4, and Table 1
    Calculate penetration
      Equations (3) through (5), if compatible with selected filter system
    Calculate required cloth area and number of compartments
      Extra area required for having compartments out of service for cleaning and maintenance
    Proceed to fan and ducting requirements

6 TESTING METHODS

There are two major test method groupings: the first is associated with the baghouse performance, while the second includes those tests that are performed on the bags or swatches of the filter medium.

6.1 Baghouse Testing

The first group includes measurements of penetration, particle size, pressure drop, leaks, and opacity. Penetration is usually measured by EPA Method 5,[13] the procedure usually specified for compliance tests. For compliance testing, the baghouse outlet particulate concentration alone is measured. The Method 5 train consists of a dry filter collector followed by a series of condensers. The condensable hydrocarbons, and so on, captured in the condenser train are usually excluded from any emission calculations. An alternative to Method 5 is the ASME Power Test Code 27 Procedure 27. Although this test method is simpler to use, the results may differ from those obtained by Method 5, if there are condensation problems or very fine particles present. Modifications of both methods are also employed. Care must be exercised in choosing the test method for either compliance or performance guarantee purposes to be certain that it is acceptable. Performance guarantees may be written in terms of both outlet emissions and/or percent removal. Simultaneous inlet and outlet emission tests are often required for performance guarantee evaluations. Generally, the duration of the tests must be much longer on the system outlet due to the dramatically lower mass emission on the outlet side. Multiple inlet tests may be required during the period of the outlet tests if significant inlet fluctuations are anticipated.

Historically, the particle size measurements have been obtained by collecting a gas stream dust sample for subsequent analysis in the laboratory. The sizing technique endorsed by the IGCI has been the Bahco Analyzer.[15] The Bahco method and numerous other lab techniques all fail to account for the problem of particle agglomeration and/or dispersion while in transit.

One attempt to solve this problem has been the use of cascade impactors[16] to measure the particle size in situ by inertial classification in the gas stream. Although this approach has been used with some success, formation of artifacts can in some applications result in misleading data. Due to differences in inlet and outlet mass loading, different impactor designs are frequently employed at the two locations, for example, a Brink unit on the inlet and an Anderson impactor on the outlet.

While pressure drop measurements can be simply obtained with a U-tube manometer or a magnehelic gauge, care should be taken in specifying where the measurement is made. Pressure drop measurements at three locations are usually advisable; the total system pressure drop (needed for fan specification), the flange-to-flange baghouse drop, and the pressure drop across the bags in each cell or module. The three measurements cited above are very helpful in tracking down the source of any major pressure drop increases. Continuous monitoring and recording of the flange-to-flange pressure drop and comparison of these data with the source process information can identify process modes or upsets that increase pressure drop. Avoidance of such irregularities will result in lower pressure drop and, consequently, lower operating costs.

There are a number of techniques that can be used to identify the location of dust leaks that may be due to bag holes or tears, faulty compartment welds, or improper seating of bags in the tube sheet. In the case of a system housing hundreds of bags and usually multiple modules or compartments, it is cost effective to narrow down the probable leak sources. If observation of the stack shows the visible dust emissions to be intermittent, one can identify the defective module by noting which compartment is being cleaned when the stack emission disappears. The initial internal inspection of this module should include a "clean side" search for dust "tracks." If a particular bag is torn or improperly seated, dust should be discernible on the clean side of the tube sheet around or above this bag. Should this approach fail to locate the leak, one can then resort to the injection of a fluorescent tracer powder at the inlet of the baghouse followed by a scanning of the "clean side" of the fabric with an ultraviolet light.

Visible emission, or plume opacity, measurements for compliance purposes, are usually based on spot checks of the plume appearance by a trained observer. Continuous opacity measurements using in-duct optical transmissometers have been successfully employed on several baghouses. The outlet aerosol from a properly performing baghouse is less than 10% opacity.

6.2   Bag Testing

The second test group, that is, those tests that are performed on the bags, include the following: permeability, MIT flex, Mullen burst, tensile and microscopic analyses.

The permeability test measures the air flow through a swatch of the filter medium at a given pressure drop. By comparing the permeability of a bag that has seen service with the original clean bag permeability, one can sometimes determine if the bags are blinded or if the pores have opened up. Similarly, comparison of Mullen burst and tensile strengths of unused and used bags can give an indication of strength deterioration. The tensile test that provides stretch, elongation, and tear data for fabrics can be used to measure separately the relative strength of warp and fill yarns in fabric samples. The Mullen burst test shows the relative overall capability of fabrics to withstand severe pulsing or pressure. A third strength test is the MIT flex test that can be employed to test for a loss in flexing properties. One variation of this test is to expose the swatch to an acidic atmosphere prior to flexing, but only when the bags are to be employed in an atmosphere that may see acid dewpoint excursions.

Microscopic analysis of the bags is also helpful in identifying deterioration of the fibers. Use of the various bag tests cited above to forecast bag life is still an art form and caution must be exercised in the interpretation of the data. These test methods do provide an excellent quality control approach for new bags as well as proving helpful tools in troubleshooting baghouses with bag life, emission, and pressure drop problems.

7   BAGHOUSE OPERATION

7.1   Optimization

Optimum pressure drop, bag life, cleaning conditions, bag style, or other parameters may be individually important in particular circumstances. All parameters are probably best considered in the aggregate as they impact optimum cost, or minimum long-term cost. Billings and Wilder[17] give capital and operating cost breakdowns as shown in Tables 7 and 8. These tables which are based on collector

Table 7   Installed Capital Cost of Fabric Filters[12]

Baghouse	33.6
Ductwork	27.3
Foundation/installation	11.8
Fan/motor	10.5
Disposal equipment	4.2
Design	4.2
Startup	4.2
Instrumentation	2.1
Freight	2.1
	100.0

Table 8   Annual Operating Cost
of Fabric Filters[12]

Power	15.6
Labor	39.0
Plant overhead	32.5
Fabric	13.0
	100.1

sizes ranging from about 1000 ft² to 100,000 ft² of fabric, were prepared before energy costs rose precipitously in the 1970s. For large baghouses, cost of the structure would be proportionately smaller while ductwork and disposal equipment costs would be larger. For annual operating costs, power would probably represent a larger percentage than shown in the table. Fabric costs would also be higher if expensive fabrics were used under short life conditions.

Baghouse structure and ductwork account for over 50% of installed cost. Therefore, there is the temptation to increase gas-to-cloth ratio in order to reduce structure size. There are penalties, however, associated with increased gas-to-cloth ratios. They include: shorter bag life because of more frequent cleaning and higher particle velocity through the fabric (more abrasion), greater potential for high-pressure drop or blinding (K_2 increases with velocity), and greater penetration.

A generalized cost versus cleaning curve can be drawn, as shown in Figure 17, that relates costs for pressure drop, bag life, and cleaning energy with cleaning frequency. Curves have been published that relate bag life to gas-to-cloth ratio (and by inference, to pressure drop) for fly ash applications on several boilers. Those curves, however, may be misleading because of variables other than gas-to-cloth ratio that affect bag life.

An indication of performance and bag life for a shake-cleaned bag is shown in Figures 18 and 19.[10] This information derived from a single-bag pilot baghouse using redispersed fly ash that was admitted infrequently. Bag life, in number of shakes, for an industrial baghouse would be far less, perhaps on the order of 5 to 10 million.

An indication of change in pressure drop with increased cleaning energy is given in Figure 20.[10] This figure, which is for a specific isolated installation, should be used with caution. The cost of pressure drop can be estimated from Eq. (6).

Long-term minimum operating cost results from reliable equipment operation. A prerequisite to reliable operation is good record keeping. Records are necessary for analysis of operational problems and for appraisal of normal operating procedures and hardware items that may be altered to reduce costs. Once the baghouse goes on-line, finding optimum operating procedures is primarily in the hands of the operator.

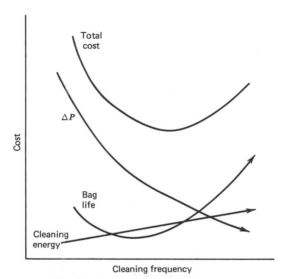

Fig. 17   Cost versus cleaning frequency.

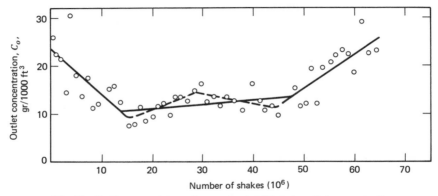

Fig. 18 Performance data under accelerated life test: outlet concentration.

7.2 Troubleshooting

If a problem develops after a unit has been brought on-stream, it usually falls into one of two major classes. Either it is a problem related to improper design or misapplication, or it is related to an unexpected process change that cannot be accommodated by the original baghouse design and operating parameters.

The first step in troubleshooting should be a review of system design and hardware specifications, followed by a thorough inspection of the system hardware to ensure that its operation is in accord with the original design. For example, a system inspection should be made to ensure that all dampers are properly positioned, all bags are dry, intact and seated properly, and that all instruments are functioning and/or are reading properly. These measures may sound obvious and they are. It is not uncommon, however, to find a critical damper reversed, or a number of bags missing or not connected at tube sheet. Just one or two bags out of a thousand can be a source of a severe emission problem.

Following a review of design specifications, and a physical inspection of the installation, the next step should be an attempt to clearly define the problem. An approach that helps to define the nature of the problem is to first identify the symptoms of the problem or malfunction. For example, the symptom may be a rapid pressure drop increase with no evidence of an emission problem. Conversely, the symptom may be a visible emission while the pressure drop remains within acceptable limits. Once having listed the symptom or symptoms of the problem, the next step is to list all probable causes for such symptoms. Table 9 provides a tabulation of common symptoms and their related causes (or ailments).

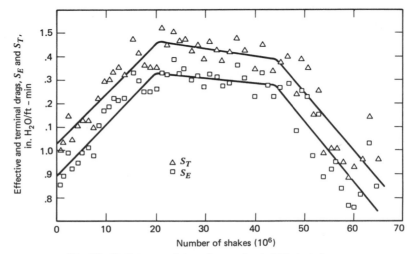

Fig. 19 Performance data under accelerated life test: drags.

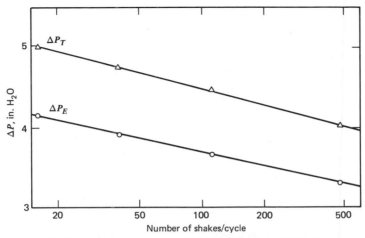

Fig. 20 Relationship between bag pressure drops and lengths of shake-cleaning period.

Table 9 Troubleshooting Guide

Symptom	Possible Cause	Remedy
High collector pressure drop	Malfunction of bag cleaning system	Check all cleaning system components
	Ineffective cleaning	Modify cleaning cycle Review with designer
	Reentrainment of dust in collector due to low density material or inleakage at discharge	Check discharge valves Lower AC ratio
	Wetting of bags	Control dewpoint excursions Dry bags with clean air
		Clean bags with vacuum or wet wash
	Too high A/C ratio either through added capacity or improper original design	Verify gas volume if possible Reduce inlet volume if possible Review with designer
	Change in inlet leading or particle distribution	Test Review with designer
Abnormally low pressure drop	Manometer line(s) plugged	Blow back through lines
		Protect sensing point from dust or water buildup
		Incorporate autopurging system in sensing lines
	Manometer line(s) broken or uncoupled	Verify with local manometer Inspect and repair
	Overcleaning of bags	Reduce cleaning energy and/or cycle time

Table 9 Cont.

Symptom	Possible Cause	Remedy
Stack emission	Broken bag	Replace, repair, or tie off bag
	Bag permeability increase	Test bag
		Check cleaning energy/cycle and reduce if possible
	Clean to dirty plenum leakage	Inspect and repair
	Change of inlet conditions	Test and review
Puffing	High-pressure drop across baghouse	See above
	Low system fan speed	Check drive system Increase speed
	Improper duct balancing	Rebalance system
	Plugged duct lines	Clean out
	Poor hood design	Evaluate temporary modifications and implement
	Improper system fan damper position	Check and adjust
Low dust discharge	Inleakage at discharge points	Inspect and repair seals or valves
	Malfunction of discharge valve, screw conveyor, or material transfer equipment	Inspect and repair
	Reentrainment of dust within collector	Lower GC ratio
	Reentrainment of dust on filter bags	Increase cleaning
Loud or unusual noises	Vibrations Banging of moving parts Squealing of belt drives	
Corrosion	Improper paint material or application	
	Improper insulation Emission of nearby equipment Dewpoint excursions Improper shutdowns	

One step that may quickly identify the problem is to compare the design specification with the current operating data. Gross differences between the two can often be at the root of operational problems. For example, the gas volume may have increased because more pickup (exhaust) points have been added or because of an increased production rate. As a result, a problem of excessive pressure drop stems from the increase in gas-to-cloth ratio. Another example is where the feedstock is changed such that a finer particle size results in blinding of the bags despite no change in gas flow. The baghouse will often tolerate process changes, for example, volume, dust load, particle size, temperature, and so on, within plus or minus 10–20% of the design values. Gross changes, however, will frequently lead to serious operational problems.

If the problem does not lend itself to early diagnosis via the above two approaches, then an in-depth analysis is required either by in-house experts or outside specialists. Again, the first step here should be to confirm the prior baghouse inlet conditions, preferably by means of stack sampling at baghouse inlet to determine gas volume, temperature, chemical description, dust mass loading, and gas flow profile. Particle size distribution may be obtained by in situ stack tests (cascade impactor) or by subsequent lab analysis of a dust sample obtained during the stack test. The latter procedure, however, often gives misleading results. Subsequent laboratory chemical and microscopic analyses of the dust and filter media for chemical and physical changes can provide additional information on the process variables. A listing of common bag and fabric analyses is provided in Table 10.

Table 10   Standard Fabric Tests

Test	Method
Weight	ASTM D1910
Thickness	ASTM D1777
Count	ASTM D1910
Permeability	ASTM D737
Tensile strength	ASTM 1682—Method IR-T
Mullen Burst	ASTM D231
MIT flex	ASTM D2176
Organic content	ASTM D578
Water repellency	ASTM D2721
Yarn weight	ASTM D578
Yarn twist	ASTM D578

Once the problem has been defined, the possible solutions, many of which are shown in Table 9, are usually easy to establish. Selecting the best option invariably involves an economic decision. For example, if the system gas flow is excessive, the options are to:

1. If possible, modify the source operation.
2. Increase the baghouse capacity by addition of another baghouse or module (if a modular system already exists).
3. Increase the baghouse capacity by use of double-walled bags, or other change in bag type.
4. If a high-temperature application, consider cooling of the gas stream to reduce volume.

As each of these approaches has its distinct economic impact, the need to make an economic decision at this point often arises.

Performance guarantees are usually provided by manufacturers of the larger systems. If it can be determined that current operations are compatible with the original specification, the solution should involve the baghouse manufacturer at this stage.

To summarize, troubleshooting a baghouse should include the following steps:

1. Review of the original design specifications.
2. Thorough inspection of the system and baghouse.
3. Comparison of the original design inputs with those existing at the time difficulties arose.
4. Identification of the problem symptoms.
5. Listing of possible causes of such symptoms.
6. Final definition of the problem.
7. Determination of alternative solutions and their associated cost.
8. Selection of the most suitable solution and means of implementation.

GLOSSARY

DRAG.   Normalized value for pressure drop is obtained by dividing pressure drop by the gas velocity. This parameter allows comparison of one dust/filter medium to another on a common basis and at various times during the filtration cycle.

DUST LOADING.   Grains of dust per cubic foot of air, or grams of dust per cubic meter.

FABRIC.   A collective term applied to cloth no matter how constructed and regardless of the kind of fiber used. In the commonest sense, it refers to a woven cloth.

FELT.   Fabric structures constructed by the interlocking action of the fibers themselves, without spinning, weaving, or knitting.

FIBER.   Fundamental unit of a textile raw material.

FILTER CAKE.   The accumulation of dust on the surface of a bag.

FILTER MEDIA.   The permeable barrier employed in the filtration process to separate the particles from the fluid stream and also to act as the substrate for dust cake development.

FILTRATION.   A process by which particles are separated from a fluid stream by means of a permeable barrier.

GAS-TO-CLOTH RATIO.   The amount of process gas entering the fabric filter dust collector divided by the amount of cloth area filtering the dust from the air. Normally, the gas flow given in cfm and the cloth in square feet. The term also defines the superficial or face velocity.

$$\text{gas-to-cloth ratio} = \frac{\text{gas volumetric flow rate}}{\text{cloth area}}$$

"GRAB" TENSILE.   The tensile strength, in lb/in., of a textile sample cut 4 by 6 in. and pulled in two lengthwise by two 1-in.^{2} clamp jaws set 3 in. apart and pulled at a constant specified speed.

GROSS GAS-TO-CLOTH RATIO.   The total inlet gas flow rate divided by the total available filter media (cloth).

$$\text{gross gas-to-cloth ratio} = \frac{\text{total inlet gas volumetric flow}}{\text{total filter cloth in collector}}$$

MULLEN BURST.   Pressure necessary to rupture a secured specimen of cloth; usually expressed in psi.

NEEDLED FELT.   A felt constructed by the use of barbed needles moving up and down, pushing and pulling the fibers usually through a supporting plain weave fabric (scrim) to form an interlocking of adjacent fibers.

NET GAS-TO-CLOTH RATIO.   The total inlet gas volume divided by only the on-stream filter media. For example, if one cell is off-stream for cleaning this will be deducted from the total available cloth before calculating the net gas-to-cloth ratio.

$$\text{net gas-to-cloth ratio} = \frac{\text{total inlet gas volumetric flow}}{\text{on-stream cloth}}$$

PERMEABILITY.   A measure of fabric porosity or openness, usually expressed as the flow of cfm of air/ft^2 of fabric that produces a 0.5 in. H$_2$O pressure differential.

PRECOAT.   Material dispersed in air stream during initial process startup to aid in establishing filter cake on bags.

PRESSURE DROP.   Resistance of filter medium and retained dust to air flow; pressure differential across the cloth.

PULSE JET.   Generic name given to all pulsing collectors.

REVERSE-AIR BAGHOUSE.   A unit employing reverse flow flushing air to dislodge the dust from the bags.

RINGS.   Metal bands sewn in the bag at various intervals to prevent bag from total collapse while cleaning.

SHAKER BAGHOUSE.   A unit wherein cleaning is accomplished by shaking the bags.

SPECIFIC RESISTANCE COEFFICIENT OF THE FILTER CAKE.   An indicator of how rapidly pressure drop increases during filtration.

SPUNBONDED.   A nonwoven fabric formed by producing, laying, and self-bonding a web of filamentous material in one continuous set of processing steps. Usually fabricated from polyesters, polyamides, or polyolefins.

TENSILE STRENGTH.   A measure of the ability of yarn or fabric to resist breaking by direct tension.

WOVENS.   Filter media fabrics constructed solely by weaving or interlacing yarns more or less at right angles into a uniform structure.

YARN.   A term for a twisted assemblage of fibers or filaments forming a strand that can be woven or otherwise formed into a textile material.

REFERENCES

1.   Bradway, R. M., and Cass, R. W., *Fractional Efficiency of a Utility Boiler Baghouse—Nucla Generating Plant*, Report No. EPA-600/2-75-013a, NTIS No. PB 240 641/AS, August 1975.

2.   Cass, R. W., and Bradway, R. M., *Fractional Efficiency of a Utility Boiler Baghouse: Sunbury Steam-Electric Station*, Report No. EPA-600/2-76-077a, NTIS No. PB 253 943/AS, March 1976.

3.   Dennis, R., and Wilder, J., *Fabric Filter Cleaning Studies*, Report No. EPA-650/2-75-009, NTIS No. PB 240 372/AS, January 1975.

4.   Williams, C. E., Hatch, T., and Greenburg, L., Determination of Cloth Area for Industrial Air Filters, *Heating, Piping and Air Conditioning* 12, 259 (1940).

5.   Dennis, R., and Klemm, H. A., *Fabric Filter Model Format Change; Volume 1. Detailed Technical Report*, Report No. EPA-600/7-79-043a, NTIS No. PB 293 551/AS, February 1979.

6.   Danielson, J. A. (ed.), *Air Pollution Engineering Manual*, 2nd edition, Report No. AP-40 (U.S. EPA), May 1973.

7. Tomaides, M., as reported in *Handbook of Fabric Filter Technology, Volume 1, Fabric Filter Systems Study,* p. 2–83, by C. E. Billings and J. Wilder, NTIS No. PB 200 648, December 1970.

8. *Industrial Ventilation, A Manual of Recommended Practices,* Committee on Industrial Ventilation, P. O. Box 16153, Lansing, MI, 1980.

9. Dennis, R., et al., *Filtration Model for Coal Fly Ash with Glass Fabrics,* Report No. EPA-600/7-77-084, NTIS No. PB 276 489/AS, August 1977.

10. Donovan, R. P., Daniel, B. E., and Turner, J. H., *EPA Fabric Filtration Studies: 4. Bag Aging Effects,* Report No. EPA-600/7-77-095a, NTIS No. PB 271 966/AS, August 1977.

11. Dennis, R., Klemm, H. A., and Battye, W., *Fabric Filter Model Sensitivity Analysis,* Report No. EPA-600/7-79-043c, NTIS No. PB 297 755/AS, April 1979.

12. Dennis, R., and Klemm, H. A., *Fabric Filter Model Format Change; Volume II. User's Guide,* Report No. EPA-600/7-79-043b, NTIS No. PB 249 042/AS, February 1979.

13. *Federal Register,* August 18, 1977, Part II, Vol. 42, No. 160, p. 41754 and Vol. 43 (160), pp. 41776–41782.

14. *Determining Dust Concentration in a Gas Stream,* ASME Pamphlet, PTC 27, 1957.

15. *Standardized Method for Particle Size Determination and Collection Efficiency,* Industrial Gas Cleaning Institute, Publication M-5.

16. Van Osdell, D. W., *Proceedings: Seminar on In-Stack Particle Sizing for Particulate Control Device Evaluation,* Report No. EPA-600/2-77-060, February 1977.

17. Billings, C. E., and Wilder, J. *Handbook of Fabric Filter Technology, Volume I. Fabric Filter Systems Study,* NTIS No. PB 200 648, December 1970.

18. Bergman, L., New Fabrics and Their Potential Application, *JAPCA* 24(12), 1187–1192 (1974).

CHAPTER 12

CONTROL OF PARTICULATES BY ELECTROSTATIC PRECIPITATION

Harry J. White

Consultant
Carmel, California

1 INTRODUCTION

Practical development of electrostatic precipitation as an effective method for control of particle emissions from smelters, cement plants, and other heavy industries originated in the early years of this century with the pioneer work of F. G. Cottrell. The electrostatic precipitation process differs basically from mechanical methods of particle separation in that the separation forces are applied directly to the individual particles rather than indirectly through forces applied to the entire gas stream. This direct and highly efficient use of forces explains the modest energy requirements and the low resistance to gas flow characteristics of the electrical method. Even the finest particles in the submicrometer range are collected effectively because of the relatively large electric forces acting on the particles. There is no intrinsic limit to the degree of cleaning obtainable since the efficiency can always be raised by increasing the exposure time of the particles in the precipitator.

Precipitators for modern conditions commonly meet efficiencies of 99.5% or higher, with many operating at 99.8–99.9% or sometimes better. The combination of high-collection efficiency, moderate energy use, ability to treat large gas flows at high temperatures, and to handle corrosive atmospheres and particles accounts for the wide use and varied applications of electrostatic precipitation.

In practice, the process involves a surprising number of scientific and technical disciplines, including physics, chemistry, chemical, mechanical, electrical and civil engineering, electronics, and aerosol technology. In addition, a knowledge of the industrial processes and plants to which precipitators are applied is vitally important to ensure technically and economically sound installations.

Despite the many fundamental and practical advances made in the process since its origin in the early part of the century, wide areas for improvement remain in dealing with high-resistivity particles, reducing the size and cost of precipitator installations, reducing energy consumption, and other similar aspects. International attention is being renewed on the research, development, and engineering phases of the precipitation process to meet more stringent demands for performance and for coping with new applications.

1.1 Operating Principles

An electrostatic precipitator is an apparatus or equipment that uses electric forces to separate suspended particles from gases. Many types and configurations of precipitators are used in practice, the most familiar being duct and pipe types used for industrial service, and the two-stage type mainly used for air cleaning and light industrial applications. All precipitators, regardless of type, are based on the same underlying principles. These principles include three major fundamental steps of electric charging of the suspended particles, collection of the charged particles in an electric field, and removal of the precipitated material to an outside receptacle.

Field experiments have shown that most industrial particles are charged during the process of formation by such means as flame ionization, friction, and grinding, but usually only to a low or

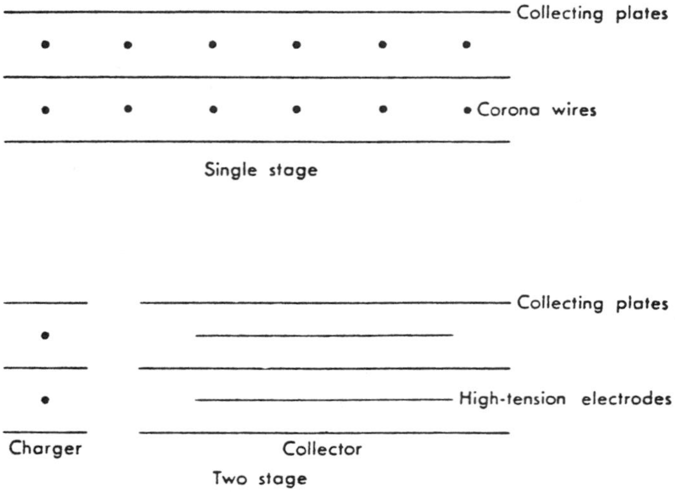

Collecting plates

•   •   •   •   •   •

•   •   •   •   •   • Corona wires

Single stage

Collecting plates

•

High-tension electrodes

•

Charger   Collector

Two stage

Fig. 1   Schematic diagrams of single-stage and two-stage precipitator arrangements.

moderate degree. These natural charges are far too low for effective precipitation. The high-voltage DC corona is the most effective means for particle charging and is universally used for electrostatic precipitation. The corona is formed between an active high-voltage electrode such as a fine wire and a passive ground electrode such as a plate or a pipe. Particle charging occurs in the gas space between the electrodes which is filled with highly concentrated gas ions usually of negative sign. Dust particles carried with the gas through this space become highly charged by bombardment of the gas ions, usually in a few-hundredths of a second or less. The level of particle charge achieved varies with the particle diameter, but typically is of the order of 300 electron charges for a 1-μm particle and 30,000 electron charges for a 10-μm particle.

Collection of the charged particles is effected by subjecting them to a continuation of the corona field or by passing them through a purely electrostatic field between smooth nondischarging electrodes.

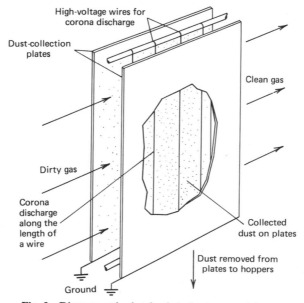

High-voltage wires for corona discharge

Dust-collection plates

Clean gas

Dirty gas

Corona discharge along the length of a wire

Collected dust on plates

Dust removed from plates to hoppers

Ground

Fig. 2   Diagrammatic sketch of single-stage precipitator.

The first type is known as a single-stage or Cottrell precipitator, while the second type is called a two-stage arrangement because the charging and collecting are carried out in separate stages. These two types are shown schematically in Figure 1. The general method of using the corona discharge for particle collection in a single-stage precipitator is shown schematically in Figure 2.

Dust particles become highly charged in the first 10 or 20 cm of travel and are driven to the grounded collecting electrodes by the intense electric field of the corona. The entire process is very fast, usually taking only a few seconds to collect virtually all the particles.

Collected dust layers on the ground electrodes are removed by rapping and fall into hoppers from which the dust is removed from the precipitator for disposal. In the case of liquid particles, such as sulfuric acid or tar, the collected particles coalesce and drip into appropriate removal chambers under the electrodes.

1.2 Applications

Earliest applications of precipitators were to smelters and cement plants, these being the outgrowth of Cottrell's pioneer work.[1] Major growth of the process since the pioneer period came initially from application to other industrial air pollution control problems, and later from new problems resulting from technical advances in numerous industrial fields which gave rise to new gas cleaning problems. Another major factor has been the increasingly stringent air pollution control legislation of the past decade which has led to expanded fields of application and required much higher performance levels.

Important examples of developing technology which have led to large new fields of application are the development of powdered-coal-fired boilers for electric power generation, the fluidized catalyst process for gasoline production in the oil refining field, and the basic oxygen furnaces in the steel industry. Economic recovery of valuable materials has also played an important role. One of the best known examples is the recovery of soda ash in kraft paper mills.

The broad range of application of precipitators in industry is indicated in Table 1. In the United States, fly ash collection in electric generation plants comprises about 75% of the total in terms of

Table 1   Precipitator Applications in Major Industries in the United States

Industry	Application
Electric power and industrial boilers	Fly ash collection from pulverized coal-fired furnaces
Nonferrous smelters: copper, lead, and zinc	Particle collection from furnace operations; gas cleaning for sulfuric acid production
Primary aluminum	Cleaning of potline gases and anode furnace emissions
Iron and steel	Open hearth furnaces Basic oxygen furnaces Sinter plants Scarfing machines Gray iron cupolas
Cement production	Rotary kilns Clinker coolers
Gypsum	Gypsum kettles
Paper mills	Black liquor recovery furnaces Sludge lime kilns
Chemical	Sulfuric acid Hot phosphorus Phosphoric acid Many others
Municipal incinerators	Cleaning stack emissions
Oil refining	Recovery of fluidized catalyst in high-octane gasoline production
Manufactured and by-product fuel gases	Detarring of gaseous products
Forest and lumber	Bark and wood chip boiler gas cleaning
Glass manufacture	Recovery of emissions from glass furnaces

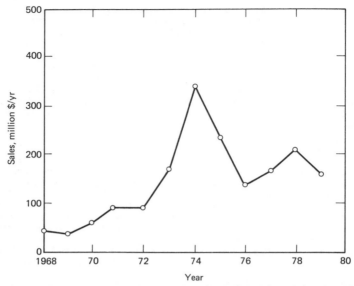

Fig. 3 Annual sales of precipitators in the United States. Compiled by the Industrial Gas Cleaning Institute for hardware only.

gas volume treated. Metallurgical, cement, and paper mill applications are each in 5–10% range. All other fields account for less than 5%, although individually they may be of great importance and represent growing areas of application. Precipitator sales data are available from several sources, but those shown in Figure 3 indicate the trend.[2] Note that these figures cover only the flange-to-flange hardware costs, and that the installed costs including foundations, dust removal systems, flue connections, and so on, will run several times higher.

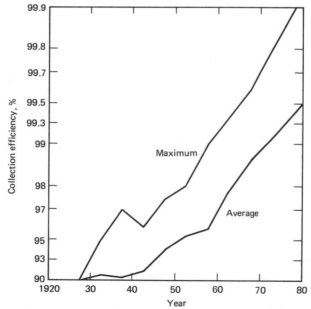

Fig. 4 Showing increasing collection efficiency trends for fly ash precipitators in the United States in recent years.

Growth of the precipitator market in the United States has been remarkable when both increases in gas flow treated and efficiency required are taken into account. Efficiency trends for fly ash precipitators since 1950 are plotted in Figure 4. Average efficiency has risen from 94% in 1950 to about 99% in 1970 and 99.8% in 1980. Maximum efficiencies for fly ash in 1980 are observed to exceed 99.95%. Since the higher efficiencies mean substantially larger precipitators, the effect on growth rate is very large.

Continued growth of precipitator use on a substantial scale is expected because of their many advantages, the broad spectrum of established applications, and the opening of new markets such as in oil shale and coal gasification. The ability of precipitators to achieve very high collection efficiencies with only modest power requirements will become increasingly important as energy costs continue to inflate. Although competing gas cleaning technologies undoubtedly will have an impact on precipitator use, the great flexibility and long record of precipitator performance in heavy industry under rugged conditions and varied applications, together with advanced precipitator technology now being developed, augur for continued and expanding use of precipitators.

2 FUNDAMENTALS

Understanding the fundamental processes that underlie electrostatic precipitation is essential to the rational design and operation of precipitators. Although early precipitators were designed principally by analogy with prior installations and by certain rule-of-thumb procedures, this approach is limited in scope and provides little guidance for advances and improvements. Intensive studies of fundamentals began in the 1930s and evolved into a mature activity by about 1960. Field experience remains a major factor, but scientific methods now dominate precipitator technology and the interpretation and application of field experience. The net result of this deeper understanding is more effective precipitator design, higher cleaning performance and reliability, and more economical and energy efficient precipitators.

2.1 Corona Discharge

The corona discharge as used in electrostatic precipitation occurs between fine wires and outer pipes, as shown schematically in Figure 5, or between wires and plates as depicted in Figure 1. Mechanism of the corona can be explained by reference to the corona discharge diagram of Figure 5.

Referring to this diagram, electrons released from or near the wire surface move in the intense electric field of the corona glow region and produce enormous numbers of additional electrons by impact ionization with gas molecules. The electrons thus produced are attracted to the outer positive polarity cylinder and attach to gas molecules in the passive region between the glow and the pipe surface. Electron attachment occurs because free electrons in the relatively low electric field in the passive region attach to the gaseous molecules instead of producing further ionization. This passive zone typically has an ion density of 10^7 to 10^8 ions/cm³.

Electron impact ionization in the glow region produces a positive ion for every electron generated. These positive ions move to the wire where they release new free electrons by impact with the wire surface or by photoelectric effect in the gas or at the wire surface. This regeneration action of the positive ions in releasing new electrons at the wire ensures that the corona discharge will be self-maintaining.

Theoretical equations can be derived for symmetrical electrode systems under clean conditions in gases such as air. However, these equations tend to be complicated and are useful mainly for research and development work. In addition, the theoretical equations assume smooth round wire electrodes and other ideal conditions. For these reasons the actual corona characteristics of precipitators are best determined by experimental or field measurements. Knowledge of corona characteristics is often essential in evaluating the operation and performance of precipitators, and it is usually advisable to provide kilovolt meters and milliampere meters with rectifier sets for this purpose.

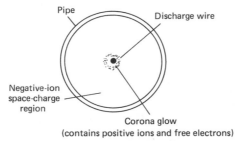

Fig. 5 Corona discharge diagram.

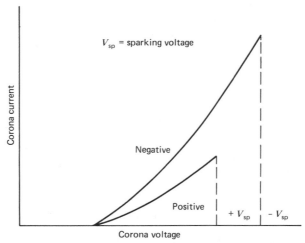

Fig. 6   Corona current–voltage characteristics for negative and positive corona in air.

The general form of the corona current–voltage curves for both negative and positive polarity corona is illustrated in Figure 6. Current flow starts at a certain minimum voltage V_0, called the corona starting voltage. The current then increases parabolically with voltage until sparkover is reached, beyond which it is impossible to go because of the spark breakdown of the corona gap. Note the superiority of negative polarity because of its much higher sparkover voltage, a fact that was originally discovered by Cottrell.

Corona characteristics depend on many factors, including electrode spacing and configuration, gas composition, pressure, and temperature, dust loading and particle size, dust deposits on the corona

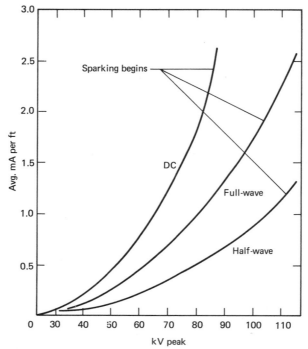

Fig. 7   Effect of voltage waveform on corona characteristics in air.

and collecting electrodes, and on the electrical resistivity of the particulates being collected. Voltage waveform also has a strong effect, as illustrated in Figure 7. Full-wave and half-wave voltages are commonly used in practice, with DC or direct current used only for special situations and laboratory investigations.

Typical corona voltages for precipitators of 10–15-cm electrode spacing range from 40 to 60 kV peak value. Corresponding corona current densities range from 0.1 to 1 mA/m², depending on dust and gas conditions.

2.2 Particle Charging

Particle charging is the first basic step in the electrostatic precipitation process. Most particles encountered in industrial gas cleaning do carry some charge acquired in the natural processes of their formation, but these charges are far too small for effective precipitation. It is therefore essential to provide a high degree of charging as a part of the collection process. In practice this is accomplished by passing the particles through the high-voltage DC corona which exists between the precipitator electrodes. Either positive or negative corona may be used, but for industrial gas cleaning negative polarity is preferred because of its greater stability and the higher operating voltages and currents possible. For air cleaning positive corona is the rule because of the lower quantites of ozone produced.

Two distinct particle charging mechanisms occur in the corona field of a precipitator. The most important is charging by ions driven to the particles by the force of the applied electric field. A secondary charging process occurs owing to the phenomenon of ion diffusion, which depends on the thermal energy of the ions but not on the electric field. The field charging process is predominant for particles larger than about 0.5 μm in diameter, the diffusion process for particles smaller than about 0.2 μm, and both are important for particles in the intermediate range between 0.2 and 0.5 μm.

Theoretical and experimental investigations of particle charging began in the 1920s and 1930s. Understanding of the basic processes was further advanced in the 1950s. In recent years particle charging studies have been pursued with renewed intensity because of the great importance of this factor in achieving advanced precipitator design and performance.

Particle charging by the field process under conditions existing in the unipolar corona is well understood and the theoretical predictions have been amply confirmed by experiment.[3] The theory has been developed for the case of an isolated spherical particle placed in a uniform field with a uniform ion density. These conditions are met in the ordinary unipolar corona for particles in the micrometer size range and for particle concentrations normally met within practice.

Consider an initially uncharged particle suddenly placed in a corona field. Particle charging begins immediately as the result of gas ions impinging on the particle, and continues until the ion-repelling field produced by the charge accumulated on the particle becomes sufficiently strong to prevent more ions from reaching the particle. The problem of determining the rate and magnitude of the charge is solved by the methods of classical electrostatics. Only the results of these calculations are covered herein.

The particle charge attained in time t is given by the following equation

$$q = 12\pi\epsilon_0 E_0 r_p^2 \left(\frac{\epsilon}{\epsilon+2}\right)\left[\frac{t}{t + (4\epsilon_0/N_0 eK)}\right] \tag{1}$$

where ϵ is the relative dielectric constant of the particle compared to that of the free space, ϵ_0 is the permittivity of free space equal to 8.85×10^{-12} F/m in SI units, E_0 is the electric field strength, r_p is the particle radius, K is the mobility of the gas ions, e is the electronic charge, N_0 is the ion density, and t is the time.

Equation (1) may be written in more succinct form by noting that the charge approaches a saturation limit q_s given by

$$q_s = 12\pi\epsilon_0 E_0 r_p^2 \left(\frac{\epsilon}{\epsilon+2}\right) \tag{2}$$

and that the quantity $4\epsilon_0/N_0 eK$ has the dimensions of time. This quantity is called the charging time constant, denoted by the symbol t_0. Using q_s and t_0, Eq. (1) can be written in the simplified form

$$q = q_s \left(\frac{t}{t + t_0}\right) \tag{3}$$

Note that the particle receives over 90% of its saturation charge in a time $10t_0$.

The order of magnitude of the charging quantities is best illustrated by means of a numerical example. Assume a conductive particle of 1-μm diameter placed in a corona field having the following properties

$$E_0 = 6 \times 10^5 \text{ V/m}$$
$$N_0 = 5 \times 10^{14} \text{ ions/m}^3$$
$$K = 2.2 \times 10^{-4} \text{ m/sec-V-m}$$

The saturation charge q_s from Eq. (2) is

$$q_s = 12\pi\epsilon_0 E_0 r_p^2 = 5.00 \times 10^{-17} \text{ C}$$

It is often useful to express the particle charge in terms of the number of electronic charges, n_s. For the above example n_s is found to be

$$n_s = \frac{q_s}{e} = \frac{5.00 \times 10^{-17}}{1.60 \times 10^{-19}}$$
$$= 313 \text{ electronic charges}$$

The charging time constant for the example is

$$t_0 = \frac{4\epsilon_0}{N_0 eK}$$
$$= \frac{4 \times 8.85 \times 10^{-12}}{5 \times 10^{14} \times 1.60 \times 10^{-19} \times 2 \times 10^{-9}}$$
$$= 0.002 \text{ sec}$$

The important particle properties for field charging are the radius r_p and the dielectric constant ϵ, while for the corona field they are the field strength E_0 and the ion density N_0. Note that the particle charge increases directly with field strength E_0 and with the square of the particle radius r_p. The time rate of charging increases with the ion density, but is independent of the field. In practice,

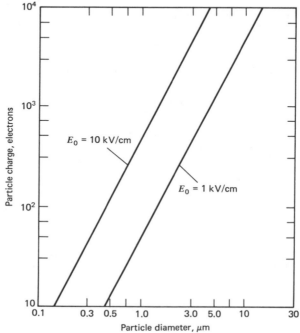

Fig. 8   Theoretical saturation particle charge as a function of particle diameter for the field charging process. Computed for conductive particles and charging fields of 1 kV/cm and 10 kV/cm.

Table 2   Saturation Charge and Charging Time Constant for the Field Charging Process

Particle Diameter, m	Charging Field, V/m	Ion Density, ions/m³	Charging Time Constant, sec	Saturation Charge, n_s
0.2	5×10^5	10^{13}	0.11	10
0.5	5×10^5	5×10^{13}	0.02	65
1.0	5×10^5	10^{14}	0.01	260
5.0	5×10^5	5×10^{14}	0.002	6,500
10.0	5×10^5	10^{15}	0.001	26,000
1.0	10^6	5×10^{13}	0.02	520
5.0	10^6	10^{14}	0.01	13,000
10.0	10^6	10^{15}	0.001	260,000

the charging fields are typically in the range 3–6 kV/cm, but may exceed 10 kV/cm in special designs. Ion densities are of the order of 10^{13}–10^{14} ions/m³, but again may be significantly higher in special cases. The dielectric constants of insulating particles of mineral origin commonly are of the order of 2 to 10, but many industrial particulates from mineral sources act like conductors for particle charging purposes.

Ion mobility varies somewhat for different gases, but usually not markedly. Differences also exist for positive and negative gases, although the differences are not large. Experiment shows that the ion mobility for atmospheric air at sea level is about 2×10^{-4} m/sec-V-m.

Orienting values of the saturation particle charge are shown in Figure 8. Additional data are listed in Table 2. Values of n_s for a charging field of 5 kV/cm are seen to range from 26 electronic charges for a 1-μm diameter particle up to 26,000 charges for a 10-μm particle. Doubling the field strength increases these values by a factor of 2. The charging time constant t_0 is 0.11 sec for the relatively low ion density of 10^{13} ions/m³ and drops to 0.001 sec for the high density of 10^{15} ions/m³. If a time $10t_0$ is allowed for essentially complete charging, it is observed that the charging time for the moderate ion density of 10^{14} ions/m³ is 0.1 sec. This time corresponds to only 10–20 cm travel of the gas through the precipitator, so that particle charging generally is attained in a negligible distance of passage through the precipitator.

2.2.1   Diffusion Charging Process

Ions present in a gas share the thermal energy of the gas molecules and in essence obey the laws of kinetic theory. The thermal motions of the ions cause them to diffuse through the gas, and to collide with any particulates present. Such ions adhere to the particles because of the attractive electrical image forces which become strong as the ions approach the particles at close range. It is evident that ion diffusion provides a particle charging mechanism which does not depend on an externally applied electric field. Such a field will aid in charging the particles, but is not necessary for the diffusion charging process. Accumulation of charge on a particle gives rise to a repelling field which tends to prevent additional ions from reaching the particle. However, in contrast to the field charging process, there is no maximum limit on the charge that can be reached by diffusion charging, because in accord with kinetic theory there is no upper limit to the thermal energy of the ions. Particle charging under these conditions depends on the thermal energy of the ions, particle size, and exposure time.

Derivation of the theoretical equation for diffusion or thermal charging can be carried out by application of kinetic theory principles.[4] The resulting equation for the particle charge in terms of electronic charges n is

$$n = \frac{4\pi\epsilon_0 kTr_p}{e^2} \ln\left(1 + \frac{e^2 \bar{v} r_p N_0 t}{4\epsilon_0 kT}\right) \tag{4}$$

where k is Bolzmann's constant, T is the temperature, and $\bar{v}$ is the average velocity of the gas ions.

To illustrate by a numerical example, let $N_0 = 5 \times 10^{14}$ ions/m³ and $T = 100°C$. For air at room pressure $\bar{v} = 467$ m/sec. Substitution in Eq. (4) gives

$$n = 1.80 \times 10^7 r_p \ln(1 + 4.08 \times 10^{10} r_p t)$$

Further, let $r_p = 1 \times 10^{-6}$ m and $t = 1$ sec. Then

$$n = 18 \times \ln(1 + 4.08 \times 10^4) = 191 \text{ electronic charges}$$

Table 3   Particle Charges for Diffusion Charging Process

Diameter, μm	Time, sec			
	10^{-2}	10^{-1}	1	10
0.1	2	3	6	8
0.2	4	8	12	16
0.4	11	19	27	35
1.0	34	55	76	96

Calculated values of particle charge by the diffusion process for representative conditions are listed in Table 3.

2.2.2   Combination Field and Diffusion Charging

Under normal precipitator conditions, field charging generally predominates for particles larger than about 1 μm in diameter, and diffusion charging for particles smaller than a few-tenths μm in diameter. For intermediate-size particles it is necessary to take into account both field and diffusion charging. The combination of these two charging processes leads to differential equations which cannot be solved by analytical methods, so that recourse must be had to approximate and numerical methods. Many approximate solutions for the simultaneous field and diffusion charging equations are to be found in the literature of electrostatic precipitation, but the most satisfactory solution is that by Smith and McDonald.[5] Graphs of particle charge calculated from the theory for field charging, diffusion charging, and combined field and diffusion charging for representative conditions are shown in Figure 9. Note the dominance of diffusion charging for particles smaller than about 0.4 μm, and of field charging for particles larger than about 0.8 μm.

For many years the most reliable and accurate experimental work on particle charging was that of Hewitt[6] published in 1957. Hewitt's experimental results are also plotted in Figure 9 for $E_0 = 3.6$ kV/cm and $N_0 = 10^{13}$ sec/m^3. Note the reasonably good agreement between Hewitt's results and the theoretical curve for combined field and diffusion charging. Recently, Hewitt's work has been checked and extended to cover a broader set of conditions by Smith et al.,[7] and also by McDonald et al.[8] using the Millikan oil-drop method for measuring charges on individual particles. An example of the close agreement between theory and experiment is shown in Figure 10 for McDonald's measurements.

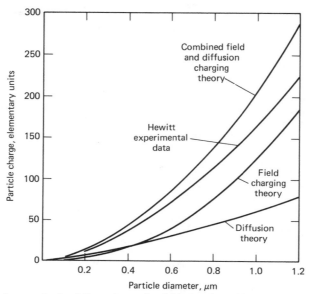

Fig. 9   Particle charge calculated from theory for field charging, diffusion charging, and combined field and diffusion charging for typical conditions.

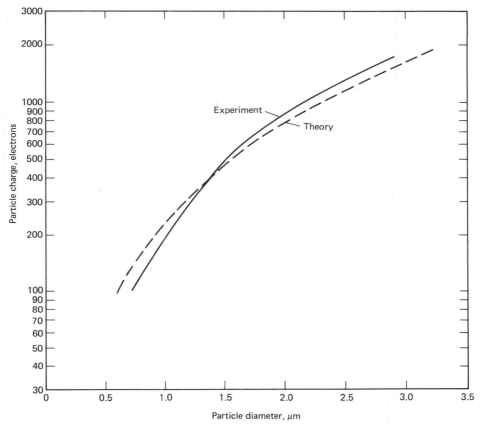

Fig. 10 Measured particle charge values versus theory for combined field and diffusion charging; charging field 3.5 kV/cm and charging current density 0.22 mA/m².

2.2.3   *Charging by Electron Attachment*

Although no theory for particle charging by attachment of free electrons present in the main body of the corona discharge has been developed, it has been known for some years from experimental investigations that such charging occurs in the negative corona at high field strengths. For example, in one recent experimental investigation[8] in which individual particle charge measurements were made using the Millikan oil-drop technique, the average measured negative particle charge was found to be 30–90% higher than predicted by the combination field and diffusion charging theory. These measurements were made in air at temperatures of 230 and 340°C using a laboratory 10-in. duct precipitator operating at 27-kV negative voltage.

Qualitatively, it is evident that free electrons move through the corona field at very high velocities and therefore tend to impact dust particles in their paths with negligible deflections. Thus electron charging should depend on the number and distribution of the electrons, and the probability of striking the suspended dust particles. The importance of free-electron charging in field precipitators has not, however, been determined, so that ionic charging mechanisms are generally assumed in precipitation theory. Limited investigations have been made, but these are preliminary and tentative.

2.2.4   *Abnormal Particle Charging Conditions*

There are several conditions that can lead to abnormally low particle charging in the precipitator. The most important of these are as follows.

1. High-resistivity particles deposited on the collection surfaces disrupt the corona by causing sparkover to occur at unusually low voltages or by forming a reverse or back corona from

the collection surface. The critical value of resistivity is about 2×10^{10} ohm-cm, above which the sparking and reverse corona phenomena occur.

2. Electric space charge suppression of the corona current caused by a high concentration of very fine particles (usually submicrometer) in the gas stream. The net result is that there is insufficient ion concentration to provide enough ion charges for the particles.

3. Reentrainment loss of particles from the collection surfaces caused by poor gas flow, too high gas velocities, or poor rapping conditions. The particles thus released from the collection electrodes tend to carry a positive charge (for negative corona) due to the pith-ball effect. These particles may not be recharged or only partially recharged. In any event, they are quite possibly carried out of the precipitator, with a substantial lowering of precipitator efficiency.

2.3 Particle Collection

Particle collection in precipitators occurs by the coulomb or electric forces acting on the particles. These forces cause the particles to move toward the collecting electrodes at velocities determined by the equilibrium between the electric forces and the drag forces acting on the particles as a result of the gas viscosity. The rate increases with the migration velocity of the particles toward the collection surfaces, so these velocities should be maximized.

Particle collection theories depend on the nature of the gas flow through the precipitator. The simplest case is that of particles carried in a gas moving in laminar flow through the precipitator. In such cases the particles move toward the collecting electrode with migration velocities that can be calculated from the laws of classical mechanics and electrostatics. The following equations apply

$$F_e = qE_p \qquad \text{Coulomb's law of electric force}$$

$$F_d = \frac{6\pi\mu r_p w}{1 + A(\lambda/r_p)} \qquad \text{Stokes–Cunningham law of drag}$$

Setting $F_e = F_d$ and solving for the migration velocity w gives

$$w = \frac{qE_p}{6\pi\mu r_p}\left(1 + A\frac{\lambda}{r_p}\right) \tag{5}$$

where q is the particle charge, E_p is the collecting field, μ is the gas viscosity, λ is the mean-free path of the ambient gas molecules, and A is a dimensionless parameter having the value of about 0.86 for atmospheric air. The dependence of w on particle size and field strength for typical conditions is depicted in Figure 11. Complete collection occurs when the slowest particle has sufficient time to move from a point near the corona electrode to the collection surface. The ideal condition of laminar flow never exists in practice, although it may be approached for certain two-stage type precipitators.

For the single-stage precipitators universally used in heavy industry, the gas moves in complex turbulent flow patterns on which is superposed electric wind effects of the corona and, in heavy dust loadings, the momentum transferred to the gas by the mass movements of the dust particles. It is evident that the locus of a particle through the precipitator is extremely complex and not subject to detailed calculations. Nor will any two particles trace out the same path. For the small particles of most interest in electrostatic precipitation, the particle migration velocities are much less than the velocity of the gas through the precipitator. Particle motions for these conditions are determined primarily by the turbulent flow patterns of the gas and only secondarily by the electric forces acting on the particles. Particle collection occurs when the particle happens to be carried close to the collection surface and enters the laminar flow boundary layer where the electric forces act to move them to the collection surface.

Particle collection for these turbulent flow conditions can be dealt with by probability theory, and leads to an exponential formula for the probability of capturing a particle passing through a precipitator field.[9] Application of this theory leads to the formula for collection efficiency η

$$\eta = 1 - e^{-(A/V)w} \tag{6}$$

where A is the collection surface area of the precipitator, V is the gas flow rate through the precipitator, and w is the particle migration velocity. It is important to note that this equation is theoretically valid only for the case of uniform size particles having migration velocities not exceeding 10 or 20% of the precipitator gas velocity. The theory has been experimentally verified in the laboratory using particle sizes of the order of 1 μm.[10] Application of the exponential precipitation equation to field conditions is constrained by various limitations such as nonuniform particle size distribution, uneven gas velocities, particle losses due to reentrainment, and the like. Nevertheless, experience shows that the equation using modified values of the parameter w can be applied with success to field conditions.

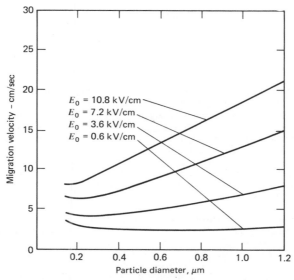

Fig. 11   Dependence of migration velocity on particle diameter and electrical conditions in atmospheric air. Adapted from Hewitt.[6]

As a matter of historical record, the exponential precipitation efficiency relationship was discovered in empirical form by Evald Anderson in 1919 from the analysis of field experiments at a cement works near Santa Cruz, California. This basic discovery was not published at the time, but is described in internal reports of the Western Precipitation Company. The exponential efficiency equation was first derived theoretically by W. Deutsch in 1922.[11] The Deutsch equation is mathematically equivalent to the Anderson equation. The joint origins of the equation have led to the designation of the result as the *Deutsch–Anderson* equation.

For nonuniform particle size distributions one can use integration methods, based on a known or assumed particle size distribution, to calculate precipitator efficiency. Let $\gamma(x)$ be the size frequency distribution function where $\gamma(x)\,dx$ is the fraction of particles having particle size between x and $x + dx$. The efficiency is then given by the equation

$$\eta = 1 - \int_0^\infty e^{-(A/V)\,w(x)}\,\gamma(x)\,dx \tag{7}$$

which can be evaluated analytically in certain cases of practical interest. The most important case is that for the log-normal distribution under conditions where field charging of the particles predominates.

2.4   Removal of Collected Particles

Removal of the collected particles from the precipitator is frequently ignored, or treated as a detail, in discussions of the electrostatic precipitation process. However, this step is also fundamental and requires as much care and sophistication as the first two steps of particle charging and collection. Liquid particles usually drain off by gravity, but the much more common case of dry particles requires the use of rapping or vibration to remove the particles from the electrodes.

Particle deposits occur on both the corona and the collection electrodes in thicknesses from a few millimeters to several centimeters. Deposits on the corona electrodes may interfere with the magnitude and uniformity of the corona current emissions. The problem is simply to keep the corona electrodes sufficiently clean. The most effective means is rapping the corona electrode supporting structure which must be sturdily constructed to effectively transmit the rapping blows to the corona electrodes.

The major problem in dust removal from the collecting electrodes is to prevent particle loss by reentrainment. Reentrainment may result from direct scouring of the collecting electrode surfaces by the gas stream, from redispersion of collected particles during rapping of the electrodes, and from sweepage of the collected dust from the hoppers. All three of these types of losses are closely related to the patterns and characteristics of the gas flow through the precipitator. Uneven gas velocity distribution and rough, turbulent flow can lead to greatly increased dust losses and poor performance. Technical means for avoiding these problems are well established. Gas flow model studies, avoidance of gas sneakage past the electrodes, and effective hopper emptying methods are among the preventive measures to be taken.

3   PRECIPITATOR DESIGN

During the pioneer era of electrostatic precipitation, precipitators were designed empirically by simulation of earlier installations which had proved successful and by certain rule-of-thumb procedures. This approach, although necessary under the conditions existing at the time, is obviously limited in scope and provides little guidance for advancements and improvements except by slow and costly cut-and-dry methods. In addition, many installations proved inadequate after completion, and not infrequently required extensive changes to bring performance up to design levels. However, since the late 1940s it has become possible to rely more and more on fundamentals, so that it is now possible to design precipitators largely from basic principles.

However, empirical input is also necessary because in the majority of cases values of some of the most important application variables, such as particle size distribution and particle resistivity, are not known with sufficient accuracy, or may even not be obtainable. For existing plants, these variables can be measured and pilot precipitator methods can be used to obtain other essential information if necessary.

Precipitator design has also acquired a new importance during the past decade, changing in character from an often routine and casual function to a more serious enterprise involving much higher performance standards and large financial stakes. These changes have been forced by the enactment of the federal clean air legislation which requires high-efficiency cleaning of industrial gases and high reliability of the gas cleaning equipment. Strong enforcement of clean air standards and regulations can require curtailment or even complete shutdown of entire production units. Hence, in this sense, precipitator design and engineering practice is now of equal importance to that of the production equipment itself.

3.1   Design Methods

The basic design problem for precipitators is determination of the principal parameters for precipitator sizing, the electrode arrangement, and the electrical energization needed to provide specified levels of performance. Ancillary factors such as rappers, gas flow control methods, dust removal systems, and performance monitoring must also be considered. Various design methods and philosophies are used in actual practice for the design of precipitators. These range from the rather crude design-by-analogy practices already referred to on the one hand, to rather sophisticated methods based on theory and fundamental principles on the other hand.

Design by analogy or simulation can be used successfully for applications such as collection of sulfuric acid mist for which conditions change little from installation to installation, but is not viable for applications such as fly ash collection where particle and gas properties tend to be highly variable because of major differences in coals, furnaces, and operation.

In the case of existing plants or where new processes are being developed, pilot precipitators are often used as a means of determining design. The main problem is the scale-up factor to be used in going from the pilot-scale to the full-size unit, since it is well known that pilot units operate much better per unit size than do those of commercial size. The scale-up effect is chiefly attributable to differences in electrical energization levels achievable and to gas flow quality, both of which are better in pilot units. Rapping and reentrainment losses may also be significantly different between pilot and full-size precipitators.

Another approach to precipitator design is to use purely statistical data to evaluate adjustable parameters to fit some empirical formula for precipitator efficiency. Values of the parameters are derived by regression analysis and usually are devoid of physical meaning. This scheme is better than simple averaging methods, but lacks the reliability and scientific integrity inherent in methods based on fundamental physical principles.

In principle, precipitator designs can be deduced from theoretical considerations alone if all the significant variables are known. However, this is seldom the case, and it is necessary to specify some of the basic design parameters primarily from field experience. Mathematical models of precipitator designs have been programmed for computer operation for cases where the particle size distribution and particle resistivity are known.[12] Semiempirical parameters are introduced to account for particle reentrainment and gas sneakage. The latter quantities are derived from field experience and measurements, mainly on fly ash precipitators. It is also necessary to account for the properties of the gas and its corona characteristics. This approach has progressed to the point where it can be used for trial designs and to check field and pilot precipitator results.[13]

In practice, precipitator design methods vary from vendor to vendor. Most precipitator suppliers consider their design methods proprietary, and design approaches range from almost purely empirical to reasonably scientific. Many precipitator users, especially in the fly ash field, issue specifications requiring not only stated performance levels, but also minimum design parameters, to assure that the performance levels will actually be achieved.

3.2 Precipitator Design and Sizing Parameters

Basic precipitator design parameters commonly used in practice are summarized in Table 4 together with values used for fly ash, the most important field of application. Values of these parameters for a given design will depend on particle and flue gas properties, total gas flow, and required collection efficiency. It is very important to stress that the performance actually attained in practice is also highly dependent on the mechanical and electrical quality of the precipitator. Experience shows that deficiencies in quality such as poor electrode alignment, warped electrodes, air inleakage, poor gas flow distribution, and mismatched or unstable rectifier sets are not infrequent and must be guarded against. In some applications allowance needs to be made for outage of one or two electrical sections of a precipitator by addition of redundant sections.

Ancillary design factors that need to be considered in addition to the basic parameters already covered are summarized in Table 5. The list is not exhaustive, but will serve as a guide to ensure attention to important practical considerations essential for a well-designed and reliable precipitator installation.

Precipitator sizing parameters include specific collection area, precipitation rate parameters, gas velocity, and aspect ratio. These will be discussed individually.

3.2.1 Specific Collection Area (SCA)

This is the most important and generally used quantity for rating precipitator size and cost. Despite the general drive toward the use of metric units, current practice for the SCA parameter is to use ft²/1000 acfm, and this practice is followed herein. Specification of the required SCA for a given installation is sufficient to determine the total collection surface and therefore the size of the precipitator required.

There are several approaches for determining the SCA needed for a given precipitator installation. Most of these reduce to choosing an appropriate Deutsch parameter w and solving for A/V from the Deutsch–Anderson equation which gives

$$\frac{A}{V} = \frac{1}{w} \cdot \ln\frac{1}{1-\eta} = \frac{1}{w} \cdot \ln\frac{1}{Q}$$

In English units A/V will be in ft²/cfs and w in ft/sec. This leads to the expression for the SCA as follows

$$SCA = \frac{A}{V} = \frac{16 \cdot 67}{w} \ln\frac{1}{Q} \text{ ft}^2/1000 \text{ acfm} \tag{8}$$

where w is in ft/sec and Q is the loss $= 1 - \eta$.

Table 4 Summary of Basic Design Parameters for Fly Ash Precipitators

Parameter	Symbol	Range of Values
Duct spacing	s	23–38 cm
Precipitation rate	w	3–18 cm/sec
Specific collection area	$\dfrac{A}{V}$	100–800 ft²/1000 cfm
Gas velocity	v	1–2 m/sec
Aspect ratio	$\dfrac{L}{H}$	0.5–1.5 length of ducts/height of ducts
Specific corona power	$\dfrac{P_c}{V}$	50–500 W/1000 cfm
Corona current per m² plate area	$\dfrac{I_c}{A}$	0.05–1.0 mA/m²
Plate area per electrical set	A_s	500–8000 m²/electrical set
No. of H.T. sections in gas flow direction	N_s	2–8
Degree of H.T. sectionalization	$\dfrac{N}{V}$	0.4–4 H.T. bus sectionalization/100,000 cfm

Table 5   Ancillary Design Factors for Precipitators

1. Corona electrodes: type and method of supporting.
2. Collecting electrodes: type, size, mounting, mechanical and aerodynamic properties.
3. Rectifier sets: ratings, automatic control system, number, instrumentation and monitoring provisions.
4. Rappers for corona and collecting electrodes: type, size, range of frequency and intensity settings, number, and arrangement.
5. Hoppers: geometry, size, storage capacity for collected dust, number, and location.
6. Hopper dust removal system: type, capacity, protection against air inleakage and dust blow-back.
7. Heat insulation of shell and hoppers, and precipitator roof protection against weather.
8. Access doors to precipitator for ease of internal inspection and repair.
9. Provisions for obtaining uniform, low-turbulence gas flow through precipitator. This will usually require a high-quality gas flow model study made by experienced people in accord with generally accepted techniques, with full report to precipitator purchaser before field construction.
10. Quality of field construction of precipitator, including adherence to electrode spacing and rigidity requirements.
11. Warranties: performance guarantees, payment schedules, adequate time allowance for performance tests, penalties for nonperformance.
12. Support insulators for high-tension frames: type, number, reliability. Air venting, if required.
13. Inlet and outlet gas duct arrangements.
14. Structure and foundation requirements.

3.2.2   Precipitation Rate Parameters

Theoretically, the precipitation rate in the Deutsch–Anderson formula is synonymous with the migration velocity w for uniform particle size, gas velocity, electric field, and the like. However, such uniform conditions do not exist in actual practice, although it is possible to account for some of the nonuniformities by modifications of the theory. But particle reentrainment, gas sneakage around the collecting zones, and similar factors cannot be accounted for by theory. Moreover, for new precipitators particle size, resistivity, and gaseous properties must be estimated.

Because of these uncertainties, most precipitator design is based on the use of an effective precipitation rate parameter w_e derived mainly from field experience rather than from theory. In this context, w_e no longer represents migration velocity, but instead is a semiempirical parameter which can be used in Eq. (8) to determine the SCAs needed for specified gas flows and collection efficiencies. Appropriate values of w_e for given applications are derived from bodies of experience accumulated from practice by design engineers and vendors. Interpretation of this experience requires assessment of the quality of the data as well as engineering knowledge and judgment.

Although the relation of w_e to collection efficiency for values of SCA can be calculated directly from Eq. (8), it is convenient to plot the relationship graphically as shown in Figure 12. The range of w_e shown in the plot, that is, from 0.05 to 0.6 ft/sec (1.5–18 cm/sec), covers most precipitator applications. The SCA required for a specific application can be read directly from the graph. For example, assume a precipitator efficiency of 99.5% and an application where a value of w_e of 0.20 ft/sec is judged suitable. From the graph the required SCA is 440 ft²/1000 acfm.

It is useful to discuss the precipitation rate parameter introduced by Matts and Ohnfeldt in 1963.[14] This parameter is usually designated by the symbol w_k to distinguish it from the Deutsch parameter. The parameter w_k was arrived at in the following way. Realizing that the Deutsch equation is limited in theory to particles of uniform diameter, investigators attempted to evaluate the integral in Eq. (7) for various particle size distributions. It is easy, for example, to find the integral for the case of the normal distribution function, but this distribution has little application in practice. On the other hand, the log-normal distribution has wide application since many particulates are found to follow this distribution reasonably well. Evaluation of the integral for the log-normal distribution had been successfully carried out by Allander and Matts[15] in 1957. Matts and Ohnfeldt used plausibility arguments to the effect that other statistical variations of precipitator operation and conditions in addition to those of particle size could be approximately accounted for by generalizing the concept of the standard deviation in the earlier Allander and Matts formula.

They also showed that it is possible to represent the rather complex integral function by the relation

$$\eta = 1 - e^{-[(A/V)\,w_k]^k} \tag{9}$$

The exponent k in this equation is a parameter having a value in the neighborhood of one-half. For our purposes we may assume that it has this value, and write

$$\eta = 1 - e^{-[(A/V)\,w_k]^{1/2}} \tag{10}$$

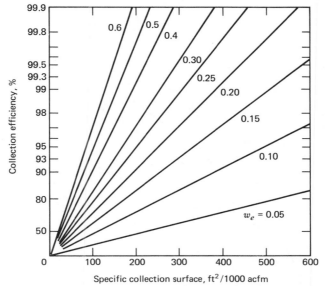

Fig. 12 Precipitator efficiency versus SCA and precipitation rate w_e.

Because of the lack of a firm theoretical base, it is advisable to regard Eq. (10) as essentially empirical, but with some fundamental elements.

It is easily shown that w_k is related to the Deutsch w in the following way

$$w_k = \frac{V}{A} \ln^2\left(\frac{1}{Q}\right) = w \ln\left(\frac{1}{Q}\right) \tag{11}$$

The main practical advantage of using w_k over w_e is that the efficiency depends much less on the SCA than is usually the case for w_e. Precipitator efficiency versus SCA and w_k is shown graphically in Figure 13.

3.2.3 Gas Velocity

Although localized gas velocities in the collection zones of precipitators usually vary over a substantial range, it is useful for design purposes to use an average value calculated from the total gas flow and the flow cross section of the precipitator. The cross section is taken as the open area between collecting plates, disregarding plate baffles. Primary importance of the average gas velocity is its relation to rapping and reentrainment losses. These losses tend to increase rapidly above some critical velocity because of the aerodynamic forces on the particles. The critical velocity for a given type of dust depends on the quality of the gas flow, plate configuration, electrical energization, precipitator size, and other factors. For reasonably good conditions, the critical gas velocity for a dust such as fly ash will vary between about 1.5 to 2 m/sec, but may be higher in some cases. Modern requirements for very high efficiencies usually make it advisable to maintain gas velocities on the low side of the range.

3.2.4 Aspect Ratio

This parameter is defined as the ratio of the effective duct length to duct height. Its importance in precipitator design stems from its relation to rapping loss. Collected dust falling from the plates is carried forward by the gas flow. If the ducts are short compared to their height, some of the falling dust will be carried out of the precipitator before it reaches the hoppers, thereby substantially increasing the dust loss. The time required for released dust clumps to fall from the top of a 12-m high duct, for example, will be several seconds. This is sufficient time for significant amounts of dust to be carried out of the precipitation zones unless the duct length exceeds at least 10 or 12 m. For orientation purposes, the aspect ratio needed for efficiencies of 99% should be at least 1.0 to 1.5.

Although it is possible to devise and use a combination parameter taking into account both the aspect ratio and the gas velocity which would be more closely related to the falling time of the dust, the aspect ratio alone suffices for most purposes.

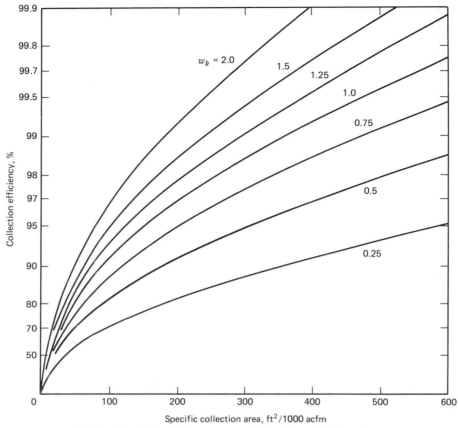

Fig. 13 Precipitator efficiency versus SCA and precipitation rate w_k.

3.3.   Electrical Energization

Electrical parameters are equal in importance to the sizing parameters in determining the performance of precipitators. The important electrical quantities for precipitator operation are the corona current density, the voltage levels that can be maintained, and the useful corona power in the precipitator.

3.3.1   *Corona Current Density*

Corona current densities should be maintained at the highest possible levels in order to maximize particle collection rates. In practice, operating current densities are limited by the following factors:

1. Gas composition, temperature, and pressure.
2. Electric space charge effects of the suspended particles.
3. Resistivity of the collected particle layers.
4. Electrode alignment and accuracy of alignment.
5. Collection area energized per electrical set.
6. Type and design of the high-voltage sets and controls.
7. Effectiveness of rappers in keeping the electrodes clean.

For most precipitator applications, the corona current density j at the collecting electrodes will be in the range of 0.05 to 1.0 mA/m^2.

3.3.2 *Corona Power*

The migration velocities of particles in a precipitator increase with the electric fields to which they are subjected. These fields in turn are dependent on the corona voltages and currents, and are closely related to the corona power used to energize the precipitator. The corona power for an electrical section of a precipitator is given, to a good approximation, by

$$P_c = \frac{1}{2}(V_p + V_m)I_c \tag{12}$$

where P_c is the corona power, V_p is the peak voltage, V_m is the minimum voltage, and I_c is the corona current for the section. In practice, the voltages may be measured by an oscilloscope or the term $\frac{1}{2}(V_p + V_m)$ approximated by the average voltage V_{av} as measured on a kilovolt resistance divider meter. For multisection precipitators, P_c is the sum of the corona powers for the individual sections.

Note that precipitator sparking represents wasted energy and should not be included in the useful corona power P_c. Nevertheless, limited or low-level sparking is usually desirable in precipitator operation to ensure that the highest practicable operating voltages are being used. This is accomplished by means of suitable automatic control systems which have optimum sparking rate as one of the control variables. Back corona or the reverse corona that occurs from the collected dust layers under conditions of high dust resistivity is also a waste of power and can seriously deteriorate precipitator performance. There are technical means for detecting and avoiding back corona which may be desirable to use in certain cases.

Specific corona power (SCP) is defined as the corona power in W/1000 acfm. In practice, the specific corona power typically ranges from about 50 to 500 W/1000 acfm. Both theory and experience show that precipitator efficiency increases markedly with increasing useful corona power. The relation between specific corona power and collection efficiency, derived from field experience with good quality fly ash precipitators, is shown in Figure 14. Note, for example, that a corona power of only about 200 W/1000 acfm is required for an efficiency of 99.5%.

3.4 **European Versus U.S. Designs**

During the decade of the 1970s significant differences between European precipitator designs and traditional U.S. designs became apparent. These differences grew out of the different air pollution control regulations which existed between many European countries and the United States. In the United

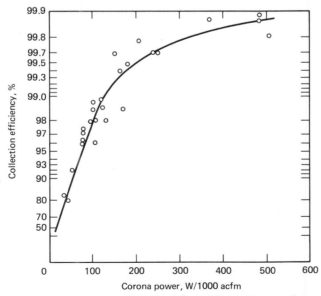

Fig. 14 Efficiency increase with corona power from field data for fly ash precipitators.

States, prior to the federal clean air legislation of 1970, air pollution control in many states was far from stringent. As a result many precipitators were designed and purchased to meet the minimum cost requirements of the marketplace. This, perhaps not surprisingly, translated in practice to marginal size precipitators, mediocre mechanical designs, and not infrequently low-quality field construction. The end result was often poor performance and unreliable operation.

On the other hand, European precipitator suppliers had developed conservatively sized sturdy designs, and well-installed equipment in response to the needs of their markets. These suppliers were therefore well positioned to take advantage of the new U.S. requirements growing out of the new federal standards.

The main features of European designs may be summarized as follows:

1. Conservative precipitator sizing.
2. Sturdy mechanical design.
3. Use of rigidly supported corona electrodes on a framework instead of suspended wires and weights.
4. High-intensity rapping of plate and corona electrodes, usually by direct-impact rotating hammers.
5. Usually designed to operate at low sparking rates.

Wire-and-weight corona electrode designs still are used in pipe-type precipitators collecting mists or tars, or operated with water flushing. They are also used for dry applications treating relatively small gas flows and for very high gas temperatures. The European designs are not well suited for these applications.

3.5 Design Examples

Design methods used in practice are discussed in Section 3.1. In any given application, the engineering design method used is likely to be a combination of the various methods available. No matter what approach is used, it is evident that experience must necessarily play a large role just as it does in other technical fields. Rather than attempting to delineate an exact design procedure, it will be more instructive to illustrate the design process by examples for actual precipitator installation.

The examples presented are taken from the field of fly ash collection which is by far the most important in practice. The first example is for a low-resistivity fly ash derived from a western low-sulfur, high-sodium coal. Even though low-sulfur western coals are notorious for high resistivity, the sodium provides adequate conductivity of the ash. For this particular case, the amount of sodium present in the fly ash is about 5% as Na_2O, an amount sufficient to reduce the resistivity to about 3×10^9 ohm-cm. The second example is for a high-resistivity fly ash from a low-sulfur western coal.

Basic designs for the two collectors are shown schematically in Figures 15 and 16, together with the design and performance data applicable to each. The design parameters for both cases were derived largely from pilot precipitator investigations, while the performance data were determined from comprehensive commissioning test programs conducted for each precipitator. These precipitators are of modern high quality, sturdy design, are relatively conservative in size and electrical energization, use high intensity rapping, and have sufficient series corona fields to minimize rapping losses and provide considerable safety margins against section outages. Both precipitators have performance efficiencies exceeding 99.8%.

Design and Performance Data

Item	Design Value	Performance Value
Gas flow	316 m³/sec	327 m³/sec
Gas temperature	155°C	145°C
Collection efficiency	99.35%	99.87%
Gas velocity	1.61 m/sec	1.68 m/sec
Specific collection area	367 ft²/1000 acfm	348 ft²/1000 acfm
Precipitation rate w_e	7.0 cm/sec	9.6 cm/sec
Precipitation rate w_k	35.3 cm/sec	64.0 cm/sec
Specific corona power	500 W/1000 acfm	480 W/1000 acfm
Corona current density	0.36 mA/m²	0.38 mA/m²
Ash resistivity	3×10^9 ohm-cm	3×10^9 ohm-cm
Inlet ash loading	18 g/m³	8 g/m³
Outlet ash loading	0.12 g/m³	0.01 g/m³
Stack emission		0.007 lb/10⁶ Btu

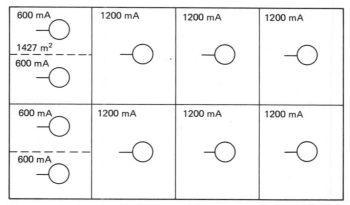

Fig. 15   Precipitator design (schematic): Example 1—low-resistivity fly ash. Precipitator consists of 80 ducts, each 10 m high by 14.3 m long by 0.25 m wide. Total collection surface area = 22,880 m².

Design and Performance Data

Item	Design Value	Performance Value
Gas flow	1100 m³/sec	897 m³/sec
Collection efficiency	99.33%	99.85%
Gas velocity	1.35 m/sec	1.10 m/sec
Specific collection area	488 ft²/1000 acfm	598 ft²/1000 acfm
Precipitation rate w_e	5.22 cm/sec	5.52 cm/sec
Precipitation rate w_k	26.1 cm/sec	36.0 cm/sec
Specific corona power		369 W/1000 acfm
Corona current density		0.18 mA/m²
Ash resistivity		5×10^{11} ohm-cm
Gas temperature		106°C
Inlet ash loading		9.3 g/m³
Outlet ash loading		0.021 g/m³
Stack emission		0.014 lb/10⁶ Btu

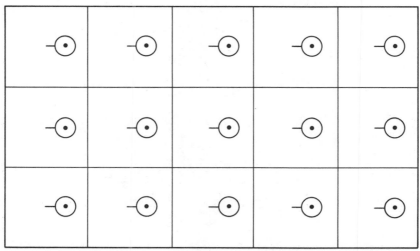

Fig. 16   Precipitator design (layout, one-half): Example 2—high-resistivity fly ash. Total precipitator consists of 30 sections, comprised of 264 ducts, each 12.5 m high by 15 m long by 0.25 m wide. Total collection surface is 105,600 m².

4 PRECIPITATOR EQUIPMENT

A broad variety of precipitator equipment is used in practice to meet requirements for the dusts and gases encountered, ambient conditions, collection efficiencies, and space and cost constraints. The purpose of this section is to convey a general understanding of the types of equipment used, rather than a detailed survey or inventory of all the kinds and variations of precipitators available. Information of the latter type is available from manufacturers and in the literature.

4.1 Precipitator Types

Precipitators are classified as single-stage and two-stage (see Figure 1). Two-stage precipitators are used for cleaning ventilating air and for certain light industrial applications. They are compact, can be mass produced, and fit the needs of the consumer market. Attention in this treatment of electrostatic precipitation is focused, however, on the single-stage units used by heavy industry for controlling air pollution from process and furnace gases. The two major types of such precipitators are the pipe type and the duct type. Pipe-type precipitators are used for small gas flows, for collection of mists and fogs, and frequently for applications requiring water-flushed electrodes. Duct-type precipitators comprise the major part of the market and are used for gas flows above 25 or 50 m³/sec (50,000–100,000 acfm). Figure 17 shows a cutaway view of a typical duct-type precipitator design used for fly ash collection.

4.2 Corona Electrodes

Corona electrodes in U.S. practice most commonly have been round wires on the order of 2.5 mm in diameter maintained taut by suspension weights on the order of 5–10 kg as illustrated in Figure

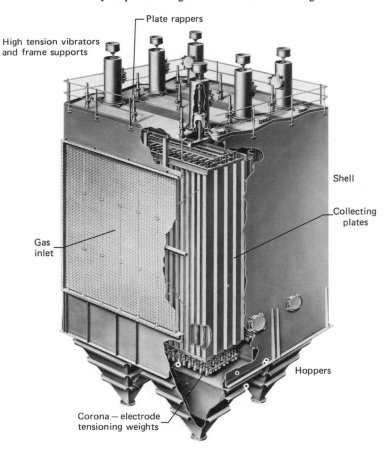

Fig. 17   Duct-type precipitator.

18. Barbed wires, twisted-square wires, ribbons, and various special electrodes have also found limited use. European designs, on the other hand, favor rigid support frames with coiled spring wires, serrated strips, needle points mounted on supporting elements, or similar schemes for the corona-emitting electrodes. Examples of two of the most common types are shown in Figures 19 and 20.

4.3 Collection Electrodes

Pipe precipitators use pipes ranging in size from about 15 cm in diameter by 3 m long for small units, up to 40 cm in diameter by 6 m long for large units. The number of pipes per precipitator varies from a few up to 100 or more, depending on application and amount of gas treated. Duct precipitators use vertically mounted collecting plates, spaced apart in rows, with corona electrodes placed midway between adjacent plates. Duct sizes are typically 10–20 m long by 10–15 m high by 20–40 cm wide. Plates are commonly 1–3 m wide by 10–15 m high. Large precipitators treating 2 or 3 million acfm (944–1416 m³/sec), for example, use several hundred ducts for efficiencies of 99.5%.

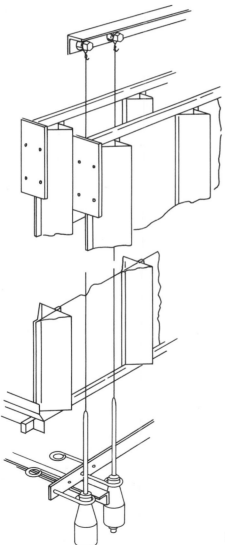

Fig. 18 Schematic diagram for wire-and-weight corona electrodes.

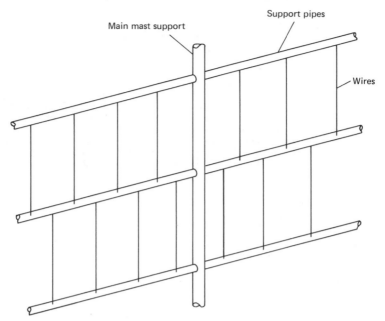

Fig. 19   Mast-type rigid corona electrode design.

Large precipitators are subdivided into sections transverse to the gas flow, and chambers or fields lateral to the gas flow. The individual sections usually are energized by separate electrical sets.

Collecting plates are usually made of 16–20 gauge steel sheets. Schematic diagrams of several commonly used duct–electrode arrangements are shown in Figure 21. Note that the plates have vanes or baffles to provide the necessary stiffness and to reduce direct impact of the gas stream on the collection surfaces.

An interesting variation in duct precipitator technology is the development of wide-duct precipitators which, although of European origin,[16] have become popular for certain applications in Japan in recent years. It has been demonstrated that in some cases duct widths can be increased by 50–100% above usual values without loss of performance. This results in substantial savings in electrode and structural costs, although at the expense of providing greater voltages for the wider ducts. There is a net overall cost saving in many cases.

4.4   High-Voltage Equipment

The high-voltage equipment provides the intense electric fields and corona currents needed for particle charging and collection. To meet service requirements, the electrical sets must also be very stable in operation and have useful operating lives of 20 yr or more. Fulfillment of these essentials requires unfiltered rectifiers with stability against precipitator sparkover, sturdy electrical and mechanical design, and automatic control. Set ratings are commonly 70–100 kV peak voltage and 100–2000 mA current capacity. Silicon high-voltage rectifiers are now almost universal. Rectifier output voltage may be either half-wave or full-wave, depending on the application.

Figure 22 shows a schematic diagram for a modern high-voltage rectifier set for precipitation service. Multiple feedback loops are provided to obtain necessary regulation and fast control of transient spark disturbances in the precipitator. A linear inductive reactor is used as an important feature in obtaining good current waveform and stable operation under sparking conditions. For monitoring purposes kilovolt meters, milliammeters, and spark-rate meters should be included as an integral part of each rectifier set.

Large precipitators may use 50 or more electrical sets to provide needed sectionalization of the high-voltage power. Such sectionalization greatly increases precipitation efficiency and also needed redundancy in case of outages.

4.5   Rappers

Dry particle deposits on the electrodes usually must be removed by rapping forces. This is necessary both to rid the precipitator of collected deposits and also to help maintain optimum electrical conditions.

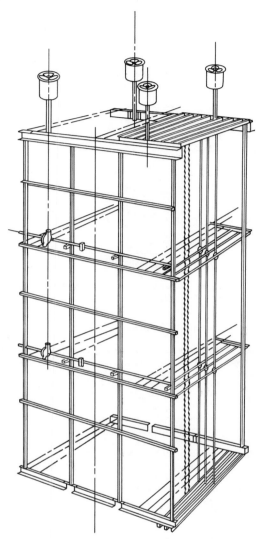

Fig. 20   Rigid support frame used with coiled spring wire corona electrodes.

The rapping forces are supplied by mechanical impulses or by vibration of the electrodes. Most modern precipitators use impulse- or hammer-type rappers, although vibrators are still used for rapping the corona electrodes in some installations. Rapping systems must be highly reliable, provide high-intensity rapping forces when needed, and be adjustable in intensity or frequency.

Two main types of rappers are used for cleaning plates, namely, the magnetic impulse type and the tumbling hammer type. Magnetic impulse rappers are mounted vertically on the top or roof of the precipitator and rap several plates in parallel through connecting rods. See Figure 23. Tumbling or rotating hammer-type rappers are mounted internally and each plate is individually rapped by the hammer striking an anvil on the bottom of the plate as illustrated in Figure 24. Impulse-type rappers are used with most U.S. designs. Both the intensity and frequency of rapping can be varied by using appropriate electronic controls. The tumbling hammer rappers are characteristic of European designs. They are especially useful when collecting high-resistivity dusts because they can provide rapping intensities of hundreds of g's (g = acceleration of gravity) at the plate surfaces, which prevents significant dust buildups. Frequency of rapping is easily controlled by changing the speed of the externally driven rotating rods on which the hammers are mounted.

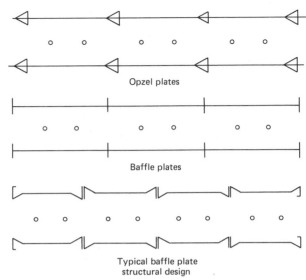

Opzel plates

Baffle plates

Typical baffle plate
structural design

Fig. 21 Schematic diagram for commonly used duct electrode arrangements.

4.6 Shell, Hoppers, and Structural Features

The precipitator shell and structure serve the dual purpose of enclosing the electrodes and supporting them in a stable, rigid manner. The shell also acts as a gas-tight enclosure which can be heat insulated and heat traced when hot gases are being treated. Structural support and foundations for the precipitator hoppers, and connecting flues, must be conservatively designed to ensure long-term stability and maintenance of electrode alignment. The support structure is especially critical for precipitators operating at temperatures above about 200°C because of the large temperature expansions and high-temperature differential stresses which can literally tear shell and hopper joints and welds apart.

Conventional practice for large precipitators is to support the collecting plates and corona electrodes from the top, so that these elements hang vertically under gravity and can expand or contract under temperature changes without binding or distorting the electrodes. Modular designs may be used for smaller or less critical precipitators, but may be subject to electrode alignment problems during construction and maintenance of alignment under temperature swings in operation.

Dust hoppers should be of adequate size and number to hold the dust collected over at least an 8- to 12-hr period. The dust removal system should be adequate to remove the dust from the hoppers

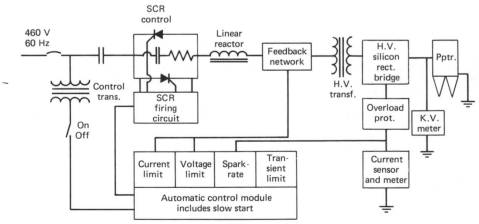

Fig. 22 Schematic diagram for modern rectifier sets.

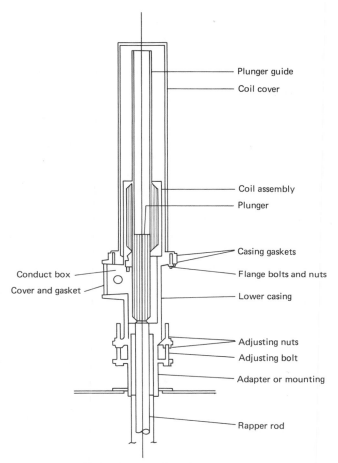

Fig. 23   Magnetic-impulse-type rappers.

on a continuous basis. For many applications it is necessary to heat insulate and trace the hoppers to prevent caking or cementing of the collected dust, a problem that can occur especially during precipitator shutdowns.

Air inleakage into hoppers must be prevented to avoid loss of dust by reentrainment caused by the incoming air. Reliable hopper dust-level indicators, several types of which are available, should be provided. Overflowing hoppers are a common cause of trouble. Not only is the dust loss greatly increased, but both the corona and collecting electrodes may be seriously damaged, necessitating shutdowns and troublesome repairs.

4.7   Ensuring High-Quality Gas Flow

Provisions for ensuring high-quality gas flow through the precipitator are highly essential for achieving high-efficiency performance, and therefore need to be dealt with at the initial design stage. Long experience has shown that precipitator gas flow systems can seldom be successfully designed by intuitive methods. Conditions of cramped space and asymmetric, irregularly shaped flues usually encountered rule out mathematical and fluid–dynamic methods. This leaves scale-model laboratory studies as the most practical and reliable approach to ensure good gas flow for precipitator installations. Model techniques are well documented and the close correlation between model study and field results has been demonstrated by many successful examples.[17]

Techniques available for controlling and correcting gas flow patterns include guide vanes to change gas flow direction, flue transitions to couple flues of different sizes and shapes, and various types of diffusion screens to reduce turbulence.[18]

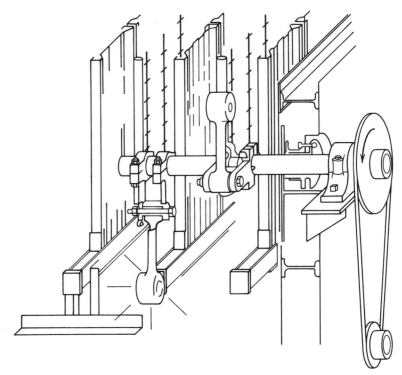

Fig. 24 Rotating-hammer-type rappers.

Minimum standards for gas flow quality have been set by the Industrial Gas Cleaning Institute. These are minimal in character and are being updated. More generally, it is useful to use the root-mean-square deviation of gas velocity as measured at the inlet face of the precipitator as a criterion for gas flow quality. Perfect or zero deviation corresponds to uniform velocity. In practice, 10–15% deviation would be considered good. Deviations greater than 20–25% are poor or unacceptable.

5 PRECIPITATOR PROBLEMS

Experience has shown that problems are encountered in a significant percentage of precipitator installations. These problems can be classified into three major areas: fundamental, mechanical, and operational, as summarized in Table 6. In some cases multiple problems occur, a circumstance that tends to complicate the diagnosis and correction of the difficulties. Problem solution should be approached by systematic and rational methods. Many unfortunate experiences have shown that guess-work is almost invariably counterproductive, and usually increases the time and expense of correction.

5.1 Fundamental Problems

Fundamental difficulties include high-resistivity particles, particle reentrainment, poor gas flow, poor rapping, inadequately designed electrode systems, insufficient high-voltage electrical equipment, and in some cases undersize precipitators. Scientific procedures exist for determining and isolating these deficiencies. For example, high resistivity can be detected and evaluated with field probe apparatus of the type discussed in Section 6, and reentrainment can usually be seen by visual observations in the outlet flue under intense illumination or by comparative size analyses and the distribution of the particulates at the outlet of the precipitator.

5.2 Mechanical Problems

Mechanical problems comprise, principally, poor alignment of electrodes, bowed or distorted collecting plates, vibrating wires or swinging high-tension frames in the case of wire-and-weight corona electrodes, excessive dust buildups on the corona or collection electrodes, hoppers full or overflowing with collected

Table 6   Commonly Encountered Precipitator Problems

Fundamental Problems
1. High-resistivity particles.
2. Reentrainment of collected particles.
3. Poor gas flow.
4. Insufficient or unstable rectifier equipment.
5. Insufficient number of corona sections.
6. Improper or incompatible rapping.
7. Gas velocity too high.
8. Aspect ratio too small.
9. Precipitator size too small.

Mechanical Problems
1. Poor electrode alignment.
2. Distorted or skewed collecting plates.
3. Vibrating or swinging corona wires.
4. Excessive dust deposits on corona electrodes and/or collecting plates (sometimes cemented on).
5. Formation of dust mountains in precipitator inlet and outlet ducts.
6. Gas turning vanes and/or gas distribution screens plugged with dust.
7. Air inleakage into hoppers, shells, or gas ducts.
8. Gas sneakage around precipitation zones and/or through hoppers.

Operational Problems
1. Full or overflowing hoppers.
2. Shorted corona sections (broken wires, etc.).
3. Rectifier sets or controls poorly adjusted.
4. Precipitator overloaded by excessive gas flow.
5. Precipitator overloaded by excessive dust concentration.
6. Process upsets (poor combustion, steam leaks, etc.).

dust, air inleakage into hoppers, and dust mountains or piles in connecting gas ducts or on turning vanes. The correction of these difficulties usually is fairly obvious, once they have been located. Again, scientific methods based on symptoms, measurements, and observations are most effective in making corrections.

5.3   Operating Problems

These problems are primarily associated with plant maintenance and operation. Overflowing hoppers, for example, can be avoided by establishing adequate hopper-emptying schedules. Most operating problems can be controlled by following procedures set forth in the operating instruction manual provided by the precipitator supplier. Precipitator operation and maintenance has received special attention in recent literature.[19,20]

6   PARTICLE RESISTIVITY[21]

Particle deposits on the collection surfaces of precipitators must possess at least a small degree of electrical conductivity in order to conduct the ion currents from the corona to ground. Minimum conductivity required is about 10^{-10} inverse ohm-cm. This is merely a trace conductivity compared to ordinary metals, but much greater than the conductivity of good insulators such as silica and most plastics. Particles having conductivities less than the critical value of about 10^{-10} inverse ohm-cm, which corresponds to a resistivity of 1×10^{10} ohm-cm, are referred to as high-resistivity particles.

The presence of high-resistivity particles is manifested by disturbed electrical operation in the form of excessive sparking and somewhat lowered operating voltages, or by excessive currents at greatly lowered voltages. The adverse effects increase with increasing resistivity. Resistivities in the range of about 2×10^{10} to 10^{11} ohm-cm make it necessary to reduce operating voltages and currents in order to prevent excessive sparking. Resistivities above about 10^{11} ohm-cm cause back corona or a back discharge from the collected dust layers, making it necessary to substantially reduce corona voltages and currents. The smaller currents and voltages are reflected in lower collection efficiencies. In practice, it is necessary to provide more collection surface, or to take measures to reduce the resistivity. Research and development work is being done to overcome the adverse effects of high resistivity.

Liquid particles and certain types of solid particles are intrinsically conductive, and therefore cannot cause difficulty because of high resistivity. Many dusts and fumes met with in industrial precipitator applications, however, are composed of silicates, metallic oxides, and similar inorganic compounds which in the pure dry state are good insulators and therefore might be expected to cause trouble in precipitators. However, moisture and chemical impurities present in the gases and particles always provide at least some of the trace conductivity required. In other cases the gas temperature may be sufficiently high to ensure adequate conductivity. Thus high resistivity is a problem in only a fraction of applications.

Electrical conduction in a bulk layer of particles takes place over the surface and through the interiors of the particles. In surface conduction, electric charges are carried in the surface moisture and chemical films adsorbed on the particles. These films usually differ both physically and chemically from the interiors of the particles, owing to adsorption phenomena. Volume conduction, or the motion of electrical charges through the interiors of the particles, depends, on the other hand, on the composition and temperature of the particles. It should also be noted that both surface and volume conduction involve ancillary factors, such as particle size and shape, thickness and compression of the particle layers, and the intensity of the electric field applied to the layer.

6.1 Measurement of Resistivity

Particle resistivity measurements are made with high-voltage conductivity cells designed for use under high temperature and rugged conditions found in the field. A number of such cells have been developed, but the accepted method in the United States is the point-plane method first used in this country in 1940. See Figure 25. In field practice, the resistivity is measured by placing the point-plane device directly in the flue gas through a suitable opening in the flue. Dust is collected on the plane by the corona discharge from the point. A movable high-voltage disk is then lowered on the dust and the voltage increased gradually until sparkover through the dust layer occurs. The resistivity is computed from the current and voltage across the dust layer at breakdown. There are several variations of the technique, but it is vital that the resistivity be measured for voltages near the breakdown value since the resistivity usually decreases with the applied electric field.

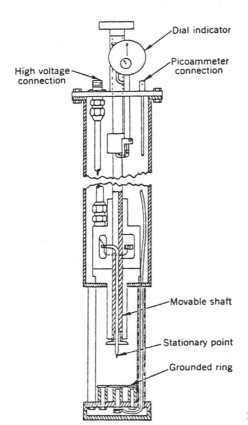

Fig. 25 Point-plane resistivity apparatus.

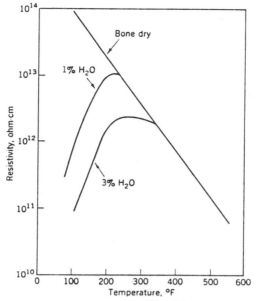

Fig. 26 Effect of moisture in reducing the resistivity of a typical fly ash.

In practice, resistivity values for industrial particles determined in the field commonly range between about 10^8 to 10^{13} ohm-cm. Such field measurements are usually called *in situ* to distinguish them from values for collected dust samples measured in the laboratory. It is not uncommon for in situ resistivity values to be 100 to 1000 times lower than the corresponding values measured in the laboratory on the same dust. It is for this reason that only in situ resistivity values should be used in evaluating field operation of precipitators. Extreme care must be taken in using reported values of resistivity which may have been determined by other methods and therefore may not be representative of the actual resistivity conditions.

Figure 26 shows the effect of moisture in reducing the resistivity of a typical fly ash as determined by laboratory measurements. Note that the resistivity for bone dry conditions increases with decreasing temperature in accord with a well-known exponential law. Even small moisture additions to the gas reduce the resistivity markedly for temperatures below about 150°C (300°F), the reduction increasing as the temperature is lowered. Moisture conditioning of cement kiln dust is illustrated in Figure 27. More generally, the effect of environmental treatment on the resistivity of a dust is brought out in Figure 28. The ash resistivity as received from the field was measured as a function of gas temperature. Heating the ash and allowing it to cool increased the peak resistivity by a factor of about 10, while washing the ash in distilled water and again going through the heating and cooling cycle further increased the peak resistivity by about 100. These results are quite typical of many industrial dusts and fumes.

6.2 Effects of High Resistivity on Precipitator Performance

High-resistivity particles can be readily recognized in precipitator operation by the disturbed electrical conditions and reduced collection efficiencies they cause. If the precipitator has been designed to take these adverse effects into account, high efficiencies and good operation can still be achieved by proper adjustments and the use of effective automatic control systems for the electrical sets.

Although no universally accepted method and technique exist for resistivity measurements, the point-plane apparatus with resistivity determined at fields near the breakdown value across the layer is widely used in the United States and other countries. Differences in measurement techniques can yield resistivity values on the same dust which differ by factors of 100 or more.

The point-plane method has been used in the United States since about 1940, and a large body of experience has been built up which shows a close correlation between resistivities measured under actual operating conditions in the field and the level of precipitator performance. Measurements made in many different precipitator applications show that there exists a critical value of resistivity of about 10^{10} ohm-cm, below which resistivity has little effect on electrical operation and precipitator performance. Resistivities between 10^{10} and 10^{11} ohm-cm, in general, increase sparking and lower operating voltages.

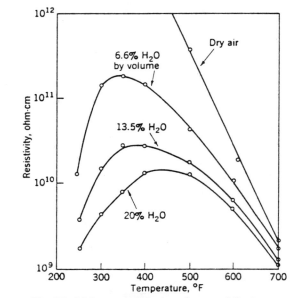

Fig. 27 Moisture conditioning of cement kiln dust.

Above about 10^{11} ohm-cm a more serious localized back corona forms which causes both particle charging and particle collection to be greatly reduced with corresponding loss of precipitator efficiency.

The adverse effects of high resistivity increase with the magnitude of the resistivity. If the resistivity is not above about 10^{11} ohm-cm, the effects are limited to moderately reduced operating voltages and currents, but for higher resistivities back corona tends to form and operating corona current densities must be greatly reduced. These effects are illustrated graphically in Figure 29. The effect of

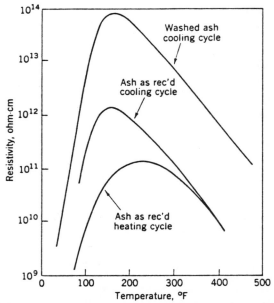

Fig. 28 Effect of environmental conditions and treatment on the resistivity of a fly ash field sample as measured in the laboratory.

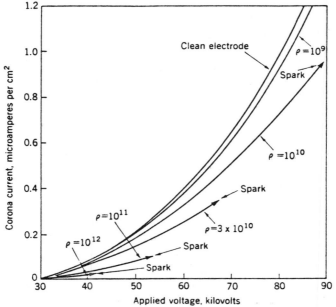

Fig. 29   Theoretically calculated effect of resistive dust layer on voltage–current characteristics for a 20-cm diameter pipe precipitator.

resistivity on precipitation rate w_e is shown in Figure 30 for fly ash precipitators having particle resistivities ranging between 10^9 and 10^{13} ohm-cm. The data are based on comprehensive field performance tests.

6.3   Methods for Overcoming High Resistivity

Major methods used in practice for overcoming high-resistivity problems are: keeping electrodes as clean as possible; providing better electrical energization; and using chemical and moisture conditioning and temperature conditioning. In addition, several advanced concepts for overcoming the adverse effects of high resistivity are being developed in the laboratory and field.

In principle, keeping the electrodes dust free should eliminate high-resistivity problems. Although it is impossible in practice to maintain completely dust-free surfaces, the dust thicknesses can be reduced by heavy rapping to the order of only 1 mm or less, which is sufficient to substantially reduce the adverse effects of high resistivity.

Improved electrical energization methods include the use of pulse voltages, greater high-tension sectionalization, fast-acting spark-quenching circuits, and effective automatic control systems.

Particle conductivity can be increased by moisture addition to the gas stream, and by addition of small amounts of certain chemical agents such as SO_3, NH_3, and Na_2CO_3. By far the most widely used chemical conditioning agent is sulfuric acid, SO_3. This has been used in nonferrous smelters since 1915, and more recently in coal-fired power plants to condition fly ash from low-sulfur coals. Use of sodium compounds for conditioning fly ash was first discovered in about 1970.[22] Sodium occurs naturally in some western coals in sufficient quantity (about 0.2% or more by weight as Na_2O) to provide adequate conductivity of the fly ash. Laboratory and field investigations show that sodium can be added in small quantities to the coal feed in power plants to provide effective conditioning both for cold-side and hot-side precipitators.

Many industrial particles of mineral origin which have high resistivities at temperatures in the 150–200°C range become sufficiently conductive for precipitation purposes at both lower and higher temperatures. In practice, it has been found that sufficient conductivity for precipitation often can be achieved by lowering gas temperature below about 130°C or by raising gas temperature above about 350°C. These lower and higher temperature levels can in some cases be achieved by changes in the plant process, or, at the design stage, by suitable location of the precipitator in the plant system.

Cooling by water sprays is especially effective because both moisture and lower temperature effects are obtained simultaneously.

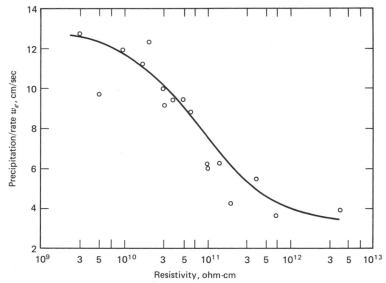

Fig. 30   Variation of precipitation rate w_e with resistivity, for fly ash precipitators, based on performance data for full-scale installations.

7   HOT-SIDE PRECIPITATORS

High-temperature precipitators, operating in the range of 300 to 400°C where dust resistivity tends to be below the critical level of 10^{10} ohm-cm, have long been used for the collection of cement kiln and certain other dusts. The idea was first applied to fly ash collection from boilers burning low-sulfur coals in the mid-1960s, by installing the precipitator ahead of the air preheaters where the gas temperature is usually above 300°C. This arrangement became known colloquially as a "hot" or "hot-side" precipitator. Its popularity increased through much of the 1970s but because of serious unforeseen problems no new installations are being designed for fly ash collection.[23]

The primary advantage of the hot-side precipitator is avoidance of high-resistivity ash which often occurs with low-sulfur coals at flue gas temperatures of the order of 150°C typical for flue gases after the air preheater. There are also serious disadvantages. Gas flows are about 50% higher at the high gas temperatures. Precipitator operating voltages are substantially reduced owing to the lower densities of hot gases. Gas viscosity increases with gas temperature, thus reducing precipitation rate. Structural and mechanical problems have occurred, such as precipitator shell failures and support structure distortions stemming from differential thermal expansions between the shell and the support structure. Other problems and high costs are associated with the necessarily long interconnecting flues needed between the precipitator and the boiler. But the major problem is the deterioration of precipitator performance which has been found to occur in a significant percentage of cases in the form of degraded electrical conditions and lowered precipitator efficiency. This deterioration of performance typically occurs over periods ranging from a few days to a few weeks or even months.

Laboratory and field research has identified the cause of this degradation as sodium depletion in the residual dust layers on the collection surfaces.[24] This depletion occurs only for low-sodium coals where the amount of sodium in the fly ash (expressed as Na_2O) is less than about 0.5–1.0%. Sodium addition to the coal ahead of the furnace feed point in the amount of 1–2% as Na_2O in the fly ash has been found adequate to restore precipitator performance and to prevent deterioration of performance thereafter.

In view of the fact that other options are available, including conditioning by sodium, sulfur trioxide, and ammonia in some cases in cold-side precipitators installed after the air preheaters, it seems unlikely that hot-side precipitators will regain their former popularity. The use of microprocessor control systems for precipitators and advanced electrical energization methods such as pulse energization also opens the way for other attractive options.

REFERENCES

1. Cottrell, F. G., Problems in Smoke, Fume, and Dust Abatement, *Smithsonian Report for 1913,* Publication 2307, 653–685 (1914).

2. Sales data provided by Industrial Gas Cleaning Institute (IGCI), Alexandria, VA. Sales shown are for flange-to-flange hardware only.

3. Pauthenier, M. M., and Moreau-Hanot, M., La Charge des Particles Spherique dans un Champ Ionisé, *Journal de Physique et le Radium* **3**, 590 (1932).

4. White, H. J., Particle Charging in Electrostatic Precipitation, *Trans. Am. Inst. El. Engrs.* **70**, 1189 (1951).

5. Smith, W. B., and McDonald, J. R., Development of a Theory for the Charging of Particles by Unipolar Ions, *J. Aerosol Science* **7**, 151 (1976).

6. Hewitt, G. W., The Charging of Small Particles for Electrostatic Precipitation, *Trans. Am. Inst. El. Engrs.* **76**, 300 (1957).

7. Smith, W. B., et al., Experimental Investigations of Fine Particle Charging—A Review, *J. Aerosol Science* **9**, 101 (1978).

8. McDonald, J. R., et al., Charge Measurements on Individual Particles Exiting Laboratory Precipitators with Positive and Negative Corona at Various Temperatures, *J. Applied Physics* **51**, 3632 (1980).

9. White, H. J., Modern Electrostatic Precipitation, *Ind. and Eng. Chem.* **47**, 932 (1955).

10. White, H. J., *Industrial Electrostatic Precipitation,* Addison-Wesley, Reading, MA, 1963, pp. 185–190.

11. Deutsch, W., *Ann. der Physik* **68**, 335 (1922).

12. EPA Report, *A Mathematical Model of Electrostatic Precipitation* (EPA-650/2-75-037), April 1975. Report prepared by Southern Research Institute, Birmingham, AL.

13. EPA Report, *A Manual for the Use of Electrostatic Precipitators to Collect Fly Ash Particles* (EPA-600/8-80-025), Section 12, May 1980. Report prepared by Southern Research Institute, Birmingham, AL.

14. Matts, S., and Ohnfeldt, P., *Efficient Gas Cleaning with SF Electrostatic Precipitators,* Bulletin of AB Svenska Flaktfabriken, Stockholm, Sweden, 1963.

15. Allander, C., and Matts, S., The Effect of Particle Size Distribution in Electrical Precipitators, *Staub* **52**, 738 (1957).

16. Heinrich, D. O., Electrostatic Precipitators with Wide Collector Spacing, Inst. El. Engrs. Colloquium on Corona Discharges, London, December 1979.

17. Burton, C. L., and Willison, R. E., Application of Model Studies to Industrial Gas Flow Systems, Paper 59-A-280, Am. Soc. Mech. Eng., Atlantic City, NJ, December 1959.

18. White, H. J., *Industrial Electrostatic Precipitation,* Addison-Wesley, Reading, MA, 1963, Chapter 8.

19. Air Pollution Control Association, *Proceedings of Specialty Conference on Operation and Maintenance of Electrostatic Precipitators,* published by APCA, Pittsburgh, PA, 1978.

20. Katz, J., *The Art of Electrostatic Precipitation,* Precipitator Technology, Inc., Munhall, PA, 1980.

21. High resistivity is treated *in extenso* in White, H. J., Resistivity Problems in Electrostatic Precipitation, *J. Air Poll. Contr. Assoc.* **24**, 314 (1974).

22. Ibid, p. 333.

23. Walker, A. B., and Gawreluk, G., Performance Capability and Utilization of Electrostatic Precipitators Past Future, *Proceedings of International Conference on Electrostatic Precipitation,* Monterey, CA, October 1981.

24. Nichols, G. B., Sodium Conditioning Tests to Combat the Time Dependent Performance Degradation in Hot Side ESP, *Proceedings of the Institute of Electrostatics Japan* **6**, 48 (1982).

CHAPTER 13

CONTROL OF PARTICLES BY MECHANICAL COLLECTORS

WILLIAM LICHT

Department of Chemical Engineering
University of Cincinnati
Cincinnati, Ohio

1 PRINCIPLES AND SCOPE OF MECHANICAL COLLECTION

1.1 Nature of Collecting Forces Utilized

The term mechanical collectors is generally taken to refer to devices in which particles are collected either by the action of gravity or by inertial effects upon the particles or perhaps by a combination of both. In the case of gravity collectors, particles simply settle out of the gas stream due to their weight. In momentum collectors, the flowing stream of particles in suspension is subjected to a sudden change of direction. The resulting inertial effect causes the particles to tend to be thrown out of the gas stream. Collectors employing centrifugal force (cyclones) are an important special case of the inertial effect.

The rate of particle removal is proportional to the size of the collecting force. Because the weight of small particles is quite low, gravity collection is generally a slow and ineffective process for particles smaller than 100 μm in size. The rate of collection can be greatly increased by employing the inertial effect. This reduces the size of equipment and extends the range of effective collecting generally down to particles of about 20 μm in size. In the case of certain cyclones it may go down to 5–10 μm.

1.2 Types of Equipment

Collection by gravity is usually carried out by simply providing a large chamber through which the gas stream moves slowly, thus giving particles an opportunity to settle into hoppers in the bottom. The distance of settling required may be reduced by placing a number of parallel horizontal trays in the chamber.

Baffles may be placed in gravity chambers to change the direction of gas flow and bring about the inertial effect to enhance that of gravity. Other designs involve the use of louvers, shutters, or impingement surfaces to create an inertial effect.

In cyclone collectors the gas is given a spinning or vortex motion to impart a centrifugal force to the particles. This is done either (a) by introducing the flow tangentially into a circular chamber, or (b) by passing the gas through radial vanes along an axial flow.

All of these types of equipment are characterized by simplicity of construction and operation. They are relatively inexpensive in comparison with other kinds of collectors. Generally there are no moving parts, and any material of construction may be used as may be required to withstand operating conditions. The energy required for operation is also relatively low. It is due only to a rather small pressure drop required to move the gas through the equipment.

1.3   Scope of Applications

These collectors are useful primarily to remove coarse particles from a gas stream. In most air pollution control problems, there is a need to collect much finer (in the neighborhood of 1 μm) particles, so other types of collectors are usually required. However, the mechanical collector may be used as a primary collector in series with another type in order to reduce the load on the latter. This is especially true for heavily loaded gas streams. The mechanical collectors can give long periods of maintenance-free operation at low energy costs.

2   SETTLING CHAMBERS

2.1   Gravity Force and Drag Force

The design of settling chambers is based on a calculation of the forces acting upon a particle and the velocity of downward motion the net force produces. According to Newton's law the net downward acceleration of a particle is determined by the resultant of the weight of the particle, as offset by buoyancy, and of the drag force opposing downward motion. In gases the buoyancy effect may be neglected. The drag force is estimated by use of the drag coefficient (C_D) as related to the Reynolds number (N_{Re}) for particle motion.

Thus, the vertical motion of a particle is determined by

$$m\frac{du_s}{dt} = gm - C_D\frac{\rho_f u_s^2}{2}\pi r_p^2 \tag{1}$$

in which

$$C_D = C_D(N_{Re}) = C_D\left(\frac{\rho_f u_s d_p}{\mu}\right)$$

The functional relationship between C_D and N_{Re} is taken from the standard graph (or descriptive equations) for individual spherical particles. In case $N_{Re} \leq 0.5$, this is represented well by Stokes law: $C_D = 24/N_{Re}$. For larger values of N_{Re} an empirical equation by Klyachko,[1] $C_D = 24/N_{Re} + 4/N_{Re}^{1/3}$, fits the data well for $0.5 < N_{Re} < 800$. For very small particles the drag force must be corrected by the Cunningham slip factor, but this case is of no importance in gravity settling.

2.1.1   Terminal Settling Velocity

It is evident from Eq. (1) that as a particle settles it will accelerate vertically until the drag force becomes equal to the weight, after which further motion is at constant speed. This speed is called the terminal settling velocity of a particle. The time required for it to be reached in gases is very short. This special value of u_s may be found by setting Eq. (1) equal to zero, replacing m by ($\rho_p - \rho_f$) $d_p^3/6$, and placing in dimensionless form

$$\frac{4}{3}\frac{gd_p^3\rho_f(\rho_p - \rho_f)}{\mu^2} - C_D N_{Re}^2 = 0$$

Thus, the Galileo number, $N_{Ga} = 4gd_p^3\rho_f(\rho_p - \rho_f)/3\mu^2$ is defined. It is proportional to d_p^3 and is independent of u_s. The condition for terminal settling is that $N_{Ga} = C_D N_{Re}^2$.

Another dimensionless number, which is proportional to u_s^3 and independent of d_p, may be defined as

$$\frac{N_{Re}}{C_D} = \frac{N_{Re}^3}{N_{Ga}} = \frac{3\rho_f^2 u_s^3}{4g(\rho_p - \rho_f)\mu}$$

The direct relationship between d_p and u_s may be given then in terms of $(N_{Re}/C_D)^{1/3}$ as a function of $N_{Ga}^{1/3}$. Empirically this can be well represented by an equation due to Koch.[2]

$$\log(N_{Re}/C_D)^{1/3} = -1.387 + 2.153\log(N_{Ga}^{1/3}) - 0.548(\log N_{Ga}^{1/3})^2 + 0.05665(\log N_{Ga}^{1/3})^3 \tag{2}$$

Thus to calculate to the terminal settling velocity for a particle of any size d_p, first calculate the value of N_{Ga} as defined above. Then calculate the value of N_{Re}/C_D from Eq. (2) and finally obtain u_s from the definition of N_{Re}/C_D given above.

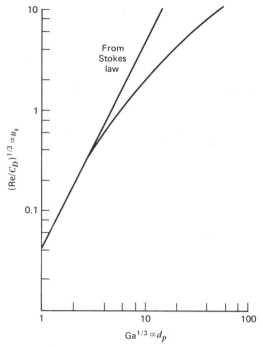

Fig. 1 Terminal settling velocity as related to particle size. Reprinted by permission of Marcel Dekker, Inc.

In the range of Stokes law this becomes simply

$$u_s = \frac{g(\rho_p - \rho_f)d_p^2}{18\mu}$$

The relation between particle size and terminal settling velocity is as shown in Figure 1.

2.2 Chamber Design

2.2.1 Single Chamber

A simple model for chamber design is obtained by assuming the gas stream to be in plug flow with a uniform velocity of u_0, and the particles to be uniformly distributed in the gas. Each particle is assumed to settle independently at its own terminal settling velocity. Figure 2 is a schematic representation of a cross section of the chamber. Particles of a given size entering at a given position (level h_c) will follow a straight line trajectory as shown. Whether a given particle settles to the bottom of the chamber before being carried out with the gas will be determined by whether $u_s \cdot h_c \leq u_0 l$. The fraction of all particles having the same u_s that will be collected is given by $h_c/H = u_0 l/u_s$.

The dimensions of the chamber (H, l, B) are determined by first selecting the smallest particle size d_p^* to be collected completely. The value of u_s^* for this particle is calculated and h_c/H is set at 1. Then $l = u_s^*/u_0$. The value of u_0 is taken to be less than the particle pickup velocity, or less than 10 ft/sec, whichever is smaller. Finally $BH = Q/u_0$.

2.2.2 Multiple Channels

The efficiency of collection by gravity settling may be increased by reducing the distance the particles have to fall. This can be accomplished by placing horizontal plates in a chamber, effectively creating a set of multiple chambers in parallel. The entire gas flow is divided up and passes through the parallel spaces between the plates. H in the above equations becomes the spacing between the plates which may be of the order of one-tenth of the chamber height. The concept is sound but the practical difficulty is that of removing the particles that settle on the plates.

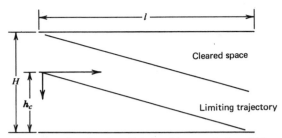

Fig. 2 Cross section of gravity chamber. Reprinted by permission of Marcel Dekker, Inc.

2.2.3 Baffled Chambers

The action of gravity may be enhanced by an inertial force if a vertical baffle is attached to the top of the chamber. As the gas flow is forced round the bottom end of the baffle, particles tend to be thrown downward by the inertial effect of this curvature in the path. No precise data are readily available for the calculation of this added collection force.

2.3 Effectiveness

The effectiveness of a collector may be predicted by calculating a grade efficiency relationship, giving efficiency of collection as a function of particle size. This coupled with the particle size distribution in the feed to the collector will predict the overall collection efficiency.

2.3.1 Prediction of Collection Efficiency

The nature of the grade efficiency relationship obtained for a simple settling chamber will depend on the assumption made to describe the flow pattern of the gas. Four cases may be considered: (a) plug flow, no mixing; (b) laminar flow with a parabolic velocity profile, no mixing; (c) turbulent plug flow with lateral (vertical) back mixing of uncollected particles; (d) turbulent plug flow with complete mixing of all uncollected particles. The grade efficiency in each case may be generalized against $(u_s l / u_0 H)^{1/2}$, which in turn may be taken to be proportional to d_p in the case of Stokes law particle behavior. The grade efficiency curves obtained for cases (a), (c), and (d) are plotted for comparison in Figure 3.

The performance of a real chamber will involve turbulence, some degree of mixing, and some distortion of plug flow. Hence a real grade efficiency curve will lie in the lower range of those shown.

Overall collection efficiency (E) is defined as the mass of particles removed divided by the mass entering the collector, per unit time. It may be calculated by integrating the grade efficiency over the particle size distribution:

$$E = \int_0^1 E_g \, df \approx \Sigma E_g \cdot \Delta f \tag{3}$$

where E_g is the grade efficiency at a given particle size d_p and f is the mass fraction less than that size in the feed stream. The technique of this calculation is illustrated below in connection with cyclone performance.

2.3.2 Actual Performance and Testing

The actual performance of settling chambers is hardly ever subjected to experimental measurement or testing. At best such devices provide only a preliminary cleaning of gas to remove the coarsest and heaviest particles. The particles escaping a settling chamber must be subsequently collected by other devices. It is the performance of these which is of critical importance in the system.

3 CYCLONE COLLECTORS

3.1 Utilization of Centrifugal Force

A particle of mass m moving in a circular path of radius r with a tangential velocity u_T is acted upon by a centrifugal force of $m u_T^2 / r$. For typical values of $u_T = 50$ ft/sec, $r = 2$ ft, this force is

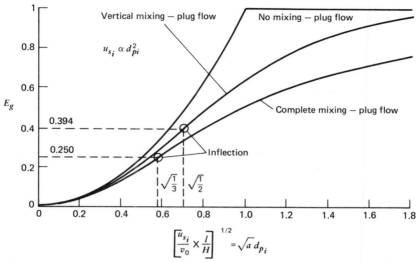

Fig. 3 Generalized grade efficiency for gravity collection. Reprinted by permission of Marcel Dekker, Inc.

$(50)^2/2 \times 32.2 = 38.8$ times that of gravity. It can therefore be used as a greatly enhanced collecting force in comparison with a settling chamber. A practical collection system is obtained by causing this particle-laden gas to flow in a spinning or circular path, confined by a cylindrical wall. Particles are collected by being thrown against this wall. Such a device is called a cyclone. It may have a grade efficiency well above 80% for particles as fine as 10 μm.

3.1.1 Types of Cyclones

There are two principal classes of cyclones: tangential entry and axial entry. Figure 4 shows a typical configuration for the common type, a tangential entry, reverse-flow cyclone. This is mounted vertically. Particles thrown against the wall are collected by sliding down into a hopper. The gas flows in a vortex which reverses its direction of rotation near the bottom of the cone and cleaned gas leaves through the exit duct D_e.

In an axial flow device the gas stream enters at the center of one end of a cylinder. It flows through vanes which impart the spinning motion. Collected particles are carried out by a peripheral stream, while the clean gas exits through a central stream at the opposite end from the entry. This device may be mounted in any position. There are a number of proprietary designs which are variations of these basic designs.

These devices are usually built of ordinary carbon steel, but any type of metal or even ceramic material may be used if necessary in order to withstand high temperatures, abrasive particles, or corrosive atmospheres. It is important that the interior surface be smooth. There are no moving parts, so operation is usually simple and relatively free of maintenance.

3.1.2 Single-Cyclone Configurations

As is shown in Figure 4, eight dimensions are required to specify a tangential entry cyclone. Seven dimension ratios, that is, $K_a = a/D$, $K_b = b/D$, and so on, will specify the configuration and one dimension, chosen as D, will specify the size. Table 1 lists ratios for several standard configurations which have been found to be practical and effective.

A designer might wish to select other configurations but ought to keep in mind certain useful guidelines:

1. $a \leqq S$ to prevent short-circuiting of incoming dust to the outlet tube.
2. $b \leqq (D - D_e)/2$ to avoid excessive pressure drop.
3. $H \geqq 3D$ to keep tip of vortex inside the cone.
4. Angle of cone ≈ 7–$8°$ for ready slippage of dust.
5. $D_e/D \approx 0.4$–0.5, $H/D_e \approx 8$–10, and $S/D_e \approx 1$, for maximum efficiency.[3]

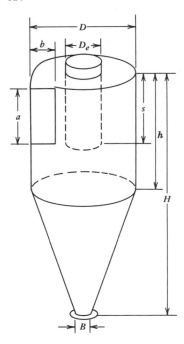

Fig. 4 Cyclone shape and dimensions. Reprinted by permission of Marcel Dekker, Inc.

3.1.3 Multiple Cyclones

Two or more cyclones of the same configuration and size may be used in parallel, the flow being divided equally between them. In this case, each cyclone can be of smaller D than if only one were used. The collection grade efficiency for each cyclone is the same, and is greater than if only one larger one were used. A large number of very small cyclones in parallel (called a "multiclone") may be used to increase efficiency for very large flows of gas. However, it may be difficult to maintain an

Table 1   Cyclone Design Configurations (Tangential Entry)

Term	Description		"High Efficiency" Stairmand[a]	Swift[b]	Shepherd and Lapple[c]	"General Purpose" Swift[b]	Peterson and Whitby[d]
D	Body diameter		1.0	1.0	1.0	1.0	1.0
a	Inlet height	K_a:	0.5	0.44	0.5	0.5	0.583
b	Inlet width	K_b:	0.2	0.21	0.25	0.25	0.208
S	Outlet length	K_s:	0.5	0.5	0.625	0.6	0.583
D_e	Outlet diameter	K_{D_e}:	0.5	0.4	0.5	0.5	0.5
h	Cylinder height	K_h:	1.5	1.4	2.0	1.75	1.333
H	Overall height	K_H:	4.0	3.9	4.0	3.75	3.17
B	Dust outlet diameter	K_B:	0.375	0.4	0.25	0.4	0.5
K	Configuration parameter		551.3	699.2	402.9	381.8	342.3
N_H	Inlet velocity heads		6.40	9.24	8.0	8.0	7.76
Surf	Surface parameter		3.67	3.57	3.78	3.65	3.20

[a] Stairmand, C. J., *Trans. Inst. Chem. Engrs.* **29**, 356 (1951).
[b] Swift, P., *Steam Heating Eng.* **38**, 453 (1969).
[c] Shepherd, G. B., and Lapple, C. E., *Ind. Eng. Chem.* **31**, 972 (1939).
[d] Peterson, C. M., and Whitby, K. T., *ASHRAE Journ.* **42** (1965).

equal flow of gas in each cyclone of the parallel array, especially if they all discharge dust into a common hopper.

Two or three cyclones may also be used in series, the emission from the first becoming the feed to the second, and so on. In this case, the size of the cyclones is made successively smaller downstream, in order to improve collection efficiency of the finer particles which have penetrated upstream. The collected product from each cyclone has a different size distribution. This arrangement may be used to separate a feed particulate into two or three size fractions.

3.2 Cyclone Design

Cyclone design consists of selecting a configuration, then determining the size, grade efficiency, pressure drop, and power requirement of each cyclone to be used. These determinations are based on given gas flow rate, composition, temperature, pressure, and grain loading, together with data on the particle size distribution in the feed. The design will also give a predicted overall efficiency of collection, emission rate, outlet grain loading, and particle size distribution. These latter items will provide the basis for design of a secondary collector following the cyclone, if one is to be used.

3.2.1 Particle Trajectory in a Spinning Gas

The design method must be based on a knowledge of the spiral path followed by a particle in the collector. This path results from the centrifugal force induced on the particle by the circular streamlines followed by the spinning gas. The tangential velocity of the gas on these streamlines is inversely proportional to a power of the radius, according to the "vortex" law: $u_T R^n = $ constant, where $n \leqq 1$. It has been found experimentally[4] that n may be estimated from

$$n = 1 - [1 - 0.67(D)^{0.14}]\left(\frac{T}{283}\right)^{0.3}$$

where

$$D = \text{diameter of cyclone, m}$$
$$T = \text{absolute temperature of gas, K}$$

The path of a particle may usually be well represented by[5]

$$t = \frac{9\mu R_1^2}{(n+1)\rho d_p^2 u_{T_1}^2}\left[\left(\frac{R}{R_1}\right)^{2n+2} - 1\right] \tag{4}$$

This expresses the time elapsed for a particle, initially at radius R_1 having tangential velocity u_{T_1}, to reach radius R. It results from Eq. (1) by replacing (gm) with the centrifugal force, assuming the radial drag force to be expressed by Stokes' law, and by taking the radial drift velocity of the particles to be constant, that is, $du_s/dt = 0$. Both of these assumptions may be somewhat in error under unusual circumstances: large particles, and/or very short times. Equation (4) deals with the horizontal motion only. Vertical motion is governed by the terminal settling velocity, as explained in Section 2.1.1 above.

3.2.2 Cyclone Dimensions

Cyclone dimensions are determined by a selection of the configuration ratios (see Section 3.1.2 and Table 1) coupled with a selection of the desired gas velocity at the inlet. For high collection efficiency the inlet velocity should be as large as possible without causing excessive rebounding or reentrainment of particles and without exceeding the saltation velocity. This limitation on u_{T_1} may be estimated according to Kalen and Zenz[6] by

$$\text{max } u_{T_1} = 22.6\frac{g\mu\rho_p(K_b)^{1.2}D^{0.201}}{\rho_f^2(1 - K_b)}$$

This is an empirical equation in which the units of feet, pounds mass, and seconds must be used. For particles of density in the range of 1–2.5 g/cm³, this gives working values of u_{T_1} in the commonly accepted range of 15–30 m/sec for typical cyclones, $K_b = 0.2$.

From the given gas volumetric flow rate Q and the selected inlet velocity, the dimensions of the cyclone are determined by the configuration ratios: $a = (QK_a/u_{T_1}K_b)^{1/2}$, $D = a/K_a b = K_b D$, $S = K_s D$, and so on.

3.2.3   Overall and Grade Efficiency

The overall efficiency of collection E may be calculated by Eq. (3), which requires a knowledge of the grade efficiency E_g as a function of the particle size. This may be determined by the model of Leith and Licht.[7]

This model is based on Eq. (4), coupled with the concepts of radial back-mixing of uncollected particles, and the definition of an effective time of collection. The resulting equation is of the form

$$E_g = 1 - \exp(-Md_p^N) \tag{5}$$

in which

$$M = 2\left[\frac{KQ}{D^3}\frac{\rho_p^{n+1}}{18\mu}\right]^{N/2} \quad \text{and} \quad N = \frac{1}{n+1}$$

A 50% cut diameter may be found from

$$d_{p50} = \left(\frac{0.6931}{M}\right)^{n+1}$$

This is equivalent to stating that E_g depends on a Stokes number and a geometric configuration parameter K for the cyclone. The configuration parameter is calculated from the dimension ratios only, and is independent of the size of the cyclone. Values of it for some standard designs are given in Table 1. For other configurations, the calculation is explained by Licht.[8]

This model is satisfactory for cyclones of common sizes (say $D > 8$ in.) and over a fairly extended range of operating temperatures. It does not apply to very small cyclones, such as ones used in respirable dust sampling, nor to operation under high pressure. There is some indication that here E_g depends on a Reynolds number for the cyclone in addition to the Stokes number involved in M.

The model gives a conservative estimate of performance such as would be obtained on fairly dilute particle/gas streams, that is, at low "grain loading," say below 10 g/m³. The limited data available indicate that efficiency improves as grain loading increases. The improvement may be estimated roughly by[9] $(100 - E_1)/(100 - E_2) = (C_{o1}/C_{o2})^{0.182}$.

For the assumed operating conditions and configuration, E_g may be calculated from Eq. (5) and combined with inlet particle-size distribution data to evaluate E by using Eq. (3). This may conveniently be done graphically, as illustrated in Figure 5. If the size distribution happens to follow the log-normal form, the integration may be done numerically. Charts are available[10] which give the value of E as a function of N, d_{p50}, and the mass median size and geometric standard deviation of the distribution.

The amount and size distribution of the uncollected emissions from the cyclone may be estimated by material balance, using the values of E_g and E obtained above. If the emission is unacceptable, the procedure may need to be repeated using a different set of assumed conditions or dimensions. The use of two or more cyclones in parallel or in series may also be considered by applying the procedure to each single cyclone individually.

3.2.4   Pressure Drop and Power Requirements

The pressure drop across a cyclone may be estimated by the method of Shepherd and Lapple.[11]

$$\Delta P = 5.12\, \rho_g u_{T_1}^2 N_H \quad \text{in cm H}_2\text{O, gauge} \tag{6}$$

where

$$\rho_g = \text{density of gas–particle stream, g/cm}^3$$
$$u_{T_1} = \text{inlet velocity, m/sec}$$
$$N_H = \text{number of inlet velocity heads}$$

N_H depends on the configuration and may be estimated by $N_H = 16K_aK_b/K_e^2$. More complex equations for N_H are available[12] but none seems to be generally superior to this one. Acceptable levels of ΔP for cyclone operation are generally less than 20 cm H₂O gauge.

The power required to be supplied by the gas moving system (fan or blower) will be given by

$$\text{power} = 0.0981 Q \cdot \Delta P \cdot \text{W} \tag{7}$$

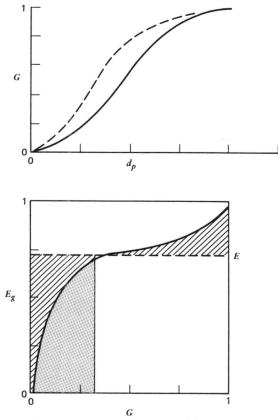

Fig. 5 Graphical determination of overall efficiency and outlet particle size distribution. Reprinted by permission of Marcel Dekker, Inc.

where

$$Q = \text{flow rate, m}^3/\text{sec}$$
$$\Delta P = \text{pressure drop, cm H}_2\text{0 gauge}$$

3.2.5 *Optimizing the Design*

In order to optimize the design, a parameter may be constructed which is proportional to the benefit/cost ratio for a given cyclone configuration. The "benefit" or accomplishment of collection may be taken to be proportional to the configuration parameter K as used in Eq. (5) and given in Table 1. Note that this is inversely related to the "cut diameter." Cost is made up mainly of two parts: (a) initial cost of construction, which is essentially proportional to the total surface area of the material of construction used, and (b) cost of operation, which is proportional to power consumption. Power consumption also reflects, roughly, the initial cost of the gas moving system.

A useful optimization parameter (OP) may therefore be conceived as

$$\text{OP} \propto K/\text{total surface area} \times \text{power}$$

Total surface area may be calculated by using $(\pi D^2 \text{Surf})$, where "Surf" is again a function of configuration only. Values are given in Table 1 for the standard configurations, and methods of calculation for other shapes are given by Licht.[13] Power is calculated by combining Eq. (6) with Eq. (7). For a given set of operating conditions then, let

$$\text{OP} = K/(D^2 \text{Surf})(Q u_{T_1}^2 N_H)$$

dropping the various constants. The group $(D^2 Q u_{T_1}^2)$ may be replaced by $Q^3/K_a K_b$, resulting in

$$OP = \frac{K \cdot K_a K_b}{N_H \cdot \text{Surf} \cdot Q^3}$$

For a given gas flow, OP depends only on the configuration parameters, K, K_a, K_b, N_H, and Surf. Using these as given in Table 1, values of OP are as follows:

Configuration	OP
Stairmand	2.35
Swift (high)	1.96
Lapple	1.66
Swift (general)	1.64
Peterson and Whitby	1.67

3.2.6 Cost Estimation

The initial cost of a cyclone collector may be estimated by the method of Neveril, Price, and Engdahl.[14] The total cost is the sum of that for (1) the cyclone body, (2) the cyclone hopper, and (3) the support. These costs are estimated individually, by correlation with the inlet area ($A = K_a K_b D^2$ ft^2) and materials of construction. The following relations are given:

1. Cost of cyclone body, for 1 ft$^2 < A < 14$ ft^2:
   Carbon steel:

$\frac{3}{16}$-in. plate:	Cost $ = 1140 + 570A$
10-gauge plate:	$ = 1026 + 433A$
14-gauge plate:	$ = 684 + 342A$

   Stainless steel:

$\frac{3}{16}$-in. plate:	$ = 1710 + 1767A$
10-gauge plate:	$ = 1482 + 1094A$
14-gauge plate:	$ = 912 + 741A$

2. Cost of dust hoppers:
   Carbon steel:

$\frac{3}{16}$-in. plate:	$ = 160 + 92A$
10-gauge plate:	$ = 148 + 77A$
14-gauge plate:	$ = 137 + 68A$

   Stainless steel:

$\frac{3}{16}$-in. plate:	$ = 228 + 296A$
10-gauge plate:	$ = 228 + 195A$
14-gauge plate:	$ = 182 + 141A$

3. Cost of support:

$ = 570 + 148A$	$1 < A < 2$ ft^2
$ = 1026 + 143A$	$2 < A < 6$ ft^2
$ = 1938 + 120A$	$6 < A < 14$ ft^2

All costs are given in 1977 dollars. They may be scaled up to current dollars by the use of the Equipment Cost Index published by *Chemical Engineering*.

3.3 Cyclone Performance

The overall collection efficiency of an operating cyclone may be determined rather easily by measuring:

1. The total cyclone catch collected and weighed over a period of time, expressed as kg/min.
2. The total grain loading, or particulate concentration in the inlet stream, expressed as kg/m^3.
3. The gas flow rate entering, expressed as m^3/min.

Then $E = $ total catch/grain loading $\times$ flow rate. The grain loading may be determined by withdrawing an isokinetic sample of the feed stream from a point well upstream of the cyclone inlet, collecting all of the suspended particulate on a filter, and metering the sampling gas flow rate. The gas flow rate may be found by a Pitot tube traverse at the same location as the sampling point. Care should be taken to select a point located in a straight run of duct preferably 8 or 10 duct diameters either upstream or downstream away from any bend or other flow disturbance.

The grade efficiency is more difficult to determine. In addition to the measurements listed above, isokinetic samples of particulate taken both upstream and downstream of the cyclone must be subjected to determination of particle size distribution. This is best done in situ by running each sampling stream directly into a cascade impactor of at least six stages. Both sample streams should be withdrawn at as nearly the same time as possible. A less desirable alternative is to collect the total particulate at each point on a filter and take the composite to the laboratory for size distribution analysis by a Bahco or a Coulter counter. Care must be taken in selecting the downstream sampling point location to avoid distortion due to the swirling flow in the exit gas stream caused by the vortex in the cyclone. If necessary straightening vanes may be inserted in the exit duct just ahead of the sampling point.

Experience has shown that the measured overall collection efficiency will be likely to agree quite well with that predicted by employing the design procedure outlined above. It is likely that the measured grade efficiency will not agree nearly as well with the predictive model. The difficulties inherent in sampling and in particle size determination are considerable. In any case, a particle size determination on the stream to be treated must be obtained before the design can be carried out.

REFERENCES

1. Fuchs, N. A., *The Mechanical of Aerosols*, Pergamon Press, New York, 1964, p. 33.
2. Licht, W., *Air Pollution Control Engineering*, Dekker, New York, 1980, p. 140.
3. TerLinden, A. J., *Tonindustrie-Zeitung* 22(iii), 49 (1953).
4. Alexander, R. McK., *Proc. Austral. Inst. Min. and Met.* (*N.S.*), **152/3,** 202 (1949).
5. Licht, W., *Air Pollution Control Engineering*, Dekker, New York, 1980, p. 177.
6. Kalen, R., and Zenz, F. A., *AIChE Symposium Ser.* **70**(137), 388 (1974).
7. Leith, D., and Licht, W., *AIChE Sympos. Ser.* **68**(126), 196 (1972).
8. Licht, W., *Air Pollution Control Engineering*, Dekker, New York, 1980, p. 243.
9. Baxter, W. A., in *Air Pollution Control—Vol. IV*, 3rd edition (A. Stern, ed.), Academic Press, New York, 1977, Chapter 3, p. 124.
10. Leith, D., *Handbook of Pollution Abatement Engineering*, Humana Press, Clifton, NJ, 1978.
11. Shepherd, G. B., and Lapple, G. E., *Ind. Eng. Chem.* **31,** 972 (1939).
12. Strauss, W., *Industrial Gas Cleaning*, 3rd edition, Pergamon Press, Oxford, 1975, p. 263.
13. Licht, W., loc. cit., p. 250.
14. Neveril, R. B., Price, J. V., and Engdahl, K. L., *J. Air Poll. Contr. Assoc.* **28,** 963 (1978).

CHAPTER 14

CONTROL COSTS

WILLIAM M. VATAVUK

U.S. Environmental Protection Agency
Research Triangle Park, North Carolina

1 INTRODUCTION

Although the technical parameters influencing the selection and design of a given type of air pollution control system may be unique, cost is the only parameter relevant to *all* systems. This is especially so where different types of systems can control a source to achieve an emission limit. Here, cost is the parameter used to select the optimum system from the alternatives available.

Cost is also an important factor in the setting of many state and federal regulations. An example of the latter are New Source Performance Standards (NSPS), established under the authority of Section 111 of the 1977 Clean Air Act Amendments. An NSPS is to[1]

> . . . *reflect the degree of emission reduction achievable through the application of the best system of continuous emission reduction which* (taking into consideration the cost of achieving such emission reduction . . .) *the Administrator determines has been adequately demonstrated for that category of sources.* (Emphasis added.)

In addition, cost may influence how a facility (or an enforcement agency) determines whether a regulation is being complied with. For instance, compliance with an opacity limit could be measured by visual observation (e.g., via EPA Method 9) or by using an optical transmissometer. The costs of these and other compliance methods may be considerably different.

There are surely many more situations where cost plays an important part in air pollution control decision making. This chapter will focus on one of the most important aspects of this growing specialty, the conceptual design and costing of particulate and gaseous pollutant "add-on" control systems. (Because the costs of fugitive emission controls, process modifications, fuel switching, and other control methods are highly dependent on the locations of the sources being controlled, they will not be covered here.)

This chapter will be divided into five parts:

1. *Methodology.* This will discuss, and present examples of, the technical and cost parameters that comprise a control system cost estimate. It will also present a simple, practical method for combining these parameters with raw cost data to develop total installed and annualized costs of the systems.

2. *Auxiliary Equipment Costs.* Here, methods for sizing and costing control system auxiliaries (such as ductwork) will be presented.

3. *Particulate Control Devices.* Sizing and costing of the most commonly used particulate controls—electrostatic precipitators, wet scrubbers, and fabric filters—will be the subject of this section.

4. *Gaseous Control Devices.* This section will deal with the gaseous (primarily volatile organic compounds) controls: thermal and catalytic incinerators, flares, gas absorbers, and carbon adsorbers. (Specialized combinations of one or more gaseous or particulate control devices, such as flue gas desulfurization systems, will not be covered. However, there are many excellent sources of cost data for these systems, such as Laseke et al.[2])

5. *Cost Escalation.* Escalating historic costs to present or future times by means of well-tabulated indices will comprise this section. Though most of the costs presented in this chapter are in current (June 1981) dollars, many of the costs available in the literature and elsewhere reflect different periods.

In closing, a few points need to be emphasized. First, most of the costs in this chapter have been presented in equational rather than graphical form. There are three reasons for this: (1) equations are more convenient to use and, if desired, computerize; (2) graphs—especially logarithmic plots— are often difficult to read and interpolate; and (3) equational cost data are much easier to escalate.

Second, because the subject of this chapter is control *costs,* the material on control system design is limited to the computation of device critical parameters and the use of these parameters in cost estimation. For information on designing specific control devices, refer to appropriate chapters in this handbook and to other references (e.g., Perry's *Chemical Engineers' Handbook).*

Finally, only *British* units are used in this chapter. This is done because cost data found in the literature have been correlated against English units, rather than SI units. This in turn reflects the widespread preference for English units in the cost engineering profession.

2   METHODOLOGY

Even before considering the parameters that influence cost estimates, we must decide how accurate our cost estimates will be. Naturally, the time and resources for making an estimate increase with the accuracy required. But for most purposes, a *study* estimate (generally accurate to within ±30%) is more than adequate. Study estimates are a reasonable compromise between *order-of-magnitude* estimates (±50%) and *firm* and *definitive* estimates (±5 to 10%, respectively). The former are usually too imprecise for serious consideration, while the latter require knowledge of site-specific variables that would likely be unavailable to corporate or governmental air pollution control engineers.

Costs will be discussed later in more detail, but for now a few basic definitions are in order. First, two kinds of costs are generally used, *installed capital* and *annualized.* Installed capital (or just capital) cost is the investment cost of the air pollution control system and includes such items as the equipment, the direct and indirect costs of installing same, and land, working capital, and other such "nondepreciable" items. Total annualized cost is defined as follows:

$$\text{TAC} = D + I - R \tag{1}$$

where

$$\text{TAC} = \text{total annualized cost, \$/yr}$$
$$D = \text{direct operating costs, \$/yr}$$
$$I = \text{indirect operating costs, \$/yr}$$
$$R = \text{recovery credits, \$/yr}$$

These costs and the items that comprise them will be discussed later. However, for now recognize that the total annualized cost is a parameter that may be used to make comparisons among competing control systems. It accounts for both the investment and the expenses necessary to operate and maintain the control system.

Note also that the TAC method is only one of several ways to present and compare control system costs. Others include the *net present worth* (or *discounted cash flow)* and *internal rate of return* analysis methods. Each of these would yield equally valid results. Nonetheless, the TAC method is preferred because of its simplicity and flexibility. It is especially useful when comparing alternative systems that have different useful lives, and where tax and inflation considerations are not a part of the analysis. The other two methods lack this versatility.

2.1   Sizing Parameters

Even before a system is sized and costed, the source to be controlled must be thoroughly studied. This involves: measuring or estimating the quantity, temperature, and composition of the emission stream(s); and determining the capacity and usage rate of the facility where the emission source is located, and, of course, the emission level to be achieved—mass rate, concentration limit, and so on. These data and other inputs are used to formulate the *regulatory alternatives,* which may include anything from "no control" to shutting down the facility. (See Table 1.)

In most cases, an add-on system is the alternative selected for controlling the source. Regardless of the number of add-ons technically capable of controlling a source, the procedure for determining their costs is the same.

First, each system must be "sized." This procedure involves combining the facility parameters in Table 1 with the systems factors listed in Table 2 to yield the equipment size measures that are inputs to the cost equations and curves. Note that most system factors are "intensive," in that their values are not so much influenced by the size of the emission source being controlled, but by its

Table 1   Facility Parameters and Regulatory Options [a]

Facility Parameters	Regulatory Alternatives
Intensive,	No control
Facility status	"Add-on" devices,
(new or existing, location)	Emission levels
Gas characteristics	Opacity levels
(temperature, pressure)	Process modifications,
Pollutant concentration and	Raw material changes
particle size distribution	Fuel switching
Extensive,	Others,
Facility capacity	Coal desulfurization
Facility use rate	Plant shutdown
Facility life	(total or partial)
Gas flow rate	

[a]Excerpted by special permission from *Chemical Engineering*, October 6, 1980. Copyright © 1980, by McGraw-Hill, Inc., New York, N.Y. 10020.

characteristics. (For instance, the choice of the system materials of construction would depend on, among other things, the corrosiveness of the gas stream being controlled.) These characteristics also influence the conceptual design of the control system, that is, the method used to capture the exhaust (hood or direct exhaust), the means for transporting the exhaust gases through the system, the nature of the equipment for precooling or precleaning the gases upstream of the control device, and the means for disposing, treating, or recycling the captured pollutants. Other considerations are whether on-site facilities, such as power generation equipment, are adequate to serve the control system.

These considerations must be addressed prior to sizing the equipment. However, regardless of the source being controlled, each system should contain, along with the device itself, the following auxiliary equipment:

Hood, or other means for exhaust capture.

Ductwork, to convey the exhaust to and from the control device.

Fan system (fan, motor, and starter), to move the exhaust through the system.

Stack, for dispersing the cleaned gases into the atmosphere.

(Methods for sizing and costing the auxiliaries and control devices will be presented in the following sections.)

Table 2   Examples of System Design Factors [a]

General	Basic Control Equipment
Material of construction:	Electrostatic precipitator,
stainless steel	Drift velocity: 16.5 ft/m
Insulated	Pressure drop: 1.0 in.
Economic life,	Fabric filter,
Electrostatic precipitator and	Air-to-cloth ratio:
fabric filter: 20 yr	7.5 to 1
Venturi scrubber: 10 yr	Pressure drop: 6.0 in.
Redundancy: none	Construction: suction
	Duty: continuous
	With dilution air port
	Venturi,
	Pressure drop: 60 in.
	Throat adjustment:
	automatic

[a]Excerpted by special permission from *Chemical Engineering*, October 6, 1980. Copyright © 1980, by McGraw-Hill, Inc., New York, N.Y. 10020.

The next type of parameter to consider is the control device *critical variable*. This is the quantity that, along with the gas volumetric flow rate, primarily determines the device size and cost. Typical values for some of these variables appear in Tables 3 and 4 for particulate and gaseous control devices, respectively. Note that, for instance, the fabric filter critical parameter, *air-to-cloth ratio*, varies not only from source to source, but among baghouse designs as well. This ratio, when divided into the baghouse volumetric flow rate, will yield the area of bags required to clean the emission stream. As Section 4.3 will show, bag area then determines the cost of the baghouse shell, as well as other features— insulation and materials, for example.

Analogously, electrostatic precipitator (ESP) *drift velocity* is the variable that, again with the gas flow rate, ultimately determines the required collecting area for the ESP. The ESP purchase cost and most of the power cost for operating it are functions of this area.

As Sections 4 and 5 will show, there are other variables that influence systems costs—some significantly, like materials of construction. However, the critical variables should be especially noted, not only because they heavily influence system costs, but because they are often correlated with cost data found in the literature and elsewhere.

2.2 Estimating Installed Costs

As noted in Section 2.1, installed capital costs are usually subdivided into *direct* and *indirect* components. Direct costs include the purchased costs of the equipment (usually F.O.B. the vendors) and related charges, such as taxes and freight. This category also includes the various installation labor and materials costs, along with costs for site preparation, buildings, and the like. Indirect installation costs cover those expenses not directly tied to the system, such as engineering and startup. Finally, there are occasionally costs for land and working capital which must be added in, but these are rarely large enough to consider when costing air pollution control systems.

There are several ways to estimate the installed cost. The most accurate method would be, naturally, to compute each equipment and installation cost individually, then to total them. Aside from being time consuming, this approach is not feasible for making study estimates, since most of the detailed, site-specific data are unavailable. At the other extreme, one could determine the cost by scaling it

Table 3   Control Equipment Parameters for Selected High-Efficiency Performance[a]

Process	Fabric Filter Air-to-Cloth Ratio			Venturi Scrubber ΔP, in. H_2O	Precipitator Drift Velocity, ft/sec
	Reverse Air	Pulse Jet	Mechanical Shaker		
Basic oxygen furnaces	1.5–2.0	6–8	2.5–3.0	40–60	0.15–0.25
Brick manufacturing	1.5–2.0	9–10	2.5–3.2	3–35	
Castable refractories	1.5–2.0	8–10	2.5–3.0		
Clay refractories	1.5–2.0	8–10	2.5–3.2	11	
Coal-fired boilers				15	0.22–0.35
Detergent manufacturing	1.2–0.15	5–6	2.0–2.5	10–40	
Electric arc furnaces	1.5–2.0	6–8	2.5–3.0		0.12–0.16
Feed mills		10–15	3.5–5.0		
Ferroalloy plants	2.0	9	2.0	40–80	
Glass manufacturing	1.5			65	0.14
Gray iron foundries	1.5–2.0	7–8	2.5–3.0	25–60	0.1–0.12
Iron and steel (sintering)	1.5–2.0	7–8	2.5–3.0		0.2–0.35
Kraft recovery furnaces				15–30	0.2–0.3
Lime kilns	1.5–2.0	8–9	2.5–3.0	12–40	0.17–0.25
Municipal incinerators					0.2–0.33
Petroleum catalytic cracking				40	0.12–0.18
Phosphate fertilizer	1.8–2.0	8–9	3.0–3.5	15–30	
Phosphate rock crushing		5–10	3.0–3.5	10–20	0.35
Polyvinyl chloride production		7			
Secondary aluminum smelters		6–8	2.0	30	
Secondary copper smelters		6–8			0.12–0.14

[a]Excerpted by special permission from *Chemical Engineering,* October 6, 1980. Copyright © 1980, by McGraw-Hill, Inc., New York, N.Y. 10020.

Table 4 Efficiency of Carbon Adsorption for Selected Solvents and Specified Operating Conditions[a]

Solvent	Average Inlet Concentration, ppm	Acceptable Ceiling Concentration, ppm	Lower Explosive Limit, % by Volume in Air	Carbon Adsorption Efficiency, lb Solvent/100 lb Carbon
Acetone	1000		2.15	8
Benzene	10	25	1.4	6
n-Butyl acetate	150		1.7	8
n-Butyl alcohol	100		1.7	8
Carbon tetrachloride	10	25	None	10
Cyclohexane	300		1.31	6
Ethyl acetate	400		2.2	8
Ethyl alcohol	1000		3.3	8
Heptane	500		1	6
Hexane	500		1.3	6
Isobutyl alcohol	100		1.68	8
Isopropyl acetate	250		2.18	8
Isopropyl alcohol	400		2.5	8
Methyl acetate	200		4.1	7
Methyl alcohol	200		6.0	7
Methylene chloride	500	1000	None	10
Methyl ethyl ketone	200		1.81	8
Methyl isobutyl ketone	100		1.4	7
Perchloroethylene	100	200	None	20
Toluene	200	300	1.27	7
Trichloroethylene	100	200	None	15
Trichlorotrifluoroethane	1000		None	8
Xylene	100		1.0	10

[a]Excerpted by special permission from *Chemical Engineering*, October 6, 1980. Copyright © 1980, by McGraw-Hill, Inc., New York, N.Y. 10020.

from the cost of another similar system of a different size. This is usually done via a power function such as

$$TIC_2 = TIC_1 \left(\frac{S_2}{S_1}\right)^n \qquad (2)$$

where

TIC = total installed cost of system 1 or 2
S = system size (e.g., as measured by the exhaust flow rate)
n = some exponent (usually between 0 and 1)

The first method is overly precise; the second, not precise enough. But in the study estimate method, the direct and indirect installation costs are computed by multiplying the total of the control device and auxiliaries purchased costs by individual factors, the values of which can be varied to suit the particular source being controlled. This method combines precision with simplicity and ease of use. Its only drawback is that a lot of data have to be compiled from existing installations before the factors can be developed.

Fortunately, this has already been done. Table 5 displays average installation cost factors for add-on equipment and illustrates how they are used. Note that the installation factors are to be applied against the base price X, not just the purchased equipment costs A and B. Note also that no factors are given for *site preparation* and *facilities and buildings*, since these costs rarely depend on the base price. However, Godfrey et al.[3] and other references provide such data in detail.

Vatavuk and Neveril[4] also list adjustment factors to apply against those in Table 5. These adjustments would account for installation conditions different from the "average" and would, it is hoped, yield more representative estimates. But again, the estimator rarely has enough information at hand to take full advantage of these adjustments.

Table 5 Average Cost Factors for Selected Air Pollution Control Equipment[a]

Direct Costs	Electrostatic Precipitator	Venturi Scrubber	Fabric Filter	Thermal and Catalytic Incinerator	Carbon Adsorber	Gas Absorber	Flare
Purchased equipment:							
Primary	A						
Auxiliary	B						
Instruments and controls	$0.10(A+B)$						
Taxes	$0.03(A+B)$	Same for all equipment					
Freight	$0.05(A+B)$						
Base price	X (total of above)						
Installation:							
Foundations and supports	0.04	0.06	0.04	0.08	0.08	0.12	0.12
Handling and erection	0.50	0.40	0.50	0.14	0.14	0.40	0.40
Electrical	0.08	0.01	0.08	0.04	0.04	0.01	0.01
Piping	0.01	0.05	0.01	0.02	0.02	0.30	0.02
Insulation	0.02	0.03	0.07	0.01	0.01	0.01	0.01
Painting	0.02	0.01	0.02	0.01	0.01	0.01	0.01
Site preparation[b]				As required			
Facilities and buildings[b]				As required			
Multiplier for direct-installation costs	0.67	0.56	0.72	0.30	0.30	0.85	0.57

Indirect Costs

Installation:							
Engineering and supervision	0.20	0.10	0.10	0.10	0.10	0.10	0.10
Construction and field	0.20	0.10	0.20	0.05	0.05	0.10	0.10
Construction fee	0.10	0.10	0.10	0.10	0.10	0.10	0.10
Startup	0.01	0.01	0.01	0.02	0.02	0.01	0.01
Performance test	0.01	0.01	0.01	0.01	0.01	0.01	0.01
Model study	0.02	—	—	—	—	—	—
Contingencies	0.03	0.03	0.03	0.03	0.03	0.03	0.03
Multiplier for indirect-installation costs	0.57	0.35	0.45	0.31	0.31	0.35	0.35

[a]Excerpted by special permission from *Chemical Engineering*, November 3, 1980. Copyright © 1980, by McGraw-Hill, Inc., New York, N.Y. 10020.
[b]Costs for site preparation and facilities and buildings are available in Ref. 3.

Another situation that influences installed costs is the "retrofit," wherein a control system is sized for, and installed in, an existing facility, rather than in a new one. Because the factors in Table 5 reflect new, "grass-roots" installations, some would have to be adjusted to suit the retrofit conditions. However, unless the size and number of auxiliaries were different in the retrofit situation, the base cost of the control system would *not* vary from the grass-roots installation. Probably the most common example of deviation in auxiliary equipment costs is attributable to the ductwork, for in many retrofits exceptionally long duct runs are necessary to tie the control system into the existing process.

Because each retrofit situation is different, no generalized factors can be developed. Nonetheless, some guidance can be given concerning the kinds of system modifications one might expect in a retrofit:

1. *Auxiliaries.* The biggest component to consider here is again the ductwork cost. In addition to requiring very long duct runs, some retrofits require extra tees, elbows, and other expensive fittings.

2. *Handling and Erection.* Again, because of a "tight fit," special care may need to be taken when unloading, transporting, and placing the equipment.

3. *Piping, Insulation, and Painting.* These costs could increase significantly in a retrofit situation. Like ductwork, large amounts of piping may be needed to "tie in" the control device; of course, the more piping and ductwork required, the more insulation and painting needed.

4. *Site Preparation.* Unlike the other categories, this cost may actually decrease, for most of this work would have been done when the original facility was built.

5. *Facilities and Buildings.* Potentially, retrofit costs here could be the largest for this category, especially so the facilities. For example, if the control system requires a great deal of power for operation (e.g., a venturi scrubber), the facility's power plant may not be able to service it. In such a case, the facility would have to either purchase additional electricity from a public utility, expand its power plant, or build another one. In any case, the cost of electricity supplied to that venturi would surely be higher than if the scrubber were installed in a new plant where adequate provision for its electrical needs could have been made.

7. *Engineering and Supervision.* Designing a control system to fit into an existing plant normally requires extra engineering, especially when the system is extraordinarily large, heavy, or utility consumptive. For the same reasons, extra supervision may be needed when the installation work is being done.

8. *Lost Production.* If the control system cannot be tied into the process during normally scheduled maintenance periods, part or all of the process may have to be shut down. The *net* (i.e., after-tax) value of the products not made during this time is a bonafide retrofit expense.

9. *Contingencies.* Due to the uncertain nature of retrofit estimates—even those based on "hard" data—the contingency or uncertainty factor should probably be increased by the estimator.

These points highlight qualitatively some of the costs that may increase in retrofit situations. However, there may be other cases where the retrofited installation cost would be *less* than the cost of installing the system in a new plant. This could occur, for instance, when one control system, say an ESP, is being replaced by another, a baghouse, for example. The ductwork, stack, and other auxiliaries for the ESP may be adequate for the new system, as perhaps would be the off-site facilities.

However, despite these seemingly random influences on retrofit installation costs, a rule-of-thumb can be given, namely, the *installed* cost of a retrofited system would be between 25 to 50% more than the cost of that system installed in a new facility. This rule-of-thumb should only be used when making very approximate, order-of-magnitude estimates. Using it with more precise costs may result in a deterioration of the estimates.

2.3 Estimating Annualized Costs

As mentioned previously, the total annualized cost is comprised of three components—direct and indirect operating costs and recovery credits. Unlike the installed cost components, which are usually "factored" from the base equipment cost, annualized cost items are computed from known data on the system size and operating mode, as well as from the facility and control device parameters.

Direct operating costs are of three kinds: *variable, semivariable,* and *fixed.* Variable costs include, for instance, utilities, the values of which are normally directly proportional to the quantity of exhaust gas the device processes per unit time. If the device processes none (as would be the case during a shutdown), these utility costs would be zero.

Costs that depend on the gas flow rate, but only partly, are semivariable. Even if the device were not operating, some portions of these costs would still be incurred. For instance, even a dormant control system would have to be periodically inspected. The labor required for this, though not as high as the operating labor when the device is running, could be a significant expense.

Some cost estimators consider certain semivariable costs as fixed expenditures. That is, their values are completely independent of the level of operation. The best example of this is maintenance which, instead of being figured directly on an hourly basis, is computed as a percentage of the system installed cost. This fixed approach, though certainly simpler to employ, introduces some inaccuracy, for it tends to overstate the maintenance costs for systems with higher capital costs. System complexities, numbers of moving parts, and other factors are usually better indicators of maintenance needs than system investment.

Except for overhead, all of the indirect operating costs are fixed, since they are computed as percentages of the investment. (Though an indirect, overhead is usually factored from operating and maintenance labor. In this chapter, however, *overhead is included in the unit labor rates.*)

The following guidelines may be used when estimating direct and indirect operating costs.

2.3.1 Operating Labor

The amount of labor required for a system depends on its size, complexity, level of automation, and operating mode (i.e., batch or continuous). The labor is usually figured on an hours-per-shift basis. Table 6 gives typical values for operating labor for the control systems discussed in this chapter. These values apply to relatively large, automated, continuously operated systems. Note that the ranges given for fabric filters, precipitators, and scrubbers generally reflect corresponding ranges in capacity. As a rule, though, data showing explicit correlations between the labor requirement and capacity are hard to obtain. But when they are, the relation is usually logarithmic, or

$$\frac{L_2}{L_1} = \left(\frac{C_2}{C_1}\right)^n \tag{3}$$

where

L_1, L_2 = labor requirements for systems 1 and 2
C_1, C_2 = capacities of systems 1 and 2 (as measured by the gas flow rate, for instance)
$n = 0.2$ to 0.25 (usually)[5]

This equation can be used to estimate labor needs when data are available for only one system capacity.

A certain amount must be added to the operating labor to cover supervisory requirements. A figure of 15% of the base labor requirement is considered reasonable.[4]

To obtain the total labor cost, merely multiply the operating and supervisory labor requirements by the respective wage rates (in \$/hr) and the number of hours per year the system is in operation (i.e., the operating factor).

Table 7 contains typical labor rates and other unit operating costs. Note that the labor rates include payroll and plant overhead.

2.3.2 Maintenance

Maintenance labor is computed in the same way as operating labor, and is influenced by the same variables. However, the labor rate is normally higher than that for operation. A 10% premium is typical.[4] (See Table 7.)

In addition, there are expenses for such maintenance materials as oil, other lubricants, duct tape, solder, and a host of small tools. These can be figured item by item, but since they are normally small expenditures, it is just as accurate to factor them from the maintenance labor. Reference 4 suggests a factor of 100%.

Table 6 Estimated Labor Hours per Shift[a]

Primary Equipment	Operation	Maintenance
Fabric filters	2 to 4	1 to 2
Precipitators	½ to 2	½ to 1
Scrubbers	2 to 8	1 to 2
Incinerators	½	½
Absorbers	½	½
Adsorbers	½	½
Refrigerators	½	½
Flares	—	½

[a]Excerpted by special permission from *Chemical Engineering*, November 3, 1980. Copyright © 1980, by McGraw-Hill, Inc., New York, N.Y. 10020.

Table 7 Typical Operating Costs for Air Pollution Control Systems[a]

Category	Cost
Direct Operating Costs	
1. Operating labor	
a. Operator	$11/hr[b]
b. Supervisor	15% of 1(a)
2. Operating materials	As required
3. Maintenance	
a. Labor	$13/hr[b]
b. Material	100% of 3(a)
4. Replacement materials	
a. Carbon	See Section 5.5
b. Catalyst	See Section 5.2
c. Filter bags	See Section 4.3
5. Utilities	
a. Electricity	$0.049/kWh
b. Fuel oil	$1.00/gal
c. Natural gas	$2.50/$10^6$ Btu
d. Cooling water	$0.10/$10^3$ gal
e. Plant water	$0.25/$10^3$ gal
f. Steam	$6.50/$10^3$ lb
g. Compressed air	$0.025/$10^3$ lb
6. Waste treatment and disposal	
a. Liquid	$1.20/$10^3$ gal
b. Solid	$25/ton
Indirect Operating Costs	
7. Overhead	60% of [1(a) + 1(b) + 3(a)]
8. Property tax	1% of capital cost
9. Insurance	1% of capital cost
10. Administration	2% of capital cost
11. Interest rate ("opportunity cost")	10%/yr

[a]Reference date, June 1980.
[b]Labor cost includes overhead. However, if desired, base labor rate can
be used and overhead computed separately, as per item 7 above.

For small- to medium-sized systems, maintenance may be calculated as simply 5% of the system installed cost—provided that the investment is less than $100,000.

2.3.3 *Replacement Parts*

This cost is computed separately from maintenance, since it can represent a substantial expenditure. Included in this category are such items as carbon (for carbon adsorbers), bags (for fabric filters), and firebrick liners (for thermal incinerators), along with the labor for installing them.

The annual cost of the replacement materials is a function of the materials cost, the life of the materials, and the interest rate (or *opportunity cost*), as in the following expression:

$$A_P = P \left[\frac{i(1 + i)^n}{(1 + i)^n - 1} \right] \tag{4}$$

where

A_P = annualized cost of replacement parts, $/yr
P = initial cost of the replacement parts, $
n = useful life of the parts, yr
i = interest rate, fraction

In this report, replacement parts are treated no differently from capital costs. (Section 2.3.8 will treat this annualization technique in more detail.)

The labor for replacement parts will vary, of course, depending on the amount of the material, its workability, the accessibility of the control device, and so forth. However, for study estimate purposes, it can be figured at *100% of A_P*, the annualized replacement parts cost.[4]

2.3.4   *Utilities*

As Table 7 shows, this cost category covers many different items, ranging from electricity to compressed air. Of these, only electricity is common to all control devices, whereas fuel oil and natural gas are generally specific to incinerators; water and water treatment, to scrubbers and absorbers; steam, to carbon adsorbers; and compressed air, to fabric filters.

Techniques for estimating utility costs for specific devices will be presented in Sections 4 and 5. However, because every system requires a fan to convey the exhaust gases, a general expression for computing fan horsepower has been developed:

$$\text{fan power (hp)} = \frac{0.746Q\,\Delta PSh}{6356\eta} \qquad (5)$$

where

Q = gas flow rate, acfm
ΔP = pressure drop, in. H_2O, gauge
S = specific gravity of gas
h = operating factor, hr/yr
η = combined fan and motor efficiency (fraction)

A similar expression can be developed for computing *pump* horsepower.

2.3.5   *Waste Disposal*

When the solid material captured by a control system can be neither recycled to the process being controlled nor sold, it must be disposed of. In most cases, this means hauling the material to a landfill, pit, or other depository. In such cases, the disposal costs would be merely the hauling costs (i.e., amortization of the trucks, drivers' wages, gasoline, etc.) plus the landfill charges. The hauling costs depend on the distance from the control device to the disposal site, the quantity of dust hauled, and other variables. Thus, most costs are computed on a $/ton-mile basis. For a typical hauling distance (5 miles), the cost would range from $10 to $75/ton-mile, based on an annual hauling load of 100–1000 tons, respectively. Since most loads are closer to the upper limit, the disposal cost would approach the lower figure—say, $25/ton-mile.

Normally, landfill charges are minimal, compared to the hauling costs. However, occasionally the waste being disposed is "hazardous," in the sense that its handling and disposal is regulated by the Resource Conservation and Recovery Act (RCRA). In such cases, the disposal costs would increase significantly, due to the complex record-keeping procedures that must be followed, the special containers required, and the like.

2.3.6   *Overhead*

This is probably the easiest cost to compute, yet the most difficult to comprehend. Much of the confusion surrounding overhead is attributable to the many different ways it is computed and to the many costs it includes, some of which appear (and often are) duplicative.

Generally, there are two categories of overhead, *payroll* and *plant*. Payroll overhead includes those expenses directly associated with operating and maintenance labor, such as: workmen's compensation, Social Security contributions, vacations, group insurance, and other fringe benefits. Though some of these are fixed, in that they must be paid regardless of how many hours per year an employee works, payroll overhead is traditionally computed as a percentage of the total annual labor cost.

Conversely, plant overhead accounts for those expenses not necessarily tied to the operation and maintenance of the control system. These would include: plant protection, employee amenities (cafeteria, locker rooms, etc.), plant lighting, parking areas, and landscaping. Some estimators compute plant overhead by taking a percentage of all labor plus replacement parts, while others factor it partly from the labor costs and partly from the total installed cost.[6] However, for most cost analyses it is sufficient to combine payroll and plant overhead into one figure and to compute it from the total of operating and maintenance labor. (See Table 7.) This is the approach preferred by Peters and Timmerhaus.[5] Consequently, the simplest way to handle overhead is to merely combine it with the individual base labor rates, as was done in Table 7.

2.3.7   *Property Taxes, Insurance, and Administration*

These three indirect operating costs are factored from the system installed (capital) cost, and typically comprise 1, 1, and 2% of it, respectively. Taxes and insurance are self-explanatory, while "administration" covers sales, research and development, accounting, and other home office expenses. For simplicity, the three items are usually lumped into a single 4% factor.

2.3.8 Capital Recovery

As discussed above, the annualization method used here is the *equivalent uniform annualized cost* method. The cornerstone of this method is the *capital recovery factor*, which, when multiplied by the system installed cost, yields a payment that accounts for the original investment and a return thereon, also known as an *opportunity cost.* That is, the capital recovery factor "spreads out" the initial investment over the life of the system, much like mortgage payments recover the principle and interest on the original note.

The capital recovery factor (CRF) is a function of the appropriate interest rate (i) and the system useful life (n, in yr):

$$CRF = \frac{i(1+i)^n}{(1+i)^n - 1} \qquad (6)$$

so that

$$CRC = CRF \cdot TIC \qquad (7)$$

where

$$CRC = \text{capital recovery charges, \$/yr}$$

Table 8 gives typical values for n, the control system life, as well as for selected replacement parts. In some cases (e.g., electrostatic precipitators) the ranges are rather significant, reflecting the wide variation in service severity.

Obtaining representative values for i, the interest rate, is a more difficult matter, however. The interest rate depends on the source of funds for the system, be it a bank loan, a stock or bond issue, or corporate reserves. (In some cases, it may be all three). The interest rate also depends on whether the cost analysis will be done in "real" or "nominal" terms—that is, with or without considering inflation—and whether the analysis is being done on a "before" or "after" (income) tax basis. Regardless of what value is selected for i, however, the same value must be used throughout. This is especially important when comparing the costs of alternative systems.

For the sake of simplicity, we can select an interest rate that would neglect income tax effects and would be essentially independent of inflation. Such a rate would be properly termed a *before-tax, real* rate of return. A commonly used value is 10%—a rate, incidentally, which the French government has formally recommended.[7]

Table 8 Guidelines for Equipment and Parts Life, yr [a]

Equipment	Type of Service		
	Low	Average	High
Electrostatic precipitators	5	20	40
Venturi scrubbers	5	10	20
Fabric filters	5	20	40
Thermal incinerators	5	10	20
Catalytic incinerators	5	10	20
Absorbers	5	10	20
Adsorbers	5	10	20
Refrigerators	5	10	20
Flares	5	15	20
Parts or Material			
Filter bags	0.3	1.5	5
Adsorbents	2	5	8
Catalyst	2	5	8
Refractories	1	5	10

[a] Excerpted by special permission from *Chemical Engineering,* November 3, 1980. Copyright © 1980, by McGraw-Hill, Inc., New York, N.Y. 10020.

3 COSTING AUXILIARY EQUIPMENT

As discussed in Section 2.1, each control device requires certain auxiliary equipment to move, contain, and, where necessary, condition the exhaust stream before it reaches the device. Because the kind of equipment needed depends primarily on the characteristics of the gas stream and the control device, designing auxiliaries is often more difficult than designing the control device itself. And occasionally (such as building evacuation systems used to control particulate from furnace operations), the cost of such auxiliaries may far exceed the cost of the primary control device.

Since the list of these auxiliaries is so long, this discussion will be limited to the most commonly used equipment: hoods, ductwork, spray chambers and mechanical collectors, fan and pump systems, screw conveyors, and stacks.

3.1 Capture Hoods

In general, two types of hoods are used to capture emissions at the source: canopy or semiclosed. The former are usually mounted at some distance from the source and, for this reason, usually entail the entrainment of large amounts of air from the surrounding area. The additional induced air, in turn, drives up the cost of the control device and auxiliaries. Conversely, because semiclosed hoods are located close to the source, little additional air is drawn in. However, because they require more material to more fully enclose the source, semiclosed hoods cost more. Hence, in those instances where either type of hood may be used, there are tradeoffs between capital and operating costs.

The dimensions of a hood (and its costs) are functions of several variables. For *cold* processes, these variables are the perimeter of the source, the distance of the hood from the source, and the hood capture velocity, which typically ranges from 50 to 200 ft/min, depending on the stability of the surrounding atmosphere. However, sizing canopy hoods for *hot* processes is much more complex, since other variables enter in—temperature of the source, heat transfer rate from source to plume, and so on. Equations for obtaining dimensions for canopy hoods appear in Ref. 8, while similar information for semiclosed hoods appears in Ref. 9.

The purchase cost ($) of a hood ($C_H$) is the sum of the material cost (C_M) and the fabrication labor cost (C_F):

$$C_H = C_M + C_F \tag{8}$$

The labor cost is a function of the hood dimensions, and may be obtained by first consulting Figure 1 or Figure 2 (for rectangular and circular hoods, respectively) and then multiplying the value read by *1.427*, to escalate the cost to June 1981 dollars.

Estimating the hood materials cost requires more effort, since it depends on the hood area, the hood dimensions, and the weight of the fabricating materials.[8] For *circular hoods*, the materials cost equation is:

$$C_M = A(F + 2.34) \tag{9}$$

where:

A = hood metal area, ft^2
$F = 1.57 - 0.0622D + 7.75 \times 10^{-4}D^2$
D = hood diameter, ft

Similarly, for *rectangular hoods,*

$$C_M = LG + 2.34A \tag{10}$$

where

L = hood length, ft
$G = 17.1$, $/ft, for $L/W = 1$
    $22.8\,W/L$, $/ft, for $2 \leq L/W \leq 8$
W = hood width, ft

(Note: the ratio L/W is always an integer value.)

The areas (A) noted in the equations may be obtained from Figures 3 and 4, respectively, for rectangular and canopy hoods. Both equations apply to hoods with metal thicknesses $\leq$ 3/16 in., which covers the majority. For larger thicknesses, substitute "2.19" for "2.34" in the equations.

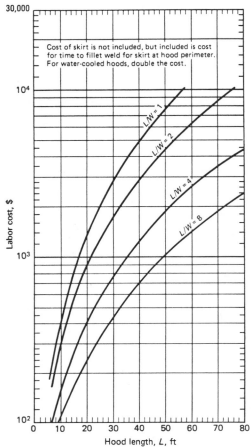

Fig. 1 Labor costs for fabricating 10-gauge carbon steel, 35° sloped rectangular canopy hoods. Excerpted by special permission from *Chemical Engineering,* December 1, 1980. Copyright © 1980, by McGraw-Hill, Inc., New York, NY 10020.

Finally, if hood skirts, booth walls, or other enclosures are needed, the material cost can be estimated by multiplying the unit price ($0.347/lb) by the product of the metal area (length × width × 1.2) and the unit weight (5.625 lb/ft²). For fabrication cost, use $0.428/lb.

3.2 Ductwork

The type of ductwork used in air pollution control systems primarily depends on the exhaust temperature, as follows: 1150°F or less, carbon steel; 1150 to 1500°F, stainless steel; above 1500°F, water-cooled or refractory-lined.[10] However, these temperatures apply to noncorrosive gases. Stainless steel can be used to convey corrosive exhausts, but only at lower temperatures.

The cost of a ductwork subsystem depends on four factors: diameter, thickness, length, and number and type of fittings and dampers used. The diameter is computed as follows:

$$D = 13.54 \left(\frac{Q}{v}\right)^{1/2} \tag{11}$$

where

$$D = \text{duct diameter, in.}$$
$$Q = \text{gas flow rate, acfm}$$
$$v = \text{duct velocity, ft/min}$$

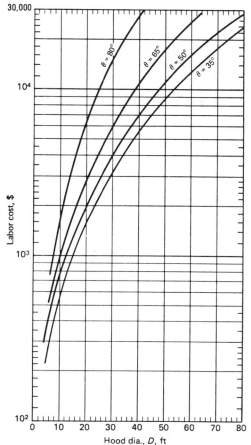

Fig. 2 Labor costs for fabricating 10-gauge carbon steel circular canopy hoods. Excerpted by special permission from *Chemical Engineering,* December 1, 1980. Copyright © 1980, by McGraw-Hill, Inc., New York, NY 10020.

The duct velocity, in turn, depends on the specific gravity (S) and diameter (d, μm) of the particles in the exhaust:

$$v = 105 \left(\frac{S}{S+1}\right) d^{0.4} \tag{12}$$

For most applications, v ranges from 2000 to 5000 ft/min. If the exhaust contains only gaseous pollutants, however, a velocity of 2000 ft/min is recommended.

The fabricated cost of straight ductwork varies linearly with diameter and parametrically with plate thickness, or

$$C_{SD} = a(t) + b(t)D \tag{13}$$

Here

$$C_{SD} = \text{straight duct cost, \$/ft}$$
$$a(t), \ b(t) = \text{parameters that vary with the duct thickness } t, \text{ in.}$$

Values for $a(t)$ and $b(t)$ in Eq. (13) are listed below for carbon and stainless steel duct, CS and SS, respectively, as functions of thickness t and diameter D. The right-hand column gives the diameters for which the equations are valid.

	CS		SS		D, in.	
t, in.	$a(t)$	$b(t)$	$a(t)$	$b(t)$	CS	SS
½	−5.51	4.30	−78.41	20.61	50–200	50–90
⅜	−4.18	3.40	−60.18	15.50	40–185	40–120
¼	−2.93	2.53	−38.30	10.76	30–150	30–150
³⁄₁₆	−2.51	1.95	−31.01	8.21	20–140	20–140
⅛	−2.08	1.56	−6.02	5.47	12–85	25–80

The prices from Eq. (13) include flanges every 40 ft of straight duct.
A similar approach is used to estimate the costs for fittings. First, for elbows

$$C_E = f(t) + g(t)D + h(t)D^2 \qquad (14)$$

where

$$C_E = \text{cost of an elbow, \$1000}$$
$$f(t),\ g(t),\ \text{and}\ h(t) = \text{parameters as above}$$

Parameter values for Eq. (14) and applicable ranges are

	SS			CS			D, in.	
t, in.	$f(t)$	$g(t)$	$h(t)$	$f(t)$	$g(t)$	$h(t)$	SS	CS
½	388	46.1	4.00	−23.4	35.3	0.685	50–180	50–180
⅜	284	33.8	3.00	−18.4	30.1	0.642	40–200	40–160
¼	191	27.5	2.08	−11.7	26.3	0.342	30–130	30–120
³⁄₁₆	177	25.1	1.50	−13.4	23.9	0.293	25–130	30–100
⅛	0	27.2	1.00	0	17.2	0.251	30–90	20–80

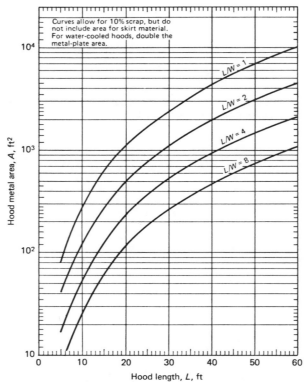

Fig. 3 Metal plate area for 35° sloped rectangular canopy hood. Excerpted by special permission from *Chemical Engineering*, December 1, 1980. Copyright © 1980, by McGraw-Hill, Inc., New York, NY 10020.

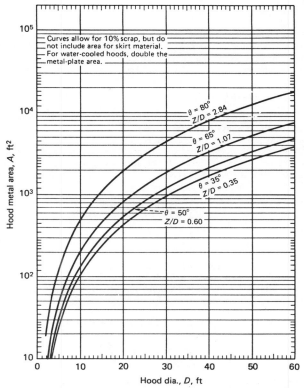

Fig. 4 Metal plate area for circular canopy hoods. Excerpted by special permission from *Chemical Engineering*, December 1, 1980. Copyright © 1980, by McGraw-Hill, Inc., New York, NY 10020.

For a given diameter and thickness, the cost of a *tee* is approximately one-third of the elbow cost. Based on its larger diameter, the cost of a *transition piece* would be about one-half the elbow cost.[10]

Obtain carbon steel *expansion joint* costs from this equation:

$$C_J = 63.32 + 1.64D \tag{15}$$

For a like diameter stainless steel expansion joint, multiply the above cost by 3.

The cost of dampers depends on their dimensions, extent of temperature control (manual or automatic), and design (rectangular or circular). For carbon steel *rectangular* dampers, the following equations provide the purchase cost:

$$C_{RD} = 0.714A, \text{ manual control} \tag{16}$$
$$= 1.848 + 2.714A - 0.1343A^2 \tag{17}$$

where

$$A = \text{damper cross-sectional area, 1000 in.}^2$$

These apply to dampers with length-to-width ratios of 1.0 to 1.3 and are valid for areas between 500 and 9500 in.²

Valid for diameters (D) between 20 and 130 in., these equations give purchase costs for carbon steel *circular* dampers:

$$C_{CD} = 2.079 - 0.0564D + 11.55D^2, \text{ manual} \tag{18}$$
$$= -3.383 + 0.2182 - 1.931 \times 10^{-4}D^2, \text{ automatic} \tag{19}$$

Finally, to obtain the purchase cost of a stainless steel damper of a given area or diameter, simply multiply the cost for one of carbon steel by 3.

3.3 Gas Conditioners

Gas conditioners cool, humidify, or clean the gas stream before it enters the primary collection device, thus enhancing its performance and, in turn, helping to minimize the cost of the entire control system. Among the more commonly used conditioners are cyclones, spray chambers, and quenchers. (Note: Other conditioning methods, dilution cooling and U-tube cooling, are discussed in Ref. 11.)

3.3.1 Cyclones

Also called *mechanical collectors,* cyclones serve to remove larger dust particles (20–30 μm and up), to ease the loading on the primary collector.

The primary costing parameter for cyclones is the inlet area (A, ft²), which is a function of the gas stream conditions, particle density, and the "critical" particle size—the size of the *smallest* particle which can be removed at 100% efficiency. If the size distribution of the particles in the exhaust is known, one can arbitrarily select a critical size (d_{cr}). From this, the inlet area can be computed, as follows:[12]

$$A = 3.34 \left[\frac{Q(\rho_p - \rho)}{\mu} \right]^{4/3} d_{cr}^{8/3} \tag{20}$$

where

$$d_{cr} = \text{critical particle size, } 10^{-6} \text{ m}$$
$$\mu = \text{gas viscosity, lb/ft-sec}$$
$$\rho_p, \rho = \text{densities of particles and gas, respectively, lb/ft}^3$$
$$Q = \text{gas flow rate, acfm}$$

Once A is known, the cyclone pressure drop (in. H_2O) can be determined:[12]

$$P = 2.36 \times 10^{-7} \left(\frac{Q}{A} \right)^2 \tag{21}$$

Both these equations hold for a "widely recommended" cyclone geometry and a "typical" number of gas stream revolutions in the unit. (For more details, see Ref. 12.)

The cost of a cyclone and its appurtenances is expressed here as a function of the inlet area. The first equation gives the cost of the cyclone, scroll, and dust hopper:

$$C_{C+S+H} = j(t) + k(t)A + l(t)A^2 \tag{22}$$

The coefficients depend on the plate thickness and materials of construction (carbon or stainless steel) and apply to inlet areas from 1 to 12 ft²:

	$j(t)$		$k(t)$		$l(t)$	
t	CS	SS	CS	SS	CS	SS
³⁄₁₆ in.	2230	3390	1210	3710	5.67	17.8
10 gauge	2030	3000	933	2380	4.15	16.6
14 gauge	1510	2000	765	1640	3.04	12.6

A set of equations provides costs for supports, which apply to *all* plate thicknesses and materials:

$$
\begin{aligned}
C_{Supp} &= 788 + 205A & (1 \leq A \leq 2) & \tag{23}\\
&= 1420 + 198A & (2 \leq A \leq 6) & \tag{24}\\
&= 2680 + 166A & (6 \leq A \leq 14) & \tag{25}
\end{aligned}
$$

Adding results from Eq. (22) to those from Eqs. (23), (24), or (25) yields the total cost for a single cyclone. Where multiple units (parallel or series) are used, the cost would be the sum of single unit costs, plus 20% for ductwork and structure.[11]

3.3.2 Spray Chambers and Quenchers

Spray chambers and quenchers also reduce the particulate loading, but their primary purpose is cooling (to reduce gas stream volume and system cost) and humidification (to enhance dust removal in primary collectors, electrostatic precipitators, especially). Both devices perform this function by introducing water into the gas stream, either by spray nozzles (spray chambers) or by outright flooding (quenchers). With spray chambers the amount of gas cooling (which occurs adiabatically) is related to the amount of water added, whereas if quenchers are used, the gas temperature is always reduced to the saturation point.

Costs for spray chambers and quenchers can be obtained from these equations:[11]

$$C_{SC} = 0.325Q + 59,500, \text{ spray chambers} \tag{26}$$
$$C_Q = 0.304Q + 11,100, \text{ quenchers} \tag{27}$$

where, as usual

$$Q = \text{gas flow rate, acfm}$$

The spray chamber cost function is based on a unit with an inlet gas velocity of 600 ft/min and a length/diameter ratio of 3/1. Included in both equations is the vessel cost (carbon steel) plus costs for support rings, platform, ladder, grating, spray system, and controls. Costs for refractory, pumps, or piping would be extra, however.

3.4   Fan and Pump Subsystems

Fans and pumps are used with control systems to transport gases and liquids, respectively. Both are equipped with certain accessories—motor, motor starter, and (if necessary) V-belt drives. In addition, fans usually require inlet and outlet dampers, to regulate gas flow rate and thereby optimize performance.

3.4.1   *Pumps*

Pumps are normally used with venturi scrubbers, spray chambers, gas absorbers, or any other device where liquid (usually water) is conveyed or circulated. One of the more commonly used designs is the vertical-turbine sump pump, which is normally designed to run at three speeds, 1170, 1750, and 3550 rpm. The optimum pump speed depends, in turn, on the capacity (in gpm):[13]

Flow Rate Range, gpm	Speed, rpm
0–1,000	3550
500–5,000	1750
2,000–10,000	1170

In addition to the capacity, costs for these pumps depend on the *head* (usually expressed in feet of water). The cost equations for these pumps are as follows:[13]

$$C_{P\,3550} = mG^n \quad \text{(for 3550 rpm)} \tag{28}$$

where

$$G = \text{pump capacity, gpm}$$
$$m, n = \text{coefficients that vary with the desired head, or}$$

Head, ft	m	n
100	614	0.155
200	756	0.142
300	949	0.120

Equation (28) applies to capacities between 100 and 1100 gpm. For 1750-rpm pumps,

$$C_{P\,1750} = p + qG \tag{29}$$

Coefficients p and q also depend on the head, or

Head, ft	p	q
100	2040	0.524
200	2550	0.545
300	3050	0.581

The applicable range here is 0–5100 gpm. Finally, two equations apply for 1170-rpm pumps:

$$C_{P\,1170} = rG^s \qquad \text{(for } 1500 \le G \le 6000\text{)} \tag{30}$$

and

$$C_{P\,1170} = t + uG + vG^2 \qquad \text{(for } 6000 \le G \le 10{,}000\text{)} \tag{31}$$

As above, the coefficients vary with the head:

	Flow Rate Range, gpm				
	$1500 \le G \le 6000$		$6000 \le G \le 10{,}000$		
Head, ft	r	s	t	u	v
100	171	0.424	22,600	−4.82	3.74×10^{-4}
200	617	0.299	14,500	−2.44	2.36×10^{-4}
300	1200	0.239	23,100	−4.78	4.19×10^{-4}

3.4.2 Fans

Centrifugal fans are the type most commonly used to convey gases through control systems. Whether the fan is to be placed before the control device (forced draft) or after (induced draft) determines the design to be used. Fans with *backwardly curved* blades are used with induced draft configurations, while *radial tip* bladed fans are employed with forced draft layouts (primarily venturi scrubbers), mainly because they can withstand the abrasiveness of heavily dust-laden streams.

The costs of backwardly curved fans are functions of the fan *class,* the delivered static pressure (in. H_2O), and the gas flow rate (acfm). The class, in turn, depends on the static pressure and the gas velocity (ft/min), as shown:[14]

Class	Applicable Pressure/ Velocity Range		
	(in. H_2O)		(ft/min)
I	5/2300	to	2.5/3200
II	8.5/3000	to	4.25/4175
III	13.5/3780	to	6.75/5260
IV	(Performance in excess of class III)		

Figure 5 illustrates how pressure, flow rate, and fan class determine backwardly curved fan purchase cost.[14] This figure applies only to fans operating at standard temperature (70°F) and pressure (sea level, or 29.92 in. Hg). For nonstandard conditions, the static pressure must be adjusted before Figure 5 can be used. The following correction factor accounts for the effects of altitude and temperature on static pressure:

$$\pi_{cf} = \frac{530}{T} \exp\left(-3.53 \times 10^{-5}z\right) \tag{32}$$

where

$$T = \text{absolute temperature of exhaust, °R}$$
$$z = \text{altitude above sea level, ft}$$

To obtain the pressure at standard conditions, simply *divide* the static pressure by π_{cf}.

If the service temperature exceeds 250°F (but is less than 600°F), add 3% to the cost from Figure 5; and for stainless steel construction, multiply the same cost by 2.5.[14]

A final point: because the Figure 5 costs are in December 1977 dollars, they must be escalated to June 1981 by multiplying the costs by 1.571.

Costs for radial-tip fans in lower pressure operation ($\le$ 20 in. H_2O) appear in Figure 6. Here again, the static pressure variable must be adjusted for conditions other than standard; and as above, the costs must be escalated to June 1981 via the factor 1.571.

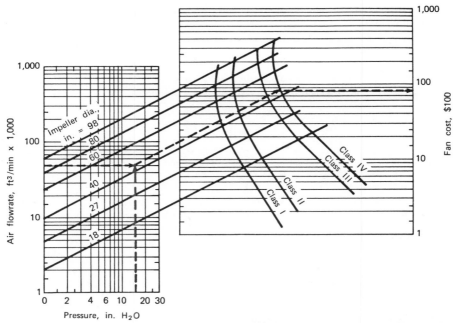

Fig. 5 Costs of backward-curved fans for transporting dust-laden gases. Excerpted by special permission from *Chemical Engineering*, May 18, 1981. Copyright © 1981, by McGraw-Hill, Inc., New York, NY 10020.

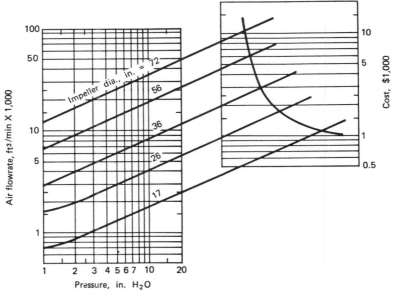

Fig. 6 Costs of radial-tip fans of 20-in. (and lower) head. Excerpted by special permission from *Chemical Engineering*, May 18, 1981. Copyright © 1981, by McGraw-Hill, Inc., New York, NY 10020.

For pressures ranging from 20 to 60 in. (often required with venturi scrubber systems) use this expression for costs:[14]

$$C_{F-RT} = a + bQ + cQ^2 \tag{33}$$

where

Q = air flow rate, cfm
a, b, c = parameters that vary with static pressure, as shown below

Static Pressure, in. H₂O	a	b	c
20	−2070	1.03×10^{-3}	-3.49×10^{-9}
40	−5690	1.06×10^{-3}	-3.33×10^{-9}
60	−3420	1.27×10^{-3}	-3.05×10^{-9}

For operating temperatures between 250 and 600°F, add 6% to costs taken from Figure 6 or Eq. (33). Finally, when stainless steel construction or rubber liners are used, multiply these costs by 2.5.

3.4.3   Motors, Starters, and Other Accessories

For most applications "drip-proof" motors are adequate. However, for surroundings containing large amounts of moisture, abrasive dust and fumes, or high vapor concentrations, "totally enclosed" motors must be specified. And where there is risk of explosion, "explosion-proof" motors should be used. The former cost from 30 to 50% more than drip-proof motors; the latter, 60 to 70% more.

For *pumps*, Figure 7 provides the costs for drip-proof motors as a function of motor size (in brake horsepower) and speed. Horsepower may be estimated as follows:

$$\text{hp} = \frac{QZS}{3960\eta} \tag{34}$$

as

Q = liquid flow rate, cpm
Z = liquid head, ft H₂O
S = specific gravity of liquid relative to water at
      standard temperature and pressure
η = pump efficiency

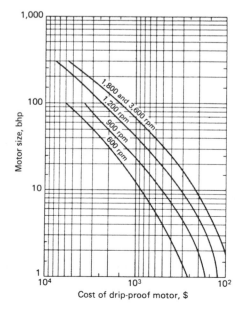

Fig. 7   Use multipliers to get costs of other types of motors. Excerpted by special permission from *Chemical Engineering*, March 23, 1981. Copyright © 1981, by McGraw-Hill, Inc., New York, NY 10020.

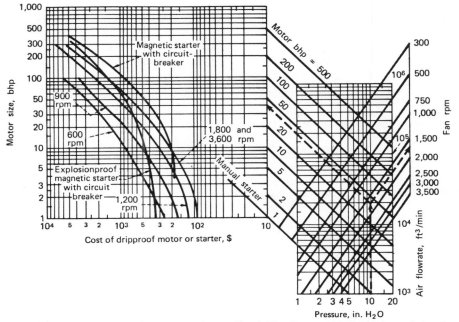

Fig. 8 Costs of motors and motor starters for centifugal fans. Excerpted by special permission from *Chemical Engineering*, May 18, 1981. Copyright © 1981, by McGraw-Hill, Inc., New York, NY 10020.

The next step is to draw a horizontal line from the ordinate of Figure 7 corresponding to the horsepower to its intersection with the curve that *comes closest* to the pump speed (e.g., a 1750-rpm pump would require an 1800-rpm motor). Next, draw a vertical line from this intersection point to the abscissa and read the motor cost. Obtain starter costs (manual or magnetic) from Figure 8 also as a function of the horsepower. Note, however, that all costs from Figures 7 and 8 must be multiplied by an escalation factor (1.366).

Obtaining *fan* motor costs requires an additional step. The right-hand grid in Figure 8 correlates centrifugal fan rpm and brake horsepower (bhp) to static pressure and air flow rate. As illustrated, a fan delivering 20,000 cfm at a static pressure of 10 in. (STP) would have a speed of 1700 rpm and a bhp of 40. At this speed, the corresponding motor speed would be 1800 rpm, as this guide shows:[14]

Fan Speed	Motor Speed
2400–4000	3600
1400–2400	1800
1000–1400	1200
700–1000	900
700	600

The motor and starter costs are obtained as above. The same escalation factor (1.366) is used as well.

Figure 9 provides a similar grid for radial-tip fans with static pressures of 20 in. or lower. With the motor size from this figure, find the motor/starter cost from Figure 8.

However, for high-pressure venturi scrubber service, radial-tip motor and starter costs can be computed from this equation:[14]

$$C_{M/S-RT} = c + dQ \qquad (35)$$

such that

Q = air flow rate
c, d = functions of *total* fan pressure at standard conditions, as follows

Total Pressure, in. H₂O	c	d
30	−9,840	0.350
50	−14,900	0.570
70	6,830	0.683

Inlet and outlet damper costs appear in Figure 10. Similar to the data in Figures 7 and 8, these costs must be escalated to June 1981 dollars via this factor: *1.427*.

Finally, where fan and motor speeds do not match, they must be coupled by means of V-belt drives. Costs for these appear in Figure 11,[14] as functions of motor size and speed. However, these too must be escalated via the factor 1.366.

3.5 Screw Conveyors

Screw conveyors are used to continuously remove captured dust from hoppers beneath dry collection devices to piles or bins for storage or to haulage vehicles for transportation to disposal sites.

Purchase costs ($) for conveyors may be obtained from these equations:[13]

$$C_{SC} \text{ (9-in. diameter)} = 903 + 94L \qquad (36)$$
$$C_{SC} \text{ (12-in. diameter)} = 1070 + 96.7L \qquad (37)$$

where

$$L = \text{conveyor length, ft}$$

Note that the costs are functions of conveyor length and diameter. The length depends on, among other things, the hopper dimensions and the distance to storage or vehicle. The diameter, however, is primarily a function of the gas flow rate. Specifically, 9-in.-diameter conveyors are adequate for flow rates up to 100,000 cfm, while the 12-in. size must be used for larger gas flows.[13]

3.6 Stacks

Stacks are necessary to disperse exhaust gases above ground level and obstacles near the control system. (So-called "tall stacks" may also be used to dilute pollutant concentrations—often to meet concentration emission standards—but these will not be covered here. For tall stack costs, see Ref. 15.)

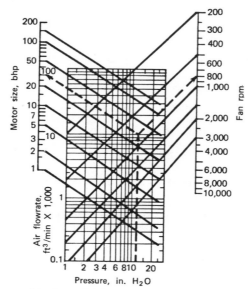

Fig. 9 Fan rpm and motor bhp for radial fans of 20-in. (and lower) static pressure. Excerpted by special permission from *Chemical Engineering,* May 18, 1981. Copyright © 1981, by McGraw-Hill, Inc., New York, NY 10020.

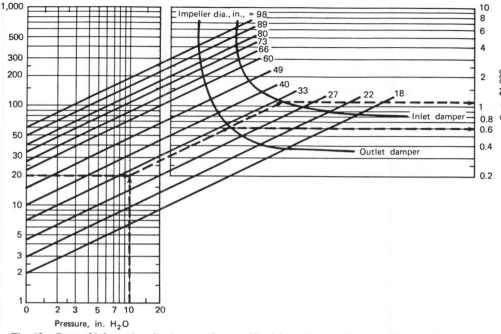

Fig. 10 Costs of inlet and outlet dampers for centrifugal fans. Excerpted by special permission from *Chemical Engineering,* May 18, 1981. Copyright © 1981, by McGraw-Hill, Inc., New York, NY 10020.

The dimensions and arrangement of the control equipment and surrounding structures dictate the stack height, while the wind velocity determines the *minimum* exit velocity and, hence, the *maximum* allowable stack diameter.

Based on this minimum velocity, the *maximum* stack diameter (in.) would be:

$$D_{S\,(max)} = 1.18 \left(\frac{Q}{v_w}\right)^{1/2} \tag{38}$$

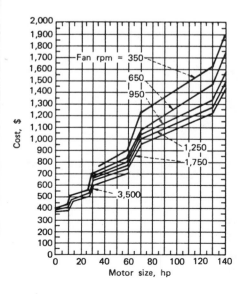

Fig. 11 Cost of V-belt drive includes OSHA drive guard. Excerpted by special permission from *Chemical Engineering,* May 18, 1981. Copyright © 1981, by McGraw-Hill, Inc., New York, NY 10020.

where

$$Q = \text{air flow rate, cfm}$$
$$v_w = \text{wind velocity, mph}$$

The stack diameter and height, along with structural considerations, determine the required plate thickness, the third parameter that influences stack cost.

In each of the following cost equations, these three variables are combined. These equations are valid for heights ranging from 20 to 90 ft and diameters from 2 to 5 ft, and are of the form[16]

$$C_{st} = e + fH \tag{39}$$

where

$H = $ stack height, ft

$e, f = $ parameters that vary with stack diameter and plate thickness, as follows

Diameter, in.	Cost Equation, $		
	¼-in. Plate	⁵⁄₁₆-in. Plate	⅜-in. Plate
24	$1230 + 70.5H$	None	None
30	$1230 + 85.3H$	None	None
36	$1230 + 93.5H$	None	None
42	$1394 + 99.9H$	$1558 + 114H$	None
48	$1477 + 113H$	$1969 + 123H$	None
54	$1641 + 120H$	$2133 + 136H$	$2626 + 146H$
60	$1805 + 125H$	$2379 + 140H$	$2954 + 148H$

In addition to the stack itself, these costs include flanges, cables, clamps, and surface coating.

4 PARTICULATE CONTROL DEVICES

The decision of whether to use an electrostatic precipitator, a venturi scrubber, or a fabric filter to control particulate emissions from a source will depend on a number of technical and cost factors. However, in many cases, there are no technical reasons preventing the use of any one device, so that the selection is solely dictated by cost. Fortunately, a wealth of cost information has been available from vendors and the literature to make such selections easier. Some of these data are presented in this section.

4.1 Electrostatic Precipitators

In general, electrostatic precipitators (ESPs) are capital intensive, as opposed to, say, venturi scrubbers, which are utility intensive. Operating and maintenance costs for ESPs are relatively low, as Table 6 shows. Missing from that table is the ESP electricity consumption for the fan motor and plate-charging transformers. Typical values are: pressure drop, ½ in. H_2O; field charging, 1.5 W/ft² of plate area.

The critical sizing parameter for an ESP is the collecting area (or plate area) A, a good approximation of which can be obtained from the well-known Deutsch–Anderson equation:

$$A = -Q(\ln[1 - \eta]) / w \tag{40}$$

where

$\eta = $ overall particulate collection efficiency, fractional

$w = $ "drift velocity," ft/min—a lumped parameter that combines the effects of the charging and collecting fields, the plate spacing, the dust resistivity, and other variables

Moreover, because dust resistivity (and the drift velocity) are better at higher humidities, the collection efficiency is often improved when spray chambers and like preconditioners are used upstream of the ESP. At the same time, increasing the moisture content may require insulating the unit, which, in turn, increases the purchase cost. Hence, the following cost equations:[17]

$$C_{ESP} \text{ (insulated)} = 218,000 + 6.77A \tag{41}$$

$$C_{ESP} \text{ (uninsulated)} = 158,000 + 4.50A \tag{42}$$

where

$$A = \text{plate area, ft}^2$$
$$C_{ESP} = \text{cost of insulated/uninsulated ESP, \$}$$

and

$$500 \leq A \leq 1{,}000{,}000$$

The materials of construction, type of field power supply, and other variables can also influence costs, but not as much as the presence or absence of insulation.

The auxiliary equipment used with an ESP consists of a capture device, ductwork, fan subsystem (usually installed downstream of the ESP), dust conveyor, stack, and preconditioners. The last can be a mechanical collector (to lighten the dust loading), a heat exchanger (for gas precooling), and, as discussed above, a spray chamber (for precooling and humidification).

Finally, Eqs. (41) and (42) give costs for *dry* ESPs only. It is much more difficult to develop costs for *wet* ESPs, since so many design variables enter in. However, for approximation purposes, estimate the purchase cost of a wet ESP at 2.0 to 2.5 times the cost of a dry ESP of the same size.

4.2 Wet Scrubbers

Of the several different types of wet scrubbers used for particulate control, venturis are generally the most efficient. This is primarily due to the high-velocity collisions between the particles and atomized water droplets in a venturi's converging section, or "throat." The pressure drop (ΔP) of the gas passing through this throat mainly affects this high collection efficiency. The required ΔP is in turn a function of the particle size distribution:[18]

$$\Delta P_v = 15.4 d^{-1.39} \tag{43}$$

where

ΔP_v = pressure drop of gas stream through the venturi throat, in. H_2O
d = diameter of *smallest* particle that can be completely (100%) removed by the venturi, μm

For instance, if one wanted to remove all particles as large and larger than 0.5 μm, the required ΔP_v would be 40 in. H_2O. (Of course, some unknown fraction of the particles *smaller* than 0.5 μm would also be removed at this ΔP_v, but for design purposes this fraction is often neglected.)

The pressure drop is also the major factor in determining venturi costs, both capital and O & M. Power consumption is extremely high, since ΔP_v can be as high as 80 in. H_2O. Therefore, the electricity usage for a venturi system is simply the product of (1) ΔP_v (plus the other pressure drops in the system), (2) air flow rate, and (3) operating factor, as Eq. (5) shows.

Although actual particulate collection occurs in the venturi, a cylindrical tank ("separator") must be installed downstream of the venturi to remove entrained liquid droplets from the gas stream. When extremely large quantities of water are present in the gas stream, the separator may be equipped with an internal gas cooler to remove the excess moisture. This cooler may be baffled, tray-type, or of other designs.

The cost of the cooler and separator, along with the costs of the venturi, elbow, pumps (for liquor recycle and slurry withdrawal), and controls are included in this base cost equation:[18]

$$C_v = 10{,}200 + 0.582Q - 1.21 \times 10^{-6}Q^2 \tag{44}$$

where

$$Q = \text{gas flow rate, acfm}$$

and

$$0 \leq Q \leq 200{,}000$$

This equation applies to $\frac{1}{8}$-in. *carbon steel* only, which is adequate for many flow rates and pressure drops. But if either parameter is unusually large, thicker plate may be required. Figure 12 provides that information. The required thickness is found where the design pressure ordinate intersects the waste-gas flow rate abscissa. Where the intersection lies between two thicknesses, always select the next highest value.

With this metal thickness and the flow rate, next obtain the "cost adjustment factor" from Figure 13, which is multiplied by the base cost [Eq. (44)] to obtain the adjusted costs.

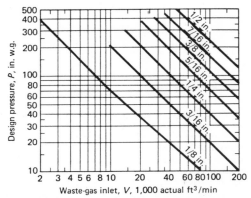

Design pressure, P, in. w.g.

Waste-gas inlet, V, 1,000 actual ft³/min

Fig. 12 Flow rate and design pressure dictate metal thickness. Excerpted by special permission from *Chemical Engineering*, November 30, 1981. Copyright © 1981, by McGraw-Hill, Inc., New York, NY 10020.

If 304 or 316 stainless steel is used instead of carbon steel, multiply this adjusted cost by a second factor, either 2.3 (304 SS) or 3.6 (316 SS).

If an adjustable throat is desired (to vary ΔP_v and, in turn, collection efficiency), add $4900 or $9100, for manual or automatic adjustments, respectively.

Finally, if the venturi is to be lined with fiberglass, simply add 15% of the *base cost* to the total [Eq. (44)]. However, if $3/16$-in. rubber is desired instead, figure the lining cost at $6.69/ft² of internal surface area. Obtain this area (A_s) from the following:[18]

$$A_s = 0.015Q \tag{45}$$

(For information on sizing and costing internal gas coolers, see Ref. 18.)

If a scrubber other than a venturi is to be installed, the following equations can provide *approximate* costs:[19]

Impingement and entrainment:

$$C_{I/E} = 1.15Q - 7.45 \times 10^{-6}Q^2 \qquad (10,000 \le Q \le 50,000) \tag{46}$$

Plate type:

$$C_{PT} = 1.82Q - 5.81 \times 10^{-5}Q^2 \qquad (3000 \le Q \le 20,000) \tag{47}$$

$$0.743Q - 4.32 \times 10^{-5}Q^2 \qquad (20,000 \le Q \le 70,000) \tag{48}$$

Centrifugal:

$$C_C = 104Q^{0.474} \qquad (3000 \le Q \le 40,000) \tag{49}$$

$$0.395Q \qquad (40,000 \le Q \le 70,000) \tag{50}$$

Moving bed:

$$C_{MB} = 187Q^{0.478} \qquad (7000 \le Q \le 70,000) \tag{51}$$

where

$$Q = \text{gas flow rate, acfm}$$

These equations apply to scrubbers fabricated of $1/4$-in. carbon steel and are valid for only the flow rates listed in parentheses.

Finally, regardless of the type of scrubber used, the same auxiliary equipment is generally employed. This includes a capture device, a quencher (for precooling, if needed), wastewater treatment facilities, a fan subsystem (radial-tip design), and stack.

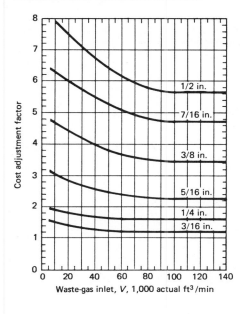

Cost adjustment factor

1/2 in.

7/16 in.

3/8 in.

5/16 in.

1/4 in.

3/16 in.

Waste-gas inlet, V, 1,000 actual ft³/min

Fig. 13 Factors adjust cost for flow rate and metal thickness. Excerpted by special permission from *Chemical Engineering*, November 30, 1981. Copyright © 1981, by McGraw-Hill, Inc., New York, NY 10020.

4.3   Fabric Filters

Unlike the other two particulate control devices discussed in this chapter, whose control efficiencies depend on the characteristics of the exhaust stream and other variables, fabric filters either operate very efficiently or not at all. Except for extremely fine particulate, their capture efficiencies are quite high—usually in excess of 99.5%.

For this reason, baghouses are fast becoming the preferred method for particulate control, even for such high-temperature sources as utility and industrial boilers. In fact, nearly 75% of all baghouse purchases made since 1977 were for these kinds of sources.[20]

Favorable cost is another reason for the growing popularity of baghouses. Nonetheless, costing a baghouse is somewhat more difficult than costing an ESP or a venturi. For one thing, baghouse cost depends on several factors:

1. Bag fabric and air-to-cloth ratio.
2. Duty—continuous or intermittent.
3. Construction—pressure or suction.
4. Design—standard or custom.
5. Cleaning mechanism (i.e., shaker, reverse-air, or pulse-jet).
6. Materials of construction.

These factors, coupled with the particulate loading and other facility parameters, determine the basic design parameter, the *net* cloth area, defined as

$$A_n = \frac{Q}{R_{a/c}} \tag{52}$$

where

$$A_n = \text{net cloth area, ft}^2$$
$$Q = \text{gas flow rate, acfm}$$
$$R_{a/c} = \text{air-to-cloth ratio, ft/min}$$

$R_{a/c}$ will depend on the cleaning mechanism, the type of bag material used, and other variables. Roughly speaking, $R_{a/c}$ can be up to 4 to 1 for shaker and reverse-air baghouses and up to 10 to 1 for pulse-jet units. Further, polyester fabrics are used for R up to 3 to 1; fiberglass, up to 2 to 1.

A_n is the area available for filtering at any given time while the baghouse is in operation. It does *not* include the area of those bags temporarily out of service for cleaning. The sum of A_n and the out of service bag area is the *gross* cloth area (A_g), which is normally factored from A_n:

$$A_g = f(A_n)A_n \tag{53}$$

where

$f(A_n) =$ a conversion factor that decreases with increasing A_n and approaches unity for extremely large A_n

(Note, however, that with *intermittent*-duty baghouses, $A_n = A_g$, since the entire unit is shut down when cleaning occurs.)

A_n, ft²	$f(A_n)$
1–4,000	2
4,001–12,000	1.5
12,001–24,000	1.25
24,001–36,000	1.17
36,001–48,000	1.125
48,001–60,000	1.11
60,001–72,000	1.10
72,001–84,000	1.09
84,001–96,000	1.08
96,001–108,000	1.07
108,001–132,000	1.06
132,001–180,000	1.05
>180,000	1.04

The total purchase cost of a baghouse is the sum of the costs of the baghouse compartment (with appropriate add-ons) and the bags. In general

$$C_{FF} = C_{comp.} + C_{bags} = [a + bA_n] + cA_g \tag{54}$$

The values of a and b inside the brackets will depend on cost factors 2 through 6 listed above. Tables 9 to 11 list these values for different cleaning mechanisms, duty types, construction, and design. Also listed are the ranges in net cloth area to which the parameters apply. The first two tables are for factory-assembled and designed units, while Table 11 applies to custom baghouses, those designed and built for the specific source.

Values for the parameter c (the bag price) appear in Table 12 for selected bag materials. Note that the bag prices (in $/ft²) vary according to the design and cleaning mechanism. Also listed in the table are typical operating temperatures for the materials.[23]

In addition to the purchase cost, one must consider the utility costs. These include electricity for the fan and cleaning mechanism motors and (with pulse-jet units) compressed air.

Table 9   Purchase Costs for Intermittent Baghouse Compartments[22]

Construction:	pressure
Cleaning mechanism:	mechanical shaker
Applicable range:	0–20,000 ft²

Component	Parameter, $	
	a	b
Basic baghouse	4780	2.63
Stainless steel add-on	3910	1.60
Insulation add-on	2900	1.20
Suction add-on	1970	0.171

Table 10 Purchase Costs for Continuous Baghouse Compartments[22]

Cleaning Mechanism[a]	Applicable Range, ft^2	Component	Parameter	
			a	b
I. Mechanical shaker	0–70,000	Basic baghouse	9,500	4.99
		SS add-on	10,500	2.71
		Insulation add-on	3,250	2.53
		Suction add-on	3,230	0.357
II. Reverse-air	0–100,000	Basic baghouse	36,600	4.28
		SS add-on	16,600	2.55
		Insulation add-on	16,000	2.37
		Suction add-on	2,410	0.457
III. Pulse-jet	0–20,000	Basic baghouse	7,660	10.8
		SS add-on	2,350	7.14
		Insulation add-on	7,010	3.42

[a]Mechanical shaker and reverse-air parameters are based on pressure construction, while those for pulse-jet units can apply to either pressure or suction construction.

As Eq. (5) shows, the fan power is proportional to the pressure drop through the baghouse and auxiliary equipment. The baghouse ΔP, in turn, is a function of the air-to-cloth ratio, the frequency and thoroughness of cleaning, dust cake properties, and other parameters. In general, ΔP varies with $R_{a/c}^2$, with a typical value of approximately 6 in. H$_2$O. One can increase $R_{a/c}$ to decrease the initial baghouse cost, but this would increase the fan power requirement more than comparably, while also possibly causing unacceptably large increases in the outlet emissions.[24]

Table 11 Purchase Costs for Custom Baghouse Compartments[22]

Construction:	pressure or suction
Cleaning mechanism:	any
Applicable range:	30,000–400,000 ft^2

Component	Parameter, $	
	a	b
Basic baghouse	145,000	3.85
Stainless steel add-on	71,400	2.00
Insulation add-on	54,200	2.00

Table 12 Bag Prices for Selected Materials

Material	Operating Temperature, °F	Price ($/ft^2)[22]				
		Shaker (< 20,000 ft^2)	Shaker (> 20,000 ft^2)	Pulse Jet	Reverse Air	Custom
Dacron	270	0.57	0.50	0.86	0.50	0.36
Orlon	248	0.93	0.71	1.36	0.86	0.50
Nylon	200	1.07	1.00	Not used	1.00	0.64
Nomex	400	1.64	1.50	1.86	1.50	0.93
Fiberglass	500	0.71	0.64	Not used	0.64	0.43
Polypropylene	200	0.93	0.78	1.00	0.78	0.50
Cotton	180	0.64	0.57	Not used	0.57	0.57

Power for motors in shaker baghouses may be estimated from the following expression:[25]

$$\text{shaker energy (kWh/yr)} = [-6.393 + 1.045 \ln A_g]h \quad (A_g > 1432 \text{ ft}^2) \tag{55}$$

$$= 0.56h \quad (A_g \le 1432 \text{ ft}^2) \tag{56}$$

where

$$h = \text{operating hours per year}$$

For reverse-air baghouse motors, figure the power as:[25]

$$\text{reverse-air energy (kWh/yr)} = 3.7 \times 10^{-4} A_g h \tag{57}$$

Finally, the auxiliaries used with a fabric filter system are the same as those used with an ESP. (See Section 4.1.)

5 GASEOUS CONTROL DEVICES

Gaseous control devices fall into two broad categories according to their function, pollutant destruction or capture. Destruction devices include incinerators (thermal and catalytic), flares, and other equipment where the pollutants (e.g., a mixture of organics) are oxidized to CO_2 and H_2O.

In capture devices, conversely, the pollutants are valuable enough to make recycling or resale economical. The second category includes the *ad*sorption, *ab*sorption, and condensation–refrigeration devices. However, where there are few technical restrictions to using destruction devices, capture devices are often limited in their applications, due to constraints of operating temperature, impurities, pollutant concentration, and like parameters.

5.1 Thermal Incinerators

Direct-fired thermal incinerators (or *afterburners*) are perhaps the simplest and most reliable gaseous control devices. They are widely used to remove a variety of combustible gases, aerosols, vapors, and even particulate from sources ranging from coating ovens to chemical plant process vents.

An afterburner is basically a narrow, refractory-lined tube equipped with either a natural gas or an oil burner. The exhaust gases are mixed with fuel and air at the inlet, ignited by a pilot flame, and proceed to the outlet, as the combustibles oxidize to CO_2 and H_2O. Next, the combustion products usually proceed to a "recuperative" heat exchanger, where some of the enthalpy in the flue gas is used to preheat the inlet gases. From here the gases are conveyed to either a stack or a secondary heat exchanger (such as a waste-heat boiler) where additional heat is recovered for steam production and other means of energy usage on-site.

Because the combustion reactions in the incinerator are essentially kinetic in nature, their extent of completion depends on the three kinetic variables: time, temperature, and turbulence, the "three T's." Turbulence is normally effected by the high length-to-diameter aspect of the afterburner—2:1 or greater. This permits intimate contact of the reactants during their residence in the combustion chamber. Residence time, the second T, is quite short, usually ranging from 0.5 to 1 sec. Although temperature is the more critical variable, residence time becomes important when higher carbon monoxide concentrations are present in the exhaust.[26]

Though the waste gas composition, oxygen content, and other variables influence combustion efficiency, the combustion temperature has the most pronounced effect. In particular, tests have shown that, to achieve efficiencies exceeding 98 and 99% in a typical pollutant stream, combustion temperatures of 1500 and 1600°F, respectively, must be maintained.[26]

Afterburner costs are normally estimated as a function of combustion chamber volume (V, ft³), where

$$V = Q\theta \tag{58}$$

and

$$Q = \text{volumetric flow rate, cfm, figured at the combustion (outlet) temperature}$$
$$\theta = \text{residence time, min}$$

Costs may be obtained from the next two equations, which reflect the normal range of combustion temperatures used.[26]
For 1400°F,

$$C_{Ab} = 228{,}000 V^{-0.703} \exp(0.0896[\ln V]^2) \tag{59}$$

For 1800°F,

$$C_{Ab} = 262{,}000\, V^{-0.721}\, \exp(0.0908[\ln V]^2) \tag{60}$$

In addition to the combustion chamber, these equations include costs for internal piping, wiring, burners, and refractory. (However, no instrumentation is included.) Also, the equations are only valid for volumes between 50 and 10,000 ft³.

The cost of recuperative heat exchangers is also figured on a size parameter, surface area, A_s (ft²). The area is computed from the standard heat transfer formula:

$$A_s = \frac{q}{U\, \Delta T_{lm}} \tag{61}$$

where

$q =$ heat lost/gained by exit/inlet gas, in Btu/hr
$\Delta T_{lm} =$ log-mean temperature difference between the countercurrently flowing streams, °F
$U =$ overall heat transfer coefficient, Btu/hr-ft²-°F

Because heat exchange is gas to gas, values for U are relatively low, ranging from 2 (cross-flow units) to 6–8 (U-tube exchangers).

Estimate recuperative heat exchanger costs from this expression.[26]

$$C_{HE} = 38{,}300\, A_s^{-0.440}\, \exp(0.0672[\ln A_s]^2) \tag{62}$$

This applies to a range of heat exchanger designs and to sizes ranging from 200 to 50,000 ft².

Despite the importance of purchase costs, by far the largest component (as much as 80%) of afterburner annualized costs is fuel. Fuel cost, in turn, is a function of the price (in $/scf or gal), the inlet and outlet temperature (after heat recovery), the exhaust flow rate and composition, and the combustion efficiency. Although, to be rigorous, one should consider the heating contribution of the combusted pollutants when figuring the fuel requirement, one can, however, conservatively estimate fuel use by neglecting this contribution and by assuming: (1) 100% efficiency of combustion, (2) negligible heat losses, and (3) no combustion air is available from the effluent. Given these assumptions, the following provides an approximation of afterburner fuel use:[4]

$$F = \frac{n_i c_{p_i}(T_i - T_o)}{\Delta H_f - \Sigma_j\, m_j c_{p_j}(T_o - T_f)} \tag{63}$$

where

$n_i =$ flow rate to incinerator, mole/hr
$c_{p_{i,j}} =$ mean heat capacity of inlet stream and combustion products, respectively, Btu/mole-°F
$T_{i,o,f} =$ temperatures of inlet stream, outlet stream, and fuel, respectively, °F
$\Delta H_f =$ gross heating value of fuel, Btu/mole
$m_j =$ moles of combustion product j/mole of fuel

Because the combustion air is obtained externally (i.e., 100% primary air), the volume flow rate after combustion is higher by the amount of the added combustion products, or

$$n_o = n_i + \Sigma_j m_j \tag{64}$$

(Note: The combustion chamber is sized according to n_o, not n_i.)

Afterburners also consume electricity for the fan motor. Power consumption is, again, proportional to the incinerator pressure drop, which depends, in turn, on the amount of heat recuperation employed. The following data provide some guidance:

Heat Recuperation, %	Pressure Drop, in. H_2O
0	2
30	6
50	8
70	10

Finally, in addition to heat exchangers, the auxiliary equipment used with afterburners includes a capture device, fan subsystem, ductwork, and stack.

5.2   Catalytic Incinerators

Catalytic incinerators are nearly identical to thermal incinerators in terms of design and operation. Like afterburners, they combust organics and other gaseous pollutants to CO_2 and H_2O with the aid of varying quantities of auxiliary fuel in a combustion chamber and may or may not recover heat from the resultant flue gases. Where catalytic units differ, however, is in the use of a catalyst to enhance the combustion kinetics, such that the combustion temperature can be reduced considerably (i.e., from 1400 to 1800°F for afterburners to 200 to 900°F).[27] This catalyst, usually a noble metal on an activated alumina base, is installed on a metal mesh pad in the combustion chamber near the outlet. Like other catalysts, it acts by supplying active sites on its surface for the oxidation reactions to occur. The only drawback is that these surfaces can become poisoned by gas stream impurities, namely, lead, halogens, and sundry particulate matter.

The quantity of catalyst required and its cost is a function of the gas flow rate and the desired destruction efficiency. The amount varies from 0.5 to 2 ft^3/100 scfm of exhaust. However, the cost per unit flow rate is essentially constant in these equations, valid for $500 \leq Q \leq 100,000$ cfm[27]

$$C_{Cat} (90\%) = 3.58 Q^{0.968} \tag{65}$$

$$C_{Cat} (99\%) = 6.60 Q^{0.946} \tag{66}$$

where

$$C_{Cat} = \text{catalyst cost, \$/cfm}$$

Note that the cost of catalyst is nearly twice as much at 99% efficiency as at 90%.

Because catalyst requirements vary with efficiency, so do the pressure drops through the incinerator. These ΔP's range from 5 in. (at 90%) to 7 in. H_2O (at 99%). However, as Section 5.1 indicates, the addition of recuperative heat recovery would increase these ΔP's.

Finally, note that eventually catalysts wear out and must be replaced. (Every *three* years is typical.) This results in an additional expenditure, computed as follows:

$$\text{catalyst replacement cost (\$)} = \text{original cost} \times \text{CRF} \tag{67}$$

where

$$\text{CRF} = \text{the capital recovery factor [see Eq. (6)]}$$

To the catalyst cost must be added the cost of the basic incinerator unit. The following equation gives prices for prepiped, prewired incinerators, including burner, blower, controls, and instrumentation:

$$C_{CI} = 166,000 Q^{-0.686} \exp(0.0622[\ln Q]^2) \tag{68}$$

such that

$$500 \leq Q \leq 100,000 \text{ scfm}$$

Note that this equation, unlike the one for afterburners, is expressed in terms of standard flow rate, not chamber volume.

The same kind of recuperative heat exchangers is used with catalytic and thermal incinerators. Equation (62) provides costs for them.

Finally, Eq. (64) can be used to compute the fuel requirement. The only difference is that T_o will be considerably lower, due to the action of the catalyst. Following are ignition temperatures for selected combustibles which will give 90% destruction in catalytic incinerators:[27]

Carbon monoxide—425°F

Ethylene—550°F

Benzene, toluene, xylene—575°F

Methyl ethyl ketone—700°F

Propane—770°F

Dimethyl formamide—800°F

Methane—920°F

5.3 Flaring

Unlike other combustion devices, flares are normally designed for intermittent flows and emergency upsets, rather than continuous exhausts. A fairly simple device, it is essentially a burner with a built-in combustion chamber which can handle a variety of combustibles (gaseous or vapor) having a range of heat contents.

There are two main flare categories: *elevated* and *enclosed ground*. With the former, the flare and auxiliaries are mounted on a stack (freestanding or supported), so as to remove it from the process area for purposes of safety (heat) and nuisance (noise). Auxiliary equipment include a stack seal (to prevent air intrusion), a knockout drum (for removal of entrained liquids), an ignition system, and, of course, the usual ductwork, capture device, and fan subsystem.

Elevated flares are of three designs—*nonsmokeless, smokeless,* and *endothermic*—the use of which depends on the type and heat content of the material burned.[28]

Unlike elevated flares, enclosed ground flares can burn both gaseous and liquid wastes and are also more versatile in terms of safety, operation, and energy usage. They consist of several burners and combust the material in a refractory chamber or an open pit. Despite the fact that ground flares need neither stack nor supporting devices, their more complex designs require higher capital costs, though they are offset somewhat by reduced utility requirements.

Flare purchase costs are a function of the design, the elevation, and, most of all, the waste gas flow rate (usually expressed in lb/hr). The general equation for flare costs is

$$C_F = aM^b \exp(c[\ln M]^2) \tag{69}$$

where

$$a, b, c = \text{parameters whose values are listed below}$$

and

$$2500 \leq M \leq 250{,}000 \text{ lb/hr}$$

Elevated flare costs also include ladders, platforms, and a stack of sufficient height to ensure a maximum grade-level radiation of 1500 Btu/hr-ft².[28]

| | Flare Type and Waste Gas Burned | | | |
| | Enclosed Ground | | Elevated | |
Parameter	High Btu (Ethylene)	Low Btu (60 Btu/scf)	High Btu	Low Btu
a	6160	14.82	411	1250
b	−0.0105	1.07	0.398	0.256
c	0.0296	−0.0314	0	0

In addition, flares have fairly high operating costs due to the substantial amounts of natural gas and (in the case of smokeless flares) steam required. The fan power costs may be high also, because of the pressure drop needed to be overcome in the ductwork, knockout drum, stack, and flare itself, which could be as high as 60 in. H_2O.

For ground flares, the following expression predicts the natural gas required for pilots and purge gas as a function of the waste gas flow rate:[28]

$$G_{GF} = 1300M^{-0.363} \exp(0.0306[\ln M]^2) \tag{70}$$

where

$$G_{GF} = \text{gas required, million Btu/yr}$$

This expression holds for all waste gas streams, regardless of heat content, since ground flares do not require supplemental ("assist") fuel.

Elevated flares do require assist gas, however. Equation (71) predicts the assist gas consumption (in million Btu/yr), as well as gas for pilots and purging when a low-Btu gas is being burned.

$$G_{EF} = 2.18M^{0.934} \tag{71}$$

Both equations have been developed assuming an 880 hr/yr operating factor and apply to the same range of gas flows as does Eq. (69).

Finally, for smokeless flares and others requiring steam injection for mixing and air introduction, the steam consumption can be estimated at 0.4 lb/lb of waste gas.

5.4 Gas Absorbers

Although widely used in the chemical process industry, gas absorption is also used to remove gaseous pollutants, particularly those present in relatively dilute concentrations (~3% by volume).

Basically, absorption involves the removal of the pollutant (solute) from the exhaust via a liquid (solvent) in which it is miscible. Absorption is most effective when the contact is intimate, such as in a tray or packed column. (Although spray towers, venturis, and other scrubbers can be used for absorption also, their removal efficiencies are relatively low.) Generally, the gas, which enters the bottom of the column, is contacted by the liquid flowing downward. The cleaned gas exits the top, while the solvent is drawn off to storage or to a stripping column where the captured pollutant is removed in a more concentrated, more manageable stream. Though both tray and packed columns are used, the latter appear to be more popular. However, the same basic parameters govern the design of either type: column diameter, pressure drop, number of transfer units, and height of a transfer unit.

The column diameter is a function of the gas and liquid flow rates and densities, packing type, and liquid viscosity. Usually determined empirically, it must be large enough to prevent "flooding," that condition when the gas velocity and resulting pressure drop become so high that liquid entrainment occurs. For guidance on determining the column diameter, see any text on unit operations (e.g., Ref. 5).

Both the tower height and the amount of packing required are determined by the number and height of transfer units, or

$$\text{column height} = Z_{og}N_{og} + \eta_e \tag{72}$$

where

$$Z_{og} = \text{height of a transfer unit, ft}$$
$$N_{og} = \text{number of transfer units}$$

such that

$$Z_{og}N_{og} = \text{depth of packing, ft}$$

and

$$\eta_e = \text{added height for vapor/liquid separation, maintenance, inspections,}$$
$$\text{etc. (typically 2–3 ft plus 25\% of the diameter)}$$

The parameter N_{og} envelops those intensive variables affecting the transfer (capture) efficiency, while Z_{og} deals with extensive variables, such as flow rate and geometry.

In general,

$$Z_{og} = \frac{\alpha G^\beta}{J^\gamma} \left(\frac{\mu_G}{\rho_G \Lambda_G} \right)^{0.5} \tag{73}$$

where

$$\alpha, \beta, \gamma = \text{packing constants (see Ref. 5)}$$
$$G, J = \text{superficial gas and liquid flow rates}$$
$$\text{through the column, respectively, lb/hr-ft}^2$$
$$\mu_G = \text{gas viscosity, lb/hr-ft}$$
$$\rho_G = \text{gas density, lb/ft}^3$$
$$\Lambda_G = \text{gas diffusivity, ft}^2/\text{hr}$$

For most air pollution applications, N_{og} can be computed as follows:

$$N_{og} = \frac{\ln\left\{\left(1 - \dfrac{mG_m}{J_m}\right)\left(\dfrac{y_1 - mx_2}{y_2 - mx_2}\right) + \dfrac{mG_m}{J_m}\right\}}{\left(1 - \dfrac{mG_m}{J_m}\right)} \tag{74}$$

where

m = slope of the solute/solvent equilibrium curve
G_m, J_m = superficial gas and liquid molar flow rates, respectively, mole/hr-ft^2
y_1, y_2 = mole fraction of solute in gas at concentrated and dilute ends of the column, respectively
x_2 = mole fraction of solute at dilute end of column

Once the column diameter, height, and packing depth are calculated, the absorber purchase cost can then be determined. In general

$$\text{absorber purchase cost} = \text{base vessel cost} + \text{ladders and platform} \\ \text{cost} + \text{packing cost} + \text{piping cost} \quad (75)$$

Mulet et al.[29] provide a succinct, accurate method for determining the costs of the first two items. First, they correlated the costs of vertical columns against shell weight to develop this expression:

$$C_c = 717 W_c^{0.183} \exp(0.0230[\ln W_c]^2) \quad (76)$$

where

$$W_c = \text{shell weight, lb}$$

and

$$4250 \le W_c \le 980{,}000$$

In addition to the shell cost this equation predicts costs for manholes, skirts, and shop prime paint. Now, W_c is computed from the column height, diameter, and thickness (T_c) as follows:

$$W_c = \pi D_c (H_c + 0.8116 D_c) T_c \rho_{pl} \quad (77)$$

where

D_c = column inside diameter, ft
ρ_{pl} = density of carbon steel plate, lb/ft^3

In turn, T_c is a complex, semiempirical function of the internal pressure, wind load, and corrosion allowance. It can range from ¼ to over 1½ in. depending on the severity of process and environmental conditions. (See Ref. 29 for a more detailed discussion of this.) But for most absorption situations operating at pressures near atmospheric, ¼–½ in. is adequate. At these thicknesses, corresponding values for $T_c \rho_{pl}$ are 10.18 to 20.35 lb/ft^2 for carbon steel.

If a material other than carbon steel is used, the column cost (C_c) must be multiplied by a material cost factor (F_m). Selected values for F_m are[29]

Stainless steel, 304: 1.7
Stainless steel, 316: 2.1
Monel 400: 3.6
Inconel 600: 3.9
Titanium: 7.7

Computing the cost for ladders and platforms ($C_{1/p}$) is much simpler:

$$C_{1/p} = 233 D_c^{0.740} H_c^{0.707} \quad (78)$$

This is valid for diameter/height pairs ranging from 3 ft/27 ft to 21 ft/40 ft, respectively.
The packing cost is figured on a volume basis, or

$$V_p = \frac{\pi D_c^2}{4} (Z_{og} N_{og}) \quad (79)$$

Although recent years have seen the introduction of several new packing designs with high surface-to-volume ratios (e.g., Tellerettes®), more traditional packings, such as Raschig rings, are still in use. Current costs for some commonly used packings are[5]

Packing Type and Material	Cost, $/ft			
	1-in.	1½-in.	2-in.	3-in.
Pall rings				
Carbon steel	24.3	16.5	15.1	—
Stainless steel	92.1	70.3	60.8	—
Polypropylene	21.9	14.8	13.6	—
Berl saddles				
Stoneware	28.1	21.7	—	—
Porcelain	34.5	25.6	—	—
Intalox saddles				
Polypropylene	21.9	—	13.6	7.0
Porcelain	19.4	14.8	13.3	12.2
Stoneware	18.2	13.3	12.2	11.0
Raschig rings				
Carbon steel	30.3	19.8	17.0	13.9
Porcelain	13.2	10.6	9.7	8.1
Stainless steel	109.0	82.6	72.9	—

Piping costs will, of course, vary from absorber to absorber, depending on the amount of liquid conveyed to and from the column, the extent of solvent recycle, column dimensions, and other factors. Because piping layouts are so site-specific, no generalized cost formulas can be given. However, Barrett[30] provides a comprehensive assortment of current piping costs, covering a range of sizes and materials. For example, for a 400-ft complex piping system (reasonably typical of an absorber piping layout), the installed piping for selected materials and diameters would be

Material	Cost, $/100 ft of piping		
	2-in.	4-in.	6-in.
Schedule 40			
Carbon steel	2,970	4,880	6,620
Aluminum	4,080	6,660	9,210
304-L stainless steel	4,660	7,690	11,200
316-L stainless steel	4,880	9,260	14,800
FRP-vinyl ester	5,440	11,000	17,600

Along with the usual auxiliaries—capture device, ductwork, fan subsystem, and stack—absorbers require pump subsystems to convey the solvent to and from the column and, if desired, to recycle a portion of the "rich" solvent to the vessel. The cost of these pumps is a function of the amount of liquid conveyed and the head to be overcome. The latter depends mainly on the column height, since other losses (friction, pressure, velocity, etc.) are relatively small. Designing for a 100-ft head will generally suffice. Costs for such pumps (vertical turbine type) appear in Section 3.

In addition to the equipment costs, there are costs for the initial solvent charge and makeup, to compensate for losses, decomposition, and so on. Since such a variety of solvents is used, no general costs can be given here. However, issues of the daily *Chemical Marketing Reporter* provide a comprehensive listing. Ironically, water is one of the most commonly used (and the least expensive) solvent. A typical cost for it is $0.25/1000 gal.

Lastly, the utility consumption of absorbers consists of electricity for the fan and pump motors. The required pump horsepower can be figured from the head and flow rates, as given above. To figure the fan horsepower, simply add the ductwork and other auxiliary pressure losses to the column pressure drop:[5]

$$\Delta P_c = \psi \frac{G^2}{\rho_g} \exp\left(2.303 \frac{\phi J}{\rho_J}\right) Z_{og} N_{og} \tag{80}$$

where

$\psi, \phi,$ = constants dependent on packing size and type (see Ref. 5)
ρ_J, ρ_g = liquid and gas densities, respectively, lb/ft³
ΔP_c = column pressure drop, in. H_2O

5.5 Carbon Adsorbers

*Ad*sorption and *ab*sorption are analogous operations, in that each involves the removal of a substance by another substance utilizing the laws of mass transfer. However, where in *ab*sorption, the substance is dissolved in a solvent, in *ad*sorption, the substance is "captured" by a porous adsorbent and held there by means of mechanical and chemical bonding. This adsorbent is usually activated carbon, though other materials (e.g., silica gel) can also be used.

The driving force in adsorption is the vapor pressure of the solute at the temperature and pressure at which the adsorption takes place. (In other words, the *partial pressure* of the solute in the exhaust stream.) The higher the vapor pressure, the higher the *saturation capacity* (SC) of the adsorbent. The equilibrium relationship between the vapor pressure and capacity at a given pressure and temperature is the adsorption *isotherm.* For example, for a toluene–carbon system at 760 mm Hg and 70°F,

$$SC = 0.307 P_v^{0.171} \tag{81}$$

where

SC = saturation capacity at stated conditions, lb toluene/lb carbon
P_v = toluene vapor pressure, mm Hg

Increasing the temperature will *decrease* the capacity, while increasing the pressure will *increase* it. Isotherms for a variety of solute/adsorbent systems are available from the literature (e.g., Ref. 31) or adsorbent manufacturers, such as Calgon Corporation.

But when designing adsorption systems, it is common practice to add a safety factor to the theoretical saturation capacity, to allow for departures from true equilibrium during operation. This factor (which is analogous to the tray efficiency correction factor in distillation) can range widely, though 0.5 is usually typical.[32] For example, using Eq. (81), if the toluene vapor pressure were 1 mm, the (theoretical) saturation capacity would be 0.307 lb toluene/lb carbon. However, for design purposes, this would be reduced to 0.154 (0.5 × 0.307). The latter value is often called the *working capacity* of the adsorbent.

Thus, the quantity of carbon required to adsorb a given amount of solute is just the quotient of the solute flow rate (e.g., in lb/hr) divided by the working capacity. But to obtain the total amount of carbon required by the adsorption system, this quotient must be multiplied by the number of carbon beds in the system. At least two beds are required. That is, while one bed is adsorbing the solute, the second bed (already saturated) is being desorbed by the countercurrent passage of saturated steam or hot air through it. The mixture of steam and solute leaving the desorbed bed is then sent to a condenser where the mixture is liquefied, after which the organic layer is decanted off and, if valuable, recycled to the process or sold. (An example is the perchloroethylene solvent captured by dry-cleaning plant adsorbers and recycled.) Note that the carbon requirement computed above has units of mass (lb) per unit time (hr). However, since adsorption is essentially a batch operation, while the solute flow rate is continuous, a time element must be introduced to obtain the total carbon requirement. This element is the *adsorption time,* which may vary from minutes to hours, though it is typically *1* hr. It is the time between when a bed is put on-line to when it reaches saturation. Another variable to note is the *regeneration time,* the time to desorb a bed. Though this is usually less than the adsorption time, it can be greater, in which case *three* beds would be required. But if the larger of the two times is used in the design, one can still use a two-bed system. That is,

$$W_c = \frac{2 M_s \theta_{A,R}}{\text{W.C.}} \tag{82}$$

where

W_c = weight of carbon required, lb
M_s = solute mass rate, lb/hr
W.C. = working capacity, lb solute/lb carbon
$\theta_{A,R}$ = adsorption or regeneration time, hr, whichever is larger

With this equation, a two-bed system with the working capacity computed above, a toluene rate of 100 lb/hr, and a 1-hr adsorption time, the total carbon required would be approximately 1300 lb.

The carbon weight is the parameter used to price adsorbers. Depending on the size, adsorbers are either "package" or "custom" units. The former are generally used with commercial and small industrial applications, the latter, for larger industrial applications. (The dividing line is 10,000 lb of carbon.)

Purchase costs (in $) for package units can be computed from the following:[21]

$$C_{Ad(p)} = 10,000 + 721 W_c^{0.481} \tag{83}$$

where

$$W_c = \text{weight of carbon in system, lb}$$

and

$$250 \leq W_c \leq 10{,}000$$

These costs include the adsorber, carbon, blower or fan, controls, and steam regenerator. For more stringent applications, 30% should be added.

This next equation gives costs for custom adsorbers:

$$C_{Ad(c)} = 3200\, W_c^{0.371} \tag{84}$$

and

$$10{,}000 \leq W_c \leq 200{,}000$$

The custom adsorber costs also assume carbon steel construction. If stainless steel and other materials are required, multiply the purchase cost by one of the F_m factors given in Section 5.4.

Note also that, in addition to the adsorber vessels, Eq. (84) only gives the cost of instrumentation and controls. The cost of the other auxiliaries (piping, condenser, etc.) and the *carbon* must be computed separately. Figure the cost of carbon at $1.35/lb,[33] a typical cost for purchase lots greater than 10,000 lb. However, if smaller quantities are bought, this price would increase by as much as 25%.

Carbon can only be regenerated a limited number of times, after which it must be replaced. This carbon replacement cost is a function of the initial carbon cost, the life, and the discount rate used. Thus,

$$\text{replacement cost} = C_c \times \text{CRF} \tag{85}$$

where

CRF = the capital recovery factor [see Eq. (6)], the value of which depends on the discount rate and the life (typically 5 yr)

However, to avoid double counting, the capital charges for the control system should be calculated based on the installed cost *less* the initial cost of the carbon, C_c. To this product should be added the carbon replacement cost computed from Eq. (85), to obtain the *total* capital charges. For instance, if the total system cost, less carbon, were $150,000, C_c were $30,000, and the lives were 10 and 5 yr, respectively, at a 10% interest rate, the total annualized capital charges would be

$$\text{total capital charges} = \$150{,}000 \times \text{CRF} (10/10\%) + \$30{,}000 \times \text{CRF} (5/10\%)$$
$$= \$24{,}420 + \$7910$$
$$= \$32{,}330$$

If this approach were used, there would be no need to compute replacement carbon cost separately, since the capital charges would already account for it.

The utilities consumed by adsorption systems include steam, cooling water (for the condensers), and power, to overcome pressure drop through the bed and the rest of the system.

A rule-of-thumb for steam usage is 4 lb/lb solute recovered. This covers the amount of steam needed to heat the captured solute to its boiling point and to vaporize it, as well as the energy to heat the carbon, vessel, and so on. The cost of the steam, however, will depend on whether it is generated by the adsorber system or purchased. (See Table 7.)

The cooling water requirement is directly proportional to the steam usage, since it represents the amount of heat evolved when the steam–solute vapor condenses. It also depends on the allowable temperature rise of the cooling water in the condenser, which, in turn, depends on the source of the cooling water, facilities available for cooling, and so on. However, 20°F is representative. At this value, the cooling water requirement would be approximately 6 gal/lb steam.

The pressure drop through the adsorber depends on the gas velocity through the bed, the bed depth, and the adsorbent properties. For most control applications, where pressures are near atmospheric, the gas velocity should be approximately 100 ft/min.[32] At this velocity, the bed depth (Z_b) would be computed as follows:

$$Z_b = \frac{\text{adsorbent volume, ft}^3}{\text{adsorbent cross-sectional area, ft}^3}$$

$$= \frac{50\, W_c}{Q \rho_c} \tag{86}$$

where

$$Z_b = \text{depth of adsorbent bed, ft}$$
$$W_c = \text{weight of adsorbent in system, lb}$$
$$\rho_c = \text{density of adsorbent, lb/ft}^3$$
$$Q = \text{gas flow rate through bed, acfm}$$

For the example of carbon weight above (1300 lb), a flow rate of 1500 cfm, and a typical carbon density of 30 lb/ft³, $Z_b = 1.44$ ft.

For a popular carbon type and size—type BPL, 4×10 mesh—the pressure drop–velocity relationship is

$$\frac{P_b}{Z_b} = 0.0484 v_b^{0.6118} \exp (0.1006[\ln v_b]^2) \tag{87}$$

where

$$v_b = \text{gas velocity through the bed, ft/min}$$
$$\frac{P_b}{Z_b} = \text{pressure drop through bed, in. H}_2\text{O/ft}$$

Thus, at 100 ft/min and a bed depth of 1.44 ft, $P_b = 1.44$ ft $\times$ 6.84 in./ft = 9.85 in. H$_2$O.

The utility and other costs would, of course, be offset by credits for any solute recycled or sold. In many cases, in fact, these recovery credits exceed the total operating costs by significant amounts, so that the adsorber appears to yield a "profit." However, this does not imply that a firm would find it as economically advantageous to purchase an adsorber as it would to build another facility or invest its funds in, say, Treasury notes. Thus, the adsorber "profit" must be balanced against other factors— the availability of capital, current market conditions for the recovered product, and other investment opportunities open to the firm. Nonetheless, if applying an adsorber to a source should yield a net credit (rather than a cost) it would probably be the preferred means for controlling that source, if control were to be eventually applied.

6   COST ESCALATION

As stated in Section 1, all of the costs in this chapter reflect June 1981 prices. These prices, however, were not obtained from data sources of that period. Instead, these costs were *escalated* from some previous time to June 1981. These escalations have been made by multiplying the original cost by a cost *index*. That is,

$$\text{cost (June 1981)} = \text{cost (prior date)} \times \frac{\text{index (June 1981)}}{\text{index (prior date)}} \tag{88}$$

This simple procedure yields surprisingly accurate costs when compared to actual current cost data. Any deviations would certainly be within the overall error band of these estimates ($\pm 30\%$).

Cost indexes are commonly used to escalate both capital and O & M costs. However, in most cases, it is merely sufficient to escalate the capital cost items, since current prices for O & M items (e.g., electricity) are usually available to the estimator. Even if they are not available, studies have shown that capital cost indexes "track" quite well with most O & M-related indexes, such as the Wholesale and Consumer Prices Indexes. (This is hardly fortuitous, since control equipment vendors often utilize the WPI and CPI data when developing their price lists.)

Among the more commonly used of the published equipment cost indexes are the Chemical Engineering (CE) Plant and the Marshall and Swift (M & S) Indexes, both of which appear in the biweekly *Chemical Engineering* magazine. Updated monthly, the CE index is a composite of several components, ranging from *construction labor* and *engineering and supervision*, to equipment category-oriented indexes, such as *fabricated equipment* and *pipes, valves, and fittings*. Both the composite index and its components are published.

The M & S index is updated quarterly. Like the CE index, it is a composite of several components, all of which are published along with the overall index. These components show equipment cost trends for various industry categories, from *cement* to *rubber* to *electric power*. (See any recent issue of a *Chemical Engineering* magazine for a complete listing of these indexes and their components.)

Because it is better suited to control equipment costing, the CE index or its components have been used in escalating the costs in this chapter. With just a few exceptions, the *fabricated equipment* component has been used in escalating these costs. The exceptions are

1. *Pipes, valves, and fittings:* piping costs.
2. *Process machinery:* carbon adsorbers (package units only).
3. *Pumps and compressors:* fans and pumps.
4. *Electrical equipment:* fan and pump motors and starters.

In addition to updating costs from the past to the present, indexes may be used to project costs into the *future.* Although there is no little risk associated with this, if the projections are done rigorously and, particularly, with adequate amounts of historical data, the results would be credible.

For example, Mascio,[34] after plotting 8 yr of various indexes against time, found that the data fit two regression curves very well. The long-term trend lines (1971–1978) were linear, while the short-term regressions (1975–1978) were exponential. The equations for the CE plant index were

$$I_{CE} = -7.79 + 10.31\theta_l \qquad \text{(long-term)} \tag{89}$$
$$= 38.33 \exp(0.170\theta_s) \qquad \text{(short-term)} \tag{90}$$

where

$$\theta_l = \text{number of years after January 1971}$$
$$\theta_s = \text{number of years after January 1975}$$
$$I_{CE} = \textit{percentage increase} \text{ in the CE plant index}$$
$$\text{from the January 1971 value (128.2)}$$

To test the accuracy of these equations, consider three later dates for which indexes are available: December 1979, August 1980, and June 1981. The actual and predicted indexes are:

	Index		
Date	Actual	Predicted	% Difference
December 1979	247.6	243.2	−1.8
August 1980	264.9	257.0	−3.0
June 1981	298.2	276.6	−7.2

Computed from the short-term equation, the predicted indexes compare reasonably well with the actual, though the percentage difference gets larger with time. Of course, this comparison would be improved if more current data were used to update the regression curves. All in all, Mascio's technique affords a convenient way to project costs into the future, provided that the projections are not made too far in advance of the data base.

REFERENCES

1. *The Clean Air Act as Amended, August 1977* (serial no. 95-11), Washington, D.C., U.S. Government Printing Office, 1977.
2. Laseke, B. A., and Devitt, T. W., Status of Flue Gas Desulfurization, *Chemical Engineering Progress,* 37–50 (February 1979).
3. Godfrey, R. S. (ed.), *Building Construction Cost Data,* 35th annual edition, R. S. Means, Duxbury, MA, 1977.
4. Vatavuk, W. M., and Neveril, R. B., Estimating Costs of Air-Pollution Control Systems, Part II: Factors for Estimating Capital and Operating Costs, *Chemical Engineering,* 157–162 (November 3, 1980).
5. Peters, M. S., and Timmerhaus, K. D., *Plant Design and Economics for Chemical Engineers,* 3rd edition, McGraw-Hill, New York, 1980.
6. Humphries, K. K., and Katell, S., *Basic Cost Engineering,* Dekker, New York, 1981.
7. Chauvel, A., et al., *Manual of Economic Analysis of Chemical Processes,* McGraw-Hill, New York, 1980.
8. Vatavuk, W. M., and Neveril, R. B., Estimating Costs of Air-Pollution Control Systems, Part III: Estimating the Size and Cost of Pollutant Capture Hoods, *Chemical Engineering,* 111–115, (December 1, 1980).
9. *Industrial Ventilation,* 10th edition, American Conference of Governmental Industrial Hygienists, Lansing, MI, 1968.

10. Vatavuk, W. M., and Neveril, R. B., Estimating Costs of Air-Pollution Control Systems, Part IV: Estimating the Size and Cost of Ductwork, *Chemical Engineering,* 71–73 (December 29, 1980).

11. *Ibid,* Part V: Estimating the Size and Cost of Gas Conditioners, *Chemical Engineering,* 127–132 (January 26, 1981).

12. Grove, D. J., Cyclones, in *Control of Particulate Emissions,* Institute for Air Pollution Training, U.S. Environmental Protection Agency, Research Triangle Park, NC, 1972.

13. Vatavuk, W. M., and Neveril, R. B., Estimating Costs of Air-Pollution Control Systems, Part VI: Estimating Costs of Dust-Removal and Water-Handling Equipment, *Chemical Engineering,* 223–228 (March 23, 1981).

14. *Ibid,* Part VII: Estimating Costs of Fans and Accessories, *Chemical Engineering,* 171–177 (May 18, 1981).

15. *Capital Costs of Free-Standing Stacks,* EPA Contract 68-02-0299, Cincinnati, Vulcan-Cincinnati, August 1973.

16. Vatavuk, W. M., and Neveril, R. B., Estimating Costs of Air-Pollution Control Systems, Part VIII: Estimating Costs of Exhaust Stacks, *Chemical Engineering,* 129–130 (June 15, 1981).

17. *Ibid,* Part IX: Costs of Electrostatic Precipitators, *Chemical Engineering,* 139–140 (September 7, 1981).

18. *Ibid,* Part X: Estimating the Size and Cost of Venturi Scrubbers, *Chemical Engineering,* 93–99 (November 30, 1981).

19. *McIlvaine Scrubber Manual,* The McIlvaine Co., Northbrook, IL, 1974.

20. Rheinfrank, L., Particulate Collection on Outside of Bags Offers Cost and Maintenance Savings on Large Volume Exhausts, *Pollution Engineering,* 48–50 (October 1977).

21. Neveril, R. B., *Capital and Operating Costs of Selected Air Pollution Control Systems,* EPA-450/5-80-002, GARD, Inc., Niles, IL, 1978.

22. Price data from Fuller Company, Catasaqua, PA, 1977.

23. Bergmann, L., Baghouse Filter Fabrics, *Chemical Engineering,* 177–178 (October 19, 1981).

24. Dennis, R., et al., *Fabric Filter Model Sensitivity Analysis: Energy/Environment R & D Program Report,* EPA-600/7-79-043c, GCA Corporation, Bedford, MA, April 1979.

25. Fraser, M. D., and Foley, G. J., Cost Models for Fabric Filter Systems, *Transactions of the 67th Annual Meeting of the Air Pollution Control Association,* June 1974.

26. Blackburn, J. W., *Emissions Control Options for the Synthetic Organic Chemicals Manufacturing Industry, Control Device Evaluation: Thermal Oxidation,* Contract 68-02-2577, U.S. Environmental Protection Agency, Research Triangle Park, NC, December 1979.

27. Key, J. A., *Emissions Control Options for the Synthetic Organic Chemicals Manufacturing Industry, Control Device Evaluation: Catalytic Oxidation,* Contract 68-02-2577, U.S. Environmental Protection Agency, Research Triangle Park, NC, June 1980.

28. Straitz, J. F., III, Flaring with Maximum Energy Conservation, *Pollution Engineering,* 47–49 (February 1980).

29. Mulet, A., et al., Estimate Costs of Distillation and Absorption Columns via Correlations, *Chemical Engineering,* 77–82 (December 28, 1981).

30. Barrett, O. H., Installed Cost of Corrosion-Resistant Piping, *Chemical Engineering,* 97–102 (November 2, 1981).

31. Grant, R. J., et al., Adsorption of Normal Paraffins and Sulfur Compounds on Activated Carbon, *AIChE Journal,* 403–406 (1962).

32. *Adsorption Handbook,* Pittsburgh Activated Carbon Division, Calgon Corporation, 1980.

33. Price data from Calgon Corporation, Pittsburgh, PA, 1980.

34. Mascio, N. E., Predict Costs Reliably via Regression Analysis, *Chemical Engineering,* 115–121 (February 12, 1979).

CHAPTER 15
FOSSIL FUEL COMBUSTION

TIMOTHY W. DEVITT

PEDCo Environmental, Inc.
Cincinnati, Ohio

1 INTRODUCTION

Annual fossil fuel consumption in the United States has totaled approximately 75×10^{15} Btu in recent years. Consumption in boilers accounts for one-third of this total, far exceeding transportation, which is the next largest usage (19×10^{15} Btu/yr). Residential usage and direct heating of processes are other major categories of fossil fuel consumption, using about 14.4×10^{15} and 8.6×10^{15} Btu, respectively. Combustion by these four categories accounts for approximately 89% of the total fossil fuel usage; the balance is used for feedstocks, raw materials, and other miscellaneous uses.

In addition to being the largest usage category, boilers consume most of the "dirty" fuels (coal and residual oil). Hence, boilers are, by virtue of the amount and type of fuel burned, the largest single source of sulfur oxide emissions, and are major sources of both particulate matter and nitrogen oxide emissions. Although the emission contributions of other combustion sources are important, this handbook focuses on boiler combustion and the three major pollutants.

The boiler population is frequently divided into three sectors: utility boilers for generating electricity; industrial boilers that generate steam or hot water for process heat, space heat, for generation of electricity; and commercial boilers for space heating. The primary fuels consumed by these boilers are natural gas, oil, and coal. The amounts and types of emissions are largely affected by quantity and type of fuel as well as by type of boiler. Sections 2 and 3 discuss the basis of combustion processes and describe the sources (boiler population). Section 4 discusses the major fuels and the resultant emissions from each.

The development and application of boiler emission control systems largely reflect the quantity and nature of the pollutants as well as the cost of controls. For example, systems for particulate emission control are widely applied in all sectors and represent a relatively common degree of technology. On the other hand, systems for control of sulfur oxides are not nearly as widespread at the present time. The technology is relatively new and expensive as evidenced by its limited applications to relatively small numbers of large utility and industrial boilers. Nitrogen oxide control systems are the most recently available pollutant control technique. They are not found at many installations, but a fair amount of research effort is being directed toward their development. Control systems and their costs are discussed in the final two sections.

2 COMBUSTION PROCESSES

2.1 Introduction

Combustible matter in fuels is composed mainly of three elements: carbon, hydrogen, and sulfur. Combustion is the rapid combination of oxygen with these fuel elements resulting in generation of useful heat. For most fuels, only carbon and hydrogen are important since the percentage of sulfur is so low as to be negligible in producing heat.

Table 1 Combustion Constants[2]

No.	Substance	Formula	Molecular Weight	lb/ft³	ft³/lb	Sp gr air = 1.0000	Heat of Combustion Btu/ft³ Gross (High)	Net (Low)	Heat of Combustion Btu/lb Gross (High)	Net (Low)	For 100% Total Air mole/mole of Combustible or ft³/ft³ of Combustible — Required for Combustion O₂	N₂	Air	Flue Products CO₂	H₂O	N₂	For 100% Total Air lb/lb of Combustible — Required for Combustion O₂	N₂	Air	Flue products CO₂	H₂O	N₂
1	Carbon[a]	C	12.01	—	—	—	—	—	14,093	14,093	1.0	3.76	4.76	1.0	—	3.76	2.66	8.86	11.53	3.66	—	8.86
2	Hydrogen	H_2	2.016	0.0053	187.723	0.0696	325	275	61,095	51,623	0.5	1.88	2.38	—	1.0	1.88	7.94	26.41	34.34	—	8.94	26.41
3	Oxygen	O_2	32.00	0.0846	11.819	1.1053	—	—	—	—	—	—	—	—	—	—	—	—	—	—	—	—
4	Nitrogen (atm)	N_2	28.01	0.0744	13.443	0.9718	—	—	—	—	—	—	—	—	—	—	—	—	—	—	—	—
5	Carbon monoxide	CO	28.01	0.0740	13.506	0.9672	321	321	4,347	4,347	0.5	1.88	2.38	1.0	—	1.88	0.57	1.90	2.47	1.57	—	1.90
6	Carbon dioxide	CO_2	44.01	0.1170	8.548	1.5282	—	—	—	—	—	—	—	—	—	—	—	—	—	—	—	—
Paraffin Series																						
7	Methane	CH_4	16.04	0.0425	23.552	0.5543	1012	911	23,875	21,495	2.0	7.53	9.53	1.0	2.0	7.53	3.99	13.28	17.27	2.74	2.25	13.28
8	Ethane	C_2H_6	30.07	0.0803	12.455	1.0488	1773	1622	22,323	20,418	3.5	13.18	16.68	2.0	3.0	13.18	3.73	12.39	16.12	2.93	1.80	12.39
9	Propane	C_3H_8	44.09	0.1196	8.365	1.5617	2524	2322	21,669	19,937	5.0	18.82	23.82	3.0	4.0	18.82	3.63	12.07	15.70	2.99	1.63	12.07
10	n-Butane	C_4H_{10}	58.12	0.1582	6.321	2.0665	3271	3018	21,321	19,678	6.5	24.47	30.97	4.0	5.0	24.47	3.58	11.91	15.49	3.03	1.55	11.91
11	Isobutane	C_4H_{10}	58.12	0.1582	6.321	2.0665	3261	3009	21,271	19,628	6.5	24.47	30.97	4.0	5.0	24.47	3.58	11.91	15.49	3.03	1.55	11.91
12	n-Pentane	C_5H_{12}	72.15	0.1904	5.252	2.4872	4020	3717	21,095	19,507	8.0	30.11	38.11	5.0	6.0	30.11	3.55	11.81	15.35	3.05	1.50	11.81
13	Isopentane	C_5H_{12}	72.15	0.1904	5.252	2.4872	4011	3708	21,047	19,459	8.0	30.11	38.11	5.0	6.0	30.11	3.55	11.81	15.35	3.05	1.50	11.81
14	Neopentane	C_5H_{12}	72.15	0.1904	5.252	2.4872	3994	3692	20,978	19,390	8.0	30.11	38.11	5.0	6.0	30.11	3.55	11.81	15.35	3.05	1.50	11.81
15	n-Hexane	C_6H_{14}	86.17	0.2274	4.398	2.9704	4768	4415	20,966	19,415	9.5	35.76	45.26	6.0	7.0	35.76	3.53	11.74	15.27	3.06	1.46	11.74

Olefin Series, *Aromatic Series*, and *Miscellaneous Gases* — combustion and physical property data.

No.	Substance	Formula	Mol Wt	lb per cu ft	cu ft per lb	Sp Gr	Heat of Combustion, Btu/cu ft Gross	Heat of Combustion, Btu/cu ft Net	Heat of Combustion, Btu/lb Gross	Heat of Combustion, Btu/lb Net	For combustion (cu ft/cu ft) O₂	N₂	Air	Flue products (cu ft/cu ft) CO₂	H₂O	N₂	For combustion (lb/lb) O₂	N₂	Air	Flue products (lb/lb) CO₂	H₂O	N₂
Olefin Series																						
16	Ethylene	C₂H₄	28.05	0.0742	13.475	0.9740	1604	1503	21,636	20,275	3.0	11.29	14.29	2.0	2.0	11.29	3.42	11.39	14.81	3.14	1.29	11.39
17	Propylene	C₃H₆	42.08	0.1110	9.007	1.4504	2340	2188	21,048	19,687	4.5	16.94	21.44	3.0	3.0	16.94	3.42	11.39	14.81	3.14	1.29	11.39
18	n-Butene	C₄H₈	56.10	0.1480	6.756	1.9336	3084	2885	20,854	19,493	6.0	22.59	28.59	4.0	4.0	22.59	3.42	11.39	14.81	3.14	1.29	11.39
19	Isobutene	C₄H₈	56.10	0.1480	6.756	1.9336	3069	2868	20,737	19,376	6.0	22.59	28.59	4.0	4.0	22.59	3.42	11.39	14.81	3.14	1.29	11.39
20	n-Pentene	C₅H₁₀	70.13	0.1852	5.400	2.4190	3837	3585	20,720	19,359	7.5	28.23	35.73	5.0	5.0	28.23	3.42	11.39	14.81	3.14	1.29	11.39
Aromatic Series																						
21	Benzene	C₆H₆	78.11	0.2060	4.852	2.6920	3752	3601	18,184	17,451	7.5	28.23	35.73	6.0	3.0	28.23	3.07	10.22	13.30	3.38	0.69	10.22
22	Toluene	C₇H₈	92.13	0.2431	4.113	3.1760	4486	4285	18,501	17,672	9.0	33.88	42.88	7.0	4.0	33.88	3.13	10.40	13.53	3.34	0.78	10.40
23	Xylene	C₈H₁₀	106.16	0.2803	3.567	3.6618	5230	4980	18,650	17,760	10.5	39.52	50.02	8.0	5.0	39.52	3.17	10.53	13.70	3.32	0.85	10.53
Miscellaneous Gases																						
24	Acetylene	C₂H₂	26.04	0.0697	14.344	0.9107	1477	1426	21,502	20,769	2.5	9.41	11.91	2.0	1.0	9.41	3.07	10.22	13.30	3.38	0.69	10.22
25	Naphthalene	C₁₀H₈	128.16	0.3384	2.955	4.4208	5854	5654	17,303	16,708	12.0	45.17	57.17	10.0	4.0	45.17	3.00	9.97	12.96	3.43	0.56	9.97
26	Methyl alcohol	CH₃OH	32.04	0.0846	11.820	1.1052	868	767	10,258	9,066	1.5	5.65	7.15	1.0	2.0	5.65	1.50	4.98	6.48	1.37	1.13	4.98
27	Ethyl alcohol	C₂H₅OH	46.07	0.1216	8.221	1.5890	1600	1449	13,161	11,917	3.0	11.29	14.29	2.0	3.0	11.29	2.08	6.93	9.02	1.92	1.17	6.93
28	Ammonia	NH₃	17.03	0.0456	21.914	0.5961	441	364	9,667	7,985	0.75	2.82	3.57	—	1.5	3.32	1.41	4.69	6.10	—	1.59	5.51
29	Sulfur[a]	S	32.06	—	—	—	—	—	3,980	3,980	1.0	3.76	4.76	1.0 (SO₂)	—	3.76	1.00	3.29	4.29	2.00 (SO₂)	—	3.29
30	Hydrogen sulfide	H₂S	34.08	0.0911	10.979	1.1898	646	595	7,097	6,537	1.5	5.65	7.15	1.0 (SO₂)	1.0	5.65	1.41	4.69	6.10	1.88 (SO₂)	0.53	4.69
31	Sulfur dioxide	SO₂	64.06	0.1733	5.770	2.2640	—	—	—	—	—	—	—	—	—	—	—	—	—	—	—	—
32	Water vapor	H₂O	18.02	0.0476	21.017	0.6215	—	—	—	—	—	—	—	—	—	—	—	—	—	—	—	—
33	Air		—	0.0766	13.063	1.0000	—	—	—	—	—	—	—	—	—	—	—	—	—	—	—	—

[a] Carbon and sulfur are considered as gases for molal calculations only.

All gas volumes corrected to 60°F and 30 in. Hg dry.

Note: This table is included by courtesy of the American Gas Association and the Industrial Press. The format and data are taken principally from *Fuel Flue Gases*, 1941 edition, American Gas Association, with modifications, especially in the four columns labeled "Heat of Combustion," using data from *Gas Engineers' Handbook*, the Industrial Press, 1965.

Every combustion process requires (1) sufficient time to complete the chemical reactions, (2) sufficient temperature to heat the fuel through its various decomposition stages and to ignite the carbon and hydrogen, and (3) sufficient turbulence to mix the oxygen and fuel elements to ensure complete combustion. The basic combustion reactions are:[1]

Carbon to carbon dioxide	$C + O_2 \rightarrow CO_2$ + heat
Carbon to carbon monoxide	$2C + O_2 \rightarrow 2CO$ + heat
Carbon monoxide to carbon dioxide	$2CO + O_2 \rightarrow 2CO_2$ + heat
Hydrogen to water vapor	$2H_2 + O_2 \rightarrow 2H_2O$ + heat
Sulfur to sulfur dioxide	$S + O_2 \rightarrow SO_2$ + heat

Ignition temperature is the temperature that must be attained or exceeded in the presence of oxygen to cause combustion. When ignition temperature is exceeded, more heat is generated by the reaction than is lost to the surroundings and combustion becomes self-sustaining. Ignition temperatures increase in order for coal, liquid fuels, and gaseous fuels.

The heat of combustion (i.e., the quantity of heat evolved by burning a unit of fuel) differs with the compound being burned. Ideally, a given weight of fuel combines with a stoichiometric weight of oxygen to form the products of combustion. Table 1 presents some of the combustion constants applicable to many hydrocarbon compounds.

2.2 Fuel Types and Heating Values

2.2.1 Solid Fuels

The primary solid fuel in use today is coal. The characteristics of coal vary greatly by deposit. Coal is classified into classes and groups based on the fixed carbon content and the heating value (on a mineral-free basis). Table 2 shows the range of coal characteristics.

Other solid fuels include wood, charcoal, tanbark, bagasse, and solid wastes. Combustion equipment types applicable to many of these fuels are presented in Table 3.

2.2.2 Liquid Fuels

Liquid fuels are reduced to small droplets (atomization) to increase the surface area available for combining with oxygen. These droplets are easily vaporized when exposed to furnace temperatures, and are in gaseous form during combustion. The degree of atomization controls the amount of excess air required to assure complete combustion. Two major classes of liquid fuel burners are (1) burners that vaporize the liquid within the burner and (2) burners that atomize the liquid so vaporization will occur in the combustion space. Properties and typical-use equipment for petroleum-derived fuel oils are shown in Table 4.

Nonpetroleum liquid fuels include coal tar, oil shale, and numerous waste streams in industrial processes of varied composition. Liquefied products from coal conversion processes and extraction from tar sands are other liquid fuels.

2.2.3 Gaseous Fuels

Gaseous fuels, very easily dispersed in air, require no fuel preparation. When gas and air are mixed prior to ignition, burning proceeds by blue flame hydroxylation. Cracking, or yellow flame burning, occurs when oxygen is added to the fuel after both have been heated. Soot and carbon black can be formed in yellow flame burning if insufficient oxygen is present or if the combustion process is stopped before completion.

Natural gas (primarily methane) is the dominant gaseous fuel used today. Heating values range from 900 to 1200 Btu/ft³, but average about 1000 Btu/ft³. The components of natural gas are depicted in Table 5.

Table 2   Typical Chemical Composition and Heating Value for Various Types of Coal[3]

Type	Carbon, %[a]	Ash, %	Sulfur, %	M Btu/lb
Anthracite	75–90	10–20	0.6–0.8	11.8–13.0
Bituminous	50–80	5–25	0.7–4.5	10.5–14.4
Subbituminous	45–65	5–25	0.7–4.5	8.5–10.2
Lignite	35–50	5–12	1.0	6.3

[a] Total carbon-ultimate analysis.

Table 3 Burning Equipment for Solid Fuels[5]

Fuel	Source	Stokers				Pulverized Fuel		Crushed	
		Underfeed	Chain Grate	Chain Grate, Jet Ignition	Spreader	U Flame, Pulverized Fuel	Horizontal Burners, Pulverized Fuel	Cyclone (Where Fusion Temperature Is Suitable)	Cell Furnaces
Coke breeze			x						
Anthracite	E. PA		x c[a]			x			
Bituminous coal:									
17–27% volatile	WV, Central PA	x			x	x	x	x	
27–35% volatile									
Strongly coking	W. PA, WV, KY, OH, UT	x		x	x	x	x	x	
Weakly coking	IN, IA, IL, CO, W. KY			x	x	x	x	x	
Pipeline slurry						x	x	x	
Lignite	ND, SD, MT, WY, TX			x	x	x	x	x	
Low-temperature fluid-coal char						x	x Aux[a]	x	
Petroleum coke, 9–14% volatile						x	x	x	
Fluid petroleum coke, 4–5% volatile						x	x Aux[a]	x	
Wood and bark[b]					x			x Aux[a]	x
Bagasse					x			x	x

Equipment indicated will usually result in a good application, but there are many factors affecting the burning of fuel, and variations of fuel properties that guide the individual selection of burning equipment.

[a] c = coarse sizes only. Aux = auxiliary fuel—coal, oil, or gas.

[b] Bark and wood are only burned on inclined grates and in Dutch oven pile furnaces.

Table 4 Properties of Fuel Oils[a]

Grade	Description	Flash Point min, °F	Water and Sediment, % by Volume	Maximum Ash, wt%	Suggested Temperature, °F	Average Heat Content, Btu/gal	Sulfur Content Range	Size Installation Using Fuel
1	A distillate oil intended for vaporizing pot-type burners and other burners requiring this grade of fuel	100	Trace	—	—	136,000	0.10–0.50	—
2	A distillate oil for general-purpose domestic heating, for use in burners not requiring no. 1 fuel oil	100	0.10	—	—	142,000	0.10–1.00	Domestic units
4	An oil for burner installations not equipped with preheating facilities	130	0.50	0.10	—	145,000	0.20–2.00	200 hp
5	A residual-type oil for burner installations equipped with preheating facilities	130	1.00	0.10	170–220	148,000	0.50–3.00	200–1000 hp
6	An oil for use in burners equipped with preheaters permitting use of high-viscosity fuel	150	2.00	—	220–260	151,000	0.70–4.00	1000 hp

Table 5 Analyses of Natural Gas[a] (Ref. 6)

City	Components of Gas, % by Volume								Miscellaneous	Heating value,[b] Btu/ft^3	Specific Gravity
	Methane	Ethane	Propane	Butanes	Pentanes	Hexanes Plus	CO$_2$	N$_2$			
Baltimore, MD	94.40	3.40	0.60	0.50	0.00	0.00	0.60	0.50	—	1051	0.590
Birmingham, AL	93.14	2.50	0.67	0.32	0.12	0.05	1.06	2.14	—	1024	0.599
Boston, MA	93.51	3.82	0.93	0.28	0.07	0.06	0.94	0.39	—	1057	0.604
Columbus, OH	93.54	3.58	0.66	0.22	0.06	0.03	0.85	1.11	—	1028	0.597
Dallas, TX	86.30	7.25	2.78	0.48	0.07	0.02	0.63	2.47	—	1093	0.641
Houston, TX	92.50	4.80	2.00	0.30	—	—	0.27	0.13	—	1031	0.623
Kansas City, MO	72.79	6.42	2.91	0.50	0.06	Trace	0.22	17.10	—	945	0.695
Los Angeles, CA	86.50	8.00	1.90	0.30	0.10	0.10	0.50	2.60	—	1084	0.638
Milwaukee, WI	89.01	5.19	1.89	0.66	0.44	0.02	0.00	2.73	0.06 He	1051	0.627
New York, NY	94.52	3.29	0.73	0.26	0.10	0.09	0.70	0.31	—	1049	0.595
Phoenix, AZ	87.37	8.11	2.26	0.13	0.00	0.00	0.61	1.37	—	1071	0.633
Salt Lake City, UT	91.17	5.29	1.69	0.55	0.16	0.03	0.29	0.82	—	1082	0.614
San Francisco, CA	88.69	7.01	1.93	0.28	0.03	0.00	0.62	1.43	0.01 He	1086	0.624
Washington, DC	95.15	2.84	0.63	0.24	0.05	0.05	0.62	0.42	—	1042	0.586

[a] Average analyses (1954 data) obtained from the operating utility company(s) supplying the city. The gas supply may vary considerably from these data, especially where more than one pipeline supplies the city. Also, as new supplies may be received from other sources, the analyses may change. Peak shaving (if used) is not accounted for in these data.

[b] Gross or higher heating value at 30 in. Hg, 60°F, dry. To convert to a saturated basis, deduct 1.73%, i.e., 17.3 from 1000, 19 from 1100.

Table 6   Composition of Combustion Air[7]

Dry Atmospheric Air

The volumetric composition of dry atmospheric air given in NACA Report 1235 (Standard Atmosphere—Tables and Data for Altitudes to 65,800 ft, November 20, 1952),[1] and the molecular weights of the gases constituting dry air are as follows:

	Volume, %	Mol. Wt.
Nitrogen	78.09	28.016
Oxygen	20.95	32.000
Argon	0.93	39.944
Carbon dioxide	0.03	44.010

(Neon, helium, krypton, hydrogen, xenon, ozone, and radon, combined, are less than 0.003%.)

Dry air with this composition has an apparent molecular weight of 28.97 lb/lb-mole and a density at 35°F and 14.7 psia of $28.97 \div 359 = 0.0807$ lb/ft³. The oxygen content is 23.14% by weight. The lb dry air/lb oxygen $= 1 \div 0.2314 = 4.32$.

Other gaseous fuels include liquefied petroleum gas (LPG), oil gases, producer gas, blast furnace gas, acetylene, hydrogen, and other industry-specific mixtures of by-product off-gases.

2.3   Stoichiometry

Almost all combustion processes utilize air rather than pure oxygen. Table 6 presents calculations of the composition of air. Since air is a mechanical mixture of gases, calculations may be performed to identify the exact proportion of air required to combust the fuel in question. The result of such calculations is presented graphically in Figure 1 for the combustion of coal.

The amount of air used in a combustion system depends on the stoichiometric amount of oxygen for complete combustion and the degree of mixing. With ideal mixing, the theoretical air-to-fuel ratio provides complete combustion but mixing is never ideal and excess air is needed to completely burn the fuel.

As shown in Figure 2, increasing excess air decreases the amount of unburned fuel and increases combustion efficiency while diluting and cooling the combustion gases. At a given point more heat is lost in the stack gases by further increasing excess air than is gained by releasing the remaining heat of combustion. This point is the point of maximum overall thermal efficiency.

Some commercial gas burners can achieve complete combustion with 0–10% excess air. Liquid fuels, less easily mixed with air, require 3–20% excess air. Coal requires 10–50% excess air for economical combustion and still leaves some unburned carbon in the ash residue. Table 7 presents typical excess air quantities used in combustion processes.

Fuel and oxygen can be mixed either as part of fuel preparation (premix) or immediately before combustion (burner mix). In general heat applications, burner mix is most common; in applications requiring precise or stoichiometric conditions, premix burning is used. Premix burning, called hydroxylation, is characterized by a small blue flame with little or no luminescence. Burner mix combustion thermally cracks the fuel and produces a yellow luminous flame. Some burning techniques employ a premix of primary air, followed by a mixture of secondary air. A high primary air rate produces a short, blue flame, whereas a low primary air rate results in a long, luminous flame.

2.4   Fuel Efficiencies

The efficiency of combustion of any fuel is the ratio of heat recovered in usable form (e.g., steam or hot water) to total heat input. Banks of heat-exchange tubes absorb heat through convection. Heat is also recovered by radiation (primarily in the flame zone area) and conduction. Multipass boiler designs and numerous tube and other heat-exchange sections are used to extract heat prior to exhausting the flue gases. Figures 3 and 4 present a view of the complexity of a pulverized-coal-fired boiler system, and the relationship of heat absorbed to heat transfer surface, respectively.

Fuel efficiencies in boiler systems range from 80 to 90% based on the type of equipment, system

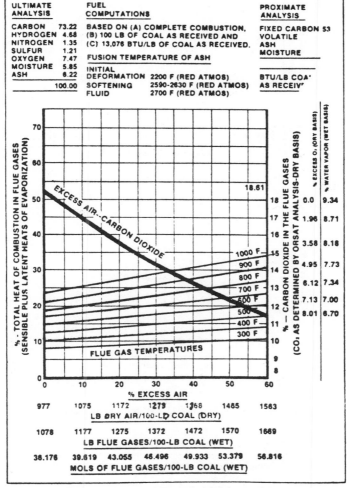

Fig. 1 Stoichiometric calculations of a bituminous coal sample.[8]

load, fuel type, excess air, fuel-to-air ratio, ductwork and boiler air inleakage, pollution control equipment, fan maintenance, and a variety of other factors. Typical utility-size boilers may optimally operate at 85–90% efficiency. Home furnaces typically convert about 80% of the fuel input energy to useful heat. The equation used to calculate the thermal efficiency of a boiler is[13]

$$\text{efficiency} = \frac{W_1(H_1 - h_1) + W_2(H_2 - h_2)}{C} \times 100\%$$

where

W_1 = actual initial evaporation, lb/hr
W_2 = actual steam reheated (if any), lb/hr
C = gross heating value of fuel burned, Btu/hr
H_1 = enthalpy in initial steam, Btu/lb
H_2 = enthalpy in reheat steam (if any), Btu/lb
h_1 = enthalpy of feed water, Btu/lb
h_2 = enthalpy in steam entering reheater (if any), Btu/lb

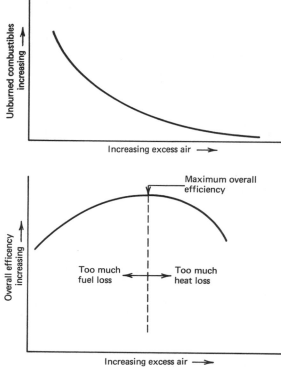

Fig. 2 Effect of excess air on combustion efficiency.[9]

Table 7 Usual Amount of Excess Air Supplied to Fuel-Burning Equipment[10]

Fuel	Type of Furnace or Burners	Excess Air % by Weight
Pulverized coal	Completely water-cooled furnace for slag-tap or dry-ash removal	15–20
	Partially water-cooled furnace for dry-ash removal	15–40
Crushed coal	Cyclone furnace—pressure or suction	10–15
Coal	Spreader stoker	30–60
	Water-cooled vibrating grate stoker	30–60
	Chain-grate and traveling grate stokers	15–50
	Underfeed stoker	20–50
Fuel oil	Oil burners, register-type	5–10
	Multifuel burners and flat-flame	10–20
Acid sludge	Cone and flat-flame-type burners, steam-atomized	10–15
Natural, coke-oven, and refinery gas	Register-type burners	5–10
	Multifuel burners	7–12
Blast-furnace gas	Intertube nozzle-type burners	15–18
Wood	Dutch oven (10–23% through grates) and Hofft-type	20–25
Bagasse	All furnaces	25–35
Black liquor	Recovery furnaces for kraft and soda-pulping processes	5–7

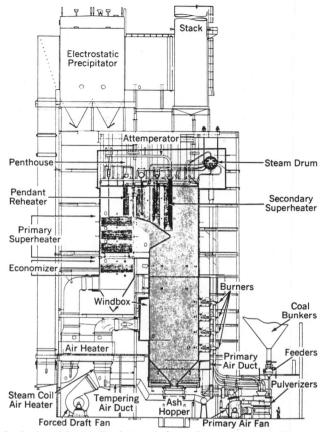

Fig. 3 Typical pulverized-coal-fired boiler heat transfer surface arrangement (Courtesy of Babcock & Wilcox).[11]

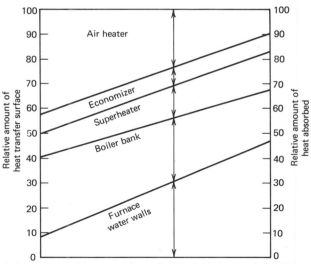

Fig. 4 Relative amount of heat transfer surface and relative amount of heat absorbed.[12]

385

REFERENCES FOR SECTION 2

1. Considine, D. M., *Energy Technology Handbook,* McGraw-Hill, New York, 1977, pp. 9–38.
2. *Steam—Its Generation and Use,* 38th edition, Babcock and Wilcox Company, Barberton, Ohio, 1975, p. 6-2.
3. Annual Book and ASTM Standards, ASTM-D388, American Society for Testing and Materials, 1977.
4. Annual Book and ASTM Standards, ASTM-D396, American Society for Testing and Materials, 1980.
5. Marks and Baumeister, *Standard Handbook for Mechanical Engineers,* ASME, New York, 1967.
6. *Gas Engineers' Handbook,* American Gas Association, 1965.
7. Singer, J. G., *Combustion—Fossil Power Systems,* 3rd edition, 1981, p. 4-2.
8. Harju, J. B., Coal Combustion Chemistry, *Pollution Engineering* 12(5), 54 (1980).
9. Center for Professional Advancement, North American Combustion Short Training Course Material, Applied Combustion Technology, 1979.
10. Ref. 2, p. 6-10.
11. Babcock and Wilcox Company.
12. Ref. 2.
13. *Useful Tables for Engineers and Steam Users,* 13th edition, Babcock and Wilcox Company, Barberton, Ohio, 1978, p. 68.

3   COMBUSTION SOURCES

3.1   Types of Boilers[1]

Boilers are classified based on their heat transfer configuration. There are three major types: water tube, fire tube, and cast iron. Water-tube and fire-tube boilers are used in a variety of applications ranging from supplying process steam to providing space heat, although water-tube units are available in much larger sizes. Cast-iron boilers, which are more limited in size, are generally used only to supply space heat.

3.1.1   *Water-Tube Boilers*

A water-tube boiler contains heat transfer tubes which are interconnected to common water channels and to steam outlets. Figure 5 is a simplified diagram of a water-tube boiler. The hot combustion gases contact the outside of the heat transfer tubes and the boiler water and steam flow inside the tubes. These boilers can generate high-pressure, high-temperature steam, and they are available as packaged or field-erected units. Capacities of packaged units range from 10,000 to 250,000 lb steam/hr; units of higher capacity are field erected. These boilers can fire all types of fuels.

Coal-fired water-tube boilers are either stoker fired or pulverized-coal fired. In a stoker unit, a conveying system both feeds coal into the furnace and provides a moving grate upon which the coal is burned. Stokers are generally used on units rated at less than 600×10^6 Btu/hr heat input. In pulverized-coal boilers, the coal is pulverized to the consistency of talcum powder and injected into

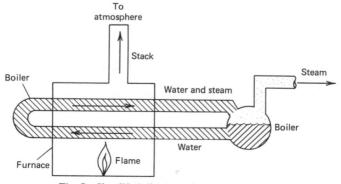

Fig. 5   Simplified diagram of a water-tube boiler.[2]

the furnace pneumatically where it is burned in suspension. The maximum capacity of individual burners in these boilers is 165×10^6 Btu/hr, but as many as 70 burners may be used (although 16 to 30 burners is more common).

Stoker-fired water-tube boilers may be further classified based on the coal feeding mechanism: underfeed, overfeed, and spreader. There are many variations on underfeed retort stokers, depending on whether the coal is fed horizontally or by gravity, whether the ash is discharged from the end or the sides, and the number of retorts. These units are best suited for bituminous coals and can be designed to generate up to 400,000 lb steam/hr. Overfeed stokers contain an endless belt of assembled links or grates that passes through the furnace. Coal is fed onto the moving assembly into the combustion zone of the furnace. Ash is discharged at the far end of the grate. Such chain-grate and traveling-grate stokers can be designed to burn all kinds of solid fuels with outputs up to 300,000 lb steam/hr. Spreader stokers combine suspension burning with a thin, fast-burning fuel bed on a grate. Feeder units distribute the fuel over the grate and forced-draft systems provide both undergrate and overgrate air. Spreader stoker capacities range up to 400,000 lb steam/hr.

3.1.2 Fire-Tube Boilers

A fire-tube boiler contains heat transfer tubes surrounded by a water basin. Figure 6 is a simplified diagram. The water basin absorbs heat both through the shell and the tubes, producing steam more efficiently than in a simple shell boiler. In most fire-tube boilers the combustion chamber is enclosed in the boiler shell. In other models the boiler shell and combustion chamber are separate. Those with internal furnaces provide better water circulation and easier ash removal. Fire-tube boilers are susceptible to structural failure when subjected to large variations in steam demand, thus they are generally used where loads are relatively constant. Capacities range up to about 20×10^6 Btu/hr heat input.

There are six basic configurations of fire-tube boilers: horizontal return tubular (HRT), Scotch marine, vertical, locomotive, short firebox, and compact. The most common are the HRT, Scotch marine, and vertical designs.

An HRT boiler contains heat transfer tubes which are horizontal to the ground. The fuel firing mechanism is located at one end, and the entire furnace is set on rollers or suspended on hangers to allow for expansion and contraction. The products of combustion make multiple passes through the water medium. HRT boilers range in size from about 0.5 to 20×10^6 Btu/hr heat input. In the smaller sizes, the combustion products make two passes through the system; larger sizes are four-pass units. All types of fuels may be used with HRT boilers, but coal firing may cause scaling and slagging.

Scotch marine boilers are small, self-contained, and portable. Capacities range up to about 12×10^6 Btu/hr heat input. As with HRT boilers, all types of fuels may be fired, but use of coal can cause scaling and slagging.

A vertical boiler contains fire tubes which rise straight up from a lower water-cooled furnace. The combustion products make only a single pass through the system. These units are also self-contained and portable and are available in sizes up to about 2.5×10^6 Btu/hr heat input. All types of fuels may be fired.

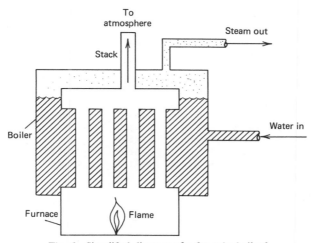

Fig. 6 Simplified diagram of a fire-tube boiler.[3]

3.1.3 Cast-Iron Boilers

A cast-iron boiler consists of several vertical sections. Water enters each section through its bottom, and steam or hot water exits through the top. The combustion products pass through a network of tubes in each section where heat is transferred to the boiler water. Cast-iron boilers are used to produce either low-pressure steam or hot water. The capacity is determined by the number of vertical sections. Available capacities range up to about 14×10^6 Btu/hr heat input. These boilers require little maintenance and can handle overloading or demand surges. The vertical sections tend to deform, however, which may allow leakage of flue gases from the joints.

Smaller cast-iron boilers used in domestic applications generally contain a furnace surrounded by a water basin which is lanced with flues. The heat of combustion is transferred from the gases to the water through these flues.

3.2 The U.S. Boiler Population[5]

Table 8 presents an estimate of the distribution of the U.S. boiler population by fuel type for 1977. The total capacity of boilers in the utility sector is over 80% of that in the industrial and commercial sectors combined. Over half of the total capacity is accounted for by coal and residual oil, the two fuels that present the most significant pollution problems. Natural gas is the dominant fuel in the industrial and commercial sectors. Replacement of a significant part of this natural gas capacity by new coal- or oil-fired units could greatly increase the levels of air pollutants generated by boilers in these sectors.

Although the total capacity of utility boilers is nearly that of those in the industrial and commercial sectors, there is a great disparity in terms of numbers. Utility boilers average nearly 1000×10^6 Btu/hr in capacity, whereas industrial boilers are only about 6×10^6 Btu on the average and commercial boilers are even smaller. The control of emissions from industrial and commercial boilers is complicated by their number, small average size, and proximity to population centers. Also, many uses in the industrial and commercial sectors cause fluctuations in load swings and other operating conditions that can influence both the rate and type of emissions.

Although they comprise only about 3% of the number of boilers, water-tube boilers represent over 75% of the total capacity. Table 9 presents the estimated 1977 industrial and commercial boiler capacity distribution by size range and type. Boilers less than 25×10^6 Btu/hr in size account for

Table 8 Estimated U.S. Boiler Population by Fuel, 1977[6]

	Number	Capacity, 10^6 Btu/hr
Utility boilers		
Coal	1,530	1,833,000
Residual oil	1,040	743,600
Distillate oil	200	57,300
Natural gas	980	1,013,700
Total utility	3,750	3,647,600
Industrial boilers		
Coal	51,180	619,000
Residual oil	123,000	846,700
Distillate oil	67,520	224,900
Natural gas	265,230	1,416,900
Total industrial	506,930	3,107,500
Commercial boilers		
Coal	163,220	196,800
Residual oil	266,100	377,300
Distillate oil	176,690	208,700
Natural gas	689,120	591,800
Total commercial	1,295,130	1,374,600
All boilers		
Coal	215,930	2,648,800
Residual oil	390,140	1,967,600
Distillate oil	244,410	490,900
Natural gas	955,330	3,022,400
Total	1,805,810	8,129,700

Table 9   Estimated U.S. Industrial/Commercial Boiler Population by Type and Size, 1977[7]

Size Range, 10⁶ Btu/hr	Capacity, 10^6 Btu/hr			
	Water Tube	Fire Tube	Cast Iron	Total
<0.4			236,100	236,000
0.4–1	8,300	179,200	303,100	490,600
1–10	51,000	412,000	357,000	820,000
10–25	112,500	320,100		432,600
Subtotal, <25	171,800	911,300	896,200	1,979,200
25–50	486,700	122,000		608,700
50–100	503,000			503,000
100–250	632,000			632,000
Subtotal, 25–250	1,621,700	122,000		1,743,700
250–500	335,400			335,400
500–1500	191,400			191,400
>1500	232,300			232,300
Subtotal, >250	759,100			759,100
Total	2,552,600	1,033,300	896,200	4,482,000

about 44% of the total capacity, slightly more than that in the 25 to 250 × 10⁶ Btu/hr size range. The smaller boilers less than 25 × 10⁶ Btu/hr are primarily cast-iron or fire-tube units, whereas those in the larger size groupings are predominantly water-tube boilers. Boiler size plays an important role in generation of pollutants. Environmental control equipment is generally required on larger boilers. The combustion process is also better controlled in larger units; this results in more complete fuel combustion and the formation of fewer air pollutants per unit of fuel burned.

REFERENCES FOR SECTION 3

1. *The Population and Characteristics of Industrial/Commercial Boilers,* EPA-600/7-79-178a, U.S. Environmental Protection Agency, Research Triangle Park, NC, 1979.
2. Ref. 1.
3. Ref. 1.
4. Ref. 1.
5. *Overview of Pollution from Combustion of Fossil Fuels in Boilers of the United States,* EPA-600/7-79-233, U.S. Environmental Protection Agency, Research Triangle Park, NC, 1979.
6. Ref. 5.
7. Ref. 5.

4   EMISSION SOURCES

Table 10 shows the estimates of particulate and sulfur oxide emissions by fuel type and combustion sector for 1978. Coal is the largest contributor of particulate and sulfur oxide emissions. Current trends in fuel displacement will result in even higher coal usage.

4.1   Solid Fuels

Coal is the dominant solid fuel burned in combustors. Other solid fuels are industry specific and are residues or by-products of industrial processes. For example, the pulp mill industry generates bark waste which is burned in boilers both to recover heat and to reduce solid waste disposal problems.

4.1.1   Emission Compositions

The burning of solid fuels produces particulate matter pollutants and several kinds of gaseous pollutants. Particulate matter is also emitted as fugitive dust during handling and storage of solid fuels.

4.1.1.1   Combustion Process Emissions.[2]   Particulates emitted from coal combustion consist primarily of carbon, silica, alumina, iron oxide, and sulfur and organic compounds. Various other elements in trace quantities are also present. The particle size distribution and quantity of particulate emissions

Table 10   1978 Estimates of Particulate and Sulfur Oxide Emissions from Stationary Point Sources[1]

Source Category	Emissions (10^3 tons)	
	Particulates	Sulfur Oxides
Fuel combustion		
Utility		
Coal	2,590	17,530
Oil	150	1,900
Gas	10	0
Industrial		
Coal	770	2,080
Oil	100	1,270
Gas	40	0
Other fuels	310	170
Commercial/institutional		
Coal	20	40
Oil	70	990
Gas	10	0
Residential		
Coal	20	70
Oil	20	240
Gas	30	0

depends on coal characteristics and boiler type. Gaseous emissions include sulfur oxides, carbon monoxide, hydrocarbons, and nitrogen oxides.

4.1.1.2   Fugitive Emissions.[3]   Fugitive dust may be emitted from several sources in the coal-fired power plant. Coal unloading is accomplished by belt conveyors which are fed from the hoppers. Fugitive emissions occur during unloading due to poor sealing of unloading equipment. At some plants, hoppers are fed by bucket-type elevators or clam shell buckets which are usually open and release fugitive emissions. Fugitive emissions from coal storage piles occur mainly due to wind erosion. Emissions may also occur from activities such as loading onto and out of the pile and vehicular traffic. Other fugitive sources are the transport of coal between various points at the plant by belt conveyors and the handling and storage of fly ash and bottom ash.

4.1.2   Emission Rates

Table 11 presents the uncontrolled emission factors for combustion processes burning solid fuels. Particulate emission rates include the fly ash present in the flue gas plus the unburned carbon.

Uncontrolled sulfur oxide emission rates are based on 95% conversion of the sulfur present in the fuel. The remaining 5% sulfur is emitted in the fly ash or combines with the slag or bottom ash in the furnace.

Nitrogen oxide emissions result from the high-temperature reaction of atmospheric nitrogen and oxygen in the combustion zone and from the partial combustion of nitrogenous compounds in the fuel. The major factors that affect NO_x production are flame and furnace temperatures, residence time of combustion gases at the flame temperature, rate of cooling of the gases, and amount of excess air present in the flame.

Combustion efficiency determines the carbon monoxide and hydrocarbon content of the flue gases. Successful combustion that results in a low level of carbon monoxide and hydrocarbon emissions requires a high degree of turbulence, a high temperature, and adequate time for combustion reactions. Thus, careful control of excess air rates, high combustion temperatures, and furnace designs permitting intimate fuel–air contact will minimize these emissions.

4.2   Liquid Fuels

Distillate fuel oils (fuel oil grades 1 and 2) are used mainly in domestic and small commercial applications where ease of handling and combustion is required. Residual oils (fuel oil grades 4, 5, and 6) are used mainly in utility, industrial, and large commercial applications.

Table 11 Emission Factors for Combustion Processes (Without Control Equipment)[a]

Furnace Size, 10^6 Btu/hr Heat Input	Particulates, lb/ton of fuel	Sulfur Oxides, lb/ton of fuel	Carbon Monoxide, lb/ton of fuel	Hydrocarbons, lb/ton of fuel	Nitrogen Oxides, lb/ton of fuel	Aldehydes, lb/ton of fuel
Bituminous Coal						
Greater than 100 (utility and large industrial boilers)						
Pulverized						
General	$16A$[a]	$38S$	1	0.3	18	0.005
Wet bottom	$13A$	$38S$	1	0.3	30	0.005
Dry bottom	$17A$	$38S$	1	0.3	18	0.005
Cyclone	$2A$	$38S$	1	0.3	55	0.005
10 to 100 (large commercial and general industrial boilers)						
Spreader stoker	$13A$	$38S$	2	1	15	0.005
Less than 10 (commercial and domestic furnaces)						
Underfeed stoker	$2A$	$38S$	10	3	6	0.005
Hand-fired units	20	$38S$	90	20	3	0.005
Anthracite Coal						
Pulverized coal	$17A$	$38S$	1	Neg[b]	18	
Traveling grate	$1A$	$38S$	1	Neg	10	
Hand-fired	10	$38S$	90	2.5	3	
Lignite						
Pulverized coal	$7A$	$30S$	1.0	<1.0	14(8)[c]	
Cyclone	$6A$	$30S$	1.0	<1.0	17	
Spreader stoker	$7A$	$30S$	2.0	1.0	6	
Other stokers	$3A$	$30S$	2.0	1.0	6	
Bark combustion with fly ash reinjection	75	1.5	2–60	2–70	10	
Bark combustion without fly ash reinjection	50	1.5	2–60	2–70	10	
Wood/bark mixture with fly ash reinjection	45	1.5	2–60	2–70	10	
Wood/bark mixture without fly ash reinjection	30	1.5	2–60	2–70	10	
Wood combustion	5–15	1.5	2–60	2–70	10	
Bagasse combustion	16	Neg			1.2	
Residential fireplaces: wood,	20	0	120	5.0	1.0	
coal	30	36S	90	20.0	3.0	
Wood stoves	4–30		260			

[a] Letters A and S represent the ash and sulfur content (in percent) of the fuel, respectively. The indicated numbers should be multiplied by these values.

[b] Neg = neglible.

[c] Use 14 lb/ton for front wall-fired and horizontally opposed wall-fired units and 8 lb/ton for tangentially fired units.

4.2.1 Emission Compositions

Distillate fuel oils are more volatile and less viscous than residual oils. They are also cleaner, having negligible ash and nitrogen contents and usually containing less than 0.3% sulfur (by weight). Heavier residual oils must be heated for pumping and proper atomization. Residual oils contain significant quantities of ash, nitrogen, and sulfur.

4.2.2 Emission Rates

Table 12 presents the particulate and gaseous emission factors for fuel oil combustion. Among residual oils, grades 4 and 5 usually result in less particulate than does the heavier grade 6. In boilers firing grade 6, particulate emissions can be described, on the average, as a function of the sulfur content of the oil.

Table 12 Particulate and Gaseous Emission Factors for Fuel Oil Combustion

	Power Plant	Type of Boiler[a] Industrial and Commercial		Domestic
Pollutant	Residual Oil, lb/10³ gal	Residual Oil, lb/10³ gal	Distillate Oil, lb/10³ gal	Distillate Oil, lb/10³ gal
Particulate	b	b	2	2.5
Sulfur dioxide[c]	157S	157S	142S	142S
Sulfur trioxide[c]	2S	2S	2S	2S
Carbon monoxide[d]	5	5	5	5
Hydrocarbons (total, as CH_4)[e]	1	1	1	1
Nitrogen oxides (total, as NO_2)	105(50)[f]	60[g]	22	18

[a] Boilers can be classified, roughly, according to their gross (higher) heat input rate, as shown below:

> Power plant (utility) boilers: $>250 \times 10^6$ Btu/hr
> Industrial boilers: $>15 \times 10^6$, but $<250 \times 10^6$ Btu/hr
> Commercial boilers: $>0.5 \times 10^6$, but $<15 \times 10^6$ Btu/hr
> Domestic (residential boilers): $<0.5 \times 10^6$ Btu/hr

[b] Particulate emission factors for residual oil combustion are best described, on the average, as a function of fuel oil grade and sulfur content, as shown below:
> Grade 6 oil: lb/10³ gal = $10(S) + 3$, where S is the percentage, by weight, of sulfur in the oil
> Grade 5 oil: 10 lb/10³ gal
> Grade 4 oil: 7 lb/10³ gal

[c] S is the percentage, by weight, of sulfur in the oil.
[d] Carbon monoxide emissions may increase by a factor of 10 to 100 if a unit is improperly operated or not well maintained.
[e] Hydrocarbon emissions are generally negligible unless unit is improperly operated or not well maintained, in which case emissions may increase by several orders of magnitude.
[f] Use 50 lb/10³ gal for tangentially fired boilers and 105 lb/10³ gal for all others, at full load, and normal ($>15\%$) excess air. At reduced loads, NO_x emissions are reduced by 0.5 to 1%, on the average, for every percentage reduction in boiler load.
[g] Nitrogen oxide emissions from residual oil combustion in industrial and commercial boilers are strongly dependent on the fuel nitrogen content and can be estimated more accurately by the following empirical relationship:

$$\text{lb } NO_2/10^3 \text{ gal} = 22 + 400(N)^2$$

where N is the percentage, by weight, of nitrogen in the oil. Note: for oils having high ($>0.5\%$, by weight) nitrogen contents, one should use 120 lb $NO_2/10^3$ gal as an emission factor.

Boiler load also affects particulate emissions in units firing grade 6 oil. At low-load conditions, particulate emissions may be lowered by 30 to 40% for utility boilers and as much as 60% for small industrial and commercial units. At too low a load condition, proper combustion conditions cannot be maintained and particulate emissions may increase drastically. Any condition that prevents proper boiler operation can result in excessive particulate formation.

Total sulfur oxide emissions are almost entirely dependent on the sulfur content of the fuel and are not affected by boiler size, burner design, or grade of fuel being fired. On the average more than 95% of the fuel sulfur is converted to SO_2, with about 1 to 3% further oxidized to SO_3. Sulfur trioxide readily reacts with water vapor (both in the air and in the flue gases) to form a sulfuric acid mist.

Fuel nitrogen conversion is the more important NO_x forming mechanism in boilers firing residual oil. Except in certain large units having unusually high peak flame temperatures, or in units firing a low-nitrogen residual oil, fuel NO_x will generally account for over 50% of the total NO_x generated. Thermal fixation, on the other hand, is the predominant NO_x forming mechanism in units firing distillate oils, primarily because of the negligible nitrogen content in these lighter oils. Because distillate oil-fired boilers usually have low heat release rates, however, the quantity of thermal NO_x formed in them is less than in larger units. Nitrogen oxide emissions from tangentially (corner) fired boilers are, on the average, only half those of horizontally opposed units. The use of limited excess air firing, flue gas recirculation, staged combustion, or some combination thereof, may result in NO_x reductions ranging from 5 to 60%.

Only minor amounts of hydrocarbons and carbon monoxide are produced during fuel oil combustion. If a unit is operated improperly or not maintained, however, the resulting concentrations of these pollutants may increase by several orders of magnitude. This is most likely to be the case with small, often unattended units.

4.3   Gaseous Fuels

Natural gas is the major gaseous fuel. Small amounts of liquefied petroleum gas (LPG) are also burned in various combustors. The primary component of natural gas is methane with varying amounts of ethane and smaller amounts of nitrogen, helium, and carbon dioxide. LPG is obtained from oil or gas wells as a by-product of gasoline refining. LPG is graded according to maximum vapor pressure with grade A being predominately propane, and grades B through E containing varying mixtures of butane and propane.

4.3.1   Emission Compositions

Emissions from natural gas and LPG combustion have components similar to coal and fuel oil combustions; however, the quantities of constituents are significantly lower.

4.3.2   Emission Rates

Tables 13 and 14 present the emission rates for natural gas and LPG combustion, respectively. Even though natural gas is considered to be a relatively clean fuel, some emissions can occur from the combustion reaction. For example, improper operating conditions, including poor mixing, insufficient air, and so on, may cause large amounts of smoke, carbon monoxide, and hydrocarbons to be produced. Moreover, because a sulfur-containing mercaptan is added to natural gas for detection purposes, small amounts of sulfur oxides will also be produced in the combustion process.

Table 13   Emission Factors for Natural Gas Combustion[6]

Pollutant	Power Plant, lb/10^6 ft^3	Industrial Process Boiler, lb/10^6 ft^3	Domestic and Commercial Heating, lb/10^6 ft^3
Particulates	5–15	5–15	5–15
Sulfur oxides (SO_2)	0.6	0.6	0.6
Carbon monoxide	17	17	20
Hydrocarbons			
(as CH_4)	1	3	8
Nitrogen oxides	700[a]	(120–230)[b]	(80–120)[c]

[a] Use 300 lb/10^6 ft^3 for tangentially fired units.
[b] This represents a typical range for many industrial boilers. For large industrial units (>100 MM Btu/hr) use the NO_x factors presented for power plants.
[c] Use 80 for domestic heating units and 120 for commercial units.

Table 14   Emission Factors for LPG Combustion[a] (Ref. 7)

Pollutant	Industrial Process Furnaces		Domestic and Commercial Furnaces	
	Butane, lb/10³ gal	Propane, lb/10³ gal	Butane, lb/10³ gal	Propane, lb/10³ gal
Particulates	1.8	1.7	1.9	1.8
Sulfur dioxide[b]	0.09S	0.09S	0.09S	0.09S
Carbon monoxide	1.6	1.5	2.0	1.9
Hydrocarbons	0.3	0.3	0.8	0.7
Nitrogen oxides[c]	12.1	11.2	(8–12)[d]	(7–11)[d]

[a] LPG emission factors calculated assuming emissions (excluding sulfur oxides) are the same, on a heat input basis, as for natural gas combustion.
[b] S equals sulfur content expressed in grains per 100 ft³ gas vapor; for example, if the sulfur content is 0.16 gr/100 ft³ (0.366 g/100 m³) vapor, the SO_2 emission factor would be 0.09 × 0.16 or 0.014 lb SO_2/1000 gal (0.01 × 0.366 or 0.0018 kg SO_2/10³ l) butane burned.
[c] Expressed as NO_2.
[d] Use lower value for domestic units and higher value for commercial units.

Nitrogen oxides are the major pollutants of concern when burning natural gas. Nitrogen oxide emissions are a function of the temperature in the combustion chamber and the rate of cooling of the combustion products. Emission levels generally vary considerably with the type and size of unit and are also a function of loading.

4.4   Factors Affecting Emissions

During the normal operation of boilers, emissions can be minimized by proper operation and maintenance procedures. Operating factors that affect the combustion process also directly affect emissions. The amount of excess air, for example, has a major impact on emissions. When less than optimum amount of excess air is supplied, incomplete combustion of fuel results and particulate emissions are increased because of the higher percentage of unburned carbon particles. Inadequate excess air will also result in increased carbon monoxide emissions.

Operating and maintenance procedures vary with boiler type. A proper operation and maintenance schedule should be developed from recommendations of the boiler manufacturer and operating experience with the unit. A properly operated and maintained unit will not only minimize the emissions, but will also provide high overall thermal efficiency.

REFERENCES FOR SECTION 4

1.   *National Emissions Estimates, 1970–78,* EPA-450/4-80-002, U.S. Environmental Protection Agency, Research Triangle Park, NC, January 1980.

2.   *Compilation of Air Pollutant Emission Factors,* 3rd edition, Office of Air and Waste Management. OAQPS, U.S. Environmental Protection Agency, Research Triangle Park, NC, Publication No. AP-42, April 1980.

3.   *Reasonably Available Control Measures for Fugitive Dust Sources,* Ohio Environmental Protection Agency, Office of Air Pollution Control, Division of Engineering, Columbus, OH, September 1980.

4.   Ref. 2.

5.   Ref. 2.

6.   Ref. 2.

7.   Ref. 2.

5   CONTROL SYSTEMS

5.1   Particulate Control

The principal particulate control devices discussed are electrostatic precipitators (ESPs), fabric filters, wet scrubbers, and mechanical collectors.

5.1.1   Electrostatic Precipitators

5.1.1.1   Present and Projected Use.   The ESP has traditionally been the control device of choice in the electric utility industry but in recent years it has had competition from fabric filters and wet scrubbers. The most recent New Source Performance Standards (NSPS) have made fabric filters more economical than ESPs at coal sulfur contents below 1% because of the large ESPs that would be required to capture high resistivity fly ash; fabric filters are unaffected by resistivity.

In the industrial sector, the ESP has not been as dormant, but is a strong competitor for control of coal- and wood-fired boilers, especially where gas flows exceed 50,000 acfm. McIlvaine predicts a market share of 25% by 1985 for ESPs.[1]

5.1.1.2   Types and Performance Capability.   There are two types of ESPs used on combustion sources, the wire weight and rigid frame. The wire weight is dominant in the industrial and utility markets, but the rigid-frame design is being increasingly applied in the utility sector, especially on units firing low-sulfur western coals, which require larger ESPs.

ESPs are classified in application as either hot side or cold side. Hot-side ESPs are installed upstream of the air preheater. These were popular on utility boilers in the 1970s because high collection efficiencies could be achieved with less plate area than cold-side ESPs due to lower resistivity of the gas at higher temperatures above 600°F. However, performance and maintenance problems at a number of installations have resulted in a return to the cold-side design.

ESPs can achieve collection efficiencies in excess of 99.5% and are quite efficient collectors of fine particulate, as shown in Figure 7, which describes ESP performance as a function of particle size for a coal-fired utility boiler.

5.1.1.3   Design Improvements.   Several design improvements are receiving attention: preionization, pulse energization, and gas conditioning.

1. *Preionization.*   The Electric Power Research Institute (EPRI) is testing this device, which imparts a change to the fly ash ahead of the ESPs. Although pilot testing has produced encouraging results, it is not yet known whether the preionizer will be commercially viable.

2. *Pulse Energization.*   This technique consists of superimposing a series of short fast-rising pulses of short duration on the normal ESP DC voltage to produce a more uniform corona discharge. This

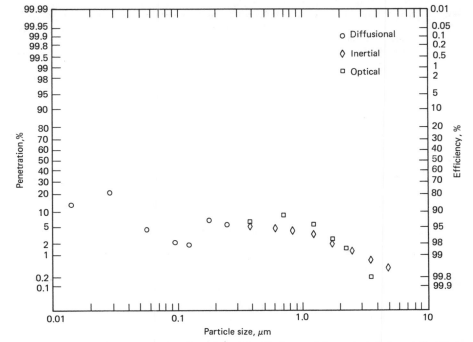

Fig. 7   Measured efficiency as a function of particle size for precipitator installation at the Gorgas Plant of Alabama Power Company.[3]

is not a new concept, having been tried about 30 yr ago; however, the need for better collection of high-resistivity dusts, advances in pulse energization technology, and the possibility of utilizing smaller ESPs with pulse energization have renewed interest in the technology. Pulse energizers have been installed on a few installations but have limited operating experience.

3. *Gas Conditioning.* Chemicals such as ammonia, SO_3, H_2SO_4, and so on, have been used to upgrade ESP performance for a number of years, and have advanced to the point where some utilities are considering their use on new units to reduce the size requirements. This reduces costs but adds the burden of having to depend on the chemical for proper performance.

5.1.2 Fabric Filters

5.1.2.1 Present and Projected Use.   In the electric utility industry, fabric filters have been increasingly used in recent years. The fabric filter share of the coal-fired utility market will probably increase gradually because utilities will tend to use ESPs until an abundance of long-term operating experience is gained with fabric filters.

In the industrial sector, fabric filters are more widely used but can be subject to fires on wood-fired boilers (except salt-laden hogged fuel), and are not recommended for use on oil-fired boilers because of fabric blinding. McIlvaine predicts that fabric filters will capture 40% of the industrial market by 1985.

5.1.2.2 Types and Performance Capability.   The three major types of fabric filters (classified by cleaning system) are mechanical shake, reverse air, and pulse jet. Utility applications in the United States generally use reverse air systems only, while all types are used for industrial applications.

Fabric filters are capable of very high collection efficiencies (99.5%+) and are the most efficient of all collectors for fine particulate. Figure 8 illustrates penetration through a fabric filter as a function of particle size on a small coal-fired utility boiler.

5.1.3 Wet Scrubbers

5.1.3.1 Present and Projected Use.   Scrubbers for particulate control have not found much application in the electric utility industry mainly because of their high operating and maintenance costs, and their fine particulate collection capability without excessive pressure loss is lower than that of ESPs or fabric filters. Rather, they have been installed in conjunction with SO_2 absorbers. This would indicate that there will be no change in future use of particulate scrubbers at utilities although some interest has been shown in use of a low-efficiency (i.e., 95%) ESP followed by a wet scrubber for both particulate and SO_2 control.

In the industrial sector, particulate scrubbers are more accepted because their small size, low initial

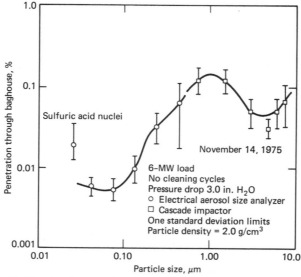

Fig. 8   Fractional penetration through Nucla baghouse (6-MW load).[15]

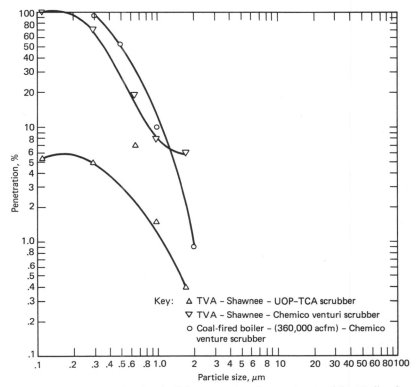

Fig. 9 Wet scrubber fractional efficiency test data from various coal-fired boilers.[8]

and operating and maintenance costs, and reasonable fine particulate collection efficiency make them attractive, especially for paper and pulp and wood products plants. McIlvaine projects a 30% market share for wet scrubbers in 1985.[5]

5.1.3.2 Types and Performance Capability. The types of scrubbers most often applied to combustion sources are the preformed spray, gas atomized, impingement and entrainment, and moving-bed-type scrubbers. Performance of wet scrubbers is basically a function of pressure drop; venturi-type scrubbers can achieve overall efficiencies in excess of 99% with reasonable fine particulate collection capability. The performance of wet scrubbers, however, is not on a par with that of ESPs or fabric filters. A typical performance curve as a function of particle size for a venturi scrubber installed on a coal-fired boiler is presented in Figure 9.

5.1.4 Mechanical Collectors

5.1.4.1 Present and Projected Use. Mechanical collectors are presently in use at some existing utility plants as precleaners ahead of other devices. There appears to be no need for these low efficiency devices on future utility installations, however.

In the industrial sector, mechanical collectors can be applicable where local regulations are not stringent (i.e., 0.3 lb/10⁶ Btu or greater), or as precleaners or cinder collectors (to prevent fires) ahead of other particulate collectors. McIlvaine predicts a 5% industrial market share for mechanical collectors (including granular-bed filters) for 1981.

5.1.4.2 Types and Performance Capability. The most commonly used mechanical collector for combustion sources is the multiple cyclone type shown in Figure 10. Collection efficiencies range from 70 to 90%, with 95 to 96% sometimes possible from two multicyclones arranged in series under ideal conditions. Fine particulate collection is poor, as indicated in Figure 11, for both single- and multiple-type mechanical collectors.

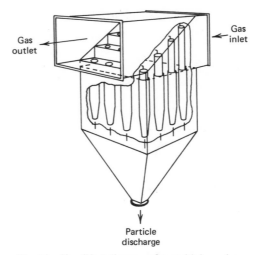

Fig. 10   Simplified diagram of a multiple cyclone.

5.2   Sulfur Dioxide Control

5.2.1   *Introduction*

Flue gas desulfurization (FGD) has become a leading means of controlling SO_x emissions in the United States; it is primarily used by the electric utility industry. Dry or wet FGD systems can be used to control SO_x emissions; wet systems, however, dominate the utility market in the United States.

Flue gas desulfurization processes are categorized as generating a saleable or throwaway product depending on whether sulfur compounds are separated from the absorbent as a by-product or disposed of as a waste.

Table 15 lists the number of domestic utility FGD systems according to status and equivalent electrical capacities as of the end of December 1981.

5.2.1.1   Power-Generating and FGD Capacity.   As indicated in Table 15, 94 coal-fired power-generating units currently equipped with operational FGD systems represent a total controlled capacity of 35,931 MW out of a total coal-fired power-generating capacity of approximately 235,000 MW.

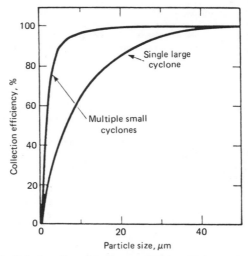

Fig. 11   Relation of particle size to collection efficiency of cyclones.

Table 15 Number and Total Capacity of FGD Systems[10]

Status	No. of Units	Total Controlled Capacity, MW[a]	Equivalent Scrubbed Capacity, MW[b]
Operational	94	35,931	32,683
Under construction	40	17,386	16,666
Planned:			
Contract awarded	17	10,035	9,819
Letter of intent	10	7,643	7,585
Requesting/evaluating bids	10	5,630	5,630
Considering only FGD systems	51	30,726	30,398
Total	222	107,351	102,781

[a] Total controlled capacity (TCC) represents the gross capacities (MW) of coal-fired units brought into compliance by FGD systems, regardless of the percentage of the flue gas treated.
[b] Equivalent scrubbed capacity (ESC) represents the effective capacities of the FGD systems (in equivalent MW), based on the percentage of the flue gas treated.

Based on the known utility commitments to FGD, the percentage of coal-fired capacity controlled by FGD will increase from its current level of 15.3% to 26.5% by the end of 1990.

Table 16 presents the projected distribution of power-generating sources (by energy source) in the electric utility industry. Table 17 presents the percentage of current and projected coal-fired and total power-generating capacities by FGD.

5.2.1.2 Process Type. Table 18 illustrates the popularity of lime- and limestone-based systems both in the past and currently. In the next section, these processes as well as others are described.

5.2.2 Process Descriptions[14,15]

5.2.2.1 Lime Process. The lime process is a wet, throwaway product SO_2 absorption process, in which an alkaline slurry formed from the lime is circulated through a scrubber/absorber tower,

Table 16 Distribution of Power-Generating Sources by Energy Source[11] (Percentage of Total)

	Coal	Nuclear	Oil	Hydro	Gas	Other	Total, MW
December 1979	39	9	25	13	13	1	603
December 1990	44	14	20	11	10	1	833

Note: Figures reflect annual losses of 0.4% of the year-end capacity attributed to retirement of older units.

Table 17 FGD-Controlled Power-Generation Capacity[12,16] (Percentage of Total)

Period	Coal-Fired Capacity	Total Capacity
August 1980[a]	11.5[b]	4.5[b]
December 1990	26.5	11.6

[a] Represents FGD-committed capacity as of August 1980.
[b] Based on FGD capacity as of August 1980 and total power-generating capacity as of December 1979. Notable changes during the 1974–1981 growth period include:

A 527% increase in the number of operational systems.
A 1026% increase in FGD operating capacity (ESC).
An increase in the average FGD capacity of the FGD-equipped unit from 170 to 348 MW.

Table 18   Distribution of FGD Systems by Process[13]

Process	FGD Capacity (ESC), MW			
	Operational	Under Construction	Planned	Total
Limestone[a]	15,903	8,929	24,060	48,892
Lime[b]	9,886	2,870	4,940	17,696
Lime/spray drying	110	2,338	1,019	3,467
Lime/limestone	20	0	475	495
Sodium carbonate	1,255	0	1,900	3,155
Magnesium oxide	0	724	0	724
Wellman–Lord	1,540	534	0	2,074
Dual alkali	1,181	421	421	2,023
Aqueous carbonate/ spray drying[c]		100	0	100
Citrate[d]	60	0	0	60
Total	29,955	15,916	32,815	78,686

[a] Includes alkaline fly ash/limestone and limestone slurry process design configurations.
[b] Includes alkaline fly ash/lime and lime slurry process design configurations.
[c] Includes nonregenerable dry collection and regenerable process design configurations.
[d] This system is operating at the St. Joseph Zinc Co., G. G. Wheaton Plant and is listed as a utility FGD system because the plant is connected by a 25-MW interchange to the Duquesne Light Company.
[e] Because the processes of all planned systems are not known, the totals in this status category are less than those in Table 19.

where it reacts with SO_2 in the flue gas. Calcium sulfite and sulfate formed by the reaction are then separated in settlers or clarifiers and filters. The sludge produced by the system can be chemically stabilized to produce an inert landfill material or can be stored in sludge ponds equipped with adequate barriers to prevent contamination of surface or groundwaters.

Lime FGD systems have demonstrated the ability to remove in excess of 90% of the inlet SO_2 at a number of utility boiler installations.

5.2.2.1a   Process Chemistry.   The following reactions take place in the absorber during SO_2 absorption by an aqueous scrubbing liquor:

$$SO_2(g) \rightarrow SO_2(aq) \tag{1}$$

$$SO_2(aq) + H_2O \rightarrow H_2SO_3 \rightarrow HSO_3^- + H^+ \tag{2}$$

$$HSO_3^- \rightarrow H^+ + SO_3^{2-} \tag{3}$$

Lime in the slurry produces calcium through the following reactions:

$$CaO + H_2O \rightarrow Ca(OH)_2(s) \tag{4}$$

$$Ca(OH)_2(s) \rightarrow Ca(OH)_2(aq) \tag{5}$$

$$Ca(OH)_2(aq) \rightarrow Ca^{2+} + 2OH^- \tag{6}$$

Sulfite ion generated [reaction (3)] combines with calcium generated [reaction (6)] to yield the insoluble calcium sulfite hemihydrate:

$$Ca^{2+} + SO_3^{2-} + \tfrac{1}{2}H_2O \rightarrow CaSO_3 \cdot \tfrac{1}{2}H_2O \tag{7}$$

In addition, sulfite ion may ultimately be converted to gypsum in the following reactions:

$$SO_3^{2-} + \tfrac{1}{2}O_2 \rightarrow SO_4^{2-} \tag{8}$$

$$Ca^{2+} + SO_4^{2-} + 2H_2O \rightarrow CaSO_4 \cdot 2H_2O(s) \tag{9}$$

Quantities of lime required by the process and sludge generated in the process are calculated from reactions (1) through (8). With assumptions of 95% lime purity and a 1.0 M stoichiometric ratio, the lime requirement is 1.05 weight per unit weight of SO_2 removed.

5.2.2.1b System Description.   The equipment for a lime FGD system is generally grouped under four major operations:

Scrubbing or Absorption.   Includes SO_2 scrubbers, hold tanks, and circulation pumps.

Flue Gas Handling.   Includes inlet and outlet ductwork, dampers, reheaters, and fan.

Lime Handling and Slurry Preparation.   Includes lime unloading and storage equipment, and lime processing and slurry preparation equipment.

Sludge Processing.   Includes clarifier and filters (if used) for sludge dewatering, sludge pumps, and sludge handling equipment.

A diagram of a typical lime FGD system is shown in Figure 12. Individual systems may deviate from that shown, depending on plant characteristics and system supplier.

5.2.2.1c Sludge Disposal.   Processing of the sludge generated by an FGD system may involve several steps. A stream is bled continuously from the scrubber to the sludge circuit. Because this stream contains a large proportion of water (90% is not uncommon), liquid–solid separation is required. The FGD sludge is thixotropic.

A typical sludge processing circuit involves solids sedimentation, dewatering, fixation, and transportation of sludge for final disposal. Clarifiers are generally used for sedimentation; the recovered water is sent back to the scrubber circuit, and the partially dewatered sludge is sent for further dewatering. When vacuum filters are used for further dewatering, the dewatered cake contains about 60% solids. Any further dewatering leads to excessive energy consumption.

After the vacuum filtration, the cake is transferred to a fixation tank, where chemicals are added to the cake and mixed thoroughly. The fixation process is a means of physically and chemically stabilizing the sludge to reduce its pollution potential and facilitate handling.

5.2.2.2 The Limestone FGD Process.   The limestone and the lime FGD processes are similar in many aspects. In the limestone process, limestone slurry is used as the absorbent, as is lime slurry in the lime process. The use of limestone, however, requires different feed preparation equipment than is used in preparing lime slurries and also necessitates other process differences; the limestone

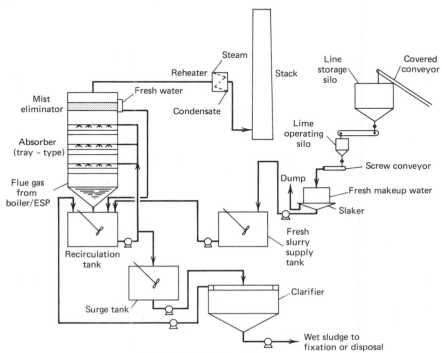

Fig. 12   Typical lime FGD system.

process, for example, requires a higher liquid to gas (L/G) ratio because the absorbent is less reactive than lime. The exact L/G ratio required is a function of SO_2 removal required, the inlet SO_2 gas stream concentration, the absorbent inlet pH, and other items. Even with such differences, the processes are so similar that it is possible to design a system that can use either lime or limestone as the absorbent.

5.2.2.2a Process Chemistry. The chemistry of the limestone process differs from that of the lime process only in the way that the calcium ion becomes available for absorption of SO_2; calcium ion generation in the limestone process takes place according to the following reactions:

$$CaCO_3(s) \longrightarrow CaCO_3(aq) \tag{10}$$

$$CaCO_3(aq) \longrightarrow Ca^{2+} + CO_3^{2-} \tag{11}$$

Other reactions are the same as those in the lime process.

The SO_2 in the flue gas stream contacts water forming the sulfite ion, which reacts with the calcium ion to yield insoluble calcium sulfite hemihydrate or gypsum, as shown in reactions (7), (8), and (9) in the lime process description.

The sludge from a limestone process consists of calcium sulfite hemihydrate, calcium carbonate, gypsum, limestone impurities, and any excess unreacted limestone. The exact proportions of calcium sulfite hemihydrate and gypsum depend on such factors as inlet SO_2 content, excess oxygen, and absorbent pH. Because the FGD sludge is always an aqueous slurry, the weight of the associated water must be included in the total sludge weight.

5.2.2.2b System Description. Most of the equipment in a limestone FGD system is similar to that in a lime-based system. The major difference is in feed and slurry preparation. In a limestone system, the feed generally must undergo size reduction before the slurry is prepared; although preground rock of less than 200-mesh particle size can be purchased and used directly for slurry preparation, this is rarely done because of the high cost. Therefore, a limestone FGD system generally incorporates equipment for size reduction.

The raw limestone can be stored in open piles without protection from weather. Because it can be stored in the open and because it is cheaper than lime, large inventories can be maintained.

The SO_2 removal efficiency of the system is dependent on the limestone stoichiometry up to a certain level, after which no increase is obtained with an increase in the limestone stoichiometry.

The previous description of the lime FGD system is applicable to the limestone system except that the feed preparation modules are different because of the difference in the properties of the two feed materials. Other aspects of the system as well as the types of scrubber/absorber are essentially similar.

Figure 13 is a diagram of a typical limestone FGD system. Individual systems may deviate from that shown, depending on plant characteristics and the system supplier.

Because limestone is less reactive than lime, some of the process parameters are different. The L/G ratio of a limestone system is higher, and the residence time in process tanks is longer than that of the lime system. Typical L/G ratios for limestone FGD systems using spray tower absorbers often are in the 60 to 80 gal/1000 acf range. Spray absorbers are the most common in limestone systems.

A low pH leads to better limestone utilization but reduces SO_2 removal efficiency. Some proposed designs take advantage of the flexibility afforded by operating multiple stages at different pH levels. A low-pH liquor contacts the entering flue gas in an initial scrubbing stage, in which part of the SO_2 is removed and the limestone dissolution is essentially completed. High-pH liquor contacts the flue gas in the second stage, where additional SO_2 removal takes place. Blowdown liquor from the second stage is bled to the first stage. Makeup limestone slurry is added in the second-stage loop.

Based on the operating experience of limestone systems, there is evidence from individual SO_2 removal test runs to show that limestone FGD systems can operate at 90% SO_2 removal or greater and that they can operate reliably (90% operability) with proper design and maintenance on both low- and high-sulfur coals.

5.2.2.2c Additives. There are two important variations of the calcium-based processes: lime/alkaline fly ash and limestone/alkaline fly ash. Some western utilities have been able to take advantage of the alkaline constituents of the ash of their low-sulfur coals by using the fly ash as the primary SO_2 absorbent. In these processes lime or limestone is used to supplement the alkaline fly ash in the slurry to maintain design pH levels. The process chemistry of these systems is nearly identical to the lime and limestone processes and will not be addressed here.

There are two important additives that have been tested in commercial lime and limestone utility FGD systems: adipic acid and sodium thiosulfate. These additives have been demonstrated to reduce scaling, improve SO_2 removal efficiency, and minimize the potential for swings in pH with inlet SO_2

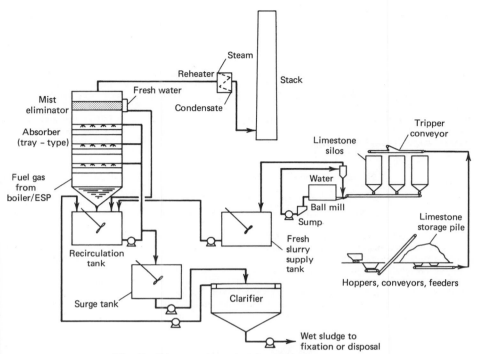

Fig. 13 Diagram of a typical limestone FGD system.

loading. The use of these additives has not yet gained wide acceptance among FGD system operations but it is expected that they will be seriously considered at troublesome sites. Another important process additive is magnesium oxide. Magnesium-promoted lime has proved to be an excellent absorbent for SO_2 control on high-sulfur coal applications. Magnesium-promoted lime processes show minimal scaling and high SO_2 removal efficiency. These systems also tend to have less sensitive process chemistry that in part is responsible for the minimal occurrence of scale accumulation.

5.2.2.3 Dual Alkali Process. Dual alkali scrubbing is an indirect lime/limestone process that removes SO_2 from exhaust gases, which avoids some of the scaling associated with direct lime/limestone scrubbing. The process normally involves absorption of SO_2 in a sodium solution in the absorber followed by regeneration of the absorbent in a separate system through reaction with a calcium-based alkaline slurry. Regenerated absorbent is recycled to the absorption loop; the calcium sulfite/sulfate is precipitated and discarded.

5.2.2.3a Process Chemistry. The alkaline solution used to absorb SO_2 may be a solution of either potassium, sodium, or ammonium compounds. In the United States, the scrubbing liquor is generally a sodium salt solution.

The basic absorbent is formed by addition of soda ash (sodium carbonate) or caustic (sodium hydroxide) to water. Several sodium species are available for reaction with the SO_2, including sodium carbonate, sodium hydroxide, and sodium sulfite. The primary products of reaction include sodium sulfite and sodium bisulfite. Following are the main absorption reactions:

$$2NaOH + SO_2 \longrightarrow Na_2SO_3 + H_2O \tag{12}$$

$$Na_2CO_3 + SO_2 \longrightarrow Na_2SO_3 + CO_2\uparrow \tag{13}$$

$$Na_2SO_3 + SO_2 + H_2O \leftrightarrows 2NaHSO_3 \tag{14}$$

$$NaOH + SO_2 \longrightarrow NaHSO_3 \tag{15}$$

Also some of the SO_3 present in the gas stream may react with sodium hydroxide to form sodium sulfate as follows:

$$2NaOH + SO_3 \longrightarrow Na_2SO_4 + H_2O \tag{16}$$

In the absorber, and through the rest of the system to a lesser degree, some sulfite is oxidized as follows:

$$2Na_2SO_3 + O_2 \rightarrow 2Na_2SO_4 \tag{17}$$

The sulfate species is inactive and is unavailable for further SO_2 removal. The extent of oxidation is a function of the oxygen and SO_2 content of the flue gas, the temperature of the gas in the absorption vessel, and the design of the absorber. As an example, the typical excess air level in high-sulfur, coal-fired utility boilers leads to oxidation levels of 10 to 15% of the SO_2 removed.

After absorption of SO_2, spent absorbing liquor is bled to the regeneration system and reacted with either lime or limestone. Regeneration with slaked lime (calcium hydroxide) takes place over several stages. The calcium hydroxide reacts with sodium bisulfite, sodium sulfite, and sodium sulfate in the following series of reactions:

$$2NaHSO_3 + Ca(OH)_2 \rightarrow Na_2SO_3 + CaSO_3 \cdot \tfrac{1}{2}H_2O\downarrow + \tfrac{3}{2}H_2O \tag{18}$$

$$Na_2SO_3 + Ca(OH)_2 + \tfrac{1}{2}H_2O \rightarrow 2NaOH + CaSO_3 \cdot \tfrac{1}{2}H_2O\downarrow \tag{19}$$

$$Na_2SO_4 + Ca(OH)_2 + 2H_2O \rightarrow 2NaOH + CaSO_4 \cdot 2H_2O\downarrow \tag{20}$$

The precipitated calcium species (sulfite and sulfate) are separated from regenerated liquor.

After removal of the calcium sulfite/sulfate from the regenerated scrubbing liquor, two additional steps may be required. The first step, needed in all cases, involves the addition of makeup sodium to replace the small amount of sodium lost in the waste solids. The sodium is usually added as sodium carbonate or sodium hydroxide. The second step involves reduction of the calcium ion concentration in the scrubbing liquor to prevent scaling in the scrubber. This step is not always needed if the calcium concentration in the return liquor is low. If excess calcium is present, however, the solution must be softened by addition of sodium carbonate as the makeup sodium. Calcium is removed by the following reaction:

$$Na_2CO_3 + Ca^{2+} \rightarrow CaCO_3\downarrow + 2Na^+ \tag{21}$$

To achieve maximum SO_2 removal with an optimum amount of energy input, various absorber types are used by the industry, the most common of which are the tray tower or packed tower types. Typically the L/G ratio is 10 to 15 gal/1000 acf, the pressure drop is 6 to 12 in. H_2O, and the spent absorbent pH is 6.0.

The sludge generated by the dual alkali process is essentially the same as that generated by lime scrubbing and is handled in the same way.

Corrosion, erosion, and scaling problems have not been important factors at dual alkali FGD installations. Full-scale versions of these systems are not expected to experience these problems either. The double alkali system has demonstrated the ability to perform well under fluctuating SO_2 inlet concentrations.

5.2.2.4 The Wellman–Lord Process.

In this process an aqueous sulfite solution is used to absorb SO_2. Sodium bisulfite is formed as the SO_2 is absorbed from the gas stream; the SO_2 is then released in a concentrated stream in the stripping step. The regenerated absorbent is returned to the absorber loop. The concentrated SO_2 stream with water vapor enters a condenser, where most of the water is removed. If necessary, the resulting SO_2 stream may be further dried in a concentrated sulfuric acid drying tower. Sulfur values from the SO_2 stream may be recovered as liquid SO_2, liquid SO_3, sulfuric acid, or elemental sulfur. The product is determined by potential use, market demand, and cost of transportation to the destination.

5.2.2.4a Process Chemistry.

In sodium sulfite/bisulfite systems, it is desirable that any fly ash or other particulate matter be removed before the absorption step to reduce the need for process purge and thereby reduce the need for makeup of fresh scrubbing solution. An electrostatic precipitator (ESP), fabric filter, wet particulate scrubber, or other device may be used for particulate removal. The gas stream normally is cooled to its adiabatic saturation temperature in a wet scrubber or presaturator.

The basic process for absorption of SO_2 by aqueous scrubbing liquor is given by the following reactions:

$$SO_2(g) \rightarrow SO_2(aq) \tag{22}$$

$$SO_2(aq) + H_2O \rightarrow H_2SO_3 \rightarrow HSO_3^- + H^+ \tag{23}$$

$$HSO_3^- \rightarrow H^+ + SO_3^{2-} \tag{24}$$

Recovery processes are based on the chemistry of the sulfite/bisulfite buffer system. After appropriate pretreatment, the flue gas containing SO_2 enters the absorber, where it is brought into contact countercurrently with a sulfite solution. The sulfite absorbs and reacts chemically with the SO_2, forming the more soluble bisulfite product.[16]

Oxygen and SO_3 in the flue gas also react with the sodium sulfite, forming the unreactive sulfate/bisulfite. The presence of the unreactive species in the system necessitates a purge from the absorber to maintain the level of reactive sulfite and to reduce the possibility of scaling.

The principal chemical reactions in the SO_2 absorber are absorption and oxidation, discussed briefly as follows:

SO_2 absorption: Sulfur dioxide and sodium sulfite react to form bisulfite.

$$SO_2 + SO_3^{2-} + H_2O \longrightarrow 2HSO_3^- \tag{25}$$

Oxidation: Some oxidation of sodium sulfite to sodium sulfate occurs.

$$2SO_3^{2-} + O_2 \longrightarrow 2SO_4^{2-} \tag{26}$$

In the sodium ion makeup reactions, sodium carbonate (soda ash) or sodium hydroxide (caustic) reacts with sodium bisulfite to regenerate the SO_2 absorbent, sodium sulfite.

$$Na_2CO_3 + 2NaHSO_3 \longrightarrow 2Na_2SO_3 + H_2O + CO_2\uparrow \tag{27}$$

$$NaOH + NaHSO_3 \longrightarrow Na_2SO_3 + H_2O \tag{28}$$

If the product is to be a concentrated SO_2 stream, the bleed stream is regenerated by use of single-effect evaporators, double-effect evaporators, or either atmospheric or vacuum steam stripping. The basic chemical reaction for regeneration of the alkali absorbent is then

$$2NaHSO_3 \xrightarrow{\Delta} Na_2SO_3 + H_2O\uparrow + SO_2\uparrow \tag{29}$$

The sodium sulfate formed must be purged at approximately the rate of formation. It can be dried for sale or disposal, or it can be neutralized and discharged as an innocuous effluent.

The concentrated SO_2 stream leaving the regeneration step can be used to produce sulfuric acid, sulfur, liquid SO_2, or some combination of these, depending on available markets.

5.2.2.4b System Description.   A sulfite/bisulfite FGD system can be considered in terms of the following general steps:

Flue gas pretreatment
SO_2 absorption
Absorbent regeneration
Sulfur product recovery
Purge treatment

Efficient flue gas pretreatment is important to these FGD systems because, in reducing particulate contamination, the requirements for regeneration and purge are reduced. The flue gas to be treated is taken after the electrostatic precipitator at a temperature of about 300°F and passed through a venturi or tray-type prescrubber, in which it is cooled to around 130°F and humidified. A tray-type prescrubber satisfactorily cools and humidifies the gas with low pressure drop, but removes less of the fly ash and chlorides. Scaling and plugging problems are virtually eliminated by use of the prescrubber and clear scrubber solutions as well as by the solubility of the absorption product, sodium bisulfite, which is more soluble than sodium sulfite.

A well-designed prescrubber may remove up to 99% of all chlorides in the flue gas; this should help maintain a low level of chloride in the scrubbing liquor and reduce the potential for stress corrosion. After neutralization with lime when necessary, the fly ash and other solids collected by the prescrubber are pumped to an ash disposal pond as about a 5% slurry.

The SO_2 absorption step in the Wellman–Lord process has been performed primarily in a tray tower absorber. The humidified gas from the prescrubber is passed upward through the absorption tower, where it meets the countercurrent flow of the aqueous absorbent solution. Utility Wellman–Lord systems have shown individual SO_2 removal test results above 90%.

When SO_2 is the product of the regeneration step, various sulfur products are possible. Production of elemental sulfur, which has been done at two full-scale utility FGD systems, requires a reducing gas (such as methane or natural gas, hydrogen sulfide, or carbon monoxide).

If sulfuric acid is a by-product, a definite market is needed to maintain proper operation of the FGD system.

The sodium sulfate, present in the decahydrate form (Glauber's salt), is continuously removed from the regenerated absorbent solution by vacuum crystallization. Subsequently, it is prepared for disposal or dried for storage and sale. The sodium sulfate is used in the pulp and paper industry and the fertilizer industry, two of several potential markets.

5.2.2.5 Dry Removal Processes. The term "dry removal process" designates any FGD process from which a dry product directly results.

Dry removal offers various advantages over wet scrubbing. Dry removal systems do not require the sludge handling equipment that many wet scrubbers need. Scaling and plugging, common problems at the wet/dry interface in wet scrubbers, are avoided in dry removal units because only a dry product contacts the walls. Whereas wet systems often use special materials of construction or coatings to prevent corrosion and erosion, the vessels and duct work of dry systems can be made of low-carbon steel. Dry removal units require less manpower to operate than wet scrubbers and can respond more quickly to fluctuations in SO_2 levels. Because dry systems operate with relatively low pressure drops through the absorption system and with smaller volumes of spent absorbent, the operating expenses that wet systems incur because of high pressure drops and greater volumes of spent absorbent can be reduced. It is estimated that dry removal units need only 25 to 50% of the energy that wet scrubbers require. Finally, dry systems consume much less water than wet systems and thus are particularly attractive in western areas of the United States where water supplies are limited.

Several disadvantages of dry scrubbing systems are higher priced absorbents, applicability primarily to low sulfur coal, lack of commercially proved systems. Most of the dry systems either in design or under construction are for low sulfur (less than 1% sulfur) coal; however, one industrial unit that is on-stream reports 85% SO_2 removal when firing a 3% sulfur coal.[17] Therefore, dry systems may well be applicable to higher sulfur coals, but this will be shown only by more operating experience and research.

5.2.2.5a Process Description. Dry systems can be designed to generate a saleable product such as the Aqueous Carbonate Process, or throwaway product such as the lime systems.

In spray dryer-based systems, the first of the major types of dry FGD systems, flue gas at air preheater outlet temperatures generally 275 to 400°F is contacted with a solution or slurry of alkaline material in a vessel of 5–10 sec residence time. The flue gas is adiabatically humidified to within 50°F of its saturation temperature by the water evaporated, liquid phase salts are precipitated, and the remaining solids are dried to generally less than 1% free moisture. These solids, along with fly ash, are entrained in the flue gas and carried out of the dryer to a particulate collection device. Reaction between the alkaline material and flue gas SO_2 proceeds both during and following the drying process. The mechanisms of the SO_2 removal reactions are not well understood. The chemical reactions are identical to those described under corresponding wet scrubbing processes (e.g., lime, sodium carbonate, etc.).

Sodium carbonate solutions and lime slurries are common sorbents. A sodium carbonate solution will generally achieve a higher level of SO_2 removal than a lime slurry at similar conditions of inlet and outlet flue gas temperature, SO_2 level, and sorbent stoichiometry. Lime, however, has become the sorbent of choice in many circumstances because of the cost advantage it enjoys over sodium carbonate and because the reaction products are not as water soluble. Liquid to gas ratios are generally in the range of 0.2–0.3 gal/1000 ft^3. The sorbent stoichiometry is varied by raising or lowering the concentration of a solution or weight percent solids of a slurry containing this set amount of water. As sorbent stoichiometry is increased to raise the level of SO_2 removal, two limiting factors are approached.

Sorbent utilization decreases, raising sorbent and disposal costs on the basis of SO_2 removed.

An upper limit is reached on the solubility of the sorbent in the solution, or on the weight percent of sorbent solids in a slurry.

One method of circumventing these limitations is to initiate sorbent recycle, either from solids that settle in the spray dryer or from material collected in the particulate collection device. This has the advantage of increasing the sorbent utilization; additionally, it can increase the opportunity for utilization of any alkalinity in the fly ash.

The second method is to operate the spray dryer at a lower outlet temperature; that is, a closer approach to saturation. This has the effect of both increasing the residence time of the liquid droplets and increasing the residual moisture level in the dried solids. As the approach to saturation is narrowed, SO_2 removal rates and sorbent utilization generally increase rapidly.

The spray dryer design can be affected by the choice of particulate collection device. Bag collectors have an inherent advantage in that the unreacted alkalinity in the collected waste on the bag surface can react with remaining SO_2 in the flue gas. Some process developers have reported SO_2 removal on bag surfaces on the order of 10%. A disadvantage of using a bag collector is that since the fabric is somewhat sensitive to wetting, a margin above saturation temperature (on the order of 25 to 35°F) must be maintained for bag protection. Electrostatic precipitator (ESP) collectors have not been demonstrated to achieve significant SO_2 removal. However, some vendors claim that the ESP is less sensitive to condensation and hence can be operated closer to saturation (less than a 25°F approach) with the associated increase in spray dryer performance.

The choice between sorbent types, use of recycle, use of warm or hot gas bypass, and types of particulate collection devices tends to be site specific. Vendor and customer preferences, system performance requirements, and site specific economic factors tend to dictate the system design for each individual application.

5.2.3 Operating Experience

In the past 5 yr, FGD has become the most commercially developed means of control of SO_2 emissions from coal-fired boilers, and operating experience has increased significantly. At the end of 1975, 20 units were either on-line (or had been), and approximately 198,000 hr of on-line experience had been accumulated. By August 1980, 85 FGD systems had been operated on utility boilers, and more than 460,000 hr of operation had been logged. This represents a 425% increase in the number of FGD systems operated and a 230% increase in total hours logged.

The operational hours above reflect the number of hours reported by the utilities. Because hours of operation often are not available for such periods as initial system startup or performance testing, the actual number of operational hours is greater than reported, as is the corresponding percentage increase.

5.3 Nitrogen Oxides Control

Among the stationary sources of NO_x emissions, fossil-fuel-fired utility boilers constitute the largest single contributing sector; in 1977 these boilers generated almost 55% of all NO_x emissions from stationary fuel combustion sources and about 31% of the total NO_x emissions in the United States. Within the utility category, coal-fired boilers account for over 80% of the NO_x emissions. Emissions of NO_x from utility boilers are expected to pose a more severe problem in the future, with a projected increase in construction of new, coal-fired boilers.

Three different NO_x emission factors were applied to the three different types of utility boilers reported to be in use: pulverized-coal-fired, wet-bottom boilers; pulverized-coal-fired, dry-bottom boilers; and cyclone boilers.

Formation of NO_x in combustion processes is caused either by thermal fixation of atmospheric nitrogen in the combustion air, which produces "thermal NO_x," or by the conversion of chemically bound nitrogen in the fuel, which produces "fuel NO_x." The relative magnitude and importance of the two modes of NO_x formation varies with the type of fuel used. In coal-fired boilers, fuel NO_x predominates. The control techniques for coal-fired boilers therefore are generally designed to reduce fuel NO_x formation, which depends primarily on oxygen concentration and not so much on temperature.

The principal options for control of NO_x from utility boilers consist of combustion modification, flue gas treatment, and fuel modifications. Among these three options, modifying the combustion conditions is the most effective and widely used technique, achieving moderate (20–60%) reductions of combustion-generated NO_x. In an effort toward greater levels of control, additional control techniques such as advanced burner/furnace designs and ammonia injection are under development.

Several combustion modification techniques may be used singly or in combination on coal-fired utility boilers. These include low excess air (LEA) firing; staged or off-stoichiometric combustion (OSC) implemented through biased burner firing (BBF), burners out of service (BOOS), or overfire air injection (OFA); low NO_x burners (LNB); flue gas recirculation (FGR); and reduced firing rate.

Two flue gas treatment (FGT) processes are receiving attention as a possible means of meeting stringent NO_x standards for utility boilers. Selective, noncatalytic reduction (SNR) of NO_x by ammonia injection has been developed and patented by Exxon Corporation. The process, called Thermal De NO_x, has been commercialized for oil- and gas-fired boilers, and recent pilot-scale tests confirm its effectiveness on coal-fired utility boilers. Disadvantages include possible ammonia emissions and a narrow temperature range of operation.

A second FGT process is known as selective catalytic reduction (SCR). In principle the process is similar to SNR, but the use of catalysts minimizes the temperature range restriction imposed by SNR. The process has been applied to commercial oil- and gas-fired boilers and is being tested on coal-fired boilers in Japan; it has not yet been installed on U.S. boilers. The SCR process offers potential for use with combustion modifications when very high NO_x removal efficiencies are required.

The techniques proposed for fuel modification include fuel switching, fuel additives, and fuel denitrification. These processes, however, do not currently appear attractive for NO_x control.

Certain boiler firing configurations and combustion modes are not amenable to NO_x reduction techniques. Cyclone boilers promote highly turbulent combustion and high flame temperature to achieve complete combustion and maintain the ash in a molten state for removal. Both of these conditions are very conducive to the formation of NO_x, and any attempt to minimize emissions could lead to a significant derating of boiler capacity and inhibit removal of slag (molten ash) from the boiler. Ash is also removed in molten form from a pulverized-coal-fired wet-bottom boiler, and attempts to reduce NO_x formation would yield results similar to those occurring in a cyclone boiler.

REFERENCES FOR SECTION 5

1. Ardell, M. Future Use of Air Pollution Control Equipment on Industrial Boilers, The McIlvaine Co., presented at the 73rd Annual Meeting of the Air Pollution Control Association, Montreal, Quebec, June 22–27, 1980.

2. Nichols, G. B., and McCain, J. D., *Particulate Collection Efficiency Measurements on Three Electrostatic Precipitators,* Southern Research Institute, EPA-600/2-74-056, October 1975.

3. Ref. 1.

4. Bradway, R. W., and Cass, R. W., *Fractional Efficiency of a Utility Boiler Baghouse, Nucla Generating Plant,* NTIS Document No. PB 245541, August 1975.

5. Ref. 1.

6. Szabo, M. F., and Gerstle, R. W., *Operation and Maintenance of Particulate Control Devices on Coal-Fired Utility Boilers,* EPA-600/2-77-129, July 1977.

7. Ref. 1.

8. Boubel, R. W., *Control of Particulate Emissions from Wood-Fired Boilers,* PEDCo Environmental, Inc., EPA Contract No. 68-02-1375, 1977.

9. Ref. 8.

10. Bruck, N., and Melia, M., *EPA Utility FGD Survey: October–December 1981* (preliminary draft), U.S. Environmental Protection Agency, Washington, D.C., Contract No. 68-02-3173, Task 71, February 1982, p. xxi.

11. Smith, M. P., et al., Recent Trends in Utility Flue Gas Desulfurization, PEDCo Environmental, Inc., presented at the 6th Symposium on Flue Gas Desulfurization, Houston, Texas, October 28–31, 1980, p. 2-4.

12. Ref. 11, p. 2-3.

13. Ref. 10.

14. Devitt, T. W., et al., *Flue Gas Desulfurization Systems Capabilities for Coal-Fired Steam Generators,* U.S. Environmental Protection Agency, Washington, D.C., EPA-600/7-78-0326, March 1978, Vol. II, pp. 3–98.

15. *Electric Utility Steam Generating Units, Background Information for Proposed SO_2 Emission Standards,* U.S. Environmental Protection Agency, Research Triangle Park, NC, EPA-450/2-78-007a, July 1978, pp. 4–95.

16. Pedroso, R., *An Update of the Wellman–Lord Flue Gas Desulfurization Process,* U.S. Environmental Protection Agency, Washington, D.C., EPA-600/2-76-136a, May 1976, p. 720.

17. *The FGS Newsletter,* Number 24, The McIlvaine Company, Northbrook, IL, April 30, 1980, p. 2.

6 CONTROL EQUIPMENT COSTS

6.1 Particulate Control Systems[1]

This section presents cost data for electrostatic precipitators, fabric filters, wet scrubbers, and mechanical collectors. Three cost curves are given for each type of control technique, representing purchased equipment costs, total capital costs, and annualized costs. All costs are estimated from information contained in Ref. 1 updated to January 1980 dollars. The accuracy of any curve relative to a specific application depends on the similarities between the assumptions used in the example and the conditions under which the system will actually be used. These curves can be considered representative for combustion sources. The curves should not be relied on, however, to provide better than ±50% accuracy.

Each of the control system cost curves includes the cost of auxiliary equipment (ductwork, fan, and fan drive) normally associated with such a system. These curves provide costs for grass-roots installation. A retrofitted installation generally costs 10–30% more than a grass-roots installation and, depending on specific difficulties at a given site, the costs can be calculated on the basis of the latter percentage.

Annualized costs are based on 8700 hr/yr operation time. Since the annualized costs vary with operating time, the annualized costs for operations of less than 8700 hr/yr will be lower than those shown in the control system cost curves. For example, the annualized costs for 2000 hr/yr operation as a percentage of the costs for 8700 hr/yr operation are approximately as follows:

Venturi scrubber	30–40%
Fabric filter	50–60%
Electrostatic precipitator	60–70%

The annualized cost includes a disposal cost of $10/ton, based on disposal of a nontoxic substance. It also includes a capital charge based on an assumed equipment life of 15 yr and an opportunity cost of 15%.

These cost curves are presented in terms of dollars versus exhaust gas volume. This relationship is based on a number of simplifying assumptions, which allow one to obtain quick, conceptual or study estimates with a minimum of effort. It must be borne in mind that these simplifications can lead to anomalous results at the extremes of the ranges, since the curves are presented in the form $y = ax^b$ and are based on regression analysis.

It should also be noted that the gas flow range of 10,000–1,000,000 cfm was chosen to represent a range of utility and industrial applications and thus does not include the largest gas flows found on a utility source or the smallest in an industrial application.

6.1.1 Electrostatic Precipitators

Figure 14 presents cost curves for systems utilizing an electrostatic precipitator housed in an insulated, carbon steel shell. The assumption is made that the uncontrolled gas stream is normally vented to a stack. Thus, the necessary fan and ductwork are considered part of the process. Costs are presented for three levels of control efficiency based on medium- and high-reactivity dust. For a given collection efficiency, high-resistivity dust requires a greater SCA (specific collection area) and the cost of the ESP is thus increased. For purposes of estimating equipment costs, plate area was calculated according to the Deutsch equation with particle drift velocities of 0.036 m/sec for high-resistivity dusts and 0.086 m/sec for low-resistivity dusts. Fly ash from low-sulfur coal combustion has high resistivity while bark boiler fly ash generally has low resistivity.

6.1.2 Fabric Filters

Fabric filters on combustion sources are commonly used across a broad range of exhaust gas volumes. Figures 15 and 16 present cost curves for carbon steel and stainless steel construction, respectively.

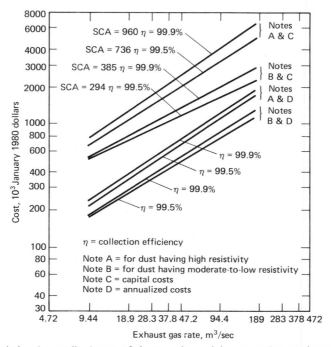

Fig. 14 Capital and annualized costs of electrostatic precipitators, carbon steel construction.[1]

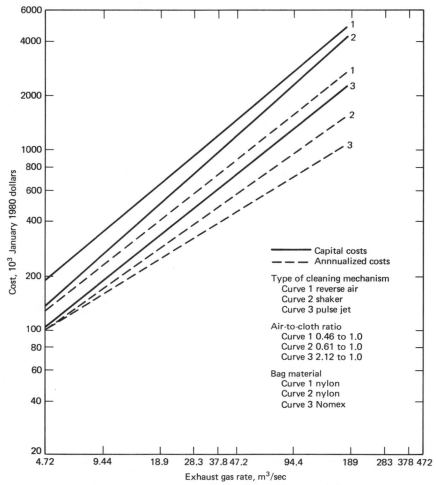

Fig. 15   Capital and annualized costs of fabric filters, carbon steel construction.[1]

Costs are presented for filters utilizing each type of bag-cleaning mechanism. The cost curves assume that the fan and drive are process equipment. The control costs include tie-in ductwork, a dust handling conveyor, and a dust storage bin. The costs of thermal insulation and heaters (necessary to prevent condensation in some applications) are not reflected in the cost curves.

6.1.3 *Venturi Scrubbers*

Venturi scrubber use on combustion sources can cover a broad range of gas flows and pressure drops; thus, Figures 17 and 18 present cost curves for carbon steel and stainless steel construction, respectively, and a variety of pressure drops. The costs include a clarifier and circulating pump for the scrubber liquor, a fan and drive, and ductwork sufficient to tie the scrubber into the process exhaust stream.

6.1.4 *Mechanical Collectors*

Capital and annualized cost curves for mechanical collector systems are shown in Figure 19. System costs include hooding to capture the exhaust at the emission point, ducting, a fan and drive, and a dust storage bin. The system cost is based on carbon steel construction. Collection efficiency for this type of system generally ranges from 80 to 90%, depending on the particle size distribution and inlet grain loading.

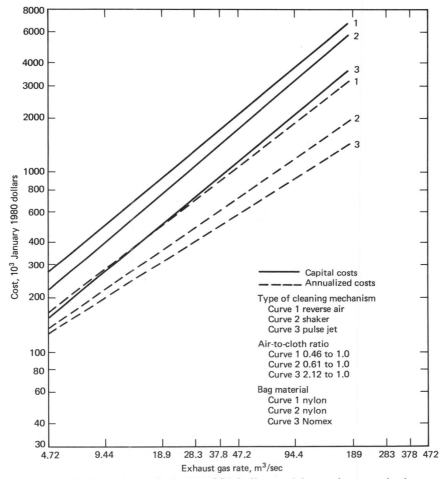

Type of cleaning mechanism
Curve 1 reverse air
Curve 2 shaker
Curve 3 pulse jet

Air-to-cloth ratio
Curve 1 0.46 to 1.0
Curve 2 0.61 to 1.0
Curve 3 2.12 to 1.0

Bag material
Curve 1 nylon
Curve 2 nylon
Curve 3 Nomex

Fig. 16   Capital and annualized costs of fabric filters, stainless steel construction.[1]

6.2   Sulfur Dioxide Control Cost

Capital and annual cost data on operational FGD systems have been obtained continuously since March 1978 by PEDCo Environmental, Inc. Costs for each system are obtained directly from the utilities and from published sources, and then itemized by individual FGD cost elements. The itemized costs are then adjusted to a common basis to enhance comparability. This adjustment includes factors for estimating costs not given by the utilities and escalating all costs to common dollars.

Table 19 summarizes both reported and adjusted costs for all operational FGD systems on which cost data were obtained for the most recent cost analysis study. This table also summarizes the results by application (new/retrofit), by sulfur content of the coal (high sulfur/low sulfur), and the results by process type.

In order that direct comparisons with conventional coal-to-electricity processes may be made, design and economic premises compatible with those commonly used in this area have been employed.

The base case for the conceptual design and detailed engineering cost studies is an approximately 500 MW (net) new utility power plant burning Illinois no. 6 coal with a sulfur content (dry) of 3.86%. This coal has a moisture content of 12%, an ash content of 8.82%, and a higher heating value of 12,771 Btu/lb (dry).

The projected operating life of the utility power plant is assumed to be 30 yr, representing 127,500 hr of generating capacity. This is an average of 4250 hr of operation per year. The projected load factor for the first year of operation is 0.8, or 7000 hr of operation.

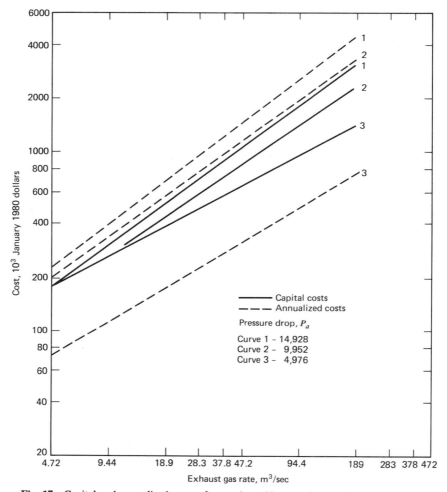

Fig. 17 Capital and annualized costs of venturi scrubbers, carbon steel construction.[1]

All of the alternative power plants have been designed to comply with the June 1979 NSPS. For particulate emissions, this is a 0.03 lb/10^6 input. For SO_2 emissions, this is an 85% removal (daily average) of sulfur in the coal as mined. The gasification/combined-cycle power systems achieve sulfur removal by using standard industrial H_2S removal techniques, such as the Stretford process with a Claus tailgas-cleanup system. Sulfur removal in the SRC process takes place during the hydrogenation of the coal that is dissolved in a coal-derived solvent. The sulfur, as hydrogen sulfide, is flashed off, separated from the recycle hydrogen, and converted to elemental sulfur.

A construction start date of 1981 and a plant startup date of 1985 have been assumed for all of the technologies. Costs have been calculated on the basis of 1980 dollars and have been scaled based on the extrapolated average annual *Chemical Engineering Cost Indices.*

6.4 Nitrogen Oxide Control Cost

The costs for retrofitting NO_x controls are dependent on the type of burner (i.e., circular or cell and other design parameters). The estimated 1980 cost for replacing a circular burner with a low-NO_x burner was estimated to be $16,000. Cell burners are much more difficult to retrofit, requiring modifications of boiler tubes, windbox, and pressure parts. The estimated total direct cost would be approximately

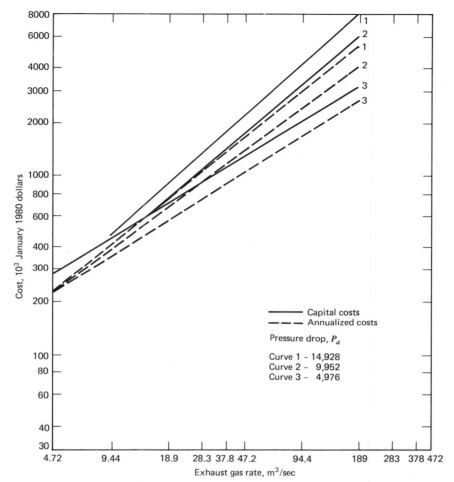

Fig. 18 Capital and annualized costs of venturi scrubbers, stainless steel construction.[1]

$110,000 per burner. Overfire air is the currently recommended NO_x control strategy for some boilers. The total direct cost of an overfire air port was estimated to be $80,000, with one port used over each column of burners.

Annualized costs consist of the annual incremental operating and maintenance cost and annual capital charges. The incremental operating and maintenance cost is the amount of increase in maintenance costs that is attributable to installation of additional equipment for NO_x control. Where overfire air ports are recommended, this increment is assumed to be 1.5% of the capital cost. Where low-NO_x burners are recommended, it was assumed that no incremental operating and maintenance would be needed. The annual capital charges include taxes, insurance, general and administrative costs, and a capital recovery rate. Table 20 gives the estimated capital and annualized costs of implementing NO_x control strategies at 36 plants.

No cost estimates are given for such items as potential derating of the boiler, replacement power costs, or possible operational problems, chiefly because retrofit NO_x control costs are so highly variable. Actual capital costs can vary by a factor of 2, and equipment costs may represent only a small portion of the total retrofit cost. Cost data are sparse because few coal-fired boilers have been retrofitted with NO_x controls. The few available cost estimates vary greatly, largely because of the inflation of material and labor costs, geographical factors, boiler size and configuration, burner design, fuel characteristics, and other site-specific parameters.

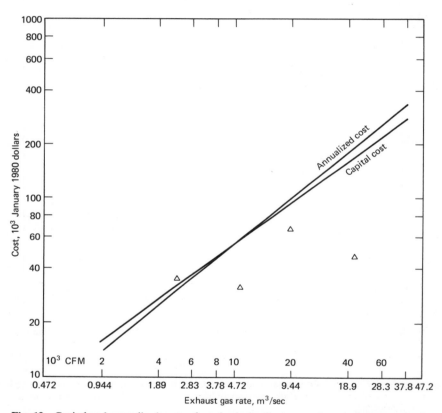

Fig. 19 Capital and annualized costs of mechanical collectors, carbon steel construction.[1]

Table 19 Categorical Results of the Reported and Adjusted Capital and Annual Costs for Operational FGD Systems[a]

| | Reported | | | | | | Adjusted | | | | | |
| | Capital | | | Annual | | | Capital | | | Annual | | |
	Range, $/kW	Average, $/kW	σ	Range, mills/kWh	Average, mills/kWh	σ	Range, $/kW	Average, $/kW	σ	Range, mills/kWh	Average, mills/kWh	σ
All	23.7–213.6	80.2	44.3	0.1–13.0	2.3	2.8	38.3–282.2	118.8	58.1	1.6–20.8	7.6	4.1
New	23.7–213.6	80.4	46.1	0.1–5.5	1.7	1.8	38.3–263.9	110.8	48.4	1.6–14.6	6.8	3.2
Retrofit	29.4–157.4	79.7	39.4	0.5–13.0	4.5	4.4	60.4–282.2	139.3	73.8	4.3–20.8	9.7	5.3
Saleable	132.8–185.0	153.1	20.6	13.0–13.0	13.0	0.0	254.6–282.2	271.6	12.1	16.7–20.8	18.1	1.9
Throwaway	23.7–213.6	75.8	41.5	0.1–11.3	2.1	2.4	38.3–263.9	110.9	47.6	1.6–17.6	7.0	3.4
Alkaline fly ash/lime	43.4–173.8	93.9	44.0	0.4–5.4	2.1	1.9	52.5–184.4	122.8	51.4	3.0–14.1	7.2	3.8
Alkaline fly ash/limestone	49.3–49.3	49.3	0.0	0.8–0.8	0.8	0.0	102.6–102.6	102.6	0.0	5.4–5.4	5.4	0.0
Dual alkali	47.2–174.8	97.8	55.3	1.3–1.3	1.3	0.0	87.8–263.9	146.7	82.9	5.0–13.9	8.7	3.8
Lime	29.4–213.6	81.8	43.7	0.3–11.3	3.2	2.7	60.4–210.0	116.5	44.2	4.0–17.6	8.1	3.6
Limestone	23.7–170.4	67.9	37.2	0.1–7.8	1.6	2.2	38.3–194.3	98.9	44.0	1.6–14.6	6.1	3.1
Sodium carbonate	42.9–100.8	69.2	26.6	0.2–0.5	0.4	0.1	87.1–150.9	110.9	26.4	5.8–7.4	6.4	0.7
Wellman–Lord	132.8–185.0	153.1	20.6	13.0–13.0	13.0	0.0	254.6–282.2	271.6	12.1	16.7–20.8	18.1	1.9

Table 20 Estimated Costs of Installing NO$_x$ Controls at 36 of the 50 Highest Uncontrolled NO$_x$ Emitters

Plant Data			NO$_x$ Control Costs				NO$_x$ Emissions		
			Capital Cost		Annualized Cost				
Name	Capacity, MW	Capacity Factor	$ × 10³	$/kW	$ × 10³	mills/kWh	Uncontrolled, 10³ tons	Controlled, 10³ tons	NO$_x$ Reduction, %
Baldwin	634.5	0.67	1,041	1.64	166	0.04	141.91	130.83	8
Gavin	2600	0.71	31,480	12.11	4,564	0.28	73.96	44.38	40
Amos	2932.6	0.84	17,821	6.08	2,584	0.12	68.56	41.14	40
Muskingum River	1054.6	0.70	8,344	7.91	1,335	0.21	62.24	49.76	20
Clifty Creek	1303.56	0.75	2,496	1.91	427	0.05	58.93	35.36	40
Monroe	3279	0.63	38,632	11.78	5,600	0.31	58.35	35.01	40
Bowen	3160	0.58	4,164	1.32	666	0.04	53.95	40.46	25
Stuart	2440.8	0.64	28,616	11.72	4,148	0.30	53.42	32.05	40
Labadie	2220	0.59	2,080	0.94	333	0.03	49.93	37.45	25
Gibson	2672	0.52	3,332	1.25	476	0.04	47.66	28.60	40
Kyger Creek	1086.5	0.76	2,080	1.91	356	0.05	46.47	27.88	40
Cumberland	2600	0.48	31,480	12.11	4,564	0.42	43.81	26.29	40
Gaston	2012.8	0.71	2,861	1.42	463	0.04	43.39	28.78	34
Homer City	1850	0.80	2,705	1.46	394	0.03	42.81	25.68	40
Conesville	1711	0.44	2,601	1.52	458	0.07	42.01	34.97	17
Belews Creek	2160	0.62	25,756	11.92	3,708	0.32	41.67	25.00	40
Cardinal	1880.46	0.71	16,169	8.60	2,361	0.20	41.48	24.89	40
Wansley	1730	0.73	2,082	1.20	327	0.03	40.98	30.74	25
Sherburne	1440	0.54	2,082	1.45	329	0.05	39.63	29.72	25
Marshall	2000	0.65	3,122	1.56	507	0.04	37.56	28.17	25
Keystone	1872	0.66	2,082	1.11	339	0.03	37.53	28.15	25
Joliet	1320	0.76	2,080	1.58	343	0.04	37.50	30.65	18
Neal	1445.2	0.52	1,977	1.37	284	0.04	36.19	25.05	31
Kingston	1723	0.65	4,680	2.72	885	0.09	35.87	26.90	25
Sammis	2303.5	0.49	16,422	7.13	3,527	0.36	35.71	21.42	40
Montour	1534	0.78	2,082	1.36	333	0.03	35.47	26.60	25
St. Clair	1547	0.24	2,184	1.41	374	0.12	34.41	24.55	29
Hatfield	1728	0.65	21,462	12.42	3,135	0.32	34.39	20.63	40
Brunner Island	1558	0.75	2,081	1.34	341	0.03	34.16	25.62	25
Will County	897	0.51	1,040	1.16	178	0.04	33.97	29.48	13
Columbia	1024	0.58	1,040	1.02	164	0.03	33.90	25.43	25
Roxboro	1813.05	0.63	4,163	2.30	674	0.07	33.48	25.11	25
Widows Creek	1980	0.46	2,600	1.31	458	0.06	31.95	21.81	32
Mansfield	1670	0.53	2,186	1.31	312	0.04	31.01	18.61	40
Mitchell	1632.3	0.67	2,082	1.28	304	0.03	30.46	18.27	40
Karn	530	0.74	988	1.86	164	0.05	29.17	25.10	14

REFERENCES FOR SECTION 6

1. Neveril, R. B., *Capital and Annual Costs of Selected Air Pollution Control Systems,* EPA-450/5-80-002, December 1978.

2. Bruck, N., and Melia, M., *EPA Utility FGD Survey: October–December 1981* (preliminary draft), U.S. Environmental Protection Agency, Washington, D.C., Contract No. 68-02-3173, Task 71, February 1982, p. A-8.

3. Ref. 2, p. A-7.

CHAPTER 16
UNCONVENTIONAL FUELS

NORMAN J. WEINSTEIN

Recon Systems, Inc.
Three Bridges, New Jersey

1 INTRODUCTION

As the great Industrial Revolution which began in the seventeenth century accelerated, it was fueled by coal and then in the nineteenth and twentieth centuries by oil and natural gas. Without the development of these fossil fuels, the revolution might have been merely an evolution. We have now become so accustomed to the availability of coal, oil, and gas, and even nuclear power, that we often think of these energy sources, along with hydroelectric power, as conventional.

Yet the Industrial Revolution did begin with a variety of other energy sources, for example, water power to directly drive machinery, animals fueled with cultivated and uncultivated crops for transportation, and charcoal derived from woodlands for iron production. Even though they are very old, these energy sources are today considered to be unconventional.

In this discussion we will deal with those unconventional (sometimes called alternative) energy sources which are combustible. Included will be all combustible materials (fuels) except conventional coal and oil. Some lower forms of fossil fuels such as peat, lignite, tar sands, and oil-bearing shale will be treated as unconventional fuels, as well as wood, industrial and agricultural wastes, and some synthetic fuels (e.g., alcohols) which can be made from a variety of conventional and unconventional raw materials.

The present and future importance of unconventional fuels is related to the replacement of diminishing supplies of oil and gas and in some instances, to effective disposal of a waste while obtaining energy as a useful by-product. The incentive for the use of unconventional fuels is therefore based on a complex set of circumstances which may involve alternative energy sources, alternative disposal methods, and relative impact on the environment.

The combustion of unconventional fuels often results in air and water pollutants and in undesirable solid wastes, but the air emission problem is usually the most serious of the environmental concerns. The water pollution and solid waste problems are often secondary, resulting from air purification systems.

The potential for air pollution is a characteristic of all combustion processes, regardless of whether the fuel is conventional or unconventional. The pollution potential is related primarily to fuel composition, fuel combustibility, and the design of the combustion equipment. Actual emissions to the ambient air depend also on the effectiveness of the air pollution control equipment used.

2 SOURCES OF UNCONVENTIONAL FUEL

Unconventional fuel sources may be classified as follows:

1. *Municipal Solid Waste.* Includes unprocessed wastes for mass burning, or those processed to remove salvageable materials such as metals and glass. The latter fuels are usually called refuse-derived fuel (RDF).

2. *Commercial and Institutional Solid Wastes.* Usually high in cellulosic material (paper, cardboard), but often includes greases and other food residues, and plastics. In the case of hospitals and other institutions, pathogenic wastes may be included.

3. *Agricultural Materials and Wastes.* Includes wood, wood processing wastes, wastes from forestry operations, wastes from paper manufacture, farm wastes, and also crops harvested primarily as fuels.

  4. *Aquatic Plants and Wastes.* Includes fish wastes and aquatic plants harvested primarily as fuels.

  5. *Sewage Treatment Plants.* Includes sludges, greases, and other solids.

  6. *Combustible Industrial and Mining Wastes.* Includes solids, liquids, and gases, some of which are considered hazardous and some nonhazardous. Waste oils and waste solvents are included in this group.

  7. *Naturally Occurring Carbonaceous and Hydrocarbon-Containing Mineral Resources.* Other than natural gas, petroleum, and coal, including peat, lignite, oil shale, and tar sands.

  8. *Synthetics.* Fuels prepared by chemical conversion (chemical reaction, microbiological conversion, thermal processing) of various raw materials from the above sources, from conventional fuels, and from other hydrocarbon sources such as hydrocarbon-rich agricultural products. Includes H_2, CO, wood char, pyrolytic fuels, synthetic natural gas (SNG), methanol, ethanol, and methane-containing gases from landfills and other anaerobic digestion or conversion of wastes, such as sewage sludges. Coke, such as is prepared from coking coals for iron ore reduction or from petroleum coking, could also be considered to be in this category.

  The term biomass is sometimes applied to all of the materials in groups 1 through 5 above, but the breakdown provided here is somewhat more useful in considering combustion and air pollution control technology. It should be noted that sewage sludge is hardly ever useful as a fuel since the energy recovered by burning dried sewage sludge rarely exceeds that required for dewatering and drying.

  Unconventional fuels suitable for energy recovery could become significant by the year 2000. For example, municipal solid waste could possibly contribute 2 quads (1 quad = 10^{15} Btu) of energy, or about 2% of the U.S. total required.[1] Wood wastes could provide 5 quads, while intensive silvaculture to produce wood as a fuel could conceivably provide up to 40 quads,[2] though serious air pollution and other problems might accompany such a drastic change in fuel source. More realistic estimates of future fuel sources show continued dependence on conventional fuels, but increasing importance of unconventional fuel sources, for example:[3]

	Quads	
	-------------	--------
	1977	2000
Domestic petroleum liquids	20.0	15.0
Imported petroleum liquids	17.4	17.0
Domestic gas	19.1	9.0
Imported gas	1.0	1.0
Coal	13.0	32.0
Hydroelectric	2.7	3.5
Nuclear	2.7	11.0
Other		
Wood		4.0
Agricultural waste		0.7
Municipal solid waste		1.0
Other		0.8
Total other	1.8	6.5
Total	77.7	95.0

The increase in coal use will include some lower grade coals such as lignite. Some conversion to coal liquids and gases may also be expected, but these are not likely to be a major energy factor in this time frame.

3   FUEL PREPARATION TECHNIQUES

There is a strong tendency toward adaptation of unconventional fuels for use in conventional combustion systems, for example, in oil- or coal-fired steam boilers. Several good reasons for this approach are: the many years of experience in the design and operation of such conventional facilities; the possibility of conversion of existing facilities; the availability of package units from manufacturers; and advantages that often exist for simultaneous firing of conventional and unconventional fuels in a single combustor or using a conventional fuel as a standby. However, as will be shown, overemphasis on the use of conventional combustion systems for unconventional fuels can lead to operating problems and projects with questionable economic viability.

Many techniques have been used to convert unconventional fuels to more useful forms, often for use in conventional combustion systems, but sometimes merely to improve ease of handling or combustion characteristics. The techniques used may result in merely a physical change, for example, size reduction or compaction; purification, for example, by chemical treatment or physical separations; or a complete change in characteristics by chemical modification to another fuel form, for example, microbiological conversion to alcohol or gasification to a fuel gas. There are so many possible fuel preparation techniques, most of which have a profound effect on air pollution control problems, that full documentation is beyond the scope of this discussion. A few examples will be cited here to suggest the range of possibilities.

Many techniques have been used to prepare municipal solid waste for use as a fuel, often known as RDF. Some of these are shown in Table 1, but even similar methodologies have resulted in widely varying success, often not fully elaborated in the literature cited.

Although not included above, solid waste combustion always results in nitrogen oxide emissions, and some sulfur oxide emissions, depending on the sulfur content of the waste. Carbon monoxide, excessive hydrocarbons, and soot will be emitted if combustion is not complete.

Shredding and other RDF preparation steps are normally designed both to recover materials of value and to allow use of conventional combustion equipment, as compared to mass burning incinerators. These techniques have often been technically and economically unsuccessful.[14]

The latter comment can also be made about current pyrolysis technology as a method for fuel preparation.[14] For example, a pyrolysis plant completed in Baltimore in 1975 produced pyrolysis gas from shredded waste which was then burned in a boiler to produce steam.[13] The same overall result, solid-waste-to-steam, could have been accomplished in a water-wall incinerator. This fact and the technical difficulties encountered in this multistep system raised serious questions about the usefulness of pyrolysis as a technique for fuel preparation. A later trial in San Diego with pyrolysis to produce primarily liquids was also unsuccessful.

In summary, there appears to be little justification for using presently available fuel preparation techniques for municipal solid waste. The same can be said as well for commercial and institutional solid wastes and for many combustible industrial wastes, unless fuel preparation leads to products valuable enough as raw materials to justify the separation steps involved.

Table 1   Methods for Municipal Solid Waste Preparation as Fuel

	Fuel Preparation Technique	Examples of Combustion Systems Used	Potential for Air Pollution
1.	Manual or source separation of white goods (refrigerators, etc.)	Special water-wall mass burning incinerator[4,5]	Particulates; HCl; trace hydrocarbons
2.	Shredding and magnetic separation of ferrous metals	Special suspension fired boiler[6,7]	Particulate emissions may be a little lower than for mass burning depending on bottom ash versus fly ash; HCl; trace hydrocarbons
3.	Shredding, magnetic separation, and separation of some nonferrous metals and glass by screening, flotation, and other techniques[8,9,10]	RDF for use in a special suspension fired boiler or as a supplement in existing pulverized fuel boilers	Lower particulate emissions; HCl; trace hydrocarbons
4.	Further grinding of RDF from 3 above to −200 mesh powder[11]	As a supplement in existing pulverized fuel boilers	Essentially same as 3 above
5.	Compaction of RDF from 3 above by pelletizing or briquetting[1]	As a supplement or to replace coal in a stoker-fired boiler[12]	Particulate emissions may be reduced compared to 3 above depending on bottom ash versus fly ash; HCl; trace hydrocarbons
6.	Pyrolysis to manufacture solid, liquid, and gaseous fuels from raw municipal solid waste or a form of RDF[13]	Various types of conventional combustion equipment	Varying pollutant emissions depending on fraction being burned and type of combustor

Table 2 Fuel Preparation Techniques

Raw Fuel	Preparation Technique	Effect on Potential for Air Pollution
1. Lignite	Milling and flotation to reduce ash content	Reduces fly ash
2. Oil shale	Retorting to separate oil from shale	Can almost eliminate potential particulate emissions, especially in lighter fuels
3. Tar sands	Thermal treatment to separate oil and sands and coking to recover light oils	Can almost eliminate potential particulate emissions especially in lighter fuels, but residual coke is a potential source of particulate emissions
4. Waste oils	Drying and filtering	Reduces fly ash, including some heavy metals

On the other hand, simple fuel preparation techniques are often useful for low-grade fossil fuels such as peat and lignite and for waste oils. Fuel preparation becomes essential for shale oil and tar sands. Some examples of useful fuel preparation techniques can be cited (Table 2).

Therefore, it is wise to carefully examine each unconventional fuel and its proposed use to determine whether special combustion equipment is required, with or without pretreatment, or whether it is best to adapt the fuel for use in conventional combustion systems.

4 TYPES OF COMBUSTION EQUIPMENT

The most familiar types of stationary combustion equipment in use today are gas-, oil-, and coal-fired steam boilers, hot water or hot oil heaters, and space heaters. Other types range from gas-fired

Table 3 Combustion Equipment and Unconventional Fuels

Type of Combustion Equipment	Types of Unconventional Fuels That May Be Considered
1. Pulverized fuel-fired boilers wall burners	Gaseous, liquid, and solid fuels with minimal ash or with ash that shows minimal tendency to slag, including powdered or shredded combustibles which can be pneumatically injected
2. Pulverized fuel-fired boilers cyclone burners	Powdered or shredded materials
3. Stoker-fired boilers	Dense lump fuels
4. Two-stage combustors (first stage—partial oxidation or pyrolysis; second stage—final incineration)	Solid wastes
5. Process heaters, e.g., rotary kilns, vertical shaft furnaces, and multiple hearth furnaces, where combustion gases come into contact with products, or are used as an inert gas with or without CO_2 scrubbing	Gaseous, liquid, or solid fuels which will not contaminate product (however, products such as cement or asphalt may be helpful in absorbing fuel contaminants to prevent air pollution)[15]
6. Thermal incinerators with convective heat recovery (e.g., stationary shaft or rotary kilns)	Gaseous, liquid, or solid wastes where designated temperature/time history is required for destruction efficiency
7. Catalytic incinerator with convective heat recovery	Gaseous or liquid wastes with minimal ash or catalyst poisons to make possible lower temperature/time history for required destruction efficiency
8. Fluidized bed (inert, chemically active, or catalytic solids) with in-bed and/or convective heat recovery	Gaseous, liquid, or solid fuels of many types
9. Molten salt	Gaseous, liquid, or solid fuels (development stage)

cooking devices to a myriad of important industrial devices such as asphalt kilns, cement kilns, ferrous and nonferrous blast furnaces, glass furnaces, and refinery heaters. Nonconventional fuels could be considered for any of these equipment types, or in modified combustion equipment for the same applications. Modifications may include added fuel preparation and handling steps, special fuel injection techniques (e.g., pneumatic suspension of solids, special air or steam atomized nozzles for high viscosity or dirty liquids, and special controls for volatile fuels), and upgraded air pollution control equipment.

Some types of combustion equipment which may be of special interest for unconventional fuels are shown in Table 3.

As will be discussed in the next section, potential air emissions are a function of both the type of combustion equipment and the composition and form of the fuel.

A more detailed list and description of combustion devices especially useful for hazardous wastes are available,[16,17] including design criteria.[16,18] However, trial burns may be necessary to demonstrate combustion efficiency, destruction efficiency, and pollutant emissions control.

5 POTENTIAL AIR EMISSIONS

The term "potential air emissions" is meant here to designate those emissions that can be detected between the combustor and air pollution control equipment, sometimes called uncontrolled emissions. It is useful to classify air pollutants resulting from combustion of unconventional fuels into two groups: particulate emissions and gaseous emissions. However, one should recognize some overlap in this classification since some gaseous pollutants may be condensed to liquid or solid particulates as the temperature of the effluent gas decreases; or by reactions between various gaseous constituents, for example, oxidation of sulfur dioxide to sulfur trioxide, or sulfuric acid in the presence of moisture.

The quantity of potential particulate emissions is a strong function of both the ash content of the fuel and the design of the combustion equipment. Major gaseous emissions, except for nitrogen oxides, are primarily dependent on the composition of the fuel. Nitrogen oxides will be emitted even if the nitrogen content of the fuel is zero, resulting from high-temperature oxidation of nitrogen in the combustion air. The presence of organic nitrogen compounds in the fuel will normally increase nitrogen oxide emissions.

Quantitative estimates of potential emissions from combustion of unconventional fuels can be made by material balance from the fuel composition (except for nitrogen oxides) and by reference to the U.S. Environmental Protection Agency's AP-42,[19] including some data for nitrogen oxides, and from other literature sources.[13,20,21,22] Material balance estimates are generally conservative (predicting higher than actual emissions) in the case of sulfur oxide and acid halide emissions, but may be high or low in the case of particulates depending on flue gas reactions and the test methods that define the quantity of particulate emissions (see Table 4).

Results for waste oil combustion[24] are provided in Table 5 as an example of the use of material balances for emission predictions. Actual emissions from waste oil combustion are dependent on the

Table 4 Air Pollutants and Quantities

Potential Air Pollutant	Estimation of the Quantity Emitted
Total particulates	Preliminary estimate from ash content of fuel. Correct for changes in ash composition during combustion, if known. Correct for split between fly ash, bottom ash, and slag held in combustion and heat recovery devices, if known. Account for particulates emitted during soot blowing, if practiced. Poor combustion may result in high-particulate carbon content, adding to total particulate emissions.
Metallic elements	Preliminary estimate from analysis for each element in fuel. Correct for split and concentration differences between fly ash, bottom ash, and slag held in combustion and heat recovery devices, if known. Distinguish between volatile compounds, e.g., certain metallic chlorides, and nonvolatile metal salts.

Table 4 *(Continued)*

Potential Air Pollutant	Estimation of the Quantity Emitted
Nitrogen oxides (NO, NO$_2$, other oxides)	Accurate estimates of NO$_x$ are difficult. Can generally be made only from experience and measurements with specific type of equipment to be used. Depends on residence time, temperature, oxygen concentration. NO predominates, with NO$_2$ usually 2–10% of total NO$_x$ in boilers. NO$_x$ lowest in low-temperature combustion equipment such as fluidized beds.
Sulfur oxides (SO$_2$, SO$_3$, H$_2$SO$_4$) ·	Preliminary estimates from sulfur content of fuel. Correct for reactions with other pollutants, e.g., formation of sulfites, bisulfites, sulfates. SO$_2$ usually predominates. SO$_2$/SO$_3$/H$_2$SO$_4$ split difficult to predict. SO$_3$ and H$_2$SO$_4$ often measured as particulates depending on effluent temperature and composition. Moisture content has major effect on dewpoint.
Halides	Preliminary estimate of chlorides and other halides from composition of fuel. Most halides converted to corresponding acids (hydrogen halides). Halide acids will be found in vapor phase except those which may condense at low temperature or which react with metallic elements (often substantial).
Hydrocarbons	Estimates difficult. Will be high under low-temperature or otherwise poor combustion conditions. Trace quantites of oxygenated hydrocarbons common, e.g., ketones and aldehydes. Trace quantities of high boiling hydrocarbons such as polynuclear aromatics (PNAs) emitted from practically all combustors whether in the fuel or not, with concentration higher under poor combustion conditions. PCBs, insecticides, and other toxic materials in commercial use could be emitted when present in fuel. Dioxins have been found in waste combustion flue gas, but a recent interim evaluation of TCDD emissions concluded that concentrations from municipal waste combustors are "far below the level of credible health risk."[23]
Carbon monoxide	Estimates difficult. Generally present in significant quantities only under poor combustion conditions, e.g., insufficient excess air.
Carbon dioxide	Usually not considered a local pollutant, but it is obvious from carbon balance calculations that increased reliance on solid fossil fuels (normally high carbon to hydrogen ratio) and synthetic fuels derived from solid fossil fuels could lead to increased global CO$_2$ concentration and higher atmospheric temperatures through the "greenhouse" effect.

Table 5 Uncontrolled Emission Factors for Combustion of Used Oil[24]

Pollutant	Emission Factors, lb/10^3 gal		Comments
	EPA AP-42 (3)	Suggested for Used Oil[a]	
Pb	Waste oil 0.0075(L)	0.0075(L)	L = ppm Pb in oil. Based on 100% emission at 7.5 lb/gal oil density.
Pb	Virgin oils 0.0042(L) (residual, distillate)	—	Based on substantially less than 100% emissions. Av L = 1.0 for residual oils, and 0.1 for distillate oils.
	Coal 1.6(L) lb/10^3 ton (bituminous, anthracite)	—	Based on 80% emissions.
Particulate	Waste oil 75(A)	75(A)	A = % ash in oil. Based on 100% equivalent emission at 7.5 lb/gal oil density.
Particulate	Virgin oils #6 10(S) + 3 #5 10 #4 7 Ind./comm. dist. 2 Domestic dist. 2.5	—	S = % sulfur in oil. Note that used oil with approx. 0.13% ash would be equivalent to #5 fuel oil.
Other metals (in particulate)	Not included	0.0075(L)	L = ppm metal in oil.
SO₂	Residual oil—157(S) Distillate oil—142(S)	150(S)	S = % sulfur in oil. Suggested factor for used oil based on 100% conversion of S to SO_2 for 7.5 lb/gal oil density.
SO₃	All virgin oils—2S	2S	S = % sulfur in oil.
NO$_x$ (total as NO₂)	Residual oils Power plant tangential—50 Power plant other—105 Ind./comm.—22 + 400(N)² Ind./comm. dist.—22 Domestic dist.—18	22	N = % nitrogen in oil. See AP-42 1.3 for further discussion of NO$_x$ emissions.
Hydrocarbons (total, as CH₄)	All virgin oils—1	1	RECON measurements ranged from 14 to 165 μg/g fuel (113 av) as compared to 1 lb/10^3 gal (approx. 133 μg/g) emission factor.
PNAs	Not included	0.0075	Corresponds to 1 μg/g. Insufficient data to determine how PNA emissions for used oils compare to virgin oils.
HCl	Not included	77(C) max.	C = % chlorine in oil.
HBr	Not included	76(B) max.	B = % bromine in oil.
P (in particulate)	Not included	75(P) max.	P = % phosphorous in oil.
CO	5	5	CO emissions vary with combustion control on all fuels. No CO emission detected by Orsat analyses in RECON tests 1–4. Determinations by Kitagawa detector tube in runs 5–9 showed 10 to 100 ppm in the flue gas or an average of about 5 lb/10^3 gal.

[a] And for used oil/virgin oil mixtures.

type of combustor used. Air or steam atomization used in industrial boilers emits higher levels of trace metals than does a vaporizing burner used in many space heaters. Hydrocarbons are usually higher from a vaporizing burner than from an air atomizing burner.

A search of the U.S. literature provides data that can be used to identify possible trace metal and other emissions from combustion of coal, oil, municipal refuse, and wood.[22] Data are available comparing emissions from coal firing with refuse firing and RDF/coal cofiring for a variety of pollutants.[25] Ultimately, only very careful testing can provide accurate emission estimates. This is especially true for unconventional fuels for which there are generally few reliable data.

6 REGULATIONS

The use of unconventional fuels is governed by federal, state, and local air pollution regulations. The variety and complexity of state and local regulations precludes a discussion of these here. Since one can assume that these regulations are generally similar to or more stringent than the federal regulations that are discussed, it is necessary that local and state regulations be investigated before proceeding with projects involving unconventional fuels.

Federal air pollution regulations that may affect the use of unconventional fuels find their basis primarily in the following legislation:

> The Clean Air Act of 1970 (as amended in 1974 and 1977)
> The Toxic Substances Control Act of 1976
> The Resource Conservation and Recovery Act of 1976

The responsibility for regulations under these acts lies primarily with the Environmental Protection Agency (EPA).

6.1 The Clean Air Act (CAA)

The Clean Air Act was adopted in 1970 and amended in 1974 and 1977 to protect public health and welfare from any actual or potential adverse air pollution effects. Regulations under CAA that may affect the use of unconventional fuels in stationary sources can be divided into the following categories:[26]

> Primary and Secondary National Ambient Air Quality Standards
> Prevention of Significant Deterioration
> "Nonattainment region" provisions, including offset policy
> New Source Performance Standards
> National Emission Standards for Hazardous Air Pollutants
> State Implementation Plans

Each of these categories is discussed further below. There are additional specific regulations for mobile sources not discussed here.

6.1.1 National Ambient Air Quality Standards (NAAQS)

Existing NAAQS limit ground level concentrations for sulfur dioxide (SO_2), total suspended particles (TSP), nitrogen dioxide (NO_2), carbon monoxide (CO), ozone, nonmethane hydrocarbons, and lead (Pb). Primary NAAQS were instituted to protect the public health while secondary NAAQS were designed to protect the public welfare. Established standards are provided in Table 6. Assuming complete combustion of unconventional fuels, SO_2, TSP, NO_2, and Pb standards may be significant, depending on fuel composition and combustion conditions as previously discussed. Ground level concentration of pollutants is determined by dispersion modeling and ambient air quality measurements. An example of modeling for Pb and other emissions can be found in "Used Oil Burning."[24]

6.1.2 Prevention of Significant Deterioration (PSD)

The PSD program was developed to preserve air quality in those areas where the air is better than NAAQS and to ensure that future growth is consistent with the preservation of clean air. As shown in Table 7, the PSD regulations set forth the maximum allowable incremental changes in existing ambient levels of SO_2 and TSP. Increments in class I areas restrict severely any industrial growth; increments in class II areas allow moderate growth; and increments in class III areas permit the most industrial growth.

PSD regulations provide in general that new major stationary sources or major modifications must obtain a permit before construction may begin. Existing facilities are not subject to PSD regulations unless major modifications are made to a major source that would result in a "significant net increase"

Table 6   National Ambient Air Quality Standards

Air Pollutant	Averaging Period	Maximum Allowable Concentrations[a]			
		Primary Standard		Secondary Standard	
		(μg/m^3)	(ppm)	(μg/m^3)	(ppm)
Sulfur oxides (as SO$_2$)	Annual arithmetic mean	80	0.03	—	—
	24-hr	365	0.14	—	—
	3-hr	—	—	1,300	0.5
Particulate matter	Annual geometric mean	75	—	60	—
	24-hr	260	—	150	—
Carbon monoxide	8-hr	10,000	9	10,000	9
	1-hr	40,000	35	40,000	35
Ozone	1-hr	235	0.12	235	0.12
Nitrogen dioxide	Annual arithmetic mean	100	0.05	100	0.05
Lead and its compounds as Pb	1 calendar quarter	1.5	—	1.5	—

[a] Other than annual and calendar quarter periods, maximum allowable concentrations may be exceeded no more than once per calendar year.

Table 7   National Standards for the Prevention of Significant Deterioration of Air Quality

Air Pollutant	Averaging Period	Maximum Allowable Increments[a]					
		Class I		Class II		Class III	
		(μg/m^3)	(ppm)	(μg/m^3)	(ppm)	(μg/m^3)	(ppm)
Sulfur dioxide	Annual arithmetic mean	2	0.001	20	0.008	40	0.016
	24-hr	5	0.002	91	0.036	182	0.071
	3-hr	25	0.010	512	0.201	700	0.270
Total suspended particulates	Annual geometric mean	5	—	19	—	37	—
	24-hr	10	—	37	—	75	—

[a] Note: 1. Increments refer to the maximum allowable increase of ambient air pollutant concentrations over baseline concentrations.

2. Baseline concentration is that ambient concentration level that exists at the time of the first permit application based on available air quality data, taking into account the effect of all projected emissions from any major emitting facility that commenced construction prior to January 6, 1975, but has not begun operation before the date of the baseline concentration determination.

3. The total ambient concentration shall not exceed the respective national secondary or primary ambient air quality standard, whichever is lower.

4. Other than annual periods, maximum allowable increases may be exceeded once per calendar year.

5. All areas are initially designated class II areas, except international parks, national wilderness areas, and national memorial parks greater than 5000 acres in size, national parks greater than 6000 acres in size, which are designated as class I areas, all other areas previously designated as class I, and nonattainment areas.

in that source's "potential to emit." A significant net emissions increase occurs when the following are exceeded:

	Tons/yr
Carbon monoxide	100
Nitrogen oxides	40
Sulfur dioxide	40
Particulate matter	25
Ozone (volatile organic compounds)	40
Lead	0.6
Asbestos	0.007
Beryllium	0.0004
Mercury	0.1
Vinyl chloride	1
Fluorides	3
Sulfuric acid mist	7
Hydrogen sulfide (H_2S)	10
Total reduced sulfur (including H_2S)	10
Reduced sulfur compounds (including H_2S)	10

Twenty-eight major sources with the "potential to emit" 100 tons/yr or more of any air pollutant are required to undergo a preconstruction review and permit process under PSD. Among these are, for example, fossil fuel-fired boilers (or combinations thereof) which have a heat input of greater than 250 million Btu/hr, municipal incinerators which are capable of charging more than 250 tons/ day, fuel conversion plants, and portland cement plants. Also required to undergo the review and permit process are sources not listed but having the "potential to emit" 250 tons/yr or more of any pollutant regulated by the CAA.

Conversion from conventional to unconventional fuels in a major stationary source that results in significant net emissions would be a major modification under the PSD program except for alternative fuel at a steam-generating unit to the extent that the fuel is generated from municipal solid waste, and except for certain other alternative fuels used by reason of other federal regulations. PSD regulations do allow the use of the "bubble" approach to offset new emissions.

As an example of the effect of PSD regulations, even moderately sized industrial boilers burning a fuel with 2.2% ash and 0.5% sulfur can have a significant uncontrolled "potential to emit":

Fuel, 10^6 Btu/hr	Total Potential to Emit, Tons/yr (Uncontrolled)		
	Particulate	SO_2	NO_x (as NO_2)
10	52	23	7
100	516	235	69
500	2581	1173	344
1500	7743	3520	1032

Therefore, new or modified industrial boilers about 100 million Btu/hr in size, burning unconventional fuels, could be required to undergo the review and permit process in areas governed by PSD, depending on ash and sulfur content.

6.1.3 Nonattainment Region Provisions

If proposed new or modified major sources lie in or impact on a nonattainment area (one that does not comply with a NAAQS), they will be subject to preconstruction review provisions of the applicable State Implementation Plan (SIP), or to a prohibition on construction if the SIP does not meet applicable requirements. Major sources are defined as those that will have "potential" emissions greater than 100 tons/yr for any applicable pollutant.

For such new sources, EPA's emission offset policy requires that:

1. All existing major sources in the nonattainment area owned by the owner of the proposed source are in compliance with applicable emission standards.
2. Proposed emissions from the new sources are more than "offset" by a reduction of emissions from other sources in the nonattainment area.

3. The emissions offset must represent a net air quality benefit.
4. The proposed sources will be subject to the lowest achievable emission rate (LAER). LAER is defined as the more stringent of either: (a) the most stringent emission limitation for this type of source in any SIP in the country, or (b) the lowest emission rate that can be achieved for this type of source with current technology.

Based on the previous discussion of "potential to emit," it is anticipated that some unconventional fuels would be governed by the offset policy, depending on ash and sulfur content and combustor size.

6.1.4 New Source Performance Standards (NSPS)

NSPS apply to new sources or to existing sources modified in a way that alters process capacity significantly and increases emissions, or are reconstructed at a cost equal to 50% of a new facility cost. Although existing sources need not meet NSPS, state standards are required in order to meet NAAQS. These are often less stringent than NSPS, sometimes more stringent, but in many instances are essentially equivalent to NSPS.

NSPS have been applied to many types of plants which could affect the use of unconventional fuels including:

Fossil fuel steam generators that have a heat input greater than 250 million Btu/hr

Solid waste incinerators with a charging rate greater than 50 tons/day

Kilns and other facilities in portland cement plants

Asphalt concrete plants

Storage vessels for petroleum liquids with a storage capacity greater than 40,000 gal

Secondary lead smelter pot furnaces of more than 550-lb capacity, blast (cupola) furnaces, and reverberatory furnaces

Incinerators that combust wastes containing more than 10% sewage sludge (dry basis) produced by municipal sewage treatment plants, or incinerators that charge more than 2205 lb/day municipal sewage sludge (dry basis)

Other chemical, metallurgical, and miscellaneous operations

Controlled pollutants vary, but include particulates, SO_2, and NO_x for steam generators; particulates for incinerators, portland cement plants, asphalt concrete plants, secondary lead smelters, and sludge incinerators; and hydrocarbons for storage vessels. Other pollutants covered by NSPS for some plants include fluorides, visible emissions, and CO. NSPS also include test methods and procedures, and may also include monitoring provisions.

Presumably cases where substitution of unconventional for conventional fuels tends to increase particulate or other emissions would cause imposition of NSPS for all pollutants. Therefore, strict adherence to NSPS might tend to inhibit substitution in steam generators larger than 250 million Btu/hr. On the other hand, if no emission increase could be expected, emissions would be governed by state and local regulations.

While the federal standards above apply only to relatively large new and modified sources, state standards often apply to much smaller sizes. Although no NSPS now exist for steam generators firing less than 250 million Btu/hr, such standards may be expected in the future to govern industrial boilers and possibly commercial boilers. The fact that there is now a NAAQS for lead suggests the possibility of future NSPS for this pollutant. Hydrochloric acid and other emissions not covered by NSPS are included in some state and local standards.

6.1.5 National Emission Standards for Hazardous Air Pollutants (NESHAP)

NESHAP govern asbestos, beryllium, mercury, and vinyl chloride emissions from specific sources.[27] These and other toxic elements such as lead, cadmium, arsenic, and nickel should be considered as potential pollutants when present in fuels.[28]

6.1.6 State Implementation Plans (SIP)

Each state must prepare a SIP for attainment and maintenance of NAAQS.[29] The SIP includes control strategies, evidence of legal authority, compliance schedules, contingency plans to prevent air pollution emergency episodes, provisions for an air quality surveillance system, procedures for review of new sources and modifications, procedures for source surveillance, copies of state rules and regulations, provisions for PSD, and analysis and plans for air quality maintenance areas (AQMA) where NAAQS are exceeded.

Thus, the SIP provides the framework through which state regulations are used to ensure meeting and maintaining NAAQS. The SIP must address all pollutants governed by NAAQS.

6.2 The Toxic Substances Control Act (TSCA)

Of primary concern under TSCA is the relationship of PCB disposal regulations to burning practices for waste oils or other unconventional fuels which may be contaminated by PCBs. Under these regulations[30]

For PCB liquids containing 500 ppm PCB or greater, disposal is permitted only in EPA-approved incinerators.

For PCB liquids containing 50–500 ppm, disposal is permitted in EPA-approved incinerators, in high-efficiency boilers rated at a minimum of 50 million Btu/hr (under rigidly controlled combustion conditions), and in EPA-approved chemical waste landfills (approved for PCBs).

Liquids containing less than 50 ppm are not considered PCBs (unless dilution was involved) and their burning is not regulated.

However, current litigation may affect future EPA PCB regulations under TSCA.

6.3 The Resource Conservation and Recovery Act (RCRA)

Insofar as burning of hazardous waste with energy recovery can be considered a beneficial use, the special requirements of 40 CFR 261.6[31] apply. Under these requirements, hazardous wastes used as an unconventional fuel are exempt from many RCRA provisions. However, state regulations may be more stringent in some instances.

7 AIR POLLUTION CONTROL EQUIPMENT

The air pollution control equipment available for meeting regulations is essentially the same as that for other fuels and processes discussed elsewhere in this book. Particulate emissions are generally controlled by centrifugal separators, electrostatic precipitators, fabric filters, venturi-type scrubbers, and sometimes combined forms of these devices. Gaseous pollutants, where control is required, will generally be removed by venturi, packed-bed, and other types of scrubbing systems.

Recently developed dry sulfur oxide removal schemes based on reactions with sorbents such as soda ash or lime have been applied to bituminous and subbituminous coals and lignite-fired boilers,[32,33] using either fabric filters or electrostatic precipitators to remove the spent sorbent. Similar sorption systems are available for use directly in fluidized beds to control sulfur oxide emissions.[34] Halide acids and other pollutants could also be controlled by sorbents, but such control is not now known to be practiced.

Hydrocarbon and CO emissions are best controlled by improvements in the basic combustion process. For control of nitrogen oxides, two-stage combustion is sometimes practiced, where the first stage is operated fuel rich and the second stage is operated fuel lean or with only air injection to complete the overall combustion process. Nitrogen oxides are also controlled by other combustion modifications, but scrubbing, ammonia reduction, catalytic burning, and other methods are available when required.[35,36,37,38]

Obviously, the use of scrubbers results in water pollution control problems, and often in solid waste problems resulting from water cleanup. Electrostatic precipitators and fabric filters require disposal means for the recovered particulates. In many instances the material to be disposed of must be considered as a possible hazardous waste under RCRA.

8 A SAMPLING OF COMMERCIAL EXPERIENCE

Unconventional fuels are being used today in many applications. The following cases are merely a small sampling of these.

8.1 Municipal Solid Waste in a Water-Wall Incinerator

The RESCO facility in Saugus, Massachusetts, one example of a successful municipal solid waste, mass-burning incinerator producing process steam,[5] started up in 1975. Other such facilities are in operation in the United States, Japan, and Europe. Some of the characteristics of this operation follow:

Current operating capacity: 1150 tons/day, 7 days/week, 24 hr/day

Steam produced: 625 psig, 785–825°F, up to 370,000 lb/hr (typical production 300,000 lb/hr)

Major equipment: 6700-ton storage pit; two overhead cranes; two Wheelabrator/von Roll water-wall combustion units with reciprocating grates, superheater, generator, and economizer sections; two 2-field electrostatic precipitators; bottom and fly ash quench tank; trommels and rotary drum magnet for metal recovery from ash

Standby boilers: two oil-fired boilers with combined capacity of 240,000 lb/hr steam

Many modifications have been made to overcome problems of corrosion, erosion, and air pollution and to otherwise improve the combustion process. "Black flake" emissions were controlled by changes in the furnace configuration. Particulate emission stack tests in 1979 and 1980 showed emissions to be roughly half of the state requirement of 0.05 gr/scf corrected to 12% CO_2.

8.2 Thermal Oxidation of Liquid Waste with Waste Heat Boiler

Described here is one example of perhaps hundreds of facilities burning organic wastes and recovering useful energy in the form of steam. In this instance, a manufacturing process produces both a combustible liquid waste and odorous vent gases containing cyanides and chlorinated hydrocarbons. The thermal oxidizer destroys the wastes, eliminates the source of odors, and produces process steam. Some of the characteristics of this system follow (see Figure 1):

Liquid waste: 764 lb/hr max, 100 ssu, 100 psig (steam or air atomization)

Waste gas: 120 lb/hr (28 scfm), 10-in. W.C.

Total flue gas: 5400 scfm max

Thermal oxidizer parameters: 1650°F, 0.6-sec residence time, castable refractory lined steel shell

Waste heat boiler: 7500 lb/hr max, 250 psig saturated steam (based on 267°F feed water), fire-tube type

Fan: 50 hp (induced draft-following scrubber)

Auxiliary fuel: normally zero, max 10,500 scfh to sustain steam production when wastes are not available

Air pollution control: Hastelloy quench section followed by a packed scrubber constructed of fiberglass reinforced polyester with water recycle and neutralization to reduce HCl to 10 ppm

Fig. 1 Hirt Combustion Engineers' Thermal Oxidation/Energy Recovery System for disposal of chlorinated liquid wastes and hydrogen cyanide fume stream (quench/scrubber system not shown).

8.3 Wood-Fired Steam Generator

A new boiler owned by the Potlach Corp. in Lewiston, Idaho uses a Combustion Engineering boiler, suspension-firing fine ($-$ ⅛ in.) low-moisture wood wastes, including sawdust and veneer, and burning high-moisture bark and sawdust on a traveling grate spreader–stoker.[39] The system includes

A shredder, surge bin, and pneumatic blower for fine waste

A bark hogger ($-$ 3 in.), hoppers, and variable-speed turn screw feeders

Boiler producing 550,000 lb/hr steam at 1400 psig, 950°F

Electrostatic precipitator reportedly operating as high as 99.5% efficiency with 0.015 gr/ft³ (dry) outlet loading, suitable for LAER in a nonattainment area

In another approach to wood burning, hot flue gas is used to predry wet wood before firing.[39] Pulp and paper makers have long practiced wood burning, but newer projects that have been announced are designed for steam and power production in industrial and utility applications.[40]

8.4 Canadian Tar Sands

Synthetic crude oils from the two current plants provide about 10% of Canadian crude oil production. Rapid growth is expected over the next 20 years.

Bitumen extracted from tar sands can be characterized by its hydrogen content as a material between conventional petroleum crude oil and coal. Its molecules are high-boiling polycyclic aromatic and naphthenic structures. The material is very viscous, unsuitable for conventional pipeline transport. Sulfur, nitrogen, iron, vanadium, and nickel contents tend to be high compared to many petroleum crude oils. Therefore, the bitumen is cracked by coking and hydrogenated to make a synthetic crude oil with properties closer to conventional petroleum crudes, but even then the properties of refined products such as gasoline and kerosene may vary from conventional products—either favorably or unfavorably.[41]

The bitumen metals tend to remain with the coke (during coking) and may be troublesome when the coke is burned as a fuel. Sulfur and nitrogen are removed during hydrogenation as H_2S and NH_3 which may cause air pollution problems, although the H_2S can be converted to sulfur for recovery.

Overall, tar sands as a source of unconventional fuel results in some air pollution problems, but these are subject to familiar solutions previously developed in the petroleum refining industry. The same conclusion could be drawn for petroleum products derived from oil shale.

8.5 Anthracite Culm as a Fuel in a Fluidized Bed Boiler

Culm, the waste product of anthracite coal mining, is fed to an atmospheric fluidized bed combustion boiler that delivers 23,400 lb/hr of 200 psig saturated steam. Heating value of the culm is 3000–4000 Btu/lb, about a third of that for coal; ash content is about 65–75%.[42]

9 THE FUTURE OF UNCONVENTIONAL FUELS

As noted previously, unconventional fuels will play an increasingly important role in our economy. However, the optimal use of fuels such as biomass and low-grade fossil fuels will require the development of useful fuel preparation techniques and energy recovery equipment specifically designed for the fuel to be used. Careful characterization of the unconventional fuels and combustion trials with extensive air emission testing will be required to design suitable air pollution control equipment.

Extreme care will be required to detect unusual and potentially hazardous emissions during fuel preparation and combustion, including those pollutants already subject to control, other heavy metals, bacteria and viruses, and toxic organics. Isolated use of unconventional fuels may have a negligible environmental impact, but extensive conversions can cause problems as shown by a recent study of potential pollution caused by residential wood combustion.[43]

REFERENCES

1. Anon., Producing and Burning d-RDF, *NCRR Bulletin,* 86–90 (December 1980).
2. Keller, R., A New Look at an Old Energy Source, *Energy Report II,* No. 5, The National Energy Research and Information Institute, May 1981, pp. 1–4.
3. Hayes, E. T., Energy Resources Available to the United States, 1975 to 2000, *Combustion,* 12–19 (April 1979).
4. Stabenow, G., Performance of the New Chicago Northwest Incinerator, *Proceedings, 1972 National Incinerator Conference,* New York, June 4–7, 1972, American Society of Mechanical Engineers, pp. 178–194.

5. Reilly, T., et al., Resource Recovery Facilities—Part 4: Adapting a Proven Technology—Saugus, MA, *Solid Wastes Management,* 30–35 (July 1981).

6. Sutin, G. L., The East Hamilton Solid Waste Reduction Unit, *Engineering Digest* 15(7), 45–51 (1969).

7. Reilly, T., et al., Resource Recovery Facilities—Part 3: Improving Systems in Shakedown—Akron, OH, *Solid Wastes Management,* 50–56 (June 1981).

8. Anon., *Evaluation of the Ames Solid Waste Recovery System—Part I—Summary of Environmental Emissions: Equipment, Facilities and Economic Evaluations,* EPA-600/2-77-205, November 1977.

9. Musselwhite, R. W., Resource Recovery in Madison, *NCRR Bulletin,* 76–81 (December 1980).

10. Rueth, N., A Solid Waste Package Deal—Energy and Materials from Garbage, *Mech. Eng.,* 24–29 (December 1977).

11. Beningson, R. M., et al., Production of Eco-Fuel-II from Municipal Solid Waste CEA/ADL Process, *Proceedings, First International Conference, Conversion of Refuse to Energy,* Montreux-Switzerland, November 1975, IEEE Catalog No. 75CH1008-2 CRE, pp. 14–21.

12. Kleinhenz, N. J., *Coal: dRDF Demonstration Test in an Industrial Spreader Stoker Boiler. Use of Coal: dRDF Blends in Stoker-Fired Boilers,* Vol. I, PB 82-100 868, Vol. II, PB 82-100 876, NTIS, Springfield, VA, October 1981.

13. Weinstein, N. J., and Toro, R. F., *Thermal Processing of Municipal Solid Waste for Resource and Energy Recovery,* Ann Arbor Science, Ann Arbor, MI, 1976, 179 pp.

14. Serper, A., Resource Recovery Field Stands Between Problems, Solutions, *Solid Waste Management,* 16 (continued on p. 86) (May 1980).

15. Berry, E. E., et al., *Experimental Burning of Waste Oil as a Fuel in Cement Manufacture,* Report No. EPS 4-WP-75-1, Environment Canada, June 1975.

16. Hitchcock, D. A., Solid-Waste Disposal: Incineration, *Chem. Eng.,* 185–194 (May 21, 1979).

17. Sittig, M., *Incineration of Industrial Hazardous Wastes and Sludges,* Noyes Data Corp., Park Ridge, NJ, 1979, 348 pp.

18. Manson, L., and Unger, S., *Hazardous Material Incinerator Design Criteria,* EPA-600/2-79-198, October 1979.

19. Anon., *Compilation of Air Pollutant Emission Factors,* Publ. No. AP-42, U.S. Environmental Protection Agency, August 1977 (plus Supplements through No. 12, April 1981).

20. Coe, W. W., Combustion: Efficiency vs No_x, *Hydrocarbon Processing,* 130–134 (May 1980).

21. England, G. C., et al., Control of NO_x Emissions, *Hydrocarbon Processing,* 167–171 (January 1980).

22. Edwards, L. O., et al., *Trace Metals and Stationary Conventional Combustion Processes,* PB 80-216 161, NTIS, Springfield, VA, February 1981.

23. Anon., EPA Says no Dioxin Risk for Incinerators, *Chem. & Eng. News,* 20 (November 30, 1981).

24. RECON Systems, Inc., *Used Oil Burned as a Fuel,* SW-892, U.S. Environmental Protection Agency, 1980, 213 pp.

25. Duckett, E. J., Health Aspects of Resource Recovery—Part II: Air Pollution, *NCRR Bulletin,* 105–112 (Fall 1978).

26. 40 CFR 50-67, July 1, 1982.

27. 40 CFR 61, July 1, 1982.

28. Jenkins, D. W., *Biological Monitoring of Toxic Trace Metals,* Vol. I, PB 81-103 475, Vol. 2, Parts I, II, III, PB 81-103 483, PB 81-103 491, PB 81-103 509, NTIS, Springfield, VA, February 1981.

29. 40 CFR 51, July 1, 1982.

30. FR 44, 31514, May 31, 1979.

31. 40 CFR 261 to 267, July 1, 1982.

32. Lane, W., Will the Spray-Drier System Meet the SO_2 Removal Requirements in Your Plant?, *Power,* 43–45 (December 1979).

33. Anon., Commercial-Scale Dry Scrubbing Holds Promise for SO_2 Control, *Power,* 26–29 (July 1980).

34. Robinson, J. M., et al., *Environmental Aspects of Fluidized-Bed Combustion,* PB 81-217 630, NTIS, Springfield, VA, August 1981.

35. Anon., *Controlling Nitrogen Oxides,* EPA-600/8-80-004, February 1980.

36. Hill, H. L., SCR Process Cuts No_x Emissions, Hydrocarbon Processing, 141–143 (February 1981).

37. Parkinson, G., Catalytic Burning Tries for NO_x Control Jobs, *Chem. Eng.*, 51–55 (June 15, 1981).

38. Yaverbaum, L. H., *Nitrogen Oxides Control and Removal*, Noyes Data Corp., 1979, 388 pp.

39. Strauss, S. D., Wood-Burning Provides Energy Stability, *Power* 126–127 (September 1981).

40. Berry, R. I., An Ancient Fuel Provides Energy for Modern Times, *Chem. Eng.*, 73–76 (April 21, 1981).

41. Steere, D. E., et al., Tar Sands Products Bring Changes, *Hydr. Proc.*, 263–268 (September 1981).

42. Anon., Steam Plant Fueled with Anthracite Culm, *Chem. & Eng. News,* 25 (November 2, 1981).

43. Meyer, H. E., *Contribution of Residential Wood Combustion to Local Airshed Pollutant Concentrations,* CONF-810674-1, Oak Ridge National Lab., TN, 1981, 12 pp.

CHAPTER 17

COAL CLEANING

JAMES D. KILGROE

Industrial Environmental Research Laboratory
U.S. Environmental Protection Agency
Research Triangle Park, North Carolina

1 INTRODUCTION AND SUMMARY

Physical coal cleaning can be used to reduce the ash and sulfur content of coal and hence the potential sulfur dioxide (SO_2) emission which results from coal combustion. This chapter summarizes information on U.S. coal resources, describes physical coal-cleaning technology, and discusses the potential for desulfurizing U.S. coals by physical techniques. It summarizes the amounts of cleaned coals which can comply with different emission standards and presents estimates of the cost of cleaning. It also discusses the characteristic variability in the sulfur content of coal and provides a method for estimating the mean coal sulfur value which will provide an expectation that a coal can comply with a specified emission regulation. Finally, sample calculations are given to illustrate some uses of coal washability and sulfur variability data.

Sulfur in coal is found in two major forms: organic and pyritic. Organic sulfur is chemically bonded to the carbon molecules of the coal and can only be removed if the bonds are broken by chemical processes. Pyritic sulfur, a mineral, occurs in discrete particles throughout the coal structure. In U.S. coals, the pyritic sulfur may comprise from 30 to 70% of the total sulfur content. Pyritic sulfur can be removed by physical methods.

Coal is physically cleaned—a process sometimes called washing—by crushing it to a point where some of the mineral impurities are released from the coal structure. The mineral and coal particles can then be separated by techniques which rely on differences in density or surface properties. The degree of desulfurization obtained by crushing and specific gravity separation depends on the sulfur level in the raw coal, type of sulfur, and the coal preparation techniques used. Commercial coal-cleaning plants are capable of reducing the potential SO_2 emissions from coal combustion by 10 to 50%, depending on the coal and process employed.

In 1979, there were more than 460 physical coal-cleaning plants in the United States. They were capable of processing over 400 million tons (363 million Mg) of raw coal per year. In 1979, approximately 344 million tons (312 million Mg) of coal was delivered to utilities from eight eastern and midwestern U.S. coal producing states. Approximately 33% of this coal was cleaned. It is estimated that these cleaning activities reduced the potential annual SO_2 emission from the combustion of these coals by over 1.7 million tons (1.5 million Mg). While physical coal-cleaning processes are principally designed for product standardization, increasing emphasis is being placed on cleaning for sulfur removal to aid in complying with SO_2 emission standards.

Emission limits for coal-fired boilers in the United States may range from less than 1.0 lb SO_2/10^6 Btu (430 ng/J) to greater than 9.0 lb SO_2/10^6 Btu (3869 ng/J). Older boilers are generally subject to less stringent standards than new boilers. Low-sulfur coals and physically cleaned coals can be used to comply with many emission regulations which incorporate moderate emission limits. New boilers require the use of best available pollution control technology. Some standards for new sources require a 70–90% reduction in the potential SO_2 emissions as related to the raw coal. Conventional physical coal cleaning by itself cannot be used as a technique to comply with these standards. However, in some instances combinations of physical coal cleaning and flue gas desulfurization (FGD) may be the most cost-effective method for complying with these emission regulations.

This chapter summarizes a wide range of information on U.S. coals, coal cleaning, and SO_2 emission regulations from coal combustion. The treatment of these topics has been greatly summarized and simplified.

Both English units and Standard International (SI) units of measure are used in the text. For simplification, figures and tables are expressed solely in English units. A table with English to SI unit conversions is provided at the end of this chapter.

2   COAL CHARACTERISTICS

Coal is a complex assemblage of materials that originate from peat and detrital material deposited in swamps or bogs. In addition to its plant-derived organic constituents (carbon, hydrogen, oxygen, and nitrogen), coal contains significant quantities of minerals and inorganic elements. The relative abundance of the organic and inorganic constituents varies significantly between coals of different types and between different locations in a single coal seam. The variability of coal properties is a function of

1. The chemical composition and structure of the coal-forming vegetal matter.
2. The chemical changes that occurred during the accumulation and burial of the coal-forming matter.
3. The deposition of inert sediments during accumulation.
4. The characteristics of the sedimentary rock above and below the coal seam.
5. The subsequent geologic history of the deposit such as the permeation of groundwater and the depth of burial.

A number of different systems are used to classify and characterize coal. Some systems depend on the gross physical and chemical properties which are related to the use for which coal is intended—combustion, coke making, or synthetic fuels. Other systems are based on the macroscopic or microscopic constituents of which the whole coal is composed.

2.1   Coal Rank

The classification of coal by rank provides information useful in assessing its properties as a fuel. Rank is related to the degree of coalification—the process of deposition, burial, and metamorphosis. Age, volatile matter, fixed carbon, bed moisture, and oxygen are indicative of rank, but no one item completely defines it. The American Society for Testing and Materials (ASTM) classification system includes four classes of coals: anthracite, bituminous, subbituminous, and lignite.[1] Each class is in turn divided into groups. For the higher ranked coals, the ordering is based on the fixed carbon and volatile matter content. Fixed carbon increases and volatile matter decreases with increasing rank for these coals. Lower ranked coals are ordered in terms of their calorific value as determined on a moist mineral matter free basis: the heating value decreases with decreasing rank.

The location of coal fields and the rank of coals in these fields are shown in Figure 1. Anthracite is found mainly in Pennsylvania, with small deposits in Virginia, Arkansas, Colorado, and New Mexico. It is used primarily for home heating and for fuel in small commercial and industrial boilers. The total amount of anthracite in the demonstrated reserve base of the United States is estimated to be 7.3 billion tons (6.6 billion Mg).[2]

Bituminous coals are found in almost every coal producing state. However, they predominate in the eastern and midwestern coal states. Bituminous coals are used mainly for electrical production, coke production, and small boiler fuels. There are more than 241.9 billion tons (219 billion Mg) of bituminous coals in the demonstrated reserve base.[2] Most of the coal which is cleaned in the United States is bituminous.

Subbituminous coals are found mainly in Montana, Wyoming, Alaska, Colorado, New Mexico, and Washington. There are more than 182.4 billion tons (165 billion Mg) of these coals in the demonstrated reserve base.[2] They are used primarily as boiler fuels.

There are more than 42.9 billion tons (38.9 billion Mg) of lignitic coals in the demonstrated reserve base.[2] The largest reserves are found in Montana, North Dakota, Texas, Alaska, and Alabama. In the past these coals were not widely used because of their low heating value and the distance of most lignite reserves from major industrial centers. Recently, mine mouth power plants have been constructed in North Dakota and Texas to utilize these resources.

2.2   Coal Constituents

Coals contain a wide variety of organic and mineralogical constituents. In the same way that inorganic rocks are composed of minerals—for example, granite is made up of feldspar, quartz, and mica—the organic coal structure consists of identifiable components called macerals. But there is a difference. Whereas a mineral is characterized by a fairly well-defined chemical composition, by the uniformity of its substance, and by the fact that most minerals are crystalline, coal macerals vary widely in chemical composition and physical properties.[3]

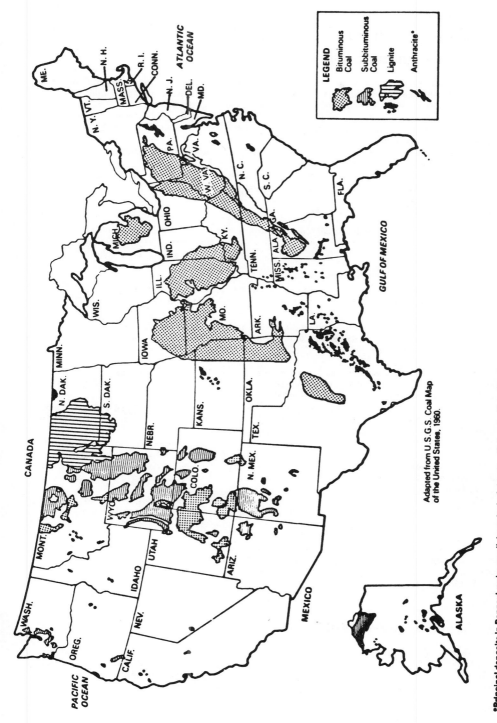

LEGEND

Bituminous Coal

Subbituminous Coal

Lignite

Anthracite*

Adapted from U.S.G.S. Coal Map of the United States, 1960.

Fig. 1   Coal fields of the United States (coal rank not distinguished in Alaska).

*Principal deposits in Pennsylvania; small deposits in Arkansas, Colorado, New Mexico, and Virginia. Anthracite - meta-anthracite deposits in Massachusetts and Rhode Island

Although microscopic examinations show these distinct maceral types to be more or less homogeneous, they often contain a large number of small mineral inclusions. Systems that are used to classify the organic microlithotypes in coal are described in Ref. 3.

2.2.1 Organically Related Constituents

Organic sulfur is believed to exist in the form of several functional groups: thiophenes, aryl sulfides, cyclic sulfides, aliphatic sulfides, disulfides, and mercaptans (thiols).[4] The total organic sulfur content in bituminous coals consists primarily of thiophenes (40–70% of total). The rest is believed to consist of aryl sulfides, cyclic sulfides, and aliphatic sulfides. Disulfides and mercaptans are not believed to contribute significantly to the organic sulfur content in bituminous coal. They may be present in lignites in larger proportions.[4]

Nitrogen in coal is believed to be associated primarily with aromatic ring structures in the form of pyridine, picoline, quinoline, and nicotine.[5]

Trace elements commonly associated with the organic structure of coal include: germanium, beryllium, boron, antimony, cobalt, nickel, copper, chromium, and selenium.[6]

2.2.2 Inorganically Related Constituents

In addition to purely organic substance, coal contains varying amounts of inorganic components: minerals and trace elements. These inorganic components are classified into three groups according to their origin:[3]

1. Inorganic matter from the original plants.
2. Inorganic/organic complexes and minerals which formed during the *first phase* of the coalification process or which were introduced by wind or water into the coal deposits as they were forming.
3. Minerals deposited during the *second phase* of the coalification process, after consolidation of the coal, by ascending or descending solutions in cracks, fissures, or cavities or by the alteration of primarily deposited minerals.

Minerals and elements associated with the original organic matter and the first phase of coalification are fine grained and intimately interspersed in the coal. This inorganic material is often called "inherent" minerals. By contrast, the minerals that formed during the second stage of the coalification process are large grained and distinct from the coal organic structure, because most of them were deposited in cracks and fissures.[3] These minerals are often called "extraneous" minerals. Extraneous minerals can be removed by conventional physical coal-cleaning techniques; inherent minerals cannot.

There is generally not a strong correlation between coal rank and the quantity and characteristics of the coal mineral matter. The mineral content of coal is generally a function of the depositional environment, the subsequent exposure of sediments, and the presence of mineralized ground or surface water. Coal rank and degree of metamorphosis are mainly functions of the temperature, pressure, and time of coalification.

Minerals associated with coal include clays, iron disulfides, carbonates, oxides, hydroxides, sulfides, phosphates, sulfates, silicates, and salts. Clays are by far the most important coal minerals. They generally account for 60–80% of the coal mineral content. Carbonates are the next most common, followed by iron disulfides, oxides, hydroxides, and salts.

Trace elements that are normally associated with the inorganic structure of coal include: zinc, cadmium, manganese, arsenic, molybdenum, and iron. A number of metals (including cobalt, nickel, copper, chromium, and selenium) are often associated with both the inorganic and the organic constituents of coal. In some instances, they are probably associated with sulfide minerals in coal. In other cases, they may be associated with organometallic compounds as chelated species, or as absorbed cations.[6]

2.2.3 Relationship of Mineral Matter to Ash

An evaluation of the mineral content of coal is a difficult and costly process. For this reason the coal ash content is used as a surrogate for quantifying the mineral content of coal. Whole coal samples are burned in a laboratory combustor, converting the minerals and inorganic elements to ash. The weight of this ash approximates the original mineral content of the coal.

2.2.4 The Sulfur and Ash Content of U.S. Coals

The sulfur and ash content of coal varies from coal field to coal field and from seam to seam within the same coal field. It also varies from location to location within a single mine and between different elevations at the same location.

The average ash content, calorific content, and total sulfur content for U.S. coals are summarized in Table 1.[7] These data are presented for six coal regions and the states within those regions. These data are from a computerized data base developed by the Department of Energy on the desulfurization potential of U.S. coals.

The sulfur and ash content of northern Appalachian, eastern-midwestern, and western-midwestern region coals is higher than the sulfur and ash content of the southern Appalachian, Alabama, and western coals. This is also reflected in the potential sulfur emissions which would result from the combustion of these coals. The relative difference between the sulfur and ash content of the various coal basins results primarily from differences in the depositional environments which controlled initial stages of coalification in the ancient peat swamps. Differences within a coal mine or bed may have resulted from localized differences in the depositional environment or from differences in the subsequent coalification process.

2.3 The Potential for Constituent Removal

Constituents can be removed from coal by either physical or chemical techniques. In most commercial physical cleaning processes, coal and mineral particles are separated by techniques that rely on differences in their size, shape, and specific gravity or their surface properties. Those that rely on specific gravity separation techniques predominate. The amount of constituents that can be removed by a given technique or process changes from coal to coal. The potential for constituent removal or cleanability can be considered to be an inherent property of each coal—a property determined by the relative amount and size distribution of the macerals and minerals of which the whole coal is composed.

Most of the published data on the cleanability of different coals are based on specific gravity separation techniques. No systematic collection of data on other techniques has been published.

2.3.1 Specific Gravity and Size Designations

Coal and mineral particles are often characterized according to their size and density. It is common practice to specify their density relative to water, that is, their specific gravity. Thus a particle with a specific gravity of 1.0 would have a density of 62.4 lb/ft³ (1.0 g/cm³), and a particle with a specific gravity of 2.0 would have a density of 124.8 lb/ft³ (2.0 g/cm³).

Large particles are generally specified by their maximum mean diameter in inches or millimeters. Small particles are generally specified by their Tyler mesh size. The Tyler mesh size is the number of screen openings per inch in a standard square screen sieve. The maximum size or top size of a 14 mesh particle is the largest particle size that will pass through the 0.0469-in. (1.2-mm)† opening. All particles smaller than 0.0469 in. (1.2 mm) will also pass through the 14 mesh screen. This size range of particles is denoted as 14 mesh × 0. In a similar fashion the size range of particles which will pass through a 14 mesh screen but which will be retained on a 100 mesh screen (0.0059-in. or 0.150-mm screen opening) is denoted by 14 × 100 mesh. In the text of this chapter the Tyler mesh size will be followed by the screen opening in millimeters, for example, 14 × 100 mesh (1.2 × 0.15 mm).

2.3.2 Washability

Pyrite, the most common iron disulfide (FeS_2), may constitute from less than 1 to more than 7% of the coal weight. It has a specific gravity of 5.0. The specific gravity of other coal minerals ranges from approximately 1.9 to 2.2, and the specific gravity of coal macerals ranges from 1.2 to 1.7.

The potential for reducing the mineral and sulfur content of a coal by specific gravity separations can be determined by laboratory tests called float–sink tests or washability tests.[8] In these tests, a coal sample is crushed and separated into various size ranges. Each size fraction is float–sink tested in organic liquids of standardized specific gravity. The weight, percent ash, heat content, total sulfur, and pyritic sulfur content of each density and size fraction are then determined. This matrix of data characterizes the washability of a given coal. It is used to determine the mineral and desulfurization potential and to specify the design and operating conditions for physical coal preparation equipment.

2.3.3 Float–Sink Data

The effects of crushing and specific gravity of separation are "qualitatively" similar for nearly all coals. The weight recovery (yield), energy recovery, ash content, and sulfur content decrease with decreasing specific gravity of separation. At low specific gravities, only those particles that are relatively free of embedded pyrite or mineral particles will float. Coal particles containing moderate amounts of mineral and pyrite particles sink along with pure mineral particles. This produces a cleaner product but at the expense of high energy losses to the sink or refuse fraction.

† English and metric equivalents for particle screen size are from: *Testing Sieves and Their Uses, Handbook 53*, 1976 edition, CE Tyler Industrial Products.

Table 1 Coal Properties in Coal Washability Data Base—State and Regional Summary[7]

Region[a]	State	Number of Samples	Heating Value, Btu/lb				Total Sulfur, %				Ash, %			
			Min.	Max.	Mean	S.D.	Min.	Max.	Mean	S.D.	Min.	Max.	Mean	S.D.
1	MD	35	11,026	14,244	12,661	797	0.71	6.13	2.58	1.26	8.0	26.90	16.41	4.36
	OH	90	8,571	14,256	12,494	930	0.67	6.55	3.55	1.39	3.57	37.43	13.61	5.56
	PA	170	7,070	14,658	12,982	1,057	0.53	9.40	2.56	1.56	4.80	51.9	14.41	6.60
	WV-N	30	9,186	13,973	12,647	1,266	1.06	6.33	3.06	1.21	6.63	37.53	15.74	8.21
	Total[b]	325	7,070	14,658	12,781	1,038	0.53	9.40	2.88	1.51	3.57	51.90	14.53	6.32
2	GA	0	0	0	0	0	0	0	0	0	0	0	0	0
	KY-E	13	11,120	14,366	12,986	1,163	0.56	4.88	1.44	1.21	2.13	23.47	11.48	7.44
	NC	0	0	0	0	0	0	0	0	0	0	0	0	0
	TN	8	12,966	14,321	13,672	484	0.66	3.61	1.17	0.99	3.00	12.87	7.90	3.56
	VA	8	11,872	14,828	13,715	1,143	0.48	1.17	0.77	0.23	1.97	22.77	10.04	7.83
	WV-S	16	11,000	14,210	13,064	901	0.59	1.90	0.90	0.35	5.30	26.07	12.73	5.84
	Total[b]	45	11,000	14,828	13,266	998	0.48	4.88	1.08	0.82	1.97	26.07	11.03	6.45
3	AL	10	12,765	15,056	13,696	914	0.59	3.76	1.33	0.95	1.73	16.27	9.50	6.38
	Total[b]	10	12,765	15,056	13,696	914	0.59	3.76	1.33	0.95	1.73	16.27	9.50	6.38
5	IL	40	10,777	13,294	11,944	683	1.14	7.82	3.86	1.39	7.63	25.03	15.29	4.45
	IN	21	11,598	13,452	12,546	531	0.67	7.17	3.67	1.54	6.37	19.57	11.89	3.46
	KY-W	37	8,698	13,676	12,339	864	2.17	5.02	4.05	0.71	5.17	37.63	13.81	5.52
	MI	0	0	0	0	0	0	0	0	0	0	0	0	0
	Total[b]	98	8,698	13,676	12,222	763	0.67	7.82	3.89	1.21	5.17	37.63	14.00	4.84

Region	State													
5	AR	3	13,900	14,293	14,123	202	1.81	3.96	3.12	1.15	7.10	8.97	8.27	1.02
	IA	17	10,165	13,230	11,601	759	2.53	12.11	6.76	2.49	7.00	26.87	16.15	4.67
	KS	8	9,550	12,857	11,544	1,212	4.02	6.65	5.04	1.03	13.40	34.50	21.48	7.79
	MO	9	9,788	12,914	11,665	934	4.22	6.16	4.97	0.64	11.60	29.57	19.04	5.06
	OK	7	12,156	14,934	13,461	923	0.57	6.00	3.10	2.33	3.97	20.20	10.09	5.68
	TX	0	0	0	0	0	0	0	0	0	0	0	0	0
	Total[b]	44	9,550	14,934	12,072	1,232	0.57	12.11	5.25	2.32	3.97	34.50	16.21	6.68
6	AZ	6	12,197	12,649	12,348	162	0.47	0.68	0.56	0.08	6.37	9.33	8.36	1.11
	CO	19	11,311	13,839	12,449	796	0.27	1.84	0.58	0.36	3.53	18.87	9.68	4.01
	MT	9	11,417	12,288	11,749	277	0.74	1.75	1.00	0.37	7.37	10.43	8.83	0.98
	NM	11	10,577	13,695	11,815	1,167	0.48	0.76	0.64	0.10	4.50	22.17	14.96	6.49
	ND	1	11,007	11,007	11,007	0	1.18	1.18	1.18	0	8.87	8.87	8.87	0
	OR	0	0	0	0	0	0	0	0	0	0	0	0	0
	SD	0	0	0	0	0	0	0	0	0	0	0	0	0
	UT	9	12,598	13,778	13,217	383	0.40	0.67	0.52	0.10	4.33	10.37	7.09	2.10
	WA	0	0	0	0	0	0	0	0	0	0	0	0	0
	WY	10	9,947	12,733	11,676	1,056	0.55	1.31	0.88	0.29	2.23	15.90	7.48	4.62
	Total[b]	65	9,947	13,839	12,201	929	0.27	1.84	0.70	0.32	2.23	22.17	9.63	4.65
	Total U.S.[b]	587	7,070	15,056	12,623	1,059	0.27	12.11	2.82	1.84	1.73	51.90	13.67	6.24

[a] 1—Northern Appalachian, 2—southern Appalachian, 3—Alabama, 4—eastern-midwestern, 5—western-midwestern, 6—western.

[b] Due to independent rounding, data may not add to totals.

Crushing the coal to finer particle sizes liberates a higher fraction of the mineral and pyrite particles from the organic coal structure. This results in improved separation of the mineral and organic components during the subsequent float–sink tests.

Washability data from a coal face sample taken from the Pittsburgh coal bed in Harrison County, Ohio, are shown in Table 2.[8] These data are for three representative subsamples which have been crushed to pass through screen openings of 1½ in. (37.5 mm), ⅜ in. (9.5 mm), and 14 mesh (1.2 mm). Each subsample was again fractionated by float–sink separations at specific gravities of 1.3, 1.4, 1.6, and 1.9.

The data in Table 2 show that, by crushing the sample to 1½ in. (37.5 mm) top size and by taking a float product at 1.6 s.g. (specific gravity), 90.3% by weight of the sample would be recovered. This float product would have a moisture-free heating value of 13,375 Btu/lb (31,110 J/g), an ash content of 7.5%, a pyritic sulfur content of 1.43%, and a total sulfur content of 2.98%. Although 90.3% of the weight content of the original sample was recovered, 96.0% of its original energy content was recovered. Upon combustion this sample would have emitted 4.5 lb $SO_2/10^6$ Btu (1935 ng SO_2/ J).

In a similar fashion data from Table 2 can be used to determine the properties that can be obtained by washing the sample at four different specific gravities after crushing to any of three different top sizes. Data for the raw coal samples are given in the rows designated as total.

The effect that the specific gravity of separation has on the ⅜-in. (9.5-mm) float product of the Harrison County coal is presented graphically in Figure 2. The weight recovery, Btu recovery, ash content, sulfur content, and SO_2 emission parameter (lb $SO_2/10^6$ Btu) all decrease with decreasing specific gravity of separation. Only the heating value increases. This increase is a consequence of the reduced ash content of the coal at low separation gravities.

The effects of size reduction and specific gravity of separation on ash, pyritic sulfur, total sulfur, and lb $SO_2/10^6$ Btu reduction are shown in Figure 3. These data are for the average reduction in

Table 2 Cumulative Washability Data for Pittsburgh Coal Bed, Harrison County, Ohio[a]

| | Sample Crushed to Pass 1½ in. | | | | | | |
| | Recovery, % | | | | Sulfur, % | | lb $SO_2/10^6$ |
Product	Weight	Btu	Btu/lb	Ash, %	Pyritic	Total	Btu
Float-1.3	35.4	39.2	13,925	3.7	0.40	1.73	2.5
Float-1.4	73.8	79.7	13,592	6.0	1.01	2.47	3.6
Float-1.6	90.3	96.0	13,375	7.5	1.43	2.98	4.5
Float-1.9	93.2	98.0	13,231	8.5	1.56	3.12	4.7
Total	100.0	100.0	12,580	13.0	2.01	3.48	5.5

| | Sample Crushed to Pass ⅜ in. | | | | | | |
| | Recovery, % | | | | Sulfur, % | | lb $SO_2/10^6$ |
Product	Weight	Btu	Btu/lb	Ash, %	Pyritic	Total	Btu
Float-1.3	52.5	57.8	13,939	3.6	0.32	1.71	2.5
Float-1.4	81.2	88.1	13,722	5.1	0.74	2.21	3.2
Float-1.6	88.9	95.1	13,534	6.4	0.94	2.44	3.6
Float-1.9	92.4	97.7	13,375	7.5	1.11	2.62	3.9
Total	100.0	100.0	12,652	12.5	1.89	3.35	5.3

| | Sample Crushed to Pass 14 Mesh | | | | | | |
| | Recovery, % | | | | Sulfur, % | | lb $SO_2/10^6$ |
Product	Weight	Btu	Btu/lb	Ash, %	Pyritic	Total	Btu
Float-1.3	23.3	26.1	14,084	2.6	0.23	1.61	2.3
Float-1.4	78.6	85.9	13,751	4.9	0.56	1.99	2.9
Float-1.6	87.3	93.9	13,534	6.4	0.77	2.21	3.3
Float-1.9	91.5	97.1	13,346	7.7	0.98	2.42	3.6
Total	100.0	100.0	12,580	13.0	2.00	3.48	5.5

[a] All values given on a moisture-free basis. The air-dried raw coal moisture content of the original sample was 2.6%. Data from Ref. 8.

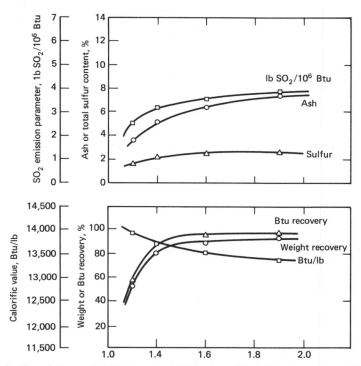

Fig. 2 Cumulative washability data for Pittsburgh coal bed, Harrison County, Ohio.[8]

these properties as determined by washability analysis of 227 coal samples from the northern Appalachian coal region. The northern Appalachian region includes the states of Maryland, western Pennsylvania, Ohio, and northern West Virginia. Although the reductions that can be obtained at low specific gravities and small particle sizes are desirable from the viewpoint of coal quality, separations at these conditions are not presently considered economically viable. At low specific gravities the coal energy losses are excessive. For example, the average Btu recovery for the 227 northern Appalachian samples corresponding to 1.3 s.g. float product at 1½-in. (37.5-mm) top size is only 37.5%. Alternatively, crushing all the coal to fine sizes and separating at specific gravities of 1.6 or higher would result in excessive coal preparation costs. The costs of cleaning fine coal are substantially higher than the costs of cleaning coarse coal.

Crushing coal to a top size of 1½ in. (37.5 mm) and separating the particles at a specific gravity of 1.6 provide a float sample that is representative of the product that can be obtained by many commercial coal-cleaning plants. Crushing coal to a top size of ⅜ in. (9.5 mm) and separating the particles at a specific gravity of 1.3 provide a float product that can be used to represent the coal quality limit that can be achieved with advanced technologies for physically cleaning fine coal.

In a manner similar to raw coal properties, the properties of cleaned coal vary from coal bed to coal bed and coal region to coal region. Table 3 presents a summary of the average physical desulfurization results obtained from laboratory washability analyses for six different U.S. Bureau of Mines coal regions. The average potential raw coal SO_2 emissions for the northern Appalachian, eastern-midwestern, and western-midwestern regions range from 4.8 to 9.0 lb $SO_2/10^6$ Btu (2064 to 3869 ng SO_2/J). After cleaning by crushing to 1½-in. (37.5-mm) top size and taking a 1.6 s.g. float product, the average potential raw coal SO_2 emissions for these three regions range from 3.2 to 6.1 lb $SO_2/10^6$ Btu (1376 to 2622 ng SO_2/J). Although the average reduction in SO_2 emissions is about 30% for these high-sulfur coal regions, SO_2 emission reductions for individual coals may range from less than 20 to greater than 50%.

The southern Appalachian, Alabama, and western regions contain mostly low-sulfur coals. The potential raw and cleaned coal sulfur emissions from these coals generally fall in the range of 1.0 to 2.0 lb $SO_2/10^6$ Btu (430 to 860 ng SO_2/J). Some individual low-sulfur coals show little or no reduction in potential SO_2 emissions when washed. Others show reductions of 25% and higher. The average potential SO_2 emission reduction for cleaning low-sulfur coals falls in the range of 11 to 12%.

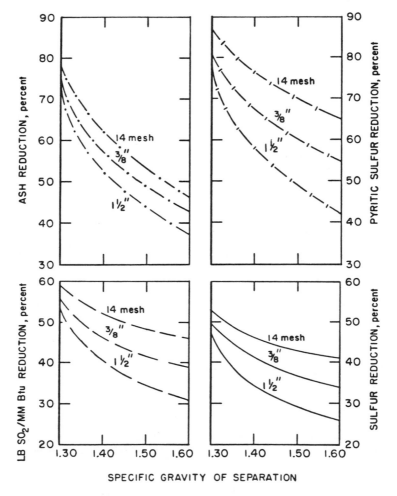

Northern Appalachian region (227 samples)

Fig. 3 The effect of crushing top size and specific gravity of separation on the reduction of ash, pyritic sulfur, total sulfur, and lb SO$_2$/10^6 Btu.

3 COAL PREPARATION TECHNOLOGY

Coal preparation or beneficiation is used to remove mineral matter from coal. Historically, the major motivation for these activities has been to reduce the ash level in steam coals and the ash and sulfur levels in metallurgical coals. Since the advent of sulfur emission regulations, many development activities have focused on coal desulfurization.

A number of different unit operations are employed in coal preparation. They include size reduction, size classification, coal cleaning, dewatering and drying, and pollution control/waste disposal.[1,9,10,11,12] Modern commercial coal preparation plants crush coal and separate the particles into a number of size ranges: coarse, medium, and fine. Each size fraction is processed in common or separate circuits using equipment suitable to the size range and cleaning objectives. Following cleaning, the coal is dewatered and sometimes thermally dried. Pollution control/waste disposal techniques are employed as needed to meet environmental regulations.

3.1 Size Reduction

Size reduction liberates mineral impurities and produces the range of sizes needed in subsequent cleaning activities. Size reduction is generally accomplished by mechanical equipment which breaks the coal

Table 3 Summary of Average Physical Desulfurization Potential of Coals by Region[8] (Cumulative Analysis of Float 1.6 Product for 1½-in. Top Size)

Coal Region	No. Samples	Face Sample		Laboratory Float–Sink Product		Percent Reduction of lb $SO_2/10^6$ Btu	
		Mean (lb $SO_2/10^6$ Btu)	S.D.	Mean (lb $SO_2/10^6$ Btu)	S.D.	Mean	S.D.
Northern Appalachian	227	4.8	2.8	3.2	1.8	31	17
Southern Appalachian	35	1.6	1.0	1.3	0.7	12	10
Alabama	10	2.0	1.5	1.7	1.3	11	8
Eastern-midwest	95	6.5	2.1	4.4	1.3	30	13
Western-midwest	44	9.0	4.5	6.1	2.5	29	13
Western	44	1.1	0.6	0.9	0.3	12	18

by impaction, compression, splitting, shearing, or attrition. Energy usage is minimized by staged crushing and bypassing fines.

Equipment used for size reduction may include single- or double-roll crushers, rotary breakers, and ring-roller mills. Primary size reduction is generally accomplished by rotary breakers or crushers. The second and third stages generally use crushers or mills.

3.2 Size Classification

Size classification is generally used for: removal of fines prior to size reduction; separation of size ranges prior to further processing or sale; dewatering, separation of fines, and rinsing; and removing rock and trash. The major size classification technique used is screening. It may be performed in either a wet or dry state. Screens may be stationary or moving. The screening surface may be a perforated plate, a woven cloth, formed bars, or nonstationary parallel bars. The most common screens used in coal preparation are perforated plate and square opening wire screens that shake or vibrate. These dynamic screens are used to stratify the coal bed and separate the smaller particles which are presented to the screen apertures. Curved stationary screens called sieve bends are used to separate fine and intermediate size particles in water slurries. Classifying cyclones are also commonly used to classify fine coal particles.

3.3 State-of-the-Art Cleaning Methods

Undesirable coal constituents can be removed by either physical or chemical techniques. In physical cleaning, the organic and mineral particles are separated by techniques that rely on differences in their size, shape, specific gravity, surface properties, electrical properties, or magnetic properties. Chemical cleaning employs chemical reactions which selectively remove mineral or organically associated constituents. State-of-the-art cleaning methods employed in the United States are limited to physical techniques which rely on specific gravity or surface property separation.

3.3.1 *Specific Gravity Separation*

The equipment most commonly used for physical cleaning relies on differences in the size, shape, and specific gravity of particles. Hydraulic washers, dense medium separators, and air concentrators are classes of equipment used in the specific gravity separation of coal. While, for the purpose of simplicity, this equipment is classified as specific gravity separation equipment, the phenomena describing separation are often complex. Separation of particles in this equipment has been described by a number of terms: stratification, hindered settling, differential acceleration, consolidation trickling, centrifugal separation, and so on. A detailed description of this equipment and its operating principles is well documented.[1,10,11]

The most popularly used hydraulic washers are jigs, hydrocyclones, and concentrating tables. The hydraulic jig is the most commonly used coal-cleaning equipment (see Figure 4). Jigs are used to remove mineral matter and mining refuse from coarse and intermediate-size coal. The jig is operated by applying periodic hydraulic pulsations to stratify a bed of coal and gangue materials. The heavy refuse materials are removed from the bottom of the bed and clean coal particles are removed from the top. Jigs can process large quantities of coal at low cost, but they do not efficiently separate coal

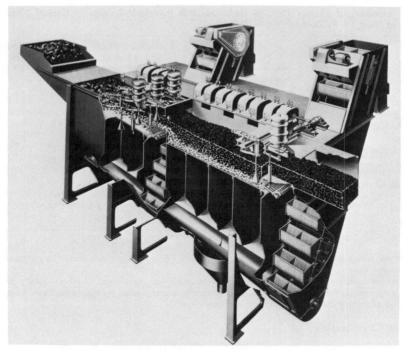

Fig. 4 McNally Mogul Washer. Source: McNally Pittsburgh, Inc.

and mineral particles whose specific gravity is nearly the same (near-gravity materials). If a large portion of the coal pyrite is contained in the near-gravity material, sulfur removal efficiencies may suffer.

Wet concentrating tables (Deister tables) are vibrating ribbed surfaces onto which a slurry of raw coal is fed. Segregation takes place along the surface so that various regions of the table contain relatively different amounts of coal and mineral particles of different sizes and specific gravities (see Figure 5). Concentrating tables are generally used to wash intermediate- to fine-size coal ranging from ⅜ in. to 200 mesh (9.5–0.075 mm). Tables can be used effectively for separating coal and pyrite particles. While the separating efficiencies of tables are good, they require more floor space than hydrocyclones of equivalent capacity.

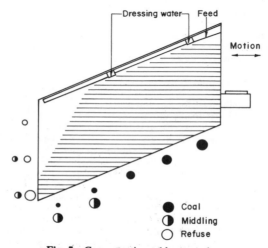

Fig. 5 Concentrating table—top view.

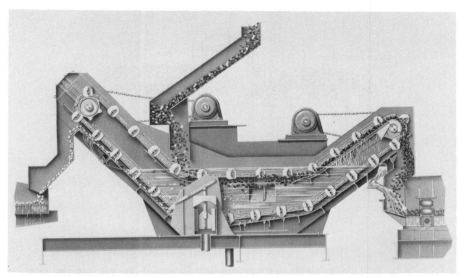

Fig. 6 McNally Lo-Flo Dense Media Vessel. Source: McNally Pittsburgh, Inc.

Hydrocyclones are also used to process medium- to fine-size coal. However, the underflow (refuse) normally contains high amounts of misplaced coal, which must be recovered by further processing. Another drawback is the large amounts of water required for proper operation. Despite these drawbacks, hydrocyclones have gained wide acceptance, as they can process large tonnages for relatively low capital investments.

Dense media separators include dense media vessels and dense media cyclones. The bulk of coal cleaned by dense media processes is cleaned in a suspension of magnetite, although organic liquids, dissolved salts in water, aerated solids, and sand/water suspension have been used on an experimental or commercial basis. Dense media vessels are static (laminar-flowing) baths in which particles are separated by virtue of differences in their specific gravity (see Figure 6). The use of magnetite (5.0 s. g.) permits practical suspension densities ranging from 81.2 to 124.9 lb/ft³ (1.3 to 2.0 g/cm³). For coal coarser than ¼ in. (6.3 mm), the separation efficiency and energy recovery are between 95 and 99% of the values expected from laboratory float–sink tests. Dense media cyclones are used to wash intermediate- and fine-size coal. A sharp separation of coal and minerals is achieved for all particle sizes above 28 mesh (0.60 mm). Below 48 mesh (0.30 mm), the efficiency of particle separation deteriorates rapidly with decreasing particle size.[13]

The increased costs of constructing and operating dense media circuits are generally offset by improvements in the product coal quality and coal recovery efficiency.

Equipment using air as a separating medium has been used for cleaning coal with sizes ranging from ⅜ in. to 48 mesh (9.5–0.30 mm). However, the use of this equipment has declined because of problems with separating wet coal and the costs of controlling dust emissions from the process.

3.3.2 Separation by Surface Properties

Cleaning processes that depend on the differences in the surface properties of particles include froth flotation and oil agglomeration.

Froth flotation is used to clean fine coals of particle size in the range of 28 mesh × 0 (0.60 mm × 0). Suspensions of the finely divided coal and mineral matter are agitated in a cell with small amounts of reagents in the presence of water and air. The reagents help to form small air bubbles which collect the hydrophobic coal particles and carry them to the surface. The clean coal froth is drawn off at the top of the cell and the hydrophilic mineral matter is wetted by water and drawn off as tailings (see Figure 7).

The effectiveness of froth flotation varies widely with different coals depending on the content of fixed carbon, volatile matter, and inherent moisture. In general, low-volatile bituminous coal is easier to float than high-volatile bituminous coal. Anthracite is more difficult to float, followed by subbituminous, with lignite the least floatable.[14] Also, the floatability of the same rank coal depends on whether it is freshly mined or weathered.[14] In addition to coal characteristics, there are many factors controlling froth flotation. The major factors affecting coal flotation are particle size, solids concentration, pH value, and the flotation reagents.[1,14,15]

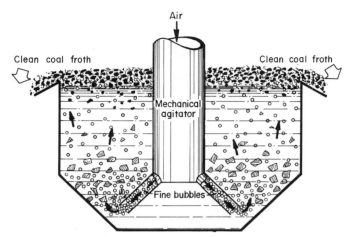

Fig. 7 Froth flotation cell.

Flotation has been used primarily to remove ash from fine coal. It has not been used extensively for the removal of pyritic sulfur from coal. Because the surface properties of coal and pyrite do not differ greatly, special depressants or other coal pretreatment may be required to improve separation.[15] A special two-stage froth flotation method developed by the Bureau of Mines has been shown to be effective on many coals.[16] In some instances, conventional flotation processes can be successful in removing pyrite. Tests at four preparation plants with conventional froth flotation circuits indicated mean ash reduction across the flotation cells ranging from 65 to 81%. Pyrite reductions for the same flotation circuits ranged from 48 to 65%.[17]

While it is apparent that froth flotation can be used for desulfurization of fine coal, the effectiveness of the process may vary widely depending on the coal and the design and operating conditions of the flotation circuit. It is, therefore, difficult to make generalizations concerning the potential of this technique for fine coal desulfurization.

3.4 Advanced Cleaning Techniques

A number of advanced techniques are under development for the cleaning of fine coals. The objectives of these techniques are to increase separation efficiencies of fine coal and mineral particles or to increase the amounts of sulfur that can be removed.

3.4.1 *Oil Agglomeration*

Oil agglomeration methods use a water-immiscible liquid, usually hydrocarbons, to preferentially wet and agglomerate coal particles. The small hydrophilic mineral particles remain in suspension and larger oil–coal agglomerates are separated. While oil agglomeration methods are reported to show promise in recovering fine coal and for the selective separation of pyrite, they are not used in the United States on a commercial basis.[15]

3.4.2 *Electrostatic and Magnetic Cleaning*

These cleaning techniques are considered important because of their potential for cleaning very fine coal. Crushing of the coal to particle sizes less than 200 mesh (0.075 mm) will liberate many of the impurities from coal; however, specific gravity and flotation separation techniques are not effective at these size ranges. While neither technique is used for commercially clean coal, substantial research has been done to evaluate these processes. Electrostatic separation is conducted on dry coal while magnetic separation may be conducted with dry coal or coal in a liquid suspension.

Early research on the electrostatic cleaning of coal indicated that cleaning is strongly and often unfavorably influenced by many variables such as the electrophysical properties of the particles, the particle size, the separator design, particle surface moisture, particle shape, and coal rank.[18] However, recent developments have been made to improve the process. Laboratory tests on four eastern and midwestern high-sulfur coals indicated sulfur reductions per million Btu ranging from 37 to 68%.[19] The heat specific ash reductions on the same coals ranged from 25 to 59%. Further evaluations are now under way to evaluate the performance and economics of the process.

Recent research on magnetic separation has used high gradient magnetic separators (HGMS) or pretreated coal for use with conventional magnetic separators.[20,21] The results of a number of research projects indicate that in general, while a large fraction of the pyrite can be removed, too much coal is lost. Economic studies show that the costs for preparing the coal for the magnetic separator, and subsequent handling of the product, are much higher than the magnetic separation step itself.[22] Further research is now in progress to evaluate the success of magnetic separation with a variety of process conditions.

3.4.3 Chemical Cleaning

Chemical cleaning processes entail grinding the coal to small particles and reacting these particles with or without chemical agents at specified temperatures and pressures. The coal's sulfur is converted to elemental sulfur or sulfur compounds which can be removed from the coal structure. Successful processes have the following characteristics, they:

1. Attack large fractions of the mineral and organic sulfur.
2. Convert the sulfur to forms that can be readily removed from the coal structure.
3. Minimize coal energy losses and processing costs.

While more than 30 chemical coal-cleaning processes have been evaluated, the most successful in terms of total sulfur reduction involve oxidative and caustic treatments.[15] Thermal decomposition mechanisms may also play an important complementary role at the elevated temperatures employed in the most successful processes.

The oxidative treatments have used air, oxygen, chlorine, nitrogen, or ferric sulfate for oxidation. The most prominent caustic treatment processes have used aqueous solutions or melts of sodium hydroxide, calcium hydroxide, or potassium hydroxide.[15,23] The most successful processes are reported to be capable of converting more than 95% of the pyritic sulfur and up to 50% of the organic sulfur to extractable forms. Reaction of more than 50% of the organic sulfur is difficult to achieve without substantial coal energy losses.[15,23] Organic sulfur exists in a number of functional forms, and the heterocyclic groups that appear to compose a large fraction of the total organic sulfur are resistive to chemical attack unless the carbon structure is decomposed resulting in a loss of chemical energy.

One of the least studied aspects of chemical coal cleaning is the removal or washing of reacted sulfur, ash, and reagents from the reacted coal structure. The removal of these constituents from the coal structure and the subsequent dewatering and drying operations may well be some of the most costly unit operations in chemical coal desulfurization.[24]

The prospects for commercial applications of most of the processes that have been evaluated to date are questionable. The revised 1978 federal regulations for new utility boilers require reductions in the potential SO_2 emission of 70–90% before a cleaned coal can qualify as a compliance fuel. Few, if any, chemical coal-cleaning processes can meet this performance requirement and compete on a cost basis with utility-size FGD units.[25,26]

If chemical coal cleaning is used commercially for sulfur emission control, the most likely application is the production of compliance coals for industrial boilers or the production of low-sulfur and low-ash fuel for use in coal–water or coal–oil mixtures.

3.5 Dewatering and Drying

Following wet cleaning methods, the coal must be dewatered by the use of vibrating screens, centrifuges, vacuum filters, or pressure filters. Thermal dryers are sometimes used to reduce the shipping weight of coal and to increase the heat content (and to prevent freezing); however, due to the cost of drying and complying with air pollution emission regulations, thermal drying is being displaced by physical drying equipment such as vacuum filters, centrifuges, or filter presses.

3.6 Pollution Control/Waste Disposal

The most concentrated source of air pollution at a coal-cleaning plant is the thermal dryer. Particulate emission from the furnace may be as high as 0.0062 to 0.025 lb/ft³ (100 to 400 g/m³) at standard temperature and pressure.[27] Stack gases from thermal dryers are usually cleaned by a combination of centrifugal collectors and high-efficiency venturi wet scrubbers.[27]

Centrifugal collectors can accommodate high particle loadings and high temperatures while collecting 95% of the particulates from the gas stream. These collectors, however, are not effective for particles smaller than 10 μm.[27] High-efficiency venturi scrubbers are 95% efficient in collecting smaller particles, and may collect some gaseous contaminants as well.

Air pollution from transport and loading operations is difficult to control. Prevention rather than cleanup is the rule. Methods include the use of closed conveyors, spraying of water or chemicals on haul roads and storage piles, and reduction of vehicle speed on unpaved areas.

Waste materials in process waters are fine coal particles and inorganic elements and compounds dissolved from the ash in the coal. The large volumes of water used in the cleaning process and the large amount of suspended solids generated dictate that process waters be clarified and recycled. The usual means of clarification is retention in large sedimentation ponds.[10]

If colloidal material from mine overburden is present, flocculants may be added to the ponds to aggregate the fine solids and facilitate clarification. Mechanical thickeners may also be used, sometimes in conjunction with flocculants. Thickeners can concentrate solids from a feed of 20% to an underflow of 60% solids, with the resultant decant suitable for recycling.

Federal standards regulate the total suspended solids, iron, pH, and (if a discharge is acidic prior to treatment) manganese of effluent discharges from coal preparation plants.[28] Toxic constituents must also be reported and controlled if above certain levels, but studies have not indicated this to be a problem at operating facilities.[29]

Contamination of surface water or groundwater can occur from coal waste piles or storage piles, if water is allowed to infiltrate them.[30] Contaminants such as iron and manganese and heavy metals such as cadmium and silver may leach from the wastes.[31,32] Because processing wastes are higher in ash (mineral) content than the cleaned coal, leachate from wastes poses the greater threat. Coal-handling operations can also pollute waters if dust is allowed to settle in streams or be washed into them by precipitation and runoff.

Solid wastes from coal cleaning can be coarse or fine and consist of waste coal, slate, carbonaceous and pyritic shale, and clay. The recommended disposal method for coarse coal-processing waste is deposition in layers on a cleared area, compaction, covering with earth, and ultimately revegetation.[10] These procedures are designed to prevent subsidence, spontaneous combustion, and infiltration of water into the waste material. Subsurface systems are also used under waste banks to prevent water buildup that could cause structural deterioration or leachate generation. Fine refuse is discharged to settling ponds as a slurry, or dewatered and added to coarse refuse piles.

Another option for disposal of coal-processing wastes is returning them to spent underground mine workings. This method is not acceptable if the mine location or the character of the waste poses a threat of contamination to an underground source of drinking water.[33] These environmental considerations and the high costs of disposal have generally prohibited use of this disposal technique in the United States.

4   STATUS AND DESIGN OF COMMERCIAL PLANTS

Physical coal cleaning is a mature technology. There were more than 460 operating coal-cleaning plants in the United States in 1979 which employed wet cleaning processes. About one-third of utility coals are cleaned prior to combustion.

4.1   Level of Cleaning

Many different plant configurations employing many different types of equipment are used to clean coal in the United States. In describing a cleaning process it is desirable to classify the preparation plant according to the complexity of the plant employed. One such approach has been to designate "levels of cleaning." Although there are many definitions of cleaning levels in the literature, it is convenient to use a system based on the number of different coal size ranges cleaned in separate circuits and the number of product streams.

Level 1.   Mine debris and noncombustible impurities are removed. Unit operations include crushing and particle sizing.

Level 2.   Coal is crushed and classified. Coarse and medium-size coal is cleaned in a single circuit. The top size of the coarse coal may range from 4 in. (100 mm) to approximately ⅜ in. (9.5 mm). The fine 28 mesh × 0 (0.60 mm × 0) raw coal is dewatered and shipped with the coarse coal. In some cases the fine coal is not sold but is discarded in slurry ponds.

Level 3.   Coal is crushed and classified. The coarse, greater than ¼ in. (0.60 mm), and medium ¼ in. × 28 mesh (6.3 × 0.60 mm) size fractions are cleaned in separate circuits. Fine coal is treated in the same manner as for level 2.

Level 4.   Coal is crushed and classified. Coarse, medium, and fine coal are cleaned in separate circuits. The coal from these circuits is combined to form a single product. Figure 8 is a schematic of a typical level 4 plant.

Level 5.   Coal is crushed. Coarse, medium, and fine coal are cleaned in a number of different circuits. The cleaned coal from these circuits is combined to form two products: a deep cleaned coal and a middling coal. The deep cleaned coal is a metallurgical grade coal or compliance grade coal that is low in ash and sulfur. The middling coal has an ash and sulfur content that is higher than for the deep cleaned coal.

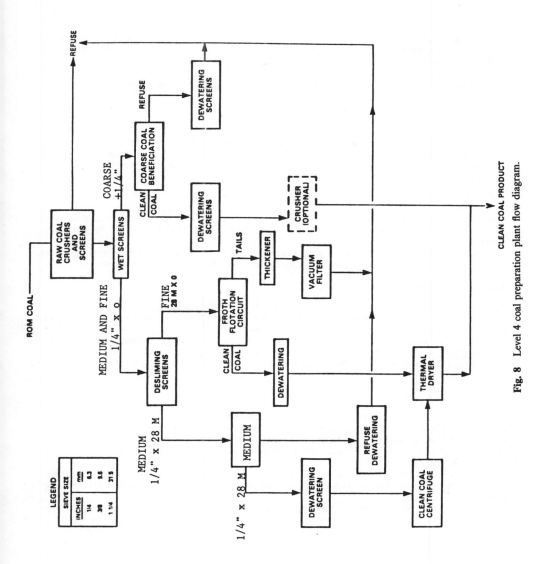

Fig. 8 Level 4 coal preparation plant flow diagram.

For a given coal the amount of mineral matter and sulfur that can be removed improves with the level of cleaning. So also does the capital and some operating costs.[9,34,35] Other operating costs, such as the amount of coal lost to refuse, may be reduced as the cleaning level increases.

In selecting the optimum plant design for any given application, performance and cost trade-offs must be made. A balance must be made between the costs of preparation and the increased value of the product coal(s).

4.2 Plant Design Trade-offs

Plant design and choice of cleaning equipment depend on the washability of the feed coal, the specifications of the desired product, and acceptable costs of cleaning. The major variables that can be considered when selecting a plant design include size reduction, specific gravity of separation, sharpness of separation, and number of cleaning circuits.

4.2.1 Size Reduction

Crushing the coal to finer sizes liberates more of the mineral particles. However, fine coal is more costly and difficult to clean than coarse coal. The increase in coal quality may not be worth the added cost.

Commercial coal preparation plants process coals with top sizes which range from approximately 4 to ⅜ in. (100–9.5 mm). No commercial plant crushes all of its coal to fine sizes before cleaning.

4.2.2 Specific Gravity of Separation

Reductions in the specific gravity of separation result in increased coal quality. This increased quality must be balanced against a reduction of yield and Btu recovery. Separation at low specific gravities is also more difficult. At low specific gravities a large weight fraction of the coal may lie in the specific gravity range near the separating point. The difficulty of separation increases as the amount of near-gravity material increases (material within ±0.10 of the specific gravity of separation). The result is increased amounts of misplaced material—refuse in the clean coal and clean coal in the refuse.

4.2.3 Sharpness of Separation

The efficiency of coal-cleaning equipment in separating particles of different specific gravities is defined by a probability or distribution curve which characterizes the sharpness of separation.[1] The amount of misplaced material decreases as the sharpness of separation increases. Thus, for a given yield the ash content and sulfur content of the product will decrease with increasing sharpness of separation. The calorific value and Btu recovery increase.

The sharpness of separation is a critical consideration for separations at low specific gravities. There is more near-gravity material and the amount of misplaced material will increase with decreasing sharpness of separation.

Float–sink tests give a perfectly sharp separation. The coal float–sink density distribution is the criterion against which equipment performance is measured. Dense media separators and concentrating tables produce sharper separations than jigs and hydrocyclones. A plant with dense media vessels and concentrating tables can produce a better quality coal than a plant with jigs and hydrocyclones. The marginal costs of installing and operating dense media vessels and concentrating tables must be offset by increases in the cleaned coal value.

4.2.4 Number of Cleaning Circuits

The sharpness of separation that can be achieved by a given type of equipment is dependent on the coal particle size. The best plant performance for a given coal top size is achieved by separating the coal with equipment that gives the sharpest separation. However, capital and operating costs increase as the number of circuits increases. A very complex plant may give a better quality product, but it may be subject to more unscheduled stoppages. Alternatively it may require a significantly higher amount of operating and maintenance labor to match the availability of a less complex plant. In many cases simplicity, reliability, and ease of operation are the main criteria on which plant design is based.

4.3 Plant Performance

The sulfur reductions and Btu enhancements (resulting from ash reductions) that are achieved at commercial coal preparation plants can result in potential reductions of SO_2 emission which range from less than 10% to more than 50%.[9] Table 4 presents a summary of data on the performance of coal preparation plants in seven eastern and midwestern coal-producing states. An analysis of this larger collection led to the conclusions that:[9]

Table 4 Coal Preparation Plant Feed and Product Data[a]

Coal Region[b]	State, County, Coal Seam	Plant No., Cleaning Level[c]	Plant Feed			Plant Product			Reduction in Emission Parameter, %
			Calorific Value, Btu/lb	Sulfur Content, %	Emission Parameter, lb $SO_2/10^6$ Btu	Calorific Value, Btu/lb	Sulfur Content, %	Emission Parameter, lb $SO_2/10^6$ Btu	
NA	WVA, Harrison, Lower Freeport (6A)	52, 3	10,466	2.74	5.36	13,900	2.64	3.80	28.0
NA	OH, Perry, Ohio No. 6	20, 2	11,059	3.94	7.14	12,459	3.01	4.79	31.8
NA	Unknown, Lower Kittanning	53, 3	12,950	1.94	3.00	14,368	1.07	1.49	48.3
SA	WVA, Wise, SWVA	21, 3	9,983	0.93	1.84	14,470	1.17	1.61	10.6
SA	WVA, Wyoming	54, 3	8,167	0.70	1.70	13,242	0.91	1.38	17.0
AL	AL, Jefferson, Blue Creek	42, 4	12,150	0.77	1.27	14,258	0.60	0.84	33.9
EMW	KY, Ohio, KY No. 9 & KY No. 14	16, 2	11,325	4.15	7.36	13,164	3.19	4.84	33.9
EMW	KY, Ohio, KY No. 9 & KY No. 17	17, 2	12,234	4.30	7.03	12,974	3.38	5.21	25.3
EMW	KY, Muhlenberg KY No. 11 & KY No. 12	19, 2	10,429	4.00	7.67	12,560	3.33	5.30	30.9
EMW	IL, Randolph, IL No. 6	18, 3	10,866	4.41	8.12	12,048	3.84	6.37	21.6

[a] Data are for weighted plant averages calculated from data in Ref. 9, Appendix C.
[b] NA—northern Appalachian, SA—southern Appalachian, AL—Alabama, EMW—eastern-midwest.
[c] Plant No. corresponds to plant No. given in Ref. 9.

1. Physical coal cleaning is an effective technology for reducing SO_2 emissions from coal combustion. The sulfur emission parameter (lb $SO_2/10^6$ Btu) of the coal is significantly reduced by coal cleaning. The average reductions achieved for different coal regions using data provided by coal companies were 33% for northern Appalachian coals, 23% for southern Appalachian coals, and 30% for eastern-midwestern coals.

2. In terms of the sulfur emission parameter, the preparation plants reduced the mean, standard deviation, and relative standard deviation (standard deviation/mean) of the product coal in almost every case. The only exceptions were plants cleaning low-sulfur southern Appalachian coal.

3. The difference in reductions of the sulfur emission parameter between cleaning levels 2, 3, and 4 is small, although cleaning level 4 (deep cleaning) always showed the greatest reduction on a regional basis.

Table 5 presents a comparison of plant performance data and washability data.[9,36] The emission and percentage reduction estimates are given as average state values. The washability data are for the 1.6 s.g. float product of coal samples which were crushed to a top size of 1½ in. (37.5 mm). The emission reduction estimates for the plant data range from 22.0 to 41.2%. Emission reduction estimates for the washability data range from 10.1 to 32.2%. The comparisons are judged to be quite good considering the facts that (1) data are matched only on a regional and state basis, and (2) there are not enough data to obtain accurate averages for each state.

However, while the plant and washability data appear to fall in the same range, the plant raw coal emission data and the plant percentage reduction data are at times substantially higher than the corresponding washability data. This probably results from the fact that the run-of-mine (ROM) heating value of the plant feed is low in relationship to the raw coal washability samples. Modern mining techniques (continuous and long-wall miners) often remove roof, floor, and parting materials, called dilution material, along with the coal. The washability data in Tables 2 and 3, and those most commonly given in the literature, were derived from coal mine face samples. These face samples differ from ROM samples in that they do not include major partings, roof, or floor materials. The roof and floor materials, which may also contain sulfur, reduce the heating value of the ROM samples and result in a higher ROM sulfur emission parameter and a higher potential reduction of this parameter when compared with face samples.

Since there is little ROM washability data it is desirable to use the face sample washability data to estimate the performance of plants designed for controlling SO_2 emissions. Studies show that face sample washability data can be used as a surrogate for estimating the performance of coal preparation plants.[9,36] In one study, data on the feed coals from 85 coal preparation plants were used along with laboratory washability data to predict the sulfur and Btu content of the cleaned coal.[36] Since the washability data were not available for the feed coal, the cleaned coal properties were predicted by multiplying the feed coal properties by the ratio of the corresponding cleaned to uncleaned coal properties for similar coals in the washability data base using a computer program called the Coal Assessment Processor (CAP).* Two washability conditions were used. The first, designated as PCC-1, was for crushing to a 1½-in. (37.5-mm) top size and taking a 1.6 s.g. float product. The second, designated as PCC-2, was for crushing to a ⅜-in. (9.5-mm) top size and taking a 1.3 s.g. float product. These predicted results were compared with the clean coal product data from the 85 coal preparation plants. Figure 9 shows the results of these comparisons.

It can be concluded from Figure 9 that laboratory washability results obtained by crushing face samples to 1½ in. (37.5 mm) and taking a 1.6 s.g. float product provide a good first approximation of the likely performance of a level 3 or level 4 coal preparation plant. The exact SO_2 emission value predicted for the cleaned coal will be somewhat uncertain because of (1) differences in ROM and face samples, and (2) plant and washability performance differences which result from differences in the sharpness of separation. Laboratory washability data corresponding to separations at lower specific gravities can be used to estimate potential improvements which can be made for different design and operating conditions. However, these estimates will be more speculative because of the increased importance of sharpness of separation criteria at low specific gravities.

4.4 Costs of Cleaning

The cost of coal cleaning is a function of level of cleaning, percentage of material discarded as refuse, coal washability, and plant size. The cleaning costs are not strongly influenced by the coal sulfur content. There is potentially a wide range of costs for coal cleaning, and the costs for each plant must be considered individually with respect to location, terrain, coal properties, preparation specifica-

* Washability data were assigned to each coal preparation plant by matching the closest coal composition (in a least squares sense) using washability data from the same state, county, and coal bed. If there were no washability samples in the same county, then state, region, and coal bed matches were used.

Table 5 Comparison of Plant and Laboratory Washability Emission Parameters (lb SO$_2$/10^6 Btu) [a]

Region and State	Laboratory Data [b]				Plant Data [c]			
	No. Samples	Raw Coal	Clean Coal	Reduction, %	No. Plants	ROM Coal	Product Coal	Reduction, %
Northern Appalachian								
Pennsylvania	170	3.9	2.4	32.2	4	3.1	1.8	35.9
Ohio	90	5.7	4.2	25.9	4	7.3	5.0	32.2
N. West Virginia	30	4.8	3.4	28.6	9	4.8	3.6	26.0
Southern Appalachian								
S. West Virginia	16	1.4	1.2	10.1	19	1.8	1.3	22.0
Alabama								
Alabama	10	1.9	1.7	10.8	4	2.2	1.2	41.2
Western and Eastern Midwest								
Indiana	21	5.9	4.3	26.4	3	8.0	5.2	35.6
W. Kentucky	37	6.7	4.5	31.5	3	7.5	5.2	29.6
Illinois	40	6.5	4.4	29.3	10	8.1	6.0	27.9

[a] State averages for the indicated number of washability samples and number of plants.
[b] Laboratory data from Ref. 36 for 1½-in. (37.5-mm) top size and 1.6 s.g. separation. Percentage reduction values were given. The raw and clean coal emission parameters are estimated using the average sulfur and heating values of coal from each state.
[c] Plant data summarized from Ref. 9, Appendix C.

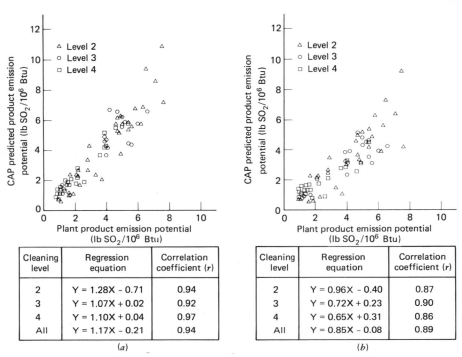

Cleaning level	Regression equation	Correlation coefficient (r)
2	Y = 1.28X - 0.71	0.94
3	Y = 1.07X + 0.02	0.92
4	Y = 1.10X + 0.04	0.97
All	Y = 1.17X - 0.21	0.94

(a)

Cleaning level	Regression equation	Correlation coefficient (r)
2	Y = 0.96X - 0.40	0.87
3	Y = 0.72X + 0.23	0.90
4	Y = 0.65X + 0.31	0.86
All	Y = 0.85X - 0.08	0.89

(b)

Fig. 9 Comparison of PCC-1 and 2 computer model results with coal preparation plant performance.

tions, and coal production rate. The costs of cleaning can be summarized in terms of the total capital requirements and annual costs.

4.4.1 Capital Requirements

The total capital requirements include direct costs, indirect costs, contingencies, land, and working capital. Direct costs include raw coal storage and handling equipment costs, preparation plant equipment cost, and facility costs. Facility costs are normally estimated to be some fixed percentage of equipment costs. Indirect costs include engineering costs, construction costs, and field expenses. Total capital requirements are generally 1.6 to 2.3 times higher than total direct costs.[26,35,37]

Table 6 presents estimates of the total capital requirements for three levels of cleaning for each of five different coals. These costs were taken from a 1981 Electric Power Research Institute (EPRI) study on the impact of coal cleaning on the cost of new coal-fired power generation.[35] The costs for treating two additional coals at three levels of cleaning are also contained in the 1981 EPRI study. They were not included here because of space limitations.

Each coal preparation plant in the 1981 EPRI study was designed to provide coal for a 1000-MW power plant. The capacity and annual production rate of each treatment facility, therefore, changes as the properties of raw and cleaned coal vary. In the 1981 EPRI study, each of the coals was treated at three different levels: screening and crushing (level 1), partial preparation (level 2), and intensive cleaning (level 4). All costs in the study were expressed in mid-1978 dollars. Note that these costs are outdated because of inflation and that they are used here to illustrate the relative importance of various coal preparation cost components.

Table 7 provides a summary of the raw and cleaned coal properties which result from preparing these five coals (A through E) to three levels each (1, 2, and 4). For each coal the ash content is decreased and the calorific value is increased as the level of cleaning increases. The sulfur contents of some coals are reduced while others remain essentially constant. However, in all cases the SO_2 emission parameter decreases with increased level of cleaning. A portion of this reduction in all cases results from an increase in the coal calorific values. In other cases it also results from a reduction in coal sulfur content.

Plant yields for the five coals decrease with increasing level of cleaning. Level 2 yields range from 79.7 to 94.3%. Level 4 yields range from 76.1 to 87.6%. Although the plant feed material was adjusted to include dilution material from mining (roof, floor, and parting material), these yields are moderately

Table 6 Coal Preparation Plant Costs for Fine Coals When Prepared at Levels 1, 2, and 4[a]

Coal: State; Seam; County	Cleaning Level	Annual Coal Rate,[b] 10³ tons/yr	Total Capital[c] Requirements, 10³ $	Unit Annual Costs, $/ton of Prepared Coal				
				ROM[d] Coal	Prepa- ration[e]	Waste Disposal	Total Prepared	Incre- mental
A. IL; No. 6; Macoupin	1	3,120	7,370	21.39	0.78	—	22.17	—
	2	2,944	17,820	23.74	1.80	0.20	25.74	3.57
	4	2,820	31,620	24.62	3.09	0.28	27.99	5.82
B. WVA; Cedar Grove; Logan	1	2,592	7,310	23.39	0.93	—	24.32	—
	2	2,296	20,550	27.67	2.89	0.35	30.91	6.59
	4	2,242	31,180	28.00	4.17	0.37	32.54	8.22
C. AL; American & Mary Lee; Walker	1	3,224	7,490	22.39	0.77	—	23.16	—
	2	2,908	17,090	25.55	1.76	0.28	27.59	4.43
	4	2,540	31,390	28.99	3.44	0.58	33.01	9.85
D. IA; Lower Ford; Panora	1	3,276	7,430	19.39	0.75	—	20.14	—
	2	3,206	15,730	20.47	1.45	0.10	22.02	1.88
	4	3,154	34,760	21.46	2.93	0.24	24.63	4.49
E. UT; Rock Canyon, Gilson, and Sunny Side; Carbon	1	3,172	7,330	19.39	0.76	—	20.15	—
	2	2,662	17,220	23.97	1.93	0.46	26.36	6.21
	4	2,558	31,640	24.41	3.50	0.51	28.42	8.27

[a] Data adapted from Ref. 35. A = case 1, B = case 3, C = case 4, D = case 6, and E = case 7 in the 1981 EPRI study. All capital and first year unit annual costs are presented in mid-1978 dollars.

[b] Prepared coal on an "as-received" basis. Rate required to produce a net 1000-MW electricity in a new power plant.

[c] Capital costs can be updated using an inflation factor developed from the Chemical Engineering Plant Cost Index. Example: ($31.620 million in mid-1982 $) × 1.46 = $46.165 million in first quarter 1982 $.

[d] ROM coal costs = cost of as-mined coal cost in $/ton ÷ (preparation plant yield/100) + UMW pension and trust fund of $1.39/ton. Coal costs can be updated by assuming a 6.2% annual inflation rate.[36] The inflation factor for first quarter 1982 dollars is 1.25. Example: ($24.62/ton in mid-1978 $) × 1.26 = $30.78/ton in first quarter 1982 $.

[e] Mean first year cost ÷ annual prepared coal rate. Capital costs can be updated using a weighted average of labor, electricity, material, and capital cost inflation factors. Example: ($3.09/ton in mid-1982 $) × 1.32 = $4.08/ton in first quarter 1982 $.

Table 7 Summary of Raw and Cleaned Coal Properties for Flue Coals When Prepared at Levels 1, 2, and 4[a]

Coal: State; Seam; Mine Type	Cleaning Level	Yield, %	Btu Recovery, %	Ash Content,[b] %	Sulfur Content, %	Calorific Value, Btu/lb	Potential Emissions, lb SO$_2$/ 10^6 Btu	Emission Reduction,[c] %
A. IL; No. 6; Deep	1	100	100	16.5	3.39	9,860	6.88	—
	2	89.5	95.1	12.0	3.26	10,480	6.22	9.6
	4	86.1	94.9	9.2	3.23	10,870	5.94	13.7
B. WVA; Cedar Grove; Deep	1	100	100	16.0	0.85	11,680	1.46	—
	2	84.6	94.1	7.5	0.83	13,130	1.26	13.7
	4	83.6	95.1	5.5	0.84	13,430	1.25	14.4
C. AL; American & Mary Lee; Deep and surface	1	100	100	27.0	1.26	9,450	2.67	—
	2	86.9	96.3	20.7	1.28	10,470	2.45	8.2
	4	76.1	95.8	12.2	1.28	11,910	2.15	19.5
D. IA; Lower Ford; Surface	1	100	100	15.1	6.90	9,450	14.60	—
	2	94.3	96.1	13.8	5.46	9,630	11.34	22.3
	4	87.6	93.4	10.2	4.66	9,840	9.47	35.1
E. UT; Rock Canyon, Gilson, and Sunny Side; Deep	1	100	100	22.9	0.64	9,650	1.33	—
	2	79.7	94.1	11.4	0.63	11,390	1.11	16.5
	4	78.2	96.0	8.4	0.62	11,850	1.05	21.1

[a] Data adapted from Ref. 35. Level 1 = screening and crushing (raw coal), level 2 = partial preparation, and level 4 = intensive preparation.
[b] Coal properties on an "as-received" basis.
[c] Two coals not summarized here were from the Pittsburgh seam in WVA and the Rosebud Seam in MT. The level 2 and level 4 emission reduction for these coals were: WVA coal 13.3 and 21.5%; MT coal 9.3 and 27.9%.

high. Based on the calorific values of the feed and product coal, the removal of ash and dilution material during cleaning may result in yields as low as 50% in some commercial plants.[9]

The Btu recovery may decrease or increase with an increasing level of cleaning (see Table 7). Btu recoveries in the level 2 plants range from 94.1 to 99.1%. In the level 4 plants the Btu recoveries range from 93.4 to 96.0. An increase in Btu recovery when cleaning from level 2 to level 4 can result from the use of equipment in the level 4 plant which improves the sharpness of separation. A reduction in Btu recovery occurs in the level 4 plants when Btu losses at lower specific gravity separations are not offset by Btu recovery gains related to the sharpness of separation.

The total capital requirements for level 1 preparation—which includes screening, crushing, coal sampling, crushed coal storage, and unit train loadout facilities—vary little from coal to coal (see Table 6). The average for the five level 1 plants is $7,386,000. The total capital requirements for the five level 2 plants range from $15,730,000 to $20,550,000. Included in these costs are provisions for raw coal screening and crushing, a raw coal storage silo, clean coal samplers, clean coal storage silos, and unit train loadout equipment. The cleaning circuit for each of the level 2 plants is nearly identical. In all cases a baum jig is used to wash 4 in. × 0 (100 mm × 0) coal. Cost differences between level 2 plants result primarily because of differences in equipment used for dewatering the clean coal and refuse. The lower capital cost plants correspond to those plants that use only screens and centrifuges for dewatering of coal and refuse. Most of the other plants employ screens, centrifuges, thickeners, and vacuum filters for dewatering of the coal and refuse. The plant with the highest capital cost also contains a thermal dryer for drying of the intermediate- and fine-size coal.

Total capital requirements for the five level 4 plants in the 1981 EPRI study ranged from $31,180,000 to $34,760,000. The level 4 plants all employ dense media vessels for cleaning of the coarse 4 × ¼ in. (100 × 6.3 mm) coal, heavy media cyclones or concentrating tables for cleaning the medium ¼ in. × 28 mesh (6.3 × 0.60 mm) coal, and concentrating tables or froth flotation for cleaning the fine 28 mesh × 0 (0.60 mm × 0) coal. Cost differences between the plants result primarily from differences in the intermediate and fine coal-cleaning equipment and associated dewatering and drying equipment. Other cost differentials arise from the variation in plant capacity.

Economies of scale can be obtained through the use of large preparation plants. Cost economies result primarily from the fact that costs per unit of capacity decrease with increasing size of equipment and facilities. Capital costs for larger or smaller plants may be estimated by use of the 0.7 power rule which states that the ratio of capital costs is proportional to the ratio of plant capacities raised to the 0.7 power. Thus, if the 609 ton/hr (544 Mg/hr) clean coal capacity of the level 4 plant cleaning in case B were to be increased to 1200 ton/hr (1088 Mg/hr), the total capital investment may be estimated to be:

$$\$31.2 \text{ million} \times \left(\frac{1200}{609}\right)^{0.7} = \$50.2 \text{ million}$$

Economies of scale are also reflected in annual costs in the form of reduced unit costs related to operating and maintenance expenses.

The 0.7 power rule evidently does not apply to modular plants. Plant modules are available at feed size ranges of 100, 150, and 200 ton/hr (91, 136, and 181 Mg/hr). Different types of modules available include those for heavy media, jig, hydrocyclone, and concentrating table processes. It is sometimes possible to cut the total capital requirement of a conventional level 2 plant expressed in $/ton-hr of feed by as much as 25% by using a modular plant.[38] Unfortunately, modular plants cannot generally achieve the levels of performance that can be attained in a conventional level 4 plant.

4.4.2 Annual Costs

Annual costs include: material costs, operating and maintenance costs, overhead expenses, and capital charges (amortization, taxes, and return on investment). Annual charges are often expressed in terms of $/ton of cleaned coal. To assess the cost differences between plants and to determine the costs of various activities, it is convenient to break down the unit annual costs for coal preparation into a number of different components:

1. *ROM Coal Cost.* This cost includes two terms: (1) the cost of mining a ton of coal divided by the preparation plant yield, plus (2) the contribution to the United Mine Workers of America (UMWA) pension/benefits fund per ton of coal shipped from the preparation plant. The first term accounts for the ROM material that ends up as clean coal and that fraction of coal that must be mined then discarded as refuse.

2. *Coal Preparation Cost.* This cost is for size reduction, size classification, coal cleaning, dewatering and drying, and pollution control. The cost of raw coal storage, clean coal storage, clean coal loadout system, and refuse disposal can also be included in the coal preparation cost. If these and other "grass roots" plant costs are included, then the capital cost of coal preparation will be two to three times the cost of a "battery limits" plant.[1] Accordingly, annual costs will also be higher. Equipment

costs, facility costs, and operating expenses included in various coal preparation plant cost estimates will be dependent on whether the plant is a mine-site plant, a central cleaning plant, or a power plant site facility. It will depend on the essential facilities and operations that are charged to the mining and power plant costs, rather than to the preparation plant account.

3. *Waste Disposal Cost.* This cost is sometimes included in a separate category.

4. *The Total Cost of Prepared Coal or Cleaned Coal.* This cost includes the ROM coal cost, the coal preparation cost, and the waste disposal cost.

5. *Cost of Mined and Discarded Material.* This cost can be defined as the sum of the ROM coal cost and the refuse disposal cost at a given level minus the level 1 ROM coal cost.

6. *Incremental Cost of Cleaning to Higher Level.* This cost is defined as the total cost of prepared coal at a given cleaning level minus the total cost of prepared coal at some lower cleaning level.

Unit annual costs for cleaning five coals at three different levels are given in Table 6. The ROM coal costs and refuse disposal costs increase with the level of cleaning due to decreasing plant yield. The preparation costs also increase with level of cleaning due to increases in capital costs and operating expenses. Level 1 preparation costs range from $0.75/ton to $0.93/ton ($0.83/Mg to $1.03/Mg). The level 2 preparation costs range from $1.45/ton to $2.89/ton ($1.60/Mg to $3.19/Mg). This wide range of preparation costs is in part due to differences in capital costs and operating expenses. It is also due in part to differences in the clean coal production rate. For low production rates, unit annual costs are spread over a smaller number of units. Unit annual costs based solely on the tonnage of material produced penalize the plants that produce a high calorific value coal. In these cases more appropriate unit measures of cost would be $/$10^6$ Btu or $/kWh. However, the $/ton costs are widely used because coal is most commonly sold on a weight basis.

Level 4 preparation costs in Table 6 range from $2.93/ton to $4.17/ton ($3.23/Mg to $4.60/Mg). The low costs correspond to a high production rate with a modest amount of coal dewatering operations. The high costs correspond to low production rates and extensive dewatering and drying.

The total cost of prepared coal in Table 6 ranges from $20.14/ton to $33.01/ton ($22.20/Mg to $36.39/Mg). In all cases the cost of prepared coal increases with level of cleaning. Increased incremental costs for cleaning from level 1 to level 2 range from $1.88/ton to $6.59/ton ($2.07/Mg to $7.27/Mg). Incremental costs for cleaning from level 1 to level 4 range from $4.49/ton to $9.85/ton ($4.95/Mg to $10.86/Mg). These incremental preparation costs are the net costs added to a ton of cleaned coal by preparation (level 1 corresponds to raw coal). The incremental costs above level 1 consist of three components: the coal preparation costs, the costs of material mined and discarded material, and the costs of waste disposal. For level 2 plants, the costs of preparation range from $1.45/ton to $2.89/ton ($1.60/Mg to $3.19/Mg), while the costs of the mined and discarded material, as calculated from Table 6, range from $1.18/ton to $5.04/ton ($1.3/Mg to $5.56/Mg). For level 4 plants the costs of preparation range from $2.93/ton to $4.17/ton ($3.23/Mg to $4.60/Mg), while the calculated costs of the mined and discarded materials range from $2.31/ton to $7.18/ton ($2.55/Mg to $7.92/Mg). Waste disposal costs, which are a portion of the costs of mined and discarded material, range from $0.10/ton ($0.11/Mg) to $0.58/ton ($0.64/Mg), depending on the coal and level of cleaning.

The cost of mined and discarded material is a measure of costs related to both the coal quality and mining practice. Coal preparation removes the extraneous minerals which are intimately associated with the coal facies. It also removes dilution material that originates in the mine roof, mine floor, and coal seam partings. Modern mechanical mining methods are not selective, and they often result in ROM coal which contains large quantities of dilution materials. It is not unusual to discard more than 40% of the material which is removed from modern underground mines. While this adds substantially to the total costs of prepared coal, the ROM coal could not be sold unless it were cleaned. Although the costs of mining and discarding dilution materials are often attributed to the "costs of cleaning," it is more appropriate to charge them to the costs of producing a marketable coal. In this fashion the "costs of cleaning" can then be defined as the incremental costs of cleaning to higher levels; that is, incremental levels that are above the base level needed to market the mined material. In most cases the base level would be level 2 cleaning. If increases in the incremental costs of cleaning are more than offset by increases in the value of the product coal, then higher levels of cleaning are warranted. The benefits of cleaning must outweigh the costs, otherwise a buyer would purchase raw coal or coal cleaned to a lower level.

4.4.3 *Cost Benefits*

Cost benefits may accrue from a reduction in transportation costs, coal-handling costs, ash disposal costs, boiler capital costs, boiler operating costs, and flue gas cleaning costs. Cost penalties may result from an increase in the amount of coal fines and the surface moisture content. Large amounts of fines and frozen coal (a consequence of additional moisture) increase handling costs.

Cost benefits that occur from reduced transportation and boiler ash disposal costs are well known. These costs are coal and site specific. Transportation and boiler ash disposal cost savings are increased as the amount of ash that is removed by cleaning is increased. Potential cost savings also increase as coal transportation and ash disposal distances increase.

The use of cleaned coal can result in the reduction of the capital costs of new steam generators. These cost reductions result from cost savings in: the firing equipment (pulverizers, fuel piping, wind boxes, etc.); the boiler design (furnace volume, heat transfer surface arrangement, tube spacing, soot blowing, etc.); the coal-handling and storage equipment (conveyors, bunkers, etc.); and ash handling and storage. Most of these steam generator capital costs generally decrease with reductions in mineral (ash) content. Reductions in the sulfur content can also result in lowered problems with fireside performance and, hence, can lead to a less conservative, lower cost design. In most cases, increasing the level of cleaning to produce a better quality coal should result in lower capital costs for new boilers.

Boiler operating costs are affected by boiler efficiency, boiler availability, and boiler maintenance costs. These factors are in turn affected by coal quality. Excessive slagging or fouling of boiler heat transfer surfaces by coal minerals results in a reduction of heat transfer efficiencies. A major factor in boiler availability is boiler tube failure. Boiler tube failures can result from many causes, including fly ash and soot blower erosion, external corrosion, iron attack, slag erosion, and slag falls. A reduction in coal sulfur and ash content can reduce fuel related tube failures and result in increased boiler availability. Plant equipment with maintenance expenses related to coal quality may include: coal breakers, boilers and accessories, soot blowers, coal transport piping and valves, boiler room scale and feeders, pulverizer mills, burners and lighters, air preheaters, ash crushing equipment, and ash disposal piping systems. Reducing the ash and sulfur content of coal can reduce boiler maintenance costs.

Flue gas cleaning costs are also reduced by coal cleaning. Particulate control device capital and operating expenses are generally reduced as the amount of ash that must be collected is reduced. Flue gas desulfurization (FGD) costs are reduced with the decreasing sulfur and ash content of the coal. In many cases, major FGD cost reductions result because (1) a portion of the flue gas can be bypassed, reducing the volumetric flow through the scrubber, (2) with bypass, flue gas reheat requirements are decreased, (3) the quantities of limestone or other chemical reagents are reduced, and (4) FGD sludge quantities that must be disposed of are reduced.

A number of studies have attempted to quantify the cost benefits of improved coal quality which can be achieved by coal cleaning.[39-47] Others have attempted to assess the impacts of coal cleaning on the costs of SO_2 emission control or the costs of power generation.[26,35,48,49] They too have attempted to quantify the cost benefits of cleaning. In all of these studies the most elusive cost benefits have been those related to boiler capital costs and boiler operating expenses.

In the 1981 EPRI study on the cost of new power generation, it was estimated that the costs of new steam generators can be reduced by coal cleaning.[35] In the seven cases studied it was estimated that coal cleaning could reduce the total capital requirement for new 1000-MW power plants by 0.3 to 5.0%. This represented capital cost savings ranging from $2.3 to $39.5 million. These savings were exclusive of savings that could be achieved by cost reductions in the FGD system or boiler/FGD waste disposal. Most of the cost savings were achieved in firing system cost reductions. The cost reductions were dependent on the coal and level of cleaning. In all cases the cost savings increased with increased level of cleaning.

Estimates of boiler operating cost benefits are based on studies that show that coal with high sulfur and ash contents generally result in decreased boiler efficiency and boiler availability, and increased boiler maintenance costs.[39,41-47] Early studies based on limited data estimated that boiler-related cost benefits may range from less than $1/ton to more than $7/ton ($1.1/Mg to $7.7/Mg) of coal burned.[39,41] Contrary to these estimates, an EPRI study on the impact of cleaned coal on power plant performance and reliability failed to find significant differences in the boiler operating costs at two plants that switched from run-of-mine to washed coal.[40] However, recent studies supported by the Tennessee Valley Authority (TVA) and the Department of Energy (DOE) have confirmed the relationship between coal quality and boiler operating costs.[43-47] These studies involved the evaluation of operating data collected over a period of 19 yr from 58 TVA boilers. Major findings of the studies were:

1. *Boiler Efficiency.* Major factors influencing boiler efficiency include coal ash content, coal moisture content, and boiler age. Sulfur content does not appreciably affect efficiency. A large change in coal, ash, and moisture content can have a significant impact on boiler efficiency. For example, increasing coal ash from 10 to 20% and moisture from 6 to 10% can reduce boiler efficiency by approximately 0.77%.[44]

2. *Boiler Availability.* Forced outage hours attributed to coal related equipment are a function of coal ash content, coal sulfur content, and boiler age. Models based on TVA data predict that switching from a very good coal (1.0% sulfur and 12.0% ash) to a very poor coal (5.0% sulfur and 24% ash) would increase the forced outage rate in a "typical" TVA plant by 850 hr/yr.[45]

3. *Maintenance Costs.* Of 10 plants analyzed, 7 showed correlations between coal quality and maintenance costs when cost data were evaluated according to the annual consumption of ash and sulfur. The degree of correlation varied significantly between the fuel-related cost centers at each plant as well as between various plants. Plantwide maintenance cost dependencies were nonlinear. The logarithm of maintenance costs increased proportionally with the annual plant consumption of ash and sulfur. Using a plantwide maintenance cost saving model it was estimated that plants in the TVA system can save between $0.05/ton and $0.10/ton ($0.06/Mg and $0.11/Mg) of coal burned for each

1% reduction in ash. Similarly, for each 1% reduction in sulfur, between $0.30/ton and $0.50/ton ($0.33/Mg and $0.55/Mg) can be saved for each ton of coal burned. Substantial cost savings can be achieved by reducing both ash and sulfur. For example, at the Widows Creek Plant, it was estimated that if ash were to be reduced from 20 to 12% and sulfur from 3.5 to 2.0%, a plantwide maintenance cost saving of $1.13/ton ($1.25/Mg) of coal burned could be achieved.[46,47]

Although the impact of coal preparation on boiler operating costs has not been demonstrated on boilers that burn raw and then cleaned coal, the preponderance of evidence indicates that the reduction in coal ash and sulfur content will *in general* have a beneficial impact on boiler operations. The difficulty arises in quantifying these cost benefits, since potential reductions in steam-generator costs are dependent on the individual coal properties and the specific boiler design. In some instances cleaning coal may increase its fouling and slagging properties because of selective removal of minerals which will result in an increase in the coal's slagging and fouling index. Fortunately, these conditions probably apply to few coals. In most cases the use of coal cleaning would not only reduce the ash loading but, more importantly, it would also reduce the slagging and fouling potential of the remaining ash.[50]

4.4.4   *Impact of Coal Cleaning on Power Generation Costs*

Perhaps the best method of illustrating the cost benefits of coal cleaning is to present data from case studies of the impact of coal cleaning on power generation.

Table 8 presents a comparison of power generation cost components as a function of level-of-cleaning for case B from the 1981 EPRI study.[35] These costs are presented in terms of the mean first-year costs given in mid-1978 dollars. Each cost component is presented in terms of three different unit measures: $/ton of as-fired coal, $/$10^6$ Btu of as-fired coal, and mills/net kWh of power generation.

When using $/ton as a measure, level 1 coal preparation (raw coal) *appears* to be the least-cost method of power generation. When using $/$10^6$ Btu or mills/kWh, level 4 coal cleaning is indicated as being the least-cost method of power generation. In this instance, $/ton of as-fired coal is an inappropriate measure for making relative comparisons between coal-cleaning cases: in cleaning at levels 2 and 4, the coal heating value is enhanced and the amount of coal needed for generation of a fixed amount of power decreases. Multiplying the unit power generation costs by the amount of coal actually burned provides the true rankings:

Level 1. $116.77/ton $\times$ 2.592 $\times$ 10^6 tons = $302.7 million
Level 2. $129.96/ton $\times$ 2.296 $\times$ 10^6 tons = $298.4 million
Level 4. $132.61/ton $\times$ 2.242 $\times$ 10^6 tons = $297.3 million

The ranks remain unchanged if we divide by any appropriate constant, in this case, the amount of raw coal mined and burned at level 1:

Level 1. $302.7 million $\div$ 2.592 $\times$ 10^6 tons = $116.78/ton raw coal ($128.75/Mg raw coal)
Level 2. $298.4 million $\div$ 2.592 $\times$ 10^6 tons = $115.12/ton raw coal ($126.92/Mg raw coal)
Level 4. $297.31 million $\div$ 2.592 $\times$ 10^6 tons = $114.70/ton raw coal ($126.46/Mg raw coal)

Table 9 presents the power plant and transportation costs which are related to the coal-cleaning costs given in Table 6. The unit costs in Table 9 are expressed in terms of relative costs in $/ton. These relative $/ton costs have been adjusted to the level 1 coal mining rate applicable to each case. This provides information on the correct cost ranking for each cleaning option and provides approximate estimates of the relative savings (costs) in $/ton. In four of the five cases in Table 9, the use of coal cleaning provides the lowest cost of power generation. In three out of the four cases, level 4 cleaning is the most cost effective and provides for lower costs than does raw coal or coal prepared at level 2. The cost reductions attainable by cleaning ranged from 0.2 to 3.2%. In the case where the use of raw coal was the most cost effective approach, level 2 preparation increased net power production costs by 0.4%. Level 4 preparation increased power production costs by 0.2%. In these last two cases the cost benefits related to reduced transportation, power generation, and FGD costs were not sufficient to offset the increased costs of prepared coal.

The potential cost savings (benefits) that are achieved by coal cleaning in the 1981 EPRI Case B coal to level 2 (B-2) and to level 4 (B-4) are presented in Table 10. These potential savings are expressed in $/ton of level 1 coal and mills/kWh. The potential transportation, power generation, FGD waste disposal, and cost savings increase with level of cleaning. The total potential saving for level 2 preparation is approximately $4.71/ton ($5.19/Mg) of level 1 coal or more precisely 1.9 mills/kWh. For level 4 preparation, the potential saving is $5.89/ton ($6.49/Mg) of level 1 coal or 2.55 mills/kWh. The percentage reduction in costs as determined by [100 $\times$ (cost saving) $\div$ (level 1 cost)] is in close agreement when calculated either from the potential savings in $/ton or mills/kWh (see Table 10).

The relative value of the transportation, power generation, flue gas desulfurization, and waste disposal cost savings will depend on site specific factors. For example, in the 1981 EPRI study the ranges of

Table 8 Comparison of Power Generation Cost Components as a Function of Level of Cleaning[35]

	Mean First-Year Costs (1978 $)								
	$/ton As-Fired Coal			$/10⁶ Btu As-Fired Coal			Mills/Net kWh		
	Level of Cleaning			Level of Cleaning			Level of Cleaning		
Cost Component	1	2	4	1	2	4	1	2	4
ROM coal[a]	23.39	27.67	28.00	1.00	1.05	1.04	9.9	10.4	10.25
Coal preparation	0.93	2.89	4.17	0.04	0.11	0.16	0.4	1.1	1.55
Coal preparation refuse disposal	—	0.35	0.37	—	0.01	0.01	—	0.1	0.15
Prepared coal (subtotal)	24.32	30.91	32.54	1.04	1.17	1.21	10.3	11.6	11.95
Coal transportation[b]	16.50	16.50	16.50	0.71	0.63	0.62	7.0	6.2	6.05
Delivered coal (subtotal)	40.82	47.41	49.04	1.75	1.80	1.83	17.3	17.8	18.0
Power generation[c]	62.15	68.55	69.76	2.66	2.61	2.60	26.25	25.75	25.5
FGD[d]	11.54	12.37	12.31	0.49	0.47	0.46	4.85	4.65	4.5
Power plant waste disposal[e]	2.26	1.63	1.50	0.10	0.07	0.05	1.0	0.6	0.5
Total	116.77	129.96	132.61	5.00	4.95	4.94	49.4	48.8	48.5

[a] Coal source: West Virginia, Logan County, Cedar Grove seam.
[b] Coal transportation by unit train and collier (1000 miles at 1.65¢/ton-mile).
[c] Power plant location: Quincy, Mass.
[d] Limestone FGD.
[e] Includes boiler ash and FGD wastes.

Table 9 Relative Unit Power Generation Cost Components in $/ton of Level 1 Coal[a,b]

Case/Coal	Cleaning Level	Prepared Coal Cost	Power Plant Site (Mileage)[c]	Transportation Cost	Power Plant Cost	FGD Cost	Waste Disposal Cost	Total Power Generation Cost
A. IL, No. 6	1	22.17	Glasford, IL	3.00	47.02	13.27	2.39	87.85
	2	24.30	(160)	2.83	46.51	12.92	2.05	88.61
	4	25.30		2.71	46.05	12.44	1.93	88.43
B. WVA, Cedar Grove	1	24.32	Quincy, MA	16.50	62.15	11.54	2.26	116.77
	2	27.39	(1000)	14.62	60.72	10.96	1.44	115.13
	4	28.15		14.27	60.34	10.65	1.30	114.71
C. AL, American & Mary Lee	1	23.16	Dade City, FL	7.75	45.47	10.33	1.90	88.61
	2	24.89	(570)	6.99	44.39	9.92	1.52	87.71
	4	26.01		6.11	43.46	9.12	1.17	85.87
D. IA, Lower Ford	1	20.14	Panora, IA	3.00	46.36	18.47	3.02	90.99
	2	21.56	(160)	2.94	46.21	17.61	2.34	90.66
	4	23.72		2.89	46.06	16.48	1.91	91.06
E. UT, Rock Canyon,	1	20.15	Delta, UT	3.00	54.86	9.68	1.12	88.81
Gilson, & Sunny Side	2	22.12	(180)	2.52	53.16	9.05	0.74	87.59
	4	22.91		2.42	52.62	8.77	0.61	87.33

[a] Data adapted from Ref. 35.
[b] All costs adjusted to annual coal rate at level 1. Annual coal rates are presented in Table 6. Example: Case A, adjusted prepared coal cost for level 4 = $27.99/ton × (2,820,000 tons/yr/3,120,000 tons/yr) = $25.30/ton.
[c] Transportation distance from preparation plant to power plant.

Table 10 Potential Coal-Cleaning Cost Savings for Case B[a]

Cleaning Level/ Cost Component	Potential Cost Savings Relative to Level 1			
	$/ton Raw Coal[b]	Percent Reduction[c]	mills/kWh	Percent Reduction[c]
Level 2				
Coal transportation	1.88	11.4	0.80	11.4
Power generation	1.43	2.3	0.50	1.9
FGD	0.58	5.3	0.20	4.1
Waste disposal	0.82	36.3	0.40	40.0
Total savings	4.71	5.1[d]	1.90	4.9[d]
Level 4				
Coal transportation	2.23	13.5	0.95	13.6
Power generation	1.81	2.9	0.75	2.9
FGD	0.89	7.7	0.35	7.2
Waste disposal	0.96	42.5	0.50	50.0
Total savings	5.89	6.4[d]	2.55	6.5[d]

[a] Adapted from data in Ref. 35.
[b] Total savings in $/ton = level 1 cost − adjusted level i cost; adjusted costs are from Table 9 where i = 2, 4.
[c] Component percent reduction = 100 (saving) ÷ (level 1 component cost).
[d] Total percent reduction = 100 (total savings) ÷ (level 1 costs for transportation, power generation, FGD, and waste disposal).

potential cost savings for cost components which provide coal-cleaning cost benefits were: transportation, 0–37%; power generation, 0–51%; flue gas desulfurization, 14–56%; and waste disposal, 13–44% (see Table 11).

Thus, coal-cleaning cost benefits are very site specific and must be used with caution, either in general coal-cleaning studies or in specific plant design studies.

5 COAL CLEANING FOR SO₂ EMISSION REDUCTION

The relationships between coal resources, coal cleaning, and SO₂ emission regulations are complex. As discussed in the introduction, the quantity and quality of coal reserves vary significantly. There are also a variety of coal-cleaning techniques that can be used, and there are many different SO₂ emission regulations pertaining to the combustion of coal. In assessing the potential uses of coal cleaning for reducing SO₂ emissions, the Clean Air Act and emission regulations which have been promulgated under its mandate will be discussed. Estimates of the gross amounts of coal available for a given

Table 11 Change in Relative Percentage of Potential Cost Savings with Coal and Power Plant Site Factors[35]

Case	Total Potential Savings, mills/kWh	Percentage of Total Potential Savings				Net Benefit (Cost), mills/kWh
		Trans- portation	Power Generation	FGD	Waste Disposal	
A	1.4	7	43	36	14	(0.1)
B	2.55	37	29	14	20	0.9
C	3.0	30	37	20	13	1.5
D[a]	0.8	—	—	56	44	0.1
E	2.15	16	51	19	14	0.7
Average	1.98	18	32	29	21	0.62

[a] Level 2 cleaning; all others are level 4. Level 4 cleaning in case D resulted in a net power generation cost increase of 0.1 mills/kWh.

emission limit and percentage reduction requirement will then be made. Next, generalizations concerning the uses and limitations of physically cleaned coal (PCC) for compliance with selected emission regulations will be made. Some information on the amount of utility coals that are cleaned will be summarized, and some observations on the cost effectiveness of coal cleaning for reducing SO_2 emissions will be made. The prime question of coal sulfur variability will then be discussed: if an emission limit is not to be exceeded, how much lower than the emission limit must the mean sulfur value of a compliance coal be? Coal washability data and a model of coal sulfur variability will be used to estimate the conditions for which a coal can be expected to comply with SO_2 emission limits over specified averaging times. Finally, sample problems will be presented and solved to show some uses of washability. In this discussion the term "compliance coal" will be used to denote any coal, either raw or cleaned, which can be used to comply with any state or federal SO_2 emission regulation.

5.1 The Clean Air Act

The Clean Air Act of 1970 (P.L. 91-604) and its amendments of 1977 (P.L. 95-95) provide authority for the Environmental Protection Agency (EPA) to regulate the discharge of air pollutants into the atmosphere.[51,52] The act requires EPA to develop and implement regulations designed to limit the quantity of pollutants in the ambient air and to limit pollutant emissions to the atmosphere from stationary and mobile sources. As such, the act contains several regulatory and enforcement actions for control of emissions from stationary sources using coal. Actions include: (1) creating National Ambient Air Quality Standards (NAAQS) and approving State Implementation Plans (SIPs) to meet these standards, (2) promulgating Federal New Source Performance Standards (NSPS), and (3) establishing national regulations for the Prevention of Significant Deterioration (PSD). A discussion of regulatory activities relating to the Clean Air Act follows.

5.1.1 State Implementation Plans

Under Section 109 of the act, NAAQS have been established by EPA for sulfur dioxide (SO_2), particulate matter, nitrogen dioxide (NO_2), carbon monoxide (CO), photochemical oxidants, hydrocarbons (HC), and lead (Pb). To implement the NAAQS, each state develops a detailed plan for attaining and maintaining the NAAQS in all regions of the state. These SIPs, which must be approved by EPA (per Section 110 of the act), include emission limits, timetables for compliance, and enforcement programs. Pursuant to the 1977 amendments, the SIPs must also (1) stipulate the implementation of reasonable *further progress* toward attainment of NAAQS, (2) include a comprehensive emission inventory, and (3) identify a quantified *growth allowance*. The primary human source of SO_2 that affects ambient air quality is the combustion of solid and liquid fuels in stationary sources. Therefore, pre- and postcombustion control of sulfur in power plant fuel is a major tool in each SIP for meeting the NAAQS for SO_2.

SIP limits are designated for attainment and maintenance of NAAQS and are, therefore, tailored to local air quality conditions. Lower emission limits are often set for plants located in urban areas, where a high density of sources degrades local air quality, than for plants located in rural (clean air) areas. Where possible, SO_2 emission standards are set at levels that will allow the use of local coal.

Table 12   Range of SIP SO_2 Emission Limits for Coal-Fired Boilers in Selected States[53]

	Emission Limit, lb $SO_2/10^6$ Btu[a]	
State	Metropolitan Areas[b,c]	Nonmetropolitan Areas[c]
Alabama	1.2–1.8	4.0
Colorado	0.4–1.2	0.4–1.2
Iowa	5.0–6.0	5.0–6.0
Kentucky	1.2–3.3	5.2–6.0
Massachusetts	0.56–1.1	1.1–2.4
Ohio	0.32–5.6	1.6–9.6
Pennsylvania	0.6–3.0	4.0
West Virginia	1.6–5.1	2.7–7.5

[a] For boilers with heat input of approximately 10^9 Btu/hr (1.05 J/hr). Many SIPs differentiate emission limits with respect to size. Large boilers generally have more stringent emission limits than small boilers.
[b] Metropolitan and adjacent areas in the same air basin.
[c] Some states specify emission limits on a county or plant basis, resulting in a range of emission limits.

For example, in Pennsylvania the SIP limit for many units located in rural areas near major coal fields is 4.0 lb $SO_2/10^6$ Btu (1720 ng SO_2/J). The SIP limit for units located near urban areas is in the 0.6 to 3.0 lb $SO_2/10^6$ Btu (258 to 1290 ng SO_2/J) range. Table 12 lists typical SIP SO_2 emission limits for a selected number of states.

Emission averaging times under SIP regulations vary from state to state. Some regulations do not specify averaging times. Others specify hourly, daily, monthly, or yearly averaging times.

5.1.2 New Source Performance Standards

All boilers sited within a state must comply with applicable SIP emission regulations. In addition, coal-fired boilers of greater than 250×10^6 Btu/hr (264 GJ/hr) must comply with either of two federal standards that apply to new boilers.[54,55] The first NSPS applies to boilers constructed between August 17, 1971, and September 18, 1978. The second NSPS applies to coal-fired power plants for which construction commenced after September 18, 1978. The first NSPS specifies an SO_2 emission limit of 1.2 lb $SO_2/10^6$ Btu (516 ng SO_2/J). The 1977 CAA amendments required EPA to promulgate emission standards for coal-fired steam generators which "reflect a degree of emission limitation and *percentage reduction* achievable through application of the best technological system of emission reduction." Accordingly, plants constructed after September 17, 1978, cannot emit more than 1.2 lb $SO_2/10^6$ Btu (516 ng SO_2/J) and must reduce potential sulfur emissions from high-sulfur coal by 90%. If emissions are less than 0.6 lb $SO_2/10^6$ Btu (258 ng SO_2/J), then a 70% reduction in potential SO_2 emission is required. Sulfur dioxide emission reduction credits which result from coal cleaning are given on the basis of a three-month average for sulfur removed at the preparation plant.

The emission limit and percentage reduction requirements of the 1978 NSPS are based on a 30-day rolling average. The 1971 NSPS currently incorporate a compliance test that requires 3 hr to complete. The standard is sometimes interpreted to mean that for every 3 hr the mean SO_2 emission cannot exceed the limit.

5.1.3 Nonattainment Areas

The 1977 amendments also contain provisions to alleviate air pollution in areas where one or more air pollutants exceed the NAAQS (nonattainment areas). Before construction permits are issued by a state, a reduction in emissions from existing sources must be guaranteed to "offset" emission of nonattainment pollutants from the new source. One means to achieve this reduction without the installation of pollution control equipment is to burn cleaned coal instead of run-of-mine (ROM) coal in existing boilers.

5.1.4 Prevention of Significant Deterioration

Provisions of the 1977 amendments specify formulation of regulations for PSD in the ambient air quality. These regulations are to preserve the air quality in regions that meet the NAAQS (attainment areas). The amendments establish allowable increases (increments) above baseline concentrations of SO_2 and total suspended particulate (TSP) for three classes of attainment areas. Increments for the other criteria pollutants must also be established by EPA.

In order of increasing PSD allowable increments, clean air areas are designated as class I, class II, or class III. National parks and national wildlife areas are mandatory class I areas that cannot be redesignated. All other areas were automatically classified as class II. A class II area within a state may be redesignated as class I or III upon request of the state's governor, approved by the EPA administrator.

Class I areas are those in which even small deteriorations in air quality are undesirable. Class II areas are those in which deterioration of air quality accompanying moderate industrial growth would not be considered significant. Class III areas are those in which deterioration not exceeding NAAQS would be permitted to allow concentrated or large-scale industrial development.

The construction of major new sources or the modification of existing sources in attainment areas is subject to the PSD provisions of the act. These sources must incorporate the best available control technology (BACT) for each pollutant subject to regulation. BACT, which is determined on a case-by-case basis, must consider the available technologies and their energy, environmental, and economic impacts. BACT must be at least as stringent as the NSPS for the category of sources. In some instances where higher levels of control are desired, combinations of technologies such as flue gas desulfurization and physical coal cleaning (FGD + PCC) may be required.

5.1.5 Control Trading

Control trading is a generic term EPA uses to describe a number of different market-based approaches which are being studied as possible methods of controlling air pollution in accordance with provisions of the act.[56] It covers offsets, bubbles, netting, and banking of pollution reduction credits for future use. The objective of control trading is to facilitate economic savings in pollution control measures which are taken to meet the PSD and nonattainment provisions of the act.

Control trading allows emission reductions at one source to be credited to emission control requirements at another source. "Bubbles" are emission trade-offs between existing sources controlled under SIP requirements. "Offsets" are reductions in emissions from existing sources which must be obtained before a new source can be constructed in a nonattainment area. "Netting" applies only to modifications that lead to emission increases at existing sources. By the use of netting, the requirements of a new source review can be avoided by reducing emissions at other nearby existing sources. While the bubble allows existing sources to reduce their costs of control, the offset and netting programs serve an analogous function for new or expanding sources by allowing them to locate in nonattainment or PSD areas without violating ambient air quality restrictions.

In many cases coal cleaning can be used for control trading transactions to provide emission reduction for other new, modified, or existing sources.

5.1.6 *Other Applications*

Provisions of the act authorize research on prospective environmental problems. Under these provisions research is being conducted to evaluate the causes and effects of acid rain. Research is also being conducted on alternative means of controlling potential human precursors of acid rain. This research includes assessment of the techniques and costs of reducing SO_2 emissions from coal combustion. Coal cleaning is a major technology under consideration for reducing SO_2 emissions from existing power plants.

5.2 SO_2 Emission Regulations

There are a number of components to an emission regulation. A regulation may include an emission limit, a percentage reduction requirement, a time period of compliance, the number of times the emission limit may be exceeded, and the technical methods required for demonstrating compliance. The emission limit may be expressed in various terms: $lb/10^6$ Btu of heat input, ppm of flue gas concentration, lb/ton of material processed, lb/hr of pollutant emitted, and so on.[53,54,55] The percentage reduction requirement can be expressed in any of the terms used in specifying the emission limits; for example, a percentage reduction in flue gas concentration or lb $SO_2/10^6$ Btu. The time of compliance or time period over which the emission limit or percentage reduction is averaged may range from instantaneously to yearly averages. In some cases it is not specified.

The technical method of demonstrating compliance may include the use of coal sampling and combustion calculations, the use of stack sampling techniques, or the use of continuous instrumental emission monitors.

All of the components of the emission regulation influence or constrain the use of a given coal and the technological methods used for producing a compliance fuel. The major constraints are a consequence of the average fuel properties, the technologies used to provide average reductions in emissions, and the time-varying coal properties that may cause the emission limit to be exceeded.

To simplify the discussions that follow, the emission limits and percentage reduction requirements shall be expressed in terms of the heat specific SO_2 emission parameter (lb $SO_2/10^6$ Btu).

5.3 Potential Supply of Compliance Coals

Estimates have been made of the amounts of coal that can be used to comply with various emission limits and percentage reduction requirements. These estimates use coal washability data and projected data on coal production in 1985.[7,36]

5.3.1 *Emission Limits*

Figure 10 illustrates the amounts of compliance coal that can potentially be produced by the state of Ohio in 1985. The figure contains curves for raw coal and three levels of cleaning. The first level of physical cleaning, designated PCC-1, corresponds to crushing coal to a 1½-in. (37.5-mm) top size and taking a clean coal float product at 1.6 s.g. This is a good estimator for the level of coal desulfurization that can be achieved in commercial coal preparation plants. The second level of cleaning, PCC-2, corresponds to crushing to ⅜ in. (9.5 mm) and taking a clean coal float product at 1.3 s.g. PCC-2, which generally overestimates the potential sulfur emission reductions that can be achieved by modern coal preparation plants, can be used to designate the upper limit of commercially available technology. The third level of cleaning, called Gravichem, is a combined physical/chemical process. It removes approximately 90% of the pyritic sulfur but none of the organic sulfur. It can be assumed to represent the upper limit of advanced physical processes for cleaning fine coal.

Figure 10 shows that for all emission limits above 2.0 lb $SO_2/10^6$ Btu (860 ng/J) the amounts of compliance coals increase with the level of cleaning. Also, the amounts of compliance coals increase

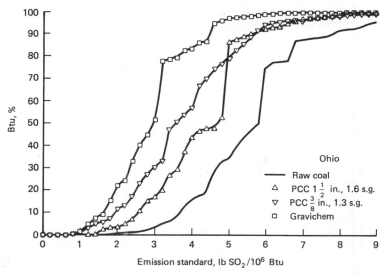

Fig. 10 Projected 1985 coal production in Ohio able to meet various emission limits (total projected 1985 coal production = 1.34 quads).

with increasing emission limit. For example, 17% of the raw coal (Btu basis) is capable of complying with an emission limit of 4.0 lb $SO_2/10^6$ Btu (1720 ng/J). By cleaning the raw coal which does not satisfy the 4.0 lb $SO_2/10^6$ Btu (1720 ng/J) emission limit at PCC-1 conditions, a total of 43% of the coal produced in Ohio can be made available for compliance with the 4.0 lb $SO_2/10^6$ (1720 ng/J) limit. The amounts available when cleaned at PCC-2 or Gravichem increase to 58 and 82%, respectively. At an emission limit of 6.0 lb $SO_2/10^6$ Btu (2579 ng/J), the availability of compliance coals under different scenarios are: raw coal, 75%; PCC-1 coal, 92%; PCC-2 coal, 95%; and Gravichem coal, 99%.

The shape of the compliance coal distribution curves for other states that produce high-sulfur coals is similar to the Ohio distribution curves.

The compliance coal distribution curves for states producing predominantly low-sulfur coals are different from those that produce high-sulfur coals. Figure 11 presents data on the potential distribution

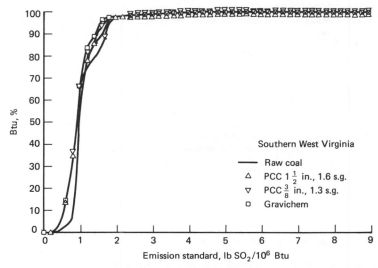

Fig. 11 Projected 1985 coal production in southern West Virginia able to meet various emission limits (total projected 1985 coal production = 2.42 quads).

Table 13   Percentage of Projected 1985 Coal Production Able to Meet Selected Emission Limits Using Raw Coal or Cleaned Coal[7]

Region	State[a]	1985 Production, 10⁶ tons	Raw Coal Emission Limit, lb SO₂/10⁶ Btu			Clean Coal[b] Emission Limit, lb SO₂/10⁶ Btu		
			1.2	2.0	4.0	1.2	2.0	4.0
1	OH	54.5	0	1	16	0	4	43
	PA	99.6	5	20	79	15	48	88
	WV-N	52.8	19	26	62	23	42	81
	Total	209.7	8	17	59	13	35	75
2	KY-E	106.2	49	85	94	58	88	96
	VA	46.4	67	87	100	67	87	100
	WV-S	88.0	75	99	99	78	97	99
	Total	249.6	61	89	97	66	90	98
3	AL	32.1	45	71	93	49	77	94
	Total	32.1	45	71	93	49	77	94
4	IL	92.6	0	7	25	1	16	39
	IN	30.6	6	11	42	10	15	51
	KY-W	58.7	0	0	7	0	0	26
	Total	181.9	1	6	22	2	11	37
5	TX	66.3	0	28	85	1	32	86
	Total	79.2	2	27	69	3	30	76
6	CO	38.0	89	100	100	95	95	100
	MT	75.0	29	100	100	81	100	100
	MN	39.2	53	72	100	54	95	100
	ND	43.8	12	67	100	66	99	100
	UT	33.6	82	99	100	82	99	100
	WY	241.9	46	97	99	54	97	99
	Total	490.6	48	93	100	65	98	99
Total United States		1243.1	34	62	74	43	68	84

[a] States producing less than 20 tons not listed but included in totals: MD, GA, TN, AR, IA, KS, MO, OK, AZ, WA.
[b] PCC estimates based on cleaning by crushing to 1½-in. top size and separating at 1.6 s.g.

of compliance coals produced in southern West Virginia. These distribution curves are similar to those from other states that produce only low-sulfur coals. The potential sulfur emissions from these coals are low either before or after cleaning. Under different scenarios for using the southern West Virginia coal for meeting an emission limit of 1.2 lb SO₂/10⁶ Btu (516 ng/J), 75% of the raw coal, 78% of the PCC-1 coal, 80% of the PCC-2 coal, and 83% of the Gravichem coal produced can be used as compliance fuels. More than 98% of all raw and cleaned coals are available in compliance with an emission limit of 2.0 lb SO₂/10⁶ Btu (860 ng/J).

Table 13 provides estimates of the 1985 coal production which can be used to comply with SO₂ emission regulations of 1.2, 2.0, and 4.0 lb SO₂/10⁶ Btu (516, 860, and 1720 ng/J). These estimates are aggregated at the state, regional, and national levels and are presented for raw coal and one level of physical cleaning (PCC-1). Approximately 34% of the uncleaned coal can be used for compliance with an emission limit of 1.2 lb SO₂/10⁶ Btu (516 ng/J). When cleaned, the potential availability of 1.2 lb SO₂/10⁶ Btu (516 ng/J) compliance coal increases to 43%. Most of these raw or cleaned compliance coals come from the low-sulfur coal producing states in the southern Appalachian, Alabama, or western coal regions. In states in which high-sulfur coals predominate, the percentage of raw compliance coals is small for emission limits less than 4.0 lb SO₂/10⁶ Btu (1720 ng/J). Cleaning of these coals substantially increases the amounts of compliance fuels for emission limits of 4.0 lb SO₂/10⁶ Btu (1720 ng/J) and above.

5.3.2   Percentage Reduction Requirements

Percentage reduction requirements impose a significant constraint on the production of compliance coals by cleaning. This constraint is more severe for low-sulfur coals than high-sulfur coals. Figure 12 illustrates the impact of percentage reduction requirements on the potential production of Ohio coals in 1985. Curves in Figure 12 are coals which have been cleaned by PCC-1 and which are subject to a percentage reduction requirement and no emission limit or an emission limit of 3.0, 4.0, 5.0, or

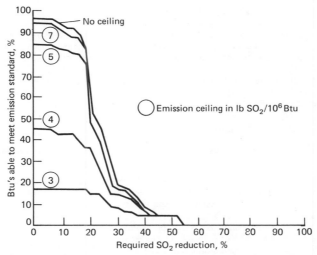

Fig. 12   Percentage of projected 1985 Ohio coal production able to meet various SO₂ emission standards defined by an emission ceiling and percentage SO₂ reduction using physical coal cleaning at 1½ in., 1.6 s.g.

7.0 lb SO₂/10⁶ Btu (1290, 1720, 2150, or 3009 ng/J). Below a percentage reduction requirement of approximately 15%, the emission limit is the primary constraint on the availability of compliance coals. For example, at a percentage reduction requirement of 10% there are only marginal reductions in the amounts of available compliance coal as compared with a standard without a percentage reduction requirement. Above 15%, the percentage reduction requirement becomes more restrictive and all curves converge for percentage reduction requirements greater than 45%. No compliance coals are available in Ohio for PCC-1 cleaning conditions when the percentage requirement is 55% or greater. While other data show that higher levels of physical cleaning increase the amount of coal at most percentage reduction requirements, advanced physical cleaning techniques are capable of producing only small amounts of compliance coals at reductions above 70%. None can be produced for reduction requirements greater than 90%.

The impact of percentage reduction requirements on the availability of compliance fuels in low-sulfur coal states is more severe. Figure 13 illustrates the effects of percentage reduction requirements

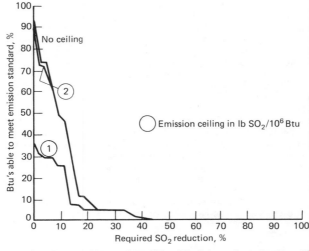

Fig. 13   Percentage of projected 1985 southern West Virginia coal production able to meet various SO₂ emission standards defined by an emission ceiling and percentage SO₂ reduction using physical coal cleaning at 1½ in., 1.6 s.g.

and emission limits on the availability of cleaned compliance coals from southern West Virginia. At an emission limit of 2.0 lb $SO_2/10^6$ Btu (860 ng/J), more than 98% of the coal can be used as compliance coal. Specifying a percentage reduction requirement causes the availability to fall off precipitously. Compliance coal availability at a 10% reduction requirement is approximately 50% of the total projected production. At a 20% reduction requirement the availability has fallen to approximately 20%. There are virtually no compliance coals available for percentage reduction requirements above 45%, even at the highest level of physical cleaning represented by removal of 90% of the pyritic sulfur. The effects of sulfur reduction requirements on the availability of compliance coals are similar for all low-sulfur coal states.

5.3.3 Generalizations on Compliance Coal Availability

In evaluating available data on the desulfurization of coal, the following general observations can be made:

1. Physically cleaned northern Appalachian and midwestern coals can be used to meet the moderate emission limits specified for old boilers. However, few of these coals can be cleaned to the 1.2 lb $SO_2/10^6$ Btu (516 ng SO_2/J) level specified by the initial new source standards (1971) for coal-fired steam generators.

2. Many southern Appalachian, Alabama, or western coals are capable of meeting a long-term emission limit of 1.2 lb $SO_2/10^6$ Btu (516 ng SO_2/J), either before or after cleaning.

3. Few physically cleaned high-sulfur coals can be used to comply with regulations specifying sulfur emission reductions exceeding 50%. The physical sulfur reduction potential of many low-sulfur coals falls in the range of 10–20%.

4. Except for a small number of coal seams, emission regulations specifying any combination of emission limits below 1.2 lb $SO_2/10^6$ Btu (516 ng SO_2/J) and sulfur reduction requirements above 30% eliminate the use of compliance coals that are produced by commercially available physical coal-cleaning techniques.

5. Advanced physical cleaning and chemical coal cleaning can be used to remove additional amounts of sulfur from coal. However, few processes are capable of 70–90% reduction in sulfur emissions required for compliance with the latest new source standards for coal-fired utility boilers.

5.4 Cleaning of Utility Coals

Utilities have traditionally used coal cleaning for ash removal. The extent to which utilities are using coal cleaning to comply with SO_2 emission regulations is uncertain. Table 14 summarizes the current extent to which electric utilities are using coal cleaning in eight eastern and midwestern coal producing states. These data provide estimates of the reduction in potential SO_2 emissions that are achieved by current practices. The information in Table 14 was developed through use of a 1979 deliveries-to-utilities data and a coal washability data base.[36] The deliveries-to-utilities data base specifies the SO_2 content of the as-delivered coal and the quantity of coal cleaned prior to delivery. These specifications were converted to the as-mined qualities and SO_2 contents presented in Table 14 by assuming that the washability characteristics of the coals washed prior to delivery were the same as those of the coals that were not washed prior to delivery. For example, a 21% SO_2 reduction can be achieved by washing Alabama coals. Therefore, it was assumed that the washed Alabama coals delivered to utilities had had their SO_2 content reduced by 21% in the washing process, and that, accordingly, the as-mined SO_2 content was equal to the as-delivered SO_2 content divided by 0.79 (i.e., 1–0.21). Coal weight prior to washing was calculated in a similar manner.

The amount of coal cleaned varies from state to state. Table 14 shows, for example, that over half of the coal from Indiana and Illinois and about one-third of the coal from Pennsylvania, western Kentucky, and Alabama was cleaned prior to delivery to utilities in 1979. This cleaning resulted in significant SO_2 reductions. On the other hand, only small percentages of the coal from Ohio, Virginia, and southern West Virginia was cleaned prior to utility use in 1979.

The SO_2 reduction that could have been achieved if all coal delivered to utilities in 1979 had been cleaned at $1\frac{1}{2}$-in. (37.5-mm) top size and 1.6 s.g. (i.e., moderate cleaning) is also summarized in Table 14. About 1.7 million (1.5 million Mg) tons more of SO_2 could have been removed in 1979 by cleaning all the utility coal from Pennsylvania, Ohio, and western Kentucky. Moderate additional reductions in SO_2 emissions could have been achieved by washing other northern Appalachian and eastern-midwest coals. By washing all of the utility coals from these eight states, the SO_2 emission from combustion of these coals would have been reduced by 18%.

5.5 Cost Effectiveness

An evaluation of the cost effectiveness of using coal cleaning for reducing SO_2 emissions must be based on requirements of the various regulations. The most cost-effective emission control method is generally that which will satisfy all emission requirements and minimize the cost of power production.

Table 14 Current (1979) SO_2 and Potential Reduction by Cleaning Utility Coal from Eight States[36]

Region and State in Which Coal Was Mined	Coal Delivered to Utilities in 1979, 10^3 tons	Utility Coal Cleaned in 1979, %	SO_2 Content of Coal		Average SO_2 Reduction by Coal Cleaning in 1979, %	Additional SO_2 Reduction by Cleaning All Coal	
			As Mined, 10^3 tons	As Delivered, 10^3 tons		10^3 tons	%
Northern Appalachia							
Pennsylvania	47,400	30	2,100	1,860	12	470	25
Ohio	38,300	11	2,750	2,670	3	740	28
Northern West Virginia	31,300	23	1,760	1,690	4	280	16
Southern Appalachia							
Southern West Virginia	17,500	9	300	290	1	30	11
Virginia	13,400	7	280	270	1	30	10
Eastern Kentucky	68,600	22	1,630	1,570	4	260	16
Eastern Midwest							
Western Kentucky	38,100	34	2,880	2,600	10	530	21
Indiana	25,300	52	1,620	1,410	13	180	13
Illinois	49,500	72	3,570	2,780	22	230	8
Alabama							
Alabama	14,600	32	460	440	5	70	17
Eight-state total	344,000	33	17,350	15,580	10	2,850	18

Three different types of regulations are of interest: SIP regulations, NSPS regulations, and other regulations such as those that are being advocated for the control of acid deposition.

5.5.1   SIP Regulations

Where SIP regulations specify moderate emission limits, the lowest cost strategies may be the use of uncleaned low-sulfur coal, the blending of low- and high-sulfur coals, the use of physically cleaned coals, or combinations of these strategies. Selection of the lowest cost strategies must include all site specific factors including delivered coal cost (raw and cleaned), transportation costs, coal/boiler compatibility, and cost benefits that may accrue from the use of cleaned coals.

These strategy studies must necessarily consider many potential coal supply and cleaning options in arriving at the least-cost compliance method. If a strategy involving coal cleaning is the least-cost option, then it is the most cost effective.

5.5.2   NSPS Regulations

NSPS may require either a 1.2 lb $SO_2/10^6$ Btu (516 ng/J) emission limit as specified by the 1971 standard for coal-fired steam generators, or an emission limit and percentage removal requirement as specified in the 1979 revisions applicable to coal-fired electric utility boilers. In the case of the 1971 standard, low-sulfur coal and physically cleaned coals will provide the least-cost option if they can be obtained at a reasonable price. However, in some instances where transportation distances are large, the use of limestone FGD may provide a lower cost option. In these cases studies have shown that combinations of physical coal cleaning and FGD are generally more cost effective than FGD alone.[48,49] In these instances coal-cleaning cost benefits more than offset the costs of cleaning.

For the revised NSPS which requires a 70–90% reduction in potential sulfur emissions, the most cost-effective strategy is dependent on the cost benefits that accrue from the use of physically cleaned coal (see Section 4.4.3). In evaluating studies that have assessed the use of physical coal cleaning and limestone FGD, the following general conclusions can be made.

1.   For percentage reduction requirements of 70% and less, coal cleaning and FGD are generally more cost effective than FGD alone.[26,48,49] Cost savings related to reduced transportation, FGD, and waste disposal expenses more than offset the costs of cleaning.

2.   For medium- and high-sulfur coals with percentage reduction requirements greater than 70%, coal cleaning and FGD are the most cost-effective strategy *only if* there are substantial cost savings in transportation and boiler related costs.[26,35,48,49] At high percentage reduction requirements, the FGD cost savings from flue bypass and reduced reheat requirements are not sufficiently large to offset a major fraction of the coal-cleaning costs.

3.   For very high sulfur coals (>5.0%) combinations of coal cleaning and limestone FGD appear to be the most cost-effective strategy. While FGD bypass cost savings are not available, the reduction in FGD costs due to reduced reagents, slurry pumping, the FGD waste disposal requirements are large enough to offset a major portion of the costs of preparation.[57]

Note that the above conclusions are based on comparisons of physical coal cleaning and wet FGD systems that employ limestone scrubbing. These same conclusions may not apply to other FGD systems. The FGD cost savings obtained by using cleaned coal may be substantially different for dual alkali scrubbers, lime spray dryers, or other FGD systems. In any event, studies considering all site specific factors that influence the costs of power generation and SO_2 emission control must be made before the most cost-effective strategy is selected.

5.5.3   Other Regulations

In studies concerning state, regional, and national strategies, such as those for reducing the emission of acid rain precursors, the use of different emission control technologies are compared in terms of $/ton of SO_2 reduction. The cost effectiveness of alternative control technologies are then compared by the use of this measure. In the case of coal cleaning, the cost effectiveness for SO_2 emission reduction can be expressed as the unit annual coal preparation cost (in $/ton of clean coal) divided by the potential reduction in SO_2 emissions expressed as the tons of SO_2 reduction per ton of clean coal. Alternatively, it can be expressed as the annual costs of preparation divided by the annual reduction in SO_2 emissions. The cost effectiveness for cleaning is, therefore, a function of the coal preparation cost, the original sulfur content and washability of the coal, and the cleaning level employed. It is most sensitive to the original sulfur level and the washability. It is also dependent on any cost benefits that may result from cleaning. Not surprisingly, coal-cleaning cost effectiveness varies from coal to coal. The cost effectiveness for cleaning high-sulfur coals is quite different from that for cleaning low-sulfur coals.

The cost effectiveness of cleaning Ohio and southern West Virginia coal is compared in Table 15.[36] These comparisons correspond to SO_2 emission control scenarios that require that all coals that

Table 15 Potential SO$_2$ Emission Reductions and Costs due to Selective Cleaning of Ohio and Southern West Virginia Coals Delivered to Utilities in 1979[36]

Coals to Be Cleaned	Quantity to Be Cleaned, 10³ tons	Total SO$_2$ Emissions After Selective Cleaning, 10³ tons	SO$_2$ Emission Reduction Achieved by Selective Cleaning		Levelized Cost of Cleaning,[c] 10⁶ 1979 $	Cost Effectiveness, $/ton SO$_2$
			10³ tons	%		
Ohio						
No coals[a]	0	2,479	0	0	0	—
Coals with SO$_2$ contents above floor of:						
7 lb/10⁶ Btu	12,353	2,101	378	15	137	360
6 lb/10⁶ Btu	20,225	1,964	515	21	201	390
5 lb/10⁶ Btu	27,063	1,832	647	26	273	420
4 lb/10⁶ Btu	31,986	1,757	722	29	302	420
All coals	34,527	1,735	744	30	322	430
Southern West Virginia						
No coals[b]	0	267	0	0	0	—
Coals with SO$_2$ contents above floor of:						
4 lb/10⁶ Btu	82	267	<1	<1	<1	1,000
3 lb/10⁶ Btu	337	263	4	2	4	1,000
2 lb/10⁶ Btu	933	260	8	3	8	1,000
1 lb/10⁶ Btu	15,677	236	31	12	123	3,970
All coals	16,013	236	31	12	126	4,060

[a] Excludes 3,787,000 tons of Ohio coal actually cleaned in 1979.
[b] Excludes 1,450,000 tons of southern West Virginia coal actually cleaned in 1979.
[c] Levelized costs 1.178 times higher than the first year annualized costs.

were not washed and which have SO_2 emission potentials greater than a given floor level be washed. As the floor level decreases the cost effectiveness of cleaning all the previously unwashed coal in $/ton SO_2 increases. This results from the fact that smaller amounts of sulfur are being removed from the lower sulfur coals while the *costs* of cleaning remain nearly constant. In Ohio, which contains high-sulfur coals, the levelized cost effectiveness of cleaning ranges from $360 to $430/ton ($397 to $474/Mg) of SO_2 removed. In southern West Virginia the cost effectiveness of cleaning ranges from $1000 to $4060/ton ($1102 to $4476/Mg) of SO_2 removed. These values for cost effectiveness are typical of those given in Ref. 36 for cleaning coals from other high- and low-sulfur coal states. Note that these values can be substantially reduced by accounting for coal-cleaning cost benefits.

5.6 Sulfur Variability

Many boiler operators attempt to comply with SO_2 emission regulations by using low-sulfur coal, blends of low- and high-sulfur coals, or physically cleaned coal. However, the sulfur and heating value of coal may vary significantly with time, and it is difficult to ensure that a "compliance coal" will indeed satisfy conditions of the emission regulation for which it was purchased.

Emission regulations generally specify an emission limit and an averaging time. Average pollutant emissions during this time period are not to exceed the limit. The ability of a coal to comply with an emission regulation depends on the statistical characteristics of the coal sulfur and ash, the requirements of the regulation, and the rate at which coal is burned (boiler size).

Sulfur emissions from coal combustion vary in time in a probabilistic manner. There is also some structure to the sulfur emissions. If the sulfur emissions were high 1 hr ago, it is likely that they are high now.

The sulfur and ash content of coal varies vertically and laterally within a coal seam. Lines of constant sulfur representing the average sulfur content at a given location can be drawn on the map of a coal mine in a manner similar to lines of constant elevations on a contour map. The sulfur content of individual coal particles at any mine location may vary randomly around the average value for that location.

Mining of the coal transforms its spatial properties into a unique time sequence of potential sulfur emissions. This time sequence of potential sulfur emissions is dependent on the mining scheme—a different mining scheme will produce a different series of potential sulfur emissions. Blending, coal preparation, coal transportation, and boiler pollution control devices will modify and attenuate the variability of the "as-mined" coal.[58]

5.6.1 *Importance of Sulfur Variability*

The importance of coal sulfur variability can best be illustrated by examining some data that show the variability of potential sulfur emissions in time. Figure 14 is a plot of the sulfur emission parameter (lb SO_2/10^6 Btu) for raw and cleaned coal data from the Kitt Mine preparation plant in West Virginia.[58]

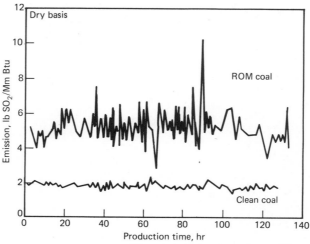

Fig. 14 Kitt Mine Coal Preparation Plant—hourly incremental data for sulfur dioxide emission parameter.

Each raw coal data point represents the average emission which would result from burning 800 tons (726 Mg) of coal. Each raw coal data point is said to represent a lot size of 800 tons (726 Mg). Each clean coal data point represents a lot size of 460 tons (417 Mg).

Under typical conditions an 80-MW boiler may burn approximately 800 tons (726 Mg) of coal in 24 hr. It is assumed that the Kitt Mine raw coal must be used to comply with an emission limit incorporating a 24-hr averaging time, the amount of coal represented by each hour of production (one data point). It is seen from Figure 14 that the maximum emission average for 1 hr of production is greater than 10 lb $SO_2/10^6$ Btu (4299 ng SO_2/J). Many of the 1-hr production averages and hence many of the 24-hr emission averages are above 5.5 lb $SO_2/10^6$ Btu (2364 ng SO_2/J). The mean raw coal sulfur emission for the entire 140 hr of production would be 5.2 lb $SO_2/10^6$ Btu (2236 ng SO_2/J). The raw Kitt Mine coal could be used to comply with a 5.2 lb $SO_2/10^6$ Btu (2236 ng SO_2/J) emission standard if the regulation specified a 140-day averaging time instead of a 24-hr averaging time.

An obvious conclusion is that emission regulations that incorporate short averaging times are more difficult to comply with than those with long averaging times. In the same manner, it is easier for large boilers to comply with a given emission limit than small boilers. In 1 hr a 500-MW boiler will burn 20 times more coal than a 25-MW boiler. A 1-hr averaging time for a 500-MW boiler would be equivalent to a 20-hr averaging time for a 25-MW boiler since equal amounts of coal would be burned during these time periods. Long-term emission averages or large lot size averages are always less than or equal to the maximum values of the smaller incremental averages which they comprise.

The variability of cleaned coal sulfur emissions is much less than that for raw coal. The maximum emissions from the hourly production of cleaned Kitt Mine coal were approximately 2.25 lb $SO_2/10^6$ Btu (967 ng SO_2/J), a factor of 4.4 less than the maximum potential emission from the raw coal.

In some instances emission limits are defined as maximum values that are never to be exceeded under penalty of law. In other cases, standards are interpreted to mean that the maximum emission as determined by the compliance test method is never to exceed the emission limit. These are stringent requirements where continuous emission monitors and/or 3-hr stack emission tests are employed to determine compliance. These difficulties can largely be mitigated by regulations which specify 24- or 30-day averages.

Some state, county, or municipal air pollution authorities enforce sulfur emission regulations on the basis of short averaging times. It is, therefore, important to understand the relationship between the maximum and average SO₂ emissions, and various types of SO₂ emission standards. To aid in enforcement and to ensure that the short-term ambient air quality standards (3-hr average) are met, it is desirable to develop methods for specifying mean coal sulfur emission parameter values which should not be exceeded. These determinations must be based on the statistical characteristics of coal sulfur emissions.

5.6.2 Definition of Statistical Terms

Mean value, variance, standard deviation, relative standard deviation, and autocorrelation are terms used to evaluate coal sulfur variability. The mean value, μ, is simply the large sample average which is estimated by the sum of all values of the variable x divided by the number of data points. The variance is a measure of the data scattered about the mean value. The variance, σ^2, is estimated by the sum of the squared distance of each point from the mean divided by an appropriate constant. The standard deviation, σ, is the square root of the variance. The relative standard deviation, RSD $= \sigma/\mu$, is a normalized form of the standard deviation.

If values in a sequence of data points are dependent on previous values, then they are said to be autocorrelated. A normalized version of the autocovariance is the autocorrelation. In more detail, ρ_j is the autocorrelation for lag j. It measures the dependence of data separated by j time intervals.

The usefulness of the autocovariance or autocorrelation is in forecasting present and future data from past events. Small autocorrelations may indicate that the relation between past and present values is small so that past data probably provide little useful information in predicting present and future events. If the autocorrelation is large, it is essential to determine the time-dependent effect through modeling in order to predict future trends. Models that use past values of x to predict present and future values of x are called autoregressive models. The number of past data points needed to discuss future effects is the order of an autoregressive model. Thus with an AR(1) model only the previous data point is useful in forecasting the present. An AR(2) model requires the two previous data points to forecast present conditions. Once these trends have been forecast by time series analysis, the variance can be used to provide an estimated measure of their accuracy. The statistical techniques of time related values are called time series analysis.[59]

5.6.3 Time Series Model of Emissions

Assuming that all of the coal sulfur is emitted as SO₂ upon combustion, the relationship between the estimated mean coal sulfur emission parameter and a limiting emission value may be expressed by

$$E_L = \mu(1 + Z_\alpha \, \text{RSD}_n) \tag{1}$$

where

E_L = time average emission limit for an averaging time, t

μ = time independent mean emission value for a given coal population. This value of μ will provide an expected exceedance rate of α (i.e., α is the average allowable proportion of the averaging periods during which the emission standard may be exceeded).

Z_α = normal percentile (percentage) point corresponding to an expected exceedance rate α. For $Z_\alpha = 1$, it is expected that emissions will on the average be exceeded 15.9% of the averaging periods. For $Z_\alpha = 1.5$, 2.0, and 2.5 the average expected exceedances are 6.7, 2.3, and 0.6%, respectively.

RSD_n = a relative standard deviation for a time sequence of n reference lots of size l such that n lots are burned during the emission averaging time of E_L. The total lot size (weight) for n reference lots is $L = nl$.

The values of parameters in Eq. (1) must be estimated using coal sampling and analysis (CSA) data. The method for estimating these parameters is summarized below. A discussion of application of the model follows the discussion of mathematical development.

5.6.4 Mathematical Development of Model

The coal statistical parameters must be estimated using CSA data. The major parameter to be estimated is RSD_n. An estimated value of a parameter will be designated by placing a "hat" over the symbol, for example, $\widehat{\text{RSD}}_n$.

RSD_n is a measure of the variability of coal lots of size L. This variability is a function of the lot size and autocorrelation structure. Previous studies have shown that the RSD_n can be expressed by

$$\text{RSD}_n = \frac{\text{RSD}_1 \cdot \omega_n}{\sqrt{n}} \tag{2}$$

where

RSD_1 = the RSD for all reference lots of size l corresponding to $n = 1$

ω_n = a function of the autocorrelation function for n reference lots with autocorrelation structure ρ_n

n = number of reference lots in l

As noted, RSD_1 is a measure of the sulfur variability between reference lots of size l and is explicitly dependent on errors in CSA data. CSA data must be used to estimate RSD_1, and limitations of current CSA data sets make it impossible to separate errors related to the CSA process from the inherent variation in coal properties. Since overestimation of the variability will lead to a conservative result, it is reasonable to use an estimation of RSD_1 which is obtained by

$$\text{RSD}_1 = \frac{1}{\overline{X}_n} \left[\sum_{i=1}^{n} \frac{(x_i - \overline{X}_n)^2}{N-1} \right]^{1/2} \tag{3}$$

where $\overline{X}_n = \Sigma \, x_i/N$ and x_i is the value of each of N independent values of x. In general, this estimation will include variability terms related to the variance between the reference lots, the variance of the coal samples within the reference lots (a representation error), and the variance resulting from the inherent variability in sample preparation and analysis procedures.

The value of ω_n can be estimated from CSA data by the use of time series models. No single time series model has been found which can describe exactly the potential sulfur emission from all coal sulfur data sets. However, the effects of averaging time (n) and autocorrelation can be approximated by an autoregressive model of order 1. For the AR(1) model the function of autocorrelation can be expressed in the simplified form:[60]

$$\hat{\omega}_n = \left[\frac{1 - \hat{\rho}_1^2 - 2\hat{\rho}_1/n + 2\hat{\rho}_1^{n+1}/n}{(1 - \hat{\rho}_1)^2} \right]^{1/2} \tag{4}$$

where $\hat{\rho}_1$ is an estimated value of the autocorrelation at lag 1 as determined from CSA data with the equation

$$\hat{\rho}_1 = \frac{\displaystyle\sum_{i=1}^{n} (x_{i+1} - \overline{X}_n)(x_i - \overline{X}_n)}{\displaystyle\sum_{i=1}^{n} (x_i - \overline{X}_n)^2} \tag{5}$$

The effects of coal sulfur variability on the mean emission parameter which is required for compliance with a given emission regulation may be determined using CSA data and Eqs. (1) through (5). To illustrate the effects of coal statistical parameters on the probable value of μ needed for compliance, it is necessary to identify some representative values of $\widehat{RSD}_1$ and $\hat{\rho}_1$.

Depending on the lot size and other considerations, $\widehat{RSD}_1$ may range from less than 0.07 to greater than 0.35.[58,61,62] In one carefully controlled study at the R&F coal preparation plant in Cadiz, Ohio, it was determined that $\widehat{RSD}_1 = 0.14$ for raw coal reference lots of 428 tons (388 Mg).[58] At the same plant, $\widehat{RSD}_1 = 0.10$ for cleaned coal reference lots of 330 tons (299 Mg). Using this information, 0.20 and 0.12 can be assumed as reasonable values of $\widehat{RSD}_1$ for raw and cleaned coal (respectively) at a lot size of 240 tons (218 Mg).

The autocorrelation structure of coal has not been studied widely. However, cleaned coal is more highly correlated than raw coal. The cleaning process reduces variability, and properties of adjacent coal lots are more likely to be similar for clean coal than for raw coal. For the R&F plant, the values of $\hat{\rho}_1$ for coal data collected every half-hour were about 0.25 for raw coal and 0.50 for cleaned coal. These values of $\hat{\rho}_1$ will be used hereafter for illustrative purposes.

Emission regulations do not normally specify a value of Z_α nor do they identify the percentage of time emissions can exceed the limit without violating the regulation. A value of $Z_\alpha = 2.5$, corresponding to compliance with the emission limit 99.4% of averaging periods, will be assumed.

5.6.5 Effects of Variability Parameters on $\hat{\mu}/E_L$

Using equations developed earlier, Eq. (1) can be expressed in terms of CSA data:

$$\hat{\mu} = \frac{E_L}{1 + Z_\alpha \widehat{RSD}_n} = \frac{E_L}{1 + Z_\alpha \dfrac{\widehat{RSD}_1}{\sqrt{n}} \hat{\omega}_n} \tag{6}$$

where $\hat{\mu}$ is the estimated value of the time independent emission parameter needed for compliance with E_L. As discussed earlier, RSD_1 is an estimate of the variability of the reference lots of size l, and $\hat{\omega}_n$ is a measure of the autocorrelation structure of coal samples which are taken to estimate properties of the reference lots.

In the development of Eq. (6), it was assumed that CSA data provide an unbiased estimate of the mean coal properties and that all of the coal sulfur is converted to SO₂ during combustion. The factors of sulfur capture and CSA biases were not incorporated in Eq. (6) because they may vary significantly from site to site. In actual application these factors must be taken into account.

5.6.6 Model Analyses

An analysis of the effects of n and $\hat{\rho}_1$ on the normalized lot size, $\widehat{RSD}_n/\widehat{RSD}_1$, shows that the averaging of reference lots over successively larger lot sizes significantly attenuates the effects of coal sulfur variability. However, these attenuating effects are reduced as autocorrelation increases. The effects of n and $\hat{\rho}_1$ on $\widehat{RSD}_n/\widehat{RSD}_1$ are shown in Figure 15. For example, with zero autocorrelation ($\hat{\rho}_1 = 0$)

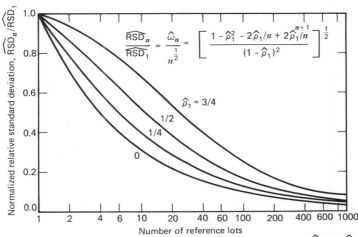

$$\frac{\widehat{RSD}_n}{\widehat{RSD}_1} = \frac{\hat{\omega}_n}{n^{\frac{1}{2}}} = \left[\frac{1 - \hat{\rho}_1^2 - 2\hat{\rho}_1/n + 2\hat{\rho}_1/n^{n+1}}{(1 - \hat{\rho}_1)^2} \right]^{-\frac{1}{2}}$$

$\hat{\rho}_1 = 3/4$

1/2

1/4

0

Normalized relative standard deviation, $\widehat{RSD}_n/\widehat{RSD}_1$

Number of reference lots

Fig. 15 Effect of autocorrelation and number of reference lots on $\widehat{RSD}_n/\widehat{RSD}_1$.

and with a lot size equal to four reference lots ($n = 4$), the relative standard deviation is decreased to 0.50 times its initial value ($\widehat{RSD}_n/\widehat{RSD}_1 = 0.5$). For $n = 4$ and $\hat{\rho}_1 = \frac{1}{2}$, the value of $\widehat{RSD}_n/\widehat{RSD}_1$ is 0.71. For $n = 600$ and autocorrelation of 0 and $\frac{1}{2}$, the values of $\widehat{RSD}_n/\widehat{RSD}_1$ are 0.04 and 0.07, respectively.

The effects of emission averaging time, boiler size, and autocorrelation on $\widehat{RSD}_n$ and $\hat{\mu}/E_L$ are shown by Table 16. The total coal lot size is equal to the emission averaging time multiplied by the rate at which coal is fired in a specific size boiler. The raw and clean coal statistical parameters used in Table 16 are estimated from coal sample test data discussed earlier. While the cleaned coal generally has a higher autocorrelation coefficient than raw coal, this is offset by a lower value of $\widehat{RSD}_1$ for clean coal. For small boilers and short averaging times, the effects of cleaning in reducing fuel sulfur variability are significant. For large boilers and long averaging times, the difference between raw and cleaned coal is small. For example, the mean coal emission parameters required to comply with a 4.0 emission limit with alternative averaging times of 24 hr and 30 days in 25- and 500-MW boilers are

	Required $\hat{\mu}$ for $E_L = 4.0$	
	24 hr	30 days
25 MW		
Raw coal	2.67	3.58
Clean coal	3.08	3.66
500 MW		
Raw coal	3.50	3.90
Clean coal	3.60	3.92

Note that the values of $\hat{\mu}/E_L$ are uniquely determined by $\widehat{RSD}_1$, $\hat{\rho}_1$, and n. The relationship between lot size and averaging time is not defined until a reference lot and a boiler size are defined. Thus the values of $\hat{\mu}/E_L$ as defined by $\widehat{RSD}_1$, $\hat{\rho}_1$, and n may apply to lot sizes, boiler capacities, and averaging periods other than shown in Table 16. For example, assume that coal data are obtained for a reference lot size of 960 tons (871 Mg) and that $\widehat{RSD}_1 = 0.20$ and $\hat{\rho}_1 = 0.25$. If this coal is to be burned in a 100-MW boiler, then for a 24-hr averaging period $n = 1$ and $\hat{\mu}/E_L = 0.667$. This is a lower value of $\hat{\mu}/E_L$ than given in Table 16 for a 100-MW boiler, a 24-hr averaging period, and a reference lot size of 240 tons (218 Mg). This lower value of $\hat{\mu}/E_L$ is a consequence of the larger reference lot size. The relationship between boiler size, averaging period, and coal lot size must be adjusted for different reference lot sizes.

5.6.7 Compliance Determinations

Earlier, a coal sulfur variability model was used to illustrate the effects of $\widehat{RSD}_1$, n, and $\hat{\rho}_1$ on the value of $\hat{\mu}/E_L$ needed for compliance. In practice, this procedure will be reversed. CSA data will be

Table 16 Effect of Coal Statistical Parameters on Fractional Coal Emission Parameter[a]

No. of Ref. Lots, n	Lot Size, tons	Boiler Size, MW[b]	Averaging Period hr (days)	Raw Coal[c]		Clean Coal[d]	
				$\widehat{RSD}_n$	$\hat{\mu}/E_L$	$\widehat{RSD}_n$	$\hat{\mu}/E_L$
1	240	25	24(1)	0.200	0.667	0.120	0.769
25[e]	600	500	3(⅛)	0.145	0.733	0.099	0.802
4	960	100	24(1)	0.120	0.769	0.086	0.823
7	1,680	25	168(7)	0.094	0.810	0.071	0.850
20	4,800	500	24(1)	0.059	0.875	0.045	0.899
30	7,200	25	720(30)	0.047	0.895	0.037	0.915
120	28,800	100	720(30)	0.024	0.945	0.019	0.955
600	144,000	500	720(30)	0.011	0.974	0.009	0.979

[a] Based on time series analysis using an AR(1) model.
[b] Coal firing rate = 0.40 ton/hr-MW.
[c] For raw coal reference lot size of 240 tons: $\widehat{RSD}_1 = 0.20$ and $\hat{\rho}_1 = 0.25$.
[d] For clean coal reference lot size of 240 tons: $\widehat{RSD}_1 = 0.12$ and $\hat{\rho}_1 = 0.50$.
[e] It is assumed that the AR(1) model can be used to estimate the RSD_n for fractional reference lot numbers >1.

used to determine if a specific coal complies with a given emission regulation. This can be accomplished as follows.

The values of $\widehat{RSD}_1$, $\hat{\rho}_1$, and $\hat{\omega}_n$ can be determined from data collected during the previous weeks of operation. Then if $\overline{X}_n$ is the average emission parameter for any given coal lot size L, $\overline{X}_n$ can be expected to comply with E_L with average exceedances of less than α if

$$\overline{X}_n \le \hat{\mu} = \frac{E_L}{1 + Z_\alpha(\widehat{RSD}_1 \hat{\omega}_n / \sqrt{n})} \tag{7}$$

Note that the model can only be used when coal data are available for coal reference lot sizes which are equal to, or less than, the amount of coal burned during the averaging period, that is, for $n \ge 1$. If the reference lot size, l, is greater than L there will generally not be enough data to estimate $\widehat{RSD}_1$ without additional model assumptions. This fact poses problems when making compliance determinations in small boilers. It would require that the coal and utility industry collect data for smaller reference lot sizes than are customary. This can be illustrated by examining the following limiting reference lot sizes which are required for boilers of various capacities and averaging times.

Boiler Capacity, MW	Limiting Reference Lot Size, tons		
	3 hr	24 hr	30 days
5	6	48	1,440
25	30	240	7,200
100	120	960	28,800
500	600	4,800	144,000
1,000	1,200	9,600	288,000

Compliance determinations by the previously outlined method will generally require additional CSA data for boilers of less than about 100 MW with averaging times less than 24 hr. The coal and utility industries seldom obtain coal data on reference lot sizes smaller than 1000 tons (907 Mg). If CSA data are to be used for compliance determinations for small boilers, then either (1) more sophisticated models must be developed to estimate the statistical properties of small lot sizes, (2) CSA data must be collected for smaller reference lots than is customary, or (3) longer averaging times must be used.

One final observation is appropriate. The variability between lot sizes decreases with increasing lot size. This is not always apparent from CSA data because the proportion of the various components which make up RSD_1 may vary with reference lot size. For small reference lots the variance of properties between lots may be large, but the representation variance (the difference between the coal sample properties and the average properties of the reference lot) may be small. Alternatively, for large reference lots, the variance between lots may be small, but the representation variance may be large. The representation variance for large reference lots can be reduced by taking many subsamples from the reference lot to form a single sample for laboratory analysis. The representation variance will then be dependent on the number (frequency) of subsamples and the manner in which the subsamples are collected, composited, and prepared for laboratory analysis. The estimated value of RSD_1 will thus be highly dependent on the CSA procedures used to collect coal data.

5.6.8 Validation and Use of Model

The coal sulfur variability model described here has been shown to provide a reasonable first estimate of the statistical behavior of coal properties in all data sets against which it has been tested. Other coal emission parameters and boiler SO₂ emission data generated by EPA and other organizations agree with the general features of this model. However, insufficient information is known about available coal data sets to allow for a general validation of the model.

In some cases the model may prove inadequate. The model described here does not include the effects of long-term (daily or weekly) autocorrelation effects that are a consequence of specific coal mining or utilization patterns. Coal from many different mines or different mine locations is often fired in the same boiler during the period of a year. These coal-use patterns represent a mixing of different coal populations. While the model was shown to apply to the mixed coal populations over a 30-day period at the R&F coal preparation plant,[60] it may not apply to mixed populations under other conditions.

Another model may be required for assessing compliance in very small boilers with short averaging times. The frequency of coal sampling required to make compliance determinations with the model presented here may be economically prohibitive.

Federal and state SO_2 emission regulations are not now structured to employ statistical methods for determining compliance with SO_2 emission regulations. The EPA and some state air pollution control authorities are currently formulating CSA procedures that are to be used for compliance determinations. The models described here to *estimate* the mean SO_2 emission parameter values needed for compliance will only be valid if the promulgated CSA procedures and the associated emission regulations are based on similar statistical techniques.

In summary, it is emphasized that the mean coal sulfur values needed for compliance must be determined on a case-by-case basis and must be consistent with regulatory requirements. A simplified model was used here to illustrate the relative importance of reference lot size, averaging time, relative standard deviation, and autocorrelation in estimating the statistical properties of emissions. This model is probably valid for short-term daily emission averages with moderately well-behaved populations. Other models may be necessary to account for longer term variations in mixed coal populations or for small lots sizes which are burned over short time periods in small boilers.

5.7 Estimates of Performance

A complete description of plant design procedures and equipment performance specifications is beyond the scope of this chapter. However, preliminary estimates of the ash and sulfur reductions that can be achieved in coal preparation plants can be made from washability data. The effects of sulfur variability on the level of cleaning needed to produce a compliance coal can also be estimated. The best approach to understanding the use of coal washability data and the effects of coal sulfur variability is to present and solve a number of problems.

5.7.1 *Problem 1*

A 500-MW utility boiler in Harrison County, Ohio, must comply with a 4.5 lb $SO_2/10^6$ Btu (1935 ng SO_2/J) emission limit. The regulation is based on a 24-hr emission averaging time. A nearby mine in Harrison County is providing cleaned coal under a long-term contract. The mean properties of this coal are those which are given by the washability data in Table 2.

The plant is currently designed and operated at level 2 conditions which are approximated by crushing to 1½ in. (37.5 mm) and separating at 1.6 s.g. This results in a product with a mean sulfur emission parameter of 4.5 lb $SO_2/10^6$ Btu (1935 ng SO_2/J). CSA data collected for cleaned coal at a reference lot size of 1200 tons (1088 Mg) show that $\widehat{RSD}_1 = 0.10$ and $\hat{\rho}_1 = \frac{1}{2}$. (a) Determine the value of the mean SO_2 emission parameter, $\hat{\mu}$, which will provide an expectation that the emission limit is met 99.4% of the time ($Z_\alpha = 2.5$). (b) What cleaning conditions must be used to produce a compliance coal? (c) Are these cleaning conditions reasonable?

(a) *Determination of $\hat{\mu}$.* It is assumed that the variability in the SO_2 emission parameter can be described by an AR(1) model and the Harrison County boiler has a heat rate of 10,000 Btu/kWh (10.55×10^6 J/kWh). To avoid an iterative solution, a nominal as-fired coal heating value of 12,500 Btu/lb (29,076 J/g) will be used to calculate the coal-firing rate and reference lot size.

The nominal coal firing rate is

$$10{,}000 \text{ Btu/kWh} \times 500{,}000 \text{ kW} \times 1/25 \times 10^6 \text{ Btu/ton} = 200 \text{ tons/hr (181 Mg/hr)}$$

In 24 hr, 4800 tons of coal are burned. The number of coal reference lots is

$$n = \frac{\text{total coal burned over averaging time}}{\text{coal reference lot size}} = \frac{4800}{1200} = 4$$

The cleaned coal function of autocorrelation is calculated using Eq. (4):

$$\hat{\omega}_n(\hat{\rho}_1, n) = \left[\frac{1 - \hat{\rho}_1^2 - 2\hat{\rho}_1/n + 2\hat{\rho}_1^{n+1}/n}{(1 - \hat{\rho}_1)^2} \right]^{1/2}$$

$$= \left[\frac{1 - (0.5)^2 - 2(0.5)/4 + 2(0.5)^5/4}{(1 - 0.5)^2} \right]^{1/2} = 1.436$$

The value of $\widehat{RSD}_n$ is

$$\widehat{RSD}_n = \frac{\widehat{RSD}_1 \hat{\omega}_n}{\sqrt{n}} = \frac{0.10 \times 1.436}{\sqrt{4}} = 0.0718$$

Finally, the value of $\hat{\mu}$ is determined

$$\hat{\mu} = \frac{E_L}{1 + Z_\alpha \widehat{RSD}_n} = \frac{4.5}{1 + 2.5(0.0718)} = 3.82 \text{ lb } SO_2/10^6 \text{ Btu (1642 ng } SO_2/J)$$

Note that the coal currently being produced by the level 2 plant does not satisfy the estimated compliance requirements.

(b) *Estimation of Cleaning Conditions.* Estimates of the required cleaning conditions are made using Table 2. Note that the coal properties in Table 2 are given on a moisture-free basis and that the moisture-free and as-fired sulfur emission parameters differ in most cases by less than 1%.

(i) Assume crushing to 1½-in. (37.5-mm) top size. A linear interpolation of data in Table 2 can be performed to determine the float specific gravity required for a 3.8 lb SO₂/10⁶ Btu (1634 ng SO₂/J) product.

The required float specific gravity is

$$1.6 - \left[(1.6 - 1.4)\frac{(4.5 - 3.8)}{(4.5 - 3.6)}\right] = 1.44$$

It is difficult to control separating conditions to 1.44. After rounding down to 1.4 to ensure a satisfactory product, the properties of the cleaned coal are found from Table 2 to be

mean emission parameter = 3.6 lb SO₂/10⁶ Btu (1548 ng SO₂/J)
Btu recovery = 79.8%
heat content = 13,595 Btu/lb (31,622 J/g)
ash content = 6.0%

(ii) Crushing to ⅜-in. (9.5-mm) top size. An inspection of Table 2 shows that a 3.8 lb SO₂/10⁶ Btu (1634 ng SO₂/J) product can be obtained at a float specific gravity between 1.9 and 1.6. The required float specific gravity is

$$1.9 - \left[(1.9 - 1.6)\frac{(3.9 - 3.8)}{(3.9 - 3.6)}\right] = 1.8$$

The coal properties at this specific gravity are:

$$\text{Btu recovery} = 97.7 - \left[(97.7 - 95.1)\frac{(1.9 - 1.8)}{(1.9 - 1.6)}\right] = 96.8$$

$$\text{heat content} = 13,375 - \left[(13,375 - 13,534)\frac{(1.9 - 1.8)}{(1.9 - 1.6)}\right]$$

$$= 13,428 \text{ Btu/lb (31,234 J/g)}$$

$$\text{ash content} = 7.5 - \left[(7.5 - 6.4)\frac{(1.9 - 1.8)}{(1.9 - 1.6)}\right] = 7.1\%$$

(c) *Reasonableness of Required Cleaning Conditions.* Crushing to a top size of 1½ in. (37.5 mm) and separating at 1.4 s.g. results in excessive coal energy losses. Crushing to ⅜ in. (9.5 mm) and separating at 1.8 s.g. provides a 96.8% Btu recovery. However, crushing to ⅜ in. (9.5 mm) may produce a large amount of fine coal and may result in high cleaning costs. Cleaning can probably be used to produce a compliance coal, but detailed plant design studies must be made to evaluate cost and performance trade-offs between the various coal top sizes and cleaning conditions that will provide a compliance coal. Studies must also be made on the feasibility of either modifying the existing level 2 preparation plant or building a new plant. The use of alternative coals must also be considered.

5.7.2 Problem 2

The utility in problem 1 is evaluating coals from alternate sources to locate potential compliance fuels. A raw coal from the Pittsburgh seam in West Virginia is available at an attractive price. The mean emission parameter of the coal is 3.8 lb SO₂/10⁶ Btu (1364 ng SO₂/J). CSA data for 1200-ton (1088-Mg) reference lot sizes show that $\widehat{RSD}_1 = 0.16$ and $\hat{\rho}_1 = 0.25$. Can this coal meet the conditions expected for a compliance fuel?

As in problem 1, four reference lots are to be fired during the averaging period. The raw coal function of autocorrelation is

$$\hat{\omega}_n(0.25, 4) = \left[\frac{1 - (0.25)^2 - 2(0.25)/4 + 2(0.25)^5/4}{(1 - 0.25)^2}\right]^{1/2} = 1.202$$

The value of $\widehat{RSD}_n$ is

$$\widehat{RSD}_n = \frac{0.16 \times 1.202}{\sqrt{4}} = 0.0962$$

Finally, the value of $\hat{\mu}$ is determined to be:

$$\hat{\mu} = \frac{4.5}{1 + 2.5(0.0962)} = 3.63 \text{ lb SO}_2/10^6 \text{ Btu (1561 ng SO}_2/\text{J)}$$

For these sampling conditions, the coal cannot be shown to meet the assumed compliance require-ments. However, by increasing the coal sampling frequency (reducing the reference lot size) it is possible that this coal can be used as a compliance fuel. For example, assume that, at a reference lot size of 480 tons (435 Mg), it is found that $\widehat{RSD}_1 = 0.17$ and $\hat{\rho}_1 = 0.35$. The required value of $\hat{\mu}$ for these conditions at $n = 10$ can be calculated to be

$$\hat{\omega}_n(0.35, 10) = 1.383$$
$$\widehat{RSD}_n = 0.0743$$
$$\hat{\mu} = 3.80 \text{ lb SO}_2/10^6 \text{ Btu (1634 ng SO}_2/\text{J)}$$

For this sampling frequency the raw West Virginia coal is seen to meet the assumed compliance requirements.

5.7.3 Practical Considerations

This chapter has considered the use and limitations of coal cleaning as a technology for complying with SO_2 emission regulations. However, selection of the least-cost option for complying with an SO_2 emission regulation at a given site is complex. Many different coals can be used. Blending of low- and high-sulfur coals, either cleaned or uncleaned, can also be considered. The use of flue gas desulfuriza-tion or other SO_2 emission reduction technologies is also possible. The final choice will depend on a careful consideration of all options, taking the major site-specific design factors into account. Once the number of options has been narrowed, preliminary design, performance, and cost studies can be conducted to develop information on which investment decisions can be made. Within this overall context, an evaluation of coal washability data and potential preparation plant costs is a necessary first step in evaluating the coal cleaning options. Information in this chapter should aid in selection of appropriate technologies for air pollution control.

6 TABLE OF CONVERSION FACTORS

Multiply English Unit	By	To Obtain SI Unit
Pount (lb)	453.59	Gram (g)
Ton (2000 lb)	0.907	Million grams (Mg) = metric tonne
Inch (in.)	0.0254	Millimeter (mm)
Foot (ft)	3.048	Meter (m)
British thermal unit	1054.88	Joule (J)
Btu/lb	2.326	J/g
lb/10^6 Btu	429.907	ng/J
\$/ton	1.1025	\$/Mg
\$/10^6 Btu	0.9480	mill/MJ = 10^{-3} \$/MJ
lb/ft^3	0.0160	g/cm^3

7 TYLER SCREEN SIZE MESH OPENINGS

Mesh Size	Sieve Opening	
	in.	mm
14	0.0469	1.18
28	0.0234	0.60
48	0.0117	0.30
100	0.0059	0.15
200	0.0029	0.075
270	0.0021	0.053
325	0.0017	0.045

REFERENCES

1. Leonard, J. W. (ed.), *Coal Preparation,* 4th edition, The American Institute of Mining, Metallurgical and Petroleum Engineers, Inc., New York, 1979.

2. Anon., *Demonstrated Reserve Base of Coal in the United States on January 1, 1979,* DOE/EIA-0280(79), U.S. Department of Energy, Energy Information Administration, Washington, D.C., May 1981.

3. Stach, E., et al., *Stach's Textbook of Coal Petrology,* 2nd edition, Gebruder Borntraeger, Berlin-Stuttgart, 1975.

4. Attar, A., Chemistry, Thermodynamics and Kinetics of Reactions of Sulfur in Coal-Gas Reactions: A Review, *FUEL* **57** (April 1978).

5. Mezey, E. J., Singh, S., and Hissong, D. W., *Fuel Contaminants, Vol. 1, Chemistry (Battelle Columbus Laboratory),* EPA-600/2-76-177a, NTIS PB 256020, U.S. Environmental Protection Agency, Industrial Environmental Research Laboratory, Research Triangle Park, NC, July 1976.

6. Gluskoter, H. J., et al., *Trace Elements in Coal: Occurrence and Distribution (Illinois State Geological Survey),* EPA-600/7-77-064 (NTIS PB 270922), U.S. Environmental Protection Agency, Industrial Environmental Research Laboratory, Research Triangle Park, NC, June 1977.

7. Wells, M. A., et al., *Coal Resources and Sulfur Emission Regulations: A Background Document, Draft Report (Teknekron Research, Inc.),* EPA Contract 68–02–3136, U.S. Environmental Protection Agency, Industrial Environmental Research Laboratory, Research Triangle Park, NC, February 1980.

8. Cavallaro, J. A., Johnston, M. T., and Deurbrouck, A. W., *Sulfur Reduction Potential of U.S. Coals: A Revised Report of Investigations (U.S. Bureau of Mines, RI 8118),* EPA-600/2-76-091 (NTIS PB 252965), U.S. Environmental Protection Agency, Industrial Environmental Research Laboratory, Research Triangle Park, NC, April 1976.

9. Buroff, J., et al., *Technology Assessment Report for Industrial Boiler Applications: Coal Cleaning and Low Sulfur Coal (Versar, Inc.),* EPA-600/7-79-178c (NTIS PB 80-174055), U.S. Environmental Protection Agency, Industrial Environmental Research Laboratory, Research Triangle Park, NC, December 1979.

10. Nunenkamp, D. C., *Coal Preparation Environmental Engineering Manual (J. J. Davis Assoc.),* EPA-600/2-76-138 (NTIS PB 262-716/AS), U.S. Environmental Protection Agency, Industrial Environmental Research Laboratory, Research Triangle Park, NC, May 1976.

11. Phillips, P. J., *Coal Preparation for Combustion and Conversion,* Electric Power Research Institute, EPRI AF-791, Palo Alto, CA, May 1978.

12. Buder, M., et al., *Environmental Control Implications of Generating Electric Power from Coal, 1977 Technology Status Report, Appendix A, Part 1, Coal Preparation and Cleaning Assessment Study: Report for Department of Energy (Bechtel Corp.),* December 1977.

13. Onursal, B., et al., *Assessment of Coal Cleaning Technology: An Evaluation of Dense-Medium Cyclone Circuits for Removal of Sulfur from Fine Coal, Draft Report (Versar, Inc.),* EPA Contract 68-02-2199, U.S. Environmental Protection Agency, Industrial Environmental Research Laboratory, Research Triangle Park, NC, December 1980.

14. Fuerstenau, D. W., *Froth Flotation 50th Anniversary Volume,* The American Institute of Mining, Metallurgical and Petroleum Engineers, Inc., New York, 1962.

15. Wheelock, T. D., and Markuszewski, R., *Physical and Chemical Coal Cleaning, Conference on the Chemistry and Physics of Coal Utilization,* Energy Research Center, West Virginia University, Morgantown, WV, June 1980.

16. Miller, K. J., *Coal-Pyrite Flotation: A Modified Technique Using Concentrated Second-Stage Pulp,* U.S. Bureau of Mines, TPR 91 (1975).

17. Onursal, B., and McCandless, L. C., *Assessment of Coal Cleaning Technology: An Evaluation of Froth Flotation Circuits for Removal of Sulfur from Coal, Draft Report (Versar, Inc.),* EPA Contract 68-02-2199, U.S. Environmental Protection Agency, Industrial Environmental Research Laboratory, Research Triangle Park, NC, July 1980.

18. Abel, W. T., et al., *Removing Pyrite from Coal by Dry-Separation Methods, Report of Investigations 7732,* U.S. Bureau of Mines, 1973.

19. Rich, S. R., *Economic Analysis and Evaluation of Sulfur and Ash Reduction Capability of the Advanced Energy Dynamics Dry Electrostatic Coal Cleaning System—Tested on 27 Bituminous Coals (Advanced Energy Dynamics, Inc.), Draft Final Report,* EPA Contract 68-02-3563, U.S. Environmental Protection Agency, Industrial Environmental Research Laboratory, Research Triangle Park, NC, June 1981.

20. Wechsler, I., Doulin, J., and Eddy, R., *Coal Preparation Using Magnetic Separation, Volume 3 (Sala Magnetics, Inc.),* EPRI CS-1517, Vol. 3, Electric Power Research Institute, July 1980.

21. Hise, E. C., Wechsler, I., and Doulin, J. M., *Separation of Dry Crushed Coals by High-Gradient Magnetic Separation,* ORNL-5571, Oak Ridge National Laboratory, October 1979.

22. Karlson, F. V., et al., *Coal Preparation Using Magnetic Separation, Volume 5: Evaluation of Magnetic Coal Desulfurization Concepts (Bechtel National, Inc.),* EPRI CS-1517, Vol. 5, Electric Power Research Institute, July 1980.

23. Contos, G. Y., Frankel, I. F., and McCandless, L. C., *Assessment of Coal Cleaning Technology: An Evaluation of Chemical Coal Cleaning Processes (Versar, Inc.),* EPA-600/7-78-173a (NTIS PB 289493), U.S. Environmental Protection Agency, Industrial Environmental Research Laboratory, Research Triangle Park, NC, August 1978.

24. Stambaugh, E. P., et al., *Process Improvement Studies on the Battelle Hydrothermal Coal Process (Battelle Columbus Laboratories),* Draft Final Report on EPA Contract 68-02-2187, U.S. Environmental Protection Agency, Industrial Environmental Research Laboratory, Research Triangle Park, NC, June 1980.

25. Kilgroe, J. D., Coal Cleaning for Compliance with SO_2 Emission Regulations, *Proceedings of the Fourth Symposium on Coal Utilization,* NCA/BCR Coal Conference and Expo IV, October 1977.

26. Tarkington, T. W., Kennedy, F. M., and Patterson, J. G., *Evaluation of Physical/Chemical Coal Cleaning and Flue Gas Desulfurization (TVA),* EPA-600/7-79-250 (NTIS PB 80-147622), U.S. Environmental Protection Agency, Industrial Environmental Research Laboratory, Research Triangle Park, NC, November 1979.

27. Lemon, A. W., Jr., et al., *Environmental Assessment of Coal Cleaning Processes: Final Report (Battelle Columbus Laboratories),* EPA-600/7-82-024 (NTIS PB 82-222910), U.S. Environmental Protection Agency, Industrial Environmental Research Laboratory, Research Triangle Park, NC, April 1982.

28. U.S. Environmental Protection Agency, Coal Mining Point Source Category: Effluent Limitation Guidelines for Existing Sources, Standards of Performance for New Sources and Pretreatment Standards (Proposed Regulation), 40 CFR Part 434, 46 FR 3136, January 13, 1981.

29. U.S. Environmental Protection Agency, *Guidelines and Standards for Coal Mining,* 40 CFR 434, June 27, 1980.

30. Wewerka, E. M., et al., *Environmental Contamination from Trace Elements in Coal Preparation Wastes: A Literature Review and Assessment (Los Alamos Scientific Laboratory),* EPA-600/7-76-007 (NTIS PB 267339), U.S. Environmental Protection Agency, Industrial Environmental Research Laboratory, Research Triangle Park, NC, August 1976.

31. Wewerka, E. M., et al., *Trace Element Characterization of Coal Wastes—Second Annual Progress Report (Los Alamos Scientific Laboratory),* EPA-600/7-78-028a (NTIS PB 284450), U.S. Environmental Protection Agency, Industrial Environmental Research Laboratory, Research Triangle Park, NC, July 1978.

32. Wewerka, E. M., et al., *Trace Element Characterization of Coal Wastes, Third Annual Progress Report (Los Alamos Scientific Laboratory),* EPA-600/7-79-144 (NTIS PB 80-166150), U.S. Environmental Protection Agency, Industrial Environmental Research Laboratory, Research Triangle Park, NC, June 1979.

33. Moore, J. C., and Kilgroe, J. D., Federal Environmental Regulations for Coal Preparation Plants, *Proceedings of the 4th International Coal Utilization Exhibition and Conference,* Houston, TX, November 1981.

34. Kilgroe, J. D., and Lagemann, R. C., The Technology and Costs of Physical Coal Cleaning for Controlling Sulfur Dioxide Emissions, paper prepared for the Third Seminar on Desulphurization of Fuels and Combustion Gases, Salzburg, Austria, May 1981.

35. Buder, M. K., et al., *Impact of Coal Cleaning on the Cost of New Coal-Fired Power Generation (Bechtel National, Inc.),* EPRI CS-1622, Electric Power Research Institute, Palo Alto, CA, March 1981.

36. Chapman, R. A., and Wells, M. A., *Coal Resources and Sulfur Emission Regulations: A Summary of Eight Eastern and Midwestern States (Versar, Inc.),* EPA-600/7-81-086 (NTIS PB 81-240319), U.S. Environmental Protection Agency, Industrial Environmental Research Laboratory, Research Triangle Park, NC, May 1981.

37. Holt, E. C., Jr., *An Engineering/Economic Analysis of Coal Preparation Plant Operation and Cost (Hoffman-Muntner Corp.),* EPA-600/7-78-124 (NTIS PB 285251), U.S. Environmental Protection Agency, Office of Energy, Minerals and Industry, Washington, D.C., July 1978.

38. *1980 Keystone Coal Industry Manual,* McGraw-Hill, New York, 1980, p. 719.

39. Phillips, P. J., and Cole, R. M., Economic Penalties Attributable to Ash Content of Steam Coals, presented at the Coal Utilization Symposium, AIME Annual Meeting, New Orleans, LA, February 1979.

40. *Impact of Cleaned Coal on Power Plant Performance and Reliability,* CS1400, Research Project 1030-6, Electric Power Research Institute, Palo Alto, CA, April 1980.

41. Isaacs, G., Ressl, R., and Spaite, P., *Cost Benefits Associated with the Use of Physically Cleaned Coal (PEDCo Environmental, Inc.),* EPA-600/7-80-105 (NTIS PB 81-113953), U.S. Environmental Protection Agency, Industrial Environmental Research Laboratory, Research Triangle Park, NC, May 1980.

42. Blackmore, G., A Discussion on Coal Economics as It Relates to Electric Utilities and Energy Independence, presented at the Consolidation Coal Company's Annual Management Meeting, August 19–20, 1980.

43. Barrett, R. E., Holt, E. C., and Cole, R. M., Examining Relationships Between Coal Characteristics and the Performance of TVA Power Plants, Part 1: Approach and Some Early Results, presented at the 1980 Joint Power Conference, Phoenix, AZ, September 29–October 2, 1980.

44. Barrett, R. E., and Frank, R. L., Examining Relationships Between Coal Characteristics and the Performance of TVA Power Plants, Part 2: Boiler Efficiency, presented at the 1981 American Power Conference, Chicago, Illinois, April 27–29, 1981.

45. Barrett, R. E., and Frank, R. L., Examining Relationships Between Coal Characteristics and the Performance of TVA Power Plants, Part 3: Boiler Availability, presented at the 1981 Joint Power Generation Conference, St. Louis, Missouri, October 4–8, 1981.

46. Holt, E. C., and Barrett, R. E., Examining Relationships Between Coal Characteristics and the Performance of TVA Power Plants, Part 4: Maintenance Costs, presented at the 1981 Joint Power Generation Conference, St. Louis, Missouri, October 4–8, 1981.

47. Holt, E. C., Effect of Coal Quality on Maintenance Costs at Utility Plants. *Mining Congress Journal* (May 1982).

48. Hoffman, L., Aresco, S. J., and Holt, E. C., Jr., *Engineering/Economic Analyses of Coal Preparation with SO₂ Cleanup Processes for Keeping High Sulfur Coals in the Energy Market (Hoffman-Muntner Corporation for U.S. Bureau of Mines, Contract JO155171),* EPA-600/7-78-002 (NTIS PB 276769), U.S. Environmental Protection Agency, Office of Energy, Minerals and Industry, Washington, D.C., January 1978.

49. Kilgroe, J., Combined Coal Cleaning and FGD, in *Proceedings: Symposium on Flue Gas Desulfurization—Las Vegas, Nevada, March 1979; Volume I (Research Triangle Institute),* EPA-600/7-79-167a (NTIS PB 80-133168), U.S. Environmental Protection Agency, Industrial Environmental Research Laboratory, Research Triangle Park, NC, July 1979.

50. Hazard, H. R., *Influence of Coal Mineral Matter on Slagging of Utility Boilers (Battelle Columbus Laboratories),* EPRI CS-1418, Electric Power Research Institute, June 1980.

51. The Clean Air Act, U.S. Environmental Protection Agency, Washington, D.C., December 1970.

52. The Clean Air Act as amended August 1977, Serial 95-11, U.S. Government Printing Office, Washington, D.C., November 1977.

53. Woodard, K. R., Quidley, D., and Hester, C., *Analysis of State and Federal Sulfur Dioxide Emission Regulations for Combustion Sources,* EPA-450/2-81-079, U.S. Environmental Protection Agency, Office of Control Programs Development Division, Research Triangle Park, NC, November 1981.

54. U.S. Environmental Protection Agency, *Standards of Performance for New Stationary Sources,* 42 CFR 466, August 17, 1971.

55. U.S. Environmental Protection Agency, *New Stationary Sources Performance Standards; Electric Utility Steam Generating Units,* 40 CFR 60, June 11, 1979.

56. Anon., *Concept Paper: Emission Reduction Banking and Trading Concepts,* OPA-113-0, U.S. Environmental Protection Agency, Office of Planning and Evaluation, Washington, D.C., June 1980.

57. Maxwell, J. D., et al., *Physical Coal Cleaning Computer Economics (Tennessee Valley Authority),* EPA Draft Project Report on Interagency Agreement 79DX0511, Industrial Environmental Research Laboratory, Research Triangle Park, NC, July 1982.

58. Cheng, B., et al., *Variability and Correlation in Coal, Draft Report (Versar, Inc.),* EPA Contract 68-02-2199, U.S. Environmental Protection Agency, Industrial Environmental Research Laboratory, Research Triangle Park, NC, October 1981.

59. Box, G. E. P., and Jenkins, G. M., *Time Series Analysis: Forecasting and Control,* revised edition, Holden-Day, San Francisco, 1976.

60. Cheng, B. H., et al., Time Series Analysis of Coal Data from Preparation Plants, working draft of paper submitted to *Journal of the Air Pollution Control Assoc.,* Versar, Inc., Springfield, VA, August 1981.

61. Sargent, D. H., et al., *Effect of Physical Coal Cleaning on Sulfur Content and Variability (Versar,*

Inc.), EPA-600/7-80-107 (NTIS PB 80-210529), U.S. Environmental Protection Agency, Research Triangle Park, NC, May 1980.

62. Warhlic, G. H., et al., *A Statistical Study of Coal Sulfur Variability and Related Factors (Foster Associates, Inc.*), U.S. Environmental Protection Agency, Office of Air Quality Planning and Standards, Research Triangle Park, NC, May 1980.

CHAPTER 18

EMISSION CONTROL IN INTERNAL COMBUSTION ENGINES

EDWARD DeKIEP

DONALD J. PATTERSON

Department of Mechanical Engineering and Applied Mechanics
University of Michigan
Ann Arbor, Michigan

1 INTRODUCTION

The pollutant emissions from internal combustion engines have become of increasing concern over the years because of health hazard and environmental damage. Table 1 lists several sources of man-made air pollution and their contribution prior to pollution control efforts and more recently. The portion attributed to human activity as a percentage of the total of man-made and natural sources is given also. On a mass basis, transportation is by far the largest source, although the majority in this category arises from relatively nontoxic carbon monoxide.

In recognition of the contribution of transportation sources to air pollution, regulations have been imposed both on motor vehicles and engines by the federal government beginning in 1968. The state of California has its own regulations which in some cases are more strict than the federal requirements. Future federal regulations may include some stationary internal combustion engines. The remainder of this chapter will be devoted to internal combustion engine emissions, their sources, methods of control, and test procedures.

1.1 Internal Combustion Engine Emissions

An estimate of the major emissions from the various transportation sources is given in Table 2. Compared to other sources, those from highway vehicles are dominant, especially in the categories of hydrocarbons, carbon monoxide, and nitrogen oxides.

Hydrocarbons is the name given to an unspecified mixture of hydrocarbon compounds. These arise from gasoline and diesel fuels and their combustion products. These fuels are made up of hundreds of different hydrocarbon compounds. To simplify matters, hydrocarbons are usually reported in terms of a single hydrocarbon compound equivalent such as methane (CH_4), propane (C_3H_8), or hexane (C_6H_{14}). For EPA certification testing hydrocarbons are taken to be CH_x, where x is a value between 1.85 and 2.33.

The engineering terminology for an unknown mixture of nitrogen oxides is NO_x. Commonly NO_x emissions are predominately nitric oxide (NO) with a small fraction (<10%) of nitrogen dioxide (NO_2). In the atmosphere, NO is converted to NO_2. Federal regulations mandate that NO_x mass emissions be reported as NO_2.

Engine exhaust emissions from a typical spark-ignited (SI) and compression-ignited (CI) engine are shown in Table 3. Values reflect no exhaust aftertreatment.

Table 1 Estimated U.S. Man-Made Air Pollution, tg/yr

Source	Particulate Matter		Sulfur Oxides		Nitrogen Oxides		Hydrocarbons		Carbon Monoxide		Source Total		Source Total, %	
	1960	1980	1960	1980	1960	1980	1960	1980	1960	1980	1960	1980	1960	1980
Transportation	0.7	1.4	0.4	0.9	4.6	9.1	10.2	7.8	63.4	69.1	79.3	88.3	48	55
Stationary source fuel combustion	4.9	1.4	12.7	19.0	6.8	10.6	0.3	0.2	3.3	2.1	28.0	33.3	17	21
Industrial processes	12.0	3.7	5.6	3.8	0.6	0.7	6.6	10.8	9.3	5.8	34.1	24.8	21	16
Solid waste disposal	0.9	0.4	0	0	0.3	0.1	1.4	0.6	5.1	2.2	7.7	3.3	5	2
Miscellaneous	1.7	0.9	0.5	0	0.4	0.2	3.1	2.4	9.7	6.2	15.4	9.7	9	6
Man-made total	20.2	7.8	19.2	23.7	12.7	20.7	21.6	21.8	90.8	85.4	164.5	159.4	100	100
Man-made as percentage of total including natural sources	50		82		20		23		9.1					

Note: One teragram equals 10^{12} g (10^6 metric tons).
Data from Refs. 1 and 2.

Table 2 Estimated U.S. Transportation Source Emission Distribution—1980, tg/yr

Transportation	Particulate Matter	Sulfur Oxides	Nitrogen Oxides	Hydro-carbons	Carbon Monoxide
Highway vehicles	1.1	0.4	6.6	6.4	61.9
Aircraft	0.1	0.0	0.1	0.2	1.0
Railroads	0.1	0.1	0.7	0.2	0.3
Vessels	0.0	0.3	0.2	0.5	1.5
Other off-highway vehicles	0.1	0.1	1.5	0.5	4.4
Total	1.4	0.9	9.1	7.8	69.1

Data from Ref. 2.

Note that hydrocarbon and carbon monoxide emissions are significantly higher for SI engines. In part, the low ppm values for CI engines arise from high dilution (lean mixtures) under part-load conditions. Table 4 summarizes federal exhaust emissions standards which must be met for automobiles, light trucks, motorcycles, and heavy-duty engines and vehicles. Automotive standards refer to model year. Note that the heavy-duty standards are expressed as g/bhp-hr. The variability in trucks and their engines has precluded heavy-duty standards tied to vehicle packages.

To reach the highest levels of control for automobiles requires at least a 96% reduction in HC and CO, a 75% reduction in NO_x, and up to a 90% reduction in particulates compared to an uncontrolled vehicle. Such a high level of control requires substantial modifications to internal combustion engines and these often require compromises in fuel economy, performance, and drivability.

1.2 Sources

The principal sources of pollutant emissions from internal combustion engines differ between spark-ignited (SI) and diesel engines (CI). This is because SI engines operate with essentially homogeneous fuel–air mixtures while CI engines operate with heterogeneous mixtures. Because of these differences the sources will be discussed in separate sections for SI and CI engines.

1.2.1 *Spark-Ignition Engines (SI)*

The pollution emissions from spark-ignited engines are classified as hydrocarbons, carbon monoxide, oxides of nitrogen, and unregulated emissions. These emissions arise mainly from surface effects and from bulk gas reactions during combustion. Blowby and scavenging emissions are lesser sources.

For a properly operating SI engine, wall quenching is the major source of unburned hydrocarbons leading to emission levels representing about 1% unburned fuel and the production of up to 200 different HC species. Quenching is an extinction of the combustion process near or at the cool chamber surfaces (single-wall) and within narrow crevices less than 1.5 mm wide (two-wall). Crevice quenching

Table 3 Typical Exhaust Constituents

	Idle, ppm	Intermediate Load		High Load	
		ppm	g/kWh	ppm	g/kWh
Hydrocarbons as $CH_{1.85}$					
SI	4,000	2,400	7.5	6,000	12
CI	200	50	0.5	100	0.3
Carbon monoxide					
SI	10,000	10,000	73	60,000	240
CI	150	700	3.8	500	5
Oxides of nitrogen					
SI	100	2,500	17	500	2.2
CI	50	1,700	14	1,400	14
CI smoke particulate concentration, g/m³	0.05	0.6		1.2	

Table 4 Federal Gasoline and Diesel Engine and Vehicle Standards

Light-Duty Passenger Car, *Federal Register*, Vol. 42, no. 124, June 28, 1977 and Vol. 44, no. 23, Feb. 1, 1979 [Under 2720 kg (6000 lb$_m$) GVW]

Year	Test Procedure	Hydro-carbons	Carbon Monoxide	Oxides of Nitrogen	Particulates	Evaporative Losses
1981	FTP	0.41 g/mile	3.4 g/mile	1.0 g/mile	—	2 g/test
1982–1984	FTP	0.41 g/mile	3.4 g/mile	1.0 g/mile	0.6 g/mile	2 g/test
1985 and later	FTP	0.41 g/mile	3.4 g/mile	1.0 g/mile	0.2 g/mile	2 g/test

Light-Duty Trucks, *Federal Register*, Vol. 42, no. 124, June 28, 1977 [Under 3860 kg (8500 lb$_m$) GVW]

| 1979–1983 | FTP | 1.7 g/mile | 18 g/mile | 2.3 g/mile | — | 2 g/test |
| 1984 | FTP | 0.8 g/mile | 10 g/mile | 2.3 g/mile | — | 2 g/test |

Motorcycles, *Federal Register*, Vol. 42, no. 3, Jan. 5, 1979 (50 cm³ and Greater Displacement)

1980–1982	FTP	5 g/mile	12 g/mile	—	—	—
1983–1984	FTP	5 g/mile	12 g/mile	—	—	6 g/test
1985 and later	FTP	5 g/mile	12 g/mile	—	—	2 g/test

Heavy Duty Gasoline and Diesel Engines, *Federal Register*, Vol. 45, no. 14, Jan. 21, 1980 and Vol. 42, no. 174, Sept. 8, 1977

		Hydro-carbons	Carbon Monoxide	Hydrocarbons plus Oxides of Nitrogen
1979–1983	13 mode (A)	1.5 g/bhp-hr	25 g/bhp-hr	10 g/bhp-hr
	(B)	No standard	25 g/bhp-hr	5 g/bhp-hr
1984+	Transient	1.3 g/bhp-hr	15.5 g/bhp-hr[a]	10.7 g/bhp-hr

[a] Idle CO limit of 0.47% for gasoline engines.

is the dominant form. Adsorption and desorption of volatile fuel hydrocarbons by the lubricating oil film on the cylinder wall may also contribute to hydrocarbon emissions from surface sources. Likewise, combustion chamber deposits may trap fuel particles which are later expelled as unburned hydrocarbons. The surface effect of wall quenching is also a source for the incomplete combustion production of carbon monoxide (CO). This is the major source of CO when the engine is operating with a lean fuel–air mixture and flame propagation is otherwise complete. Wall and crevice quenching are sources for unregulated emissions of various partial combustion products including aldehydes and alcohols.

The bulk gas is normally a minor source of hydrocarbon emissions and theoretically complete combustion of the fuel is thermodynamically favored. However, rich or lean mixture operation especially when exhaust residuals are high will cause incomplete flame propagation. Bulk gas hydrocarbon emissions are associated with the startup, warmup, and deceleration modes of operation, and when present may represent 10–100% of the fuel, thus dominating the HC emissions. Incomplete combustion in the bulk gases is the major source for carbon monoxide with rich mixtures due to lack of oxygen. Incomplete flame propagation is a major source of CO for mixtures near the lean combustion limit and is a major source of aldehydes and other oxygenates.

The oxides of nitrogen are formed in the bulk gases and their amounts depend on temperature, time, and fuel–air ratio. The three chemical equations below which are termed the Zeldovich mechanism are thought to govern NO formation.

$$O_2 = 2O \tag{1}$$
$$O + N_2 = NO + N \tag{2}$$
$$N + O_2 = NO + O \tag{3}$$

NO is formed in the hot products after the passage of the flame. Fuel–air mixtures just lean of chemically correct produce the highest NO_x concentrations because of the availability of some oxygen coupled with relatively high combustion temperatures. NO_x levels may reach as high as 1 mole % of the combustion products. In an engine combustion chamber, concentrations of NO_x are highest in the portion of the charge that is burned first because of the increased time for reaction and the temperature stratification associated with progressive burning in the confined combustion chamber. Blumberg and Kummer[3] describe a theoretical computer model for the prediction of NO_x formation in spark-ignition engines which uses a modification of the Zeldovich mechanism.

Blowby past the piston rings which is vented to the atmosphere will result in hydrocarbon emissions. Blowby can amount to 20–25% of all hydrocarbon emissions from an uncontrolled vehicle. Small amounts of carbon monoxide and NO_x are also present in blowby gases.

Scavenging losses are the predominate source of hydrocarbon emissions for carbureted two-stroke, spark-ignited engines. Scavenging losses can also contribute to hydrocarbon emissions in four-stroke engines with high overlap camshafts or with turbo- or supercharging. In these cases, part of the fuel–air charge passes from the intake port or valve into the exhaust, thereby escaping combustion completely.

Because of the high volatility of gasoline, evaporation from the fuel system is a source of hydrocarbon emissions.

As the engine becomes older, emissions can increase due to the engine's deteriorating condition. As piston ring wear increases, the amount of blowby will increase. Exhaust valve leakage may develop into a major source of hydrocarbon emissions. An ignition system that is in poor condition may permit misfiring with a resulting large HC emission increase.

1.2.2 Diesel Engines (CI)

The main pollutant emissions of diesel engines are the same as those of spark-ignition engines (hydrocarbons, carbon monoxide, oxides of nitrogen, and unregulated emissions) and, in addition, carbon particulates (black smoke). It is necessary to consider the combustion process of the CI engine to understand the sources of emissions. This section will describe separately the combustion process and emission sources for the two types of CI engines: the direct injection (DI) and indirect injection (IDI).

1.2.2.1 Direct Injection Engines. In DI engines the fuel is sprayed under high pressure of as much as 100,000 kPa into the compressed air in the cylinder. The spray breaks up into droplets which are dispersed by the swirling air, Figure 1. Consequently, the fuel–air ratio varies greatly throughout the spray. Henein[4] has divided the spray into four regions: lean flame region (LFR), lean flame out region (LFOR), spray core, and spray tail. Examination of these regions shows how pollutant emissions arise in direct injection engines.

The lean flame region is so named because the mixture is lean. Ignition starts at the downstream edge of this region in premixed zones and combustion is complete throughout. Thus emissions of hydrocarbons and carbon monoxide are negligible. However, the bulk of nitrogen oxides are formed here due to high temperatures and plentiful oxygen. Moreover, since the combustion is initiated in this region a relatively longer time is available for NO_x formation by the mechanism discussed previously in regard to SI engines.

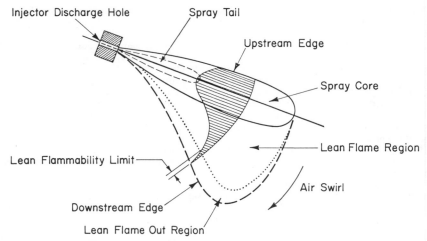

Fig. 1 Schematic diagram for a fuel spray injected into swirling air. The length of lines in the shaded region of the spray is proportional to the *F/A* ratio.

The lean flame out region is located at the far downstream edge of the spray. The fuel–air ratio in the LFOR is too lean to support complete combustion. The size of the LFOR varies and depends on the temperature and pressure in the combustion chamber, the air swirl, and the volatility of the fuel. The LFOR is reduced in size with increase in temperature and pressure which allow combustion in leaner mixtures. The emissions formed in this region include unburned hydrocarbons, carbon monoxide, and other intermediate oxidation products, including aldehydes.

The spray core contains larger fuel droplets than the LFOR. As the temperature increases due to combustion in the LFR, the core droplets partially or completely evaporate. The combustion in this region depends on the local fuel–air ratio which is affected most by the interaction between the spray core and the air swirl in the cylinder. At part load the spray core region contains adequate oxygen for complete combustion and is therefore a source of NO_x. Near full load the spray core is fuel rich and incomplete combustion produces hydrocarbons, carbon monoxide, oxygenated compounds, and carbon particulates. Very little NO_x is produced under full-load operation in the core.

The spray tail is the last part of the fuel to be injected. Large droplets are usually formed due to decreased fuel injection pressure and increased cylinder pressure. Penetration and mixing with air are poor. The high temperature of the surrounding gases causes the fuel from the spray tail to quickly evaporate and decompose. The resulting emissions are hydrocarbons, carbon monoxide, aldehydes, and carbon particulates.

Some injection systems produce "afterinjection" which occurs after the main injection and arises from hydraulic fuel line transients. The amount of fuel injected is usually small, but the injection occurs late in the expansion stroke, with little atomization and penetration. The fuel is evaporated and decomposed, resulting in the formation of hydrocarbons and carbon particulates.

Some of the fuel spray may impinge on the walls. Combustion of this fuel depends on the rate of evaporation and mixing of the fuel with oxygen. If oxygen for combustion is unavailable in the surrounding gases, the fuel will evaporate, decompose, and form unburned hydrocarbons, carbon particulates, and partial oxidation products.

1.2.2.2 Indirect Injection Engines. Many of the pollutant-forming mechanisms of the DI engine apply to emission formation in indirect injection engines as well.

In indirect injection (IDI) engines (also called divided chamber engines) fuel is sprayed into the air in an ante- or prechamber which may contain 40–70% of the total combustion volume. The inward flow of air into the prechamber during the compression stroke produces turbulence for the mixing of injected fuel and air. Combustion begins there and forces the gases through a narrow passage into the main chamber. The emitted jet turbulence causes further mixing and combustion within the main chamber.

As the load is increased by injecting more fuel, the oxygen concentration in the prechamber decreases. Thus the extent of burning in the prechamber is reduced and this increases the amount of hydrocarbons and carbon monoxide formed there. Countering this, the rate of oxidation of these emissions in the main chamber increases due to its higher average temperature. At high loads, near the smoke limit, the oxidation reactions in the main chamber are insufficient to offset the increased hydrocarbon and carbon monoxide emissions from the prechamber. This results in increased HC and CO emissions and smoke particulates.

Emissions from IDI engines are lower than those from DI engines. This is attributed to the highly effective oxidation reaction in the main chamber.

1.2.2.3 General Remarks. Uncontrolled gasoline and diesel engines emit relatively large quantities of unburned hydrocarbons and oxides of nitrogen. Carbon monoxide and hydrocarbons are low enough in well-designed diesel engines that special controls are not required. On the other hand, CO emissions from gasoline engines usually require control even when mixtures are lean. The diesel has a serious problem with smoke particulates and has a characteristic odor arising from a combination of unburned fuel components and partially reacted products. Section 2.2 outlines some control techniques for diesel engine emissions.

Unlike SI engines, diesel hydrocarbon evaporative losses from the fuel system and tank are not a problem. The low volatility of diesel fuel prevents significant fuel tank evaporative losses at normal ambient temperatures and the fuel injection system itself is sealed.

In addition, crankcase hydrocarbon emissions are low compared to the SI engine because of both low fuel volatility and little fuel on or near the walls which can blow-by into the crankcase. Hare and Baines[5] have found that hydrocarbon emissions from blowby are only 3–4% of exhaust levels.

2 INTERNAL ENGINE CONTROLS

This section discusses some engine operating and design variables which influence exhaust emissions and may be used in various combinations to control engine-out emission. These operating and design variables are classified as "internal engine controls" and are distinguished from "external engine controls" because they only affect the combustion process within the cylinder.

2.1 SI Engine

The operating and design variables that affect emissions in spark-ignited engines include the following:

1. Fuel–air ratio
2. Load or power level
3. Speed
4. Spark timing
5. Combustion chamber deposit buildup
6. Surface temperature
7. Exhaust back pressure
8. Valve overlap
9. Intake manifold pressure
10. Surface to volume ratio
11. Displacement per cylinder
12. Compression ratio
13. Exhaust gas recirculation
14. Combustion chamber design
15. Stroke-to-bore ratio

2.1.1 *Operating Variables*

The air–fuel ratio affects hydrocarbon and carbon monoxide emissions as shown in Figure 2. HC and CO emissions increase as the mixture becomes richer. The CO increase is due to incomplete combustion caused by insufficient oxygen in the mixture. The HC increase arises primarily from a combination of crevice and oil adsorption mechanisms. Lean mixtures produce lower HC and CO emissions as a result of more complete combustion. As the lean combustion limit is approached, however, the HC and CO emissions increase due to incomplete flame propagation which leaves pockets of partially burned charge. Figure 2 does not extend to the incomplete combustion region. Oxygen and high temperatures are required to produce oxides of nitrogen. Consequently, rich mixtures produce low NO_x emissions and mixtures just lean of chemically correct produce high NO_x emissions. At very lean mixtures, NO_x emissions are reduced due to low combustion temperatures. Shown in Figure 2 also is brake specific fuel consumption (bsfc). This is the mass of fuel required to produce a unit of work and is a measure of efficiency. Low bsfc means high efficiency. Lean mixtures provide low bsfc.

At a fixed speed and mixture ratio, the load or power level has little effect on engine-out HC and

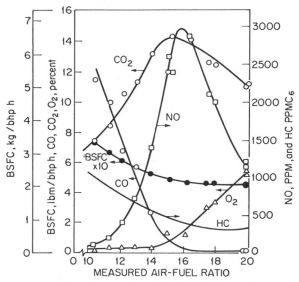

Fig. 2 Emissions and specific fuel consumption as a function of air–fuel ratio for a gasoline engine.

CO emissions when best economy spark timing is employed. As power output is increased, the volume flow through the engine and thereby the mass emissions of both HC and CO increase. Increases in power output increase peak combustion temperatures, thereby increasing the concentration of NO_x emissions.

Increasing engine speed tends to reduce HC emission concentration. This arises because of the improved combustion process within the cylinder resulting from increased turbulent mixing. Unburned hydrocarbons near the walls have less time to become entrained in the bulk gases and a greater fraction is left within the cylinder. Exhaust temperatures are increased with increased engine speeds and this promotes further combustion and reduction of CO and HC if sufficient oxygen is present. At higher speeds, combustion chamber temperatures are increased. For rich mixtures where combustion and NO_x formation are rapid, NO_x concentrations increase with engine speed. For lean mixtures where combustion and NO_x formation are slow, NO_x emission concentrations are reduced with engine speed increase because formation time is reduced.

Spark timing can be retarded to reduce HC emission concentration. This reduction results primarily from an increase in exhaust temperature, which promotes postcombustion chamber oxidation. Spark retard has little effect on CO concentrations for rich mixtures and has an effect similar to that on HC at lean mixtures. Because of the increase in working fluid mass flow necessary to maintain constant power with a retarded spark, HC mass emissions do not decrease as much as their concentration levels. An advance in spark timing increases cycle temperatures which increase NO_x concentration.

Combustion chamber deposit buildup primarily results from the tetraethyl lead antiknock fuel additive. Lubricant additives that leave ash upon combustion also contribute to deposits. Deposits, especially those having a porous structure, increase the surface area and may act as a sponge, trapping fuel droplets which are later exhausted. Deposit buildup increases effective compression ratio which in turn increases HC emissions. Emissions of CO are unaffected by deposit buildup for rich mixtures but are affected similarly to HCs for lean mixtures. Use of unleaded gasolines significantly reduces combustion chamber deposits and their impact on HC emissions. This has also extended spark plug life and reduced the possibility of HC emissions from misfiring spark plugs.

An increase in cylinder temperature caused by increased coolant temperature increases the emissions of NO_x. By increasing surface temperatures by 75°C, Myers and Alkidas[6] were able to increase NO_x emissions by 73%. They also noted no significant effect of surface temperature on CO emissions. Higher surface temperatures lowered HC emissions. In one test an increase in coolant temperature of 100°F decreased HC emissions one-third.[7]

2.1.2 Design Variables

The amount of exhaust gas remaining in the combustion chamber affects both HC and NO_x emissions. The last portions of the exhaust gas to leave the cylinder contain a large quantity of unburned hydrocarbons. As more of this portion is retained in the cylinder and burned during the next cycle, HC emission concentration will be reduced. Both increased exhaust backpressure and valve overlap affect this. If the amount of retained residual is increased to the point where incomplete flame propagation occurs, HC emission concentrations will be increased greatly. Increasing the amount of residual retained in the cylinder reduces flame speed and the maximum temperature of the cycle. This results in lower NO_x emission concentration.

Valve overlap increase will increase the amount of residual exhaust gas retained in the cylinder. Siewert[8] investigated the effect of changing intake and exhaust opening and closing times on HC, CO, and NO_x emissions. He found the best reductions in emissions occurred with valve timing such that the last portion of the exhaust gas was drawn back into the cylinder during the intake stroke. This was done by using either early intake valve opening, which allowed exhaust gas to enter the intake manifold, or by late exhaust valve closing which kept the exhaust valve open longer at the beginning of the intake stroke. Large amounts of valve overlap produce a rough idle due to the poor combustion of the residual diluted mixture. If the mixture ratio is richened to smooth the idle, then HC and CO concentrations will be increased. If the throttle opening is increased to overcome the increase in dilution by the residual gas, the mass emission of HC and CO may be increased. References 9–11 describe attempts to make use of high valve overlap for emission reduction and to eliminate the problem of rough idle by using variable valve timing. This has not come into general use due to the cost and complexity of these devices.

The surface-to-volume ratio (s/v) of the combustion chamber affects HC emissions.[12,13] For a given clearance volume reducing the surface area of the chamber (including crevices) will reduce HC emissions.

Combustion chamber shape has been used to reduce HC emissions by designing for a low surface-to-volume area as shown in Figure 3. Engine s/v values range from 5 (low) to 10 (high), depending on design.

The amount of residual exhaust gas in the cylinder can be augmented through the use of exhaust gas recirculation (EGR). A system for the recirculation of exhaust gas is shown in Figure 4. The flow is controlled by an EGR control valve which is activated in response to exhaust backpressure, carburetor venturi, or manifold vacuum. The valve restricts flow at light loads where residual is normally high and EGR would produce rough idle and off-idle operation. After passing the EGR valve, the exhaust gases are drawn into the intake manifold plenum, mixed with the fresh intake charge, and drawn into the cylinder. The effect of EGR on emissions is the same as an increase in exhaust backpressure or valve overlap. The emissions of NO_x are reduced due to reduced flame speed and lower maximum cycle temperature. Depending on mixture ratio and load, 5% EGR can reduce NO_x more than 50%, and 15% EGR by more than 75%.[14] EGR in excess of 15% often produces incomplete combustion and increases HC and CO emissions. EGR is normally eliminated at full load in order to permit maximum power.

Older automobiles used a road draft tube to relieve the crankcase pressure caused by blowby gases. This allowed the blowby gases to escape directly into the atmosphere. Since 1963 automobiles have used positive crankcase ventilation (PCV) systems which, in their present closed form, completely eliminate emissions due to blowby. A closed PCV system is shown in Figure 5. The system operates as follows. Crankcase gases are purged through the PCV valve which restricts the flow, depending on the intake manifold vacuum. The valve is designed to restrict flow at light loads which would otherwise lean the mixture excessively, and produces a rough idle. Filtered air enters the crankcase to assist in the purging process. As the engine wears and more blowby arises, less filtered air is introduced.

2.1.2.1 Control Systems. From the foregoing brief discussion, it is clear that changes in design and operating variables affect gasoline engine emissions in complex and usually interrelated ways. A variety of devices have been applied to control spark advance and carburetion variables during critical portions of vehicle operation.

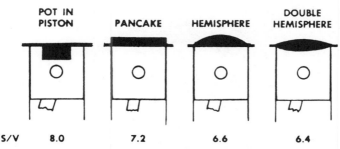

Fig. 3 Influence of combustion chamber shape on surface/volume ratio. Calculations are for 101-mm bore, 5.7-liter V-8, 9:1 compression ratio.

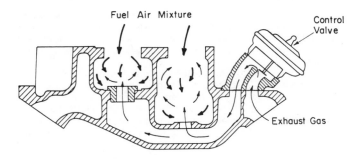

Intake Manifold Section

Fig. 4 Exhaust recirculation system schematic. The control valve modulates the amount of exhaust gas recirculated into the intake manifold.

Examples are:

CONTROLLED CARBURETOR AIR TEMPERATURE. Employs exhaust heat to control carburetor air temperature to about 40°C and allows leaner carburetor settings.

TRANSMISSION-CONTROLLED SPARK. Eliminates vacuum advance during shifting which retards timing.

SPARK DELAY VALVE. Delays advance of spark during load transient.

PORTED SPARK. Provides no vacuum advance at idle or off-idle throttle setting.

THERMAL SPARK SWITCH. Applies maximum distributor vacuum during cold engine operation or if engine overheats.

THROTTLE CRACKER. Slows throttle closing during deceleration—limits residual dilution.

IDLE SOLENOID. Limits minimum throttle position—controls residual dilution at idle.

ELECTRIC CHOKE. Increases rate of choke removal during cold start.

FUEL INJECTION. Provides more precise fuel control, especially under cold starting and deceleration conditions and reduces need for starting and transient fuel enrichment. Injection is either into each intake port or above the throttle (central fuel injection).

COMPUTER ENGINE CONTROLS. Provides optimized values for mixture strength, spark timing, and control settings for emission control devices such as PCV and EGR valves. Values arise from sensors responsive to ambient and engine conditions coupled with preprogrammed logic and learned responses.

2.2 CI Engine

The operating and design variables that affect emissions in compression-ignition engines include the following:

1. Fuel–air ratio
2. Injection timing

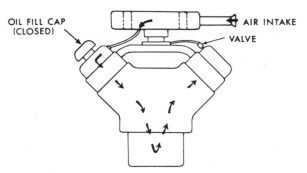

Fig. 5 Closed positive crankcase ventilation system.

3. Inlet air temperature
4. Fuel composition (including additives)
5. Turbocharging
6. Air swirl
7. Combustion chamber design
8. Injector and spray characteristics
9. Exhaust gas recycle
10. Crankcase ventilation system

Since these variables can affect emissions differently for direct and indirect injection engines, these two types of engines are discussed separately. Also, since there are many different diesel engine designs of each type, one engine may not respond to a change in an operating or design variable like another engine would. The following discussion is intended to present a general overview of the engine response to changes in operating or design variables. The reader is referred to Ref. 4 for more details.

2.2.1 Operating Variables

2.2.1.1 Fuel–Air Ratio. In diesel engines the load is changed by varying the amount of fuel injected. This results in changes in fuel spray distribution, amount of fuel that impinges on the walls, cylinder pressures, temperatures, and duration of the injection.

In direct injection engines, an increase in the amount of fuel injected (high load) increases the injection duration and if injection timing and rate are constant, more fuel is injected later in the cycle. The last part of fuel injected has a shorter time for reaction and less oxygen available. These tend to lower the rate of hydrocarbon and soot burnup. On the other hand, because more fuel is burned, the cycle temperatures are increased. This tends to increase the rate of burnup. The ensuing discussion explores how these opposite effects combine to change HC and soot emissions as load is varied. The reader is referred to Figure 1.

At light loads the fuel spray does not reach the walls and fuel concentration in the spray core is low. Most of the HC and aldehyde emissions are from the LFOR. As the fuel–air ratio is increased, the HC emissions from the LFOR decrease as a result of increased temperatures. At high loads more fuel is deposited on the walls and its concentration in the spray core increases. As a result, more unburned hydrocarbons arise in these regions. At medium loads there is sufficient oxygen and as a result of higher temperatures, the hydrocarbon and soot emission concentrations are reduced compared to their amounts at light loads. At full load the increase in fuel–air ratio results in more unburned hydrocarbons at the walls and in the spray core. The contribution of the LFOR to the total is small. In spite of high temperatures the reactions are limited due to lack of oxygen. Thus, hydrocarbon and soot emission concentrations increase at full load. The NO emissions are formed mainly in the LFR (premixed mixture combustion) and their concentration increases with an increase in fuel–air ratio. In some engines, the NO mass emissions tend to be in proportion to the fuel burned.

In indirect injection engines more fuel is injected into the prechamber as the load is increased. This will occur later in the cycle, if injection timing and rate are constant. The oxygen concentration is decreased in the prechamber as more fuel is injected and burned. These factors limit the extent of burning in the prechamber especially for the last portion of the fuel to be injected. Thus, the increase in fuel–air ratio increases HC, CO, and soot formation in the prechamber. However, as the fuel–air ratio is increased, the temperature in the main chamber increases and the resulting increase in the rate of oxidation reactions there produces lower HC and CO emission concentrations. At light loads, NO is formed mainly in the prechamber where most of the fuel is burned. As the load is increased, the mixture in the prechamber becomes rich and the amount of NO formed there is greatly reduced. As the temperature in the main chamber increases with increases in load, it might be expected that the formation of NO in the main chamber would increase. This is not the case however, since as the load increases, the amount of unburned compounds discharged from the prechamber increases and subsequent burning reduces oxygen concentration in the main chamber. Therefore, as the load increases NO concentrations commonly decrease.

2.2.1.2 Injection Timing. Advancing the injection timing increases the unburned HC emissions as well as NO emissions. Early injection results in a longer ignition delay. This permits the fuel vapor and small droplets to be spread over a larger volume resulting in a larger LFOR. Early injection also increases the amount of fuel that impinges on the walls since the air density is not as great in the earlier part of the cycle. Reference 15 states that advancing the injection timing provides additional burning time in the expansion stroke and yields higher combustion temperatures. These conditions may be expected to increase the rate of combustion of carbon particulates. Retarding injection timing usually reduces NO emissions at the expense of increasing emissions of HC, CO, and smoke. This result arises from the reduced premixed burning associated with a short delay period. The reader is referred to Ref. 16 for details.

2.2.1.3 Intake Air Temperature. An increase in inlet air temperature results in higher cycle temperatures. This reduces penetration and increases atomization, evaporation, and diffusion[4] with fuels of normal volatility. These factors produce rich mixtures near the nozzle and increase smoke emissions. However, with fuels of lower volatility, the increased temperatures aid oxidation reactions and result in lower smoke emissions. Any increase in cycle temperatures will produce higher NO_x emissions.

2.2.1.4 Fuel Composition (Including Additives). The cetane number of the fuel has an effect on particulate emissions. Reference 17 reports that reducing the cetane number of the fuel lowers particulates. A possible explanation is that since lower cetane number fuels have longer ignition delay periods, more fuel is injected and mixed before ignition. Particulate formation is less since the amount of fuel burned under heterogeneous conditions is less. On the other hand, the increase in the premixed combustion portion increases NO_x. If the ignition delay is very long, the LFOR is increased which increases HC emissions. Otherwise, HC may increase or decrease with cetane number change.

Barium additives reduce smoke but have little or no effect on power, odor, or gaseous emissions. Reference 18 reports a maximum reduction of 40% in smoke over the entire load range. These findings are disputed by Ref. 17 which notes that reductions in smoke may not be accompanied by a similar reduction in mass emissions.

Although barium additives reduced smoke levels, extended use of the additive was found in Ref. 19 to increase smoke levels due to the formation of heavy deposits in the combustion chamber. The mechanism by which barium acts to reduce smoke is not understood. There are potential drawbacks to the use of barium since combustion products may be toxic.

2.2.2 Design Variables

2.2.2.1 Turbocharging. Turbocharging increases cycle temperatures and therefore increases oxidation reactions. Higher exhaust temperatures enhance reactions in the exhaust system. These factors reduce HC emissions. Intercooling can be used with turbocharging to reduce cycle temperatures and thereby reduce NO emission.

2.2.2.2 Swirl. The amount of swirl in a direct injection engine affects the mixing process. Too much swirl can result in a larger LFOR or an overlap of sprays. Thus, for a given fuel injection pattern too much swirl will increase HC emissions and smoke. Too little swirl may cause the same emission increases due to poor mixing. The amount of swirl can be changed by varying the intake port shape or the ratio of the bowl diameter to its depth. Deep bowl pistons tend to produce more swirl than shallow bowl pistons.

2.2.2.3 Combustion Chamber Design. Combustion chamber design plays an important part in emission formation. The piston bowl geometry of a direct injection diesel and the prechamber geometry of an indirect injection diesel have a great effect on the emissions produced. Middlemiss[20] has performed an extensive investigation of these effects for a direct injection diesel. The parameters investigated were: bowl throat diameter, compression ratio, bowl flank angle, and lip shape. The effect of a "pip" or bump in the center of the bowl was also examined. Radovanovic and Djordjevic[21] have performed an extensive investigation of these effects for an indirect injection diesel. The geometric parameters they varied were: prechamber rear cell geometry, cone angle of nozzle tip, passageway diameter, downward exit angle, spread angle, number of passageways, and prechamber volume ratio. These parameters can be optimized to obtain the desired emission reductions.

2.2.2.4 Injector and Spray Characteristics. The injector and its spray characteristics have an effect on emissions. The sac volume between the needle seat and nozzle holes contains fuel which seeps out late in the expansion stroke. This fuel has little chance to react and therefore adds to smoke and HC emissions. Since this fuel is a major source for HC emissions, reducing the sac volume can greatly reduce HC emissions. The nozzle design has an effect on smoke concentration as well. Larger diameter orifices result in less atomization and thus increased smoke. On the other hand, Hames et al.[22] found that an increase in orifice size or a reduction in the number of orifices decreased NO concentration throughout the engine speed and load range, probably through a decrease in premixed burning.

2.2.2.5 EGR Systems. Exhaust gas recirculation is used to reduce NO emissions. The reductions in NO are due to the increased heat capacity of the charge which lowers temperatures. Emissions of HC, CO, and NO decrease with EGR while smoke increases. Usually, exhaust gas recirculation can be used up to 20% to reduce NO with little effect on HC or CO but with an increase in smoke. Above 20%, EGR increases smoke sharply for part load and above 25% EGR, smoke increases sharply at full load.[4] To reduce NO emissions without increasing other pollutants, the amount of EGR should be reduced as the load is increased.

2.2.2.6 Crankcase Ventilation Systems.   Where greater emission control is needed, a positive crankcase ventilation system may be employed. Since diesel blowby is largely air, it accounts for only a small portion of the emissions from diesel engines. As a result, many diesel engines presently use a road draft type of crankcase ventilation system. EPA is currently considering diesel blowby standards.

3   EXTERNAL ENGINE CONTROLS

External engine controls are classified as devices used to reduce emissions outside of the combustion chamber. They are different for gasoline and diesel engines and therefore will be discussed separately.

3.1   Gasoline Engine

External devices are commonly used to control emissions from gasoline engines. These are thermal reactors, catalytic reactors, feedback systems, and evaporative controls.

3.1.1   Thermal Reactors

The purpose of thermal reactors is to oxidize hydrocarbons and carbon monoxide through noncatalytic homogeneous gas reactions. They are oxidizing devices and therefore do not remove NO_x. Such reactors maintain the exhaust gases at elevated temperatures (up to 900°C) for a period of time (average of 100 msec) so that oxidation reactions will continue in the exhaust gases after they have left the cylinder. Thermal reactors are often large, well-insulated baffled containers which replace the conventional exhaust manifold. Total reactor volume may be 1.5 to 2 times engine displacement. On Vee engines two small reactors are usually used. Operating temperatures must be between 800 and 900°C in order to fully oxidize CO and thus reactors must be mounted close to the engine. At lower temperatures, hydrocarbons may be partially oxidized to CO and aldehydes, thus increasing those emissions.[23] High-temperature materials are required, usually stainless steels, and consideration for thermal expansion in the design is essential. If the engine malfunctions, much higher temperatures may arise which can damage the reactor.

Thermal reactors are often baffled internally. Usually a large mixing volume (backmix reactor) is used to collect and mix the exhaust from each port. This volume is followed by one or more passages in which the gases move in a plug flow manner. The mixing volume is essential if exhaust port air injection is used. Reactors with elaborate baffling minimize "breakthrough" and can provide up to 100% conversion of HC and CO under warmed up, high-temperature operation. However, elaborate internal structures slow the rate of warmup. Under rich engine operation, a substantial amount of HC and CO oxidation may occur in a conventional exhaust manifold when air is injected into the exhaust port. Thus the manifold itself may be considered to be a small reactor. In automotive systems where warmup rate is important, a compromise is made between the reactor mass and its steady-state efficiency.[24]

Thermal reactors are of two types based on the rich or lean air–fuel ratio of the engine's intake charge. A secondary air injection system is required with rich reactors to provide the necessary air for combustion. An oxygen level in the reactor of 1 or 2% is often optimum and this may require an air volume flow of 20–30% of the engine air flow itself. Such a high flow requires a substantial pump and a significant amount of power. The basic difference in rich and lean reactors is that the inlet concentration of hydrogen and CO is normally several percent for rich reactors and a small fraction of 1% for lean reactors. One percent of hydrogen or CO burns to give a temperature rise of about 80°C in exhaust gases. Thus, the rich reactor can "light off" and reach a relatively high operating temperature. The operating temperature of the lean reactor is determined primarily by the exhaust temperature itself. Commonly lean reactors do operate at lower temperatures. This results in lower conversion efficiency compared to rich reactors. Lower temperatures give lean reactors the advantage of improved reactor durability and reduced underhood heat rejection. Lean reactors have lower cost and complexity than rich reactors because a secondary air injection system is not required. Rich engine operation increases fuel consumption, consequently rich reactors are of little interest except for providing rapid "light off" during engine starting and warmup. In that capacity they may be used as a "light off" reactor for a catalyst.

The secondary air injection system plays a major part in the conversion efficiency of the rich reactor.[25] Conversion efficiency may be substantially improved if the secondary air is proportioned to the engine air flow. Moreover, the degree of mixing of the secondary air with the exhaust gas has a great effect on conversion efficiency.[26,27] Open-ended air injection tubes were found to provide poor mixing because the increasing exhaust pressure interrupted air flow during the exhaust event. As a result, alternating slugs of exhaust gas and secondary air entered the reactor with mixing occurring only at the interface between the slugs. The use of a sparger tube with radial holes in place of the open-ended tube improved mixing and the completion of reaction. If secondary air is injected continuously, the reactor is cooled and efficiency reduced during the portions of the engine cycle when the

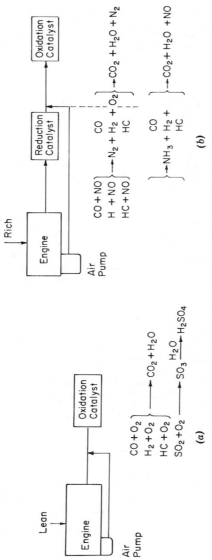

Fig. 6 (*a*) Oxidation catalyst schematic. (*b*) Dual catalyst schematic.

exhaust valve is closed. To overcome this deficiency a timed injection system might be considered. Preheating the injected air with exhaust heat minimizes this cooling effect also.

Herrin[28] has examined some variables that may increase the operating temperature of a lean reactor. These were insulation, increasing engine speed, retarding spark timing, richening the fuel–air ratio, advanced exhaust valve opening, and exhaust port liners.

The development of effective catalytic reactors has reduced interest in thermal reactors. This is because catalysts operate efficiently at lower temperatures and thus engine efficiency compromises are minimized. However, thermal reactors are relatively tolerant to lead antiknock compounds and this is an advantage for the thermal reactor.

3.1.2 Catalytic Reactors

Catalytic reactors are mounted in the exhaust system often slightly removed from the engine and, depending on design, are used to remove not only HC and CO but also NO_x. For automotive vehicles platinum (Pt) and palladium (Pd) are used as catalysts for HC and CO oxidations. Rhodium (Rd) is used as a catalyst to reduce oxides of nitrogen. Typically 2–4 g total of noble metals are used. Base metal catalysts may be effective with alcohol fuels but their catalytic activity drops rapidly in use with conventional hydrocarbon fuels. Two types of support structures, pellets (gamma alumina) or monoliths (cordierite or stainless steel), are used. To cordierite supports a wash coat of gamma alumina is added prior to application of the catalytic metals themselves.

In order to function properly, oxidizing catalysts require some oxygen and reducing catalysts require some CO, HC, or H_2. Figure 6 shows typical systems and reactions. Depending on catalyst selectivity some ammonia may form in the reduction process which is subsequently reoxidized to NO yielding a low overall conversion efficiency for NO_x. Sulfuric acid may be an undesirable by-product. For a near chemically correct mixture, both oxidizing and reducing species coexist in the exhaust thereby permitting use of three-way catalysts.

Figure 7 shows the conversion efficiency as a function of air–fuel ratio for a typical noble metal catalyst. Dual-bed catalyst systems (such as Figure 8) permit simultaneous rich and lean operation which can achieve at best nearly 100% conversion of each pollutant. The three-way catalyst simultaneously achieves all the reactions shown in Figure 7 in a single bed (Figure 9), but at efficiencies that are below optimum. Oxidation systems are effective with HC and CO, but not with NO_x.

Dual systems are larger, heavier, more expensive, and more efficient compared to single-bed systems. Compared to three-way catalysts, the richer engine operation required to achieve a chemically reducing environment for the dual catalyst system lowers fuel economy. Lean catalyst operation permits optimum fuel economy but requires total NO_x control within the engine itself. Depending on emission control requirements, the control of NO_x with three-way or dual systems may provide the best fuel economy solution since use of internal controls alone may reduce fuel economy even more.

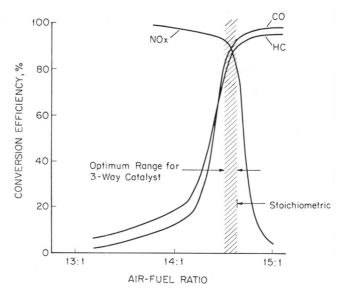

Fig. 7 Catalytic converter efficiency as a function of air–fuel ratio for a typical warmed up precious metal catalyst. Figure from Ref. 29.

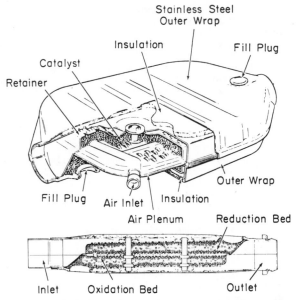

Fig. 8 Cross section of a dual pelleted converter. Exhaust enters the upper portion in which reduction reactions remove NO_x. Together with injected air it then passes into the lower portion where HC and CO are oxidized. Figure from Ref. 29.

Catalytic converter efficiency is lowered by the presence of metallic compounds in the exhaust arising from fuel and lubricant additives as well as wear metals. This is known as catalyst poisoning. Tetraethyl lead (TEL) antiknock additive especially lowers catalyst efficiency and therefore virtually lead-free gasolines are required. Use of metal substrates may offer increased tolerance to TEL.

Noble metal catalysts are relatively efficient at temperatures exceeding 250°C. Thus, compared to thermal reactor use, engine efficiency can be optimized to a greater extent. Lower temperatures reduce container material and thermal expansion requirements. Compared to thermal reactors, catalysts may

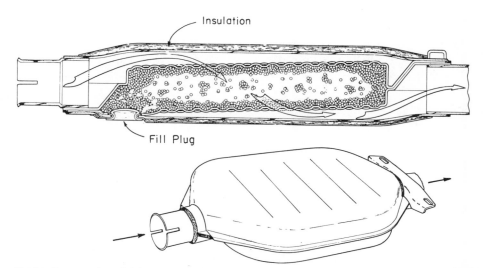

Fig. 9 Cross section of either an oxidizing or three-way converter of the pelleted type. Figure from Ref. 29.

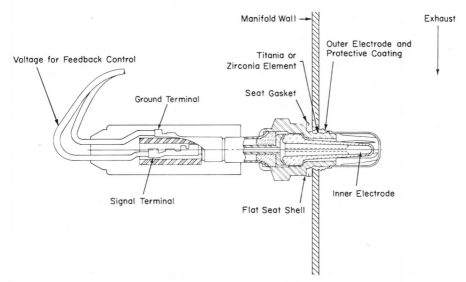

Fig. 10 Exhaust oxygen sensor. Figure from Ref. 29.

warm up more slowly due to their large internal mass. This limitation is important where efficient emission control is required upon initial engine startup.

3.1.3 *Fuel Feedback Control Systems*

Fuel feedback control systems can be used to precisely control the fuel–air ratio which is essential for correct three-way catalyst operation. For this a sensor is placed in the exhaust stream which measures the oxygen concentration. The sensor consists of a zirconia (ZrO_2) or titania (TiO_2) element coated with platinum. Figure 10 shows a zirconia sensor installation.

At a sufficiently high temperature, this electrochemical cell produces a voltage of about 1 V, depending on exhaust oxygen content (see Figure 11). This characteristic is used to sense when the fuel–air ratio is near stoichiometric. The voltage is provided to a signal processor which, in turn, provides a corrective signal to a fuel injector or carburetor system. Either the observed voltage is compared to a desired reference value and the fuel control adjusted, or the fuel control is caused to oscillate rapidly between a rich and lean limit. In this latter case, the amount of time the fuel control is constrained at the lean limit establishes the average mixture strength provided. A proportional and integral control strategy is used to maintain the fuel–air ratio within the optimum range as the engine throttle position and speed change. Since the sensors do not operate at low temperatures, an open loop option is provided for starting until sensor temperature reaches about 350°C. A system schematic is shown in Figure

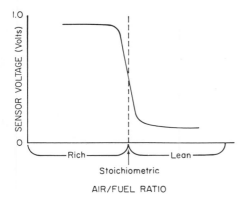

Fig. 11 Exhaust oxygen sensor voltage characteristic.

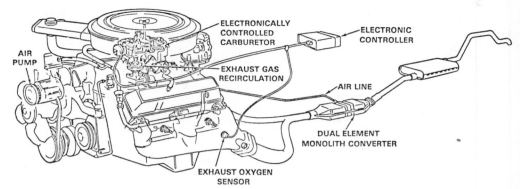

Fig. 12 Closed loop dual catalytic system employing a monolithic converter. Figure from Ref. 29.

12. A unique advantage of fuel feedback control is its ability to adapt carburetion to the stoichiometric point variability associated with commercial gasoline fuels.

In addition to the exhaust gas oxygen sensor, Ref. 30 describes other sensors that may be used to help control emissions. These include engine crank position, air temperature, throttle position, and coolant temperature sensors. Information from these devices processed by the electronic controller can be used to accurately set engine variables of fuel rate, air rate, EGR, and spark timing. The reader is referred to Refs. 31–34 for additional detail.

3.1.4 *Evaporative Controls*

A system for the control of hydrocarbon emissions caused by evaporation from a gasoline fuel system is presented in Figure 13. During hot soaking conditions, fuel vapors emitted from the fuel tank or carburetor float bowl are collected and stored on the surfaces of charcoal granules in a canister located near the engine. A restricted central takeoff vapor point in the tank dome allows vapors to vent regardless of tank attitude. The dome prevents liquid fuel arising by thermal expansion from reaching the canister.

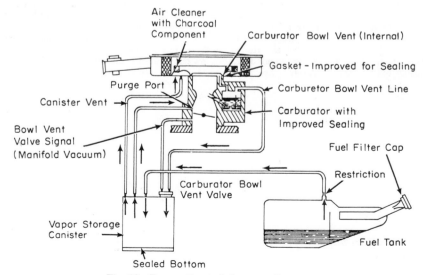

Fig. 13 Evaporative emission control system.

The system shown in Figure 13 is a "closed" system in that vapors displaced during the hot soak period are trapped by a charcoal element in the air cleaner. Purging is accomplished through the purge port employing filtered air inducted through the canister vent line. A valve actuated by a ported vacuum signal controls the carburetor bowl venting.

Under running conditions the vapor losses are drawn into the engine and burned. This occurs when the throttle blade tip exposes the purge port to vacuum. Thus purging is eliminated at idle and very light load operation modes in which vapor addition can significantly increase tailpipe HC emission and even cause drivability problems.

On a recycle basis, activated charcoal can store 30–200 g fuel/100 g charcoal. Typically, vehicles may emit 100–200 g fuel/day depending primarily on ambient and vehicle temperature changes, fuel volatility, fuel tank vapor volume, and carburetor fuel bowl volume.[12] A typical vehicle canister may employ 1000 g charcoal. If the canister is required to store the additional vapors expelled during the filling of the fuel tank, a significantly larger volume of charcoal will be required.

Diesel fuel evaporation losses are very small (low volatility) and no controls have been proposed.

3.2 Diesel Engine Particulate Traps

There are no external emission control devices in use at this time for diesel engines. Emissions of HC and CO are low enough from the combustion process that an external device to further oxidize them in the exhaust system may be unnecessary. Also, low exhaust temperatures virtually preclude external devices for use on a continuous basis. However, external emission control devices for particulates are currently under development. The remainder of this section will deal with such devices.

The basic problem with the removal of diesel soot particulates from exhaust gases is that the particles are very small in that one-half are less than 0.5 μm in diameter, and they are very light (0.05 g/cm³). Typically emitted particles total from 0.1 to 0.5% of the fuel by weight. Consequently, a conventional filter would quickly plug.

Reference 35 describes one device which uses screens in a first stage to agglomerate particles. This is followed by inertial separation in a second stage. The trapped particles must be removed periodically to avoid excessive backpressure increase.

An alternate method uses a filter to trap particles and the filter is periodically regenerated. One such trap is shown in Figure 14.[36] This device uses a stainless steel mesh followed by a ceramic filter. The ceramic filter has alternate channels blocked so that the exhaust flow must pass through the porous walls. Particulates can be oxidized if exhaust temperatures are raised to 600°C or more and held there for several minutes. Such a temperature is substantially above the normal diesel exhaust level. One possibility for achieving the high exhaust temperatures is to throttle the engine intermittently.[37] The addition of catalytic materials to the fuel or trap is being studied as a means of lowering ignition temperatures. While these traps look promising, much development is required before any commercial application.

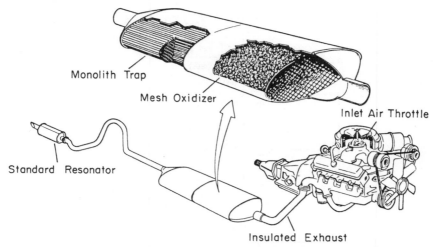

Monolith Trap

Mesh Oxidizer

Inlet Air Throttle

Standard Resonator

Insulated Exhaust

Fig. 14 Prototype diesel particulate trap-oxidizer. Figure from Ref. 36.

4   EMISSION TEST PROCEDURES

Worldwide there are many different emission test procedures for motor vehicles and stationary engines. For light- and medium-duty vehicle exhaust and evaporative emissions in the United States, the Federal Test Procedure (FTP) is used. This procedure, which is sometimes termed the "city cycle," is discussed briefly below. Heavy-duty engine emissions are tested using either the 13-mode engine dynamometer procedure which is explained in Ref. 38 or a transient cycle technique still under development[39,40] which measures particulates as well as gaseous compounds. A brief description of non-U.S. vehicle procedures can be found in Ref. 41.

The FTP exhaust and evaporative emission test is specified for both spark-ignited and diesel-engine powered light-duty cars and trucks up to 3856 kg (8500 lb$_m$) gross vehicle weight (GVW). Figure 15 is a diagram showing the steps in the test. Results are expressed as g/mile for each exhaust pollutant and as g/test for evaporated fuel.

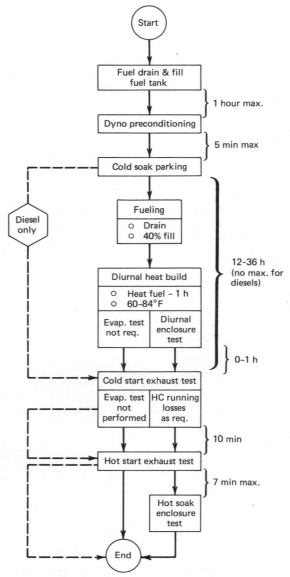

Fig. 15   U.S. FTP test sequence—light-duty vehicles. Figure from Ref. 42.

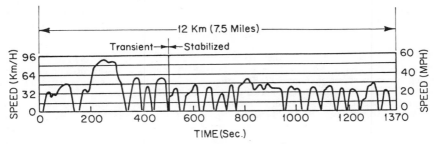

Fig. 16 U.S. FTP driving cycle.

4.1 Light-Duty Vehicle Exhaust Emissions

Exhaust emissions of unburned hydrocarbons, carbon monoxide, and oxides of nitrogen are measured by operating the motor vehicle on a chassis dynamometer. Carbon dioxide is also measured to compute fuel economy. The dynamometer load that is imposed on the vehicle is determined either by a formula using the vehicle's frontal area, aerodynamic characteristics, tire type and weight, or by a direct on-the-road coast-down procedure. The dynamometer power setting is made at 80 km/hr (50 mph). An inertia weight is used to simulate acceleration and deceleration loadings. Incremental inertia weight classes are specified with increments of 228 kg (500 lb$_m$) above 2500 kg (5500 lb$_m$), 114 kg (250 lb$_m$) between 1818 kg (4000 lb$_m$) and 2500 kg, and 57 kg (125 lb$_m$) below 1818 kg GVW. Detail may be found in Ref. 42.

For certification a special test gasoline is used, termed "Indolene." Its specifications include hydrogen-to-carbon ratio, research octane number, density, vapor pressure, distillation range, lead, sulfur, phosphorus, and hydrocarbon class maxima. Diesel fuel specifications include cetane number, gravity, flash point, and viscosity in addition to distillation range and sulfur content. The diesel fuel is a type 2-D grade. Lubricants are limited to those specified for use by the manufacturer with the restrictions that they must be generally available and reasonably priced.

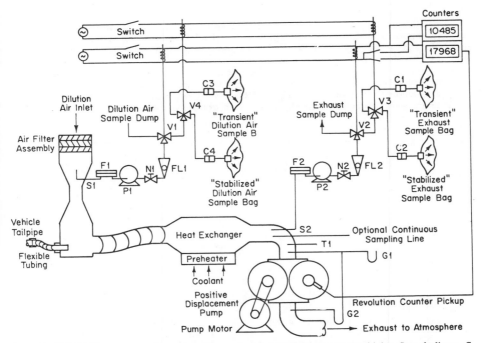

Fig. 17 U.S. FTP constant volume sampling system for gasoline powered vehicles. Sample lines—S, filters—F, pumps—P, flow control valves—N, flow meters—FL, connectors—C, diverter valves—V, thermocouple—T, manometer—G.

In the FTP test for exhaust emissions the vehicle is operated over the 1372-sec driving cycle illustrated in Figure 16. A professional driver rather than a machine is used to operate the vehicle. Various driving rules are specified. If the driver does not match the driving cycle within the limits specified in Ref. 42, the test is voided.

The cycle is run with a vehicle that has been stabilized for 12 hr (cold soak) at 20–30°C (68–86°F) and is run at the temperature of the cold soak. A schematic of the variable dilution CVS sampling system is shown in Figure 17. This dilution system dilutes the exhaust by as much as a factor of 10. Three 5-ft³ teflon bags are used to collect the exhaust samples.

Gases are collected during cranking, starting, and for 5 sec after the ignition is turned off. One bag collects emissions during the "cold transient" portion of the cold start exhaust test which extends up to 505 sec. After 505 sec, the exhaust is collected in a second bag for the "cold stabilized" portion of the remainder of the cold start exhaust test up to 1372 sec. After the driving cycle is completed the engine is turned off. After 10 min the engine is restarted and the first 505 sec of the driving cycle are repeated. This is termed "hot transient" portion. The third bag is used to collect emissions during that test portion. The stabilized portion of the hot test (505–1372 sec) is not repeated since it is assumed to be the same as the stabilized part of the cold start test.

Within each bag, the concentration of HC, CO, and NO_x is measured by flame ionization, nondispersive infrared, and chemiluminescent analyzers, respectively. The mass of each is determined from ideal gas law calculations and a knowledge of the total volume of exhaust plus dilution air provided by the CVS system during the test. Diesel particulate filters are weighed. In the determination of g/mile of each compound, the cold start results are weighted 43% and the hot restart results 57%.

Emission regulations require that each vehicle type meet the emission standards at 80,000 km (50,000 miles). Fleet averaging is not permitted. Vehicles are tested when 6400 km (4000 miles) have been accumulated on a mileage accumulation cycle and their emissions at 80,000 km (50,000 miles) are projected by application of a deterioration factor. To develop this deterioration factor, selected vehicles, termed durability vehicles, are driven to 80,000 km (50,000 miles) during which period several emission tests are performed. The best-fit straight line between 6400 and 80,000 km (4000 and 60,000 miles) is used to determine deterioration factors.

The FTP test is sufficiently complicated that repeatability is a problem. Figure 18 below indicates the variables in the FTP test and their relative contribution to test variations. Reference 43 gives a detailed discussion of exhaust emission test variability.

4.2 Evaporative Emissions

The evaporative emissions test applies only to gasoline-fueled vehicles since diesel fuel is not volatile. The test is known as the SHED test, an acronym for sealed housing evaporative detection. There are two portions to the test. The diurnal portion is performed prior to the FTP. The hot-soak portion is performed after the FTP.

In the diurnal test the car is placed in the sealed enclosure with fresh chilled test fuel in the tank which is filled 40% with liquid fuel. A heating blanket is placed on the fuel tank and the fuel is heated at a constant rate from 15.6 to 28.9°C (60 to 84°F) in the period of 1 hr. A blower purges the enclosure when testing is not in progress. The hydrocarbon concentration in the enclosure is

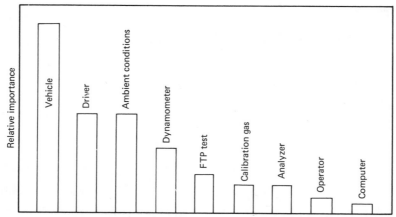

Fig. 18 Relative importance of several factors contributing to variability in the U.S. FTP emission test.

measured at the beginning and end of the test by a flame ionization detector. Typically 50 g of vapor are emitted. Results are very sensitive to front-end fuel volatility and fuel tank vapor volume.

The hot-soak enclosure test relies on vehicle heat for the evaporation of fuel. Within 7 min after the hot-start exhaust test is finished, the vehicle is put in the sealed enclosure one more time. The hydrocarbon concentration in the enclosure is measured at the beginning and end of the test. Emissions during the hot-soak evaporation test come primarily from carburetor bowl fuel boiling since temperatures there may rise to 80°C. Typically 12 g of vapor are emitted, depending on the fuel distillation characteristics, peak bowl temperature, and volume.

U.S. federal evaporative standards limit total losses to 2 g/test. Reference 42 gives further details of the evaporative procedure and the required calculations.

REFERENCES

1. Myers, P. S., Automobile Emissions—A Study in Environmental Benefits Versus Technological Costs, *SAE Paper 700182* (1970).
2. *National Air Pollutant Emission Estimates 1940–1980,* EPA Report 450/4-82-001, 1982.
3. Blumberg, P., and Kummer, J., Prediction of NO Formation in Spark Ignited Engines, *Comb. Sci. & Tech.* **14**, 73–95 (1971).
4. Henein, N. A., Analysis of Pollutant Formation and Control and Fuel Economy in Diesel Engines, *Prog. Energy Combus. Sci.,* Vol. 1, Pergamon Press, 1976, pp. 165–207.
5. Hare, C. T., and Baines, T. M., Characterization of Diesel Crankcase Emissions, *SAE Paper 770719* (1977).
6. Myers, J. P., and Alkidas, A. C., Effects of Combustion-Chamber Surface Temperature on the Exhaust Emissions of a Single-Cylinder Spark Ignition Engine, *SAE Paper 780642* (1978).
7. Wentworth, J. T., Effect of Combustion Chamber Surface Temperature on Exhaust Hydrocarbon Concentrations, *SAE Paper 710587* (1971).
8. Siewert, R. M., How Individual Valve Timing Events Affect Exhaust Emissions, *SAE Paper 710609* (1978).
9. Freeman, M. A., and Nicholson, R. C., Valve Timing for Control of Oxides of Nitrogen (NO_x), *SAE Paper 720121* (1978).
10. Schiele, C. A., Denagel, S. F., and Bennethum, J. E., Design and Development of A' Variable Valve Timing (VVT) Camshaft, *SAE Paper 740102* (1974).
11. Anon., Variable Camshaft Boosts Engine Efficiency, *Automotive Engineering* **81**(12), 11–13 (1973).
12. Patterson, D. J., and Henein, N. A., *Emissions from Combustion Engines and Their Control,* Ann Arbor Science Publishers, Ann Arbor, MI, 1972.
13. Sheffler, C. E., Combustion Chamber Surface Area, A Key to Exhaust Hydrocarbons, *SAE PT-12,* 60–70 (1963–1966).
14. Benson, J. D., Reduction of Nitrogen Oxides in Automotive Exhaust, *SAE Paper 690019* (1969).
15. Bryzik, W., and Smith, C. O., Relationships Between Exhaust Smoke Emissions and Operating Variables in Diesel Engines, *SAE Paper 770718* (1977).
16. Pischinger, R., and Cartellieri, W., Combustion System Parameters and Their Effect upon Diesel Engine Exhaust Emissions, *SAE Paper 720750* (1972).
17. Kittelson, D. B., Dolan, D. F., Diver, R. B., and Aufderheide, E., Diesel Exhaust Particulate Size Distributions—Fuel and Additive Effects, *SAE Paper 780787* (1978).
18. Apostolescu, N. D., Matthew, R. D., and Sawyer, R. F., Effects of a Barium-Based Fuel Additive on Particulate Emissions from Diesel Engines, *SAE Paper 770828* (1977).
19. Marshall, W. F., and Hurn, R. W., Factors Influencing Diesel Emissions, *SAE Paper 680528* (1968).
20. Middlemiss, I. D., Characteristics of the Perkins "Squish lip" Direct Injection Combustion System, *SAE Paper 780113* (1978).
21. Radovanovic, R. S., and Djordjevic, M. S., Diesel Precombustion Chamber Design and Its Influence on the Engine Performance and Pollutants, *SAE Paper 790496* (1979).
22. Hames, R. J., Merrion, D. F., and Ford, H. S., Some Effects of Fuel Injection Systems Parameters on Diesel Exhaust Emissions, *SAE Paper 710671* (1971).
23. Sakai, Y., Nakagawa, Y., Tange, S., and Maruyama, R., Fundamental Study of Oxidation in a Lean Thermal Reactor, *SAE Paper 770297* (1977).
24. Patterson, D. J., et al., Warmup Limitations on Thermal Reactors, *SAE Paper 730201* (1973).
25. Pozniak, D. J., and Siewert, R. M., Continuous Secondary Air Modulation—Its Effect on Thermal Manifold Reactor Performance, *SAE Paper 730493* (1973).

26. Herrin, R. J., The Importance of Secondary Air Mixing in Exhaust Thermal Reactor Systems, *SAE Paper 750174* (1975).

27. Kadlec, R. H., et al., Limiting Factors on Steady State Thermal Reactor Performance, *SAE Paper 730202* (1973).

28. Herrin, R. J., Lean Thermal Reactor Performance Characteristics—A Screening Study, *SAE Paper 760319* (1976).

29. Zemke, B. E., and Gumbleton, J. J., General Motors Progress Towards the Federal Research Objective Emission Levels, *SAE Paper 800398* (1980).

30. Reddy, J. N., Application of Automotive Sensors to Engine Control, *SAE Paper 780210* (1978).

31. Seiter, R. E., and Clark, R. J., Ford Three-Way Catalyst and Feedback Fuel Control System, *SAE Paper 780203* (1978).

32. McDonald, W. R., Feedback Carburetor Control Electronic Design for Improved System Performance, *SAE SP-441 and Paper 780654* (1978).

33. Esper, M. J., Logothetis, E. M., and Chu, J. C., Titania Exhaust Gas Sensor for Automotive Applications, *SAE Paper 790140* (1979).

34. Sensors for Automotive System, *SAE SP-458* (1980).

35. Springer, K. J., and Stahman, R. C., Removal of Exhaust Particulate from a Mercedes 300D Diesel Car, *SAE Paper 770716* (1977).

36. Winsor, R., Diesel Particulates, *Combustion Engine Performance, Economy and Emissions,* Univ. of Mich. Engr. Sum. Conf., July 7–11, 1980.

37. Wade, W. R., Diesel Particulate Trap Regeneration Techniques, *SAE Paper 810118* (1981).

38. Heavy-Duty Engines for 1979 and Later Model Years, *Federal Register,* Part III, Vol. 42, No. 174, Sept. 8, 1977.

39. Control of Air Pollution from New Motor Vehicles and New Motor Vehicle Engines: Particulate Regulations for Heavy-Duty Diesel Engines, *Federal Register,* Part III, Vol. 46, No. 4, Jan. 7, 1981.

40. Gaseous Emission Regulations for 1984 and Later Model Year Heavy-Duty Engines, *Federal Register,* Part II, Vol. 45, No. 14, Jan. 21, 1980.

41. Simanaitis, D. J., Emission Test Cycles Around the World, *Automotive Engineering* **85**(8) (1977).

42. Control of Air Pollution from New Motor Vehicles and New Motor Vehicle Engines, *Federal Register,* Part III, Vol. 42, No. 124, June 28, 1977.

43. Junija, W. K., Horchler, D. D., and Haskew, H. M., A Treatise on Exhaust Emission Test Variability, *SAE Paper 770136* (1977).

CHAPTER 19

SOURCE CONTROL—
MUNICIPAL SOLID WASTE
INCINERATORS

ALBERT J. KLEE

Industrial Environmental Research Laboratory
U.S. Environmental Protection Agency
Cincinnati, Ohio

1 INCINERATION OF MUNICIPAL SOLID WASTE

1.1 Characteristics of Municipal Solid Waste

Municipal solid waste (MSW) is a somewhat imprecise collective term for a variety of materials, primarily household, commercial, and institutional wastes, street sweepings, and construction or demolition refuse, but it may also include limited quantities of bona fide industrial solid wastes as well. As might be expected, both the quantity and composition of MSW varies significantly with geographical locale; it also, however, varies continuously, but predictably, within any given community by day, week, or month. A brief comparison between an Iowa coal and MSW from St. Louis, MO, is shown in Table 1. As can be seen, the MSW is 179% more variable than the coal with regard to higher heating value, 292% more variable in carbon content, and 247% more variable in ash content.

Several of the more significant major constituents of MSW are shown in Table 2. Generally speaking, MSW has a higher ash, oxygen, moisture, metals, and chlorine content than coal, and a lower heating value, sulfur, carbon, and hydrogen content. The ranges given for these constituents in Table 2 again reflect the great variability of MSW composition. This variability affects MSW incineration emissions directly, through changing composition per se, and indirectly, through changing process requirements.

1.2 Large Systems for the Incineration of Municipal Solid Waste

For quantities greater than 50 tons/day, the rectangular furnace with continuous feed is by far the most common configuration for the incineration of MSW, although a limited number of rotary kiln types have also been constructed for this purpose. (Older designs included "teepee" or conical burners, open pit, and upright cylindrical types as well.) Often the rectangular furnace has two or more cells set side by side, sharing a common secondary combustion chamber and residue disposal hopper. In all but the rotary kiln types, the MSW rests on grates while burning. Usually the grates are movable, in order to (a) move the refuse through the system, (b) agitate the refuse to promote combustion, and (c) remove ash and residue from the furnace. These mechanical systems are called stokers and can be of various types, for example, traveling grate, reciprocating grate, reverse reciprocating, rocking grate, vibrating, and inertial grate stokers. The principal residue is that discharged from the stoker, but there are also siftings that fall through the grate openings and fly ash collected from the flue gas. Stoker residue is typically quenched and cooled in a water trough before it is conveyed to storage hoppers. From these, delivery is made to trucks which ultimately carry the material to a landfill. Siftings and fly ash can either be removed dry or flushed out with water.

Table 1   Relative Variability of a Coal and a Municipal Solid Waste (Figures Shown Are Coefficients of Variation, Expressed as Percentages)

	Iowa Coal	St. Louis MSW
Higher heating value	5.2	14.5
Carbon	2.7	10.6
Ash	5.8	21.7

Sources: Refs. 8 and 16.

Air that enters the furnace through and around the grates is called "underfire" air; that which enters the furnace above the grates through the sides or roof of the furnace is called "overfire" air. The latter is used both to mix and burn the combustible gases driven off from the burning MSW to ensure optimal combustion, and to cool the burning gases to protect the furnace walls (a water spray is sometimes used to cool the flue gases as well). To control the flow of air, forced draft fans are used to remove the flue gases. In older furnaces the draft required to remove the flue gases was provided by tall stacks or chimneys, but modern air pollution control devices generally require induced draft fans ahead of the stack. The secondary combustion zones in MSW incinerators are chambers connected by passages from the primary combustion chamber, but are seldom sharply defined. They serve to improve the degree of combustion.

There are basically two types of furnace walls; refractory and water wall. The older incinerators in the United States are almost exclusively refractory lined since they generally have neither sophisticated air pollution control equipment nor heat recovery systems (exceptions include incinerators located at Chicago, Southwest, IL; Merrick, Hempstead, NY; and Miami, 20th Street, FL). The water walls, which are simply walls lined with structural steel tubes through which water is circulated for the generation of steam, have a significant cooling effect on furnace temperatures. The amount of excess air required for cooling the furnace is substantially reduced and the need for flue gas cooling is eliminated. The quantity of air supplied to the incinerator is a major determinant as to the size of the equipment. In incinerators without water walls in the combustion zone, the quantity of air required to maintain the process at customary conditions is two to three times that theoretically necessary for complete combustion, or 6.4–9.6 tons of air per ton of refuse. To reduce this air requirement, one must either extract heat, as in the water-wall type, or else operate at a high temperature where the ash tends to

Table 2   Ranges of Major Constituents of Municipal Solid Waste ("Typical" Figures Are Given in Parentheses)

Higher heating value, Btu/lb	2800–6300 (4500)
Proximate analysis (wt%, as received)	
Moisture	8–40 (28)
Volatile matter	37–65 (45)
Fixed carbon	0.6–15 (9)
Ash	11–39 (21)
Ultimate analysis (wt%, as received)	
Carbon	15–37 (26)
Oxygen	15–35 (21)
Hydrogen	2–5 (3.5)
Nitrogen	0.5–0.7 (0.6)
Total chlorine	
Water soluble chlorine	0.03–0.41 (0.08)
Organic chlorine	0.01–0.31 (0.09)
Sulfur	0.06–0.28 (0.10)
Metals (wt%, as received)	
Fe	0.2–6.6 (1.1)
Al	0.5–2.4 (0.8)
Cu	0.01–0.22 (0.04)
Pb	0.02–0.21 (0.06)
Ni	0.01–0.05 (0.02)
Zn	0.02–0.22 (0.06)

Sources: Refs. 8, 16, 18, and 32.

fuse into a slag. Slagging, nonpyrolytic incinerators for MSW have not, however, been developed to a satisfactory point at this time.

In addition to these systems for the incineration of MSW on a large scale, there are several other systems of interest as well. One modification of the water-wall incinerator is used at Hamilton, Ontario (Canada) where the solid waste is dumped into a "live" bottom pit (a pit carpeted with 4-ft-wide conveyors) from which it is conveyed to shredders after manual sorting of objects that are difficult to mill (e.g., rugs, mattresses, tires, etc.). This eliminates the need for an overhead crane system. The milled material is transported to a bin from which it is conveyed to the furnaces. Approximately 60% of the milled material is reported to burn in air suspension, with the remainder burning on the traveling grate system. Another type, used in Hempstead, New York, incorporates hydropulping or wet shredding of the solid waste into an aqueous slurry from which the heavy fraction consisting of ferrous metals, nonferrous metals, and glass is removed. The ferrous metals are then removed by magnetic separation and the aluminum separated, leaving the glass which subsequently is sorted by color. The light fraction is processed through extractors to reduce the moisture content to about 45%, and then bunkered and subsequently pneumatically fed into two air-swept spreader–stoker water-wall boilers (typical of those used in the wood industry) which burn approximately 50% of the material in suspension.

The Torrax system is a slagging pyrolysis process in which the gasifier, or vertical shaft furnace which resembles a cupola, is the central piece of equipment. The refuse is initially dried by the gases moving upward through the shaft. As the refuse proceeds farther down the shaft, it is heated and pyrolyzed. The pyrolysis gas has many components, the most important of which are hydrogen, carbon monoxide, and methane. The carbon char remaining from the pyrolysis process proceeds to the base of the shaft (known as the hearth) where it is combusted. The inerts remaining form a slag which is tapped off. The preheated air enters the hearth with auxiliary fuel through tuyeres. This air supports the combustion in the hearth. A similar system, which is no longer on the market since it did not prove to be economically competitive, is the Purox process, differing mainly in that (a) pure oxygen is used instead of air, and (b) the refuse is coarse ground and processed for the recovery of ferrous metals. The gases leaving the Purox converter are cleaned of their oil mist and excess water vapor by passing through a recirculating water scrubber system and an electrostatic precipitator. The liquid hydrocarbons and any entrained solids are separated from the scrubber water and recycled to the furnace for disposal. The net water product discharge from the scrubber system is cleaned of organics and sent to the sewer.

1.3 Basic Concepts in the Incineration of Municipal Solid Waste: Large Systems

As is indicated in Table 2, the primary combustible elements in MSW are carbon and hydrogen, with much lower but significant amounts of sulfur, chlorine, and nitrogen. The net result of effective combustion is the conversion of the carbon in the MSW to carbon dioxide, and the hydrogen to water. Sulfur is converted to sulfur oxides (primarily SO_2), some nitrogen is converted to nitrogen oxides, and organic chlorides are converted to hydrogen chloride. These last-named compounds, along with particulate emissions, constitute the major air pollutants from the incineration of MSW.

Incineration may be thought of as occurring in three overlapping stages that may take place in different areas of the furnace. First, heat from the combustion process is used to drive off moisture from the internal and external surfaces of the MSW; second, the MSW is further heated, causing physical and chemical changes called "pyrolysis"; and third, oxygen in the air reacts with the combustible materials both in the solid portions of the MSW itself and in the gases driven from the solids during pyrolysis, emitting large quantities of heat. Proper incinerator design requires that adequate residence time, temperature, and turbulence be provided for each stage of combustion to ensure optimal contact of oxygen with the combustible materials under conditions where combustion is essentially complete. (These principles also apply to municipal sewage sludge and industrial waste incinerators, to be discussed later in this chapter. All incinerators are designed around these three "T's" of time, temperature, and turbulence. Any unit that does not follow these principles can be expected to emit pollutants and suffer other problems as well.)

Time is designed into the incinerator through the space in the combustion chambers of the furnace. The volume is made large enough to retain the gas flow long enough to allow for complete combustion of the MSW and the derived volatile gases from pyrolysis. Temperature is the most important "T" since heat is used as the driving force to sustain combustion. In many instances, supplemental heat is supplied by an auxiliary burner to preheat the incinerator or support the combustion of those materials with a large amount of moisture or with a low heat value. Combustion at 890°C (1600 to 1800°F) will normally produce a sterile residue and an effluent gas free of odors, assuming adequate furnace design and operations that do not exceed design feed rate. In some units, turbulence is designed into the incinerator by a series of baffles or constrictions. These changes in direction (due to the baffles) and increases or decreases in velocity (due to the constrictions or expansions) serve to mix the products of combustion with the air (oxygen) necessary for combustion. Separation or stratification of the combustion gases would occur if turbulence were not included in the design. Under these conditions, some of the gases would go through the incinerator unburned. As MSW is decidedly not uniform material, this is an important consideration.

The air supplied to the incinerator has several important functions. It supplies the oxygen necessary for combustion; it carries combustion products and vaporized water from the incinerator; it aids in establishing turbulent conditions in the furnace and in cooling vital furnace elements; and finally, the air and the resulting flue gas absorb heat from the combustion reactions and carry it from the combustion zone. Excess air is added to the incinerator to ensure complete combustion and to regulate the incinerator temperature. The excess air requirements vary for different moisture contents of the MSW; refuse with higher moisture content has lower excess air requirement. Many incinerators are designed to operate at a temperature of 1000°C (1800°F), the reason being to eliminate odors [the minimum temperature required to eliminate odors under perfect conditions is 830°C (1500°F)] and to protect the refractory lining [common refractory materials start to break down above 1110°C (2000°F)]. A typical refuse will burn with a temperature of 1000°C (1800°F) when provided with about 135% excess air. The more excess air supplied, the lower the operating temperature in the incinerator.

1.4 Small-Scale Systems for the Incineration of Municipal Solid Waste

Controlled (starved) air MSW incinerators for quantities less than 50 tons/day are factory engineered and constructed, and shipped to the sites as prepackaged systems. They are installed in municipal incinerator plants in identical modules to achieve the desired plant capacity. Normal variations in MSW can be processed in these incinerators without special treatment (such as shredding) before charging. They operate on a batch feed basis, rather than continuously, within a 24-hr cycle; normally they are charged for 7–8 hr, then burned down with the use of auxiliary fuel for approximately three more hours, and then allowed to cool overnight. Ash residue is removed by an operator each morning, before the start of the next 24-hr cycle.

The design of these units is based on the controlled air principle. Two refractory-lined and insulated chambers, designed "primary" and "secondary," are used for the combustion process, and a refractory-lined stack is used to duct the exhaust gases to the atmosphere. The MSW is charged into the primary chamber where ignition takes place. Once preset conditions of temperature and time have been achieved, the primary chamber auxiliary fuel burners (which use either natural gas, oil, or propane) are turned off automatically.

A reducing atmosphere (insufficient oxygen for complete combustion) is maintained in the primary chamber and the gas velocity is kept very low to minimize the entrainment of particulates in the effluent gas stream. This gas, which is high in fuel value, is introduced into the secondary chamber through a zone of turbulent air flow where the combustion process continues. An oxidizing atmosphere (excess air) is maintained in the secondary chamber to assist in the further oxidation of the combustible material in the gas. The effluent from the secondary chamber flows into the stack through an air induction section. The inducer serves to cool the hot gas stream prior to its exit from the system. (The primary chambers are also usually supplied with water spray systems to assist in controlling temperatures.) The unit operates at a slightly negative pressure, which ensures that any infiltration is directed into the system.

In these controlled air incinerator designs, particulates, including metallics, are considerably reduced in the gas stream by the high temperatures in the secondary chamber (afterburner), due to the use of auxiliary fuel, and without the use of mechanical pollution control systems or devices. Hence, the energy conversion systems that have been developed for use with these incinerators have utilized the hot gas stream exiting from the afterburner chamber as the heat source to operate a packaged boiler installed adjacent to the incinerator.

With regard to apartment house incinerator designs, these commonly are either single- or double-chamber designs using a fixed grate. Large amounts of underfire and overfire air are used in the main chamber, and a secondary gas burner is located in the second chamber if one is used. The flue gas is sometimes treated in a wet scrubber.

1.5 Air Pollution Regulations Applicable to Municipal Solid Waste Incinerators

For thermal processing facilities burning wastes where more than 50% is MSW, which commenced construction after December 23, 1971, and which charge more than 50 tons/day of solid waste, the very specific federal regulation for particulate emission concentrations must be met. These standards prohibit the discharge into the atmosphere of particulate matter, the concentration of which is in excess of 0.08 gr/scf (70°F, 1 atm) of flue gas on a dry basis corrected to 12% carbon dioxide by volume, maximum 2-hr average. The particulate emissions are to be measured in accordance with U.S. Environmental Protection Agency Method 5, *Determination of Particulate Emissions from Stationary Sources* (Environmental Protection Agency Standards of Performance for New Stationary Sources, 36 FR 24880ff, December 23, 1971), which measures only "dry catch" particulates. A requirement for recording burning rates, hours of operation, and any particulate emission measurements that are made is also included.

On a state level, similar regulations exist for both new and existing facilities but the actual numerical standard, the definition of "particulate" (for the purpose of air quality standards, "particulate" generally means that which is filterable from the air and which remains on the filter after conditioning at 59 to 95°F), and the test method may be significantly different. The state of Maryland, for example, has a particulate standard of 0.03 gr/scf at 12% CO_2. State regulations may also have prohibitions against particulate emissions that are visible. These regulations may set a specific standard such as a limit on opacity (typically 20%) or an equivalent smoke density (e.g., Ringelmann No. 1) or simple prohibition of any visible particles.

In addition to regulations based on concentration or rate of emissions from the incinerator and the visual appearance of the emissions, allowable particulate emissions are also determined by the air quality in the region affected by the incinerator. The 1970 amendments to the Clean Air Act required the EPA to set two levels of national ambient air quality standards; primary standards to protect human health, and stricter secondary standards to protect public welfare (i.e., to prevent damage to crops, livestock, buildings, etc.). The EPA has issued primary and secondary standards for seven major air pollutants: sulfur oxides (SO_x), total suspended particulates, carbon monoxide (CO), hydrocarbons, nitrogen oxides (NO_x), ozone, and lead. These standards are shown in Tables 3 and 4. The 1977 amendments specified that all areas of the country must meet primary standards by December 31, 1982, with a possible 5-yr extension for those areas having a particular problem with CO and ozone. Secondary standards are to be met in an unspecified "reasonable" time period. While not specific to MSW incinerators, the air quality standards nevertheless must not be exceeded because of insufficient control in new and existing installations.

Table 3   National Ambient Air Quality Standards for Particulate Matter

Primary Standards	Secondary Standards
(a) 75 $\mu g/m^3$, annual geometric mean. (b) 260 $\mu g/m^3$, maximum 24-hr concentration not to be exceeded more than once per year.	(a) 60 $\mu g/m^3$, annual geometric mean (as a guide to be used in assessing implementation plans to achieve the 24-hr standard) (b) 150 $\mu g/m^3$, maximum 24-hr concentration not to be exceeded more than once per year.

Source: Ref. 15.

Table 4   Federal Ambient Air Standards for Gaseous Pollutants

	Primary Standard	Secondary Standard
Carbon monoxide	(a) 10 mg/m³ (9 ppm) maximum 8-hr concentration[a]	Same as primary
	(b) 40 mg/m³ (35 ppm) maximum 1-hr concentration[a]	Same as primary
Hydrocarbons	160 $\mu g/m^3$ (0.24 ppm) maximum 3-hr concentration (6–9 A.M.)[a]	Same as primary
Ozone	235 $\mu g/m^3$ (0.12 ppm)	Same as primary
Sulfur oxides (as SO_2)	(a) 80 $\mu g/m^3$ (0.03 ppm) annual arithmetic mean	1300 $\mu g/m^3$ (0.5 ppm) maximum 3-hr concentration
	(b) 365 $\mu g/m^3$ (0.14 ppm) maximum 24-hr concentration[a]	
Nitrogen dioxide	100 $\mu g/m^3$ (0.05 ppm) annual arithmetic mean	Same as primary
Lead	1.5 $\mu g/m^3$ maximum arithmetic mean averaged over a calendar quarter	Same as primary

[a] Not to be exceeded more than once per year.
Source: Ref. 15.

Another factor is the promulgation of "major source" criteria. Briefly, this limitation requires any "major source" seeking to locate in a "nonattainment area" to "improve" the air quality in that area. A major source is defined by the following:

Pollutant Emitted	Limitation, tons/yr
Particulate matter	100
Sulfur oxides	100
Nitrogen oxides	100
Nonmethane hydrocarbons (organics)	100
Carbon monoxide	1000

At this writing there is considerable controversy over these regulations and Congress is constantly redefining these to allow for economic growth. In general, however, it would be expected that the emission of ozone and carbon monoxide would be well under statutory limits of the "major source" requirements as far as MSW incinerators are concerned.

No emission regulations exist that limit mercury emissions from MSW incinerators but the EPA does have a national emission standard for facilities that process sludge from municipal wastewater treatment plants thermally. This limit has been set at 3200 g/24 hr. Even at the highest emission rates known for MSW incinerators, only about 1530 g/24 hr would be emitted, well below the standard.

At the present time, a single national emission standard for small-scale MSW incinerators does not exist. Potential undesirable emissions from small-scale incinerators include odors, noxious gases, and particulates. Odor emissions can be readily eliminated if the combustion products are processed at a sufficiently high temperature, for example, 890–1000°C (1600–1800°F). Modern small-scale incinerators rarely produce excessive amounts of gaseous emissions and generally have no problem meeting existing gaseous emission standards. The major emission problem associated with small-scale incinerators is generally accepted to be connected with particulate emissions. Generally, particulate emission standards for these incinerators fall under one of the following two categories:

1. The requirements imposed by the EPA on all federally owned and financed facilities, regardless of their location. Individual federal agencies may write more restrictive requirements than the EPA standards.

Table 5 EPA and Veterans Administration Regulations for Small-Scale Incinerators

EPA	VA
1. *Visible Smoke Standards* Except during startup or stoking, visible smoke emission to the atmosphere shall not exceed no. 1 on the Ringelmann scale or Public Health Service Smoke Inspection Guide.	1. *Visible Smoke Standards* Visible smoke emission shall not exceed no. 0 on the Ringelmann scale, nor more than a Ringelmann no. 1 during any 3 min of continuous operation in any given hour.
2. *Particulate Standards* (a) Incinerators, capacity 91 kg (200 lb)/hr and over Incinerators having burning rates equal to or greater than 91 kg/hr shall not exceed 0.46 g of particulate matter per standard cubic meter of dry flue gas (0.2 gr/scf) corrected to 12% CO_2 and shall not include particles larger than 60 μm. (b) Incinerators, capacity under 91 kg (200 lb)/hr Incinerators having burning rates of less than 91 kg/hr shall not emit more than 0.69 g of particulate matter per standard cubic meter of dry flue gas (0.3 gr/scf) corrected to 12% CO.	2. *Particulate Standards* Particulate emission shall not exceed 0.23 g/scm of dry flue gas (0.1 gr/scf) adjusted to 12% CO_2 when burning Public Health Service standard waste or equivalent and not more than 0.46 g/scm (0.2 gr/scf) measured in the same manner when burning 100% type 4 waste. In addition, the VA emissions standards must be achieved without the use of scrubbers. As with the EPA requirements, if state or local codes are more stringent than the VA requirements, the stringent local codes must be met.

Source: Ref. 15.

2. The requirements imposed by state and local governments on all facilities other than those of the federal government.

It should be noted that state or local regulations for a particular federal facility take precedence if they are stricter than the EPA regulations. The EPA regulations for small-scale incinerators include both a smoke standard and a particulate standard as shown in Table 5. The Veterans Administration (VA) has written the most restrictive standard of any federal agency, and this is also shown in Table 5.

As with large incinerators, a significant variation among state and local emission standards exists throughout the country. In addition, a wide variety of source test methods and procedures are specified as part of the certification process. Thirty-two states and the District of Columbia currently require that measured particulate emissions be reported either on a 12% CO_2 basis or on the basis of mass of particulate per mass of waste charged. The most frequently specified standards are: 0.46 and 0.23 g/scm of dry flue gas (0.2 and 0.1 gr/scf) corrected to 12% CO_2; and 0.2 and 0.1 kg of particulate per 100 kg of waste charged (0.2 and 0.1 lb/100 lb). Exact conversion of the kg/100 kg waste charged standard to g/scm at 12% CO_2 depends on waste composition and the operating conditions of the incinerator. Using the conversion factors shown in Table 6, based on a representative set of operating conditions and a waste composition that is typical of municipal waste, gives the approximate conversions 0.2 kg/100 kg = 0.488 g/scm at 12% CO_2, and 0.1 kg/100 kg = 0.244 g/scm at 12% CO_2. Small-scale incinerators with particulate emission less than 0.46 g/scm (0.2 gr/scf) at 12% CO_2 will generally meet emission standards in 21 states, while those with emissions less than 0.23 g/scm (0.1 gr/scf) at 12% CO_2 will meet the standards in a total of 32 states. State emission standards change rapidly, however, and state air quality control offices should be consulted for current information when needed.

1.6 Particulate Emissions: Large Incinerators

Perhaps no other aspect of the incineration of MSW receives the attention now given to air pollution emissions. Particulate emissions are by far the greatest concern. Such emissions are reported using a wide range of units, the most common being lb/ton (short ton) of MSW as fired. Often, however, regulations are written to limit particulate emission concentrations to a specified value based on operating the incinerator with an amount of air that would result in a carbon dioxide content of 12% by volume in the flue gas (excluding the contribution of the auxiliary fuel, if any), when measured on a moisture-free basis. Other bases have been used also; Table 7 provides factors for conversions among several

Table 6 Conversion Factors for the Approximate Measurement of Particulate Matter

	lb/ton refuse (as received)	lb/1000 lb flue gas at 50% excess air	lb/1000 lb flue gas at 12% CO_2	gr/scf[a] at 50% excess air	gr/scf[a] at 12% CO_2	g/scm at ntp[b] and 7% CO_2
lb/ton refuse (as received)	1	0.089	0.10	0.047	0.053	0.067
lb/1000 lb flue gas at 50% excess air	11.27	1	1.12	0.52	0.585	0.74
lb/1000 lb flue gas at 12% CO_2	10.0	0.89	1	0.46	0.52	0.66
gr/scf[a] at 50% excess air	21.31	1.93	2.16	1	1.12	1.42
gr/scf[a] at 12% CO_2	18.85	1.71	1.92	0.89	1	1.26
g/scm at ntp[b] and 7% CO_2	15.00	1.36	1.53	0.704	0.79	1

[a] 60°F, 1 atm.
[b] 32°F, 1 atm.
Source: Ref. 20.

Table 7 Conversion Factors for the Measurement of Particulate Matter

$A = \dfrac{N_2}{N_2 - 3.788(O_2 - 0.5CO)}$

i.e., the ratio of actual air used in the combustion (exclusive of secondary burner requirements) to the stoichiometric air required for burning the incinerator fuel.

B = the heating value of the as-fired incinerator fuel, Btu/lb.

E = the excess air as a decimal; $E = A - 1$.

F = the volume of air required to burn the as-fired incinerator fuel at 0% excess air, scf/lb.

G = the volume ratio of combustion products to required air.

H = fuel being burned, lb/hr.

L = percent moisture in the fuel.

M = dry molecular weight of the flue gas corrected to reflect primary fuel data; $M = 0.16CO_2 + 0.04O_2 + 28$.

$P = G + E$.

N_2, O_2, CO_2, and CO are percentage concentrations in the dry flue gas as given by Orsat data (corrected to eliminate the effects of secondary burners).

Multiply This \ To Get	gr/scf at 12% CO₂	lb/100 lb dry refuse	lb/hr	gr/scf	lb/1000 lb dry gas at 50% E	lb/hr per lb/hr charged	lb/1000 lb dry gas	gr/scf at 7% O₂	gr/scf at 50% E	lb/10⁶ Btu	lb/1000 lb dry gas at 12% CO₂
gr/scf @ 12% CO₂	—	$\dfrac{CO_2LFP}{840}$	$\dfrac{CO_2HFP}{84,000}$	$\dfrac{CO_2}{12}$	$\dfrac{CO_2A}{0.326M}$	$\dfrac{CO_2FP}{84,000}$	$\dfrac{4.6CO_2}{M}$	$\dfrac{7CO_2}{12O_2}$	$\dfrac{CO_2A}{18}$	$\dfrac{CO_2FP}{0.084B}$	$\dfrac{55.2}{M}$
lb/100 lb dry refuse	$\dfrac{840}{CO_2LFP}$	—	$\dfrac{H}{100L}$	$\dfrac{70}{LFP}$	$\dfrac{258A}{LMFP}$	$\dfrac{1}{100L}$	$\dfrac{3865}{MLFP}$	$\dfrac{490}{O_2LFP}$	$\dfrac{46.7A}{LFP}$	$\dfrac{10,000}{LB}$	$\dfrac{46,380}{CO_2LMFP}$

lb/hr	$\frac{84{,}000}{CO_2HFP}$	$\frac{100L}{H}$	—	$\frac{7000}{HFP}$	$\frac{258A}{HFP}$	$\frac{1}{H}$	$\frac{386.5}{HFP}$	$\frac{49{,}000}{O_2HFP}$	$\frac{4670A}{HFP}$	$\frac{10^6}{HB}$	$\frac{4640}{CO_2HFP}$
gr/scf	$\frac{12}{CO_2}$	$\frac{LFP}{70}$	$\frac{HFP}{7000}$	—	$\frac{36.8A}{M}$	$\frac{FP}{7000}$	$\frac{55.2}{M}$	$\frac{7}{O_2}$	$\frac{A}{1.5}$	$\frac{143FP}{B}$	$\frac{662}{CO_2M}$
lb/1000 lb dry gas @ 50% *E*	$\frac{0.326M}{CO_2A}$	$\frac{LMFP}{258A}$	$\frac{HFP}{258A}$	$\frac{M}{36.8A}$	—	$\frac{MFP}{258{,}000A}$	$\frac{1.5}{A}$	$\frac{73.5M}{O_2}$	$7M$	$\frac{MFP}{0.258BA}$	$\frac{18}{CO_2A}$
lb/hr per lb/hr charged	$\frac{84{,}000}{CO_2FP}$	$100L$	H	$\frac{7000}{FP}$	$\frac{258{,}000A}{MFP}$	—	$\frac{386{,}000}{MFP}$	$\frac{49{,}000}{O_2FP}$	$\frac{4670A}{FP}$	$\frac{10^6}{H}$	$\frac{4.64 \times 10^6}{CO_2MFP}$
lb/1000 lb dry gas	$\frac{M}{4.6CO_2}$	$\frac{MLFP}{3865}$	$\frac{HFP}{386.5}$	$\frac{M}{55.2}$	$\frac{A}{1.5}$	$\frac{MFP}{386{,}500}$	—	$\frac{M}{7.89O_2}$	$\frac{MA}{82.8}$	$\frac{MFP}{0.386B}$	$\frac{12}{CO_2}$
gr/scf at 7% O_2	$\frac{12O_2}{7CO_2}$	$\frac{O_2LFP}{490}$	$\frac{HFP}{49{,}000}$	$\frac{O_2}{7}$	$\frac{O_2}{73.5}$	$\frac{O_2FP}{49{,}000}$	$\frac{7.89O_2}{M}$	—	$\frac{O_2A}{10.5}$	$\frac{O_2FP}{0.049B}$	$\frac{94.7O_2}{CO_2M}$
gr/scf at 50% *E*	$\frac{18}{CO_2A}$	$\frac{LFP}{46.7A}$	$\frac{HFP}{4670A}$	$\frac{1.5}{A}$	$\frac{1}{7M}$	$\frac{FP}{4670A}$	$\frac{82.8}{MA}$	$\frac{10.5}{O_2A}$	—	$\frac{FP}{0.00467BA}$	$\frac{994}{CO_2MA}$
lb/10^6 Btu	$\frac{0.084B}{CO_2FP}$	$\frac{LB}{10{,}000}$	$\frac{HB}{10^6}$	$\frac{0.007B}{FP}$	$\frac{0.258BA}{MFP}$	$\frac{H}{10^6}$	$\frac{0.386B}{MFP}$	$\frac{0.049B}{O_2FP}$	$\frac{0.00467BA}{FP}$	—	$\frac{4.63B}{CO_2MFP}$
lb/1000 lb dry gas at 12% CO_2	$\frac{M}{55.2}$	$\frac{CO_2FPML}{46{,}380}$	$\frac{CO_2HFP}{4640}$	$\frac{CO_2M}{662}$	$\frac{CO_2A}{18}$	$\frac{CO_2MFP}{4.64 \times 10^6}$	$\frac{CO_2}{12}$	$\frac{CO_2M}{94.7O_2}$	$\frac{CO_2MA}{994}$	$\frac{CO_2MFP}{4.63B}$	—

Source: Ref. 24.

of the most important forms, and Table 8 shows the particulate definitions used by the various states. Unfortunately, the conversion factors depend on the heating value and other characteristics of the fuel and the operating conditions of the particular incinerator. The conversion factors shown in Table 6, however, can be used for quick and approximate purposes.

Uncontrolled particulate emissions vary with the design and operation of the equipment as well as with the nature of the refuse, but typically average somewhere between 17 and 35 lb/ton. It should be noted that while the particulate emissions from a water-wall furnace may be similar to that from a refractory type on a per ton of refuse basis, water-wall emissions may be significantly higher on a concentration basis, due to lower air flows. Some of the major variables are: (a) ash content of the MSW, (b) underfire and overfire air flows, (c) burning rate, (d) furnace temperature, (e) grate agitation, and (f) combustion chamber design. Three mechanisms are believed to be mainly responsible for particulate emissions:

1. Mechanical entrainment of particles from the burning waste bed.
2. The cracking or pyrolysis gases.
3. The vaporization of metal salts or oxides.

The first of these mechanisms is favored by refuse with a high percentage of fine ash, by high underfire air velocities, or by other factors that induce a high gas velocity through the bed. The second is favored by refuse with a high volatile content producing pyrolysis gases with a high carbon content, and by conditions above the fuel bed that prevent the burnout of the coked particles formed by the cracking of the volatiles. The third mechanism is favored by high concentrations of metals that form low boiling point oxides, and by high temperatures within the bed.

The composition of particulate emissions from MSW incinerators is dependent on design and operation, as well as refuse ash composition. A poorly designed or operated incinerator may emit carbon particles (soot), and the inorganic (mineral) ash will contain a significant quantity of combustibles. Typically, incinerators show a range of from 6 to 40 percent in the combustible content of their

Table 8 Particulate Definitions Used by the States

Alabama 1	Louisiana 1	Oregon 4
Alaska 1	Maryland 4	Pennsylvania 1
Arizona 5	Massachusetts 1	Rhode Island 1
Arkansas 1	Michigan 7	South Carolina 10
California 4	Minnesota 4	South Dakota 2
Colorado 1	Mississippi 1	Tennessee 2
Connecticut 1	Missouri 1	Texas 3
Delaware 3	Montana 1	Vermont 2
Florida 9	Nebraska 1	Virginia 1
Georgia 1	Nevada 3	Washington 8
Hawaii 2	New Hampshire 1	West Virginia 3
Idaho 2	New Jersey 1	Wisconsin 11
Illinois 1	New York 3	Wyoming 2
Indiana 5	North Carolina 3	Washington, D.C. 1
Iowa 1	North Dakota 4	Puerto Rico 2
Kansas 1	Ohio 2	Federal 1
Kentucky 1	Oklahoma 6	

Key
1. gr/scf at 12% CO_2 (2.29 g/scm)
2. lb/100 lb dry refuse charged (kg/100 kg)
3. lb/hr (0.454 kg/hr)
4. gr/scf (2.29 g/scm)
5. lb/1000 lb gas at 50% excess air (g/kg)
6. lb/hr per lb/hr charge (0.454 kg/hr per 0.454 kg/hr)
7. lb/1000 lb gas (g/kg)
8. gr/scf at 7% O_2 (2.29 g/scm)
9. gr/scf at 50% excess air (2.29 g/scm)
10. lb/10^6 Btu (1.80 g/10^6 cal)
11. lb/1000 lb gas at 12% CO_2 (g/kg)
12. lb/ton fuel charged (kg/2 metric tons); this is not legal but is widely used.

Source: Ref. 24.

particulates. The inorganic content of fly ash samples from incinerators of both the older refractory and the newer water-wall type are listed in Table 9. However, the composition of the suspended particles emitted is quite different from that of the collected fly ash, with concentrations of many elements differing by an order of magnitude between the two materials (Table 10). Contribution of MSW incinerators to urban aerosols, therefore, must be estimated with the use of elemental concentrations in the suspended particles and not in the fly ash. Two such suspended particle analyses are shown in Table 11, one from an incinerator with an efficient pollution control system (mechanical separators plus an electrostatic precipitator) and the other from one with an inefficient system (water spray baffle system). With regard to suspended particles, elements that have predominantly small-particle distributions are Na, Cs, Cl, Br, Cu, Zn, As, Ag, Cd, In, Sn, Sb, W, and Pb. These elements usually have greater than 75% of their mass in particles of diameter less than 2 μm. Elements that have predominantly large-particle distributions are Ca, Al, Ti, Sc, and La. Vanadium, Cr, Mn, Fe, Co, and Se have mixed size distributions. MSW incinerators can be the major source of Zn, Cd, and Sb in many cities, and it appears also that major fractions of Ag, Sn, and In in urban air can also come from incinerators. Despite the high concentrations of Pb in incinerator material, the contribution of leaded gasoline is now a much greater source (as well as Br). Marine aerosols and leaded gasoline are usually considered the major Cl source, but appreciable amounts are emitted on particles from MSW incinerators, and there may be considerably more in the gas phase.

Particle size distribution and specific gravity of the particulate matter are properties that are essential design data for most particulate removal devices. The smaller particles require more sophisticated

Table 9 Concentrations of Elements in Municipal Solid Waste Incinerator Fly Ash (as μg/g unless % Indicated)

Element	Washington, D.C.[a]	Alexandria, VA[a]	Braintree, MA[b]	Hamilton, Ontario[b]
Potassium, %	—	—	1.0	8.3
Silicon, %	—	—	1.0	5.9
Calcium, %	5.0	4.3	0.8	4.8
Sodium, %	1.5	1.5	0.6	4.2
Aluminum, %	13.5	10.9	0.9	3.4
Zinc, %	2.4	1.1	1.0	2.7
Iron, %	2.5	5.2	0.4	1.0
Lead, %	—	0.4	1.0	0.6
Titanium, %	2.9	3.2	0.6	0.4
Magnesium, %	1.3	1.3	0.4	0.1
Cadmium	185	42	1100	1400
Manganese	2100	4300	770	1200
Boron	—	—	120	110
Tin	2400	1430	4400	1100
Copper	950	980	1800	700
Nickel	—	740	—	670
Chromium	780	1330	270	600
Mercury	—	—	—	300
Strontium	280	610	970	120
Barium	3200	2800	540	50
Molybdenum	—	—	43	40
Zirconium	—	—	31	30
Antimony	580	270	1000	30
Bismuth	—	—	1900	30
Vanadium	140	135	39	20
Cobalt	27	35	38	10
Silver	220	85	165	10
Gallium	—	—	38	10
Arsenic	59	40	120	10
Phosphorus	—	—	4400	—

[a] Refractory furnace.
[b] Water-wall furnace.
Sources: Refs. 14 and 23.

Table 10 Differences Between Fly Ash and Suspended Particles for Selected Elements Emitted from Municipal Solid Waste Incinerators

	Alexandria, VA[a]		Washington, D.C.[b]	
	Fly Ash	Suspended Particles	Fly Ash	Suspended Particles
Zinc, %	1.1	12.0	2.4	14.0
Silver, $\mu g/g$	85	390	220	1000
Cadmium, $\mu g/g$	42	1100	185	1900
Indium, $\mu g/g$	0.7	6.5	1.0	3.6
Tin, %	0.14	1.07	0.24	1.08
Lead, %	0.4	9.7	7.8	6.7

[a] Water spray baffle system.
[b] Mechanical separator and ESP.
Source: Ref. 14.

(and expensive) equipment to meet a specific emission limit. Table 12 presents data on the particle size and other properties of particulate matter gathered in the area between the combustion chamber and the gas cleaning system for three refractory-type incinerators (installations 1, 2, and 3) and two modern water-wall types (installations 4 and 5). The data for the last does not differ much in size distribution from one of the refractory types. Loading, however, like size distribution in general, varies widely. Most factors that affect particle loading, it should be noted, also affect the size of the particles emitted.

Electrical resistivity is an important property of particulates in the design of the electrostatic precipitators commonly used in modern incinerators. High resistivity reduces collection efficiency, while low resistivity may result in reentrainment of the particles into the gas stream after collection. Resistivity is a function of the basic particle characteristics, and the composition and temperature of the flue gas stream. The presence of moisture and very low concentrations of certain chemicals in the flue gas, such as sulfur trioxide and ammonia, may strongly influence particle resistivity and thus precipitator efficiency. Because the particle electrical resistivity as a function of ESP operating temperature will vary with the composition of the waste burned, the desirable range of resistivity (10^6 to 2×10^{10} ohm-cm) clearly will influence the choice of electrostatic precipitator operating temperature.

1.7 Gaseous Emissions: Large Municipal Solid Waste Incinerators

The overwhelming quantity of stack emissions from MSW incinerators consists of relatively innocuous gaseous combustion products, viz., carbon dioxide and water, and unused oxygen and inert nitrogen from the combustion air. The relative and absolute quantities of each are determined primarily by (a) the composition of the refuse, (b) the amount of excess air used (deliberately as well as through leakage), and (c) the amount of air and/or water used for flue gas cooling and cleaning. These gases are generally of little concern, except for water vapor plume formation and insofar as certain air pollution standards are based on a specific gas composition (e.g., 12 volume % CO_2 or 7 volume % O_2). Water plumes are not regulated but practical considerations, such as icing of nearby roads or buildings, fogging over roadways, or psychological reaction to a visible emission may necessitate at least some degree of control. Table 13 shows typical gas compositions for several types of incinerators and air pollution control combinations.

Poor combustion can result in emissions of carbon monoxide, hydrocarbons, oxygenated hydrocarbons, and other complex compounds. Some of these emissions are the source of odors that are associated with older incinerator operations. (Handling and storage of MSW, as well as furnace leakage, often can create odors that may be detected both within the working areas of the facility and in nearby residential or commercial areas.) Table 14 lists some measurements that have been made on older incinerators with either scrubbers or spray chambers, and two modern water-wall types. As can be seen, the wet collection devices can remove small amounts of some of these materials.

The emission of CO is a function of proper employment of the combustion air as it relates to: (a) distribution within the ignition chamber, and (b) the levels of oxygen available in the combustion zone. The hydrocarbon content of incinerator flue gases, like that of CO, is usually very low. There does not seem to be any significant difference in the total amount of hydrocarbons produced by varying: (1) the fuel charging rate, (2) the stoking interval, (3) the distribution of combustion air, or (4) the pounds of fuel per charge. Aldehydes are generated in MSW incinerators in minute quantities. The amount generated has been shown to be related to the temperature of the furnace gases, which in turn is related to the amount of excess air supplied and the amount of underfire air. In general, as

Table 11   Concentrations of Inorganic Constituents of Municipal Solid Waste Incinerator Suspended Particles (Concentrations in μg/g unless % Indicated)

	A	B
Na, %	9.8 ± 2.8	6.5 ± 1.3
Cs	3.1 ± 1.7	5.5 ± 2.2
Mg, %	0.68 ± 0.25	0.36 ± 0.07
Ca, %	2.3 ± 1.1	1.1 ± 0.3
Sr	<1000	<1000
Ba	890 ± 570	990 ± 410
Cl, %	20 ± 5	14 ± 3
Br	2600 ± 140	920 ± 520
I	<100	~40
Al, %	1.6 ± 0.8	2.1 ± 1.0
Sc	1.0 ± 0.8	1.6 ± 0.4
Ti, %	0.29 ± 0.14	0.36 ± 0.20
V	22 ± 12	40 ± 20
Cr	490 ± 350	870 ± 370
Mn	1500 ± 1400	410 ± 110
Fe, %	0.90 ± 0.33	0.71 ± 0.14
Co	12 ± 7	5.4 ± 2.5
Ni	200 ± 80	170 ± 70
Cu	2000 ± 1200	1500 ± 500
Zn, %	12 ± 6	14 ± 5
As	210 ± 100	310 ± 160
Se	32 ± 21	39 ± 29
Ag	390 ± 360	1000 ± 800
Cd	1100 ± 400	1900 ± 700
In	6.5 ± 5.4	3.6 ± 0.7
Sn, %	1.07 ± 0.15	1.08 ± 0.11
Sb	2400 ± 2400	2400 ± 1100
La	3.8 ± 2.6	4.3 ± 1.1
Ce	11 ± 9	18 ± 2
Sm	~0.8	0.62 ± 0.15
Eu	~0.5	<0.3
Lu	<0.2	~1
Yb	~2	~0.2
Th	1.8 ± 1.4	~3
Hf	0.96 ± 0.57	~1
Ta	<1	~4
W	17 ± 11	22 ± 8
Au	0.77 ± 0.71	0.70 ± 0.06
Pb, %	9.7 ± 2.6	7.8 ± 1.1

A = water spray baffle system.
B = mechanical separator and ESP.
Source: Ref. 14.

the amount of excess air is increased, which decreases the furnace temperature, the quantities of aldehydes emitted increase. Table 14 shows ranges and typical concentrations emitted by MSW incinerators of these compounds.

Minor quantities of sulfur oxides, ammonia, and halide gases are produced from the sulfur, nitrogen, and halide (chlorine, bromine, and fluorine) content of the MSW. Nitrogen oxides emitted may result from the nitrogen content of the waste or high-temperature oxidation of nitrogen in the air. Nitrogen dioxide (NO_2) emissions increase with increasing amounts of excess air; wet cleaning processes tend to increase slightly the amount of NO_2 produced. The amount of underfire air has a significant effect on NO_2 production, an occurrence that has been explained as being a result of the variance in oxygen consumption and its residence time with underfire air. Another variable found to be significant in the formation of nitrogen oxides is feed rate; no distinguishable effects on NO_2 by (1) pounds of fuel

Table 12 Physical Properties of Particulates Leaving Furnaces

	Installation				
	1	2	3	4	5
Specific gravity, g/cm³	2.65	2.70	3.77	—	—
Bulk density, lb/cf	—	30.87	9.4	—	—
Loss on ignition at 750°C, wt%	18.5	8.15	30.4	—	—
Size distribution, wt%					
0.3 μm	—	—	—	14.6	12.0
2 μm	13.5	14.6	23.5	23.9	42.2
4 μm	16.0	19.2	30.0	26.7	55.6
6 μm	19.0	22.3	33.7	30.4	64.0
8 μm	21.0	24.8	36.3	34.9	70.0
10 μm	23.0	26.8	38.1	39.2	76.3
15 μm	25.0	31.1	42.1	—	87.2
20 μm	27.5	34.6	45.0	—	—
30 μm	30.0	40.4	50.0	—	—
Particulate emission rate, lb/ton refuse	12.4	25.1	9.1	27.5	11.3
Particulate emission rate, lb/hr	156	241	46.1	415	60
Particulate emission rate, gr/scf at 12% CO_2	0.55	0.69	0.38	1.46	0.36
Incinerator capacity, ton/day	250	250	120	360	120

Sources: Refs. 3 and 29.

Table 13 Gas Compositions for Selected Refractory and Water-Wall Municipal Solid Waste Incinerators

	Installation[a]				
	1	2	3	4	5
Refuse heating value (HHV), Btu/lb	5246	4360	4900	4391	—
Refuse firing rate, ton/hr	13.6	16.7	15.1	4.9	9.2
Excess air, %	—	71.7	84	400	180
Stack exhaust temperature, °F	269	411	430	388	152
Volume % CO_2, dry basis	8.6	10.5	10.4	4.2	4.8
Volume % O_2, dry basis	11.5	9.0	9.5	16.1	13.0
Volume % N_2, dry basis	79.9	80.5	80.1	—	82.2
Volume % H_2O	13.4	13.3	—	6.3	28.4
Flow, acfm	97,870	84,700	—	35,561	74,800

[a] Installations 1, 2, 3, and 4 are water-wall types; no. 5 is of refractory design. Installation no. 1 uses shredded solid waste as a feed.
Sources: Refs. 3, 6, 12, 23, and 28.

Table 14 Carbon Monoxide and Organic Gas Emissions from Municipal Solid Waste Incinerators ("Typical" Figures Are Given in Parentheses)

Constituent	Refractory	Water Wall
Carbon monoxide, ppm	0–3000 (440)	0–501 (209)
Hydrocarbons, ppm	0–240 (30)	9–1250 (358)
Oxygenated hydrocarbons		
Aldehydes, ppm	0.3–9.2 (5.3)	0.4–21 (9)
Organic acids, ppm	0.1–1.6 (1.1)	6–19 (13)

Sources: Refs. 3, 5, 6, 12, 13, and 23.

Table 15 Sulfur Oxide, Ammonia, Mercury, Nitrogen Oxide, and Halide Emissions from Municipal Solid Waste Incinerators ("Typical" Figures Are Given in Parentheses)

Constituent	Refractory	Water Wall
Nitrogen oxides, ppm	3–92 (35)	0–200 (76)
Sulfur oxides (as SO_2), ppm	15–40 (26)	114–236 (117)
Ammonia, ppm	(29)	—
Chlorides (as HCl), ppm	12–732 (227)	47–259 (166)
Fluorides (as HF), ppm	0.6–2.3 (1.1)	1.0–2.4 (1.6)
Mercury, ppm	—	0.015–1.230 (0.200)

Sources: Refs. 13, 23, and 32.

charged, (2) stoking interval, (3) percent secondary air, or (4) percent underfire air have been found. Sulfur oxide emissions from MSW incinerators are practically negligible because the sulfur content in refuse generally averages only about 0.1%. For comparison, the sulfur content of coals fired in U.S. power plants averages approximately 1.0–2.5%. If the stack gas concentration of SO_2 is measured in parts per million by volume, the amount of diluting excess air supplied to the furnace will directly affect the concentration of SO_2. Incinerators equipped with auxiliary burners for either waste heat recovery boilers or for burning low calorific value refuse often use high-sulfur content fuels such as coal and oil. In this type of incinerator, SO_2 emissions are considerably higher and can be of some concern.

All of these emissions are expected to be directionally, but not necessarily quantitatively, related to the concentration of the source element in the waste burned. Table 15 shows typical quantities emitted. Hydrogen chloride has been of particular concern to health scientists because of increasing emissions due to increased disposal of polyvinyl chloride (PVC) and other halide-containing plastics and aerosols. It is believed that almost all of the chlorine in PVC (greater than 50% chlorine) is converted to HCl but that some of the HCl reacts with particulate matter and is removed by particulate control equipment. However, gaseous incinerator emissions are not, at least at the present, of primary concern as a source of air pollution. When compared to other gaseous emission sources, the contribution of MSW incinerators presently is relatively small. It is for this reason that MSW incinerator air pollution control equipment is adapted to the removal of particulate matter rather than gases.

1.8 Emissions from Slagging Municipal Solid Waste Incinerators

Table 16 summarizes the information available on the Torrax and Purox slagging MSW incinerators. The characteristics of the pyrolysis gases produced by these two systems differ considerably since the latter uses pure oxygen as the combustion gas. For the Torrax unit, the SO_2 and NO_x emissions are well below the EPA new source requirements for stationary combustion sources of 1.2 and 1.3 lb/

Table 16 Emissions from Slagging Municipal Solid Waste Incinerators

	Torrax	Purox
Pyrolysis gas		
H_2, volume %, dry	15.4	26
CO, volume %, dry	14.8	40
CH_4, volume %, dry	2.6	5
C_2H_6, volume %, dry	1.5	5
CO_2, volume %, dry	14.3	23
O_2, volume %, dry	2	0
N_2, volume %, dry	49.4	1
SO_2, lb/10^6 Btu	0.23	—
NO_x, lb/10^6 Btu	0.24	—
Particulate loading rate, lb/10^6 Btu	4.80	—
Particulate emission rate, gr/scf at 12% CO_2	—	0.02–0.05

Sources: Refs. 22 and 25.

10^6 Btu, respectively, and the controlled particulate emissions from the Purox unit are well below the EPA emission limitation on MSW-fired resource recovery units of 0.08 gr/scf at 12% CO_2.

1.9 Emissions from Small-Scale Municipal Solid Waste Incinerators

The data available indicate that controlled air incinerators produce less particulate emission than the apartment house types. In all but one test, the controlled air results fell below the 0.46 g/scm (0.2 gr/scf) standard, and most of the results were below the 0.23 g/scm (0.1 gr/scf) standard. This is achieved without additional treatment of the flue gas such as scrubbing, electrostatic precipitation, or the use of a cyclone separator. It is further noted that if the emission standard were to be arbitrarily reduced to 0.184 g/scm (0.08 gr/scf) at 12% CO_2, about 40% of the controlled air designs would not be able to comply without the addition of particulate control devices. The types of apartment house incinerators tested do not meet the applicable (federal facility) small-scale standard of 0.46 g/scm (0.20 gr/scf) at 12% CO_2. Other emission data from apartment house incinerators are shown in Table 17.

Although for the controlled air incinerator data available there is no correlation between the particulate emissions and the type of solid waste burned, Table 18 provides additional data from three controlled air incinerators having waste burning capacities of less than 50 tons/day in which only municipal solid waste is burned. The CO_2 content of the flue gases is lower than that produced by large water-wall MSW incinerators; the CO emissions are low, as are SO_x, NO_x, chloride, fluoride, and hydrocarbons (for the one incinerator for which these last five constituents were measured). Stack emissions from the tested incinerators ranged from 0.030 to 0.154 gr of particulate matter per standard cubic foot of dry gas corrected to 12% CO_2. It would appear that, unless sufficient quantities of auxiliary fuel are burned to maintain high temperatures in the secondary chamber, these units have some difficulty in meeting the federal requirement for large incinerators of 0.08 gr/scf at 12% CO_2 but do meet the less strict standards for small incinerators (federal facilities) of 0.10 gr/scf at 12% CO_2.

1.10 Control of Emissions from Municipal Solid Waste Incinerators

Considering the particle size data presented in Table 12, it is apparent that to achieve even a modest 90% efficiency, all the particles larger than 1–3 μm must be removed. This requirement effectively eliminates the simple air pollution control systems traditionally used on MSW incinerators (e.g., settling chambers, wetted baffle spray systems, cyclones, and low-energy scrubbers), although it may sometimes be advantageous to use one of these simple devices as a first-stage collector. The ineffectiveness of these simpler systems is illustrated in Table 19 where particulate emissions for seven older incinerators utilizing such devices are listed. None of the incinerators meets the 0.08 gr/scf at 12% CO_2 standard.

Electrostatic precipitators (and perhaps certain types of scrubbers) appear to be the only commercially available devices that have the capability to meet the current emission standards for MSW incinerators. Newer forms of these devices, including charged droplet scrubbers and chromatographic systems, may have advantages over more conventional equipment, but these have not been demonstrated commercially for MSW incinerator applications. The chromatographic system is in commercial operation treating the effluents from secondary aluminum reverberatory furnaces, container glass, and fiberglass melters. These "dry" systems simultaneously remove particulates and acid gases from the hot effluent similiar to that of an MSW incinerator emission. The gases are quenched in the normal manner in a wet bottom temperature control chamber to a temperature of the order of 400°F (204°C). The quench is accomplished, however, with a proprietary procedure using a slurry of an inexpensive reagent. The acid gases are neutralized to the extent of 70–90% at this juncture, thus protecting the downstream equipment from corrosion. The quenched gases are duct contacted with nepheline syenite (a natural ore) in a proprietary system. Two phenomena occur. The acid gas neutralization is completed, and

Table 17 Typical Emissions from Apartment House Incinerators (Figures Shown Are lb/ton of Refuse Burned)

Constituent	Unmodified Flue Fed	Roof Afterburner	Basement Afterburner	Multiple-Chamber Basement Type
Particulates	33	4.5	6.3	4.8
Organic acids	9.5	—	5.6	3.9
Nitrogen oxides	6	—	10.1	2.1
Aldehydes	1.5	—	2.5	0.5
Hydrocarbons	2.0	—	—	2.7

Source: Ref. 9.

Table 18 Emission Data from Small Controlled Air Municipal Solid Waste Incinerators

		Installation			
	I	II	III	IV [a]	V
Heating value of refuse (HHV), Btu/lb	2,408	3,857	4,353	4,353	4,777
Moisture % of refuse	49.9	34.0	20.8	20.8	23.0
Air cooling method	Water spray and inducer	Water spray and inducer	Water spray and inducer	Water spray and boiler	Water spray and boiler
Design capacity, ton/day	8.5	12.5	10.5	10.5	25
Actual burn rate, lb/hr	1,221	2,142	1,527	1,654	1,890
Stack exhaust temperature, °F	967	984	1,333	264	259
Vol. % CO_2 (dry basis)	5.6	6.5	6.6	3.1	3.3
Vol. % O_2 (dry basis)	12.4	12.1	13.5	15.9	17.1
Vol. % N_2 (dry basis)	81.1	80.8	79.9	81.0	—
Vol. % H_2O	12.8	13.4	4.5	6.9	6.7
Flow, acfm	9,362	12,927	9,046	8,373	22,267 [b]
CO, ppm	0.4	0	25	10	28.4
SO_2, ppm	—	—	—	—	5
NO_x, ppm	—	—	—	—	68
Chloride, ppm	—	—	—	—	284.7
Fluoride, ppm	—	—	—	—	2.4
Hydrocarbons, ppm	—	—	—	—	1.1
Opacity, %	—	—	—	—	24
Auxiliary fuel (oil) consumed, lb/ton of refuse	114	—	—	—	—
Auxiliary fuel (gas) consumed, lb/ton of refuse	—	63.1	0.5	1.7	0.002
Particulate emission rate, (dry catch), lb/ton refuse	1.87	1.87	0.50	0.36	3.60
Particulate emission rate, (dry catch), gr/scf at 12% CO_2	0.065	0.084	0.037	0.030	0.154

[a] Installation IV is installation III with boiler in operation.
[b] Flow for two units feeding into common stack.
Source: Ref. 11.

Table 19 Particulate Emission Data from Older, Non-ESP-Equipped Municipal Solid Waste Incinerators

Incinerator	Air Cleaning Method	Test Date	Particulate Emissions					Stack Temperature, °F	Gas Flow Rate, acfm
			gr/scf at 12% CO_2	lb/1000 lb at 50% Excess Air	lb/hr	lb/ton of Waste Charged	CO_2, %		
A	Wetted column water scrubber	1968	0.55	1.06	122	10.4	4.6	455	69,800
B	Flooded baffle-wall water scrubber	1968	1.12	—	186	14.5	3.5	585	131,000
C	Flooded baffle-wall water scrubber	1968	0.46	0.85	173	8.8	5.0	485	120,000
D[a]	Baffle-wall and water spray	1968	0.73	1.19	238	8.6	5.0	305	186,000
E[a]	Baffle-wall and water spray	1968	0.72	1.18	—	12.5	3.9	365	165,000
F	Dry cyclone and wet baffle-wall	1969	1.35	2.70	386	20.4	3.2	500	130,000
G	Scrubber	1975	0.17	0.29	48	3.2	10.4	430	97,870

[a] Rotary kiln type.
Sources: Refs. 1 and 2.

the submicrometer particulate is collected by the nepheline syenite. The agglomerated particulate, now greater than 10 μm, is collected in a baghouse.

The ionizing wet scrubber system is a feasible add-on unit provided that an existing wet scrubber is in operation and the emissions are less than 0.4 gr/scf at 12% CO_2. The gaseous effluent from the existing wet scrubber, preferably a low-energy unit, contains submicrometer particulates and residual acid gases. The ionizing wet scrubber causes ionizing of the gas. The particulates are collected in an irrigated packed bed by ion image attraction. The residual acid gases are absorbed in this section. Thus, the action is similiar to that of an ESP, but the wet contact on a nonmetallic surface circumvents the problems of oil buildup, corrosion, and absence of acid gas recovery. The characteristics of these systems, compared to ESPs and multiclones, are shown in Table 20.

Table 21 shows a partial listing of ESP installations together with design parameters; Table 22 shows the emissions and efficiencies achieved during actual tests of these same incinerators, together with additional data of actual tests on other incinerators as well. It should be noted that almost all new MSW incinerators built since 1969, other than controlled air units, have utilized ESPs for particulate emission control. The chromatographic and ionizing wet scrubber systems are too new to evaluate their effectiveness at this writing.

Carbon monoxide and hydrocarbon emissions, including odorous compounds, are effectively controlled by a well-designed combustion chamber, and careful control of operating conditions. Odors from waste handling and storage can be controlled by drawing the combustion air into the plant through these areas so that the odor-bearing gases will be burned in the incinerator. Similiarly, leakage from the furnaces can be prevented by maintaining a slight negative pressure within. Inorganic gaseous emissions do exist in relatively low concentrations but in the absence of specific regulations, control will not be mentioned further except to say that water scrubbers will remove significant quantities of HCl, SO_2, and NO_2 (but not nitric oxide), and that ESPs will remove only very small quantities of these compounds.

The only realistic methods of water vapor plume control are to increase the stack gas temperature or to reduce stack gas water concentration. The former can be accomplished by a stack burner which injects very hot combustion gases directly into the stack, by mixing with a warmed gas from another source or by heat exchange with a source of heat (e.g., steam or furnace flue gases). The first two may also achieve a reduction in water concentration by dilution. These methods increase auxiliary fuel requirements, of course, and the increased gas flow may require a larger stack. Heat exchange with hot furnace flue gases requires a large amount of heat exchange surface. If the heat exchange surface is placed so as to remove heat directly from the combustion zone, it is theoretically possible to decrease the excess air normally required for temperature control in conventional incinerators. Indirect heat exchange does not decrease moisture content; it could, in fact, even increase it if excess air is reduced. Plume control by reducing stack gas moisture content poses more problems than it is worth. Thus, for water vapor plume control, intermittent control with the use of fuel for those periods when serious local environmental conditions prevail (such as road icing) is probably the best choice. Proper siting, of course, avoids many of these undesirable results in a good number of cases.

Finally, even with modern air pollution control devices, stacks are generally still required to disperse residual pollutants in order to avoid high ground level concentrations. Sophisticated dispersion modeling

Table 20 Characteristics of Emission Control Systems Compared

	ESP	Multi-clones	High-Energy Wet Scrubber	Low-Energy Wet Scrubber	Chromato-graphic	Ionizing Wet Scrubber
Particulate emissions, gr/scf at 12% CO_2	0.02–0.15	0.2–0.4	0.05–0.4	0.1–0.5	0.03	0.06–0.08
Acid gas emission, ppm	20–200	20–200	5–20	2–20	3	2
Maintenance	High	Moderate	High	Low	Low	Low
Capital cost, $/ton/day of capacity	4400–8800	500–1000	2000–5000	1000–4000	4300	3400

Source: Ref. 29.

Table 21 Partial Listing of Electrostatic Precipitator Installations at Municipal Solid Waste Incinerators, Including Design Parameters

Installation	Capacity, ton/day	Furnace Type[a]	Gas Flow, acfm	Gas Temperature °F	Gas Velocity, ft/sec	Residence Time, sec	Plate Area, acfm/ft²	Power Input, kVa	Pressure Drop, in. H$_2$O	Efficiency, wt%
Montreal	4 × 300	WW	112,000	536	3.5	3.3	6.2	35	0.5	95.0
South Shore, NY	1 × 250	R	136,000	600	5.5	3.3	6.8	33	0.5	95.0
Dade City, FL	1 × 300	R	286,000	570	3.9	4.0	5.7	48	0.4	95.6
Chicago, IL	4 × 400	WW	110,000	450	2.9	4.6	5.5	40	0.2	96.9
Braintree, MA	2 × 120	WW	32,000	600	3.1	4.5	5.7	19	0.4	93.0
Hamilton, Ontario	2 × 300	WW	81,000	585	3.5	5.4	3.9	70	0.5	98.5
Washington, DC	6 × 250	R	130,000	550	4.1	3.9	4.9	77	0.4	95.0
Harrisburg, PA	2 × 360	WW	100,000	410	3.5	5.1	5.0	40	0.2	96.8

[a] WW = water wall, R = refractory.
Source: Ref. 4.

Table 22 Actual Emissions and Efficiencies Achieved with Electrostatic Installations at Municipal Solid Waste Incinerators

Installation	Date of Test	Furnace Type[a]	Design Gas Flow, acfm	Actual Gas Flow, acfm	Stack Temperature, °F	Inlet Particulate, gr/scf at 12% CO_2	Outlet Particulate, gr/scf at 12% CO_2	ESP % Weight Efficiency
Des Carriers, Montreal	1970	WW	112,000	82,500	540	2.43	0.013	99.45
	1971	WW	112,000	—	540	1.24	0.008	93.5
South Shore, NY	1970	R	136,000	96,054	600	0.89	0.056	93.7
Dade City, FL	1970	R	286,000	199,000	570	0.19	0.027	86.0
Braintree, MA	1971	WW	32,000	40,000	—	0.60	0.108	82.1
	1977 (boiler #1)	WW	32,000	—	—	—	0.042	—
	1978 (boiler #1)	WW	32,000	—	—	—	0.042	—
	1978 (boiler #2)	WW	32,000	—	—	—	0.094	—
	1978 (boiler #1)	WW	32,000	37,520	388	0.34	0.083	73.6
Chicago, IL	1971	WW	110,000	—	—	1.06	0.026	97.5
Quebec City, P.Q.	1974	WW	100,000	—	500	—	0.095	—
Hamilton, Ontario	1976	WW	81,000	97,870	516	6.71	0.680	90.0
Washington, DC	1972	R	130,000	128,696	550	0.30	0.058	80.0
Harrisburg, PA	1973	WW	100,000	—	525	0.00	0.043	95.6
Philadelphia, PA	1974	R	219,300	—	550	—	0.041	—
Merrick, Hempstead, NY	1974	R	148,000	—	600	0.10	0.016	83.7
Saugus, MA	—	WW	—	—	—	—	0.050	—
Nashville, TN	—	WW	—	—	—	—	0.024	—
Norfolk, VA	—	WW	—	—	—	—	0.040	—

[a] WW = water wall, R = refractory.
Sources: Refs. 4 and 10.

techniques, usually computerized, can predict air quality resulting from utilizing various stack heights and diameters, and can even aid in stack location to avoid "downdraft" effects due to buildings and hills. If water vapor plumes are experienced, stacks also function to disperse them before impinging on surfaces where icing or condensation can be harmful.

2   INCINERATION OF MUNICIPAL WASTEWATER SLUDGE

2.1   Characteristics of Municipal Wastewater Sludge

The major factors that influence the characteristics of municipal wastewater sludge (MWS) are:

1. Influent waste characteristics
2. Degree of treatment
3. Unit processes selected
4. Design of unit processes
5. Operating mode

The first exerts a profound influence on the quality of the sludge generated and it, in turn, is affected by many factors such as industrial contributions, presence of garbage grinders, water supply characteristics, stormwater inclusion, and so on. The solids in domestic sewage are of two types: suspended and soluble with the former composed of approximately 60% settled and 40% colloidal. Settled and floating solids discharged to a municipal sewer system will, for the most part, be removed in primary treatment and sent directly to the sludge disposal system. In primary treatment without coagulants, about 50–60% of the suspended solids and 30–35% of the biochemical oxygen demand (BOD) are removed. In the secondary treatment, most of the soluble BOD (up to 90–95%) is removed and converted to biological solids. Although the composition of sewage sludge varies widely, depending on a complexity of factors, Table 23 provides an idea of some average characteristics of sewage solids. Primary sludges are higher in caloric value than biological sludges because of their high grease content; thus it is more economical to burn undigested solids than digested solids.

2.2   Treatment of Municipal Wastewater Sludge Prior to Incineration

Typical pretreatment steps include grit removal, blending, thickening, conditioning, and dewatering. Grit removal is necessary to protect pumps and other mechanical equipment against plugging and wear. Also, as can be seen from Table 23, it increases the heat value of the sludge by increasing its volatile content. When different types of sludges are handled, blending of the sludges improves the operation of the thickening, dewatering, and incineration operations. Sludge thickening, which is usually accomplished by processes such as gravity thickening, flotation thickening, or centrifugation, is designed to reduce the volume of the sludge to be handled. Sludge conditioning methods primarily aim at the reduction of bound and surface water, generally through heat treatment, chemicals, or use of polymers.

Sludge dewatering processes include: centrifugation, vacuum filtration, plug presses, and filter presses. Reduction of sludge moisture content of up to 75% is achieved in these dewatering processes in order that the fuel requirements for incineration are minimized. It should be mentioned that, when chemicals are used in the pretreatment steps, the weight of the sludge increases about 10% and, because of their inert nature, the heat content of the sludge is reduced accordingly. Secondary sludges are dewatered with difficulty, producing a high-moisture sludge cake containing solids of low heating value. The addition of large quantities of secondary sludge (resulting from improved wastewater treat-

Table 23   Average Characteristics of Sewage Sludge

Material	Combustibles, %	Ash, %	Btu/lb
Grease and scum	88.5	11.5	16,750
Raw sewage solids	74.0	26.0	10,285
Fine screenings	86.4	13.6	8,990
Ground garbage	84.8	15.2	8,245
Digested sewage solids and ground garbage	49.6	50.4	8,020
Digested sludge	59.6	40.4	5,290
Grit	30.2	69.8	4,000

Source: Ref. 2.

ment) to the feed of a sludge incinerator, will reduce the combustibility of the total sludge load, thereby increasing the auxiliary fuel requirement. Dewatered primary treatment plant sludge requires little auxiliary fuel since the sludge is readily dewatered and the solids have a significant heating value.

2.3 Municipal Wastewater Sludge Incinerators

The principal types of sludge incineration systems are:

1. Multiple-hearth furnace
2. Fluidized bed
3. Flash drying used in conjunction with fossil fuel or refuse-fired furnace
4. Cyclonic reactors
5. Wet oxidation (Zimpro process)
6. Atomized suspension technique
7. Coincineration with refuse via pyrolysis

The most widely used type of incineration system is the multiple-hearth furnace. Such units are simple, durable, and have the flexibility of burning a wide variety of materials even with wide fluctuations in feed rate. The furnace consists of a circular steel shell surrounding a number of solid refractory hearths and a central rotating shaft to which rabble arms are attached. Each hearth has openings that allow the sludge to be dropped to the next lower hearth. The central shaft and rabble arms are air cooled. Rabbling is very important to combustion because it breaks up large sludge particles, thereby exposing more surface area to the hot furnace gases that induce rapid and improved combustion. Partially dewatered sludge is continuously fed to the upper hearths which form a drying and cooling zone. In the drying zone, vaporization of some free moisture and cooling of exhaust gases occur by transfer of heat from the hot gases to the sludge. Intermediate hearths form a high-temperature burning zone whose principal function is to burn volatile gases and solids. The lowest hearth of the combustion zone is the place where most of the total fixed carbon is burned. The bottom hearth of the furnace functions as a cooling and air preheating zone where ash is cooled by giving up heat to the shaft cooling air which is returned to the furnace in this zone. Typical temperatures are 330°C (600°F) at the bottom, 890–1000°C (1600–1800°F) in the middle hearths, and 555°C (1000°F) on the top hearths. The waste gases from combustion are heated, via auxiliary fuels, to deodorizing temperature so as to guard against odor nuisance. Exhaust gases leaving the incinerator at the top are generally scrubbed in a wet scrubber to remove fly ash.

The next most widely used incineration system is the fluidized bed. Some of the advantages of the fluidized bed are: (a) ideal mixing of the sludge with the combustion air, (b) no moving parts, (c) it can be operated 4–8 hr a day with little reheating when restarting since the same bed acts as a heat reservoir, and (d) no mechanical ash system is required since the ash is removed from the reactor by the upflowing combustion gases. The bed material is composed of silica sand. (The principle of the fluidized bed is, of course, that when particles are suspended in an upward-moving stream of gases, the mixture of particles and gases behaves much like a fluid.) Sufficient air is used to keep the sand in suspension but not to carry it out of the reactor. The intense and violent mixing of the solids and gases results in uniform conditions of temperature, composition, and particle size distribution throughout the bed. Heat transfer between the gases and the solids is extremely rapid because of the large surface area available. The sand bed retains the organic particles until they are reduced to mineral ash and the violent motion of the bed comminutes the ash material, preventing the buildup of clinkers. The resulting fine ash is constantly stripped from the bed by the up-flowing gases.

Flash drying is the instantaneous removal of moisture from solids by introducing them into a hot gas stream. The system comprises four distinct cycles which can be combined in different arrangements to give the system maximum flexibility to meet specific requirements. The first cycle is the flash-drying cycle, the second is the fuel-burning cycle, the third is the effluent gas or induced draft cycle, and the fourth is the fertilizer-handling cycle. Some of the advantages arising out of flexibility in operation are: (a) sludge can be dried or incinerated to suit the plant's immediate requirements; (b) the final moisture content can be automatically controlled very closely since a relatively small amount of sludge is in the system at one time; (c) the system can be started and shut down in a short period of time and no standby fuel is required when sludge is not being processed. The lack of fertilizer market for dried sewage has eliminated the major advantage of this system, that is, the flexibility of drying or burning. As an incineration unit, the flash-drying system has the major disadvantage of complexity, potential for explosions, and potential for air pollution by fine particles. Even though air pollution controls are readily applicable to the flash-drying and incineration systems, in comparable situations it is not equal to other furnace designs.

Cyclonic reactors are ideally suited for sludge disposal in the smaller sewage treatment plants because of their simplicity in installation and flexibility in operation. The mechanism in cyclonic reactors is that high velocity air, preheated with combustion gases from a burner, is introduced tangentially into the cylindrical combustion chamber. Concentrated sludge solids are sprayed radially toward the intensely heated walls of the combustion chamber. This feed is immediately caught up in the rapid cyclonic flow of hot gases and combustion takes place so rapidly that no material adheres to the walls. The ash residue is carried off in the cyclonic flow and passes out of the reactor. Basically, the performance of the cyclonic reactor depends on (a) the cyclonic flow pattern, (b) the dispersion of the feed, and (c) the temperature of the combustion chamber walls. Cyclonic reactors have high-efficiency operation and this is achieved by the cyclonic action.

The wet oxidation process is based on the discovery that any substance capable of burning can be oxidized in the presence of liquid water at temperatures between 140 and 390°C (250 to 700°F). They are not, strictly speaking, comparable to incinerators since their primary use is for sterilization. Fly ash or dust are not produced as the oxidation takes place in the presence of water. Sulfur dioxide and nitrogen oxides are not formed, and odor control is assured by use of gas incineration devices. The major disadvantage associated with the wet oxidation process is the cost of construction and operation. Also, odor problems can develop from the off-gases and from lagooning of the ash-containing effluent. Though air pollution caused by the stack gases can be controlled by catalytic burning at high temperatures, this is an unknown added expense. Another suggested disadvantage of wet combustion systems is the need for high-quality supervision and frequent maintenance due to use of sophisticated equipment and controls.

The atomized suspension technique is designed for high-temperature/low-pressure thermal processing of wastewater sludges. In this system, sludges are reduced to an innocuous ash, and bacteria and odors are destroyed. The process generally includes the following steps:

1. Thickening of the feed sludge to 8% and higher solids content.
2. Grinding the sludge to reduce the particle size to less than 25 μm.
3. Spraying the sludge into the top of a reactor to form an "atomized suspension."
4. Drying and burning the sludge in the reactor.
5. Collecting and separating the ash from the hot gases.

This system has the following advantages: (a) versatility in sludge handled, (b) small space requirement, (c) rapid conversion of raw sludge to innocuous ash, steam, and CO_2, and (d) no nuisance conditions.

2.4 Air Pollution Regulations Applicable to Municipal Wastewater Sludge Incineration

The EPA standards applicable to sludge incineration are shown in Table 24. There are no EPA standards for oxides of nitrogen and sulfur because the quantities and concentrations of these compounds are low compared to other sources. The emission standards for particulate matter vary from state to state. Prior to the 1970s, the practice usually was to control these emissions to 0.85 lb of fly ash per 1000 lb of flue gas, adjusted to 50% excess air (1 lb/10^6 Btu), and most incinerator manufacturers advertised to limit the particulate matter to 0.20–0.28 lb/1000 lb of stack gas at 50% excess air. However, in the event of increased air pollution standards, there is a trend toward the use of electrostatic precipitators or high-efficiency scrubbers.

2.5 Emissions from Municipal Wastewater Incinerators

Tables 25 and 26 show particulate emission data from 10 MWS incinerators, two being of the fluidized-bed type and the remainder of the multiple-hearth design. Four of the incinerators have two- or three-tray impinger-type scrubbers for particulate removal; one incinerator has a cyclonic dust collector

Table 24 Federal Standards Applicable to Sewage Sludge Incineration

Pollutant	Standard
Particulates	0.65 g/kg dry sludge feed
Opacity	20%
Beryllium	10 g/24 hr
Mercury	3200 g/24 hr
Lead	Ambient air standard is 1500 ng/m³
Cadmium	No standard but a 100 ng/m³ in ambient air is being discussed as a possible future standard

Source: Ref. 15.

Table 25 Total Particulate Emissions from Sewage Sludge Incinerators with Impingement Scrubbers

Plant	A	B	C	D	K[a]
Emissions (g/kg dry sludge feed)	0.72	0.40	0.67	0.74	5.9
Scrubber efficiency	83	92	92	99	78
Pressure drop					
cm water	25	48	42	15	—
in. water	10	19	16	6	—
Year of construction or latest scrubber modification	1974	1972	—	1972	1966
Incinerator type	MH[b]	MH[b]	MH[b]	MH[b]	MH[b]
Sludge rate (kg/hr dry sludge)	1280	1398	2443	2711	778

[a] No scrubber, only a dry cyclone.
[b] MH = multiple-hearth furnace.
Source: Ref. 31.

Table 26 Total Particulate Emissions from Sewage Sludge Incinerators with Venturi and Impingement Scrubbers

Plant	E	F	G	H	J
Emissions (g/kg dry sludge feed)	0.13	0.59	2.6	0.87	0.036
Scrubber efficiency (%)	93	93	96	96	97
Pressure drop					
cm water	51	51	—	—	41
in. water	20	20	—	—	16
Year of construction or latest scrubber modification	1960	1976	1973	1976	1970
Incinerator type	MH[a]	MH[a]	FB[b]	MH[a]	FB[b]
Sludge rate (kg/hr dry sludge)	1597	936	93	640	1340

[a] MH = multiple hearth furnace.
[b] FB = fluidized bed incinerator.
Source: Ref. 31.

and the other five incinerators have venturi scrubbers followed by two- or three-tray impingement scrubbers. Capacities range from 90 to 2700 kg/hr (dry basis).

The data of these tables show that three of the five incinerators equipped with both venturi and impingement scrubbers met the federal standard (see Table 26), whereas only one of the five incinerators equipped solely with the impingement scrubbers met the standard (Table 25). The scrubber on this incinerator (plant B) had the highest pressure drop of the four scrubbers shown in Table 25 (plant K had no pressure measuring device). The incinerator with the dry cyclone did not meet the standard. The results do not provide a rigorous comparison between the effectiveness of the combination of the two types of scrubbers versus the impingement scrubbers alone, but they indicate that the combination gives better performance.

The failure of any of these incinerators to meet the emission standard on a given day does not indicate that they could not be tuned up to this performance level. In fact, most of the incinerators at one time passed the emission tests for total particulate discharge. Sludge flow rate, moisture content, and particularly scrubber pressure drop and water rate influence the results substantially.

Table 27   Percentage of Metal in Feed That Leaves Scrubber

Plant	Scrubber Type[a]	(Metal Lost/Metal in Feed) (100)			
		Cd	Fe	Ni	Pb
A	I	29.4	0.006	0.38	3.8
C	I	35.9	0.02	0.14	3.5
D	I	14.2	0.05	0.35	13.8
E	V–I	9.1	0.01	0.12	4.5
F	V–I	6.0	4.16	0.08	5.6
H	V–I	16.0	0.12	0.28	14.7
J	V–I	0	0.02	1.1	0
K	C	14.7	0.40	1.5	16.9

[a] I = impingement.
V–I = venturi and impingement.
C = dry cyclone.
Source: Ref. 31.

Table 28   Percentage of Particulate Metal That Is in the 0.1–1.0 μm Fraction

Plant	C		H	
Scrubber type	I[a]		V–I[b]	
Scrubber location	Inlet	Outlet	Inlet	Outlet
Metals				
Cd	86	99	68	97
Fe	24	83	5	19
Ni	38	89	10	18
Pb	42	99	68	44

[a] I = impingement.
[b] V–I = venturi and impingement.
Source: Ref. 31.

For these 10 incinerators, the solids collected were analyzed for the principal heavy metals. Consequently, the relative removal of any particular metal can be calculated at any stage of the process. The mass of metal lost in the gases leaving the scrubber relative to the mass of that metal in the feed is presented in Table 27 for cadmium, iron, nickel, and lead. It is evident that the recovery of lead and cadmium is much poorer than for total particulates, iron, and nickel. A factor contributing to the high loss of lead and cadmium relative to iron and nickel is the high proportion of lead and cadmium in the finer fractions of the particulates leaving the incinerators. Like all collection devices, scrubbers collect fine particles less effectively than coarse particles so lead and cadmium removals are lower than overall efficiencies might indicate. The percentage of the mass of each of the four metals in the 0.1–1.0 μm fraction relative to the total mass of metal in all four of the fractions is presented in Table 28 for two of the plants. In each case, cadmium and lead are heavily concentrated in the fine fractions of the particulates entering the scrubber.

The total discharged per day of a pollutant is important, but it is equally important to be able to estimate the local impact of a stack discharge on ground level concentrations. These concentrations can be calculated from plume models. Using the data of Tables 25 to 28, the average annual ground concentration levels resulting from the stack discharge for each test site were calculated for cadmium, iron, nickel, and lead and compared with ambient concentrations. The maximum calculated average ground level concentrations at each site for cadmium were less than ambient concentrations in most cases. For lead, they were normally less than 2% of the ambient value. The calculated maximum average annual lead concentrations averaged about 2% of the federal standard of 1500 ng/m³. For cadmium, they were about 1.5% of the 100 ng/m³ of air, which is the level being discussed if a cadmium standard is recommended in the future. These low levels appear unlikely to be a cause for concern.

2.6   Control of Emissions from Municipal Wastewater Incinerators

Particulate matter typically is controlled by centrifugal dust collectors or wet scrubbers. Centrifugal collectors remove 75–80% of the particles and are suitable for exhaust gas temperatures of 650–700°F.

Water scrubbers are less sensitive to loadings and gas temperatures and they collect the condensable portion of the exit gases. In general, the nature of the emitted particulate matter from sludge incinerators does not lend itself to centrifugal collection and most systems utilize wet scrubbers of a variety of types including venturi, baffle plate, packed tower, and impingement models. These scrubbers have the added advantage of absorbing significant amounts of gases including sulfur oxides and odorous organics.

Odors can be eliminated at their source or can be prevented from reaching the atmosphere by control. The basic requirements for preventing odor are good plant design and operation. Septicity of sludge can be prevented by providing adequate sludge hoppers and flexibility in pumping schedules. Odors, when emitted, can be controlled by any one of the following five methods with certain limitations: (a) combustion, (b) chemical oxidation, (c) adsorption, (d) dilution, and (e) masking.

While the above methods have some usefulness in the control of odors, the control of odors from sludge incinerators is generally limited to two techniques. The main and most successful approach is to incorporate a means of ensuring that all gases arising from the system are raised to and held at a sufficiently high temperature and for a sufficient time period to obtain satisfactory oxidation of all organics. It is generally considered that if the gases are held at 780°C (1400°F), oxidation will occur in a matter of seconds. Thus, if the gases are held at 780°C (1400°F) for the usual gas phase detention time (10–60 sec), no odors should be present in the gas exhaust. However, through poor design, operation, and/or maintenance, these conditions are frequently not achieved and a serious odor problem can and does arise.

A less frequently utilized control technique is to take off-gases to a secondary incineration chamber. This may be of the flame type in which the gases are passed through a natural gas or oil (usually the former) flame to increase the organic destruction. As an alternative, the gases may be passed over a catalytic system where the same oxidation takes place but at a lower temperature since the catalytic surface lowers the oxidation energy "hump." In both cases, the chemistry is identical to that described relative to incineration and additional air may be added to ensure satisfactory combustion.

3 INCINERATION OF INDUSTRIAL WASTES

3.1 Industrial Incineration Systems

This section deals with the state of the art of the incineration of industrial wastes. Six technologies are described: rotary kilns, liquid injection incinerators, and fluidized bed incinerators; multiple-hearth, multiple-chamber, and catalytic combustors.

3.1.1 *Rotary Kiln*

The rotary kiln is a cylindrical, horizontal, refractory-lined shell that is mounted at a slight incline. Rotation of the shell causes mixing of the material with the combustion air. This rotation provides turbulence and agitation to maximize burnout. Examples of the kinds of materials destroyed in rotary kilns are: polyvinyl chloride, PCB wastes in capacitors, nitrochlorobenzene tars, chlorotoluene production wastes, and steam still bottoms from aniline and alkylated phenol production.

Rotary kilns are designed to be operated at temperatures in excess of 1400°C (2500°F), making them well suited for the destruction of toxic compounds that are difficult to degrade thermally. They are adaptable for use with a wet gas scrubbing system, and if most of the heat is supplied by auxiliary source the waste can be fed directly into the incinerator without any preparation such as preheating, mixing, and so on. (If most of the heat comes from the wastes themselves, and if the heat contents vary widely from one waste to another, mixing is desirable to avoid thermal shock.) There is also continuous ash removal which does not interfere with the waste oxidation. On the other hand, the capital cost is high, especially for low feed rates. The particulate loadings are high and the thermal efficiency is low.

3.1.2 *Liquid Injection*

Liquid waste combustors (liquid injection combustors) are flexible units that can be used to dispose of virtually any combustible liquid waste. The heart of the liquid injection system is the waste atomization device or nozzle (burner). Because a liquid combustion device is essentially a suspension burner, efficient and complete combustion is obtained only if the waste is adequately divided or atomized and mixed with the oxygen source. Atomization is usually achieved either mechanically using rotary cup or pressure atomization systems, or via gas fluid nozzles using high-pressure air or steam. Atmospheric emissions from the combustion of chemicals are usually controlled by an afterburner system and a scrubber.

Liquid injection combustors have widespread applicability for chemical waste incineration. Typical chemicals destroyed in such a system include phenols, cyanide, and chrome plating chemicals, thinners, solvents, dodecyl mercaptan chemicals, hexachlorocyclopentadiene, and fluorinated herbicide chemicals.

3.1.3 Fluidized Beds

The fluidized bed incinerator is a simple device consisting of a refractory-lined vessel containing inert granular material. Gases are blown through this material at a rate sufficiently high to cause the bed to expand and act as a theoretical fluid. The gases are injected through nozzles that permit flow up into the bed but restrict downflow of the material. Waste feed enters the bed through nozzles located either above or within the bed. Preheating of the bed to startup temperatures is accomplished by a burner located above and impinging down on the bed.

Fluidized beds have general applicability for the disposal of combustible solids, liquids, and gaseous wastes. They are simple in design and require no moving parts in the combustion zone. It is difficult, however, to remove residual materials from the bed, and incineration temperatures are limited to a maximum of about 830°C (1500°F). The operating costs are relatively high, particularly power costs.

3.1.4 Catalytic Combustion

Catalytic incineration is applied to gaseous wastes containing low concentrations of combustible materials and air. Usually noble metals such as platinum and palladium are the catalytic agents. The catalyst is supported in the hot waste gas stream in a manner that will expose the greatest surface area to the waste gas so that the combustion reaction can occur on the surface, producing nontoxic effluent gases of carbon dioxide, nitrogen, and water vapor. Since most waste gases from ordinary industrial processes are at low temperatures up to 170°C (300°F), a preheat burner is required to bring these gases up to the reaction temperature. The waste gases are preheated before exposure to the catalyst, usually to about 330–555°C (600–1000°F). Most of the combustion occurs during flow through the catalyst bed which operates at maximum temperatures of 555–890°C (1000–1600°F).

This is a proven technology for the incineration of gaseous material such as that generated from the production of vinyl chloride monomer manufacturing wastes. The process is more economical than the direct flame form of incineration but the catalyst systems are susceptible to materials that reduce or suppress their activity.

3.1.5 Multiple-Hearth Furnaces

A typical multiple-hearth furnace is a vertical cylinder that includes a refractory-lined steel shell, a central shaft that rotates, a series of solid flat hearths, a series of rabble arms with teeth for each hearth, an air blower, fuel burners mounted on the walls, an ash removal system, and a waste feeding system. Sludge and/or granulated solid combustible waste is fed through the furnace roof by a screw feeder or belt and flapgate. The rotating air-cooled central shaft with air-cooled rabble arms and teeth plows the waste material across the top hearth to drop holes. The waste then falls to the next hearth and then the next until discharged as ash at the bottom. The waste is agitated as it moves across the hearths to make sure fresh surface is exposed to hot gases.

This technology has moderate applicability to the incineration of chemical wastes. It is suited to such wastes as isophthalic acid and terephthalic acid still bottoms, solid residues from the manufacture of aromatic amines, and reactor bottoms from PVC manufacture. The retention time in multiple-hearth incinerators is usually higher than in other incinerator configurations. Fuel efficiency is high, and large quantities of water can be evaporated. On the other hand, due to the longer residence times of the waste materials, temperature response throughout the incinerator when the burners are adjusted is usually very low, and maintenance costs are high because of the many moving parts.

3.1.6 Multiple Chamber

Multiple-chamber incinerators are divided into three separate zones: (1) an ignition or primary combustion chamber, (2) a downdraft mixing chamber, and (3) an up-pass secondary combustion chamber. Solid wastes are either manually or automatically fed into the incinerator through charging doors onto grates at the bottom of the ignition chamber. Here, the wastes are dried, ignited, volatilized, and partially oxidized into gases and particulates. As more material is charged to the system, the pile of burning material is pushed farther along the hearth toward the ash pit. The amount of ash entrained as particulate, versus the amount leaving the system as bottom ash, is primarily a function of the underfire–overfire air ratio.

Multiple-chamber incinerators are used for solids, usually refuse. These units are generally marginal to inadequate for acceptable destruction of solid chemicals because of poor solids mixing and temperature control. Some of the materials that have been burned in such units include phenolic resins, polyvinyl chloride, and rubber.

3.1.7 Emerging Technologies

Among the emerging technologies that can be included within the general rubric of incineration include: pyrolysis, wet air oxidation, microwave plasma destruction, starved air combustion, molten salt, coincineration, and coburning incinerators. The use of these devices for the incineration of chemicals, however, is not widespread.

3.2 Emissions from Industrial Waste Incineration: General

Table 29 shows some typical major reaction products resulting from the incineration of specific industrial wastes. Besides the incinerator itself, auxiliary equipment may also emit air pollution. Fugitive dust can be produced by material handling operations. Steam strippers for water pollution control and quench water might cause volatile compounds to be emitted into the air; chemicals added to scrubber water might conceivably react with chemicals in the waste-producing gaseous emissions, and so on.

Emissions from chemical incinerator stacks include suspended particulates and various gases. These emissions are confined and can thus be ducted into various types of air pollution control equipment. Fugitive emissions are pollutants arising from sources other than stacks. They can be generated from a wide variety of sources associated with the total incineration process. Because fugitive emissions are not confined, controlling emissions can be more difficult than controlling emissions in stacks. They have to be collected to be controlled by conventional pollution control equipment, controlled in situ with water sprays or other measures, or the source may be elminated (by tightening leaky flanges, changing the process, etc.).

If the combustion of pure hydrocarbons went to completion, only CO_2 and H_2O would be produced. Industrial wastes, however, are often complex organic compounds containing a variety of elements in several structures, and combustion efficiency may not be particularly high. Emissions produced as a result of incineration can contain some of the original compounds in the wastes, new products formed as a result of the partial and complete combustion process, and trace elements. In some cases, hazardous wastes may be converted to even more dangerous forms by incineration.

Incineration processes generally require fuels and air to produce efficient combustion reactions. Combustion products generally include CO, CO_2, H_2O vapor, particulates, NO_x, hydrocarbons, and SO_2 (from the sulfur in the fuel and waste). Depending on such factors as the combustion efficiency of the incinerator and the chemical composition of the waste, other compounds can be produced. The incomplete combustion of waste may result in emissions of CO, aldehydes, amines, organic acids, carbon, partially degraded products of the original waste material, partially oxidized products, and some unreacted waste. The products of complete combustion may include dangerous substances such as As_2O_3 and P_2O_5.

Examination of the lists of chemical waste streams indicates that approximately 75% of those suitable for disposal by incineration fall into one of the following three classes:

1. C—H and C—H—O compounds, yielding CO_2 and H_2O on complete combustion.
2. C—H—N and C—H—O—N compounds, yielding CO_2, H_2O, and NO_x on complete combustion.
3. C—H—Cl and C—H—O—Cl compounds, yielding CO_2, H_2O, and primarily HCl on complete combustion.

Table 29 Chemical Descriptions of Some Industrial Wastes and Major Reaction Products

Chemical Description	Example Compounds or Types of Wastes	Major Reaction Products
Organic salts + inorganic salts	Na_2SO_4, phthalates	Inorganic oxides, inorganic carbonates
Halogen- (Cl, Br, F) containing compounds	PCBs, chlorinated hydrocarbons, pesticides and herbicides	Halogen acids (HCl, HBr, HF) and gases (Cl_2, Br_2)
Sulfur-containing compounds	Pesticides, herbicides, explosives	SO_2/SO_3, trace H_2S
Nitrogen-containing compounds	CN-, HNO_3 production wastes, NH_3	CN-, NO_x
Phosphorus-containing compounds	Explosives, pesticides and herbicides	H_3PO_4 gas, P_2O_5
Organics containing heavy metals/organometallic compounds	Paints, pigments, pesticides and herbicides	Pb, Fe, Zn, Ni, Hg in various forms

Source: Ref. 19.

Within each class, of course, the combustion characteristics of waste streams may differ from one another substantially. Nevertheless, the final combustion products are primarily dependent on the elemental composition of the waste streams and auxiliary fuel only.

In addition to the three classes of waste streams discussed above, wastes containing organofluorides, organobromides, organophosphorus compounds, organosulfur compounds, organosilicates, and organic wastes containing significant amounts of inorganic salts and oxides are also disposed of by incineration. Thus, incinerator effluent gases can also contain hydrogen fluoride (HF), hydrogen bromide (HBr), bromine (Br), phosphorus pentoxide (P_2O_5), sulfur dioxide (SO_2), sulfur trioxide (SO_3), and a multitude of mineral salts and oxides. Also to be considered are gaseous contaminants from incomplete combustion of the wastes and auxiliary fuel as has previously been mentioned.

Particulate emissions from industrial waste incineration include particles of mineral matter, mineral matter containing unburned combustible contaminants, soot, and heavy tar, as well as noncombustible aerosols such as sulfuric acid (H_2SO_4) or hydrochloric acid (HCl) mists.

3.3 NO_x Emissions

Nitrogen oxides (NO_x) is the collective term used principally for the two gaseous oxides of nitrogen, nitric oxide (NO) and nitrogen dioxide (NO_2). Nitric oxide is formed by the reaction of N_2 and O_2 in the presence of heat:

$$N_2 + O_2 \rightarrow 2NO$$

The further reaction of nitric oxide with oxygen forms nitrogen dioxide:

$$2NO + O_2 \rightarrow 2NO_2$$

In combustion or incineration processes, N_2 and O_2 for the formation of NO_x will come from the fuel to run the system (i.e., gas, coal, oil), the excess air injected into the system and/or the waste material incinerated. Variables such as the furnace or flame temperature, concentrations of O_2 and N_2, the chemical makeup of the waste material, the pressure differences in the furnace, the catalytic action of trace elements and compounds, and the amount of excess air present will determine the amount of NO_x produced in the system.

For example, in the incineration of waste materials containing nitrogen, such as still bottoms and scrubber wastes from organic chemical nitration processes, NO_x will be produced. Atmospheric nitrogen present in the excess air injected in the system and nitrogen in the combustion fuel and in the waste will be combined with oxygen to create oxides of nitrogen. Oxygen will be in the air and in certain types of the waste material.

Nitric oxide is formed primarily in the portions of flames possessing the highest temperature. For example, at a residence time of 0.5 sec and at temperatures above 1860°C (3350°F), the NO concentration is well over 500 ppm for both oil- and gas-fired combustion units.

The amount of excess air injected into the unit will affect NO_x production. In horizontally and tangentially fired oil combustion units, for example, NO_x emissions decrease as the percentage of oxygen in the flue gas is reduced. NO_x reductions of 28% in a tangentially fired unit and 36% in a horizontally fired unit have been observed as the flue gas oxygen content was lowered from about 4 to 2%.

In relation to the various other sources of NO_x emissions, NO_x emissions from industrial waste incineration have so far been only a small fraction of the national emissions. However, the possibility of an incinerator in a specific area producing very high levels of NO_x, for a short time span, cannot be eliminated a priori.

3.4   Effect of Industrial Waste Incineration on Emissions of Other Criteria Pollutants

The type and amount of criteria pollutants produced by chemical waste incineration depends largely on the composition of the waste material prior to incineration. As the reader will recall, the following is a list of EPA criteria pollutants:

1.   Suspended particulates
2.   Sulfur dioxide
3.   Lead
4.   Carbon monoxide
5.   Photochemical oxidants
6.   Nonmethane hydrocarbons
7.   Oxides of nitrogen

Emissions of criteria pollutants can be produced directly by incineration (except for photochemical oxidants, which are generally formed downwind by reaction of hydrocarbons and oxides of nitrogen in the presence of sunlight). For example, lead can be emitted by the incineration of paint production sludge containing lead-based organics. Large amounts of suspended particulates, CO, and nonmethane hydrocarbons can be caused by the incomplete combustion of organic waste material. Sulfur dioxide and nitrogen oxides can be formed in the incineration process by reactions of waste compounds containing sulfur and nitrogen with air.

Suspended particulates arise by both physical and chemical means. Dust may be emitted directly from the mechanical agitation of specific waste products and from residues and deposits in the stack, secondary burner units, flue passageways, and so on. Fumes may form from incomplete combustion.

3.5 Regulations Relevant to Incineration of Industrial Wastes

Most of the existing regulations have been mentioned in Sections 1 and 2 of this chapter. However, among the most important provisions are those in proposed regulations, particularly under the Resource Conservation and Recovery Act (RCRA). These "strawman" regulations contain the following:

1. Limits the particulate concentration in incinerator flue gas as a function of the percent CO_2 or excess air. It would allow the permit writer to specify the hydrogen halide and sulfur dioxide removal efficiency required for pollution control equipment.

2. Allows the permit writer to specify levels of carbon monoxide and incomplete products of combustion in exhaust gas, along with the destruction efficiency required for principal toxic components.

3. Requires monitoring the exhaust gas during test burns for the principal hazardous components, incomplete products of combustion, hydrogen halides, CO, CO_2, O_2, and particulates.

4. Requires visual hourly observations of plume color and opacity for normal appearance and daily inspection for leaks, spills, and fugitive emissions. Carbon monoxide concentrations, O_2 concentrations, and air pollution control device operating parameters must also be monitored during operational burns.

5. Limits the concentration of air pollution above a landfill, land farm, or impoundment to the OSHA standard. If two or more gases with OSHA standards are involved, the sum of the following expression E_m must not exceed unity:

$$E_m = \frac{C_i}{L_i} + \cdots + \frac{C_n}{L_n}$$

where E_m is the equivalent exposure for the mixture, C_i is the concentration of a particular contaminant, and L_i is the OSHA exposure limit for that contaminant.

Under the Clean Air Act, incinerators burning beryllium-containing waste must either:

1. Not emit more than 10 g of beryllium over a 24-hr period
2. Demonstrate that the 30-day mean ambient Be concentration in the vicinity of the plant will not exceed 0.01 $\mu m/m^3$

Under the Toxic Substances Control Act, a 99.9% combustion efficiency is required for incinerators used for the destruction of PCBs. For units burning solid PCBs, no more than 0.0001 g of PCB per kilogram burned may be emitted. Minimum combustion temperatures and retention times are specified; HCl control is required; and safety systems must be installed to shut off PCB feed if certain specifications are not met.

3.6 Air Pollution Control Devices

3.6.1 *General Considerations*

Control equipment commonly considered suitable for the reduction of particulate and gaseous emissions from industrial combustion sources includes wet scrubbers, electrostatic precipitators, and fabric filters. In industrial waste incineration facilities designed to handle a variety of solid and liquid wastes, the use of wet venturi and packed bed scrubbers has been predominant. Additionally, new control technologies such as wet electrostatic precipitators, electrostatically augmented scrubbers, and dry collection on fabric filters with chemical additives are sometimes employed in special applications and as secondary

Table 30   Advantages and Disadvantages of Selected Emission Control Devices

Device	Advantages	Disadvantages
Cyclone	Continuous wet or dry dust collection Suitable for high-temperature operation Low to moderate pressure drop Applicable to high or low dust loadings	Generally inefficient for particles $< 5 \ \mu m$ Sensitive to changes in flow rate Not capable of removing gaseous pollutants
Electrostatic precipitator	Dry dust collection Low pressure drop and operating cost Efficient removal of fine particles Suitable for high-temperature operation	Relatively high capital cost Sensitive to changes in flow rate Particle resistivity affects removal and economics Not capable of removing gaseous pollutants Fouling potential with tacky particles
Fabric filter	Dry dust collection High efficiency at low to moderate pressure drop Efficient removal of fine particles	Gas temperatures cannot exceed 290°C Fabrics may be susceptible to chemical attack Filter may be fouled by acid mist, condensation, or tacky particles Not capable of removing gaseous pollutants
Sorbent-filled baghouse	Simultaneous removal of gaseous and particulate pollutants High gas and particulate removal efficiency Dry collection of particulate and sorbent material	Potential nozzle plugging Potential secondary dust problem during disposal
Gas-atomized spray	Simultaneous gas absorption and dust removal Suitable for high temperature, high moisture, and high dust loading applications Cut diameter of 0.5 μm is attainable Collection efficiency may be varied	Corrosion and erosion problem Dust is collected wet Moderate to high pressure drop Requires downstream mist eliminator
Preformed spray	Simultaneous gas adsorption and dust removal Suitable for high temperature, high moisture, and high dust loading applications Collection efficiency may be varied	High efficiency may require high pump discharge pressures Dust is collected wet Nozzles are susceptible to plugging Requires downstream mist eliminator
Plate-type scrubbers and packed bed	Simultaneous gas adsorption and dust removal High removal efficiency for gaseous and aerosol pollutants Low to moderate pressure drop	Low efficiency for fine particles Not suitable for high temperature or high dust loading applications Requires downstream mist eliminator
Ionizing wet scrubber	Simultaneous gas adsorption and dust removal Low energy consumption No dust resistivity problems Cut diameter of 0.5 μm is attainable	Not suitable for high dust loadings Dust is collected wet
Wet electrostatic precipitator	Simultaneous gas absorption and dust removal Low energy consumption No dust resistivity problems Efficient removal of fine particles	Low gas absorption efficiency Sensitive to changes in flow rate Dust collection is wet

Table 31 Configurations of Selected Gas Cleaning Devices Applicable to Industrial Waste Incineration Facilities

Gas Stream Characteristics	Particulate Control Device[a]	Mist Eliminator	Absorption Device	Mist Eliminator
No significant gaseous pollutants present[b] — High or low particulate loading	Cyclone Electrostatic precipitator Fabric filter	Not required	Not required	Not required
	Gas-atomized sprays[c] Preformed spray	Required	Not required	Not required
	Wet electrostatic precipitator	Not required	Not required	Not required
Significant gaseous pollutants present[b] — High or low particulate loading	Cyclone Electrostatic precipitator	Not required	Gas-atomized spray Preformed spray Plate-type scrubber Packed bed scrubber	Required
			Sorbent-filled baghouse Ionizing wet scrubber	Not required
	Gas-atomized sprays[c] Preformed spray	Optional	Plate-type scrubber (optional)[c] Packed bed scrubber (optional)[c]	Required
			Sorbent-filled baghouse (optional) Ionizing wet scrubber (optional) Wet electrostatic precipitator (as a polishing device)	Not required
Low particulate loading	Sorbent-filled baghouse	Not required	Not required	Not required
	Plate-type scrubbers[c] Packed bed scrubbers[c]	Required	Not required	Not required
	Ionizing wet scrubber	Not required	Not required	Not required

[a] Particulate control may require combinations of the indicated devices depending on loading, size distribution, and economics.
[b] The term significant gaseous pollutants is meant to include HCl, Cl_2, HF, HBr, Br_2, P_2O_5, and/or SO_x.
[c] Existing industrial waste incineration facilities generally employ these device configurations.
Source: Ref. 19.

Table 32 Summary of NO$_x$ Control Techniques

Technique	Principle of Operation	Status of Development	Limitations	Applications	
				Near-term	Long-term
Combustion modification	Suppress thermal NO$_x$ through reduced flame temperature, reduced O$_2$ level; suppress fuel NO$_x$ through delaying fuel–air mixing or reduced O$_2$ level in primary flame zone	Operational for point sources; pilot-scale and full-scale studies on combined modifications, operational problems, and advanced design concepts for area sources	Degree of control limited by operational problems	Retrofit utility, industrial boilers, gas turbines; improved designs; new utility boilers	Optimized design area, point sources
Flue gas/ noncombustion tail gas treatment	Additional adsorption of NO$_x$ to HNO$_3$; conversion NO$_x$ to NH$_4$NO$_3$; reduction of NO$_x$ to N$_2$ by catalytic treatment	Operational for existing and new nitric acid plants meeting NSPS; pilot-scale feasibility studies for conventional combustion systems	New wet processing developing experience in applications; old catalytic processes have high costs interference by fuel sulfur or metallic compounds	Noncombustion sources (nitric acid plants)	Possible supplement to combustion modifications; simultaneous SO$_x$/NO$_x$ removal

Fuel switching	Simultaneous SO_x and NO_x control by conversion to clean fuels; synthetic gas or oil from coal; SRC; methanol; hydrogen	Synthetic fuel plants in pilot-scale stage; commercial plants due by mid-1980s	Fuel cost differential may exceed NO_x, SO_x, control costs with coal	Negligible use	New point sources, (combined cycle) Convert area sources (residential)
Fuel additives	Reduce or suppress NO by catalytic action of fuel additives	Inactive; preliminary screening studies indicated poor effectiveness	Large makeup rate of additive for significant effect; presence of additive as pollutant	Negligible use	Not promising
Fuel denitrification	Removal of fuel compounds by pretreatment	Oil desulfurization yields partial denitrification	Effectiveness for coal doubtful; no effect on thermal NO_x	Negligible use	Supplement to combustion modification
Catalyst combustion	Heterogeneously catalyzed reactions yield low combustion temperature, low NO_x	Pilot-scale test beds for catalyst screening; feasibility studies	Limited retrofit applications; requires clean fuels	Small space heaters	Possible use for residential heating, small boilers, gas turbines
Fluidized bed combustion	Coal combustion in solid bed yields low temperature, low NO_x	Pilot-scale study of atmospheric and pressurized systems; focus on sulfur retention devices	Fuel nitrogen conversion may require control (staging); may require large makeup of limestone sulfur absorbent	Negligible use	Utility, industrial boilers beginning 1980s; possible combined cycle, waste fuel application

Source: Ref. 19.

devices to further reduce emission levels. The selection of appropriate control technologies should be based on both the physical and the chemical characteristics of the incinerator effluent system. As a general guideline, the following considerations apply:

1. *Electrostatic Precipitators.*   Effective for the collection of fine particles, but unable to capture noxious gases and performs poorly for particles with high resistivity.
2. *Fabric Filters.*   Also effective for the collection of fine particles and unable to capture noxious gases, but have problems with particles that are hydroscopic, tacky, or have the tendency to solidify on fabric material.
3. *Wet Scrubbers.*   Absorption towers such as packed bed scrubbers are most suitable for the removal of noxious gases but are not recommended for gas streams with high particulate loadings; venturi scrubbers have high collection efficiencies for particles, and are fairly effective in removing noxious gases that are highly soluble or reactive with the scrubber solution.

3.6.2 Decision Guide

Applicability of air pollution control devices or systems of devices to a specified gas-cleaning duty will be contingent on the operating characteristics of the devices, the physical–chemical characteristics of the stream to be treated, and the emission standards defining the extent of the required treatment. The numerous control devices currently available may be categorized as either wet or dry control devices, and each has distinct operating characteristics that may prove advantageous under certain applications. Dry control devices include cyclones, electrostatic precipitators, and fabric filters or bag-houses. Wet control devices include gas-atomized sprays (venturi, orifice, flooded disk), preformed sprays (eductor venturi, spray tower), plate-type scrubbers, and packed bed scrubbers.

The principal advantages and disadvantages associated with selected control devices are briefly summarized in Table 30. Dry devices offer the distinct advantage of direct dust collection without sludge and wastewater treatment. Dry control devices are generally effective for removing fine particles (less than 1 μm in diameter) at low to moderate pressure drop (less than 2 kPa). On the other hand, they cannot remove gaseous pollutants unless sorbent materials are injected into the gas prior to treatment. Injection of sorbent material has been primarily used in conjunction with baghouses, and this approach is referred to as "sorbent-filled baghouse" in the table. Caution must be exercised during dry dust disposal operations to prevent secondary emission problems. Wet control devices afford simultaneous removal of dust and gaseous pollutants but require sludge treatment prior to waste disposal. Conventional wet control devices operate at moderate to high unit pressure drop to attain particle cut diameters (diameter of particles collected with 50% efficiency) in the 0.3–1 μm range.

Depending on the physical–chemical characteristics of the incinerator gas stream, there are a variety of devices or systems of devices that may be applied to industrial waste incinerator facilities. Potential configurations are presented in Table 31 for broad ranges of incinerator gas characteristics. Gaseous pollutants including HCl, Cl_2, HF, HBr, Br_2, P_2O_5, and/or SO_x may be present in sufficiently high concentrations to require control. Specific concentrations requiring control are dependent on applicable local, state, and federal regulations. Low particulate loadings indicate levels typically associated with liquid injection incinerators while high particulate loadings indicate levels associated with solid waste incinerators or fluid bed incinerators. Fluid bed incinerators may yield high particulate emissions due to entrainment of the bed material regardless of whether liquid or solid wastes are burned.

Mist eliminators are indicated as separate items that are required downstream of any conventional wet scrubber; however, they may be considered as optional when a wet scrubber is followed by another wet scrubbing device. Ionizing wet scrubbers and wet ESPs generally contain an integral mist eliminator and do not require separate devices.

Most hazardous waste incineration facilities currently employ either venturi scrubbers or sequential venturi and plate-type or packed bed scrubbers. For these systems, a gas quench is optional since the venturi may be utilized to effect gas cooling. Such systems are capable of handling a variety of incineration gas compositions and dust loadings. Several liquid injection incineration systems utilize a gas quench and either plate-type or packed bed scrubbers. These are dedicated type systems in that they are designed specifically for low dust loadings. Configurations that are not capable of removing noxious gases as well as high dust loadings should be used only for dedicated incinerators having reasonably constant emission levels.

3.6.3 Control of NO_x Emissions

Methods for controlling NO_x emissions from waste incineration units are complicated by trade-offs. For example, attempting to increase the efficiency of destruction by increasing the temperature and retention time can cause an increase of NO_x emissions. Catalysts that, when added, speed up the decomposition of NO_x, may react with other components of the wastes, forming toxic compounds.

An overall summary of possible NO_x control techniques is given in Table 32. Using oxygen instead of air for combustion in an incinerator is a possibility not listed in the table.

REFERENCES

1. Achinger, W. C., and Daniels, L. E., An Evaluation of Seven Incinerators, *Proceedings of the 1970 National Incinerator Conference*, ASME, Cincinnati, OH, May 1970, pp. 32–61.

2. Balakrishnan, S., Williamson, D. E., and Okey, R. W., *State of the Art Review on Sludge Incineration Practice*, Water Pollution Control Research Series, 17070DIV, U.S. Dept. of the Interior, Federal Water Quality Administration, 1970.

3. Bozeka, C. G., Nashville Incinerator Performance Tests, *Proceedings of the 1976 National Waste Processing Conference*, ASME, Boston, MA, May 1976, pp. 215–227.

4. Bump, R. L., The Use of Electrostatic Precipitators on Municipal Incinerators in Recent Years, *Proceedings of the 1976 National Waste Processing Conference*, ASME, Boston, MA, May 1976, pp. 193–201.

5. Carotti, A. A., and Smith, R. A., *Gaseous Emissions from Municipal Incinerators*, Publication SW-18c, U.S. Environmental Protection Agency, 1974.

6. Ellison, W., Control of Air and Water Pollution from Municipal Incinerators with the Wet-Approach Venturi Scrubber, *Proceedings of the 1970 National Incinerator Conference*, ASME, Cincinnati, OH, May 1970, pp. 157–166.

7. Environmental Protection Agency Standards of Performance for New Stationary Sources, *Federal Register 36*, 24,880ff (December 23, 1971).

8. Fiscus, D. E., et al., *St. Louis Demonstration Final Report: Refuse Processing Plant; Equipment, Facilities, and Environmental Evaluations*, EPA-600/2-77-155a, U.S. Environmental Protection Agency, 1977.

9. Fogiel, M. (ed.), Incineration, in *Modern Pollution Control Technology*, Research and Education Association, New York, 1978.

10. Freeman, H., Pollutants from Waste to Energy Conversion Systems, *Environmental Science and Technology*, 12(12), 1252–1256 (November 1978).

11. Frounfelker, R., *Small Modular Incinerator Systems with Heat Recovery*, SW 177c, U.S. Environmental Protection Agency, November 1979.

12. Gilardi, E. F., and Schiff, H. F., Comparative Results of Sampling Procedures Used During Testing of Prototype Air Pollution Control Devices at New York City Municipal Incinerators, *Proceedings of the 1972 National Incinerator Conference*, ASME, New York, June 1972, pp. 102–110.

13. Golembiewski, M., et al., *Environmental Assessment of a Waste to Energy Process*, Revised Final Report, EPA Contract No. 68-02-2166, April 1979.

14. Greenberg, R. R., et al., Composition and Size Distributions of Particles Released in Refuse Incineration, *Environmental Science and Technology* 12(5), 566–573 (1978).

15. Greenwood, D. R., Kinsbury, G. L., and Cleland, J. G., *A Handbook of Key Federal Regulations and Criteria for Multimedia Environmental Control*, EPA-600/7-79-175, U.S. Environmental Protection Agency, 1979.

16. Hall, J. L., et al., *Evaluation of the Ames Solid Waste Recovery System, Part III: Environmental Emissions of the Stoker Fired Steam Generators*, EPA-600/7-79-222, U.S. Environmental Protection Agency, 1979.

17. Hoffman, R., *Small Modular Incinerator Systems with Heat Recovery*, SW 177c, U.S. Environmental Protection Agency, November 1979.

18. Hollander, H. I., et al., A Comprehensive Municipal Refuse Characterization Program, presented at the Ninth ASME National Waste Processing Conference, Washington, D.C., May 11–14, 1980.

19. Klee, A. J., State of the Art Review of the Air Pollution Aspects of the Incineration of Industrial Waste, *Proceedings of the APCA Speciality Meeting, Waste Treatment and Disposal Aspects: Combustion and Air Pollution Control Processes*, Charlotte, NC, February 9–11, 1981.

20. Niessen, W. R., and Sarofim, A. F., Incinerator Air Pollution: Facts and Speculation, *Proceedings of the 1970 National Incinerator Conference*, ASME, Cincinnati, OH, May 1970, pp. 167–181.

21. Page, F. J., Torrax—A System for Recovery of Energy from Solid Waste, *Proceedings of the 1976 National Waste Processing Conference*, ASME, Boston, MA, May 1976, pp. 109–116.

22. Paige, S. F., *Environmental Assessment: At-Sea and Land-Based Incineration of Organochlorine Wastes*, EPA-600/2-78-087, U.S. Environmental Protection Agency, 1978.

23. Reid, R. S., and Heber, D. H., Flue Gas Emissions from a Shredded Municipal Refuse-Fired Steam Generator, *Proceedings of the 1978 National Waste Conference*, ASME, May 1978, pp. 167–178.

24. Rinehart, R., Complete Conversions Among the Regulatory Incineration Particulate Emission Definitions, *Proceedings of the 1976 National Waste Processing Conference*, ASME, Boston, MA, May 1976, pp. 185–191.

25. Rivers, J. R., et al., The Purox System, *Proceedings of the 1976 National Waste Processing Conference,* ASME, Boston, MA, May 1976, pp. 125–132.
26. Rubel, F. N., *Incineration of Solid Wastes,* Noyes Data Corporation, Park Ridge, NJ, 1974.
27. Smith, L. T., et al., Emission Standards and Emissions from Small-Scale Solid Waste Incinerators, *Proceedings of the 1976 National Waste Processing Conference,* ASME, Boston, MA. May 1976.
28. Stabenow, G., Performance of the New Chicago Northwest Incinerator, *Proceedings of the 1972 National Incinerator Conference,* ASME, New York, June 1972, pp. 178–194.
29. Teller, A. J., New Systems for Municipal Incinerator Emission Control, *Proceedings of the 1978 National Waste Processing Conference,* ASME, Chicago, IL, May 1978, pp. 179–187.
30. Walker, A. B., and Schmitz, F. W., Characteristics of Furnace Emissions from Large, Mechanically Stoked Municipal Incinerators, *Proceedings of the 1966 National Incinerator Conference,* ASME, May 1966, pp. 64–73.
31. Wall, H. O., and Farrell, J. B., Particulate Emissions from Municipal Wastewater Sludge Incinerators, *Proceedings of the Mid-Atlantic States Section Semi-Annual Conference of Air Quality Impacts of Ocean Disposal Alternatives,* APCA, Newark, NJ, April 27, 1979.
32. Weinstein, N. J., and Toro, R. F., *Municipal Scale Thermal Processing of Solid Wastes,* Publication SW-133c, U.S. Environmental Protection Agency, 1977.

The following are suggested as general information sources regarding the incineration of wastes.
1. Publications of the U.S. Environmental Protection Agency, in particular: (a) Office of Solid Waste (OSW Publications Distribution, EPA, 26 West St. Clair, Cincinnati, OH 45268), and (b) Office of Research and Development (Mailing List Manager, Washington, D.C., 20460).
2. Publications of the American Society of Mechanical Engineers (United Engineering Center, 345 East 47th Street, New York, NY 10017), particularly their *Proceedings of the National Waste Processing Conferences* (formerly *National Incinerator Conferences*).
3. Publications of the Air Pollution Control Association (P.O. Box 2861, Pittsburgh, PA 15230), including the *APCA Reprint Series, Proceedings of Specialty Conferences,* and the *Journal of the Air Pollution Control Association.*

CHAPTER 20

SOURCE CONTROL— FERROUS METALLURGY

MURRAY S. GREENFIELD

Dofasco, Inc.
Hamilton, Ontario
Canada

1 INTRODUCTION

The world's steel industry produced over 740 million tons of steel in 1979, mainly in two types of plants:

1. Large integrated plants utilizing coke ovens, blast furnaces, and steel making, and producing from 2 to 10 million tons/yr at an individual location. Over 75% of the world's steel was produced by integrated plants in 1980.
2. Electric furnace operations characterized by smaller plants usually making less than 1 million ingot tons/yr from cold scrap or sponge iron.

In areas with abundant supplies of low-cost natural gas, complexes producing over 2 million tons/yr of steel, using direct reduced iron, have been installed.

The growth rate for the electric arc furnace route in the 1970s was 5.3% per annum. The growth rate for steel made by both integrated and electric furnace routes worldwide was 1.8% per annum.

The integrated route is very capital intensive and requires expenditures of $228/ton (1979 costs) for the coke oven, blast furnace, basic oxygen process furnace route versus $56/ton (1978) for the scrap electric arc furnace route. No new integrated plant sites were established in the United States in the 1970s. Considerable modernization of facilities, including the installation of environmental control facilities, has been carried out. The AISI estimated that air pollution abatement costs (including operating, maintenance, and capital recovery), as a percentage of production costs, were 2.1% for integrated plants in 1980. Capital costs for air and water pollution control equipment at a new integrated site can reach 10–18% of the total capital costs.

The major air pollution problem from the steel industry is suspended particulate matter. Emissions associated with the combustion of sulfur-containing fuels are a lesser problem. Coke ovens, foundries, and ferroalloy furnaces can be significant sources of organic emissions.

Great strides have been made in the installation of environmental control equipment in the steel industry. The United States and other countries are now at a point where minor sources once considered insignificant, such as coke-quenching and basic oxygen furnace (BOF) clean gas stack emissions, are now the first and third highest contributors to particulate emissions from the U.S. steel industry as shown in Table 1.

The regulatory requirements of the various government jurisdictions are beyond the scope of this chapter. Many jurisdictions limit particulate emissions from wet scrubbers to 100 mg/Nm³ and from baghouses and ESPs to 50 mg/Nm³.

Table 1 Listing of Annual Particulate Emissions from Various U.S. Iron and Steel Unit Operations for the Year 1980[1]

Process	Particulate Emissions, ton/yr
Coke quenching	34,500
Blast furnace cast house	22,700
BOF stack	20,000
Material stockpiles	16,300
Roadway travel	16,300
Coke combustion stack	16,300
BOF charge and tap	14,500
Coke pushing	8,900
Sinter, miscellaneous fugitives	8,700
Sinter windbox	8,200
EAF charge, tap, slag	7,600
Coal preparation	7,400
OH stack	7,300
Coke door leaks	7,100
EAF stack	6,600
Sinter discharge end	5,700
Blast furnace top	3,700
Teeming	3,700
Ore screening	3,300
BOF, miscellaneous fugitives	2,200
Coke topside leaks	2,100
Reheat furnaces	2,000
Blast furnace combustion	2,000
OH roof monitor	2,000
Coal charging	1,800
Open area	1,100
Machine scarfing	670
BOF, hot metal transfer	650
OH, miscellaneous fugitives	640
Soaking pits	570
EAF, miscellaneous fugitives	540
OH, hot metal transfer	190

1.1 Air Pollution Control Costs

Reference 3 has developed cost functions for control technologies in the form of an equation:

$$\text{cost} = A \,(\text{production capacity})^B = Ax^B$$

Table 2 outlines uncontrolled and controlled emission factors, control technologies, and coefficients for the equation. Costs produced by the equations are for 1977 dollars and are estimates ±35%. Further details on the use of the equations can be found in Ref. 3. The technology outlined in this chapter, described as "best available technology," may not coincide with the assumptions made in the preparation of Table 2.

2 SINTERING PROCESS

2.1 Introduction

In 1978, 30 million tons of sinter were used worldwide to feed blast furnaces, while pellets accounted for only 10 million tons. The sintering process agglomerates fine iron oxides into sizable chunks for blast furnace feed. Dolomite and limestone added to the sinter feed is calcined, lowering the blast furnace coke rate. The first step for sinter production is the mixing of feed raw materials to obtain a consistent mix which is fed onto the horizontal sinter strand. The individual pallets forming the strand travel over windboxes that are under suction, drawing air through the bed of materials. Sufficient fuel in the form of coke breeze, FeO, or FeS is present in the mix so that after the sinter has passed under the ignition hood, a self-sustaining flame front forms which heats the bed. By the time the sinter has reached the discharge end, the flame front has passed completely through the bed. The sinter fuses as it cools.

2.2 Nature of Emissions

2.2.1 Sinter Feed and Product Handling

Figure 1 notes the large number of transfer points for the raw material feed and the sinter discharge. If sufficient moisture is contained in the sinter plant raw materials, dusting may not be a problem. Dusting is a problem at all sinter product transfer points, including hot screens, crash deck, and cold screens.

Dust loadings from the product discharge of a sinter plant vary from 9.2 to 13 g/Nm^3. Particle size range is 10% less than 10 μm, 40% from 10 to 100 μm, and 50% greater than 100 μm.

2.2.2 Windbox Exhaust

The suction in the windbox is maintained at 760–1500 mm W.C. to create the gas flow through the sinter bed. A typical exhaust flow is 2500 m^3/ton of total raw mix feed. As the flame front moves through the sinter, the temperature in the bed rises to about 1450 to 1500°C. Compounds with low vaporization temperatures, such as $ZnCl_2$ and alkali salts, can be driven off with the exhaust gases. Below the flame front there exists a temperature gradient that can either volatilize or partially combust organic materials. The exhaust temperature from each windbox gradually increases toward the discharge end of the machine, as shown in Figure 2, and the average windbox temperature varies from 105 to 200°C. Typical gas analysis and particle size and analysis are shown in Tables 3, 4, and 5, respectively.

2.3 Factors Affecting Emissions

2.3.1 Sinter Feed and Product Handling

The sinter product is screened and the <5 mm fraction, amounting to approximately 275 kg/ton of sinter feed, is normally mixed with feed materials. If the return fines are hot, drying can occur in the mixing granulating drum, resulting in dusting. The production of a poor quality sinter with considerable fines would increase the quantity of material circulating, and therefore, total emissions.

In most plants it is the practice to lay down a hearth layer, 1½ layers thick, directly on top of the pallets. The sinter mix is then added on top of the hearth layer. The hearth layer consists of material ranging in size from 10 to 15 mm. Due to the dry, dusty nature of the hearth layer materials, emissions are produced at each transfer point including feed to the strand.

A major factor affecting sinter discharge emissions is the quality of the sinter, as poor quality sinter is weak and dusts more. Sinter plants may screen the product with cold screens, hot screens, or a combination of hot and cold screens. Due to thermal currents, emissions are higher from hot screening. If the sinter is not screened before being added to the cooler, fines can be entrained.

The most popular method for cooling sinter is the carousel or circular cooler. Emissions from the cooler are dependent on the strength of the sinter, the fines present, the velocity of the gases through the bed, and the depth of the bed. Some sinter plants incorporate "on-strand cooling." An extra long strand cools the sinter and eliminates the cooler dusting and associated collection requirements.

Certain plants recycle the hot gases from either the cooler or the windbox onto the top of the sinter strand in order to obtain heat recovery, improve sinter quality, or reduce emissions. Any particles in the recirculated gas stream may be filtered by the upper surface of the bed. Since the top of the bed outside the ignition hood is below the fusion temperature, the fines may not be fused and might cause increased particulate emissions from the sinter product handling system.

2.3.2 Windbox Exhaust

SO_2 emissions originate from the S in the coke breeze, S in the raw materials, and the use of sulfur-containing fuels under the ignition hood. The S content in coke varies from 0.6 to 1.8%. Most ore sources are low in S, and the use of natural gas or desulfurized coke oven gas will eliminate sulfurous emissions from the ignition hood. Reference 11 states that a basicity of 3 to 4 in the lower part of the bed reduces SO_2 emissions by a factor of 10 and that the fuel N_2 in the coke breeze contributes to NO_x formation.

CO is produced by the incomplete combustion of the coke. Despite a lower O_2 content in the air supply to the bed, Ref. 5 reports a reduction in CO emissions from 22.75 to 18.00 kg/ton by recycling 25% of the waste gas back onto the bed. With waste gas recycle, the air supply to the bed is at a higher temperature.

Table 2 Capital Costs for Iron and Steel Plant Environmental Control Installations as a Function of Production Unit Capacity[3] (Values of A and B for the Equation $y = Ax^B$)

Process or Operation	Basis for Emission Measurement	Emission Rate[a]		Control	New Installations		Existing Installations		Units of X (2000 lb = 1 ton)
		Uncontrolled	Controlled		A	B	A	B	
Raw Materials									
Ore handling and storage	Hot metal produced	0.28	0.06	Water spray dust suppression	234,030.0	0.054	294,225.2	0.050	Total plant, annual tons of hot metal capacity
Coal handling and storage	Coal used	0.06	0.015	Water spray dust suppression	219,765.5	0.047	262,700.5	0.045	Total plant, annual tons of coke capacity
Coal crushing and transfer	Coal used	0.2	0.002 or 0.01 g/m³	Baghouse	3,284.3	0.326	3,358.9	0.331	Total plant, annual tons of coal capacity
Sintering									
Sinter windbox	Sinter produced	2.2 SO$_x$ 0.9 HC 0.12	0.15 or 0.04 g/m³	Scrubber None None	17,172.7	0.413	17,692.2	0.419	Sinter plant, annual tons of sinter capacity
Sinter discharge	Sinter produced	3.5	0.05 or 0.02 g/m³	Baghouse	23,262.5	0.321	24,923.1	0.323	Sinter plant, annual tons of sinter capacity
Sinter building fugitives	Sinter produced	0.35	0.0035 or 0.02 g/m³	Baghouse	17,460.9	0.199	17,010.2	0.207	Sinter plant annual tons of sinter capacity
Coking									
Wet coal charging	Coke produced	0.57 SO$_x$ 0.015 HC 1.8	0.01	Stage charging— new larry car	8,620.6	0.396	9,461.1	0.396	One battery, annual tons of coke capacity

Process			Control					Basis
Coke pushing	2.85	0.021	Enclosed hot car	385,888.2	0.194	423,541.2	0.194	One battery, annual tons of coke capacity
Coke quenching	4.3	0.5	Baffles and clean water	17.5	0.684	19.3	0.684	Total plant, annual tons of coke capacity
Door emissions	0.355	0.035	Door maintenance and auto cleaning	376,483.8	0.0	451,801.0	0.000	One battery, annual tons of coke capacity
Topside leaks	0.245	0.021	Good maintenance	0.0	0.000	0.0	0.000	One battery, annual tons of coke capacity
Underfire stack	0.5	0.075 or 0.06 g/m³	Dry ESP	4,392.3	0.439	4,833.1	0.446	One battery, annual tons of coke capacity
Coke handling	0.015	0.001	Baghouse	864.5	0.464	931.4	0.466	Total plant, annual tons of coke capacity
Coke oven gas	SO_x 6.65	0.5	Desulfurization	9,888.6	0.481	12,802.9	0.481	Total plant, annual tons of coke capacity
Coal preheater	0.065	0.0125	Scrubber	568.9	0.504	623.0	0.504	One battery, annual tons of coal capacity

Iron Making

Process			Control					Basis
Cast house emissions	0.35	0.021	RACT and runner covers	158,839.0	0.250	156,588.9	0.269	One blast furnace, annual tons of hot metal capacity
Slag pouring	0.14	0.007	Hood and scrubber	4,484.4	0.495	5,287.8	0.496	Total plant, annual tons of hot metal capacity
Slag crushing and screening	0.12	0.013 or 0.01 g/m³	Baghouse	10,181.1	0.224	10,829.0	0.226	Total plant, annual tons of hot metal capacity

Table 2 (*Cont.*)

Process or Operation	Basis for Emission Measurement	Emission Rate[a]		Control	New Installations		Existing Installations		Units of X (2000 lb = 1 ton)
		Uncontrolled	Controlled		A	B	A	B	
Steel Making									
Open hearth hot metal transfer	Hot metal used	0.175	0.0035 or 0.02 g/m³	Same as RACT	35,925.1	0.243	39,837.9	0.246	One OH shop, annual tons of steel capacity
Open hearth stack	Steel produced	8.7	0.175	Same as RACT	995.6	0.632	916.7	0.657	One OH shop, annual tons of steel capacity
Open hearth building fugitives	Steel produced	0.15	0.015	Same as RACT	0.0	0.000	0.0	0.000	One OH shop, annual tons of steel capacity
Open hearth slag crushing and screening	Steel produced	0.11	0.105	Same as RACT	25,338.9	0.000	25,338.9	0.000	Total plant, annual tons of steel capacity
BOF hot metal transfer	Hot metal used	0.18	0.0035 or 0.02 g/m³	Baghouse	33,307.1	0.246	35,835.6	0.247	One BOF shop, annual tons of steel capacity
BOF stack	Steel produced	25.5	0.02 or 0.03 g/m³	Closed hood—scrubber	6,812.5	0.489	15,887.1	0.464	One BOF shop, annual tons of steel capacity
BOF charging, tapping, and sampling	Steel produced	0.5	0.04	Furnace enclosure	6,505.6	0.450	8,578.4	0.443	One BOF shop, annual tons of steel capacity
BOF slag pouring	Steel produced	0.06	0.005	Baghouse	1,199,378.0	0.025	1,232,843.8		One BOF shop, annual tons of steel capacity
BOF slag crushing and screening	Steel produced	0.085	0.005	Baghouse	2,341.0	0.320	2,158.5	0.332	Total plant, annual tons of steel capacity

Source	Basis			Control device					Basis
Electric furnace emissions including fugitives	Steel produced	—	—	—	—	—	—	—	—
Carbon steel		15.0	0.455	Direct evacuation and canopy hood	1,308.2	0.642	1,438.9	0.643	One EAF shop, annual tons of steel capacity
Alloy steel		7.5	0.98	Canopy hood	1,022.2	0.663	1,172.5	0.665	One EAF shop, annual tons of steel capacity
Electric furnace slag	Steel produced	0.035	0.005	Baghouse	1,287.4	0.516	1,493.5	0.513	One EAF shop, annual tons of steel capacity
Electric furnace slag crushing and screening	Steel produced	0.05	0.005	Baghouse	86,711.7	0.079	93,357.3	0.080	Total plant, annual tons of steel capacity
Conventional casting	Steel produced	0.03	0.03	None	—	—	—	—	—
Continuous casting	Steel produced	0.06	0.005	Baghouse	1,261,960.0	0.024	1,457,337.0	0.025	One casting machine, annual tons of steel capacity
Soaking pits using 100% oil at 1.0% sulfur	Steel produced	0.1 SO_x 0.72	0.015 0.72	ESP	574.7	0.581	632.5	0.586	Group of pits, annual tons of steel capacity
Automatic scarfing	Steel scarfed	0.12	0.015	Wet ESP	529,826.1	0.128	573,959.3	0.129	One scarfing machine, annual tons of steel capacity
Reheat furnaces using 100% oil at 1.0% sulfur	—	0.21	0.03	ESP	1,541.0	0.558	1,740.0	0.561	Group of furnaces, annual tons of steel capacity

[a] All figures in kilograms of suspended particulate per tonne of production unless otherwise noted.

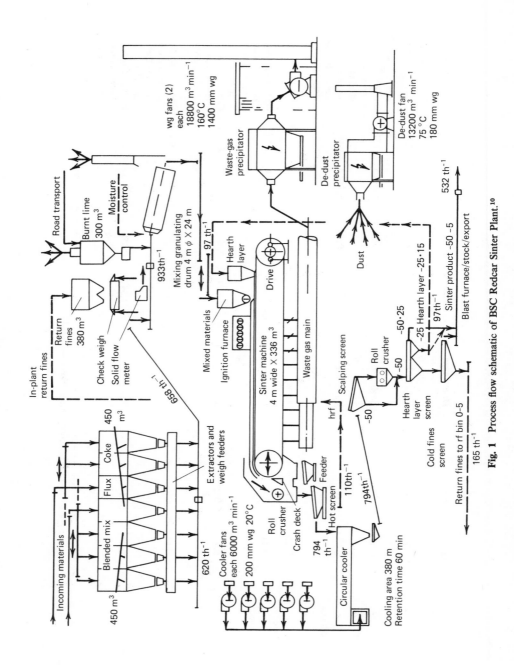

Fig. 1 Process flow schematic of BSC Redcar Sinter Plant.[10]

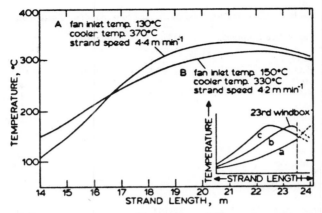

Fig. 2 Exhaust gas temperatures from various windboxes along sinter strand.[10]

Alkalis and hydrocarbons present in the feed may vaporize from the bed. Some plants add $CaCl_2$ to the mix to increase alkali removal from the sinter strand. Hydrocarbons that volatilize ahead of the flame front may not be combusted but instead they subsequently condense in the exhaust gas, proving difficult to clean. A fairly complete listing of organic emissions measured from two sinter plants can be found in Ref. 4.

The addition of blast furnace top gas solids containing up to 0.1 to 0.2% Na and K, respectively, to the sinter mix, would increase windbox fine particulate. The addition of oily millscale from hot strip mills to the sinter mix increases hydrocarbon formation. Coke breeze may also be a source of hydrocarbons. BOPF slag fines containing fluorides can increase HF emissions.

Table 3   Analysis of Sinter Plant Windbox Gases

CO_2	5–7%	
O_2	12–16%	
CO	0.7–14%	
N_2	78%	
H_2O	150 g/Nm³	
SO_2	0.2–3.0 g/Nm³	(1–12 kg/ton)
NO_x	0.4 g/Nm³	(0.3–0.8 kg/ton)
HCl	4 mg/Nm³	
Hydrocarbons		(0.02–0.2 kg/ton)

Table 4   Particle Size of Uncontrolled Sinter Plant Windbox Particulate

Particle Size, μm	wt% Undersize	
	A^a	B^b
0.3	8	2
1	19	6
5	44	16
7	60	20

[a] A—Plant using blast furnace top gas solids.

[b] B—Plant not using blast furnace top gas solids.

Table 5 Chemical Analysis of Sinter Plant Windbox Particulate

Particulate Component	Plant F	Plant G	Plant H
Fe_2O_3	33.9	11.7	28.0
CaO	7.1	10.9	15.0
MgO	5.3	0.4	2.0
K_2O	5.2	0.6	8.1
SiO_2	4.8	2.4	4.6
Al_2O_3	2.6	4.3	2.5
Na_2O	1.6	0.8	0.0
ZnO	0.4	0.1	0.0
MnO	0.2	0.1	0.0
Chlorides	8.5	3.0	8.8
Sulfates	7.5	16.5	2.1
Hydrocarbons	7.4	36.9	0.0
Other	1.6	0.0	0.0
Loss on ignition	13.9	12.3	28.9
Total	100.0	100.0	100.0

Poorly maintained dump valves in the bottom of the windbox and leaky pallet edge seals increase the air volume to be cleaned per ton of sinter produced.

Some newer plants feature on-strand cooling, where about 60% of the strand length is used for sintering and the remaining 40% of the strand length is used for cooling. The windboxes from the two sections may be segregated since the particulate sizing in the cooling section should be coarser and less difficult to clean than that from the sintering section.

2.4 Best Available Control Technology

2.4.1 Sinter Feed and Product Handling

All transfer and handling points for sinter product return fines and hearth layer require hooding and cleaning. Reference 6 outlines hooding design and flow requirements for collection at various sinter handling and transfer points. Sinter feed and product handling collection points are usually ducted to a common baghouse or ESP. Duct velocities must be high to avoid the dropout of the coarse particles collected. Coarse particles moving at high velocities cause considerable abrasion at elbows, and so on.

Considerable steam is generated where hot return fines are added to the sinter mix. Baghouses are unsuitable due to the high gas moisture content. Medium energy wet scrubbers with short duct lengths are normally used.

2.4.2 Windbox Exhaust

Gas recycle systems, as described in Ref. 12, can be used to reduce the quantity of gas to be cleaned by 20%. The exhaust volume that can be recycled is limited by the requirement to purge the CO_2 and maintain adequate O_2 levels to combust the bed.

The quantity of oil in the feed affects the gas cleaning device for sinter plants. According to Ref. 7, the presence of more than 3% oil in the waste dust can produce a pyrophoric material that will self-ignite and cause damage to dry cleaning devices such as baghouses and dry ESPs. Dry ESPs, wet ESPs, wet scrubbers, and baghouses are used to clean windbox gases.

Baghouses are in use at three sinter plants in the United States for windbox cleaning. In cold weather, precautions must be taken during startups to avoid wetting the bags from acid dewpoint condensation. Fluoride attack may also be a problem with fiberglass bags.

High windbox gas temperatures and a low level of moisture and SO_3 in the gas increase resistivity into the range of 10^{11}–10^{12} ohm-cm. Due to the lack of cooling, uncondensed hydrocarbons may be carried through baghouses and ESPs to form a detached plume.

An advantage of wet cleaning devices is their ability to condense and remove hydrocarbons. SO_x and NO_x removal systems have been experimented with on certain European and Japanese sinter plants. Washing windbox gases with NH_3 from the coke plant to obtain $(NH_4)_2SO_4$ has been proposed. Reference 8 describes the 650,000 Nm^3/hr catalytic NO_x removal system in operation at Kawasaki Steel, China Works, Japan. The catalyst is an iron silica alumina type and a 1.1:1 ratio of NH_3 to NO_x at 375°C is used to obtain 90% NO_x removal. Catalyst fouling is reported to be a problem.

Pl—Pd catalysts can effectively oxidize CO in the presence of SO_x. Pd—Co catalysts were effective at temperatures of 160°C in the absence of SO_x.[11]

3 SLOT OVENS—COKING

3.1 Introduction

The modern by-product coke oven operation involves the destructive distillation of coal to produce blast furnace or foundry coke and a medium Btu gas. The coking coal is usually a mixture of coals from different mines that are blended at the coke plant. Often, the coals have been cleaned at the mine to lower S and ash. The 6–10% moisture-blended coal may be directly charged into the oven, or it may be preheated to about 200°C and then charged hot into the oven. Individual ovens are 0.36–0.51 m wide, 12–17 m long, and 3.5–7.0 m high.

During the heating process, the coal loses about 30% of its weight to gases and tars. The gases are cleaned in a by-product plant and then distributed for use as a fuel. Upon completion of the coking cycle, the doors of the oven are removed and a ram pushes the coke out of the oven into a receiver car. The coke is usually quenched with water. In the United States, 99% of the coke is produced in slot coke ovens and the remainder in beehive ovens. In 1979, world coke production and U.S. coke production was 341 million tons and 48 million tons, respectively. A summary of coke oven emissions appears in Table 6. Particle size distributions from different operations are shown in Table 7.

Table 6 Uncontrolled Particulate and Organic Emission Factors from Coke Oven Operations (kg/ ton)

Emission Source	TSP[a]	BSO[b]	BaP[c]	Benzene
Larry car charge (wet coal)	0.5	0.55	0.001	0.25
Coke pushing	1	0.04	2×10^{-5}	0.003
Quench, clean water	0.85	8.5×10^{-4}	7×10^{-5}	1.5×10^{-5}
Doors	0.2	0.25	0.0015	0.01
Topside leaks	0.1	0.13	0.0005	0.0025
Combustion stack (old)	0.65	0.003	3×10^{-5}	6
Coke handling	0.5	0.0	0.0	0.0
Coal preheat	3.5	0.525	1.95×10^{-4}	0.007
Coal preparation	0.25	0.0	0.0	0.0
Coal storage	0.08	0.0	0.0	0.0
Pipeline charge (dry coal)	0.008	0.0095	1.75×10^{-5}	0.004
Redler conveyor (dry coal)	0.05	0.003	5.5×10^{-6}	0.0025
Hot larry car (dry coal)	0.085	0.0095	1.75×10^{-5}	0.004
By-product	0.0	0.15	0.0	0.1
Combustion stack (new)	0.065	3×10^{-4}	3×10^{-6}	0.0
Quench, dirty water	1.6	3.2×10^{-3}	1.55×10^{-4}	1.3×10^{-4}

[a] TSP—Total suspended particulate.
[b] BSO—Benzene soluble organics.
[c] BaP—Benz (a) pyrene.

Table 7 Particle Size of Uncontrolled Particulate Emissions from Various Coke Oven Operations

| Particle Size, μm | wt% Undersize | | | |
	Coal Preheater Inlet	Coal Charging	Coke Pushing	Flue Stack
0.25	38	—	35	—
0.5	40	20	52	23
1.0	41	25	58	50
3	71	40	72	88
5	85	42	94	—
10	—	50.0	98.9	—

3.2 Nature of Emissions

3.2.1 Coal and Coke Handling

The unloading, handling, and stockpiling of coal and coke at the various transfer points can result in the loss of the smaller size particles from the materials being handled.

3.2.2 Coal Preheating

Coal is preheated to about 200°C in flash dryers prior to being charged to the ovens. Charging preheated coal to the ovens reduces the coking time from 15 to 18 hr with wet coal to about 12 to 15 hr. Preheating improves coke stability. A typical preheater consists of a combustion chamber fed by a coke oven gas burner, rated at 85 million MJ/hr for a coal feed of 90 tons/hr. About one-third, or 500 Nm³/min, of the gas from the cyclones is cleaned and exhausted to the atmosphere, while the other two-thirds are recycled back to the combustion chamber. Particulate inlet loadings to the exhaust scrubber vary from 2.6 to 4.8 kg/ton.[13] A typical preheater exhaust gas analysis is: O_2—6%, CO_2—11%, CH_4—0.2%, total hydrocarbons—0.23%, water vapor—49%, NO_x—250 to 760 ppm, and SO_x—700 to 1100 ppm. Reference 13 documents further details of the organic content before and after the exhaust gas scrubber.

3.2.3 Coal Charging

Preheated coal may be charged into an oven by pipeline, conveyor, or larry car. Wet coal is usually charged with a larry car. The lids from the oven are removed and the larry car discharge boots positioned over the open ports. When the coal is fed from the hoppers into the oven, it contacts the hot incandescent oven walls and large volumes of steam and volatile matter are generated. Normally, during the 2- to 9-min charging cycle, the oven is kept under suction but, once the charging cycle is finished, the suction is turned off and the lids are replaced.

3.2.4 Coking Cycle

The heat to coke the coal is transferred from flues through the silica brickwork walls to the coal inside the oven. Pyrolization products include water, hydroaromatic compounds, parafins, olefins, phenolic and nitrogen compounds, and gases listed in Table 8.

Table 8 Chemical Analysis of Major and Minor Components of Coke Oven Gas

Major Components of Coke Oven Gas	
Compound	Percent Volume
H_2	46.5–65
CH_4	22.6–32.1
CO	4.0–10.2
CO_2	1.1–2.8
N_2	0.7–8.5
O_2	0.2–0.8
C_mH_n	1.7–5.2
H_2S	0.5–4.5
NH_3	1.3–9.0

Minor Components of Coke Oven Gas	
Component	Concentration, g/m³
HCN	0.1–4.0
Dust	1.8–36
BaP	0.2–0.6
Benzene	21.4–35.8
Toluene	1.5–3.0

About 320 Nm³ of coke oven gas is generated per ton of wet coal charged to an oven. When the center of the coke mass has reached 850 to 1000°C, the doors are removed from the ends of the oven and the coke is ready to be pushed out.

3.2.5 Coke Pushing

A ram pushes the coke from the oven into a receiver car. Although most of the volatiles from the coal have been evolved during the coking cycle, a minor amount is evolved during the push, as shown in Table 6.

3.2.6 Wet Coke Quenching

The hot coke is transported to the quench tower where about 1.7 m³/ton of water is sprayed on the coke for 90–120 sec to put the fire out. Quenching coke produces 864 m³ of water vapor per ton of coke which rises up through the tower and entrains considerable air. The temperature range of the stack gases during a quench varies from 60 to 90°C. In certain plants, it is the practice to use contaminated excess standpipe ammonia liquor to quench the coke. Table 9 presents particulate emission measurements for several quenching situations with baffle-type entrainment separators.

3.2.7 Dry Coke Quenching

The first step in the dry coke quenching operation is to hoist the hot coke container to the top of the quenching chamber where the coke is dumped. About 1400 Nm³ of gas (containing 14% CO, 10–12% CO_2, and 4% H_2) is circulated per ton of coke, cooling the coke from 960 to 150°C. For each ton of coke quenched, 0.5 ton of steam is generated. Dust is produced as the hot coke is charged into the quenching chamber and as the cool coke is discharged from the quench chamber. Three times the emissions at the coke screening station were measured from dry quenched coke as compared to wet quenched coke.[19]

3.2.8 Flue Stack

To coke 3–5% moisture coal requires 2600 to 3300 J/ton of coke. The products of combustion pass through the flues, through the regenerators, and out the battery stack. The underfiring system of a battery consists of a split regenerator below each oven. The hot flue gases are passed through one side of the regenerator for 20 to 30 min. The flow is then reversed, combustion air is heated by the regenerator brickwork, and the flue gases heat the opposite regenerator.

Suction inside the battery flues is provided by 60- to 80-m high chimneys. The pressure inside the oven varies from top to bottom and coke oven gases can be drawn through leaks in the brickwork from the oven to the flue. At the point of leakage into the flue, the gas mixture will be enriched. The characteristics of the flue gas from batteries experiencing flue leakage are shown in Table 10. The use of blast furnace gas increases the flue gas volume to about 2500 Nm³/ton.

3.3 Factors Affecting Emissions

3.3.1 Coal and Coke Handling

Coals that have been passed through a cleaning operation are finer and much dustier than noncleaned coals. The addition of pulverized coke breeze or petroleum coke to the coal blend also increases coal-handling emissions. Coals with less than 6% moisture are dustier than wetter coals. Coal conveyors are normally shrouded or covered to reduce windblown dust.

3.3.2 Coal Preheating

The coal is fluidized in the flash dryer section, leading to collision and resultant particle breakdown. Higher coal feed rates and preheat temperatures require higher combustion chamber gas volumes and temperatures, increasing organic emissions. As the coal feed rate increased from 82 to 109 tons/hr, the chloroform soluble organic emissions increased from 0.321 to 0.584 kg/ton of coal processed.[13]

3.3.3 Coal Charging

Before the coal is charged into the oven, a water or steam aspirator is turned on and educts a gas flow from the oven up through the standpipe. If the flow is strong enough and the openings around the charging boots sealed, the oven can be maintained under vacuum. The flow educted depends on the nozzle standpipe configuration, as well as the nozzle size and header pressure used, as documented in Ref. 23. Ovens built in the 1950s and 1960s were generally 4 or 5 m in height, while ovens built in the 1970s and 1980s are 6 or 7 m high. Due to stack effect, taller ovens have a higher gas pressure

Table 9 Coke Plant Quench Tower Particulate Emissions Collected in Various Parts of Sampling Train at Different Steel Plants (kg/ton of Coal)

Test Location	Cyclone with Nozzle	Front Half without Cyclone	Front Half Total	Back Half	Full Train	Makeup Water TDS,[a] mg/l		Cyclone Size, μm	Velocity, m/sec	Spray Method	Baffle Type
Clean Water											
Lorain	0.14	0.59	0.73	0.075	0.8	520	0.46	10	9.5	Coherent stream	Single row 45°
Dofasco no. 2	0.02	0.115	0.135	0.12	0.255	175	0.165	10	2.4	Regular	Double row 20°
Gary no. 3	0.135	0.03	0.165	0.02	0.185	468	0.14	10	—	Regular	Carl still plastic
Gary no. 5	0.12	0.04	0.16	0.035	0.195	512	0.13	11	2.1	Regular	Single row 45°
Dofasco no. 1	0.015	0.13	0.145	0.068	0.213	350	0.17	10	4.0	Coherent stream	Double row 45°
Dirty Water											
Lorain	0.57	0.795	1.365	0.395	1.76	5370	5.69	10	9.8	Coherent stream	Single row 45°
Gary no. 3	0.145	0.07	0.215	0.22	0.435	1786	0.45	9	3.7	Coherent stream	Carl still plastic
Gary no. 5	0.18	0.14	0.32	0.65	0.97	1917	0.58	12	2.4	Regular	Single row 45°
Dofasco no. 1	0.157	0.474	0.631	0.349	0.980	6000	2.9	10	4.0	Coherent stream	Double row 45°

[a] TDS—Total dissolved solids.

**Table 10   Coke Oven Battery Stack
Gaseous Analysis, Temperature and
Moisture Contents**

Gas temperature—260°C (average)

Gas moisture content—13.0% (average)

Gas composition (dry basis)
  O_2—4.6 to 15.4% (average 10.9%)
  CO_2—2.5 to 7.0% (average 4.8%)
  CO—1.18 g/Nm³ (1000 ppm)
  NO_x—25 to 155 ppm
  SO_x—50 to 700 ppm

at the charging boots and require a more effective charging control system. When the chuck door is open, air in-draft may significantly reduce oven suction. If the gas passage at the top of the oven between the charging port and the standpipe or jumper pipe is blocked, emissions from the charging boots can result. Any buildup of tar and coal in the standpipe may reduce the suction in the oven during a charge.

3.3.4   Coking Cycle

Coking cycle emissions include leaks from doors, standpipe caps, and lids. The pressure in the collecting main is maintained from 2 to 9 mm W.C. positive pressure. At the beginning of the coking cycle, the coal is densely packed and the resistance to gas flow is high. Gas pressure adjacent to the door below the coal surface may be high, leading to leaks through any gaps in the door jamb interface. As the coal heats up, it develops fissures which reduce gas flow resistance and the gas temperature at the top of the oven rises. A pressure differential develops between the top and bottom of the oven. Sufficient positive pressure in the standpipe is required to maintain +2 mm W.C. positive pressure at the base of the oven to avoid negative pressure drawing in the air which could damage the brickwork. With 6- to 7-m high ovens, collecting main pressures are maintained at 9–12 mm W.C., as compared to 5–6 mm W.C. with 4-m high ovens.

  To control gas leakage, 26 m of door seals and 10 m of lid seals must be maintained per oven. The most popular door seals are the rigid metal seals. Hardened deposits on the door seal or the door jamb may cause gaps, and therefore, are removed periodically. Other factors leading to gaps between the door seal and the jamb include insufficient latch force, damaged or incorrectly aligned door seals, and excessive bowing of the door or jamb. After the door is replaced, there may be numerous gaps but, if the gaps are less than 0.2 mm, they will seal in less than 1 hr.[24]

3.3.5   Coke Pushing

As mentioned previously, most of the volatiles are evolved during the coking cycle. Due to oven heating problems, however, certain sections of the ovens may not be heated sufficiently. These so-called "green" sections produce large clouds of black smoke when pushed out of the oven. Reference 25 notes that the average plume temperature, measured from a boom above the coke receiver car, was 110°C for green pushes and 47°C for clean pushes. The same study reported particulate emissions of 1 kg/ton for green pushes and 0.35 kg/ton for clean pushes.

3.3.6   Wet Coke Quenching

Coke stability is a measurement of coke strength and the higher the stability or strength of the coke, the lower the emissions should be from the quenching operation. Another factor that may influence quench tower emissions is the coke quenching rate. The use of spray nozzles to quench the coke produces small water droplets which may not penetrate the steam blanket, resulting in a slow cooling rate. On the other hand, open water pipes produce a coherent water stream that penetrates the steam blanket, achieving a higher cooling rate. More fracturing of the coke and increased quench tower emissions can result.

3.3.7   Dry Coke Quenching

The unloading of green coke into the quenching chamber would be expected to produce higher emissions, as compared to well coked-out coke. The higher the positive pressure maintained at the top and bottom of the quench chamber, the greater the volume of gases escaping during the charging and discharging operations.

3.3.8 *Flue Stack*

Coke oven gas can be drawn into the flues through any gaps in the brickwork between the oven and the flue. Certain coke practices can result in carbon sealing the cracks in the brickwork. Sealing the cracks with refractory mixes, silica welding and silica dusting, control the leakage into the flues.

3.4 Best Available Control Technology

3.4.1 *Coal and Coke Handling*

If the coal handled is above about 6% moisture, dusting during handling is usually not a problem. If the coal is dry, control techniques include hooding and gas cleaning, foam or water sprays, or enclosed conveyor systems.

3.4.2 *Coal Preheating*

The most popular coal preheater exhaust gas cleaning systems installed in the United States are venturi scrubbers with pressure drops ranging from 400 to 1000 mm W.C. The organic content or "back-half" catch, from a 400 mm W.C. venturi scrubber outlet installed on a preheater exhaust, was reported at 400–1200 mg/Nm³. Cooling of the gases reduces the organic or noncondensable emissions.

3.4.3 *Coal Charging*

Technologies for control of coke oven charging emissions include:

1. Staged charging.
2. Scrubbers mounted on the larry car.
3. Connection of the larry car charging boot shrouds to fixed ducts and gas cleaning devices.

Staged charging consists of emptying the outside coal hoppers first, replacing the lids, and then charging the inside hoppers. In order to achieve low emission levels using staged charging, the following conditions must be achieved:

1. Adequate standpipe suction and low-leakage charging boots to maintain a negative pressure inside the oven during the entire charge.
2. Suction at both ends of the oven from a double collector main or jumper pipe arrangement (i.e., Figure 3).

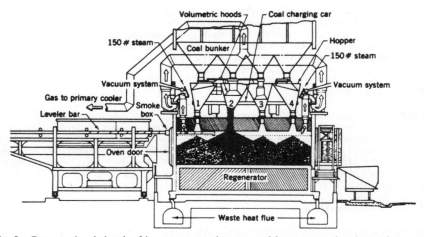

Fig. 3 Cross-sectional sketch of larry car emptying one coal hopper at a time into coke oven.[23]

3. Larger outside hopper holding capacity to reduce leveling times and leveler bar seals to reduce indraft during the leveling operation.

If the flow drawn by the standpipe is excessive, coal carryover from the oven into the collecting main can occur, reducing tar quality and increasing collector main cleanout frequency. Excessive standpipe gas flows also increase the O_2 content of the coke oven gas, reducing the Btu content. A number of U.S.-based coke oven facilities have been able to reduce the duration of emissions with staged charging to less than 15 sec/charge, which corresponds to a particulate emission of 0.76 g for a 30-ton coal charge.

Particulate emissions from a larry car scrubber with gas flow capacities from 60 to 320 Nm^3/car varied from 83 to 201 g for coal charges which varied from 30 to 40 tons.[27]

At a number of Japanese plants, the emissions from the larry car charging boots have been shrouded and exhausted by fixed ducts and gas cleaning devices at a rate of 1000 Nm^3/min. A gas cleanliness of 25 mg/Nm^3 and a 3-min charging time produces an emission of 75 g/charge.

3.4.4 Coking Cycle

Control of emissions from the coking cycle consists of limiting leaks from lids and doors. Magnetic lid lifters rotate the charging hole lid as it is replaced back into the machined charging hole casting. The rotating action usually cleans the two machined surfaces sufficiently so that leakage is not a problem. A luting mixture is poured around the edge of the manually replaced lids, forming a gas-tight seal.

Various types of machined seals are used to reduce leakage from standpipe caps. Often it is necessary to pour luting mixtures around the cap to reduce visible leakage. Specially designed water-sealed standpipe caps are gaining wider acceptance. A water reservoir is provided around the top of the standpipe. The cap is equipped with an extended skirt that sits in the reservoir, providing a seal. The depth of skirt submerged must be sufficient to seal against the positive pressure inside the collecting main. A continuous water supply and overflow is necessary to maintain the water seal.

To control door emissions, metal seals backed by spring-loaded plungers to follow the contour of the jamb are used. If adequate maintenance of the seal is carried out, the total number of leaking doors can be limited to 10%. In Japan, hoods drawing 500 Nm^3/min at each door have been installed to collect door leakage. The effectiveness of door hoods is reduced by strong winds. To increase the hood collection efficiency, strips have been fastened to the door to form a channel between the edge of the door and the buck stays.

3.4.5 Coke Pushing

A variety of systems have been installed for collection of emissions from coke pushing and the systems can be grouped into three classifications:

1. Sheds.
2. Mobile hoods connected to fixed ducts and gas cleaning.
3. Mobile hoods connected to mobile gas cleaning devices.

Sheds consist of a building running along the length of the coke side of the battery inside which the coke-receiver car travels. The volume requirement for a shed installed over 56, 6-m high ovens is approximately 8400 Nm^3/min. Entry and exit doors, oven-to-shed interface, and the various intersections of the roof and walls must be adequately sealed to attain a high collection efficiency.

For traveling coke-receiver cars, the typical open area at the start of a push is about 20 m^2. For a one-spot car, it is about 7.5 m^2. Sufficient air must be in-drafted to cool the hot plume so that ductwork damage does not occur. The temperature rise in the collection system for a typical 9-ton push is limited to 144°C for an air flow of 2550 Nm^3/min.[17] Baghouses, venturi scrubbers, and wet ESPs are in use for cleaning the gases from coke oven pushing operations. Precoat materials can be used to prevent the tars in pushing plumes from plugging baghouse filter fabrics.[18]

3.4.6 Wet Coke Quenching

Approximately 80% of quench tower particulate is less than 15 μm. Mist eliminators are best suited for removal of particles above 50 μm. Ninety percent of the 10-μm particles can be removed by using water spray nozzles with an L/G of 4×10^{-4} m^3 of water per m^3 of gas.[28] The water spray nozzles feature small diameter holes and require clean water.

Offset towers are appropriate when sprays are used to cool or clean the gases to eliminate spray water falling on the coke, thereby causing variable coke moistures.

In order to generate sufficient draft to allow the use of efficient mist eliminators, a minimum tower height of 30 m is recommended.

3.4.7 *Dry Coke Quenching*

Emission points from dry coke quenching include:

1. Coke unloading into the chamber.
2. Discharging coke from the chamber.
3. Excess chamber exhaust gas.

An exhaust volume of 3000 Nm^3/min is used at the coke discharge transfer point into the dry quencher.[19] The exhaust gas from a 70 ton/hr capacity unit is cleaned in a baghouse and then burned in an incinerator, producing 20 m^3 of gas per ton of coke quenched.[19]

3.4.8 *Flue Stack*

New ovens, if properly constructed, are rarely troubled by gas leaking from the ovens to the flues. As the ovens age, gaps are created by shifting brickwork and can cause flue stack opacities above 20%. Cracks are most prevalent near the ends of the oven and can easily be reached, but repair of cracks toward the center of the oven requires specialized machinery. Baghouses and ESPs have been installed to clean flue gases.

Reference 21 reports on the installation of a flue gas ESP for the removal of the 80–90% carbon dust which has a resistivity varying from 0.0035 to 0.0058 ohm-cm at 20°C. Particles with such a low resistivity may lose their charge when they contact the collecting plate and may be reentrained. A baghouse for flue stack cleaning was installed at Kaiser Steel, Fontana, California. Stack outlet loadings were below 50 mg/Nm^3 as long as inlet temperatures were below 350°C.[20] Cooling of the gases to below 175°C apparently will reduce emissions.[20] Due to the pyrophoric nature of the dust, burning of the bags has been a problem.

4 BLAST FURNACE

4.1 Introduction

Worldwide total iron production from blast furnaces is about 300 million ton/yr. A blast furnace consists of a refractory-lined steel shaft in which the charge is fed into the top through a gas seal. Air heated from 871 to 1100°C is blown through tuyeres into the bosh near the bottom of the furnace. The combustion of coke provides the CO to reduce the iron oxides to iron and provides the heat to melt the iron and gangue. As the burden moves downward through the furnace, it is heated by the upward flow of gases that exit at the top of the furnace. The molten iron and slag drain to the hearth of the furnace and a taphole is drilled into the hearth to drain the slag and iron accumulation into a trough. The trough is equipped with a skimmer and dam at the outlet end, allowing the iron to be drawn off separately from the slag. A typical charge required to produce a ton of hot metal consists of 1.55 tons of iron ore, 0.5 ton of coke, and 0.05–0.15 ton of limestone and dolomite. Higher productivities have been achieved through the use of high-purity iron units, sized raw materials, higher blast temperatures, humidity control, higher top pressures, fuel, and O$_2$ injection.

4.2 Nature of Emissions

4.2.1 *Top Gas*

The top gas (characteristics shown in Tables 11 and 12) contains sufficient energy content to be used as a process fuel. Submicrometer particles form when low boiling point materials such as Na, K, and Zn compounds vaporize in the lower hotter sections of the furnace and condense in the upper section of the furnace. The gas leaving the furnace contains only trace quantities of H$_2$S and SO$_2$ due to the efficiency of the iron oxide and flux in removing sulfurous gases. Volatile organics added with feed materials will vaporize and exit in the top gas before reaching the hot tuyere zone.

4.2.2 *Casting*

A taphole is drilled to drain the pool of iron and slag accumulated in the furnace. Once the molten iron has been drained down to the level of the taphole, a combination of slag–metal mixture is cast from the furnace. The taphole is plugged when hot blast flows from it. The molten iron and slag drained from the taphole is directed into a trough which is equipped with a dam and skimmer arrangement to separate the iron and slag into two separate streams.

The iron stream drains through a runner system into a cylindrical-shaped railroad car for transportation to the steel-making operation. The slag runner drains into a slag pit or slag pots. Casting particulate analysis is shown in Table 12. Particulate emissions are produced either by the direct vaporization of

Table 11 Particulate Content, Energy Content, Gas Flow, Temperature and Analysis of Blast Furnace Top Gas

Temperature	175–250°C
Gas quantity	1000–2000 Nm³/ton of iron
Gas analysis	CO—23 to 40% CO_2—15 to 22% H_2—1.5 to 6.0% N_2—remainder
Energy content	3.7 MJ/m³
Dust content	Up to 30 g/Nm³

Table 12 Chemical Analysis of Blast Furnace Particulates from Top Gas, Casting Emissions and Calcium Carbide Desulfurization (wt%)

Constituent	Top Gas	Casting Emissions	Desulfurization Using CaC_2
Fe	47	53	25
C	12	7	23
SiO_2	12	2	10
Ca	2.9	1	32
Mg	1.2	0.1	1.3
Mn	0.38	1.4	1.14

compounds or the bursting of CO bubbles at the metal–atmosphere interface, ejecting particulate into the air. Since the ratio of Mn in the dust versus Mn in the melt measured at Dofasco is only 1.6, droplet ejection must be the major mechanism. Further discussion of the bubble bursting phenomena can be found in Ref. 29. The particle size analysis collected from a total building ventilation system is shown in Table 13.

SO_2 and CO, collected at a local hood mounted over a tilting iron runner, amounted to 0.018 kg/ton and 0.026 kg/ton, respectively. Particulate collected at the same location amounted to 0.14 kg/ton. Total particulate collected by a local collection system was reported at 0.65 kg/ton with 0.14 kg/ton less than 15 μm.[36]

4.2.3 Slag Cooling

Small amounts of the S in the slag may be oxidized to SO_2 during the cast. A common practice is to drain the molten slag into a pit adjacent to the furnace. Water may be sprayed onto the slag to

Table 13 Particle Sizing Distribution from Blast Furnace Cast House and Hot Metal Desulfurization

Aerodynamic Particle Diameter, μm	Blast Furnace Cast House[a]	Desulfurizing Hot Metal
	Fraction Smaller Than:	
0.25	0.026	
0.50	0.083	0.015
0.75	0.110	0.025
1.0	0.177	0.03

[a] Total building ventilation.

increase the cooling rate or produce an expanded slag. The S in the slag cast from the furnace is present as CaS.[30] If water contacts hot slag, then H_2S emissions are produced. H_2S concentrations inside a cover, mounted over a section of the slag pit being quenched, peaked at 250 ppm H_2S by volume.[31]

4.2.4   Hot Metal Desulfurization

Blast furnaces can be operated to produce an iron containing from 0.02 to 0.03% S, which is utilized directly in steel-making operations. Certain plants have increased blast furnace productivity and lowered operating costs by producing iron containing from 0.04 to 0.1% S which is externally desulfurized by injecting CaC_2, Mg, or soda-based reagents. Desulfurization station particulate emission characteristics are shown in Tables 12 and 13. The particulate emission factor for a system injecting 3.2 kg CaC_2 per ton of metal was 0.5 kg/ton.

4.3   Factors Affecting Emissions

4.3.1   Top Gas

If all or part of the raw materials are screened in the stockhouse prior to being added to the furnace, dust entrained by the top gas is considerably reduced. Buildup or secretions may occur on the walls of the furnace from the burden of the furnace to "hang up." When the burden finally moves, it may produce a roll in which the material on one side of the furnace is moving downward while the material on the other side of the furnace is moving upward. This results in a furnace top pressure surge. In order to prevent damage to the furnace, the excess gas is vented out the safety release valve or bleeder valve. Estimates of emissions from slips and bell leaks can be found in Refs. 33 and 34.

4.3.2   Casting

At high C contents, decarburization of the Fe—C melt is the result of a surface area reaction and the rate of reaction is limited only by the rate of O_2 transport in the gas phase. During blast furnace casting, O_2 transport is low since there is no forced circulation of the air directly above the molten iron. Slag frequently covers the metal, reducing the O_2 transport. Turbulent conditions break up the slag layer, increasing O_2 transport by exposing more metal directly to the air. Runners with a high width-to-depth ratio would increase O_2 transport emissions. The taphole is usually drilled at a 2–20° angle to the horizontal. A higher angle increases the trajectory of the metal as it exits the taphole, leading to increased turbulence in the trough. The diameter of the taphole determines the casting rate and resultant turbulence.

Coal tars and other organics are sometimes used as binders in blast furnace runner and trough refractories, resulting in organic emissions when the refractories are heated.

With high top pressure furnaces after completion of the cast, large quantities of fume may be evolved if there are problems in plugging the taphole.

4.3.3   Slag Cooling

A typical practice is to drain the slag from successive casts into a pit established at the end of the cast house. The outer layer of slag cools down between casts, but the inner layer can remain molten. A typical cooling rate at 1.4 m below the surface is 2.8°C/hr.[31] When the pit is full, water is frequently sprayed onto the slag to allow the pit to be dug safely. To increase the cooling rate and avoid the use of water, pits with a high surface area and shallow depth are favored.[31] H_2S emissions can be reduced by allowing a three-day cooling period with no casting before the pit is watered and dug.[31]

4.3.4   Hot Metal Desulfurization

Turbulence in the ladle is required for satisfactory desulfurization, but excess turbulence will increase emissions. Failure of the lance above the iron level can result in unreacted reagent being picked up by the gas cleaning system until the flow is stopped. Insufficient freeboard between the top of the iron and the top of the ladle car may cause molten materials to be ejected from the ladle car and increase emissions. On the other hand, insufficient metal in the car to allow required lance submergence may also result in increased emissions.

A popular injection material is salt-coated Mg granules containing about 85% Mg. Eighty percent of the injected salt is captured by the slag on top of the ladle and 20% of the salt escapes as a fume.[32] The particle size of salt-coated material would be that of a vaporized fume due to the evaporation of the alkali salts and the evaporation and oxidation of nonreacted metallic Mg.

4.4 Best Available Control Technology

4.4.1 *Top Gas*

Due to the energy content, blast furnace top gas is used as a fuel in blast furnace stoves, coke ovens, reheat furnaces, boilers, and so on.

In order to minimize maintenance problems with burners, high-efficiency cleaning of the gas has been a general practice.

Older furnaces generally run at top pressures varying from 0.25 to 0.5 atm, while new furnaces may run at top pressures of up to 3 atm. Low top pressure furnaces often use wet ESPs following the venturi scrubber to achieve the desired gas cleanliness. Outlet particulate loadings of 5 mg/Nm³ can be achieved with venturi scrubbers using the pressure drop available from high top pressure furnaces.

4.4.2 *Casting*

The major difficulty with the collection of blast furnace casting fumes is the long length of runner over which the fumes are generated, especially with multiple ladle spot furnaces. Modern furnaces often use tilting runners which allow the runner length to be cut in half, decreasing the area from which fumes are evolved. A modern fume collection system shown in Figure 4 consists of local collection

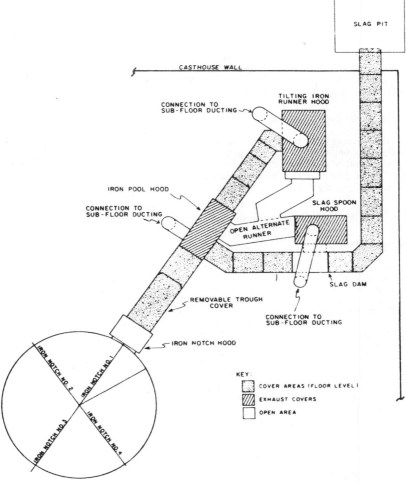

Fig. 4 Sketch showing runner cover and local suction hood locations for blast furnace cast house fume control system.[36]

Table 14   Blast Furnace Cast House Design Suction Volumes for Local Exhaust Capture Systems Equipped with Runner Covers

Location	Ventilation Rate
Taphole hood	250,000 Nm³/hr
Taphole runner cover	190,000 Nm³/hr
Dam-skimmer	184,000 Nm³/hr
Tilting iron runner	310,000 Nm³/hr

at the taphole, skimmer, dam, and tilting runner with below floor ducting to avoid interference with overhead cranes. Well-sealed covers over the runners limit O_2 transfer and resultant fume production. Steam and combustible gases may be distributed over the metal surface to limit O_2 transfer.

Volume requirements for a local hooding and runner cover system are shown in Table 14.

Kobe Steel of Japan has installed a wet ESP on the roof of the cast house. Natural draft carries the fumes through the precipitator, saving the installation of ductwork and fan.

Since the gas temperature from local hoods does not usually exceed 90°C, satisfactory gas cleaning devices for casting emissions include baghouses and ESPs.

4.4.3   Slag Cooling

As mentioned previously, the use of water for quenching slag results in H_2S emissions. A 50% reduction in H_2S can be attained by adding 100 ppm of $KMnO_4$ to the quench water.[31] A total H_2S reduction of 88% was attained by a combination of a three-day cooling period before the quench as well as 100-ppm addition of $KMnO_4$ to the quench water.[31]

4.4.4   Hot Metal Desulfurization

A collection volume of 1000 Nm³/min for a hood located within 0.3 m of the ladle car will satisfactorily collect emissions from the injection of 80 kg/min of a CaC_2—$CaCO_3$ mixture. The average temperature of the gases would be approximately 80°C and could peak much higher if a lance break occurred. Wet scrubbers, baghouses, or ESPs are satisfactory gas cleaning devices. Provisions for the safe collection of unreacted CaC_2 or Mg reagents are required.

5   OXYGEN BLOWN STEEL MAKING

5.1   Introduction

An integrated mill produces steel from molten iron. More steel is produced by BOPFs than by electric arc furnaces and open hearths combined. O_2 blown steel-making techniques to be covered in this section include top blown furnaces in which the O_2 is injected through a lance from above the bath, bottom blown furnaces in which the O_2 is injected through tuyeres below the surface of the molten bath, and a combination of top and bottom blown systems. The oxidation of the C, Si, and Mn in the molten iron generates sufficient heat to allow the addition of 15–35% scrap to the vessel.

All three systems use an open mouth converter vessel, lined with basic refractory as shown in Figure 5. The furnace is mounted on trunions that allow the vessel to be rotated through 360° in either direction. Production steps to produce steel consist of:

1. Charging, in which the scrap and molten iron are charged into the furnace.
2. O_2 blowing, during which approximately 3.15 m³/min of O_2 per ton of steel is blown into the vessel.
3. The turndown, during which the vessel is tilted sufficiently to allow metal samples to be taken from the bath and analysis checked.
4. The reblow, during which further refining of the metal takes place if the specified metal characteristics have not been attained.
5. Tapping, during which the molten steel is poured from the vessel into a ladle.

BOPF emission factors are summarized in Table 15.

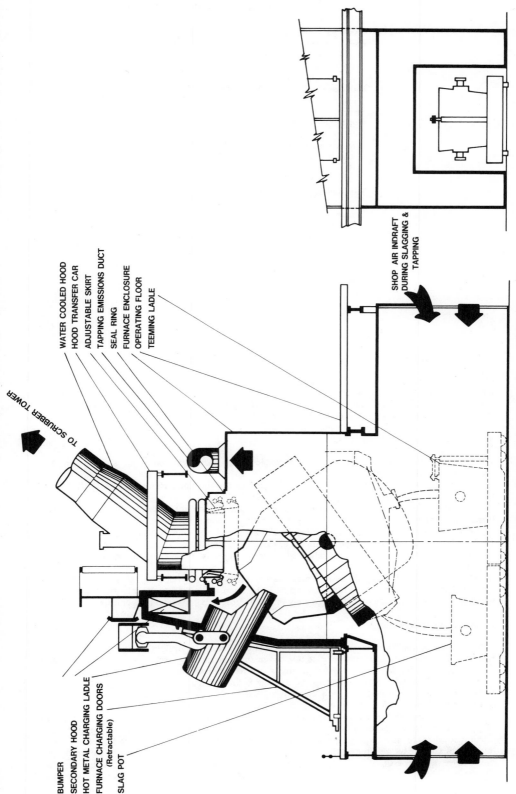

BUMPER
SECONDARY HOOD
HOT METAL CHARGING LADLE
FURNACE CHARGING DOORS (Retractable)
SLAG POT

TO SCRUBBER TOWER

WATER COOLED HOOD
HOOD TRANSFER CAR
ADJUSTABLE SKIRT
TAPPING EMISSIONS DUCT
SEAL RING
FURNACE ENCLOSURE
OPERATING FLOOR
TEEMING LADLE

SHOP AIR INDRAFT DURING SLAGGING & TAPPING

Fig. 5   Sketch showing enclosure around BOP furnace.[26]

Table 15 Uncontrolled Particulate Emission Factors from Basic Oxygen Process Furnace Operations

Process	Emission Factor, g/ton of Steel
Hot metal transfer	89.5
Hot metal charge	
Top blown	189[a]
Bottom blown	310[a]
Turndown	
Tapping	
Top blown	146
Bottom blown	460
Teeming	35

[a] Based on 85% furnace yield and 80% of charge being hot metal.

5.2 Nature of Emissions

5.2.1 Hot Metal Transfer

Hot metal transfer is the pouring of the molten iron from the torpedo car into the furnace charging ladle. The transfer of undesulfurized iron into the ladle produces emissions similar to those from blast furnace casting. The transfer of spent desulfurizing reagent along with the metal will increase the analysis of Ca and Mg, depending on the reagent.

5.2.2 Raw Materials Charging

The furnace is tilted sideways to about a 45° angle to the horizontal and scrap is dumped into the furnace from a charging bucket.

At certain installations, the furnace may be tilted to the upright position and an O_2 fuel burner used to preheat the scrap.

After the scrap charge, the furnace is positioned to receive the hot metal. With the bottom blown furnace, N_2 is blown through the tuyeres to prevent plugging during the hot metal charge. Contaminants in the scrap such as oils and Zn are ignited by the hot metal and considerable heat is evolved. Table 16 lists the size of the heat plume measured at different installations. Table 17 indicates particle size data for BOF charging emissions. Additional chemical analysis of charging emissions can be found in Refs. 39 and 43.

During hot metal charging, peak temperatures of 800°C with velocities of 3 to 4 m/sec were measured in the vicinity of the furnace mouth. Integration of the measurements at a height of 3 m above the furnace produced a calculated flow of 215 m³/sec at an average temperature of 165°C.[48]

5.2.3 Oxygen Blowing

After the charge has been completed, the furnace is tilted into the upright position and the O_2 flow is initiated. In the bottom blown furnace, O_2 blown through tuyeres reacts exothermally with the C, Si, or Mn, producing local hot spot temperatures of over 2100°C. Vaporized iron forms inside rising bubbles of CO, subsequently cooling to the average bath temperature and condensing. Vaporization fume is very fine, the smallest particle being about 0.02 μm while the majority are less than 0.2 μm in diameter. Considerable Mn will be evaporated due to the high vapor pressure.

The mechanism for fume generation from the top blown furnace is much different. At the beginning of the O_2 blow, the condition of high lance, low bath velocity, low bath temperature, and high metal–C content all result in a large quantity of fine metal spray being generated by the O_2 jet impact and bursting of CO bubbles.[40] During this period, the slag volume is small and unable to entrap much of the spray.[40] Fume particles formed as a result of metal spray oxidation are spherical in shape with the majority being less than 1 μm in diameter. For a 140-ton heat, emissions of particles less than 5

Table 16 Secondary Ventilation System Ventilation Rate, Heat Evolution and Gas Temperature for Various Steel Plants[52]

Plant	BOF			Charge			Tapping		Reladling			Other		Total	
	No.	M.T.	Sec.	m³/min	°C	Gcal/min	m³/min	°C	m³/min	°C	Gcal/min	m³/min	°C	m³/min	°C
Stelco LED	2	230	40	10,000	200	0.316	10,000	—	6,000	150	0.151	—	—	16,000	135
Fukuyama	2	300	40	10,000	200	0.316	5,000	150	6,000	150	0.151	Skimming 4,000	—	16,000	150
OITA	2	300	300	11,300	200	0.357	8,400	80	10,100	150	0.254	Desul., Deslag 9,600	0	14,500	87
Kimitsu No. 2	2	220	—	11,200	200	0.354	—	—	3,700	60	0.033	Desul., Deslag 8,350	0	12,600	130
Inland	2	200	—	Canopy (7,800)	95	(0.127)	—	—	Too small 3,500	95	0.057	—	—	11,300	120
Stelco Hilton	3	114	65	6,120	315	0.260	Vessel hood		Separate filter 3,000	120	0.062	—	—	—	—
Youngstown	2	240	—	4,250	15	(0.220)a	—	—	4,500	15	0.167a	—	—	—	—
Italsider, Taranto	3	350	240	8,300	90	0.244a	—	—	—	—	—	Desul.		16,600 (for two vessels)	90
Bethlehem	—	—	—	—	—	—	—	—	3,000	120	0.062	—	—	—	—
Kaiser,															
Fontana	2	200	120	12,750	200	0.403	—	—	4,250	200	0.134	—	—	17,000	200

a Assumed values.

Table 17   Particle Size Distribution of Uncontrolled Emissions from BOPF Operations

Particle Size, μm		wt% Undersize	
	Charging[41]	Top Blowing[37]	Tapping[41]
65	88	68	99
5	4	55	90
0.3		23	
0.1	0.06		9

Table 18   Analysis of Gases Evolved from Top Blown and Bottom Blown BOPFs[49]

	BOF	Q-BOP
CO	79	67
CO_2	16	13
H_2		3
H_2O		3
N_2	25	14

μm decreased from 2.3 g/min at the beginning of the blow to 0.4 g/min at the end of the blow. As the blow proceeds, the decarburization rate increases, increasing the gas velocity and total particulate exiting from the furnace from 20 kg/min at the 5-min mark to 300 kg/min at the 16-min mark of the 23-min blow.[40]

For the bottom blown furnace, a hydrocarbon is added with the O_2 to cool the tuyeres and increase tuyere life. If the hydrocarbon is natural gas, the required volume is about 10% of the O_2 flow. H_2 and CO are the products of the cracking, which increases the gas volume to be handled. The decarburization rate for top blowing is slow initially and peaks at 1.6 to 1.8 times the theoretical average decarburization rate. The decarburization rate for bottom blowing is smoother, with the peak at 1.2 to 1.4 times the theoretical average decarburization rate. An average gas analysis from the two converter processes is shown in Table 18.

5.2.4   Tapping

When the steel in the converter has been adjusted to specifications, the taphole is unplugged and the vessel tilted over into the tapping position. Depending on the final requirements of the steel, additives such as Cu, Ni, Mo, ferromanganese, deoxidizers (such as Al, ferrosilicon, etc.) can be added to the ladle during tapping. In the case of the bottom blown furnace, the tuyere N_2 flow is left on. A typical tapping time would be about 5 min. Particle size of tapping emissions is finer than either BOF charging or blowing emissions, as shown in Table 17.

During tapping, measured gas volumes close to the furnace mouth were 100 Nm³/sec at 165°C.[48]

After the steel is tapped, the furnace is rotated and the slag poured into a slag pot on the opposite side. The slag is then usually transported to the slag-handling site where it is cooled and screened. Emissions are similar to aggregate-handling operations.

5.2.5   Teeming and Continuous Casting

The steel ladle is carried to either the teeming aisle where steel is drained into molds or to the continuous caster. The emissions are similar to tapping emissions.

5.3   Factors Affecting Emissions

5.3.1   Hot Metal Transfer

The iron pouring rate and the height from which the metal falls into the charging ladle will affect turbulence and resultant emissions. The quantity and type of spent desulfurizing reagent transferred with the metal will also affect emissions. During slag skimming, any iron that is skimmed out of the ladle into the pit will increase emissions. The unreacted desulfurizing reagent frequently contains considerable C that will burn when agitated.

5.3.2 Raw Materials Charging

Contaminants coating the scrap may include Zn, Pb, Sn, Cd, oils, and other organics, depending on the source of the scrap. No. 2 bundles consist of old black and galvanized steel sheet scrap, hyraulically compressed to charging box size. Increasing the quantity of No. 2 bundles charged to a heat from 2270 to 8600 kg/heat increased emissions from 0.14 to 0.5 kg/ton, respectively.[49]

5.3.3 Oxygen Blowing

The major difficulty with the collection of emissions from the bottom blown furnace is the splash and fume generated by the gas flow necessary to avoid plugging the tuyeres when the furnace has been rotated away from the primary hood. Secondary hooding, which will be discussed in the next section, must be designed to capture emissions from a full range of furnace positions.

The slag layer inside a top blown coverter plays a large part in limiting emissions. Minimum emissions are obtained with a stable foamy slag that covers the metal and entraps the metal spray being generated in the jet impact zone. Low bath temperatures lower the slag temperature and increase slag viscosity, hindering the formation of a stable foamy slag. During the period of rapid decarburization, the slag is deoxidized and the concentration of iron oxides in the slag decreases, lowering the melting point.[40] The slag loses its capacity to foam, and decreases in thickness, resulting in a higher fume rate usually called slopping. Varying MnO, FeO, P, and SiO_2 levels affect the slag viscosity, and therefore, the emissions. An excessively foamy slag may overfill the vessel, requiring the lifting of the suppressed combustion skirt. Certain shops have installed microphones to measure slag thickness. Slag thickness is controlled by changing lance heights and/or N_2 and Ar flows through porous plugs in the bottom of the furnace.

5.3.4 Tapping

As the metal is poured out of the converter, various alloying agents are added to the steel. Fines present in the alloys can be entrained by the heat plume and enter the air. Light elements such as S and C may float on the surface of the bath and burn.

5.3.5 Teeming and Continuous Casting

Organic coatings may be sprayed inside the ingot mold to improve ingot surface quality and lengthen mold life. The products of combustion are dependent on the type of coating. To produce killed steels, additives are added during the pour to produce a reducing atmosphere inside the mold during the filling and an insulating layer on top of the metal surface.

Additives are also used to form a layer on top of the continuous casting tundish. Periodically, the tundish must be dumped, but the emissions from this source are small. The slabs produced from a continuous casting machine are cut to certain lengths. Particulate emissions during the cutting are dependent on the type of torch and fuel used.

5.4 Best Available Control Technology

5.4.1 Hot Metal Transfer

A close fitting hood design is important to minimize collection volume requirements. Baghouses are the most frequently used devices for cleaning hot metal transfer emissions.

5.4.2 Raw Materials Charging

The plume produced from charging scrap into the furnace is small compared to that produced by charging hot metal. Local charging hoods should be positioned so that they are directly over the hot metal charging ladle mouth over the duration of the charge. At the end of the hot metal charge, the hot metal ladle should deflect the emissions into the hood.

5.4.3 Oxygen Blowing

The more traditional type of hood used to collect O_2 blowing emissions is the full combustion hood which features a gap of approximately 0.45 m between the top of the furnace and the top of the hood. This hood draws in sufficient air to burn the CO evolved from inside the furnace. Most new O_2 steel-making installations utilize the suppressed combustion hood, which includes a movable skirt

Table 19   Gas Volume, Temperature, Heat Content, and Gas Cleaning
Energy Requirements for Various Combustion Factors[49]

	Combustion Factor		
	1	0.30	0.10
Gas volume (Nm³/hr)	270,000	148,000	113,000
Theoretical gas temperatures at inlet hood (°C)	2,400	2,200	1,800
Heat to be removed in hood (Mkcal/hr)	224	82	42
Fan power (kW) (high-energy scrubber)	4,100	2,200	1,640

that lowers to seal the gap between the hood and the top of the furnace. The amount of air drawn
in is limited to approximately 10–15% of the amount required to stoichiometrically combust the CO.
The ratio of the air supplied to the ratio of the stoichiometric quantity to combust the gas is called
the combustion factor.

Equations for calculating theoretical waste gas volumes can be found in Ref. 47. The gap between
the furnace and the hood determines the fan volume required to maintain negative hood pressures.
Devices in use to clean BOF blowing emissions include scrubbers, wet precipitators, dry precipitators,
and baghouses. Dry precipitators require gas conditioning during the first few minutes of the blow
when particle size is small and the gas temperature is too low to saturate the gas. A baghouse was
in operation at Crucible Steel's BOF shop and utilized a dry bottom evaporative cooler to maintain
the required inlet temperatures to the fiberglass baghouse. The system operated successfully until the
integrated facilities were replaced with electric arc furnaces.

Some partial combustion systems operate at a combustion factor of 0.3 and do not employ a
movable skirt. Although most of the CO gas is combusted at the flare, there is an interval at the
beginning of the blow and an interval at the end of the blow during which the CO content is neither
burnt at the mouth of the furnace nor at the flare, generating an estimated 9 kg of CO per ton of
steel produced. The use of a skirt allows the CO emissions to be minimized by operating in the full
combustion mode with the skirt up during the beginning and end of the blow when the CO content
is low. Fast closing and opening of the skirt reduces CO emissions to an estimated 0.5 kg/ton.

The major advantage of the closed hood is the lower volume of gases that have to be cleaned
and cooled, as shown in Table 19. One disadvantage is the extra maintenance entailed with the movable
skirt.

The suction inside a movable hood is usually maintained at about 0.5 mm W.C. Rapidly varying
gaseous evolution rates, as experienced during slopping, can cause positive pressures inside the hood
and emissions from under the skirt. Secondary hoods, installed to collect these puffs, usually attain
varying efficiencies. Enclosures, such as shown in Figure 5, hold the emissions until they are gradually
withdrawn back into the hood.

Enclosure is also used for capturing emissions during the tilting of the bottom blown furnace.

5.4.4   Tapping

Because of the requirement to rotate the furnace, fixed local hoods cannot be installed directly over
the ladle to capture the emissions. The use of an enclosure, including wings on the ends of the transfer
car, can attain highly efficient collection of tapping emissions.

5.4.5   Teeming and Continuous Casting

Emissions from nonleaded teeming are rarely controlled due to the low cost effectiveness of emissions
control. Emissions from the pouring of leaded heats require the installation of hooding and cleaning
devices.

During continuous casting, shrouds purged with inert gas are installed around the steel stream to
reduce oxidation. The shrouds also help to reduce emissions. The surfaces of the metal in the tundish
and mold are usually covered with insulating materials to maintain metal temperature and improve
steel quality, but they also tend to reduce emissions.

6   ELECTRIC STEEL MAKING

6.1   Introduction

An electric furnace is a squat cylindrical steel shell vessel lined with refractory, as shown in Figure
6. Three carbon electrodes are inserted through holes in the roof and a current is passed between

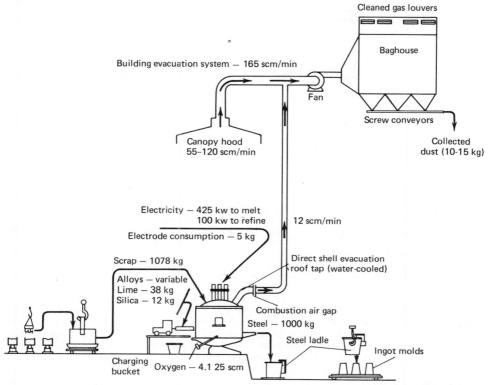

Fig. 6 Sketch showing process weights of various feeds and products from electric arc furnace production (based on 1000 kg of steel produced).[58]

the electrodes and the steel, providing the energy to melt the charge. The typical charge consists of 100% cold scrap, but molten iron, preheated scrap, or prereduced pellets are also used. O_2 may or may not be blown into the bath to increase the refining rate.

In 1980, 27.9% of U.S. steel production was made in electric furnaces. Current energy consumption for electric arc furnaces using scrap amounts to 10,400 mg/ton versus 37,000 mg/ton for integrated plants using blast furnaces, coke ovens, and BOF steel-making units.

A typical production sequence for an EAF consists of charging the furnace, melting the charge with full power, followed by up to three more charge–melt sequences. Once all the scrap is melted, the power is usually reduced during the refining cycle. Prereduced pellets are normally added continuously with full power on.

6.2 Nature of Emissions

6.2.1 *Raw Materials Charging*

The roof is usually swung aside to allow the scrap to be dumped into the furnace from a bucket carried by the crane. Temperatures at the upper cross section of the charging bucket vary from 60 to 300°C and velocities vary from 1.7 to 3.2 m/sec for regular scrap, and up to 6 m/sec for charging oily scrap.[57] Emissions from charging vary from 0.25 to 0.5 kg/ton.

6.2.2 *Melting*

After the charge, the electrodes are lowered to within 2–3 cm of the scrap and an intermediate power level is used for about 15 min to bore into the scrap. The power is then increased to its maximum to melt the scrap as fast as possible. About 425 kW are used to melt and 100 kW to refine a ton of steel. Depending on the required C level and oxidation requirements, gaseous O_2 may be blown through a lance to increase the oxidation rate. Dust evolution rates during the melting period are 1 to 2 kg/ton-hr and during O_2 lancing, 10 kg/ton-hr.[57] Particle size during melting is shown in Table 20.

Table 20   Particle Size Distribution of Uncontrolled Electric Arc Furnace Melting, Tapping, and AOD Operations

Particle Size, μm	wt% Undersize		
	Melting	Tapping	AOD
0.3	2	20–45	—
0.5	7	20–60	15
1.0	18	28–75	37
2.5	50	30–80	66
5	60	40–85	76
10	65	—	81

For production of nonalloy grades of steel, it is only necessary to use one oxidizing slag for the entire heat. For other grades of steel, it may be necessary to slag off the oxidizing slag by back-tilting the furnace to the point that the slag–metal interface is just below the level of the charging door. The slag is raked out of the furnace, followed by the addition of materials such as lime, fluorspar, silica sand, and silicon carbide to form a reducing slag. For certain heats, it may be necessary to carry out a number of slag-offs. Ferromanganese, Cr, and other oxidizable alloying agents are added during the reducing period. Analysis of the dust collected during the various phases of the operation is shown in Table 21.

CO is generated by the reaction of C electrodes or C in the bath with iron oxide or O_2. The emission of CO varies from 0.26 to 3.3 kg/ton of steel produced. In any event, the CO emission may not exceed the C loss from the electrodes plus the C lost from the bath. For a direct evacuation system, the volume of gases leaving the furnace amounts to 8–10 times the O_2 blow rate before the combustion gap, and 12 to 15 times after the combustion gap.[59] Exhaust gas temperatures leaving the furnace vary from 650 to 1700°C. Emissions may escape around the electrode ports or doors.

6.2.3   Tapping

When the metal has been adjusted to the desired specifications, the taphole is cleaned out and the furnace is tilted to pour the metal into an appropriately positioned ladle. The ladle may be sitting on a ladle car or carried by a crane. Temperatures at 1.5 m above the rim of a 1.8-m diameter 40-ton ladle varied from 150 to 300°C.[57] Gas volumes produced per m² of tapping ladle area varied from 3.5 m²/sec at the ladle rim to 40 m²/sec at the roof truss level.[60] The particle size of tapping emissions is shown in Table 20. Tapping emissions vary from 0.02 to 0.50 kg/ton.

6.2.4   AOD Refining

The amount of Cr to be added to the scrap in an electric arc furnace is limited to about 14%, since higher amounts will oxidize and be lost to the slag. To produce high Cr heats in an electric arc furnace, a double slag practice is carried out with the addition of high-cost, low C ferrochromium during the reducing period.

To allow the use of low-cost, high C ferrochromium and high Cr scrap, the charge is first melted

Table 21   Chemical Analysis of Uncontrolled Particulate from Electric Arc Furnace Melting, Oxidizing, Oxygen Lancing, and Reduction Operations

Phase	Dust Composition, %								
	SiO_2	CaO	MgO	Fe_2O_3[a]	Al_2O_3	MnO	Cr_2O_3	SO_3	P_2O_5
Melting	9.77	3.39	0.45	65.75	0.31	10.15	1.32	2.08	0.60
Oxidizing	0.76	6.30	0.67	66.00	0.17	5.81	1.32	6.00	0.59
Oxygen lancing	2.42	3.10	1.83	65.37	0.14	9.17	0.86	1.84	0.76
Reduction	Tr.	35.22	2.72	26.60	0.45	0.70	0.53	7.55	0.55

[a] The iron content was determined as total iron and converted to Fe_2O_3.

in an electric arc furnace, then poured into the AOD where the Cr is added. The AOD furnace is a pear-shaped vessel equipped with two or three side-mounted tuyeres through which various mixtures of N_2—O_2 or A—O_2 mixtures are blown. Vessel capacities range from 5 to 100 tons. Typical quantities of O_2, A, and N_2 used in an AOD are 500 m³/ton, 400 m³/ton, and 300 m³/ton, respectively. The O_2 removes the C, while the A reduces Cr oxidation to a very low level.

Emissions are produced during the charging, blowing, and tapping stages. The tuyeres of an AOD project horizontally toward the opposite wall, whereas the tuyeres of a bottom blown vessel usually point toward the mouth. Particle sizing for AOD emissions is shown in Table 20. The emission factor is 8 kg/ton.

6.3 Factors Affecting Emissions

6.3.1 *Raw Materials Charging*

Significant quantities of oils, nonmetallics, nonferrous metals, and coated steels may increase emissions during charging. Emissions from dirty scrap during the first charge are not as serious as during subsequent charges because the furnace bottom may not be hot enough to ignite the scrap. Dust loadings averaging 139 mg/Nm³ were measured from a canopy hood during the first bucket charge and an average of 316 mg/Nm³ during the second bucket charge.[56] Systems have been developed for the continuous addition of prereduced pellets or scrap to the furnace through a fifth hole in the roof, allowing the capture of charging emissions with melting emission control systems.

6.3.2 *Melting*

Not all of the scrap contaminants are burnt off during the charging. Nonferrous metals such as Zn "fume off" during the melting period. Electric arc furnace dust may contain as high as 37% ZnO. As the electrodes melt holes in the scrap charge, bridges of cold scrap may be formed which can fall suddenly into the molten pool, creating a "cave in," accompanied by a large burst of gas. Gas bursts may also be formed during the addition of lime or iron ore with full power. The rate of fume formation in electric arc furnaces is dependent on the decarburization rate, as shown in Figure 7.

6.3.3 *Tapping*

Factors increasing tapping emissions include fines in the alloy additions, a rough nozzle producing metal spray, and a greater than average distance between the furnace spout and the ladle. Emissions also increase with increased alloy additions to the ladle.

6.3.4 *AOD Refining*

Emission rates increase at gas flow rates in excess of 39 Nm³/ton-hr.[61] To avoid plugging the tuyeres, a gas flow must be maintained when metal is in the furnace. Refining times vary depending on the deslagging and number of analysis checks. As the cycle time increases, the emissions increase. Most AOD furnaces use a swingaway hood which allows the vessel to remain fairly upright during a charge.

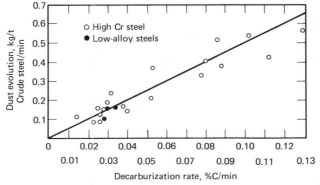

Fig. 7 Graph showing relationship between electric arc furnace dust evolution and decarburization rate.

6.4 Best Available Control Technology

6.4.1 *Raw Materials Charging*

The best available control technology depends on the type of scrap to be charged. The most common control technique used for collection of charging emissions is the canopy hood, installed above the charging bucket in the building roof trusses. A three-section canopy hood, with control dampers to direct the air to the two hoods over the furnace during the scrap charge, and two hoods over the ladle during tapping, is often used. Ductwork should extract air from the hood face proportional to the bell-shaped velocity profile at the mouth of the hood, so as to minimize exhaust volumes.[62] As shown in Figure 8, the required hood face area can be determined by projecting a 12–20° angle from the top edge of the furnace ladle or charging bucket to the bottom edge of the hood. A 20° angle allows for some movement of the plume due to cross drafts.

To attain high collection efficiencies at minimum air volumes, enclosures have been built around the furnace. Enclosures, as shown in Figure 9, are equipped with doors sized large enough for the scrap bucket. A 1.5-m gap in the top of the enclosure is allowed for crane cables carrying the charging bucket. An air curtain can be used to seal the gap. Higher enclosures enclosing the crane trolley have also been used. The evacuation rate applied to the enclosure during charging is up to five times that of the direct evacuation system during melting. In comparison, the evacuation rate for a canopy hood for scrap chargings is up to 10 times the volume requirement of a direct evacuation system.[63]

6.4.2 *Melting*

The most practical and effective technique for the collection of furnace off-gases is the direct evacuation or roof tap as shown in Figure 6. Advantages include lowest gas flow, lowest power requirements, and highest CO combustion rate. Volumes to be ventilated from the furnace must be sufficient to prevent leakage at electrode openings, doors, and so on. CO is combusted by air drawn in at the combustion gap. To effectively capture emissions with direct evacuation, the furnace must be maintained

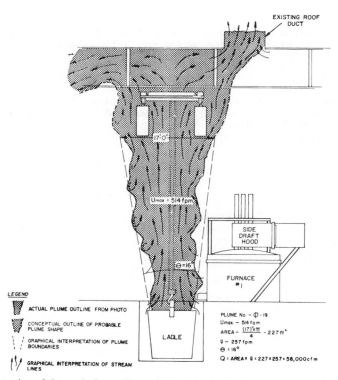

Fig. 8 Illustration of the angle formed by a rising expanding plume evolved from a steel ladle.[56]

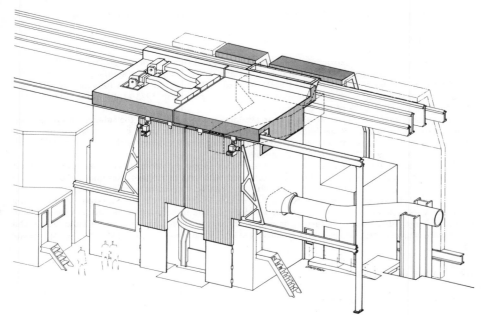

Fig. 9 Sketch of enclosure around an electric arc furnace.[56]

at a negative pressure. The production of specialty or high-alloy heats requires positive furnace pressure to attain reducing conditions inside the furnace. Alternatives for high-alloy shops include side-draft hoods, full-roof hoods, enclosures around the furnace, canopy hoods, or total building ventilation.

Exhaust gases from the roof tap may be cooled by hairpin coolers, evaporative coolers, or dilution air. The dilution air may be provided from an overhead canopy hood.

A gas residence time of 0.3–0.5 sec is required for combustion of the CO from the direct evacuation tap before gas cooling to avoid the possibilities of an explosion in the baghouse or ESP. Sufficient air is allowed in at the combustion gap to provide 100% excess air. Scrubbers are also used for cleaning the gases from a direct evacuation tap.

6.4.3 Tapping

Canopy hoods may also be used for tapping emission control, but the exhaust volumes required are very high and collection efficiency may be low due to cross drafts. A local hood flow of 4000 Am³/min at 112°C is used for a pouring rate of 65 tons/min.[55]

Enclosures around the furnace are gaining popularity since the volume requirements for charging and tapping emission control are lower than overhead canopy hoods. Baghouses and ESPs may be used for cleaning tapping emissions.

6.4.4 AOD Refining

The normal practice for collection of AOD charging emissions is a canopy hood above the crane. The hood must be designed to collect emissions when the furnace is tilted forward and backward. The face area of the canopy hood can be determined by projecting a 13–20° angle from the mouth of the AOD in the two extreme positions of tilt to the bottom of the hood.

A 12.1 × 24.4 m canopy hood, evacuated by a 9800 m³/min fan, is used to capture emissions from a 90-ton AOD.[64]

During the blowing part of the cycle, a hood may be swung over the mouth of the AOD to collect blowing emissions. The hood may be tied directly into separate ducting or it may channel the flow into the roof canopy. The use of separate ducting reduces the volume requirements. High excess air ratios minimize hood overheating and gas cooling problems.

Hot gas from an AOD would require cooling by dilution or hairpin coolers before cleaning in a baghouse. ESPs and scrubbers can also be used.

7 ELECTRIC SUBMERGED ARC FOR FERROALLOY PRODUCTION

7.1 Introduction

Various ferroalloys such as ferrochrome, ferromanganese, and ferrosilicon are produced in the electric submerged arc furnace and are utilized in C steel making, alloy steel making, stainless steel production, and foundries. Production in the United States peaked in 1965 at 2.3 million tons and by 1977 was reduced to 1.466 million tons. Consumption of ferroalloys peaked in 1973 at 2.7 million tons, and in 1977 imports supplied 44% of the U.S. supply.

The primary manufacturing process for the production of ferroalloys is the submerged arc furnace. Other production and refining methods are vacuum and induction furnaces, exothermic alumino-silico-thermic processes, and electrolytic manufacturers. The energy requirements for the production of ferroalloys vary from 2 to 9 MW-hr/ton of alloy produced.

The submerged arc furnace has a flat bottom and is supported on an open foundation to permit air cooling to dissipate heat. The furnace is equipped with one taphole or multiple tapholes for the removal of slag and metal. Certain specialty furnaces can be rotated or oscillated to aid in slag and metal removal. Basic raw materials used are iron oxides, Si, Mn ore, Cr ore, limestone, and reducing agents such as coke or low volatile coal and wood chips. The three graphite electrodes extend 1–2 m into the charge. Metal from the furnace is tapped through a short launderer into a ladle. The metal in the ladle is usually cast into cast iron moulds from which the metal is then crushed to the appropriate size.

7.2 Nature of Emissions

7.2.1 Charging and Tapping

There are two techniques used to charge a submerged arc furnace. With mix seal furnaces, materials are charged through openings around electrodes. Fumes escape from the furnace through the mix seals. With the closed furnace, mechanical seals are used between the cover and electrodes and the mix is fed into a mix column that feeds through a spout into the furnace, controlling charging emissions. Particulate emissions, lost through mix seals, are shown in Table 22.

Approximately once every 2.5 hr, the taphole is drilled out and the furnace drained through a launderer into a ladle. A typical tap time might be 15 min and produce emissions of 0.3 kg/MW-hr.

7.2.2 Submerged Arc Furnace Refining

Furnace sizes vary from 10 to 96 MW. Production and emission factors for uncontrolled open furnaces are shown in Table 23.

Lump ore and coke descend through the burden and begin to react as the slag layer is reached. The current passes between the electrodes and slag via the coke layer and the coke reacts with the oxides to form metal plus CO. The CO may reduce oxides such as Fe and Mn, forming CO_2 which rises up through the burden. Fines in the burden may be entrained by the passage of the gases. Modern submerged arc furnaces are designed for resistance heating and the current density is not high enough to cause arcing at the electrode tips. Some furnaces operate with current densities high enough to create arcing at the electrode tips, increasing fume formation. Particulate emission factors and analysis of uncontrolled fume from submerged arc furnaces are shown in Tables 23 and 24, respectively. Particulate size analysis of ferroalloy emissions is 88, 84, and 79% finer than 10, 3, and 1 μm, respectively.[66] C is added in the form of coke, anthracite coal, and wood chips. As the burden moves down through the furnace, the organic content in the charge is evolved. Detailed organic and heavy metal measurements from ferroalloy facilities can be found in Ref. 66.

Table 22 Uncontrolled Particulate Emission Factors from Ferroalloy Furnaces for Various Ferroalloy Products[71] (kg/MW-hr)

Product	Open Furnaces	Semiclosed Furnaces
FeSi (50%)	40.4	3.6–20
FeSi (65–75%)	47	16.3–23
SiMn	22.7	4–16
FeMn	28	4.5–34
HC FeCr	28	13–16

Table 23   Chemical Analysis of Particulate from Various Types of Ferroalloy Furnace Operations[71]

Fume Characteristics	Furnace Product			
	FeSi (50%)	SiMn[a]	FeMn	HC FeCr
Furnace type	Open	Covered	Open	Covered
Chemical analysis (%):				
SiO_2	63–88	15.68	25.48	20.96
FeO		6.75	5.96	10.92
MgO		1.12	1.03	15.41
CaO		—	2.24	—
MnO		31.35	33.60	2.84
Al_2O_3		5.55	8.38	7.12
LOI[c]		23.25	—	—
Total Cr as Cr_2O_3		—	—	29.27[b]

[a] Manganese fume analyses in particular are subject to wide variations, depending on the ores used.

[b] Fumes from open furnace contain less chrome oxide.

[c] LOI is loss on ignition at 1000°C.

7.3   Factors Affecting Emissions

7.3.1   Charging and Tapping

Closed furnaces with mechanical seals are designed to operate at slight positive pressures, normally in the range of 0.1–0.3 mm W.C. to avoid air in-draft into the furnace. Excessive air in-draft into the hood could result in explosive conditions. Similarly with mixed seal furnaces, air in-draft at the seals must be avoided, resulting in particulate leakage around the electrodes.

Tapping emissions are influenced by the type of product being made. The tapping of silvery iron produces airborne graphite flakes.[69]

Table 24   Uncontrolled Organic Emission Rates from Various Ferroalloy Operations[66]

Facility	Reducing Agent, kg/MW-hr	Seal Type	Uncontrolled Organic Emission Rate, kg/MW-hr	Operating Power, MW	O_2 Content of Off-Gases, %
A-1	213 buckwheat coke	Mix seal	0.06	11.4	11.8
A-2	190 buckwheat coke 11 millscale	Open	0.35	15.8	
B-1	108 rosa pea coal 70 wood chips	Open	0.25	48.4	18.2
B-2	81 jewel coal 36 Cleveland coke	Mix seal	1.5	48	
C-1	68 quinwood coal 37 rosa pea coal 149 wood chips	Mix seal	1.27	15.5	1.4
C-2	9 pea coke 57 rosa nut coal 45 rosa pea coal 43 wood chips	Mix seal	0.59	16.8	

7.3.2   *Submerged Arc Furnace Refining*

With the production of certain ferroalloys, crusting or sintering of the mix occurs and openings in the hood are necessary to allow rods to be inserted and moved around to break up the crust. Furnaces equipped with doors for stoking are referred to as open furnaces.

There has been an evolution in open hood designs, from the canopy hood and the high hood with movable sections which require shutdowns to change electrodes, to the low hood design which allows electrode changing under load. The required hood extraction rate depends on the openings. In the so-called semiclosed furnace, sidewalls extend to the edge of the furnace body and doors are installed to allow access for stoking, thereby reducing the volume of gas to be cleaned, as shown in Figure 10. Sufficient air is allowed into the hood to combust the CO.

Fines or dense materials in the feed can promote bridging and nonuniform descent of the charge. Gas channels develop with a momentary burst of gases when the bridge collapses. A porous burden favors uniform gas distribution and avoids fines entrainment, due to gas short-circuiting through sections of the burden. Lack of a coke layer can lead to arcing in furnaces designed for resistance heating.[67] The temperature in an arc can exceed 3600°C.

Uncontrolled POM emissions are 100 to 900 kg/yr-MW for open furnaces and 1230 to 11,080 kg/yr-MW for closed furnaces.[66] Feed materials containing organics include coal, wood chips, millscale, and oily turnings. According to Table 24, uncontrolled organic emissions from open furnaces can be appreciable. Combustion of organics in an open furnace requires closely controlled combustion conditions.

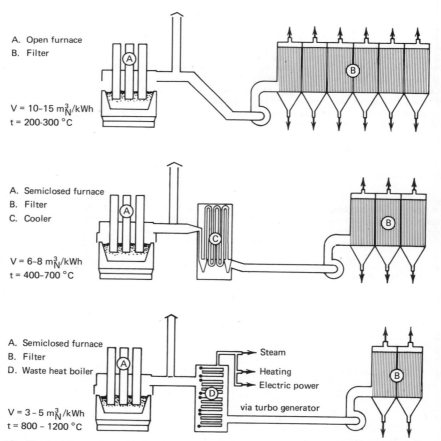

Fig. 10   Hood capture volumes and gas temperatures from open and semiclosed ferroalloy furnaces.[70]

7.4 Best Available Control Technology

7.4.1 *Charging and Tapping*

No controls are required for collection of charging emissions from open, semiclosed, and completely sealed furnaces. Control gates in the mix column provide a gas seal so that fumes cannot escape the furnace from the charging operation.

A 5600 Am³/min system is used for collection of emissions from the following locations:[69]

1. Metal transfer from the tap ladle into the bull ladle.
2. Ladle additions.
3. Raking kish off the bull ladle onto conveyors.
4. Cleaning ladles and handling graphite.
5. Crushing and sizing systems.
6. Transporting conveyors.

Baghouses are normally used for cleaning tapping emissions due to the low temperatures involved.

7.4.2 *Submerged Arc Furnace Refining*

The volume requirements for a semiclosed furnace are shown in Figure 10. Semiclosed furnaces draft air to obtain combustion underneath the hood, burning the organics and CO evolved from the charge. If efficient combustion conditions can be obtained, then baghouses and dry ESPs are feasible. If efficient combustion cannot be attained, the use of wet devices will achieve higher organic removal.

Closed furnaces require exhaust volumes of only 0.3 Nm³/kWh. The gas temperature from a closed furnace is lower than from the open furnace since the CO has not been combusted. Due to the potentially explosive nature of the gas and the high organic content, wet scrubbers are recommended for closed furnaces. After cleaning, the gases could be used as a fuel or passed through a high-combustion efficiency flare.

8 FERROUS FOUNDRIES

8.1 Introduction

In 1978, there were 4438 foundries in the United States of which 2097 were iron foundries and 631 were steel foundries. The production of ferrous castings involves melting the specified ferrous metal in a furnace and then pouring it into a mold normally made from sand. The ferrous segment can be broken down into two groupings: iron castings with C contents varying from 1.5 to 5%, and steel castings with C contents less than 1.5%.

As the steel must be poured at temperatures of approximately 1600°C, the molds must be made from 99.9% pure silica sand to withstand the temperatures. Typical temperatures for pouring grey iron castings are approximately 1500°C, which allows the use of less pure sands. Iron and steel foundries use similar mold-making and core-making practices. In an iron foundry, iron scrap, steel scrap, or pig iron is melted in either a cupola furnace, induction furnace, or electric furnace. Induction furnaces or electric arc furnaces are used for the steel foundry.

8.2 Nature of Emissions

8.2.1 *Nonmelting Particulate*

Particulate emissions from the various processes are shown in Table 25. The source indicates that the data have a low reliability.[72] Gates and risers feed molten metal to the casting during the cooling period. The gates and risers are subsequently cut off and form part of the scrap charge to the melting facilities. Emissions may be created by the handling of the sand-coated scrap. The sand, separated from the cooled casting, is normally recycled back to the muller where additives and moisture are added. The hot returned sand gives off dust when handled.

Reclamation equipment to remove the accumulated buildup of clay and carbonaceous material on the sand grains is used to extend the working life of the sand. Certain sand reclaimers use fluidization with the fine particulate swept away by exhaust gases. In general, molding sand is prepared by mixing organic or inorganic binders and water to clean silica sand. The dry sand and dry additives can produce emissions during mixing. If sufficient water is added, the mix may be nondusting when handled.

Techniques used to remove excess metal include flame-cutting torches and grinding. Methods used to remove excess sand and improve the appearance of the casting include shotblasting and sandblasting.

Table 25 Uncontrolled Particulate Emissions from Various Foundry Operations[72]

Emission Source	Estimated Emission Factor	Fine Particles, $\% < 5 \ \mu m$
Raw materials input		
Scrap	0.05 kg/ton scrap[a]	30
Handling and transfer		
Spent sand	1.1 kg/ton melt[a]	30
Coke	0.1 kg/ton coke	30
Melting and casting		
Induction furnace	0.8 kg/ton charge	80
Inoculation	1.2 kg/ton inoculated	80
Pouring	2.0 kg/ton poured	95
Cooling	2.1 kg/ton poured	95
Finishing		
Shakeout	1.52 kg/ton cast[a]	50
Mold and core preparation		
Mulling	1.0 kg/ton melt[a]	30
Molding	0.02 kg/ton melt[a]	30
Waste handling		
Waste sand transfer	0.15 kg/ton sand	30
Storage		
Sand	0.12 kg/ton sand	30
Slag	0.09 kg/ton slag	30

[a] Weight less than 50 μm.

8.2.2 Cupola Furnace

A cupola consists of a water-cooled or refractory-lined cylindrical furnace in which pig iron, scrap, limestone, and coke are charged. Air is blown into the bottom of the furnace through tuyeres, combusting the coke and melting the iron. The average uncontrolled CO emission factor is estimated at 70 kg/ ton of metal charged. Particulate chemical analysis and gas analysis are shown in Tables 26 and 27, respectively. Particulate size analysis of cupola particulate is 93, 85, 50, and 20% less than 5, 3, 1, and 0.5 μm, respectively.[73]

Iron from the cupola may be tapped into a floor hearth where the slag is skimmed off and then transferred into a ladle or is tapped directly into the ladle. As the launderer is short, turbulence is low and emissions also tend to be low. Approximately 15% of the iron castings produced in the United States are ductile iron, which requires an inoculating agent to change the graphite from a flake to spheroidal form. Inoculating promoters include Mg, Ca, Ce, Yb, Ba, Nd, and Pr. Other ladle alloys added to increase strength include Cr, Ni, and Mn. Inoculants can be added to the molten metal in the cupola spout, floor hearth, transfer ladle, pouring ladle, and mold. A further discussion of inoculating emissions can be found in Ref. 74.

8.2.3 Pouring, Cooling, and Shakeout

The mold is made by packing sand around a pattern and then separating the pattern from the mold. The sand must have sufficient strength to retain the outline of the pattern until the metal has solidified. Cores are special sand shapes placed inside the mold to form depressions or cavities inside the casting. Green sand is the most popular molding sand and is produced by adding bentonite and various other organic and inorganic binders and water to sand. Cores must be stronger than molds.

When the molten metal is poured into the mold, organic binders may evolve gases and vapors. The gases continue to be evolved during the cooling of the casting and the shakeout. Further details regarding emissions from pouring, cooling and shakeout of castings are available in Refs. 72, 74, and 75.

Table 26 Chemical Analysis of Uncontrolled Particulate Emissions from Various Foundry Cupola Operations[81] (kg/ton of Metal Melted)

Foundry	Total	Combustible	Non-Combustible	Bi	Pb	Zn	Fe	Mn	Ca	SiO_2[a]
A	22.5 ± 6.7	2.8 ± 0.9	19.2 ± 5.5	0.14 ± 0.08	0.07 ± 0.06	0.06 ± 0.06	1.7 ± 0.8	0.33 ± 0.14	0.13 ± 0.08	15.8
B	7.8 ± 1.94	1.6 ± 1.0	0.63 ± 1.3	0.07 ± 0.07	0.13 ± 0.10	0.07 ± 0.03	1.28 ± 0.87	0.14 ± 0.05	0.31 ± 0.06	3.27
C	9.0 ± 3.67	3.1 ± 1.2	6.1 ± 2.4	ND[b]	0.39 ± 0.11	0.28 ± 0.08	0.60 ± 0.30	0.08 ± 0.02	0.17 ± 0.07	3.88
D	4.1 ± 1.32	0.79 ± 0.29	3.2 ± 0.8	ND	0.16 ± 0.07	0.19 ± 0.06	0.34 ± 0.10	0.05 ± 0.02	0.07 ± 0.04	2.0
E	17.1 ± 4.70	4.3 ± 1.3	1.5 ± 4.3	ND	0.61 ± 0.08	0.84 ± 0.06	1.5 ± 0.07	0.17 ± 0.05	0.41 ± 0.22	9.4
F	2.4 ± 0.58	0.62 ± 0.09	1.78 ± 0.41	ND	0.16 ± 0.04	0.10 ± 0.01	0.2 ± 0.04	ND	ND	1.2

[a] SiO_2 emission rate was obtained by subtracting weight of O tied up in other metal oxides from fraction which consisted of elements too light for accurate fluorescence analyses. The assumption that it is primarily SiO_2 (quartz) is supported by X-ray diffraction.
[b] ND—None detected.

Table 27 Cupola Gas Flow Rate, Temperature, Dilution Factor, and Gas Analysis Measurements from Various Cupola Operations[81]

Foundry[a]	Flow Rate, m³/min	Temperature, °C	CO, % Volume	CO_2, % Volume	O_2, % Volume	NO_x, ppm	SO_x, ppm	Dilution Factor[b]
A-1	322	163	1.54	5.1	13.5	—	—	3.8
B-1	518	529 ± 21	0.03 ± 0.01	7.6 ± 0.5	12.3	12	710	4.1
C	325	646	0.38	8.2	13.0	33	480	N/A
D-2	150	732	0.73	9.3	10.6	22	290	1.9
E-2	493	171	0.50 ± 0.35	2.6 ± 0.1	18.7	—	280	8.3
F	165	385				14	290	N/A

[a] -1 and -2 refer to chronological sequence of daily runs. A, C, and D refer to the operation of automatic charge door.
[b] Dilution factor = exhaust rate/blast rate. N/A = not available.

8.3 Factors Affecting Emissions

8.3.1 *Nonmelting Particulate*

Emissions produced during the handling of waste sand from the shakeout to the sand mixing station are dependent on the sand temperature, moisture content, and silt content. Hot sand loses residual moisture, leading to more dusting. Fine C in the form of sea coal or wood flour is often added to improve the surface quality of the casting but dusting tendencies are increased. If the sand reclamation operation includes fines removal, dusting during subsequent handling may be reduced.

8.3.2 *Cupola Furnace*

Considerable loose sand may accompany foundry scrap returns. Loose charge dust is entrained up the stack as soon as the charge bucket is dumped.[76] The dust emission rate, measured during the 30 sec immediately following the introduction of a charge into the cupola, was as high as 7 times the average emission rate and 15 times the off-peak rate. Lower emission rates are attained when fine coke breeze is removed from the charge.[14] High stack combustibles are associated with the presence of dirt and grease in the scrap.

Increasing blast rate increases particulate loading, but has no measurable effect on particle size distribution.[77] Blast temperature affects particle loading and particle size distribution. Melting metal at higher rates produces a higher loading of small particles. Foundries melting primarily low-quality domestic and auto scrap emit up to four times as much submicrometer dust per ton of metal melted as foundries melting cleaner, higher quality ferrous scrap.[76]

Average exhaust gas temperatures are controlled by factors such as air-to-coke and metal-to-coke ratios, burden depth, permeability, and dilution by charge door air.[76] Further, for a given cupola, variations in average stack velocity are produced by control of the blast rate and by changes in the permeability of the charge material. The use of a hot blast system decreases the coke consumption, and therefore, the total volume of gases to be cleaned.

Exhaust gases from the cupola may be exhausted from above or below the charge door. Less air is educted with takeoff below the door. Smaller doors also reduce the dilution.

Various techniques are used to inoculate the iron to produce ductile iron. Inoculant emissions are lowest when the material is introduced and held below the metal surface. Ni—Mg alloys are available that sink reducing particulate emissions.

8.3.3 *Pouring, Cooling, and Shakeout*

Organic decomposition products during the pouring, cooling, and shakeout are dependent on the type of organic binder used, the temperature reached by the sand, and the time the mold sits until it is shaken. When the metal is poured into the mold, gases such as CO and CH_4 permeate through the sand and may ignite at the surface of the mold.

The higher molecular weight materials volatilize close to the molten metal and recondense inside the mold. During the shakeout, however, the condensed organics on the sand may boil off when they contact the surface of the hot casting. The cooler the casting surface, the fewer the organics that should be evolved during the shakeout. The use of inorganic binders reduces organic emissions during the pouring, cooling, and shakeout cycle.

8.4 Best Available Control Technology

8.4.1 *Nonmelting Particulate*

Collection systems for dry handling transfer points are covered by Ref. 78. Baghouses or low-energy wet scrubbers are normally used for cleaning sand-handling exhaust gases.

The Schumacher process mixes the dry shakeout sand with sufficient quantities of moist sand in an inundator to produce a nondusting mixture which can be handled without dusting, as described in U.S. Patent No. 3,461,941. The process increases the sand-to-metal ratio from about 6:1 to about 12:1.

8.4.2 *Cupola Furnace*

As shown in Figure 11, cupola installations with below charge door gas extraction can be provided with a separate combustion chamber equipped with automatically controlled air injection to burn the CO. Since the takeoff is below the top of the charge, the gas temperature and CO content extracted from the furnace are relatively constant. On the other hand, above the door gas extraction is characterized by variations in gas analysis and temperature whenever charge materials are added to the burden.

To reignite the gases after a charge, gas jets can be provided just below or opposite the charging

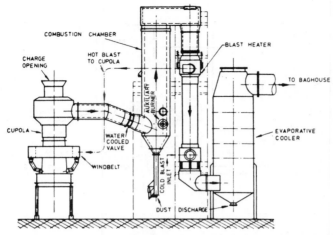

Fig. 11 Sketch of cupola with gas extraction below the charge door and use of exhaust gases for heating and blast.[82]

door. The fan volume is adjusted to draw in sufficient air through the door to attain a 4–8% O_2 content in the stack gases after combustion. Air drawn through the door is not mixed well with the stack gases and long duct runs are required to ensure sufficient residence time for combustion. To achieve low CO emissions, sufficient gas must be added through the igniters to reach temperatures of 700°C, especially during a charge. With coke-to-metal ratios of less than 11%, more auxiliary gas is necessary to maintain the required combustion temperatures.

ESPs, baghouses, and wet scrubbers can be used for cupola exhaust cleaning. Proper operation of the afterburning system is required with an ESP to avoid potential explosions. Cooling techniques that can be used with a baghouse include evaporative cooling and hairpin convection coolers. Movable hoods for cupola tapping and inoculation are illustrated in Refs. 80 and 72, respectively.

8.4.3 *Pouring, Cooling, and Shakeout*

In a typical foundry, pouring and cooling of molds may take place over a large area, making it expensive to provide collection. In more automated foundries, it is possible to centralize the pouring to a hooded area. Molds may also be cooled in a covered tunnel, exhausted at 7 to 9 m³/min-m of hooding length with a flow of 60 m/min through openings.[79]

To capture emissions from a shakeout less than 2 m wide requires a side-draft hood exhausted at a rate of 150 m³/min-m³ of grate area, in conjunction with hooding and exhausting the sand hopper to obtain a down-draft flow through the grate of 12 m/min.

During the pouring, cooling, and shakeout of green sand molds, considerable moisture is evaporated and the variable moisture and temperature of the exhaust gases can cause baghouse plugging. Scrubbers obtain hydrocarbon removal.

9 DIRECT REDUCTION

9.1 Introduction

Direct reduction (DR) processes convert iron ore materials into sponge iron or metallized iron at temperatures well below the melting point of iron. The fuel requirements in the DR process can be satisfied by a large variety of fuels, including coal, natural gas, oil, or gases produced from any of these fossil fuels. The product produced from DR is normally added to the electric arc furnace, but may also be used as a blast furnace feed to increase productivity. In this chapter, DR processes using coal and natural gas will be outlined.

9.2 Nature of Emissions

9.2.1 *Gaseous Direct Reduction*

The most popular reducing gas for DR is natural gas that is catalytically reformed with steam on a Ni catalyst to produce H_2 and CO. The gas to be reformed must contain less than 1 ppm S to avoid catalyst fouling.

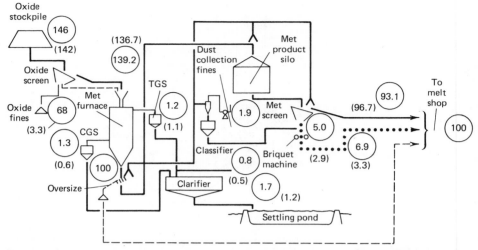

Fig. 12 Flow diagram of process weights of raw material feeds, product discharges, and dust evolved from Midrex direct reduction process. ○ Module I—typical mix feed. ( ) Module II—using 100% pellets. TGS—top gas scrubber. CGS—cooling gas scrubber.[84]

Iron oxide pellets descend through a shaft furnace from 4.88 to 5.49 m in diameter where they are heated and reduced by a rising flow of hot-reducing gas. To prevent infiltration of air or escape of reducing gas from the furnace, the feed pipe at the top of the furnace is kept full of solids and supplied with a small amount of inert gas. The hot-reducing gas is fed through entry ports one-third up from the bottom of the reducing section and exits at the top of the furnace. The top gases are cleaned and recirculated back to the reformer. Mixing valves control the composition of the feed gas, which is made up of natural gas and clean process off-gas.

Table 28 Gas Temperature and Analysis from Midrex and Krupp Direct Reduction Processes

	Midrex	Krupp
Exhaust Gases	3000 Nm³/ton Fe	2750 Nm³/ton Fe
Temperature	400°C	850°C
Composition		
CO_2	12.9%	25%
O_2	4.0%	0.7%
N_2	65.7%	—
H_2O	17.4%	—
CO	—	0.7%
Dust	4 mg/Nm³ or	
	12 g/ton Fe	55 mg/Nm³

Particulate Emission

Total   3 to 5 ton/hr
for a 400,000 ton/yr module with:
          0.5 kg/hr
from the stock as indicated above
          2.5 kg/hr
from dust collection systems
          1.0 kg/hr
from storage and load-out collection system
          1.0 kg/hr
from screening dust collection system

As the metallized material passes down through the cone section of the shaft, it is cooled to about 50°C before being discharged. As shown in the mass balance illustrated in Figure 12, 0.86% of the feed is entrained and removed by the top gas scrubber and about the same amount is removed by the cooling gas scrubber. Dust pickup is provided at the screening of the discharge, the discharge to the product silo, and the screening of the final product. Typical gas analysis of the exhaust gas discharge is shown in Table 28.

9.2.2 Solid Reductant Systems

Approximately 16% of the world's sponge iron capacity is produced in rotary kiln reactors. A number of different rotary kiln technologies are used for producing sponge iron, including the Krupp and SL/RN process. Process coal, iron ore, and dolomite or limestone are fed into the charge end of the kiln. Air is blown in along the length of the kiln to provide for the combustion of the volatiles from the coal, CO, and H_2 generated in the kiln. Gas flow is countercurrent to the solids flow and exits the kiln at about 525°C. Characteristics of exhaust gases from the Krupp process are shown in Table 28.

Typical coal requirements are 0.53 ton of fixed C per ton of Fe charged and air rates of 1720 m^3/ton of feed. The solid mixture is heated to about 1000°C. There are two distinct oxygen conditions inside the kiln. Where the coal and ore are in intimate contact, the atmosphere is composed of CO and H_2. Above the bed, an O_2 potential is maintained, allowing for combustion of organics. Oxidation of organics depends on the time and temperature for combustion. The temperature of the gases at the burner end is 1200°C and is reduced to 500°C at the product discharge end. Conditions required for the destruction of PAH are temperatures of 900°C for at least 0.3 sec. Although average residence times in the reactor vary from 5 to 9 sec, volatiles evolved at the feed end of the kiln would not likely be burned due to the low temperature and short residence time. Volatiles evolved in the middle of the kiln would more likely be combusted. SO_2 content in the stack gases would be approximately 320 ppm.

9.3 Factors Affecting Emissions

9.3.1 Gaseous Direct Reduction

To avoid fouling the catalyst, gases must be cleaned of particulate and S. As shown in Figure 2, a mix feed using a combination of pellets and coarse ores produces more fines than high quality pellets.

9.3.2 Solid Reductant Systems

The use of more volatile coal can increase organic emission from a kiln DR process. Additional lime fed into the furnace will reduce SO_2 emissions from the exhaust. High ash coal and fine ore would tend to increase particulate emissions.

9.4 Best Available Control Technology

9.4.1 Gaseous Direct Reduction

High particulate or S contents in the gas plug the catalyst so that emissions are minimal. Dust collection is required at all transfer points.

9.4.2 Solid Reductant Systems

Due to the presence of volatiles and CO in exhaust gases, afterburning of the gases is normally carried out. Cleaning of the gases with precipitators or baghouses requires gas cooling. For systems not equipped with afterburners, wet scrubbers would be recommended.

REFERENCES

1. McCrillis, R. C., Iron and Steel Inhalable Particulate Sampling Program for Iron and Steel: An Overview Progress Report, *Third Symposium on Iron and Steel Pollution Abatement Technology,* Chicago, IL, November, 1981.
2. Arthur D. Little, Inc., *Environmental Policy for the 1980's: Impact on the American Iron and Steel Industry, Summary,* 1981.
3. PEDCo Environmental Inc., *Development of Air Pollution Control Cost Functions for the Integrated Iron and Steel Industry,* Cincinnati, Ohio, EPA-450/1-80-001, July 1979.
4. Westbrook, C. W., *Level I—Assessment of Uncontrolled Sinter Plant Emissions,* prepared for U.S. EPA, PB 298-055, May 1979.

5. Current, G. P., *Sinter Plant Windbox Recirculation and Gravel Bed Filter Demonstration, Phase 2—Construction, Operation and Evaluation,* prepared for U.S. EPA, EPA-600/2-79-203.

6. Ricketts, D. B., and Bock, R. W., Dust Collection—Material Handling Operation Sinter Plants, *AIME Ironmaking Proceedings* **37** (1978).

7. Bandi, W. R., and Pignocco, A. J., Control of Combustion in Sinter Plant Precipitators, *AIME Ironmaking Proceedings* **38** (1979).

8. Hirooka, N., Equipment for Removal of Nitrogen Oxides for No. 4 Sinter Plant at Chiba Works, *Second Symposium on Environmental Control in the Steel Industry,* IISI, Chicago, IL, 1979.

9 Varga, J., *Control of Reclamation (Sinter) Plant Emissions Using Electrostatic Precipitators, Battelle Columbus Labs,* prepared for U.S. EPA, NTIS PB 249-505, January 1976.

10. Campbell, D. A., and Rist, D., Redcar Sinter Plant, *Ironmaking & Steelmaking* **7**(3) (1980).

11. Livshits, E. Y., et al., Main Ways of Reducing Harmful Emissions in Sinter Production **5**, 442–444 (1980).

12. Faigen, M. R., Application of Windbox Recycle and Dry Electrostatic Precipitation to the Cleaning of Sinter Plant Windbox Gases, APCA Annual Meeting, Paper 81-59.5, June 1981.

13. Sutherland, T. K., et al., *Environmental Assessment of a Coal Preheater,* prepared for U.S. EPA, EPA-600/2-80-082, May 1980.

14. Smith, H. M., and Spiker, C. A., Use of Suction Type Baghouses Control Emissions from Coke Oven Stacks, Kaiser Steel Corporation, *Proceedings of the 39th AIME Ironmaking Conference,* Washington, D.C., March 1980.

15. Kemner, W. F., *Cost Effectiveness Model for Pollution Control at Coking Facilities,* prepared for U.S. EPA, EPA-600/2-79-185, August 1979.

16. Lownie, H. W., Jr., and Hoffman, A. O., *Study of Concepts for Minimizing Emissions from Coke Oven Door Seals, Battelle Columbus Labs,* prepared for U.S. EPA, EPA-650/2-75-064, July 1975.

17. Conners, A., and Mullen, J., Developments in Coke Oven Emission Control, *Iron and Steel Engineer* (June 1980).

18. Bratina, J. E., Fabric Filter Applications on Coke Oven Pushing Operations, *Proceedings, Control of Air Emissions from Coke Plants,* APCA, April 1979.

19. Westbrook, C. W., and Coy, D. W., *Environmental Assessment of Dry Coke Quenching vs. Continuous Wet Quenching,* prepared for U.S. EPA, EPA-600/2-80-106, May 1980.

20. *Coke Oven Battery Stacks—Background Information for Proposed Standards,* U.S. EPA, Research Triangle Park, NC, May 1980.

21. Lafitte, H., Coke Plant Combustion Stack Gas Cleaning, 70th Annual Meeting of the APCA, Paper 77-6.4, Toronto, Ontario, June 1977.

22. Buonicore, A. J., et al., *Air Pollution Emissions Characterization of a Coal Preheater,* prepared for U.S. EPA, Contract No. 68-02-2819, Task No. 4.

23. Munson, J. G., et al., Emission Control in Coking Operations by Use of Stage Charging, *JAPCA* **24**(11) (1974).

24. Giunta, J. S., and Anderson, R. G., U.S. Steel Development of Coke-Oven-Door System Technology, *Proceedings of Control of Air Emissions from Coke Plants,* APCA, April 1979.

25. Jacko, R. B., et al., Plume Parameters and Particulate Emissions from the By-Product Coke Oven Pushing Operation, 71st Annual Meeting of APCA, Paper 78-9.4, June 1978.

26. Nicola, A. G., Fugitive Emission Control in the Steel Industry, Pennsylvania Engineering Corporation, reprinted from *Iron and Steel Engineer* (July 1976).

27. *Draft Coke Oven Emissions from By-Product Coke Oven Charging, Door Leaks and Topside Leaks on Wet-Coal Charged Batteries—Background Information for Proposed Standards,* Office of Air Quality and Standards, U.S. EPA, Washington, D.C.

28. Calvert, S. C., Yung, S. C., and Drehmel, D. C., Spray Charging and Trapping Scrubber for Fugitive Particle Emission Control, *JAPCA* **30**(11) (November 1980).

29. Pehlke, R. D., et al., *BOF Steelmaking, Volume 2—Theory,* Process Technology Division, Iron and Steel Society, AIME.

30. Stoehr, R. A., and Pezze, J. P., Effect of Oxidizing and Reducing Conditions on the Reaction of Water with Sulfur Bearing Blast Furnace Slags, *JAPCA* **25**(11) (1975).

31. Rehmus, F. H., et al., Control of H_2S Emissions During Slag Quenching, *JAPCA* (October 1973).

32. Smillie, A. M., and Huber, R. A., Operating Experience at Youngstown Steel with Injected Salt Coated Magnesium Granules for External Desulfurization of Hot Metal, *AIME Steelmaking Proceedings,* Detroit, Michigan, 1979.

33. *Pollution Effects of Abnormal Operations in Iron and Steelmaking—Volume III, Blast Furnace Ironmaking Manual of Practice,* prepared for U.S. EPA, NTIS PB 284-051/AS.

34. *Blast Furnace Slips and Accompanying Emissions as an Air Pollution Source,* prepared for U.S. EPA, NTIS PB 261-065/AS.

35. Spawn, P., and Craig, R., Status of Casthouse Control Technology in the U.S., Canada and West Germany in 1980, *GCA Technology Division, 2nd Symposium on Iron and Steel Pollution Abatement Technology,* Bedford, MA, November 1980.

36. Maslany, T. J., Blast Furnace Control Technology Update, Fall of 1981, *Symposium of Iron and Steel Pollution Abatement Technology for 1981,* Chicago, IL.

37. Ensor, D. S., Investigation of Opacity and Particulate Mass Concentrations from Hot Metal Operations, *Symposium of Iron and Steel Pollution Abatement Technology for 1981,* Chicago, IL.

38. Pope, R., and Steiner, J., *Particulate Mass and Particle Size Measurements for the Hot Metal Desulphurization Plant at Kaiser Steel,* Fontana, California, Draft Report, U.S. EPA, Contract No. 68-02-3159.

39. Westbrook, C. W., BOF and Q-BOP Hot Metal Charging Emission Comparison, Research Triangle Institute, Research Triangle Park, NC, *Second Symposium on Iron and Steel Pollution Abatement Technology,* Philadelphia, PA, 1980.

40. Goetz, F., Mechanism of BOF Fume Formation, Thesis, McMaster University, Hamilton, Ontario, April 1980.

41. Prater, B. E., and Colls, J. J., Visual Impact of BOF Plant Fugitive Emissions, *International Iron and Steel Institute Symposium,* June 1979.

42. *Development of Technology for Controlling BOP (Basic Oxygen Process) Charging Emissions,* prepared for U.S. EPA, NTIS PB 277-011.

43. *Level 1—Assessment of Uncontrolled Q-BOP Emissions,* prepared for U.S. EPA, NTIS PB 80-100399.

44. Tanaka, I., et al., Technology of BOF Waste Gas Handling Systems, *AIME Steelmaking Proceedings,* Pittsburgh, PA, 1977.

45. *Pollution Effects of Abnormal Operations in Iron and Steel Making—Volume VI, Basic Oxygen Process Manual of Practice,* prepared for U.S. EPA, NTIS PB 284-054/AS, June 1978.

46. Parish, A., and Pengelly, A., The Future: Is a New Technology Emerging?, *Basic Oxygen Steelmaking—A New Technology Emerges,* The Metals Society, 1979.

47. Rowe, A. D., et al., Waste Gas Cleaning Systems for Large Capacity Basic Oxygen Furnaces, *Iron and Steel Engineer* (January 1970).

48. Pilkington, S., Collection of Secondary Fume in BOF Steelmaking, *Engineering Aspects of Pollution Control in the Metals Industry,* The Metals Society.

49. Pearce, J., Gas Cooling and Cleaning: Environmental Protection Related to Primary and Secondary Extraction, *Basic Oxygen Steelmaking—A New Technology Emerges,* The Metals Society, 1979.

50. Greenfield, M. S., et al., BOF and BF Fugitive Fume Emission Control Economics, *Second Symposium on Environmental Control in the Steel Industry,* IISI, Chicago, IL, 1979.

51. *Basic Oxygen Process Furnace—Background Information for Proposed Standards,* U.S. EPA, September 1981.

52. Schuldt, A., et al., BOF Secondary Fume Collection at Stelco's Lake Erie Works, *CIM Conference of Metallurgists,* Hamilton, Ontario, 1981.

53. Baker, D. E., and Barkdoll, M. P., Retrofitting Emission Controls on the Electric Arc Furnace Facility at Knoxville Iron Co., *Iron and Steel Engineer* (August 1981).

54. Goodfellow, H. D., Solving Fume Control and Ventilation Problems for an Electric Melt Shop, APCA 73rd Meeting, Montreal, Quebec, 1980.

55. Lindstrom, R. N., Fugitive Emission Rate and Characterization for Electric Arc Steelmaking Furnaces, APCA Paper 79-39.3, 1979.

56. Brand, P. G. A., Current Trends in Electric Arc Furnace Emission Control, *Iron and Steel Engineer* (February 1981).

57. Marchand, D., Possible Improvement to Dust Collection in Electric Steel Plants and Summary of All Planned and Existing Collection Systems in the Federal Republic of Germany, *Ironmaking & Steelmaking* (4) (1976).

58. *Pollution Effects of Abnormal Operations in Iron and Steelmaking—Volume V—Electric Arc Furnace, Manual of Practice,* NTIS PB 284-053/AS.

59. Flux, J. H., The Clean Melting Shop, *Ironmaking & Steelmaking* (4) (1976).

60. Hutten-Czapski, L., Efficient and Economical Dust Control System for Electric Arc Furnace, *Symposium on Iron and Steel Pollution Abatement Technology for 1980,* sponsored by the U.S. EPA, Philadelphia, PA.

61. Argon-Oxygen Practices at J & L Steel Corporation, *AIME Electric Furnace Proceedings* **32**, Pittsburgh, PA, 1974.

62. Marchand, D. S., Fume Collection on Electric Arc Furnaces BFI Research and Developments During the Past 20 Years, *Proceedings of Air Pollution Control in the Iron and Steel Industry Specialty Conference*, APCA, April 1981.

63. Eisenbarth, M. J., Fume Extraction Hoods in the Iron and Steel Industry, *Proceedings of Air Pollution Control in the Iron and Steel Industry Specialty Conference*, APCA, April 1981.

64. Nowcsics, D. M., Construction and Optimization of Republic's AOD Facility at Canton, *Iron and Steel Engineer* (October 1980).

65. Culhane, F. R., and Conley, C. M., Air Pollution Control—Electric Arc Furnaces, *Proceedings of Annual Industrial Air Pollution Control Conference*, 1974.

66. Westbrook, C. W., and Daugherty, D. P., *Environmental Assessment of Electric Submerged Arc Furnaces for Production of Ferroalloys*, EPA 68-02-2630.

67. Urquhart, R. C., The Role of the Coke Bed in Electric Furnace Production of Ferroalloys, *AIME Electric Furnace Proceedings* **35** (1977).

68. Rentz, O., et al., Reducing Fume Emissions by Improving Furnace Operations by Feed Pretreatment, *AIME Electric Furnace Proceedings* **30** (1972).

69. Bailey, R. T., Installation of Sealed Covers for Fume Control on Submerged Arc Furnaces at Foute Mineral Company's Keokun, Iowa Plant, *AIME Electric Furnace Proceedings* **34** (1976).

70. Krogsrud, H., et al., Recent Achievements in the Development of the Modern Ferrosilicon Furnace—The Elkem Split Furnace Body, *AIME Electric Furnace Proceedings* **35** (1977).

71. Dealy, J. D., and Killin, A. M., *Engineering and Cost Study of the Ferroalloy Industry*, prepared for EPA, EPA-450/2-74-008, May 1974.

72. *Fugitive Emissions from Iron Foundries, Midwest Research Institute*, prepared for EPA, PB 110976, August 1979.

73. Chmiewlenski, R., and Calvert, S., Cupola Emission Control by F/C Scrubbing, APCA Paper 80-66.1, Montreal, Quebec, 1980.

74. *Environmental Assessment of Iron Casting*, prepared for EPA, NTIS PB 80-187545.

75. Scott, W. D., et al., Chemical Emissions from Foundry Molds, *AFS Transactions* **85** (1977).

76. Warda, R. D., and Buhr, R. K., A Method for Sampling Cupolas, Physical Metallurgy Division, Internal Report PM-M-72-24, *AFS Transactions* **81** (1973).

77. Davis, J. W., and Draper, A. B., Effect of Operating Parameters in Cupola Furnaces on Particle Emissions, *AFS Transactions* **86** (1978).

78. *Design of Sand Handling and Ventilation Systems*, American Foundrymen's Society, 1972.

79. *Molding, Coremaking, and Patternmaking*, American Foundrymen's Society, 1972.

80. Sushil, T., Fairbanks Foundry Gets Added Benefits from Environmental Control, *Modern Casting* (April 1980).

81. Warda, R. D., and Buhr, R. K., A Detailed Study of Cupola Emissions, *AFS Transactions* **81** (1973).

82. Escher, H., Recuperative Hot Blast Cupolas with Emission Control in Australia, Japan and the USA, American Foundrymen's Society, *AFS Transactions* **84** (1976).

83. Guseman, J. R., and Hendrickson, L. G., Fuels for the Direct Reduction of Iron Ore, *Proceedings of the 39th Ironmaking Conference*, AIME, Washington, D.C., 1980.

84. Faucher, A., et al., Review of the Sidbec-Dosco Direct Reduction Operations, *Proceedings of the 38th Ironmaking Conference*, AIME, Detroit, MI, 1979.

85. Stephenson, R. L., *Direct Reduced Iron, Technology and Economics of Production and Use*, AIME, Iron and Steel Society, 1980.

CHAPTER 21

AIR POLLUTION CONTROL IN THE ALUMINUM INDUSTRY

PATRICK R. ATKINS

Aluminum Company of America
Pittsburgh, Pennsylvania

1 INTRODUCTION

The nonferrous metals industry is dominated by aluminum. In 1980, aluminum production represented over 50% by weight of all of the light metals produced in the United States. Figure 1 shows that the growth rate of aluminum, magnesium, zinc, lead, and copper has increased rapidly in the last decades but those metals are reaching a level of maturity. It is estimated that the aluminum industry will continue to grow in the 1980s and 1990s at a rate of approximately 3 to 5%—somewhat faster than other nonferrous metals. This growth will be used to sustain aluminum's position in present markets and to provide metal for some new applications. Rapidly increasing capabilities for recycling aluminum will produce a recycled metal stream capable of fulfilling many of the needs for aluminum in the future. Thus, it can be expected that primary aluminum production will continue to grow slowly and that fabricated aluminum facilities will grow more rapidly. Production techniques will increase in sophistication as the demand for aluminum products is filled by primary metals and recycled metals. Technological developments in the industry will lead to improved manufacturing methods, pollution control techniques, and product quality and product utilization in a variety of end uses.

2 THE DOMESTIC ALUMINUM INDUSTRY

The United States aluminum production capacity includes two bauxite mines, five alumina production facilities, and 33 primary aluminum smelters owned by 13 separate companies. These facilities are located throughout the United States, but tend to be concentrated in areas near water transportation in the case of alumina refineries and near electric power facilities in the case of aluminum reduction plants. Approximately one-third of the primary aluminum capacity is located in the Pacific Northwest, with concentrations of reduction facilities also in the Tennessee Valley area, the Southwest (Texas, Arkansas, and Louisiana), and the Ohio River Valley.

Virtually all aluminum made in the free world is produced by refining bauxite to aluminum oxide (alumina) by the Bayer process and reducing that alumina in Hall–Heroult electrolytic cells to win the aluminum metal from the oxygen. The Bayer process utilizes a strong caustic solution of sodium hydroxide to dissolve alumina from aluminum-rich ores called bauxite found mainly in equatorial regions. The bauxite material is crushed and digested in a heated caustic solution. The solution is then filtered to remove insoluble residue and the pregnant liquor is cooled so that precipitation of aluminum oxide occurs. This precipitated material is removed from the process stream by filtration and calcined to produce aluminum oxide. A schematic diagram of the Bayer process is shown in Figure 2.

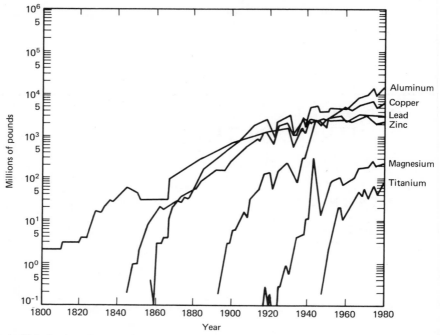

Fig. 1   U.S. Consumption—nonferrous metals (millions of pounds per year.) *Source:* Aluminum Company of America.

In the Hall–Heroult process, aluminum oxide is dissolved in an electrolytic bath composed mainly of cryolite, sodium aluminum fluoride, and aluminum fluoride. The electrolytic cells consist of open steel vessels that are lined with carbon. The carbon lining serves as the cathode for electrical conductance. Consumable carbon anodes are used to provide a carbon source to react with the oxygen liberated in the electrolysis. The cells are electrically connected in series in long buildings called potrooms. A completed electrical circuit normally consists of two rows of cells in the same or adjacent buildings. This configuration is called a potline. A schematic diagram of the Hall–Heroult process is shown in Figure 3.

Potlines may contain as few as 50 cells or more than 200. Potroom buildings may be over 600 m in length, and a plant may have from 2 to 28 such buildings.

There are basically two types of electrolytic cells in common use today. The Soderberg cell, Figure 4, utilizes a continuous anode made by adding carbon paste consisting of a mixture of petroleum and coal tar pitch to an anode casing. As the anode is consumed in the cell, the carbon material drops lower and lower in the casing. The heat from the electrolytic reduction process bakes the paste material into a continuous carbon block through which electrical current can pass. Either horizontal or vertical pins are inserted in the anode to carry current from the source into the cell. The advantage to this type of cell is that a separate carbon anode facility is not required and the capital costs of an aluminum reduction plant are reduced. The disadvantages are that energy consumption is higher and the fumes from the reduction cells are more difficult to collect and treat.

The prebake cell, Figure 5, utilizes anodes that are produced in a separate facility. These anode blocks are formed and baked separately to produce hard, dense carbon blocks that are attached to copper or aluminum stubs and inserted into the pots periodically. Improved hooding systems and cleaner off-gas streams are possible with prebake anode cells.

Although both Soderberg and center-worked prebake cells are in widespread use today, most new facilities utilize the prebake concept. In the last decade, only two greenfield aluminum smelters have been constructed in the United States and potlines have been added to only six other U.S. plants. Both greenfield smelters and five of the six potline additions have been of the center-worked, prebake variety.

3   REGULATORY REQUIREMENTS

At the present time, there are no general regulatory requirements that apply to bauxite handling. However, the United States Environmental Protection Agency is currently developing emission factors

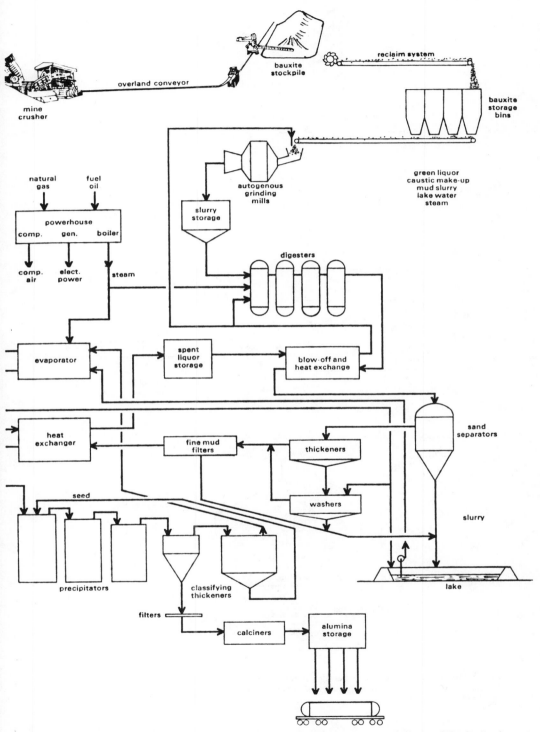

Fig. 2 Refinery schematic diagram. *Source:* Wagerup Alumina Project, Environmental Review and Management Programme, May 1978.

599

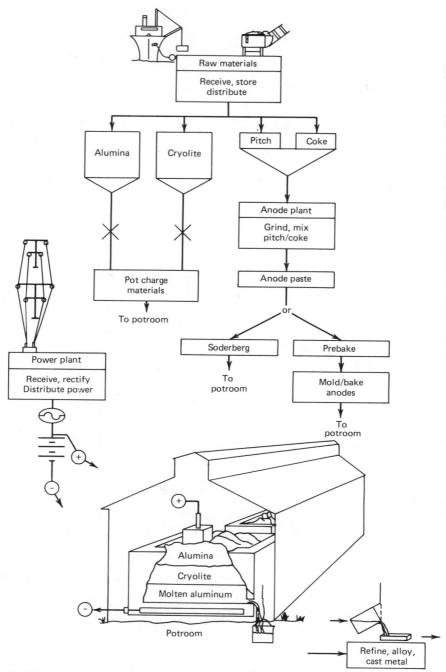

Fig. 3 Schematic of a primary aluminum reduction plant. *Source: Primary Aluminum: Guidelines for Control of Fluoride Emissions from Existing Primary Aluminum Plants,* EPA-450/2-78-0496, December 1979.

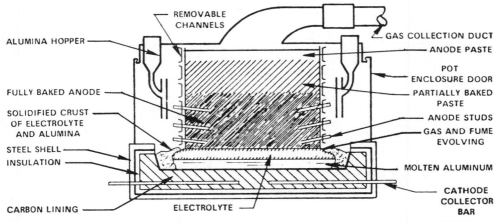

Fig. 4 Details of horizontal stud Soderberg reduction cell. *Source: Primary Aluminum: Guidelines for Control of Fluoride Emissions from Existing Primary Aluminum Plants,* EPA-450/2-78-0496, December 1979.

for ore processing facilities that would include bauxite handling facilities. Bauxite is a term that applies to a wide variety of ores. Therefore, it is impossible to describe bauxite in a succinct fashion so that general regulations could apply.

Some bauxites are extremely dusty at moisture contents as high as 16% while others become very sticky and difficult to handle at moisture contents as low as 6%. In the case of bauxites from Jamaica, for example, at 15% moisture the material can be dusty and at 18% moisture the material can be extremely sticky. Therefore, the emission factors and applicable control systems change markedly as the moisture content changes by this small amount. It is recommended at the present time that opacity be used as the main criteria for judging whether or not additional control is needed on bauxite unloading facilities and transfer points.

Alumina refineries are not currently subject to new source performance standards. Since most of the processing in the Bayer system occurs when the materials are wet, air pollution control requirements are minimal except at the beginning of the process stream (bauxite) and at the end of the process where the product (alumina) is calcined, transported for storage, and/or loading for shipment. At the present time, the EPA is reviewing calciner technology to develop new source performance standards for ore calcining. These regulations will include alumina calciners. Since both kiln-type calciners and fluid-bed, flash calciners are used in the alumina industry, both will be included in new source performance standards. It is anticipated that new source performance standard limits will be established in terms of allowable mass of emission per ton of product produced.

The New Source Performance Standards for Aluminum Reduction Plants were promulgated in 1976. A lawsuit was initiated by several aluminum companies during the 90-day period following promulgation and after 2 yr of negotiations the modified New Source Performance Standards were published in 1978. The original NSPS required the following:

1. Total fluoride emissions must be limited to 0.9 kg of fluoride per ton of aluminum produced from prebake potlines and 0.05 kg of fluoride per ton of aluminum (equivalent from anode baking furnace systems).
2. Total fluoride emissions from Soderberg operations must be limited to 1.0 kg/ton of aluminum produced.
3. Opacity from the carbon baking furnace stack must be less than 20%.
4. Opacity from all other sources in the reduction plant must be less than 10%.

The current NSPS are summarized in Table 1.

In 1980, the EPA issued a guidance document for the development of emission limitations for existing aluminum plants. Each state in which an aluminum reduction plant is located is required to develop fluoride emission limitations applicable to those plants. The guidelines document states that plant age, the cost of retrofit, the environmental conditions around the plant, and future land uses should be incorporated into any decisions concerning emissions to be placed on existing aluminum plants. The guidelines specify that hooding efficiency ranges for existing prebake plants should be 90–95% for fluorides and that removal efficiency for fluorides by primary control systems should be

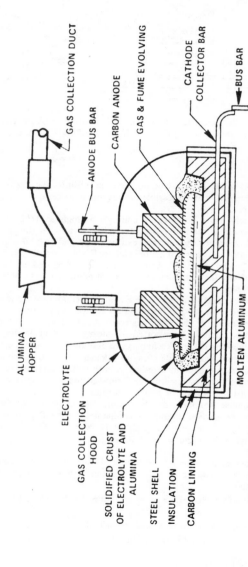

Fig. 5 Details of prebake reduction cell. *Source: Primary Aluminum: Guidelines for Control of Fluoride Emissions from Existing Primary Aluminum Plants,* EPA-450/2-78-0496, December 1979.

GAS COLLECTION DUCT

ANODE BUS BAR

CARBON ANODE

GAS & FUME EVOLVING

CATHODE COLLECTOR BAR

BUS BAR

ALUMINA HOPPER

ELECTROLYTE

GAS COLLECTION HOOD

SOLIDIFIED CRUST OF ELECTROLYTE AND ALUMINA

STEEL SHELL

INSULATION

CARBON LINING

MOLTEN ALUMINUM

Table 1   New Source Performance Standards

Subpart S—Standards of Performance for Primary Aluminum Reduction Plants

Sec. 60.192   Standards for Fluorides (Revised by 45 FR 44206, 6/30/80)

(a)   On and after the date on which the initial performance test required to be conducted by Sec. 60.8 is completed, no owner or operator subject to the provisions of this subpart shall cause to be discharged into the atmosphere from any affected facility any gases containing total fluorides, as measured according to Sec. 60.8 above, in excess of:

    (1)   1.0 kg/Mg (2.0 lb/ton) of aluminum produced from potroom groups at Soderberg plants; except that emissions between 1.0 kg/Mg and 1.3 kg/Mg (2.6 lb/ton) will be considered in compliance if the owner or operator demonstrates that exemplary operation and maintenance procedures were used with respect to the emission control system and that proper control equipment was operating at the affected facility during the performance tests;

    (2)   0.95 kg/Mg (1.9 lb/ton) of aluminum produced from potroom groups at prebake plants; except that emissions between 0.95 kg/Mg and 1.25 kg/Mg (2.5 lb/ton) will be considered in compliance if the owner or operator demonstrates that exemplary operation and maintenance procedures were used with respect to the emission control system and that proper control equipment was operating at the affected facility during the performance test; and

    (3)   0.05 kg/Mg (0.1 lb/ton) of aluminum equivalent for anode bake plants.

Sec. 60.193   Standard for Visible Emissions (Revised by 45 FR 44206, 6/30/80)

(a)   On and after the date on which the performance test required to be conducted by Sec. 60.8 is completed, no owner or operator subject to the provisions of this subpart shall cause to be discharged into the atmosphere:

    (1)   From any potroom group any gases which exhibit 10% opacity or greater, or

    (2)   From any anode bake plant any gases which exhibit 20% opacity or greater.

Source: Environment Reporter, Bureau of National Affairs, Washington, D.C., March 1982.

95–98.5%. Several states have adopted these guidelines while others have developed specific fluoride emission limitations based on the equipment efficiency criteria. Since the exact limitation is based on the assumptions of uncontrolled emission rate to which the hooding and collection efficiencies are applied, the range of allowable fluoride emissions is broad.

The existing source guidance criteria are summarized in Table 2.

4   BAUXITE HANDLING DUST CONTROL

The EPA is currently attempting to develop emission factors for metallic ore handling. A number of ore handling and processing facilities have been sampled to determine uncontrolled emission rates. However, because of the wide range of ore types and the moisture content of these ores, it is difficult to develop a series of emission factors that will adequately describe the range of emissions that can occur. This is particularly true of bauxite ores. Since bauxite represents a broad spectrum of materials with varying particle sizes and mineral content, the emission potential from bauxite loading and unloading systems is highly variable. The ores can be extremely dusty or they can be relatively sticky and difficult to remove from conveyor belts and control devices.

Particle size, surface area, and moisture content appear to be important characteristics that affect the emission potential. Jamaican ores, for example, with very finely divided particles may be dusty with moisture contents in excess of 16%, while ores from Trombetas, Brazil, and Surinam may be quite sticky at the same moisture content. With many bauxites, the dividing line between dusty ore and sticky ore may be less than a 1% change in moisture content. This characteristic makes it difficult to design liquid spray suppression systems to reduce emissions from some dusty bauxites. The ore is also inconsistent in particle size, varying from extremely fine to large and lumpy. Thus, control equipment and handling devices must be designed to be effective through the entire range of particle sizes.

Bauxite handling facilities are usually very large systems that are used intermittently when bauxite ships arrive at a loading facility. The need for bauxite dust control should therefore be carefully evaluated on a case-by-case basis. Since some bauxites can be highly dusty, while others will generate almost no dust at all, the sources of bauxite should be established and control equipment designed accordingly. In all cases, care must be taken to ensure that the collection device is compatible with the wide range of materials that exhibit quite different properties when exposed to moisture contents that may only differ by several percentage points.

Bauxite dust, when relatively dry, can be and is controlled by the use of conventional dust collection equipment such as baghouses and mechanical collectors. The dry material, although somewhat abrasive,

Table 2 State Guidelines for Control of Fluoride Emissions from Existing Primary Aluminum Plants

Cell Type	Recommended Efficiencies for Proposed Retrofits			Guideline Recommendations
	Primary Collection	Primary Removal	Secondary Removal	
VSS	80	98.5	75[a]	All plants now have best achievable hooding and primary removal. Install secondary control, but only if justified, depending on severity of fluoride problem.
HSS	90[a]	98.5[a]		All U.S. plants but one now have the best achievable primary collection efficiency. Secondary control does not appear to be justified.
CWPB	95[b]	98.5[b]		Best control is best hooding and primary removal equipment. Install where needed. Secondary control does not appear to be justified in most locations.

[a] If an existing HSS plant can achieve 85–90% collection and 95–98.5% removal efficiency, retrofit does not seem justified unless a local fluoride problem exists.
[b] If an existing prebake plant can achieve 90–95% collection and 95–98.5% removal efficiency, retrofit does not seem justified unless a local fluoride problem exists.
Source: Primary Aluminum: Guidelines for Control of Fluoride Emissions from Existing Primary Aluminum Plants. EPA 450/2-78-0496, December 1979.

is relatively easy to capture and the sizing of dust collection equipment is routine. The degree of agitation that the material receives and the distance of freefall dictates the dust loadings that will occur when a given volume of air is removed from a transfer facility. Inlet concentrations to control equipment can range from 5 to 15 g/m³. The great difference in particle size has made it necessary and wise to size dust collection equipment on the conservative side.

Past history has shown that equipment installed to capture only large particles is incapable of handling a wide variety of particle sizes that exist when bauxite ores are loaded or unloaded. Pulse-type bag collectors are normally used at small transfer points and storage tanks. Large shaker-type bag filter systems have been used successfully on ship unloaders where clamshell buckets remove the bauxite from sea-going vessels and deposit it in the loading hopper. Outlet concentrations of 0.05 g/m³ are achievable and can usually be guaranteed by equipment suppliers.

A major problem associated with equipment collection from bauxite handling facilities is the impact of moisture on the material to be collected. If wet dust is drawn into a duct and bag collector, the material will cake and eventually plug the duct, blind the bags, and stick to belts. When this occurs, a major effort is required to clean the entire system and make the collection workable at reasonable pressure drops. Large baghouse installations handling bauxite dust must be designed to overcome problems associated with high relative humidity situations that occur when slightly moist bauxite is handled.

Wet capture techniques generally have not been used successfully to control bauxite dust. The high energy consumption, equipment sizing, and, most of all, disposing of the wet material in an effective manner have steered engineers away from using the wet-type collectors on bauxite handling facilities.

5 ALUMINA REFINERY DUST CONTROL

There are a number of situations that occur between the mining of bauxite and the shipping of aluminum ingot where dust can be generated and should be controlled. The original bauxite material is changed from a highly variable mineral ore that can be extremely dusty or wet and sticky to a dry, dusty product with an average particle size near 80–100 μm and approximately 10% by weight less than 44 μm. The major point sources that must be controlled are the alumina calciners and lime burning facilities. Only calciner control will be discussed here. Lime kiln control will be discussed elsewhere in this book. In addition to the major point sources, a number of small sources, including loading and unloading facilities and transfer points, should be equipped with control equipment.

5.1 Alumina Calciner Emission Control

The major point source for alumina refining operations is the calcining system. Aluminum hydrate from the Bayer process is removed from the precipitation step by filtration and the wet cake is fed to calciners to reduce the moisture content from 45 to 2% or less. Calciners consist of rotary kilns or newer fluid-bed calciner systems. Both types of systems produce off-gas streams ladened with dry alumina dust (10–25 g/m³) and approximately 40% moisture.

The most effective control systems for these emissions are electrostatic precipitators. In some kiln systems, the hot end (product discharge end) is controlled by cyclones, multiclones, or electrostatic precipitators. The "cool end" of the kiln is usually equipped with an electrostatic precipitator. Fluid-bed calciners are complex systems which rely on product movement through a series of heating chambers and cyclones to maintain appropriate temperature levels during the drying process. All fluid-bed calciners are equipped with electrostatic precipitators as the final dust removal device.

A typical precipitator system on a modern flash calciner is designed to handle 2000 to 3000 m³/min at a gas temperature of 120–170°C with an inlet concentration of 11.4 to 24 g/m³. The design outlet concentration is 0.08 g/m³. The precipitator pressure drop is 1.25 cm of water and the specific collection area of the plates is 2154 m². A typical precipitator will contain three to four separate fields, each energized and rapped separately.

The high moisture content of the gas stream, plus the presence of any sulfur emitted by the fuel burned, assists in the conditioning of the alumina particles to ensure efficient collection. Since alumina particles in high-moisture gas streams exhibit unique electrical properties, precipitator design should be based on a thorough study of the gas stream to be treated. Precipitator life is good, but a comprehensive preventive maintenance program is required to prevent wire breakage and plate fouling.

5.2 Nuisance Dust Collection

The alumina product is conveyed either by belt conveyors or air slide conveyors from the refining process to storage facilities. It must then be transported by rail, truck, or ship to reduction plants for conversion to aluminum. Each time the alumina is handled, care must be taken to control dust emissions generated by agitating the material. (The captured material is normally returned directly to the process.) Small, pulse-type dust collectors are usually placed at transfer points or on storage tanks to keep the fine alumina particles from escaping to the atmosphere. Collectors of this type probably account for several percent of the total weight of the alumina that is handled. The particle size of the collected material averages approximately 10 μm. Dust collectors installed at transfer points may be exposed to inlet concentrations on the order of 3 to 8 g/m³. Care should be taken in designing the dust collector system to ensure that the exhaust velocity to the baghouse system is not so large that it removes dust material from the conveying system that would not normally be airborne. High-velocity offtakes are less desirable than tightly hooded transfer points with more modest air removal rates. Vendors are usually willing to guarantee 0.02 g/m³ emission rates.

Ship unloading of alumina is a particularly difficult control problem. Some ship unloaders are of the pneumatic variety where the alumina is removed from the ship by a vacuum system and deposited in an on-shore storage facility. However, in many cases, available vacuum unloading systems are too small or too energy intensive to adequately serve the large ocean-going vessels that deliver alumina to the U.S. ports. In these situations, clamshell unloaders on gantry cranes are utilized.

Two major problems exist at these loaders. First, dust is generated in the hold of the ship because the alumina is usually warmer than the surrounding environment. Since alumina has excellent insulating qualities, the heat is retained from the calcining operation imparted to the alumina at the refining plant. This heat creates turbulence in the hold which allows dust to be emitted from the large open hold as the clamshell unloading operation is under way and when the hold cleanup operation occurs. The problem is exacerbated by the turbulence created when winds blow across the large open holds. To date, there is no workable solution to this problem since the configuration of holds is variable and the size of the openings must be large enough to accommodate the clamshell unloading systems.

The second dust situation occurs when the clamshell bucket dumps its load into the receiving hopper. These areas are normally controlled by large baghouse facilities that create a face velocity of approximately 50–70 m/min at the hopper entrance. With appropriate baffling and aprons around the hopper entrance to direct air flow and spilled alumina, the dusting situation on the dock facility can be adequately controlled. The baghouse systems for such facilities are quite large but effective, easily capable of maintaining outlet emissions below 0.02 g/m³.

6 REDUCTION PLANT EMISSION CONTROL

Fluoride emissions from aluminum reduction plants have been of major concern since the late 1920s. As aluminum plants became larger in production capacity, the fluoride emission rates became significant enough to cause discernible effects on sensitive vegetation and grazing animals. Initial attempts to control these emissions utilized a variety of wet scrubber techniques. Today, improved wet scrubbers, wet electrostatic precipitators, and dry scrubbers are utilized. Since 1970, all of the aluminum reduction facilities constructed in the United States have been equipped with dry scrubbers, and all existing plants retrofitted with new control systems have also used dry scrubbers, with the exception of four horizontal stud Soderberg plants. Therefore, the major emphasis of this discussion will be on dry scrubbers.

6.1 Primary Hooding

Aluminum reduction plants are large, consisting of 2 to 28 separate buildings up to 600 m in length. Plants can contain over 900 separate reduction cells, each of which must be worked individually. These cells are usually hooded to provide a mechanism for the fluoride-bearing off-gases to be collected for treatment. These gases are called primary emissions. (See Figure 6.)

The level of hooding efficiency that can be achieved is finite since the cells must be opened periodically for anode work, bath adjustments, metal removal, and other process operations. In addition, perfect hooding is difficult to maintain as the cell ages. Current hooding technology for modern, prebake cells consists of closely fitting side shields and hinged end skirts, see Figure 7. The side shields should be substantial enough to withstand the normal wear associated with daily removal and replacement in the hot environment around the electrolytic cells. Some new hooding systems are also equipped with latches that draw each shield close to the adjacent shield to minimize the possibility of gaps in the hooding system. Latches are not needed if hooding placement is proper. The hooding systems without latches can achieve similar efficiency levels through management attention, but the latched system increases the probability of good hood placement and higher hooding efficiency.

The exhaust volumes associated with cell operations also play an important part in the hooding efficiency obtained. A number of criteria can be used to establish exhaust volume levels. Some rely on an exhaust rate based on the meters of cell perimeter or the square meters of cell surface area. Others attempt to estimate the square centimeters of cell opening that exist with all hoods in place and establish a 1 to 2 m/sec average inlet velocity through these openings. Most often, exhaust volumes are determined through smoke bomb tests and experience with other operating systems.

Many exhaust systems are equipped with dual exhaust volume capabilities. During normal operations, when all the cells controlled by a common exhaust system, usually 8 to 32 cells, have all hoods in place, each is exhausted at the same rate. When one to three of the cells are opened to be worked,

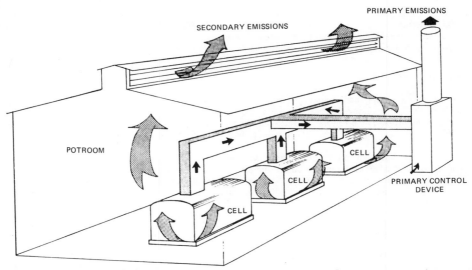

Fig. 6 Potroom fluoride emission balance. *Source: Primary Aluminum: Guidelines for Control of Fluoride Emissions from Existing Primary Aluminum Plants,* EPA-450/2-78-0496, December 1979.

the dual volume system will allow the exhaust rates of these cells to be increased by 50–200% at the expense of the closed cells. The total volume from the system is not increased but simply redistributed by opening dampers in the cells that are being worked. The slight reduction in inlet head loss causes the exhaust volume to increase significantly in those cells to overcome potential losses that would occur when side shields are removed.

Large, modern prebake cells are provided with exhaust rates of 6000–9000 am³/hr. Vertical stud Soderberg cells are exhausted at approximately one-tenth that rate because of the difficulties associated with hooding those types of cells. Horizontal stud cells are usually exhausted at rates higher than prebake cells because the hooding system consists of large enclosures which fit over the entire cell rather than individual side shields.

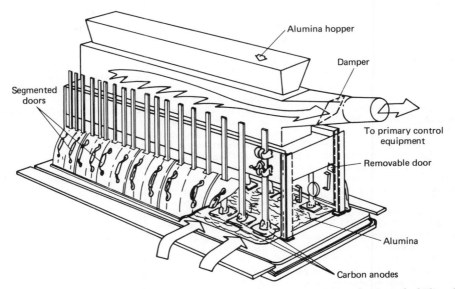

Fig. 7 Typical prebake cell hooding. *Source: Primary Aluminum: Guidelines for Control of Fluoride Emissions from Existing Primary Aluminum Plants,* EPA-450/2-78-0496, December 1979.

According to the EPA, hooding efficiencies of 85–90% can easily be achieved with horizontal stud Soderberg cells. Vertical stud cells can achieve 75–80% hooding efficiency. Modern, prebake cells can achieve 90–95% hooding efficiency on existing plants while new cells with modern hooding systems included as part of the cell design can achieve 97% efficiency or more.

6.2 Secondary Ventilation

The gases that escape the hooding systems, 20–25% for vertical stud Soderbergs to 3–10% for modern, prebake cells, are removed from the cell room by natural draft ventilation or in a few instances, power ventilators. These ventilation rates may be as high as 1 million m^3/ton of aluminum produced, while the primary exhaust from the hooding systems is on the order of 80,000–125,000 m^3/ton of aluminum produced.

These ventilation gases are dilute, often containing less than 1 mg of total fluoride per cubic meter, and 2 to 10 mg/m^3 of total particulate matter with a mass median diameter in the 3 to 10 μm range. These dilute gases and small particle sizes discourage the use of control systems to treat these emissions. Normally control attention is focused on the hooding and primary control systems rather than the cell building ventilation air.

6.3 Primary Control Systems

Primary control systems for aluminum reduction plants must have high reliability and should operate with high efficiency. The EPA has determined that 98.5% total fluoride removal efficiency should be achieved in control systems added to existing reduction plants. Most state regulations require 95–98.5% removal efficiency. New facilities often achieve removal efficiencies of 99% or more. To achieve these levels of efficiency, the control systems must be effective for gaseous fluoride as well as fine particulate and fume.

Low-energy wet scrubbers have often been used. Spray towers, moving bed scrubbers, and spray screens are in common use. Calcium or sodium compounds are added to the scrubber water to enhance the gaseous fluoride removal, producing compounds that can be removed as sludges:

$$(CaO + 2HF = CaF_2 + H_2O)$$
$$\text{lime} \qquad\qquad\qquad \text{sludge}$$

or further treated to produce compounds that can be reused in the process:

$$HF + Na_2CO_3 = NaF + NaHCO_3$$

and

$$6NaF + 4NaHCO_3 + NaAlO_2 \cdot 2H_2O = Na_3AlF_6 + 4Na_2CO_3 + 4H_2O$$
$$\text{sodium aluminate} \quad \text{cryolite}$$

The major problems to be avoided are system scale-up in the scrubber water recirculation system, nozzle pluggage in the scrubber system, scrubber pluggage due to particulates in the off-gas system (and hydrocarbons in Soderberg cell off-gases), and corrosion. All wet scrubbers should have an extensive preventive maintenance program designed and implemented at startup.

Many wet scrubber systems are preceded by cyclones, multiclones, or electrostatic precipitators to remove some of the particulate matter from the off-gases. Such systems operate with relatively low head loss and low efficiency. Since the mass median diameter of the particulates is less than 10 μm in such systems, high efficiency is impossible to achieve with mechanical collectors and difficult to achieve and maintain with simple electrostatic precipitators.

In a limited number of situations, wet electrostatic precipitators are used for both gaseous and particulate fluoride removal and particulate control. Such systems are capable of handling the off-gases from Soderberg cells with increased hydrocarbons present. Again, an effective preventive maintenance program is needed to ensure reliable operations of these systems.

Although wet scrubber systems are still in use in the United States, all of the aluminum reduction facilities built since 1972 have been equipped with dry scrubber systems, and seven existing plants have been converted to dry scrubber systems during that time. Such systems have the advantage of closed-cycle operation where no contaminated water or sludges are produced, and the collected materials can be returned directly to the cells. Since the dry scrubber system returns material directly to the process, it should be designed as part of the process itself, rather than as an "end-of-the-pipe" treatment process, if total success is to be achieved.

Dry scrubber systems utilize alumina as the scrubbing medium. After use the material is transported to the cells and used as the ore feed for the process. The collected fluorides and particulates are also returned to the cell and become part of the electrolytic bath. Appropriate adjustments must be made to the bath chemistry to maintain the proper balance in the cells. The returned fluoride and alumina dust are valuable constituents which may offset the cost of operation of the dry scrubber system. However, trace contaminants such as iron and silica are also returned to the cells and ultimately reach an equilibrium concentration in the metal produced. If high purity metal is desired, some cells may be operated on pure alumina, while the aluminum used to scrub the off-gases is directed to other cells. Such ore management is possible for a limited number of cells, but because the sorption capacity of the ore for fluoride is finite, care must be taken in designing the ore management system to ensure that the scrubber systems receive the required amount of feed.

The factors that dictate the efficiency of the ore to adsorb fluoride include surface area of the ore, sodium content of the ore, the relative humidity of the off-gas stream, and the contact time between the ore and the fluoride gases.

A surface area of 45 m²/g is normally recognized as the minimum that should be utilized for dry scrubbing applications. At surface areas in this range, alumina can adsorb approximately 4% of its weight in gaseous fluoride. However, to achieve this high removal a long contact time is required. There is evidence that the sodium content of the alumina develops active sites that are more efficient at sorbing fluoride. Therefore, the sodium content of the alumina plays a role in the ability of the alumina to retain fluoride. The relative humidity of the gases in contact with the alumina plays a role in fluoride adsorption. There is an optimum relative humidity level for each situation. However, since off-gas concentration, surface area, and contact time also are important, it is difficult to quantify the impact of relative humidity. In most situations, the relative humidity is considered an uncontrolled variable that cannot be used as a design criteria to increase fluoride removal.

Perhaps the most important variable in the design of a dry scrubber system is the contact time. It appears that a contact time of 1 to 3 sec is appropriate for good fluoride adsorption. Shorter contact times can be used if good distribution of alumina particles throughout the gas stream is accomplished.

The two major types of dry scrubber systems differ in the way that the contact time parameter is achieved. The injected alumina system, Figure 8, relies on contact between the absorbing medium, alumina, and the fluoride-bearing gases in a disbursed phase system. Proper distribution of the alumina throughout the gas stream is critical for efficient fluoride removal. Alumina is injected into the gas stream through various types of jets and nozzles and the contact occurs as the alumina is carried along in the gas stream. Some systems utilize venturi sections to ensure good turbulent mixing and to aid in the dispersion of the alumina particles in the gas stream.

The alumina off-gas mixture enters a disengaging section where much of the scrubbing medium is removed by gravity. The gas stream is then directed to a bag filter or electrostatic precipitator system to capture the finer fraction of the ore. The scrubbing medium is returned to the cell room feed system or recycled to storage tanks where it can again be passed through the injection system. Most injected systems rely on some recycle to ensure adequate removal of the fluoride gases. In some cases, this recycle may amount to 200% of the ore feed and appears to be related to how effectively the ore is distributed throughout the gas stream during the injection step.

The injected alumina scrubber systems have the advantage of accomplishing the dry scrubbing at lower head loss requirements than fluid-bed systems. The disadvantages of the system include the fact that recirculation is often required and the attrition of the alumina and abrasion of the equipment is increased.

The other major type of dry scrubber is the fluid-bed system, Figure 9. In this type of system, the off-gas stream is contacted with the scrubbing ore by passing the gas through an expanded bed of alumina. The bed moves along a horizontal perforated plate and the gas stream enters through the perforations, fluidizing the ore. Fresh alumina is fed into one end of the reactor system and the reacted ore is removed at the opposite end. The flow of alumina across the bed is controlled by the inlet feed rate. A baghouse system is placed above the fluid bed to collect the particulate matter that is carried from the bed to the gas stream. Pulse-jet bag systems or shaker-bag systems may be used for this application. In the case of the former system, the reactor operates continuously. With shaker-bag systems, the air flow must be shut off to each individual reactor unit for 20–30 min every few hours to clean the bags.

The advantage of the dry scrubber system is that even if alumina feed is lost to the dry scrubber system for a period of several hours, the fluid bed retains the capacity to continue to remove fluoride with high efficiency. Breakthrough tests have indicated that, under normal conditions, fluid-bed systems may have up to 8 hr of "safety factor" built into the system should alumina feed be lost. In the fluid-bed systems, no recirculation of the ore is required so alumina handling can be simplified. A major disadvantage of the system is that the fluid bed itself and the perforated plates required to support the bed generate higher head losses than the injected alumina systems. Therefore, there is an energy penalty associated with the operation of fluid-bed dry scrubber systems.

In both types of dry scrubber systems that utilize baghouses for final particulate collection, bag lives on the order of 15–24 months have been reported for the pulsed-air systems and 48–60 months for the shaker-bag systems. Most bag failure problems are associated with poor design of the entrance conditions that allow alumina-laden air streams to impact directly on the bag systems. Careful design

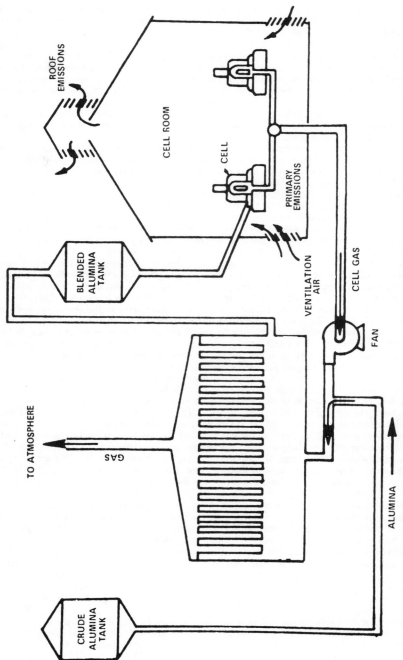

Fig. 8 Flow diagram for injected alumina dry scrubbing process. *Source: Primary Aluminum: Guidelines for control of Fluoride Emissions from Existing Primary Aluminum Plants,* EPA-450/2-78-0496, December 1979.

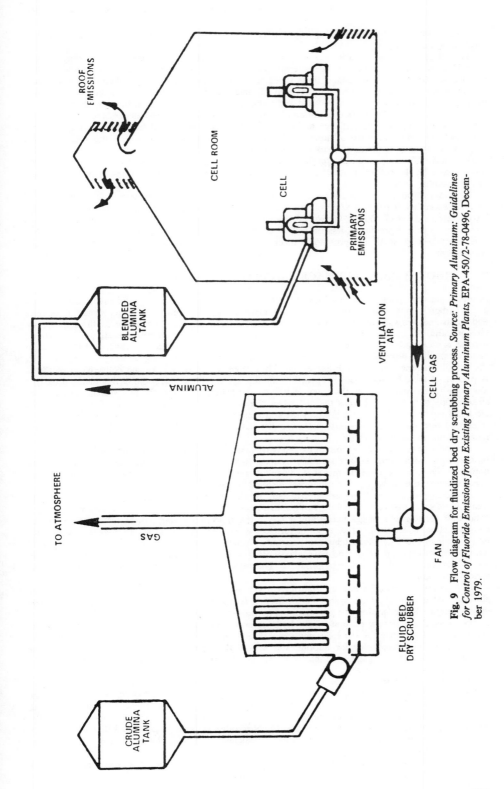

Fig. 9 Flow diagram for fluidized bed dry scrubbing process. *Source: Primary Aluminum: Guidelines for Control of Fluoride Emissions from Existing Primary Aluminum Plants,* EPA-450/2-78-0496, December 1979.

of entrance systems plenums and bag supports can significantly impact the bag life achieved in dry scrubber systems.

The off-gas temperatures from reduction cells will vary as the effectiveness of the hooding systems vary. However, in most cases, the off-gas temperatures are in the range of 110 to 125°C and necessitate the use of bags that can withstand these types of temperatures. Efforts to condition the gas stream by wet or dry cooling systems have not proved effective and efforts to bleed in ambient air for cooling purposes resulted in increased power cost and reduced exhaust volumes from the cells themselves.

Since reduction cells operate continuously, dry scrubber systems must be designed to also treat the off-gases continuously. Scrubber maintenance, fan maintenance, and bag cleaning requirements necessitate periodic shutdown of units, so the scrubber systems are constructed with multiple units in parallel. Typical scrubber systems may consist of 5 to 20 units of 100,000 m³/hr capacity in a bank to serve all or a portion of a potline. The removal of any single unit from the system will have little impact on the collection rate from the cells and will have no impact on the removal efficiency of the operating units.

Most dry scrubber systems are capable of achieving fluoride removal efficiencies in excess of 99%. Inlet gases containing 100 mg/m³ of fluoride are exhausted at concentrations less than 1 mg/m³. Particulate loadings of 100 to 200 mg/m³ are also reduced to similar levels. Dry scrubber emissions are normally low, representing less than 10% of the total plant discharge. The EPA recognized that these emissions are small and provided a mechanism to waive the monthly sampling requirement for these systems.

7 CARBON BAKING FURNACE EMISSION CONTROL

A major emission point from aluminum reduction facilities utilizing the prebake anode configuration is the carbon baking furnace. In this operation, the anode blocks are formed using a mixture of coke and pitch and are subsequently baked at high temperatures to form a block with the appropriate strength and electrical resistivity characteristics required for reduction cell operations. The baking process is accomplished in furnaces that consist of pits separated by flues in which fuel is burned to generate the heat necessary to perform the baking operation (see Figure 10). During the baking cycle, the volatile materials in the pitch binder are driven from the anode blocks and collected in the flues. A portion of these hydrocarbons is burned in the flue system. However, some of this material escapes along with particulate matter that may leak from the baking pits into the flue system. In addition, since anode butts removed from the cell when the anode becomes too small to effectively conduct electrical current to the process are crushed and reused in new anodes, some fluoride contamination is carried into the baking operation. During the baking cycle, this fluoride is also volatilized and escapes from the furnace system.

A variety of control devices to remove the particulate, hydrocarbons, and fluorides from the baking process have been utilized. Wet scrubber systems have been used for a number of years, but the nature of the collected material makes it difficult to remove the sludge from the system. In addition, once removed, the sludge poses a significant disposal problem.

In recent years, several types of electrostatic precipitators, both wet and dry, have been used to treat the off-gases from carbon baking furnaces. In some types of bake furnace operations where oxygen conditions are low and off-gas temperatures are in the appropriate range, these systems have proved to be effective. However, in other types of furnaces where oxygen conditions are higher and off-gases may be above 250°C, such systems can pose safety problems because of the fire potential associated with electrostatic precipitators. The materials collected by these systems are normally sold to reprocessors for conversion to fuels or other by-products. The material cannot be reused directly in the anode forming process.

Dry scrubbers have also been utilized for carbon baking furnace emission control. Both injected-alumina systems and fluid-bed systems are currently in operation. In the injected systems, precoolers may be required to condition the gas stream. The temperature is lowered, usually with water sprays, to the point that a major portion of the hydrocarbons will condense and can be adsorbed by the alumina particles. In these systems, distribution of the alumina in the gas stream is critical in determining the efficiency of the collector for fluoride and hydrocarbon removal. Baghouse systems follow the injection device and disengaging section. Alumina recycle capabilities are designed into the system to improve collection efficiency.

In the fluid-bed-type systems used to control carbon baking furnace emissions, the gas conditioning and scrubbing normally take place simultaneously. An appropriate amount of water is added to the reactor with the off-gases to be treated to ensure that appropriate collection of hydrocarbons is achieved. Both hydrocarbons and fluoride are collected by the alumina bed. A baghouse system, usually of the pulsed type, is placed above the fluidized bed to collect alumina entrained in the exhaust gas stream.

The alumina from either type of system can be calcined to remove the hydrocarbons collected or reused directly in the reduction process. Evidence indicates that if calcining occurs at or below 600°C, the adsorbed fluoride will not be driven off the alumina. Recent experience indicates that calcining may not be necessary if hydrocarbon levels in the furnace off-gases are low.

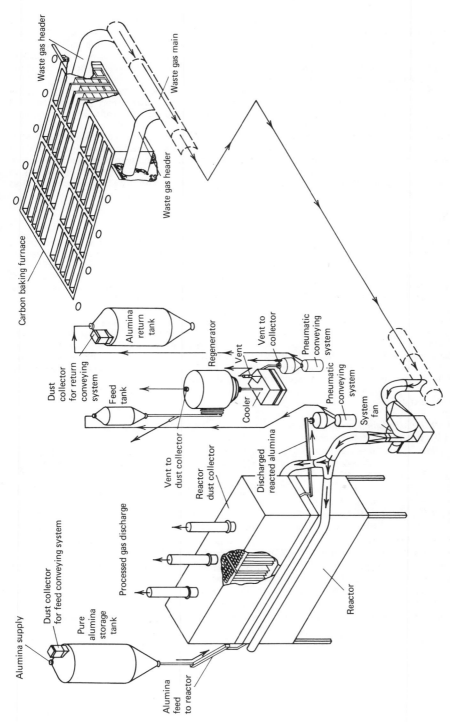

Fig. 10 Schematic of carbon baking furnace and fluidized bed dry scrubber system. *Source:* Aluminum Company of America.

Alumina supply

Dust collector for feed conveying system

Pure alumina storage tank

Processed gas discharge

Alumina feed to reactor

Reactor

Reactor dust collector

Discharged reacted alumina

Vent to dust collector

System fan

Pneumatic conveying system

Cooler

Feed tank

Dust collector for return conveying system

Vent to collector

Vent

Pneumatic conveying system

Regenerator

Alumina return tank

Carbon baking furnace

Waste gas header

Waste gas main

Waste gas header

The off-gases from carbon baking furnaces may contain several hundred milligrams per cubic meter of condensable hydrocarbons and 100 mg/m³ of fluoride. In order to meet New Source Performance Standards in the United States, the hydrocarbon emissions from new furnaces must be reduced to the point that the opacity of the plume is less than a Ringelmann 1. This requires reduction to the 10 to 20 mg/m³ range. Fluoride emissions from new furnaces must be reduced to 0.05 kg/ton of aluminum which is equivalent to approximately 5 mg/m³.

Attempts have been made to utilize coke material as the scrubbing medium in carbon baking emission control systems. However, low fluoride capture and material transport problems have led researchers away from this concept. Variations of the idea, which include dual-bed scrubbers, have been described but not yet implemented.

Recent advances in furnace firing techniques have resulted in significant energy savings as well as improved combustion of the hydrocarbon waste products. These advances have made it possible to achieve the Ringelmann 1 opacity limits without control systems. However, the stringent fluoride emission standards may not be achievable through process management. This area requires further study to determine the minimum fluoride levels that can be achieved through improved butt cleaning practices and process improvement.

8   COSTS

Control system costs for primary aluminum production facilities are highly variable since there are a number of methods utilized to achieve acceptable emission levels. Dust control systems for bauxite unloading facilities represent 10–12% of the total unloading system costs. Electrostatic precipitators for calciner dust control at refineries may be 16–20% of the total calciner cost. The International Primary Aluminium Institute has determined that fluoride emission control systems, exclusive of the hooding systems, range in cost from 2 to 11% with an average cost of 6.6%. Carbon baking emission control systems are 8–10% of the cost of a baking facility.

In all cases, there is some return of usable material to the process from well-designed and operated control equipment. The value of these returned materials is difficult to quantify, since the impacts of these returned materials on process operations and product quality are not fully understood. However, in most cases, it is generally believed that the recovered material value is enough to offset the operating costs of the control systems. In some instances, the recovered material value may also be large enough to amortize a portion of the capital required to install the systems.

REFERENCES

1. *Air Pollution Control in the Primary Aluminum Industry,* EPA-450/3-73-004A and EPA-450/3-73-004B, Singmaster & Breyer, New York, July 1973.

2. *Primary Aluminum: Guidelines for Control of Fluoride Emissions from Existing Primary Aluminum Plants,* EPA-450/2-78-049b, Office of Air Quality, Planning and Standards, Research Triangle Park, NC, December 1979.

3. Environmental Aspects of Aluminium Smelting, United Nations Environment Programme, *Industry & Environment Technical Review Series* 3 (1981).

4. *UNEP/UNIDO Workshop on the Environmental Aspects of Alumina Production,* Secretariat Report, United Nations Environment Programme, Industry and Environment Office, Paris, November 1981.

5. *Fluoride Emissions Control: Costs for New Aluminium Reduction Plants,* International Primary Aluminium Institute, Environmental Committee Report, April 1975.

6. Cook, C. C., Swany, G. R., and Colpitts, J. W., Operating Experience with the Alcoa 398 Process for Fluoride Recovery, *J. Air Pollution Control Association* 21(8) (1971).

7. Cochran, C. N., Sleppy, W. C., and Frank, W. B., Fumes in Aluminum Smelting: Chemistry of Evolution and Recovery, *J. of Metals* (September 1970).

8. Rush, D., Russell, J. C., and Iverson, R. E., Air Pollution Abatement on Primary Aluminum Potlines: Effectiveness and Cost, *J. Air Pollution Control Association* 23(2) (1973).

9. *Environment Reporter,* Bureau of National Affairs, Washington, D.C., March 1982.

CHAPTER 22
SOURCE CONTROL—
MINERAL PROCESSING

RICHARD W. GERSTLE

PEDCo Environmental, Inc.
Cincinnati, Ohio

1 INTRODUCTION

The mineral processing industry includes a very broad cross section of processes that convert naturally occurring materials into usable products. The raw materials are mined or dug from the ground in open pit, strip, dredge, and underground mining operations. Due to the relatively dry nature of many of the raw materials the mining and subsequent handling of the minerals causes emissions of particulate matter. The main atmospheric emission from the processing is also particulate. However, if thermal processing or melting occurs, the emission of volatile matter and gases may also occur, depending on the composition of the raw materials.

This chapter briefly describes the processing operations common to many minerals and then describes in more detail common mineral processes; their emission characteristics and emission controls.

2 MINERAL PROCESSING OPERATIONS

All mineral processing operations begin with mining or dredging of the raw materials. Atmospheric emissions vary widely with the type of mine and the mining methods, the moisture content of the material, and its friability. Thus a deep mine, such as a coal mine, might have little atmospheric emission, while an open pit or strip mine that requires drilling and blasting to loosen the material would have much higher emissions; especially during dry weather.

Table 1 summarizes the particulate emission sources associated with the mining and initial processing of most minerals.[1] These sources include relatively enclosed processes such as crushing and screening and open sources such as loading, stockpiles, and roadways. All of these operations are not encountered in all mineral mining operations. Overburden removal is an operation in almost all surface mining and entails removal of topsoil, subsoil, and other strata overlying the deposit to be mined. For some types of surface mining such as open pit mining and stone quarrying, overburden removal may be only a one-time or occasional operation rather than continuous. For these types of mines, the deposit to be removed is of the same magnitude or larger than the overburden volume and the location of the mining activity is relatively fixed. Therefore, the overburden is removed permanently and may be transported off-site for disposal.

In excavating overburden, three kinds of equipment are used in typical surface mining operations:

1. Draglines
2. Shovels
3. Mobile tractors, including bulldozers, scrapers, and front-end loaders.

The subsequent processing of the raw material varies with the mineral and desired end use. Figure 1 illustrates the various typical mineral processing steps. Drilling and blasting especially are used only for very solidified minerals such as rock, metallic ore, and some coals. The initial handling of the raw materials is accomplished by mechanical equipment such as draglines, front-end loaders, and conveyor belts. This equipment is supplemented by large off-the-road trucks and rail cars. Haul roads, most temporary unpaved roads between the active mining areas, loading and unloading areas, waste

Table 1   Particulate Emission Sources from Mineral Mining and Initial
Processing

Process or Enclosed Sources	Fugitive or Open Dust Sources
Drilling	Overburden removal
Crushing	Blasting
Screening	Loading, hauling, and dumping
Conveyor transfer points	Haul roads
	Storage piles
	Conveying

disposal areas, and equipment service areas are also sources of dust emissions. Because of the size of the trucks and crawler-mounted equipment that use these roads, they are normally constructed at least 40 ft wide. In newer strip mine operations that use 100- to 200-ton capacity trucks, the roads may be as wide as 100 ft.

Each time dry raw materials are handled, dust is emitted. Raw materials are unloaded directly into a feeder or hopper, crushed, and sized prior to any further processing. This is accomplished by utilizing a feeder or short conveyor which feeds the material at a constant rate from a hopper or unloading area to a crusher. Feeders come in a wide variety of sizes and configurations including vibrating, belt, and apron types with capacities up to many thousands of tons per hour. The feeder may be combined with a size separator or "grizzly" to remove extra large pieces of raw material that might damage the crusher. The feeder discharges into a primary crusher where the material is reduced in size through a series of crushers and separators. Feeders are generally open devices and sources of dust. Material is reduced in size by utilizing primary crushers of the jaw or gyratory type, set to act upon material from about 2 to 3 ft to about 6 in. in size and to pass smaller sizes. A great variety of sizes and types of crushers exist and their use depends on the type of material, the feed rate, and the initial and final material size. Many crushers, though noisy, are fairly well enclosed and emit relatively small amounts of dust compared to materials-handling processes.

Depending on the desired final size, the discharge from the primary crusher may be fed to a secondary crusher for further size reduction. A screening step frequently separates the two crushing processes with the fines going directly to storage. Secondary crushers are also fairly enclosed and consist, for example, of hammermills, cone crushers, and gyratory crushers.

Sizing in the minerals industry is almost universally accomplished by screening with conveying to separate storage piles or to further crushing. Inclined vibrating screens are generally used. For the smaller sized material, covered screens are used to reduce dust emissions.

Thermal processing of raw material is required when products such as lightweight aggregate, cement clinker, lime, or glass are produced. Drying of other minerals such as coal is also required at times. Thermal processing generally occurs in inclined rotary kilns where the raw materials are fed in a counter flow manner to the hot flue gases from a burner. Pulverized coal, oil, or natural gas are all used as fuels in rotary kilns. Fluidized bed reactors are also used especially in calcining limestone to produce lime. All of the kilns and fluidized bed reactors are major sources of particulate and require particulate controls. Furnaces are used to further process raw materials to produce glass or clay products. Gaseous emissions in the form of fluorides, SO_2, and NO_x may also occur depending on the fuel used and the raw material composition.

3   PROCESS DESCRIPTIONS

3.1   Crushed Stone, Gravel, and Aggregate

Crushed stone, as the name implies, is mined in an open quarry or pit and crushed to the desired product sizes. Gravel on the other hand is excavated from pits and then sized by screening into various classifications. Gravel is found naturally in relatively small sizes, usually with sand, and does not require crushing. Aggregate is a mixture of crushed stone, gravel, sand, and other inert materials such as slag and is most commonly used with cement to form concrete.

3.1.1   Process Description[2]

In the production of crushed stone, quarrying operations include drilling, blasting, and excavating: processing operations include crushing and screening. Gravel production includes the excavating steps followed by screening and washing operations.

As shown in Figure 2, rock obtained from the quarry is dumped into a hoppered feeder, usually a vibrating grizzly type, and fed to the primary crusher for initial size reduction. Jaw or gyratory

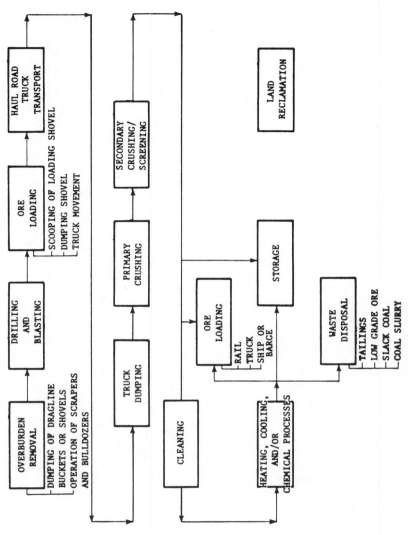

Fig. 1 Example flow diagram for typical mineral mining and processing.

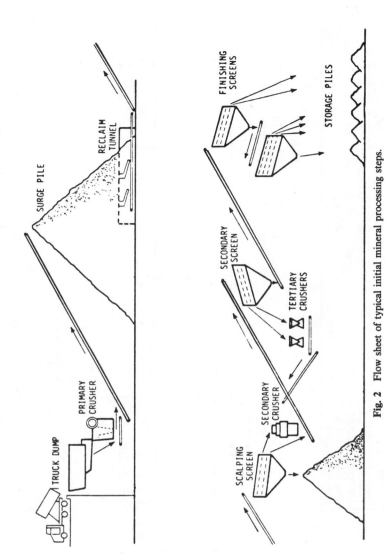

Fig. 2 Flow sheet of typical initial mineral processing steps.

crushers are often used, but impact crushers are also used when low-abrasion rock types (like limestones) are crushed and when high size reduction ratios are desired. The crusher product (approximately 3–12 in. in size) and the grizzly throughs are discharged onto a belt conveyor and transported to a surge pile or silo for temporary storage.

The material is then reclaimed by a series of vibrating feeders under the surge pile and conveyed to a scalping screen that separates the process flow into three size fractions (oversize, undersize, and throughs) prior to secondary crushing. The oversize is discharged to the secondary crusher for further reduction. The undersize, which requires no further reduction at this stage, bypasses the secondary crushers, thus reducing its crushing load. The throughs, which contain unwanted fines and screenings, are removed from the process flow and stockpiled. Secondary crushers are usually gyratory or cone type, but impact crushers are used at some installations.

If required by the product specifications, material from the secondary crushing stage, approximately 1 in. or less in size, is transported to a secondary screen for further sizing. Oversized material from this screen is conveyed or discharged directly to tertiary cone crushers or hammermills. The product from the tertiary crushers is generally conveyed back to the secondary screen. The throughs from this screen are then discharged to a conveyor and elevated to a screen house or tower containing multiple screen lines for final sizing. At this point, end products of desired gradation are discharged directly to finished-product bins or are stockpiled in open areas by conveyors or trucks.

Gravel is generally unloaded through a large bar screen or grizzly and stored in an open surge pile prior to screening into various size fractions. The screening is performed with water to wash the gravel and reduce dust. Since gravel is almost always found in the presence of sand, extensive washing steps are required to produce a "clean" product. This is most commonly accomplished by utilizing vibrating screens with water or by jigs and heavy media separation equipment.

3.1.2 Emissions

Atmospheric emissions from stone and gravel operations will naturally vary widely from plant to plant and within a given plant from day to day. Plant throughput, layout, and the moisture content of the raw material are key variables affecting the overall emissions. Table 2 summarizes the approximate emission rates from stone and gravel process operations including mining activities. These values can only be used as very general estimates of long-term average emission rates. Other researchers have reported lower emission rates for the process operations, but higher haul road emissions of 52 lb/vehicle mile for dry roads.[5] Over 50% of the emissions are usually greater than 30 μm near the source and particles there tend to settle out rapidly.

Visible emissions occur frequently, but are generally less than 20% opacity at well-controlled plants.

Table 2 Particulate Emission Factors for Stone-Processing Operations

	Uncontrolled Emission Factor[a]	
	Total, lb/ton	Suspended
Process Operations[3]		
Primary crushing	0.5	0.1
Secondary crushing and screening	1.5	0.6
Tertiary crushing and screening	6.0	3.6
Recrushing and screening	5.0[b]	2.5[b]
Screening, conveying, and handling	2.0	—
Total process operation emission	15.0	
Fugitive Operations[4]		
Drilling and blasting	0.16	
Haul roads	0.92–2.5 lb/vehicle mile	
Storage (5 active days/week)	0.33	

[a] Based on primary crusher throughput.

[b] Based on recrushing and screening throughput. Assuming 20% of the primary crusher throughput undergoes recrushing, the emission factor may be expressed as 1 lb/ton of primary crusher throughput for total emissions and 0.5 lb/ton for suspended.

3.1.3 *Emission Controls*

Dust suppression techniques are most commonly used to control dust emissions from stone and gravel processes. These techniques are especially useful for open or fugitive sources. These suppression techniques mainly involve the utilization of water, and chemical wetting or foaming agents which reduce the water's surface tension and provide for better adhesion of the water to the dust. Some chemical wetting and foaming agents can have a deleterious effect on concrete, and compatibility with subsequent products should be checked.

Enclosures and covers are also widely used for conveying, crushing, and screening operations. The enclosed process is then vented to a fabric filter system. These systems can operate very efficiently and the key to good overall control is containment and capture of the dust through a hood and vent system.

Table 3 summarizes the particulate control operations available to crushed stone and gravel operations.

3.2 Asphaltic Concrete

Though they produce a variety of asphalt paving materials, hot-mix asphalt plants are similar. Variations in products are determined by the aggregate size and type and the asphalt characteristics. The Asphalt Institute classifies hot-mix asphalt paving by the amounts of coarse aggregate, fine aggregate, and mineral dust or filler. Aggregate up to 2.5 in. in size is used in some paving and consists of crushed stone, gravel, or crushed slag. The finer aggregates include sand. The asphalt is petroleum derived and is a solid at ambient temperature, but is used as a liquid at 275–325°F.

Table 3 Emission Sources and Control Options[6]

Operation or Source	Control Options
Drilling	Liquid injection (water or water plus a wetting agent). Capturing and venting emissions to a control device
Blasting	No control
Loading	Water wetting
Hauling (emissions from roads)	Water wetting Treatment with surface agents Soil stabilization Paving Traffic control
Crushing	Wet dust suppression systems Capturing and venting emissions to a control device
Screening	Same as for crushing
Conveying (transfer points)	Same as for crushing
Stockpiling	Stone ladders Stacker conveyors Water sprays at conveyor discharge Telescoping chutes
Conveying	Covering Wet dust suppression
Windblown dust from stockpiles	Water wetting Surface active agents Covering Windbreaks
Windblown dust from roads	Oiling Surface active agents Soil stabilization Paving Sweeping

3.2.1 *Process Description*[7]

In a typical batch type plant, as shown in Figure 3, the aggregate containing 3–5% moisture is hauled by vehicle or conveyor from the storage piles and placed in the appropriate hoppers of the feed unit. The material is metered from the hoppers onto a conveyor belt and transported into a gas- or oil-fired rotary dryer. Because a substantial portion of the heat is transferred by radiation, dryers are equipped with flights that are designed to tumble the aggregate and promote uniform drying and heating.

The hot aggregate then drops into a bucket elevator and is transferred upward to a set of vibrating screens where it is classified by size into as many as four different grades. The classified hot materials then drop into one of four large bins. The operator controls the aggregate size distribution by opening individual bins and allowing the classified aggregate to drop into a weigh hopper until the desired weight is obtained. After all the material is weighed, the sized aggregates are dropped into a mixer and mixed dry for about 30 sec. The asphalt is pumped from heated storage tanks, weighed, and then injected into the mixer. The hot mixed batch is then dropped into a truck and hauled to the job site. Typical batch size is 4000 lb with production in the range of 100–150 tons/hr.

Approximately two-thirds of the dryers are oil fired with the balance utilizing natural gas. Only about 6% of the plants are of the continuous mix type, with the balance using mostly the batch process described above. About 20% of the asphalt batch plants can be moved from site to site.

3.2.2 *Emissions*

Particulate emissions from the rotary dryer are by far the major source of atmospheric emissions. Fugitive particulate emission for aggregate storage and handling, and plant roads also occur as do occasional "tarry" odors from the asphalt storage and mixing area.

Rotary dryer emissions vary with the aggregate size and type of fuel. Uncontrolled emissions average approximately 45 lb/ton. This particulate is relatively coarse with more than 50% larger than 20 μm,[8] but varies with the aggregate used. Gaseous emissions of sulfur and nitrogen oxides are much less than particulate, generally amounting to less than 0.1 lb/ton. Aldehyde emissions from the dryer have been reported at 0.025 lb/ton.[9] Emissions from the mixer are low, and except for potential odor problems are not usually significant.

3.2.3 *Emission Controls*

Dryer emissions are controlled by either scrubbers or fabric filter systems preceded by a cyclone collector. These devices can reduce emissions by over 99%. The New Source Performance Standard sets an

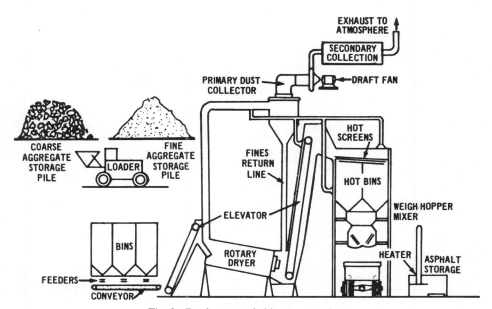

Fig. 3 Batch type asphaltic concrete plant.

emission limit of 0.04 gr/dscf from any source within the plant. Visible emissions cannot exceed 20% opacity.[10] This emission limit can be achieved with either a low energy scrubber or fabric filter.

3.3 Concrete Batching

Concrete batching is accomplished by numerous plants, both mobile and permanent, which carefully mix measured portions of sand, cement, and aggregate to form a mixture for concrete. In one type of plant this mixture is added with water into mixer trucks which mix the batch by rotation during transit to the work site. The concrete mixture is also at times transported in a dry state to the work site and water is added and mixing occurs there. At large construction projects, the concrete batching operations may include proportioning and mixing the ingredients with water at a central facility and transporting the finished wet mix directly to the pouring site.

3.3.1 Process Description[11]

A variety of designs exist for concrete batch plants, but they all include elevated storage bins, bucket elevators, weigh hoppers, and ancillary equipment. Sand and aggregate are normally delivered by truck and transferred into the bins by bucket elevator. Cement is generally delivered by trucks equipped with compressors for pneumatic delivery. Hopper bottom trucks may also be used for gravity feed into an underground bin and screw conveyor and bucket elevator delivery system.

From the storage bins, sand and aggregates are weighed out into a weigh hopper and cement is added to a separate weigh hopper. The weighed batches are then dropped into another single hopper for discharge into a truck. In wet batch plants, water is also added simultaneously to the truck.

3.3.2 Emissions

Particulates are the only atmospheric emissions from concrete batch plants and they occur from unloading, loading, and handling the materials used in concrete (especially cement). The sand, gravel, and crushed stone aggregates, having been washed prior to shipping to the batch plant, are relatively dust free and moist. As such, they are only minor dust sources. The cement, however, due to its dry condition and fine particle size (10–20% less than 5 μm, depending on grade), tends to easily become airborne. Lightweight or expanded aggregate, if used, also tends to create dust problems unless it is wetted prior to use.

The cement dust is emitted during unloading at the receiving site, from leaks in the conveying system and from bin vents. Discharge of the batch into a truck also causes dust at dry batch plants where no water is added to the truck. While emissions vary widely from plant to plant, depending on how well the conveyors, bins, and loading operations are sealed, the average emission rate of particulate is about 0.2 lb/yd³ of concrete (0.1 lb/ton).[12]

3.3.3 Emission Controls

Reduction of emissions is accomplished by enclosing transfer operations and venting them and the bin vents to fabric filtration systems. These systems along with the judicious use of water to spray aggregate, outside storage piles, and plant road areas can reduce total emissions by about 90%.

3.4 Portland Cement Production

Portland cement is composed mainly of calcium- and silica-based materials with smaller amounts of alumina and iron compounds. Approximately 3200 lb of raw materials are required to produce 1 ton of cement. The production process involves quarrying and crushing, grinding and blending, clinker production, drying and calcining, and finish grinding and packaging.

3.4.1 Process Description[13,14]

The most common combination of raw materials is limestone and clay or shale, carefully blended to yield the final chemical proportions required in the cement. Limestone and shale are blasted from quarries, usually close to the cement facility. The raw material is transported to the primary crusher by dump trucks, rail cars, and conveyor belts. The other raw materials are obtained from clay, shale, bauxite, silica sand, and iron ore.

Primary crushing reduces rock as large as 4–5 ft in size to pieces 6 to 10 in. across by utilizing gyratory, jaw, and roll crushers. After the rock is broken by the primary crushers, it is carried by conveyors to the secondary crushers, usually of the "hammer mill" type. Rock crushed by the hammer mill is usually less than ¾ in. across. The crushed raw material then undergoes a fine grinding process, which further reduces the rock size in preparation for processing in the kiln. As shown in Figure 4,

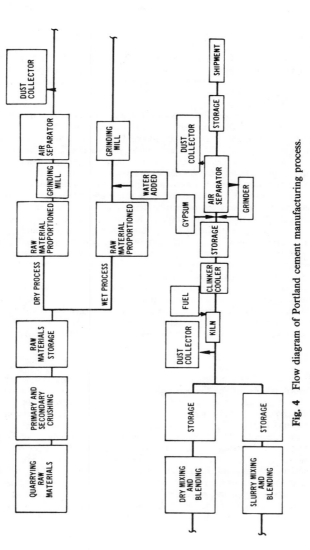

Fig. 4 Flow diagram of Portland cement manufacturing process.

the subsequent processing used after secondary crushing depends on whether cement is produced by the wet or dry process.

In the wet process, raw feed is fed with water to the grinding operation which consists of ball and rod mills. The resulting slurry is discharged from the mill and stored in open tanks where additional mixing takes place. The wet slurry is then pumped directly to the kilns or dewatered by vacuum filters, or thickeners to a moisture content of 20–50%.

In the dry process hot gases for drying the feed materials are provided by direct firing of separate furnaces or by flow of exhaust gases from the kiln. Ground raw materials are proportioned as they enter the grinding mills. The milled rock must be finely ground and thoroughly mixed to produce clinker with a uniform composition. In closed-circuit grinding, air separators return the coarse fraction to the mill for further grinding, while the fine fraction goes to a series of silos, bins, or storage buildings.

Dry feed or wet slurry is calcined or "burned" in a kiln to form portland cement clinker. The rotary kiln used for calcining is a steel cylinder with a refractory lining, slightly inclined downward from the feed end. Raw materials are fed into the high end through a kiln feeder. Fuel in the form of pulverized coal, oil, or natural gas is burned at the lower end of the kiln. On the order of 3 to 5 million Btu are required to calcine 1 ton of cement depending on the heat recovery system utilized.[15] The hot combustion gases flow through the rotating kiln countercurrent to the incoming raw materials. Kilns range in size, depending on capacity, up to about 25 ft in diameter and 700 ft in length. As the kiln revolves, the raw materials roll and slide downward toward the lower end, gradually becoming exposed to more and more heat. Water is evaporated in the upper part of the kiln. Hanging chains within the kiln promote heat transfer. In the middle section of the kiln, carbon dioxide and combined water are driven from the raw materials, and the original limestone, silica, iron ore, and clay are changed into new compounds, such as calcium silicates, aluminates, and ferrites. The lower third of the kiln is the burning zone, maintained at temperatures of approximately 2700°F, where the material becomes incandescent. The clinker forms as round marble-size, glass-hard balls and drops from the burner end of the kiln. The clinker is then cooled with air, mixed with gypsum ($CaSO_4$), and ground to the desired product size. The finished cement is shipped in bulk form or packaged.

3.4.2 Emissions

Particulate matter is the primary emission in the manufacture of portland cement. Emissions also include the combustion products from the fuel used to supply heat for the kiln and drying operations, including oxides of nitrogen and sulfur. Smaller amounts of reduced sulfur compounds such as H_2S or organic sulfur compounds may also be emitted when low oxygen levels occur in the kiln. Sources of dust at cement plants include: (1) quarrying and crushing, (2) raw material storage, (3) grinding and blending (dry process only), (4) clinker production, (5) finish grinding, and (6) packaging. The largest potential source of emissions by far is the kiln operation which consists of three units: the feed system, the fuel-firing system, and the clinker-cooling and handling systems.

Table 4 summarizes the emission data from kilns and ancillary grinding and drying processes. Kiln emissions are the major source of particulate amounting to an average of 245 and 228 lb/ton of cement for the dry and wet systems, respectively. The wet process tends to yield lower uncontrolled particulate emissions, especially from the initial grinding operation. The approximate size distribution of the particulate emitted from the kiln is shown in Table 5 and indicates that about 40% of the

Table 4   Emission Factors for Cement Manufacturing Without Controls, lb/ton of Cement[16,17]

Pollutant	Dry Process		Wet Process	
	Kilns	Dryers, Grinders, etc.	Kilns	Dryers, Grinders, etc.
Particulate,[a] lb/ton	245	96	228	32
Sulfur dioxide[b]				
Mineral source, lb/ton	10		10	
Gas combustion, lb/ton	Neg.		Neg.	
Oil combustion, lb/ton	4.2S[c]		4.2S	
Coal combustion, lb/ton	6.8S		6.8S	
Nitrogen oxides, lb/ton	2.6		2.6	

[a] These emission factors include emissions from fuel combustion.
[b] Total sulfur dioxide is the sum of the SO_2 from the mineral (clinker) and from the fuel. This factor takes into account SO_2 retained by the clinker.
[c] S is the percent sulfur in fuel.

Table 5   Size Distribution of Dust Emitted from Kiln Operations Without Controls[16]

Particle Size, μm	Kiln Dust Finer than Corresponding Particle Size, %
60	93
50	90
40	84
30	74
20	58
10	38
5	23
1	3

particulate is less than 10 μm in size. The relatively fine and dry nature of the cement product after finish grinding requires careful control to reduce losses. Clinker cooler emissions are approximately 30 lb/ton of raw material when using a cyclone for emission control. However, emissions as low as 1.0 lb/ton have been measured on clinker coolers or small (12 ton/hr) kilns.

In addition, the many materials handling and sizing processes all may emit particulate unless well sealed and controlled. Fugitive emissions from plant roads, stockpiles, and quarrying operations must also be considered in estimating the total plant emissions.

3.4.3   Emission Controls

Fabric filter systems are widely used to reduce particulate emissions from materials handling, sizing, product grinding, and product loading operations. When the operation is well enclosed to capture the particulate, the filtration system can reduce emissions by over 99%. These fabric filter systems use cotton or Dacron bags and automatic shakers or pulse-jet cleaning systems. Air-to-cloth ratios up to 6:1 may be used. For storage piles and unpaved roads, water sprays with wetting agents have proved useful in reducing airborne emissions. Some chemical wetting agents may be deleterious to concrete and should be used with caution.

Control of particulate emissions from the kiln is complex due to the large volume of hot dirty gas. Fabric filters or electrostatic precipitators usually preceded by cyclone collectors are utilized to control particulate emissions. Due to the high temperature, heavy grain loading, and moisture content of the kiln off-gases, a variety of designs have been implemented to reduce emissions and recover heat. Depending on product specifications and the alkalinity of the collected dust, some of this dust may be returned to the kiln feed stream.

With either control system, the gases must be cooled to about 500°F or less by water sprays, air dilution, waste heat boilers, or some combination. Due to the presence of sulfuric acid, the acid dewpoint for the gas stream could be as high as 270–300°F. To prevent corrosion or possible fabric plugging and blinding, the exhaust gas temperature should not approach the dewpoint. Fabric filter systems will also reduce SO_2 emissions by about 50% due to reactions with the alkaline dust. Electrostatic precipitators operate in the 300–500°F range and are less susceptible to operating problems from the high moisture content in the flue gas. Corrosion will still be a problem if operated below the dewpoint.

Clinker cooler emissions can be effectively reduced by fabric filter systems or electrostatic precipitators. A portion of the clinker cooler exhaust air is frequently sent to the kiln as combustion air, and to other drying processes within the plant to utilize the available heat in this air.[18] Gravel bed filters have also been used successfully to reduce emissions from clinker coolers, but they have not received wide application.

The Federal New Source Performance Standards in the United States limit particulate emission from kilns to 0.3 lb/ton of dry kiln feed and for clinker coolers, to 0.10 lb/ton of feed. Visible emissions from the kiln cannot exceed 20%; visible emissions from the clinker cooler and all other processes cannot exceed 10%.[19]

3.5   Lime

The manufacture of lime involves the calcining of limestone ($CaCO_3$) or dolomite ($CaCO_3 \cdot MgCO_3$) to release carbon dioxide and form lime (CaO) or dolomitic lime ($CaO \cdot MgO$). Most lime facilities are located at or near a limestone quarry to reduce transportation costs. Transfer of the quarried limestone to the crushing and screening site is usually accomplished by large off-highway trucks. Limestone and/or dolomite is discharged to a feeder and primary crusher and then to secondary crushing/

screening operations. This initial part of the process is identical to stone processing. Limestone is calcined in either vertical or more commonly in rotary kilns. The limestone feed material size for vertical kilns is 6–8 in., and consequently only primary crushing is required. However, some vertical kilns do require smaller (3–5 in.) sized feed material. Inclined rotary kilns require the smaller size feed provided by secondary crushing. Fluidized bed reactors are also used to calcine limestone and require a finely ground feed material which remains suspended in the combustion air stream.

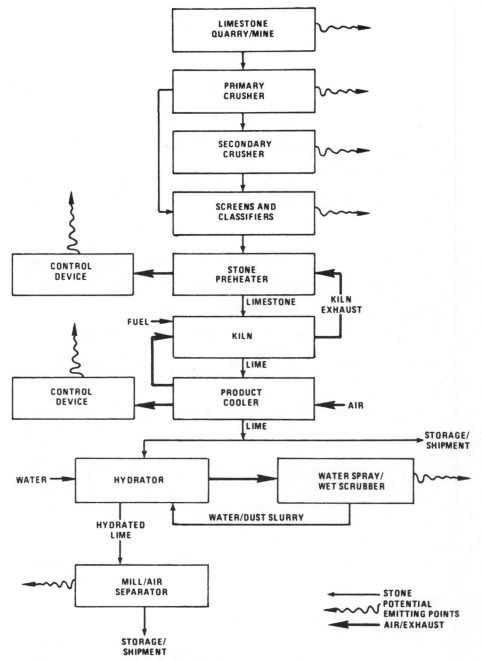

Fig. 5 Generalized lime manufacturing process flow.

3.5.1 Process Description

As shown in Figure 5 initial processing involves limestone crushing and sizing. The main process in any lime plant is however the calcining operation which occurs in fuel-fired kilns or fluidized-bed reactors. Most common is the rotary kiln which accounts for about 90% of production and is identical to a cement kiln (though usually smaller). Crushed limestone is preheated and partially dried, usually by the hot kiln flue gas, and fed to the upper end of the slowly rotating (about 1 rpm) kiln and passes down the kiln countercurrent to the hot flue gases. As in a cement kiln, the feed passes through three distinct zones consisting of drying, preheating, and calcining. Pulverized coal, oil, or gas combustion are used to provide heat.

The vertical kiln is an upright steel cylinder lined with refractory material. All vertical kilns operate similarly and have four distinct zones from top to bottom, namely, stone storage, preheating, calcining, and cooling and discharge zone. The limestone is charged at the top and calcined as it descends slowly to the bottom of the kiln where it is discharged. A primary advantage of vertical kilns over rotary kilns is the higher average fuel efficiency. The primary disadvantages of vertical kilns are their lower production rates and the fact that coal cannot be used without degrading the quality of the lime produced.

Regardless of kiln type, the temperature in the feed end of the kiln is kept below 1000°F while the operating temperatures in the preheating and calcining zones are generally in the 2000–2400°F range, with higher temperatures found in shorter kilns.[20]

The calcined lime (CaO) is then screened, milled, and shipped. About 10% of the lime produced is converted to hydrated or slaked lime (CaOH) by reacting lime with water.

3.5.2 Emissions

As with cement, particulate is the primary problem associated with lime manufacture. This occurs not only from the quantity that could potentially be emitted, but also its obvious white color. Products from fuel combustion, including sulfur, carbon, and nitrogen oxides, are also emitted and depend on the fuel type and kiln efficiency. Copious amounts of carbon dioxide are emitted, approximately equivalent by weight to the amount of lime produced. Table 6 summarizes uncontrolled particulate emission data from lime production.[21] In addition to the emissions shown, sulfur dioxide is emitted at a rate of from 10 to 50% of that expected from the combustion of sulfur in the fuel. The balance of the sulfur dioxide combines with the lime.

The emissions associated with quarrying, hauling, crushing, screening, conveying, and storing raw materials also occur as described in Section 3.1.

3.5.3 Emission Controls

Most of the more common particulate control devices have been used to successfully reduce particulate emissions from lime kilns and product coolers. Fabric filters, electrostatic precipitators, and venturi scrubbers, usually preceded by cyclone collectors are the most common devices. A number of gravel bed filters have also been installed. All of these devices can operate at collection efficiencies of 99% or more. As in the case of cement kilns, the exhaust gas must be cooled prior to control if fabric filters or ESPs are used. However, gravel bed filters and scrubbers can operate at high temperatures.

Emissions from lime hydrators are controlled with water sprays and scrubbers with the water recycled back to the hydrator. Emissions from product dryers are controlled by cyclones followed by fabric filters or by scrubbers with pressure drops of about 30 in. of water.

In addition to the kiln control systems, controls are utilized to reduce particulate emissions from raw materials and product handling systems.[22,23] Raw material crushing and sizing operations are controlled by wet suppression and by handling and venting through a fabric filter system. Storage piles must be sprayed to prevent airborne dust.

Table 6 Uncontrolled Emissions from Lime Processes, lb/ton of Lime

Process	Particulate	Nitrogen Oxides	Carbon Monoxide
Rotary kiln	340	3	2
Vertical kiln	8	No data	No data
Product cooler	40[a]	0	0
Hydrators	0.1[b]	0	0

[a] These emissions are usually vented to the kiln.
[b] After a water spray system.

Lime storage silo vent air and pneumatic transport system air are usually controlled by fabric filters. Lime packaging and bulk truck, rail, and ship/barge loading operations are also frequently controlled by aspiration through fabric filters. Gravity-feed fill spout mechanisms with outer concentric aspiration ducts to fabric filters greatly reduce dust emissions at loading sources.

Most of the material collected by fabric filters at a lime plant is returned to the process. However, when this collected material cannot be returned (e.g., kiln flue dust), disposing of it to lime by-product storage or waste areas by discharge and transport in open trucks, can be an intermittent emission source. Wet suppression and enclosure of the unloading operation and covering the trucks are used to reduce these emissions.

3.6 Coal Cleaning

As mined, coal contains many impurities and noncombustible minerals. To enhance the coal quality, meet customer specifications, and reduce transportation charges, many of these impurities are removed at the mine site by coal cleaning processes. The nature and amount of impurity vary with the type of mine (deep or strip), the mining methods, the type of coal (lignite, bituminous, anthracite), and the mine location. In its natural state, coal contains clay, rock, shale, pyrites, and other minerals referred to as ash. The actual mining operation adds other impurities in the form of dirt, mine rock, wood, and tramp iron. Coal cleaning plants mainly utilize the difference in specific gravity between the combustible coal fraction and the heavier impurities to physically separate the fractions. Chemical coal processing to produce synthetic fuels is also accomplished at some plants. This is an entirely different process and is not included in this section.

3.6.1 Process Description[24]

Physical coal cleaning as performed at most coal cleaning plants involves crushing and screening to free entrapped impurities. Impurities that occur as finely divided mixtures in coal are more difficult to remove than the coarser materials, for example, pyritic sulfur is much more difficult to remove than rock or shale. The degree of separation achieved thus depends on how fine the coal is crushed. The finer the particle size, the more impurities are freed for subsequent separation. Scalping separators and magnets remove wood and iron prior to size reduction.

After the crushing and sizing operations, classifying screens separate coal particles by size and route them to the cleaning processes which may be wet, dry, or a combination of both.

Wet cleaning systems utilize centrifugal or gravity separation in a liquid medium to separate heavier fractions from coal. None of the wet cleaning methods cause atmospheric emissions. However, the auxiliary processes of handling and drying can be major emission sources. After the coal is wetted by the cleaning process, it is dried mechanically by dewatering screens followed by centrifugal dryers. Where customer demand is for low surface moisture (3–6%), secondary drying is required, usually by thermal drying. New plants utilize fluid-bed dryers, in which hot combustion gases from a coal-fired furnace are passed upward through a moving bed of fine wet coal. The coal is dried as the fine particles come into contact with the hot gases. Particulate emissions occur predominately from ultrafine (−200 mesh) coal particles which are entrained and carried from the dryer by the combustion gases. Flash and multilouver dryers are also in common use.

All dry coal cleaning systems use pulsating air columns to separate coal from impurities. In these processes, dirty crushed coal enters the air table where it is stratified into a bed by pulsating air. The heavier impurities settle in a layer beneath the coal and as the bed travels forward, these impurities drop into pockets or wells and the upper layer containing the lighter coal fraction is reclaimed.

3.6.2 Emissions

Emissions from coal cleaning plants consist mainly of particulate matter and combustion products, and fine coal particles from thermal drying. Particulate emissions from crushing, screening, storage, transfer, grinding, conveying, or loading operations are highly variable and increase as the surface moisture in the coal decreases.

Thermal dryers in a coal plant utilizing wet cleaning incorporate cyclones as an integral part of the coal cleaning process. An average emission factor for fluid-bed dryers with only cyclone control is 3 gr/dscf or about 10 lb/ton of coal processed.[25] Uncontrolled particulate emissions from flash dryers and multilouvered dryers are approximately 16 and 25 lb/ton, respectively.[26]

Gaseous combustion products are emitted from thermal dryers as shown in Table 7. Sulfur oxides vary with the sulfur content of the coal while the other emissions are a function of burner design and excess air. Total emissions depend on the amount of heat required to properly dry the coal.

Particulate emissions from dry coal cleaning systems vary widely with the coal's moisture content. For a relatively dry coal, air classifier emissions after a cyclone control device are approximately 0.7 lb/ton. With a wetter coal, emissions are reduced to about 0.18 lb/ton.[28]

Table 7 Gaseous Emissions from Thermal Dryers[27]

Compound	Concentration, ppm	Emission Rate, lb/10⁶ Btu[a]
NO_x	40–70	0.39–0.68
SO_x	0–11	0–0.09
HC[b]	20–100	0.07–0.35
CO	<50	<0.30

[a] Based on heat input to dryer.
[b] Expressed as methane.

3.6.3 Emission Controls

Primary control systems in the form of cyclone collectors are an integral part of all coal drying systems since they reduce product loss. For wet coal cleaning systems that utilize a thermal dryer, a wet scrubber is the only type of control device that is used to further reduce emissions. Scrubbers can reduce emissions to the 0.02 to 0.04 gr/dscf range or about 0.06 to 0.15 lb/ton of product. Dry air separation systems are controlled by fabric filter systems which can reduce emissions to less than 0.02 gr/dscf or about 0.08 lb/ton. Visible emissions of less than 10% opacity can be achieved. With either type of control, collected materials may be returned to the product or burned on site to generate heat.

The Federal Standard of Performance for new coal cleaning plants limits dryer emissions to 0.031 gr/dscf and air classifier emissions to 0.018 gr/dscf.[29]

3.7 Glass Manufacturing

Glass, an inorganic product produced by melting raw materials, shaping to the desired form and cooling without crystallization, is used to produce containers, pressed and blown shapes, and flat glass. Soda–lime glass is by far the major type of glass produced. Other types of glass include borosilicate (heat resistant), lead, and opal glass. Raw materials for manufacturing glass are relatively plentiful and the primary ingredient is sand (SiO_2). In the case of soda–lime glass, soda ash (Na_2CO_3) and limestone ($CaCO_3$) are the other major raw materials. Cullet or broken glass is also recycled as a raw material.

3.7.1 Process Description

Regardless of which product is produced, the manufacture of glass consists of four basic processes; preparation and blending of raw materials, melting in a furnace, product forming, and product finishing. In the preparation and blending step, raw materials stored in hoppers are carefully weighed, and mixed and stored until fed to the melting furnace.

Continuous regenerative-type furnaces with direct firing of gas or oil are used to melt the batch. The temperature in the melting zone of the furnace varies with the type of glass, but is in the 2700–2900°F range. Approximately 5 million Btu are required per ton of glass melted in a large efficient furnace. Electric furnaces are also used in the pressed and blown glassware industry, but they account for less than 5% of production. As the fresh material enters the furnace, it floats on the molten glass already in the furnace. As the feed material melts, it passes toward the cooler opposite end of the furnace, flows through an opening to the refining section of the furnace, and then flows from the furnace to the forming process.

In the float process for flat glass production, the molten glass passes directly to a sealed chamber containing molten tin. By careful temperature control, the glass solidifies on the surface of the tin bath, and forms a flat surface.

The continuous regenerative furnaces commonly used in the glass industries have capacities in the 50 to 450 ton/day range. Brick checkerwork is utilized to recover heat by passing flue gas over the brick checkerwork, and then passing combustion air over the heated bricks while flue gas is preheating another section of brickwork. The flue gas and combustion air streams are alternately switched to preheat and then cool the brick checkerwork every 15 to 20 min.

3.7.2 Emissions

Particulate from raw material handling and especially from the melting furnace are the major atmospheric emissions for glass production. The raw material handling emissions amount to about 0.04–0.06 lb/ton of material processed for flat and container glass and up to 4 lb/ton for pressed blown

Table 8   Uncontrolled Emissions from Glass Melting Furnaces,[32] lb/ton of Glass[a]

Emission	Type of Glass Produced		
	Flat	Container	Pressed and Blown
Particulate	0.8–3.2	0.9–1.9	1.0–25.0
SO_x	2.2–3.8	2.0–4.8	1.1–11.0
NO_x	5.6–10.4	3.3–9.1	0.8–20.0
CO	<0.1	0–0.5	0.1–0.3
HC	<0.1	0–0.4	0.1–1.0

[a] Representative of fuel-fired furnaces only.

glass.[30] Particulate emissions from the furnace vary from about 1 to 3 lb/ton of glass as shown in Table 8 (except for pressed and blown glass). The particulate is largely submicrometer in size due to the condensation of volatile species as they leave the furnace and the entrapment of larger sized particulate in the brick checkerwork. Geometric mean particle diameters in the 0.1–0.13 μm range have been measured with geometric standard deviations of 1.5 and 1.7, respectively.[31] The fine nature of the emissions also causes the opacity of the resulting plume to be in the 20–50% range. These uncontrolled particulate emissions vary with the type of product and are affected mainly by the volatile constitutents in the feed such as borates. Melting furnace emissions as shown in Table 8 include gaseous compounds from the fuel combustion. Sulfur oxides occur from fuel sulfur and also from decomposition of sulfates in the glass batch, especially sodium sulfate. Nitrogen oxides occur largely from thermal fixation of atmospheric nitrogen and are seen to be the largest pollutant by weight.

Fluorides are emitted from furnaces producing opal, borosilicate, and lead products. They occur from the fluospar and other fluoride-containing minerals and can be as much as 20 lb/ton. Arsenic may also be emitted if present in the raw materials.

Hydrocarbon emissions from glass forming and finishing for container and pressed or blown products occur due to the lubricating oils on the mechanical equipment used to shape and handle the hot glass. These volatile emissions can amount to up to 9 lb/ton of glass.

3.7.3 Emission Controls

Process modifications dealing with feed composition and size and reduction in operating temperature can reduce particulate emissions. The utilization of electric furnaces or the use of supplemental electric melting on fuel-fired furnaces (known as boosting) can also reduce emissions since the upper surface of the melt is at a lower temperature and hotter melting zones occur beneath the surface.

The fine nature of the particulate coupled with the hot gas temperature and a fairly dilute particulate loading makes the application of control devices fairly complex. If scrubbers are utilized to reduce emissions, only the high-energy type are useful due to the very fine nature of the particulate. Electrostatic precipitators and fabric filters have been utilized to reduce emissions from both charging and melting operations. Table 9 presents particulate emission data for gas-fired glass furnaces obtained during an EPA study. Container and flat glass particulate emissions were in the range of 0.12–0.42 lb/ton. Pressed and blown glass emissions ranged up to 1.0 lb/ton.

Methods for reducing NO_x emissions have not been applied to glass furnaces. While some of the techniques used in boilers could be utilized in glass furnaces, their effectiveness is difficult to predict.

3.8   Brick and Clay Processing

The manufacture of brick and related clay products involves (1) mining, grinding, screening, and blending of the raw materials, (2) forming, cutting, or shaping, and (3) drying and firing of the final product. Raw materials for brick and clay products are naturally occurring minerals that are usually mined in open pit operations. Some of the finer sized clays are, however, deep mined. These clays are hydrates of alumino-silicates with impurities such as feldspar, sand, limestone, and carbonaceous matter. Clay forms a plastic moldable mass when wet and becomes hard and brittle when dry.

Bricks are a predominate product of the clay processing industry; other products include clay pipe, roof tiles, pottery, and other ceramic products.

3.8.1 Process Description[34,35]

Bricks and related clay products are manufactured by either the dry press, stiff mud, or soft mud method. In either method, the raw materials must be mined, crushed, and sized. Grinding (either

Table 9 Particulate Emissions from Glass Melting Furnaces after Control Devices[33]

Type of Product and Glass	Control Device[a]	Emissions, lb/ton
Container glass (soda–lime)	ESP	0.12
Container glass	VS	0.42 (av of 4 furnaces)
Pressed and blown (borosilicate)	ESP	1.0 (av of 4 furnaces)
Pressed and blown (fluoride/opal)	ESP	0.34
Pressed and blown (soda–lime)	FF	0.2
Pressed and blown (lead)	ESP	0.24 (av of 6 furnaces)
Flat glass	ESP	0.3 (estimated)

[a] ESP = electrostatic precipitator.
VS = venturi scrubber.
FF = fabric filter system.

wet or dry) is also frequently used in preparing fine clays. Ash, coke, sawdust, or sand are added to the clay at times to produce a specific product characteristic. As the processing name implies in the dry press process, very little, if any, water is added to the clay and it is pressed into the desired shape using pressures in the 500–1500 psi range. In the stiff mud process, clay is mixed in a pug mill and enough water is added to provide some plasticity. The mixture is then extruded through a die to form the desired shape. In the soft mud process, up to 20–30% water is added to the clay and the desired shapes are formed in molds.

Before firing, the formed clay shapes are almost completely dried by exposure to waste heat from the baking kilns. Many types of kilns are used for firing clay products; however, the tunnel and periodic kilns are the most popular. Kilns are usually gas or oil fired, but coal is occasionally used. Firing in a kiln causes six distinct changes in the clay, namely, evaporation of free water, dehydration, oxidation, vitrification, flashing, and cooling. In the tunnel kiln, the clay shapes are stacked on small carts on rails and moved slowly through the heated tunnel. The hottest zone is in the middle of the tunnel with temperatures of about 2000°F. A total firing time of 50–100 hr is required for 9-in. brick. For refractory and fire brick kiln temperatures up to about 3000°F are utilized. This type of kiln is a continuous operation with new clay shapes entering one end, and finished product leaving the opposite end. The periodic kiln is a batch process where the kiln is loaded, heated, cooled, and unloaded. Hot combustion gases circulate around the clay products through a series of flues and are finally vented to the atmosphere.

3.8.2 Emissions

Particulate matter in the form of fine dry clay and dirt particles from the materials handling and processing steps is the main emission problem in processing clay. Kiln emissions occur from fuel combustion and the ignition or volatization of impurities in the clay. These emissions naturally vary with the type of fuel, the nature of the impurities, and the operating temperature.

Particulate emissions from the material handling steps include both contained and fugitive losses, and vary widely depending on the clay's moisture content, the amount of processing, and the degree of process enclosure. Estimates of uncontrolled particulate from drying and grinding operations are in the range of 70 to 96 lb/ton of finished product.[36]

Uncontrolled tunnel kiln emissions are shown in Table 10. Emissions while firing coal are the highest since the combustion products are emitted along with the clay curing products. Emissions from any single plant could vary from these average emission factors depending on the combustion efficiency, and kiln design, and fluoride content of the clay. Particulate and hydrocarbon emissions may be much higher when sawdust or other combustible matter is mixed with the clay.

3.8.3 Emission Controls

Many brick and clay processing plants operate with little air pollution control equipment if the clay is wet and the kiln is gas or oil fired. For drier fine clays where handling and processing causes

Table 10   Average Uncontrolled
Emissions from Tunnel Kilns, lb/ton of
Product[37]

	Type of Fuel		
	Gas	Oil	Coal
Particulate	0.04	0.6	1.4^a
SO_2	Neg.	$4S^b$	$7.2S^b$
NO_x	0.15	1.1	0.9
CO	0.04	Neg.	1.9
HC	0.02	0.1	0.6
Fluorides	1.0	1.0	1.0

$^a A$ is % ash in coal.
$^b S$ is % sulfur in fuel.

emissions, water spray systems along with covered conveyers and enclosed screens are utilized to reduce particulate emissions.

Kiln emissions require control when particulate or fluoride emissions are excessive. Wet scrubbers can effectively reduce these emissions. Treatment of the waste water to precipitate fluorides is required at times.

REFERENCES

1. Based on information from *Technical Guidelines for Control of Industrial Process Fugitive Particulate Emissions,* U.S. Environmental Protection Agency, No. EPA-450/3-77-010, Section 2.6, March 1977.

2. Kothari, A., and Gerstle, R., *Air Pollution Control Techniques for Crushed and Broken Stone Industry,* U.S. Environmental Protection Agency, No. EPA-450/3-80-019, May 1980.

3. *Compilation of Air Pollutant Emission Factors,* 3rd edition, U.S. Environmental Protection Agency, Publication No. AP-42, August 1977, Section 8.20.

4. See Ref. 1, pp. 2-33 and 2-241.

5. Bennette, J. H., and Gordon, R. J., Assessment of Fugitive Emissions for Sand and Gravel Processing Operations, Paper 80-12.3, presented at the 73rd Air Pollution Control Association Annual Meeting, Montreal, Canada, June 1980.

6. Ref. 2, p. 3-2.

7. Ref. 3, Chapter 8.1.

8. *Air Pollution Engineering Manual,* U.S. Public Health Service, No. 999-AP-40, 1967, p. 328.

9. Kahn, I. S., and Hughes, T., *Source Assessment: Asphalt Hot Mix,* U.S. Environmental Protection Agency, No. EPA-600/2-77-107 n, December 1977.

10. *Federal Register* **38,** June 6, 1973, p. 15406.

11. Ref. 8, p. 334.

12. Ref. 3, Section 8.10.

13. Kulujian, N. K., *Inspection Manual for the Enforcement of New Source Performance Standards: Portland Cement Plant,* U.S. Environmental Protection Agency, January 1975.

14. Ref. 3, Section 8.6.

15. *Pit and Quarry Handbook,* Chicago, IL, 1975–1976, p. B-200.

16. Kreichelt, T. E., et al., *Atmospheric Emissions from the Manufacturer of Portland Cement,* U.S. DHEW, PHS Publication No. 994-AP-17, 1967.

17. Ref. 3, Section 8.6.

18. Simard, R., and Cote, D., Exhaust Air Recirculation: A Solution for Emission Control of Clinker Coolers, presented at the 73rd Air Pollution Control Association Annual Meeting, Montreal, Canada, June 1980.

19. *Federal Register* **39,** June 14, 1974, p. 20790; **39,** Nov. 12, 1974, p. 39872; **40,** Oct. 6, 1975, p. 46250.

20. Ref. 3, Section 8.15.

21. Based on information in Ref. 3, Section 8.15.

22. Minnick, J. C., Control of Particulate Emissions from Lime Plants—A Survey, *J. Air Pollution Control Association* **21**(4), 195–200 (1971).

23. Krohn, D. J., U.S. Lime Division's Dust Abatement Efforts Whip Pollution Problem, *Pit and Quarry* **66**, 87–92 (May 1974).

24. *Background Information for Standards of Performance: Coal Preparation Plants, Volume 1: Proposed Standards,* U.S. Environmental Protection Agency, Publication No. EPA-450/2-84-021a, October 1974.

25. *Background Information for Standards of Performance: Coal Preparation Plants, Volume 2: Test Data Summary,* Publication No. EPA-450/2-74-021b.

26. Ref. 3, Section 7.9.

27. Ref. 24, p. 12.

28. Ref. 24, p. 10.

29. *Federal Register* **41**, January 15, 1976, p. 2231.

30. Schorr, J. R., et al., *Source Assessment: Glass Container Manufacturing Plants, Battelle–Columbus Laboratories,* U.S. Environmental Protection Agency, Publication No. EPA-600/2-70-209.

31. Stockman, J. D., The Composition of Glass Furnace Emissions, *J. Air Pollution Control Association* **21**, 713–715 (Nov. 1971).

32. Ref. 3, Section 8.13.

33. *Glass Manufacturing Plants, Background Information: Proposed Standards of Performance,* U.S. Environmental Protection Agency, Publication No. EPA-450/3-79-005a, 1979.

34. Ref. 3, Section 8.3.

35. Hardison, L. C., *Air Pollution Control Technology and Costs in Seven Selected Industries,* Industrial Gas Cleaning Institute, Stamford, Conn., Office of Air and Water Programs, Contract 68-02-0289, Rept. EPA-450/3-73-010, IGCI Rept. 47-173, Dec. 1973.

36. Ref. 3, Sections 8.3 and 8.7.

37. Ref. 3, Section 8.3.

CHAPTER 23

SOURCE CONTROL—
FERTILIZER

LEE BECK

U.S. Environmental Protection Agency
Research Triangle Park, North Carolina

1 INTRODUCTION

Fertilizers are divided into two major categories: organic and inorganic. Organic fertilizers consist of various manures and composts which, though sometimes packaged and sold commercially, are not highly processed and consequently do not usually present a significant potential for air pollution. On the other hand, inorganic or chemical fertilizers are typically manufactured using basic raw chemicals that are combined, granulated, and dried in process equipment that requires large volumes of air. The exhausts from these processes contain chemical gases and dusts which can threaten air quality if not collected. This chapter deals exclusively with the manufacture of inorganic fertilizers.

Plants need three basic nutrients for survival: nitrogen, phosphorus, and potassium. These nutrients are normally applied in a mixture specifically suited for the crop, and are always applied as a chemical combination of ingredients rather than in elemental form. There are several popular manufacturing processes yielding different chemical compounds for each of the three nutrients. For example, nitrogen can be manufactured as ammonium nitrate, ammonium sulfate, urea, monoammonium phosphate, diammonium phosphate, or a number of other compounds. Because of the large number of processes used in the manufacture of inorganic fertilizers, this chapter presents one example of a typical process used to manufacture a fertilizer in each of the three major groups. The emphasis is on methods for controlling emissions of air pollutants. If the reader requires specific information on a process not used as an example, he is given adequate information in Section 4 to direct him to a source that can satisfy his requirements.

2 PROCESSES, EMISSIONS, AND CONTROL TECHNIQUES

2.1 Urea

Urea is an example of a popular nitrogenous fertilizer. It is the most widely used nitrogenous fertilizer, supplying almost 1 million Mg of nitrogen in 1979.[1]

2.1.1 Process Overview[2]

The process for manufacturing urea involves a combination of up to seven major unit operations. The basic arrangement of these operations is shown in the block diagram given in Figure 1. These major operations are:

1. Solution synthesis (solution formation)
2. Solution concentration
3. Solids formation
     Prilling
     Granulation
4. Solids cooling
5. Solids screening

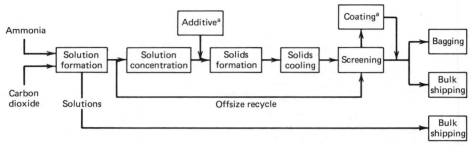

Fig. 1 Urea manufacturing.[a] These processes are optional depending on individual manufacturing practices.

6. Solids coating
7. Bagging and/or bulk shipping

The combinations of processing steps are determined by the desired end products. Plants producing urea solutions alone are comprised of only the first and seventh unit operations, solution formation, and bulk shipping. Facilities producing solid urea employ these two operations and various combinations of the remaining five operations, depending on the specific end product being produced.[3]

2.1.2 Types of Urea Plants[4]

All urea plants produce an aqueous urea solution as depicted in the process flow diagram shown in Figure 1. In these plants, ammonia and carbon dioxide are reacted to form ammonium carbamate. The carbamate is then dehydrated to yield a 70–77% aqueous urea solution. The solution can be sold as an ingredient in nitrogen solution fertilizers or can be further concentrated to produce solid urea. There are three methods of concentrating urea solution: vacuum evaporation, atmospheric evaporation, and crystallization. Vacuum and atmospheric evaporation produce a urea melt containing from 99–99.9% urea at a nominal temperature of 140.5°C (285°F). Crystallization is used primarily when product requirements dictate an extremely low biuret concentration in the final product.

Urea solids are produced from the urea melt by two basic methods: prilling and granulation. In prilling there are two types of prill towers: fluidized bed and nonfluidized bed. Each of these is capable of producing both agricultural grade and feed grade urea prills. The major difference between these towers is that a separate solids cooling operation may be required when producing agricultural grade prills in a nonfluidized bed prill tower. Because the smaller feed grade prills exhibit better heat transfer properties, additional cooling external to the nonfluidized bed tower is not required when producing feed grade urea. The fluidized bed supplies the required cooling in a fluidized bed prill tower.

The other methods of solids formation used in the urea industry are drum and pan granulation. In drum granulation, solids are built up in layers on seed granules in a rotating drum granulator/cooler approximately 4.3 m (14 ft) in diameter. Pan granulators also form the product in a layering process, but the equipment used is different from the drum granulator. There is only one pilot scale pan granulator operating in the domestic industry, providing 61,000 Mg/yr (67,000 tons/yr) of urea granules.

2.1.3 Emissions and Emission Control Techniques[5]

Emissions from urea processes include particulate matter, ammonia, and formaldehyde. Table 1 presents uncontrolled emission factors for each of the major processes in the urea industry.

Ammonia is emitted during the urea synthesis (solution production) and solids production processes. Ammonia emissions range from 14.40 kg/Mg of urea produced (28.80 lb/ton) for synthesis processes to 0.0255 kg/Mg (0.051 lb/ton) for a rotary drum prill cooler. Ammonia in these low concentrations is usually not considered detrimental to air quality.

Formaldehyde has been added to the urea melt in recent years for the purpose of reducing urea dust emissions and to prevent solid urea melt in concentrations of 0.5% or less prior to solids formation. The use of formaldehyde as an additive has resulted in formaldehyde emissions which range from 0.0095 kg/Mg (0.0190 lb/ton) of urea produced for a fluidized bed prill tower producing agricultural grade urea, to 0.0020 kg/Mg (0.0040 lb/ton) of urea produced for a fluidized bed prill tower producing feed grade urea solids.

Particulate matter is the primary emission being addressed in this section. Table 1 includes a summary of uncontrolled particulate emissions from all urea processes. These particulate emissions

Table 1 Uncontrolled Emissions from Urea Facilities

Process	Plant	Particulate		Ammonia		Formaldehyde	
		kg/Mg	(lb/ton)	kg/Mg	(lb/ton)	kg/Mg	(lb/ton)
Solution formation and concentration	A	0.00241	(0.00482)	12.89	(25.77)	—	
Solution formation and concentration	B	0.0150	(0.0317)	4.01	(8.02)	—	
Solution formation and concentration	D	0.0052	(0.0104)	14.40	(28.80)	—	
Drum granulation	A	148.8	(297.6)	1.08	(2.15)	0.00359	(0.0072)
Drum granulation	B	63.6	(127.2)	1.07	(2.13)	0.00555	(0.0111)
Nonfluidized bed prill tower (agricultural grade)	E	1.90	(3.80)	0.433	(0.865)	—	
Fluidized bed prill tower (feed grade)	D	1.80	(3.60)	2.07	(4.14)	0.0020	(0.0040)
Fluidized bed prill tower (agricultural grade)	D	3.12	(6.23)	1.42	(2.91)	0.0095	(0.0190)
Rotary drum cooler	C	3.72	(7.45)	0.0255	(0.051)	—	

range from 148.8 kg/Mg (297.6 lb/ton) of urea produced for a rotary drum granulator to 0.00241 kg/Mg (0.00482 lb/ton) of urea produced for a synthesis process.

With the exception of bagging operations, urea emission sources are typically controlled with wet scrubbers. The preference toward scrubber systems as opposed to dry collection systems is primarily due to the ease of recycling dissolved urea collected in the device. Scrubber liquors are recycled back to the solution concentration process, eliminating potential waste disposal problems and recovering the urea collected.

Fabric filters (baghouses) are not suitable for controlling emissions from many sources because the hygroscopic nature of urea particulate combined with the moisture content of the gas streams could cause blinding of the bags. Dry cyclones offer lower collection efficiencies than scrubbers in urea particulate applications. Electrostatic precipitators are not currently in use in any urea industry applications.

Fabric filters are used in the control of fugitive dust generated in bagging operations where humidities are lower and blinding is not a problem. Many bagging operations are uncontrolled. However, if a control device is used, baghouses are the typical method of control.

Emissions from prill towers are typically controlled using scrubbers of the spray tower, packed tower, entrainment, or fibrous filter design.

The spray tower currently used in urea prill tower applications is designed to operate at 0.25–0.5 kPa (1- to 2-in. W.G.) pressure drop with liquid-to-gas ratios of 0.134–0.268 l/m³ (1–2 gal/1000 ft³). Spray nozzle pressure is 689–1380 kPa (100–200 psig). Although efficiency curves for various particle sizes are not available, the manufacturer claims exit loadings of 0.0115–0.0344 g/m³ (0.005–0.015 gr/dscf) are achievable in urea prill tower applications.[6] The manufacturer has also reported that opacities of 20% or less are achievable.[7]

Two nonfluidized bed prill towers were tested by the United States Environmental Protection Agency (EPA) and reported in a technical document describing emission control techniques for the urea manufacturing industry.[8] One of the plants used a fibrous filter scrubber which achieved an average particulate control efficiency of 98.3%. Particulate emissions at the inlet to the scrubber averaged 1.88 kg/Mg of urea produced (3.76 lb/ton of urea produced), and outlet emissions averaged 0.027 kg/Mg (0.054 lb/ton). The second plant tested used a packed tower to control particulate emissions. Inlet emissions were not measured at this plant, but outlet emissions averaged 0.188 kg/Mg (0.376 lb/ton).

Two fluidized bed prill towers were also tested by EPA during the study. Both used entrainment scrubbers to control emissions, and both demonstrated particulate emission control efficiencies of about 87%. Inlet emissions for the one plant were 3.12 kg/Mg (6.24 lb/ton). The other plant averaged inlet emissions of 1.80 kg/Mg (3.59 lb/ton). Outlet emissions for the two plants were 0.392 kg/Mg (0.785 lb/ton) and 0.240 kg/Mg (0.479 lb/ton), respectively.

EPA also tested two drum granulators which controlled particulate emissions using entrainment scrubbers. The control efficiencies for these two plants were 99.9 and 99.8%, and outlet particulate emissions for the two plants averaged 0.115 kg/Mg (0.230 lb/ton) and 0.122 kg/Mg (0.244 lb/ton).

Two additional sources of particulate emissions from urea manufacturing operations are cooling and bagging. Controlled mass emission data submitted by several plants operating coolers indicate controlled emissions that range from 0.01 to 0.1 kg/Mg (0.02 to 0.2 lb/ton) with an average emission of 0.035 kg/Mg (0.07 lb/ton). These emission levels are generally confirmed by predicted controlled emissions using uncontrolled EPA emission data and control device performance curves. A single-tray-type scrubber operating at a pressure drop of 0.375 kPa (1.5-in. W.G.) would remove approximately 98.9% of the particles in the cooler exhaust. An entrainment scrubber operating at an overall 1.25–1.50 kPa (5- to 6-in. W.G.) pressure drop would perform at approximately 99.6% efficiency. These high efficiencies are possible because of the large particles in cooler exhausts. Using these efficiencies and EPA uncontrolled mass emission measurements, controlled emissions of 0.043 and 0.016 kg/Mg (0.086 and 0.031 lb/ton) are estimated for coolers controlled with tray-type and entrainment scrubbers, respectively.

Mass emission test data are not available for fabric filters controlling emissions from a urea bagging operation. However, regardless of the type, baghouses can attain collection efficiencies greater than 99% even on submicron particle sizes.[9] Testing conducted by EPA on baghouses used to control emissions in the nonmetallic mineral industry demonstrated efficiencies of 99.8% or better with no visible emissions (0% opacity).[10]

2.1.4   Costs[11]

A 726 Mg/day (800 ton/day) urea plant with a prill tower and a cooler would expect to incur capital costs of $1,695,000 (first-quarter 1980 dollars) for the particulate control devices. This cost assumes a fibrous filter scrubber for the prill tower exhaust and a tray-type scrubber for the cooler exhaust. Both control devices would operate at a particulate control efficiency of 98%. The total annualized cost for the plant would be $1,079,100 (first-quarter 1980 dollars). These annualized costs represent the yearly cost of operating and maintaining the control systems, and includes utilities, operating labor, maintenance, taxes, insurance, administrative overhead, and other ongoing costs.

The capital cost of an entrainment scrubber operating with a 99.8% control efficiency for a 726 Mg/day granulation plant would be $1,112,000 (first-quarter 1980 dollars). The annualized cost in this case would be a yearly credit of $1,314,000 (first-quarter 1980 dollars). The reason for the credit is the value of the recovered particulate (product) which would otherwise be lost.

2.2   Granular Triple Superphosphate

Granular triple superphosphate, commonly called GTSP, is manufactured by reacting phosphate rock with phosphoric acid. It is one of the most common phosphate fertilizers, with others being normal superphosphate (NSP), run-of-pile triple superphosphate (ROP-TSP), superphosphoric acid (SPA), mono-ammonium phosphate (MAP), and diammonium phosphate (DAP).

DAP is the most popular phosphate fertilizer with consumption totaling about 1.5 million Mg (almost 1.7 million tons) in 1979. GTSP was second in popularity, with about 500,000 Mg consumed in 1979.[12] GTSP is used as an example of a phosphate fertilizer in this chapter since DAP is technically a mixed fertilizer, containing nitrogen and phosphorus, whereas GTSP contains only the phosphate component.

2.2.1   Process Description[13]

The direct-slurry process for production of granular triple superphosphate begins with the reaction of phosphate rock and phosphoric acid. The reaction proceeds as indicated:

$$Ca_3(PO_4)_2 + 4H_3PO_4 + 3H_2O \longrightarrow 3[CaH_4(PO_4)_2 \cdot H_2O]$$

A schematic diagram of the process is shown in Figure 2. The reactor slurry is pumped to the granulator (similar to the drum granulator described in Section 2.1.2) where it is mixed with undersize material from the product screens. From the granulator, the granules flow to a rotary dryer. After drying, the product is cooled, screened, and conveyed to the storage building.

2.2.2   Emissions and Emission Control Techniques

The primary pollutant emitted from GTSP plants is fluorides. The fluorides are present in both gaseous and particulate form; consequently, scrubbers are usually chosen to control emissions. As shown in Figure 2, the scrubbing medium is generally not clean water. Emissions from the production train (granulator, dryer, cooler, mills, and screens) are generally precleaned using venturi scrubbers which utilize either gypsum pond water or phosphoric acid as the scrubbing liquor. The use of phosphoric acid allows recovery of the particulate, since the spent scrubbing liquor can be fed to the reactor. Phosphoric acid contains fluoride, however, which makes it a source of fluoride emissions when it is used as a scrubber liquor. This necessitates the use of a second scrubber placed in series with the first. This final cleaning device is typically a packed scrubber which uses water from the gypsum pond to control gaseous emissions from the production train and, if phosphoric acid scrubbing is

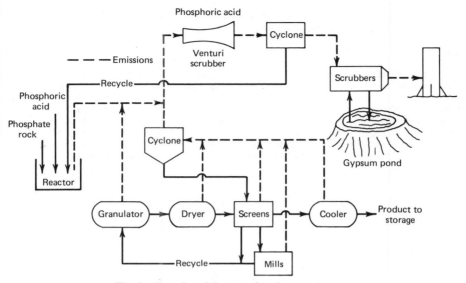

Fig. 2 Granular triple superphosphate production.

used, emissions from the acid scrubbing liquor. The gypsum pond is a wastewater cooling and settling pond present at practically every phosphate fertilizer complex because of the large volumes of water needed in the manufacture of wet-process phosphoric acid, the starting material for almost all phosphate fertilizers.

Fluoride emission regulations applicable to phosphate fertilizer processes usually relate the amount of fluoride emissions to the amount of P_2O_5 fed to the process. The reason for this is that phosphate fertilizer processes do not typically measure their production rate accurately. They do, however, have to closely monitor the input of raw materials to assure product quality. The state of Florida, where most of the nation's phosphate fertilizer is produced, and the EPA have chosen units of "grams of fluoride per megagram of P_2O_5 input to the process" for emission regulations applicable to phosphate fertilizer processes.

EPA conducted three emission tests at two GTSP plants during their program for development of emission regulations for this source. One plant was tested twice on different dates separated by about eight months. The tests yielded average results of 90 and 30 g of fluoride per Mg of P_2O_5 input to the process (g/Mg P_2O_5), respectively. The operator of the plant sampled emissions on the same day as the first EPA test and reported 60 g/Mg P_2O_5. Tests by EPA at another plant averaged fluoride emissions of 105 g/Mg P_2O_5.

The performance of the emission control systems at GTSP plants cannot be quantified accurately because data on emissions at the inlet to the control devices were not measured during the EPA tests. However, an EPA document estimates uncontrolled fluoride emissions from GTSP plants to be approximately 10,500 g fluoride/Mg P_2O_5.[14] If the plants tested by EPA during the above study had comparable amounts of uncontrolled fluorides, the overall fluoride control efficiency of the control systems would range from 99.0 to 99.7%.

Triple superphosphate (both GTSP and ROP-TSP) is unique with respect to other fertilizer manufacturing operations in that gaseous fluorides continue to evolve from the product for three to five days following production. Consequently, the storage building is also a source of fluoride emissions.

Since storage emissions are mostly gaseous, packed scrubbers using gypsum pond water are typically used for control. EPA sampled emissions from two such installations.[15] Tests conducted at the first storage building revealed average fluoride emissions of 236 g/hr. The storage building contained about 1297 Mg P_2O_5 (21% full) during the tests. Related to the amount of GTSP in storage, the emissions were 0.18 g/hr per Mg of P_2O_5 stored. Two emission tests were conducted by EPA at the second storage building. The tests were separated by about three months. The first time the building averaged 4121 Mg P_2O_5 in storage (about 30% full), and the fluoride emissions averaged 121 g/hr, or 0.02 g/hr per Mg P_2O_5. The second test averaged 1906 Mg P_2O_5 in storage (about 15% full). Fluoride emissions during this test averaged 435 g/hr, or 0.25 g/hr per Mg P_2O_5 stored.

Estimates of the control device efficiency during the EPA tests are not available. Emissions at the inlet to the scrubbers were not measured. However, fluoride control efficiencies for the types of packed scrubbers used by the two installations typically exceed 95%.[16]

2.2.3 Costs

EPA estimated costs associated with controlling fluoride emissions from a new GTSP plant.[17] The costs are based on a 363 Mg P_2O_5/day plant which uses separate venturi scrubbers to control emission streams from the reactor–granulator, the dryer, and the cooler. Emissions from the reactor–granulator scrubber are ducted to a cyclone mist eliminator for final cleanup. Emissions from the other two scrubbers are ducted to separate spray–cross-flow packed scrubbers. Emissions from the storage building are also ducted to a packed scrubber, and gypsum pond water is used as the scrubbing liquor for each control device. The EPA costs were given in terms of 1973 dollars. These costs, updated to January 1981 by using the Chemical Engineering Plant Cost Index,[18] are as follows:

	Capital Costs ($ × 1000)	Annualized Costs ($ × 1000)
Reactor–granulator	371.2	170.1
Dryer	480.8	305.6
Cooler	504.2	311.5
Storage	329.0	129.6
Total	1685.0	916.8

The annualized costs include operating labor, maintenance, utilities, depreciation, interest, taxes and insurance, and administrative costs.

In a separate study, EPA estimated retrofit control costs for GTSP manufacturing plants.[19] The plant parameters were identical to those presented above except that the reactor–granulator venturi scrubber was assumed to be followed by a spray–crossflow packed scrubber instead of a cyclonic mist eliminator. Unlike the new plant costs presented above, the retrofit costs assume that the venturi scrubbers are actually part of the process since they recover usable product, so costs are presented for four packed scrubbers only (three for the production train and one for storage emissions). The 1974 costs updated to January, 1981 total $911,500 for capital costs and $425,100 for annualized costs.

2.3 Muriate of Potash[20]

The term "potash" is generally used to describe chemical compounds which contain the element potassium. In the fertilizer industry, the term is used in reference to potassium oxide (K_2O). Although potassium oxide is not actually produced, all potassium values are by convention expressed in equivalent weight of potassium oxide. Major potash products include muriate of potash or potassium chloride (KCl); sulfate of potash or potassium sulfate (K_2SO_4); and sulfate of potash magnesia, which is termed langbeinite or potassium magnesium sulfate ($K_2SO_4 \cdot 2MgSO_4$). The muriate is by far the most popular product for fertilizer applications, accounting for around 80% of U.S. potash production. Around 95% of U.S. potash consumption is for agricultural purposes; the remaining 5% is used in a wide range of chemical manufacturing applications, including ceramics, explosives, glass, medicines, and textiles.

Because of its popularity, muriate production will be used as an example of a potassium-based fertilizer.

2.3.1 Process Description[21]

Muriate of potash production involves separating potassium chloride from sodium chloride and clays. There are two parallel operations in muriate production. The fine and dissolved potash is recovered in a crystallization operation, while the large particles of potash are recovered by a flotation operation.

The operations involved in the production of muriate of potash from sylvinite (crude KCl) ore are illustrated in Figure 3. The sylvinite ore is crushed and slurried with saturated brine solution, then agitated in scrubbers. After scrubbing, the clay and the fine and dissolved potash are separated with a cyclone from the coarse ore which contains the large particles of potash. The fine and dissolved potash in the cyclone overflow is recovered through a four-step recrystallization process. First, a better separation of large and fine particles is made in a hydroseparator where the fine potash is dissolved

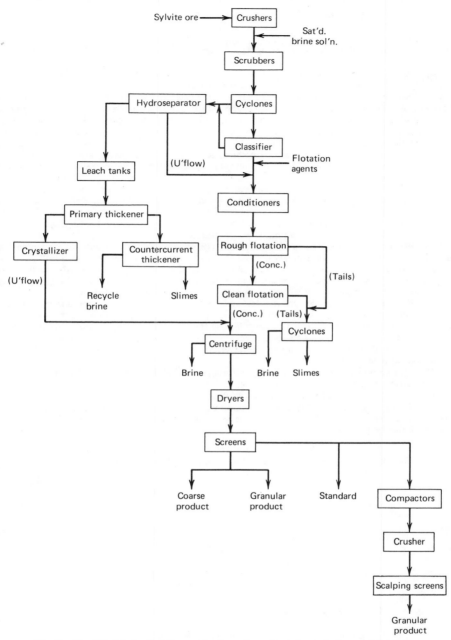

Fig. 3 Simplified flow sheet for production of muriate of potash from sylvinite ore.

from accompanying clays in heated leach tanks. The clays are then separated from the dissolved potash in thickeners. Finally, the dissolved potash is recrystallized in crystallizers. In a parallel processing operation, the larger potash particles are recovered from the coarse ore by flotation. First the coarse ore is slurried with recycled brine. Then further separation of fine clays from the coarse ore is made in a classifier. Flotation agents are added to the coarse ore and the potassium chloride particles are floated from the remaining clays and sodium chloride. This flotation product and the recrystallized product are combined, centrifuged, dried, and screened into the desired product sizes (granular, coarse, standard, soluble, and fine standard), and the undersized fines are compacted, recrushed, and screened to granular potash.

2.3.2 Emissions and Emission Control Techniques[22]

Atmospheric emissions result from the following muriate processing operations: crushing, screening, conveying, drying, compacting, product storage, and loading.

Emissions from crushing, screening, and conveying are caused by entrainment of fine dust particles in exhaust air during mechanical processing of the potash. Emissions from crushing, generally done in hammermills, are the result of impact and attrition. Screening causes fines to become airborne as the potash product is mechanically vibrated to separate ore or product sizes. Emissions from conveying result from the entrainment of fines as the product moves along the conveyor, and at transfer points where material is discharged or received. Crushing, screening, and conveying are minor sources of emissions, and are typically uncontrolled or controlled by using low-efficiency cyclones. No emission data are available for these sources.

Dryers are the major source of atmospheric emissions. Rotary dryers are commonly used and fines are entrained as the muriate tumbles through the rotating dryer. Also, the exit gas velocity for direct-fired rotary dryers is generally high, 0.5 m/sec (100 ft/min), which results in additional entrainment of fine particles. Particulate control equipment used on dryers in the United States includes dry cyclones, wet scrubbers, and to a limited extent, baghouses.

A comparison of the effectiveness of dry cyclones, venturi scrubbers, and baghouses for controlling particulate emissions from potash dryers is illustrated in Table 2 below. An "uncontrolled" emission rate is presented in this comparison, but it should be noted that the value of entrained potash makes some recovery necessary for economical dryer operation. Therefore, a dry cyclone probably represents a minimum level of control which would be necessary in any competitive installation.

Note that the dryer controlled by the dry cyclone (estimated at 94% control efficiency) is less than twice the size of the dryer controlled by the venturi scrubber, yet its emission rate is more than an order of magnitude larger. The dryer controlled by the venturi scrubber in turn is less than twice the size of the dryer controlled by the baghouse, but its emissions are again greater than an order of magnitude larger than from the baghouse. However, as discussed earlier, baghouses are hardly ever used to control potash dryers.

There are four major steps in compacting, crushing, screening, and conveying. The emissions from crushing, screening, and conveying operations were discussed earlier in this section.

The compacting step itself is a minor source of particulate emissions. When the fine muriate is compressed into slates some particles become airborne, but most of the particles are larger than 1 μm and can be captured relatively easily. Table 3 is a summary of data on equipment used for controlling emissions from compactors.

Table 2 Controlled Emissions from Three Rotary Gas-Fired Potash Dryers

Dryer Capacity		Controlled Emissions		Uncontrolled Emissions	
TPY Product	Control Equipment	Short tons/yr	Metric tons/yr	Short tons/yr	Metric tons/yr
350,000	Dry cyclone	500	450	8000	7000
200,000	Venturi scrubber	15	14		
120,000	Baghouse	0.2	0.2		

Table 3 U.S. Potash Plant Compacting Area Control Equipment Summary

Control Equipment	Controlled Emissions,[a] metric tons/yr	Efficiencies,[a] %
Baghouse	7 (E)	
Dry cyclone and venturi scrubber	13 (M)	
Joy venturi scrubber	30 (M)	99+ (E)
Krebs Elbair scrubber		99− (M)
Dry cyclone	790 (M)	95 (M)
Dry cyclone	660 (M)	98 (E)
Dry cyclone	135	91 (E)

[a] E—estimated, M—measured.

Information presented in the above table was obtained from inspection reports and permit applications from the following sources:

South Coast Air Quality Management District, Eastern Zone, San Bernardino, California.

Environmental Improvement Agency, Air Quality Division, Santa Fe, New Mexico.

U.S. Environmental Protection Agency, Region VII, Denver, Colorado.

Muriate of potash is stored in enclosed storage buildings which can be a source of fugitive particle emissions at some plants.

Emissions from bulk loading or bagging are caused by displaced air which entrains potash particles. One muriate plant reports 3.0 g/sec (24 lb/hr) on an annual basis of particulate emissions for bagging operations.

2.3.3 Costs

As noted earlier, dryers are the major source of emissions from muriate plants. Crushing, compacting, and bagging are typically controlled by cyclones, if controlled at all. Only dryer emission control costs are presented in this section.

For cost purposes, a direct-fired rotary dryer with an annual (7446 operating hours) production of 301,000 Mg is assumed. The dryer exhaust is controlled by using a cyclone followed by a venturi scrubber. The total control efficiency for the system is 99.88%, with controlled emissions of 0.09 g/Nm^3 and 0.06 g/kg feed. For such a system, the installed equipment cost for the control system, including the stack, is about $734,000. The total direct operating cost of the system (utilities, labor, maintenance, etc.) would be about $325,000 and the total annualized costs would be about $516,000. All costs are given in January 1981 dollars, updated from 1979 calculated costs.

3 SUMMARY

Chemical fertilizers are produced by manufacturing chemical compounds which contain one or more of the primary plant nutrients: nitrogen, phosphorus, and potassium. The processes for producing compounds containing each of the primary nutrients are many and varied. Examples of commonly used fertilizers are urea (containing nitrogen), granular triple superphosphate (containing phosphorus), and muriate of potash (containing potassium). Particulate matter is the most predominant air pollutant from processes manufacturing these compounds. The major sources of the particulates are crushing and classification of raw materials, granulation, drying, and cooling of the fertilizer compounds, which are usually produced by combining materials in a reactor, and bagging of the products. Some of the processes also emit gaseous pollutants such as ammonia, fluorides, and formaldehyde, but these pollutants are typically controlled along with the particulate matter if wet collection devices (scrubbers) are used. In addition to scrubbers, fabric filters (baghouses) are commonly used to control emissions from fertilizer manufacturing processes. Emission control costs are frequently partially offset by collection of material which can be added back to the process to increase product yields.

4 SOURCES FOR ADDITIONAL INFORMATION

The publications referenced in this chapter contain more detail on each of the processes described. Many of the publications also contain information pertinent to fertilizer manufacturing operations not discussed in this chapter. A partial listing of additional publications which may provide useful information is as follows:

Sittig, M., Fertilizer Industry—Processes, Pollution Control and Energy Conservation, Park Ridge, NJ, Noyes Data Corporation, 1979, 204 pp.

Ammonium Nitrate Manufacturing Industry—Technical Document, U.S. Environmental Protection Agency, Research Triangle Park, NC, Publication No. EPA-450/3-81-002, January 1981, 302 pp.

Augenstein, D. M., Air Pollutant Control Techniques for Phosphate Rock Processing Industry, U.S. Environmental Protection Agency, Research Triangle Park, NC, Publication No. EPA-450/3-78-030, June 1978, 224 pp.

Sauchelli, V., Chemistry and Technology of Fertilizers, New York, Reinhold Publishing Corporation, 1960.

Published information is very useful to the engineer searching for information. The data contained in the publications have generally been validated and the source of information can be easily referenced. Also, the information is generally not subject to misinterpretation, a frequent problem with verbally transmitted information. However, published information has its disadvantages. Technical publications are frequently not readily available at public or university libraries and can take months to obtain.

Also, with the fast pace of today's technology, a publication only a few years old can be out of date. Consequently, telephone contact with experts in the field can provide quick, up-to-date information specific to the concerns of the engineer. Large fertilizer manufacturing companies such as W. R. Grace & Company, Agrico Chemical Company, Mobil Chemical Company, Swift Chemical Company, and J. R. Simplot Corporation usually have technical expertise which is current, and a contact with the right person can provide quick, valuable information. The fertilizer industry is very competitive though, and frequently an engineer employed by a large fertilizer company is unable to discuss technical matters over the telephone because of company or departmental policies which discourage free exchange of information. This potential problem can frequently be overcome by sending a letter to the individual which states the reason the information is needed, the intended use of the information, and very specifically, what information is needed.

Trade organizations are another valuable source of information, especially for industrywide statistics on growth, production, and so on. They can also frequently refer the engineer to other sources of information. The largest trade organization for fertilizer manufacturing companies is The Fertilizer Institute, 1015 18th Street, N.W., Washington, D.C. 20036, and an example of a trade association of phosphate fertilizer manufacturers is Florida Phosphate Council, P.O. Box 1626, Lakeland, Florida 33802.

Governmental agencies probably provide the most readily available source of information. As public servants, government employees are required to provide information to whomever requests it, except for information which a company claims is proprietary. The government employee may also have a broader knowledge than the industry employee regarding industry statistics or the applicability of pollution control equipment, since the company employee may not be familiar with information not pertinent to his company's operations. The government employee can frequently also suggest experts in the field and can sometimes provide insight regarding the availability of information from industry sources. A partial list of federal agencies which have information regarding fertilizer manufacturing operations is as follows:

Tennessee Valley Authority
National Fertilizer Development Center
Muscle Shoals, AL 35660

United States Environmental Protection Agency
Office of Air Quality Planning and Standards
Research Triangle Park, NC 27711

Crop Reporting Board
United States Department of Agriculture
Washington, D.C.

Bureau of Mines
United States Department of the Interior
Washington, D.C.

There are also many state and local pollution control agencies which have information on control of pollution from fertilizer manufacturing operations. It is wise to first determine where the industry is predominantly located, then contact the state and/or local pollution control agencies in that area. For instance, phosphate ore reserves, and consequently phosphate fertilizer manufacturing companies, are predominantly located in Florida, Idaho, North Carolina, and Tennessee. These states are likely to have more extensive information than other states in which there are no phosphate reserves.

REFERENCES

1. Bridges, J. D., *Fertilizer Trends 1979*, Tennessee Valley Authority, Muscle Shoals, Alabama, Bulletin 4-150, January 1980, p. 12.
2. *Urea Manufacturing Industry—Technical Document*, U.S. Environmental Protection Agency, Research Triangle Park, NC, Publication No. EPA-450/3-81-001, January 1981, p. 3-2.
3. Ibid., p. 3-3.
4. Ibid., pp. 3-2 through 3-4.
5. Ibid., p 3-4.
6. Ibid., p. 4-9.
7. Ibid.
8. Ibid., pp. A-1 through 8-81.
9. Ibid., p. 4-43.

10. Ibid.

11. Ibid., pp. 7-1 through 7-24.

12. Bridges, J. D., op. cit., p. 21.

13. *Background Information for Standards of Performance,* U.S. Environmental Protection Agency, Research Triangle Park, NC, Publication No. EPA-450/2-74-019a, October 1974, 148 pp.

14. *Final Guideline Document: Control of Fluoride Emissions from Existing Phosphate Fertilizer Plants,* U.S. Environmental Protection Agency, Research Triangle Park, NC, Publication No. EPA-450/2-77-005, March 1977, p. 5-11.

15. *Background Information for Standards of Performance,* op. cit.

16. Ibid.

17. Ibid.

18. "Economic Indicators," *Chemical Engineering* **88**(7) (April 20, 1981).

19. *Final Guideline Document: Control of Fluoride Emissions from Existing Phosphate Fertilizer Plants,* op. cit., p. 6-65.

20. Blythe, G. M., and Trede, K. N., *Screening Study to Determine Need for Standards of Performance for the Potash Industry,* Radian Corporation, Austin, TX, Publication No. DCN 78-200-187-35-08, October 1978, 110 pp.

21. Ibid.

22. Ibid.

23. *Preliminary Costs—Chapter 8 for the Potash Industry,* Radian Corporation, Durham, NC, Radian Contract No. 230-372-04, July 1979, pp. 8-1 through 8-17.

CHAPTER 24

SOURCE CONTROL—
CHEMICAL

T. R. BLACKWOOD
B. B. CROCKER

Monsanto Company
St. Louis, Missouri

1 SCOPE

1.1 Definition of the Chemical Industry

The chemical industry can be loosely defined as a process industry in which crude raw materials of a mineral or petroleum nature are connected by way of chemical intermediates into semifinished or final products. For this chapter, consideration of the chemical industry has been restricted to the manufacture of solvents, acids, bases, and chemical intermediates. Note that the manufacture of agricultural nutrients has been covered in Chapter 23 and petrochemicals is covered in Chapter 25. To illustrate the complexity of the industry, over 500 industrial organic chemicals are derived from nearly 400 processes using one of the following 10 feedstocks: benzene, butylene, cresol, ethylene, methane, naphthalene, paraffin, propylene, toluene, and xylene. These organic intermediates along with the nearly 100 inorganic acids, bases, and salts made in the chemical industry can eventually be found in over 70,000 substances such as synthetic fibers, plastics, synthetic rubber, pesticides, dyes, pigments, food products, and pharmaceuticals. The chemical and allied products industry has been estimated to manufacture between 500,000 and 600,000 synthetic compounds, although some of these are by-products. Often, a chemical by-product can be turned from a waste material to productive use by processing changes that alter its physical or chemical properties. Because of its diversity, the chemical industry is one of the most complex for emission control.

Typical processes have been selected to show the major types of problems encountered with respect to APC regulation. Process alternatives for pollution control are discussed rather than attempting definitive coverage. Often control requirements are as unique as the process itself, calling for individual engineering in process design. As an example, acid mists at a phosphoric acid plant had previously been removed with a venturi scrubber. Although 98.4% removal was achieved, a visible plume persisted and emission standards were not met. The addition of a mesh collector to convert fine acid mist to larger liquid particles increased efficiency to 99.9% and eliminated the plume. This method of improving efficiency is unique to phosphoric acid because of the physical and chemical interactions of water and phosphoric acid.[1] Many processes have no continuous atmospheric discharge and air pollution is limited to fugitive emissions (from leaks or spills) and to cleaning and maintenance operations. Many modern day air pollution control methods are adaptations of chemical process technology either in the equipment design or the chemistry of separation. Frequently, pollution problems form incentives to change processes, eliminating the source or converting it to a gainful product. These can be unique to a particular plant. At a midwestern chemical plant, high-purity magnesium nitrate and manganese nitrate are produced by dissolving electrolytic metal in nitric acid. To prevent rapid pH change of the solution, an excess of nitric acid is used. Too rapid acid addition or poor temperature control can release nitrogen oxides in a highly visible cloud. While control with absorption equipment could reduce the emissions, close control of the temperature and acid addition rate is more economical and effective. It is, however, difficult to determine how often this type of control is applied, since company trade secrets may be at risk if an application is divulged.

1.2 Major Sources Classed by Production

The quantity of a chemical that is produced cannot be equated to pollution impact; however, many regulatory agencies look at an industry from its potential to pollute. Table 1 is a summary of the major chemicals produced in the United States in 1978 and 1979.[2] Sulfuric acid is by far the largest commodity chemical produced in the United States, but a large portion of the production is as a by-product of nonferrous smelting and oil refining. Whenever a high concentration of sulfur can be found in a process gas, sulfuric acid can be produced, and usually at a competitive price. This has altered the marketing of sulfuric acid because recovery of sulfur from stack gas can, in some cases, offset the cost of air pollution control equipment. However, process gas sulfur recovery is not simple and cannot usually give acid of a quality needed by most industries. Disposal of such acid can be a problem and is one of the many reasons recovery from low-concentration sulfur sources has not been widely practiced. Sulfuric acid production has grown 4.5%/yr since 1971 with most of the growth in sulfur recovery in the early seventies. Other major inorganics have grown 3% or less per year since 1971. As a comparison, the largest petrochemical produced, ethylene dichloride, has grown about 5.7%/yr since 1971. Most of the petrochemical growth is due to increased usage of plastics derived from chemicals such as vinyl chloride, styrene, and terephthalic acid. Ethylene dichloride, ethylbenzene, and p-xylene are used to manufacture these intermediates, respectively.

1.3 Major Sources Classed by Pollutant

Looking at the chemical industry's contribution to each of the regulated pollutants gives a different view. Tables 2 through 6 are compilations of criteria pollutant emissions for major chemical process sources.[3] For comparison, each table also shows the major pollutant source.

The lime industry, which is estimated to release 31.2 Gg (34,000 tons) of particulate matter per year, produces less than 0.5% of the emissions from unpaved roads. Most particulate emissions in the chemical industry are from crushing or grinding and are not produced by chemical reaction. This is important for control since chemical reaction generally produces finer particles which are more difficult to collect. The mass of fine particle mists is usually very small. Such emissions are best controlled by chemical reaction or physical means rather than mechanical or electrical forces.

Sulfuric acid accounts for about 56 Gg (62,000 tons) of sulfur dioxide emission, less than 0.5% of the major source, combustion of fossil fuels. Chemical industry steam generators, furnaces, and kilns combined produce far more sulfur dioxide than acid plants. While sulfur recovery is practical for some sources, the particulate matter and nitrogen oxides in the power plant carbon dioxide-rich flue gas have played havoc with recovery process development. While contact sulfuric acid plants

Table 1   The 20 Major Chemicals Ranked by U.S. Production in 1978 and 1979[2]

Chemical	Production, Tg	
	1978	1979
Sulfuric acid	37.31	38.13
Lime	17.61	17.61
Ammonia	15.61	16.45
Sodium hydroxide	9.72	11.25
Chlorine	10.03	11.00
Phosphoric acid	8.50	9.20
Nitric acid	7.20	7.78
Sodium carbonate (synthetic and natural)	7.53	7.50
Ammonium nitrate	6.54	7.08
Urea	4.94	6.14
Ethylene dichloride	4.99	5.37
Ethylbenzene	3.81	3.87
Vinyl chloride	3.15	3.42
Styrene	3.26	3.40
Methanol	2.92	3.36
Terephthalic acid	2.70	3.30
Xylene	2.78	3.13
Hydrochloric acid	2.53	2.70
Ethylene oxide	2.27	2.40
Ethylene glycol	1.77	2.09

Table 2 Major Sources of Particulate Emissions in the Chemical Industry[3]

Source	Gg/yr
Unpaved roads[a]	99,900.0
Lime	312.4
Gypsum	99.4
Phosphate rock	81.9
Alumina, aluminum oxide	56.3
Nylon	28.1
Potash	23.6
Ammonium nitrate	18.9
Soap and detergents	18.4
Polyvinyl chloride	16.6
Rubber processing	12.8
Calcium carbide	12.2
Ammonium sulfate	11.0
Ammonium phosphate	10.2

[a] Largest particulate source (not in chemical industry).

Table 3 Major Sources of Sulfur Oxides in the Chemical Industry[3]

Source	Gg/yr
Electric generation from fossil fuels[a]	14,135.00
Sulfuric acid	56.56
Methyl parathion	9.67
Barium chemicals	7.26
Ammonia	3.34
Ethyl parathion	3.16
Carbon disulfide	1.66
Phthalic anhydride—O-xylene	1.65

[a] Major source of sulfur oxides (not in chemical industry).

Table 4 Major Sources of Nitrogen Oxides in the Chemical Industry[3]

Source	Gg/yr
Ammonia	43.99
Nitric acid	27.05
Propylene	12.60
Toluene	6.54
Adipic acid	3.55

produce 99% acid, recovery processes give 70–80% acid, the peak range for sulfuric acid corrosion. The need for exotic metals to handle the dilute acid produced has added economic problems to recovery plant selection. In addition, the dilute process gas needs supplemental heat to drive the recovery reaction.[4]

Nitrogen oxides (NO_x) are also produced in the combustion of fuels, primarily due to the high level of temperature and nitrogen concentration in the combustion air of most processes. Production of synthetic ammonia (discussed in Chapter 23) is the largest source of nitrogen oxides, followed by nitric acid. With the increased emphasis on combustion as a method of disposal of hydrocarbons

Table 5　Major Sources of Hydrocarbons in the Chemical Industry[3]

Source	Gg/yr	
Plastics processing	462.90	
Maleic anhydride from benzene	156.80	
Phthalic anhydride—naphthalene	83.75 ⎫	148.80
Phthalic anhydride—O-xylene	65.05 ⎭	
Ethylene oxide	84.85	
Polyvinyl chloride	78.65	
Carbon black	70.20	
Rubber processing	58.54	
Acrylonitrile	56.09	
Polyethylene resin—low density	23.96 ⎫	52.18
Polyethylene resin—high density	28.22 ⎭	
Ethylene dichloride—ethylene chlorination	29.82 ⎫	51.29
Ethylene dichloride—oxyhydrochlorination	21.47 ⎭	
Dimethyl terephthalate	49.44	
Ethyl benzene	44.68	
2-Ethyl-1-hexanol	41.35	
Cyclohexanone	31.24	

Table 6　Major Sources of Carbon Monoxide in the Chemical Industry[3]

Source	Gg/yr
Carbon black	2,082.00
Maleic anhydride from benzene	121.60
Acrylonitrile	68.20
Cyclohexanone	34.58
Dimethyl terephthalate	28.71
N-Butyl alcohol	17.32
Methanol	17.09
Ammonia	14.41
Acetic anhydride	14.23

and toxic chemicals, the amount of nitrogen oxides may increase but probably not due to a specific chemical like ammonia.

While inorganic chemicals dominated the first three criteria pollutants, industrial organic chemicals account for most of the hydrocarbons and carbon monoxide from the chemical industry. Plastics processing is not an integral part of the chemical industry, but many products (solvents and resins) escape during formulation. Although some manufacturers have been able to recover and reuse solvents, most must be destroyed since they are of low quality. Maleic anhydride, which is used in coatings, polyester resins, pesticides, and preservatives is the largest source of volatile organic compounds (VOCs). Its manufacture is also the second largest source of carbon monoxide. This is because one-third of the carbon in the benzene molecule is converted to carbon dioxide. To prevent the formation of large quantities of maleic acid, the oxygen level has to be kept low. Chemical equilibrium dictates that a large portion of the carbon dioxide will shift to carbon monoxide.

A comparison of the major production sources (Table 1) to the major sources of pollutants shows that for inorganic production, the equating of production and pollution impact is a fairly reasonable assumption; however, there are some notable exceptions such as sodium hydroxide (insignificant pollution) and alumina (very low production). In contrast to the organic chemicals, major hydrocarbon and carbon oxide sources are generally not in proportion to the production statistics.

2　CONTROL FOR INORGANIC PROCESSES

The major criteria pollutants produced by inorganic processes are particulates, sulfur oxides, and nitrogen oxides. When steam or heat is required for the process, the combustion emissions can overshadow the chemical process emissions. Examples of criteria pollutants needing control in the inorganic chemical industry are:

Source	Pollutant
Lime kiln	Particulate
Sulfuric acid plants	Sulfur oxides
Nitric acid plants	Nitrogen oxides
Ammonia	Nitrogen oxides

There are three industries within the inorganic chemical industry for which new source performance standards have been developed. Newly constructed or modified sources must comply with the standards summarized in Table 7.[5] In addition, control of noncriteria pollutant emissions needs to be considered. For most inorganic sources, these emissions are inorganic compounds with appreciable vapor pressure (i.e., Pb, HF, SiF_4, HCl, and other halogen acids). Absorptive and adsorptive methods are generally used for control. Submicrometer particles can be formed in vapor condensation or during reaction. Such particles are very difficult to remove and the process should be studied with a view to preventing their formation.

Control of fugitive emissions can be one of the more difficult problems in inorganic processing. These emissions are usually of large particle size compared to process emissions, but repeated material handling can produce significant problems. Minimizing the fall of material onto storage piles or into process equipment can reduce fugitive emissions. Alternative handling methods (such as pneumatic conveying) may reduce both these problems and handling costs.

2.1 Control of Particulates

Most lime (calcium oxide) is produced from limestone rock. The typical composition of selected products is shown in Table 8. About 85% of the lime is produced in rotary kilns, while the remainder is produced in vertical kilns or fluidized beds.

Fuel is burned in the kilns to decompose the calcium carbonate to carbon dioxide and lime at about 1173 K (1650°F). The rotary kiln takes smaller limestone [6–65 mm (0.25–2.5 in.) in diameter], while vertical kilns handle larger lumps [typically 150–205 mm (6–8 in.) in diameter]. Vertical kilns are restricted to low throughput. A typical vertical kiln would have a steel-encased refractory shell 3–7 m (10–23 ft) in diameter and 10–23 m (30–75 ft) high. The top of these kilns can be sealed with a charging mechanism and the dust-laden gases conducted to suitable treatment equipment. The limestone travels countercurrently to the hot gas in both types of kilns, and the exhaust gases leave at between 570 and 1270 K (550 and 1800°F).[7] The lower temperature is more common.

Rotary kilns are from 2 to 4 m (6 to 12 ft) in diameter and as long as 120 m (400 ft).[8] Rotary kilns are not as energy efficient as vertical kilns; however, a high production rate can be achieved. This greatly reduces manpower requirements and is the main reason for selection of a rotary kiln.

Although a rotary kiln can produce lime on a continuous and rapid basis, emissions are generally higher than for vertical kilns, 90 to 100 kg/ton (180 to 200 lb/ton) versus 3.5 to 4 kg/ton (7 to 8 lb/ton).[6] Abrasion of the limestone and lime in the kiln produces a large amount of airborne particulate matter. As the CO_2 boils off the limestone, additional dusting is created by attrition and decomposition of the stone. This dust is very difficult to control. It is of mixed composition, difficult to wet, hot, dry, and prone to development of electrostatic charges. When pulverized coal is used as the fuel for the kiln, fly ash, tars, and carbon can also be present. Typically, 5–15% of the weight of the lime produced is blown from a rotary kiln.[7] Dry cyclonic collectors knock out large particles (above 10 μm) which comprise about 60–85% of the dust emissions. The emissions are primarily lime and are usually returned to the process. For final cleanup of the gas, fabric filters, electrostatic precipitators,

Table 7 Standards of Performance for New Sources[5]

Source	Pollutant	Emission Level
Nitric acid plants	Opacity	10%
	Nitrogen oxides	1.5 g/kg
Sulfuric acid plants	Opacity	10%
	Sulfur dioxide	2 g/kg
	Acid mist	0.075 g/kg
Lime plants		
Rotary kilns	Opacity	10%
	Particulate	0.15 g/kg
Lime hydrators	Particulate	0.075 g/kg

Table 8 Composition of Commercial Lime[6]

Component	High Calcium Lime, %	Dolomite Lime, %
CaO	93.25–98.00	55.5–57.50
MgO	0.30–2.50	37.60–40.80
SiO_2	0.20–1.50	0.10–1.50
Fe_2O_3	0.10–0.40	0.05–0.40
Al_2O_3	0.10–0.50	0.05–0.50
H_2O	0.10–0.90	0.10–0.90
CO_2	0.40–1.50	0.40–1.50

wet scrubbers, and granular bed filters have been employed. Of these devices, only granular filters have been found to be unacceptable: 100 mg/Nm³ (0.05 gr/scf) versus the usual emission concentration of 40 mg/Nm³ (0.02 gr/scf). Test results of the performance of gravel bed filters have shown emissions that are about twice as large as the Federal New Source Performance Standard of 0.15 g/kg (0.3 lb/ton).[5,9] Users of baghouses have reported controlled emission levels of 2 mg/Nm³ (0.001 gr/scf) or lower.[10] Bag life can be as long as 2 yr.[7] It is good practice to replace all bags when bags start to fail; otherwise, bag failure will result in a significant loss of production time.

Since lime dust has high electrical resistivity, the injection of water is usually required to use electrostatic precipitators effectively. The type of fuel and minor impurities in the limestone can influence the choice and efficiency of the collector. Fabric filters are usually more efficient and costly than wet scrubbers; but, the wet scrubber operating costs are usually higher. The choice depends mainly on local economics, the presence of toxic metals in the dust, or uses of the wet or dry products from the collector. Typical control efficiencies by method for lime kilns are given in Table 9.[6] Often the weak lime solution from a wet scrubber can be used in hydrators or slaking operations.

Slaking operations also produce emissions as the lime is reacted with water to produce hydrated lime (calcium hydroxide). Emissions are usually controlled by wet sprays or a wet scrubber, and the scrubbing liquid is used in the process to conserve product. The major sources of emission in the slaking of lime are milling and handling of the slaked lime and packaging. Fugitive emissions for the sources are about 2.5 g/kg of lime. Table 10 shows the emissions and control methods for hydration and packing.[6]

Table 9 Typical Lime Emissions and Control Methods[6]

Source or Operation	Particulate Emissions, g/m³	Collection Efficiency, %	Control Method
Vertical lime kiln	0.70–2.29		None
Rotary kiln	0.002	99.99	Glass bag filter
Rotary kiln	0.05–0.18	99.7–97.5	Four-stage cyclonic scrubber
Rotary kiln	9.80	70.0	High-efficiency cyclones
Rotary kiln	0.50	95.0	Single-stage precipitator
Rotary kiln	0.25–0.57	96–97	Venturi scrubber
Rotary kiln	0.7–0.9	97.5	Impingement scrubber
Calcimatic kiln	0.05	99.2	Glass bag filter

Table 10 Emissions and Control Methods for Hydration/Packing[6]

Source or Operation	Particulate Emissions, g/m³	Collection Efficiency, %	Control Method
Hydration	0.02–2.15		Water sprays in stack
Hydration	0.02–0.16		Wet scrubber
Hydrate milling	No visible dust	99+	Bag filter
Hydrate loader and packer	0.02	99+	Bag filter

2.2 Control of Sulfur Oxides

By far the most common source of sulfur oxides at a chemical plant is the power boiler. Because of the low mass of emissions, most chemical plants do not employ controls for sulfur oxides on boilers. In the future, this may change with newer plants needing sulfur control. Most industrial power plants utilize low sulfur coal, when available, to reduce sulfur emissions. Removing sulfur oxides from the stack has not been considered feasible, based on the experience of electric utilities in removing sulfur oxides. In comparison to the emissions of electric utilities, industrial power plants are very small sources of emissions.[3] The major source of sulfur dioxide in the chemical industry is the manufacture of sulfuric acid. Although there are several processes for making sulfuric acid, the most common is the contact process in which sulfur is oxidized at 670–880 K (750–1130°F) over a catalyst to sulfur trioxide.[11] The acid gas is then absorbed in 98.5% sulfuric acid at about 370 K (200°F) in a packed tower.[10]

The primary method of sulfur dioxide control in sulfuric acid plants is interpass absorption of sulfur trioxide. Figure 1 shows a typical dual absorption acid plant.[12] Chemical equilibrium prevents conversion of all the SO_2 to SO_3 when appreciable reaction products are present. Since near-equilibrium is achieved over modern catalysts, interpass absorption is required to remove the acid gas to permit further conversion over an additional bed of catalyst.[11] Interpass absorption is generally applied between the third and fourth catalyst bed. New plants utilizing interpass absorption can achieve sulfur oxide levels of about 200 to 550 ppm, depending on feed gas concentration, steadiness of operation, and catalyst condition or age. For removal below these levels, costly supplemental control measures such as chemical scrubbing (i.e., caustic in a packed tower) would be required after interpass absorption plants.

Acid mist emitted from sulfuric acid plants must also be controlled. It is generally accepted that the most efficient and economical equipment for this purpose is packed fiber bed mist eliminators, since typical particle sizes are small (in the range of 2 or 3 μm and less).[13] Acid mist can also be reduced by drying the process gas fed to the sulfur trioxide convertor. Organic matter in sulfur increases mist generation by combustion of hydrocarbons to form water. For this reason, organic matter is preferably kept below 0.1% in the raw sulfur.[12] Temperature control in the absorption process can minimize mist formation by avoiding excessive gas velocity or shock cooling. Glass fiber eliminators can reduce emissions by 92 to 99.9%.[10] In conjunction with interpass absorption, the acid mist can be held below 1.5 mg/Nm³ (0.0008 gr/scf) when the above controls and two-stage coarse-fiber mist elimination is applied.[14] If fine fibers are used for the mist eliminator, only one stage is required to achieve 99+% removal of mist.[13,15]

Typical sulfur oxide conversion in interpass absorption plants is in the range of 99.7 to 99.8+%. Many existing single-absorption plants achieve conversions of about 98%, equivalent to 2000 ppm SO_2. However, some plants have significantly higher emissions because they were originally designed

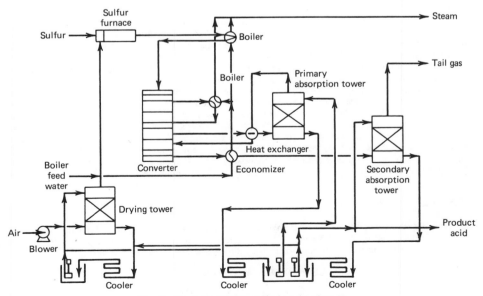

Fig. 1 Dual absorption sulfuric acid plant.[12]

for lower conversions or are operated at rates significantly above design. The major operational factors affecting emissions are plant load, converter temperature, and catalyst age.

Utilizing the theoretical aspects of SO_2 oxidation and practical chemical engineering, higher conversion can be achieved by several methods. Specific recommendations depend on the plant arrangement and operating conditions. Some general suggestions are: (1) add new catalyst to replace deteriorated material or to increase contact time in the first reaction bed; (2) install an additional catalyst bed with intercooling and absorption in series with the present reactor; (3) reduce operating rates, preferably using lower SO_2 concentrations and/or higher oxygen concentrations; and (4) repair deteriorated or faulty equipment to eliminate SO_2 leaks between process streams and/or to achieve optimum gas temperatures at each catalyst bed. Further suggestions are given in Ref. 16.

Add-on scrubber systems or other treatment processes have occasionally been used to reduce SO_2 emissions from single-absorption plants (or to treat tail gases from interpass absorption plants in Japan). There are several processes available but the most frequently employed is ammonia scrubbing.[17-20] To be viable, ammonia value must be recovered, preferably by utilizing spent scrubbing liquor in fertilizer production.

2.3   Control of Nitrogen Oxides

Nitric acid production is one of the larger chemical industry emission sources of nitrogen oxides. If uncontrolled, it can be a major source of high NO_x concentrations in the immediate area of the plant site. Ammonia is the starting raw material for nitric acid and its synthesis produces additional NO_x. The NH_3 admixed with air in the presence of a platinum–rhodium catalyst is oxidized to NO which is then mixed with additional air and slowly oxidized to NO_2 or its dimer, N_2O_4. The latter is absorbed into water, producing weak nitric acid and freeing gaseous NO. The overall reaction, comprising several steps, can be represented by the equation

$$3NO_2 + H_2O \rightleftharpoons 2HNO_3 + NO \tag{1}$$

Complete absorption of the NO_x in the process is impossible because 1 mole of NO is constantly being released for every 3 moles of NO_2 absorbed. The NO, in turn, must be reoxidized to NO_2 (a very slow reaction) before it can be reabsorbed. More air must be added for the oxidation, and the reaction kinetics become slower and slower as the reactants become progressively diluted with more nitrogen. (The substitution of oxygen for air would aid reaction but has not been found to be economically practical.) In addition, the absorption reaction is exothermic, and the increased temperatures produced tend to reverse the above reaction equation.

The absorption is generally carried out in a stainless steel countercurrent plate column with water being added at the top and the NO_2 entering at the bottom. The water-to-gas ratio is set to produce about 60% nitric acid at the tower bottom. Cooling coils on the plates remove the absorption and reaction heat with the aid of additional air introduced at intervals up the column. The most efficient absorption is obtained by operating at low temperatures and throughput, and long residence time to provide for reoxidation of the NO.

Because gas volume contracts as the reaction proceeds, completion of the reaction is aided by increased pressure. In many plants, the absorption is carried out at 140–520 kPa (20–60 psig), but a substantial number of plants operate at atmospheric pressure. Concentrated nitric acid (98%), if desired, can be produced by distilling the 60% acid with concentrated sulfuric acid in a separate operation by direct absorption at pressures up to 5.275 MPa (750 psig), coupled with refrigerated cooling. The chemistry of NO_x absorption has been discussed by Sherwood and Pigford.[21]

The major emission source is the vent gas from the weak acid absorption tower. Aside from better design and operation of the tower, control methods are limited to (1) catalytic reduction, (2) flame reduction, (3) catalytic oxidation and concentration, (4) adsorption, and (5) absorption and chemical reaction. Of these, only catalytic reduction has been widely demonstrated and practiced commercially.

Many state regulations limit NO_x emissions from existing nitric acid plants to 2.8 kg/metric ton of 100% acid produced (5.5 lb/ton acid), which is equivalent to an emission concentration of 400 ppmv. The EPA New Source Performance Standards (NSPS) limit emissions to 1.5 kg/metric ton (3.0 lb/ton) or about 215 ppmv. Florida and Colorado have applied NSPS levels to existing plants as well.

2.3.1   Catalytic Reduction

Catalytic reduction of NO_x with a fuel (hydrogen, natural gas, and LPG) have typically been used to achieve NO_x levels of 200–400 ppmv. The fuel is mixed with the tail gas before passing it through a nonselective catalytic reduction unit where the following catalytic reactions occur:

$$CH_4 + 4NO_2 \rightarrow 4NO + CO_2 + 2H_2O \tag{2}$$

$$CH_4 + 2O_2 \rightarrow CO_2 + 2H_2O \tag{3}$$

$$CH_4 + 4NO \rightarrow 2N_2 + CO_2 + 2H_2O \tag{4}$$

The reaction kinetics of the first reaction (which gives plume decolorization but not NO_x destruction) is the fastest, while the last reaction is the slowest. Further, the fuel will preferentially react with O_2 as long as any appreciable quantity is present, so that Eq. (3) must be essentially completed before the last reaction can proceed. This increases the fuel requirements and may cause too high a temperature for the reaction products, resulting in melting of the catalyst and damage to the process turboexpanders.

In the early 1970s, it was popular to complete only Reaction (2), resulting in plume decolorization to comply with plume appearance regulations. Such an approach is unacceptable today when NO_x destruction is required. Many early decomposers used honeycomb catalysis masses to give increased surface area, but these were more easily damaged from overtemperature. Pelleted catalysts (precious metals on alumina support) have proved more rugged. Adlhort et al.[22] and Newman[23] have discussed fitting catalytic decomposers into nitric acid plants. Gillespie et al.[24] have discussed many of the reasons for the unsatisfactory operation of the early honeycomb decomposers. In addition to problems from the composition of the tail gas, other problems were: poor mixing of fuel and tail gas, poor control of fuel/oxygen ratio resulting in carbon deposition on the catalyst, and catalyst poisoning with sulfur compounds from the fuel. Better instrumental controls have improved process reliability. Reed and Harvin[25] have reported similar observations.

The fuel/tail gas mixture must be preheated to a suitable ignition temperature depending on fuel choice [420 K (300°F) for hydrogen; 750 K (890°F) for methane] before entering the catalytic unit. Obviously, hydrogen would be the preferred fuel but it is seldom used because of its cost. Exhaust NO_x concentrations as low as 100 ppmv can be obtained with proper control, but more expensive two-stage catalytic beds may be required.

An alternate selective process for catalytic reduction uses ammonia as a fuel. It is advantageous in oxygen-rich gases because NO_x can be reduced without appreciable oxygen removal. Ammonia reduces the nitrogen oxides according to the following reaction equations:

$$2NH_3 + 3NO \rightarrow \tfrac{5}{2}N_2 + 3H_2O \tag{5}$$

$$4NH_3 + 3NO_2 \rightarrow \tfrac{7}{2}N_2 + 6H_2O \tag{6}$$

$$4NH_3 + 3O_2 \rightarrow 2N_2 + 6H_2O \tag{7}$$

Reactions (5) and (6) proceed at much faster rates than Reaction (7) so that good selectivity is possible. Supported precious metal catalysts (platinum preferred) give the best results. Although ammonia is an expensive fuel, considerably less is required, so the process has considerable merit. Unfortunately, the reduction must be carried out within a very narrow temperature range of 475–515 K (400–470°F) to obtain satisfactory efficiency and operation. At lower temperatures, ammonium nitrate salts can be formed with possible damage to the downstream turbine expanders. Above 515 K (470°F), oxides of nitrogen are produced by reaction between NH_3 and O_2:

$$4NH_3 + 5O_2 \rightarrow 4NO + 6H_2O \tag{8}$$

This narrow temperature limitation has restricted appreciable commercialization of this method.

2.3.2 Catalytic Oxidation, Concentration, and Adsorption Methods

A number of other control methods for nitric acid tail gas have been investigated without any appreciable commercialization. One approach is to oxidize the NO in the tail gas catalytically to NO_2, concentrate the NO_2 stream, and return it to the nitric acid absorption tower inlet.

Fornoff[26] has proposed the separation and concentration of NO_2 with molecular sieves. Since the tail gas contains appreciable moisture, this appears an unlikely choice at first glance. However, Fornoff provides sieves adequately sized to hold both the moisture and the NO_2. The major drawback appears to be the high cost of the large amount of adsorbent required.

Mayland and Heinze[27] have proposed absorption of NO_2 in airblown, bleached nitric acid (acid stripped of NO_2). While this could be used as an NO_2 stream concentrator in lieu of Fornoff's molecular sieves following vapor phase catalytic oxidation of NO, they have developed a catalytic tower packing which oxidizes the NO to NO_2 directly in the absorber.

2.3.3 Absorption and Chemical Reaction Methods

Swanson et al.[28] report the successful retrofitting of a 220 metric ton/day (250 TPD) Allied Chemical nitric acid plant with an "extended absorber" giving outlet NO_x concentrations of 100–300 ppmv. Although normal nitric acid plant technology was employed, the absorbers were pushed to their limit. The tail gas from the original 15-m (50-ft) absorber was introduced into a 3.65 m i.d. × 33.5 m high (12 ft diameter × 110 ft) plate countercurrent absorber. The procedure involves water being added to the top of the tower with 0.19 m³/sec (3000 gpm) of well water at 286 K (55°F), providing cooling on each plate. The weak acid produced was used as the feed to the original absorber, and hot gases from the ammonia converter were used to reheat the vent gas from the extended absorber. An additional cooler–condenser was added to precool the converter gases with well water prior to entering the original absorber. The extended absorber operates at a pressure of 724 kPa (90 psig).

Kelley et al.[29] report modification of an Illinois Nitrogen Corp. 320 metric ton/day (350 TPD) nitric acid plant with a novel urea scrubber. Tail gases from the nitric acid absorber were passed through a cooler chilled by evaporating liquid NH_3 from the converter. The cooled gas then entered a two-section saddle-packed absorption tower. Chilled nitric acid feed water was recirculated over the lower packed section and chilled urea solution over the upper packed section. NO and NO_2 absorbed in the urea section produce nitric and nitrous acids. Some of the urea hydrolyzed, producing ammonium ions which reacted with the nitric acid producing ammonium nitrate in solution. The nitrous acid and urea reacted to nitrogen, carbon oxides, and water. A spent urea blowdown stream containing the ammonium nitrate was and is still used for fertilizer production. NO_x effluent concentrations are in the 100–200 ppmv range with test results as low as 60 ppmv.

Other absorption methods have been reported in the literature using sodium hydroxide, sodium carbonate, sodium sulfite, and ammoniun hydroxide solutions, as well as calcium carbonate slurries. Nitrate and nitrite salt solutions result. Removal efficiencies up to 90% have been reported but disposal of the water soluble scrub solutions have generally resulted in insurmountable problems unless they could be used in a product such as fertilizer. A scrubbing process developed in Japan[30] destroys the NO_x with sodium sulfite according to the following equations:

$$2Na_2SO_3 + 2NO \rightarrow 2Na_2SO_4 + N_2 \tag{9}$$

$$4Na_2SO_3 + 2NO_2 \rightarrow 4Na_2SO_4 + N_2 \tag{10}$$

Destruction efficiency was 98% in laboratory tests.

Literature data on absorption of NO and NO_2 in alkaline solutions are contradictory. NO alone is poorly absorbed. Mixtures of NO and NO_2 appear to be absorbed efficiently when the NO_2/NO ratio equals or exceeds unity. When NO is the predominant species, prescrubbing with an oxidizer or hydrogen peroxide has produced the desired conversion to NO_2. However, the operating cost for chemical removal of NO_x can be expensive in addition to disposal problems with the waste products.

NO_x can also be readily absorbed in concentrated sulfuric acid if the presence of the NO_x is not objectionable for future uses of the sulfuric acid. Another drawback is the dilution of the sulfuric acid, which occurs if the gas stream is humid.

2.4 Control of Hydrocarbon Emissions

Inorganic chemical manufacture usually produces little, if any, hydrocarbon emissions. Some manufacturing operations produce odorous compounds which may or may not contain organic molecules. When possible, these emissions may be scrubbed and treated as liquid waste. However, a variety of vapor incinerators similar to the type used for solvent emission control may be applied. Combustible odorous materials in very dilute concentrations in exhaust streams may be controlled by passing the gas through a plant steam boiler or kiln.

Incineration of liquid or solid waste chemicals results in the generation of varying concentrations of submicrometer particulate matter. When the waste contains alkali metal compounds, the portion of submicrometer material is extremely high. Control of such emissions is extremely difficult, and each situation must be considered on a case by case basis. However, the general control principles discussed in Section 3.1 apply.

Table 11 Selected Noncriteria Pollutants from the Chemical Industry[31]

Source	Pollutant	Emission Rate (Mg/yr)
Ammonia	Ammonia	31,858
Ammonium nitrate	Ammonia	2,136
Ammonium phosphates	Ammonia	210
Ammonium sulfate	Ammonia	1,687
Hydrochloric acid	Hydrogen chloride	753
Phosphorus	Hydrogen fluoride	2,905
Phosphoric acid	Hydrogen fluoride	1,734
Phosphoric acid	Phosphoric acid mist	606
Potassium hydroxide	Chlorine	64
Potassium hydroxide	Mercury	3
Urea	Ammonia	4,464

2.5   Control of Noncriteria Pollutants

Noncriteria pollutants present a more difficult situation for emission control design since emission standards have not been defined. The following topics have been intended to be representative of control methods for noncriteria pollutants. Particulates and hydrocarbons are criteria pollutants but the specific chemical composition of the emissions strongly affects the degree of control that needs to

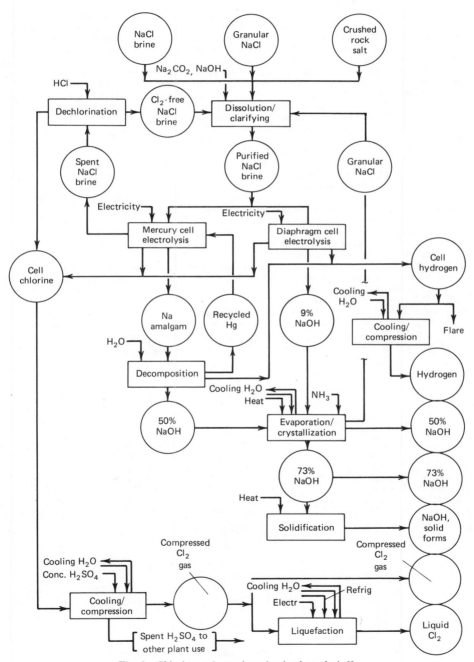

Fig. 2   Chlorine and caustic soda via electrolysis.[32]

be applied. A recent study of approximately 800 types of stationary sources identified 389 noncriteria pollutants covering a wide range of hazard potential. Such pollutants are generally very low in mass emission rate compared to criteria pollutants as shown in Table 11. While some of these emissions appear obvious, there are notable exceptions such as potassium hydroxide manufacture. Since the data base (Ref. 31) is incomplete, there may be other noncriteria sources which have not been quantified. Because of recognized hazards of some noncriteria pollutants, a much higher degree of control is often required to protect life and property. Selected examples for noncriteria pollutant control are chlorine, hydrofluoric acid, and phosphorus.

Chlorine is produced in conjunction with sodium hydroxide in an electrolytic cell. There are several process variations,[32] but chlorine emission control is the same because only two major emission points exist (see Figure 2). Leakage of chlorine gas around the cell has been estimated to be less than 0.1 g/kg (0.05 ton). Control is affected by control of cell room ventilation and caustic scrubbing. Product chlorine from the cell is compressed and condensed as liquid chlorine. Chlorine in the remaining vent gas can be recovered economically in a packed-bed absorption tower with water or carbon tetrachloride as long as the chlorine concentration is in excess of 1% (typically 10% or higher). At chlorine vent concentrations below 0.5%, alkaline scrubbing with caustic or lime is needed. The effluent chlorine concentration is limited by chemical kinetics and not equilibrium. Levels of less than 0.2 ppm can be expected with caustic solutions in packed towers operating at 303 K (90°F).[32]

Hydrofluoric acid (HF) is produced by the reaction of fluorspar (CaF_2) with sulfuric acid as shown in Figure 3. Typically, the reaction takes 30–60 min at 523 K (480°F) in a rotary kiln. Fugitive emissions from the discharge of the calcium sulfate are minimal since an airlock is employed on the waste end of the kiln. The HF is purified in a series of absorption and distillation steps to remove excess sulfuric acid, sulfur dioxide, and carbon dioxide. Silicon tetrafluoride and HF can be scrubbed with water to make a marketable 30–35% solution of H_2SiF_4. When high-purity HF is desired, sulfuric acid can be used to give 99+% pure gas. Sulfur dioxide released is about 10 g/kg (5 lb/ton) and is usually not controlled.[33] It has been suggested that a caustic scrubber could be used for more complete cleanup, although operating cost would be difficult to justify.[10]

An elemental phosphorus plant has a large number of small particulate sources. Various ore sizes are blended and charged to a rotary kiln where a 50–150 mm (2–6 in.) clinker is produced. Crushing and sizing reduces the clinker to a desired size nodule for the electric furnace. The cumulative effect of all of these fugitive emission sources has not been evaluated or estimated. Cyclones followed by low-pressure drop spray water scrubbers [6–8 mm Hg (3–4 in. H_2O)] are used to remove the particulate and fluorides from the kiln effluent. This type of system can control emissions to as low as 0.0046 g/Nm^3 (0.002 gr/scf), fluoride, 0.007 g/Nm^3 P_2O_5 (0.003 gr/scf).[10] Control of fugitive dust emissions from nodule crushing and screening operations and burden conveying and stocking is accomplished by hooding, ventilation, and removal in bag filters, low-energy wet scrubbers, and venturi scrubbers [18–30 mm Hg (10–16 in. H_2O)]. For furnace slag tapping, venturi scrubbers operating at 55–115 mm Hg (30–60 in. H_2O) pressure drop are required to remove the P_2O_5 and silicon metal fumes.

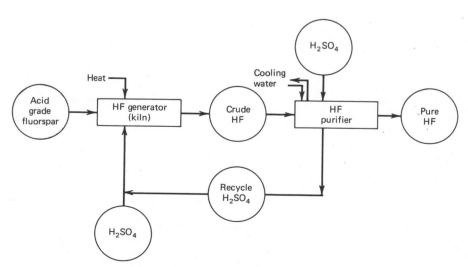

Fig. 3   Flowsheet for the production of hydrogen fluoride.[33]

Table 12 Wet Electrostatic Precipitator Performance on
P₂O₅ Aerosol[34]

Average effective migration velocity	22.3 cm/sec
Average specific collection area	8.1 m²/m³-sec (41.2 ft²/1000 acfm)
Average P₂O₅ removal efficiency	84%
Particle size distribution	Greater than 90% by weight below 1.0 μm in diameter

Emissions of fluorides can be expected to be as low as 0.069 g/Nm³ (0.03 gr/scf) while P₂O₅ may reach 0.0023 g/Nm³ (0.001 gr/scf).[10] It has been reported that wet-electrostatic precipitators can achieve similar control of emissions at a significantly lower pressure drop.[34] The results of 22 pilot tests (summarized in Table 12) have demonstrated high-control efficiency on P₂O₅ aerosol. This is typical of some control problems where significant improvements can be made both on energy conservation and emission reduction. Phosphorus produced in the furnace is separated from carbon monoxide by condensation. The CO produced is normally burned as fuel in the nodulizing kilns and remaining P₂O₅ scrubbed out in the kiln flue-gas scrubbers.

3 CONTROL FOR ORGANIC PROCESSES

Control problems for the organic chemical industry focus mainly on hydrocarbon and carbon oxide emissions as seen from Tables 2 through 7. Recent emphasis on hazardous chemicals also pose special problems for control. Flares are used to control both hydrocarbons and carbon monoxide when it is not economical to recover the heat or chemical compound. When the combustible substance is classified as hazardous, normal combustion practices are insufficient since regulations often require control of emissions to levels of 2–10 ppmv. This does not rule out incineration with high turbulence, high temperature, and long exposure time. Usually absorption and adsorption are more cost effective, especially for low volumes of emissions.[35]

The only New Source Performance Standard that applies to the organic chemical industry is on the storage of petroleum liquids; however, specific chemicals may be subject to stricter standards. For storage of volatile organics in tanks of more than 150,000 l (40,000 gal), hydrocarbons must be controlled. EPA recommends the use of floating roof tanks or vapor recovery systems for vapor pressures of 10–76 kPa (1.5–11 psi). Vapor recovery systems are recommended above 76 kPa (11 psi).[5]

When feasible, containment by sealing the process is the preferred method of control. There are many instances where this is impractical or too expensive. The presence of a hazardous emission is not necessarily related to the chemical being produced. Toxic emissions may result from the reactants, from undesired by-products, or trace contaminants in reactants even though the product is nontoxic. The emissions may be present in very small quantities but be highly volatile. Thus, during storage, the quantity of emission would be small but the concentration very large. Often the vent is isolated from the rest of the plant so control for such small quantities is extremely costly.

Emissions may occur from many points in a process, being caused by process upsets, incomplete separation, leaking pump seals, spills, and vessel decontamination during maintenance. When the points are close together, the emissions may be combined and sent to a single treatment facility. The selection of a specific control will depend on the size of the gas stream and the contaminant concentration as well as the required removal efficiency.

The following section discusses selected industrial sources that exemplify the types of controls applied to the organic chemical industry.

3.1 Control of Hydrocarbons

Ethylene dichloride manufacture typifies the complexities of chemical process and control equipment selection.

Ethylene dichloride (EDC) is produced by either direct chlorination [Eq. (11)] or oxychlorination [Eq. (12)] of ethylene; however, most production is used captively to make vinyl chloride so the two methods are used in combination to utilize the by-products of vinyl chloride manufacture.[36] In the direct chlorination process, ethylene is reacted in the liquid phase with chlorine in the presence of ferric chloride catalyst.

$$HC_2 = CH_2 + Cl_2 \xrightarrow{FeCl_3} ClCH_2CH_2Cl \tag{11}$$

$$CH_2 = CH_2 + \tfrac{1}{2}O_2 + 2HCl \xrightarrow{CuCl_2} ClCH_2CH_2Cl + H_2O \tag{12}$$

$$ClCH_2CH_2Cl \rightarrow CH_2 = CHCl + HCl \tag{13}$$

The reaction [Eq. (11)] is fast, complete and exothermic. The towerlike reactor serves also as the fractionator to provide the proper molar ratio of reactants in the presence of the catalyst. The gas stream from the reactor is passed through a caustic scrubber to remove unreacted gases while the uncondensed gases (primarily ethylene and chlorine) are returned to the reactor. The liquid stream from the reactor contains crude EDC which is purified in a series of distillations. After the EDC is purified, it is cracked to form vinyl chloride [Eq. (13)]. The HCl produced is compressed and fed to a fluid-bed oxychlorination unit to produce more ethylene dichloride, as shown by Eq. (12).

The overall manufacturing scheme, showing air emission points, is given in Figure 4. The major sources of emission are the reflux condenser vent, the EDC stripper vent, and the distillation vents. The vent on the reflux condenser releases hydrocarbons, chlorine, and HCl as follows:[36]

Compound	Emission Factor (kg/metric ton EDC)
Ethane	3.0
Ethylene	7.5
Methane	3.0
Chlorine	0.5
HCl	0.5

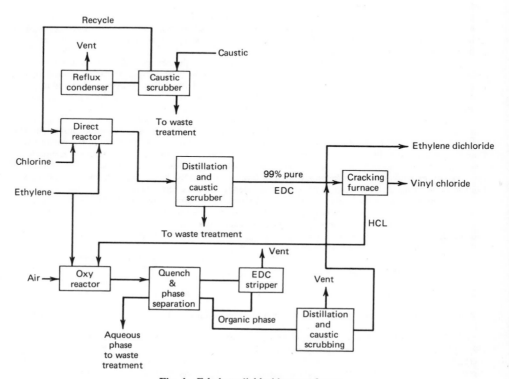

Fig. 4 Ethylene dichloride manufacture.

As a comparison, the process vent gas on the oxychlorination reactor has the following emission factors:[36]

Compound	Emission Factor (kg/metric ton EDC)
Carbon monoxide	5.5–12
Methane	0–47
Ethylene	1.9–7.4
Ethane	0–35
EDC	0.65–7
Ethyl chloride	0–7
Aromatic solvent	0–7

In this same process, the distillation vent releases only 1.3% as much mass of hydrocarbons as the process vent so this source is of little consequence. In addition, the composition of each stream will vary considerably depending on operating conditions, temperature control, catalyst activity, and scrubber efficiency.

The hydrocarbons from vent lines can be incinerated and the HCl in the combustion gas removed by caustic scrubbing in a packed tower or cross-flow scrubber. Heat recovery from incineration can be considered, but the additional investment in capital can seldom be justified because of heat losses and chloride corrosion problems. The presence of a large amount of chloride in incinerator feed gas requires higher than normal temperatures to get complete combustion. This higher temperature increases the nitrogen fixation which can make selection of an incineration system less desirable. Supplemental fuel is required for most vent streams to achieve effective combustion. With the spiraling energy costs, incineration has become an unfavorable choice for destruction of gases with a heating value of less than 10 MJ/m^3 (300 Btu/ft^3).[37] As an alternative to incineration, the hydrocarbons from vents can be treated by absorption or adsorption.

The application success of a particular method depends on the concentration, quantity, and operating conditions. Adsorption concentration (Section 3.6) can be used to reduce the volume of the gas to be incinerated and increase its fuel value. As an example, the vent stream hydrocarbon could be adsorbed on activated carbon and then the carbon regenerated with recycled hot flue gas from the hydrocarbon combustion.

Control of emissions from the manufacture of acrylonitrile (AN) through incineration can generally be considered economical. Acrylonitrile is manufactured exclusively from propylene and ammonia by the SOHIO process. The process, shown in Figure 5,[38] has three primary emission sources: (1) absorber vent, (2) fractionating column vents, and (3) storage tank breathing vents.

The absorber vent may be controlled with direct-fired or catalytic incinerator, or flare. Most plants use thermal or catalytic incineration with heat recovery. Since hydrocarbon emissions result from incomplete conversion in the reactor, most chemical companies have concentrated their efforts for control on development of more selective catalysts.[39] Such catalysts would increase AN production while lowering the amount of by-products vented. The fractionating column vents are generally flared because the volume is small, and the concentrations can vary widely.

Storage tanks have been reported by the EPA to be one of the largest sources of emission. Fixed roof tanks, for the most part, are being replaced with floating roof tanks, internal floating covers, and interconnected gas holder tanks. Vapor recovery systems have also been applied to fixed roof tanks. Vapors generated in the tank are displaced through a piping system to an external surge tank called a vapor saver. Inbreathing of saturated vapor instead of air prevents additional losses of hydrocarbons. Several storage tanks can be manifolded into a single vapor saver, gas holder, or vapor recovery system. The chemical industry puts severe restrictions on this type of approach due to the potential of explosions, vessel rupture, or contamination.

3.2 Control of Carbon Monoxide

Carbon black is manufactured by reduction of a hydrocarbon feedstock in the furnace black process (90% of production) or the thermal black process. The thermal black process produces a large particle-size carbon black whereas the furnace black process can produce a large range of particle sizes. Because of the larger particle size, utilization of natural gas as the feedstock, and recycle of the off-gases, atmospheric emissions from the thermal black process are insignificant.[40] A puff of carbon black can sometimes be seen during switching between cracking and heating cycles but the mass is small.

The typical furnace black process is shown in Figure 6. Burner operation in these furnaces is considered a trade secret but usually involves changing residence time in the cracking zone of the reactor, and control of temperature profiles in the flame front. The primary emissions are carbon

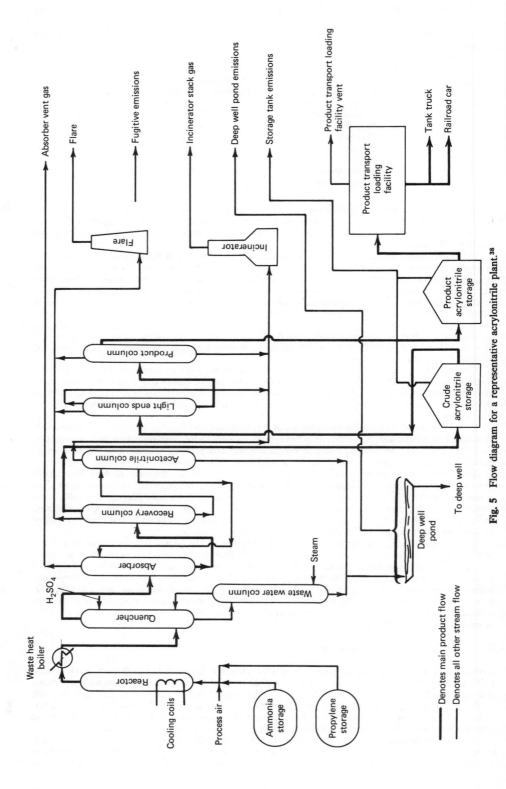

Fig. 5 Flow diagram for a representative acrylonitrile plant.[38]

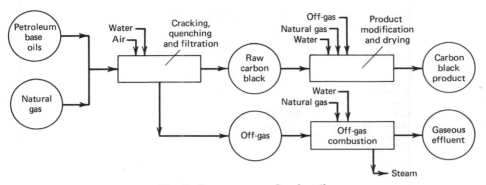

Fig. 6   Furnace process flowsheet.[40]

monoxide and hydrogen sulfide, with small quantities of particulates and hydrocarbons. Table 13 gives
a typical summary of overall emission factors for a furnace black plant.[41] Smaller particle size carbon
blacks give larger mass emissions of carbon monoxide and the residual carbon black is more difficult
to control. Particulate emissions are very low but there is increasing pressure to lower them further
due to the hazardous nature of carbon black. Table 14 gives concentrations of potentially hazardous
polynuclear aromatics (PNA) for a carbon black plant.[42] Hydrogen sulfide is expected to increase
due to a trend to higher sulfur oils in the future.

Table 13   Typical Atmospheric Emissions Factors for Carbon Black Plant[41]

Component	Range Mole, %	Typical kg/kg of Product	Mole, %
Hydrogen	2.7–7.5	0.11	6.7
Carbon dioxide	1.5–3.3	0.91	2.5
Carbon monoxide	3–7	1.3	5.5
Hydrogen sulfide	0.005–0.1	0.03	0.1
Sulfur oxide	0–0.015[a]	—	0
Methane	0.1–0.35	0.03	0.2
Acetylene	0.05–0.5	0.03	0.2
Nitrogen	32.5–40	8.2	35.5
Oxygen	0–2.5	0.07	0.3
Nitrogen oxide	8–100 ppm[a]	—	44 ppm
Particulate[b]	0.048 gr/scf[c]	0.002	0.048 gr/scf[c]
Water	42–50	7.3	49
Total		18.0	100

[a] Most data are near low side of range.
[b] After fabric filter with 99.78% control.
[c] Including water vapor.

**Table 14   Concentration of PNA in
Particulates in Air near Carbon Black
Plant[42]**

Compound	Concentration, μg/1000 m³ Air
Benzo(a)pyrene	0.035–0.58
Pyrene	0.15–0.33
Benzo(e)pyrene	0.18–0.76
Perylene	0.03
Benzo(ghi)perylene	0.37–0.16
Coronene	0.16–0.59
Fluoranthene	0.07–0.43

Table 15 Emission Factors for Major Pollutants from Furnace Carbon Black Process[40]

Pollutant	Emission Factor, g/kg of Product		
	Cracking Furnace	Product Modification	Off-Gas Combustion
CO	800–3000	—	5
Hydrocarbons	70–100	—	4
Hydrogen sulfide	5–13	—	<0.3
Particulate	1–3	0.04	1
Sulfur dioxide	—	—	18

Fabric filters are used to control particulates from the furnace. The hot gas must be cooled to below 530 K (—°F) to protect the bags, which are usually fiberglass. Silicon-graphitized and Teflon® coatings have been used to improve filter performance. Typical air-to-cloth ratios are about 0.025 m^3/sec-m^2 (5 ft^3/min-ft^2) of cloth area. Bag life varies widely from 9 to 30 months with reverse-jet air cleaning. Pressure drop is kept high, at about 140–250 mm (10–18 in.) water, to obtain maximum efficiency. Proper temperature control is important since condensation of corrosive gases can occur when the gas temperature falls below 505 K (450°F).

Some plants alter the product after manufacture. Product modification involves wetting and pelletizing the raw carbon black, and drying the pellets to remove moisture. Some particulate emissions result and are controlled with fabric filters similar to the type used on the cracking furnace. Emissions are very low and consist primarily of particulate as shown in Table 15.

The off-gas from the cracking furnace is burned in a CO boiler, incinerator, or combination. In the typical carbon black plant, only 40–60% of the heat can be utilized, so the use of a boiler will depend on the plant location. Emissions are substantially reduced as shown in Table 15, although sulfur oxides are increased. The combustor is usually operated at about 1250 K (1800°F). Flares operate at somewhat lower temperatures.[40] The development of a high-temperature fabric filter [875 K (1100°F)] would reduce energy requirements by eliminating the need for additional water quench in the furnace to protect the fabric filter.

3.3 Hazardous and Toxic Pollutants

When a potential emission is suspected of being extremely toxic or containing a cancer-suspect material, exceptional measures are needed in the control strategy. Whereas the control efficiency of 90–99% may be sufficient for most hydrocarbons in vents, hazardous and toxic chemicals may require removal down to levels of a few ppmv. In addition, most hazardous and toxic materials have been in the previously unregulated category (i.e., polychlorinated bisphenol).

Two basic approaches are available for removing hazardous and volatile organics from vent streams. The pollutant may be recovered in concentrated form for use in the process or used for process heat. The presence of halogen atoms in a recovered material usually makes combustion uneconomical. The other approach is to destroy the toxic material before it reaches the atmosphere. Five control methods constitute the most common methods for controlling hazardous pollutants. These are: absorption, adsorption, condensation, chemical reaction, and incineration. It is possible to combine two or more of these methods to achieve a desired goal. The selection of the best method will depend on effluent quantity, pollutant concentration, required efficiency, desired ultimate disposal, economic factors, and chemical and physical characteristics of the stream.[35]

For substances such as benzene, acrylonitrile, and styrene, absorption is generally more desirable. Figure 7 shows how adsorption, absorption, and incineration compare on a relative economic basis, taking into account the value of recovered material (benzene). These economics can also be expected to be typical of control costs for organics with low water solubility. Absorption with no. 2 fuel oil is ideal since the heating value of the fuel oil is improved with little capital expenditure, and the vapor pressure of the fuel oil is so low that hydrocarbon emissions are negligible. Generally, absorption is most economical at initial organic concentrations above about 300 ppmv. When absorption is not desirable, adsorption is the best choice, especially for large gas flow volumes. However, when the organic combustion heat is significant, incineration can be a better choice. For concentrations below 200 ppmv, incineration would be the best choice for most organics.[35] Consideration must also be given to the presence of solid particulates, acid gases, and oxygen as these factors complicate the equipment selection. Particles tend to clog absorber packing and most adsorption beds. Acid gases can be undesirable in combustion or absorption equipment. The presence of air or oxygen can result

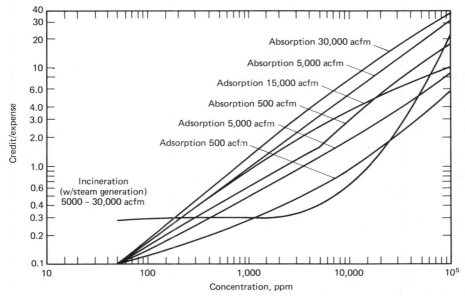

Fig. 7 Ratio of credit to expense for Bz as a function of treatment method, flow rate, and effluent concentration.[35]

in an explosive atmosphere which will require additional control or even negate the application of the method.

3.3.1 Control of Benzene

Benzene is a major raw material to the chemical industry and is a major noncriteria emission, as shown in Table 16. Benzene emissions can be controlled by preventing the release of the pollutant or by treating the waste stream. Different control methods have been developed for different types of emissions. The following are generally applicable to any process containing benzene. Major sources that need to be considered are fugitive releases, catalyst regeneration, vents, and storage areas.

Fugitive hydrocarbons are best controlled through good housekeeping, inspection, and maintenance. For pump and compressor seals, specific devices such as double mechanical or external liquid seals have been employed to help control emissions.[43] Other sources require major equipment additions such as installing an integrated vapor recovery or flare system for all vents and relief valves.

Catalyst regeneration is very common in processes using benzene. During the regeneration, benzene may be released along with other hydrocarbons. Special controls are not usually applied because of the infrequency of catalyst regeneration. Incineration of the regenerator gas to control overall hydrocarbons is usually practiced.[44]

Table 16 Major Sources of Benzene in Chemical Manufacture[31]

Source	Emission Rate, mg/yr
Maleic anhydride	18,524
Nitrobenzene	3,392
p-dichlorobenzene	892
o-dichlorobenzene	603
Fumaric acid	300
Acetone and phenol from cumene	183
Acrylonitrile	174
Chlorobenzene	84

Table 17   Control Devices and Reported Control Efficiencies for the Maleic Anhydride Industry[45]

	Emission Points		
Company	Flaking, Pelletizing, and Packaging	Product Recovery Scrubber	Storage Tank Vents
Amoco Chemicals Corp.	NR[a]	NR	NR
Ashland Oil, Inc.	—[b]	Scrubber[c]	Floating roof tanks[c]
Denka Chemical Corp.	Scrubber[c]	Thermal incinerator 93%, 95%	Floating roof tanks[c]
Koppers Co.	Scrubber[c]	Thermal incinerator 97%[e]	Return vents[c]
Monsanto Co.	Scrubber[c]	—[f]	Scrubber[c]
Reichold Chemicals, Inc.	—[b]	Carbon absorber 84%[d]	Scrubber, conservation vents[c]
Tenneco, Inc.	—[f]	—[f]	Scrubber, conservation vents[c]
U.S. Steel Corp.	Scrubber[c]	Catalytic incinerator 85%[d]	Floating roof tanks[c]

[a] Not reported.
[b] Plant does not have the emission point.
[c] Control efficiency not reported.
[d] Hydrocarbon control efficiency.
[e] Carbon monoxide control efficiency.
[f] No control.

Benzene can be controlled from vents by routing the emission to an integrated vapor recovery or incineration system. Wet scrubbing in an absorber, a combination scrubbing-incineration system, or an adsorption system are frequently used to achieve the high efficiency of control necessary for benzene.[44]

Controlling storage losses not only results in an emissions reduction, but prevents the loss of product or reactant and thereby improves the overall economics of the process. Most organic storage tanks being constructed in the chemical industry today have some type of vapor conservation or collection system. Open tanks can be converted to floating roof tanks in some cases, and pressure storage of chemicals can often be practiced for small storage quantities. Emissions from these types of storage, then, are virtually eliminated.

The efficiency of any method of control depends on the operating conditions, quality, and quantity of the waste, and state of repair of the system. One of the major sources of benzene in the chemical industry is the manufacturing of maleic anhydride, which is reported to release 18.5 Gg (20,000 tons) per year of benzene.[3] Table 17 lists producers of maleic anhydride along with control technologies and reported efficiencies.[45]

Incineration can achieve an overall hydrocarbon removal efficiency in excess of 99%. The incineration usually occurs in direct flame or a catalytic afterburner or boiler. When the amount of hydrocarbons present in the exhaust stream is very high, such as the product recovery scrubber in maleic anhydride, heat generation may be feasible, and can offset some of the cost of the control equipment. Adsorption onto activated carbon is usually the most economical choice over incineration without heat recovery when concentrations are below 200 ppm. Benzene control can approach 100% with adsorption.[44] As can be seen from Table 17, scrubbing is very popular for the removal of benzene in the maleic anhydride process. Absorption involves the transfer of the vapor phase components into a liquid solvent. After the benzene is removed, it may be reacted in the solvent or more easily disposed of. When the concentration of benzene in the gas is low, large quantities of absorbent and long contact times are required for adequate removal. For this reason, scrubbers are often used in series with incinerators or adsorbers for effective treatment.[44]

3.3.2   Control of Chlorinated Linear Alkyls

Several chlorinated ethanes are suspected toxic substances, and are typical of pollution problems with chlorinated organics. These include vinyl chloride, methyl chloroform (1,1,1-trichloroethane), 1,1,2-trichloroethane, and vinylidene chloride (1,1-dichloroethylene). These substances are all derived from ethylene dichloride. The major products are vinyl chloride (for manufacture of PVC) and methyl chloroform (a powerful solvent). Vinylidene chloride (VDC) and vinyl chloride are also used to produce transparent heat-shrink wraps which have excellent resistance to gases and water for the protection of food. The following reactions show the process of manufacture of these chlorinated ethanes.

$$\underset{\text{EDC}}{\text{ClCH}_2\text{CH}_2\text{Cl}} + \underset{\text{chlorine}}{\text{Cl}_2} \rightarrow \underset{\text{1,1,2-trichloroethylene}}{\text{CH}_2\text{ClCHCl}_2} + \underset{\substack{\text{hydrogen} \\ \text{chloride}}}{\text{HCl}} \tag{14}$$

$$\text{ClCH}_2\text{CH}_2\text{Cl} + \text{HCl} + \tfrac{1}{2}\text{O}_2 \rightarrow \text{CH}_2\text{ClCHCl}_2 + \underset{\text{water}}{\text{H}_2\text{O}} \tag{15}$$

$$\text{CH}_2\text{ClCHCl}_2 + \underset{\substack{\text{sodium} \\ \text{hydroxide}}}{\text{NaOH}} \rightarrow \underset{\text{VDC}}{\text{CH}_2=\text{CCl}_2} + \underset{\substack{\text{sodium} \\ \text{chloride}}}{\text{NaCl}} + \text{H}_2\text{O} \tag{16}$$

$$\text{CH}_2 + \text{CCl}_2 + \text{HCl} \rightarrow \underset{\text{methyl chloroform}}{\text{CH}_3\text{CCl}_3} \tag{17}$$

The manufacture and purification of vinylidene chloride is typical of the complexity and difficulty of control for substances of this type. Figure 8 gives a flow diagram for this process.[46] Emissions occur in vents on the separator, finishing columns, recycling tower, and storage tanks. VDC emissions are ducted from the vents to activated carbon adsorbers and VDC removed. Regenerating spent carbon[47] is not economical. VDC can be recovered by refrigerated solvent scrubbing using acetone, methyl ethyl ketone, ethylene dichloride, butyl acetate, or heptyl butyl ketone.

Flammable liquid wastes containing these types of chlorinated chemicals are usually disposed of by incineration. Often the liquid wastes are of high enough heating value to be self-sustaining. Best emission control is obtained with supplemental fuel.[47] Because of the high toxicity of vinyl chloride and VDC, the following control measures are also required to reduce air emissions and protect workers and equipment in the manufacturing operations:[47]

Pump, compressor, and agitator seals are controlled by installing double mechanical seals and maintaining a liquid between the seals at sufficient pressure to cause the liquid to leak into the pump should the seal fail. The gas in equipment that is to be opened for maintenance or inspection can be vented to a control device by purging the equipment with an inert gas such as nitrogen, or displacing the contents with water.

Purging of lines for loading, unloading, and sampling of these chemicals with the purge gas vented to a control device such as an air incinerator or carbon adsorption unit is also good practice.

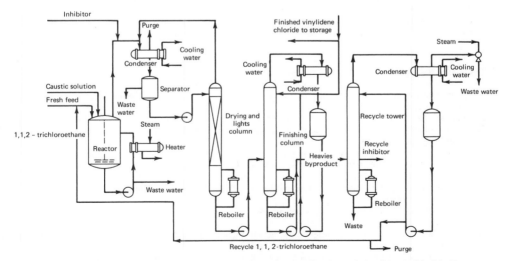

Fig. 8 Flow diagram showing the production and purification of vinylidene chloride.[46]

Fugitive emissions are especially difficult to control because every joint, flange, or relief valve is a potential source. The most common method of control is through a formal program of leak detection and repair.[43] Fixed and portable monitoring devices are used to locate problem areas for repair of the leaks. A good program in this area can reduce fugitive emissions by 90%.[47]

For vinyl chloride, tests have shown that captive and point sources in monomer and polymer plants can be controlled by absorption in an organic solvent, adsorption onto activated carbon, or incineration to less than 10 ppm. Based on other studies[35,43,44,47] this would appear to be optimistic for most hazardous organic chemicals.

3.4 Control of Nitrogen Oxides

A number of other chemical operations give rise to the release of nitrogen oxides. Organic nitration processes manufacturing materials such as nitrocellulose, nitrobenzene, TNT, and nitrofertilizers are probably the most common. Emissions also arise from metal etching and dissolving operations. The total emissions so produced are not significant nationally, but can produce objectionable local concentrations. In the past, treatment has often been comprised of absorption in water in stoneware-packed towers producing a dilute nitric acid. Because the vent gases were at atmospheric pressure and often quite dilute, efficient absorption was usually not achieved. Any of the control methods discussed in Section 2.3 could be utilized for efficient control.

Where the NO_x generation can be controlled in an enclosed vessel, a concentrated stream of NO_x which may be amenable to flame reduction can be obtained. Flame reduction is a three-stage process. In the first stage, O_2 and NO_x in the vent gas are mixed with excess fuel (natural or LP gas) and burned at a high temperature of 1700 K (2000°F) to form CO_2, N_2, and H_2O. Next, the gases are cooled by indirect heat exchange to 1030 K (1400°F) or by direct injection of steam. In the third stage, air is admitted to burn the excess fuel at low temperatures slightly above 1030 K (1400°F) to prevent atmospheric nitrogen fixation. Oxidation at even lower temperatures with a catalyst is desirable. The interrelationship of fuel type, excess fuel needed for NO_x reduction, and reaction temperature must be considered for a successful design, which may include heat interchangers or energy recovery. Michalek[49] has briefly discussed many NO_x control methods.

4 MINIMIZING EMISSION CONTROL COSTS

Many things can be done in the process industries to reduce the cost of air pollution control. One of the first is to consider changes to the process to prevent generation of an emission initially. This can involve utilization of different raw materials or treatment to remove a substance that becomes airborne.

Many natural ores, minerals, and clays contain small amounts of fluorides or other chemical species which become volatilized when processed thermally. Pretreatment to remove the objectionable components prior to processing is an alternative to cleaning of process vent gases.

Removal of fines which are easily airborne from a granular material can greatly reduce fugitive emissions from conveying and drying operations. If sufficiently valuable, the separated fines can be agglomerated or pelleted before further processing.

Substitution of low-sulfur fuels for high-sulfur fuels and carrying on combustion at lower temperatures and lower excess air can reduce SO_x and NO_x emissions without control equipment.

Other process modifications that can minimize control costs are to seal up the process to prevent interchange of process gases with the atmosphere; reduce the quantity of gas to be treated; or render

Table 18 Principles of Good Hood Design[58]

1. Completely enclose the emission point within the hood. If impractical, locate the exhaust port as close to the emission source as possible.
2. Position the hood to receive the "throw" of the source without need to deflect its direction.
3. Utilize density differences between the emission and ambient air or natural convective currents to aid capture.
4. Make atmospheric openings in the hood as small as possible and located where they will not encounter the natural movement of the emission. For access openings, use hinged panels opened only when access is needed.
5. Access panels should be hinged, fitted with soft gaskets, and quick-opening or self-sealing latches to encourage sealing.
6. Use the minimum face velocity across openings needed to prevent contaminant escape and overcome external air turbulence.
7. Locate atmospheric openings where external air currents from drafts, machinery, moving objects, or personnel will not be encountered, or provide baffles for shielding. Do not locate openings where air movement on one side can cause outflow through an opposing opening.

Table 19 Fundamental Means of Reducing or Eliminating Pollutant Emissions to the Atmosphere[59]

I. Eliminate the source of the pollutant
   1. Seal the system to prevent interchanges between system and atmosphere
      a. Use pressure vessels
      b. Interconnect vents on receiving and discharging containers
      c. Provide seals on rotating shafts and other necessary openings
   2. Change raw materials, fuels, etc., to eliminate the pollutant from the process
   3. Change the manner of process operation to prevent or reduce formation of, or air entrainment of, a pollutant
   4. Change the type of process step to eliminate the pollutant
   5. Use a recycle gas, or recycle the pollutants rather than using fresh air or venting
II. Reduce the quantity of pollutant released or the quantity of carrier gas to be treated
   1. Minimize entrainment of pollutants into a gas stream
   2. Reduce number of points in system in which materials can become airborne
   3. Recycle a portion of process gas
   4. Design hoods to exhaust the minimum quantity of air necessary to ensure pollutant capture
III. Use equipment for dual purposes, such as a fuel combustion furnace to serve as a pollutant incinerator

the pollutants more amenable to collection at lower cost. The employment of the principles of good hood design will help.[50-58] Table 18 lists seven principles of good hood design. Treatment and recycling of gases used in processes such as drying rather than using fresh atmospheric air can reduce the size of control equipment. Transfer of thermal energy to process materials by indirect heat transfer can also eliminate the need for control devices. Keeping gaseous pollutants as concentrated as possible maximizes the driving force for mass transfer and minimizes the number of transfer units needed in the collection device. Table 19 summarizes general process means of reducing control costs.

REFERENCES

1. *The Chemical Industry and Pollution Control,* National Industrial Pollution Control Council, Washington D.C., 1971.
2. *Chemical and Engineering News* **58**(23), 36 (June 9, 1980).
3. Eimutis, E. C., Quill, R. P., and Rinaldi, G. M., *Source Assessment: Overview Matrix for National Criteria Pollutant Emission (1978),* U.S. Environmental Protection Agency, EPA-600/2-78-004r, July 1978.
4. Semrau, K. T., Control of Sulfur Oxide Emissions from Primary Copper, Lead, and Zinc Smelters, *J. Air Pollution Control Assoc.* **21**(4), 185 (1971).
5. Tabler, S. K., Federal Standards of Performance for New Stationary Sources of Air Pollution, *J. Air Pollution Control Assoc.* **29**(8), 803–811 (1979).
6. Doumas, A. C., Shepherd, B. P., and Muehlberg, P. E., The Lime Industry, in *Industrial Process Profiles for Environmental Use,* U.S. Environmental Protection Agency, EPA-600/2-77-023r, February 1977.
7. Lewis, C. J., and Crocker, B. B., The Lime Industry's Problem of Airborne Dust, *J. Air Pollution Control* **19**(1), 31–39 (1969).
8. Chalekode, P. K., Blackwood, T. R., and Archer, S. R., *Source Assessment Crushed Limestone,* U.S. Environmental Protection Agency, EPA-600/2-78-004e, April 1978.
9. Simonini, J. W., Gravel Bed Filter, in *Air Pollution Control and Design Handbook, Part 2,* (N. Cheremisinoff and R. A. Young, eds.), Dekker, New York, 1977, pp. 924–927.
10. Cuffe, S. T., Walsh, R. T., and Evans, L. B., Chemical Industries, in *Air Pollution, 3rd edition, Volume IV, Engineering Control of Air Pollution* (A. C. Stern, ed.), Academic Press, New York, 1977, pp. 735–812.
11. Donovan, J. R., Palermo, J. S., and Smith, R. M., Sulfuric Acid Converter Optimization, *CEP* **74**(9), 51–54 (1978).
12. Tucker, W. G., and Burleigh, J. R., SO_2 Emission Control from Acid Plants in *Sulfur and SO_2 Developments* Section of Chemical Engineering Progress Technical Manual, AIChE, New York, p. 102.
13. Duros, D. R., and Kennedy, E. D., Acid Mist Control, *CEP* **74**(9), 70–77 (1978).
14. York, O. H., and Poppele, E. W., Two Stage Mist Eliminators for Sulfuric Acid Plants, in Ref. 11, pp. 85–90.

15. Brink, J., *Can. J. Chem. Eng.* **41**, 134 (1963).

16. Donovan, J. R., Smith, R. M., and Palermo, J. S., The Role of Catalyst Engineering in Sulfuric Acid Plant Operations, *Sulfur* (July/August 1977).

17. Scheidel, C. F., Sulfur Dioxide Removal from Tail Gas by the Sulfacid Process, preprint 6E, 61st Annual AIChE Meeting, Los Angeles, Dec. 1–5, 1968.

18. Davis, J. C., SO₂ Absorbed from Tail Gas with Sodium Sulfite, *Chem. Eng.* **78**, 43 (Nov. 29, 1971).

19. Uno, T., Atsukawa, M., and Muramatsu, K., The Pilot Scale R&D and Prototype Plant of MHI Line-Gypsum Process, *Proc. 2nd International Lime/Limestone Wet-Scrubbing Symposium,* EPA-APTD-1161, 833–849 (1972).

20. Kronseder, J. E., Cost of Reducing Sulfur Dioxide Emissions, *CEP* **64**(11), 71 (1968).

21. Sherwood, T. K., and Pigford, R. L., *Absorption and Extraction,* McGraw-Hill, New York, 1952, pp. 368–383.

22. Adlhart, O. J., Hindin, S. G., and Kenson, R. E., Processing Nitric Acid Tail Gas, *CEP* **67**, 73–78 (Feb. 1971).

23. Newman, D. J., Nitric Acid Plant Pollutants, *CEP* **67**, 79–84 (Feb. 1971).

24. Gillespie, G. R., Boyum, A. A., and Collins, M. F., Nitric Acid: Catalytic Purification of Tail Gas, *CEP* **68**, 72–77 (April 1972).

25. Reed, R. M., and Harvin, R. L., Nitric Acid Plant Fume Abaters, *CEP* **68**, 78–79 (April 1972).

26. Fornoff, L. L., A New Molecular Sieve Process for NO$_x$ Removal and Recovery from Nitric Acid Plant Tail Gas, *AIChE Symp. Ser.* **68**(126), 111–114 (1972).

27. Mayland, B. J., and Heinze, R. C., Continuous Catalytic Absorption for NO$_x$ Emission Control, *CEP* **69**, 75–76 (May 1973).

28. Swanson, C. G., Jr., Prusa, J. V., Hellman, T. M., and Elliott, D. E., NO$_x$ Absorber—A Winning System for Energy Conservation and Product Recovery, *Poll. Eng.* **10**, 52–53 (October 1978).

29. Kelly, H. D., Block, C., and Kuncl, K. L., NO$_x$ Emissions Plummeted Far Below Minimum EPA Standards with No Increase in Fuel Needs, *Chem. Processing* **40**, 22–24 (Jan. 1977).

30. Ando, J., and Tohata, H., Nitrogen Oxides Abatement Technology in Japan—1973, *Environmental Tech. Ser.,* EPA-R-2-73-284, U.S. EPA, Research Triangle Park, June 1973.

31. Eimutis, E. C., Quill, R. P., and Rinaldi, G. M., *Source Assessment: Noncriteria Pollutant Emissions,* U.S. Environmental Protection Agency, EPA-600/2-78-004t, July 1978.

32. Muehlberg, P. E., Shepherd, B. P., Redding, J. T., Behrens, H. C., and Parsons, T., Brine and Evaporite Chemicals Industry, in *Industrial Process Profiles for Environmental Use,* U.S. Environmental Protection Agency, EPA-600/2-77-023o, February 1977.

33. Doorenbus, H. E., and Parsons, T., The Fluorocarbon-Hydrogen Floride Industry, in *Industrial Process Profiles for Environmental Use,* U.S. Environmental Protection Agency, EPA-600/2-77-023p, February 1977.

34. Jaasund, S. A., Control of Fine Particle Emissions with Wet Electrostatic Precipitation, in *Second Symposium on Transfer and Utilization of Particulate Control Technology,* U.S. Environmental Protection Agency, EPA-600/9-80-039b, September 1980, pp. 452–467.

35. Crocker, B. B., Removal of Hazardous Organic Vapors from Vent Gases, in *Proceedings, Control of Specific (Toxic) Pollutants, Gainesville, FL, Feb. 13–16, 1979,* APCA, Pittsburgh, 1979, pp. 360–376.

36. Liepins, R., Mixon, F., Hudak, C., and Parsons, T., The Industrial Organic Chemicals Industry, in *Industrial Process Profiles for Environmental Use,* U.S. Environmental Protection Agency, EPA-600/2-77-023f, February 1977.

37. Blackwood, T. R., *Feasibility Study—Small Scale Combustion Test Unit for EPA,* Report on Contract 68-03-2550 Task 6 for U.S. Environmental Protection Agency, IERL-Co, Cincinnati, Ohio, June 1977.

38. Hughes, T. W., and Horn, D. A., *Source Assessment: Acrylonitrile Manufacture,* U.S. Environmental Protection Agency, EPA-600/2-77-107j, September 1977.

39. Tierney, D. R., and Blackwood, T. R., *Status Assessment of Toxic Chemicals: Acrylonitrile,* U.S. Environmental Protection Agency, EPA-600/2-79-210a, December 1979.

40. Gerstle, R. W., Richards, J. R., Parsons, T., and Hudak, C., The Carbon Black Industry, in *Industrial Process Profiles for Environmental Use,* U.S. Environmental Protection Agency, EPA-600/2-77-023d, February 1977.

41. *An Investigation of the Best Systems of Emission Reduction for Furnace Process Carbon Black Plants in the Carbon Black Industry,* Emission Standards and Engineering Division, Office of Air Quality and Planning and Standards, U.S. Environmental Protection Agency, April 1976.

42. Hangebrauck, R. P., Von Lehmden, D. J., and Meeker, J. E., *Sources of Polynuclear Hydrocarbons in the Atmosphere,* Public Health Service, 999-AP-33, 1967.

43. Wallace, M. J., Controlling Fugitive Emissions, *Chem. Eng.* 18(86), 78–92 (August 27, 1979).

44. Ochsner, J. C., Blackwood, T. R., and Zeagler, L. D., *Status Assessment of Toxic Chemicals, Benzene,* U.S. Environmental Protection Agency, EPA-600/2-79-210d, December 1979.

45. Lewis, W. A., Rinaldi, G. M., and Hughes, T. W., *Source Assessment, Maleic Anhydride,* report on contract 68-02-1874 submitted to U.S. Environmental Protection Agency, IERL-RTP, Research Triangle Park, NC, June 1978.

46. Shelton, L. G., Hamilton, D. E., and Fisackerly, R. H., Vinyl and Vinylidine Chloride, in *Vinyl and Diene Monomers, Part 3, Volume 24* (E. C. Leonard, ed.), Wiley-Interscience, New York, 1971, pp. 1205–1282.

47. Tierney, D. R., Blackwood, T. R., and Piana, M. R., *Status Assessment of Toxic Chemicals: Vinylidene Chloride,* U.S. Environmental Protection Agency, EPA-600/2-79-210c, December 1979.

48. Hushon, J., and Korneich, M., Air Pollution Assessment of Vinylidene Chloride, U.S. Environmental Protection Agency, MTR-7230, May 1976.

49. Michalek, R., Removal of Nitrogen Oxides from Industrial Non-Combustion Sources, *Poll. Eng.* 8, 30–33 (March 1976).

50. *Industrial Ventilation,* 15th edition, Amer. Conf. of Governmental Ind. Hygienists, Lansing, MI, 1978.

51. AHSRAE *Guide and Data Book—1976 Systems Handbook,* Industrial Ventilation Section, ASHRAE, New York, 1976.

52. Brandt, A. O., Exhaust System Design, *Heating and Ventilating,* May 1945.

53. *Fundamentals Governing the Design and Operation of Local Exhaust Systems,* ANSI Z9.2—1971, Amer. Nat. Stds. Inst., Inc., New York, 1972.

54. *Practices for Ventilation and Operation of Open-Surface Tanks,* ANSI Z9.1—1977, Amer. Nat. Stds. Inst., Inc., New York, 1977.

55. Alden, J. L., and Kane, J. M., *Design of Industrial Exhaust Systems for Duct and Fume Removal,* 4th edition, Industrial Press, New York, 1970.

56. Dalla Valla, J. M., *Exhaust Hoods,* 2nd edition, Industrial Press, New York, 1952.

57. Hemeon, W. E. L., *Plant and Process Ventilation,* Industrial Press, New York, 1954.

58. Crocker, B. B., Capture of Hazardous Emissions, *Proceedings, Control of Specific (Toxic) Pollutants, Feb. 13–16, 1979, Gainesville, FL,* Air Pollution Control Assn., Pittsburgh, 1979, pp. 415–433.

59. Perry, R. H., *Engineering Manual,* 3rd edition, McGraw-Hill, New York, 1976, pp. 10–50.

CHAPTER 25

AIR POLLUTION CONTROL IN PETROLEUM REFINERIES

LEIGH SHORT

Environmental Research & Technology, Inc.
Houston, Texas

1 INTRODUCTION

At the present time (1982), there are some 250 petroleum refineries in the United States, varying in size from several hundred thousand barrels per day of crude oil throughput to a few hundred barrels per day throughput. With the lessening demand for petroleum products, the number of operating refineries has been decreasing steadily for the last 2 yr.

There is a wide variation in the process units within a refinery, ranging from a "topping unit" at the simplest extreme to a fully integrated refinery with perhaps a dozen different processes in one refinery. Additionally, some refineries may operate on sweet (low-sulfur) crude while others may operate using sour crude as a feed stock (there is a worldwide trend toward the use of more high-sulfur crude oils because the supply of readily available low-sulfur crudes is diminishing). Both the parameters of refinery complexity and crude oil sulfur content have a substantial impact on the potential air emission from a refinery and, hence, on the air pollution control techniques used.

Before describing the air pollution control techniques, a brief description will be given of the various refinery processes and air emission sources within a refinery.

The following is a list of the major processes normally found in a petroleum refinery (by no means are all processes found in all refineries).

1. Distillation—atmospheric and vacuum. This includes not only the crude unit, common to essentially all refineries, but distillation equipment is also a part of many of the processes listed below.
2. Catalytic cracking. Most commonly fluid bed (FCCU).
3. Thermal cracking and visbreaking.
4. Hydrocracking.
5. Catalytic reforming.
6. Alkylation.
7. Isomerization.
8. Solvent refining or extraction.
9. Hydrotreating.
10. Product treating, for example, sweetening.
11. Wax, grease, and lube oil production.
12. Asphalt manufacture.
13. Coking.
14. Refinery gas treating, for example, amine absorption.
15. Gas processing (LPG).
16. Sulfur plant (Claus plant).
17. Sulfuric acid manufacture.

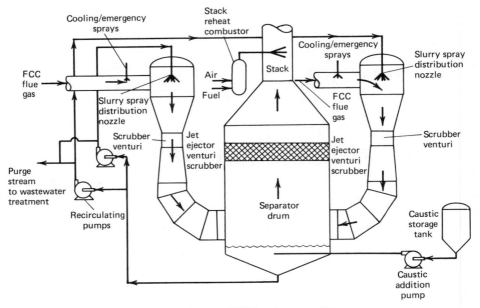

Fig. 1 Schematic of Exxon FCC jet ejector scrubbing system.

Figure 1 is a block flow diagram for a refinery which illustrates the process flow sequence.

In addition to the above processes, there are several generic types of equipment which are potential sources of air polution. These include:

Crude and product storage

Product blending facilities

Process heaters and boilers

Oil–water separators (API separator)

Blowdown and gas flaring systems

Fugitive hydrocarbon emissions will also potentially occur at each valve, flange, pump or tank seal, compressor seal, and so on. The measurement and control of fugitive emissions are discussed at the end of this chapter.

Table 1 summarizes the major sources of air emissions within a refinery, excluding fugitive emissions from valves, flanges, pumps, tanks, and the API separator.

This chapter does not present techniques for estimating emission quantities. These methods are available in EPA publication AP-42 data, and in some cases, via estimating techniques included in the CFR.

The need for air pollution control stems from one of three reasons:

1. *Safety and Health Protection.* For example, hydrogen sulfide is an extremely toxic gas and is fatal at very low concentrations.
2. *Economic Reasons.* For example, allowing some solvents to escape to the atmosphere would simply be too expensive.
3. *Environmental Regulations.* State, federal, or local.

2 REGULATORY FRAMEWORK

The basic air pollution control regulations for petroleum refineries stem from the Clean Air Act and standards of performance promulgated by the EPA, to meet the requirements of the act (40 CFR 60 subpart J, K, Ka). In addition, states have promulgated the State Implementation Plans (SIPs) which incorporate the federal regulations and provide for additional control measures and/or inspection measures. This is particularly true in states such as California, where regulations are also imposed by air pollution control districts—each of which may have slight variations in the impacts on petroleum

Table 1   Summary of Major Sources of Air Emissions Within a Refinery

Process Units	Emission Type
Crude unit including desalter	None.
Vacuum unit	Noncondensable vapors from steam barometric condenser hot wells. If surface condensers are used, this source will be essentially eliminated.
Catalytic cracking	SO_2, NO_x, particulates, carbon monoxide, hydrocarbons, aldehydes, ammonia. These emissions are released from the regenerator and/or CO boiler and not the reactor.
Catalytic reforming	Generally none, although some hydrocarbons may release during regeneration. If regeneration is continuous, minor amounts of CO may be released. The quantities of NH_3 and/or H_2S which may be formed are typically recovered in the sour water or refinery fuel gas stream.
Catalytic hydrocracking	Emissions from catalyst regenerations which may release significant quantities of CO over relatively short time frames. As in the case of catalytic reformers, any NH_3/H_2S formed is captured either in the refinery fuel gas system or the sour water system.
Alkylation	None.
Light ends recovery	Essentially none.
Catalytic hydrotreating	Similar to catalytic hydrocracking.
Hydrogen manufacture	None.
Solvent deasphalting	None.
Residual oil hydrodesulfurizing	None, other than catalyst regeneration.
Asphalt blowing	None, assuming the vent gases, which are highly odorous, are vented before release.
Coking	Wind-blown coke dust and emissions from the storage containers for the water used in cutting the coke.
Storage and blending	All storage tanks containing light hydrocarbons, including crude, have the potential to emit hydrocarbons to the atmosphere. Breathing losses can result from weathering or during emptying and filling. Also, the seals of the floating roof tanks will leak to a certain extent.
Process heaters	Particulates, SO_x, CO, hydrocarbons, aldehydes, NO_x. The quantities will, of course, depend on the source of fuel—gas, oil, residual oil, etc., the physical properties of each, and the burner design (for NO_x emissions, particularly).
Wastewater treatment	API separator (hydrocarbons).
Pressure relief and flare systems	Hydrocarbons.

refineries. The federal EPA has also issued control technology guidelines (CTGs) which specify control technique guidelines or control technique options. An important example of a CTG is the one issued for fugitive emissions control, which has now been adopted (or incorporated) in several state SIPs.

The standards of performance of subpart J, 40 CFR 60 are applicable to the following facilities in petroleum refineries:

(a) *The provisions of this subpart are applicable to the following affected facilities in petroleum refineries: fluid catalytic cracking unit catalyst regenerators, fuel gas combustion devices, and all Claus sulfur recovery plants except Claus plants of 20 long tons per day (LTD) or less. The Claus sulfur recovery plant need not be physically located within the boundaries of a petroleum refinery to be an affected facility, provided it processes gases produced within a petroleum refinery.*

(b) *Any fluid catalytic cracking unit catalyst regenerator or fuel gas combustion device under paragraph (a) of this section which commences construction or modification after June 11, 1973, or any Claus sulfur recovery plant under paragraph (a) of this section which commences construction or modification after October 4, 1976, is subject to the requirements of this part.*

The standards are summarized in Table 2.

The standards of performance of subpart K, Vessels for Storage of Petroleum Liquids, is summarized in Table 3. This standard applies to vessels with a capacity greater than 40,000 gal (except at a drilling

Table 2 Summary of Standards

	Performance Standard
Particulates	a. 1.0 lb/1000 lb of coke burn off in the regenerator
	b. If the gases from the regenerator pass through a waste heat boiler (CO boiler), particulate emissions may exceed (a) but not exceed 0.1 lb/million Btu of heat input
Carbon monoxide	0.050% by volume
Sulfur dioxide	a. Maximum of 230 mg/dscm (0.10 gr/dscf) of H_2S in fuel gas. This standard does not apply to process upsets or relief valve leakage sent to a flare. The standard also does not preclude higher H_2S concentrations if the SO_2 emissions from combustion are controlled
	b. Claus plant emissions
	i. 0.025% by volume sulfur dioxide at 0% oxygen (dry basis) if emissions controlled by oxidation or reduction control followed by incineration
	ii. 0.030% volume sulfur dioxide if emissions controlled by reduction not followed by incineration

Table 3 Standards of Performance of Subpart K

Construction Time	Standard
Before May 18, 1978	i. Vessel to be equipped with floating roof and vapor recovery system. (Vapor pressure 1.5–11.1 psia)
	ii. Vessel to be equipped with a vapor recovery system (vapor pressure greater than 11.1 psia)
After May 19, 1978	i. If the true vapor pressure of the stored liquid is 1.1 to 11.1 psia, the vessel shall have (1) an external floating roof consisting of a pontoon type or double deck type cover or (2) a fixed roof with a floating type cover and closure device, or (3) a vapor recovery system
	ii. If the vapor pressure exceeds 15 psig and there are no atmospheric emissions, this standard does not apply

or production facility). It separates the requirements for vessels built after May 18, 1978 and applies if the true vapor pressure of the stored liquids exceeds 1.5 psia.

Note that in the above federal regulations there are no standards for NO_x or fugitive emissions. These standards and/or guidelines are contained in the SIPs, or in local (county, air district) ordinances.

As an example of the complexity and wide variation in air pollution regulations at the *state* level, the American Petroleum Institute each year publishes the fuel sulfur regulations, federal, state, and local for all 50 states, all of which are different.

2.1 Regulatory Control Technology Definitions

The type of control technology selected will depend on the degree of removal efficiency, the particular circumstances of the local region (whether or not it is attainment for a specific air pollutant), and whether or not the source is defined as major.

Within the United States a source is defined as major, and therefore, requiring air pollution controls if the emission rates exceed a "de minimis" value. The de minimis rates are shown in Table 4. Very few refineries of any size or complexity fail to exceed these rates and, hence, require controls.

Again, within the United States, one of two types of control is required—best available control technology (BACT) or lowest achievable emission rate (LAER) technology. The essential distinctions between the two are:

1. BACT is applicable when the air pollution region is attainment—that is, ambient values of the pollutant do not exceed some specified value. In specifying BACT, one must consider the economic costs of the various competing technologies.

Table 4 De Minimis Emission Rates

Pollutant	Emission Rate, tons/yr
Carbon monoxide	100
Nitrogen dioxide	10
Total suspended particulates	10
Sulfur dioxide	10
Ozone	10 of volatile organic compounds
Lead	1
Mercury	0.2
Beryllium	0.004
Asbestos	1
Fluorides	0.02
Sulfuric acid mist	1
Vinyl chloride	1
Total reduced sulfur:	
Hydrogen sulfide	1
Methyl mercaptan	1
Dimethyl sulfide	1
Dimethyl disulfide	1
Reduced sulfur compounds:	
Hydrogen sulfide	(See above)
Carbon disulfide	10
Carbonyl sulfide	10

2. LAER is applicable when a region is nonattainment—that is, ambient values of the pollutant exceed some specified value. In the specification of LAER technology, economics are not considered.

The level of control (removal efficiency) is typically specified via a new source performance standard (NSPS) as described previously. These standards do not specify technology but rather the degree of control required. A refiner is free to select whatever technology is appropriate provided it meets BACT or, if necessary, LAER guidelines.

3 SOURCES OF EMISSIONS

Table 1 listed the primary sources of process and stack emissions in a refinery. As this table illustrates, the major sources of process emissions are sulfur recovery, fluid catalytic cracking regenerators, and process heaters and boilers. (Storage tanks and the API separator are sources of fugitive emissions, discussed separately in Section 5.)

3.1 Sulfur Recovery (Amine Absorption and Claus Unit)

The amount of sulfur in crude oil obviously has a direct bearing on the potential SO_2 and H_2S emissions. Crude oil sulfur contents vary widely from low sulfur (sweet) to high sulfur (sour). In general, the majority of the sulfur in crude oil is present in the heavier fractions and, hence, most of the potential sulfur emissions will be associated with processes that treat these fractions. Typically, the sulfur content in the oil is converted to H_2S in a hydrotreating/hydrocracking process, or directly to SO_2 via combustion. The H_2S is absorbed and recovered in an amine unit and then converted to elemental sulfur, in a Claus plant, with only minimal SO_2 releases to the material. The tail gas from a Claus plant is one of the main potential sources of SO_x emissions in a refinery (the FCC regenerator being the other).

The tail gas from a Claus unit contains H_2S, SO_2, CS_2, COS, and sulfur. The emission rates and concentrations will depend on the efficiency of the Claus plant. A typical three-stage Claus unit, approximately 95% efficient, will produce a tail gas containing about 7000–12,000 ppm sulfur compounds. This tail gas will also contain small quantities of CO formed by reaction between the hydrocarbons and the CO_2 in the Claus unit feed stream.

Table 5 shows typical compositions of the tail gas from a Claus unit (before SO_2 removal), as well as typical feed compositions.

3.2 FCCU Catalyst Regeneration

Catalysts are used in several refining processes—catalytic reforming, fluid catalytic cracking, catalytic hydrocracking, and so on. These catalysts will become coated with carbon and metals which must

Table 5 Typical Compositions of Feed Stream and Tail Gas for a 94% Efficient Claus Unit

Component	Sour Gas Feed Volume, %	Claus Tail Gas Volume, %
H_2S	89.9	0.85
SO_2	0.0	0.42 [b]
S_8 vapor	0.0	0.10 as S_1
S_8 aerosol	0.0	0.30 as S_1
COS	0.0	0.05
CS_2	0.0	0.05
CO	0.0	0.22
CO_2	4.6	2.37
O_2	0.0	0.00
N_2	0.0	61.04
H_2	0.0	1.60
H_2O	5.5	33.00
HC	0.0	0.00
	100.0	100.00
Temperature, °F	104	284
Pressure, psig	6.6	1.5
Total gas volume [a]	—	3.0 × feed gas volume

[a] Gas volumes compared at standard conditions.

[b] NSPS requires an emission of less than 250 ppmv (0.025%) SO_2, 0% O_2, dry basis, if Claus unit tail gas is oxidized as the last control step, or, 300 ppmv SO_2 equivalent reduced compounds (H_2S, COS, CS_2) and only 10 ppm H_2S as SO_2, 0% O_2, dry basis, if the tail gas is reduced as the last control step.

be removed (regenerated) to restore activity. During regeneration the carbon is oxidized to CO, CO_2 and the hydrocarbons will be burned incompletely.

In some applications regeneration occurs only once or twice per year. Though there may be significant emissions (NO_x, CO, CO_2, particulates) during regeneration, the total quantity released is small.

However, fluid catalytic cracking regeneration is a continuous process, and the uncontrolled emissions are perhaps the most important air pollution source within a refinery. Other gases from FCC regeneration will contain particulates, SO_x, CO, hydrocarbons, NO_x, aldehydes, and ammonia. A recent EPA report (EPA-600/2-80-075d) summarized the estimates of these emission rates, shown in Table 6. The data in this table show dramatically the impact of a CO boiler in lessening the air emissions.

Table 7 presents similar data for emission concentrations from an FCC unit, but this table includes concentrations of the noncontrolled pollutants such as aldehydes.

3.3 Boilers and Process Heaters

Refinery boilers (for process steam production) and heaters are fired with the most readily available and economic fuel—usually some combination of purchased natural gas, refinery fuel gas, and refinery fuel oil(s). The latter is typically a residual fuel oil. Usually one-half or more of the heating needs will be provided by refinery fuel gas.

The emissions from boilers and heaters depend on the fuel type, but typical emission values are listed in Table 8.

3.4 Coking and Calcining

The primary air emissions from coking and calcining are H_2S, organic sulfur compounds, and coke particulates not removed in the air pollution control equipment. The primary emission sources are summarized below:

Sulfides
    Delayed coking: Steam from the steam-out operation and accumulator fuel gas

    Fluid coking: Accumulator fuel gas reactor scrubber, burner, or boiler emissions

    Calcining: Gases from the stacks and gases from the kiln

Table 6 Emission Rates from a "Typical FCC" Unit (Capacity, 50,000 BPSD)

Significant Emission Sources	Gas Volume scfm	Gas Composition, ppm (Wet Basis Except for Particulates)						Pollutant Emission Rate, lb/hr				
		Part., gr/dscf	SO_x[a]	NO_x[b]	CO	HC	O_2	Part.	SO_x	NO_x	CO	HC
Point A[c]												
Traditional regeneration	87,050 (4.3% H_2O)	0.055	1.360	200	114,000	500	Nil	39.4	1,200	127	44,000	110.3
High-temperature regeneration	92,500 (3.4% H_2)	0.043	1.070	200	500	10	9,800	32.8	1,000	134.4	205.0	2.3
Point B[d]												
Traditional regeneration	110,950	0.043	1.070	200	10	10	10,000	39.4	1,200	162	4.9	2.8

[a] Expressed as SO_2.
[b] Expressed as NO_2.
[c] Lies between electrostatic precipitator and CO boiler.
[d] Lies downstream of CO boiler.

Table 7   Emission Rates from FCCU Regenerators, Before and After CO Boiler

Chemical Species[a]	Composition of Flue Gas from FCCU Without CO Boiler	Emissions from FCCU Equipped with CO Boilers: Regenerator Flue Gas Composition	
SO_2, ppmv	130–3300	2700	14–871[b]
SO_3, ppmv	NA[c]	NA[c]	0.7–13.5
NO_x (as NO_2), ppmv	8–394	500	94–453
CO, vol.%	7.2–12.0	0–14	0.0
CO_2, vol.%	10.5–11.3	11.2–14.0	13.5–16.1
O_2, vol.%	0.2–2.4	2.0–6.4	3.2–7.0
N_2, vol.%	78.5–80.3	82.0–84.2	77.0–82.7
H_2O, vol.%	13.9–26.3	13.4–23.9	9.2–22.7
Hydrocarbons, ppmv	98–1213	NA[c]	0–46
Ammonia, ppmv	0–675	NA[c]	0–15
Aldehydes, ppmv	3–130	NA[c]	0–20
Cyanides, ppmv	0.19–0.94	NA[c]	0–19
Particulates, gr/scf	0.08–1.39	0.017–1.03	0.012–0.304
Temp., °F	1000–1200	458–820	386–727

[a] All considerations on dry basis.
[b] Based on sampling of six stacks.
[c] Not available.

Table 8   Typical Emission Values

Pollutant	Fuel	
	Natural Gas, lb/10^6 scf	Fuel Oil, lb/10^3 gal
Hydrocarbons (as CH_4)	3	1
Particulates	5–15	—[a]
SO_x as SO_2	0.6[b]	157S[c]
CO	17	5
NO_x as NO_2	120–130[d]	60[e]

[a] A function of fuel oil grade and sulfur content. For grade 6: lb/10^3 gal = 10S + 3; for grade 4: 7 lb/10^3 gal; for grade 5: 10 lb/10^3 gal.
[b] Based on average sulfur content of natural gas of 2000 gr/10^6 scf.
[c] S equals percent by weight of sulfur in fuel.
[d] Use first number for tangentially fired units, second for horizontally fired units.
[e] Strongly dependent on the fuel nitrogen content.

Particulates	
Delayed coking:	Emissions from the fired heater and fugitive dusts from the coking units
Fluid coking:	Reactor scrubbers, burner or boiler emissions, product coke recovery
Calcining:	Fugitive emissions from conveyor belts, stacks, and the kiln

Typical emission rates are 30 lb CO/1000 barrels of feed and 520 lb particulates/1000 barrels of feed.

3.5 Other Process Emission Sources

Representative hydrocarbons and other emissions occurring from the remaining process units in a refinery are shown below (reference EPA-600/2-80-75d):

Process	Emissions
Vacuum distillation	50 lb HC/1000 barrels vacuum unit charge (barometric condensers only)
Asphalt blowing	40–80 lb hydrocarbons/ton asphalt treated
Blowdown	580 lb hydrocarbons/1000 barrels refinery feed (uncontrolled)

Data on compressor engines are given in Table 9.

Table 9   Compressor Engines

	Pollutant, lb/10³ ft³ Gas Burned			
Engine Type	NO_x as NO_2^a	CO	HC as CH_4^b	SO_x as SO_2^c
Reciprocating	3.4	0.43	1.4	$2S$
Gas turbine	0.3	0.12	0.02	$2S$

[a] At rated load. In general, NO_x emissions increase with increasing load and intake air temperature. They generally decrease with increasing air–fuel ratios and absolute humidity.
[b] Overall less than 1% by weight is methane.
[c] S = refinery gas sulfur content (lb/1000 scf): factors based on 100% combustion of sulfur to SO_2.

4   CONTROL TECHNOLOGY

Air pollution emissions are typically controlled through one or more of the following methods.

1. Process modifications—to prevent or minimize the pollutant formation.
2. Add-on devices—for example, low NO_x burners.
3. Electrostatic precipitators, cyclones, and so on.
4. Chemical or physical processes—for example, absorption, adsorption, combustion, catalytic convertors, and so on.
5. Mechanical design features—for example, double versus single seals, closed vent systems which collect and flare emissions.

Table 10 summarizes the more important control technologies for specific applications within petroleum refineries.

4.1   Fluid Catalytic Cracking—Air Pollution Controls

The purpose of catalytic cracking (FCCU) is to process a feed of relatively heavy gas oil, to produce valuable, lighter fractions (predominantly gasoline). This is done by splitting apart the hydrocarbon chains of the heavier fractions into smaller (and lighter) hydrocarbon fractions, in the presence of a catalyst.

In an FCCU, the hot catalyst is mixed with fuel and recycled in a riser tube where the cracking reactions take place. Above the riser the reactor products are injected with steam, causing the products to be stripped from the catalyst. The products rise as vapor through the reactor, are cooled, and separated by distillation. The spent catalyst flows to the regenerator. Here air is used to regenerate the catalyst in a controlled combustion reaction, where carbon (coke) is burned off the catalyst. The catalyst is then recycled for reuse.

The flue gas from the regenerator then flows through an internal cyclone to remove catalyst fines and to a CO boiler and to the atmosphere, often via an electrostatic precipitator. Note that high-temperature regeneration, a process developed recently and in operation in some refineries, will eliminate the need for a CO boiler, and dramatically alter the air pollution emissions from an FCCU. This high-temperature regeneration process is discussed below.

Table 10   Air Pollution Control Technologies in Petroleum Refineries

Pollutant	Major Sources	Control Techniques
SO_2	FCC unit	None required at present but some companies use caustic scrubbing. Also, flue gas desulfurization processes developed for the utility industry are being evaluated for their applicability.
	Claus plant tail gas	SCOT process, Wellman–Lord process, plus others.
	Boiler	Low-sulfur fuel or flue gas desulfurization (not now used).
H_2S	Hydrocracking, hydrotreating, etc.	Amine scrubbers and the recovered H_2S is sent to the Claus plant for conversion to sulfur.
CO	FCC unit	High-temperature regeneration or CO boiler to recover the waste heat.
Particulates	FCC unit delayed coker	Cyclones or electrostatic precipitators.
NO_x	Process furnaces	Low NO_x burners, flue gas recirculation, and in rare instances, selective or nonselective catalytic reduction.

4.1.1   Particulates

The existing NSPS for particulate emissions is 1 kg/1000 kg of coke burned (0.038 gr/scfd). This control level can be readily achieved by use of cyclones followed by an electrostatic precipitator, that is, the NSPS is consistent with BACT. Medium- to high-energy level scrubbers may also meet this standard, but it is doubtful if low-energy scrubbers can do so.

A multicyclone is most commonly used as the primary particulate control device. The off-gases are forced tangentially into a cylindrical vessel and the particulates are forced to the outer wall, primarily by centrifugal force, where they fall to the bottom of the cyclone and are removed. Cyclones will generally remove particles of 40 μm and larger—hence, a cyclone alone will not consistently meet the NSPS standards. In this event, an electrostatic precipitator is often located on the CO boiler stack to remove these residual particulates (a third-stage high-energy cyclone or water scrubber could be used but this is not common).

In the electrostatic precipitator, very fine particulates are charged by a high-voltage direct current corona. A large platelike electrode is used to collect these charged particles, which are then removed by rapping or washing. Electrostatic precipitators will typically achieve 99% plus removal of catalyst fines, thereby meeting the performance standards.

4.1.2   Carbon Monoxide

There are two routes to control of carbon monoxide emissions—use of a CO boiler, and high-temperature regeneration (the latter may involve use of a promoted cracking catalyst system). As stated earlier, the NSPS for carbon monoxide is 500 ppm, which can be met by either technique. In fact, the CO content of flue gases leaving a CO boiler is often less than 50 ppm. High-temperature regeneration is not capable of achieving such a low value—CO concentrations are usually much nearer to, but less than, 500 ppm when using this control technique (see Table 7).

Since high-temperature regeneration (HTR) is a relatively new technique, until very recently most FCC units used a CO boiler. HTR does offer significant advantages over conventional regeneration. These are summarized below.

1. More complete catalyst regeneration. Both the catalyst activity and selectivity are improved; less coke is produced per unit of feed, which reduces the SO_x emissions, because the sulfur on the catalyst going to the regenerator is related to the carbon on the catalyst.

2. Lower catalyst inventory. The complete regeneration increases the catalyst activity, which means that a lower catalyst-to-oil ratio is possible. In turn, the unit's capacity or severity can be increased if bottlenecks are removed from the rest of the process.

3. Better heat recovery within the regenerator. The major oxidation reactions are as follows:

Reaction	Heat Released, Btu/lb Carbon
$C + \frac{1}{2}O_2 \rightarrow CO$	4,440
$CO + \frac{1}{2}O_2 \rightarrow CO_2$	10,160
$C + O_2 \rightarrow CO_2$	14,160

Traditional regenerators do not allow complete CO oxidation, and, hence, a large quantity of heat is either lost or must be recovered in a CO boiler.

4. Low CO emissions. It is possible to meet the 500 ppm CO NSPS without a CO boiler.

Due to these advantages, some "traditional" units have been revamped for high-temperature regeneration. Generally, revamping requires the replacement of cyclones, the plenum chamber, cyclone diplegs, the regenerator grid and seals, and the catalyst overflow weir, all of which must be type 304 stainless steel rather than carbon steel in order to withstand the higher temperatures resulting from HTR.

Recent data indicated that essentially complete catalyst regeneration at lower temperatures can be achieved through the use of CO oxidation promoter catalysts. These catalysts are identical in terms of cracking performance to their nonpromoter counterparts; however, they contain metals that catalyze the regenerator reaction

$$CO + \frac{1}{2}O_2 \rightarrow CO_2$$

Hence, a lower regenerator temperature can be used to meet the CO NSPS without upgrading the metallurgy of the regenerator internals for HTR. The promoter catalysts are considerably more expensive, however, which means that there is a tradeoff between capital expenditure and increased catalyst cost. Regeneration of the catalyst is not quite as effective at the lower temperature; hence, the selectivity of the catalyst is slightly poorer in that more coke is produced.

4.1.3 FCCU SO_x Emissions

At present, there is no NSPS for SO_x control of FCC regenerators (the EPA has indicated it would like to promulgate such a regulation). Hence, these possible control measures discussed here are not now mandatory in most cases. There are, however, some control techniques for SO_x which are specific to FCC units—feed desulfurization and combustion promoters. In addition, Exxon has developed a jet ejector liquid scrubber process for simultaneous particulate and SO_x control.

Upson has presented data showing that use of combustion promoters will significantly decrease regenerator SO_x emissions—his data show a reduction of 250 to 90 ppm SO_2 in the regenerator off-gas with the use of a promoter. Upson postulates that in the presence of the promoter, the free alumina on the catalyst reacts in the regenerator to form nonvolatile aluminum sulfate, which is then reduced to H_2S in the reactor and leaves with the reactor products. This H_2S will be recovered in the refinery H_2S (amine) system and sent to the Claus plant.

Regenerator $Al_2O_3 + 3S + \frac{9}{2} O_2 \rightarrow Al_2(SO_4)_3$
Reactor $Al_2(SO_4)_3 + 12 H_2 \rightarrow Al_2O_3 + 3H_2S + 9 H_2O$

The Exxon jet liquid scrubber has been developed for simultaneous removal of SO_x and particulates. A schematic of this system is shown in Figure 1.

In this scrubber, the particulates are removed in much the same manner as in a conventional venturi scrubber, except that in a conventional scrubber energy is provided by a fan on the gas side, in the jet ejector scrubber the energy is provided by a pump on the liquid side. Also in the scrubber absorption of SO_2 takes place:

Absorption $2NaOH + SO_2 \rightarrow Na_2SO_3 + H_2O$
              $Na_2SO_3 + SO_2 + \frac{1}{2} H_2O \rightarrow 2NaHSO_3$
Oxidation $Na_2SO_3 + \frac{1}{2} O_2 \rightarrow 2Na_2SO_4$

The purge stream from the scrubber will contain soluble salts and insoluble materials (particulates). This liquid stream will require treatment before discharge.

Performance data for this system are shown below:

Range of Pollutant Loadings to Scrubber	Collection Efficiency, %
SO_x	
200–500 ppm	95–99
Particulate	
0.1–0.3 gr/scf	85–95
Condensables	
0.1–0.2 gr/scf	90

Table 11   Desulfurization Processes[a]

Process	Licensors
1. Gulfining	Gulf Research & Development
2. Hydrofining	British Petroleum and Exxon
3. Hydrodesulfurization	Institut Français du Pétrole
4. Isomax	Chevron Research or UOP
5. Trickle hydrosulfurization	Shell Development
6. HGO-unicracking	Union Oil Co. of California
7. Unifining	Union Oil Co. of California and UOP
8. Gofining	Exxon R&E

[a] Source: *The Oil and Gas Journal.*

The feed to the FCC unit may be desulfurized as an alternate to treating the flue gas. This type of technology is well known and consists of mixing the feed stock with a hydrogen-rich gas in the presence of a catalyst. The sulfur in the feed stock is converted to H_2S, which may then be recovered. Use of this process does also alter the characteristics of the liquid feed (gas oils). The reader is referred to the literature describing these processes for specific details. Table 11 provides a summary of available desulfurization processes.

4.2   Coking

Coking is a thermal cracking process used to convert reduced crudes, tars, vacuum distillation bottoms, and the like to produce gas, naphtha, gasoline, gas oils, and coke. There are two commonly used coking processes—delayed coking and fluid coking. In addition, some refineries may calcine coke.

4.2.1   Delayed Coking

In this process, feed enters the coker fractionator to remove "light" material such as gas oils, naphthas, and other light fractions. The fractionator bottoms combine with a recycle stream, are heated in a reaction heater to 900–1000°F (steam is also injected here), and fed to a coke drum, where the coking reaction takes place. In this drum the vapors evolve from the drum, but liquids experience cracking and polymerization to produce coke.

To remove coke from the drum, steam is injected to remove hydrocarbon vapors. This mixture is cooled to form a stream that may be separated into three parts:

1. Water with coke particles. This is added to the coke removal stream (described below)
2. Hydrocarbon liquid—added to the refinery slop oil system
3. Noncondensables—these either go to a flare or to a refinery fuel gas system

After the coke drum is further cooled by filling with water and draining, the coke is cut out with a high-pressure water jet.

In the delayed coking process, steam and hydrocarbons may be emitted to the atmosphere when the drum is opened. Some steam and hydrocarbon vapors may be produced by vaporization of a portion of the cooling water and/or release of trapped pockets of hydrocarbons. These potential emissions may be minimized by venting the quenching steam to a vapor recovery or blowdown system. The organics may be flared, although some SO_x emissions will result. When the drum cools to 212°F, a water flood will further minimize emissions, although hydrocarbon vapors can be released from this water. (See Section 5.2 on tanks for control measures to minimize this problem.)

4.2.2   Fluid Coking

In this process the coking reaction is carried out in a fluidized bed. A portion of the feedstock is converted to vapors, which leave in the reactor overhead. The remainder of the charge stock is deposited on the fluidized bed of coke particles. This coke is removed and fed to a burner vessel, where some of the coke is burned to maintain heat balance, and the coke product is removed through an elutriator.

The gas and distillate products leave the reactor through cyclones and enter a scrubbing section above the reactor, where particulates are removed and partial condensation occurs. The vapors leaving are separated into naphtha and gas oil products, while the high boiling distillate is recycled to the reactor and coked.

In the fluid coking process, the gas from the burner contains CO and sulfur containing hydrocarbons. A boiler may be used to recover this heat, but again all sulfur compounds are converted to SO_2 and ultimately are released to the atmosphere.

4.2.3 Coke Calcining

Coke calcining is the process whereby green coke is converted to the more valuable needle coke, which is lower in sulfur content. Green coke is fed from one of the feeding bins to an inclined, rotary kiln calciner. The calcining reaction takes place at a temperature of 2400–2700°F. The coke travels countercurrent to the combustion gases from the burner. The calcined coke leaving the kiln is cooled and stored.

Gases from the cooler discharge end hood and conveyors pass through a dust removal chamber and fan to a stack. Gases from the hotter end of the kiln pass through a separate dust removal chamber, an incinerator, and from there to a main (hotter) stack.

A calcining unit may require both particulate and sulfur oxide control measures. The particulate control equipment used is a scrubber, filter, or an electrostatic precipitator. Amine absorption is used to control the H_2S emissions.

4.3 Sulfur Recovery

Sulfur in the refinery vapor streams will be predominantly in one of two forms, H_2S or SO_2. The H_2S will be recovered from fuel gas streams and the like by use of an amine unit. In this process, the H_2S is absorbed in monoethanolamine (usually), or some other amine. The H_2S is then recovered from this MEA solution via regeneration with steam. Figure 1 is a schematic flow diagram of an MEA absorption system. The H_2S is then typically routed to a Claus sulfur recovery unit for conversion to sulfur. In the Claus process, some H_2S is oxidized to form SO_2 and water. Additional H_2S reacts with this SO_2 to form elemental sulfur and water. There is an NSPS for control of SO_2 leaving the Claus plant.

Four processes have demonstrated an ability to meet SO_2 NSPS compliance levels. These are the Beavon process, the SCOT process (Shell Claus off-gas treating), the Wellman–Lord process, and the Institut Français du Pétrole (IFP) process.

4.3.1 Beavon Process

The Beavon process developed jointly by the Ralph M. Parsons Company and the Union Oil Company of California is licensed by the Union Oil Company. A schematic diagram of the process is shown in Figure 2. The Claus plant tail gas is heated to reaction temperature by mixing with the hot combustion products formed in the reducing gas (CO, H_2) generator. The combined gas stream is then passed through a cobalt–molybdenum catalyst bed where all sulfur compounds are converted to H_2S by hydrogeneration and hydrolysis according to the following reactions:

$$S + H_2 \rightarrow H_2S$$
$$SO_2 + 3H_2 \rightarrow H_2S + 2H_2O$$
$$CS_2 + 2H_2O \rightarrow 2H_2S + CO_2$$
$$COS + H_2O \rightarrow H_2S + CO_2$$
$$CO + H_2O \rightarrow CO_2 + H_2$$

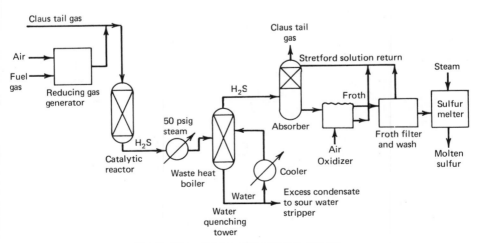

Fig. 2   Flow diagram of the Beavon process.

After exiting the catalytic reactor, the gas stream is cooled in a waste heat boiler and/or quench tower. After cooling, the gases are fed to a Stretford absorber. The H_2S is absorbed in an aqueous solution of sodium carbonate (Na_2CO_3), sodium metavanadate ($NaVO_3$), and anthraquinone 2,7-disulfonic acid (ADA). The H_2S is initially absorbed in the alkaline solution per the following reaction:

$$H_2S + Na_2CO_3 \rightarrow NaHS + NaHCO_3$$

Sodium bisulfide (NaHS) and sodium bicarbonate ($NaHCO_3$) further react with $NaVO_3$ to precipitate sulfur.

$$NaHS + NaHCO_3 + 2NaVO_3 \rightarrow S + Na_2V_2O_5 + Na_2CO_3 + H_2O$$

The reduced vanadium is subsequently oxidized back to the pentavalent state by blowing in air in the presence of the ADA which works as an oxidation catalyst:

$$Na_2V_2O_5 + \frac{1}{2}O_2 \rightarrow 2NaVO_3$$

In the same operation, the finely divided sulfur appears as a froth which is skimmed off, filtered, and washed. The product sulfur is then separated from the washwater and melted. The Stretford solution is recirculated to the absorber from the oxidizer and the sulfur filter.

The treated tail gas exits the system from the absorber overhead. It is odorless, contains only a trace of H_2S, and small amounts of COS and CS_2 which are not converted in the catalytic reactor. The concentration of all sulfur compounds combined is less than 100 ppm. Incineration of the off-gas is not required.

4.3.2 Shell Claus Off-Gas Treating (SCOT) Process

The SCOT process was developed by Shell International Research in the Netherlands, and is licensed in the United States by Shell Development Company. The flow diagram of this process is illustrated in Figure 3. The sulfur compounds in the Claus plant tail gas are converted to hydrogen sulfide as in the Beavon process; that is, the Claus off-gas is heated along with the addition of reducing gas (H_2, CO) and passed through a reactor containing a cobalt–molybdenum catalyst. The gas is cooled utilizing a waste heat boiler for steam generation and a water quenching tower. Excess condensate from the quench is routed to a sour water stripper.

Gas from the quench tower is countercurrently contacted in an absorption column with a di-isopropanolamine (DIPA) solution which absorbs all but trace amounts of the H_2S plus about 30% of the CO_2. The sulfide-rich DIPA is sent to a conventional stripper where the H_2S is removed from

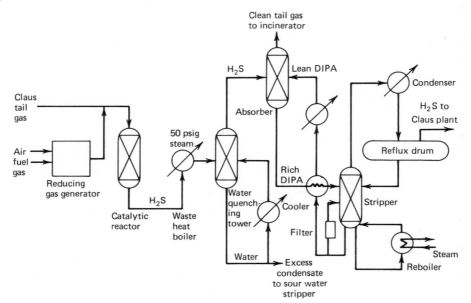

Fig. 3 Flow diagram of the SCOT process.

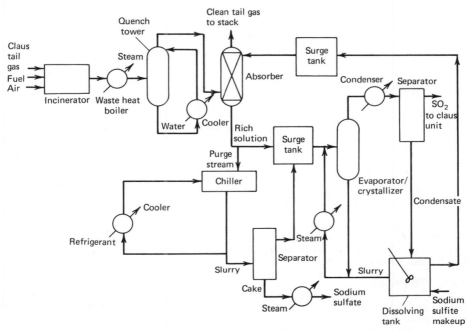

Fig. 4   Wellman–Lord SO$_2$ recovery process flow diagram.

the solution. The overhead gas, mostly H$_2$S, is recycled to the Claus unit inlet. The lean amine solution is returned to the absorber. The absorber outlet gas, containing some unabsorbed H$_2$S and residual organic sulfides, is sent to the original Claus tail gas incinerator before discharge to the atmosphere. Typical concentrations of SO$_2$ are in the 200–500 ppm range.

4.3.3   *Wellman–Lord Process*

This process was developed by Davy-Powergas, Inc. The process is illustrated in Figure 4. The Claus tail gas stream is incinerated, the gas is then cooled in a waste heat boiler and direct-quench tower before entering the absorber. In the absorber, the gas stream is scrubbed with a lean Na$_2$SO$_3$ to form sodium bisulfite (NaHSO$_3$) and sodium pyrosulfite (Na$_2$S$_2$O$_5$) for the following reactions:

$$SO_2 + Na_2SO_3 + H_2O \rightarrow 2NaHSO_3$$
$$2NaHSO_3 \rightarrow Na_2S_2O_5 + H_2O$$

The rich bisulfite solution is fed to an evaporator/crystallizer regeneration system. Heat supplied to the evaporator causes the decomposition of the bisulfite and pyrosulfite:

$$2NaHSO_3 \rightarrow Na_2SO_3 + SO_2 + H_2O$$
$$Na_2S_2O_5 \rightarrow SO_2 + Na_2SO_3$$

The SO$_2$ and water vapor pass overhead from the evaporator to a primary condenser, where 80 to 90% of the water is removed. The condensate is stripped with steam to remove dissolved SO$_2$. Stripper overhead and condenser overhead vapors are mixed and sent to a secondary condenser. Condensate is returned to the stripper and the vapor stream is recycled to the Claus plant. The recycle is 90–95% SO$_2$.

The main problem facing the Wellman–Lord process is the formation of nonregenerable sulfate and thiosulfate by the following side reactions:

$$Na_2SO_3 + \tfrac{1}{2}O_2 \rightarrow Na_2SO_4$$
$$SO_2 + 3Na_2SO_3 \rightarrow 2Na_2SO_4 + Na_2S_2O_3$$
$$Na_2S_2O_5 \rightarrow SO_2 + Na_2SO_3$$
$$2Na_2SO_3 + SO_3 \rightarrow Na_2SO_4 + Na_2S_2O_5$$

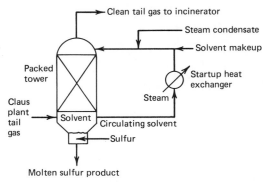

Fig. 5   IFP-1500 Claus tail gas treatment.

This creates the need to purge a portion of the solution from the system, necessitating chemical replacement and posing a waste disposal problem. Chemical requirements have been minimized by a system which selectively removes the sulfate by crystallization. The concentrated purge stream can then be dried for sale or disposal.

4.3.4   *IFP Process*

The IFP process was developed by the Institut Français du Pétrole for the removal of sulfur compounds from Claus plant tail gas. A schematic diagram is given in Figure 5.

This process converts mixed hydrogen sulfide/sulfur dioxide streams to sulfur and water by a liquid-phase Claus reaction using a proprietary catalyst. The process is primarily used to clean up Claus unit tail gas. The technology is an extension of the Claus reduction process but is carried out in the liquid phase.

The tail gas, at Claus unit exit pressure, is injected into the bottom of a packed tower which provides the necessary surface area for gas–liquid contact. A low-vapor pressure polyethylene glycol solvent containing a proprietary carboxylic acid salt catalyst in solution circulates countercurrent to the gas.

The catalyst forms a complex with H_2S and sulfur dioxide which in turn reacts with more of the gases to regenerate the catalyst and forms elemental sulfur. The reaction is exothermic and the heat released is removed by injecting and vaporizing steam condensate. Temperature is maintained at about 250–270°F, sufficient to keep the sulfur molten but not high enough to cause much loss of sulfur or glycol overhead. The sulfur is drawn off continuously through a seal leg. Overhead gases are incinerated. (Outlet SO_2 concentration is 1000–2000 ppm.)

4.4 NO_x Control

NO_x is formed during combustion from either nitrogen in the air or via reaction with nitrogen in the fuel. There is no generally applicable NSPS for NO_x in refineries. However, there is in California and some other states, substantial interest in NO_x control, and in some air pollution control districts regulations of NO_x are in effect, either for process heaters or compressor exhausts.

There are NSPS for NO_x for large (greater than 250 MBtu/hr) boilers. These are shown below.

Fuel Type	NSPS	
	lb NO_x/MBtu Input to Boilers	NO_x ppm (3% excess O_2 dry)
Gaseous fuel	0.2	150
Liquid fuel	0.3	225
Solid fuel	0.5 (subbituminous)	375
	0.6	450

It is possible that the boiler in a refinery will fall under this regulation, that is, that the heat release will be greater than 250 MBtu/hr.

There are two basic routes to NO_x control:

1. Combustion modification
2. Flue gas treatment

Combustion modification includes methods to inhibit the thermal and fuel NO_x formation. This may be done either by advanced (low NO_x) burner design or by flue gas recirculation. Flue gas recirculation has proved effective for oil and gas, but has not generally proved as effective as coal. Low NO_x burner designs are available which will provide NO_x emission rates of 0.2–0.4 lb/MBtu-hr. These burners are manufactured by Foster Wheeler, Riley Stoker, Babcock and Wilcox, among others.

Flue gas treatment generally consists of either catalytic reduction (selective or nonselective), or other processes such as activated carbon, copper oxide, or electron beam processes. Of all these methods, only selective catalytic reduction appears to be near or have commercial acceptance. This method is relatively widely used in Japan, but has not been much used in the United States.

A recent report (EPA-600/7-81-120) presents a detailed comparative evaluation of the advanced low NO_x burner designs and two flue gas treatment processes, the Exxon and Hitachi Zosen DeNO$_x$ processes. This report describes in detail the status of each process and compares the capital and operating costs for the three processes.

5 FUGITIVE EMISSIONS

In June 1978, EPA issued an OAQPS guideline series (EPA-450/2-78-036) "Control of Volatile Organic Compound Leaks from Petroleum Refinery Equipment." This guideline, based on an extensive sampling program conducted in 1976 and 1977 at several United States refineries, provided the framework for what state regulations now exist for fugitive emissions control. The control methodology in this guideline contains the assumption that reasonably available control technology (RACT) is the norm that can be applied to existing petroleum refineries. RACT is defined as the lowest emission limit that a particular source is capable of meeting by the application of control technology that is reasonably available considering technological and economic feasibility. It may require technology that has been applied to similar, but not necessarily identical, source categories. It is not intended that extensive research and development be conducted before a given control technology can be applied to the source.

5.1 Emission Factors and Leak Frequencies

Table 12 presents a summary of the emission factors and leak frequencies for petroleum refinery equipment—valves, flanges, pump seals, compressor seals, and drains. These factors are further broken down by process unit type in Tables 13 through 18. In Table 12, "Gas streams" refers to streams containing hydrocarbons that are normally gaseous at ambient conditions and to streams at elevated temperatures such that all liquid is in the vapor phase. Leaking sources are defined as those with hydrocarbon concentrations at the source surface greater than 200 ppm or actual measured leaks greater than 0.00001 lb/hr.

Table 12   Estimated Vapor Emission Factors for Nonmethane Hydrocarbon Petroleum Refineries

Source Type	Estimated Percent Leaking[a]	Emission Factor Estimated, lb/hr-source
Valves		
Gas streams	29.3	0.047
Light liquid/two-phase streams	36.5	0.023
Heavy liquid streams	6.7	0.0007
Flanges	3.1	0.00058
Pump seals		
Light liquid streams	63.8	0.255
Heavy liquid streams	22.6	0.045
Compressor seals	76.5	1.1
Drains	19.2	0.070
Relief valves	41.2	0.35

[a] Leaking sources in this report are defined as sources with screening values greater than or equal to 200 ppmv or sources with measured leak rates greater than 0.00001 lb/hr.

Table 13   Summary of Emission Data by Process Unit—All Valve Types

Unit Identification	Percent Leaking[a]	Maximum Measured Leak, lb/hr	Emission Factor Estimated, lb/hr-source
Atmospheric distillation	22.3	0.123	0.0023
Fuel gas/lt ends processing	34.3	2.264	0.046
Catalytic cracking	25.8	2.321	0.47
Catalytic reforming	56.2	0.286	0.029
Alkylation	37.4	7.15	0.031
Vacuum distillation	0	—	Neg
Catalytic hydrotreating/ refining	24.2	0.338	0.0051
Aromatics extraction	33.3	0.054	0.0053
Delayed coking	10.5	0.091	0.0019
Dewaxing/treating	15.2	0.101	0.011
Sulfur recovery	0	—	*
Hydrocracking	32.5	0.381	0.057
Hydrogen production	18.4	0.087	0.0013
Hydrodealkylation	38.9	0.048	0.013
Other	0	—	*
All units combined	27.8	7.15	0.023

[a] Leaking sources in this report are defined as sources with screening values greater than or equal to 200 ppmv or sources with measured leak rates greater than 0.00001 lb/hr.

Table 14   Summary of Emission Data by Process Unit—Pumps

Unit Identification	Percent Leaking[a]	Maximum Measured Leak, lb/hr	Emission Factor Estimated, lb/hr-source
Atmospheric distillation	43.6	0.983	0.022
Fuel gas/lt ends processing	53.2	10.45	0.186
Catalytic cracking	40.3	2.50	0.081
Catalytic reforming	78.0	3.00	0.177
Alkylation	78.9	14.00	1.34
Vacuum distillation	12.0	0.084	*
Catalytic hydrotreating/ refining	37.7	0.779	0.033
Aromatics extraction	58.1	2.06	0.195
Delayed coking	27.0	0.190	0.020
Dewaxing/treating	40.0	3.42	0.056
Sulfur recovery	—	—	—
Hydrocracking	50.0	0.231	0.053
Hydrogen production	0.0	—	*
Hydrodealkylation	80.0	0.151	*
Other	0.0	—	*
All units combined	48.5	14.00	0.169

[a] Leaking sources in this report are defined as sources with screening values greater than or equal to 200 ppmv or sources with measured leak rates greater than 0.00001 lb/hr.

Table 15 Summary of Emission Data by Process Unit—Compressors

Unit Identification	Percent Leaking[a]	Maximum Measured Leak, lb/hr	Emission Factor Estimated, lb/hr-source
Atmospheric distillation	100.0	3.76	*
Fuel gas/lt ends processing	61.0	19.17	1.05
Catalytic cracking	77.4	13.1	0.66
Catalytic reforming	90.3	0.911	0.18
Alkylation	100.0	1.07	*
Vacuum distillation	—	—	—
Catalytic hydrotreating/ refining	41.7	0.242	*
Aromatics extraction	100.0	0.640	*
Delayed coking	100.0	3.43	*
Dewaxing/treating	80.0	0.481	*
Sulfur recovery	—	—	—
Hydrocracking	66.7	0.292	*
Hydrogen production	—	—	—
Hydrodealkylation	50.0	0.00086	*
Totals	76.5	19.17	1.07

[a] Leaking sources in this report are defined as sources with screening values greater than or equal to 200 ppmv or sources with measured leak rates greater than 0.00001 lb/hr.

Table 16 Summary of Emission Data by Process Unit—Flanges

Unit Identification	Percent Leaking[a]	Maximum Measured Leak, lb/hr	Emission Factor Estimated, lb/hr-source
Atmospheric distillation	1.47	0.0328	0.0001
Fuel gas/lt ends processing	7.43	0.1817	0.00085
Catalytic cracking	0	—	Neg
Catalytic reforming	7.76	0.413	0.0034
Alkylation	3.03	0.0204	0.00014
Vacuum distillation	0	—	Neg
Catalytic hydrotreating/ refining	3.67	0.0738	0.00042
Aromatics extraction	6.67	0.0113	*
Delayed coking	0	—	Neg
Dewaxing/treating	1.67	0.0036	0.00003
Sulfur recovery	0	—	*
Hydrocracking	6.06	0.0867	0.0028
Hydrogen production	5.26	0.0016	*
Hydrodealkylation	0	—	*
Totals	3.05	0.413	0.00058

[a] Leaking sources in this report are defined as sources with screening values greater than or equal to 200 ppmv or sources with measured leak rates greater than 0.00001 lb/hr.

Table 17 Summary of Emission Data by Process Unit—Drains

Unit Identification	Percent Leaking[a]	Maximum Measured Leak, lb/hr	Emission Factor Estimated, lb/hr-source
Atmospheric distillation	21.7	1.13	*
Fuel gas/lt ends processing	14.0	1.21	0.026
Catalytic cracking	27.8	0.0223	*
Catalytic reforming	19.0	0.379	*
Alkylation	12.1	0.239	0.027
Vacuum distillation	100.0	0.019	*
Catalytic hydrotreating/ refining	18.5	0.048	*
Aromatics extraction	2.7	0.048	*
Delayed coking	27.3	0.058	*
Dewaxing/treating	19.2	3.16	*
Sulfur recovery	0	—	*
Hydrocracking	25.0	0.331	*
Hydrogen production	25.0	0.698	*
Hydrodealkylation	20.0	0.007	*
Other	0	—	*
Totals	19.2	3.16	0.070

[a] Leaking sources in this report are defined as sources with screening values greater than or equal to 200 ppmv or sources with measured leak rates greater than 0.00001 lb/hr.

Table 18 Summary of Emission Data by Process Unit—Relief Valves

Unit Identification	Percent Leaking[a]	Maximum Measured Leak, lb/hr	Emission Factor Estimated, lb/hr-source
Atmospheric distillation	14.3	0.034	*
Fuel gas/lt ends processing	75.0	3.33	*
Catalytic cracking	20.0	2.51	*
Catalytic reforming	28.6	0.20	*
Alkylation	50.0	4.04	*
Vacuum distillation	0	—	*
Catalytic hydrotreating/ refining	28.6	0.175	*
Aromatics extraction	25.0	0.00091	*
Delayed coking	25.0	2.411	*
Dewaxing/treating	—	—	*
Sulfur recovery	—	—	*
Hydrocracking	0	—	*
Hydrogen production	0	—	*
Hydrodealkylation	—	—	*
Totals	41.2	4.04	0.35

[a] Leaking sources in this report are defined as sources with screening values greater than or equal to 200 ppmv or sources with measured leak rates greater than 0.00001 lb/hr.

Table 19   Percentage of Sources Emitting 90% of Total Source Mass Emissions

	Percentage of Total Mass Emissions	Percentage of Total Number of Sources	Minimum Screening Value
Valves			
Gas streams	90	5	70,000
Light liquid/			
two-phase streams	90	15	5,300
Heavy liquid streams	90	16	50
Pump seals			
Light liquid streams	90	27	6,500
Heavy liquid streams	90	27	200
Compressor seals	90	45	4,200
Flanges	90	45	90
Drains	90	6	500
Relief valves	90	18	1,400

To orient the reader toward the significance of this data and the distribution of "leaking" equipment, the data presented in the tables above were analyzed statistically. Results of this analysis are presented in Table 19.

It is evident from this data that most of the mass emissions come from a relatively small number of the sources—that is, to effectively control emissions one should ideally be able to identify the "large" leak sources. Control strategies promulgated consist of an inspection, maintenance, and reporting system designed to locate and repair the identified leaks. If the leak source is such that it can be repaired or replaced without shutting down the equipment, the repair time allowed is a few days. If this is not the case, repairs are to be made at the next available unit shutdown.

State regulations have generally defined a "leak" to be a specified concentration of hydrocarbon at or near the surface of the source. EPA has suggested a VOC concentration of 10,000 ppm at the surface of the source, which is equivalent to 1000 ppm, 5 cm from the source. California has imposed a more stringent definition of 1000 ppm.

5.2   Storage Tanks

OAQPS has also issued a guideline for the control of volatile organic emission from petroleum storage tanks (EPA-450/2-78-047) of the external floating roof type. Figure 6 is a representation of this type

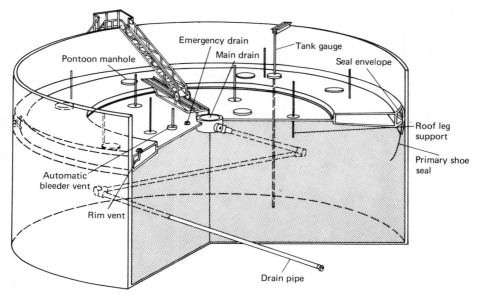

Fig. 6.   External floating roof tank (pontoon type).

Table 20   Example of Tank/Seal Type and Wind Speed

Tank/Seal Type	Hydrocarbon Emissions,[a] lb/yr	
	Long Beach, CA	Galveston, TX
Welded tanks		
Mechanical shoe seal		
Primary	13,842	31,939
Shoe-mounted secondary	5,311	10,371
Rim-mounted secondary	919	1,605
Liquid-mounted resilient		
Fluid-filled seal		
Primary only	5,056	8,828
Weather seal	3,058	5,048
Rim-mounted secondary	1,067	1,332
Vapor-mounted resilient		
Filled seal		
Primary only	60,358	217,504
Weather seal	37,658	128,346
Rim-mounted secondary	17,401	74,117
Riveted tanks		
Mechanical shoe primary only	14,995	34,599
Shoe-mounted secondary	9,295	18,151
Rim-mounted secondary	2,772	6,765

[a]Calculated Using API 2517 (February 1980).

of tank. When the fit between the seal and tank wall is well maintained, most losses by the seal are attributable to the wind. The American Petroleum Institute has published calculational procedures (API 2517) for predicting the tank losses, with varying types of seal and construction. Table 20 presents the results calculated using the API procedures for two locations, with varying wind speeds. Note that for all categories and construction types, higher wind speeds result in larger hydrocarbon losses.

The OAQPS guideline recommends retrofitting a rim-mounted secondary seal as the control technology. EPA presents data which show that use of this technique will reduce VOC emission significantly. Figure 7 shows their estimated tank losses as a function of vapor pressure of the stored liquid.

Equally, if not more important than seal type(s) is good and constant maintenance. As in the case of valves and fittings VOC emissions can be limited by good maintenance.

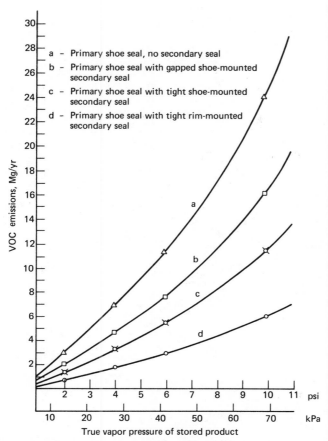

Fig. 7 Emissions from 30.5 m (100 ft) diameter welded gasoline tank with primary shoe seal at 16.1 kph (10 mph) average wind speed.

CHAPTER 26
SOURCE CONTROL—FOOD

RICHARD F. TORO

Recon Systems, Inc.
Three Bridges, New Jersey

1 INTRODUCTION

The food industry is very large and diverse, as reflected by the myriad of products available in American supermarkets and specialty shops. Even though economic trends are toward growth of the "agri-business conglomerates," the processing operations still remain quite independent and specific to the product. This chapter will deal with processing only, and not the production of food on farms, orchards, and ranches.

As in many other air pollution source industries, the food industry emits solid, liquid, and gaseous pollutants. However, aside from particulate, the emissions generally are not of the regulated variety. Sulfur oxides, carbon monoxide, hydrocarbons, and nitrogen oxides are not common emissions except of course from auxiliary systems such as incinerators, steam boilers (including processing residue waste fuel fired systems), and glass and can manufacturing. This chapter will not deal with such systems since they are covered elsewhere in this work.

The air pollution problems of the food industry tend to be more of the nuisance variety, specifically, visible and odorous emissions.

For this reason, the industry has not been a primary regulatory target, although all states have visible emission (opacity) standards, and some have attempted odor emission standards. Probably all utilize the common law doctrine of nuisance and the "permit to construct and operate" system to control these emissions.

The degree of controls necessary depend to a major degree on the regulatory jurisdiction, and the proximity to impacted neighbors. It is possible with modern dispersion modeling to predict the impact and the degree of emission reduction required.

Solid particulate emissions are controlled by the usual devices such as cyclones, baghouses and other dust collectors, and scrubbers. However, special care is required to select, size, and operate these, since much food of course is organic, and such particulates present potential safety (explosions) and operating (hygroscopic filter blinding) problems.

Liquid and semiliquid particulate emissions are controlled by aerosol collectors, fume incinerators, and to some degree by wet scrubbers or electrostatic precipitators.

Gaseous emissions are controlled by fume incinerators, chemical scrubbers, carbon adsorption, and where allowed, by dilution or tall stacks, or masking.

2 OPERATIONS CAUSING EMISSIONS

A wide variety of processing operations handle dry bulk products such as sugar, salt, grains, flour, coffee, tea, starch, and so on. As with any bulk material handling, dust control measures must be utilized. Some examples of this category are alfalfa dehydrating, feed and grain mills, and elevators.

Some commercial cooking, roasting, and smoking operations cause visible and/or odorous emissions. Vegetable oil expellers, coffee bean roasting operations, meat and fish smoke houses, fish frying, peanut roasters, and commercial hamburger broiling are example operations in this category. Although sometimes visibly "masked" by steam evolving from the same process, the emissions are often quite distinguishable in that they "trail" on after the steam has dissipated.

Odors are often associated with visible emissions, but there are some food industry sources which emit odors without any major visibility. Examples of this category include tomato cooking, flavor manufacture (spray dryers), rendering, and fish processing.

It is crucial to effective and economic control of emissions to understand the nature of the emissions relative to their state of matter, as well as to their behavior in the control systems under consideration and the atmosphere subsequently. Time–temperature relationships enter into this understanding. If any doubt exists, source and pilot testing should be carried out.

3 NATURE OF THE EMISSIONS

Particulate (dust) emissions from uncontrolled bulk handling systems vary in quantity depending on the degree of "agitation" and the size and amount of "fines." Some typical data are summarized in Table 1.

Particle sizing of the dusts, being mechanically generated, do not usually involve the difficult submicrometer materials. More often, the ranges encountered are 2–5 μm on the low end, and 50–500 μm on the high end. For this reason, cyclones and mechanical collectors were often used with sufficient removal efficiency. Even today some applications can be effectively controlled with cyclones.

Perhaps of most significance is the tendency of many of the dusts to be hygroscopic, absorbing moisture from the atmosphere, and becoming sticky and agglomerated. While this does not usually affect removal efficiency, cleaning of the removal device becomes difficult, and if severe enough will cause deterioration of the flow handling capability of the system, and/or irreversible loss of the media.

As with any organic dust, food processing dusts also have a high explosion tendency and precautions must be taken.

"Smokes" and liquid or semiliquid emissions are highly visible because they are submicrometer in size, ranging from about 0.05–5 μm. For this reason, such sources are problems even though they may be in compliance with regulatory particulate mass emission standards. Table 2 illustrates some emission quantities and opacity (% obscuration of vision) data.

These smoke and liquid emissions have significant vapor pressures associated with them, and therefore also have secondary effects to be considered.

Low vapor pressure smokes will easily begin condensing on ductwork, stacks, and any heat recovery devices, causing severe maintenance costs as well as possible fire hazards. However, since they are easily condensed, they can be removed by physical methods at relatively high temperatures. Secondary odors are minimal in this case.

Higher vapor pressure smokes cause less of these problems, but must be cooled to lower temperatures before physical methods of removal can be sufficiently effective. Often even high efficiency removal of the smoke is insufficient to eliminate secondary odors and an additional removal system is required.

Data on the differentiation between these condensable emissions are not readily available, so pilot testing and modeling of residual emission impacts often are needed.

True odors not associated with condensable emissions are gases, and cannot be removed by physical methods. Combustion, adsorption, absorption, and chemical reaction methods are necessary. Aldehydes, ketones, organic acids, sulfides, and mercaptans are often the major causes of these odors, but odors are usually complex mixtures with unpredictable effects such as synergism and masking. It therefore is usually not productive to have a detailed chemical analysis of the emissions, unless there is some possibility of affecting the emissions by, for example, removal of raw material components.

Visible emissions can be caused by solid particulate dust; carbonaceous, liquid, or semiliquid smokes; fats, oils, and greases; acids, or combinations thereof. It is essential to determine the exact nature of the source, or misapplication of a control system can occur with catastrophic results.

Table 1   Example Emission Factors for Food Processing Operations Involving Bulk Handling of Solid Materials

Sources	Uncontrolled Emissions, lb/ton of Product per Step
Alfalfa dehydrating plants, various cyclones	2.6–10
Grain elevators, various steps	0.3–3.0
Feed mills, various steps	0.1–3.0
Wheat or rye mills	1.0–70
Soybean mills	1.6–7.2
Grain handling, beer and whiskey	3.0
Fish meal dryer	0.1
Starch manufacturing	8.0

Table 2   Example Particulate Emissions and Opacity Data from Food Processing Operations Involving Heating

	Uncontrolled Emissions	
Sources	Particulates,[a] gr/dscf	Opacity, %
Coffee roasting—batch	0.20	20–80
Coffee roasting—continuous	0.18	60
Peanut oil expeller	0.14	40–50
Peanut roaster	0.29	40
Bacon smoking—batch	0.08	40
Weiner smoking—continuous	0.54	60
Sausage smoking—batch	0.05	50
Hamburger broiling	0.50	60

[a] Based on measurements made with EPA Method 5, but including the impinger (back half) catch.

Emission testing of food processing sources requires some advance knowledge of the nature of the emissions, and careful selection of sampling and analytical techniques, many of which are not well-known standardized methods. Table 3 provides some guidance in this matter.

4   PERTINENT AIR POLLUTION CONTROL REGULATIONS

There are no specific federal standards for food processing plants. However, various state and perhaps regional and municipal regulations do apply. These include:

1. Limits on opacity (visible emissions)
2. Requirements for permits to construct and operate
3. Nuisance prohibitions
4. Limits on odor emission quantities
5. Limits on particulate emission quantities

Table 3   Example Testing Techniques for Food Processing Sources

Pollutant	Sampling Method	Analytical Method
Solid particulates and dust	EPA Method 5	Gravimetric
Solid and liquid particulates or liquid	EPA Method 5 plus impinger catch	Gravimetric, organic extractions, chemical or instrumental analysis
Particle (large) sizing	EPA Method 17	Microscope, BAHCO, Coulter counter
Particle (fine) sizing	Single-point isokinetic-cascade impactor	Gravimetric
Liquid or condensable emissions	N.J. Air Test Method 3, EPA Methods 6, 8, 25	Extraction/gravimetric, chemical or instrumental analysis
Noncondensable gases and odors	Grab or integrated sample into inert gas bags or containers	Mass spec, gas chromatography, ASTM odor units
Condensable and noncondensable emissions	N.J. Air Test Method 3, EPA Method 25	Extraction/gravimetric, chemical or instrumental analysis, odor unit

Table 4   Example Opacity Data from Controlled Sources

Sources	Particulates,[a] gr/dscf	Opacity, %
Coffee roasting—batch	0.0061	0–5
Coffee roasting—continuous	0.0042	0
Peanut oil expeller	0.0032	0
Peanut roaster	0.0058	5
Bacon smoking—batch	0.0018	0
Weiner smoking—continuous	0.0083	5
Sausage smoking—batch	0.0013	0
Hamburger broiling	0.0066	0

[a] Based on measurements made with EPA Method 5, but including the impinger (back half) catch.

Opacity is defined as the percent obscuration of light transmission and is the difference between 100% light transmission and the percent transmission through the stack gases. It is a scientific and objective parameter, but is mostly measured somewhat subjectively by regular (e.g., every six months) calibration (memorization) of human eyes. Numerous "techno-legal" challenges to this technique have been brought, but it still remains in the regulations of most if not all states. The main difference between the old Ringelmann scale (1–5) and the opacity scale (0–100%) is the latter's suitability for white, blue, and grey as well as black smoke. Roughly speaking, one Ringelmann unit equals 20% opacity.

A typical opacity regulation will prohibit release of any stack gas with an opacity greater than a specific amount. In some cases, this amount will be the same for all and any stacks and will range from 0–20% opacity in highly populated and industrialized states to perhaps 30–40% in others. Some variations in allowable opacity exist in an attempt to correlate opacity and mass emissions so that compliance with one coincides with the others. A short exclusion period is usually allowed (e.g., 5–10% of the time) to allow for unexpected or even planned upsets (e.g., soot blowing).

A solid particulate emission of less than 0.01–0.02 gr/dscf often will meet an opacity standard of 10–20%. However, a liquid or condensable emission of 0.005 or less may be needed. This is due to the fact that liquids and condensables often are heat or combustion generated and therefore are submicrometer in size. Submicrometer particles produce much more opacity for the same mass. Table 4 lists some opacity data.

Permits to construct and operate require submission to the regulatory agency of some of the details of the source operation, more details of the air pollution control system, potential and expected emissions without controls, and expected emissions with controls. The applicable regulations usually require the air pollution control system to be "state of the art" and the controlled emissions on the permit to reflect that.

The controlled emissions as noted on the permit application become an additional standard to be met once the permit is issued. Therefore overly optimistic estimates are detrimental, since it is very difficult and in some cases impossible to get permits modified to less stringent emissions. Table 5 lists some provisions of a permit regulation.

Table 5   Example Provisions of a State Permit Regulation

No person shall construct, install, or alter any equipment or control apparatus without first having obtained a "Permit to Construct, Install or Alter Control Apparatus or Equipment" from the Department of Environmental Protection. Such certificates shall be valid for 5 years unless revoked by the Department, and these certificates may be renewed not less than 90 days prior to their expiration date.

The Department may issue a temporary certificate after receiving an application for a permit. The temporary permit is valid for 90 days and is issued to allow for evaluation of the source and control equipment while in operation. It may call for stack tests.

Any person in possession of a permit shall maintain said certificate readily available on the operation premises.

All conditions and provisions of the permit shall be fulfilled. The components connected or attached to the equipment or control apparatus shall function properly and be used in accordance with the permit issued.

Table 6 Example Codification of Nuisance Doctrine

Notwithstanding compliance with other subchapters of the chapter, no person shall cause, suffer, allow, or permit to be emitted into the outdoor atmosphere substances in quantities which shall result in air pollution as defined.

Air pollution is defined as the presence in the outdoor atmosphere of one or more air contaminants in such quantities and duration as are or tend to be injurious to human health or welfare, animal or plant life or property, or would unreasonably interfere with the enjoyment of life or property throughout the state and in such territories of the state as shall be affected thereby and excludes all aspects of employer–employee relationship as to health and safety hazards.

Insecticides, rodenticides, fungicides, herbicides, nematocides, and defoliants are not covered.

Nuisance is a common law doctrine which prohibits disturbing of neighbors by, for example, visual, noise, odor, dust, or corrosive gas emissions. In some states it has been codified as an air pollution control regulation, but in any case it is a difficult issue for an industrial source to deal with, since it is subjective and potentially conspiratorial. It is also an extremely local matter, and this work cannot effectively deal with it other than to note its importance to food processing operations. Table 6 presents one such codified regulation.

Odors are most often dealt with under the codified or common law nuisance doctrine. In simple terms, odors that cause a nuisance are prohibited and those that do not are not. It is clear then, that odors that do not impact on the human nose or are diluted before then are of no particular concern. This impact may occur at ground level, or at elevated points such as on a mountain or highway overlooking a processing plant, or a high-rise building downwind. Since such problems are enforced after the fact, odor quantification and dispersion modeling may be advisable.

A few states have implemented specific odor control regulations which define the number of odor units which may be emitted. In effect, these regulations are attempting to predict what total quantity of stack gas and odor concentration combination will be below a nuisance level. Table 7 presents one such codified regulation.

Particulate regulations are widespread and well known since they apply to virtually all manufacturing processes. They exist as concentration (e.g., gr/dscf), mass (e.g., lb/hr, lb/ton) emission or removal efficiency standards, but since they are covered in detail elsewhere in this work, they will not be dealt with again in this chapter.

5 CONTROLLING EMISSIONS

As with any process, changes to the raw materials, fuels, and operating systems should be considered wherever possible. One example of this is the use of liquid smoke flavoring for meats to avoid controlling the emissions from smoke houses. However, this approach is so entwined with proprietary processes, product quality, marketing and economics, that it is beyond the scope of this work.

The use of add-on air pollution control systems is the most widely used approach for complying with the regulations outlined earlier. Table 8 provides an overview of the often used devices for the different types of emissions from food processing operations.

Table 7 Examples of Specific Odor Control Regulation

Violation shall be any discharge of air contaminants in excess of the following odor emission limits:

Well-defined stack 50 ft or taller	150 ocu
All other sources	25 ocu
All sources	1,000,000 ocu/min

Violation shall also be any discharge which causes odors in excess of the following limits:

Residential, recreational, institutional, retail sales, hotel, educational zones	1 ocu
Light industrial zones	2 ocu
Other zones	4 ocu

An ocu (odor concentration unit) is defined as the number of standard cubic feet of odor-free air needed to dilute each cubic foot of contaminated air so that at least 50% of the test panel does not detect any odor in the diluted mixture. The ocu/min is determined by multiplying the total source gas standard cubic feet per minute by the number of ocu.

The ocu are determined by ASTM D-1391-57 and *JAPCA* **19**(2), 101–105 (1969).

Table 8 Air Pollution Control Devices for Food Processing Applications

Emission	Often Used Control Device
Solid particulate and dust	Cyclones, baghouses, scrubbers
Liquid particulate and associated odors	Fiber beds, two-stage electrostatic precipitators, HEAF™,[a] combined with precooler/ condenser if necessary
Semiliquid, tarry particulate, or solid–liquid mixtures and odors therefrom	HEAF™, wet electrostatic precipitators, washed two-stage electrostatic precipitators, combined with precooler/condenser if necessary; vapor phase incinerator
Vapor phase and noncondensable emissions and odors	Chemical scrubbers/absorbers, carbon adsorbers, vapor phase incinerators
Combustible particulate and gases causing odors and visible emissions	Vapor phase incinerator, combinations of above devices

[a] HEAF™ is a proprietary product utilizing high velocity impaction/filtration.

Individual profiles of add-on devices follow in Tables 9–22, showing principle of operation, relevant design and operating parameters, performance characteristics, and cost parameters.

Appropriate materials of construction are determined by the characteristics of the pollutant and other components of the stack gas, any externally added materials, temperatures, humidity, and whether the materials collected will be added to the product.

Combinations of add-on devices is not unusual. Pilot testing for problems not well understood is strongly recommended.

Table 9 Cyclones for Collection of Solid Particulates and Dust

Principle of Operation

Dust-laden air is passed through the specially shaped device generating within itself a high velocity vortex. Centrifugal force causes the denser particulates toward the walls, where they spiral down into the collection receptacle. The lighter, relatively clean gases are exhausted from the outlet.

Relevant Design and Operating Parameters

Inlet velocity 10–150 (usually 40–60) ft/sec
Pressure loss 1–50 (usually less than 10) in. H_2O gauge

Housing Materials of Construction

Mild steel for nonconsumables

Epoxy coated for animal consumption

304SS, 2B finish for human consumption; see USDA 3A "Sanitary Standards for Dry Milk and Dry Milk Products"

Rubber lining for abrasion resistance

Performance Characteristics

Collection efficiency complex exponential function of inlet velocity, air density and viscosity, particle mass, size, shape and roughness, and cyclone design

Typically, greater than 90% collection efficiency on 3–10 μm particle sizes and 98% on larger sizes. Ineffective on fine particles and often unable to meet air emission standards. Large dust handling capability. Used as product collector.

Cost Parameters

Equipment cost without accessories \$0.20–\$0.40/cfm in mild steel. Epoxy coated about 1.2 times the cost. 304SS, 2B finish about twice the cost. Rubber-lined mild steel about twice the cost. Power consumption typically 1.0–2.5 kWh/hr/1000 acfm. Low maintenance.

Table 10   Baghouses and Other Filtering Systems for Collection of Solid Particulates and Dusts

Principle of Operation

Dust-laden air is drawn through a bank of filter tubes suspended in a housing. The filter cake builds up on the outside of a tubular bag and periodically a blast of compressed air, introduced on the clean side of the bag, flexes the bag and forces release of filter cake; dust then falls to bottom into hopper. Compressed air pulse controlled by variable timer and/or pressure drop sensor.

Relevant Design and Operating Parameters

Air/filter ratio depends on dust nature, concentration, and particle size, and gas temperature and composition. Typically less than 6–8 cfm/ft² of filter area.

Filter media typically polyester or polypropylene. Interwoven grounding strips for static electricity control may be required.

Compressed air pulse typically 0.005 cfm/ft² of filter area, 80–100 psig pressure. Pulse rate variable from one every 3–180 sec or longer.

Housing Materials of Construction

Mild steel for nonconsumables; epoxy coating for animal consumption; 304SS, 2B finish for human consumption; see USDA 3A "Sanitary Standards for Dry Milk and Dry Milk Products."

Performance Characteristics

Filtration efficiency typically 99.9% on particles 2 μm and larger

Pressure drop typically 2–8 in. H_2O gauge

Water or other liquids in air will foul the filters causing excessive pressure drop and eventual blockage of flow

Cost Parameters

Equipment cost without accessories $0.75–$1.50/cfm in mild steel. Epoxy coated about twice the cost. 304SS, 2B finish about five times the cost.

Power consumption 0.7–2.5 kWh/hr/1000 acfm

Bag life typically 1–4 yr

Maintenance not unusual unless preventative maintenance not practiced; then severe

Table 11 Scrubbers for Collection of Solid Particulates and Dusts

Principle of Operation

The contaminated gas stream and the scrubbing liquid are drawn through a restricted passage (throat, venturi, slot) at increased pressure drop causing intimate contact, droplet formation, and impaction of the particulates onto the liquid droplets. The particulates are completely wetted, and become much larger in diameter, enabling collection on secondary self-draining mist eliminator.

Relevant Design and Operating Parameters

Restricted passage velocity 100 ft/sec to sonic velocity

Pressure loss 1–100 (usually 15–60) in. H_2O gauge

Liquid/gas ratio 3–30 gal/acf

No temperature limit so long as materials of construction and water flows adequate

Materials of Construction

Suitable materials determined from temperature, nature of contaminants, scrubbing media. Seldom used for product recovery, therefore life and reliability of materials main criteria.

Performance Characteristics

Exhaust gas stream saturated with water at adiabatic saturation temperature often resulting in visible water vapor plume

Collection/removal efficiencies:

  99% on 2-μm and larger particulate

  10% on 0.1-μm and smaller particulate

  Direct function of pressure drop on 0.1–2 μm particulate

Particularly effective on water soluble or easily "wetted" particulate

Inherently safe on explosive or corrosive particulate

Cost Parameters

Equipment purchase cost $1–$10/acfm

Water consumption 1–10 gal/1000 acfm

Power consumption 1–10 kWh/hr/1000 acfm exhaust

Maintenance—normal

Table 12 Fiber Beds for Collection of Liquid Particulate and Associated Odors

Principle of Operation

The contaminated gas stream is passed horizontally through a vertical fiber bed shaped as a cylinder with an open core where the gases exit. Particles are collected by Brownian movement or diffusion causing collision with a fiber, where they coalesce into liquid films which drain from the bed. Soluble solids can be also collected with continuous or intermittent flushing with liquid.

Relevant Design and Operating Parameters

Gas velocity through fiber bed, typically 20–30 ft/min

Pressure drop 2–20 in. H_2O, typically 6–10 in.

Very fine glass fibers compressed into hollow cylinder

Temperature cryogenic to 800°F

Solid particulate fouls the bed

Materials of Construction

Fiber bed typically compressed fiberglass. Some work on Teflon and other fibers. Housing and cage materials function of contaminants nature, temperature, and flush liquid if any.

Performance Characteristics

100% removal of particles 3 μm and larger

94–99.95% removal of particles smaller than 3 μm

Self-draining on liquids with viscosity up to about 5000 cP

Effective on opacity problems

Turndown tends to increase efficiency

Cost Parameters

Equipment purchase cost

$5–$20/cfm with steel housing

$15–$30/cfm with stainless steel housing

Operating costs—0.5–3 kWh/hr/1000 cfm

Maintenance normal unless insoluble solids or viscous materials build up in bed; then severe

Table 13   Two-Stage Electrostatic Precipitators for Collection of Liquid Particulate and Associated Odors

Principle of Operation

Contaminated gas stream is passed through high-voltage field where particulates are electrically charged and then to grounded plate section where particles are removed from stream. Liquid particulates form film and drain from plates.

Precooling, prefiltering, and after filtering sometimes required

Heat recovery sometimes practiced

Relevant Design and Operating Parameters

Voltage typically 10–15 kV

No. of stages—one to three in series, two typical

Residence time—typically about 0.25 sec per stage

Velocity—typically around 500 ft/min

Temperatures must be below contaminant condensation point, above contaminant hardening point, and above water dewpoint, usually less than 150°F

Fire protection important if contaminant combustible

Materials of Construction

Aluminum components, steel or coated steel housings, and stainless sumps are often used

Performance Characteristics

Almost always completely eliminates visible emissions and associated odors, often resulting in sufficient or substantial odor reduction

95–99% collection of submicrometer particulate matter

Maintenance can be dramatic if preventative care not practiced—soak cleaning required typically every 2–6 weeks

Cost Parameters

Equipment purchase cost $2–$6/cfm

Power consumption—in the range of 0.05 kWh/hr/1000 acfm

Maintenance—significant

Table 14 Washed Two-Stage Electrostatic Precipitators for Collection of Semiliquids, Tars, Solid–Liquid Mixtures, and Associated Odors

Principle of Operation

Contaminated gas stream is passed through high-voltage field where particulates are electrically charged and then to grounded plate section where particles are removed from stream. The nondraining particulates are then periodically cleaned from the plates by a variety of manual or automatic washing systems.

Precooling, prefiltering, and after filtering sometimes required although difficult due to fouling

Relevant Design and Operating Parameters

Voltage—typically 10–15 kV

No. of stages—one to three in series, two typical

Residence time—typically about 0.25 sec per stage

Velocity—typically around 500 ft/min

Temperatures must be below contaminant condensation point and above water dewpoint—usually less than 150°F

Cleaning—automatic or manual by spraying, flushing, or soaking in appropriate cleaning solutions

Materials of Construction

Aluminum components, steel, or coated steel housings, and stainless sumps are often used. Effect of cleaning solution must be accounted for.

Performance Characteristics

Almost always completely eliminates visible emissions and associated odors, often resulting in sufficient or substantial odor reduction

95–99% collection of submicrometer particulate matter

Maintenance can be dramatic if preventative care not practiced—often cleaning required as much as daily or more

Cost Parameters

Equipment purchase cost—$3–$8/cfm

Power consumption—in the range of 0.05 kWh/hr/1000 acfm

Maintenance—significant

Table 15   The HEAF℠ System for Collection of Liquid Particulate and Associated Odors

Principle of Operation

Contaminated gas stream is drawn through a dense fiber mat at high velocity. The liquid aerosols impact on the fibers in the mat and are agglomerated into larger liquid droplets. Once the droplets reach size which overcomes surface tension they are released and collected in a secondary self-draining mist eliminator.

Relevant Design and Operating Parameters

Gas velocity through mat 1500–1900 ft/min

Automatic advance of fiber mat on pressure or flow control

Differential pressure 25–30 in. H_2O

Temperature ambient to less than 300°F

Excessive solid particulate (more than 0.10 gr/dscf) results in too short mat life

Materials of Construction

Mat is fiberglass. Housing and mist eliminator dependent on nature of contaminants.

Performance Characteristics

Almost always completely eliminates visible emissions and associated odors, often resulting in sufficient or substantial total odor reduction.

96–100% collection of submicrometer particulate matter

Cost Parameters

Equipment purchase cost—\$3–\$12/cfm

Mat consumption 0.5–3 ft²/hr/1000 acfm (at adiabatic dewpoint); \$0.083/ft²

Power consumption 5.0–6.0 kWh/hr/1000 acfm (at adiabatic dewpoint)

Maintenance—normal

Other

Proprietary process of Andersen 2000, Atlanta, Georgia

Table 16 The HEAF ® System for Collection of Semiliquid, Tars, Solid–Liquid Mixtures, and Associated Odors

Principle of Operation

Contaminated gas stream is drawn through a dense fiber mat at high velocity. Solid and sticky particulates are filtered, and submicrometer oils are impacted, agglomerated, and reentrained as large droplets which are then collected in a secondary self-draining mist eliminator.

Relevant Design and Operating Parameters

Gas velocity through mat 1500–1900 ft/min

Automatic advance of fiber mat on pressure or flow control

Differential pressure 25–30 in. H_2O

Temperature ambient to less than 300°F

Excessive solid particulate (more than 0.10 gr/dscf) results in too short mat life

Materials of Construction

Mat is fiberglass. Housing and mist eliminator dependent on nature of contaminants.

Performance Characteristics

Almost always completely eliminates visible emissions and associated odors, often resulting in sufficient or substantial total odor reduction

96–100% collection of submicrometer particulate matter

Cost Parameters

Equipment purchase cost—$3–$12/cfm

Mat consumption 0.5–3 ft²/hr/1000 acfm (at adiabatic dewpoint); $0.083/ft²

Power consumption 5.0–6.0 kWh/hr/1000 acfm (at adiabatic dewpoint)

Maintenance—normal

Other

Proprietary process of Andersen 2000, Atlanta, Georgia

Table 17 Wet Electrostatic Precipitator for Collection of Semiliquids, Tars, Solid–Liquid Mixtures, and Associated Odors

Principle of Operation

Contaminated gas stream is passed through high-voltage field where particulates are electrically charged, and then to grounded plate section where particles are removed from stream. The nondraining particulates are then continuously cleaned from the plates by a variety of washing systems.

Precooling, prefiltering, and after filtering sometimes required although difficult due to fouling

Relevant Design and Operating Parameters

Voltage—typically 30–40 kV

Stages—typically one

Velocity—typically less than 4–5 ft/sec

Residence time—typically about 1 sec

Temperature—limited by boiling point of flush liquid, typically less than 150°F

Cleaning—continuous flushing by liquid films which acts as grounded collecting surface

Materials of Construction

Electrodes may be high alloy. Collecting surface and housing may be fiberglass, or suitable metal.

Performance Characteristics

Almost always completely eliminates visible emissions and associated odors, often resulting in sufficient or substantial odor reduction

Exhaust gas stream saturated with water at adiabatic saturation temperature often resulting in visible water vapor plume

95–99% collection of submicrometer particulate matter

Maintenance can be dramatic if preventative care not practiced

Cost Parameters

Equipment purchase cost—very high, probably greater than $20/cfm

Power consumption—in the range of 0.04 kWh/hr/1000 cfm

Not practical for less than about 10,000 cfm and usually not considered for food processing applications

Table 18   Chemical Scrubbers/Absorbers for Removal of Vapor Phase and Noncondensable Emissions and Odors

Principle of Operation

Vapors and gas are absorbed from the contaminated air stream into water or chemical solutions. Scrubbers use high-surface area mass transfer packing to increase the rate of absorption. Typically acid gases are absorbed in alkaline solutions, while odors are absorbed into chemicals which convert them to soluble low vapor pressure compounds.

Relevant Design and Operating Parameters

Gas flow—concurrent to liquid

Gas velocities—low

Residence time—long

Packing depth 3–10 ft per bed

Differential pressure 2–6 in. H_2O

Liquid rates 15–20 gal/acf (at exhaust)

Materials of Construction

Typically corrosion resistant metals, thermoplastics or fiberglass reinforced polyester

Performance Characteristics

Exhaust gas stream saturated with water at adiabatic saturation temperature often resulting in visible water vapor plume

Typically 99+% on acid gases
     95% on odorous gases

Cost Parameters

Equipment cost $1–$5/acfm

Water consumption 0.1–1 gal/1000 acf (at exhaust)

Chemical consumption is function of nature and quantity of contaminants

Power consumption 0.5–1.5 kWh/hr/1000 cfm

Maintenance—normal

Table 19   Carbon Adsorbers for Removal of Vapor Phase and Noncondensable Emissions and Odors

Principle of Operation

The contaminated gas stream is passed through a bed of granular activated carbon. The carbon selectively attracts and holds the molecules of contaminants up to the point of having 5–50% of its weight adsorbed. At this point, the carbon is either regenerated in its holding vessel (or switching multiple vessels), or is replaced and sent to a central regenerator, or is disposed of (single vessel). If contaminants have value, regeneration can return them.

Relevant Design and Operating Parameters

Gas velocity typically less than 100 ft/min

Temperature typically limited to less than 100°F

Relative humidity less than 100%

Carbon loading—1–40% by weight at saturation; 1–20% by weight in cycling system

Materials of Construction

Usually minimum of coated steel; 304, 316 stainless, E-brite and Hastelloy also used.

Performance Characteristics

85–99% removal of contaminants

Not effective on true gases

85–99% recovery of contaminants in regenerable system

Condensables and particulates foul carbon, often irreversibly

Fire protection may be required

Cost Parameters

Equipment purchase cost per cubic foot per minute

   $6–$20 for single-bed system in coated steel

   $7–$25 for single-bed system in 316L stainless steel

   $10–$30 for switching multiple-bed system in coated steel

   $12–$40 for switching multiple-bed system in 316L stainless steel

Operating cost—extremely variable. Pressure drop horsepower and carbon consumption or regeneration charge main items for single bed. Steam, cooling water, power, and wastewater disposal main items for regenerable system.

Table 20   Vapor Phase Incinerators for Destruction of Combustible Vapor Phase and Noncondensable Emissions and Odors

Principle of Operation

The contaminated vapor stream is heated to the contaminant destruction temperature by combustion of auxiliary fuel or the contaminants themselves, and then retained at that temperature for sufficient residence time to accomplish destruction. At high concentrations of contaminants, process may be energy sufficient requiring only minimal auxiliary fuel for maintaining stable operation. Heat recovery usually essential.

Relevant Design and Operating Parameters

Operating temperatures 900–1600°F. Catalysts reduce required temperature, but life must be long, thus limiting practical use. Residence time at temperature 0.3–1.0 sec. High oxygen ($> 15\%$) content of waste stream can be utilized as combustion air. Intimate mixing of fuel and contaminated vapors. Combustible content 0–5% of LEL requires fuel; 25–50% of LEL usually self-sustaining with heat recovery; 50–80% of LEL usually self-sustaining. Heat recovery by preheat; steam, hot air, hot water, or hot oil generation; combination thereof. Gaseous (e.g., HCl, SO_2) or particulate removal devices may be required after vapor incinerator.

Materials of Construction

Typically stainless burners, refractory lined steel housing with dewpoint protection

Performance Characteristics

Destruction efficiency 85–99+%.

Energy efficiency 40–85% (higher in special cases)

Automatic unattended operation

Cost Parameters

Equipment cost: $6–$8/scfm without heat recovery

$12–$14/scfm with 48% heat recovery

$21–$24/scfm with 70% heat recovery (preheat)

$14–$16/scfm with 70% heat recovery (waste heat boiler)

Fuel consumption = 1.15 (total flow, scfm) (temperature increase, °F) = Btu/hr, to be supplied by auxiliary fuel and/or combustible content of contaminated stream and/or heat recovery.

If used, catalyst life is a major cost element

Maintenance—normal

Table 21   Vapor Phase Incinerators for Destruction of Combustible Particulates and Gases

Principle of Operation

The contaminated vapor stream is heated to the contaminant destruction temperature by combustion of auxiliary fuel or the contaminants themselves, and then retained at that temperature for sufficient residence time to accomplish destruction. At high concentrations of contaminants, process may be energy sufficient requiring only minimal auxiliary fuel for maintaining stable operation. Heat recovery usually essential.

Relevant Design and Operating Parameters

Operating temperatures 1200–1800°F. Catalysts usually foul too quickly. Residence time at temperature 0.5–1.0 sec. High oxygen (15%) content of waste stream can be utilized as combustion air. Intimate mixing of fuel and contaminated vapors. Combustible content 0–5% of LEL requires fuel; 25–50% of LEL usually self-sustaining with heat recovery; 50–80% of LEL usually self-sustaining. Heat recovery by preheat (may not be possible because of particulate fouling), steam, hot air, hot water, or hot oil generation; combination thereof. Gaseous (e.g., HCl, SO_2) or particulate removal devices may be required after vapor incinerator.

Materials of Construction

Typically stainless burners, refractory lined steel housing with dewpoint protection

Performance Characteristics

Destruction efficiency 85–99+%

Energy efficiency 40–85% (higher in special cases)

Automatic unattended operation

Cost Parameters

Equipment cost: $7–$9/scfm without heat recovery

$13–$15/scfm with 48% heat recovery

$22–$25/scfm with 70% heat recovery (preheat)

$15–$17/scfm with 70% heat recovery (waste heat boiler)

Fuel consumption = 1.15 (total flow, scfm) (temperature increase, °F) = Btu/hr, to be supplied by auxiliary fuel and/or combustible content of contaminated stream and/or heat recovery

If used, catalyst life is a major cost element

Maintenance—normal

Table 22   Evaporative Coolers and Condensers

Principle of Operation

Contaminated air stream is cooled by injection of water. Condensable compounds will condense for removal in secondary collector. Water injected by air atomized or high-pressure nozzles to create large surface area for rapid evaporation. Wet bottom recirculates water; dry bottom limits cooling to above adiabatic saturation temperature.

Relevant Design and Operating Parameters

Pressure drop 1–3 in. H_2O

Liquid rate 0.01–2 gal/1000 scf

Contact time 0.5–1.5 sec

Velocity 5–10 ft/sec

Water pressure 80–400 psig

Air/water pressure 10–18/30–50 psig

Materials of Construction

Mild steel construction for dry bottom

Stainless or nonmetallic construction for wet bottom

Performance Characteristics

Achieves cooling to desired temperature to as low as adiabatic saturation temperature. Also condenses contaminants in that range. Cooling to below saturation possible with additional indirect cooling or large quantities of very cold water.

Cost Parameters

Equipment cost: $0.5–$2/acfm in mild steel

$1–$4/acfm in stainless steel

Operating costs—apply water and pressure drop power cost factor to usage above

Maintenance—significant, for example, spray nozzle plugging

REFERENCES

1. Brady, J. D., High Velocity Filtration for Control of Emissions from Food Processing Operations, Air Pollution Control Association, 72nd Annual Meeting Paper 79-44.1, Cincinnati, OH, June 1979.

2. *Pollution Problems in Selected Food Industries,* National Industrial Pollution Control Council—Sub Council Report, U.S. Government Printing Office, 437-388/2, 1971.

3. *Compilation of Air Pollutant Emission Factors,* U.S. Environmental Protection Agency, Research Triangle Park, NC, AP-42, August 1977 and supplements.

4. New Jersey Air Test Method 3 (Draft); New Jersey Administrative Code 7:27.

5. *U.S. Environmental Protection Agency Test Methods,* U.S. Code of Federal Regulations, Title 40.

6. Danielson, J. A., *Air Pollution Engineering Manual,* 2nd edition, U.S. Environmental Protection Agency, Research Triangle Park, NC, AP-40, May 1973.

7. New Jersey Administrative Code 7:27.

8. Minnesota Pollution Control Agency, Rule APC 9.

9. American Society for Testing Materials Method D-1391-57, Philadelphia, PA.

10. Benforado, D. M., Rotella, W. J., and Horton, D. L., Development of an Odor Panel for Evaluation of Odor Control Equipment, *Journal of the Air Pollution Control Association* 19(2), 101-115 (February 1969).

11. Sanitary Standards for Dry Milk and Dry Milk Products, U.S. Department of Agriculture, 3-A Committee.

12. Private communications with, and literature from: Andersen 2000, Inc., Atlanta, GA; Fisher Klosterman, Louisville, KY; Flex Kleen Corp, Chicago, IL; Hirt Combustion Engineers, Montebello, CA; Monsanto Envirochem Systems, St. Louis, MO; United Air Specialists, Cincinnati, OH; Vic Mfg Co., Minneapolis, MN.

CHAPTER 27

SOURCE CONTROL—
FOREST PRODUCTS

JAMES A. EDDINGER

U.S. Environmental Protection Agency
Research Triangle Park, North Carolina

The production of pulp and paper from wood is categorized as the forest products industry. Manufacturing of paper and paper products is a complex process which is carried out in two distinct phases: the pulping of the wood and the manufacture of the paper. Wood pulp is prepared either mechanically or chemically. Mechanical pulp is produced by grinding or shredding wood to free the fibers. In chemical pulping processes, the wood fibers are freed by dissolving the binding material (lignin) in chemical solutions. Mechanical pulping and the paper-making process itself produce negligible air pollution, except for the boilers that supply a significant portion of the in-plant energy needs and the paper coating process. Emission and applicable control technologies for these boilers are characterized in Chapter 15.

There are four major chemical pulping techniques: (1) kraft or sulfate, (2) sulfite, (3) semichemical, and (4) soda. Of the four, the kraft or sulfate process produces over 80% of the chemical pulp produced annually in the United States and will, therefore, be emphasized.

1 KRAFT PULPING PROCESS

1.1 Process Description

Pulp wood can be considered to have two basic components—cellulose and lignin. The fibers of cellulose which comprise the pulp are bound together in the wood by the lignin. To render cellulose usable for paper manufacture, any chemical pulping process must first remove the lignin.

The kraft process for producing pulp from wood is shown in Figure 1. In the process, wood chips are cooked (digested) at an elevated temperature and pressure in "white liquor," which is a water solution of sodium sulfide (Na_2S) and sodium hydroxide (NaOH). The white liquor chemically dissolves lignin from the wood. The remaining cellulose (pulp) is filtered from the spent cooking liquor and washed with water. Usually, the pulp proceeds through various intermittent stages of washing and possibly bleaching, after which it is pressed and dried into the finished product (paper).

The balance of the process is designed to recover the cooking chemicals and heat. Spent cooking liquor and the pulp wash water are combined to form a weak black liquor which is concentrated in a multiple-effect evaporator system to about 55% solids. The black liquor can then be further concentrated to 65% solids in a direct-contact evaporator, which evaporates water by bringing the liquor in contact with the flue gases from a recovery furnace, or in an indirect-contact evaporator. The strong black liquor is then fired in a recovery furnace. Combustion of the organics dissolved in the black liquor provides heat for generating process steam and converting sodium sulfate (Na_2SO_4) to Na_2S. To make up for chemicals lost in the operating cycle, salt cake (sodium sulfate) is usually added to the concentrated black liquor before it is sprayed into the furnace. Inorganic chemicals present in the black liquor collect as a molten smelt at the bottom of the furnace.

The smelt, consisting of sodium carbonate (Na_2CO_3) and sodium sulfide, is dissolved in water to form green liquor which is transferred to a causticizing tank where quicklime (CaO) is added to convert the sodium carbonate to sodium hydroxide. Formation of the sodium hydroxide completes the regeneration of white liquor, which is returned to the digester system. A calcium carbonate mud precipitates from the causticizing tank and is calcined in a lime kiln to regenerate quicklime.

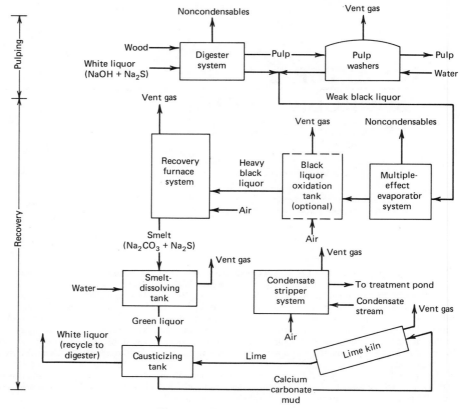

Fig. 1 Kraft pulping process.

1.2 Sources and Their Emissions

The atmospheric emissions from the kraft process include both gaseous and particulate materials. The major gaseous emissions are malodorous reduced sulfur compounds, such as hydrogen sulfide (H_2S), methyl mercaptan (CH_3SH), dimethyl sulfide (CH_3SCH_3), and dimethyl disulfide (CH_3SSCH_3); oxides of sulfur (SO_x); and oxides of nitrogen (NO_x). The particulate matter emissions are primarily sodium sulfate (Na_2CO_3) from the recovery furnace, sodium salts, and calcium compounds from the lime kiln and sodium compounds from the smelt tanks.

H_2S and the organic sulfides, when taken as a group, are called total reduced sulfur (TRS). They are extremely odorous and are detectable at a concentration of only a few parts per billion. Thus, odor control is one of the principal air pollution problems in a kraft pulp mill.

1.2.1 TRS Emission Sources

The major potential sources for the reduced sulfur gas emissions to the atmosphere include digester blow and relief gases, vacuum washer hood and seal tank vents, multiple-effect evaporation hot-well vents, recovery furnace flue gases, smelt-dissolving tanks, slaker vents, black liquor oxidation tanks, lime kiln exit vents, and wastewater treatment operations. Summaries of values on typical gas flow rates, variations in malodorous sulfur gas concentrations, and emission rates per unit production for the kraft process units are presented in Table 1. Most kraft mill flue gas streams contain appreciable amounts of water vapor.

1.2.1.1 Recovery Furnace System.
TRS emissions are generated both in the furnace and in the direct-contact evaporator. The furnace-generated TRS concentration is as high as several hundred parts per million (ppm) and as low as 1 ppm, depending on the furnace design and operation. Recovery furnace emissions are affected by the relative quantity and distribution of combustion air, rate of

Table 1 TRS Emissions from an Uncontrolled Kraft Pulp Mill[1]

Source	Typical Exhaust Gas Flow Rate, m³/a (acfm)	TRS Emission Range		Average TRS Emission Rate	
		ppm	g/kg ADP (lb/ton ADP)	ppm	g/kg ADP (lb/ton ADP)
Recovery furnace	212 (450,000)	18–1,303	0.75–31 (1.5–62)	550	7.5 (15.0)
Digester system	3 (6,200)	1,525–30,000	0.24–5.3 (0.47–10.5)	9,500	0.75 (1.5)
Multiple-effect evaporator system	1 (2,200)	92–44,000	0.015–3.2 (0.03–6.3)	6,700	0.5 (1.0)
Lime kiln	37 (79,200)	3–613	0.01–2.1 (0.02–4.2)	170	0.4 (0.8)
Brown stock washer system	71 (150,000)	—	0.005–0.5 (0.01–0.9)	30	0.15 (0.3)
Black liquor oxidation system	14 (30,000)	3–335	0.005–0.37 (0.01–0.73)	35	0.05 (0.1)
Smelt-dissolving tank	27 (58,100)	5–811	0.007–1.9 (0.013–3.70)	60	0.1 (0.2)
Condensate stripper system	2 (4,000)	—	—	5,000	1.0 (2.0)

solids (concentrated black liquor) feed, spray pattern and droplet size of the liquor fed, turbulence in the oxidation zone, smelt bed disturbance, and the combination of sulfidity and heat content value of the liquor fed. The impact of these variables on TRS emissions is independent of the absence or presence of a direct-contact evaporator.

TRS emissions generated in the direct-contact evaporator depend largely on the concentration of sodium sulfide in the black liquor. Acidic gases such as carbon dioxide in the flue gas can change the black liquor equilibrium, resulting in the release of increased quantities of hydrogen sulfide and methyl mercaptan.

1.2.1.2 Digester System. The noncondensable gases from the relief system and the blow tank vent contain TRS as high as 30,000 ppm. Both streams are sometimes referred to as digester "noncondensables." TRS compounds formed in the digester are mainly methyl mercaptan, dimethyl sulfide, and dimethyl disulfide. Operating variables that affect digester TRS emissions include the black liquor recycle rate, cook duration, cooking liquor sulfidity (percentage of sodium sulfide to total alkali, Na_2S and NaOH, in white liquor), and residual alkali level.

1.2.1.3 Multiple-Effect Evaporator System. The noncondensable gases from a multiple-effect evaporator (MEE) system consist of air drawn in through system leaks and reduced sulfur compounds that were either in the dilute black liquor or formed during the evaporation process. TRS emissions from the MEE system are as high as 44,000 ppm.

The type of condenser used can influence the concentration of TRS emissions. Certain types of condensers (e.g., direct-contact) allow the noncondensable gases and the condensate to mix, which results in a limited quantity of hydrogen sulfide and methyl mercaptan gases dissolved in the water. This reduces the TRS concentration from the system, but increases the sulfide level in the condensate. Sulfidity and pH of the weak black liquor also have an effect on the TRS concentration from the multiple-effect evaporators. Higher levels result in higher TRS emissions. TRS levels increase with decreasing pH levels.

1.2.1.4 Lime Kiln. TRS emissions can be generated in the lime kiln proper and in the downstream scrubber which is normally installed to control particulate emissions.

TRS emissions originating in the lime kiln are affected by the oxygen content of the exhaust, the kiln length to diameter ratio, the lime mud sulfide content, cold-end exit gas temperature, and simultaneous burning of sulfur-bearing materials contained in the lime mud (e.g., green liquor dregs, the impurities resulting from clarifying the green liquor).

If digester and evaporator condensates are used as lime kiln scrubber water, reduced sulfur compounds can be stripped into the exit gas stream. If the scrubbing liquor contains sodium sulfide, as it does in some installations, H_2S may be released in the scrubber as a result of the equilibrium shift caused by the absorption of CO_2 in the liquor.

1.2.1.5 Brown Stock Washer System. TRS emissions from the brown stock washers arise primarily from the vaporization of the volatile reduced sulfur compound. TRS compounds emitted are principally dimethyl sulfide and dimethyl disulfide.

Brown stock washer TRS emissions are affected by the wash water source, water temperature, degree of agitation and turbulence in the filtrate tank, and blow tank pulp consistency. TRS emissions will increase significantly if contaminated condensate from the digester and evaporator systems are used for washing. Higher temperatures and agitation result in increased stripping of TRS during the washing.

1.2.1.6 Black Liquor Oxidation System. TRS emissions from the oxidation system are created by the stripping of the reduced sulfur compounds from the black liquor by air passing through the liquor. Uncontrolled TRS emissions are principally dimethyl sulfide and dimethyl disulfide. Oxidation systems that use only molecular oxygen have the advantage of emitting virtually no off-gases because the total gas stream reacts with the sparge system.

Primary factors affecting TRS emissions from black liquor oxidation systems are the inlet sulfide content, the temperature of the black liquor, residence time, and the air flow rate per unit volume. TRS emissions tend to increase for higher liquor temperatures and greater air flow rates because of greater volatility of the gases and stripping action of the air, respectively. TRS emissions also tend to increase with increasing sulfide concentrations in the incoming black liquor and with increasing residence time.

1.2.1.7 Smelt-Dissolving Tank. Because of the presence of a small percentage of reduced sulfur compounds in the smelt, some of these odorous materials escape the tank with the flushed steam.

Several factors affect the TRS emissions. Among these are the water used in the smelt tank, turbulence of the dissolving water, scrubbing liquor used in the particulate control device, pH of scrubbing liquor, and sulfide content of the particulate collected in the control device. The use of contaminated condensate in the smelt tank or the scrubber can result in the stripping of TRS compounds into the gas stream. Turbulence can increase the stripping action. Increased H_2S formation can occur with an increase in sulfide content of the scrubbing liquid and a decrease in pH of the scrubbing liquor.

1.2.1.8   Condensate Stripping System.   The kraft process produces two main condensates, namely, digester and evaporator condensates. Both contain compounds that are volatile, chemical oxygen demanding, reduced sulfur containing, and odorous. These compounds can generate both air and water pollution. A general course of action is to decrease the amount of condensates, collect them, and reuse them; if they are not reusable, the procedure then is to strip them of their contaminating compounds and oxidize these compounds to less harmful forms.

The main components of typical kraft mill contaminated condensates responsible for the odor are hydrogen sulfide, methyl mercaptan, dimethyl sulfide, and dimethyl disulfide. The components will exist in kraft mill condensates in varying amounts depending on place of origin (digester or evaporator), pulp raw material (wood species), operating practices (such as sulfidity and cooking time), type of equipment (continuous or batch), and condition of equipment (such as capacity and age).

The condensate can be stripped in a multistage column with a countercurrent flow of air or steam. The stripping efficiency is greater than 95%.[2] Uncontrolled TRS emissions are estimated to be about 1 g/kg ADP (5000 ppm) from a condensate stripping system.

1.2.2   Particulate Emission Sources

The major potential process sources of particulate emissions from the kraft chemical recovery system are the recovery furnace, the smelt-dissolving tank, and the lime kiln. The recovery furnace is the largest potential particulate emission source. The major chemical constituent in the recovery boiler particulate emissions is Na_2SO_4, with smaller quantities of Na_2CO_3 and sodium chloride (NaCl) also present. The smelt-dissolving tank vents and lime kiln exhaust gases are also sources of varying quantities of particulate matter consisting primarily of carbonate, hydroxide, sulfate, and chloride salts of calcium and sodium. Particle sizes from these sources can range from 0.1 μm to greater than 1000 μm in diameter for uncontrolled emissions and from 0.1 to 10 μm in diameter where these sources have high-efficiency particulate control devices.

A summary of typical ranges in particulate concentrations and emission rates from kraft pulp mill process sources is presented in Table 2.

1.2.3   Other Emission Sources

Both oxides of sulfur (SO_x) and oxides of nitrogen (NO_x) can be emitted in varying quantities from specific sources in the kraft recovery system. The major source of sulfur dioxide (SO_2) emissions is the recovery furnace, because of combustion of sulfur-containing black liquor. Under certain conditions, somewhat similar quantities of sulfur trioxide (SO_3) can be released to the atmosphere, particularly when residual fuel oil is added as an auxiliary fuel. Lesser quantities of SO_2 can also be released from the lime kiln and smelt-dissolving tank. Trace quantities of sulfur oxides may also be released from other kraft mill sources. Oxides of nitrogen can be formed in any fuel combustion process by the reaction between oxygen and nitrogen at elevated temperatures. Nitrogen oxide emissions from kraft pulp mill process sources, such as the recovery furnace and lime kiln, are normally lower than for most other fuel combustion processes. This is primarily due to the large quantities of water present in black liquor and lime, which act as a heat sink to suppress the flame temperature. Larger quantities of oxides of nitrogen can be formed, however, when auxiliary fuels such as natural gas or fuel oil are added to the recovery furnace.

A summary of concentrations and emission rates for oxides of sulfur and oxides of nitrogen for specific kraft pulp mill sources is presented in Table 3. The extreme variations in operating conditions that occur in the industry, including operating combustion temperature and type of fuel, account for the broad ranges in these data.

Table 2   Typical Concentrations and Emission Rates for Particulate Matter from Kraft Pulp Mill Sources[2,3]

	Concentration,		Emission Rate,	
	g/dscm	(gr/dscf)	g/kg	(lb/ton)
Recovery furnace	3.10–22.79	(1.35–9.95)	22.1–283	(44.1–565)
Lime kiln	5.85–33.96	(2.55–14.81)	6.4–49.9	(12.7–99.8)
Smelt-dissolving tank	0.89–13.60	(0.39–5.94)	0.1–11.9	(0.19–23.7)

Table 3   Typical Emission Concentrations and Rates for SO_x and NO_x from Kraft Pulp Mill Combustion Sources[2]

Emission Source	Concentration, ppm by Volume			Emission Rate, kg/Mg		
	SO_2	SO_3	NO_x (as NO_2)	SO_2	SO_3	NO_x (as NO_2)
Recovery furnace						
No auxiliary fuel	0–1,200	0–100	10–70	0–40	0–4	0.75
Auxiliary fuel added	0–1,500	0–150	50–400	0–50	0–6	1.2–10
Lime kiln exhaust	0–200	—	100–260	0–1.4	—	10–25
Smelt-dissolving tank	0–100	—	—	0–0.2	—	—

1.3   Control Techniques

The various control techniques that have been or can be applied to the emission sources are discussed in this section. The emission sources are the recovery furnace, digester system, multiple-effect evaporator system, lime kiln, brown stock washer system, black liquor oxidation system, smelt-dissolving tank, and condensate stripper system. The applicability and effectiveness of the control techniques when retrofitted on existing facilities are also discussed.

1.3.1   TRS Control Techniques

1.3.1.1   Recovery Furnace System.   TRS emissions from a recovery furnace system can originate in the recovery furnace itself, or in the direct-contact evaporator if this type of evaporator is used.

Several operating and design variables that have some effect on, or relationship to, the generation of TRS emissions in a recovery furnace have been identified. These include the quantity and manner of introduction of combustion air, the rate of solids (concentrated black liquor) feed, the degree of turbulence in the oxidation zone, the oxygen content of the flue gas, the spray pattern and droplet size of the liquor fed to the furnace, the degree of disturbance of the smelt bed, and the combination of sulfidity and heat content value of the liquor fed. The effect of these variables is independent of the absence or presence of a direct-contact evaporator.

The age of existing furnaces has been reported to be a significant indicator of the furnace's ability to control TRS emissions. Generally, the age reflects an absence or lack of refinement in controls and instrumentation that assist the operator in maintaining close control of the process. Also, older furnaces may not incorporate recent manufacturer's improvements, such as new means of introducing air, flexibility in distributing air in the furnace, and means to change air velocity at injection ports. Furthermore, a major design change was made to recovery furnaces around late 1964. This change consisted of installing a membrane between the wall tubes located in front of the furnace's wall insulation. This design change made the furnace airtight. The wall insulation on furnaces without this membrane wall concept tends to deteriorate. This allows air to leak into the furnace. This in turn affects the combustion in the furnace and reduces significantly the capability of the operator to control TRS emissions. These older recovery furnaces could be modified to incorporate these new design features but the modifications would be extremely expensive.[1] However, changes in operating procedures can more easily be made.

There are two control techniques to reduce TRS emissions from the direct-contact evaporator: black liquor oxidation and conversion to a noncontact evaporator. Black liquor oxidation inhibits the reactions between the combustion gases and black liquor that normally generate hydrogen sulfide. This is accomplished by oxidizing the Na_2S to $Na_2S_2O_3$ in the black liquor before it enters the direct-contact evaporator. In converting to a noncontact evaporator, the direct contact between furnace gases and black liquor is eliminated, and hydrogen sulfide formation is prevented.

There are several modes of operation of black liquor oxidation systems. The black liquor is sometimes oxidized before being concentrated in the multiple-effect evaporators (weak black liquor oxidation), sometimes following evaporation (strong black liquor oxidation) and sometimes both, before and after. Air is the normal oxidizing agent, but molecular oxygen is also used when available on site. Air sparging reactors are the most common units, but packed towers and bubble tray towers are also used.

In modifying an existing recovery furnace with a direct contact evaporator to a noncontact design, a black liquor evaporator (concentrator) and a second feed water economizer are necessary. In addition, elimination of the existing direct-contact evaporator will result in an increased particulate concentration discharge from the furnace system into the particulate control device. To maintain particulate emissions at the original level, it may be necessary to replace the existing collector with a new higher efficiency precipitator or install an additional secondary collector.

Cross recovery liquors are somewhat different than straight kraft liquor, and, therefore, it is possible that the TRS emissions from a cross recovery furnace are not controllable to the same degree as are those from the straight kraft furnace. There are three reasons why TRS emissions may be higher from cross recovery furnaces. The first relates to the sulfur content of the liquor which is higher with this process than in straight kraft processes. In cross recovery operations, the heat content of the black liquor is lower than found in straight kraft mills. This is because the NSSC process gives higher pulp yields than the kraft process and, as a consequence, the spent liquor associated with the NSSC process contains less organic content. Therefore, its Btu value is lower as compared with kraft black liquor. The third reason pertains to the restriction on excess oxygen available in cross recovery furnaces to oxidize the relatively large quantities of volatile sulfur compounds given off as a consequence of the heavy sulfur loading and lower furnace operating temperatures. If enough excess oxygen is supplied to completely oxidize all volatile sulfur compounds, a sticky dust problem will develop which can plug up the precipitator and render furnace operation impossible.

Based on a study conducted on one cross recovery furnace, cross recovery furnaces that experience green liquor sulfidities in excess of 28% and liquor mixtures of more than 7% NSSC on an air dry ton basis cannot achieve the same TRS levels as straight kraft recovery furnaces.[4]

A recently developed control technique for recovery furnaces is alkaline adsorption with carbon activated oxidation of the scrubbing solution. Pilot plant studies indicate that this technique can reduce TRS emissions from 20 to 2500 ppm to between 1 and 10 ppm.[5] Reduction in particulate and SO_2 emissions are also reportedly achieved. This technique could be used to control TRS emissions on those older existing furnaces or cross recovery furnaces which do not have the combustion control capability for low TRS emissions. This technique could prevent the need to replace or reduce the load on older existing furnaces that are not capable of achieving the necessary TRS regulations.

1.3.1.2 Digester and Multiple-Effect Evaporator Systems. The digesters and multiple-effect evaporators will be considered together because noncondensable gases discharged from these two sources are normally combined for treatment. Most commonly, the gases are burned in the lime kiln. However, special gas-fired incinerators are also used, either as backup for the kiln when it is shut down, or as the full-time control device.

Retrofitting an existing mill to handle and incinerate these noncondensable gases is apparently no significant problem. Generally, it is simply a matter of ducting the gases to the kiln or incinerator and installing necessary condensers and gas holding equipment. The noncondensable gases are added to the primary air to the kiln.

The blow gases from batch digesters are generated in strong bursts that normally exceed the capacity of the lime kiln. For this reason, special gas handling equipment has been developed to make the gas flows more uniform. Adjustable volume gas holders, with movable diaphragms or floating tops, receive the gas surges, and a small steady stream is bled to the kiln. Although the noncondensable gases form explosive mixtures in air, possible explosion hazards have been minimized by the development of appropriate gas holding systems, flame arrestors, and rupture disks in the gas holding ducts, and flame-out controls at the lime kiln. Incineration of these gases in existing process equipment such as the lime kiln is particularly attractive since no additional fuel is required to achieve effective emission control.

Scrubbers are used at a few existing mills. White liquor, the usual scrubbing medium, is effective for removing hydrogen sulfide and methyl mercaptan, but not dimethyl sulfide or dimethyl disulfide. Several mills scrub the noncondensable gases before incineration to: (1) recover sulfur, (2) condense steam, and (3) remove turpentine vapors and mist, thereby reducing the explosion hazards.

Combustion of noncondensable gases in a lime kiln or gas-fired incinerator provides nearly complete destruction of TRS compounds. During a test on a separate incinerator burning noncondensables from a digester system and a multiple-effect evaporator system, the residual unburned TRS ranged between 0.5 and 3.0 ppm, and averaged 1.5 ppm (dry gas basis).[6] During the tests, the incinerator was operating at 1000°F (measured) with a calculated retention time for the gases of at least 0.5 sec.

Scrubber efficiencies are much lower than properly operated incinerators because only hydrogen sulfide and methyl mercaptan react with the alkaline medium. The composition of noncondensable gases is highly variable, but on the average hydrogen sulfide and methyl mercaptan comprise about half the TRS compounds. Since caustic scrubbing is only effective in controlling hydrogen sulfide and methyl mercaptan, alkaline scrubber efficiencies are roughly only 50%.

1.3.1.3 Lime Kiln. TRS emissions, principally hydrogen sulfide, can originate from two areas in the lime kiln installation, the lime kiln proper and a scrubber that serves as the particulate control device. TRS emissions from the lime kiln installation are controlled by maintaining proper process conditions. The most important parameters that were identified in an industry study include the temperature at the cold end (point of exhaust discharge) of the kiln, the oxygen content of the gases leaving the kiln, the sulfide content of the lime mud fed to the kiln, and the pH and sulfide content of the water used in a particulate scrubber.[7] If contaminated condensate is used as the scrubbing medium, the exhaust gases could strip out the dissolved TRS and increase the TRS emissions from the lime kiln installation. Scrubbing the exhaust gases with a caustic solution can reduce the TRS emissions from a lime kiln.

The amount of retrofitting necessary to achieve proper process conditions depends on the design of the existing kiln installation. If the existing kiln does not achieve sufficient oxygen levels, increased fan capacity or changes to the scrubber system may be necessary to increase the air flow through the kiln. Molecular oxygen can also be used to replace a portion of the combustion air to increase oxygen levels. Additional lime mud washing capacity may also be necessary to reduce the sulfide content of the mud and thereby reduce TRS emissions. This may require replacement of existing centrifuges with more efficient vacuum drum filters, and the addition of another mud washing stage. Furthermore, a mill presently using condensate that contains dissolved reduced sulfur compounds for a scrubbing medium would have to either install a condensate stripper to remove the dissolved TRS prior to the scrubber or replace the condensate with fresh water.

1.3.1.4 Brown Stock Washer System.

Nearly all existing kraft mills vent the brown stock washing system gases directly to the atmosphere without control. However, at least three mills in the United States and Canada, and one in Sweden, utilize the gases as combustion air in a recovery furnace. The furnace systems handling these gases are newer furnace systems which were designed to burn the washer gases. No existing recovery furnace (not designed for burning these gases) has yet been used to incinerate the washer gases.

Since the gas volume from the washer drums is large, about 112 m³/Mg (150 cfm/tpd), the most likely equipment for combustion is a recovery furnace or power boiler. The gases, due to their large volume, would have to supplement the recovery furnace's combustion air requirements. Even if the washers were enclosed with tight hoods, the gas volume would be too large to burn in a lime kiln.

The vent gases from the filtrate tank are considerably smaller in volume, about 4.5 m³/Mg (6 cfm/tpd). This stream is sufficiently small for combustion in a lime kiln, or to be blended with the hood vent gas and burned in a recovery furnace.

Incineration of the washer gases in a recovery furnace will not affect furnace operation, provided the moisture content of the gases is not too great. High moisture content can increase gaseous sulfur emissions and produce unsafe operating conditions. Bed (furnace) temperature decreases almost linearly with increased content of vaporized water in the combustion air because of sensible heat losses. With decreased bed temperatures, SO_2 emissions increase at a rapid rate and reduced sulfur compounds become increasingly difficult to control. Water entrained in the combustion gases can create extremely dangerous conditions such as smelt–water explosions.

One furnace manufacturer recommends that the washer gases be incinerated only in the secondary or tertiary air zones of the furnace.[1] This would keep the moist washer gases away from the smelt bed. Burning the gases only in the secondary or tertiary zones may affect the flexibility of the recovery furnace, however, since the operator would not have the ability to vary the air flow rate to each zone.

High moisture content would result in an increase in gas flow and reduce the capacity of the recovery furnace.

An alternative to incineration of brown stock washer gases is chemical scrubbing. White liquor (caustic) scrubbing, as previously mentioned, is only effective in controlling hydrogen sulfide and methyl mercaptan. However, the TRS emissions from a brown stock washer system are principally dimethyl sulfide and dimethyl disulfide. A more effective system is reportedly a chlorination-caustic scrubbing system. In this system, the chlorine absorbs and oxidizes the dimethyl sulfide and dimethyl disulfide. This technique has demonstrated TRS emissions of less than 5 ppm.[1] Another technique is chlorine gas injection. This technique has demonstrated TRS emissions of less than 5 ppm and a control efficiency of 80%.[1]

1.3.1.5 Black Liquor Oxidation System.

The vent gases from nearly all existing black liquor oxidation (BLO) systems are emitted directly to the atmosphere without control.

One control technique is incineration. Incineration has proved highly effective in controlling similar streams in some mills, for example, the vent gases from pulp washing systems, the noncondensable gases from digesters and multiple-effect evaporators, and vent gases from condensate strippers. Similar to the pulp washing system, incineration in the recovery furnace or power boiler is most likely, since the BLO gas volume is usually too large to be handled by an existing kiln. This would result in no significant fuel penalty.

Because of the high moisture content of the BLO gases, it would be necessary to use condensers to reduce the moisture content before burning, especially if the moist washer gases are burned in the same furnace. Incineration of these moist gases in the furnace would probably cause increased corrosion problems in the forced-draft fan ductwork and the forced-draft fan itself. This would probably necessitate the replacement of this equipment with corrosion-resistant equipment. A larger forced-draft fan may be necessary to handle the increased mass flow due to the high moisture content of the gases, even after using condensers.

The recovery furnace operation should not be adversely affected by burning the BLO gases, even in combination with the washer gases, provided the moisture content is sufficiently reduced and the gases are burned high in the furnace. Since the BLO gases are deficient in oxygen, one furnace manufacturer suggests burning them in the secondary or tertiary air zone but states that the gases should still contain sufficient oxygen to preclude adversely affecting the furnace operation. As mentioned previously, the operational flexibility of the furnace is reduced because a portion (BLO gases and washer gases) of the total combustion air must always be introduced into the secondary and tertiary air zones and cannot be used in the primary air zone when air in this zone is needed to adjust furnace operation.

A second control technique is the use of molecular oxygen in oxidation systems instead of air. At least two mills in the United States now oxidize black liquor by pumping oxygen directly into the black liquor lines. There are no vent gases from this closed system. The economic feasibility of such a system depends largely on the price and availability of oxygen.

Another technique is chlorine gas injection. This technique is used at one mill on the vent gases from the primary oxidation system. Tests conducted demonstrated TRS emissions of less than 5 ppm and a control efficiency of 95%.[1]

1.3.1.6 Smelt-Dissolving Tank.

Smelt-dissolving tank TRS emissions are governed by process conditions; that is, the presence of reduced sulfur compounds either in the smelt or the water. The principal control option available is the choice of water in the smelt-dissolving tank or the particulate control device. Clean water, low in dissolved sulfides, is preferable, although low emissions have been reported with nearly all process streams. If TRS emissions are high and no particulate control device (scrubber) is used, a wet scrubber (e.g., packed tower) can be used to control the TRS emissions. This scrubber would also result in controlling particulate emissions.

1.3.1.7 Condensate Stripping System.

In at least four United States mills, dissolved sulfides and other volatile compounds are stripped from the digester and evaporator condensates prior to discharge to treatment ponds. One mill, which uses steam as the stripping medium, discharges the gases from the stripper column to a lime kiln. Two mills use air as the stripping medium. One of these incinerates the stripper gases in a separate incinerator, while the other incinerates the gases in the recovery furnace. One mill, which uses steam, is presently scrubbing the stripper gases with white liquor, but this technique is not as effective as incineration.

As mentioned previously, incineration has proved to reduce TRS levels from digester and multiple-effect evaporator systems to less than 5 ppm. Since the vent gas from condensate strippers contains the same TRS compounds present in the digester and multiple-effect evaporator gases, TRS emissions in the condensate stripper gases after incineration can be reduced to 5 ppm (0.01 g/kg ADP).

1.3.2 Particulate Control

Particulate emissions from the kraft process occur primarily from the recovery furnace, lime kiln, and smelt-dissolving tank. These emissions consist mainly of sodium salts but include some calcium salts from the lime kiln. The dust collected in the kraft industry, especially from indirect-contact recovery systems, is more corrosive and sticky than that encountered in other industries. This leads to some special problems with the particulate emission control equipment. Recovery furnace exhaust gases and particulate emissions have different characteristics depending on whether they are generated from a direct-contact (conventional) or an indirect-contact (noncontact) system, as shown in Table 4.

The recovery furnace, lime kiln, and smelt-dissolving tank are the primary sources of particulate emissions to which control devices are applied. Electrostatic precipitator (ESP) and scrubber systems are employed on recovery boilers, with ESP systems being used most frequently. Lime kilns generally utilize scrubber systems, but occasionally ESP systems are used; demister pads and other low-energy scrubber systems are generally applied to control particulate emissions from smelt-dissolving tanks.

Table 4 Particulate Property Values from Conventional and Low-Odor Recovery Processes[8]

	Conventional	Noncontact
Temperature, °K	410–435	445–505
Moisture content, %	30	7–20
Particle size, μm	6–10	Less than 6
Density, kg/m³	320–400	80–160
Tenacity	Reasonable	Difficult
Resistivity	Low	High
Sulfur content	Low	High

Fabric filters are not used in kraft mills because of the high moisture content of the exhaust gases and the fact that mechanical collectors are not efficient enough by themselves, due to the size distribution of the particulate matter, to provide the degree of control required.

1.3.2.1 Recovery Boilers. Application and design of ESP systems for recovery boiler emission control depend on whether the system is to be applied to a recovery system that uses a direct-contact or an indirect-contact evaporator. For a recovery system using a direct-contact evaporator, the evaporator itself may serve to reduce the mass of particulate emissions by as much as 50%. This, along with the fact that the particles emitted by an indirect-contact recovery boiler are generally smaller and less dense than particles emitted by a direct-contact recovery boiler, means that ESP systems for indirect-contact evaporators must be designed with more conservative sizing estimates. Table 5 lists the design parameters for a typical ESP applied to an indirect-contact recovery boiler. Further information on the design parameters for ESP systems may be found in Chapter 12.

Problems encountered in applying ESP systems to recovery boilers include corrosion, plugging, and wire breakage. These problems are apparently due to operating the equipment at conditions for which it was not designed (i.e., higher gas volumes, higher inlet loadings, or lower inlet temperatures). In order to prevent corrosion, the manufacturers install insulation or heated shells to maintain the gas temperature above the gas dewpoint throughout the precipitator.

Scrubbers applied to kraft recovery boilers are generally of the venturi type and are multiple stage (i.e., venturi scrubbers connected in series). These systems are not widely used, and removal efficiencies as high as 95% by weight are obtainable with pressure drops around 3 kPa.[9]

1.3.2.2 Lime Kilns. Wet scrubbers are frequently applied to lime kilns, with venturi and impingement designs being the most prevalent. Typical operating characteristics of particulate liquid scrubbers on kraft lime kilns are summarized in Table 6. Average collection efficiencies for venturi and impingement scrubbers range from 92 to 95% removal by weight.

1.3.2.3 Smelt-Dissolving Tanks. Showered mist eliminators are used almost exclusively on smelt-dissolving tanks. Showered mist eliminators consist of fine wire pads approximately 30 cm thick. Removal efficiencies are roughly 70–80% by weight. Demister pads used in series with a packed tower or scrubber attain efficiencies of 92–96%.

1.4 Air Pollution Regulations

New Source Performance Standards (NSPS) for new and modified kraft pulp mills were promulgated by the United States Environmental Protection Agency on February 23, 1978.[10] The standards apply to particulate and TRS emissions from the recovery furnaces, lime kilns, and smelt-dissolving tanks and to TRS emissions from digesters, multiple-effect evaporators, brown stock washers, black liquor oxidation systems, and condensate stripping systems. With the exception of the smelt-dissolving tank, TRS standards are 5 ppm for all sources except the lime kiln, which is 8 ppm, and recovery furnaces

Table 5 Typical Indirect-Contact Kraft Recovery Boiler Electrostatic Precipitator System Design Parameters[8]

Compartments	2
No. of fields	6–8
Collection plate area	1.2–1.5 m²/am³/min
Residence time	Minimum 10 sec
Gas velocity	Maximum 1.1 m/sec
Power input	40–90 W/1000 am³/min
	700–900 mA/1000 am³/min
	164–490 mA/1000 m electrode
Electrode rappers	550–1220 m/unit
Collection plate rappers	140–230 m²/unit
Rake speed	2–3 cm/sec
Rake torque	Minimum 75 Nm/m² (60 in.-lb/ft²)
Screw speed	20–40 rpm
Screw torque	Minimum 55 Nm/m (150 in.-lb/ft)
Dust density	80–130 kg/m³
Inlet concentration	11–18 g/Nm³
Dust volume	1.6–3.7 m³/Mg of pulp
Dust compartment depth	1.8–3 m
Efficiency	99.5–99.8% removal by weight

Table 6   Operating Characteristics of Particulate Liquid
Scrubbers on Kraft Lime Kilns

Parameter	Scrubber Type	
	Venturi	Impingement
Liquid-to-gas ratio, l/m³	1.73–3.21	0.54–2.0
Slurry solids, wt%	10–30	1–2
Pressure drop, kPa	2.5–3.75	1.25–1.75

burning a combination of kraft and semichemical liquor, where emission standards are 25 ppm. The TRS standard for the smelt-dissolving tank is 0.0084 g/kg. Particulate standards are 0.10 g/dscm and 35% opacity for recovery furnaces, 0.15 g/dscm for gas-fired lime kilns, 0.30 g/dscm for oil-fired lime kilns, and 0.1 g/kg for smelt-dissolving tanks.

Guidelines issued by EPA on May 22, 1979,[11] which mandate that each state have a TRS emission control program for existing mills, are similar to many that exist in a number of states. The guidelines apply to recovery furnaces, lime kilns, smelt-dissolving tanks, digesters, multiple-effect evaporators, and condensate strippers. No guidelines were issued for brown stock washers and black liquor oxidation systems.

The TRS guidelines for recovery furnaces make a distinction between furnaces purchased with TRS emission control as a criterion in contrast to those purchased during the period when TRS emission control was not a prime consideration. The guidelines for straight kraft recovery furnaces are 5 ppm for furnaces designed for low TRS emissions and 20 ppm for all other straight kraft furnaces. The TRS guideline for cross recovery furnaces is 25 ppm. The guidelines for digesters, evaporators, smelt-dissolving tanks, condensate strippers, and cross recovery furnaces are the same levels as those set for new kraft pulp facilities. The guideline for existing lime kilns is, however, 20 ppm.

Typical state regulations for particulate emissions from existing sources are 2.0 kg/Mg ADP (4.0 lb/ton ADP) for recovery furnaces, 0.4 kg/Mg ADP (1.0 lb/ton ADP) for lime kilns, and 0.25 kg/Mg ADP (0.5 lb/ton ADP) for smelt-dissolving tanks.

1.5   Best Control Technology

The New Source Performance Standards discussed above reflect best control technology. The control techniques required to meet these levels are

Recovery furnace	Electrostatic precipitator plus process controls and black liquor oxidation or noncontact evaporation
Lime kiln	Venturi scrubber with caustic addition plus process controls
Smelt-dissolving tank	Scrubber plus use of clean water
Digester systems	Incineration
Multiple-effect evaporators	Incineration
Condensate strippers	Incineration
Brown stock washers	Incineration or chlorination
Black liquor oxidation system	Incineration or chlorination

The treatment of the black liquor oxidation system, which is composed essentially of the organic sulfur compounds, is practiced in only one instance. This moisture-laden gas stream, which is lacking 25–35% of the original oxygen present in air, has not been considered a suitable source for use as combustion air in furnaces or boilers. This probably accounts for the use of gas-phase chlorination at the only source being treated at this time.[12]

1.6   Control Costs

The control costs for the individual sources are presented in Table 7. The costs are for the best control system discussed above. Costs are given for all sources except for TRS control on the smelt-dissolving tank. The control technique for reducing TRS emissions from the smelt-dissolving tank is to use fresh water (or water that is essentially free of dissolved TRS compounds) in the smelt-dissolving tank scrubber. The costs associated with the use of fresh water is expected to be small. Costs given for incineration are based on incinerating the gases in either the recovery furnace or the lime kiln. Costs given for TRS control on the lime kiln assumed that a scrubber is already installed for particulate control. The costs for TRS control on the indirect-contact recovery furnace are based on the heat loss of the indirect-contact furnace compared to the direct-contact furnace. All costs in Table 7 reflect

Table 7 Summary of Control Costs per Facility[13]

Mill Size, Mg/d (tpd)		454 (500)			907 (1000)			1360 (1500)		
Facility	Control Technique	Capital Costs, $	Annualized Costs, $/yr	Unit Annualized Costs, $/Mg	Capital Costs, $	Annualized Costs, $/yr	Unit Annualized Costs, $/Mg	Capital Costs, $	Annualized Costs, $/yr	Unit Annualized Costs, $/Mg
Particulates										
Recovery Furnace										
(a) Direct-contact	ESP	2,251,650	569,170	3.82	4,003,000	990,000	3.31	5,723,000	1,400,000	3.12
(b) Indirect-contact	ESP	3,612,000	799,000	5.35	6,255,000	1,375,000	4.61	8,600,000	1,892,000	4.23
Smelt-dissolving tank	Scrubber	136,820	38,780	0.26	216,000	65,050	0.22	274,000	87,410	0.20
Lime kiln	Venturi scrubber	186,000	114,770	0.77	258,000	213,280	0.72	335,000	313,000	0.71
	ESP	479,000	289,000	1.94	691,000	514,400	1.73	852,000	730,000	1.63
TRS										
Recovery furnace										
(a) Direct-contact	Process control and BLO	617,600	220,500	1.48	899,000	328,400	1.10	1,204,000	454,000	1.01
(b) Indirect-contact	Process control	467,500	545,700	3.66	733,350	1,052,000	3.52	927,000	1,564,000	3.49
Digester (batch) and multiple-effect evaporators	Incineration	202,000	43,780	0.30	275,200	63,000	0.21	380,000	87,700	0.20
Brown stock washers	Incineration	339,000	67,000	0.45	550,000	110,000	0.36	735,000	147,000	0.33
Black liquor oxidation system vents (direct-contact furnace only)	Incineration	313,000	84,400	0.56	477,000	139,000	0.46	625,500	192,000	0.43
Lime kiln	Process controls and caustic scrubbing	0	51,600	0.34	0	112,600	0.37	0	175,000	0.40
Condensate stripper	Incineration	23,500	9,000	0.06	32,800	11,260	0.03	40,700	12,800	0.03

April 1981 prices for fabricated equipment. These costs were updated to April 1981 using the *Chemical Engineering* plant cost index. The costs that were updated were originally in terms of (fourth quarter) 1975 dollars.

2 SULFITE PULPING

2.1 Process Description

The production of sulfite pulp is similar to kraft pulping but the basic difference is that in place of the sulfide-containing caustic solution used to dissolve the lignin in the wood, a salt of sulfurous acid is employed. To buffer the solution, a bisulfite of magnesium, ammonium, calcium, or sodium is used. Calcium-base systems are used only in older mills and are being replaced with new magnesium- or ammonium-base mills due to problems with disposal of spent liquor and recovery systems.

Digestion is carried out under high pressure and elevated temperature in the presence of a sulfurous acid–bisulfite cooking liquor. When cooking is completed, the digester is either discharged at high pressure (blowing) into a blow pit or its contents are pumped out at a lower pressure into a dump tank (dumping). The spent sulfite solution is drained and is either treated and disposed, incinerated, or sent to a plant for recovery of heat and chemicals. The choice of whether or not chemical recovery is desirable is dictated by the base employed in the cooking liquor. In calcium-base systems, chemical recovery is not practical, due to the low solubility of calcium salts and the formation of scale and calcium sulfate ash in the recovery process; therefore, the spent liquor is either discarded or incinerated. In ammonium-base operations, heat and sulfur can be recovered from the spent liquor through combustion and subsequent SO_2 absorption, but the ammonium base is consumed in the process. In sodium- or magnesium-base operations, heat, sulfur, and base recovery are feasible. If recovery is practiced, the spent liquor is concentrated to 55–60% solids and then sprayed into a furnace and burned.

The major units in the sulfite pulping process are the digester, the blow pit or dump tank, the washers and screens, the evaporators, the recovery system, and the liquor preparation system. These process units are shown in Figure 2. Each of these units is a potential emission source for atmospheric pollutants.

2.2 Sources and Their Emissions

The primary emissions from the sulfite pulping process are SO_2 and particulate matter. In special cases of burning alkaline sulfite liquor in recovery furnaces under reducing conditions, H_2S emissions may also occur. Otherwise, there are practically no organic reduced sulfur compounds produced in the sulfite process.

Various process sources within the sulfite mill can emit SO_2. The main sources are the digester blow pits, multiple-effect evaporators, and liquid burning or chemical recovery systems. Minor process sources include pulp washers and the acid preparation plant. Typical values of SO_2 emission rates are listed in Table 8.

The recovery furnace is the significant process source of particulate matter in a sulfite pulp mill. Potential particulate matter emissions depend greatly on the degree of recovery of sulfite waste liquor, as well as on the degree of control of particulate matter.

2.2.1 Digesters

Two types of digesters are used in wood pulping operations: batch and continuous. The batch digester is the most common type used in the sulfite process.

Batch digesters have two potential sources of atmospheric emissions: relief gases removed from the unit during the cooking cycle or prior to pulp discharge and blow gases expelled during the pulp discharge. Depending on the pH of the cooking liquor, digester relief gases may contain high concentrations of sulfur dioxide. Acid sulfite cooking liquors usually have a pH less than 2.0 and are more likely to release large amounts of gaseous SO_2 in the digester. For a bisulfite liquor (pH between 2 and 5), the amount of gaseous SO_2 evolved during the cook is much less.

During the cooking cycle, relief gases are continuously removed in order to maintain a constant digester pressure. In addition, a large amount of relief gas may be vented from the system prior to the digester blow, depending on the type of pulp discharge being practiced. Both types of relief gas contain significant amounts of SO_2. Since the purpose of the preblow digester relief is to reduce blow emissions, this relief is usually collected for appropriate disposal.

Digester blow emissions can be very large. The emissions will depend on the type of pulp discharge practiced. For a high-pressure blow, the pulp/liquor slurry is passing from a pressurized to atmospheric regime with the initial liquor temperature above the boiling point. Consequently, the liquor flashes and produces large amounts of steam. Similarly, the sulfur dioxide, which was more soluble at high pressure, is desorbed and liberated with the steam. Cooking liquors with high sulfite concentrations (low pH) produce greater SO_2 emissions.

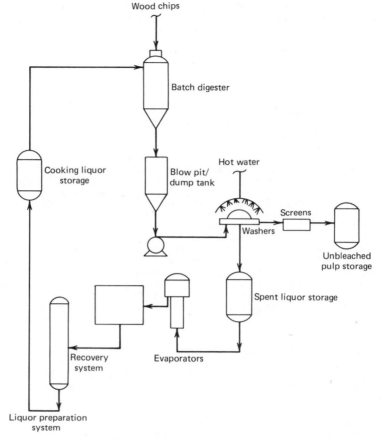

Fig. 2   Generalized sulfite process.

Emissions resulting from a digester blow are short and infrequent. A blow will last only a few minutes with the peak flows and concentrations lasting for 5–6 min. Since the blow occurs only at the end of a cooking cycle, the emissions for each digester should occur no more than once every 4 to 8 hr. Pulp mills, however, will have several digesters operating on a staggered sequence so that one digester or another will be blowing every hour or two.

Digester blow emissions are reduced considerably in a low-pressure blow. Much of the SO_2 is removed in the relief gases when the digester pressure is lowered. The emissions are reduced even

Table 8   Typical SO_2 Uncontrolled Emission Rates from Sulfite Pulp Mill Sources[2]

Emission Source	Emission Rate, kg/ton (lb/ton)
Blow pit	
Hot blow	30–75 (60–150)
Cold blow	2–10 (4–20)
Evaporators	1–30 (2–60)
Recovery process	80–250 (160–500)
Washers	0.5–1 (1–2)
Acid preparation	0.5–1 (1–2)

further if a cool rinse liquor is added. The rinse liquor lowers the slurry temperature and therefore decreases the amount of liquor that flashes when the pulp is discharged.

Since pulp pump-out systems are completely enclosed, digester blow emissions are virtually eliminated. These systems are expensive and time consuming, however, and have received only limited use.

The only source of emissions from continuous digesters is pressure relief gases. Since the discharge of pulp is uninterrupted, there is no intermittent digester blow and, therefore, no emissions associated with this step in the pulping operation. Continuous digesters are, for the most part, completely closed systems and normally use a cooking liquor in the pH range of 4 to 6. These features mean that continuous digesters usually have negligible emissions.[2]

2.2.2 Blow Pits/Dump Tanks

Blow pits and dump tanks are used as temporary storage vessels for the large amounts of pulp and liquor which are intermittently discharged from batch digesters. Blow pits and dump tanks may receive pulp from one or several digesters. Following discharge to these vessels, the pulp is drained of spent sulfite liquor and then transferred to the washers. In some cases, acid sulfite pulp may also be neutralized. The spent liquor is either directed to a recovery system for recovery of heat and/or chemicals or treated for disposal or sale.

The emissions most commonly associated with blow pits and dump tanks result from the discharge of pulp from batch digesters. These emissions, therefore, are considered to be from the digester and are discussed in Section 2.2.1.

In the event that a blow pit or dump tank is also used for pulp washing, small amounts of sulfur dioxide may be emitted. When the cool wash water contacts the hot pulp, steam is produced and some sulfur dioxide may vaporize. These SO_2 emissions would be very low.

2.2.3 Washers and Screens

From the blow pits and dump tanks the pulp is directed to a system of washers and screens. The washers rinse the pulp with hot water to remove remaining spent sulfite liquor. Several types of washers are currently in use, but the most common is a series of rotary drum vacuum filters, arranged for multistage countercurrent washing. Wash water may be either evaporator condensate or fresh water. The effluent wash water is sent to spent liquor storage.

The purpose of screens is to prepare a pulp of uniform consistency by removing uncooked knots, fiber bundles, and other oversize material.

Washers and screens are usually considered to be very minor sources of sulfur dioxide emissions.[14] However, if other SO_2 emission sources are well controlled, then washers and screens may be a significant fraction of the total mill emissions.

2.2.4 Evaporators

The purpose of the evaporators is to increase the total solids content of the spent sulfite liquor from about 15% to as much as 55–65%. The liquor is then either sold as a by-product or combusted in a furnace for recovery of heat and/or chemicals.

The evaporation of spent sulfite liquor releases sulfur dioxide. The more acidic the cooking liquor, the more SO_2 is released during the evaporation process. For a multiple-effect vacuum evaporation plant, similar to those found in kraft pulping operations, there are two potential emission sources. The vacuum system vent is the main emission point for evaporator gases, and the hot well is the main emission point for evaporator condensates. The hot well is a shell and tube surface condenser used to reduce the amount of evaporator off-gas by indirect condensation. The hot well condensate may contain appreciable amounts of dissolved SO_2. This condensate is usually combined with the evaporator system condensate. Both points are considered minor, but do contribute to the total sulfur dioxide emissions from the mill.

2.2.5 Recovery Systems

There are different recovery systems for different bases, but there can also be different recovery system designs for a single base. Usually these designs vary as to the amount and type of heat recovery or the method of SO_2 removal. Heat can be recovered by producing steam, heating process water, or evaporating spent liquor. Sulfur dioxide is usually removed by scrubbing the furnace flue gas in venturi scrubbers, packed bed absorbers, or tray towers.

Recovery systems are designed to incinerate spent sulfite liquor and to produce in a recoverable form the chemicals necessary to produce fresh cooking liquor. Except for sodium-base recovery systems that generate a usable sodium smelt, all the chemicals are recovered from the recovery boiler/furnace flue gas. These chemicals may be either in a gaseous state, such as sulfur dioxide, or in a particulate form, such as magnesium oxide. Depending on the efficiency of a recovery system, trace amounts of these chemicals may pass through the system and be discharged to the atmosphere from the flue gas exhaust stack. In addition, the recovery systems are not specifically designed for the removal of other chemical species such as sulfates, which would also be discharged from the flue gas exhaust stack.

2.2.6   Liquor Preparation Systems

The purpose of the liquor preparation system (also referred to as acid preparation) is to provide fresh cooking liquor for the digesters. For magnesium, ammonium, or sodium-base systems, liquor preparation is very closely associated with the recovery system. The weak acid produced by SO_2 removal in the recovery system is used for the cooking liquor preparation. In many cases the recovery system and the liquor preparation system are so closely integrated that it is difficult to consider them as two distinct systems.

Cooking liquor preparation consists of fortifying weak acid liquor from the recovery system with additional sulfur dioxide. Liquid sulfur is burned to produce SO_2 which is then absorbed by the weak acid liquor in an acid fortification tower. Then, depending on the desired strength (pH) of the cooking liquor, it can be further fortified in a low-pressure and/or high-pressure accumulator. Accumulators are sprayed chambers where steam is condensed and SO_2 from process unit relief gases is absorbed for subsequent reuse in the cooking liquor.

The major emission source in the liquor preparation system is the acid fortification tower. The weak acid from the recovery system does not absorb all the sulfur dioxide passing through the tower. In some cases the excess SO_2 is sent to the recovery system for removal. In many instances, however, the acid fortification tower is vented directly to the atmosphere.

The accumulators may also discharge vent gases containing SO_2 directly to the atmosphere. Many pulp mills have arranged a system of cascading SO_2 vent gases, that is, digester to high-pressure accumulator to low-pressure accumulator to acid fortification tower. Still, several other mills have no such system and the accumulators vent directly to the atmosphere.

2.3   Control Techniques

2.3.1   Control of Sulfur Dioxide Emissions

2.3.1.1   Digester Relief Vent.   Usually one of two process changes is made to effectively control emissions from the digester relief vent. The relief gases may be directed to a high- or low-pressure accumulator so that the SO_2 may be recycled to fresh cooking liquor. After passing through the accumulators, the relief gases may be sent to the recovery system so that the SO_2 is absorbed in the weak acid liquor which is eventually used for the production of fresh cooking liquor.

Another means of controlling SO_2 emissions from the digester relief vent is to direct the relief gases to a scrubber specially designed for that purpose. This approach may not be practical for some mills since it can be difficult to design a scrubber that will efficiently remove appreciable amounts of SO_2 from intermittent quantities of gas. On the other hand, for a large mill which may be operating several digesters simultaneously, the feed gas to the scrubber is much more uniform in composition and flow rate.

2.3.1.2   Digester Blow.   Emissions from the digester blow can be reduced either by collecting and scrubbing the blow gases or by modifying the pulp discharge operation to prevent liquor flashing.

Controlling SO_2 emissions with a scrubber offers a definite advantage in that it requires little or no alteration to the existing digesters and blow tanks. The entire area would need to be completely enclosed for the collection of blow gases. However, this should not be a problem unless the plant layout imposes unusually strict space limitations. The major problem with this control method is proper scrubber design. A blow will last only a few minutes with the peak flows and concentrations lasting only 5 to 6 min. Therefore, a scrubber must be designed for a maximum gas and vapor flow which is many times the average gas flow and likewise for peak concentrations of SO_2. SO_2 recovery efficiencies are reported to be as high as 97% for this type of control.[2]

Recently, many pulp mills have been modifying the pulp discharge operation in an effort to reduce sulfur dioxide emissions. The basic principle is to prevent liquor flashing by reducing the temperature and pressure of the pulp/liquor slurry prior to discharge. Two of the most common modifications are the low-pressure blow or dump system and the pump-out system.

The low-pressure blow or dump system involves lowering the digester pressure to nearly the atmospheric level and then discharging the pulp into a dump tank located directly beneath the digester. In some instances all or part of the hot spent cooking liquor may be drained from the digester and replaced by cooler water from the pulp washers. This lowers the pulp temperature and reduces the steam flashing on discharge.

There are several significant problems associated with the low-pressure blow or dump system. First, there is the problem of relief gases. In order to lower the digester pressure to near atmospheric, digester gases containing large amounts of SO_2 must be withdrawn and sent to appropriate treatment units. If the gases are vented to the recovery system or the liquor preparation system, such large intermittent doses may upset the efficiency of that particular system. Secondly, there is the problem of sunken dump tanks. Most of the sulfite mills currently operating were originally built with blow pits that are not located directly beneath the digester. Therefore, using a low-pressure dump would require major equipment modifications, assuming there was allowable space. Finally, there is the problem of reduced capacity. This type of pulp discharge is more time consuming than a high-pressure blow and consequently increases the time requirement for use of a digester.

In the pump-out system, the digester pressure is relieved to nearly the atmospheric level and then the pulp/liquor slurry is pumped from the digester into a closed storage vessel. Since the system is completely enclosed, all relief and vent gases can be collected and SO_2 emissions from the discharge of pulp are kept at a minimum. In some instances all or part of the spent cooking liquor may be withdrawn and replaced with water from the pulp washers to cool the pulp.

The pump-out system has problems very similar to those associated with the dump system. Although all relief and vent gases are collected, there may still be the problem of proper treatment. Equipment modification is not nearly as drastic for the pump-out system. The pulp storage tanks, which serve the same purpose as a dump tank, do not need to be located directly beneath the digesters. The pumps can be sized to transfer the pulp/liquor slurry to storage tanks located at any level in the plant. As before, there will also be a loss of pulping capability due to the increased time required for pulp discharge.

2.3.1.3 Recovery System Exhaust.

The recovery system is inherently a pollution control system. It is designed to reduce water pollution by providing an alternate means of disposal for spent sulfite liquor. The recovery system is made economically attractive by its ability to recover expensive chemicals necessary for the preparation of fresh cooking liquor. With the exception of sodium base liquors, all recovery systems are designed to recover sulfur as sulfur dioxide from the recovery furnace flue gas. Depending on the efficiency of SO_2 removal, large quantities of sulfur dioxide can be emitted from the recovery system exhaust stack.

The current method of recovering SO_2 is to pass the flue gas through a multistage system of absorbers or venturis and scrub the gas with an alkaline solution. Theoretically, if the scrubbers were made large enough and enough alkaline solution were used, SO_2 exhaust emissions could be reduced to zero. Practically, however, this would result in a scrubber slurry effluent of too great a quantity and too dilute a concentration to be of any use in cooking liquor preparation.

Hence, the best means of controlling SO_2 emissions from the recovery system exhaust is the proper operation of a well-designed scrubber system. This does not preclude the addition of another scrubber to an already existing recovery system. Depending on SO_2 control measures implemented at other emission sources in the mill, the recovery system scrubbers may become overtaxed. For example, if digester relief and blow gases are vented to the recovery system, an additional scrubber may be necessary to avoid a system upset from the sudden surge in SO_2-laden gases.

The scrubber system design will depend primarily on the type of alkaline scrubbing medium. Calcium hydroxide and magnesium hydroxide are considered rather cumbersome slurries because of their tendency to scale. Therefore, when using these bases, venturi scrubbers are preferred for their induced slurry turbulence. On the other hand, ammonium hydroxide and sodium hydroxide are somewhat easier to handle and may be effectively used in towers packed with an inert matrix or fitted with a number of perforated plates.

2.3.1.4 Liquor Preparation System.

Sulfur dioxide emissions from the liquor preparation system can be appreciable since the purpose of this system is to produce fresh cooking liquor by fortifying weak acid liquor with SO_2. The major emission source is the acid fortification tower which may vent large amounts of SO_2 directly to the atmosphere. These emissions can be reduced by proper equipment design and operation to ensure almost complete absorption of SO_2 by the weak acid liquor. Any tower off-gas containing significant amounts of SO_2 can then be directed to the recovery system for removal or to an additional scrubber.

In addition to the acid fortification tower, the accumulators also have off-gas vents that are sometimes released directly to the atmosphere. These emissions are best controlled by cascading the vent gases from one process unit to another. For instance, relief gases from the digester are directed to the high-pressure accumulator in which portions of SO_2 are redissolved in fresh cooking liquor. The high-pressure accumulator will, in turn, vent to the low-pressure accumulator, which will subsequently vent to the acid fortification tower. At each step along the cascade of units, some SO_2 is redissolved in the freshly prepared cooking liquor. Off-gas from the acid fortification tower is ultimately sent to the recovery system for removal of remaining SO_2.

2.3.1.5 Other Emission Sources.

Two major control options are available for reducing the emissions from stock washers and screens. In both options, vapors from the washers and screens are collected in hoods and removed from the work areas by fans. If the hoods are close fitting, then the total gas flow rate will be low and the vapors will contain considerable amounts of SO_2. In such cases, it is usually practical to send the vapors to the SO_2 scrubbers in the recovery system. On the other hand, if the total gas flow rate is high, then the SO_2 vapors will be relatively dilute. Under these circumstances, the control is to pass the gas through a water wash column (often referred to as a nuisance tower) for SO_2 removal before venting to the atmosphere.

The SO_2 emissions from the evaporation systems can be treated and recovered by several methods. One method is to scrub the gases with an alkaline solution of the base and then return the solution to acid preparation. This procedure, however, is difficult with calcium and, to a lesser degree, with a magnesium base. A widely practiced method of eliminating evaporator SO_2 emissions is to return the evaporator gases to the acid preparation plant to recover the SO_2.

2.3.2   Control of Particulate Emissions

The only source of particulate emissions is the recovery system exhaust stack. Emission control equipment applied to sulfite pulp mill recovery furnaces is generally dependent on the chemical base used in the cooking liquor. Magnesium-base recovery furnaces are frequently controlled with multicyclones and venturi scrubbers. Often multicyclones are followed in series by one or more venturi scrubbers, and these systems operate to eliminate sulfur emissions in addition to particulate matter emissions. In one plant, the multicyclone consists of 7640 tubes, 7.6 cm in diameter. Efficiencies of these multicyclones range from 96 to 98% in removing magnesium oxide.[14]

Ammonium-base sulfite recovery furnace emissions are controlled by low-pressure drop tray scrubbers followed by glass-fiber packed filter units or mist eliminators. The multiple tray scrubber is designed primarily for sulfur dioxide absorption; however, some particulate matter removal is also achieved. A typical flue gas scrubber applied to an ammonium-base sulfite recovery boiler operates with a pressure drop of 2.75–3.85 kPa, and removal efficiencies range from 85 to 95%.[14]

A venturi scrubber followed by a cross-flow packed bed scrubber is in use in the only sodium-base recovery system in the United States. The system serves several functions by reducing particulate and reduced sulfur gaseous emissions as well as recovering heat from the recovery furnace exhaust gases. The venturi scrubber operates at a pressure drop of 1.75 to 2.5 kPa; the total system pressure drop varies between 2.5 and 3 kPa. Removal efficiencies are about 97%.[14]

2.4   Applicable Regulations

At present, ten states regulate sulfur emissions and six states regulate particulate emissions from the sulfite wood pulping process. Only four states (Alaska, New Hampshire, Oregon, and Washington) regulate both sulfur and particulate emissions. Five of the states that have sulfite mill regulations do not have any sulfite mills.

For the most part, these regulations are designed to limit SO_2 emissions from the digesters (or blows), recovery systems, and the mill in general (i.e., washer vents, storage tanks, etc.). SO_2 limitations are normally expressed as g SO_2/kg of air-dried pulp (lb/ton) or ppmv SO_2. The most restrictive regulations limit total SO_2 emissions from the mill to 4.5 g/kg of pulp (9.0 lb/ton). To meet these regulations, the mill would have to reduce uncontrolled emissions approximately 89%.

Particulate regulations apply only to recovery system exhaust stacks. The particulate limitations are typically expressed as g particulate/kg of air-dried pulp (lb/ton). The most restrictive regulations limit particulate emissions from the recovery system exhaust stack to 1.0 g/kg of pulp (2.0 lb/ton). To meet these regulations, the mill would have to reduce uncontrolled emissions approximately 44%.

In states that have not established specific regulations for the sulfite pulping operations, these operations will probably be required to meet the emission limitations designed for process units. These regulations normally determine the allowable emissions from an equation based on the process weight rate. The process weight rate is usually defined as the total weight of all materials introduced into the process, divided by that period of time (in hours) during which such materials are introduced into the process. Liquid and gaseous fuels, uncombined water, and combustion air are excluded from the weight of materials introduced into the process. The process weight rate is usually calculated in kg/hr (ton/hr) and the allowable emissions are typically determined in g/hr (lb/hr).

Some states also have opacity regulations, usually based on the Ringelmann number. With the exception of ammonium-base systems, opacity requirements should not be a problem. However, for an ammonium-base liquor, the recovery system exhaust stack may emit particulates of ammonium sulfate and ammonium sulfite that produce a blue haze.

2.5   Best Control Techniques

Since each mill is unique in its application of the sulfite pulping process, no single best control system can be defined for the control of atmospheric emissions. The exact emission control strategy for a particular mill will depend on the age and type of equipment currently in use, the type of cooking liquor, the availability of space to retrofit or add equipment, and the associated capital and operating costs. Some generalizations concerning best control systems are discussed below.

From an economic standpoint, probably the best means of controlling emissions from the digester relief vent is to direct these gases to the recovery system or liquor preparation system. This method of control requires minimal process alteration and capital investment. However, care must be taken not to overload the design capacities for either the recovery system or the liquor preparation system. If these systems are unable to handle the digester relief gases, the next best control method is the installation of a scrubber. Reported emissions for the two control methods are comparable.

Digester blow emissions can be controlled either by collecting blow gases for scrubbing or by reducing blow gases by discharge modifications. Scrubbing usually requires complete enclosure of the digester blow area and installation of a new scrubber. Pulp discharge modifications require considerable equipment alterations for pressure relief, cooling liquid addition, and possible pulp pump-out. In addition, pulp discharge modifications increase digester turnaround time and consequently decrease mill production. While neither control system is economically attractive, both systems achieve similar emission reductions.

The best means of controlling SO_2 emissions from the recovery system exhaust is the proper operation of a well-designed scrubber system. For recovery systems that accept digester relief and blow gases, an additional scrubber may be necessary to avoid a system upset from the sudden surge in SO_2-laden gases.

The best means of controlling particulate emissions from the recovery system stack is to install a demister prior to flue gas exhaust. Demisters may also aid in increased SO_2 removal.

The best means of controlling emissions from the liquor preparation system is a proper design and operation. Such a system would consist of an acid fortification tower capable of almost complete absorption of SO_2 by the weak acid liquor. This result may be achieved by placing two towers in series. The system would also be designed for cascading of process unit vent gases to prevent any atmospheric release of SO_2.

The SO_2 emissions from the liquor preparation system in a calcium-base mill are usually higher than other sulfite mills. These mills are continuously preparing fresh cooking liquor without the benefit of a weak acid recycle stream from the recovery system. Therefore, it may be necessary to install a caustic scrubber for treating off-gas from the acid preparation tower in calcium-base mills.

3 NEUTRAL SULFITE SEMICHEMICAL (NSSC) PROCESS

3.1 Process Description

The pulping of wood by the neutral sulfite semichemical (NSSC) process actually involves two processes, a chemical action to partially free cellulosic fibers by lignin sulfonation and a mechanical action to completely free the fibers by friction and compression. Although NSSC pulping has been practiced for over 40 yr, the operations and equipment are not yet standardized to the extent of modern kraft pulping.

A generalized flow diagram for the NSSC process is shown in Figure 3. Wood chips, usually hardwood, are cooked in batch or continuous digesters with a nearly neutral solution of sulfite containing a small amount of alkaline agents such as carbonate, bicarbonate, or hydroxide. After cooking, the partially pulped chips pass through a blow tank for liquor drainage and through presses for additional liquor removal. The cooked chips are further processed in disk refiners for mechanical defibering and through washers and screens for rinsing and removal of unpulped material. The spent liquor is either recovered in systems similar to those used in the sulfite process or combined with the liquor used in the kraft process, a process commonly referred to as cross recovery, as a source of makeup chemicals.

The cooking liquor may be prepared either by adding fresh chemicals (sulfite and carbonate) to water or spent liquor or by absorbing sulfur dioxide generated in a sulfur burner in a sodium carbonate solution.

3.2 Sources and Their Emissions

Various process sources within the NSSC mill can emit SO_2. The main sources are the digesters, evaporators, recovery systems, and liquor preparation systems. The presses, refiners, washers, and screens are not considered potential sources of atmospheric emissions because at these points in the pulping process most of the spent liquor has been removed. The recovery system is also a source of particulate emissions.

3.2.1 Digesters

Unlike the sulfite process, the most common type of digester in the NSSC process is the continuous digester. Operating conditions for these digesters are slightly more intense than for batch digesters. Temperatures and pressures may range from 160 to 185°C (320 to 365°F) and from 690 to 1100

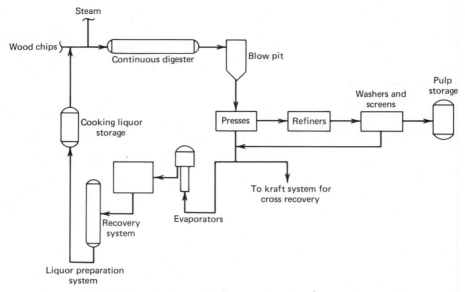

Fig. 3   Generalized neutral sulfite semichemical (NSSC) process.

kPa (100 to 160 psi), respectively. Digestion time is considerably less for the NSSC process, usually on the order of 30 min to 3 hr. Pulping may be conducted in either the liquid or vapor phase. Although sodium is the most common cooking liquor base, magnesium- and ammonium-base liquors can also be used.

Depending on the operating pressure, the partially pulped chips are discharged in a number of ways. Pressurized digesters will blow the chips through some type of pressure control valve into a blow tank. A digester operating at atmospheric pressure will normally have the chips scraped into a conveyor serving an orifice for discharge to a blow pit.

The continuous digesters have two potential sources of atmospheric emissions: relief gases and blow gases. Both types of emissions might contain sulfur dioxide, but only in trace amounts. The relatively high pH of the cooking liquor (usually between 7 and 8) indicates that high SO_2 concentrations in these gases are very unlikely. Indeed, source test data for blow emissions at one NSSC mill estimated SO_2 emissions at 0.045 g SO_2/kg of pulp production (0.09 lb/ton).[14]

For plants associated with kraft mills, digester relief and blow gases may also contain hydrogen sulfide, if green liquor is used in the cooking process.

3.2.2   Blow Tanks

For continuous digesters, blow tanks act as a hold tank where spent liquor can drain from the freshly discharged pulp slurry. The blow tanks are equipped with a leach caster with revolving arms and plows that work downward to sweep the cooked chips to an outlet. Pressurized digesters will generally blow into a cyclone where steam is separated. The cooked chips drop from the cyclone into tanks with live bottoms or into some other type of conveyance device for transport to the presses.

Blow tank emissions result from the gases evolved during the discharge of partially pulped chips from the digesters. These emissions are expected to be very low and are discussed in Section 3.2.1.

3.2.3   Disposal of Spent Cooking Liquor

In the NSSC process, spent cooking liquor may be disposed of in one of two ways: transfer to an associated kraft pulping process or evaporation and incineration for recovery of chemicals in systems very similar to those used in the sulfite process. Those mills that are operated in association with a kraft process will mix the spent sulfite liquor with the kraft cooking liquor as a means of sodium and sulfur chemical makeup. In this case, the NSSC process is considered to be free of emissions normally associated with the disposal of spent cooking liquor as the chemicals are now part of another process.

For those mills that recover heat and chemicals from the spent cooking liquor, the systems are almost identical to those designed for the sulfite process. The liquor is evaporated and incinerated, and the cooking liquor chemicals are recovered, when possible, to regenerate a fresh cooking liquor.

Potential emission sources from chemical recovery will include the evaporator gases and the recovery system exhaust. Of these two sources, the recovery system exhaust is the more significant. In many cases the evaporator gases are vented to the recovery system, so there is only a single source. The exhaust stack may be the source of SO_2 and particulate emissions.

3.2.4 *Liquor Preparation System*

The methods of preparing NSSC cooking liquor are comparable to those used in sulfite pulping. The major exception is the difference in liquor pH levels. Since the NSSC cooking liquor is designed for neutral pH levels (7 to 8), acid fortification equipment is unnecessary. Cooking liquor can be freshly prepared by contacting a sodium hydroxide solution with SO_2 from a sulfur burner. The contacting equipment can be a packed bed absorber or tray tower. If cross recovery is practiced, then fresh cooking liquor must be continuously prepared.

However, if a recovery system is employed for spent liquor disposal, then cooking chemicals can be recycled. If the cooking base is ammonia, fresh ammonium hydroxide makeup is required because of the combustion of ammonia in the recovery system.

The major emission source in cooking liquor preparation is the packed bed or tray tower used for absorbing SO_2. These absorbers can be operated at low SO_2 loadings and high efficiencies to produce a neutral sulfite cooking liquor. One mill reports operating a packed bed absorber at 99.9% efficiency and measuring typical SO_2 concentrations of 5 to 15 ppm in the off-gas.[14]

3.3 Control Techniques

Little information is available on emission rates and control techniques for NSSC mills. However, those unit operations that are similar to kraft or sulfite sources (absorption tower, blow tank, evaporators, and recovery furnaces) can use control systems similar to those already described.

3.4 Applicable Regulations

Currently three states regulate SO_2 emissions and four states regulate particulate emissions from the NSSC pulping process. Only two states (Alabama and Louisiana) regulate both sulfur and particulate emissions. Unlike the sulfite regulations (Section 2.4), which are specifically designed for that process, the NSSC emission limitations are closely associated with those limitations designated for the kraft process. Those regulations probably represent an effort by the states to regulate two integrated processes as one. No states have implemented regulations specifically designed for NSSC pulping. However, the state of Washington is currently gathering data to determine the need and possible form for specific NSSC emission regulations.

As with sulfite mills, states that have not established specific regulations for NSSC pulping operations will probably require these operations to meet the emission limitations designed for process units.

3.5 Best Control Techniques

As discussed above, little information is available on control techniques for NSSC mills. Best control techniques for NSSC sources should be similar to the best control techniques already described for those kraft and sulfite sources which are similar to NSSC sources.

4 PAPER COATING

4.1 Process Description

Paper is coated for a variety of decorative and functional purposes, using water-borne, organic solvent-borne, or solventless extrusion-type materials. Because the organic solvent-borne coating process is a source of hydrocarbon emissions, it is an air pollution concern.

In organic solvent paper coating, resins are dissolved in an organic solvent or solvent mixture and this solution is applied to a web (continuous roll) of paper. As the coated web is dried, the solvent evaporates and the coating cures. An organic solvent has several advantages: it will dissolve organic resins that are not soluble in water, its components can be changed to control drying rate, and organic base coatings show superior water resistance and better mechanical properties than some types of water-borne coatings. In addition, a large variety of surface textures can be obtained using solvent coatings.

Most organic solvent-borne coating is done by paper converting companies that buy paper from the mills and apply coatings to produce a final product. The paper mills themselves sometimes apply coatings but these are usually water-borne coatings consisting of a pigment such as clay and a binder such as starch or casein. These water-borne coatings are not normally sources of organic emissions.

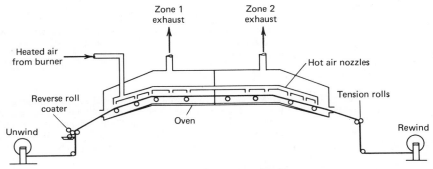

Fig. 4 Typical paper coating line.

Figure 4 shows a typical paper coating line. Components include an unwind roll, a coating applicator, an oven, various tension and chill rolls, and a rewind roll. Coatings may be applied to paper in several ways. The main applicators are knives, reverse rollers, or rotogravure devices. A knife coater consists of a blade that scrapes excess coating from the paper. The position of the knife can be adjusted to control the thickness of coating. The reverse roll coater applies a constant thickness of coating to the paper, usually by means of three rollers, each rotating in the same direction. A transfer roll picks up the coating solution from a trough and transfers it to a coating roll. A "doctor" roll removes excess material from the coating roll, thereby determining the thickness of the coating. A rubber backing roll supports the paper web at the point of contact with the coating roll, which is rotating in a direction opposite to that of the paper. Rotogravure is usually considered a printing process— the coating is picked up in a recessed area of the roll and transferred directly to the substrate.

4.2 Sources and Their Emissions

The coating applicator and the oven are the main areas of organic emissions in the paper coating facility. Most solvent emissions from coating paper come from the dryer or oven. Ovens range from 6.1 to 61 m (20 to 200 ft) in length and may be divided into two to five temperature zones. The first zone, where the coated paper enters the oven, is usually at a low temperature (~43°C). Solvent emissions are highest in this zone. Other zones have progressively higher temperatures that cure the coating after most of the solvent has evaporated. The typical curing temperature is 121°C, although in some ovens temperatures of 204°C are reached. This is generally the maximum because higher temperatures can damage the paper. Exhaust streams from oven zones may be discharged independently to the atmosphere or into a common heater, and sent to some type of air pollution control device. The average exhaust temperature is about 93°C.

Oven heaters are either direct or indirect fired. With direct-fired heaters, combustion products contact the coated web inside the oven. The burners themselves may be inside the oven chamber. More commonly, the burners are mounted external to the oven. In this case, heated air (along with products of combustion) is blown directly from the burner to the oven chamber.

Indirect-fired oven heaters are arranged so that products of combustion do not enter the oven chamber. A heat exchanger of some type is used to transfer heat from the burner to the oven chamber. Because combustion products do not enter the oven chamber in the direct-fired heater, there is no chance for contamination of the paper coating, and dirtier fuels can be burned.

Steam produced in gas or oil-fired boilers is sometimes used to heat ovens in the paper industry because paper coating ovens operate at fairly low temperatures. Typically, the steam is piped to the oven, and fresh air drawn into the oven is heated by passing it over the steam coils.

Most paper coaters try to maintain air flow through their ovens so the solvent concentration will be 25% of the lower explosive level (LEL), although many ovens are run at much lower solvent concentrations. As the energy shortages intensify, coaters are making greater efforts to minimize dilution air and thus raise solvent concentrations.

Precise methods are available for calculating the amount of dilution air needed to maintain the exhaust solvent concentration at a given LEL level. However, most of the paper-converting industry uses the estimation method of assuming 283 m³ (10,000 ft³) of fresh air, referred to 21°C, per gallon of solvent evaporated in the oven.[15] This method will give a solvent concentration of approximately 25% of LEL for most solvents, but the range may vary from 10 to 32% of the LEL for some solvents.

The exhaust flow rates from paper coating ovens vary from 2.4 to 16.5 m³/sec (5000 to 35,000 scfm) depending on size. Average exhaust rates are 4.7–9.4 m³/sec (10,000–20,000 scfm).

In a typical paper coating plant, about 70% of all solvents used are emitted from the coating line. The emphasis in this chapter is on control of the coating line. However, about 30% of plant emissions are from the other sources. These include solvent transfer, storage, and mixing operations. In order to control solvent emissions from these areas, provisions must be made to ensure that solvent-containing vessels have tight fitting covers and are kept closed. Another often overlooked source of solvent loss is use of solvents for cleaning various coatings and sludges from the coating line. This must be done before every color change. Areas of the coating line that are frequently cleaned with solvent can be hooded so that solvent fumes are captured and sent to a control device. Dirty cleanup solvent can be collected, distilled, and reused. Solvent soaked wiping rags should be kept in closed containers.

Almost all emissions of the solvent from the coating line itself can be collected and sent to a control device. Many plants report that 96% of solvent introduced to the coating line is recovered. The oven emissions can be exhausted directly to a control device.

Part of the solvent remains with the finished product after it has cured in the oven. For example, certain types of pressure-sensitive tapes have 150 to 2000 ppm by weight of solvent in the adhesive mass on the finished tape. Some coaters estimate that 2 or 3% of solvent remains in the product.

Solvent emissions from an individual coating facility will vary with the size and number of coating lines. A plant may have only one or as many as 20 coating lines. Uncontrolled emissions from a single line may vary from 0.006 to 0.13 kg/sec (50 to 1000 lb/hr), depending on the line size.[16] The amount of solvent emitted also depends on the number of hours the line operates each day.

4.3 Control Techniques

4.3.1 *Low-Solvent Paper Coatings*

A variety of low-solvent coatings have been developed for coating paper. These coatings form organic resin films that can equal the properties exhibited by typical solvent-borne coatings for some uses.

Water-borne coatings have long been used in coating paper to improve printability and gloss. The most widely used types of water-borne coatings consist of an inorganic pigment and nonvolatile adhesive. Such older water-borne coatings are useful but cannot compete with organic solvent coatings in properties such as weather, scuff, and chemical resistance. Newer water-borne coatings have been developed in which a synthetic insoluble polymer is carried in water as a colloidal dispersion or an emulsion. This is a two-phase system in which water is the continuous phase and the polymer resin is the dispersed phase. When the water is evaporated and the coating cured, the polymer forms a film that has properties similar to those obtained from organic solvent-based coatings.

Plastisols and organisols are low-solvent coatings. Plastisols are a colloidal dispersion of a synthetic resin in a plasticizer. When the plasticizer is heated, the resin particles are solvated by the plasticizer so that they fuse together to form a continuous film. Plastisols usually contain little or no solvent, but sometimes the addition of a filler or pigment will change the viscosity so that organic solvents must be added to obtain desirable flow characteristics. When the volatile content of a plastisol exceeds 5% of the total weight, it is referred to as an organisol.

Although organic solvents are not evaporated from plastisols, some of the plasticizer may volatilize in the oven. This plasticizer will condense when emitted from the exhaust stack to form a visible emission. Companies that use plastisols often have a small electrostatic precipitator to remove these droplets from the oven exhaust.

Hot melt coatings contain no solvent; the polymer resins are applied in a molten state to the paper surfaces. All the materials deposited on the paper remain as part of the coating. Because the hot melt cools to a solid coating soon after it is applied, a drying oven is not needed to evaporate solvent or to cure the coating.

One disadvantage with hot melt coatings is that materials that char or burn when heated cannot be applied by hot melt. Other materials will slowly degrade when they are held at the necessary elevated temperatures.

Hot melts may be applied by heated gravure or roll coaters and are usually applied at temperatures from 66 to 232°C (150 to 450°F). The lower melting point materials are generally waxy type materials with resins added to increase gloss and hardness. The higher melting point materials form films that have superior scuff resistance, transparency, and gloss. One particular advantage of hot melts is that a smooth finish can be applied over a rough textured paper. This is possible because the hot melt does not penetrate into the pores of the paper.

A type of hot melt coating, plastic extrusion coating, is a solventless system in which a molten thermoplastic sheet is discharged from a slotted dye onto a substrate of paper, paperboard, or synthetic material. The moving substrate and molten plastic are combined in a nip between a rubber roll and a chill roll. A screw-type extruder extrudes the coating at a temperature sometimes as high as 316°C (600°F). Low- and medium-density polyethylene are used for extrusion coating more than any other type resins.

Silicone release coatings, usually solvent-borne, are sometimes used for pressure-sensitive, adhesive-coated products. Two low-solvent alternatives are currently on the market. The first is a 100% nonvolatile coating which is usually heat cured, but may be radiation cured. This is a prepolymer coating which is applied as liquid monomers that are cross-linked by the curing process to form a solid film. The second system is water emulsion coatings.

Products are being developed that will allow solvent recovery from solvent-borne silicone coatings using carbon adsorption. Currently, there are difficulties with recovering solvent from silicone coatings because some silicone is carried into the adsorber where it fouls the carbon and lowers collection efficiency.

These low-solvent coatings are capable of achieving 80% to greater than 99% reduction in organic emissions over typical solvent-borne coatings.

4.3.2 Incineration

Thermal (noncatalytic) incinerators may be used to control organic vapors from paper coating operations. Catalytic incinerators, widely used for printing operations, have rarely been applied to control paper coating operations using roll coating or blade coating but certainly are applicable.

Incinerators, if properly operated, can be over 95% efficient in controlling organic vapors which are directed to the incinerator. The overall control for the entire plant will be less because of the emissions that escape.

4.3.3 Carbon Adsorption

Carbon adsorption units can be over 90% efficient in controlling organic solvent vapors that are drawn into the carbon bed.

Carbon adsorption is used for collecting solvents emitted from paper coating operations. Most operational systems on paper coating lines were installed because they were profitable. Pollution control has usually been a minor concern. Carbon adsorption systems at existing paper coating plants range in size from 9.0 to 28.3 m³/sec (19,000 to 60,000 scfm). Exhausts from several paper coating lines are often manifolded together to permit one carbon adsorption unit to serve several coating lines.

Carbon adsorption is most adaptable to single solvent processes. Many coaters using carbon adsorption have reformulated their coatings so that only one solvent is required. Toluene, probably the most widely used solvent for paper coating, is readily captured in carbon adsorption systems.

The greatest obstacle to the economical use of carbon adsorption is that in some cases, reusing solvent may be difficult. In many coating formulations, a mixture of several solvents is needed to attain the desired solvency and evaporation rates. If this solvent mixture is recovered, it sometimes cannot be reused in formulating new batches of coatings. Also, if different coating lines within the plant use different solvents and are all ducted to one carbon adsorption system, then there may be difficulty reusing the collected solvent mixture. In this case, solvents must be separated by distillation.

4.4 Applicable Regulations

Most organic solvent emission regulations are patterned after what is now Rule 442 of the South Coast (California) Air Quality Management District. Regulations applicable in the 16 states that contain about 85% of all surface coating industries are essentially the same as Rule 442.[16] Indiana has the most stringent regulation in that it limits organic solvent emissions to 1.4 kg/hr (3 lb/hr) or 6.8 kg/day (15 lb/day) unless such emissions are reduced by at least 85%, regardless of the reactivity or temperature of the solvent. Organic solvents that have been determined to be photochemically unreactive or that contain less than specified percentages of photochemically reactive organic materials are exempt from this regulation.

California Air Resources Board has recently adopted a model rule for the control of volatile organic compounds (VOC) emissions from paper and fabric coating operations. This model rule limits VOC emissions from the coating line to 120 g solvent/liter (1.0 lb/gal) of coating minus water through the use of add-on control equipment, unless the solvent content of the coating used is no more than 265 g/liter (2.2 lb/gal) of coating minus water.

4.5 Best Control Techniques

The two proven add-on control devices for controlling organic solvent emissions from paper coating lines are incinerators and carbon adsorbers. Both have been retrofitted onto a number of paper coating lines and are being operated successfully.

The main constraint to the use of incinerators is the possible shortage of natural gas. However, in many cases, the combination of afterburner and oven will use no more fuel than the oven alone if proper heat recovery is used. Incinerators can be operated on LPG or distilled fuel oil if natural gas is not available.

The major drawback to the use of carbon adsorption is that in some cases solvent mixtures may not be economically recoverable in usable form. If the recovered solvent has no value, it is more economical to incinerate and recover heat than install a carbon adsorber. However, if the recovered solvent can be used as fuel, carbon adsorption compares favorably in operating cost with an incinerator. If the solvent can be recovered as usable solvent, use of carbon adsorption represents an economic advantage to the paper coater.

It is more difficult to estimate costs for low-solvent coatings, because the costs will vary depending on the type of coating. Low organic solvent coatings will usually cost less in dollars per pound of coating solids applied than will conventional organic solvent coatings with some type of add-on control device.

An emission level of 0.35 kg/liter of coating (minus water) is the recommended guideline by the U.S. Environmental Protection Agency for all coatings put on paper, pressure sensitive tapes regardless of substrate (including paper, fabric, or plastic film), and related web coating processes on plastic film such as typewriter ribbons, photographic film, and magnetic tape.[15] Also included in this guideline are decorative coatings on metal foil such as gift wrap and packaging. These limits can be achieved in all cases using incineration and in many cases with coatings that contain low fractions of organic solvents.

4.6 Control Costs

Costs will vary for low-solvent systems depending on the type of low-solvent coating and the particular end use. There can be large costs involved in initially developing the coatings, purchasing new application equipment, and learning to use the new systems. Cost comparisons between various low-solvent coatings are not as easy to make as are cost comparisons between various types of add-on control systems.

Chapter 14 provides cost data for controlling emissions using an incinerator. Exhaust rates from typical paper coating ovens range from 3.8 to 9.4 m³/sec (8000 to 20,000 scfm) at exhaust temperatures of 79 to 149°C (175 to 300°F).

The cost of using carbon adsorption to control hydrocarbon emissions is also outlined generally in Chapter 14. A carbon steel adsorber is used except for certain solvents such as ketones and ethyl acetate which require that the vessel be made of special alloys. These solvents form acids when exposed to steam, and can corrode carbon steel. Stainless steel alloys are normally used in these cases.

If a distillation unit must be included, the installed cost of the carbon adsorption system will increase significantly. The installed cost of the distillation unit will depend on the number of distillation columns, the complexity, and physical properties of the solvents to be separated.

REFERENCES

1. *Kraft Pulping—Control of TRS Emissions from Existing Mills,* Environmental Protection Agency, Research Triangle Park, NC, Publication No. EPA-450/2-78-003b, March 1979.

2. *Environmental Protection Control, Pulp and Paper Industry, Part I Air,* Environmental Protection Agency, Technology Transfer, Cincinnati, OH, Publication No. EPA-625/7-76-001, October 1976.

3. *Atmospheric Emissions from the Pulp and Paper Manufacturing Industry,* Cooperative NCASI-U.S. EPA Study Project, Environmental Protection Agency, Research Triangle Park, NC, Publication No. EPA-450/1-73-002, September 1973.

4. *A Report on the Study of TRS Emissions from a NSSC-Kraft Recovery Boiler,* Container Corporation of America, March 9, 1977.

5. Teller, A. J., and Amberg, H. R., *Considerations in the Design for TRS and Particulate Recovery from Effluents of Kraft Recovery Furnace,* Preprint, TAPPI Environmental Conference, May 1975.

6. *Malodorous Reduced Sulfur Emissions from Incineration of Noncondensable Off-Gases,* Environmental Protection Agency, Research Triangle Park, NC, EPA Test Report No. 73-KPM-1A, 1973.

7. *Suggested Procedures for the Conduct of Lime Kiln Studies to Define Emissions of Reduced Sulfur Through Control of Kiln and Scrubber Operating Variables,* NCASI Special Report No. 70-71, January 1971.

8. *Control Techniques for Particulate Emissions from Stationary Sources,* Volume 2, Preliminary Draft, Environmental Protection Agency, Research Triangle Park, NC, July 1980.

9. *The Electrostatic Precipitator Manual,* The McIlvaine Company, Northbrook, IL, April 1977.

10. *Federal Register,* **43**(37) (February 23, 1978), Part V.

11. *Federal Register* **44**(100) (May 22, 1979).

12. Blosser, R. O., Trends in Odor Control Technology: How They Affect Kraft Pulp Mills, *Pulp and Paper,* September 1979.

13. *Standards Support and Environmental Impact Statement, Volume 1: Proposed Standards of Performance for Kraft Pulp Mills,* Environmental Protection Agency, Research Triangle Park, NC, Publication No. EPA-450/2-76-014a, September 1976.

14. *Screening Study on Feasibility of Standards of Performance for Two Wood Pulping Processes,* Environmental Protection Agency, Research Triangle Park, NC, Publication No. EPA-450/3-78-111, November 1978.

15. *Control of Volatile Organic Emissions from Existing Stationary Sources—Volume II: Surface Coating of Cans, Coils, Paper, Fabrics, Automobiles, and Light-Duty Trucks,* Environmental Protection Agency, Research Triangle Park, NC, Publication No. EPA-450/2-77-08, May 1977.

16. *Guidance for Lowest Achievable Emission Rates from 18 Major Stationary Sources of Particulate, Nitrogen Oxides, Sulfur Dioxide, or Volatile Organic Compounds,* Environmental Protection Agency, Research Triangle Park, NC, Publication No. EPA-450/3-79-024, April 1979.

CHAPTER 28

CONTROL OF FUGITIVE EMISSIONS

DAVID V. BUBENICK

GCA Corporation
Bedford, Massachusetts

1 BACKGROUND

Historically, due to difficulties in measuring fugitive particulate emissions and therefore their environmental impact, little attention had been given to controlling their sources. However, in view of the widespread failure to attain the National Ambient Air Quality Standards (NAAQS) for particulate matter in many areas, EPA recognized that a rigorous examination of the nature of these sources and their control strategies was in order.

Fugitive particulate emissions are divided into two basic categories depending on their origin. They are fugitive process emissions (FPEs) and fugitive dust emissions (referred to as open or area source emissions). To avoid confusion regarding terminology, FPEs are generally assumed to include particulates arising from industry-related operations. In contrast to ducted (controlled point source) emissions, FPEs escape to the atmosphere through windows, doors, or vents. FPEs may result from incompletely controlled point sources, such as furnace charging and tapping; from the handling, transfer, and storage of materials; and from poor equipment maintenance and environmentally careless process operation (e.g., leaking furnaces or oven doors). Outdoor industrial processes such as rock crushing may also constitute significant sources of FPEs.

In contrast, area fugitive dust emissions are generally related to particulate matter arising from open (or area) sources as the result of reentrainment and resuspension by wind, vehicular traffic, construction, or agricultural activities. Fugitive dust emissions include particulate matter resuspended from road surfaces, exposed surfaces at construction sites, and tilled cropland, as well as windblown particulate matter. Strictly speaking, process-related area sources may also be considered to fall within the category of area fugitive dust emissions.

In order to appreciate the rationale for the selection and application of control measures currently in use or being developed for fugitive particulate emissions, it is important to recognize the magnitude of the problem. This topic is briefly reviewed in the following section with emphasis on source identification, emission characterization, and emission factors.

2 MAGNITUDE OF THE PROBLEM

Traditional control strategies that focus on the reduction of particulate emissions from conventional point sources have not been successful in bringing about attainment of the NAAQS for total suspended particulates (TSP).* Recent reports[1,2] have shown, for example, that more than 60% of the nation's 247 Air Quality Control Regions (AQCRs) were not meeting TSP standards. Estimated fugitive particulate (dust) emissions were found to significantly exceed particulate emissions from point sources in over 90% of the 150 AQCRs that were out of compliance. In nearly 40% of the 150 AQCRs, fugitive dust emissions exceeded point source emissions by a factor of 10.

Open sources account, by far, for the largest share of fugitive particulate materials. Although numerous open sources have been identified, four broad categories may be responsible for as much as 98% of open source emissions.[3] These include unpaved roads, agricultural activities (tilling and wind erosion of cropland), construction activities, and paved roads. Annual fugitive dust emissions

---

* The NAAQS for TSP are 75 $\mu g/m^3$ (annual geometric mean) and 260 $\mu g/m^3$ for a 24-hr period.

Table 1   Estimated Totals of Anthropogenic Open Source Fugitive and Point
Source Particulate Emission Rates for the United States[3]

Source	Estimated Emission Rate		Percent of Total
	10^9 kg/yr	10^6 ton/yr	
Unpaved roads	290	320	74.4
Agricultural activities	43	47	10.9
Construction activities	25	27	6.3
Paved roads	7	8	1.9
Other open sources	7	8	1.9
All point sources	18	20	4.6
Total	390	430	100.0

from unpaved roads alone have been estimated at 290 × 10^9 kg (320 × 10^6 tons).[3] As shown in Table 1, this figure exceeds annual point source particulate emissions by a factor of 16.

In AQCRs where concentrated industrial activity is absent, fugitive dusts are probably responsible for most of the observed TSP standard violations. However, fugitive dusts also contribute significantly to ambient TSP levels in industrialized urban areas. An analysis of 300 filters from high-volume samplers, collected in 14 major cities in the United States, concluded that soil-like mineral matter predominated over combustion products and other types of particulate matter.[4,5] The annual open source contribution to TSP levels in industrialized urban areas may be greater than 25 $\mu g/m^3$.[4,5,6]

Industrial fugitive particulate emissions arise from poorly controlled point sources as well as from plant-site open sources such as haul roads. A number of industrial categories have been identified as significant fugitive emission sources, including pyrometallurgical operations and processes involving the crushing and grinding of mineral products.[1,7,8] Typically, there are many fugitive emission sources within a given industry (e.g., 20 separate sources have been identified for iron foundries).[7]

Even though their total is much less than open source emissions, fugitive emissions from industrial sources may still exert an effect on ambient TSP based on the extremely high values measured in predominately industrial areas.[4,5] In some industries, fugitive emissions may actually exceed those from point sources, further demonstrating their importance.[8]

Because of the relatively recent interest in fugitive emissions from open and industrial sources, the data base describing size properties and emission rates has not yet been adequately defined. Nevertheless, information currently available suggests that both categories contribute significant amounts of inhalable particulate matter (particles less than 15 μm diameter) to ambient TSP. Figure 1 shows that for open sources in the Phoenix, Arizona area, all emissions consist of particles finer than 100 μm.[9] Depending on the source, from 45 to 100% of emissions may consist of inhalable particulate matter (IPM). However, since open source emissions tend to be from ground level sources, the IPM fraction of emissions reaching a receptor (either a sampling instrument or a person) is likely to be increased because of the preferential settling of large particles.

Size characteristics for process fugitive emissions have been assessed by Midwest Research Institute (MRI) for several industries.[10,11] In the absence of direct measurement, additional data may be obtained by assuming that process fugitive emissions will be similar to previously measured stack emissions, although caution in using this approach is suggested. A recent review of such data by Research Triangle Institute (RTI)[12] for the pyrometallurgical industries demonstrates that for nonferrous sources, nearly all of the fugitive emission mass consists of IPM. For ferrous metal sources, the size range is broader but most emissions still fall within the IPM category.

Emission factors are required for the development of fugitive emission control strategies and the determination of control technology needs. The general approach to this problem is to develop source-specific emission factors that can be used to estimate the magnitude of emissions. For an open source such as an unpaved road, the fugitive dust emission rate can be related to vehicle speed and kilometers traveled, silt content of the road surface, and local precipitation history.[13] With appropriate modifications to the last two parameters, this emission factor can be applied to different geographical areas.

For industrial processes, emission factors relate the mass of emissions to the amount of process input or output. Although fugitive emission factors for many open source categories[3] and industrial processes[7,8] have been published, the reliability rating for the majority of these factors is not high. Engineering judgment and extrapolation from similar processes have played a large part in fugitive emission factor development, and few of the emission factors are supported by extensive test data.

Open and process source fugitive emissions have been identified as significant contributors to ambient TSP levels. Although data have been collected to support this observation, fugitive emission rates and chemical and physical characteristics need to be further defined. This task is especially important for the determination of the fugitive particulate influence on IPM levels, should such a requirement be implemented.

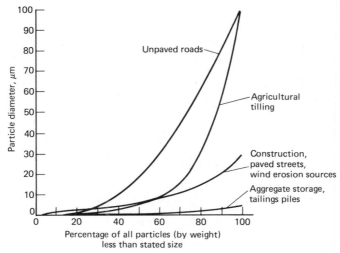

Fig. 1 Size distribution of particulate emissions from various fugitive dust sources in the Phoenix, Arizona area.

3   CURRENT CONTROL MEASURES FOR PROCESS FUGITIVE EMISSION SOURCES

The status of industrial process fugitive particulate emission (FPE) controls has been characterized by the terms "emerging" and "site-specific." Control of FPE sources has received attention only within the last 3 or 4 yr. Sources of FPEs are numerous and quite varied, not only from industry to industry, but also within the same plant. Consequently, control equipment design parameters must be developed almost on a case-by-case basis. For example, in the iron and steel industry, FPEs from small blast furnace cast houses have been controlled by exhausting all building air to a baghouse, while larger cast houses rely on local hooding to attain control at a more reasonable cost. In the future, it is unlikely that any single method of FPE control will predominate, and it can be expected that process systems, in general, will continue to require individual control strategies.[12]

The primary problem in FPE control is collection (capture) of the emissions. Once collected, these diffuse FPEs are transferred via a ventilation system to a gas cleaning device where particulates can be removed. Collection systems for FPEs typically consist of secondary hooding at the local source of emissions, large canopy-type hoods suspended above the source, or complete building evacuation. Each method has some drawbacks, especially when retrofitted to an existing process operation. Large air flows are often necessary, requiring large-diameter, expensive ductwork and fans that draw significant power. Often, the building structure itself must be reinforced because it is not strong enough to handle the additional load of hooding and ductwork. Ductwork routing and ventilation hood placement are often difficult due to space restrictions in a retrofit application. Also, secondary hooding or enclosures may restrict personnel and/or equipment access.

An important control strategy for FPEs consists of minor process modifications such as the substitution of clean scrap for charging in an electric arc furnace (EAF). Construction of wind screens around a bucket-loader transfer operation is another process modification that helps to reduce process-related dust emissions.

Some degree of FPE control can sometimes be achieved with little or no investment in additional equipment by adherence to good housekeeping, operation, and maintenance practices. Prompt repair of exhaust hood leaks, maintenance of coke oven door seals, proper handling of baghouse dusts, quick cleanup of spills, and other elements of good manufacturing practice all serve to reduce FPEs.

3.1   Pyrometallurgical Sources

FPEs and process-related area sources account for about 80% of current particulate emissions from steel mills.[11] Retrofitting of FPE controls to existing steel mill sources is often difficult due to space limitations, a lack of process technology, the need for large air flows, and required worker access to the process operation itself.

In a recent MRI study of steel mill emissions,[11] it was found that although substantial progress has been made in developing fugitive particulate emission controls for steel mill sources, several major control problems exist. Retrofitting proposed control systems to existing operations has often proved difficult and there is a serious deficiency with respect to data describing uncontrolled emission rates and composition, control device efficiency, and control costs. Available controls for paved and unpaved in-plant roads were found to be at least a factor of 20 times more cost effective than, for example, use of canopy hoods for control of electric arc furnaces. These two sources were found to constitute the largest amounts of both total particulate and fine particulate emissions from steel mills.

The coke battery represents one of the most difficult-to-control FPE sources in the steel industry due to the large number of potential sources, their diffuse nature, and the harsh coke plant environment which frequently causes equipment malfunctions. Although door leaks are caused by a number of factors, they are primarily attributable to warped door jambs, poor sealing of spring-loaded knife edges, and poor cleaning practices. Older door designs have not proved very successful in eliminating door leaks. Consequently, IERL-RTP and the American Iron and Steel Institute jointly funded a Battelle research program that determined factors causing door leaks through field research, and then developed an improved retrofit door seal designed to minimize door jamb warpage.[14]

Control of basic oxygen furnace (BOF) secondary particulate emissions is often hindered by a lack of demonstrated control technology that can be retrofitted at reasonable cost. To aid the development work under way in the United States, IERL-RTP has studied foreign technology to determine applicability of techniques and devices to domestic steel mills. In an assessment of western European and Japanese control technology for 12 BOF plants,[15] the best local hooding systems observed were estimated to have a capture efficiency in the range of 90–100%.

For secondary emission control systems, fabric filters were preferred over scrubbers and electrostatic precipitators (ESPs). While partial building evacuation systems for furnace emissions were not found in the majority of cases, there are situations where this may be advantageous over local hooding. The use of a roof-mounted ESP with no fans attached (viz. no pressure drop due to air movement) may be a cost-effective alternative to complete reliance on local hooding connected to fabric filters. In another survey of two iron and steel plants in Japan, it was found that natural draft, roof-mounted ESPs with special light-weight, widely spaced plates to control FPEs from roof monitors in steel-making shops, appear to have promising applications in the U.S. industry.[16]

Iron foundry FPEs have long been recognized as a significant problem due to a lack of available controls for secondary emissions from several operations. A recent MRI study[17] found that better emission data were needed and that currently available controls for EAFs and shakeouts needed more evaluation. New control technology development was recommended for pouring, cooling, and core and mold preparation operations.

3.2 Rock and Mineral Processing Operations

Fugitive emissions from the nonmetallic minerals industry (i.e., stone quarrying, crushed limestone, quartz, sandstone, quartzite, granite, and crushed stone) have been assessed in detail.[18] Additionally, significant quantities of fugitive particulate emissions were also identified with drilling, blasting, haulage of material, and beneficiation. Beneficiation operations may include crushing, screening, grinding, and transport of material at a processing plant located either indoors or outdoors.

The controls used by the nonmetallic industries vary according to the process and its location. The only viable control options currently available for drilling are based on wet suppression. Water or water with surfactants is injected into a drilling hole so that the dust created will adhere to the water droplets. Foam injection systems may be used in place of water alone to increase the suppression capability while using the same amount of water.[19] Beneficiation operations employ wet suppression as well as hood capture and ventilation of emissions to a central control device. Wet suppression is used in virtually all nonmetallic industries although blinding of screens can occur in certain applications such as lime and limestone products preparation.

Waste dumps for asbestos cement manufacturing plants are often located in high population density areas. Asbestos fibers display carcinogenic properties when inhaled and lodged in human lungs. An engineering study conducted by IIT Research Institute found that an 87% reduction in air emissions from this fugitive source could be achieved by (1) bagging of fine waste, (2) application of a soil-vegetative cover on the inactive dump site, and (3) temporary chemical stabilization for active piles, at an annual cost of $18,000 for a typical plant.[20]

The taconite industry, which provides 75% of the ore charged to blast furnaces for iron making, is rapidly expanding as older high-grade ore deposits become depleted. Control methods currently employed include wet suppression and chemical stabilization, which are applied to such open sources as mining areas and tailing piles, while hooding is used to collect FPEs from crushers.

Coal storage piles at industrial plants represent common sources of windblown particulate matter that may contain a number of trace elements, gaseous hydrocarbons, and carbon monoxide. Most piles are controlled by wet suppression or a coating of tar derivatives.[21] However, these practices may generate additional pollutants, such as those released upon combustion of the tar coating in the boiler. Storage of coal in a pit or silo, where possible, appears to be a more economical and effective means to reduce particulate emissions. The effectiveness of several coal pile treatment methods is summarized in Table 2.[21]

Table 2   Cost Effectiveness of Coal Pile Treatment Methods[21]

Treatment	Treatment Costs, $/kg × 10³	Direct $ Savings, $/kg × 10³	Total Benefits Value, Equivalent, $/kg × 10³
Compaction	0.09	0.303	0.67
Latex crust[a]	0.126	0.717	1.20
Filled latex coating[a]	0.174	1.15	1.82
Hot-melt coating[a]	0.144	4.15	2.09

[a] Includes compaction.

MRC investigations of fugitive emissions from urea and carbon black manufacturing operations[22,23] showed the prilling tower to be the largest source of emissions from a urea plant. Due to the design of the prilling tower and the particle size of the urea, the scrubber typically used is only about 50% efficient. As with urea, the fine particle size of carbon black makes it hard to contain, thus requiring an extensive housekeeping system to reduce fugitive emissions.

Fugitive dusts are also produced during the transport of fine materials such as coal and construction aggregate, due to the movement of vehicles as well as wind erosion. Use of wetting agents, tarpaulins, and chemical stabilization helps prevent fugitive emissions from trucks and rail cars. Tarpaulin covers are preferred over wetting because they do not affect the material being transported. Chemical stabilization is used in the transportation of materials such as coal. Based on a 100-car train, the cost of latex crusting compounds is approximately $0.078/10³ kg ($0.071/ton) of material transported (e.g., coal).[24]

4   CURRENT CONTROL MEASURES FOR AREA FUGITIVE DUST SOURCES

The data on both emission quantities and their impacts on TSP levels support a strong relationship between fugitive dust emissions and nonattainment in many AQCRs. As noted earlier, fugitive dust sources may account for well over half of the ambient TSP in many AQCRs.

Wet suppression with water with or without wetting agents can provide temporary control for some agricultural and construction sources, unpaved roads, materials handling operations, and stockpiles. Since water alone is a poor suppressant due to its high surface tension, chemical wetting agents (surfactants) are added to increase wetting effectiveness. Use of wet suppression, however, is often not feasible due to water shortage, source size, temporary nature of control, and in many cases, the need to keep the dust-generating material dry. Foam systems have recently been shown effective for rock drilling activity and conveyor transfer systems.

Physical stabilization is simply the covering of a surface with a material that prevents wind disturbance of that surface. Such materials include rock, soil, crushed or granulated slag, bark, wood chips, and straw. Paving of dirt roads is a commonly used approach. Less widespread methods involve covering with elastomeric films, wax, tar, and oil.

Chemical stabilization, sometimes used on agricultural fields, unpaved roads, and waste heaps, relies on binding materials to form a protective crust that shields surface dusts from the wind. The effectiveness of chemical stabilization for a given application depends on the level of activity at that source. For example, required application (rate and quantity) to unpaved roads would be a function of the amount of traffic. Also the effectiveness of continuous spraying of aggregate piles will depend on such factors as the fraction of fines in the mix, the type of stone, and the activity of the pile.

Vegetative stabilization is restricted to inactive sources that will support growth after a layer of soil is applied. Coal piles are commonly stabilized by vegetation after acid neutralization, and mining overburden and gangue represent no particular problem. However, because some tailings often have a deficiency in plant nutrients, a hardy plant species must be considered. In general, selection of plant species is site and source specific and will depend on soil nutrients, pH, and pile slope.

4.1   Paved Roads

The composition of urban road dust is extremely variable due to both natural processes and human activities. The primary contributors to deposition include:[25]

1. Motor vehicles.
2. Sanding and salting.
3. Pavement wear.
4. Litter.

    5. Biological debris.

    6. Wind and water erosion from adjacent areas.

    7. Atmospheric pollution fallout.

Once materials have been deposited on paved surfaces, they are removed by one or more of the following mechanisms:

    1. Mechanical redispersion to the atmosphere.

    2. Aerodynamic entrainment (wind erosion) to the atmosphere.

    3. Displacement of adjacent surfaces.

    4. Rainfall runoff to a catch basin.

    5. Street cleaning.

Rainfall and wind erosion, of course, are natural phenomena, but due to their sporadic nature, they cannot be considered as reliable dust control methods. However, it has been determined that a rainfall of 1.27 cm (0.5 in.) can remove up to 50% of road particulate, while heavier rainfalls can wash away as much as 90%.[25]

Reentrainment and displacement are related to atmospheric conditions, vehicle speed and size, and traffic mix and volume. Street cleaning methods include sweeping, flushing, resurfacing, and coating. Street sweepers of the broom, vacuum, and regenerative air types have been used with generally unsatisfactory reductions in emissions. Less than 20% removal efficiency for particles below 140 μm was noted in one study.[26] Broom sweepers can actually increase emissions by stirring up dust, moving it to the center of the road, and breaking up coarse particles into sizes that can be more readily reentrained. Flushing, by way of contrast, can suppress fine particulate matter as well as control the larger particles. Its disadvantage is that it must be carried out on a continual basis because its effect diminishes as the pavement and dust dry.

Another approach to control fugitive dust from paved roads is to prevent truck spills and the carryout by motor vehicles of materials from unpaved areas such as construction sites and industrial plants. The comparative cost effectiveness for various control methods, which often may be implemented at minimal cost, is shown in Table 3.[27]

4.2 Unpaved Roads

Several programs conducted by MRC,[28] MRI,[11] Battelle,[29] RTI,[1] and Harvard University[3] have investigated conventional control methods for unpaved roads. These include both physical and chemical stabilization, wetting, and oiling. The most cost-effective control as reported by Cooper et al.[3] is paving, with a standardized (1977 base) total annual cost of $4756/yr-km ($7653/yr-mile) at 86% control efficiency. Cost and life expectancy estimates will depend on surface type to be used, conditions of existing surfaces, traffic density, climate, and the costs of aggregate and labor. The efficiency of paving depends directly on the actual surface to be treated. Materials may range from semipermanent bituminous single-chip seal to a permanent asphalt concrete layer several centimeters deep.

Wet suppression with water, plain or mixed with a wetting agent (surfactant), can provide temporary control for unpaved roads. Since water alone is a poor suppressant due to its high surface tension,

Table 3   Cost Effectiveness of Control Measures[27]

Control Measure	Annual Cost, $ (1977)	Average Reduction in Concentration, μg/m^3	$ per μg/m^3 Reduction	Area Affected, km^2
Flushing	76,400[a]	4.6	17,000	38.8
	130,700[b]	7.6	17,200	38.8
Combined flushing/sweeping	249,800	4.6	54,300	38.8
Modification of street sanding procedures	18,300	1.2	15,200	38.8
Control of mud carryout from construction sites	8,400	15.0	560	0.26
Control of mud carryout from industrial areas	268,700	17.0	15,800	41.4

[a] Flushing 3 days/week, 8 months/yr.

[b] Flushing 5 days/week, 8 months/yr.

the addition of the surfactant is necessary to increase its effectiveness. Although data on cost and effectiveness are sparse, one study indicates annual costs ranging from \$25,730 to \$39,030/km (\$41,400 to \$62,800/mile).[3]

Dust can also be controlled by application of oil to the road surface as infrequently as once per month. While this method is generally effective, the side effects often outweigh the benefits derived. From a safety standpoint, trucks rolling over an oily road have a greater tendency to slide off the road. Oiling also may be environmentally unattractive due to heavy runoff that is estimated to be 70–75% of the oil applied.[3] Road maintenance may also increase due to the development of potholes associated with oil application treatments.

Chemical stabilizers serve to protect road surfaces from wind and vehicle entrainment of dust by forming a protective crust with the road dirt. In order to ensure the effectiveness of control, the application program developed must be conscientiously followed. The stabilizer is frequently mixed with water in a ratio of 1:4 to 1:7, then applied to the road surface where it agglomerates dust particles to form a binder. Well-maintained road surfaces become "as hard as concrete" and dust free. Estimated efficiencies of 50% and greater have been reported with cost-effectiveness estimates of \$305/km-yr-% (\$491/mile-yr-%) reduction.[3]

4.3 Agricultural Activities

Fugitive dust emissions from agricultural activity often exert strong local effects as the emissions may contain pesticide residues. Wet suppression has a low control efficiency since the soil is continually turned over. Additional problems may include a short supply of water in some regions and the inability of cultivating machinery to carry enough water. In general, control methods are related to good soil conservation practice and efficient farming techniques. Vegetative stabilization is the most obvious control technique supplemented by windbreaks and wet suppression.

4.4 Construction Activities

Construction activity has the potential to generate large amounts of dust which, in some cases, may contain hazardous emissions from certain rock and soil types. Wind erosion of stripped land is also a potential source of fugitive dust. The status of control technology for construction activity is similar to that for agricultural activity. Wetting has been applied to excavating activity but continual working of the soil precludes effective control. Stabilizing with a binder is an effective control method that is applicable to short-term heaping of excavated material. The control of fugitive dust from vehicle travel is approached with the same techniques employed for unpaved roads.

4.5 Summary of Effectiveness of Controls[1,30]

Effectiveness of controls for area source emissions, which depends on a number of factors, is somewhat variable. Generally, reductions of emissions by 50% have been reported for unpaved roads, 40% for agricultural tilling (chemical stabilization effectiveness), and 30% for construction activities (wetting effectiveness). For unpaved roads, only physical stabilization (i.e., paving) has proved to provide adequate control. Vegetative stabilization of road shoulders has also been used. Control of agricultural emissions is often achieved by wetting, although generally this source lacks effective controls. Other technologies for this source are either untried or have been judged poor. Physical and vegetative stabilization as applied to construction activities has proved acceptable; wetting is not acceptable; and chemical stabilization has not been evaluated.

With regard to other sources, wetting can be fairly effective for stabilization of dust from materials handling operations. Physical and vegetative stabilization has been used with relative success on stockpiles, whereas conventional windscreens are rated very poor. Exhaust ventilation techniques have been beneficial in collecting mining operation emissions, for example, while wetting was generally found to be unacceptable. Stabilization technologies have not been rated for this source.

Numerous other control methods are available for various sources of fugitive dust emissions. Some of the more important techniques include speed reduction on unpaved roads, street cleaning of paved roads, reduction of fall distances for materials handling, and enclosure, hooding, and ducting. For example, reducing the speed of vehicles from 64 to 40 km/hr (40 to 25 mph) traveling over unpaved roads has proved successful in reducing reentrainment of road dust by over 70%.[31]

5 NEW CONCEPTS FOR FUGITIVE PARTICULATE EMISSION CONTROL

As outlined above, current technology involves improvement in capturing fugitive process emissions (FPEs) and prevention of open source fugitive emissions. Because of the important impact of fugitive sources on meeting NAAQS, EPA is involved in a program of control technology research and development. Table 4 lists the on-going IERL-RTP projects that highlight "new concepts" being developed. This term does not imply that such technology has not been previously investigated but rather that

Table 4   Ongoing Fugitive Emission Control Programs Sponsored by U.S. EPA's Industrial Environ-mental Research Laboratory, Research Triangle Park (IERL-RTP)

Project	EPA Contractor
Fugitive process emissions	
The SCAT system for pyrometallurgical sources of emissions	A.P.T., Inc.
Field test of the charged fogger	TRC
Preliminary evaluation of the charged fogger	University of Arizona
Assessment of the use of fugitive emission control devices	Research Triangle Institute
Fugitive dust emissions	
Paved and unpaved road emissions	Midwest Research Institute
Improved street sweepers for paved roads	A.P.T., Inc.
Road carpet for unpaved roads	Monsanto Research Corporation
Windscreening and chemical additives for area sources	TRC
Charged fog for construction site activity and front-end loaders	AeroVironment, Inc.

the technology has not been fully explored and will remain new until it is accepted into conventional practice.

One of the major difficulties in controlling FPEs is that they are often dispersed as low-concentration/high-volume emissions. If the particulate were more concentrated, it could be collected by a high-efficiency hood and ducted to a conventional control device. Similarly, building evacuation would be practical if the particulate were highly concentrated in a small building volume. As an alternative, EPA is actively pursuing the use of chemically and electrostatically treated water spray droplets in cases where the FPEs cannot be hooded or evacuated to a control device.

Fugitive particles entrained in a gas stream may be collected with charged or uncharged water sprays by mechanisms such as diffusion, inertial impaction, interception, and electrophoresis. Larger water droplets would enable separation of the dust/water droplets from the gas stream by such methods as gravitational settling or entrainment separation. Figure 2 is a functional diagram of the process anticipated for controlling fugitive emissions. The process steps represented in this diagram could occur concurrently, sequentially, or separately, depending on the type of equipment. Two basic approaches include the spray charging and trapping (SCAT) scrubber system and the charged fogger.

5.1   Spray Charging and Trapping (SCAT) System

In contrast to typical FPE collection systems that use secondary hooding or total building enclosure and evacuation, the SCAT system controls fugitive emissions by diverting the FPEs into a charged spray scrubber located near the source. This is a relatively simple and inexpensive method for controlling FPEs in that it minimizes the apparatus required to contain, convey, and control the FPEs.

Figure 3 shows one of the many possible designs that are being investigated by A.P.T., Inc.[32] Air curtains consist of one or more high-velocity air streams, flowing as a sheet, that are produced by one or more air jets. These high-velocity air streams push and entrain FPEs plus some additional air and carry them away from the source. Downstream, water is sprayed cocurrently into the gas stream

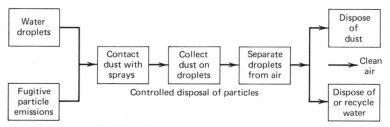

Fig. 2   Functional diagram for the major process steps involved in controlling fugitive particulate emissions with a charged droplet system.

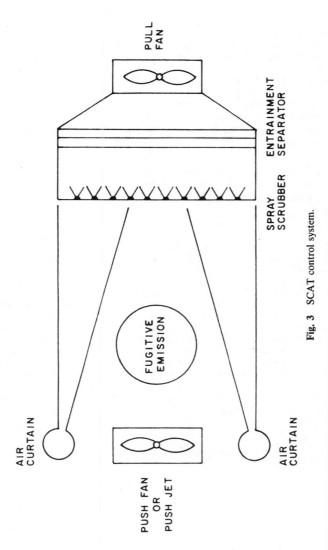

PULL FAN

ENTRAINMENT SEPARATOR

SPRAY SCRUBBER

FUGITIVE EMISSION

AIR CURTAIN

AIR CURTAIN

PUSH FAN OR PUSH JET

Fig. 3   SCAT control system.

to remove the entrained dust. After a sufficient contacting distance to effect capture of the gas stream particles, a low-pressure drop entrainment separator is used to separate the water spray drops from the gas stream. A final water treatment step would consist of filtering or separating the collected dust particles from the effluent stream enabling both water recycle and disposal of the dust in a way to minimize its redispersion.

The basic SCAT features, outlined briefly, are:[32]

1. Minimum use of solid enclosure (hooding).
2. Air curtain(s) and/or air jet(s) applied to divert, contain, and convey the FPE.
3. Charged sprays of water or aqueous solutions to collect FPEs and to aid in moving and containing the air being cleaned.
4. Trapping of collected dust and disposal so as to prevent redispersion.
5. Ability to divert crosswinds and contain hot buoyant plumes.
6. Minimum size of scrubber and entrainment separator section.
7. Minimum consumption of water.
8. Portability.

Since the design of the SCAT system is basically simple, capital investment should be low. It is expected that the use of air curtains and/or barriers would be cost effective. Air curtains have the potential for deflecting wind, thereby minimizing total air volume to be treated. Some potential SCAT applications include metal pouring and foundry operations; raw material charging into furnaces, roasters, and converters; dump sites; raw material storage, loading, and unloading; sand, gravel, and asphalt batching; transfer points on conveyor belts; and coke oven pushing operations.

Experiments have been performed on a 224 m³/min (8000 acfm) bench-scale scrubber to verify the theory and demonstrate the feasibility of collecting fugitive particles with charged sprays.[33] The effects of charge level, nozzle type, droplet size, gas velocity, and liquid-to-gas ratio were determined experimentally. A prototype SCAT system has been built and tested in crosswind conditions and on a hot, buoyant smoke plume. It has been shown that the system is effective with crosswinds up to 24 km/hr (15 mph) and with buoyant plumes of temperatures up to 470°C (880°F). It has been found that at aerodynamic particle diameters of 5 μm and below, the use of charged water droplets can significantly increase SCAT particle collection efficiency as illustrated in Figure 4.[33] The present SCAT prototype uses only a single spray bank and, depending on nozzle diameter, a liquid-to-gas ratio as low as 0.4 l/m³ (3 gal/1000 scf). Higher efficiencies can be achieved by using more water and additional

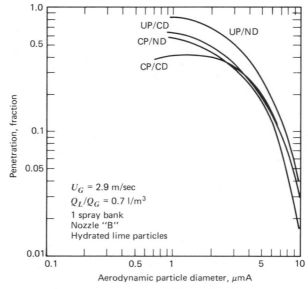

Fig. 4 Experimental spray scrubber penetration: uncharged particles/neutral drops (UP/ND); charged particles/neutral drops (CP/ND); uncharged particles/charged drops (UP/CD); charged particles/charged drops (CP/CD)

spray banks. Both capital and operating costs for the SCAT system are expected to be lower than for hooded systems.

5.2 Charged Fog Spray

Various types of wet scrubbers rely on water droplets to scour dust from industrial gas streams, and water sprays have historically been used to reduce ambient dust levels from mining and materials handling operations. In contrast to conventional water sprays, charged fog sprays consist of droplets that carry a charge of static electricity and are generally of finer size. It has been found that many respirable materials are negatively charged and most of the finer ($<1 \mu$m) particles are always negatively charged.[34] By exposing these particles to electrostatically charged water fog, the electrostatic effect augments fog/dust contact and the wetted particulates would be expected to agglomerate and settle out.

Initial studies were carried out under EPA sponsorship at the University of Arizona to determine technical feasibility, probable applications, and expected performance of the charged fogger. Potential advantages of this system are:

1. The quantity of water involved would be very low, thereby conserving water resources in the arid southwest. Limited water use would permit the application of fog to water-sensitive materials; for example, flour, cement.
2. A system of this type would be suitable for control of moving dust sources (e.g., trucks, sweepers, and front-end loaders) where conventional methods cannot be applied.
3. Charged water fog might be used for preagglomeration of dusts before final collection by other means. One example might be cyclones, which are generally not efficient in collecting dust below 10 μm in diameter. If the respirable dust were agglomerated before it entered the cyclone, it might be removed by the centrifugal action.

In laboratory experiments conducted at the University of Arizona,[35] the charged fogger was used to enhance particulate and gaseous collection primarily due to preagglomeration effects. In a series of tests to improve cyclone performance, greatest reduction of dust (viz., 71%) occurred with a positively charged fogger with low water requirements. Preagglomeration using the charged fogger accounted for up to 52% particulate collection in the 100–150°C (212–302°F) temperature range. Work is also under way involving the use of additives to reduce surface tension and to improve dust/water contact.

In a study performed by RTI[36] on the applicability of the charged fog spray for primary lead and copper smelters, several important conclusions were reached. Using uncharged sprays, a minimum collection efficiency was noted for 2-μm diameter particles, whereas no such minimum appears to exist for charged sprays. Therefore, some improvement in collection of respirable dust can be expected from charging. Charged sprays appear to be best suited to localized sources of dust, suspended in a low velocity or stationary gas stream. The charged fog sprays are not expected to be satisfactory for high temperatures or very turbulent, open areas encountered in many of the major sources of primary smelter fugitive emissions such as converter leakage or furnace taphole emissions. Finally, at practical water application rates, the nature of charged fog sprays limits them to a maximum of 50–60% collection efficiency compared to 95% for building evacuation.

Table 5 summarizes the capital costs, utility requirements, and control efficiencies determined in the RTI study.[36] While both capital investment and energy consumption are lower compared to building evacuation, the reduction of total particulate and elemental lead emissions is also lower. This is due to the larger number of sources covered by a building evacuation system. For equivalent degrees of control, it appears that charged fog sprays require approximately 10% of the capital and utility requirement for building evacuation.

No independent test data regarding system efficiencies have been reported to date. The Ritten Corporation manufactures several versions of their Electrostatic Fogger I® for various small-scale indoor applications such as spray painting. Their Fogger II® is meant for indoor/outdoor applications and uses compressed air for droplet atomization. For large outdoor applications, such as particulate collection from transfer points and grain elevators, the Fogger III® has been developed using hydraulic atomization.

A modified version of the "Spinning Cup Fog Thrower" (SCFT), originally developed by AeroVironment and the University of Arizona, is being investigated for its potential to control inhalable fugitive particulate emissions from open area sources.[37] The new SCFT, which is being considered for use on front-end loaders and street sweepers, is essentially a modified Ray Oil Burner.® In this device centrifugal forces coupled with a high-velocity air stream generate water droplets from water flowing into an atomizing cup. These droplets are then charged by directly connecting a high-voltage power supply to the inflowing water and electrically isolating the water until it becomes fog. A field test program is currently under way to verify the SCFT's overall performance capabilities under a variety of experimental conditions of wind, relative humidity, SCFT water flow rate, fog pattern, and the nature and amount of charge on the droplets.

Table 5   Summary of Charged Fog Spray Comparison with Building
Evacuation[36]

Item	Lead Smelting	Copper Smelting
Reduction in fugitive total particulate emissions		
By application of charged sprays	30%	20%
By application of building evacuation	45%	40%
Reduction in fugitive elemental lead emissions		
By application of charged sprays	40%	35%
By application of building evacuation	75%	65%
Estimated capital investment		
For application of charged sprays	$311,000	$366,000
For application of building evacuation	$8,683,000	$6,808,000
Electrical requirements		
For application of charged sprays	417 kW	450 kW
For application of building evacuation	9,000 kW	6,000 kW

5.3   Road Carpets

The problem of reentrained dust from unpaved roads can be solved if the surface of the dust can be stabilized. Such measures will include application of water, oil, or chemical surface agents. With respect to unpaved roads, application of these materials has been shown to be less cost effective than paving.[38] Advances in fabric development now provide an alternative to the use of stabilization for unpaved roads. In this approach a civil engineering fabric is laid over the soil to help support and contain the overburden aggregate. It helps to spread the concentrated stress resulting from heavy-wheeled traffic over a wide area, siphons away groundwater, and contains fine soil particles in the roadbed that can otherwise contaminate the road ballast.

The fundamental concept behind the use of a civil engineering fabric or "road carpet" for the control of fine particle emissions from unpaved roads is prevention of vortex reentrainment by isolating vehicular traffic from entrainable particles. As a vehicle passes over a poor load-bearing road surface, air compression and expansion of the road results in a draft that "pumps" dust into the air. In addition, comminution (caused by slippage between the road and vehicle tires) contributes a major portion of the reentrained emissions from unpaved roads.

With a road carpet, large aggregate is prevented from settling, while the newly deposited fines ($<70~\mu$m) are filtered by gravitation and hydraulic action down through the fabric away from vortex reentrainment. The important characteristic of being able to transport water along the plane of the fabric aids in subsurface stabilization.

At least four major manufacturers are marketing civil engineering fabrics in the United States, including Celanese, DuPont, Monsanto, and Philips Fibers. As shown in Table 6,[39] most manufacturers offer a similar range of fabric properties according to fabric thickness. Usually such fabrics are sold for use in road stabilization, support, drainage, erosion control, or reinforcement. They have been used under railroad tracks, rip-rap, and paved and unpaved roads. These fabrics are specifically designed to withstand loads from hauling vehicles and railroad trains, are rot resistant, and have estimated useful lifetimes of 12 yr. The average approximate 1980 cost for civil engineering fabrics is $1.00/yd^2 ($0.836/m^2).

The amortized costs compiled by MRC for conventional control measures for temporary roads compared to the road carpet are summarized in Table 7.[40] The road constructed with fabric is less expensive than one that is watered or oiled. Due to inflation, the high maintenance aspect of watering and oiling may reduce their cost competitiveness relative to the road carpet. Treatment with oil (or chemicals) carries the added disadvantage of being carried from the road by runoff and dust transport, resulting in ecological harm caused by the chemicals, oil, or its heavy metal constituents.[41] Use of the road carpet precludes any health or safety hazard or any other unfavorable environmental impact.

Prototype testing has been conducted for the EPA by MRC of Dayton, Ohio, who successfully carpeted a haul road at a quarry near Dayton. A more substantial field demonstration was undertaken on a light-duty vehicle haul road at Fort Carson, Colorado.[42] Ambient particulate concentrations were measured in November 1979 and April 1980 at upstream and downstream locations. These stations were equipped with high-volume and dichotomous samplers in the November testing series and high-volume and size-selective high-volume samplers in April. Ambient concentrations downwind of the fabric/aggregate road were found to be lower by an average of 44%, with a range of 30–70%, compared

Table 6   Properties of Civil Engineering Fabrics[39]

	Brand Name			
	Bidim®	Mirafi®	Supac®	Typar®
Composition	Polyester	Polypropylene	Polypropylene	Polypropylene
Structure	Spun bonded	Woven	Heat bonded	Spun bonded
Thickness, mils	60–190	25	50	15
Weight, oz/yd²	—	4	5.3 (up to 16)	4
Grab strength, lb	115–610	200	150 (Cross direction)	130
Burst strength, psi	225–850	325	300	—
Water permeability, cm/sec	0.03	—	0.05	—

To Convert to Metric Equivalents

Multiply	By	To Convert to
mils	0.00254	cm
oz/yd²	25.9	g/m²
lb	453.6	g
psi	703.1	kg/m²

to a control road cross section without the road carpet. Wind speeds in November were noted to be lower than during April, which corresponded to greater emission reductions during November than during April.

Data for the control of particles specifically in the inhalable size range (diameters less than 15 μm) obtained from the Fort Carson testing apparatus are shown in Table 8.[42] Concentrations of respirable dust generated from vehicles passing over the fabric/aggregate road are shown to be reduced from 26 to 53%, with an average of 43%, compared to the control road. Again, the reduction was slightly greater at lower wind speeds; viz, the November 1979 tests 1 through 4 had an average inhalable particulate reduction of 45% compared to the April 1980 tests 5 through 8 which had an average reduction of 40%.

5.4   Improved Street Sweepers

A number of studies have been conducted to evaluate street cleaning programs and sweeper effectiveness.[25,26,27] However, these studies have almost exclusively centered around existing street cleaning methods and practices. Results have shown primarily that such existing practices are relatively

Table 7   Amortized Costs of Dust Control on Unpaved Roads (1979)[40]

Type of Road[a] and/or Dust Control	Cost, $/km
Unpaved road on firm soil (no control); $27,400 over 25 yr (17% interest)	4,750
Watering and ballast replacement	6,750
Oiling and ballast replacement	6,200
Ordinary road with watering and ballast replacement	11,500
Ordinary road with oiling and ballast replacement	10,950
Fabric unpaved road; $30,000 over 12 yr (17% interest)	6,000
Ballast replacement	1,600
Fabric road with ballast replacement	7,600

[a] Road width = 10 m; road length = 1.6 km.

Table 8   Effect of Bidim® Road Carpet on Inhalable Particulate Concentration[a] at Fort
Carson Army Base, Colorado Springs, Colorado[42]

Test[b]	Ambient Concentration, Uncontrolled Condition, $\mu g/m^3$	Ambient Concentration After Installation of Road Carpet, $\mu g/m^3$	Concentration Reduction, %
1	636	299	53
2	396	272	31
3	144	73[c]	49
4	906	473	48
5	760	420	45
6	1040	632	39
7	1283	944	26
8	2450	1243	49

[a] Defined as <15 μm in size.
[b] Tests 1 through 4 were conducted in November 1979 and tests 5 through 8 were conducted
in April 1980.
[c] After snowfall.

ineffective. At present, street sweepers are not designed to collect inhalable particles that are dispersed
during street cleaning. EPA has identified the research needs in this area as focusing on improvements
to existing equipment with emphasis on greater removal efficiencies.

A.P.T., Inc. has undertaken a program to develop a special-purpose street sweeper.[43] The objective
of this study is to retrofit existing vacuum sweepers with an air pollution control system designed to
collect 90% of the inhalable particles dispersed by the sweeper. The control system uses air jets,
hooding, and spray scrubbing to capture and retain particles for subsequent disposal. Analysis of
conventional street sweeper operation indicates that the gutterbroom and leakage from the "pickup
hood" are the major sources of inhalable particulate emissions from the regenerative type vacuum
sweepers. Therefore, the particulate control approach consists of containing emissions from these sources
as illustrated in Figure 5.[43]

The recirculating air sweeper design is used so that only a portion of the total air flow requires
scrubbing. This minimizes the size and water requirements of the scrubber and creates a positive
influx of air to the pickup hood, thereby eliminating leakage. The gutterbroom emissions are controlled
by interactive hoods that are specifically designed to collect the dust dispersed by the gutterbroom
and convey it to the hopper, while not restricting the gutterbroom's ability to sweep the street surface
and gutter.

Preliminary test results are presented in Figure 6, showing street dust density before and after
sweeping as a function of distance from the curb.[43] It is observed that street dust is generally concentrated
in the gutter area. Removal efficiencies calculated from the measured data shown in Figure 6 range
from 75 to 97%, which is a substantial improvement over the performance capability of the best
conventional vacuum sweepers currently available. The major breakthrough has been the development
of an effective gutterbroom head. Additional power requirements are minimal since the sweeper provides
sufficient blower power, pumps, and auxiliary power for the required service. The improved street
sweeper could also be used for special problems such as sweeping roads at heavy industrial sites and
after sanding and salting operations used for general snow and ice control on any paved road.

5.5   Windscreens

Screen fencing has historically been used to reduce glare along highway medians and to reduce snow
drifting on roads. Screening systems can also be designed to reduce wind velocity, creating a significant
lee zone and helping to control the migration of fugitive dust. In Sweden windscreens are used to
confine and reduce fugitive dust emissions from iron ore, coal, coke, and gypsum storage piles. Windscreens have also been installed for glare and snow control in the United States. Where this system
can be sited in the prevailing upwind direction from open storage areas, it provides continuous protection
without the potential undesirable side effects attendant with wet suppression using water and/or chemicals, or the impracticality in certain cases of confinement by covering or enclosure.

The design and siting of a wind fence for optimum wind reduction for a given application must
include the following factors:[44]

1.   Windscreen positioning.
2.   Windscreen height and width.

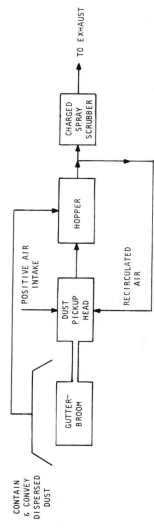

Fig. 5 Schematic diagram of improved vacuum (regenerative air) type sweeper.

759

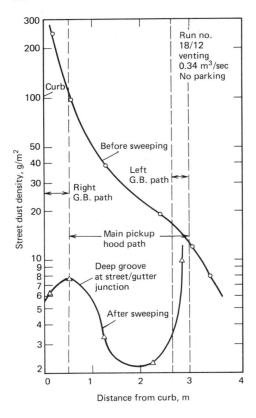

Fig. 6 Street dust distribution before and after sweeping.

3. Windscreen curvature.
4. Windscreen permeability.
5. Size of the windscreen elements or openings.
6. Terrain roughness.
7. Turbulent intensity in the incident wind field and in the air flow downwind of the screen.

The heart of the system is an extremely tough and durable knitted polyester screen secured by weather-resistant fasteners that enable the screen to be adapted to a variety of support structures. Generally, a 50% screen porosity has been found to be an optimum balance between windspeed reduction and created turbulence. Typical field data,[45] reported for a fence height of 0.5 m (1.64 ft), indicate that an effective windscreen should give a mean velocity reduction of 50% up to 10 fence heights downstream and a maximum reduction of 70 to 80% 1 to 5 fence heights downstream.

Tests were recently performed by TRC on two commercially available polyester windscreens, manu-factured by Julius Koch, Inc., to reduce open field dust entrainment at the State airport in Quonset, Rhode Island.[44] Results for the 65% permeable windscreen at an average incident wind of 3.0 m/sec (9.84 ft/sec) showed a wind velocity reduction of 70% in the adjacent area downwind of the windscreen. A reduction of 40% was observed as much as 14 fence heights downwind. With increasing incident wind speed, the area experiencing a 40% reduction is reduced significantly. The lower fence efficiency in this case is attributed to an increase in the turbulence intensity within the incident wind field.[44]

Tests conducted with the 50% permeable fence for an incident wind of 4.2 m/sec (13.8 ft/sec) showed a reduction in wind velocity of 60%. This result is comparable to the efficiency of the 65% permeable fence tested at an incident wind speed of 3.0 m/sec (9.84 ft/sec). In this case, however, there is no apparent zone of stagnation separating the "bleed flow" through the windscreen from the "displacement flow" that passes over the fence.[44]

Based on equations developed by TRC,[46] the effectiveness of windscreens in controlling fugitive emissions from a 54,000 metric ton (59,525 ton) coal storage pile using an 8-m (26.2-ft) high windscreen has been assessed in a case study. Under a set of typically encountered wind speeds and coal pile geometry and properties, windscreens were found to offer the potential for reducing fugitive emissions by 80%.

6 CONTROL FLEXIBILITY AND FACTORS AFFECTING COMPLIANCE WITH AN IPM STANDARD[12]

As discussed earlier, the most important factor in reducing the emission rate of process fugitive particulates is not control device removal efficiency but rather the effectiveness of hooding and ducting systems to collect FPEs from various process sources. While the energy and capital requirements for ventilation control of fugitive particulates are considerable, their severity is not sensitive to particle size, and an inhalable particulate matter (IPM) standard will not significantly increase the difficulty of ventilation design. Increasing the number of emission sources that are ventilated to a control device is really the only means of affecting overall removal efficiency since control devices are fixed in design. The technology is not now available to accurately predict the relative reduction corresponding to various sources, and therefore any attempts to meet standards by partial control would be difficult at best.

The importance of an ambient IPM standard based on a 15 μm upper size limit as opposed to TSP does not appear to be critical if the level of control required is the same for both standards. This is due to the fact that most of the FPEs are less than 15 μm in size. Therefore control techniques presently being used to meet a TSP standard would also be adequate to meet an IPM standard. However, if an IPM reduction level much lower than the TSP level were adopted, a greater degree of control would be required. Since no controls for fugitive emissions are designed to be highly efficient, the cost of control under a more strict IPM standard would be considerable.

The achievement of any ambient suspended particulate standard for FPEs depends not only on the above considerations but also on the location and method of compliance testing. For example, fugitive emissions measured at a plant boundary would be expected to contain fewer large particles because these would tend to settle out inside the plant boundaries. Fugitive emission measurements will also depend on the height of the emission source, meteorological conditions, topography, and particle size properties. Measurement methods require adaptation or modification for nearly every specific site/source combination and their accuracy can vary considerably. A change in the sampling method or location or in the size basis of ambient standards could alter the compliance status of a site for either standard.

Open sources differ from process sources of fugitive particulate emissions because of their sheer magnitude and because control efforts are aimed more at their prevention than collection. Table 9 shows how the contribution of fugitive emissions compares with point sources for each of the AQCRs that did not meet either the primary or secondary TSP standards.[27] As an illustration of the magnitude of fugitive source emissions, if they were cut by slightly less than 90%, 58 of the AQCRs would still have area source emissions exceeding point sources. If one were confident that a reduction of a unit weight of road dust emissions, for example, could be expected to have the same impact as a reduction of a unit weight of point source emissions, then increased open source control has the potential for becoming an attractive option compared to implementing more stringent controls on industrial sources.[3] This, of course, assumes that benefits associated with each of these approaches are equal.

In recognition of the fact that even a small reduction in fugitive dust emissions can result in a meaningful reduction in TSP, several important research and test programs have been undertaken, as listed in Table 4. The road carpet concept is an important development applied to the reduction of emissions from unpaved roads. Assuming that the road carpet is 50% efficient for all particle sizes (it is at least 50% efficient in the inhalable particulate size range), if 3% of the emissions from unpaved roads were controlled by road carpets, the resultant total reduction in fugitive dust emissions would equal the total amount of emissions coming from all controlled point sources. This is certainly a significant reduction relative to stated compliance levels and standards for point sources.

Table 9 Magnitude of the Emission Problem Associated with Fugitive Dust[27]

	Number of AQCRs	Percent of Total
1. Point source > area source	9	6.0
2. Area > point[a]	139	92.7
3. Data missing	2	1.3
Total	150	100.0

[a] There are approximately 97 AQCRs in which area source emissions exceed point source emissions by at least a factor of 5, and 58 AQCRs in which area source emissions exceed point sources by a factor of 10 or more.

7 ACKNOWLEDGMENTS

The work on which this chapter is based was sponsored by the Particulate Technology Branch, Office of Research and Development, Industrial Environmental Research Laboratory (Research Triangle Park), U.S. Environmental Protection Agency under Contract No. 68-02-3168, Technical Service Area 2, Work Assignment No. 3. The contents of this chapter do not necessarily reflect the views and policies of the U.S. Environmental Protection Agency, nor does mention of trade names, commercial products, or organizations imply endorsement by the United States government.

The authors wish to acknowledge the contributions of John A. Dirgo, Peter D. Spawn, and Eugene F. Bergson of GCA/Technology Divisio.ı in the preparation of the original report under the above contract.

REFERENCES

1. Weant, G. E., and Carpenter, B. H., *Particulate Control for Fugitive Dust,* Research Triangle Institute, EPA-600/7-78-071 (NTIS PB 282269), April 1978.

2. Kalinowski, T., Myers, R., Ellenbecker, M., and Spengler, J., *The Impact of EPA's Interpretive Ruling on Industrial Growth and the Siting of New Plants,* Harvard School of Public Health, March 1977.

3. Cooper, D. W., Sullivan, J. S., Quinn, M., Antonelli, R. C., and Schneider, M., *Setting Priorities for Control of Fugitive Particulate Emissions from Open Sources,* Harvard School of Public Health, EPA-600/7-79-186 (NTIS PB 80-108962), August 1979.

4. Lynn, D. A., Deane, G. L., Galkiewicz, R. C., and Bradway, R. M., *National Assessment of the Urban Particulate Problem: Volume I. Summary of National Assessment,* GCA/Technology Division, EPA-450/3-76-024 (NTIS PB 263665), July 1976.

5. Bradway, R. M., and Record, F. A., *National Assessment of the Urban Particulate Problem: Volume II. Particle Characterization,* GCA/Technology Division, EPA-450/3-76-025 (NTIS PB 263666), July 1976.

6. McCutchen, G., Regulatory Aspects of Fugitive Emissions, in *Symposium on Fugitive Emissions: Measurement and Control (May 1976, Hartford, CT),* The Research Corporation of New England, EPA-600/2-76-246 (NTIS PB 261955), September 1976.

7. Jutze, G. A., Zoller, J. M., Janszen, T., Amick, R. S., Zimmer, C. E., and Gerstle, R. W., *Technical Guidance for Control of Industrial Process Fugitive Particulate Emissions,* PEDCo Environmental, Inc., EPA-450/3-77-010 (NTIS PB 272288), March 1977.

8. Zoller, J., Bertke, T., and Janszen, T., *Assessment of Fugitive Particulate Emission Factors for Industrial Processes,* PEDCo Environmental, Inc., EPA-450/3-78-107 (NTIS PB 288859), September 1975.

9. Richard, G., and Safriet, D., Developing Control Strategies for Fugitive Dust Sources, in *Symposium on the Transfer and Utilization of Particulate Control Technology: Volume 4. Fugitive Dusts and Sampling, Analysis and Characterization of Aerosols,* Denver Research Institute, EPA-600/7-79-044d (NTIS PB 295229), February 1979.

10. Constant, P., Marcus, M., and Maxwell, W., *Sampling Fugitive Lead Emissions from Two Primary Lead Smelters,* Midwest Research Institute, EPA-450/3-77-031 (NTIS PB 276356), October 1977.

11. Bohn, R., Cuscino, T., and Cowherd, C., *Fugitive Emissions from Integrated Iron and Steel Plants,* Midwest Research Institute, EPA-600/2-78-050 (NTIS PB 281322), March 1978.

12. Drehmel, D. C., Daugherty, D. P., and Gooding, C. H., State of Control Technology for Industrial Fugitive Process Particulate Emissions, in *Symposium on the Transfer and Utilization of Particulate Control Technology: Volume 4. Fugitive Dusts and Sampling, Analysis and Characterization of Aerosols,* Denver Research Institute, EPA-600/7-79-044d (NTIS PB 295229), February 1979.

13. *Compilation of Air Pollutant Emissions Factors,* 3rd edition, including Supplements 1–7, EPA Publication AP-42 (NTIS PB 275525) Parts A and B, August 1977, and Supplements 8 (NTIS PB 288905), May 1978; 9 (NTIS PB 295614), 1979; and 10, July 1980.

14. Phelps, R. G., AISI-EPA-Battelle Coke Oven Door Sealing Program, *J. Air Pollut. Control Assoc.* **29**(9) (1979).

15. Coy, D. W., and Jablin, R., Review of Foreign Air Pollution Control Technology for BOF Fugitive Emissions, presented at the *Symposium on Iron and Steel Pollution Abatement Technology,* Chicago, IL (October 30, 31, and November 1, 1979).

16. Drehmel, D. C., Gooding, C. H., and Nichols, G. B., *Particulate Control Highlights: Recent Developments in Japan,* EPA/IERL-RTP, EPA-600/8-79-031a (NTIS PB 80-148802), November 1979.

17. Wallace, D., and Cowherd, C., *Fugitive Emissions from Iron Foundries,* Midwest Research Institute, EPA-600/7-79-195 (NTIS PB 80-110976), August 1979.

18. Chalekode, P. K., Blackwood, T. R., and Archer, S. R., *Source Assessment: Crushed Limestone, State of the Art,* Monsanto Research Corporation, EPA-600/2-78-004e (NTIS PB 281422), April 1978.

19. Metzger, C. L., Dust Suppression and Drilling with Foaming Agents, *Pit and Quarry Magazine,* March 1976.

20. Harwood, C. F., and Ase, P. K., *Field Testing of Emissions Controls for Asbestos Manufacturing Waste Piles,* Illinois Institute of Technology Research Institute, EPA-600/2-77-098 (NTIS PB 270081), May 1977.

21. Kromrey, R. V., Naismith, R., Scheffee, R. S., and Valentine, R. S., Development of Coatings to Reduce Fugitive Emissions from Coal Stockpiles, in *Third Symposium on Fugitive Emissions Measurement and Control* (October 1978, San Francisco, CA), The Research Corporation of New England, EPA-600/7-79-182 (NTIS PB 80-130891), August 1979.

22. Search, W. J., and Reznik, R. B., *Source Assessment: Urea Manufacture,* Monsanto Research Corporation, EPA-600/2-77-1071 (NTIS PB 274367), November 1977.

23. Serth, R. W., and Hughes, T. W., *Source Assessment: Carbon Black Manufacture,* Monsanto Research Corporation, EPA-600/2-77-107k (NTIS PB 273068), October 1977.

24. Richard, G., and Safriet, D., *Guideline for Development of Control Strategies in Areas with Fugitive Dust Problems,* TRW, Inc., EPA-450/2-77-029 (NTIS PB 275474), October 1977.

25. Brookman, E. T., and Drehmel, D. C., Future Areas of Investigation Regarding the Problem of Urban Road Dust, in *Second Symposium on the Transfer and Utilization of Particulate Control Technology: Volume 4. Special Applications for Air Pollution Measurement and Control,* Denver Research Institute, EPA-600/9-80-039d (NTIS PB 81-122228), September 1980.

26. Sartor, J. D., Boyd, B., and Van Horn, W. H., How Effective Is Your Street Sweeping, *APWA Reporter* 39(4) (1972).

27. Axetell, K., and Zell, J., *Control of Reentrained Dust from Paved Streets,* PEDCo Environmental, Inc., EPA-907/9-77-007 (NTIS PB 280325), July 1977.

28. Blackwood, T. R., Chalekode, P. K., and Wachter, R. A., *Source Assessment: Crushed Stone,* Monsanto Research Corporation, EPA-600/2-78-004l (NTIS PB 284029), May 1978.

29. Reed, A. K., Meeks, H. C., Pomercy, S. E., and Hale, V. Q., *Assessment of Environmental Aspects of Uranium Mining and Milling,* Battelle-Columbus Laboratories, EPA-600-/7-76-036 (NTIS PB 266413), December 1976.

30. Weant, G. E., and Carpenter, B. H., Fugitive Dust Emissions and Control, in *Symposium on the Transfer and Utilization of Particulate Control Technology: Volume 4. Fugitive Dusts and Sampling, Analysis and Characterization of Aerosols,* Denver Research Institute, EPA-600/7- 79-044d (NTIS PB 295229), February 1979.

31. Jutze, G., and Axetell, K., *Investigation of Fugitive Dust, Volume 1: Sources, Emissions, and Control,* PEDCo Environmental, Inc., EPA-450/3-74-036a, June 1974.

32. Yung, S. C., Calvert, S., and Drehmel, D. C., Spray Charging and Trapping Scrubber for Fugitive Particulate Emission Control, in *Second Symposium on the Transfer and Utilization of Particulate Control Technology: Volume 4. Special Applications for Air Pollution Measurement and Control,* Denver Research Institute, EPA-600/9-80-039d (NTIS PB 81-122228), September 1980.

33. Yung, S. C., Curran, J., and Calvert, S., *Spray Charging and Trapping Scrubber for Fugitive Particle Emission Control,* A.P.T., Inc., EPA-600/7-81-125 (NTIS PB 82-115304), July 1981.

34. Hoenig, S. A., *Fugitive and Fine Particle Control Using Electrostatically Charged Fog,* University of Arizona, EPA-600/7-79-078 (NTIS PB 298069), March 1979.

35. Hoenig, S. A., *Use of Electrostatically Charged Fog for Control of Fugitive Dust Emissions,* University of Arizona, EPA-600/7-77-131 (NTIS PB 276645), November 1977.

36. Daugherty, D. P., and Coy, D. W., *Assessment of the Use of Fugitive Emission Control Devices,* Research Triangle Institute, EPA-600/7-79-045 (NTIS PB 292748), February 1979.

37. Mathai, C. V., Rathbun, L. A., and Drehmel, D. C., An Electrostatically Charged Fog Generator for the Control of Inhalable Particles, presented at the *Third Symposium on the Transfer and Utilization of Particulate Control Technology,* Orlando, FL, March 9–12, 1981.

38. Evans, J. S., Schneider, M., Cooper, D. W., and Quinn, M., Setting Priorities for the Control of Particulate Emissions from Open Sources, in *Symposium on the Transfer and Utilization of Particulate Control Technology: Volume 4. Fugitive Dusts and Sampling, Analysis and Characterization of Aerosols,* Denver Research Institute, EPA-600/7-79-044d (NTIS PB 295229), February 1979.

39. Levene, B., and Drehmel, D. C., Civil Engineering Fabrics Applied to Fugitive Dust Control Problems, in *Proceedings: Fourth Symposium on Fugitive Emissions, Measurement and Control,* TRC—Environmental Consultants, EPA-600/9-80-041 (NTIS PB 81-174393), December 1980.

40. Blackwood, T. R., *Assessment of Road Carpet for Control of Fugitive Emissions from Unpaved Roads,* Monsanto Research Corporation, EPA-600/7-79-115 (NTIS PB 298874), May 1979.

41. Ochsner, J. C., Chalekode, P. K., and Blackwood, T. R., *Source Assessment: Transport of Sand and Gravel,* The Research Corporation of New England, EPA-600/2-78-004y (NTIS PB 289788), October 1978.

42. Tackett, K. M., Blackwood, T. R., and Hedley, W. H., *Evaluation of Road Carpet for Control of Fugitive Emissions from Unpaved Roads,* a problem-oriented report prepared by Monsanto Research Corporation for the U.S. Environmental Protection Agency, EPA Contract No. 68-02-3107, October 1980.

43. Calvert, S., Brattin, H., Bhutra, S., and Parker, R., Improved Street Sweeper for Controlling Urban Inhalable Particulate Matter, presented at the *Third Symposium on the Transfer and Utilization of Particulate Control Technology,* Orlando, FL, March 9–12, 1981.

44. Carnes, D., and Drehmel, D. C., The Control of Fugitive Emissions Using Windscreens, presented at the *Third Symposium on the Transfer and Utilization of Particulate Control Technology, Orlando, FL, March 9–12, 1981.*

45. Raine, J. K., and Stevenson, D. C., Wind Protection by Model Fences in a Simulated Atmospheric Boundary Layer, *J. of Industrial Aerodynamics* **2** (1977).

46. Blackwood, T. R., and Wachter, R. A., *Source Assessment: Coal Storage Piles,* Monsanto Research Corporation, EPA-600/2-78-004k (NTIS PB 284297), May 1978.

CHAPTER 29

SAMPLING AND ANALYSIS OF SOURCE GASES

KENNETH T. KNAPP

U.S. Environmental Protection Agency
Research Triangle Park, North Carolina

1 INTRODUCTION

While ancient peoples recognized the noxious effects of sulfur gases emitted from volcanos, it was not until the last half of the current century that people accepted the fact that human activities produced most of the sulfur oxide pollution and something should be done to control these emissions. The sulfur oxide pollution results mainly from the combustion of fossil fuels and the use of sulfur-containing ores in metal production. Human activities also account for the major amount of the oxides of nitrogen and several other gases emitted into the atmosphere.

Because the amounts of these gases emitted into the atmosphere need to be limited, standards for the amounts emitted by various industries have been set. To ensure that these standards are met and that needed data to set new standards are available, standard reference methods for measuring source emitted pollutants have been developed. In this chapter, these and other methods for measuring source emitted gaseous pollutants are discussed. The first section describes the reference methods. Nonreference methods are discussed in the next section. In the third section, a discussion of continuous monitors for gaseous emissions is given. The final section has a discussion of measurement methods for mobile source gaseous emissions.

2 STANDARD REFERENCE METHODS

Standard reference methods have been developed to measure source emitted gases in those industries for which emission standards have been set. A list of these methods is given in Table 1 along with the first and latest *Federal Register* (FR) references. The first nine standard reference methods, numbered sequentially from 1 to 9, were first listed as promulgated methods in FR 36, 24877, 12/23/71. Methods 1 through 4 are the ancillary methods for sample and velocity traverses, velocity and volumetric flow rates, gas analyses for O_2, CO_2, and dry molecular weight, and moisture. Method 5 is for measuring particulate emissions. Methods 6, 7, and 8 are for measuring SO_2, oxides of nitrogen (NO_x), and sulfuric acid and SO_2 at sulfuric acid plants. Method 9 is for the measurement of visible emissions, plume opacity.

The principle of Method 6 is to extract the stack gases through a probe into a sampling system composed of a midget bubbler and three midget impingers which collect the SO_2 to be measured. The midget bubbler contains an 80% isopropanol and water solution which collects the sulfuric acid and any SO_3 but not the SO_2. The first two midget impingers contain a 3% hydrogen peroxide solution which collects and oxidizes the SO_2. The resulting sulfate is measured by the barium–thorin titration method. The remaining components of the sampling train are an empty midget impinger, silica gel drying tube, needle valve, pump, rate meter, and dry gas meter. Thermometers are located in the gas flow stream after the empty impinger and before the dry gas meter. The barometric pressure must be known, therefore a barometer is recommended unless the data are available from a weather station. From the data taken, the concentration of SO_2 in mg/dscm (dry standard cubic meters) is obtained. Free ammonia interferes with the thorin indicator and when present an alternate method must be used.

The principle of Method 7 is to collect the stack gases in an evacuated flask which contains a dilute sulfuric acid–hydrogen peroxide absorbing solution. The absorbing solution oxidizes the nitrogen

Table 1   EPA Standard Methods[a] for Source Emitted Gaseous Pollutants

Method No.	Gases Measured	Lower Detection Limit	First Reference[b]	Latest Reference[b]
6	SO_2	3.4 mg/m³	36; 247, 12/23/71	42; 160, 8/18/77
7	NO_x (NO + NO_2)	2 mg/m³	36; 247, 12/23/71	42; 160, 8/18/77
8	H_2SO_4 and SO_2	1.2 mg/m³ (SO_2) 0.05 mg/m³ (as SO_3)	36; 247, 12/23/71	42; 160, 8/18/77
10	CO	20 ppm	39; 47, 3/23/74[c]	—
11	H_2S	8 mg/m³ (6 ppm)	43; 6, 1/10/78[c]	—
13A	HF (colorimetric)	Not determined	40; 152, 8/6/75	45; 121, 6/20/80
13B	HF (specific ion electrode)	0.02 µg/ml	40; 152, 8/6/75	45; 121, 6/20/80
14	HF (roof monitor)	Uses Methods 13A or B	41; 17, 1/26/76	45; 127, 6/30/80
15	H_2S, COS, CS_2	~0.5 ppm/compound	43; 51, 3/15/78[c]	—
16	TRS (H_2S, DMS, DMDS, MeSH)	50 ppb/compound	44; 9, 1/12/79[c]	—
19	SO_2 (removal efficiency)	—	44; 113, 6/11/79[c]	—
20	SO_2, NO_x, O_2 (gas turbines)	—	44; 176, 9/10/79[c]	—

[a] Standards for New Source Performance Standards under Title 40, Part 60.
[b] *Federal Register* references—volume; number, and date.
[c] Only reference.

oxides (NO_x), except nitrous oxide N_2O, to nitrogen dioxide which is measured by the phenoldisulfonic acid (PDS) procedure. The NO_x containing stack gases are drawn into the evacuated round bottom flask whose volume has been measured to within 2%. The flask with the sample is shaken for at least 5 min and then allowed to stand at least 16 hr. The contents of the flask are transferred with distilled water rinsing to a sample bottle. The pH of the solution is adjusted to between 9 and 12 with sodium hydroxide solution. The bottle is sealed for shipping. In the lab, a 25-ml aliquot is taken and analyzed by the PDS method. The results are reported as concentration of NO_x (measured as NO_2) corrected to standard conditions, mg/dscm.

Method 8 is similar to Method 6 but has several changes. The midget bubbler and midget impingers are replaced by Greenburg–Smith impingers. A filter is placed between the first and second impingers. The first impinger is filled with 100 ml of 80% isopropanol. The second and third impingers are filled with 100 ml of 3% hydrogen peroxide. The last impinger is filled with about 200 g of silica gel. The sample is extracted from the source isokinetically as in Method 5. The probe and front-half of the filter holder is washed with 80% isopropanol. The contents of the first impinger, these washings, and the filter are combined and analyzed together by the barium–thorin titration method. This measures the sulfuric acid, SO_3, and sulfate. The second and third impinger solutions are analyzed by the same titration method. The results are the SO_2 content. Method 8 has the same problem as Method 6 with free ammonia. The method was developed for measuring the sulfur oxide emissions from sulfuric acid plants and does not give good results at other types of plants, especially where the emissions have a high particulate loading.

Method 10 is for the measurement of carbon monoxide by nondispersive infrared analyzer (NDIR). The emissions are drawn through a probe into a condenser to remove excess water and then into the analyzer. If an integrated sample is desired, a bag sample is taken after the condenser and later fed into an analyzer.

Method 11 is for the determination of H_2S. The sample is extracted from the source with a sampling train containing five midget impingers. The first impinger contains 3% hydrogen peroxide; the second is empty; and the third, fourth, and fifth contain a cadmium sulfate absorbing solution. The H_2S concentration is determined by iodometric titration of aliquots from the combined contents of the last three impingers. The results are reported as concentration of H_2S at standard conditions, mg/dscm.

Methods 13A, 13B, and 14 are for measuring hydrogen fluoride emissions. The sampling trains for the three methods are essentially the same. Method 14 is for sampling the roof monitors of the pot room of an aluminum plant and has a complicated manifold for transporting the gases to the sampling train. Methods 13A and 13B differ in the analyzing procedure used. Method 13A uses the SPADNS zirconium lake colorimetric method while 13B uses the fluoride specific ion electrode. Either Method 13A or 13B is used to determine the fluoride content of samples collected by Method 14. These methods measure total fluoride, that is, HF and fluoride ions.

Method 15 is for the determination of H_2S, COS, and CS_2. An emission sample is extracted, diluted with clean dry air, and analyzed by gas chromatography (GC) with a flame photometric detector (FPD).

Method 16 is a semicontinuous method for measuring total reduced sulfur emissions from various points in a kraft pulp mill. This method, like Method 15, determines the sulfur compound by a GC with a FPD. As in Method 15, the sample is extracted, diluted with clean dry air, and analyzed by GC. The method is used to measure H_2S, methyl mercaptan, dimethyl sulfide, and dimethyl disulfide.

Method 19 is for the determination of SO_2 removal efficiency and particulate, SO_2 and NO_x emission rates from electric utility steam generators. This method gives a choice in the method to be used for measuring the SO_2. The sulfur content of the fuel along with the heat content can be measured and from these data, the amount of SO_2 feeding into the sulfur emission control system can be determined. Alternately, the SO_2 before and after the control device can be measured by a monitor or an emission measuring method specified in the regulations. The SO_2 removal efficiency is determined from the data obtained from these measurements. The NO_x content is determined according to the method specified in the regulations.

Method 20 is for the determination of NO_x, SO_2, and O_2 in the emissions from stationary gas turbines. The NO_x and O_2 content are determined in a sample stream from the turbine exhaust by instrumental analyzers. The SO_2 can be determined by either Method 6 or an instrument.

Besides the standard methods for the regulated gaseous emissions from stationary sources, reference methods have been promulgated for certain source emitted hazardous gas pollutants. These methods are Method 101 and Method 102 for gaseous and particulate mercury and Method 106 and Method 107 for vinyl chloride. These methods are given in CFR 40, part 61 and FR 38, 8826, 4/6/73 for methods 101 and 102 and FR 41, 46569, 10/21/76 for methods 106 and 107. This information is summarized in Table 2.

3   SPECIAL METHODS

In many cases, unregulated pollutants need to be measured for which no standard methods exist or from sources for which the standard methods were not intended. In these cases, special nonstandard methods are required. Methods for several of these pollutants have been developed. An important example of such methods are those that measure sulfuric acid from power plants, portland cement plants, and pulp and paper mills.

Much development work has been spent on measurement methods for stationary source emitted sulfuric acid, primary sulfuric acid. In all these methods, two major problems must be overcome. These problems are separating the sulfuric acid from SO_2 and particulate sulfate and measuring very small amounts of sulfuric acid. Several approaches to solve these problems have been published. The best results are obtained from the procedure described by Cheney and Homolya.[1] This procedure has a heated filter or quartz wool plug to remove the particulate sulfate and a controlled temperature condenser to separate the SO_2 from the acid. Several modifications of Method 6 which uses 80% isopropanol (IPA) to separate the SO_2 from the acid have been reported. However, this approach generally gives high results due to small amounts of SO_2 collecting in the IPA. Control condensation systems for measuring sulfuric acid are now available commercially. Several other types of sulfuric acid measurement instruments are under development and should be available within a few years.

Sampling and analysis methods for various organic compounds have been developed; however, a detailed discussion of them is beyond the scope of this chapter. For details of methods for common organic gaseous pollutants consult the book by Ruch.[2] For methods on the collection and analysis of polycyclic organic matter (POM) consult the papers by Jones et al.[3] and Bennett et al.[4]

For methods to measure other source emitted gaseous pollutants, consult the books by Ruch[2] or those by Driscoll[5] and Stern.[6]

Table 2   EPA Standard Methods[a] for Source-Emitted Gaseous Hazardous Air Pollutants

Method No.	Gases Measured	Reference[b]	Page
101	Hg in air streams (gaseous and particulate)	38; 66, 4/6/73	8835
102	Hg in hydrogen streams (gaseous and particulate)	38; 66, 4/6/73	8840
106	Vinyl chloride, source emissions	41; 205, 10/21/76	46569
107	Vinyl chloride, waste	41; 205, 10/21/76	46571

[a] National Emission Standards for Hazardous Air Pollutants, Title 40, Part 61.
[b] *Federal Register* references—volume; number, and date.

4 CONTINUOUS MONITORS

The state of the art for analytical instruments and air pollution monitors continue to improve. Many new and improved monitors are now available. However, because of several major problems, many of the laboratory instruments and ambient air pollution monitors cannot be used as stationary source emission monitors. The main problems are related to the harsh environment of most industrial sources. The gaseous emissions to be measured are generally hot and highly corrosive. The plants are also dusty, dirty, and have varying degrees of vibrations at the locations of the monitors. In addition, the plant personnel handle the equipment much more harshly than the general laboratory technician. These conditions dictate that the stationary source monitors must be made rugged and out of corrosion resistant material.

Most of the gaseous pollution monitors are based on measuring some property of the individual molecule. One of the more commonly used properties of molecules for analyzers is their electromagnetic spectra. The individual spectrum of molecules ranges from the internal electron transitions of the individual atoms yielding radiation in the x-ray region to the bonding electron movements of the infrared spectrum. Since many compounds absorb in the UV-visible region, spectral band overlap of several compounds can be a problem. However, enough spectral bands free of interference exist for several of the more common pollutants for good measurement. Monitors based on UV-visible absorption are available for the major pollutants SO_2, NO, and NO_2.

Another type of spectral monitor is one based on the property of certain gases to emit light as a result of a chemical reaction or interaction between the gas of interest and another gas supplied by an instrument. This approach is known as chemiluminescence and instruments to measure oxides of nitrogen, and O_3 have been developed for both source and ambient measurements. Few sources have O_3 emissions and therefore no source O_3 monitor is available.

The absorption of the infrared spectral lines is used in monitors for SO_2, NO, CO_2, and CO. These monitors are based on the infrared absorption differences between two cells. One cell is divided into two parts separated by some gas tight partition. In one part is a known amount of the gas to be measured. In the other part the source gases are drawn through. The other cell contains only the gas of interest in the same amount as the first cell. Infrared radiation is passed through both cells and the amount of radiation absorbed is detected alternately. The difference in the detector signal is determined and is a measure of the gas of interest. This approach is referred to as nondispersive infrared, NDIR.

The electrical and magnetic properties of gases are also used in pollution monitors. For hydrocarbons and other organic compounds, standard gas chromatographic hydrogen flame ionization detectors are used. A list of the types of monitors available and the gases they determine is given in Table 3. The manufacturers of the monitors will supply detailed information on the principle of operation, availability, cost, and much more.

Monitors can be used to measure source emitted gaseous pollutants for many reasons. The EPA has adopted the use of source monitors to measure the performance of pollution control equipment in several industries. A list of the industries and gases for which regulations have been set is given in Table 4. This table also lists the reference for the specifications and calibration procedures required for the monitors. For a more complete discussion of these requirements, the reader should consult

Table 3 Types of Continuous Monitors for Gases

Gases	Type of Detectors	Examples[a]
SO_2, NO, NO_2	UV-visible photometric	DuPont Model 400
SO_2	Fluorescence	TECO Model 40
SO_2, CO_2	Gas filter correlation infrared	Leeds and Northrup 7864
NO_2, NO	Chemiluminescence	Beckman
SO_2	Plume photometric	Meloy
SO_2, NO, CO_2	Nondispersive infrared	MSA
SO_2, NO, NO_2, CO, O_2	Polarographic	Dynascience
SO_2, NO	Second derivative UV	Lear–Siegler
SO_2	Amperometric	Barton ITT
O_2	Electrocatalytic	Teledyne
O_2	Paramagnetic	Scott
Hydrocarbon	Flume ionization	Beckman Model 402

[a] Mention of these examples does not constitute an endorsement. Several good instruments exist for most categories.

Table 4   Continuous Monitor Requirements[a] for Source-Emitted Gaseous Air Pollutants

Source	Gases Measured	Reference[b]	Page
Fossil fuel-fired steam generator	SO_2, NO, O_2 or CO_2	40; 194, 10/6/75	46256
Nitric acid plants	NO_2	40; 194, 10/6/75	46258
Sulfuric acid plants	SO_2	40; 194, 10/6/75	46258
Petroleum refineries	CO, SO_2	40; 194, 10/6/75	46259
Petroleum refineries, Claus sulfur recovery plant	CO, H_2S and/or SO_2	43; 51, 3/15/78	10869
Primary copper smelters	SO_2	41; 10, 1/15/76	2339
Primary zinc smelters	SO_2	41; 10, 1/15/76	2340
Primary lead smelters	SO_2	41; 10, 1/15/76	2341

[a] National Emission Standards for Hazardous Air Pollutants, Title 40, Part 61.
[b] *Federal Register* references—volume; number, and date.

either those listed *Federal Register* references or the Code of Federal Regulations 40, Part 60, Appendix B.

Several remote systems for measuring source emitted gases have been developed. One such system is the EPA's Remote Optical Sensing of Emissions system known as the ROSE system. The system uses a high-resolution Fourier transform infrared interferometer system (FT-IR) to measure the gases. The ROSE system has three modes of operation: measuring the absorption through the atmosphere of the IR radiation from some remote light source, path lengths up to about 2 km can be used; measuring the absorption of a light source aimed through ports in a stack; and measuring the IR emissions from hot stack plumes and flares. Another remote system is the ultraviolet television system (UVTV) which measures SO_2 in plumes. The Visiplume No. 121 is such a system that is commercially available. The system works on the principle of viewing the plume at two different wavelengths; one where SO_2 absorbs and one where it does not absorb. The differences yield the SO_2 content in the plume. Both signals are displayed on the TV screen which shows the position of the SO_2 plume. For a more complete discussion on remote methods consult the two papers by Herget.[7,8]

5   MOBILE SOURCES

The gaseous emissions from motor vehicles that are regulated are CO, oxides of nitrogen, and hydrocarbons. Since the fuels burned in these vehicles have only trace amounts of sulfur, SO_2 is not regulated. In addition to the regulated gases, CO_2 is measured. The sampling and analysis of these gases are carried out with a specifically designed system. The emission gases from the vehicles are passed into a dilution tunnel to which the gas monitors are connected. The system and procedure used are described in several volumes of the *Federal Register,* the latest being volume 45, March 5, 1980. These procedures can also be found in the Code of Federal Regulations 40, Part 86. Detailed specifications and calibration procedures for each of the gas monitors are given in these references. The carbon monoxide and dioxide concentrations are measured by NDIR analyzers. Chemiluminescence analyzers are used to measure the oxides of nitrogen. The hydrocarbons are measured with a flame ionization monitor. When diesel-fueled vehicles are being tested, a heated flame ionization detector is used.

For studies where gases other than those regulated are being analyzed, procedures have been developed. Many of the procedures are described in a paper by Dietzmann and Black.[9] They give procedures for the following unregulated gases; H_2S, organic sulfides, SO_2, individual hydrocarbons, aldehydes, HCN, ammonia, amines, nickel carbonyl, and phenols. The gases are collected from the dilution tunnel in either bags, impingers, or cold traps. The organics and HCN are determined by gas chromatography. Ion chromatography is used for ammonia and SO_2. Nickel carbonyl is measured by a chemiluminescence analyzer.

In general, testing the emissions from motor vehicles requires a complex setup which must be operated by experienced people. For more details, either consult the CRF 40, Part 86, the *Federal Register,* volume 45, or a text such as volume III of the series *Air Pollution* by Stern.[6]

REFERENCES

1. Cheney, J. L., and Homolya, J. B., *Environ. Sci. Technol.* **13**, 584 (1979).
2. Ruch, W. E., *Quantitative Analysis of Gaseous Pollutants,* Ann Arbor-Humphrey Science Publication, Ann Arbor, MI, 1970.
3. Jones, P. W., Giammar, R. D., Strup, P. E., and Stanford, T. B., *Environ. Sci. Technol.* **10**, 806 (1976).
4. Bennett, R. L., Knapp, K. T., Jones, P. W., Wilkerson, J. E., and Strup, P. E., in *Polynuclear Aromatic Hydrocarbons* (P. W. Jones and P. Leber, eds.), Ann Arbor Science Publishers Inc., Ann Arbor, MI, 1979, p. 419.
5. Driscoll, J. N., *Flue Gas Monitoring Techniques,* Ann Arbor Science Publishers Inc., Ann Arbor, MI, 1974.
6. Stern, A. C., *Air Pollution,* volume III, Academic Press, New York, 1977.
7. Herget, W. F., and Conner, W. D., *Environ. Sci. Technol.* **11**, 962 (1977).
8. Herget, W. F., and Brasher, J. D., *Applied Optics* **18**, 3404 (1979).
9. Dietzmann, H. E., and Black, F. M., *SAE Technical Paper Series,* 790816, September 1979.

CHAPTER 30

SAMPLING AND ANALYSIS OF AMBIENT GASES

H. H. WESTBERG

Washington State University
Pullman, Washington

1 INTRODUCTION

The purpose of this chapter is to outline the principal methods used to measure gaseous species in the atmosphere. In general, detailed analytical procedures are not provided; however, references are included which can be consulted for this type of information. Ambient air measurements require the utilization of proper sampling and detection procedures. The accuracy of a particular measurement can be seriously compromised by a deficiency in either of these two facets of an analytical scheme. In most instances, ambient gaseous concentrations are very low which compounds the chances for error. Analytical procedures that require a minimum number of operations are generally preferred. Currently, all of the gaseous species classified as criteria pollutants by the U.S. Environmental Protection Agency can be monitored by automated methods on a real-time basis. Succeeding sections of this manuscript include a brief discussion of the commonly employed sampling methods and a survey of the various detection principles for the gaseous pollutants of greatest interest to atmospheric scientists.

2 SAMPLING PROCEDURES

A sampling technique that involves the use of a pump to draw an air stream through a loop, cell, or reaction chamber is generally employed in methods designed for real-time analysis. Sample storage problems are eliminated using this approach since the time between collection and analysis is usually less than 1 min. The analyst must only be concerned that sample integrity be maintained for the short period of time that is required to transfer the gas from the ambient atmosphere to the detector. Transfer lines must be properly selected such that reactive gases are not removed and contaminates are not introduced into the system. The use of stainless steel and/or Teflon tubing is generally preferred. When sample collection and analysis on a real-time basis is impractical, various alternate sample collection methods can be employed. Whole air methods which involve the capture of a volume of air in either a plastic bag or rigid container are useful for obtaining samples of nonreactive trace gas species. Adsorption, absorption, and cryogenic procedures are also available which permit the simultaneous collection and concentration of ambient gases.

2.1 Whole Air Methods

Air samples can be collected in either plastic bags or rigid containers. Materials used for constructing bags normally include Teflon, Tedlar, Mylar, or Scotchpak. Teflon appears to be the most inert; however, it is also the most difficult to seal. Bags are not acceptable for the collection and storage of reactive gases such as oxidants, sulfur compounds, and the oxides of nitrogen. They have been used extensively for the collection of vapor phase organic compounds. Some contamination problems have been reported due to emissions from the bag material itself. Thus, care must be exercised to check for contamination whenever plastic bags are employed for sample collection. Plastic bags must also be carefully checked for leaks. Poorly sealed seams and pinhole punctures are the main sources of leakage problems. Bags are advantageous from the standpoint of being lightweight and easily transportable. They are preferred over rigid containers for collecting whole air samples over an integrated time period (hours).

Rigid containers are normally constructed of glass, stainless steel, or aluminum. A sample is collected by drawing air into the evacuated container or by displacement of existing air within the vessel. The evacuated container technique is useful when utilization of a pump for displacement is impossible or impractical. Glass vessels can be evacuated to very low pressures and sealed under reduced pressure. Then by simply breaking the sealed neck, ambient air will fill the container. The main problem associated with this technique is resealing the vessel in order to maintain sample integrity between the time of collection and analysis. Rigid metal containers can be utilized in a similar manner. Installation of a good valve will allow the canister to be evacuated and maintained at reduced pressure. The container can then be filled by opening the valve and allowing the system to attain atmospheric pressure. Samples collected in this manner are well protected from contamination prior to analysis.

Displacement sampling into a rigid container involves displacing residual air in the vessel with the ambient air of interest. This is usually accomplished by flushing the container with 10 to 20 times its volume of ambient air and then sealing at either atmospheric pressure or at a positive pressure. The use of containers that permit the collection of pressurized samples provides two advantages— more sample is available for analytical purposes and there exists less likelihood of sample contamination. A metal container can be cleaned by purging with high-purity nitrogen or air and sealed at a positive pressure. Maintenance of positive pressure prior to sample collection ensures that no leakage into the vessel has occurred. Similarly, if the sample container is filled to a positive pressure during the collection step, contamination during transport and storage will not occur because leakage will be from inside out rather than outside into the container. Extreme care must be exercised when using the evacuated container technique to ensure the absence of small leaks in welds, seals, or valves. If leaks do exist, sample contamination will occur.

2.2   Cryogenic Methods

Cryogenic trapping techniques are commonly employed to obtain sufficient quantities of carbon-containing species for analytical purposes. The standard procedure involves passage of an air stream through a trap maintained at subambient temperature. A variety of coolant solutions are available which provide subambient temperatures as low as $-196°C$ (liquid N_2). Liquid oxygen and liquid argon are popular coolants because they are cold enough to retain all except the most volatile organic species and yet they allow nitrogen in the air stream to pass on through the trap. Organic compounds concentrated in this manner can be transferred to a gas chromatograph for qualitative and quantitative analysis.

Cryogenic collection devices have also been described which trap a whole air sample at liquid nitrogen temperatures. This technique differs from the flow-through system described previously in that nitrogen, oxygen, and all trace species are retained. Prior to analysis for any of the trace constituents, nitrogen and oxygen must be removed either by distillation or cryogenic pumping. Some loss of the more volatile species (e.g., C_1—C_4 hydrocarbons) is bound to occur during the distillation step. Consequently, the whole air cryogenic procedure is probably best suited for obtaining sufficient quantities of trace organics for qualitative identification.

2.3   Absorption Methods

Sample collection via absorption is a commonly used technique in which a gaseous species is dissolved in a liquid medium. An air stream is dispersed through the absorption medium by means of a fritted glass tube placed below the surface of the liquid. This method has been traditionally used for the collection of sulfur dioxide and nitrogen dioxide samples. It also currently is the method of choice for collecting aldehyde samples in urban atmospheres. Factors that must be characterized or controlled when employing absorption sampling include (1) collection efficiencies, (2) loss of absorption medium due to evaporation, and (3) pollutant decomposition in collection liquid prior to analysis.

2.4   Adsorption Methods

This technique involves the passage of an air stream through a packed bed of charcoal, molecular sieve, or some other adsorbent material which serves to retain gaseous pollutants. The retained species are generally desorbed by heating or solvent extraction. The method has been used to collect organic species present at ultratrace levels in the ambient atmosphere. The technique is very useful from a qualitative standpoint since it provides a means of separating minute quantities of trace gases from large volumes of air. However, it is extremely difficult to obtain reliable quantitative information. Difficulties arise both from the collection and desorption steps. Atmospheric water vapor will deactivate many adsorbents which results in varying collection efficiencies. Complete desorption is not always possible due to isomerization or decomposition on the adsorbent surface. The emission of artifact compounds is also a problem with some solid adsorbents.

SUGGESTED READING

Farwell, S. O., Gluck, S. J., Bamesberger, W. L., Schutte, T. M., and Adams, D. F., Determination of Sulfur-Containing Gases by a Deactivated Cryogenic Enrichment and Capillary Gas Chromatographic System, *Analytical Chemistry* **51**, 609 (1979).

Hanst, P. L., Spiller, L. L., Watts, D. M., Spence, J. W., and Miller, M. F., Infrared Measurement of Fluorocarbons, Carbon Tetrachloride, Carbonyl Sulfide and Other Atmospheric Trace Gases, *J. Air Pollution Control Association* **25**, 1220 (1975).

Harsch, D. E., Evaluation of a Versatile Gas Sampling Container Design, *Atmospheric Environment* **14**, 1105 (1980).

Lonneman, W. A., Bufalini, J. J., Kuntz, R. L., and Meeks, S. A., Contamination from Fluorocarbon Films, *Environmental Science and Technology* **15**, 99 (1981).

Rasmussen, R. A., A Quantitative Cryogenic Sampler; Design and Operation, *American Laboratory*, July 1972.

3 ANALYTICAL PROCEDURES FOR MONITORING HYDROCARBONS

Since most hydrocarbon species contribute to the production of photochemical oxidants, considerable effort has been devoted in recent years to determining the types and quantities of hydrocarbons present in ambient atmospheres. This research work has been aided to a great extent by the rapid improvement in gas chromatographic techniques. A great deal of information is currently available which characterizes the hydrocarbon composition of urban as well as rural air masses.

3.1 Methane

Since methane is present in ambient atmospheres at concentrations in excess of 1.3 ppm, detectability is not a major problem. A gas chromatograph (GC) with flame ionization detector is the preferred analytical technique for determining methane concentrations. Gas chromatographs are commercially available which are specifically designed for ambient methane analysis. The system consists of a pump which continuously purges a sample loop with ambient air. At predetermined intervals, the contents of the sample loop are transferred to the GC column via a gas sampling valve. Ordinarily, the column system consists of a precolumn which passes methane and certain other highly volatile species but retains the higher hydrocarbons. Methane passes into the analytical column (mole sieve) while the residual materials on the precolumn are discarded by backflushing. Methane is resolved on the analytical column and then routed into the detector. In automated systems methane concentrations can be determined at 5-min intervals. Precision is generally less than 1% and accuracy in the ±5% range.

Highly reliable methane calibration standards can be purchased from commercial specialty gas suppliers. The National Bureau of Standards provides methane standards that are referenced to an absolute methane standard in their possession.

SUGGESTED READING

Rasmussen, R. A., and Khalil, M. A. K., Atmospheric Methane (CH_4): Trends and Seasonal Cycles, *J. Geophysical Research* **88**, 9826 (1981).

Villalobos, R., Role of Gas Chromatography in Air Pollution Monitoring, *Analytical Methods Applied to Air Pollution Measurements,* Ann Arbor Science, Ann Arbor, MI, 1974.

3.2 Total Hydrocarbons

Analysis for total hydrocarbons can be accomplished by transferring a measured volume of ambient air directly into a flame ionization detector. The usefulness of this measurement is limited by two factors. Since methane is normally present at much higher levels, it is the predominant species recorded by this technique. However, due to its relative inert nature in the atmosphere it is not of critical concern as a pollutant. Nonmethane hydrocarbons are of more importance due to the fact that they are much more rapidly oxidized in the atmosphere. Individual hydrocarbon response for the flame ionization detector is somewhat variable which leads to calibration problems. Since a large number of hydrocarbons with varying chemical structures are normally present in ambient samples, it is difficult to prepare representative standards. Originally, methane was used as the calibration gas undoubtedly due to the fact that methane normally constituted the major species present. Recently, propane has found favor as the calibration gas of choice since it is felt it will respond in a manner more representative of the overall hydrocarbon burden.

SUGGESTED READING

Villalobos, R., Role of Gas Chromatography in Air Pollution Monitoring, *Analytical Methods Applied to Air Pollution Measurements,* Ann Arbor Science, Ann Arbor, MI, 1974.

3.3 Total Nonmethane Hydrocarbons

Analysis for total nonmethane hydrocarbons is performed with an automated gas chromatograph equipped with a flame ionization detector. A measured volume of air is transmitted directly to the flame ionization detector which provides a measure of the total hydrocarbon burden. Simultaneously, a second portion of the sample is routed through a GC column system that passes methane but not the higher molecular weight hydrocarbons. The nonmethane fraction is then determined by subtracting the methane concentration from the total hydrocarbon measurement. Most of the continuous monitors are of such complexity that a well-trained technician is required to ensure proper operation.

As indicated previously, severe calibration difficulties exist for the total hydrocarbon measurement due to the fact that the FID response varies with the type of hydrocarbon (i.e., paraffin, aromatic, etc.). The balance gas used to prepare standards can also affect response characteristics. Generally, a weight/response relationship is determined using either methane or propane with the balance gas being hydrocarbon free air.

Nonmethane hydrocarbon data obtained in relatively clean atmospheres are always subject to large uncertainties. This is due to the subtractive methodology used to determine the nonmethane hydrocarbon concentration. The lower detection limit for nonmethane hydrocarbons using this method is in the 0.1–0.2 ppm range. However, any values below 0.5 ppm are probably subject to considerable error.

SUGGESTED READING

Lonneman, W. A., Ambient Hydrocarbon Measurements in Houston, *Ozone/Oxidants Interactions with the Total Environment II,* Air Pollution Control Association, Pittsburgh, PA, 1979.

3.4 Individual Hydrocarbons in the C_2—C_{12} Molecular Weight Range

Individual hydrocarbons are determined using gas chromatographic techniques. Routine analyses generally employ a flame ionization detector. In order to facilitate hydrocarbon identification, a gas chromatograph–mass spectrometer combination has been employed. At the present time, there is not a single GC column that will resolve all species in the C_2—C_{10} molecular weight range. Two or three column systems must be used to identify and provide quantitative information for hydrocarbons in ambient atmospheres. The two component systems generally employ a column containing a chemical bonded packing (e.g., *n*-Octane/Porasil C) for C_2—C_5 hydrocarbons and a glass capillary with a nonpolar coating (e.g., SE-30) for hydrocarbons in the C_5—C_{12} range. The three column systems are designed to provide quantitative information for each of the three main hydrocarbon groups—paraffins, olefins, and aromatics.

Since individual hydrocarbons are usually present in ambient atmospheres at low ppb levels, a concentration step prior to analysis is necessary. This is best accomplished cryogenically by inserting a freezeout loop and gas sampling valve in the GC carrier gas line. A measured volume of the ambient sample is passed through the freezeout loop which is cooled with liquid oxygen. Organic species condense in the loop and are retained while most of the oxygen and nitrogen pass through. The contents of the loop are transferred to the GC column via the gas sampling valve. Concentration of the organics in 500 ml of air provides a lower detection limit of about 0.1 ppb C for individual species.

Since an ambient sample may contain more than 100 individual hydrocarbons, it is impractical to determine a GC weight/response relationship for each species. The usual practice is to determine this relationship for one or more of the most prominent types of hydrocarbons and assume that the others respond in an identical manner. Preparation of accurate hydrocarbon standards in the low ppb range is not an easy task. Injection of a measured volume of a liquid hydrocarbon into a known gas volume followed by successive dilutions provides one method for preparing low concentration standards. Certified hydrocarbon gas standards in the low ppm range can be purchased from specialty gas suppliers. The useful lifetime of the purchased standards varies depending on the type of hydrocarbon. Periodic comparisons with standards prepared by the successive dilution method should be performed.

Hydrocarbon standards that are traceable to a NBS standard are not readily available. NBS provides a certified propane standard in the 3-ppm range. Since this is two or three orders of magnitude higher than levels recorded in ambient atmospheres, procedures for running the standard and ambient samples must be different. A much smaller volume of the standard must be used and instrument range settings must be changed. Also, propane is not an ideal standard for GC systems designed to measure hydrocarbons in the C_5—C_{12} molecular weight range.

Individual hydrocarbon analysis is best classified as a research technique at the present time. The cost of instrumentation and the technical expertise required to perform the analyses restrict its use as a routine monitoring procedure. Despite these restrictions the method is rapidly gaining acceptance because it provides more reliable nonmethane hydrocarbon data and has the additional advantage of supplying species information. The paraffinic, olefinic, and aromatic content of an air mass needs to be known for modeling exercises that employ air quality simulation models.

SUGGESTED READING

Holdren, M. W., Westberg, H. H., and Zimmerman, P. R., Analysis of Monoterpene Hydrocarbons in Rural Atmospheres, *J. Geophysical Research* **84**, 5083 (1979).

Lonneman, W. A., Kopczynski, S. L., Darley, P., and Sutterfield, F. D., Hydrocarbon Composition of Urban Air Pollution, *Environmental Science and Technology* **8**, 229 (1974).

Singh, H. B., *Guidance for the Collection and Use of Ambient Hydrocarbon Species Data in Development of Ozone Control Strategies,* EPA-450/4-80-008, U.S. Environmental Protection Agency, Research Triangle Park, NC, 1980.

Westberg, H. H., Rasmussen, R. A., and Holdren, M. H., Gas Chromatographic Analysis of Ambient Air for Light Hydrocarbons Using a Chemically Bonded Stationary Phase, *Analytical Chemistry* **46**, 1852 (1974).

3.5 Oxygenated Hydrocarbons

Many types of oxygenated hydrocarbons are present in ambient atmospheres. Aldehydes have received considerable attention because of their role in oxidant production. Lower molecular weight ketones (e.g., acetone, methylethylketone, etc.) are used as solvents and consequently enter the atmosphere as evaporative emissions. Higher molecular weight ketones such as camphor are emitted by vegetation. Organic acids such as formic, acetic, and so on have been identified in urban atmospheres. Esters, alcohols, and ethers are all used in the solvent industry and certainly enter the atmosphere by evaporative processes.

Atmospheric concentrations of the oxygenated species are generally very low. This coupled with their polar nature makes analysis difficult. Techniques that involve absorption in an appropriate solution have proved most useful for aldehydes. Ketones, esters, and ethers can be monitored using gas chromatographic procedures. Formic acid has been measured by long-path IR.

3.5.1 Aldehydes

There are several analytical methods available for monitoring formaldehyde. The chromatropic acid procedure involves passing an ambient air stream through a water impinger for time periods up to 24 hr. An aliquot of the bubbler solution is combined with a mixture of chromatropic acid and sulfuric acid. The resulting colored solution is read at 580 nm in a spectrophotometer. A similar method involves passing the air stream through a solution of 3-methyl-2-benzothiazolone hydrazone hydrochloride (MBTH) solution which produces an azine. When oxidized with ferric chloride–sulfuric acid solution a blue color is developed which can be measured at 628 nm. The lower sensitivity limit for these methods is about 0.1 ppm.

Recently, HPLC analysis has been shown to provide a means of measuring ambient aldehyde concentrations. An air stream is passed through a solution containing 2,4-dinitrophenyl hydrazine. Hydrazones of the various low molecular weight aldehydes are separated on the HPLC column and detected by UV.

SUGGESTED READING

Intersociety Committee, *Methods of Air Sampling Analysis,* 2nd edition, American Public Health Association, Washington, D.C., 1977.

Kuntz, R., Lonneman, W., Namie, G., and Hull, L. A., Rapid Determination of Aldehydes in Air Analyses, *Analytical Letters* **13**, 1409 (1980).

Lowe, D. C., Schmidt, U., Ehhalt, D. H., Frishkorn, C. G. B., and Nürnberg, H. W., Determination of Formaldehyde in Clean Air, *Environmental Science and Technology* **15**, 819 (1981).

3.5.2 Ketones

Very few ketone measurements in ambient air have been reported. In most cases, the concentrations are expected to be very low. Gas chromatography and/or HPLC should provide the most sensitive and selective methods for determining ambient ketone concentrations. All ketones will respond in a flame ionization detector; however, the relative response will vary depending on the carbon–oxygen ratio. Chromatography on fused silica or glass capillary columns provides good results for the majority of ketones in the C_3—C_{10} molecular weight range. HPLC techniques would be similar to those described for aldehydes. A carbonyl derivative can be formed by absorption in a derivatizing solution followed by analysis with the HPLC system.

3.5.3   Organic Acids

Organic acids are very polar species and measurement procedures that require a chromatographic step are expected to be troublesome. Resolution of acids on all except very specialized gas chromatographic columns is poor. Formic acid has been measured in ambient Los Angeles air by FT-IR spectroscopy.

SUGGESTED READING

Hanst, P. L., Wilson, W. E., Patterson, R. K., Gay, B. W., Chaney, L., and Burton, C. S., *Proceedings of the 167th Meeting of the American Chemical Society, Environmental Chemistry No. 55*, 1974.

3.5.4   Alcohols and Phenols

Gas chromatography has been employed to measure ambient methanol concentrations in urban atmospheres. A GC procedure has also been reported for phenols. Quantitative limits are not well defined for alcohol or phenol analysis.

SUGGESTED READING

Bellar, R. A., and Sigsby, J. E., Direct Gas Chromatographic Analysis of Low Molecular Weight Substituted Organic Compounds in Emissions, *Environmental Science and Technology* **4**, 150 (1970).

Hoshika, Y., and Muto, G., Gas-Liquid-Solid Chromatographic Determination of Phenols in Air Using Tenax-GC and Alkaline Pre-Columns, *J. Chromatography* **157**, 277 (1978).

3.5.5   Esters and Ethers

Organic esters and ethers in ambient air can be measured by gas chromatography. GC techniques utilized for pure hydrocarbons can be applied for these two classes of oxygenated species.

4   ANALYTICAL PROCEDURES FOR MONITORING HALOCARBONS

Ambient halocarbon concentrations are generally measured using gas chromatographic techniques. Effluent from the GC column is passed into an electron capture detector or a mass spectrometer for quantitative analysis. The GC–electron capture (GC–EC) method is by far the most widely used procedure. Commercially available instruments can be automated such that halocarbon measurements can be made on nearly a real-time basis (~15-min intervals). With the GC–mass spectroscopy (GC–MS) procedure, samples are collected in a rigid metal container and taken to a permanent laboratory for analysis. Instrumentation costs are much greater for GC–MS systems. The main advantages of a GC–MS analysis is that it provides a better detection method than the conventional GC–EC for some of the halocarbons.

Fluorocarbon-11, fluorocarbon-113, chloroform, methyl chloroform, and carbon tetrachloride can be measured via a one column GC–EC analysis by employing a ¼-in. stainless steel column packed with SF-96 on Chromosorb W. An injection of 5 ml provides lower detection limits (twice noise level) as follows: F-11, 0.7 ppt; CH_3CCl_3, 6 ppt; CCl_4, 2 ppt. Fluorocarbon-12 is best separated on a Porasil B column and is commonly analyzed in conjunction with nitrous oxide. The chlorinated hydrocarbons CH_3Cl and CH_2Cl_2 do not respond as well as other halocarbons in the EC detector and consequently are often measured using GC–MS procedures. Columns that require low flow rates so that the entire column effluent can be transferred to the ion source of the mass spectrometer have proved to be the most useful. Using a $\frac{1}{16}$-in. column packed with Durpak *n*-Octane/Porasil C and a cryogenic concentration procedure (100 ml ambient air), lower detection limits for CH_3Cl and CH_2Cl_2 are about 35 and 10 ppt, respectively.

It has recently been demonstrated that doping of the carrier gas in GC–EC systems enhances the detectability of halocarbons with one or two chlorine or bromine atoms. Thus, CH_3Cl, CH_2Cl_2, and CH_3Br can be measured with a GC–EC system that has been modified by adding oxygen to the carrier gas. Ambient methyl chloride concentrations in the sub-ppb range can be determined using this technique. Vinyl chloride has also been measured by the doped GC–EC technique. In this case, nitrous oxide was used as the doping gas.

SUGGESTED READING

Cronn, D. R., Rasmussen, R. A., Robinson, E., and Harsch, D. E., Halogenated Compound Identification and Measurement in the Troposphere and Lower Stratosphere, *J. Geophysical Research* **82**, 5935 (1977).

Goldan, P. D., Fehsenfeld, F. C., Kuster, W. C., Phillips, M. P., and Sievers, R. E., Vinyl Chloride Detection at Sub-Parts-Per-Billion Levels with a Chemically Sensitized Electron Capture Detector, *Analytical Chemistry* **52**, 1751 (1980).

Grimsrud, E. P., and Miller, D. A., Oxygen Doping of Carrier Gas in Measurement of Halogenated Methanes by Gas Chromatography with Electron Capture Detection, *Analytical Chemistry* **50**, 1141 (1978).

Grimsrud, E. P., and Rasmussen, R. A., The Analysis of Chlorofluorocarbons in the Troposphere by Gas Chromatography–Mass Spectrometry, *Atmospheric Environment* **9**, 1010 (1975).

Singh, H. B., Salas, L. J., and Cavanagh, L. A., Distribution, Sources and Sinks of Atmospheric Halogenated Compounds, *J. Air Pollution Control Association* **27**, 332 (1977).

5 ANALYTICAL PROCEDURES FOR MONITORING CO AND CO₂

Ambient carbon monoxide concentrations can range from about 50 ppb to nearly 50 ppm, depending on the environment. Three different analytical methods have evolved for monitoring CO. The method of choice is dependent on the expected CO concentration. In urban atmospheres where levels are generally above 1 ppm, nondispersive infrared (NDIR) analyzers are commonly employed. Gas chromatographic techniques are applicable for urban monitoring, as well, and are better suited for determining CO concentrations in the low and sub-ppm range. Global background levels of carbon monoxide (<100 ppb) are best determined by a photometric method that involves liberation of mercury vapor from mercuric oxide and subsequent excitation of the vapor with UV light.

Ambient carbon dioxide concentrations are considerably higher than carbon monoxide and can be easily measured by either the nondispersive infrared or gas chromatographic method.

5.1 Carbon Monoxide

The federal reference method for CO analysis is nondispersive infrared spectrometry. The instrumentation consists of an IR light source and two cells. The signals from the reference and sample cells are electronically balanced with an inert gas (N_2, He) in the reference cell and CO free air in the sample cell. When an air sample containing carbon monoxide is admitted to the sample cell, IR radiation is absorbed which reduces the temperature and pressure in the sample cell. This displaces a diaphram the degree of which can be related to the CO concentration. This monitoring technique is very sensitive to temperature and humidity changes. The lower detection limit is approximately 1 ppm.

The CO detection limit can be lowered by about a factor of 10 if gas chromatographic techniques are employed. A measured volume of ambient air is transferred from a sample loop to a GC column capable of separating carbon monoxide from methane and CO_2. After elution from the GC column, the CO is passed through a reduction catalyst which converts it to CH_4. The CH_4 then passes through a flame ionization detector for quantitative analysis. The conversion efficiency to methane and CO absorption losses during chromatography dictate the lower detection limit of this technique. Generally, this varies between 0.100 and 0.300 ppm.

Free tropospheric carbon monoxide concentrations are generally below 0.100 ppm. The preferred method for measuring CO in this type of clean environment is to employ the photometric detection of mercury vapor liberated by the reaction of HgO with CO.

SUGGESTED READING

Seiler, W., The Cycle of Atmospheric CO, *Tellus* **26**, 117 (1974).

Villalobos, R., Role of Gas Chromatography in Air Pollution Monitoring, *Analytical Methods Applied to Air Pollution Measurements,* Ann Arbor Science, Ann Arbor, MI, 1974.

5.2 Carbon Dioxide

The nondispersive infrared and GC techniques described for CO analysis can be applied for ambient CO_2 analysis. Ambient carbon dioxide concentrations are generally high enough to allow the use of a thermal conductivity detector instead of the FID. This facilitates the analysis because a catalytic reduction step is not required. The CO_2 can be separated from other volatile gases on a molecular sieve column and analyzed directly with the thermal conductivity detector. The lower limit of detection is approximately 250 ppm. NDIR and GC–FID procedures provide much lower detection limits.

6 ANALYTICAL PROCEDURES FOR MONITORING OZONE

Through the past 30 yr, several procedures have been developed for measuring ambient oxidant concentrations. One of the first involved exposing a stretched rubber membrane to ambient air and examining the degree of cracking due to oxidation. More quantitative procedures were developed during the 1960s which involved passing an ambient air stream through a solution containing iodide ion. The I^- was oxidized to I_2 which could be determined spectrophotometrically. An automated method which employed a variation of this principle was developed for continuous monitoring. In solution, iodine combines with I^- to form the triiodide ion (I_3^-) which can be converted back to iodide ion electrically. Ozone concentrations in the air stream can be related to the amount of electrical current required. These early monitoring methods have been replaced by more sensitive and specific techniques and consequently are now seldom used.

The Federal Reference Method for ozone is an automated procedure based on the chemiluminescent measurement principle. An ambient air sample containing ozone is mixed with ethylene in a reaction chamber. Electronically excited formaldehyde is formed which emits light at $\sim$435 nm when it loses excess energy. Intensity of the chemiluminescent signal is proportional to the ozone concentration. Commercially available instruments have a lower sensitivity limit of $\sim$0.005 ppm.

UV photometry is rapidly gaining acceptance for ambient ozone monitoring. The EPA classifies it as an equivalent method. Concentration of ozone is determined from a measurement of the amount of UV light (254 nm) absorbed as the sample flows through an absorption cell. Commercially available instruments have a lower detection limit of about 0.005 ppm. Thus, the important performance specifications for UV and chemiluminescent measurements are nearly identical. Each method has certain advantages and disadvantages. The UV method requires only electrical power while the chemiluminescent method requires power plus a reactive gas (ethylene). Since ethylene is used in excess, the exhaust gases must be properly vented or combusted. This is especially true if hydrocarbon measurements are made at the same location. The chemiluminescent instruments are generally more rugged and consequently will withstand rougher treatment. Chemiluminescent instruments have a shorter response time than the UV monitors. These latter two factors are inconsequential for normal ground level sampling programs but can be very important when selecting the proper instrument for airborne monitoring. To date, chemiluminescent instruments have been used almost exclusively in aircraft sampling programs.

SUGGESTED READING

U.S. Environmental Protection Agency, Revisions to the National Ambient Air Quality Standards for Photochemical Oxidants and Calibration of Ozone Reference Methods, *Federal Register* **44**, 8202 (1979).

7 ANALYTICAL PROCEDURES FOR MONITORING INORGANIC NITROGEN COMPOUNDS

Analytical procedures for monitoring the oxides of nitrogen, ammonia, and nitric acid will be summarized in this section. Nitric oxide and nitrogen dioxide contribute to photochemical smog production. Consequently, urban areas that experience photochemical oxidant problems routinely monitor ambient NO_x levels. Nitrogen dioxide has been classified as a criteria pollutant with the National Air Quality Standard set at 0.05 ppm (annual average). Automated monitoring methods are available which provide reliable NO_x data in urban areas. However, commercial NO_x monitors do not have sufficient sensitivity for measuring global background levels of the nitrogen oxides.

7.1 Nitric Oxide

Chemiluminescence is the preferred detection technique for monitoring ambient NO levels. Commercial instruments are available which are automated and provide a continuous record of ambient NO concentrations. Nitric oxide in the ambient sample is mixed with ozone in a reaction chamber. Excited NO_2 molecules are formed which emit light in the 600–2400 nm range upon their return to ground state. The light generated is proportional to the concentration of NO present in the air sample.

The reaction of unsaturated hydrocarbons with ozone is the only potential interference and this can be eliminated by removing chemiluminescent light emissions below 600 nm with an optical filter. The lower sensitivity limit varies from about 1 to 5 ppb depending on the instrument and application. Since nitric oxide levels in clean background regions are expected to be less than 1 ppb, commercially available instruments are not acceptable for this type of measurement. Several research groups have or are currently trying to improve the sensitivity of these instruments so that NO concentrations of 0.1 ppb can be accurately measured.

SUGGESTED READING

Helas, G., and Warneck, P., Background NO$_x$ Mixing Ratios in Air Masses over the North Atlantic Ocean, *J. Geophysical Research* **86**, 7283 (1981).

7.2 Nitrogen Dioxide

The Federal Reference Method for NO$_2$ consists of a quantitative conversion to NO followed by measurement of the NO by chemiluminescence. Thermal and chemical methods have been used to convert nitrogen dioxide to nitric acid. NO$_2$ can be quantitatively converted to NO when heated to between 600 and 800°C in the presence of a metallic surface. The chemical conversion requires temperatures in the 200–400°C range and the presence of a chemical element or alloy. Molybdenum is commonly employed in commercially available instruments. Interferences from other nitrogen-containing species in the atmosphere can cause serious problems with the NO$_2$ analysis. Ammonia is converted to NO at temperatures above 600°C while PAN, organic nitrates, and amines can be converted to NO at temperatures in the 200–400°C range. The ammonia interference is eliminated by the use of converter temperatures below 600°C but other nitrogen compounds can interfere. Fortunately, most of these compounds are present in very low concentrations and in most cases make an insignificant contribution to the measured NO$_2$.

The automated instruments are engineered such that NO$_2$ concentrations are measured by difference. The instrument records the NO concentration and the NO$_x$ (NO + NO$_2$) concentration as separate outputs. NO$_2$ is then determined by subtracting the NO from NO$_x$.

7.3 Nitrous Oxide

Nitrous oxide has received interest in recent years because of the effect it can have on chemistry of the lower stratosphere. Emission sources are primarily natural. Since nitrous oxide is relatively inert in the troposphere, it is not an important contributor to urban pollution problems.

Nitrous oxide is measured by gas chromatography using an electron capture detector. Automated systems have been described which employ a fixed volume sample loop, micro gas-sampling value, and a value-minder for injecting samples into the gas chromatograph. N$_2$O is separated on a Porasil column at 50°C. Minimum detectable limit is a few ppb which is well below the N$_2$O tropospheric background concentration of approximately 300 ppb.

SUGGESTED READING

Pierotti, D., and Rasmussen, R. A., The Atmospheric Distribution of Nitrous Oxide, *J. Geophysical Research* **82**, 5822 (1977).

Singh, H. B., Salas, L. J., and Shigeishi, H., The Distribution of Nitrous Oxide (N$_2$O) in the Global Atmosphere and the Pacific Ocean, *Tellus* **31**, 313 (1979).

7.4 Ammonia

Manual and automated methods have been developed based on the principle of absorption of ammonia in an acidic medium and formation of a colored complex followed by a colorimetric determination. The usual method involves passing a stream of air through an impinger containing dilute (1 N) sulfuric acid where the ammonia is collected as ammonium sulfate. Colored complexes can be formed by adding Nesslers reagent or sodium hypochlorite and certain aromatic compounds. Ammonium compounds in particulate matter will interfere unless removed by a prefilter. Certain organic and metal ions will interfere with the analysis if present in mg/m^3 concentrations. Formaldehyde is also reported to interfere. Lower sensitivity limits are less than 1 μg/m^3; however, accuracy limits are uncertain. Collection efficiency is known to vary with the ambient concentration.

An automated method has been reported which operates on the principle described above. An air stream passes through a tube coated with citric acid. Ammonia is adsorbed on the acid surface while particulate material passes through the tube and is vented. The adsorbed ammonia is washed from the walls and determined colorimetrically using Nesslers reagent. Precision results with a calibration gas containing 43 μg/m^3 ammonia are reported to be 5%. Ambient ammonia concentrations varying from about 2 μg/m^3 to greater than 15 μg/m^3 were recorded in the Netherlands using this automated technique.

SUGGESTED READING

Bos, R., Automatic Measurement of Atmospheric Ammonia, *J. Air Pollution Control Association* **30**, 1222 (1980).

7.5 Nitric Acid

Gaseous nitric acid can be measured using two different techniques. These include coulometry and chemiluminescence with the latter method being the more versatile of the two.

7.5.1 Chemiluminescent Method

Ambient nitric acid concentrations can be continuously monitored with a modified version of commercially available chemiluminescent oxides of nitrogen analyzers. The NO channel, which in conventional instruments does not include a converter, is modified by the addition of a nylon filter and a converter containing a molybdenum catalyst. The nylon filter quantitatively removes nitric acid. Thus, the nitric acid plus NO_x concentration is recorded in one channel and only NO_x in the other. The instrument's electronics subtract the two channels which provides a measure of the concentration of nitric acid in the sampled air. The lower detection limit is dependent on the noise level of the chemiluminescent instrument. At a time constant of 60 sec, this is generally in the 2–5 ppb range.

SUGGESTED READING

Joseph, D. W., and Spicer, C. W., Chemiluminescence Method for Atmospheric Monitoring of Nitric Acid and Nitrogen Oxides, *Analytical Chemistry* **50**, 1400 (1978).

7.5.2 Coulometric Method

A Mast microcoulomb meter can be adapted for measuring acids by changing reagents, replacing Tygon transfer lines with Teflon, and incorporating a provision for protecting the reagent from the atmosphere. Ozone, the primary source of interference is removed by titrating with ethylene prior to entering the instrument. The apparatus is designed to measure the total acids content and the acid concentration with nitric acid removed. As with the chemiluminescent method described above, nitric acid is selectively removed by passing the air stream through a nylon filter. The nitric acid concentration is obtained by difference. The detection principle involves the reaction between iodate and iodide to form iodine in acid solution. The detection limit of the instrument is about 2 ppb.

SUGGESTED READING

Miller, D. F., and Spicer, C. W., Measurement of Nitric Acid in Smog, *J. Air Pollution Control Association* **25**, 940 (1975).

8 ANALYTICAL PROCEDURES FOR MONITORING ORGANIC NITROGEN COMPOUNDS

Organic nitrogen-containing compounds of primary interest include peroxyacylnitrates, amines, and organic nitrates and nitrites. These species are difficult to detect due to low ambient concentrations and their reactive nature. Monitoring methods are best classified as research techniques which generally involve a gas chromatographic method.

8.1 Peroxyacylnitrates

Peroxyacetylnitrate (PAN) is determined by gas chromatography. Automated systems have been described which include: (1) a sample loop attached to a gas-sampling valve, (2) a valve-minder for automatic injection, (3) carbowax column for separating PAN, and (4) an electron capture detector. Continuous operation of this system can provide three PAN readings per hour. Using a 5-ml sample loop, a lower detection limit of about 0.5 ppb can be achieved. Calibration is a major problem with PAN analysis due to difficulty in preparing and storing standards.

SUGGESTED READING

Lonneman, W. A., Bufalini, J. J., and Seila, R. L., PAN and Oxidant Measurement in Ambient Atmospheres, *Environmental Science and Technology* **10**, 374 (1976).

8.2 Amines

A measure of the total primary and secondary amine concentration in a volume of air can be obtained by passing the sample through an absorbing solution of hydrochloric acid in isopropanol. A colored complex is formed by adding ninhydrin and the absorbance is measured with a spectrometer at 575 nm.

9 ANALYTICAL PROCEDURES FOR MONITORING INORGANIC SULFUR COMPOUNDS

Numerous techniques have evolved over the years for measuring H_2S and SO_2 concentrations in ambient atmospheres. Ambient SO_2 concentrations vary from less than 1 ppb in rural areas to more than 1 ppm in industrialized regions. In the United States, the National Ambient Air Quality Standard for SO_2 has been set at an annual arithmetic mean of 20 ppb and 100 ppb maximum 24-hr concentration not to be exceeded more than once per year.

Sulfur dioxide and hydrogen sulfide can be determined simultaneously using gas chromatographic techniques. An air sample contained in a sample loop is transferred to a GC column and then into a flame photometric detector. This method provides maximum sensitivity and selectivity but requires considerable technical competence on the part of the analyst. When attempting to measure these two gases at concentrations in the low or sub-ppb range, adsorptive losses in the sample loop and on the GC column can be significant.

9.1 Sulfur Dioxide

The Federal Reference procedure for monitoring SO_2 is the manual pararosaniline method. There are recognized difficulties with this procedure when it is used in field situations. Consequently, the vast majority of field SO_2 measurements are made by automated techniques which have been classified as equivalent methods by the EPA.

9.1.1 Colorimetric Method

Sulfur dioxide in an air sample is absorbed in a solution of potassium tetrachloromercurate (TCM). The complex that is formed reacts with pararosaniline and formaldehyde to form a colored sulfuric acid derivative which is measured spectrophotometrically. The lower limit of detection is about 10 ppb in an air sample of 30 l. Precautions must be taken when using this method to eliminate interferences from oxides of nitrogen, ozone, and heavy metals. For best results, the absorbing solution must be maintained at subambient temperatures during collection and for extended storage periods.

9.1.2 Conductometric Method

An air stream containing SO_2 is passed through a dilute sulfuric acid–hydrogen peroxide solution. The conductivity of the solution changes due to absorption of pollutants. It is assumed that the conductivity change is due to oxidation of SO_2 to sulfuric acid. This method is nonspecific since any oxidizable pollutant will provide a response. It is primarily employed in field situations where SO_2 is the major pollutant present. The lower sensitive limit of conductometric instruments is about 50 ppb.

9.1.3 Coulometric (Amperometric) Method

This is an automated procedure in which sample air containing SO_2 is scrubbed through a liquid reactant in a detector cell. Bromine is commonly used for the titration of SO_2. The bromine concentration is reduced by reaction with SO_2 which also lowers the redox potential in the cell. This is sensed by an electrode which, via an amplifier, controls an electric current into the solution. The current replenishes bromine by electrolysis of potassium bromide and provides a direct measure of the SO_2 content of the air. This procedure is specific for SO_2 only when other sulfur compounds or species capable of oxidizing bromide to bromine do not exceed 5% of the sulfur dioxide. The optimum operating range for amperometric SO_2 detection is 0.01–2 ppm.

9.1.4 Pulsed Fluorescent Method

Sulfur dioxide molecules are energized by an ultraviolet pulsed light source. Loss of excess energy results in a fluorescent light emission that is specific to SO_2. The emitted light is proportional to the concentration of SO_2 in the sample. Commercially available instruments based on this principle have become increasingly popular in the last few years. Problems due to aromatic hydrocarbon interferences were encountered when pulsed fluorescent instruments were located in close proximity to well traveled roadways. A hydrocarbon "cutter" was developed to eliminate this interference; however, the cutter's performance must be monitored periodically. The lower detectable limit of pulsed fluorescent instruments is about 2 ppb.

SUGGESTED READING

Smith, W. J., and Buckman, F. D., A Performance Test for the Aromatic Hydrocarbon Cutter Used in Pulsed Fluorescent Sulfur Dioxide Analyzers, *J. Air Pollution Control Association* **31**, 1101 (1981).

9.1.5   GC–FPD Method

Sulfur dioxide concentrations in ambient air can be measured using gas chromatographic techniques. An air sample is injected into an appropriate GC column where the SO_2 is separated from other pollutants.

9.2   Hydrogen Sulfide

9.2.1   Colorimetric Method

An air stream containing hydrogen sulfide is aspirated through an alkaline suspension of cadmium hydroxide. The cadmium sulfide precipitate is reacted with N,N-dimethyl-p-phenylene diamine and ferric chloride in an acid solution. This produces a methylene blue that is measured with a spectrophotometer. The collection and storage apparatus must be shielded from light to avoid photodecomposition of the cadmium sulfide. STRactan 10 is normally added to the cadmium hydroxide slurry prior to sampling to minimize photodecomposition. A minimum detectable limit of about 1 ppb can be achieved with a sampling rate of 1.5 l/min for 2 hr.

9.2.2   Lead Acetate Tape Method

If air containing hydrogen sulfide is drawn through a lead acetate impregnated tape, a dark spot results. The hydrogen sulfide concentration is proportional to the optical density of the spot. Fading of the spots during sampling and prior to optical analysis seriously restricts the quantitative utility of this method. It is best employed as a qualitative test for the presence of hydrogen sulfide.

9.2.3   Silver Nitrate Impregnated Filter

Hydrogen sulfide at low concentrations can be selectively adsorbed on a silver nitrate impregnated filter. At concentrations in the low and sub ppb range, H_2S is quantitatively removed from an air stream passing through the impregnated filter at flow rates as high as 60 l/min. A lower detection limit of about 10 ppt has been reported. Sulfide is leached from the filter with a cyanide solution, combined with fluorescein mercuric acetate and the sulfide concentration is measured with a fluorometer. At the present time, this procedure appears to be the method of choice for measuring hydrogen sulfide in background (clean) air environments.

SUGGESTED READING

Natusch, D. R. S., Sewell, J. R., and Tanner, R. L., Determination of Hydrogen Sulfide in Air—An Assessment of Impregnated Paper Tape Methods, *Analytical Chemistry* **46**, 410 (1974).

Jaeschke, W., New Methods for the Analysis of SO_2 and H_2S in Remote Areas and Their Application to the Atmosphere, *Atmospheric Environment* **12**, 715 (1978).

10   ORGANIC SULFUR COMPOUNDS

Gas chromatographic procedures have been described for measuring a large number of sulfur-containing gases which are present in the atmosphere at low ppb levels. Mercaptans, organic sulfides, and organic disulfides as well as carbonyl sulfide and carbon disulfide can be separated on an SE-30 glass capillary column. For optimum performance, the glass column must be deactivated to reduce adsorption losses. The analytical system consists of a cryogenic loop for concentrating the trace sulfur species, a precolumn capillary trap and a gas chromatograph equipped with a flame photometric detector. This technique provides a very sensitive and selective method for measuring the concentration of organic sulfur gases in ambient environments.

An automated GC method has been described for methyl mercaptan and dimethyl sulfide. Air is drawn through a sample loop attached to a multiport switching valve. The valve is automated such that at preset times the contents of the loop are swept onto the GC column. A flame photometric detector is employed for quantitating the sulfur-containing species. The lower sensitivity limit is about 5 ppb; however, accuracy and reproducibility are dependent on a large number of factors including construction materials in the gas handling system, stability of gas flows, and quality of the PM tube.

Total mercaptans can be determined by passing an air stream through an aqueous solution of mercuric acetate–acetic acid. The absorbed mercaptans form a red complex when reacted with an acidic solution of N,N-dimethyl-p-phenylenediamine and ferric chloride. The absorbance is measured with a spectrophotometer at 500 nm. Care must be taken with this procedure to eliminate interferences from hydrogen sulfide and other sulfur-containing compounds. This is accomplished by careful selection of color formation conditions. This method is useful for measuring total mercaptan concentrations in the range from 2 to 100 ppb.

SUGGESTED READING

Farwell, S. O., Gluck, S. J., Bamesberger, W. L., Schutte, T. M., and Adams, D. R., Determination of Sulfur-Containing Gases by a Deactivated Cryogenic Enrichment and Capillary Gas Chromatographic System, *Analytical Chemistry* **51,** 609 (1979).

Intersociety Committee, *Methods of Air Sampling and Analysis,* 2nd edition, American Public Health Association, Washington, D.C., 1977.

CHAPTER 31

PARTICULATE SAMPLING AND ANALYSIS

LESLIE E. SPARKS

U.S. Environmental Protection Agency
Research Triangle Park, North Carolina

1 INTRODUCTION

There are several reasons for sampling particulate matter from smoke stacks. Among the most common of these are sampling to determine compliance with air pollution regulations, sampling to determine whether or not a particulate control device meets design requirements, sampling to develop emission factors, sampling to provide trouble shooting information, and sampling to develop a data base for future reference. In many of these cases large amounts of money depend on the outcome. If the emissions exceed the allowable limit, a plant may be fined, shut down, or forced to operate at reduced capacity. If a device fails to meet design requirements, the vendor may be liable for considerable dollar amounts. Therefore, it is extremely important that the sampling and data analysis that follows the sampling effort be well done.

High-quality particulate data are very difficult to obtain—both because the sampling methods generally leave something to be desired and because the conditions that exist in typical industrial situations are not the best in the world. In spite of these problems, good high-quality data can be obtained if sufficient care is taken. The general requirement is that the same care must be taken as would be taken in an analytical laboratory. This means that all equipment must be calibrated, all procedures written down and checked, and a good quality assurance program be in place. It also means that the people involved in particulate sampling must be well trained and well supervised. Given the stakes that are riding on particulate data, all of these requirements should be insisted upon by anyone who needs or uses particulate data.

The purpose of this chapter is to provide information needed to ensure that the sampling and data analysis are indeed well done. The emphasis of the discussion will be on sampling and analysis associated with mass tests using EPA Test Method 5 and EPA Test Method 17 and particle size distribution tests conducted with cascade impactors. All steps in the sampling and data analysis process will be covered. Considerably more space will be given to cascade impactors than to EPA Test Method 5, mainly because cascade impactors are more difficult to use and data reduction is difficult. Computer programs for reducing mass train and cascade impactor data are presented along with quality control recommendations.

2 STEPS IN PARTICULATE SAMPLING

All of the common particulate measurement techniques, for both total mass and mass as a function of particle diameter, are extractive techniques. This means that a representative sample of the aerosol being sampled must be withdrawn from the bulk aerosol flow and transported to the measuring device. The measurement device actually only separates the particles from the gas. Final measurement is done by weighing the captured particulate matter on a suitable balance.

An overview of the steps required for particulate sampling is given in Table 1. Note that each of these steps must be carried out with extreme care.

Table 1   Steps in Particulate Sampling

1. Gas velocity and composition determination
2. Sample extraction
3. Sample transport
4. Sample collection
5. Sample analysis
6. Data reduction and data analysis

Note that although information on particle mass concentration is desired, the gas flow rate, gas velocity profile, and the composition of the major components of the flue gas stream must be determined first.

3   VELOCITY AND VOLUMETRIC FLOW RATE

Gas velocity and volumetric flow rate are determined by using EPA Method 1 and EPA Method 2.[1] EPA Method 1 specifies the number and location of traverse points and EPA Method 2 specifies the way to make the measurement. An S-type pitot tube is used to make the measurement. The equipment used to make the measurement is shown in Figure 1.

In a traverse the duct or stack is divided into a number of equal areas. The traverse point is located in the center of each area. The sampling site should be located at least eight stack or duct diameters downstream and two diameters upstream from a flow disturbance. If such a site is not available, a location at least two duct diameters downstream and half a diameter upstream from a disturbance may be selected. For a rectangular duct the equivalent diameter D_e should be used.

$$D_e = \frac{2LW}{L + W} \tag{1}$$

L and W are the dimensions of the two sides of the rectangular duct.

The number of traverse points is a function of both the duct diameter and the sampling location. Figure 2 can be used to determine the number of traverse points necessary for a mass sample. Figures 3 and 4 show how the duct should be divided. To use the figure determine the number of duct diameters upstream and downstream from a flow disturbance. Determine the minimum number of traverse points for each case from Figure 2. Select the higher of the two, or a greater value, so that for circular ducts the number of traverse points is a multiple of 4 and for rectangular ducts it is one of the numbers given by Table 2. The location of the traverse points for circular ducts is given in Table 3.

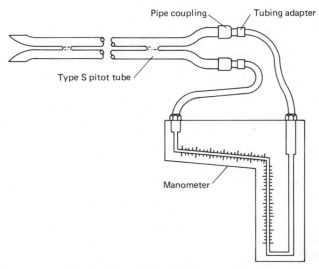

Fig. 1   Pitot tube manometer assembly.[1]

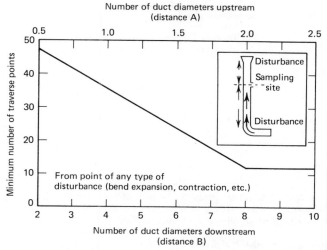

Fig. 2   Minimum number of traverse points for velocity and Method 5 traverse.[1]

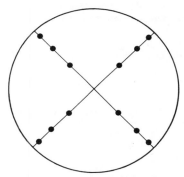

Fig. 3   Cross section of circular stack divided into 12 equal areas showing location of traverse points at centroid of each area.

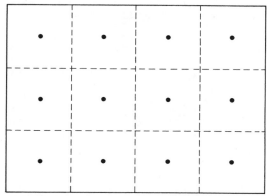

Fig. 4   Cross section of rectangular stack divided into 12 equal areas with traverse points at centroid of each area.

Table 2   Cross-Sectional Layout for Rectangular Stacks[1]

No. Traverse Points	Matrix Layout
9	3×3
12	4×3
16	4×4
20	5×4
25	5×5
30	6×5
36	6×6
42	7×6
49	7×7

Velocity measurements should be conducted at each of the traverse points using the procedures given in EPA Method 2. The data should be recorded on a form such as that shown in Figure 5.

EPA Method 2 also gives details on equipment calibration which is necessary for high quality measurement.

The stack velocity V_s is calculated from

$$V_s = k_p C_p \, [(\Delta p)^{0.5}]_{av} \left(\frac{T_g}{P_s M_s} \right)^{0.5}$$ (2)

where k_p is the pitot tube constant = 34.97 m/s g/g-mole (mm Hg)/°K(mm H_2O), C_p is the pitot tube coefficient, $[(\Delta p)^{0.5}]_{av}$ is the average square root of the velocity head in the stack which is not

Table 3   Location of Traverse Points in Circular Stacks (Percent of Stack Diameter from Inside Wall to Traverse Point)

Traverse Point Number on a Diameter	Number of Traverse Points on a Diameter											
	2	4	6	8	10	12	14	16	18	20	22	24
1	14.6	6.7	4.4	3.3	2.5	2.1	1.8	1.6	1.4	1.3	1.1	1.1
2	85.4	25.0	14.7	10.5	8.2	6.7	5.7	4.9	4.4	3.9	3.5	3.2
3		75.0	29.5	19.4	14.6	11.8	9.9	8.5	7.5	6.7	6.0	5.5
4		93.3	70.5	32.3	22.6	17.7	14.6	12.5	10.9	9.7	8.7	7.9
5			85.3	67.7	34.2	25.0	20.1	16.9	14.6	12.9	11.6	10.5
6			95.6	80.6	65.8	35.5	26.9	22.0	18.8	16.5	14.6	13.2
7				89.5	77.4	64.5	36.6	28.3	23.6	20.4	18.0	16.1
8				96.7	85.4	75.0	63.4	37.5	29.6	25.0	21.8	19.4
9					91.8	82.3	73.1	62.5	38.2	30.6	26.1	23.0
10					97.5	88.2	79.9	71.7	61.8	38.8	31.5	27.2
11						93.3	85.4	78.0	70.4	61.2	39.3	32.3
12						97.9	90.1	83.1	76.4	69.4	60.7	39.8
13							94.3	87.5	81.2	75.0	68.5	60.2
14							98.2	91.5	85.4	79.6	73.9	67.7
15								95.1	89.1	83.5	78.2	72.8
16								98.4	92.5	87.1	82.0	77.0
17									95.6	90.3	85.4	80.6
18									98.6	93.3	88.4	83.9
19										96.1	91.3	86.8
20										98.7	94.0	89.5
21											96.5	92.1
22											98.9	94.5
23												96.8
24												98.9

Plant _____

Date _____

Run no. _____

Stack Diameter, in. _____

Barometric Pressure, in. Hg. _____

Static pressure in stack (P_g), in. Hg. _____

Operators _____

Schematic of stack
cross section

Traverse point number	Velocity head, in. H_2O	$\sqrt{\Delta_p}$	Stack temperature (T_S), °F
	Average:		

Fig. 5   Form for recording velocity traverse data.[1]

the same as the square root of the average velocity head in the stack, T_g is the absolute temperature of the gas, P_s is the pressure in the stack, and M_s is the molecular weight of the stack gas. A computer program to do the calculations is presented later.

The volumetric flow rate determined by Method 2 is usually within ±10% of the true flow rate.

4   GAS COMPOSITION

EPA Test Method 3[1] is used to determine the dry molecular weight of the gas being sampled and the CO_2 and O_2 concentrations of the gas. The measurement is made using an Orsat analyzer. The Orsat analyzer is used to determine the CO_2, CO, and O_2 with N_2 determined by difference in a stack gas. If the stack gas is likely to contain other gases, then steps must be taken to ensure that the concentrations of the major gases are measured.

The operation of the Orsat analyzer is fully explained in the manufacturer's instructions which should be carefully followed.

5   SAMPLE REMOVAL

The first step in the sampling process is obtaining a representative sample of the aerosol being sampled. If this step is not carried out properly, the data will be of poor quality regardless of how well the rest of the sampling is done.

The major factors that make it difficult to obtain a representative sample are:

1. Stratification of the gas flow and/or stratification of the particulate matter in the gas.
2. Misalignment of the probe so that it is not parallel to the velocity of the bulk gas flow.
3. Failure to sample at the isokinetic sampling rate.

With proper sampling procedures, all of these difficulties can be dealt with.

5.1 Stratification

Stratification is handled by sampling at many points in the gas. The procedures for carrying out this multipoint sampling are straightforward when mass emission data are needed. With a mass train, such as the EPA Test Method 5 train, this multipoint sampling is handled by traversing the duct and sampling for specified times at each of the traverse points. As will be discussed later, such traversing is impossible with cascade impactors. Thus multipoint sampling means that single impactor runs must be conducted at each of the traverse points.

The number of traverse points depends on the location of the sampling location. The nearer the sampling point is to a flow disturbance, the more traverse points are required. For mass sampling the number and location of the sampling points are the same as the number and location of the velocity traverse points discussed in Section 3.

Ideally, the same number of sampling points as required for mass sampling should also be used for cascade impactor sampling. However, as is discussed in the section on impactors, traversing a cascade impactor is generally impossible if isokinetic sampling is to be maintained. This means that multipoint sampling with cascade impactors involves a complete cascade impactor run at each of the sampling points. A large number of traverse points would require a large number of complete cascade impactor runs. Such a large number of samples would be impossible to obtain. Thus for cascade impactor sampling a compromise needs to be made. In most cases cascade impactor sampling is done at one or two representative locations in the duct. This inability to traverse a cascade impactor train is one of the major reasons that mass concentrations measured by cascade impactor do not agree with mass concentrations measured by EPA Test Method 5.

Note that the above procedure takes care of problems due to gas flow stratification or nonuniformity. It does not take care of problems that may be caused by particle stratification. Because particles have inertia, and the larger particles have more than the smaller particles, particles can become stratified even if the gas flow is good. Such stratification should be checked for and an appropriate sampling strategy worked out if particle stratification exists.

The problem is especially serious if particle size data are desired for particles larger than 5–10 μm in diameter. Particles larger than this are likely to be adversely affected by flow disturbances, whereas smaller particles will not be affected. Thus the particle size distribution can be quite accurate for the smaller diameter particles and totally wrong for the larger diameter ones.

The stratification problem is especially serious if a full traverse is impossible. Some idea of the possible errors introduced by stratification and single point sampling can be obtained from Table 4.

Table 4   Comparison of Mass Determined
with Single-Point Impactor Measurement
with Full Traverse Mass Train[2]

Plant	Run No.	C_i/C_m
1	1	7.11
	2	2.56
	3	2.27
	4	0.74
	5	0.94
	6	0.95
	7	1.31
2	1	0.66
	2	0.65
	3	0.57
3	1	0.62
	2	0.81
	3	0.50
4	1	0.94
	2	0.55
	3	0.52
	4	0.50

Note that the data were obtained by two different methods, cascade impactors and mass trains, so some of the differences may be due to method differences and not stratification.

Brooks[3] has estimated the error due to stratification is on the order of -40% to $+60\%$ at the inlet of a particulate control device and -60% to $+150\%$ at the outlet of a particulate control device. The only way to reduce this error is by multipoint sampling.

5.2   Isokinetic Sampling

If the gas velocity in the sampling probe is not the same as the gas velocity in the bulk gas, the flow streamlines will be distorted. Because particles have finite size and mass, they will not completely follow the distorted flow streamlines caused by the velocity mismatch. Thus, sampling errors will result because of the velocity mismatch. Figure 6 shows the flow streamlines and the particle trajectories for the three cases of matched velocity, sampling velocity greater than the gas velocity, and sampling velocity less than the gas velocity.

The sampling velocity that matches the gas velocity is called the isokinetic velocity and the sampling flow rate that is achieved with the isokinetic velocity is the isokinetic sampling rate. In order to sample at the isokinetic rate the sampling volumetric flow rate and probe nozzle diameter must be carefully selected.

The sampling nozzle that will give the isokinetic sampling rate for given gas flow may be found from Figure 7.

The errors due to nonisokinetic sampling are a strong function of particle diameter. Stairmand[4] reported the data shown in Table 5. Note that the error due to nonisokinetic sampling becomes small for the 1-μm diameter particles.

Davies[5] reported that the following equation can be used to estimate the error due to nonisokinetic sampling with sharp-edged probes.

$$\frac{C_s}{C_a} = \frac{V_a}{V_s} - \frac{0.5(V_a/V_s - 1)}{K_\rho + 0.5} \tag{3}$$

where C_s is the sampled concentration, C_a is the actual concentration, V_a is the actual velocity, V_s is the sample velocity, and K_ρ is the inertial impaction parameter.

$$K_\rho = \frac{d_p^2 \, \rho \, V_a C'}{9\mu D} \tag{4}$$

where d_p is the particle diameter, ρ is the particle density, μ is the gas viscosity, D is the orifice diameter, and C is the Cunningham correction factor.

Figure 8 is a plot of Eq. (3) for the indicated conditions of gas velocity and particle diameter. Again note that for particles with diameters less than about 5 μm, the error due to nonisokinetic sampling is acceptably small.

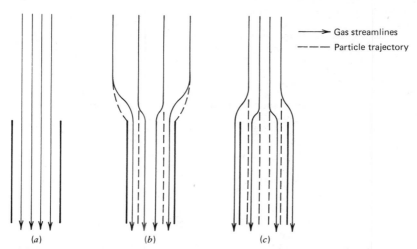

Fig. 6 Gas streamlines and particle trajectories for (a) isokinetic sampling, (b) subisokinetic sampling, and (c) superisokinetic samplíing.

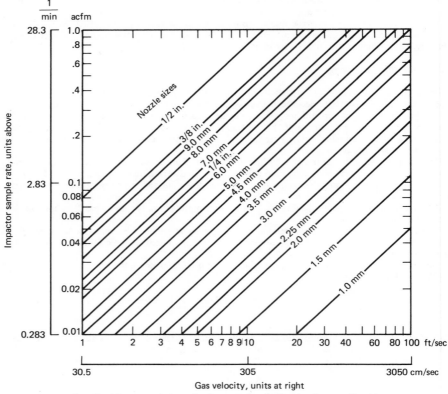

Fig. 7 Nomograph for selecting nozzles for isokinetic sampling.[14]

When the duct is traversed, it is necessary to adjust the sampling rate to maintain isokinetic sampling. This means that the gas velocity at the sampling location must be measured. The EPA Test Method 5 train has a pitot tube located at the sampling point so the operator knows at all times what the gas velocity is. Other sampling trains do not have a pitot tube so the gas velocity at each of the sampling points must be determined with a separate velocity traverse, for example, using EPA Test Method 2.

One should note that all of the discussion of isokinetic sampling and all of the experiments are based on sampling in laminar or very low turbulence flows. This is not the situation that exists in industrial installations. In most industrial situations the flow is highly turbulent. The velocity component parallel to the sampling nozzle can be expected to fluctuate over about ±10% with similar variations in the angle of the flow with respect to the nozzle.

This means that isokinetic sampling in most situations is an averaging process. In all cases the flow turbulence will be large with respect to the sampling nozzle. So in a turbulent gas flow isokinetic sampling can only be approximated.

Table 5   Effects of Nonisokinetic Sampling[4]

Particle Diameter, μm	V_a/V_s	C_s/C_a
100	2	1.99
10	2	1.54
1	2	1.013
100	0.67	0.67
10	0.67	0.82
1	0.67	0.996

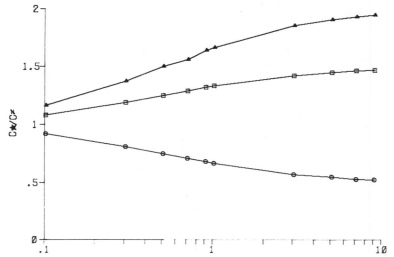

Fig. 8 Effect of anisokinetic sampling on measured concentration.

With the EPA Test Method 5 train the sampling velocity is set to match the stream velocity as measured by a pitot tube. Brooks[3] has shown that for this mode of operation:

Particle mass emission errors at a point will not be larger than the errors in the measured stream velocity as long as the sample velocity is maintained equal to the measured velocity and the sampling probe is properly aligned with the stream.

Note that this statement is for mass measurements at a single point.

5.3 Probe Alignment

Probe misalignment can be a serious problem. Even under isokinetic conditions the measured concentration will be less than the true concentration. Figure 9 shows the effect of probe misalignment on the measured concentration for several conditions. Probe misalignment is not a serious problem if the probe is aligned within 5° of the true gas velocity.

Probe misalignment and failure to sample at or near the isokinetic sample rate are operator errors. These problems are best handled by careful instruction of the sampling personnel. The leader of the sampling team must be sure that everyone in the team understands the need for care. In this way errors due to anisokinetic sampling and probe misalignment can be avoided.

In the final analysis, the error introduced by sample extraction can be held to less than a normalized standard deviation of 5% of the total mass if proper care is taken.

6 SAMPLE TRANSPORT

As soon as the sample is extracted from the gas flow, it must be transported through the sampling probe to the collector. In some sampling trains the distance between the sampling nozzle and the collector is fairly short so probe losses are minimized. In other trains, such as the EPA Test Method 5, the distance between the nozzle and the collector can be very long. In such cases probe losses can be high.

The probe loss problem depends on the sampling method used. If an instack filter or impactor is used, the sample line loss is minimal. However, if an external filter or impactor is used, the sample line loss can become large.

If the measurement is for total mass, sample line loss is not too serious a problem if the sample line is carefully cleaned and if the material contained in the sample line catch is included in the

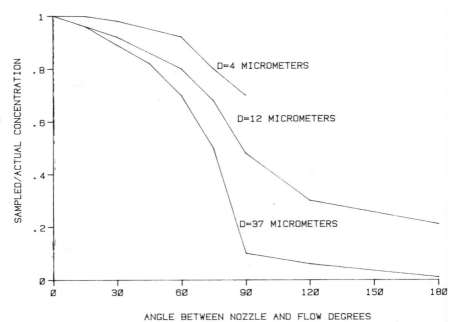

Fig. 9 Effect of probe misalignment on measured concentration under isokinetic conditions after Fuchs.[29]

total mass catch. However, as can be seen from Figure 10, if the particle size distribution data are desired, probe losses can distort the measured size distribution, especially for particles with diameters greater than 10 μm.

7 MASS SAMPLING

Sampling to determine particle mass concentration is generally required to determine compliance with air pollution regulations. Such sampling must follow procedures specified by the air pollution regulations. Failure to follow the specified procedures can invalidate the entire test.

There are two basic mass sampling techniques—one uses an instack filter to capture the particles (EPA Method 17 is an example) and the other uses an out-of-stack filter in a heated filter holder (EPA Method 5 is an example). The two methods may or may not give the same result. Particulate matter is defined by the sampling method and the temperature at which the particles are collected. Thus if the instack and out-of-stack filters are at different temperatures, they may not see the same material as particles. For example, in many combustion sources there are significant quantities of sulfuric acid. If the filter is maintained at a low temperature, the sulfuric acid will condense and be captured as particles. On the other hand, if the filter is maintained at a high temperature, the sulfuric acid will not condense and any sulfuric acid particles formed in the stack will be evaporated. The concentration of particulate matter measured at the two temperatures will be very different.

The choice of which method should be used is given in the air pollution regulations. The two EPA approved sampling methods, EPA Method 5 and EPA Method 17, are discussed below.

7.1 EPA Test Method 5

Particulate testing for determination of compliance with EPA mass emission standards must usually be conducted with EPA Test Method 5, "Determination of Particulate Emissions from Stationary Sources."[1] The exact procedures that must be followed for EPA Test Method 5 are described in the appropriate air pollution regulations. Most states also require some form of the EPA Method 5 train for compliance testing.

Note that with EPA Method 5 particulate matter is defined as whatever is captured in the probe and the filter (the so-called front half of the train). Because the filter is maintained at a temperature different from the temperature of the stack, material that is caught on the filter may or may not exist as particulate matter in the stack. This is a major conceptual difference between Method 5 and EPA Method 17 which uses an instack filter.

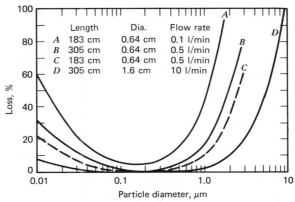

Fig. 10 Probe losses due to settling and diffusion for spherical particles having a density of 2.5 g/cm³ under laminar flow.[2]

With the EPA Method 5 train an isokinetic sample is removed from the duct at specified traverse points. The idea is to obtain an integrated mass sample over the cross section of the duct. The location and number of traverse points is discussed in Section 3 and depends on the location of the sampling ports relative to flow disturbances.

7.1.1 *Description of Sampling Train*

The design specifications of the particulate sampling train used by EPA (see Figure 11) are described in APTD-0581. Commercial models of this train are available. A brief discussion of the train follows. The nozzle is made of stainless steel (316) with sharp tapered leading edge. The probe is Pyrex glass with a heating system capable of maintaining a minimum gas temperature of 120°C at the exit end

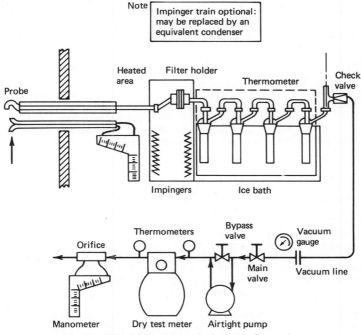

Fig. 11 EPA Method 5 particulate sampling train.

Table 6   Sources of EPA Method 5 Particulate Sampling Equipment

Company	Address
Aerotherm-Acurex	485 Clyde Avenue
	Mountain View, CA 94042
Andersen Samplers, Inc.	4215 Wendell Drive
	Atlanta, GA 30336
Joy Manufacturing Co.	Commerce Road
	Montgomeryville, PA 18936
Lear Siegler, Inc.,	One Inverness Dr. East
Environmental Technology	Englewood, CO 80110
Division	
Misco International	1021 S. Noel Avenue
Chemicals, Inc.	Wheeling, IL 60090
RCA, Inc., Div.	See address above
Andersen Samplers, Inc.	
Scientific Glass and	7246 Wynnewood
Instruments, Inc.	Houston, TX 77001

during sampling to prevent condensation from occurring. When length limitations (greater than about 3 m) are encountered at temperatures less than 315°C, Incoloy 825 or equivalent may be used. Probes for sampling gas streams at temperature in excess of 315°C must have been approved by the administrator prior to their use.

A type S pitot tube or equivalent is attached to the probe to monitor stack gas flow rate. A Pyrex filter holder is used. The filter holder is in a heated box.

The filter is fiberglass and should be at least 99.7% efficient for 0.3 μm diameter dioctyl phthalate particles. The filter should also be unreactive. Good practice requires that a blank be run by sampling filtered flue gas to determine if the filter reacts with the flue gas. The filters are not reused. The filter support media should be glass frit.

Four impingers are connected in series with glass ball joint fittings. The first, third, and fourth impingers are of the Greenberg–Smith design modified by replacing the tip with a 1-cm i.d. glass tube extending to 1 cm from the bottom of the flask. The second impinger is of the Greenberg–Smith design with the standard tip. A condenser may be used in place of the impingers, provided that the moisture content of the stack gas can still be determined. The user may either construct his own sampling train or use one of the commercial models available. Some of the vendors of EPA Method 5 equipment are listed in Table 6.

7.1.2   Equipment Preparation

When the equipment reaches the sampling site, it should be carefully uncrated and inspected. The probe should be carefully checked to be sure that it is clean. All glassware should be inspected for damage and damaged glassware replaced. All pumps are leak tested.

An initial velocity traverse should be made to determine the maximum, minimum, and average flow. This information is used to guide the selection of proper nozzle size for isokinetic sampling.

When the proper nozzle size is known, the sampling train can be moved to the stack and sampling can begin.

7.1.3   Sampling Procedures

Sampling begins with the probe properly positioned at the first traverse point. The pump is started and the flow is adjusted as needed to give isokinetic conditions. The sampling rate should be adjusted as needed to keep up with process variations. When the time for sampling at the point is over, the probe is moved to the next point and isokinetic conditions are reestablished. A data sheet similar to that shown in Figure 12 should be filled out at 5-min intervals during the sampling.

As soon as sampling is completed, the pump is turned off and the probe is carefully removed from the stack. Once the probe is removed from the stack, the train can be disassembled.

The probe and filter should be moved to a clean area before they are disassembled. While the filter and probe are cooling, the amount of water condensed in the impinger train should be measured and recorded.

When the filter holder is cool enough to be handled, the filter should be removed using tweezers. Care must be taken to ensure that all of the filter is removed and that the collected particulate matter is not disturbed. The filter should be placed in a petri dish and then desiccated before it is weighed.

Plant _____

Location _____

Operator _____

Date _____

Run no. _____

Sample box no. _____

Meter box no. _____

Meter ΔH @ _____

C factor _____

Pitot tube coefficient, Cp _____

Barometric pressure _____

Assumed moisture, % _____

Probe extension length, m(ft.) _____

Nozzle identification no. _____

Average calibrated nozzle diameter, cm (in.) _____

Filter no. _____

Leak rate, m³/min, (clm) _____

Static pressure, mm Hg (in. Hg) _____

Schematic of stack cross section

Transverse point number	Sampling time (θ). min.	Vacuum mm Hg (in. Hg)	Stack temperature (T_s), °C(°F)	Velocity head (ΔP_s), mm H$_2$O (in. H$_2$O)	Pressure differential across orifice meter, mm H$_2$O (in. H$_2$O)	Gas sample volume, m³ (ft³)	Gas sample temperature at dry gas meter — Inlet, °C(°F)	Gas sample temperature at dry gas meter — Outlet, °C(°F)	Temperature of gas leaving condenser or last impinger, °C(°F)
Total						Avg	Avg		
Average							Avg		

Fig. 12 Data form for EPA Method 5 data.

797

The probe is cleaned using the procedures in the *Federal Register*.[1] The probe catch is dried and then weighed. The amount of particulate matter collected is the sum of the probe catch and the filter catch (the front half of the train). Because the probe catch can be a significant part of the total catch, great care must be taken in cleaning the probe.

Most EPA regulations require at least three replicates, so the above procedures must be repeated three times to complete the sampling.

7.2 EPA Method 17

The second EPA approved method for mass sampling is EPA Method 17.[6] The primary difference between EPA Method 5 and Method 17 is that an instack filter (shown in Figure 13) is used with Method 17 instead of the heated out-of-stack filter used with Method 5. This means that particulate matter collected by EPA Method 17 is defined as whatever exists as particulate matter under stack conditions. In some cases the measurements made with Method 17 will agree with measurements made with Method 5; and in some cases the two measurements will not agree.

The procedures used with EPA Method 17 are basically the same as those used with EPA Method 5. Isokinetic samples are withdrawn at preselected traverse points. The sampling rate is adjusted to ensure that isokinetic sampling is maintained whenever the traverse point is changed or whenever process conditions change. Data reduction is the same as for Method 5.

The choice of which sampling method should be used depends on the reason for taking the data. In general all sampling should be done using the method required for compliance testing.

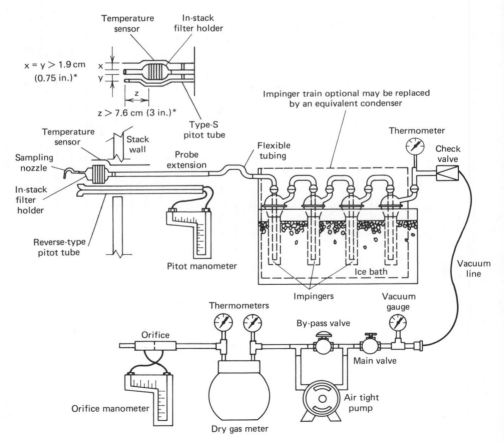

*Suggested (interference-free) spacings

Fig. 13 EPA Method 17 particulate sampling train.[6]

7.3 Mass Efficiency Measurement

Both the inlet and outlet mass concentrations are required to determine the mass collection efficiency of a particulate control device. Because of process variations, the inlet and outlet measurements should be made simultaneously. When the device has a very high efficiency, the inlet and outlet mass concentrations can vary by several orders of magnitude. Such large differences in concentration can make simultaneous sampling difficult. But with proper planning, simultaneous inlet and outlet mass sampling is possible.

7.4 Mass Measurement Calculations

The calculations necessary to reduce mass train data are:

1. Calculation of dry sample volume at standard conditions.
2. Calculation of moisture content of the gas.
3. Calculation of the dry molecular weight of the flue gas.
4. Calculation of the average stack velocity.
5. Calculation of the stack volumetric flow rate.
6. Calculation of the particle concentration.
7. Calculation of mass collection efficiency of the device.

The dry sample volume at standard conditions is

$$V_{std} = \frac{V_m T_{std}}{T_m P_m / P_{std}} \tag{5}$$

where V_m is the meter volume, T_{std} is the absolute standard temperature, T_m is the absolute meter temperature, P_{std} is the absolute standard pressure, and P_m is the absolute pressure at the meter. Any set of consistent units can be used.

The moisture content of the gas B_w is

$$B_w = \frac{V_w}{V_{std} + V_w} \tag{6}$$

where V_w, the volume of water collected, is

$$V_w = V_l R T_{std} \frac{\rho_w}{P_{std} + M_w} \tag{7}$$

V_l is the amount of liquid collected, R is the ideal gas constant, ρ_w is the density of water, and M_w is the molecular weight of water.

The molecular weight of the dry gas is given by

$$M_g = 0.44(\%CO_2) + 0.32(\%O_2) + 0.28(\%N_2 + \%CO) \tag{8}$$

The molecular weight of the wet gas is

$$M_s = M_g (1 - B_w) + 18 B_w \tag{9}$$

The average gas velocity in the stack is given by Eq. (2).

The volumetric flow rate through the stack is

$$Q_s = \frac{A_s V_s (1 - B_w) T_{std} P_s}{T_s P_{std}} \tag{10}$$

A_s is the cross-sectional area of the stack. The concentration of particulate matter at standard dry conditions is given by

$$C = \frac{m_p}{V_{std}} \tag{11}$$

where m_p is the mass of particulate matter collected.

If the efficiency of a particulate control device is to be calculated, the concentrations at both the inlet, C_{in}, and the outlet, C_{out}, must be known. The penetration Pt through the device is C_{out}/C_{in}. The efficiency E is $1 - $ Pt.

Basic language computer programs are presented later to do all of these calculations.

8   PARTICLE SIZE DISTRIBUTION MEASUREMENTS

The most important physical property of an aerosol, at least for air pollution, is its particle size distribution. Many of the important properties of the aerosol, such as its collectability and light scattering properties, are determined by the particle size distribution.[7]

Because the particle size distribution is so important, there is a growing need for good measurements of the particle size distribution from air pollution sources.

Particle size distributions are measured by many different techniques. Each technique covers a limited range of particle diameters as is shown in Table 7. Measurements with cascade impactors will be covered in detail in later sections. Only brief descriptions of the other techniques will be given.

8.1   Particle Diameter Conventions

The definition of particle size must be firmly established before further discussion is possible. The terms particle size or 5-μm size particles have no quantitative meaning. For spherical particles, size can mean diameter or radius. For nonspherical particles, the term size can mean almost anything. In most air pollution work, it is generally assumed that the particles are spheres.

There are several different types of particle diameter that are often used in air pollution. The most common diameters are the physical diameter d, the classical aerodynamic diameter d_C, and the impaction aerodynamic diameter d_A. Rabe[8] has presented a lucid discussion of the various diameters, and their use. The defining equations for the various diameters are shown in Table 8.

The impaction aerodynamic diameter is the simplest of the diameters to use, because it does not require solving for the Cunningham correction factor. Calculation of the other diameters requires solving for the Cunningham correction factor.

$$C(d) = 1 + 2A(l/d) \tag{12}$$

$$A = 1.246 + 0.42 \ \exp[-0.87d/(2l)] \tag{13}$$

where l is the mean free path of air molecules in micrometers which equals $0.0653(760/P)$ $(T/296.2)$, T is the absolute temperature in °K, and P is the absolute pressure in mm Hg. 0.0653 μm is the mean free path of air at 23°C and 760 mm Hg.

The equivalent PSL diameter is also used to describe the diameter measured by light scattering instruments. The equivalent PSL diameter is the particle diameter that has the same light scattering properties of a PSL sphere of the stated diameter. This diameter is used because light scattering instruments are sensitive to the refractive index of light which is generally unknown, because these instruments are calibrated with PSL spheres.

The choice of which diameter should be used depends on the ultimate use of the data. For example, scrubbers are sensitive to the impaction aerodynamic diameter. Thus, d_A should be used in scrubber studies. Electrostatic precipitators are sensitive to the physical diameter of the particle, so d is the appropriate diameter for electrostatic precipitator studies. Whatever diameter is used should be specified.

The equations in Table 8 can be used to convert from one diameter to another.

8.2   Cascade Impactors

The most common instrument used to measure the particle diameter of flue gases is the cascade impactor. The cascade impactor is an inertial instrument which measures the aerodynamic impaction

Table 7   Techniques for Measuring Particle Size Distribution

Technique	Particle Diameter Range
Cascade impactors	$0.3 \ < d > 15 \ \mu$m
Cyclones	$0.3 \ < d > 15 \ \mu$m
Light scattering	$0.5 \ < d > 10 \ \mu$m
Diffusion batteries	$0.01 \ < d > 0.1 \ \mu$m
Mobility analyzer	$0.001 < d > 0.1 \ \mu$m

Table 8 Defining Equations for Various Particle Diameters

Name of Diameter	Symbol	Equation
Classical aerodynamic	d_C	$d_C = d\,[\rho\,C(d)/C(d_C)]^{0.5}$
Impaction aerodynamic	d_A	$d_A = d\,[\rho\,C(d)]^{0.5}$
Physical diameter	d	$d = d_A/[C(d)]^{0.5}$

$C(d)$ = the Cunningham correction for a particle of diameter d
$C(d_C)$ = the Cunningham correction for a particle of diameter d_C
ρ = particle density

diameter d_A. Most commercial cascade impactors provide data over the range $0.5 < d_A > 12\ \mu\text{mA}$. [The usual units for impaction aerodynamic diameter are μmA and $\mu\text{mA} = \mu\text{m(g/cm}^3)^{0.5}$.]

Some special cascade impactors that operate at low pressure are available to provide data for $d_A < 0.5\ \mu\text{mA}$. Although many commercial cascade impactors claim to provide data for $d_A > 12\ \mu\text{mA}$, laboratory calibration data[10] show that the upper limit is close to $12\ \mu\text{mA}$. Probe and nozzle losses also limit the practical upper limit of cascade impactors to about $12\ \mu\text{mA}$.

8.2.1 Theory of Cascade Impactors

A typical cascade impactor has several stages. The larger particles are collected on the upper stages and the finest particles are collected on a backup filter. A typical impactor stage is shown in Figure 14. The particles are accelerated through the jet onto the impaction surface. Because of their inertia, some of the particles will not be able to follow the streamlines and will be collected. The collection efficiency of an ideal cascade impactor stage is shown in Figure 15.

The collection efficiency of a cascade impactor impaction stage is a function of the inertial impaction parameter, K_p.

$$K_p = \frac{d^2 C(d)\rho_p V_j}{9\mu D_i} \tag{14}$$

where V_j is the gas velocity in the jet, μ is the viscosity of the gas, and D_i is the diameter of the jet. Equation (14) can be rewritten in terms of the impaction aerodynamic diameter as

$$K_p = \frac{d_A V_j}{9\mu D_j} \tag{15}$$

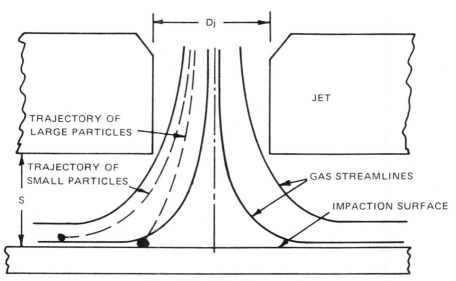

Fig. 14 Typical impactor stage.[2]

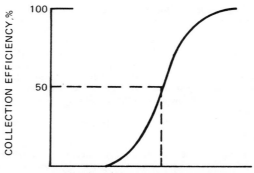

Fig. 15   Ideal impactor efficiency curve.[2]

For the case where the stage efficiency is 50%, that is, the cut point,

$$K_{p\,50} = \frac{d_{A\,50}\,V_j}{9\mu\,D_j}$$
(16)

Although there are theoretical equations for $K_{p\,50}$, experience has shown that the $K_{p\,50}$ must be determined by empirical calibration.

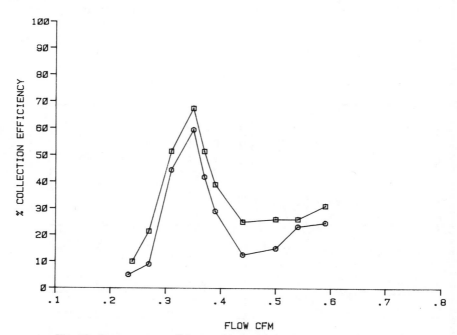

Fig. 16   Impactor stage efficiency with particle bounce bare metal substrate.

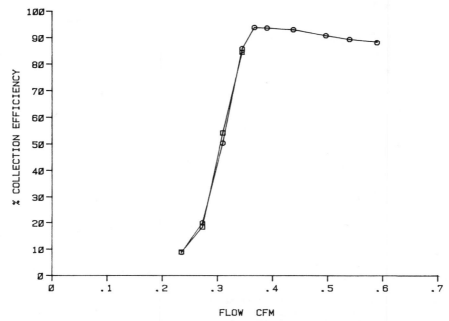

```
MRI - STAGE 4 - 2.93 MICRONS
APIEZON L   - .018 GRAMS ON DISC
OVEN AT 300 DEG F
```

Fig. 17   Impactor stage efficiency with grease substrate to reduce bounce.

8.2.2   *Nonideal Behavior of Cascade Impactors*

The most important nonideal factors affecting cascade impactors are:

1. Particle bounce.
2. Wall losses.
3. Nonideal efficiency curves.

The most serious of the nonideal factors is particle bounce. The basic assumption of cascade impactor use is that particles much larger than the cut point are collected with near 100% efficiency. As can be seen from Figure 16, this can be a poor assumption. The implication of this figure is that if a particle is not collected on the proper stage, it will probably be collected on the final filter. In cases of severe bounce, McCain and McCormack[9] suggest that the final filter should be neglected in the data reduction process.

Particle bounce is made worse if the impactor stage is overloaded with particulate matter. A general rule is that no more than 10 mg should be collected on a stage. Data reported by Esmen and Lee[11] show that if the stage loading is much larger than 10 mg, particle bounce can be a major problem.

The best way to deal with particle bounce is to use a substrate that will prevent bounce. The ability of a suitable substrate to reduce bounce is shown in Figure 17. As will be discussed in Section 8.3.4, substrates can create problems of their own. This means that extreme care must be taken in the selection of substrates.

Cushing et al.[10] have studied the wall losses of many commercial cascade impactors. Some of their results are shown in Figures 18 and 19. These are typical of the wall losses of well-designed cascade impactors. Note that the wall losses are quite low for particles with diameters less than 10 μm. The conclusion of these data is that wall losses are not a problem in well-designed cascade impactors unless data on large particles are desired.

The fact that cascade impactors have nonideal efficiency curves is best dealt with by careful laboratory calibration of cascade impactors before their use. Failure to calibrate the cascade impactors can result in significant and avoidable errors.

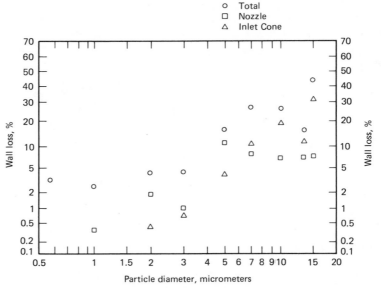

Fig. 18   Wall losses for Andersen stack sampler.[10]

8.2.3   *Cascade Impactor Calibration*

Cascade impactors, the same as other instruments, should be calibrated before they are used and periodically recalibrated with use. Impactors should also be recalibrated whenever visual inspection of the cascade impactor shows that the hole size or shape has changed. Failure to calibrate the impactor can introduce significant errors into size distribution measurements.

The calibration procedure reviewed here is from *Cascade Impactor Calibration Guidelines* by S. Calvert, C. Lake, and R. Parker, EPA Report EPA-600/2-76-118, also available as Appendix B to *Procedures for Cascade Impactor Calibration and Operations in Process Streams,* EPA Report EPA-600/2-77-004. This procedure is designed to calibrate the lower stages of a cascade impactor. Calibration of the upper stages is significantly more time consuming and will be reviewed later in this discussion.

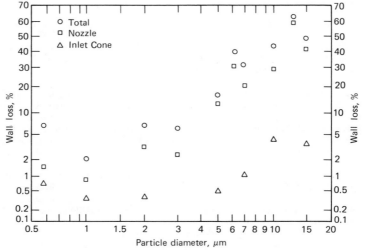

Fig. 19   Wall losses for University of Washington impactor.[10]

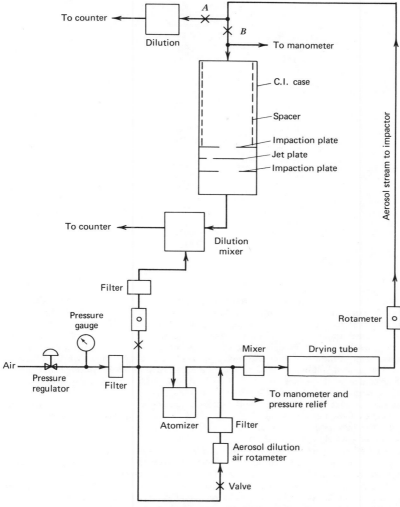

Fig. 20 Cascade impactor calibration setup.[13]

The primary objective of the calibration procedure is to determine the cut point of each impactor stage. The particle size distribution of an aerosol can be calculated from impactor data if the cut points of the stages are known and if particle bounce is minimized.

Because the calibration is intended to determine the cut point of the impactor, calibration efforts for each stage should be concentrated on obtaining data using gas flow rates and particle diameters typical of normal impactor operation. In general, particle diameters from 0.3 to 3 μm are adequate to cover the lower several stages of most commercial cascade impactors.

The experimental equipment for the calibration is shown in Figure 20. The major pieces of equipment are:

1. Aerosol generator for monodisperse particles.
2. Drying tube to remove residual water from particles.
3. Flow measuring equipment.
4. The cascade impactor stage being calibrated.
5. A particle counter to measure the particle concentration entering and leaving the cascade impactor stage.

Table 9   PSL Particles for Impactor Calibrations

Average Diameter, μm	Standard Deviation, μm	Density, g/cm^3
0.087	0.0046	1.05
0.091	0.0058	1.05
0.109	0.0027	1.05
0.176	0.0023	1.05
0.234	0.0026	1.05
0.255	0.0022	1.05
0.312	0.0022	1.05
0.357	0.0056	1.05
0.364	0.0024	1.05
0.481	0.0018	1.05
0.600	0.0030	1.05
0.721	0.0057	1.05
0.801	0.0035	1.05
0.822	0.0043	1.05
1.011	0.0054	1.05
1.101	0.0055	1.05
2.02	0.0135	1.0275
5.7	1.5	1.027

The particles used to generate the aerosol are polystyrene latex (PSL) spheres available from Dow Chemical Corporation. The available particle diameters are shown in Table 9. Note that although large diameter PSL spheres are available, they are generally not suitable for cascade impactor calibration. Problems in generating the aerosol and transporting the aerosol to the cascade impactor stage being calibrated generally make it difficult to use PSL spheres much larger than 3 μm in diameter.

The first step in the calibration is to make up a dilute solution of the PSL by diluting the stock solution. The diluted solution is used to generate an aerosol which is collected and then examined under a microscope. If doubles run more than 2 to 5%, additional dilution is required. This step must be followed for each diameter of PSL used in the calibration.

Once the required PSL dilutions are determined, the calibration can proceed. Each stage should be calibrated over the range of about $0.1 < K_p < 0.3$ to ensure adequate data. The K_p can be varied by varying the flow rate over a small range and by using more than one particle diameter. Both means of varying K_p should be used.

The penetration of particles through the impactor stage is measured using a particle counter. The particle counter is first used to measure the number of particles entering the impactor stage and then used to measure the number of particles leaving the stage. The measurement is repeated and then the flow rate or the particle diameter is changed to give a new K_p and the process repeated. When the desired range of K_p for a given stage has been studied, the process is repeated for the next stage and so on until all the stages of the impactor are calibrated.

The penetration for each K_p is determined from Pt = C_{out}/C_{in} where Pt is the penetration, C_{out} is the particle concentration leaving the impactor stage, and C_{in} is the particle concentration entering the stage. A plot of Pt versus K_p is made for each stage. The $K_{p\,50}$ is determined from the plot.

The calibration of the large particle stages of a cascade impactor is not possible using the above procedure. Cushing et al.[10] have conducted calibrations of the large particle diameter stages of many commercially available impactors. Their procedure consists of generating a monodisperse aerosol of known particle diameter and concentration. The aerosol is then sampled with an impactor. The concentration of particles collected is determined and plots of penetration versus K_p are made.

Typical calibration data for an impactor are shown in Table 10.

8.2.4   Cascade Impactor Substrates and Substrate Problems

As mentioned in Section 8.2.2, cascade impactors must be used with a suitable substrate to reduce the adverse effects of particle bounce. Failure to use a proper substrate can result in totally wrong data which is worse than no data at all.

The common substrates are either greases or filter mats. Each type of substrate has its advantages and disadvantages. Also the type of substrate used affects the collection efficiency curve of the cascade impactor stage.

The effect of substrate on collection efficiency is shown in Figure 21 which compares a grease substrate with a fiberglass filter. Note that the grease gives more ideal behavior than does the filter.

Table 10 Values of Stage Constants for MRI and University of Washington Impactors

Stage	MRI[a] C_i	University of Washington C_i
1	5.4069×10^{-2}	7.4123×10^{-2}
2	6.8611×10^{-2}	9.3954×10^{-2}
3	2.1484×10^{-2}	1.3376×10^{-2}
4	3.8666×10^{-3}	1.7759×10^{-3}
5	1.0147×10^{-3}	1.8378×10^{-3}
6	3.5065×10^{-4}	4.2206×10^{-4}
7	2.2899×10^{-4}	1.0104×10^{-4}

[a] MRI stage constants for stages 4, 5, 6, and 7 are based on EPA/IERL-RTP/Particulate Technology Branch impactor calibration data. University of Washington stage constants and MRI stage constants for stages 1, 2, and 3 are based on Southern Research Institute calibration data.

The fact that the substrate affects the collection efficiency of the impactor means that the impactor must be calibrated using the same substrate or substrates as will be used in the field test. Otherwise, the calibrations are not useful.

The major problems with substrates are weight losses due to handling, weight gains or losses due to reaction with flue gas, and deterioration of the substrate so that it no longer functions.

Weight losses can be a problem with filter substrates if they are not handled properly. Periodic checks of handling procedures should be made by assembling and then disassembling an impactor. If the substrates show any weight change, the handling procedure needs to be reviewed.

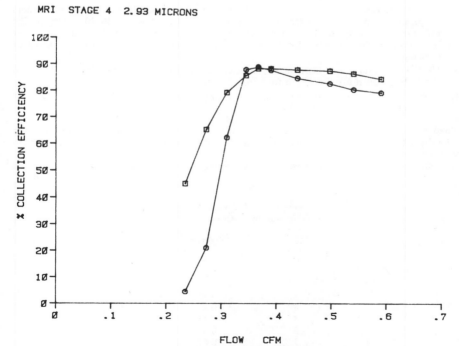

Fig. 21 Effect of substrate on impactor efficiency curve.

Grease substrates are not subject to weight losses due to handling. However, too much grease can result in weight loss due to blowoff. That is, the excess grease is blown off the impactor collection stage. Blowoff is checked for by running blanks.

Substrate weight changes due to reactions with the flue gas can be a major problem with field use of cascade impactors. Both grease and filter substrates are affected by this problem. An example of the severity of the problem of weight gain is shown in Table 11.[12] Note that the weight gain is a significant fraction of normal stage loading which can be less than 1 mg on the lower stages.

Deterioration can be caused by high temperature and/or reactions with the flue gas. If any doubts exist as to the effectiveness of a substrate, use another one. The best test for deterioration is to subject a blank to the flue gas environment and then use the blank as the substrate in a calibration run. If the substrate has deteriorated, the calibration curves would show severe particle bounce.

The only way to find the proper substrate for a given field test is to run blanks. The first step in running a blank is to assemble the cascade impactor just as if a normal run were to be made. Then a filter holder with a double filter is placed in front of the cascade impactor nozzle. Finally, the cascade impactor with filter in front is inserted into the flue gas and normal sampling procedures followed. At the end of the run, the blank cascade impactor is handled the same as a normal impactor. It is disassembled, substrates allowed to dry, and then the substrates are weighed. Any weight changes observed in the blank are due to gas substrate reactions or improper handling.

Because the normal weight change for a cascade impactor stage is at best a few milligrams, especially for the lower stages, any blank weight change is a serious matter. At times the blank weight change can exceed the weight gain expected from particles.

If a suitable substrate cannot be found, then substrate conditioning should be tried. Unfortunately, conditioning is a slow process.

The following procedures for conditioning fiberglass filter substrates are taken from Harris.[14]

Extensive laboratory and filed experiments have been performed to find a suitable filter to be used as impactor substrates. To date the only suitable material is Reeve Angel 934AH available from Whatman, Inc., 9 Bridewell Place, Clifton, NJ 07014. When this material is treated according to the procedure outlined below, mass gains due to reactions with flue gas can be kept to a minimum. Glass fiber backup filters should exhibit the same behavior and should be treated in the same manner.

Procedure for Acid Washing Substrates

1. Submerge the glass fiber substrate in a 50–50 mixture (by volume) of distilled water and reagent grade concentrated sulfuric acid at 100–115°C for 2 hr.

2. When the substrates are removed from the acid bath, they should be allowed to cool to room temperature. Next they are rinsed in flowing distilled water (flow rate 10–20 cm³/min). The pH of the rinse water should be monitored. Rinsing can stop when the pH of the rinse water is the same as the pH of the distilled water. The importance of complete washing cannot be overemphasized.

3. Next the substrates are rinsed in reagent grade isopropanol for several minutes. This step should be repeated five times with fresh isopropanol.

4. Dry the substrates at room temperature and then bake them at 50°C for 2 hr, at 200°C for 2 hr, and finally at 370°C for 3 hr.

Shred two of the substrates and place the pieces in distilled water for about 10 min and then measure the pH. If the pH is lower than that of distilled water, bake the filters at 370°C for 4 hr to drive off residual sulfuric acid.

Even after acid washing, excessive weight gains have been found in some situations. In such cases,

Table 11 Substrate Weight
Gain Due to Reaction[12]

Stage	Weight, mg
0	1.21
1	0.54
2	0.46
3	0.52
4	0.37
5	0.46
6	0.37
7	0.47
8	0.45
Filter	1.03

the acid washed substrates should be subjected to in situ conditioning. In situ conditioning is accomplished by exposing the filters to filtered flue gas for 6 to 24 hr.

After the substrates have been completely conditioned, they should be desiccated and then weighed for use.

Field sampling should not proceed until a suitable substrate is found.

Sometimes, no suitable substrate can be found. In such a case, the blank weight gain (or loss) can be subtracted from the particulate weight gains, provided that the standard deviation of the substrate weight changes is small.

Blank runs to determine a suitable substrate should be run as part of the presite survey. Also at least one blank run should be conducted during each day of testing. The results of the blank run should be reported as part of the impactor data report.

Note that the final filter and mass filters are also subject to weight changes due to flue gas reactions and should be included in the blank runs.

8.2.5 *Impactor Procedures*

The procedures for using cascade impactors are not as well specified as are the procedures for mass sampling. The procedures presented here are based on the experience of many individuals, especially Dr. S. Calvert, B. E. Daniel, Dr. D. S. Ensor, D. B. Harris, G. Ramsey, and Dr. W. B. Smith. Additional details on the procedures for using cascade impactors can be found in several EPA publications[15,16,17] listed in the references.

A typical cascade impactor train is shown in Figure 22.

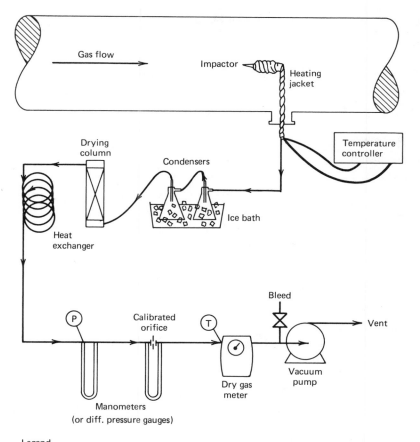

Legend

(P) Pressure measurement point

(T) Temperature measurement point

Fig. 22 Typical sample train with heated instack impactor.[14]

8.2.5.1 Presite Survey. The first step in conducting a cascade impactor test program is to gather as much information as is required to adequately prepare for the test. Minimum data are the identification of special or unusual problems, identification of potential substrates, and the initiation of work on tasks which must be completed prior to the testing (e.g., installation of ports). A form for the presite survey is given by Harris[14] and is reproduced as Table 12.

An impactor run and a blank run should be conducted as part of the presite survey. This impactor run will provide information on likely loading and will be useful in selecting sampling equipment. The blank run will enable selection of substrates.

A site for the impactor setup and disassembly should be located during the presite survey. Also a suitable location for weighing must be located. The weighing site must be free from vibration, which means that many locations in the plant are not suitable. If a suitable location in the plant is not available, the weighing can be conducted in a motel room.

Based on the information collected during the presite survey, the decisions shown in Table 13 can be made.

8.2.5.2 Field Test Procedures. As soon as the equipment is on site, it should be carefully unpacked and inspected for damage. The balance should be set up and field calibrated. A blank run should be made to ensure that the substrate selected is adequate.

A velocity traverse should be conducted to map the flow and locate suitable points to run the impactors. Note that because the flow rate of the impactor cannot be adjusted without affecting the cut points, it is impossible to traverse the duct and maintain the isokinetic sampling rate. In practice this means that impactor sampling is done at a very few points in the duct. Two points is a minimum.

Table 12 Preliminary Survey for Particulate Sizing

Plant Data   Date: _____

Company name: _____

   Address: _____ City: _____ State: _____

Name of contacts: _____ Title: _____

        _____ Title: _____

        _____ Title: _____

Telephone number: _____

Process description: _____

_____

_____

(Operating schedule): _____

_____

_____

(Batch or continuous): _____

_____

_____

(Rate/variability): _____

_____

_____

Air Pollution Control Equipment

Description: _____

_____

(Operating schedule): _____

_____

(Rate/variability): _____

_____

_____

Sketch of Sampling Sites (with approximate dimensions, ports located, upstream and downstream equipment if important).

Table 12 *(Continued)*

Conditions at Sampling Sites

Pressure _____

Temperature _____

Gas rate _____

Gas composition _____

Particulate loading _____

Precutter required? _____

Approx. size dist. _____

Weight gain/loss
by substrates, filter _____

Weight gain/loss
by grease _____

Particulate condition—
hard, sticky, etc. _____

Wet or dry _____

Port size/fitting type _____

Condensation? _____

Notes _____

_____

_____

1. Electricity source
   a. Amperage per circuit _____
   b. Location of fuse box _____
   c. Extension cord lengths _____ Quantity _____
   d. Adapters needed _____
   e. Electrician _____
2. Safety equipment needed

   a. Hard hats _____   d. Safety shoes _____

   b. Safety glasses _____   e. Alarms _____

   c. Goggles _____   f. Other _____
3. Ice
   a. Vendor _____
   b. Location _____
4. Solvents
   a. Vendor _____
   b. Location _____
5. Sampling ports
   a. Who will provide? _____ Welder: _____
   b. Size opening _____
6. Scaffolding
   a. Height _____
   b. Length _____
   c. Vendor _____

   Address _____

   Telephone _____

Table 12   (*Continued*)

7. Distilled water
    a.   Vendor _____

    b.   Location _____
8. Test site facilities
    a.   Parking _____

    b.   Restroom _____

    c.   Laboratory facilities _____

    d.   Cleanup area _____
9. Motels:
    a.   _____ Phone _____ Rate _____

    b.   _____ Phone _____ Rate _____

    c.   _____ Phone _____ Rate _____
10. Restaurants:
    a.   Near plant _____

    _____

    b.   Near motel _____

    _____

11. Airport convenient to plant _____ Distance _____
12. Comments: _____

_____

_____

_____

_____

_____

_____

Survey by: _____

A mass sample should be obtained to help determine the sampling times. Once the velocity traverse and mass sample are completed, the decisions about impactor type, nozzle size, flow rate, and sampling time can be made. Figure 7 can be used to determine the nozzle size and sampling flow rate for isokinetic sampling. Figure 23 can be used to estimate the sampling time. Experience has shown that a 50-mg sample will generally provide weighable quantities of particulate matter on all stages while preventing stage overload on any stage.

Once the decisions concerning impactor type, nozzle size, flow rate, and sample time are made, the impactor train can be assembled. Assemble the train carefully making sure that the substrates (which should have been preweighed) are not damaged during assembly. Handle the substrates with tweezers. Make sure that the impactor is assembled properly and that the O rings and substrates are not cut when the impactor is put together.

The general guidelines shown in Table 14 should be followed. More specific procedures are given in Table 15.

8.2.5.3   Sample Handling and Weighing.   After the impactor run has been completed, the impactor should be transported to the disassembly area. The impactor must be disassembled with care to avoid cutting the substrates and disturbing the particulate deposit.

The impactor should be disassembled one stage at a time. Place the substrate for the stage being disassembled in a petri dish or other container (the same container that it was originally weighed in). Carefully brush any particulate on the bottom of the jet stage into the container. Repeat the procedure for each of the stages.

Desiccate the substrates and then weigh them, recording the results on the data sheet. The balance used for weighing should have a sensitivity of at least 0.05 mg. The weighing procedure should be worked out beforehand and in consultation with the manufacturer of the balance. Proper weighing is of utmost importance.

Table 13 Impactor Decision Making

Item	Basis of Decision	Criteria
Impactor	Loading and size estimate	a. If concentration of particles smaller than 5.0 μm is less than 0.46 g/am³ (0.2 gr/acf), use high flow rate impactor (≈0.5 acfm). b. If concentration of particles smaller than 5.0 μm is greater than 0.46 g/am³ (0.2 gr/acf), use low flow rate impactor (≈0.05 acfm).
Sampling rate	Loading and gas velocity	a. Fixed, near isokinetic. b. Limit so last jet velocity does not exceed: 60 m/sec greased 35 m/sec without grease.
Nozzle	Gas velocity	a. Near isokinetic, ±10%. b. Sharp edged; minimum 1.4 mm ID.
Precutter	Size and loading	If precutter loading is comparable to first stage loading, use precutter.
Sampling time	Loading and flow rate	a. Refer to Section 5.5. b. No stage loading greater than 10 mg.
Collection substrates	Temperature and gas composition	a. Use metallic foil or fiber substrates whenever possible. b. Use adhesive coatings whenever possible.
Number of sample points	Velocity distribution and duct configuration	a. At least two points per station. b. At least two samples per point.
Orientation of impactor	Duct size, port configuration, and size	Vertical impactor axis wherever possible.
Heating	Temperature and presence of condensible vapor	a. If flue is above 177°C, sample at process temperature. b. If flue is below 177°C, sample at 11°C above process temperature at impactor exit external heaters.
Probe	Port not accessible using normal techniques	a. Only if absolutely necessary. b. Precutter on end in duct. c. Minimum length and bends possible.

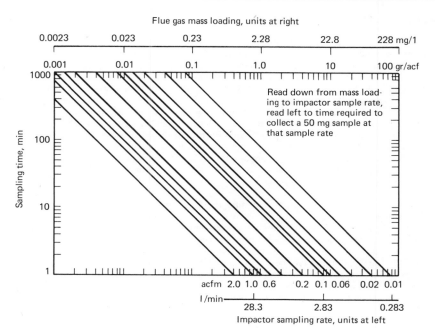

Fig. 23 Time required to get 50-mg sample.[14]

Table 14   Impactor Guidelines

The following procedures should be followed to ensure that cascade impactor data are useful.
1. All impactors should be calibrated following the guidelines in Calvert et al.[13]
2. Bare metal substrates should not be used except in special situations where the particles are sticky.
3. Blank impactors should be run to determine the suitability of a substrate.
4. Grease substrates are preferable to filter substrates.
5. Runs with a weight gain of over 10 mg on any stage should be regarded as suspect.
6. All impactors should be visually inspected before use and any suspect impactors recalibrated.
7. All impactors should be leak tested at temperatures similar to those encountered in field testing.
8. A complete data log should be maintained so that the conditions of each individual run can be reconstructed later.

8.2.6   Impactor Data Reduction

Data reduction for cascade impactors consists of three steps if particle size distributions are desired and six steps if particle collection efficiency as a function of particle diameter is calculated. The steps are:

1. Calculation of dry sample volume at standard pressure and temperature.
2. Calculation of the gas flow rate through the impactor at stack conditions.
3. Calculation of impactor cut points.
4. Calculation of cumulative size distribution.
5. Calculation of differential size distribution.
6. Calculation of collection efficiency as a function of diameter.

The equations for the first two steps are presented in Section 5.2 and will not be repeated here. The mathematics for calculation of the impactor cut point, that is, the diameter that is collected with 50% efficiency, are straightforward.

$$d_{50A} = \frac{C_i}{Q_i \times 10^{-8}\,\mu}\,\mu\text{mA} \tag{17}$$

Table 15   Procedures for Field Use of Impactors[18]

1. Obtain a velocity traverse and stack pressure reading using EPA2.
2. At the selected test point in the duct, compute the isokinetic sampling velocity.
   a. The inlet with a high dust concentration will normally require a small nozzle and low sampling rates. Also short sampling times will be required.
   b. The outlet of a particulate control device will normally have lost dust concentrations and will require a larger nozzle and higher sampling flow rates. Also long sampling times will be required to ensure that weighable quantities of particulate are collected on the stages.
3. The impactor train should be carefully assembled. All fittings and joints should be wrapped with Teflon tape. The whole impactor train must be leak tested before and after the test. This is accomplished by plugging the inlet and turning on the sampling pump. An acceptable leak rate is less than about 2% of the sampling rate at a vacuum of greater than 250 mm Hg.
5. The impactor should be preheated for at least 45 min in the stack. During the preheat the nozzle should be pointed downstream.
6. Fill in the data sheet.
7. Start the run by turning the impactor into the flow, opening the probe valve, and setting the desired flow rate.
8. During the run record the test train parameters in systematic time intervals. (For example, in a 1-hr run recording should be made every 10 min. In a 10-min run recordings should be made every minute.)
9. End the test by shutting off the pump and probe valve. Carefully remove the impactor from the port. Perform the post-test leak test.
10. Remove the hose and drain the impinger.
11. After the probe has cooled somewhat, remove the impactor and transport it to the disassembly area.

where C_i is the empirical stage constant, Q_i is the flow rate for the ith stage in l/min, μ is the gas viscosity in poise, and d_{50A} is the aerodynamic cut point in μmA. C_i for some commercially available impactors are given in Table 16. These values of C_i are based on laboratory calibration of the various impactors. The references for the calibrations are also shown in the table. If empirical calibration data for the specific impactors being used are not available, the stage constants in this table should be used. The use of empirical stage constants for the specific impactors being used is strongly suggested and the use of the data in Table 16 is only a stopgap until the impactors are calibrated.

The impaction aerodynamic diameter can be converted to the physical diameter by trial-and-error solution of the following equations:

$$d = \frac{d_A}{[\rho C(d)]^{0.5}} \, \mu m \tag{18}$$

$$C(d) = l + 2A \frac{l}{d} \tag{19}$$

$$A = 1.246 + 0.42 \exp\left(\frac{0.87d}{2l}\right) \tag{20}$$

$$l = 0.653\left(\frac{760}{P \text{ mm Hg}}\right)\left(\frac{T°K}{296.2}\right) \tag{21}$$

where l is the mean free path of the gas and ρ is the particle density in g/cm³.

Because there is a finite pressure drop across an impactor, the flow rate across each stage must be corrected for this pressure drop. The pressure drop for each stage must be estimated because the pressure drop across the whole impactor is what is measured. Because the gas flow through the impactor is accelerated and decelerated at each stage, it is reasonable to assume that the pressure drop is due to loss of velocity pressure. Therefore, the pressure drop at each stage is related to the square of the jet velocity. So for the ith stage,

$$DP_i = kV_i * V_i \tag{22}$$

where k is a constant. For many of the commercial impactors, the total pressure drop is low and can be neglected for a first cut at data reduction.

The next step in the data reduction process, after the cut points are calculated, is to construct a cumulative less than stated diameter size distribution. This is done by taking the individual stage weights, and is calculated as follows:

$$M_t = \sum_{i=0}^{N} M_i \tag{23}$$

$$f_i = \sum_{j=0}^{j=i-1} M_j M_t^{-1} \tag{24}$$

Note that because this is a cumulative less than size distribution $j = 0$ corresponds to the backup filter, $j = 1$ corresponds to the stage with the smallest cut point, and $j = N$ corresponds to the stage with the largest cut point.

Table 16 Impactor Stage Constants[a]

Stage	MRI C_i	Univ of Wash C_i	Brink Grease C_i	Brink Glass C_i
1	0.054069	0.074123	0.004053	0.003562
2	0.068611	0.093954	0.0008918	0.001262
3	0.021484	0.013376	0.0040528	0.0003711
4	0.0038666	0.0017759	0.0002577	0.00018747
5	0.0010147	0.0018378	0.000046122	0.00009852
6	0.00035065	0.0004221	0.000035934	0.00005547
7	0.00022899	0.00010104	0.000011452	0.000011452

[a] All data for stages 1, 2, and 3 based on work done by Southern Research Institute. Data for MRI stages 4, 5, and 6 from Particulate Technology Branch IERL-RTP. All other data from work done by Southern Research Institute.

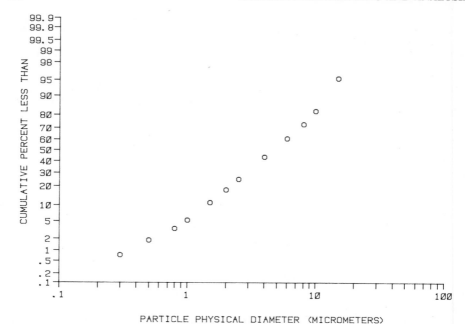

Fig. 24   Typical impactor data plotted on log probability paper.

When the f_i are calculated they can be plotted for display. Experience has shown that in most cases the data tend to give a straight line if they are plotted on log probability paper. See Figure 24.

The next step in the data analysis process is to calculate the differential size distribution curve. This curve provides information on the amount of material in each particle diameter increment. Differentiation of data is a very difficult problem. It is not something that should be done lightly. At present the best way to differentiate the cumulative curve is by use of a spline fit to the data.[19,20]

The final step in the data reduction process is the calculation of the collection efficiency as a function of particle diameter. This is a straightforward calculation performed by dividing the outlet differential size distribution by the inlet size distribution.

A major problem that occurs when the efficiency is calculated is that the inlet and outlet impactors do not give exactly the same cut points, especially at the upper and lower bounds of the size distribution. This means that either the data have to be extrapolated or that efficiency calculations be limited to the overlap region. Sparks[20] has shown that even if the efficiency calculation is limited to the overlap region, serious errors may be introduced into the efficiency calculation if the inlet and outlet impactor cut points are not well matched. The problem of errors due to imperfect match of the inlet and outlet impactors can be overcome if the data are transformed to log normal space and if a spline fit is used to perform the differential calculations and the interpolation between data points.

It is impossible to do all of these calculations by hand in a reasonable time. Fortunately, the recent development of powerful programmable calculators and small microcomputers makes it possible to mechanize all the data reduction. Programs for the Texas Instruments TI-59 Calculator,[20] large computers,[21] and for the Radio Shack TRS-80 microcomputer[22] have been published. Computer programs for the TRS-80 microcomputer are presented at the end of this chapter.

8.2.7   Data Presentation

After all the calculations have been completed and checked, it is time to write the report and present the data. There are numerous ways to present the data, but those suggested here have been found useful and informative.

The first thing to remember is to define what particle diameter is being used. Each table and figure should be labeled to show what particle diameter is being used.

The first set of data to present is the cumulative size distribution curves. The most useful way to present them is as plots of cumulative fraction or percent less than the stated diameter plotted on log probability paper. The total mass concentration should be indicated in the title of the curve.

The next set of data to present is the dM/d log d curves. These curves should be plotted as dM/d log d versus particle diameter on log–log paper. If there is a large amount of data, it is best

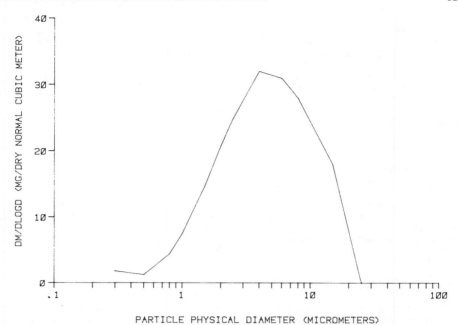

Fig. 25 Typical differential size distribution dM/d log d vs log d for physical diameter.

to present average inlet and outlet cumulative and dM/d log d curves in the body of the report. The curves for each run can be presented in an appendix.

Finally, the penetration as a function of diameter should be presented. These data should be presented as curves of penetration versus diameter. The curves should be plotted on log–log paper.

Examples of all these curves are shown in Figures 24 to 26.

All of the raw data, including stage weights, flow rates, gas temperatures, and so on, should be

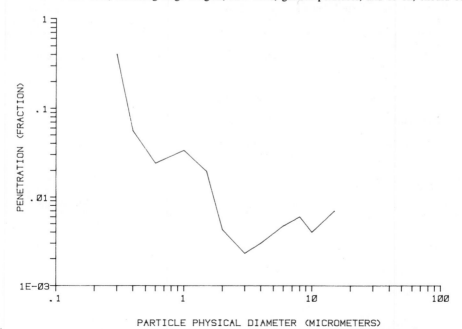

Fig. 26 Typical penetration versus physical diameter curve for impactor data.

presented in an appendix to the report. These data are necessary if someone wants to recalculate the data if, for example, it is found that the impactor calibrations are wrong.

Unless there is an obvious problem with a data set, all of the data should be presented in the appendix. If a data set has too high stage weights, or if a problem with the data set is noted in the field notebook, it should be rejected and not included in the data analysis or the appendix.

Careful attention to how the data are presented can make the data more useful. Of course if the data are to be presented in a paper for publication, it is impossible to include all of the data in the paper. But somewhere there should be a permanent record of all the data.

8.2.8 *Manufacturers of Impactors*

Some manufacturers of cascade impactors are listed in Table 17.

8.3 Other Ways to Measure Particle Size Distribution

Although the cascade impactor is the most common instrument used to measure particle size distribution in flue gases, there are other instruments that are sometimes used. Cyclone sampling trains have been used to collect large amounts of particulate matter in the same diameter range as cascade impactors. Diffusion batteries and mobility analyzers have been used by several groups to obtain data on particles with diameters smaller than 0.5 μm.

All of the instruments mentioned above are special-purpose instruments and will not be discussed in detail here.

8.3.1 *Cyclone Sampling Trains*

Small diameter cyclones have been used to collect large amounts of particulate matter in the same range of particle diameters as covered by cascade impactors. The main advantage of cyclones is that large quantities of material may be collected for chemical or other analysis. The main disadvantage of cyclones is the uncertainty of the cut points. Smith and Wilson[23] have reported on the development of cyclone sampling trains and should be consulted for additional information.

8.3.2 *Diffusion Batteries*

A diffusion battery sizes particles based on their diffusion coefficients. Diffusion batteries may be of parallel plate design, parallel tube design, or of screen design. Particles with high diffusivity are collected in a short time and particles with lower diffusivities are collected in longer times.

A diffusion battery can be constructed as a series of individual diffusion collectors along the line of a cascade impactor. In this case a detector, usually a condensation nuclei counter, is used to measure the penetration through the battery. The smaller particles are collected in the first diffusion stage and the larger particles are not collected at all.

Smith et al.[25] describe a parallel diffusion battery where the flow rate is varied. The penetration

Table 17 Commercial Cascade Impactor Sampling Systems

Manufacturer	Name of Impactor	Substrates
Andersen 2000, Inc., P. O. Box 2076, Atlanta, GA 30320	Andersen Stack Sampler	Glass fiber
Meteorology Research, Inc., Box 637, Altadena, CA 91001	MRI Cascade Impactor	Stainless steel, grease, fiberglass, Mylar, and others
Pollution Control System Corp., 321 Evergreen Bldg., Renton, WA 98055	University of Washington	Stainless steel, fiberglass, grease
Sierra Instruments, Inc., P. O. Box 909, Village Square, Carmel Valley, CA 93924	Sierra Mod 226	Glass fiber
Zoltex Corp., 68 Worthington Dr., St. Louis, MO 63043	Brink	Glass fiber, aluminum, grease

for each flow rate is measured using a condensation nuclei counter. A curve of penetration versus physical diameter is shown in Figure 26.

Diffusion batteries are used in extractive sampling and generally require that the stack aerosol be diluted before sizing.

8.3.3 *Electrical Mobility Analyzer*

The electrical mobility analyzer developed by Liu et al.[25] operates by placing a known electrical charge on the particles and then measuring their mobility. The effective diameter range of the instrument is 0.03–1 μm.

9 ERROR ANALYSIS

9.1 Errors in Mass Sampling

Brooks[3] gives a complete analysis of the errors likely in mass sampling. He concluded that system accuracies of ±16% are possible with commercial state of the art mass trains and a 16-point traverse. In general the major contribution to the error is the mapping uncertainty associated with stratification in the sample plane.

The level of accuracy estimated by Brooks requires that all steps in the sampling be carried out with extreme care. State of the art equipment was defined as a mass train "which would consist of the highest accuracy components readily available commercially. For example it would include a high accuracy differential pressure transducer."

While 16% may be within the state of the art, most commercial mass trains do not, and indeed are not required to consist of the highest accuracy components available. Therefore, Brooks' estimate should be considered as the minimum error possible.

The error with a 16-point traverse with normal commercial mass sampling trains is probably about ±25%. The largest sources of error are due to deviations from isokinetic sampling, misalignment of the probe, mapping errors, and errors in measuring the gas velocity in the duct.

Table 18 lists most of the major sources of error in mass sampling.

In a field test, source variations probably introduce more uncertainty into the measurements than do the measurement techniques themselves.

Table 18 Sources of Measurement Errors for Mass

1. Errors due to probe
   a. Anisokinetic sampling velocity
   b. Probe not parallel to gas flow
   c. Dust collected on walls of probe or sample line
   d. Condensation in probe or sample line
2. Errors due to sample collector
   a. Efficiency less than 100%
   b. Condensation in collector
3. Errors in sample flow
   a. Meter inaccuracy
   b. Leaks
4. Errors in weighing
   a. Hygroscopic loss or regain of sample or filter media
   b. Loss of sample in handling
   c. Balance without adequate precision
5. Errors in integrating traverse
   a. Variations over time or space in dust load or gas velocity
   b. Integration error
   c. Too few traverse points
6. Errors due to personnel
   a. Reading and recording errors
   b. Failure to know and comply with requirements
   c. Calculation errors
7. Errors in data reduction
   a. Errors due to poor choice of data reduction method
   b. Calculation errors
   c. Not enough data

Table 19   Magnitude of Errors Associated with Impactors[25]

Parameter Affected	Source of Error	Magnitude of Error	Comments
d_{50}	Construction, tolerances of jets, secondary flow	Up to 50% for uncalibrated impactor especially for >5 μm	Laboratory calibration can reduce to less than 10%
D_{50}	Measurement of Q, T, P	Usually less than 10%	
Size distribution	Particle bounce	May be large from stage to stage	Sensitive to loading, substrates, and design
Size distribution and mass	Reactions of flue gas with substrates	Up to 100% or more, depending on gas and substrates	Minimize by proper substrate selection and use of blanks
Size distribution and mass concentration	Weighing errors	0.0–0.05 mg with properly used electrobalance	Can be checked in field; error in data depends on loading

9.2   Errors in Impactor Sampling

There are two questions that can be asked of cascade impactor data. The first is how well does the measured particle size distribution agree with the size distribution of the aerosol that was sampled. The second is how well does the measured particle size distribution agree with the particle size distribution of the whole flue gas particle mixture. The second question can be rephrased as how representative of the entire flue gas was the small gas stream that actually entered the cascade impactor.

The errors likely in the size distribution of the aerosol actually sampled are listed in Table 19.[25] In a well-conducted field program, the total error in the particle size distribution to all of these errors can be held to ±25%. This does not mean that the size distribution leaving a stack can be measured to ±25% because of errors in obtaining a representative sample of the flue gas.

If a single-point measurement is made, the error due to stratification is usually ±50% and can be as high as several orders of magnitude. This means that the major source of error in well-conducted impactor sampling is due to stratification or mapping error, caused by too few traverse points. Therefore it is extremely important that the sampling points for cascade impactor measurements be carefully selected. In most situations the error due to stratification can be reduced to acceptable limits by sampling at two or three carefully selected points. At the outlet of a control device, where the concentration of large particles is low, it may be better to forget about isokinetic sampling and to conduct a multipoint traverse. If the deviations from isokinetic sampling are less than 30%, the error due to anisokinetic sampling will probably be acceptable. At the inlet of a control device, several runs should be conducted at various points across the duct.

The accuracy of the measured size distribution depends on the particle diameter range to be covered. If the particle diameter range is limited to aerodynamic diameters between 0.5 and 10 μmA, the particle size distribution for the entire flue gas can probably be defined within ±25% with a careful 2-point sampling program. If information on particles with diameters much larger than 10 μmA is desired, considerable sampling will be needed to define the size distribution within ±100%.

10   QUALITY ASSURANCE

As has been mentioned several times, particulate measurements are very difficult. Laboratory precision and care must be taken in very uncomfortable environments. Thus some form of quality assurance program is necessary to ensure that the proper care is taken. This quality assurance program must be designed and carried out with the same care that a quality assurance program for a precision laboratory would receive.

The first and most important part of the quality assurance program is to make sure that the people doing the field work know and understand the need for careful precise work. If they do, they will avoid many of the errors common in field programs. And that is what a good quality assurance program is all about—preventing the taking of bad data.

The quality assurance program should cover calibration of all equipment on some periodic schedule. EPA has established calibration requirements for all the equipment used in the EPA approved methods for stack sampling. Also equipment should be inspected before and after each use. If the equipment appears to be in less than top condition, it should be replaced or repaired and then recalibrated.

All substrates and filter materials for both mass and impactor sampling should be tested using blanks to ensure that there are no reactions between the flue gas and the substrate or filter. At least one blank should be run during each day's testing. If there is a reaction, a different substrate or filter should be used.

A careful and complete log should be kept of everything that happens in the field. It should be possible to find out everything about the field test by reviewing the log. Strict accountability should be maintained to ensure that the log is maintained. The field supervisor should review the log at least daily.

Each run, including the blank runs, should have its own data sheet. The data sheet should follow the run as the filters or impactors move through the data reduction process. Again a strict chain of custody must be maintained. The data sheet should indicate when the run started, when it ended, and any important things that happened during the run. It should also show who the operators of the sampling equipment were.

The balance should be checked daily with a reference substrate to make sure that it is weighing correctly. Occasional blanks should be passed through the system to see what the weighing error may be. A small percentage of the filters and substrates should be reweighed as a weighing audit. The results of the audit should be checked daily so that corrective action can be taken. It does no good to find that the weighing was bad when the weighing audits are reviewed after you come out of the field.

Preliminary data reduction should be carried out in the field. The main objective of the preliminary data reduction is to make sure that the data appear reasonable. For example, if the preliminary data reduction shows that the inlet mass loading is less than the outlet mass loading, corrective action can be taken. If the error is not found until final data reduction back at the laboratory, the entire test may be of poor quality. The sooner a problem is found, the sooner it can be corrected. Take as much data as possible.

One of the most important features of a quality assurance program is to have written procedures for each operation. Everyone involved in the test should have a copy of the procedures and must understand them. The field supervisor has the responsibility for making sure that everyone has the procedures.

The check sheets in Figure 27 can be used to help in the quality assurance job.

General Administrative

Plant name _____ Date _____

Plant address _____

Source to be tested _____

Plant contact _____ Phone _____

Observers _____ Affiliation _____

_____ _____

_____ _____

_____ _____

Reviewed test program? _____ Comments _____

Reviewed test program meeting notes? _____ Comments _____

Reviewed correspondence? _____ Comments _____

Test team company name _____ Phone _____

Supervisor's name _____ Address _____

Other members _____ Title _____

_____ _____

_____ _____

_____ _____

_____ _____

_____ _____

Fig. 27 Check sheet for quality assurance.[27]

General/Sampling Site

Stack/duct cross section dimensions _____ Equivalent diameter _____

Material of construction _____ Corroded? _____ Leaks? __

Internal appearance: Corroded? _____ Caked particulate? _____ Thickness __

Insulation? _____ Thickness _____ Lining? _____ Thickness __

Nipple? _____ I.D. _____ Length _____ Flush with inside wall? _____

Straight run before ports _____ Diameters _____

Straight run after ports _____ Diameters _____

Photos taken? _____ Of what? _____

Drawing of sampling location:

Minimum information on drawing: stack/duct dimensions, location and description of major disturbances and all minor disturbances (dampers, transmissometers, etc.), and cross-sectional view showing
dimensions and port locations.

General/Sampling System

Sampling method (e.g., EPA 5) _____

Sampling train schematic drawing:

Modifications to standard method _____

Pump type: Fibervane with in-line oiler __X__ Carbon vane __X__ Diaphragm __X__

Probe liner material _____ Heated? _____ Entire length? _____

Type "S" pitot tube: _____ Other _____

Pitot tube connected to: Inclined manometer _____ Or magnetic gauge _____

    Range _____ Approx. scale length _____ Divisions _____

Orifice meter connected to: Inclined manometer _____ Or magnehelic gauge _____

    Range _____ Approx. scale length _____ Divisions _____

Meter box brand ____X____ Sample box brand ____X____

Recent calibration of orifice meter-dry gas meter? _____ Pitot tubes? _____

    Nozzles _____ Thermometers or thermocouples? _____ Magnehelic gauges? _____

Number of sampling points/traverse from Fed. Reg. _____ Number to be used _____

Length of sampling time/point desired _____ Time to be used _____

X—Not required by regulations

Train Assembly/Final Preparations   Run # _____

(Use one sheet per run if necessary)

Filter holder clean before test? _____ Filter holder assembled _____

Correctly? _____ Filter media type _____ Filter clearly identified? _____

Filter intact? _____ Probe liner clean before test? _____

Nozzle clean? _____ Nozzle undamaged? _____

Impingers clean before test? _____ Impingers charged correctly? _____

Ball joints or screw joints? _____ Grease used? _____

Fig. 27 (*Continued*)

Kind of grease _____ Pitot tube tip undamaged? _____

Pitot lines checked for leaks? _____ Plugging? _____

Meter box leveled? _____ Pitot manometer zeroed? _____

Orifice manometer zeroed? _____ Probe markings correct? _____

Probe hot along entire length? _____ Filter compartment hot? _____

Temperature information available? _____ Impingers iced down? _____

Thermometer reading properly? _____ Barometric pressure measured? _____

If not, what is source of data? _____ ΔH@ from most recent calibration _____

ΔH@ from check against dry gas meter _____

Nomograph check:

If ΔH@ = 1.80, T_m = 100°F, % H_2O = 10%, P_s/P_m = 1.00, C = __X__ (0.95)

If C = 0.95, T_s = 200°F, DN = 0.375, Δp reference = __X__ (0.118)

Align Δp = 1.0 with ΔH = 10; @ Δp = 0.01, ΔH = __X__ (0.1)

For nomograph setup:

Estimated meter temperature __X__ °F. Estimated value of P_s/P_m __X__

Estimated moisture content __X__ %. How estimated? __X__

C factor __X__ Estimated stack temperature __X__ °F.

Desired nozzle diameter __X__

Stack thermometer checked against ambient temperature? _____

Leak test performed before start of sampling? _____ Rate _____ cfm @ _____ in. Hg.

Sampling (Use one sheet for each run if necessary)   **Run #** _____

Probe-sample box movement technique:

Is nozzle sealed when probe is in stack with pump turned off? _____

Is care taken to avoid scraping nipple or stack wall? _____

Is an effective seal made around probe at port opening? _____

Is probe seal made without disturbing flow inside stack? _____

Is probe moved to each point at the proper time? _____

Is probe marking system adequate to properly locate each point? _____

Are nozzle and pitot tube kept parallel to stack wall at each point? _____

If probe is disconnected from filter holder with probe in the stack on a negative pressure source, how is particulate matter in the probe prevented from being sucked back into the stack?

_____

If filters are changed during a run, was any particulate lost? _____

Meter box operation:

Is data recorded in a permanent manner? _____ Are data sheets complete?

_____

Average time to reach isokinetic rate at each point _____

Is nomograph setting changed when stack temperature changes significantly? _____

Are velocity pressures (Δp) read and recorded accurately? _____

Is leak test performed at completion of run? _____ cfm @ _____ in. Hg.
General comment on sampling techniques _____

If Orsat analysis is done, was it: From stack? _____ From integrated bag? _____

Was bag system leak tested? _____ Was Orsat leak tested? _____

Check against air? _____

If data sheets cannot be copied, record: approximate stack temperature _____ °F

Nozzle dia. _____ in.   Volume metered _____ acf

First 8 Δp readings _____ _____ _____ _____ _____ _____ _____ _____

Fig. 27 (*Continued*)

Sample Recovery

General environment-cleanup area _____

Wash bottle clean? _____ Brushes clean? _____ Brushes rusty? _____

Jars Clean? _____ Acetone grade _____ Residue on evap. spec. _____ %

Filter handled OK? _____ Probe handled OK? _____ Impingers handled OK? _____

After cleanup: Filter holder clean? _____ Probe liner clean? _____

   Nozzle clean? _____ Impingers clean? _____ Blanks taken? _____

Description of collected particulate _____

Silica gel all pink?   Run 1 _____ Run 2 _____ Run 3 _____

Jars adequately labeled? _____ Jars sealed tightly? _____

Liquid level marked on jars? _____ Jars locked up? _____

General comments on entire sampling project:

Observer's name _____ Title _____

Affiliation _____ Signature _____

Fig. 27   (*Continued*)[27]

11 DATA MANAGEMENT

In most field tests a large number of data sheets will be filled out. It is important that a data management system be worked out before the test begins. A single person should be responsible for checking all the data sheets for completeness at the end of each day. The same person should also be responsible for assembling all the data sheets into a single notebook.

A strict chain of custody of the data sheets should be established. The chain of custody must be enforced from the field tests through the time that the data sheets are filed in final storage after the report is written. A log book should be maintained to record who has the data when.

The sheet in Figure 28 can be used to help maintain the chain of custody.

12 COMPUTER PROGRAMS

The availability of small inexpensive microcomputers makes data reduction of particulate sampling data much easier than is possible by hand or even by a large computer. Denver Research Institute (DRI) has an excellent complete data reduction package for the TRS-80™ microcomputer.[23] This package includes graphic capability. The package is available from the National Technical Information Service or directly from DRI.

The computer programs presented here are very basic programs which will do all the necessary calculations for both mass train and impactor train sampling. The programs are written to ensure maximum portability between microcomputers and are designed for a minimal microcomputer system of cassette tape (although disk is highly recommended), minimum random ascess memory (RAM), and a line printer. For the most part the programs are straightforward translations of the various equations in the text. The only program that will be extensively commented on is the impactor data reduction program.

The programs are written in TRS-80 Model I Level II Basic. The use of special features of Level II Basic has been minimized to ensure that the programs can be used on other computers with minimum changes.

All of the programs accept data input in either English or metric units. The user must specify the type of units before beginning data entry. The units used for the various input data are given in the program.

To use the programs, simply provide the data asked for. Do not worry if you make a mistake entering data. You will be given a chance to review the data before the calculations begin.

Plant _____

Date sampled _____   Test number _____

                                                Run number _____

Sample Recovery

Container Code   Description

_____   _____

_____   _____

_____   _____

_____   _____

_____   _____

Person engaged in sample recovery
  Signature _____

  Title _____

  Location at which recovery was done _____

  _____

  Date and time of recovery _____

Sample(s) recipient, upon recovery if not recovery person
  Signature _____

  Title _____

  Date and time of receipt _____

  Sample storage _____

Laboratory person receiving sample
  Signature _____

  Title _____

  Date and time of receipt _____

  Sample storage _____

Analysis

Container code	Method of analysis	Date and time of analysis	Signature of analyst
_____	_____	_____	_____
_____	_____	_____	_____
_____	_____	_____	_____
_____	_____	_____	_____
_____	_____	_____	_____

Fig. 28 Check sheet for chain of custody.[27]

The impactor program is based on the TI-59 program presented by Sparks[20] and in part on the program in the DRI[23] package. The important feature of the program is that a spline fit is used to interpolate between data points and to differentiate the cumulative distribution curve to produce the $dm/d \log d$ distribution.

The spline fit gives the smoothest curve possible that passes through the data points. The curve is mathematically well defined. In general, the spline fit is the best way to differentiate sparse data. Before the spline is fit to the data, the cumulative distribution is transformed into log normal space. Sparks[20] has shown that use of the spline fit to data transformed into log normal space results in the smallest error in calculating penetration as a function of particle diameter as measured by cascade impactors. See Lawless[19] or Sparks[20] for details of the spline fit and the log normal transformation.

The use of the programs is best demonstrated by example problems given below.

Example 1: Calculation of Molecular Weight

Input data: % CO $= 1$
               % $CO_2 = 16$
               % O_2 $= 4$
               % N_2 $= 79$

Run program Orsat/BAS.
Answer: 30.72 g/g-mole.

Example 2: Fraction of Water Vapor in Gas

Input data: Dry gas meter volume $= 0.0283$ m^3
               Pressure at meter $= 760$ mm Hg
               Temperature at meter $= 30°C$
               Initial impinger volume $= 10$ ml
               Final impinger volume $= 12.5$ ml

Run program MOIST/BAS.
Select metric units and input data.
Answer: Fraction water vapor in gas $= 0.108707$.

Example 3: Calculation of Stack Velocity and Flow Rate

Input data: Pitot tube coefficient $= 0.87$
               Dry molecular weight $= 30.48$ g/g-mole
               Fraction water vapor $= 0.1$
               Stack temperature $= 150°C$
               Absolute stack pressure $= 760$ mm Hg
               Average delta p $= 111.484$ mm H_2O

Run program VEL/BAS.
Select metric units and input data.
Answer: Gas velocity $= 18.5608$ m/sec
           Flow rate $= 4.64 \times 10^6$ DNcm/hr

Example 4: Calculation of Mass Concentration

Input data: Volume of gas sampled $= 2.830$ m^3
               Temperature at meter $= 28°C$
               Pressure at meter $= 760$ mm Hg
               Pressure drop across orifice $= 111$ mm H_2O
               Sampling time $= 100$ min
               Fraction of water in gas $= 0.0245$
               Weight of particulate $= 100$ mg
               Stack temperature $= 150°C$
               Stack pressure $= 750$ mm Hg
               Gas velocity in stack $= 5$ m/sec
               Cross-sectional area of nozzle $= 0.000126$ m^2

Run program METH5/BAS.
Select metric units and input data.
Answer: Particle concentration $= 0.0359$ g/DNm^3
           Percent of isokinetic $= 110.20$

Note that the program will inform you if the sampling rate is within the allowable range of 90 to 110% of isokinetic. (DNm^3 is dry normal cubic meters.)

Example 5: Calculation of Mass Efficiency

Input data: Outlet mass concentration $= 0.03$ g/DNm^3
               Inlet mass concentration $= 3$ g/DNm^3

Run program EFFIC/BAS.
Input data
Answer: Mass penetration $= 0.01$
           Mass efficiency $= 99\%$

ORSAT PROGRAM

```
10    REM ORSAT PROGRAM BY LES VERSION 1.0 FILE NAME ORSAT/BAS
20    REM WRITTEN IN TRS-80 MODEL I DISK BASIC
30    CLEAR 100:REM CLEAR STRING SPACE
40    CLS
50    INPUT"ENTER TEST ID";ID$
60    INPUT"ENTER REMARKS";R$
70    PRINT STRING$(63,"=") :REM PRINTS = ACROSS SCREEN
80    PRINT"ENTER DATA ASKED FOR. PRESS ENTER AFTER EACH DATA ENTRY"
90    PRINT STRING$(63,"=")
100   PRINT"ENTER PERCENT CO (";CO;")";
110   INPUT CO
120   PRINT"ENTER PERCENT CO2 (";C2;")";
130   INPUT C2
140   PRINT"ENTER PERCENT O2 (";O2;")";
150   INPUT O2
160   PRINT"ENTER PERCENT N2 (";N2;")";
170   INPUT N2
180   REM NOW CHECK TO SEE IF DATA ARE OK
190   CLS
200   PRINT"PERCENT CO ";CO
210   PRINT"PERCENT CO2 ";C2
220   PRINT "PERCENT O2 ";O2
230   PRINT "PERCENT N2 ";N2
240   PRINT STRING$(63,"=")
250   INPUT"ARE THESE CORRECT Y OR N ";Y$
260   IF LEFT$(Y$,1)="Y" THEN 280 ELSE IF LEFT$(Y$,1)<>"N" THEN
PRINT "ANSWER Y OR N PLEASE": GOTO 250
270   PRINT"PLEASE CORRECT DATA":FOR I=1 TO 100:NEXT I: GOTO 40
280   REM NOW CALCULATE DRY MOLECULAR WEIGHT
290   MW=.44*C2 + .32*O2 + .28*(N2+CO)
300   PRINT"DRY MOLECULAR WEIGHT IS ";MW
310   PRINT STRING$(63,"=")
320   INPUT"DO YOU WANT HARDCOPY Y OR N";Y$
330   IF Y$<>"Y" THEN END
340   LPRINT"TEST ID ";ID$
350   LPRINT"REMARKS ";R$
360   LPRINT" CO2 ";C2;" CO ";CO;" O2 ";O2;" N2 ";N2
370   LPRINT"DRY MOLECULAR WEIGHT IS ";MW
380   END
```

MOISTURE DETERMINATION PROGRAM

```
10    REM DETERMINATION OF MOISTURE IN STACK GASES
20    REM VERSION 1.0 BY LES FILE NAME MOIST/BAS
30    REM WRITTEN IN TRS-80 LEVEL II BASIC
40    CLEAR 100
50    CLS
60    PRINT STRING$(63,"=")
70    INPUT"ENTER TEST ID ";ID$
80    INPUT"ENTER REMARKS ";R$
90    PRINT"DETERMINATION OF MOISTURE IN STACK GASES"
100   INPUT"ENTER E FOR ENGLISH UNITS OR M FOR METRIC ";U$
110   IF U$="E" THEN U$="ENGLISH UNITS " ELSE IF U$<> "M" THEN
PRINT "ANSWER E OR M PLEASE":GOTO 100
120   IF U$="M" THEN U$="METRIC UNITS"
130   CLS
140   PRINT STRING$(63,"=")
150   PRINT"ENTER DATA ASKED FOR"
160   PRINT STRING$(63,"=")
170   PRINT"ENTER DRY GAS METER VOLUME (";VM;")";
180   INPUT VM
190   PRINT"ENTER TEMPERATURE AT THE METER (";TM;");
200   INPUT TM
```

```
210    PRINT"ENTER PRESSURE AT THE METER (";PM;")";
220    INPUT PM
230    PRINT"ENTER INITIAL VOLUME OF IMPINGER CONTENTS ML
(";VI;")";
240    INPUT VI
250    PRINT"ENTER FINAL VOLUME OF IMPINGER CONTENTS ML
(";VF;")";
260    INPUT VF
270    CLS
280    PRINT STRING$(63,"=")
290    PRINT"CHECK DATA FOR CORRECTNESS"
300    PRINT U$
310    PRINT" DRY GAS METER VOLUME "TAB(40);VM
320    PRINT"PRESSURE AT METER "TAB(40);PM
330    PRINT"TEMPERATURE AT METER";TAB(40);TM
340    PRINT"INITIAL IMPINGER VOLUME ML";TAB(40);VI
350    PRINT"FINAL IMPINGER VOLUME ML";TAB(40);VF
360    Y$=""
370    INPUT"ARE THESE CORRECT Y OR N";Y$
380    Y$=LEFT$(Y$,1)
390    IF Y$="Y" THEN 420 ELSE IF Y$<>"N" THEN PRINT"ANSWER Y OR
N PLEASE ":GOTO 370
400    PRINT"REENTER INCORRECT DATA"
410    FOR J=1 TO 200:NEXT J:GOTO 110
420    REM DO THE CALCULATIONS
430    IF LEFT$(U$,1)="E" THEN TM=TM+460 ELSE TM=TM+273
440    IF LEFT$(U$,1)="E" THEN VW=.0474*(VF-VI) ELSE
VW=(VF-VI)/1000:REM CONVERT ML TO EITHER CUBIC FT OR CUBIC
METERS
450    IF LEFT$(U$,1)="E" THEN VE=17.71*VM*PM/TM ELSE VE=
VM*PM/760*293/TM
460    BW=VW/(VE+VW)+.025
470    CLS:PRINT:PRINT :PRINT STRING$(63,"=")
480    PRINT "FRACTION OF WATER IN GAS = ";BW
490    PRINT:PRINT STRING$(63,"=")
500    Y$=""
510    INPUT"DO YOU WANT HARDCOPY Y OR N";Y$
520    IF Y$<>"Y" THEN END ELSE PRINT"BE SURE PRINTER IS ON"
530    LPRINT"TEST ID ";ID$
540    LPRINT"REMARKS ";R$
550    LPRINT" VOLUME OF METER ";VM
560    LPRINT"TEMPERATURE OF METER ";TM;" PRESSURE OF METER ";PM
570    LPRINT"AMOUNT OF WATER COLLECTED ";VW
580    LPRINT"FRACTION OF WATER IN GAS ";BW
590    END
```

VELOCITY DETERMINATION PROGRAM

```
10     REM FILE NAME VEL/BAS
20     REM DETERMINATION OF STACK GAS VELOCITY AND FLOW RATE
30     REM TYPE S PITOT TUBE
40     REM VERSION 1.0 FOR TRS-80 MODEL I DISK BASIC
50     REM REQUIRES LEVEL II BASIC AND 16 K MINIMUM
60     CLEAR 100
70     CLS
80     PRINT STRING$(63,"=")
90     PRINT"DETERMINATION OF STACK GAS VELOCITY AND FLOW RATE"
100    PRINT STRING$(63,"=")
110    U$=""
120    INPUT"ENTER TEST ID";ID$
130    INPUT"ENTER REMARKS";R$
140    INPUT"ENTER E FOR ENGLISH UNITS OR M FOR METRIC UNITS";U$
150    If U$="E" THEN KP=85.48:FL=-1:GOTO 180
160    IF U$<>"M" THEN PRINT"ANSWER E OR M PLEASE":GOTO 140
```

```
170   KP= 34.97      : FL=1
180   CLS
190   REM DATA INPUT
200   IF FL=-1 THEN F$="ENGLISH UNITS" ELSE F$="METRIC UNITS"
210   PRINT STRING$(63,"=")
220   PRINT"ENTER DATA ASKED FOR IN ";F$
230   PRINT STRING$(63,"=")
240   INPUT"PITOT TUBE COEFFICIENT ";CP
250   INPUT"ENTER DRY MOLECULAR WEIGHT ";MW
260   INPUT"FRACTION WATER VAPOR";BW
270   INPUT"ENTER STACK TEMPERATURE ";TS
280   INPUT"ENTER AVERAGE DELTA P ";DP
290   INPUT"ENTER STACK ABSOLUTE STACK PRESSURE ";PS
300   INPUT"ENTER CROSS SECTIONAL AREA OF STACK ";A
310   CLS
320   PRINT"REVIEW INPUT DATA FOR ERRORS"
330   PRINT STRING$(63,"=")
340   PRINT"PITOT TUBE COEFFICIENT ";CP
350   PRINT"DRY MOLECULAR WEIGHT ";MW
360   PRINT"FRACTION OF WATER VAPOR ";BW
370   PRINT"STACK TEMPERATURE ";TS
380   PRINT"AVERAGE DELTA P ";DP
390   PRINT"ABSOLUTE STACK PRESSURE ";PS
400   PRINT"CROSS SECTIONAL AREA OF STACK ";A
410   PRINT STRING$(63,"=")
420   Y$=""
430   INPUT"ARE THESE CORRECT Y OR N ";Y$
440   Y$=LEFT$(Y$,1)
450   IF Y$="Y" THEN 490 ELSE IF Y$<>"N" THEN PRINT "ANSWER Y
OR N PLEASE ":GOTO 430
460   PRINT"REENTER INCORRECT DATA ":FOR J=1 TO 200:NEXT J
470   CLS
480   GOTO 140
490   REM CALCULATE VELOCITY
500   IF FL=-1 THEN TS=TS+460 ELSE TS=TS+273
510   IF FL=-1 THEN TD=528 ELSE TD=293 :REM STANDARD
TEMPERATURE
520   IF FL=-1 THEN PD=29.92 ELSE PD=760
530   IF FL=-1 THEN V$="FPS" ELSE V$="MPS"
540   IF FL=-1 THEN Q$="DSCFH" ELSE Q$="DRY NORMAL CUBIC METER
PER HR"
550   MS=MW*(1-BW)+18*BW:REM WET MOLECULAR WT
560   VS=KP*CP*DP[.5*(TS/PS/MS)[.4
570   QS=3600*(1-BW)*VS*A*(TD/TS)*(PS/PD)
580   CLS
590   PRINT STRING$(63,"=")
600   PRINT"STACK VELOCITY = ";VS;V$
610   PRINT"FLOW RATE = " ;QS;Q$
620   PRINT
630   PRINT STRING$(63,"=")
640   INPUT"DO YOU WANT HARDCOPY Y OR N"; Y$
650   IF Y$<>"Y" THEN END ELSE PRINT"BE SURE PRINTER IS ON"
660   LPRINT"TEST ID ";ID$
670   LPRINT"REMARKS ";R$
680   LPRINT"VELOCITY = ";VS;V$
690   LPRINT "FLOW RATE = ";QS;Q$
700   END
```

METHOD 5 PROGRAM

```
10    REM FILE NAME METH5/BAS
20    REM DETERMINATION OF PARTICULATE EMISSIONS
30    REM VERSION 1.0
40    REM WRITTEN IN TRS-80 LEVEL II BASIC BY LES
```

```
50    REM **********************************
60    CLEAR 200
70    DIM F$(12),X(12)
80    F$(7)="WEIGHT OF PARTICULATE COLLECTED ####.### MG"
90    CLS
100   PRINT STRING$(63,"=")
110   PRINT"DETERMINATION OF PARTICULATE EMISSIONS"
120   INPUT"ENTER TEST ID ";ID$
130   INPUT"ENTER REMARKS ";R$
140   PRINT STRING$(63,"=")
150   INPUT"ENTER E FOR ENGLISH UNITS OR M FOR METRIC UNITS";U$
160   IF U$="E" THEN 740 ELSE IF U$="M" THEN 850 ELSE
PRINT"ANSWER E OR M PLEASE"
170   GOTO 150
180   REM NOW GET THE INFORMATION
190   F$(6)="FRACTION OF WATER VAPOR IN GAS #.####"
200   CLS
210   PRINT STRING$(63,"=")
220   FOR J=1 TO 11
230   PRINT USING F$(J);X(J);
240   INPUT X(J)
250   NEXT J
260   CLS
270   PRINT STRING$(63,"=")
280   PRINT"CHECK DATA FOR ERRORS"
290   PRINT STRING$(63,"=")
300   FOR J=1 TO 11
310   PRINT USING F$(J);X(J)
320   NEXT J
330   Y$=""
340   INPUT"ARE THESE CORRECT Y OR N";Y$
350   Y$=LEFT$(Y$,1)
360   IF Y$="Y" THEN 390 ELSE IF Y$<>"N" THEN PRINT"ANSWER Y OR
N PLEASE ":GOTO 340
370   PRINT"REENTER INCORRECT DATA ":FOR J=1 TO 200:NEXT J
380   GOTO 180
390   REM NOW CALCULATE IT
400   REM FIRST CALCULATE METER FLOW AT STANDARD DRY CONDITIONS
410   IF LEFT$(U$,1)="E" THEN M=17.64 ELSE M=0.3858
420   VD=X(1)*M*(X(3)+X(4)/13.6)
430   TM=X(2)
440   IF LEFT$(U$,1)="E" THEN TM=TM+460 ELSE TM=TM+273
450   VD=VD/TM
460   IF LEFT$(U$,1)="E" THEN K1=0.0154 ELSE K1=.001
470   IF LEFT$(U$,1)="E" THEN C$="GR/DSCF" ELSE C$="G/DN CUBIC
METER"
480   C=K1/VD*X(7)
490   CLS
500   PRINT STRING$(63,"=")
510   C$="PARTICLE CONCENTRATION ###.#### "+C$
520   PRINT USING C$;C
530   PRINT STRING$(63,"=")
540   REM NOW CALCULATE % ISOKINETIC
550   IF LEFT$(U$,1)="E" THEN K4=0.0945 ELSE K4=4.320
560   IF LEFT$(U$,1)="E" THEN TS=X(8)+460 ELSE TS=X(8)+273
570   I=K4*TS*VD/(X(9)*X(10)*X(11)*X(5)*(1-X(6)))
580   IF I>110 OR I<90 THEN Z$="ISOKINETIC RATE OUT OF RANGE"
ELSE Z$="ISOKINETIC RATE IN RANGE"
590   PRINT USING"PERCENT OF ISOKINETIC ####.#### ";I
600   PRINT Z$
610   Y$=""
620   INPUT"DO YOU WANT HARDCOPY Y OR N";Y$
630   IF Y$<>"Y" THEN END ELSE PRINT"BE SURE PRINTER IS ON"
640   LPRINT"TEST ID ";ID$
650   LPRINT"REMARKS ";R$
```

```
660    LPRINT"INPUT DATA"
670    FOR J=1 TO 11
680    LPRINT USING F$(J);X(J)
690    NEXT J
700    LPRINT USING C$;C
710    LPRINT USING"PERCENT OF ISOKINETIC ####.## ";I
720    LPRINT Z$
730    END
740    REM SET UP FOR INPUT
750    F$(1)="VOLUME OF GAS THROUGH GAS METER ####.## CUBIC
FEET"
760    F$(2)="TEMPERATURE AT THE METER ####.## DEG F"
770    F$(3)="PRESSURE AT THE METER ####.## INCHES HG"
780    F$(5)="SAMPLING TIME ####.## MINUTES"
790    F$(4)="PRESSURE DROP ACROSS THE ORIFICE ##.### INCHES
WATER"
800    F$(8)="STACK TEMPERATURE ####.## DEG F"
810    F$(9)="STACK PRESSURE ####.## INCHES HG"
820    F$(10)="GAS VELOCITY IN STACK ####.### FT/S"
830    F$(11)="CROSS SECTIONAL AREA OF NOZZLE ###.##### SQUARE
FT"
840    GOTO 180
850    REM SET UP FOR METRIC DATA ENTRY
860    F$(1)="VOLUME OF GAS SAMPLED ####.### CUBIC METERS"
870    F$(2)="TEMPERATURE AT THE METER ###.## DEG C"
880    F$(3)="PRESSURE AT THE METER ####.## MM HG"
890    F$(4)="PRESSURE DROP ACROSS THE ORIFICE ###.### MM WATER"
900    F$(5)="SAMPLING TIME ####.## MINUTES"
910    F$(8)="STACK TEMPERATURE ####.## DEG C"
920    F$(9)="STACK PRESSURE ####.## MM HG"
930    F$(10)="GAS VELOCITY IN STACK ####.### M/S"
940    F$(11)="CROSS SECTIONAL AREA OF NOZZLE ##.###### SQUARE
METERS"
950    GOTO 180
```

EFFICIENCY PROGRAM

```
10    REM FILE NAME EFFIC/BAS
20    REM CALCULATE MASS EFFICIENCY
30    REM FOR TRS-80 LEVEL II BASIC BY LES
40    CLEAR 100
50    CLS
60    PRINT STRING$(63,"=")
70    PRINT"MASS EFFICIENCY CALCULATION"
80    INPUT"ENTER RUN ID ";ID$
90    INPUT"ENTER REMARKS ";R$
100   INPUT"ENTER OUTLET MASS CONCENTRATION ";CO
110   INPUT"ENTER INLET MASS CONCENTRATION ";CI
120   CLS
130   PRINT STRING$(63,"=")
140   PRINT USING"OVERALL PENETRATION = ###.#### ";CO/CI
150   PRINT USING"OVERALL EFFICIENCY = ###.##### %
";(1-CO/CI)*100
160   Y$=""
170   INPUT"DO YOU WANT HARDCOPY Y OR N";Y$
180   IF Y$<>"Y" THEN END ELSE PRINT"BE SURE PRINTER IS ON"
190   LPRINT"TEST ID ";ID$
200   LPRINT"REMARKS ";R$
210   LPRINT"INLET MASS CONCENTRATION ";CI
220   LPRINT"OUTLET MASS CONCENTRATION ";CO
230   LPRINT USING"OVERALL MASS PENETRATION = ##.####### ";CO/CI
240   LPRINT USING"OVERALL MASS COLLECTION EFFICIENCY =
###.##### ";(1-CO/CI)*100
250   END
```

REFERENCES

1. EPA, Standards of Performance for New Sources: Revision to Reference Methods 1–8, *Federal Register* **42**(160), 41776–41782 (1977).

2. Smith, W. B., Cushing, K. M., and McCain, J. D., *Particulate Sizing Techniques for Control Device Evaluation,* EPA Report EPA-650/2-74-102, Oct. 1974.

3. Brooks, E. F., *Total Particulate Mass Emission Sampling Errors,* Report for EPA Contract 68-02-2165, Task 26, Research Triangle Park, NC, not published.

4. Stairmand, C. J., *Trans. Instn. Chem. Engrs.* **29**, 15 (1951).

5. Davies, C. N., *Dust is Dangerous,* Faber and Faber, London, 1954.

6. EPA, Method 17 Determination of Particulate Emissions from Stationary Source (In-Stack Filtration Method), *Federal Register* **43**(37), 7584–7596 (1978).

7. Sparks, L. E., Importance of Particle Size Distribution, *Proc. Second Symposium of Particulate Control Technology,* EPA Report EPA-600/9-80-639, 1980.

8. Rabe, O. G., Aerosol Aerodynamic Size Conventions for Inertial Sampler Calibration, *J. Air Poll. Control Assoc.* **9**, 856 (1976).

9. McCain, J. D., and McCormack, Non-ideal Behavior in Cascade Impactors, in *Proc. Advances in Particle Sampling and Measurement (Daytona Beach, FL, October 1979),* EPA Report EPA-600/9-80-004, Jan. 1980.

10. Cushing, K. M., Lacy, G. E., McCain, J. D., and Smith, W. B., *Particulate Sizing Techniques for Control Device Evaluation: Cascade Impactor Calibrations,* EPA Report EPA-600/2-76-280, Oct. 1976.

11. Esmen, N. A., and Lee, T. C., Distortion of Cascade Impactor Measured Size Distribution Due to Bounce and Blow Off, *American Industrial Hygiene Assoc. J.* **41**, 410 (1980).

12. Gooch, J. P., Marchant, G. H., Jr., and Felix, L. G., *Particulate Collection Efficiency Measurements on an ESP Installed on a Coal Fired Boiler,* EPA Report EPA-600/2-77-011, Jan. 1977.

13. Calvert, S., Lake, C., and Parker, R., *Cascade Impactor Calibration Guidelines,* EPA Report EPA-600/2-76-118, 1976.

14. Harris, D. B., *Procedures for Cascade Impactor Calibration and Operation in Process Streams,* EPA Report EPA-600/2-777-004, Jan. 1977.

15. Smith, W. B., Cushing, K. M., and McCain, J. D., *Procedures Manual for Electrostatic Precipitator Evaluation,* EPA Report EPA-600/7-77-059, June 1977.

16. Idem, *Particulate Sizing Techniques for Control Device Evaluation,* EPA Report EPA 650/2-74-102, October 1974.

17. Lundgren, D. A., and Balfour, D., *Use and Limitations of In-Stack Impactors,* EPA Report EPA-600/2-80-048, Feb. 1980.

18. Ensor, D. S., personal communication, 1981.

19. Lawless, P. A., *Analysis of Cascade Impactor Data for Calculating Particle Penetration,* EPA Report EPA-600/7-78-189, September 1978.

20. Sparks, L. E., *Cascade Impactor Data Reduction with the SR-52 and TI-59 Programmable Calculators,* EPA Report EPA-600/7-78-226, Nov. 1978.

21. Johnson, J. W., Clinard, G. I., Felix, L. G., and McCain, J. D., *A Computer Based Impactor Data Reduction System,* EPA Report EPA-600/7-78-042, 1978.

22. Wasmundt, K. C., and Tegtmeyer, S., *TRS-80 Cascade Impactor Data Reduction System,* Electronic Division, Denver Research Institute, Denver, Colo., March 1980.

23. Smith, W. B., and Wilson, R. R., Jr., *Development and Laboratory Evaluation of a Five-Stage Cyclone System,* EPA-600/7-78-008, 1978.

24. Wilson, R. R., Jr., Cavanaugh, P. R., Cushing, K. M., Farthing, W. E., and Smith, W. B., *Guidelines for Particulate Sampling in Gaseous Effluents from Industrial Processes,* EPA-600/7-79-028, 1979.

25. Liu, B. Y. H., Whitby, K. T., and Pui, D. Y. H., A Portable Electrical Analyzer for Size Distribution Measurement of Sub-Micron Aerosols, *J. Air Pollut. Control Assoc.* **24**, 1067 (1974).

26. Ensor, D. S., personal communication, 1980.

27. U. S. Environmental Protection Agency, *Handbook Industrial Guide for Air Pollution Control,* EPA-625/6-78-004, 1978.

28. Smith, W. B., and McCain, J. D., Particle Size Measurement in Industrial Flue Gases, in *Air Pollution Control, Part III* (W. Strauss, ed.), Wiley, New York, 1972. (Much of the data in the various EPA reports are covered here.)

29. Fuchs, N. A., *The Mechanics of Aerosols,* Pergamon Press, New York, 1964.

Note that all of the EPA reports are available from the National Technical Information Service, Springfield, VA.

CHAPTER 32
PLUME OPACITY

MICHAEL J. PILAT

Department of Civil Engineering
University of Washington
Seattle, Washington

1 INTRODUCTION

1.1 History

The smoke charts developed by Professor Ringelmann (1898) in Paris, France, consisted of black grids on white paper to produce apparent shades of gray. These standard charts, having 20, 40, 60, or 80% of the white obscured by the black grid lines, were compared to the shade or density of the smoke plumes. The Ringelmann charts were introduced into the United States by W. Kent (1897) who had learned of the method from Donkin of London and apparently the charts were being used extensively in Europe at that time. In 1899 Kent proposed to the American Society of Mechanical Engineers that the Ringelmann chart be accepted as a standard method of smoke density measurement for power plants. The Technological Branch of the Federal Geological Survey (later to become the United States Bureau of Mines) used the Ringelmann chart in their studies of smokeless combustion beginning in 1904 in St. Louis and in 1908 made the charts available for public distribution. In 1910 the Massachusetts Legislature officially recognized the charts in the smoke ordinance for Boston. Marks (1937) correctly recognized that the Ringelmann smoke plume evaluation procedure is really an estimate of the fraction of light transmitted through the plume. In 1947 the state of California adopted the Health and Safety Code Section 24242 which presented the Ringelmann chart as a standard for reading the densities and opacities of stack emissions. Also the code allowed the evaluation of nonblack plumes using the Ringelmann chart. The Los Angeles Air Pollution Control District adopted rule 50 which included the Ringelmann chart and developed a method for training air pollution inspectors to read plume opacity. The training method involved comparing the plume opacity readings with measurements by a light transmissometer.

1.2 Visual Emissions Evaluation

The procedure for the visual determination of the opacity of emissions from stationary sources has been specified by the EPA in Method 9, as presented in the *Federal Register* (1971, 1974). As reported by Ensor (1970), Figure 1 shows schematically an observer viewing the plume at the top of the stack against a contrasting background, at right angles to the plume direction, and with sun to the observer's back. The EPA Method 9 procedure states that the observer shall:

1. Stand at a distance sufficient to have a clear view of the stack emission.
2. Keep the sun in the 140° sector to the observer's back.
3. Make the readings at approximately right angles to the plume direction and if the emission outlet is rectangular, at right angles to the longer axis of the emission outlet.
4. Make readings at the point of greatest opacity in that portion of the plume where condensed water vapor is not present.
5. View the plume against a contrasting background, if possible.
6. Observe the plume momentarily at 15-sec intervals (do not look continuously at the plume).
7. Record the opacity observations to the nearest 5% at 15-sec intervals on a record sheet. A minimum of 24 opacity observations shall be recorded over the approximately 6-min time period.

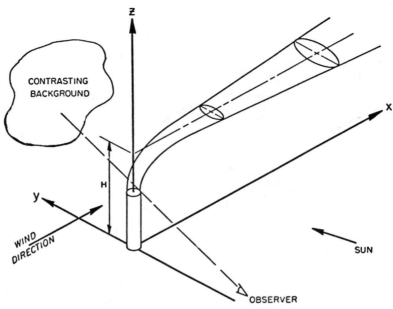

Fig. 1 Diagram of plume opacity observation.

1.3 Training of Opacity Observers

The training of opacity observers involves the reading of the opacity of black-and-white plumes produced by a smoke generator. Griswold et al. (1958) reported that the Los Angeles APCD used an instack light transmissometer measured opacity to compare with the observer readings. Coons et al. (1965) reported that the San Francisco Bay Area APCD used the opacity readings made by a group of observers to standardize the instack light transmissometers of the smoke generators. The EPA requirements for an individual to receive certification as a qualified observer include the demonstration of the ability to assign opacity readings to 25 different black plumes and 25 different white plumes with an error not to exceed 15% opacity on any one reading and an average error not to exceed 7.5% in each category (white or black). The certification is valid for a period of 6 months after which the individual must repeat the plume reading examination.

1.4 Precision of Opacity Observations

The variability of opacity readings by experienced and certified observers is of concern regarding the reliability of opacity observations. Midgett (1977) reports that observers tend to read slightly high at lower opacities (around 5%), read with good accuracy in the 10–15% opacity range, and read lower (negative bias) at the higher opacities (around 35%). Hamil et al. (1975) reported the results of the analysis of data obtained in the testing of EPA Method 9 at three different sites; a training smoke generator, a sulfuric acid plant, and a fossil fuel-fired steam generator. The results showed a composite between-observer standard deviation of 2.42% opacity. The maximum difference that could be expected between two independent observers is 6.92% opacity at the 95% confidence level. Henz (1970) pointed out that the allowance of ±7.5% opacity amounts to an error tolerance of ±75% at a maximum allowable opacity of 10%. Hague et al. (1977) reported on the testing of 100 certified opacity observers (55 from government agencies and 45 from private industry). Regression analysis of the observer scores showed poor correlation with such variables as age, vision correction, and years of experience but some correlation with the frequency of opacity observations. Conner and McElhoe (1982) compared the opacities observed on 26 different occasions at 13 different plants with opacities measured by instack transmissometers. The opacities ranged from 2 to 57%. The 13 plants included 6 coal-fired power plants, 3 oil-fired power plants, 2 portland cement plants, a kraft recovery boiler at a paper mill, and a rock dryer at a phosphate fertilizer plant. The comparison showed that in the 26 tests, 19 were within ±7.5% opacity, 4 had observer capacity more than 7.5% opacity lower than the transmissometer opacity, and 3 had the observer opacity more than 7.5% opacity higher. Two of the higher observer measurements were at an oil-fired power plant and the third was at a coal-fired power plant. A condensable plume was observed at the oil-fired power plant and with 2.5% sulfur in the oil, some sulfuric acid in the emissions is expected. In no case did observer measurements exceed the transmissometer measurements by more than 10% opacity. It should be noted that the transmissometers measured the opacity in the stack and the observers measured the opacity at the stack exit (not at the location of the greatest opacity, which in the case of condensing sulfuric acid gases may be downwind some 50–200 ft).

2 OPTICAL EFFECTS OF SMOKE PLUMES

2.1 Visual Contrast

A smoke plume has two visual effects; first is that the plume becomes visible and second it can obscure the visibility of objects behind the plume. The visual effects depend both on the plume opacity and on the environmental lighting conditions (i.e., location of sun, presence of clouds, etc.). The visual luminance contrast C of an object viewed against an extensive and uniform appearing background is given by

$$C = \frac{B_o - B_b}{B_b} = \frac{B_o}{B_b} - 1 \tag{1}$$

where B_o is the object luminance and B_b the background luminance. Our vision (ignoring color effects) depends on the eye's perception of the difference in luminance between points in the field of view. The contrast threshold is about 0.003 as reported by Blackwell (1946). A perfectly black object has an object luminance of 0.0 (i.e., $B_o = 0$) and thus the contrast C is −1.0. Contrasts in general range from around −1.0 to +10.

The plume contrast C_p, plume opacity O_p, plume luminance B_p, and plume airlight B_a can be related, as reported by Connor and Hodkinson (1967). The plume airlight is the sum of the plume scattered light from the sun and from the sky and is given by

$$B_a = B_{sun} + B_{sky} \tag{2}$$

The plume luminance B_p is given by

$$B_p = B_a + B_b T \tag{3}$$

where B_b is the background luminance and T is the fraction of background light transmitted through the plume toward the observer. The various optical parameters involved in plume contrast are illustrated in Figure 2.

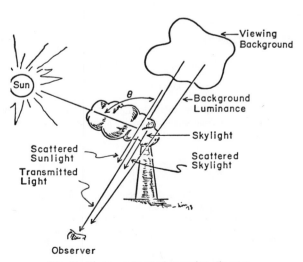

Fig. 2 Plume contrast observation diagram.

Substitution of Eq. (3) for B_o in Eq. (1) gives

$$C_p = \frac{B_a}{B_b} + (T - 1) = \frac{B_a}{B_b} - O_p \tag{4}$$

$$\text{plume contrast} = \frac{\text{plume air light luminance}}{\text{background luminance}} - \text{plume opacity}$$

This equation illustrates the three main optical effects of plumes:

1. The plume contrast C_p indicates the visual appearance of the plume against its background.
2. The plume opacity O_p is an intrinsic property of the plume which is independent of the illumination and viewing angles.
3. The plume air light luminance/background luminance ratio indicates the magnitude of light scattered to the observer. B_a/B_b is dependent on the plume light scattering properties, the angle between the sun and the observer, the presence of clouds, and the background illumination.

The relationship of the three parameters of opacity, plume air light, and background luminance are shown in Figure 3, as reported by Ensor and Pilat (1970) for data published by Connor and Hodkinson (1967). With the opacity equal to the B_a/B_b ratio, the plume is not visible. In the diagonal zone of Figure 3 where the plume contrast C_p is ± 0.02, it can be assumed that the plume contrast is less than that needed for visual detection of the plume. Above the diagonal zone the plume contrast is greater than 0.02 and the plume will appear lighter than the background. Below the diagonal line (C_p less than -0.02), the plume will appear darker than its background.

The plume air light/background luminance ratio is dependent on the angle between the sun and the observer and the location of clouds. However, the plume opacity (or fraction of light transmitted through the plume) is unaffected by lighting angles and can be calculated from the aerosol properties.

2.2 Calculation of Plume Contrast

Jarman and de Turville (1966) reported a relationship between the plume transmittance, the plume to background contrast, and the screening power of smoke plumes. Jarman and de Turville (1969) presented a method for the theoretical prediction for plume disappearance for an observer standing under the plume looking skyward based on light extinction and plume dispersion theories. The physical dimensions of a visible plume (length and width) have been used to calculate atmospheric dispersion coefficients for stack plumes, as described by Gifford (1959). Halow and Zeek (1973) presented a method of calculation of the plume contrast using information regarding the particle size distribution, the particle refractive index, and the sun angle. Graphs of the ratio of the plume air light luminance/background luminance ratio B_a/B_b versus the mass mean particle diameter were provided. Note that this method calculates the plume contrast, not the plume opacity (this may be somewhat confusing in this article which implies that Ringelmann number, opacity, and plume contrast are equal when in fact they are not).

2.3 Calculation of Plume Opacity

The plume opacity is directly related to the fraction of light transmitted through the plume by

$$O_p = 1.0 - T \tag{5}$$

The opacity of a plume has been related by Pilat and Ensor (1970) to the stack diameter L and the particle mass concentration W using an equation in the form of the Bouguer law (Lambert–Beer law)

$$\ln T = -\frac{WL}{K\rho} = -bL \tag{6}$$

where ρ is the particle density, K the ratio of the specific particulate volume to the light extinction coefficient of the particulate emission aerosol, and b the light extinction coefficient of the aerosol. The parameter K is dimensionally similar to the particle volume surface characteristic diameter described by Herdan (1960). Also, the parameter K can be thought of as the diameter of a monodisperse aerosol which extincts light in the same manner or with the same efficiency as the real polydisperse aerosol. For K in the units of cm^3/m^2, the unit of the equivalent monodisperse aerosol particle diameter is nanometers. The parameter K can be obtained from field measurements of the light transmittance T,

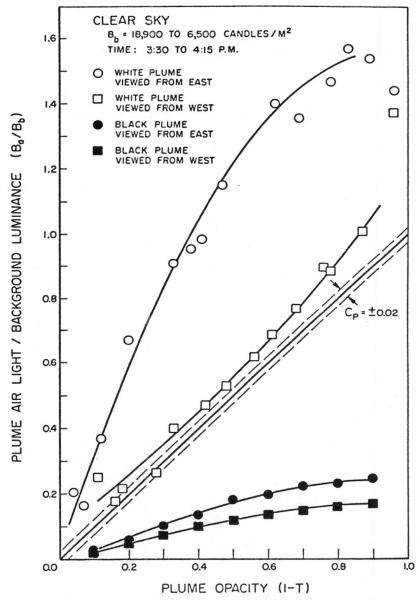

Fig. 3 Plume light scattering properties as a function of opacity.

the stack diameter L, the particle mass concentration W, and the particle density ρ, and calculated by solving for K in Eq. (6), as shown below

$$K = -\frac{WL}{\ln T\rho} = \frac{W}{b\rho} \tag{7}$$

Because of the difficulties in measuring the particle density ρ, measurements of $K\rho$ (g/m²) are at times more convenient, as reported by Crocker (1975). $K\rho$ is the particle mass concentration/light extinction ratio given by

$$K\rho = \frac{W}{b} \tag{8}$$

Also the parameter K can be theoretically calculated using the Mie equations, the particle size distribution, and the particle refractive index. With the particle mass concentration given by

$$W = \frac{4}{3}\pi N\rho \int_0^\infty r^3 f(r)\,dr \tag{9}$$

where N is the total particle number concentration and $f(r)\,dr$ the particle fraction frequency distribution. The $\ln T$ is given by

$$\ln T = -NL \int_0^\infty \pi r^2 Q_e f(r)\,dr \tag{10}$$

where Q_e is the light extinction efficiency factor. Substituting into Eq. (7) with Eq. (9) for W and Eq. (10) for $\ln T$ gives

$$K = \frac{4\int r^3 f(r)\,dr}{3\int Q_e r^2 f(r)\,dr} \tag{11}$$

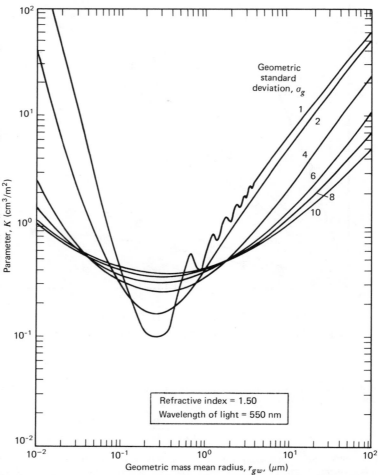

Fig. 4 Parameter K as a function of the log-normal size distribution parameters for a white aerosol.

The above equation is general and any particle size distribution function can be substituted for $f(r)$ dr. Assuming a particle size distribution which is log normal, the parameter K was calculated as a function of the particle geometric mass–mean radius and geometric standard deviation and reported by Pilat and Ensor (1970) and Ensor and Pilat (1971a). Figure 4 presents K versus the particle mass–mean radius for a white aerosol (refractive index of 1.5) and Figure 5 for a black aerosol (refractive index of $1.96 - 0.66i$). Thus by knowing the particle size distribution and refractive index, the parameter K can be obtained from one of the graphs. Then with the stack diameter, particle mass concentration, and the particle density, the opacity can be calculated using Eqs. (5) and (6).

For particle size distributions which are not log normal, Thielke and Pilat (1978) presented a calculation procedure which involves dividing the particle size distribution into increments and using parameter Ks for monodisperse particles corresponding to the mean size of the increment. The light transmittance is calculated using the equation

$$\ln T = -WL\left[\frac{f_1}{K_1\rho_1} + \frac{f_2}{K_2\rho_2} + \cdots \frac{f_n}{K_n\rho_n}\right] \tag{12}$$

where f_n is a fraction of the total particle mass concentration and K_n the specific volume/extinction coefficient ratio corresponding to the nth particle size increment.

A calculation procedure for theoretically predicting the magnitude of light transmittance downwind of an emission source was presented by Ensor et al. (1973). This procedure is based on a combination of the Mie light extinction theory and the Gaussian (Pasquill–Gifford) plume diffusion equations. Emission source information needed to estimate the downwind opacity includes the meteorological conditions, stack height, stack gas temperature and exit velocity, aerosol emission mass flow rate, size distribution, and particle refractive index.

2.4 Opacity Related to Particle Mass Concentration

Over the years, many have been interested in the relationship of the opacity to the particle mass concentration. Hurley and Bailey (1958) reported attempts to correlate the measured light extinction

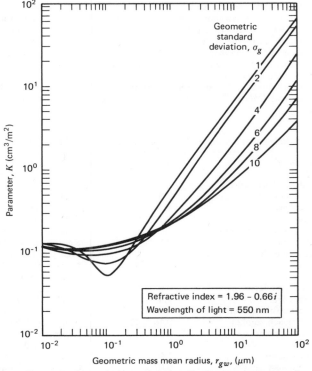

Fig. 5 Parameter K as a function of log-normal size distribution parameters for a black aerosol.

and particle concentration in stacks. The inability to control or measure the particle size distribution was reported by Engdahl (1951) and by Mitchell and Engdahl (1963) as a problem in developing a relationship between opacity and the particle mass concentration. Conner and Hodkinson (1967) reported good correlation between the light transmittance and particle mass concentration for black smoke. Larssen et al. (1972) reported good correlation of instack opacity (measured with a Bailey smoke meter) and particle mass concentration for a pulp kraft recovery boiler. Beutner (1974) reported correlations between optical density and the particle mass concentration for cement plants, a lignite-fired boiler, and a coal-fired boiler using a Lear Siegler RM4 transmissometer for the instack opacity measurements. Optical density is related to opacity and light transmittance by

$$\text{optical density} = -\log_{10} T = -\log_{10}(1 - O_p) \tag{13}$$

Cristello and Walther (1975) reported opacity–particle mass concentration data for a kraft recovery boiler and two hog-fuel boilers; Thielke and Pilat (1978) reported data for a kraft recovery boiler, a hog-fuel boiler, and a coal-fired boiler; Conner et al. (1979) reported data for portland cement plants and oil-fired boilers; Conner and White (1981) reported data for a coal-fired power plant; and Pinkerton and Blosser (1981) summarized data obtained by the National Council for Air and Stream Improvement at kraft pulp mills.

In general the measured magnitude of $K\rho$, the particle mass concentration to light extinction coefficient ratio, ranges from about 0.05 to 5 g/m². This variation is due to the differences in the particle size distribution and particle refractive indices from source to source. In general, small particle mass concentration/extinction coefficient ratios imply particles of a light absorbing nature (carbon) and small mass mean diameter (less than 0.05 μm). Large magnitudes of $K\rho$ imply the aerosol size distribution is dominated by larger particles and the mass mean diameter is larger than the optically active region (larger than 5 μm). The magnitudes of $K\rho$ for some emission sources are presented in Table 1.

2.5 Opacity of Condensing Water Vapor Plumes

The condensation of water vapor in the stack gases can cause the opacity to be substantially greater than if the gases were dry. Condensation after the stack gases exit the stack results in "detached plumes" which allow reading of the opacity before the water vapor condensation occurs. For wet plumes where condensation has occurred before exiting the stack, the plume opacity can only be read downwind at that location where the water has evaporated. For some wet plumes on a dry day, this location of the complete evaporation of the water is easily identified by the abrupt change in the opacity from a bright white plume to a residual gray haze. With hygroscopic aerosol particles such as some salts or sulfuric acid, the water evaporates slowly and the reading of the opacity of the dried plume is difficult. Yocom (1963) reviewed the problem of wet plumes and suggested that the gases be withdrawn from the stack, heated above the dewpoint, and passed through a transmissometer to measure the opacity after the water was evaporated.

The condensation of sulfuric acid along with the sorption of water vapor into the acid can significantly affect opacity. Nader and Conner (1974) reported that the out-stack opacity of emissions from oil-

Table 1 Particle Mass Concentration/Light Extinction Coefficient Ratios

Emission Source	$K\rho$, g/m²	Reference
Pulverized coal-fired boiler	3.8, 0.9	Conner (1974)
Pulverized coal-fired boiler	3.3	Ashton (1974)
Asphaltic concrete	8.5	Reisman et al. (1974)
Sludge incinerator	0.17	Reisman et al. (1974)
Catalytic cracker regenerator	1.7	Reisman et al. (1974)
Secondary brass smelter	0.8	Reisman et al. (1974)
Kraft recovery boiler	2	Pilat and Lutrick (1976)
Kraft recovery boiler	1, 3.5, 5.3	Larssen et al. (1972)
Kraft recovery boiler	1.7	NCASI (1972)
Kraft recovery boiler	0.6	Ensor and Pilat (1971a)
Cement plant kiln	1.6	Conner (1974)
Orchard heater (black smoke)	0.05	Ensor and Pilat (1971a)
Coal stoker (black smoke)	0.16–0.22	Ensor and Pilat (1971a)
Oil-fired power plant	0.11	Ensor and Pilat (1971a)
Veneer dryer	0.4	Ensor and Pilat (1971a)
Coal carbonization	1.9	Crocker (1975)
Jet engine exhaust	0.07	Grems and Chang (1976)

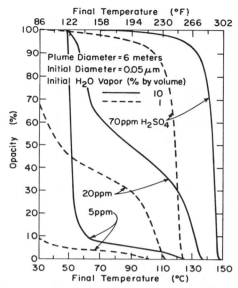

Fig. 6   Opacity as function of final temperature at 10^7 nuclei/cm^3.

fired boilers increased with increasing sulfuric acid concentrations, whereas instack opacity was not affected. Pilat and Wilder (1980) presented calculated opacities for the cooling of stack gases from 300° to 30°C with the results shown in Figure 6. The calculation assumptions include heterogeneous heteromolecular condensation of the H_2SO_4 and H_2O onto monodisperse nuclei of 0.05-μm diameter and a plume diameter of 6 m. The results show that the calculated opacities are significantly affected by the initial H_2SO_4 and initial H_2O concentrations and the final gas temperature. With stack gas temperatures in excess of 125°C, initial H_2SO_4 stack gas concentrations of 5 ppm or less results in calculated opacities of less than 20% for the 6-m plume diameter. (Note that this calculated opacity is almost entirely due to the sulfuric acid and there was little opacity due to the monodisperse nuclei particles, as shown in Figure 6.) However, for sulfuric acid concentrations in excess of 20 ppm, substantial increases in opacity are predicted. Pilat and Wilder (1983) have expanded this analysis to consider polydisperse particle size distributions.

3   INSTRUMENTAL MEASUREMENT OF OPACITY

3.1   Instack Transmissometers

Light transmissometers have been used to measure instack opacity for many years. Crider and Tash (1964) reported on the use of transmissometers to measure plume opacity. Conner and Hodkinson (1967) discussed the relationship of opacity, plume contrast, light transmissometers, and telephotometers. Ensor and Pilat (1971a) presented calculated effects of the light detector acceptance angle on the magnitude of the opacity measured by a light transmissometer. Peterson and Tomaides (1972) presented measured data of the effects of the light projection angle and the detector acceptance angle on the light transmittance. Conner (1974) discussed two design criteria for light transmissometers for opacity measurements: (1) the spectral response should be in the visible range, and (2) the transmissometer should have light collimating optics (i.e., have limited detector angle of view and limited light projection angle). Nader et al. (1974) discussed the proposed EPA transmissometer design and performance specifications prior to their promulgation in 1975. Avetta (1975) and Conner et al. (1979) reviewed the EPA transmissometer specifications and concluded that they were adequate for transmissometer opacity monitoring at the steel plant oxygen furnace and portland cement plants tested.

In 1971, the EPA promulgated Standards of Performance for New Stationary Sources which included requirements for the use on instack transmissometers for monitoring the opacity of some sources. The performance specifications for the transmissometers were issued by the EPA on October 6, 1975 in the *Federal Register*. The specifications included a calibration error of equal to or less than 3% opacity, zero drift for a 24-hr period of equal to or less than 2% opacity, a response time maximum of 10 sec, and an operation test period of 168 hr to measure the performance of the instrument. The spectral response of the instrument must have a mean and peak value between 500 and 600 nm with less than 10% of the spectral peak outside of the 400–700 nm wavelength range. The detector angle of view and the light projection angle cannot be greater than 5°.

McKee (1974) discussed the use of instack transmissometers to measure the opacity in place of visual evaluations, as permitted by the state of Texas. It was reported that the instrumental method appeared to avoid some of the difficulties inherent in the use of opacity regulations based on visual observations. Such difficulties included problems with visual observations during nighttime or cloudy and rainy days and the subjectivity of opacity evaluations made by human observers.

Steig and Pilat (1983) compared the instack opacities measured simultaneously by a portable and a cross-stack transmissometer at a coal-fired power plant. Good correlation was found between the portable opacity (measured with a Lear Siegler RM-41P) and the cross-stack opacity (measured with a Lear Siegler RM4) for the 1-hr average opacities. However, the 1-min average cross-stack opacities were consistently greater than the portable opacities, probably caused by the cross-stack transmissometer seeing puffs of particles from the electrostatic precipitator plate rapping. The tests were performed on a large rectangular duct with horizontal gas flow such that the cross-stack transmissometer had a one-way light path length of 6.15 m compared to only 1.27 m for the portable transmissometer.

3.2 Remote Measurement

Opacity measurements by remote means for the plume outside the stack have been made by telephotometers, sun photometers, and lidar. The use of laser radar (lidar) for remote measurement of the opacity of smoke plumes has been reported by Johnson et al. (1973). The techniques for using lidar to determine the transmittance though the plume involves the measurement of the light backscatter from a location in front of the plume and a location just beyond the plume. The lidar-determined transmittance is given by

$$T_L = \left(\frac{A}{B}\right)^{0.5} \tag{14}$$

where A is the magnitude of the light scattered from beyond the plume and B the magnitude of light scattered from in front of the plume. Nader and Conner (1978) reported lidar measured opacities for an oil-fired power plant along with observer and instack transmissometer measured opacities. The lidar and observer opacities compared reasonably well and were both considerably larger than the instack transmissometer opacities, probably because of sulfuric acid condensation in the stack gases after emission from the stack. Bethke (1977) reported on improvements to the electronics of the lidar system.

4 OPACITY REGULATION ISSUES

4.1 Regulation of Particulate Emissions

Opacity is the most common method used in the United States to evaluate or measure the emission of any air pollutant from a stationary source. The opacity evaluation does not require entry into a plant, does not require complex stack sampling, and does not require much time. However, the frequency of use and the frequency of plume opacity violation notices along with the accompanying fines have stimulated objections to opacity regulations over the years. However, both opacity and the Ringelmann number have apparently withstood these legal and scientific objections. Marks (1937) presented objections to the Ringelmann chart method including the complaint that the method is too strict for large diameter stacks and too lenient for small diameter stacks and the method is too dependent on the position of the sun and the observer. There have been a number of papers by attorneys objecting to opacity regulations. Henz (1970) reviewed the Ringelmann charts and questioned the relationship between opacity and the amount of pollutants being emitted into the atmosphere. Hawthorne and Rankin (1974) presented a number of objections to the use of opacity including the lack of precision and the question of the regulation being constitutional. Mastel (1976) reviewed the validity of opacity standards and issues such as: (1) does opacity accurately reflect the amount of pollution in the smoke, (2) opacity laws are unconstitutional because they are so arbitrary and capricious, and (3) the competency of the witness. Mastel claimed that the validity and admissibility of EPA and state visible emission standards are well established throughout the United States, having withstood countless challenges. Giblin (1972) presented the legal aspects of the basis for using opacity as an enforceable emission standard and identified the basic elements in a suit based on opacity as:

1. Establishment of the basic legal validity of opacity. In order to overturn a regulation such as opacity, it is necessary to demonstrate that it is so arbitrary and capricious that no reasonable administrator would have adopted it. This is very difficult to show because of the numerous air pollution control governmental agencies throughout the United States that have adopted opacity regulations and the many court cases that have upheld it.

2. Presentation of the methods by which opacity observers are trained. This establishes the qualifications of the certified observer.

3. Presentation of the testimony of the opacity observer and documentation of the proper plume observation methods.

4.2 Plume Contrast Versus Plume Opacity

Opacity observers are trained during visible emission schooling to read plume opacity, not plume contrast (or plume appearance). But these two parameters, related by Eq. (4), have been confused in the literature with the plume contrast being at times referred to as the plume opacity, when in fact they are not equal. Weir et al. (1976a) reported that previously unrecognized uncontrolled variables such as the sun angle, time of day, and geographic location of the power plant (related to the altitude of the sun) greatly influence smoke plume opacity. This is an error as these variables may influence the plume contrast but they have no effect on plume opacity. Wolbach and Key (1976) challenged the information presented in the Weir et al. article with regard to the relationship of opacity to particle mass concentration, the inability of trained observers to accurately assess plume opacity, the precision of opacity observations, and the effect of geographic location. Weir et al. (1976b) replied and claimed to have experimental data in Nevada and California relating sun altitude and opacity (claiming that opacity increases with increasing sun altitude assuming all other variables are constant). Goodwin (1977a) stated that the EPA did not agree with the analysis presented by Weir et al. and pointed out that this analysis procedure did not evaluate plume opacity but rather evaluated plume appearance. Goodwin noted that neither the Weir et al. (1976a) analysis nor the Halow and Zeek (1973) equations apply to plume opacity. Faith (1977) reported that opacity regulations provide an inexpensive way for regulatory agencies to control both large and small sources and noted that the violators usually pay the fines without contesting the violations because they know it is useless to argue. Weir (1977) responded to Goodwin's comments and presented further arguments regarding the effect of sun altitude. Key (1977) replied that Weir's statements on the effect of sun angle or altitude on opacity can be refuted by data collected by the Texas Air Control Board. Goodwin's (1977b) reply to Weir's (1977) response stated that Weir had conveyed a totally false characterization of the accuracy of visual opacity observations by distorting previous statements, utilizing misleading data presentation, utilizing data selected out of context, and basing conclusions on postulations. ES&T refused to publish Goodwin's reply.

The motivation for such a controversy is apparently the objection by those industries having emission plumes which periodically exceed opacity emission standards and accordingly receive opacity violation notices and the accompanying fines. Hence, these industries desire to have the opacity regulations revised so that the emissions no longer would be in violation. On the other side of the controversy is the preference of the regulatory agencies to retain opacity as a means of ensuring that particulate control systems required by particle mass or concentration standards are properly operating and maintained at all times. Recognizing that plume opacity is easily observed by both agency observers and the general public, it would appear that opacity standards will continue to be used for the regulation of particulate emissions for many years to come.

REFERENCES

Ashton, T. (1974), *Opacity Monitors for Stack Effluents,* Pacific Power & Light Co., Portland, Oregon.

Avetta, E. J. (1975), *Instack Transmissometer Evaluation and Application to Particulate Opacity Measurement,* EPA Report No. EPA-650/2-75-008.

Bethke, G. W. (1973), *Development of Range Squared and Off-gating Modifications for a Lidar System,* EPA Report No. EPA-650/2-73-040.

Beutner, H. (1974), Measurement of Opacity and Particulate Emissions with an On-stack Transmissometer, *JAPCA* **24,** 865–871.

Blackwell, R. H. (1946), Contrast Thresholds of the Human Eye, *J. Opt. Soc. Am.* **36,** 624–643.

Conner, W. D. (1974), *Measurement of the Opacity and Mass Concentration of Particulate Emissions by Transmissometry,* EPA Report No. EPA-650/2-74-128.

Conner, W. D., and Hodkinson, J. R. (1967), *Optical Properties and Visual Effects of Smoke Plumes,* USPHS Pub. No. 999-AP-30.

Conner, W. D., Knapp, K. T., and Nader, J. S. (1979), *Applicability of Transmissometers to Opacity Measurement of Emissions: Oil-Fired Power and Portland Cement Plants,* EPA Report No. EPA-600/2-79-188, NTIS PB 80-135-239.

Conner, W. D., and McElhoe, H. B. (1982), Comparison of Opacity Measurements by Trained Observer and Instack Transmissometer, *JAPCA* **32,** 943–946.

Conner, W. D., Smith, C., and Nader, J. (1968), Development of a Smoke Guide for Evaluation of White Plumes, *JAPCA* **18,** 748–750.

Conner, W. D., and White, N. (1981), Correlation Between Light Attenuation and Particulate Concentration of a Coal-Fired Power Plant Emission, *Atmospheric Environment* **15**, 939–944.

Coons, J. D., James, H. A., Johnson, H. C., and Walker, M. S. (1965), Development Calibration and Use of a Plume Evaluation Training Unit, *JAPCA* **15** 199–203.

Crider, W. L., and Tash, J. A. (1964), Status Report: Study of Vision Obscuration of Nonblack Plumes, *JAPCA* **14**, 161–167.

Cristello, J. D., and Walther, J. E. (1975), Continuous On-stack Monitoring of Particulate Proves Feasible, *Pulp & Paper* **49**, 122–124.

Crocker, B. B. (1975), Monitoring Particulate Emissions, *Chem. Engr. Progress* **71**, 83–89.

Engdahl, R. B. (1951), Correlations of Solids Content in Gas, *Mech. Engr.* **73**, 243–244.

Ensor, D. S. (1972), Smoke Plume Opacity Related to the Properties of Air Pollutant Aerosols, PhD dissertation, Dept. of Civil Engr., Univ. of Washington, Seattle, WA.

Ensor, D. S., and Pilat, M. J. (1970), The Relationship Between the Visibility and Aerosol Properties of Smoke-Stack Plumes, presented at the 2nd Int. Air Poll. Conf., Washington, D.C., Dec. 1970.

Ensor, D. S., and Pilat, M. J. (1971a), Calculation of Smoke Plume Opacity from Particulate Air Pollutant Properties, *JAPCA* **21**, 496–501.

Ensor, D. S., and Pilat, M. J. (1971b), The Effect of Particle Size Distribution on Light Transmittance Measurement, *Am. Ind. Hyg. Assoc. J.* **32**, 287–292.

Ensor, D. S., Sparks, L. E., and Pilat, M. J. (1973), Light Transmittance Across Smoke Plumes Downwind from Point Sources of Aerosol Emissions, *Atmospheric Environment* **7**, 1267–1277.

Faith, W. L. (1977), Opacity, *Env. Sci. & Tech.* **11**, 327.

Federal Register (1971), Standards of Performance for New Stationary Sources, 36(247), 24878–24895, Dec. 23, 1971.

Federal Register (1974), Opacity provisions for New Stationary Sources Promulgated and Appendix A, Method 9—Visual Determination of the Opacity of Emissions from Stationary Sources, 39(219), 39872–39876, Nov. 12, 1974.

Federal Register (1975), Standards of Performance Promulgated for Emission Monitoring Requirements and Revisions to Performance Testing Methods, 40(194), 46240–46271, Oct. 6, 1975.

Giblin, P. (1972), Opacity as a Readily Enforceable Standard, presented at the APCA Annual Mtg., Miami, FL.

Gifford, F. (1959), Smoke Plumes as Quantitative Air Pollution Indices, *Int. J. Air Poll.* **2**, 42–50.

Goodwin, D. R. (1977a), Opacity, a Federal View, *Env. Sci. & Tech.* **11**, 10–11.

Goodwin, D. R. (1977b), letter to S. Miller, Managing Editor, *Env. Sci. & Tech.*, June 16, 1977.

Grems, B., and Chang, D. (1976), Plume Opacity and Particulate Emissions from a Jet Engine Test Facility, presented at the 15th Annual Air Quality Conf., Purdue University.

Griswold, S., Parmelee, W., and McEwan, L. (1958), Training of Air Pollution Inspectors, presented at the APCA Annual Mtg., Philadelphia, PA.

Hague, R., Deieso, D., and Flower, F. (1977), Factors Affecting Accuracy in Visual Emissions Evaluation, presented at the APCA Annual Mtg., Toronto, Canada.

Halow, J., and Zeek, S. (1973), Predicting Ringelmann Number and Optical Characteristics of Plumes, *JAPCA* **23**, 676–684.

Hamil, H., Thomas, R., and Swynnerton, N. (1975), Evaluation and Collaborative Study of Method for Visual Determination of Opacity Emissions from Stationary Sources, EPA Report No. EPA-650/4-75-009.

Hawthorne, R., and Rankin, J. (1974), Visual Plume Readings—Too Crude for Clean Air Laws, *Natural Resources Lawyer* **7**, 457–477.

Henz, D. J. (1970), The Ringelmann Number as a Irrebutable Presumption of Guilt—An Outdated Concept, *Natural Resources Lawyer* **3**, 232–240.

Herdan, G. (1960), *Small Particle Statistics*, Butterworths, London.

Hurley, T., and Bailey, P. (1958), The Correlation of Optical Density with the Concentration and Composition of the Smoke Emitted from a Lancashire Boiler, *J. Inst. Fuel* **31**, 534–540.

Jarman, R., and de Turville, C. (1966), The Screening Power of Visible Smoke, *Int. J. Air Water Pollution* **10**, 465–467.

Jarman, R., and de Turville, C. (1969), The Visibility and Length of Chimney Plumes, *Atmospheric Environment* **3**, 257–280.

Johnson, W., Allen, R., and Evans, W. (1973), *Lidar Studies of Stack Plumes in Rural and Urban Environments*, EPA Report No. EPA-650/4-73-002.

Kent, W. (1897), *Engineering News*, Nov. 11, 1897.

Key, J. (1977), Opacity Controversy, *Env. Sci. & Tech.* **11**, 842.

Larssen, S. (1971), Relationship of Plume Light Transmittance to the Properties of the Particulates Emitted from Kraft Recovery Furnaces, MSE thesis, Dept. of Civil Engr., Univ. of Washington, Seattle, WA.

Larssen, S., Ensor, D., and Pilat, M. (1972), Relationship of Plume Opacity to the Properties of Particulates Emitted from Kraft Recovery Furnaces, *TAPPI* **55**, 88–92.

Marks, L. (1937), Inadequacy of the Ringelmann Chart, *Mech. Engr.* **59**, 681–685.

Mastel, M. (1976), *The validity of Visible Emissions and Opacity Standards,* EPA Enforcement Branch, Jan. 1976.

McKee, H. (1974), Texas Regulation Requires Control of Opacity Using Instrumental Measurements, *JAPCA* **24**, 601–604.

Midgett, R. (1976), *The EPA Program for Standardization of Stationary Source Emission Test Methodology—A Review,* EPA Report No. EPA-600/4-76-044.

Mitchell, R., and Engdahl, R. (1963), A Survey for Improved Methods for the Measurement of Particulate Concentration in Flowing Gas Streams, *JAPCA* **13**, 9–14.

Nader, J., and Conner, W. (1978), Impact of Sulfuric Acid on Plume Opacity, *Workshop Proceedings on Primary Sulfate Emissions from Combustion Sources.* EPA Report No. EPA-600/ 9-78-020b, pp. 121–136.

Nader, J., Jaye, F., and Conner, W. (1974), *Performance Specifications for Stationary Source Monitoring Systems for Gases and Visible Emissions,* EPA Report No. EPA-650/2-74-013, NTIS PB 230-934.

NCASI (1972), *The Relationship of Particulate Concentration and Observed Plume Opacity at Kraft Recovery Furnaces and Lime Kilns,* NCASI Tech. Bull. No. 82.

Peterson, C., and Tomaides, M. (1972), *Instack Transmittance Techniques for Measuring Opacities of Particulate Emissions from Stationary Sources,* EPA Report, NTIS No. PB 212-741.

Pilat, M., and Ensor, D. (1970), Plume Opacity and Particulate Mass Concentration, *Atmospheric Environment* **4**, 163–173.

Pilat, M., and Ensor, D. (1971), Comparison Between the Light Extinction Aerosol Mass Concentration Relationship of Atmospheric and Air Pollutant Emission Aerosols, *Atmospheric Environment* **5**, 209–215.

Pilat, M., and Lutrick, D. (1976), *Relationship Between Instack Opacity and Particle Properties,* Progress Report, EPA Grant 80072.

Pilat, M., and Wilder, J. (1983), Opacity of Monodisperse Sulfuric Acid Aerosols, *Atmospheric Environment* **17** (in press).

Pinkerton, J., and Blosser, R. (1981), Characterization of Kraft Pulp Mill Particulate Emissions—A Summary of Existing Measurements and Observations, *Atmospheric Environment* **15**, 2071–2078.

Reisman, E., Gerber, W., and Potter, N. (1974), *Instack Transmissometer Measurement of Particulate Opacity and Mass Concentration,* EPA Report No. EPA-650/2-74-120, NTIS PB 239-864.

Ringelmann, M. (1898), Méthode d'Estimation des Fumés Produits par les Foyer Industriels, *La Revue Technique* **268** (June 1898).

Steig, T. W. (1981), Opacity Stratification at a Coal-Fired Boiler Using Cross-Stack and Portable Transmissometers, MSCE thesis, Dept. of Civil Engr., Univ. of Washington, Seattle, WA.

Steig, T., and Pilat, M. (1983), Comparison of Opacities Measured by Portable and Cross-Stack Transmissometers at a Coal-Fired Power Plant, *Atmospheric Environment* **17**, 1–9.

Thielke, J., and Pilat, M. (1978), Plume Opacity Related to Particle Mass Concentration and Size Distribution, *Atmospheric Environment* **12**, 2439–2447.

Weir, A. (1977), Clearing the Opacity Issue, *Env. Sci. & Tech.* **11**, 561–563.

Weir, A., Jones, D., Papay, L., Calvert, S., and Yung, S. (1976a), Factors Influencing Plume Opacity, *Env. Sci. & Tech.* **10**, 539–544.

Weir, A., Jones, D., Papay, L., Calvert, S., and Yung, S. (1976b), Opacity, *Env. Sci. & Tech.* **10**, 965.

Wilder, J. M. (1981), Theoretical Droplet Size Distribution and Opacities of Condensed Sulfuric Acid Aerosols, MSCE thesis, Dept. of Civil Engr., Univ. of Washington, Seattle, WA.

Wolbach, C., and Key, J. (1976), Opacity, *Env. Sci. & Tech.* **10**, 847–848.

Yocom, J. (1963), Problems in Judging Plume Opacity—A Simple Device for Measuring Opacity of Wet Plumes, *JAPCA* **13**, 36–39.

CHAPTER 33

ODOR SAMPLING
AND ANALYSIS

GREGORY LEONARDOS

Environmental Odor Consultant
Arlington, Massachusetts

1 INTRODUCTION

Odor is the most "visible" form of air pollution in that it can be perceived through the sense of smell that we all possess. As many as 50% of all citizen complaints to local pollution control agencies are associated with odors.[1]

Odor can be defined as sensations resulting from the interaction of chemical species (i.e., odorants) present in the air with the olfactory area located in the upper reaches of the nose and registering in the brain. Trained observers sniff the air to facilitate the odor-bearing air to reach the olfactory area. Not all chemicals that impinge on the olfactory area however are odorous. Chemicals that elicit odor usually are volatile and have some degree of polarity although some low-volatility materials are powerful odorants. It is not possible at this time to predict odor of a given chemical from a consideration of its chemical structure. An odor may result from the interaction of one or more chemical species. Commonly odor is due to a mixture of many chemical species. As an example, several thousand chemicals were isolated and identified as being present in the odor of diesel exhaust.[2] Chemicals that elicit the odor response and that are the stimuli of odor are termed odorants. The odor sense is concentration dependent. Response to odorant chemicals may vary over six to eight orders of magnitude of concentration. Studies on the odor properties of commercially important chemicals indicated recognition odor threshold concentrations of 0.000001 ppm for o-iodophenol and o-bromophenol; the recognition odor threshold concentration for methylene chloride was 214 ppm.[3] Table 1 summarizes the recognition odor threshold concentrations of selected representative odorants. Other reference sources listing odor threshold concentrations for chemicals may be found in the References.[4,5,6]

The recognition odor threshold concentrations reported in the table represent the lowest concentration at which (all) four trained panelists could recognize the odor quality and at all higher test concentrations. The threshold concentration (or dilution) is not a fixed point but represents a range. At low concentrations of odorant in air, there is no recognition of the presence of odorant. At the threshold concentration range there is an increasing likelihood that the observers, on repeated presentations, will indicate the presence of odor. As the concentration is increased further, trained panelists can assign an intensity rating according to the category scale presented in Table 2 or other suitable scale (i.e., magnitude estimation, reference standard, etc.). The threshold (0.2) intensity generally coincides with 50% panel response while 100% panel response falls in the 0.5–1 (very slight to slight) intensity range. The intensity increases as a function of the logarithm of the concentration.

The foregoing two dimensions of odor were detectability and intensity. The other two dimensions of odor are quality (character of the odor) and hedonics (like–dislike or annoyance). Trained observers can be used for evaluating three of the four dimensions of odor detectability, intensity, and quality. From a consideration of intensity and quality obtained by trained observers, inferences can be made as to the hedonic dimension as measured by the general population. There are very little data available in the literature concerning the hedonic dimension.

The quality dimension of odor can be described by using the chemical name, or an associative term that is descriptive. As an example, the odor quality of H_2S can be described by the chemical name or by "eggy sulfide," trimethylamine as "fishy." A mixture of two or more odorants may give an odor quality that is dissimilar to the two components when smelled alone. Table 1 provides typical

Table 1 Recognition Odor Thresholds and Odor Quality Descriptors of Selected Chemicals

Chemical	Recognition Odor Threshold,[a] ppm V/V	Odor Description[b]
Acetone	100.0	Sweet, pungent
Acrolein	0.21	Burnt sweet, pungent
Amine, trimethyl	0.00021	Fishy
Ammonia	47.0	Pungent
Benzene	4.7	Solvent
Benzyl sulfide	0.0021	Sulfide
Butyric acid	0.001	Sour
Dimethyl sulfide	0.001	Vegetable sulfide
Diphenyl sulfide	0.0047	Burnt rubbery
Ethyl acrylate	0.00047	Hot plastic, earthy
Hydrogen sulfide	0.00047	Eggy sulfide
Methylene chloride	214.0	
Methyl ethylketone	10.0	Sweet
Nitrobenzene	0.0047	Shoe polish, pungent
o-iodophenol	0.000001	Medicinal
o-bromophenol	0.000001	Medicinal
Phosgene	1.0	Haylike
Pyridine	0.021	Burnt, pungent, diamine
Sulfur dioxide	0.47	
Trichloroethylene	21.4	Solvent

[a] Based on 100% panel response.[3]
[b] Other than chemical name.

odor descriptors for selected odorants. Quality descriptors used to describe diesel exhaust odor are burnt smoky and unburnt fuel related (oily-kerosene).

2 DEFINING ODOR POLLUTION PROBLEMS

This section describes procedures that the author has used to successfully define odor pollution problems for a variety of industrial and municipal facilities. Odors can be released into the community from a variety of source types. Odors may be released from stacks or vents, from fugitive emissions such as leaks, spills, flares, and so on, and large area sources (wastewater treatment systems such as trickling filters, sludge handling processes, etc., and feed lots, among others). Ventilation exhausts from buildings housing odor-generating processes can also be a source of odors that lead to community complaint.

To successfully abate an odor pollution problem, all sources that are significant contributors of odor in the ambient must be identified, measured to establish control target levels, and then controls implemented.

A facility that emits odor from a single source requires less effort than a facility with a multiplicity of potential odor sources. However, even the "single-source" facility requires an accurate assessment of the odor contribution of the source to define the degree of control that is necessary. Even with the single-source facility, a community (ambient) survey and an in-plant review should be carried out to confirm that the presumed single source is indeed the only odor contributor and that other unsuspected sources are not overlooked.

Table 2 Intensity Scale

3.0	Strong
2.5	Moderate to strong
2.0	Moderate
1.5	Slight to moderate
1.0	Slight
0.5	Very slight
0.2	Threshold, just recognizable
0	No odor

There are essentially three steps involved in defining the odor problem prevailing at a given facility— the ambient survey, the on-site review, and a source sampling and evaluation program.

2.1 Ambient Odor Surveys

The essential first step in determining those sources which must be controlled is to conduct odor surveys in the area surrounding the plant location. This may be carried out concurrently with the on-site review. The major goal of the off-site survey is to determine the quality and intensity of odors in the areas adjacent to and beyond the facility and to note their frequency and duration. These surveys should be carried out by a team of observers who have been trained to improve their capability in describing odor quality in associative descriptive terms and intensity through the use of appropriate odor reference standards. The surveys are carried out by the trained observers under a variety of climatic conditions, at different times of the day, and over a reasonable time period. Each survey may last for as little as 15 min up to several hours. During a survey, the observers follow a predetermined route taking into account the prevailing wind direction and velocity. The surveys are carried out in a slow moving automobile and on foot. A small (one- to three-member) team that is mobile is preferred to stationing observers at preselected points downwind. The mobile observers are not subjected to a constant exposure to odor which may induce fatigue and thus affect perception of odor. A trained mobile team of observers has the advantage of covering a large area (up to several miles) with a small number of panelists. The survey team follows the predetermined route constantly sniffing the air. During the survey, the team should cover the area in a grid pattern going from nonodorous areas to odorous areas. Upon detecting odor, the quality and intensity are recorded as well as the time, location, and the area affected by the odor. Tables and odor maps summarizing the observations are then prepared for each survey period and then analyzed. Generally, it has been found that odors present in the community that exceed the 1.0–1.5 intensity range, irrespective of odor quality, may lead to community complaint. At these intensity levels and above, even normally pleasant odors may lead to citizen complaint. Some odor types, especially those that are universally considered to be malodorous, may be considered to be at complaint level if they are at the just recognizable level.

2.2 On-Site Review

The trained observers should be utilized to review all aspects of a given facility's operations and processes. The in-plant review can be carried out in conjunction with the ambient surveys, either before or after. An advantage is that the observers can relate the odor qualities observed off-site with the particular source within the plant that emits a similar odor quality. All potential sources should be evaluated by the team with particular emphasis being given to developing description or odor character notes that adequately describe the odor quality and upon which the observers can agree. All stacks and vents should be characterized for odor quality and intensity. The team should be on the lookout for fugitive odor emissions from the site. Fugitive emissions can occur from leaks from faulty valves or couplings, spills, or poor housekeeping practices such as storing volatile odorous materials in open containers. Transfer operations such as raw material unloading and finished product loading should be evaluated. As an example of the impact of a fugitive odor emission, we have noted the odor of ethyl acrylate approximately 1 mile away from a leaking coupling that was releasing drops of ethylacrylate. Transfer operations involving highly volatile and odorous chemicals or materials with thresholds in the 10–100 ppb range or less can create odor problems if improperly handled. Another advantage of observing transfer operations is to ensure that personnel assigned are adequately trained to minimize release of odors.

Large area sources such as wastewater treatment systems (and their various elements), solid wastes, and so on, can be evaluated by the odor team by driving or walking around the source noting the quality and intensity at various downwind locations from the source.

The on-site review should cover all aspects of operations, processes, and practices carried out within the property or fence line. The ambient or off-site survey should be conducted at the property and/or fence line and beyond into the community. Thus the on-site (or in-plant) review is an extension of the ambient community survey.

Upon completion of the ambient and on-site surveys, a preliminary ranking of the major odor contributors can be made. Odors with a given odor quality observed off-site may arise from one or more sources on-site. It is important to identify all the contributors (sources) of a given complaint level odor. The descriptive capability of the observers is very useful in identifying the odors detected off-site with the respective source or sources within the plant.

2.3 Source Sampling and Evaluation

Upon identification of the major sources, samples may then be taken from these sources to quantify the odor emission level and to establish the degree of control necessary. Much of the past work on odor pollution measurement has been based on the dilution to threshold concept. In this approach, a stack or vent emitting odor-bearing air is sampled in glass or bags and then diluted in an apparatus (olfactometer) with various amounts of odor-free air to the "threshold" of an observer (i.e., the highest dilution at which the observer can detect the presence of odor). One or more observers may be on the panel. The dilutions to threshold cannot be regarded as being a measure of odor intensity. It is a measure of the detectability dimension of odor (i.e., the presence or absence of odor upon dilution). In addition to having no relationship to the odor intensity, no information is given as to odor quality.

The dilution to threshold concept has not been related to community odor nuisance and has been incorporated into only three state odor pollution control regulations. Various olfactometers for measuring thresholds that have been developed are briefly described in a later section of this chapter.

In the laboratories of Arthur D. Little, Inc., a source sampling–laboratory odor evaluation technique has been developed for relating source odor measurements to odor measurements in the community (as described in the preceding sections).[7]

For each of the sources identified in the off-site and on-site surveys as being potential major odor contributors, samples are taken utilizing a polymeric absorbent (Chromosorb 102®) for subsequent evaluation in a dynamic testroom.

2.3.1   Chromosorb 102 Source Sampling Technique

This procedure has been found to be useful for sampling such point sources as stacks and vents from a variety of industries. These include coffee spray driers, agglomerators, spent grounds scrubbers, green bean driers, diesel exhaust, jet engine exhaust, turbine exhaust, sewage sludge processing exhaust, primary sedimentation tank exhaust, aerated grit chamber exhaust, sewage sludge incinerator exhaust, automobile paint bake ovens, and refractory brick oven exhausts, among others. It has also successfully sampled odors emanating from large area sources such as trickling filters. The only instance where Chromosorb 102 was unsuccessful was in sampling trimethylamine in air mixtures generated in the laboratory (Tedlar bags also were not successful) as well as low molecular weight oxygen or sulfur-bearing species (less than C_2—C_3).

Ten grams of prewashed Chromosorb 102 is placed in a 1×6 in. stainless steel sorbent trap. The sampling train consists of a stainless steel probe, heated filter (if necessary to prevent condensation), odor sorbent trap, a metal bellows pump, and a dry gas meter to record the flow volume sampled. For continuous sources, collections may range from a 1- to 4-hr period depending on the odor strength of the source. For sporadic emissions, the sampling system can be operated selectively to reflect source operation. Sampling through the sorbent traps is at the rate of approximately 10 l/min.

The odor-bearing species are then eluted quantitatively in the laboratory from the sorbent with purified pentane collecting the first 5 or 10 ml fraction in a volumetric flask. The samples are then stored in a refrigerator until they are evaluated.

2.3.2   Odor Evaluation Procedure

A panel of four trained analysts is used to evaluate the sample in a total immersion dynamic testroom[7] that can present a range of dilutions of trapped odor components in air. As many as 12 separate dilutions may be presented. Each dilution is presented in replicate and the observer is not told the order of, or when, a dilution is being presented. Each sample dilution is presented over a 15-sec period followed by a 1–2 min exposure to background dilution air. Each observer is asked to report when odor is perceived, indicating the odor quality and intensity.

2.3.3   Dose–Response Curves

For each sample dilution, the odor intensities reported are averaged. Table 3 summarizes the panel odor data obtained from a sewage sludge incinerator scrubber exhaust and the regression analysis. In addition, a probability of detection factor for each sample dilution (PD factor) is calculated by taking into consideration the number of positive odor recognitions as a function of the total number of presentations and serves to identify the threshold dilution range. A dose–response curve is then derived using linear regression analysis for each sample from the intensity–dilution data, according to the following relationship:

$$\text{intensity} = a \log_{10} \text{dilution} + b \tag{1}$$

where a is the slope and b is the intercept.

Based on the relationship of Eq. (1), an extrapolated threshold dilution (0.2) can be derived from the regression analysis of the panel data. The source odor strength (SOS) is defined as the reciprocal of the extrapolated threshold dilution (another intensity level could be used). Figure 1 summarizes dose–response curves for four sources at a wastewater treatment plant.

Table 3   Panel Odor Data for Sludge Incinerator Scrubber Exhaust[8]

$\dfrac{\text{L. exh.}}{\text{L. air}}$	Dilutions	Average Panel Intensity	Probability of Detection
0.0026	1:390	2.13	1
0.0017	1:620	1.78	1
0.0092	1:1090	1.41	1
0.00054	1:1850	1.09	1
0.00042	1:2550	0.78	1
0.00025	1:3900	1.19	1
0.00016	1:6200	0.75	1
0.000092	1:10900	0.41	1
0.000054	1:18500	0.25	0.63
0.000043	1:23500	0.06	0.13

Regression analysis:	intensity $= a \log_{10}$ dilution $+ b$
Correlation coefficient:	0.950
(a) slope:	1.017
(b) intercept:	4.53
Extrapolated threshold dilution: (intensity $= 0.2$)	$0.000055\ \dfrac{\text{L. exh.}}{\text{L. air}}$ or 1:18200
Source odor strength:	18200

2.3.4   Analysis of Odor Source Data

The source odor strength which is based on the threshold dilution ratio does not alone provide sufficient information for determining which of many sources at a given facility may contribute to ambient odor. Other important factors are the volumetric flow rate of the emission source which can be used with the source odor strength to calculate the odor emission rate (Q). The odor emission rate can then be used on the atmospheric dispersion estimates and a consideration of the effective release point to determine a maximum ambient dilution ratio (χ) which with Eq. (2),

$$\text{ambient odor intensity} = a \log_{10} \chi + 0.2 \qquad (2)$$

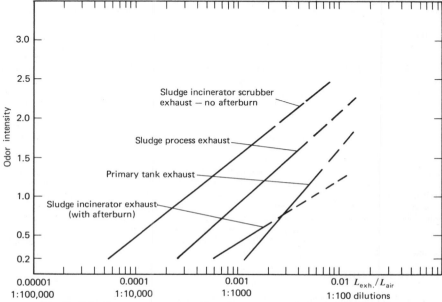

Fig. 1   Odor dose-response curves for primary sewage treatment plant exhausts.

can be used to calculate the ambient odor intensity for the source for a particular meteorological condition. a is the slope as determined for each source from Eq. (1). Table 4 summarizes the odor data obtained from four sources at a primary treatment wastewater treatment plant.[8] The predicted ambient odor intensities as shown in Table 4 were in good agreement with observed intensities observed off-site during the sampling period for the particular source. Utilizing the criteria discussed in Section 2.1., the first two sources listed do require further control as their intensity is above the slight (1) range. The primary tank exhaust is at a borderline intensity and its odor character would be a major factor (as well as frequency and duration) in deciding whether further control is necessary. Simply elevating the release point may be sufficient to reduce the odor to below perceptible levels. The afterburner has reduced emissions from the sludge incinerator exhaust so that odor would not be perceptible in the ambient. Although this source had a higher source odor strength than the primary tank exhaust, the former had a higher release point and a lower dose–response curve slope resulting in a lower ambient odor intensity.

3 ODOR CONTROL REGULATIONS

There are no federal regulations in the United States for the control of odors, nor is it likely that any will be promulgated by the Environmental Protection Agency in the foreseeable future.[9] State and local odor control regulations may be classified into nine general types.[1] These are:

No specific regulations.

Air pollution/nuisance regulations.

Criteria specified for determining objectability of an ambient odor.

Scentometer measurements in the ambient air not to exceed specified dilution to threshold levels.

Source required to use the highest and best practicable or reasonable and suitable odor control system.

Source emission standards specifying "odor concentration units" that are not to be exceeded, usually based on the syringe dilution technique.

Regulations that serve as a statement of policy for handling odor problems.

Both source and ambient standards specified.

Sixteen states have no specific regulations, while seventeen state agencies rely on the air pollution/nuisance regulation. Typically these are stated as "No person shall cause . . . any discharge of gases, vapor, or odors in such a manner that a nuisance or air pollution is created" (Maryland).

Seven states rely on the scentometer to measure odor in the ambient air. Generally, an ambient odor limit not to exceed seven dilutions to threshold is specified.

Only three states have a source emission standard using the syringe dilution technique. These source standards range from 25 or less to 150 "odor concentration units" (essentially dilutions to threshold or a dilution factor). States that include these source standards in their regulation are Minnesota, Illinois (only for inedible rendering plants), and Connecticut.

The Bay Area Air Pollution Control District (California) has recently promulgated both an ambient odor and a source odor standard using a dynamic olfactometer. The ambient limitation is four dilutions to threshold. The source standard limitation is based on elevation of the emission point above grade (Table 5).

It should be pointed out that although the same olfactometer is used to measure odor in the ambient and at the source, no relationship has been established between the two measurements for a given situation.

The variety of existing odor control regulations and procedures followed by state and local pollution control agencies make it imperative that the applicable regulations be ascertained. References 1, 9, and 10 provide recent reviews.

There is considerable dissatisfaction with the scentometer and the ASTM syringe dilution techniques as the basis for odor measurement in regulations.[1,9,10] The TT-4 Odor Committee of the Air Pollution Control Association declined to endorse the revised ASTM D1391 procedure. A consensus of the committee indicated its opposition to the use of the scentometer in regulations and recommended the use of improved devices based on dynamic dilution.

The intent of virtually all odor control regulations is to control odors that create a nuisance or annoyance to the community. Data on this dimension of the odor experience do not exist. At the present time, to estimate the annoyance dimension, both the odor quality and intensity should be measured rather than relying on measures of detectability.

4 SAMPLING ODORS

The choice of the sampling system is dependent on the chemical and physical properties of the chemical components that are responsible for the odor. These chemical components are often not known, particu-

Table 4 Odor Emissions from Sewage Treatment Plant

Source	Source Odor Strength	Vol. Flow Rate, m³/sec	Odor Emission Rate Q	Release Point (ΔH)	Max. χ Value	Slope[a]	Ambient Odor Intensity
Sludge incinerator scrubber exhaust (no afterburn)	18,200	9.77	176,000	20 m	123	1.02	2.3
Sludge process exhaust	3,800	7.36	27,700	10 m	8.4	1.18	1.1
Primary tank exhaust	800	7.36	6,000	10 m	1.8	1.54	0.6
Sludge incinerator exhaust (with afterburn)	1,700	9.77	16,500	30 m	0.48	0.80	0

[a] For D stability condition, 10 mph wind velocity.

Table 5   Bay Area Source Emission Limits

Emission Release Point above Grade, ft	Dilution Ratio
Less than 30	1,000
30–60	3,000
60–100	9,000
100–180	30,000
Greater than 180	50,000

larly in processes where oxidation or degradation can occur. Odorants can have molecular weights up to 300 to 400, and have varying degrees of polarity. The presence of moisture can interfere with effective sampling. Due to the complexity of various odors, it is highly unlikely that a single sampling technique will be useful for all odor assessment problems. A major factor that should be considered in sampling is the possibility of component condensation and/or adsorption onto the walls of the container.

4.1   Sampling Stack Odor Emissions

A number of techniques have been used for sampling well-defined sources such as stacks or vents. These have included grab sampling in glass or stainless steel bulbs, sorbent sampling, cryogenic traps, and bag sampling with and without predilution.

4.1.1   Grab Sampling

Glass bulbs (250–1000 ml) have been used for sampling emissions particularly in conjunction with the Mills modification of the ASTM syringe dilution technique.[11] The glass bulb is filled with odorous air by the vacuum created by displacement with a mercury reservoir. Because of the toxicity of mercury, the possible reaction of odor components with the mercury, and adsorptive losses due to the high surface area to volume ratio, this method is rarely used at the present time.

Another procedure is to use an evacuated or partially evacuated stainless steel cylinder. This sampling technique has similar disadvantages as sampling with glass bulbs, and would be useful in situations where the odor emission consists entirely of odorant materials that are gaseous at normal conditions.

4.1.2   Bag Sampling

Currently extensive reliance is being placed on bag sampling of odor emissions. This has come about with the development of dynamic olfactometers and their need for a larger odor sample. Sampling bags have been constructed with a variety of plastic materials which include polyethylene, Teflon®, Saran®, Mylar®, and Tedlar®. Difficulties that may occur and that should be considered in using bags include: (1) the bag material should have little or no intrinsic odor; (2) the bag material should have sufficient strength for handling; (3) there should be minimal diffusion of the odorants through the walls; and (4) there could be adsorption and condensation of odorants and/or moisture on the container walls during sampling and storage. The storage shelf life of odorous samples in bags is limited (bagged samples are usually transported and evaluated within 24 hr after sampling) and can be different for different samples.

Studies[12] on the relative merits of several plastics as bag materials for sampling odors have found that although Tedlar is a good material for storing vaporous odor compounds, sample condensation and adsorption could occur. Odor losses in Tedlar bags could be minimized by predilution during sampling. Additional studies[13] comparing Tedlar bag sampling and sorbent sampling of paint bake oven odor emissions indicated that bag sampling can be a valid technique for sampling stack emissions if sufficient predilution is employed during sampling and where it is known that volatile species are the primary odorants. Bag sampling should be acceptable in situations where the odorous emissions are known to be due to high volatility species such as hydrogen sulfide, and low molecular weight aldehydes and sulfur bearers (methyl and ethyl mercaptans). Tedlar bags were unsuccessful in quantitatively sampling trimethylamine in air.

The Committee on Odors from Stationary and Mobile Sources of the National Research Council[14] concluded that the most advisable course is to pretest the storage of the particular type of odorous sample in the proposed bags. This recent reference is recommended for further information and discussion of this topic.

4.1.3 *Sorbent Sampling*

Charcoal, silica gel, and gas chromatographic materials have been used for the collection of trace organics and odors. Although activated carbon has excellent collection efficiency and capacity, quantitative recovery of odorants can be difficult to achieve. Humid environments limit the use of silica gel.

Chromosorb 102® (a styrene-divinylbenzene resin) has been successfully used to collect emissions from a wide variety of stack sources for both sensory (odor) and/or chemical (isolation and identification studies) analyses. Other macroreticular resins that have been used include XAD® and Poropak® (styrene-based resins) and Tenax GC® (a polyphenol ether resin).

Chromosorb 102 is particularly useful for sampling emissions containing odorants in the intermediate to high molecular weight range. Certain volatile and/or highly polar species may not be quantitatively collected and recovered. Hydrogen sulfide and low molecular weight mercaptans (ethyl or less) are not collected, nor is trimethylamine. The Chromosorb is prepared by washing 50 g of sorbent successively with 750 ml of methanol and then pentane, followed by air drying. High purity solvents should be used. The stainless steel trap is packed with 10 g of sorbent. The odorant species are quantitatively eluted countercurrently with high-purity pentane. The eluate is stable at refrigerator temperatures for about 2 weeks.

For sensory evaluation, the sample eluate can be injected with a syringe pump into the dilution air stream of a total immersion dynamic test chamber or injected into a bag for evaluation with a dynamic olfactometer.

An advantage of the Chromosorb sampling procedure for chemical analysis is that large volumes of the emission can be collected, thus preconcentrating the odorous species. The eluates can be used with GC-MS for identifying the chemical nature of the species present in greatest abundance.

4.2 Sampling Ambient Odors

In principle, sampling of ambient odors should be similar to the sampling of a source sample. However, odors in the ambient air have undergone considerable dilution in the atmosphere and their concentration is consequently reduced. Sampling with bags and glass bulbs have been attempted; however it has not been shown that condensation and adsorption of odorous components on the walls of the container has not occurred. Intrinsic bag odors may also interfere with the evaluation.

This author has found that it is more expeditious to have trained observers measure odor directly in the ambient air (as described in Section 2.1) rather than to attempt to sample ambient odors for laboratory evaluation.

5 OLFACTOMETERS

Olfactometers do not measure odor, rather they are devices for presenting a known dilution of a sampled odorous emission in air to a human. One or more panelists form the panel. A variety of olfactometers have been developed primarily for measuring the detection or recognition threshold dilution (i.e., the dilutions or odor-free air required for the odorous sample to be just detectable or recognizable) for each panelist. The threshold dilution is a dimensionless dilution factor (Z). In the past, the odor threshold dilution was often expressed as odor concentration units per cubic foot or as odor units.

5.1 Scentometer

The scentometer was developed for measuring the dilution threshold of odors present in the ambient air.[15] The scentometer has two glass nosepieces which are inserted into the nostrils. These are connected to a plastic box which contains activated carbon. Several calibrated holes are drilled in the box. By closing all the holes with his fingers, the operator will take in air that has passed through the activated carbon. By opening selected holes, the observer can obtain nominal dilution to threshold ratios of 2, 7, 31, and 170. The odorous ambient air enters the box through the open finger holes where it is diluted with carbon-filtered air during inspiration. The device is specified for use in several odor control regulations to determine if an odor problem exists.

The method has been criticized on several grounds.[10] The device was not devised for use for measuring stack odor emissions.

5.2 ASTM D 1391 Syringe Dilution Test

Source odor emissions are taken by either glass bulb sampling or with bags. 100-ml syringes are utilized for preparing dilutions of the odor-bearing air with dilution air. The panelist then ejects the gas content of the syringe into the nose. In a recent revision (Ref. 4), a major change has been to present two syringes to each panelist for comparison. One syringe contains the diluted odor sample while the other syringe contains dilution air only. The TT-4 Odor Committee of the Air Pollution

Control Association declined to endorse the revised ASTM D 1391 method, preferring olfactometric methods based on dynamic dilution.

5.3 Dynamic Olfactometers

In contrast to the syringe dilution technique which may be considered as a static dilution device, dynamic olfactometers present the sample dilution in a continuous flow. Some may alternate background dilution air with the sample dilution which are both sniffed through the same port. Other devices may have two or more sniffing ports—one port containing the sample dilution while the other port contains only dilution air. There are many differences in the designs and protocols used with various dynamic olfactometers. There are three dynamic olfactometers that are available commercially. These include the Hemeon Olfactometer,* Dynamic Triangle Olfactometer,† and Misco Olfactometer.‡ These commercially available olfactometers range in flow from less than 1 l/min (Dynamic Triangle) up to 150 l/min (Hemeon). The Dynamic Triangle Olfactometer is used in conjunction with the forced choice triangle method for detection threshold measurements. It has been reported for use with stack emissions and ambient air odor. The Hemeon device presents a variety of dilutions of sampled odor in air with the observer rating the intensity (0–4) and the threshold is estimated by extrapolation to zero odor by plotting the logarithm of the dilution ratio versus the intensity rating. The Misco Olfactometer is reported for use in both the ambient air and on stack samples. The device was developed by the California Public Health Service. It has been reported that the Bay Area Air Pollution Control District uses a device similar to Misco's for determining compliance with its new (1978) odor control regulation. However, there is no description of the device in the literature.

A number of laboratories have developed their own olfactometers. Arthur D. Little, Inc., has a total immersion static testroom and a total immersion dynamic testroom available for use in odor pollution measurement studies. TRC–Environmental Consultants, Inc., has a laboratory and a mobile-van-based dynamic olfactometer.

The TT-4 Odor Committee of the Air Pollution Control Association sponsored a round robin testing program to compare odor results obtained by various laboratories. The first tests with butanol, toluene, and mineral spirits showed a wide variation in the odor thresholds reported. Additional round robin testing is presently under way under TT-4 committee sponsorship to compare results obtained by various olfactometers.

6 FURTHER READING

A recent review of odor pollution measurement, control, and regulation can be found in Ref. 14 and is recommended for those requiring in-depth information. The TT-4 Odor Committee of the Air Pollution Control Association is particularly active and is a good source of information as is the *APCA Journal.*

REFERENCES

1. Leonardos, G., A Critical Review of Regulations for the Control of Odors, *J. Air Poll. Cont. Assoc.* **24**(5) (1974).

2. Levins, P. L., et al., *Chemical Analysis of Diesel Exhaust Odor Species*, Society of Automotive Engineers, Publication 740216, March 1974.

3. Leonardos, G., The Profile Approach to Odor Measurement, *Proceedings: Mid-Atlantic States Section, Air Pollution Control Association Semi Annual Technical Conference on Odors: Their Detection, Measurement and Control*, May 1970, pp. 18–36. Pittsburgh, PA.

4. Hellman, T. M., and Small, F. H., Characterization of the Odor Properties of 101 Petrochemicals Using Sensing Methods, *J. Air Poll. Cont. Assoc.* **24**, 979–982 (1974).

5. Leonardos, G., et al., Odor Threshold Determinations of 53 Odorant Chemicals, *J. Air Poll. Cont. Assoc.,* **19**(2) (1969).

6. Fazzalari, F. A., ed., *Compilation of Odor and Taste Threshold Values Data*, ASTM DS 48A, American Society for Testing and Materials, 1978.

7. Sullivan, F., and Leonardos, G., *Determination of Odor Sources for Control*, Annals of the New York Academy of Sciences, Vol. 231, September 1974.

8. Leonardos, G., Odor Measurement at Waste Water Treatment Plants, presented to Genesee Valley Chapter, New York Water Pollution Control Association, May 16, 1979.

9. Wahl, G. H., Jr., *Regulatory Options for the Control of Odors*, U.S. Environmental Protection Agency, EPA-450/5-80-003, Feb. 1980.

* Hemeon Associates, Pittsburgh, PA.
† IIT Research Institute, Chicago, IL.
‡ Misco Company, Berkeley, CA.

10. Prokop, W. H., Developing Odor Control Regulations: Guidelines and Considerations, *J. Air Poll. Cont. Assoc.* **29,** 9 (1978).

11. Mills, J. R., et al., Quantitative Odor Measurement, *J. Air Poll. Cont. Assoc.* **13,** 467–475 (1963).

12. Schuetzle, D., et al., Sampling and Analysis of Emissions from Stationary Sources—I. Odor and Total Hydrocarbons, *J. Air Poll. Cont. Assoc.* **25,** 925 (1975).

13. Leonardos, G., et al., A Comparison of Polymer Adsorbent and Bag Sampling Techniques for Paint Bake Oven Odorous Emissions, *J. Air Poll. Cont. Assoc.* **30,** 22 (1980).

14. *Odors from Stationary and Mobile Sources,* National Academy of Sciences, Washington, D.C., January 1979.

15. Huey, N. A., et al., Objective Odor Pollution Control Investigations, *J. Air Poll. Cont. Assoc.* **10,** 441 (1960).

16. American Society for Testing and Materials, ASTM D 1391, Standard Test Method for Measurement of Odor in Atmospheres (Dilution Method), Philadelphia, ASTM, 1978.

CHAPTER 34
ATMOSPHERIC DISPERSION

NORMAN E. BOWNE

TRC Environmental Consultants, Inc.
East Hartford, Connecticut

1 REGULATORY NEED FOR MODELING

Atmospheric dispersion modeling fills two basic gaps in information that regulatory agencies must have to make decisions. These areas of knowledge that are frequently missing are detailed patterns of existing air quality, because numbers of monitors are limited, and assessment of the impact on air quality by proposed new sources or changes in existing sources. However, the degree of sophistication required for a specific dispersion modeling activity depends on the application; therefore, some background in the Clean Air Act (CAA) and its relationship to dispersion modeling activities is necessary.

1.1 The Clean Air Act and Amendments

The fundamental purpose of the Clean Air Act is to achieve and maintain air quality for the protection of public health and welfare. To this end, the Environmental Protection Agency (EPA) was directed to establish National Ambient Air Quality Standards (NAAQS). "Primary" ambient standards were designed to protect public health and "secondary" standards to protect public welfare. Each state was then required to submit an implementation plan (SIP) for EPA's approval, outlining the state's strategy for attaining and maintaining the National Ambient Standards within the time frames established by the act. Thus, while primary responsibility for implementing the act lies with the state, EPA has an enormous influence on the state implementation process.

In addition to mandating the establishment of National Ambient Air Quality Standards, the act also directs EPA to establish two types of national *emission* standards. Section 111 of the act requires EPA to establish performance standards for new and modified stationary sources. These standards are intended to keep new pollution to a minimum as the states battle pollution from existing sources under the SIPs. The second type of emission standard concerns "hazardous" air pollutants from both *new* and *existing* sources. Under Section 112 of the act, EPA can establish standards for such emissions if they will cause an increase in serious illness or mortality. Both the new source and hazardous emission standards, known respectively as New Source Performance Standards (NSPS) and National Emissions Standards for Hazardous Air Pollutants (NESHAP), will be taking an increased significance in the years ahead.

The 1977 Clean Air Act Amendments (P.L. 95–95) attempt to resolve many of the controversial issues to which the 1970 Clean Air Act Amendments spoke ambiguously or not at all. Clearly, the two most significant provisions of 95–95 are the new requirements governing areas which failed to attain the National Ambient Standards by the statutory deadline (nonattainment areas) and the new provisions for "preventing significant deterioration" of air quality in areas where the air is cleaner than the National Ambient Standards (PSD). The new provisions require dirty areas to become clean and clean areas to stay clean.

The 1977 amendments also attempt to resolve a number of problems which have been the subject of much administrative and judicial consideration since the enactment of the 1970 amendments. These new provisions cover such issues as pollution from federal facilities, the permissibility of pollution dispersion techniques, new enforcement strategies, federal regulations of previously unregulated pollutants and pollution sources, new provisions for accommodating the environmental impacts of increased coal utilization, and provisions for prohibiting interstate air pollution.

Attainment and maintenance of the standards is the primary responsibility of the states. Section 110 of the act requires each state to submit a state implementation plan (SIP) to EPA which includes, among other things, measures necessary to attain and maintain the ambient standards within the time frames established by the act. Under the 1970 legislation, all states were to have attained the primary standards by July 1, 1975, with the exception of relatively few areas where an extension to mid-1977 was granted. Secondary standards were to be attained within a "reasonable time," defined by most SIPs to be the same as the primary standard attainment date.

1.2 Continuous Emission Controls and Tall Stacks

The issue of whether dispersion techniques such as tall stacks and intermittent controls such as production cutbacks were an acceptable alternative to continuous emission controls such as stack gas scrubbers or low sulfur fuels generated considerable litigation under the 1970 Clean Air Act. EPA's answer, which received solid judicial approval, was that the statute required continuous controls as the ultimate control method in all cases and that supplementary controls such as tall stacks and production cuts were permissible only as an interim measure where continuous controls could not by themselves assure achievement of the requisite standards before the statutory deadline.

The 1977 amendments basically codify the EPA position developed during litigation. The act now defines the "emission limitations" which must be contained in all SIPs as requirements that limit pollutant emissions on a *continuous* basis. This means that the degree of emission limitation required under the State Implementation Plan shall not be affected by:

1. So much of the stack height of any source as exceeds "good engineering practice."
2. Any other dispersion technique or intermittent supplemental control varying with atmospheric conditions.

This limitation applies only to tall stacks which were constructed after December 30, 1970, and dispersion techniques which were implemented after this date.

The term good engineering practice (GEP) is defined to mean the height necessary to ensure that emissions from the stack do not result in excessive concentrations of any air pollutant in the immediate vicinity of the source as a result of atmospheric downwash which may be created by the source itself, or as a result of nearby structures or terrain features. The height allowed by this new definition shall not exceed $2\frac{1}{2}$ times the height of the source unless the owner affirmatively demonstrates that a greater height is necessary to prevent excessive pollutant downwash. Modeling of downwash problems is addressed in Section 6.

1.3 Nonattainment Offsets

By the July 1975 deadline for attaining the primary ambient standards, at least 160 of the nation's 247 Air Quality Control Regions (AQCR) had monitored violations. Lacking congressional guidance, EPA was forced to develop its own strategy for allowing new growth in such areas. That strategy requires the states to prevent the construction or modification of any source which would "interfere with" the attainment and maintenance of the ambient standards. Clearly, *any* new pollution in a nonattainment area would "interfere with" attainment of the standards. Thus, a literal interpretation of the EPA strategy would have prohibited essentially *all* new growth in and around nonattainment areas. Such an interpretation was obviously unacceptable from a political and economic standpoint and was, therefore, rejected by EPA. The alternative, as embodied in the interpretative ruling, was to allow new growth in such areas only if stringent conditions were met. These conditions are designed to ensure that new source emissions will be reduced to the greatest degree possible, that more than equivalent emission reductions (i.e., offsets) will be obtained from existing sources, and that the effect of these actions will be a net air quality benefit.

Atmospheric dispersion modeling is the tool used to determine the impact of the new source and the degree of offset that is required.

The most controversial requirement of the offset ruling is the provision from which it derives its name. Under the offset requirement, new sources will have to arrange for legally enforceable emission reductions from existing sources in the area (whether or not under the same ownership) which *exceed* the amount of emissions that will be emitted by the proposed source. Thus, if a proposed source subject to the ruling will emit SO_2 at a rate of 25 lb/hr, no permit may be issued unless the owner can secure offsetting reductions from existing area SO_2 sources in *excess* of 25 lb/hr.

1.4 Prevention of Significant Deterioration (PSD)

While the objective of the SIP and nonattainment provisions is to *achieve* and *maintain* the ambient standards, the purpose of PSD is to prevent significant deterioration of air which is already cleaner than the ambient standards.

Under EPA's regulatory scheme, clean areas of the nation can be designated under any of three "classes." Specified numerical "increments" of air pollution increases for SO_2 and particulate matter were permitted under each class up to a level considered to be "significant" for that area. Class I increments permitted moderate deterioration; class III increments permitted deterioration up to the NAAQS. EPA initially designated all clean areas of the nation as class II. States, Indian tribes, and officials having control over federal lands (Federal Land Managers) were given authority to redesignate their lands to class I or III status under specified procedures.

The central mechanism for implementing PSD is the detailed preconstruction review and permit procedure. There are two basic types of requirements involved in the preconstruction review. This first relates to air quality analysis. The second concerns the application of Best Available Control Technology (BACT) to proposed new sources. Special requirements are imposed where a source would impact a federal class I area.

Even before a PSD permit application may be filed, a detailed source impact assessment must be made available to the public. The required impact analysis must assess the source's impact on air quality, visibility, soils, and vegetation and include a description of area climate and meteorology. Air quality impacts must be analyzed using specified EPA models unless the EPA model is shown to be inappropriate.

No sources subject to the new preconstruction requirements may be constructed in *any* area of the United States unless the owner demonstrates that his proposed emissions (in conjunction with the effects of emission growth and pollution control in the impacted area after January 6, 1975) will not exceed the allowable increments in any PSD area.

New sources must be analyzed not only for their air quality impact in their immediate area, but also for their impact on distant areas. For instance, construction of a power plant proposed for location in a class III area might be prohibited if its emissions were found through air quality modeling to violate the increments of a class II area several miles away. Generally, the farther a source is located from a particular area, the less significant its air quality impact on that area will be. However, because computer modeling techniques can estimate small impacts on increments for sources hundreds of miles from the impacted area, care must be exercised in analyzing such long-range impacts.

EPA permits the use of preliminary screening techniques to determine if a full-modeling analysis is necessary for each preconstruction review. A "screening technique" is a simplified analytical method which is easy to implement and incorporates relatively conservative estimates of the air quality impact of a source. For instance, if the results of the screening indicate a maximum ambient impact of less than one-half of the allowable increment for the source, the source could be assumed to be approvable and additional ambient analysis may not be required.

The law imposes an obligation on the regulatory authority to require the owner of an approved major source to conduct such postconstruction ambient air monitoring as may be necessary to determine the source's impact on air quality. Due to the inaccuracy of existing measurement methods and the general unavailability of reliable baseline air quality data in many areas, EPA may not specifically require postconstruction monitoring to determine how much of the allowable increment has been "used up." In actual practice, assessment of the available increment is normally accomplished through an accounting procedure whereby *atmospheric modeling* of individual sources is used to keep track of the available (or "unused") increment as sources and emissions are increased or decreased.

1.5 Visibility

One of the most significant and potentially far-reaching provisions of the recent Clean Air Act Amendments of 1977 is "the prevention of any future and the remedying of any existing impairment of visibility in mandatory class I federal areas† where such impairment results from man-made air pollution."

An important problem is the need to quantify the relationship between human-made air pollution and visibility impairment in federal class I areas. The technical ability to identify regional scale sources of visibility degradation is limited. This lack of knowledge impedes the development of effective and enforceable control requirements. Until these complex modeling and optical issues are resolved, the technical basis of visibility related emission control regulations will be open to challenge.

Visibility modeling, as it exists at the time of preparation of this manuscript, is discussed briefly in Section 6; but it is an area of rapid evolution at the present time and not properly a part of an engineering treatment.

2 THE ATMOSPHERE IN MOTION

Pollutants in the atmosphere are carried along by two components of atmospheric motion: the mean wind field which transports the pollutants from one point to another and the turbulent motions which disperse the pollutants about some central position. In this section, we will touch briefly on some of

† "Mandatory federal class I areas" include all international parks, all national wilderness areas and national memorial parks over 5000 acres, and all national parks over 6000 acres. Only those parks which were in existence on August 7, 1977 are included.

the basic atmospheric motions—especially the peculiarities of those motions near the ground, the influence of friction of the ground surface, the effect of differential heat input at the surface, and how these motions affect turbulence and the transporting portion of the wind field. Most of the motions and turbulence that are important for engineering considerations occur in the planetary boundary layer.

The planetary boundary layer is that portion of the atmosphere that occurs between the ground and the free atmosphere where the motions are basically unaffected either by surface friction or by surface heating or cooling. Figure 1 represents the region of the atmospheric motions in the planetary boundary layer and gives some rough estimates of their heights. In the first 50 to 150 m above the ground, we encounter the surface boundary layer where the wind motion is affected significantly by surface roughness and vertical temperature gradient which is in turn affected by the temperature of the ground below it. In the area above 100 m, up to a height of 500 to 1000 m in general, we encounter a transition layer. The wind motions in the transition layer are affected by the surface friction, the density or temperature gradient, and the earth's rotation. Above the planetary boundary layer in the free atmosphere, the wind is generally parallel to the pressure gradient with some effect from the surface rotation of the earth; however, the upper boundary may vary. For example, in the case where cold air is advected over a warm surface such as might occur on a very warm spring day with strong sunshine, the convective currents from the earth's surface may extend to a height of as much as 3000 m. In addition, convective activities within thunderstorms can unbalance the concept of the planetary boundary layer. Most of the atmosphere dispersion models that are currently employed do not deal with such intense motions, and we will confine our discussion to the more normal boundary layer conditions.

2.1 Surface Friction Effects

The major effect of surface friction is the influence on the wind profile—that is, the change of wind speed and direction with height above the ground. Our first consideration will be given to shearing stress and the wind profile.

If we consider the flow of a gas caused by two parallel plates in a laminar flow, the shearing stress per unit area is proportional to the velocity gradient. This relationship is made precise by the introduction of a coefficient proportionality characteristic of the gas called dynamic viscosity.[1] In general, then, the shearing stress per unit area can be written as

$$\tau = \mu \frac{du}{dh} \tag{1}$$

where τ is the shearing stress per unit area, and μ is the dynamic viscosity which is independent of u and h. Of course in this case, u is the speed of the gas, and h is the distance from the nonmoving plane to the one which is moving. Frequently in terms of fluid motion, it is more convenient to use the kinematic viscosity defined as

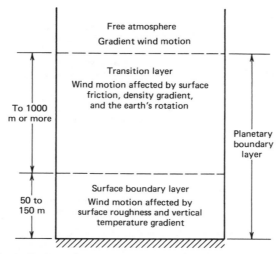

Fig. 1 Regions of atmospheric motions above the ground surface.

$$\nu = \frac{\mu}{\rho} \tag{2}$$

where ρ is the density of the fluid. In the atmosphere, we do not have laminar flow. Therefore, we are dealing primarily with the lower boundary which is fixed at the surface of the earth and some upper boundary which is normally driving the wind; and we develop a velocity profile. There are a number of profiles related to wind speed. A basic theoretical profile is given as

$$\frac{\bar{u}(z)}{u^*} = \frac{1}{K} \ln z + \text{constant} \tag{3}$$

where K is the von Karman constant, and u^*, the friction velocity, is defined as

$$u^* = \sqrt{\frac{\tau}{\rho}} \tag{4}$$

Within the layer of air near the ground, viscosity predominates; but as we depart from the ground, the shearing stress is partly turbulent and partly viscous; and when we are far enough from the ground, the shearing stress is primarily defined by the turbulent motions. The layer in which viscous stress is important is very thin.

With increase in wind speed over a given surface or as the surface becomes increasingly rough at a constrant wind speed, the stage is reached in which the purely viscous stress of the surface is outweighed by the effect of pressure forces associated with eddying wakes from the roughness elements. The flow is then said to be aerodynamically rough, and the viscosity ceases to influence the profile. Now the velocity profile is found to depend on a length which is characteristic of the surface roughness. In this case, we write the wind profile[2] as

$$\frac{\bar{u}(z)}{u^*} = \frac{1}{K} \ln \frac{z + z_0}{z_0} \tag{5}$$

The new term that has been introduced here, z_0, is the roughness length. It is generally an order of magnitude smaller than the actual height of the roughness elements. Some of the typical values of roughness lengths[3] may be less than 1 cm for such surfaces as smooth mud flats, snow, or even short mown grass—increasing to greater than 1 cm for grass that is 4 or 5 cm in length and increasing to nearly 10 cm for grass that may be 60 or 70 cm high. For larger areas where there are shrubs, trees, hedges, and other such obstacles to the wind flow, the roughness length might increase to as much as ½ m. In dealing with urban areas or cities, the roughness length may increase to as much as 2 m.

Another method of representing the wind speed profile as a function of height that is used in air quality models to relate surface winds to wind speed at stack height may be described as

$$\bar{u}(z) = \overline{u(r)} \left(\frac{z}{r} \right)^\alpha \tag{6}$$

This power law expression may be found in many textbooks.[4] The r stands for some reference height, and z is the height of interest for which we would like to determine the wind speed. We noticed that in the power law the mean wind speed at some height z is equal to the mean wind speed at the reference height times the ratio of the height raised to some power α. Simiu and Scanlon[4] have published values of α for open terrain, suburban terrain, and for centers of large cities and have also indicated the height above the terrain, for which the profile is deemed to be appropriate. However, most air quality models use other values presented later.

The representation of the wind speed profile may be found in Figure 2 which shows the wind speed change with height, and it may be represented in this manner by either Eq. (5) or (6) with the appropriate substitution of either a roughness length and a friction velocity or the appropriate power law function. Another aspect of the influence of friction on the lower layers of the atmosphere is illustrated in Figure 3 which illustrates the *Ekman spiral*. The Ekman spiral was first found studying ocean currents but has been theoretically applied to the atmosphere. It demonstrates the effect on both direction and speed of the wind. A gradient wind at the top of the boundary layer is influenced by the friction effects of the earth in a northern hemisphere causing a turning of the wind direction. The figure illustrates the vectors and the surface projection of the vectors to indicate the turning of the wind in this case. Theoretically, the Ekman spiral indicates about a 45-degree direction difference between the wind at the surface and the wind at the top of the boundary layer; however, actual observations indicate that direction changes frequently are as low as 6 degrees and usually only as high as 30 degrees. Therefore, the Ekman spiral is important theoretically but has much less importance from a practical standpoint.

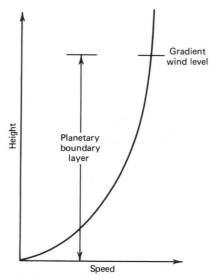

Fig. 2 Wind speed change with height above the ground.

2.2 Surface Temperature Effects

Heat convection or convection of any property for that matter denotes the transfer from one place to another by the motion of the air. Convection of heat is of concern in air pollution because it establishes vertical transfer. *Free* convection occurs when the motion or that part of the motion that carries the heat is set up strictly by buoyancy itself. There is no air flow and the vertical motion is set in progress by the heating of the air immediately adjacent to the ground. *Forced* convection is motion that is otherwise imposed. In fully forced convection buoyancy does not affect the motion or the heat transfer coefficient. It is only when the ground surface is much warmer than the air that passes over it that buoyancy does affect the motion; and according to climatology, this occurs only on the order of 10–15% of the time. However, during this 10–15% of the time, the turbulent structure of the atmosphere becomes markedly different; and the dispersion of material released into the atmo-

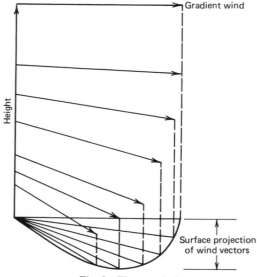

Fig. 3 Ekman spiral.

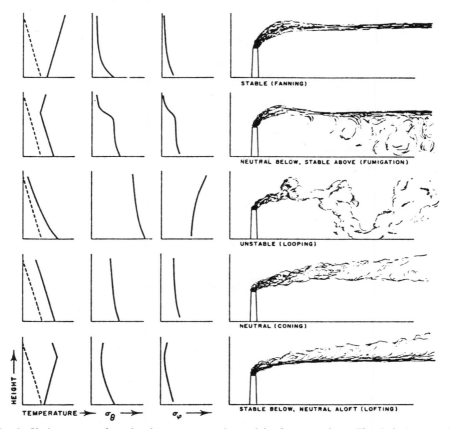

Fig. 4 Various types of smoke-plume patterns observed in the atmosphere. The dashed curves in the left-hand column of diagrams show the adiabatic lapse rate, and the solid lines are the observed profiles. The abscissas of the columns for the horizontal and vertical wind-direction standard deviations (σ_θ and σ_ϕ) represent a range of about 0° to 25°.

sphere is greatly enhanced in the vertical. Typically, it is under conditions such as these that the highest concentrations close to elevated sources are observed. The plume is said to loop or take on a serpentine kind of shape when viewed from the side.

The vertical temperature structure near the ground is influenced by the heating and, in turn, influences the density structure of the atmosphere. These are related to the behavior of material released into the atmosphere which has been illustrated in *Meteorology and Atomic Energy*[3] and which is reproduced in Figure 4. Within this figure, various types of smoke plume patterns observed in the atmosphere are pictured on the right. On the left, the dashed curve represents the adiabatic lapse rate; and the solid curves represent appropriate profiles of temperature and turbulent characteristics represented by horizontal and vertical wind direction standard deviations. The turbulent fluctuations of the winds are discussed later because they are very important to the behavior of pollutants in the atmosphere.

2.3 Local Wind Flows

There are local effects that can influence both trajectory and turbulence of the wind field. One of the local flows that can influence the trajectory of material is a valley flow. Figure 5 is a schematic representation of valley distortions of the wind flow by topographic obstacles, one of which is a valley in an area of generally flat terrain, and the second is a valley or a pass through a mountain range. It is easy to see that pollutants that are released into the areas influenced by these flows can take considerably different paths than if they were released in areas of uniform wind flow. Further, it can be seen that material released in the valley is channeled to stay within the valley. Under conditions of light gradient winds aloft, the differential heating of the valley can cause the winds to blow down valley at night and blow up valley during the day and in rare cases effectively trap the pollution within the valley.

A second type of disturbance to local wind flow is that of a sea breeze which is caused by the

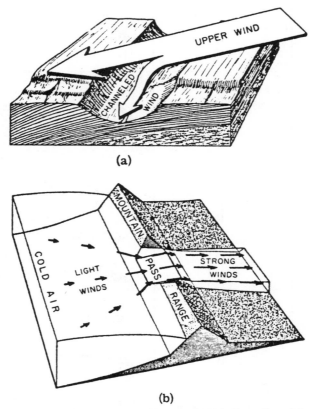

(a)

(b)

Fig. 5 Distortions of the wind flow by topographic obstacles. (*a*) Channeling of the wind by a valley. (*b*) The effect of a mountain pass on the wind flow (from *Meteorology and Atomic Energy*.[3])

temperature difference between the land and water surfaces. Figure 6 gives a simplified illustration of a sea breeze circulation that can be established near the shoreline. While the material that is transported within this simplified circulation might recirculate and enhance the pollution, it is more likely that the actual circulation pattern will not be so simple. The air flow may continue to blow parallel to the ground and achieve enhanced diffusion at the interface between the air that has blown over the water and that which has been influenced by the ground surface. This could set up enhanced mixing at a specific point that does not move and, therefore, result in high ground-level concentrations, as in Figure 7.

The third influence of importance for atmospheric pollution is that of a city. Again, a city will frequently act as a heat island—that is, an area of warmer surface temperatures than the surrounding countryside. This is true both in summer and winter—in summer because of the absorption of solar radiation and in winter because of the heat generated within the buildings and by the activities of the people that live there. There tends to be a gently upward motion with air being deflected at the surface to replace the air that has been advected upward by the additional heat; therefore, there is a confluence of pollutants into the city, a gentle rising motion, and a downward motion of the pollutants outside the city. This motion on top of the general flow may produce higher ground-level concentrations than would otherwise have been expected near the city.

2.4 Turbulence

Turbulence is that aspect of atmospheric motion which disperses the material that has been injected into the fluid. Mechanical turbulence is caused by roughness elements at the surface, be they short grass, which has very little effect, up to buildings which may have significant effect. However, turbulence is constantly being generated and decaying within the atmosphere. Back in Figure 4, we noted that profiles of the standard deviation of the wind fluctuations were pictured along with plume behavior. These profiles of standard deviation represent the turbulent structure of the atmosphere as measured by an anemometer. To further illustrate, refer to Figure 8, taken from *Meteorology and Atomic Energy*,

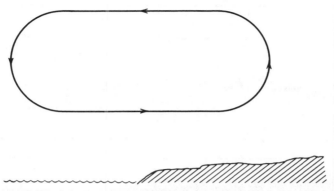

Fig. 6 Sea breeze circulation cell. Typical stream line pattern of a well-developed sea breeze or lake breeze cell. View is looking south with water on left during midafternoon. Lyons and Olsson[27] with permission.

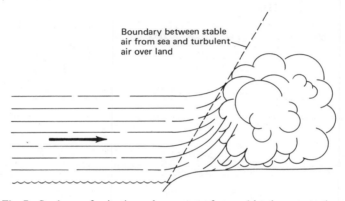

Boundary between stable air from sea and turbulent air over land

Fig. 7 Sea breeze fumigation enhancement of ground-level concentration.

1968, which provides some illustration of the appearance of turbulence on wind charts and perhaps will aid in determining the type of turbulence and picturing the appearance of the record of the anemometer to compare to the picture of the plume itself. In the (*a*) part of Figure 8, there is a record of wind direction and speed at the height of 30 m. Remember that as height is increased away from the surface, the influence of surface roughness elements is less noticeable. In the (*a*) portion, we note the change from night to day starting at the right-hand side of the figure. The primary features to note are the light wind speeds and the relatively small variability in both the speed and direction tracing of the recorder pen between 4 and 5 A.M. and how, between 8 and 9 A.M., the fluctuations are larger. This is an indication of the increase of the turbulence during this time period. Also, there is a slight increase in wind speed. Part (*b*) is a record from 16-m height starting at noon and going until 6 P.M. In the case between noon and 2 P.M., it is obvious that there was a great deal of turbulence, not only mechanical but also convective which is indicated by the wind direction trace exhibiting marked changes over periods of 15 to 30 min. At approximately 2 P.M., a cold front passed bringing about a steady wind direction at relatively high speeds, that is, in excess of 20 mph; and all of the turbulence is now mechanical. Part (*c*) shows the flow at the top of a tower (120 m) during very stable conditions created by surface temperature inversion which separates the influence of the surface roughness elements from the wind at this level. During the period from 7 P.M. to after 10 P.M., it is obvious that there is very little turbulence in the wind flow at 120-m altitude. After approximately 10:20 P.M., some disturbance had taken place which now has caused a change in the flow pattern. Finally, the (*d*) portion shows the influence of the variable wind as it changes during an intense inversion close to the ground. There is no wind speed trace, because, in general, the anemometer will not respond to such light winds and that explains part of the step function which appears in the wind direction trace, which is illustrated. There was not enough wind to cause the vane to move. However, it does indicate the highly varying wind directions that occur under these kinds of conditions.

It has been found by some investigators that the wind direction fluctuations are nearly normally distributed and, therefore, can be represented by a standard deviation. This standard deviation has

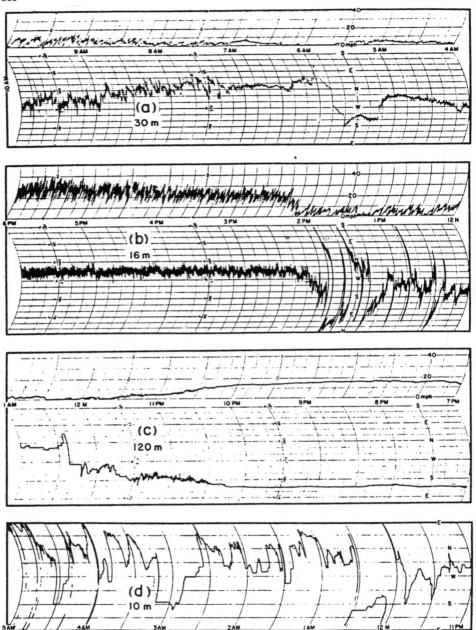

Fig. 8 Typical horizontal wind speed and direction traces at various heights. Chart speed, 3 in./hr. (*a*) Night-to-day change. (*b*) Light wind, unstable flow followed by a cold frontal passage. (*c*) Upper-level flow during an intense inversion. (*d*) Low-level flow during an intense inversion.

been related to the spread of material in a diffusing plume and that forms the basis of the Gaussian dispersion models which are discussed later in this report. In a similar manner, the vertical fluctuations are nearly normally distributed and can be represented by the vertical standard deviation. However, measurements of the vertical fluctuations are rarely made. Therefore, statistical information regarding the magnitude of the variable under various conditions is not well defined.

2.5   Summary of Importance to Dispersion Modeling

Material that is released into the atmosphere travels with the wind and is dispersed by the turbulent nature of that wind flow. The preceding sections have illustrated some of the nature of the mean wind flow. That is, the wind speed increases as a function of distance from the surface up to the top of the planetary boundary layer; and wind direction slowly changes with height although not as much as indicated theoretically. It was noted that the flow may also be affected by temperature differences. That is, land and sea breezes can be set up which influence the direction of travel of material emitted into the air. Furthermore, topography can channel or change the direction of flow. In addition to changing the direction, these types of interferences also alter the turbulent pattern of the atmosphere which influences the rate at which the material is dispersed. We noted that there are two kinds of turbulence within the atmosphere. Mechanical turbulence is caused by the roughness elements that interfere or trigger eddies within the atmosphere. In general, the mechanical eddies are not as large as those that are triggered by heating of the surface in which convective bubbles break off. The sizes of the eddies are reflected in the appearance of the wind direction and speed traces that were shown. It was noted that when mechanical turbulence was present with high wind speeds, the turbulence was relatively uniform, indicating a smaller maximum eddy size. The traces illustrated the significant influence of buoyancy indicating a much slower turbulent change and, therefore, a much larger eddy size. The size of the eddies in this case will determine the maximum size of the cloud of dispersing material and the larger the cloud, the smaller the concentration in the middle of that cloud.

Assume that most of the models described below can be given a measure of the wind speed, the wind direction, and some measure of the turbulence of these motions as reflected in a stability classification. One of the major features of most of the models is that they assume that the given condition persists over all the time and space for which the sample calculation is then being made. Most models that are currently utilized have a basic output step of approximately 1 hr, especially those that we call the Gaussian models. Therefore, they have within them the assumption that the wind direction, wind speed, and turbulence remain constant over that 1-hr period and over all space into which the material that is being modeled encompasses. Obviously in the real atmosphere, this does not occur, because the heating is not everywhere the same. The roughness elements that are affecting the turbulence close to the ground are not everywhere the same; and as air masses are advected from one location to another, the characteristics of the basic underlying turbulence within them can change.

3   DISPERSION THEORIES

Small particles or molecules of gases released into the atmosphere will separate rapidly from one another under the influence of turbulent eddies. The problem of turbulent diffusion in the atmosphere has not yet been uniquely formulated in the sense that a single basic physical model can explain all the significant aspects of the problem. There are two alternative approaches presently utilized. The first approach which is discussed is that of the gradient transfer theory. The gradient transfer approach employs the basic differential equation that attempts to describe the advective and turbulent motions of the atmosphere and its influence on the dispersion of material released into it. The second approach is the statistical theory.

Diffusion at a fixed point in the atmosphere is proportional to the local concentration gradient under the gradient transport approach. It considers properties of the atmospheric motion relative to a spatially fixed coordinate system. On the other hand, the statistical diffusion theories consider motion following the atmosphere, that is, usually following the center of the dispersing material. These theories may be classified as either continuous motion or discontinuous motion, depending on whether the particle motion is modeled to occur continuously or as discrete steps. Since both theories are supposed to be representing the atmosphere, there must be a close connection between them, and indeed, there is.

3.1   Gradient Transport (K Theory)

The mathematical statement that represents the gradient transport approach is taken from classical electric or heat conducting physics. Equation (7) shows that the rate of change of some quantity q in the one-dimensional form is related to the rate of change of the gradient and a constant, which in the atmosphere is the eddy diffusivity coefficient.

$$\frac{d\bar{q}}{dt} = K\frac{\partial^2 \bar{q}}{\partial x^2} \tag{7}$$

Boundary conditions for a point source are

$$\bar{q} \to 0 \text{ as } t \to \infty \qquad -\infty < x < \infty$$
$$\bar{q} \to 0 \text{ as } t \to 0 \qquad \text{for all } x \text{ except } x = 0$$

where $\bar{q} \to \infty$ such that

$$\int_{-\infty}^{\infty} \bar{q}\, dx = Q$$

In the three-dimensional form to represent space with respect to some point on the earth's surface, the equation is written as

$$\frac{d\bar{q}}{dt} = \frac{\partial}{\partial x}\left(K_x \frac{\partial \bar{q}}{\partial x}\right) + \frac{\partial}{\partial y}\left(K_y \frac{\partial \bar{q}}{\partial y}\right) + \frac{\partial}{\partial z}\left(K_z \frac{\partial \bar{q}}{\partial z}\right) \tag{8}$$

In this case, we find that the local rate of change with time of concentration of material q is related to the gradients in the three directions as well as the diffusivity coefficients and their change in the three directions. The boundary conditions that are applied to fundamental problems such as a point source are indicated. We noted that Q is the source strength, that is, the total release of material represented by q. The solutions to the equations have been obtained by a number of people and a number of methods because of the classical nature of the equation for its use in heat conduction. A fundamental solution of the equation is known to be a Gaussian function as illustrated below:

$$\frac{\bar{q}}{Q} = \frac{1}{at^{1/2}} \exp\left(-\frac{bx^2}{t}\right) \tag{9}$$

$$a = (4K\pi)^{1/2} \quad \text{and} \quad b = (4K)^{-1}$$

where x is the distance from the center of the diffusing cloud, and because the term is squared, we have a symmetric function. The coefficients a and b are arrived at using the continuity equation which gave us the total emission strength. Therefore, for an instantaneous point source emitted at time ($t = 0$) the solution to Eq. (7) for an instantaneous point source of $\bar{q}$ with total strength Q is given as

$$\frac{q}{Q} = \frac{1}{(4\pi Kt)^{1/2}} \exp\left(-\frac{x^2}{4Kt}\right) \tag{10}$$

The solution that was just illustrated for one-dimensional form of an instantaneous point source can be extended to three dimensions, where the turbulence is not even in all three dimensions, as

$$\frac{q}{Q} = (4\pi t)^{-3/2}\, (K_x K_y K_z)^{-1/2} \exp\left[-\frac{1}{4t}\left(\frac{x^2}{K_x} + \frac{y^2}{K_y} + \frac{z^2}{K_z}\right)\right] \tag{11}$$

Notice here that there is differentiation made with respect to the dispersion coefficient K, depending on the direction of interest, that is, in the x, y, or z coordinates. Appropriate integration of the equations permits us to examine sources other than just an instantaneous point source. For example, integration with respect to space yields an equation for instantaneous volume sources such as an explosion. Integration with respect to time gives the continuous point source solution.

As noted in the preliminary aspects of this subsection, one of the difficulties with the K theory is that it is assumed that the eddy diffusivity is constant. That is not generally the case in the planetary boundary layer which has a pronounced shear of the mean wind and large variations in the vertical temperature gradients due to different surface heating. Various workers have defined the diffusivity K as a function of stability and height above ground. For example, the vertical K may be represented by a power law function similar to that which may be used to represent wind speed as a function of distance from the surface. Some have also worked out changes for values of the lateral or K_y or K_z values to account for the change of wind direction with height in the planetary boundary layer. Seldom do these techniques yield simple analytical solutions.

A simple limiting assumption in the K theory, which has never really been demonstrated to be true, is that the diffusion which represents the flux of a quantity is proportional to the gradient of that quantity. However, the validity of the K theory has been demonstrated or at least judged to be demonstrated based on the success achieved in representing dispersion in the atmosphere. At the present time, the K theory is utilized primarily for areas such as cities which exhibit weak gradients and have many area and line sources.

Equation (12) represents the form that is most frequently used as the mathematical description of the advection and dispersion of material in the atmosphere. The equation is written in one dimension for simplification.

$$\frac{\partial c}{\partial t} = -\frac{\partial}{\partial x}(uc) + \frac{\partial}{\partial x}\left(K \frac{\partial c}{\partial x}\right) \tag{12}$$

where c represents the concentration of the substance of interest. It replaces our general q which could mean heat, material, gas, or other physical property; u is the mean wind speed in the x direction, K is the eddy diffusivity or turbulent measure as used in the previous equations, x is the distance from the source, and t is time. The first term on the right side of the equation represents the transport due to the average winds. The second term represents the diffusion due to the turbulence. Note that the transport term in Eq. (12) was not present in Eq. (11). Equation (11) represents the dispersion of material about the center of a moving puff.

As was also noted, we can obtain the Gaussian concentration equation from the K-theory equations. However, there are some assumptions that must be made. These are:

1. The solution is not time dependent.
2. The wind speed is not a function of position either laterally or vertically.
3. The diffusivities are not functions of positions; although we recognize that in the true atmosphere, they probably are.
4. Diffusion in the x direction is insignificant when compared with the mean flow or transport of material in the x direction.

As we demonstrated in generating Eqs. (10) and (11) the K-theory equation is transformed to a Gaussian or normal distribution. That distribution is discussed in the next section.

3.2 Statistical (Gaussian) Theory

It is difficult to measure the K-theory dispersion coefficients in the atmosphere; therefore, model developers long ago turned to the normal distribution as a means of representing dispersion of gas and small particles in the atmosphere.

The coordinate system that is utilized to make calculations with the Gaussian dispersion models assumes that the x direction is in the direction of plume travel extending horizontally in the direction of the mean wind. The y axis is in the horizontal plane, perpendicular to the x axis; and the z axis extends vertically. The plume travels along or parallel to the x axis as illustrated in Figure 9. Additionally, the shape of the plume in Figure 9 shows how the Gaussian distribution is employed within the plume. The oval shapes are cross sections of the simplified plume, while the concentration profiles are represented by the cross sections that are shown through the figure.

The fundamental dispersion equation represented by the Gaussian theory is*

$$C(x, y, z; H) = \frac{Q}{2\pi\sigma_y\sigma_z u} \exp\left[-\frac{1}{2}\left(\frac{y}{\sigma_y}\right)^2\right]\left\{\exp\left[-\frac{1}{2}\left(\frac{z-H}{\sigma_z}\right)^2\right] + \exp\left[-\frac{1}{2}\left(\frac{z+H}{\sigma_z}\right)^2\right]\right\} \quad (13)$$

where C, the concentration at any coordinate position (x, y, z) given an emission Q at an effective height (H), is represented by the Gaussian distribution shown on the right-hand side. H is the plume centerline for an elevated source which is the sum of the physical stack height and the plume rise. The following assumptions are made:

1. The plume spread has a Gaussian distribution in both the horizontal and vertical planes with standard deviations of plume concentration distribution in the horizontal and vertical of σ_y and σ_z, respectively.
2. The mean wind speed affecting the plume is u.
3. The uniform emission rate of pollutants is Q.
4. Total reflection of the plume takes place at the earth's surface, that is, there is no deposition or reaction at the surface which is why there is a double exponential in the z and H terms.
5. The normal units utilized in the Gaussian model are concentration in grams per cubic meter, with Q the emission rate in grams per second, u in meters per second; σ_y, σ_z, H, and all coordinates are in meters.

The basic assumptions of this model are that the turbulence is everywhere the same; σ_y and σ_z are functions of distance from the source; the wind speed is constant throughout the layer where the plume is being transported; and the wind direction remains constant during travel.

There are many equations which may be used in special circumstances that are much simpler than the full equation. When concentrations are calculated only at ground level and perfect reflection is assumed, Eq. (14) may be used.

* Note: $\exp(-a/b) = e^{-a/b}$ where e is the base of natural logarithms and is approximately equal to 2.7183.

$$C(x, y, 0; H) = \frac{Q}{\pi \sigma_y \sigma_z u} \exp\left[-\frac{1}{2}\left(\frac{y}{\sigma_y}\right)^2\right] \exp\left[-\frac{1}{2}\left(\frac{H}{\sigma_z}\right)^2\right] \tag{14}$$

It eliminates the second exponential term and the z coordinate in the exponentials as well because z is assumed equal to zero at ground level. If the user is interested only in the concentration along the centerline of the plume where that is the maximum concentration, then the following may be used.

$$C(x, 0, 0; H) = \frac{Q}{\pi \sigma_y \sigma_z u} \exp\left[-\frac{1}{2}\left(\frac{H}{\sigma_z}\right)^2\right] \tag{15}$$

In this form the exponential in y and σ_y have been removed because we are interested only when y is zero in the center of the plume. Finally, if the plume is not elevated, another form may be used.

$$C(x, 0, 0; 0) = \frac{Q}{\pi \sigma_y \sigma_z u} \tag{16}$$

Here the concentration along the centerline at ground level is a function only of the emission rate, the wind speed, and the dispersion of the material in the downwind direction.

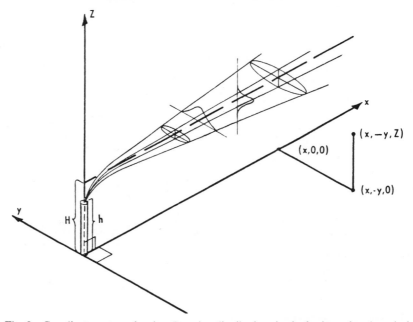

Fig. 9 Coordinate system showing Gaussian distributions in the horizontal and vertical.

The inclusion of these equations and other modifications in air quality dispersion models utilized by the Environmental Protection Agency will be demonstrated later in the description of the individual models.

4 COEFFICIENTS FOR DISPERSION MODELS

4.1 K-Theory Coefficients

Coefficients used by various model developers have been related to height above ground, Richardson number, or concentration gradient. Richardson number is a measure of the thermal stability, or instability, of the atmosphere and is defined by

$$\mathrm{Ri} = \frac{g}{T} \frac{\partial \theta / \partial z}{(\partial u / \partial z)^2} \tag{17}$$

where g is acceleration of gravity, T is the average temperature of the air in the layer ∂z in degrees Kelvin, $\partial\theta/\partial z$ is the potential temperature lapse rate $(\partial T/\partial z + 0.0098°C/m)$ and $\partial u/\partial z$ is the change of wind speed with height.

For example Pandolfo[5] and Liu[6] propose a form for the horizontal coefficient K,

$$K = [k(z + z_0)(1 + \alpha \text{Ri})]^2 \frac{\partial u}{\partial z} \quad \text{for} \quad \text{Ri} \geq 0$$

$$K = [k(z + z_0)]^2 \frac{\partial u}{\partial z}[1 - \alpha \text{Ri}]^{-3} \quad \text{for} \quad -0.048 < \text{Ri} < 0 \tag{18}$$

$$K = h(z + z_0)^2 \left| \frac{g}{Tm}\left(\frac{\partial T}{\partial z} + 0.0098\right)\right|^{1/2} \quad \text{for} \quad -0.048 > \text{Ri}$$

where k is von Karmen's constant (~ 0.4), z is height, and z_0 is surface roughness, usually small compared to Z, α is Obukhov constant and h is Priestley constant.

A simpler form was proposed by others, Slade,[3] where

$$K = \frac{\sigma_y d\sigma_y}{dt} \tag{19}$$

$$\sigma_y = \sigma_\theta(\bar{u}t)^{0.91} \tag{20}$$

$$\bar{u} = (u^2 + v^2 + w^2)^{1/2} \tag{21}$$

and σ_θ is the standard deviation of the angular fluctuation of a wind vane; u, v, and w are the orthogonal components of the wind (u is west–east, v is north–south, and w is vertical).

Vertical coefficients show similar variations in method of specification. Models developed by Pandolfo[5] and Liu[6] use the same coefficient in the vertical as for horizontal dispersion. The inherent assumption is that a plume is spreading as a cone and is completely symmetrical.

The vertical coefficient associated with Eq. (19) is given by

$$K = \frac{K_v z}{H} \quad 0 < z < H$$

$$K = K_v \quad H < z \tag{22}$$

where z is height above ground, H is the height of the turbulent mixed layer, and K_v is calculated from the vertical profile of temperature and change in time (see Sutton[1]):

$$K_v = \frac{\partial T/\partial t}{\partial^2 T/\partial z^2} \tag{23}$$

The usual method of applying K theory to engineering problems is to employ a computer code developed for dispersion problems and the calculation of the coefficients is part of the program. Finite difference and numerical integration techniques employed in the solution of K-theory models are beyond the scope of this handbook.

4.2 Statistical Models

A comprehensive set of dispersion coefficients designed to be used in the Gaussian plume models was first published by Pasquill[7] in 1959. They were modified to those illustrated in Figures 10 and 11 from Turner.[8] The lateral dispersion coefficients σ_y represent the horizontal spread of the plume perpendicular to the direction to travel. The vertical dispersion coefficients (σ_z) represent the spread of the plume in the vertical. The dispersion curves that are illustrated in these figures are based on experimental data collected at locations in the United States and Great Britain, but all data were collected for sources where the tracer material was released close to the ground, that is, generally less than 10 m and for time periods of approximately 10 min. Therefore, strictly speaking, the dispersion coefficients presented here should be used for sources near the ground for averaging time periods of 10 min. A discussion of times will be taken up in the next section. Portions of the curves in Figure 11 reflect speculation and extrapolation. Data used by Pasquill extended to about 1 km from the source and Pasquill extrapolated the curves based on his understanding of the climate of Southern England.

Notice that the lines are labeled A, B, C, D, E, and F. These terms are used to represent different classifications of stability of the atmosphere. Atmospheric stability is a representation of the turbulence that is contained within the transporting wind field affecting the dispersal of the material. A stability classification scheme was adopted by Pasquill to differentiate between periods of relatively little turbu-

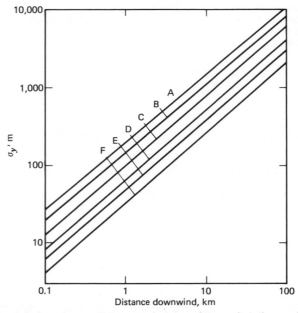

Fig. 10 Horizontal dispersion coefficient as a function of downwind distance from the source.

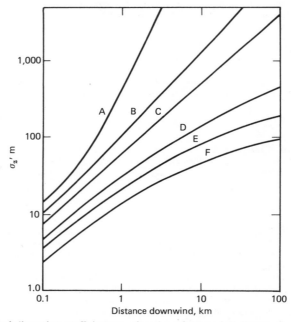

Fig. 11 Vertical dispersion coefficient as a function of downwind distance from the source.

lence, such as represented by the curves labeled *F*, and periods of high turbulence, such as in light winds on a sunny afternoon as is the case for curves labeled *A*.

Pasquill defined his stability categories according to incoming solar radiation, and in Table 1 the stability categories are listed to provide a key. During the daytime when incoming solar radiation is strong and the wind speed is light, the stability category is *A*. With strong incoming solar radiation and wind speeds greater than 6 m/sec, the stability category would be *C*, the assumption being that

Table 1 Key to Stability Categories

Surface Wind Speed (at 10 m), m/sec	Day[a]			Night[a]	
	Incoming Solar Radiation			Thinly Overcast or	
	Strong	Moderate	Slight	$\geq \frac{4}{8}$ Low Cloud	$\leq \frac{3}{8}$ Cloud
<2	A	A–B	B		
2–3	A–B	B	C	E	E
3–5	B	B–C	C	D	E
5–6	C	C–D	D	D	D
>6	C	D	D	D	D

[a] The neutral class, D, should be assumed for overcast conditions during day or night.

Table 2 P–G Stability Category Versus Vertical Wind Direction Fluctuation, σ_ϕ

P–G Stability Category	Standard Deviation of Vertical Wind Direction,[a] σ_ϕ
A	>12°
B	10°–12°
C	7.8°–10°
D	5°–7.8°
E	2.4°–5°
F	<2.4°

[a] These values should be adjusted for surface roughness by multiplying σ_ϕ by $(z_0/15 \text{ cm})^{0.2}$, where z_0 is the average surface roughness length within a 1- to 3- km radius of the source.

the strong winds would not permit thermal convection to develop to the degree that it does with light winds. Similar considerations are given at night with skies that are less than half covered by clouds, where the more stable cases occur with light winds and relatively cloud-free skies. (All tables are from *Guidelines on Air Quality Models*, revised 1980 by the EPA.[9])

Additional methods of determining stability classifications have been proposed. Table 2 lists the stability classifications as a function of vertical fluctuations if a vane is available from which the standard deviation of the vertical wind direction may be calculated for periods up to 30 min to 1 hr. Table 3 represents the Pasquill stability categories as a function of horizontal wind fluctuation. The range of standard deviation for a period of 30 min to 1 hr describes the horizontal dispersion category that should be used.

Finally, Table 4 may be used at night when a sigma theta (σ_θ) value is available and a σ_ϕ value is not. All these classifications in Tables 2, 3, and 4 may be found in Appendix C to the Environmental Protection Agency's *Guideline on Air Quality Models*.[9]

Equations that approximately fit the Pasquill–Gifford curves may be used to calculate σ_y and σ_z. The equations used to calculate σ_y in the ISC model are of the form

$$\sigma_y = 465.11628x \tan(TH) \tag{24}$$

$$TH = 0.017453293(c - d \ln x) \tag{25}$$

Table 3 P–G Stability Categories Versus Horizontal Wind Direction Fluctuations, σ_θ

P–G Stability Category	Range of Standard Deviation, Degrees
A	$\sigma_\theta \geq 22.5$
B	$22.5 > \sigma_\theta \geq 17.5$
C	$17.5 > \sigma_\theta \geq 12.5$
D	$12.5 > \sigma_\theta \geq 7.5$
E	$7.5 > \sigma_\theta \geq 3.8$
F	$3.8 > \sigma_\theta$

Table 4   Nighttime P–G Stability Categories Based on σ_θ

If the σ_θ Stability Class Is	And if the 10-m Wind Speed, u, Is		Then the Stability Class for σ_z Is
	m/sec	mph	
A	$u < 2.9$	$u < 6.4$	F
	$2.9 \leq u < 3.6$	$6.4 \leq u < 7.9$	E
	$3.6 \leq u$	$7.9 \leq u$	D
B	$u < 2.4$	$u < 5.3$	F
	$2.4 \leq u < 3.0$	$5.3 \leq u < 6.6$	F
	$3.0 \leq u$	$6.6 \leq u$	D
C	$u < 2.4$	$u < 5.3$	E
	$2.4 \leq u$	$5.3 \leq u$	D
D	Wind speed not considered		D
E	Wind speed not considered		E
F	Wind speed not considered		F

Table 5   Parameters Used to Calculate σ_y

$$\sigma_y = 465.11628x$$
$$\tan(TH)$$
$$TH = 0.017453293(c - d \ln x)$$

Pasquill Stability Category	c	d
A	24.1670	2.5334
B	18.3330	1.8096
C	12.5000	1.0857
D	8.3330	0.72382
E	6.2500	0.54287
F	4.1667	0.36191

where the downwind distance x is in kilometers. The coefficients c and d are listed in Table 5. The equation used to calculate σ_z is of the form

$$\sigma_z = ax^b \tag{26}$$

where the downwind distance x is in kilometers and the coefficients a and b are given in Table 6.

5   TIME AVERAGING

The Gaussian dispersion coefficients were determined from data acquired for about a 10-min period. Most air quality modeling requirements are for average concentrations for periods of approximately 1 hr or longer. The air quality guidelines for sulfur dioxide, for example, require estimates for 3-hr or 24-hr periods. One method proposed to obtain concentration estimates for periods longer than the 10 min appropriate to the dispersion coefficients is described in Eq. (27), where the concentration at a time of about 60 min is equal to the concentration for a 10-min period times a ratio.

$$C(t) = C(t_0)\left(\frac{t_0}{t}\right)^P \tag{27}$$

P is usually given a value of 0.17 to 0.20. It should be noted, however, that in the application of the EPA models that they do not normally account for this difference. Therefore, the dispersion coefficients are applied to 1-hr periods even though they represent 10-min dispersion. This is to maintain a conservative approach in the estimation of concentrations.

It should be pointed out that while the lateral spread of the cloud will increase as a function of time, usually the vertical spread of the cloud does not; and the values of vertical dispersion obtained for 10-min periods are appropriate to 1-hr periods as well. One model developed by the Air Resources Board of Texas utilizes an equation similar to Eq. (27) to increase the σ_y term and modifies concentration in that manner.

Table 6 Parameters Used to Calculate σ_z

Pasquill Stability Category	x	$\sigma_z = ax^b$	
		a	b
A^a	0.10–0.15	158.080	1.05420
	0.16–0.20	170.220	1.09320
	0.21–0.25	179.520	1.12620
	0.26–0.30	217.410	1.26440
	0.31–0.40	258.890	1.40940
	0.41–0.50	346.750	1.72830
	0.51–3.11	453.850	2.11660
	>3.11	b	b
B^a	0.10–0.20	90.673	0.93198
	0.21–0.40	98.483	0.98332
	>0.40	109.300	1.09710
C^a	>0.10	61.141	0.91465
D^a	0.10–0.30	34.459	0.86974
	0.31–1.00	32.093	0.81066
	1.01–3.00	32.093	0.64403
	3.01–10.00	33.504	0.60486
	10.01–30.00	36.650	0.56589
	>30.00	44.053	0.51179
E^a	0.10–0.30	23.331	0.81956
	0.31–1.00	21.628	0.75660
	1.01–2.00	21.628	0.63077
	2.01–4.00	22.534	0.57154
	4.01–10.00	24.703	0.50527
	10.01–20.00	26.970	0.46713
	20.01–40.00	35.420	0.37615
	>40.00	47.618	0.29592
F^a	0.10–0.20	15.209	0.81558
	0.21–0.70	14.457	0.78407
	0.71–1.00	13.953	0.68465
	1.01–2.00	13.953	0.63227
	2.01–3.00	14.823	0.54503
	3.01–7.00	16.187	0.46490
	7.01–15.00	17.836	0.41507
	15.01–30.00	22.651	0.32681
	30.01–60.00	27.074	0.27436
	>60.00	34.219	0.21716

a If the calculated value of σ_z exceeds 5000 m, σ_z is set to 5000 m.
b σ_z is equal to 5000 m.

Three- or 24-hr average concentrations are usually built up by superposition and addition with subsequent averaging of 1-hr plumes plotted on a horizontal plan with an appropriate wind direction and wind speed for each hour.

As longer times become involved (such as months, seasons, or years), the atmosphere provides an averaging of the concentration and a different equation is used. Equation (28) demonstrates an equation where the concentration is given as a function of distance in a sector of angular width theta (θ).

$$C(x,\ \theta) = \sum_S \sum_N \left\{ \frac{2Qf(\theta,\ S,\ N)}{\sqrt{2\pi}\ \sigma_{zs} u_N \left(\frac{2\pi x}{16}\right)} \exp\left[-\frac{1}{2}\left(\frac{H_u}{\sigma_{zs}}\right)^2\right]\right\} \tag{28}$$

The concentration calculation is made where Q is the average emission rate for the period in question; σ_z is the σ_z at distance x for each stability classification; and $f(\theta, S, N)$ represents the frequency of occurrence of the wind direction, stability class, and wind speed class. u_N in the denominator is the wind speed for class N and $2\pi x/16$ assumes that the 360-degree circle has been divided into 16 wind direction sectors. In the exponential term, height $-H_u$ is a representation of the effective plume height with wind speed u. The terms within the brackets are summed over all wind speed classes and stability classifications.

6 SPECIAL PROCESSES

There are a number of special processes that must be considered in connection with the Gaussian dispersion models. These include:

1. Plume rise.
2. Washout of material.
3. Dry deposition of material.
4. Conditions of trapping when there is an upper lid, as well as the earth's surface to contain the vertical dispersion of pollutants.
5. Fumigation, which occurs when a stable layer has been held aloft in a stable layer of air (such as, at night) and is suddenly broken up due to heating of the ground.
6. Building wakes.

These are discussed individually in the following sections.

6.1 Plume Rise

Plume rise equations developed by Briggs[10] are used in the EPA models and by most other model developers. Briggs' plume rise equations are based on theory of the entrainment of air into a jet or buoyant plume with coefficients determined empirically from observations of a wide variety of sources. The equations of plume rise developed by Briggs are given below.

$$\Delta h = 1.6F^{1/3}\, U^{-1}\rho^{2/3} \qquad \rho \leq 3.5X^* \tag{29}$$

and

$$\Delta h = 1.6F^{1/3}\, U^{-1}(3.5X^*)^{2/3} \qquad \rho > 3.5X^*$$
$$X^* = 14F^{5/8} \quad \text{if} \quad F \leq 55 \tag{30}$$
$$X^* = 34F^{2/5} \quad \text{if} \quad F > 55$$

where

$$\begin{aligned}
\Delta h &= \text{plume rise, m}\\
F &= gV_sR_s^2\,[(T_s - T_a)/T_s]\\
g &= \text{acceleration due to gravity, m/sec}^2\\
V_s &= \text{average exit velocity of gases of plume, m/sec}\\
R_s &= \text{inner radius of stack, m}\\
T_s &= \text{average temperature of gases in plume, K}\\
T_a &= \text{ambient air temperature, K}\\
U &= \text{wind speed at stack height, m/sec}\\
\rho &= \text{distance from source to receptor, m}
\end{aligned}$$

The Δh represents the plume rise in meters, and Δh is to be added to the physical stack height to determine the final height of the plume; F is the vertical flux of the buoyant plume and is determined from the equation provided. The distance to the point downwind of the stack where the plume is no longer rising is given as $3.5X^*$ and is incorporated into the plume rise equation.

Other plume rise equations have been developed; however, none have proved as popular, nor as generally useful as those given by Briggs. A comprehensive summary of plume rise equations was given by Moses,[11] where he compared approximately 17 different equations and developed a set of probability estimates of actual plume rise. It should be noted that the plume rises given by the Briggs' equations are average values, and departures of as much as 50% of the Δh term are possible.

6.2 Chemical Reactions

Gaussian models are not equipped to handle chemical reactions nearly as well as can be accomplished in the K-theory models. However, if the chemical reaction is simple, it is frequently handled by multiplying the concentration by an exponential term that is equivalent to a half-life. Therefore, the concentration is decreased because of chemical reaction at a rate equal to the transport times the half-life as shown in Eq. (31).

$$C = C_0 \exp\left(\frac{-\lambda x}{u}\right) \tag{31}$$

C_0 is the nonreactive concentration, λ is the decay rate or half-life of the reacting materials, x is the distance traveled, and u is the transporting wind speed.

6.3 Washout

When an obstacle such as a raindrop falls, it sweeps out a volume of air. This volume of air must separate to allow the obstacle to pass, and particles in the volume will tend to follow the air. Owing to their inertia and electrical attraction and to molecular diffusion, however, some fraction of these particles will cross the streamlines of the air and intersect the falling drop.

The scavenging process is thought to obey an exponential decay process (Slade[3]):

$$C = C_0 \exp(-\Lambda t) \tag{32}$$

where C_0 is concentration at the beginning of scavenging, t is time (sec), and Λ is the washout coefficient with units of $\sec^{-1}$.

In the special case of horizontally uniform precipitation over a continuous plume, the maximum washout possible at one particular distance with a rain that starts as soon as or prior to the release is given by

$$w_{max} = \frac{Q}{ex\sigma_y\sqrt{2\pi}} \exp\left(-\frac{y^2}{2\sigma_y^2}\right) \tag{33}$$

where w is the maximum washout rate (g/m²-sec) and other terms are as previously defined.

The washout rate for precipitation just beginning over a plume described by a Gaussian model is

$$w = \frac{\Lambda Q}{u\sigma_y\sqrt{2\pi}} \exp\left(-\frac{y^2}{2\sigma_y^2}\right) \tag{34}$$

The term Λ is a function of particle size, drop size, and precipitation rate but $\Lambda/\bar{u}$ may take on values of 10^{-3} to 10^{-5} (m^{-1}).

6.4 Deposition of Material

Two types of deposition may be important in air pollution problems. If particles are large enough, gravitational settling leads to deposition. Direct contact of gases and small particles with surfaces may cause deposition onto those surfaces.

The earth's gravitational field plays an important role in the deposition of particulate matter on the earth's surface. The rate of descent of the particle depends on a balance between the aerodynamic drag force and the gravitational force exerted by the earth. For a smooth spherical particle, neglecting the effect of slip flow, this balance may be expressed as

$$\rho_a v_g^2 C_D = \frac{8}{3} rg\rho \tag{35}$$

where the notation is as given below. Equation (35) cannot be solved directly for the fall velocity because the drag coefficient is an empirical function of the Reynolds number, Re, and therefore also of velocity. For the Reynolds number range between 10^{-4} and 10, the relation $C_D = 24/\text{Re}$ may be used, and since $\text{Re} = 2\rho_a v_g r/\mu$, Eq. (35) reduces to the familiar Stokes equation

$$V_g = \frac{2r^2 g\rho}{9\mu} \tag{36}$$

where v_g is fall velocity (m/sec), g is acceleration of gravity, ρ_a is atmospheric density (g/m³), ρ is particle density (g/m³), C_D is drag coefficient, r is particle radius (m), and μ is atmospheric dynamic viscosity which has a value of 1.77×10^{-2} g/m-sec at 10°C; see Sutton[1] for a table of values. The effect of shape on fall velocities is to reduce the velocity by about two-thirds from that of a smooth sphere. Smooth ellipsoidal particles theoretically vary in fall velocity by factors ranging from 0.5 to 1.04.

A plume with a single average fall velocity would realize a deflection in height of the centerline proportional to the settling velocity and the general equation for concentration according to Overcamp[12] is

$$C(x, y, z) = \frac{Q}{2\pi u \sigma_y \sigma_z} \exp\left(-\frac{y^2}{2\sigma_y^2}\right)$$
$$\times \left\{ \exp\left[-\left(z - h + \frac{V_g x}{u}\right)^2 \middle/ 2\sigma_z^2\right]\right. \tag{37}$$
$$\left. + a_0(x_g) \exp\left[-\left(z + h - \frac{V_g x}{u}\right)^2 \middle/ 2\sigma_z^2\right]\right\}$$

where the usual Gaussian terms are as previously defined and $a_0(x_g)$ is a reflection factor. For a gas or particles that are perfectly reflected at the ground, a_0 is unity. The reflection factor is only valid at the ground and is given by

$$a_0(x) = 1 - \left[\frac{2\sqrt{d}}{v_g + v_d + (uh - v_g x)\sigma_z^{-1}(d\sigma_z/dx)}\right] \tag{38}$$

where v_d is the deposition velocity or effective retention of gas or particles by the surface. See Sehmel[13] for a review of deposition and tables of deposition velocities. At point x_g, $a(x, z) = a_0(x_g)$ and the distance x_g is obtained by iteratively solving the following:

$$z + h - \frac{v_g x}{u} = \left(h - \frac{v_g x_g}{u}\right)\frac{\sigma_z(x)}{\sigma_z(x_g)} \tag{39}$$

A form of the Overcamp model is used in the EPA's Industrial Source Complex Model.

A feature of the model described here and the model described in Slade[3] is that material is assumed to be redistributed immediately upon deposition from the bottom of the cloud. Horst[14] has proposed a surface depletion model to correct for this but it has not been adapted to EPA models yet because the computation time is large.

6.5 Trapping

Trapping is a condition that permits rapid mixing of a plume beneath an inversion, but the stable air of the inversion blocks the escape of the material. This condition usually is not effective for high elevated plumes, but may lead to rather high concentrations for low-level plumes during morning conditions when an inversion is burning off slowly and the material is mixed rather uniformly between the ground and an elevated inversion. Equation (40) represents one method of estimating the concentration from a plume trapped in such a situation.

$$C(x, 0, z H) = \frac{Q}{2\pi u \sigma_y \sigma_z}\left(\exp\left[-\frac{1}{2}\left(\frac{z - H}{\sigma_z}\right)^2\right]\right.$$
$$+ \exp\left[-\frac{1}{2}\left(\frac{z + H}{\sigma_z}\right)^2\right]$$
$$+ \sum_{N=1}^{N=J}\left\{\exp\left[-\frac{1}{2}\left(\frac{z - H - 2NL}{\sigma_z}\right)^2\right]\right.$$
$$+ \exp\left[-\frac{1}{2}\left(\frac{z + H - 2NL}{\sigma_z}\right)^2\right] \tag{40}$$
$$+ \exp\left[-\frac{1}{2}\left(\frac{z - H + 2NL}{\sigma_z}\right)^2\right]$$
$$\left.\left. + \exp\left[-\frac{1}{2}\left(\frac{z + H + 2NL}{\sigma_z}\right)^2\right]\right\}\right)$$

Terms are defined as before; the new term L is the height of the inversion layer causing the trapping. The summation term that is appended to the equation represents the trapping term, and it is summed from N equals 1 to N equals J. In most cases, a value of J equal to 3 or 4 will suffice to provide a stable solution for the concentration.

6.6 Fumigation

Fumigation occurs when an elevated plume is caught in an atmosphere that is being rapidly heated from below and a plume that has been dispersing slowly in a stable layer of the atmosphere is suddenly brought to the ground by the enhanced vertical mixing. This is usually a condition that exists for only a short period of time, and an equation for the concentration is provided as

$$C(x, y, 0) = \frac{Q\left[\int_{-\infty}^{p} \frac{1}{\sqrt{2\pi}} \exp(-0.5p^2)\, dp\right]}{\sqrt{2\pi}\sigma_{yF} u h_i} \exp\left[-\frac{1}{2}\left(\frac{y}{\sigma_{yF}}\right)^2\right] \tag{41}$$

where

$$p = \frac{h_i - H}{\sigma_z}$$

and where the fumigation concentration requires an integral evaluation. Values for the integrals in brackets can be found in most statistical tables, because it is a normal distribution. This factor accounts for the portion of the plume that is mixed downward. If the inversion is eliminated up to effective stack height, half of the plume is presumed to be mixed downward; the other half remains in the stable air above. Equation (41) can be approximated when the fumigation is near its maximum by

$$C_F(x, y, 0) = \frac{Q}{\sqrt{2\pi}\, u\, \sigma_{yF} h_i} \exp\left[-\frac{1}{2}\left(\frac{y}{\sigma_{yF}}\right)^2\right] \tag{42}$$

$$h_i = H + 2\sigma_z = h + \Delta H + 2\sigma_z \tag{43}$$

The lateral dispersion coefficient has had an F added to it. This is because some difficulties are encountered in estimating a reasonable value for the horizontal dispersion, since in mixing the stable plume through a vertical depth, some additional horizontal spreading occurs. If this spreading is ignored and σ_y for stable conditions is used, a probable result would be estimated concentrations higher than actual concentrations. An approximation has been suggested by Turner[8]

$$\sigma_{yF} = \sigma_y(\text{stable}) + \frac{H}{8} \tag{44}$$

which increases the lateral dispersion coefficients by an amount equal to the height of the plume divided by 8. The computations of fumigation conditions are seldom made except in extremely special cases and are not usually incorporated in the EPA air quality models; however, trapping which was discussed above is included in those models.

6.7 Building Wakes

One of the problems that is frequently encountered in the dispersion of material from low-level sources is that the initial dispersion is greatly enhanced by the wake of the structures immediately surrounding the release point. There have been a number of methods described to account for building wakes, but the one that currently enjoys EPA favor is that developed by Huber[15] and based on wind tunnel studies.[16]

The EPA's ISC model incorporates correction for building wake effects. The ISC model programs account for the effects of building wakes by modifying σ_z for plumes from stacks with plume height to building height ratios greater than 1.2 (but less than 2.5) and by modifying both σ_y and σ_z for plumes with plume height to building height ratios less than or equal to 1.2. The plume height used in the plume height to stack height ratios is the same plume height used to determine if the plume is affected by the building wake. The ISC model defines buildings as squat ($h_w \geq h_b$) or tall ($h_w < h_b$). The building width h_w is approximated by the diameter of a circle with an area equal to the horizontal area of the building. The ISC model includes a general procedure for modifying σ_z and σ_y at distances greater than $3h_b$ for squat buildings or $3h_w$ for tall buildings. The air flow in the building cavity region is both highly turbulent and generally recirculating. The ISC model is not appropriate for estimating concentrations within such regions.

The modified σ_z equation for a squat building is given by

$$\sigma_z' = \begin{cases} 0.7h_b + 0.067[x - 3h_b]; & 3h_b < x < 10h_b \\ \sigma_z\{x + x_z\}; & x \geq 10h_b \end{cases} \tag{45}$$

where the building height h_b is in meters. For a tall building, Huber[16] suggests that the width scale h_w replace h_b in Eq. (45). The modified σ_z equation for a tall building is then given by

$$\sigma_z' = \begin{cases} 0.7h_w + 0.067[x - 3h_w]; & 3h_w < x < 10h_w \\ \sigma_z\{x + x_z\}; & x \geq 10h_w \end{cases} \tag{46}$$

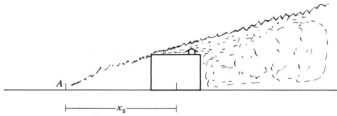

Fig. 12 Illustration of distance x_z to virtual source A that yields a cloud of material equivalent to size of cloud released into a building wake.

where h_w is in meters. It is important to note that σ'_z is not permitted to be less than the point source value given by Eq. (26), a condition that may occur with the A and B stability categories.

The vertical virtual distance x_z is added to the actual downwind distance x at downwind distances beyond $10h_b$ (squat buildings) or $10h_w$ (tall buildings) in order to account for the enhanced initial plume growth caused by the building wake. The vertical distance x_z is illustrated in Figure 12. It is the distance to an imaginary point source that would lead to a dispersed cloud size comparable to the size of the building wake. Thus, x_z for squat building is obtained as follows:

$$x_z = \left(\frac{1.2h_b}{a}\right)^{1/b} - 0.01h_b \tag{47}$$

where the stability-dependent constants a and b are given in Table 6. Similarly, the vertical virtual distance for tall buildings is given by

$$x_z = \left(\frac{1.2h_w}{a}\right)^{1/b} - 0.01h_w \tag{48}$$

For a squat building with a building width to building height ratio h_w/h_b less than or equal to 5, the modified σ_y equation is given by

$$\sigma'_y = \begin{cases} 0.35h_w + 0.067[x - 3h_b]; & 3h_b < x < 10h_b \\ \sigma_y\{x + x_y\}; & x \geq 10h_b \end{cases} \tag{49}$$

with the lateral virtual distance x_y given by

$$x_y = \left(\frac{0.35h_w + 0.5h_b}{p}\right)^{1/q} - 0.01h_b \tag{50}$$

The stability-dependent coefficients p and q are given in Table 7.

For building width to building height ratios h_w/h_b greater than 5, the presently available data are insufficient to provide general equations for σ_y. For a building that is much wider than it is tall and a stack located toward the center of the building (i.e., away from either end), only the height

Table 7 Coefficients Used to Calculate Lateral Virtual Distances

Pasquill Stability Category	$x_y = \dfrac{\sigma_{yo}^{1/q}}{p}$	
	p	q
A	209.14	0.890
B	154.46	0.902
C	103.26	0.917
D	68.26	0.919
E	51.06	0.921
F	33.92	0.919

scale is considered to be significant. The modified σ_y equation for a squat building is assumed the same as Eq. (49) with the lateral virtual distance x_y given by

$$x_y = \left(\frac{0.85 h_b}{p}\right)^{1/q} - 0.01 h_b \tag{51}$$

The modified σ_y equation for a tall building is given by

$$\sigma_y' = \begin{cases} 0.35 h_w + 0.067[x - 3h_w]; & 3h_w < x < 10h_w \\ \sigma_y\{x + x_y\}; & x \geq 10h_w \end{cases} \tag{52}$$

$$x_y = \left(\frac{0.85 h_w}{p}\right)^{1/q} - 0.01 h_w \tag{53}$$

7 ATMOSPHERIC DISPERSION MODELS

The following sections will describe atmospheric dispersion models recommended in the EPA *Guideline on Air Quality Models.*[9] All of these models are Gaussian dispersion models and primarily differ in their application, rather than in basic conceptual design or calculation. In fact, EPA has recently issued a new set of computer codes which make these models consistent with each other, so that if the same conditions are used (such as, a single elevated stack), the same answer will be achieved for like conditions. Complete descriptions of the models are included in the appendix of the EPA Modeling Guideline's document. In this section, we will only briefly describe the models, primarily concentrating on their use.

7.1 CDM—Climatological Dispersion Model

CDM is a climatological steady-state Gaussian plume model for determining long-term (seasonal or annual) arithmetic average pollutant concentrations at any ground-level receptor in an urban area. An expanded version (CDMQC) includes a statistical model to transform the average concentration data from a limited number of receptors into expected geometric mean and maximum concentration values for several different averaging times. Data requirements are the average emission rates and heights of emissions for point or areas sources including the stack gas temperature, velocity, and inside diameter. Meteorological data requirements include a stability wind rose, average afternoon mixing height, average morning mixing height, and average air temperature for the period of time involved. If the option is to be used to estimate short averaging time concentrations, a measured standard geometric deviation of concentrations in the field of receptors is required.

Output provides 1-month to 1-yr arithmetic average concentrations or with the short-term option, 1- to 24-hr concentrations may be estimated with some rather stringent assumptions.

7.2 RAM Gaussian Plume Multiple-Source Air Quality Model

RAM is a steady-state Gaussian plume model for estimating concentrations of stable pollutants for averaging times of an hour to a day from point and area sources. Level or gently rolling terrain is assumed. Calculations are performed for each hour. Both rural and urban versions are available; however, the rural version is not recommended for regulatory applications. Input requirements are the emission data which include: the location, emission rate, physical stack height, stack gas velocity, stack inside diameter, and temperature. Meteorological requirements are for hourly surface weather data including cloud ceiling, wind direction, wind speed, temperature, and opaque cloud cover. Daily mixing height is also required.

The output of the model is hourly and up to 24-hr concentrations at each receptor. There is a limited individual source contribution list and cumulative frequency distributions based on 24-hr averages. Up to 1 yr of data at a limited number of receptors may be obtained.

7.3 ISC—Industrial Source Complex Model

The ISC model is a steady-state Gaussian plume model which can be used to assess pollutant concentrations from a wide variety of sources associated with an industrial source complex. This model can account for settling and dry deposition of particles, downwash, area, line, and volume sources. Plume rise is a function of downwind distance and separation of point sources. Limited terrain adjustment is performed. ISC operates in both long-term and short-term modes. Input requirements are the emissions data, the location, the emission rate, pollutant decay coefficient, elevation of the source, the stack height, stack exit velocity, stack inside diameter, stack exit temperature, particle size distribution with corresponding settling velocities, surface reflection coefficient, and dimensions of adjacent buildings.

Meteorological data for the short-term version are hourly weather data including ceiling height, wind direction, wind speed, temperature, opaque cloud cover, and daily mixing height. For the long-term versions, stability wind rose, average afternoon mixing height, average morning mixing height, and average air temperature are required.

The output is concentration or deposition for any averaging time; and the highest and second highest values and highest 50 concentration points are provided.

7.4 MPTER—Multiple-Point Gaussian Dispersion Algorithm with Terrain Adjustments

MPTER is a multiple-point source algorithm with terrain adjustments. This algorithm is useful for estimating air quality concentrations of relatively nonreactive pollutants. Hourly estimates are made using the Gaussian steady-state model. Input requirements are emission rate, physical stack height, stack gas exit velocity, stack inside diameter, and stack gas temperature. Meteorological data used for hourly surface weather data include cloud ceiling, wind direction, wind speed, temperature, opaque cloud cover, and daily mixing height.

The output is hourly average and up to 24-hr average concentration at each receptor, highest through the fifth highest concentrations at each receptor for the period. The highest and second highest values are flagged. A limited source contribution table is printed. Note that this model may be used for rolling terrain; however, the terrain may not be higher than the physical stack top. It will handle only point sources and should only be applied in rural areas.

7.5 Single-Source Model (CRSTER)

CRSTER is a steady-state Gaussian plume model applicable to rural or urban areas and uneven terrain. The purpose of the model is: (1) to determine the maximum concentrations for certain averaging times between 1 and 24 hr over a 1-yr period due to a single-point source of up to 19 stacks; (2) to determine the meteorological conditions that cause the maximum concentrations; and (3) to store concentration information useful in calculating frequency distributions for various averaging times. A concentration for each hour of the year is calculated, and midnight-to-midnight averages are determined for each 24-hr period. A subroutine for this model is available which will produce running average concentrations for 3-, 8-, and 24-hr averaging times. The input requirements are the emission rate, physical stack height, stack gas exit velocity, stack inside diameter, and stack gas temperature. Meteorological data requirements are hourly surface weather data including cloud ceiling, wind direction, wind speed, temperature, opaque cloud cover, and mixing height.

The output is the highest and second highest concentration for the year at each receptor for averaging times of 1, 3, and 24 hr, plus a user selected averaging time which can be 2, 4, 6, 8, or 12 hr. The annual arithmetic average at each receptor is provided, and for each day, the highest 1- and 24-hr concentrations over the receptor field are given. Hourly concentrations for each receptor are written to magnetic tape.

This is the basic model used by EPA for consideration of single-point sources, particularly, new electric generating plants. It assumes that all point sources are at the same location. Even though it will accept up to 19 sources, they are considered collocated in the middle of a polar coordinate grid. The receptor locations are restricted to every 10 degrees and five-user specified radial distances. There is consideration of terrain height, but the terrain height must be below the top of the stack.

The models cited thus far, CRSTER and MPTER, calculate concentration for elevated terrain by reducing the assumed plume height above the ground by an amount equal to the elevation of the receptor above the elevation of the stack base. The computer codes will not perform calculations if the terrain is above the top of the stack. The Valley Model was designed to provide calculations for higher terrain.

7.6 Valley Model

A special-purpose model was developed to consider worst-case impact in complex terrain. The model is called Valley and is a modified Gaussian model. The horizontal crosswind distribution is assumed to be uniform rather than a normal distribution. The width of the sector of uniform concentration is assumed to be 22.5 degrees. During stable conditions, the plume centerline is assumed to approach within 10 m of the terrain obstacle at the same height. The model is strictly empirical, and it violates certain physical principles, but it is used in the regulatory process.

Input requirements are emission rate, stack diameter, stack height, stack gas temperature, and exit rate. Meteorology may be an actual stability wind rose for annual average concentrations or hypothetical for calculation of highest 24-hr concentration.

Output from the computer program consists of concentrations at 112 receptor points consisting of seven radial distances on 16 azimuth rays.

7.7 Other Gaussian Models

A number of other Gaussian models have been developed. Most of these computer codes have been viewed as providing a better or more efficient solution for a particular class of problems. If they employ the Pasquill–Gifford dispersion curves, the usual algorithm for limited mixing, and the Briggs plume rise method, then the answer for any given set of conditions should be the same as that given by the EPA models described above. Some models incorporate changes of lateral dispersion as a function of time or use different plume rise or mixing depth methods. Bowne et al.[17] and Londergan et al.[18] conclude that most are not any more accurate than the EPA versions and in some cases are even less accurate. Londergan concluded that the method of specifying dispersion rates (σ_y and σ_z) was far more important than the small differences between most of the Gaussian models.

7.8 Numerical Models

This class of models is more "complex" than the Gaussian models described above. Certainly the computer codes are more complex. The representation of physics of the atmosphere may not be complex. All models rely on numerical solutions to the K theory described in Section 3.

The Eulerian formulation has a fixed coordinate system, or grid, that covers the entire region of interest. The concentration in each square, or cell, of the grid is calculated by solving the equation using numerical methods and a computer. This approach is most useful for situations in which there are multiple sources and predicted concentrations are needed for the gridded region.

The Lagrangian formulation utilizes a coordinate system that moves with the transporting wind. This approach is most useful for single sources or a single-area source or for modeling long-distance transport of material. It does not provide concentrations over an extensive grid as in the Eulerian formulation. While the Lagrangian method saves computer time, it leads to results that can be difficult to interpret because the coordinate system is distorted. A subset of Lagrangian models called trajectory models avoids the distortion by transporting a single moving cell but creates other problems by neglecting or simplifying the diffusion process.

One solution method that is frequently seen is the hybrid technique called particle-in-cell. The source emissions are divided into individual Lagrangian cells, each of which is tracked over a fixed coordinate system. The concentration in each fixed grid square is calculated by counting the number of cells present in each square.

All of the numerical models in general use are well beyond description in a handbook. At least one model of each type described above is listed with reference below.

LIRAQ[19] developed by Lawrence Livermore National Laboratories and the Airshed Model[20] used by EPA in some applications are examples of Eulerian models. The Reactive Plume Model (RPM) described by Stewart and Yocke[21] uses a Lagrangian technique and ADPIC[22] is a particle-in-cell model.

8 DATA FOR USE IN MODELS

It is essential that the most appropriate data available be used in modeling analysis. Estimated concentrations are directly proportional to emission rate and wind direction, but the sensitivity of concentration to other variables is not always straightforward.

Data bases and related procedures for estimating input parameters are an integral part of the modeling process. Unfortunately, few of the variables required as input to a model are measured directly or are routinely available. Preprocessors are used by EPA models to convert the available source and meteorological data to a form that the air quality prediction model can accept. It is important for different emission conditions and a wide range of meteorological conditions based on several years of data to be considered in evaluating source impact in some cases. In others, only rudimentary information will suffice to determine if the highest concentration from a source is well below a limit.

8.1 Source Data

Sources can be classified into the general categories of point, line, and area sources. Point sources are individual chimneys. Line sources are most often roads but may be long buildings. Area sources are groups of buildings, roadways, and storage areas that have emissions over a limited height range and all emissions are added together to provide an average value for the area.

Table 8 lists the variables required and the usual units employed in EPA models.

If a worst-case impact is being predicted, then the usual practice is to use design emission rates. Some elevated sources may actually yield higher ground-level concentrations at some load less than 100% because of suppressed plume rise with lower stack flow. Such occurrence should be checked for large point sources by making calculations at 50 and 75% of load.

Table 8 Source Information for EPA Air Quality Models

	Point	Line	Area
Location	x, y (km)	x, y (km)	x, y (km)
End point	NA	x, y — x, y	NA
Corner point	NA	NA	x, y (km)
Size	NA	NA	(km²)
Emission rate	(g/sec)	(g/sec-m)	(g/sec-m²)
Stack/height of emission			
Height	(m)	(m)	(m)
Stack diameter	(m)	(m)	NA
Gas temperature	(K)	(K)	NA
Gas flow rate	(m/sec)	(m/sec)	NA
or gas volume	(m³/sec)	(m³/sec)	NA

8.2 Meteorological Data

The meteorological data used in any modeling process should be representative of transport and dispersion conditions for the area of interest. The representativeness of the data is dependent on proximity of the meteorological monitoring site to the area under consideration; the complexity of the site; the exposure of the meteorological measuring equipment; and the period of time during which the data are collected. Representativeness can be affected by distance, height difference between measured wind and transport wind, and terrain differences such as hills and valleys or large bodies of water.

The meteorological data required to describe transport and dispersion in most models consist of wind direction, wind speed, atmospheric stability, temperature, and mixing height. On-site data are preferred but seldom available. Necessary input data can be derived from routine measurements by National Weather Service (NWS) stations. The NWS data are available as individual observations or in summarized form from the National Climatic Center (NCC), Asheville, North Carolina. Models require either hourly meteorological data or annual (seasonal) stability wind roses. Both may be purchased from NCC.

Preliminary analyses may be accomplished by obtaining Local Climatological Data (LCD) for a NWS station from NCC for less than one dollar. Summarized data will provide prevailing wind directions, average wind speeds, temperature, and general cloud cover. If pollution concentrations can be demonstrated to be small with respect to a standard or increment, it may not be necessary to perform detailed or lengthy analyses.

If a source is significant, then EPA desires an analysis that makes use of 5 yr of meteorological data. Annual average concentrations utilize wind frequency distribution and average mixing heights and the model calculation requires no more time for 5 yr than for 1 yr. However, short-term standards and 1-, 3-, 8-, or 24-hr averages require calculations for every hour for 5 yr with the existing models.

8.3 Air Quality

Some models permit the use of existing air quality data either as a local background to which the modeled impact is added to yield total concentrations or as a means to "calibrate" a seasonal or annual model.

Air quality data are available from the U.S. Environmental Protection Agency, State Environmental Protection Agencies, and in some cases, local agencies. Some air quality data may be available from private local organizations, such as electric utilities or large manufacturing firms. The state agency is the best organization to approach for air quality data, because their records are more timely than those of EPA and they frequently have information that is not in EPA's national data bank.

9 ACCURACY OF MODELS

Comparisons of predictions of annual average concentrations with observed annual concentrations indicate that models can agree with observations within 20–30% according to Pasquill.[23] Annual averages exhibit much smoother concentration patterns than shorter averaging periods and as long as there is no undetected bias in local wind directions or stability conditions, that type of accuracy may be expected.

As time periods are reduced, the agreement between model predictions and observed concentrations is reduced. Londergan et al.[18] and Londergan and Borenstein[24] have examined the ability of EPA models to predict 1-hr average concentrations for near-ground sources and a nonbuoyant elevated source. Bowne et al.[17] have examined the ability of the EPA CRSTER model to predict 1-hr average concentrations for an elevated buoyant source. Models predicted concentrations of material from near ground sources within a factor of 2 of the observed concentration 65% of the time.

Table 9 Comparison between Observed and Predicted Plume Transport Direction (Number of Sampling Arcs with Angular Discrepancy in Each Indicated Range)

Study Name		Magnitude of Angular Discrepancy, degrees							Total Sampling
		0–3	3–7	7–11	11–15	15–21	21–29	>30	Arcs per Study
Prairie Grass		134	50	10	3	1	1	0	190
Hanford 67 (elevated)[a]		133	65	44	16	12	10	8	288
Green Glow Hanford 67 (2 m)[a]	(Combined)	68	71	17	7	3	9	4	179
N.R.T.S.		3	8	4	3	3	3	1	25
Ocean Breeze Dry Gulch	(Combined)[b]	35	50	47	46	30	31	77	316

[a] Hanford 67 results for experiments with release height of 2 m reported with Green Glow results.
[b] Results for Ocean Breeze and Dry Gulch based on NWS wind direction values.

9.1 Low-Level Sources

Londergan used tracer data, some of which was originally used to develop the rural dispersion curves in Figures 10 and 11, to test the accuracy and precision of several models. He compared observed and predicted plume centerline location, plume width, crosswind integrated concentration, and maximum concentrations.

Direction of transport of the center of the plume has some error even when the travel distance is short (800 m at most) and wind direction is measured at the source as shown in the Prairie Grass data in Table 9. Prairie Grass, Hanford 67, and so forth, refer to field measurement programs to determine rates of dispersion of tracer material in the atmosphere.

Other tests with poorer agreement covered greater distances of travel. The Ocean Breeze/Dry Gulch tests did not have anemometers at the release point and approximately 73% of the cases had modeled centerline positions that differed by more than 7 degrees. An error of 7 degrees under C stability would lead to a concentration error of a factor of 2 at 2 km at a specified point.

Plume widths precited by the EPA stability classification system using sky cover and wind speed were narrow compared to observed plume widths. The difference was greatest for stable conditions. The median plume width predicted was about 75% of that observed. Unstable elevated tests had predicted plume width equal to those observed. The median ratio of predicted to observed widths was 1.0.

Crosswind integrated concentrations (CWIC) may be compared to check the vertical dispersion coefficient. Londergan concluded that the Prairie Grass data yielded good agreement between predicted and observed CWIC values, but few of the other tests yielded information of value. Most of the data indicated a ratio of predicted to observed CWIC of 2 to 5 or higher. However, it was concluded that this represented loss of the particle tracer material, rather than a bias in the vertical dispersion rates. Prairie Grass tests were conducted with a gas tracer.

Maximum concentrations were compared using both the standard EPA stability classification method and the method where plume width is based on fluctuations of the wind direction. The results clearly indicated the superiority of using data describing the wind fluctuations to describe plume width. Unfortunately, wind fluctuation data are seldom available. An example from the Prairie Grass data which was used by Pasquill in generating his original dispersion curves is shown in Table 10.

Table 10 Comparison of Observed and Predicted Maximum Concentration at Prairie Grass (Percentage of Arcs with Ratio of Predicted to Observed Maximum Concentration in the Indicated Ranges)

Horizontal Dispersion Method	<0.2	0.2–0.5	0.5–0.67	0.67–1	1–1.5	1.5–2	2–5	>5
EPA (Turner[8])	1	8	6	20	22	15	21	7
Wind variability (Pasquill[25])	2	8	11	31	29	8	10	2

9.2   Elevated Buoyant Sources

Preliminary results from the Electric Power Research Institute's Plume Model Validation Project were described by Bowne et al.[17] Emissions from a tall stack were studied by comparing observed ground-level concentrations of sulfur dioxide and tracer gas to concentrations calculated from an EPA model. The tests were conducted at Commonwealth Edison Company's Kincaid Generating Station near Springfield, Illinois, in very flat terrain.

Preliminary evaluations of the comparison of the EPA CRSTER model to observed concentrations indicate the model predicts the maximum 1-hr concentration to be expected over a season or a year rather well. However, the ground-level pattern of observed concentration was 60% wider than predicted by the model. The Texas Episodic Model (TEM) predicted a ground pattern wider than observed and maximum concentrations less than observed because of the plume width prediction.

These results indicate the appropriate use for the Gaussian model for elevated, buoyant sources is to estimate the maximum concentration that may be observed over some period of time such as a month or a year, but it should not be used to predict concentration for a fixed location at a specific time.

10   APPLICATIONS

Sample calculations are described below which use equations, tables, and figures introduced earlier in this chapter. The calculations are developed for ever increasing complexity.

10.1   Simple Sources

The first example makes use of Eq. (16) and Figures 10 and 11. We assume a continuous release of a gas from a ruptured tank in an open, flat area. The release rate Q is 100 g/sec, and we want to determine the maximum (center of the plume) concentration 300 m away.

Meteorological conditions are: temperature, 15°C, wind speed, 3 m/sec, and the standard deviation of wind direction, 18 degrees. Inspection of Table 3 indicates the stability class is B. The value of σ_y at 500 m from Figure 10 is 84 m at 500 m downwind; the value of σ_z from Figure 11 is 51 m at 500 m from the source. EPA normally utilizes micrograms per cubic meter ($\mu g/m^3$) as a measure of concentration and the same convention is employed. Then,

$$C(x, 0, 0, 0) = \frac{Q}{\pi \sigma_y \sigma_z u} = \frac{100 \times 10^6}{\pi \cdot 84 \cdot 51 \cdot 3} = 2476.7 \ \mu g/m^3$$

at 500 m using the information developed.

Next, assume the gas is released from a stack 25 m high. The stack conditions are: diameter, 1 m, gas temperature, 330 K, exit velocity, 10 m/sec. All meteorological conditions are the same as the previous example. The first step is to solve for plume rise using Eq. (29) or (30). From these equations,

$$F = g V_s R_s^2 [(T_s - T_a)/T_s]$$
$$= 9.8 \cdot 10 \cdot 0.5^2 [(330 - 288)/330] = 3.1$$
$$X^* = 14 F^{5/8} = 14 \cdot 3.1^{5/8} = 14 \cdot 2.028 = 28.4$$

The distance to our receptor is 500 m and exceeds $3.5X^*$, which has a value of 99.4 m; therefore, Eq. (30) is appropriate.

$$\Delta h = 1.6 F^{1/3} u^{-1} (3.5 X^*)^{2/3}$$

$$\Delta h = \frac{1.6(3.1^{1/3})(99.4^{2/3})}{3} = \frac{1.6 \cdot 1.458 \cdot 21.455}{3} = 16.7 \text{ m}$$

The centerline concentration is given by Eq. (15) for the source used above. The value of H is stack height of 25 m plus the plume rise of 16.7 m or 41.7 m. The concentration is given by

$$C(x, 0, 0, H) = \frac{Q}{\pi \sigma_y \sigma_z u} \exp\left[-\frac{1}{2}\left(\frac{H}{\sigma_z}\right)^2\right]$$

$$= \frac{100 \times 10^6}{\pi \cdot 84 \cdot 51 \cdot 3} \exp\left[-\frac{1}{2}\left(\frac{41.7}{51}\right)^2\right]$$

$$= \frac{100 \times 10^6}{\pi \cdot 84 \cdot 51 \cdot 3} \times 0.71586 = 1773 \ \mu g/m^3$$

Next, add to the problem the requirement to find the concentration 150 m perpendicular to the centerline position at 500 m downwind from the source. Equation (14) is appropriate:

$$C(x, y, 0; H) = \frac{Q}{\pi \sigma_y \sigma_z u} \exp\left[-\frac{1}{2}\left(\frac{y}{\sigma_y}\right)^2\right] \exp\left[-\frac{1}{2}\left(\frac{H}{\sigma_z}\right)^2\right]$$

$$= \frac{100 \times 10^6}{\pi \cdot 84 \cdot 51 \cdot 3} \exp\left[-\frac{1}{2}\left(\frac{150}{84}\right)^2\right] \exp\left[-\frac{1}{2}\left(\frac{14.7}{51}\right)^2\right]$$

$$= \frac{100 \times 10^6 \times 0.71586 \times 0.203}{\pi \cdot 84 \cdot 51 \cdot 3} = 359.9 \ \mu g/m^3$$

All of the steps for determining concentration from a simple source were demonstrated. The effects on ground-level concentration for emitting through a stack were shown in the second calculation, and finally the effect of moving toward the edge of the plume was demonstrated in the third step.

10.2 Building Wakes

The problem of building wake influence on dispersion from a stack or vent on the roof of the building is a very common concern. The conditions to be used for calculation are set forth below in Table 11. The steps to determine concentration are:

a. Calculate plume height (final rise).
b. Compare plume height (H) to building height (h_b),
if $H \leq 1.2 \ h_b$, then both σ_y and σ_z are modified;
if $1.2h_b > H \leq 2.5h_{gb}$, then σ_z is modified;
if $H > 2.5h_b$, then neither σ_y or σ_z is modified and the source is treated as a stack without building wake.
c. Determine appropriate equations and calculate σ_y' and σ_z'.
d. Calculate concentration.

Following the steps outlined above,

$$F = 9.8 \times 2 \times 0.75^2\left(\frac{290 - 285}{290}\right) = 0.2$$

$$X^* = 14 \times 0.2^{5/8} = 5$$

$$\Delta h = 1.6 \frac{F^{1/3}}{u} 5^{2/3} = 2.7$$

$$H = 32 + 2.7 - 34.7$$

$H < 1.2h_b$ (34.7 < 36); therefore both σ_y and σ_z are modified. Equation (45) gives σ_z' at 200 m from

$$\sigma_z' = 0.7h_b + 0.067(X - 3h_b)$$
$$= 0.7 \times 30 + 0.067(200 - 90) = 28.4 \ m$$

Equation (49) gives σ_y' at 200 m from

$$\sigma_y' = 0.35h_w + 0.067(X - 3h_b)$$
$$= 0.35 \times 100 + 0.067(200 - 90) = 42.4$$

Table 11 Conditions for Calculation of Building Wake Effect on Concentration

Building length	100 m	Vent gas temperature	290 K
Building height	30 m	Vent gas velocity	2 m/sec
Vent height above roof	2 m	Vent diameter	1.5 m
Emission rate	100 g/sec		
Stability class	D		
Air temperature	12°C	Wind speed	4 m/sec
Distance to receptor area	200 m		

The concentration is given by

$$C(x, 0, 0; H) = \frac{Q}{\pi \sigma_y' \sigma_z' u} \exp\left[-\frac{1}{2}\left(\frac{H}{\sigma_z'}\right)^2\right]$$

$$= \frac{100 \times 10^6}{\pi \cdot 42.4 \cdot 28.4 \cdot 4} \exp\left[-\frac{1}{2}\left(\frac{34.7}{28.4}\right)^2\right]$$

$$= 3132.8 \ \mu g/m^3$$

When multiple sources or horizontal profiles of concentration are required, the computation burden becomes large and computer programs are used to provide the necessary calculations. Computational burden increases for other factors such as trapping which requires solution of the equations above plus a triple summation of four exponential terms. Similar difficulties are encountered for deposition and washout.

REFERENCES

1. Sutton, O. G., *Micrometeorology*, McGraw-Hill, New York, 1953.
2. Lumley, J. L., and Panofsky, H. A., *The Structure of Atmospheric Turbulence*, Wiley, New York, 1964.
3. Slade, D. (ed.), *Meteorology and Atomic Energy*, U.S. Atomic Energy Commission, TID-24190, Clearinghouse for Federal Scientific and Technical Information, 1968.
4. Simiu, E., and Scanlon, R. H., *Wind Effects on Structures*, Wiley, New York, 1978.
5. Pandolfo, J. P., and Jacobs, C. A., *Tests of an Urban Meteorological-Pollutant Model Using CO Validation Data in the Los Angeles Metropolitan Area*, Vol. 1, 1973; prepared for U.S. Environmental Protection Agency under Contract No. 68-02-0223, Center for Environment and Man, Inc., Hartford, CT.
6. Liu, M. K., Mandkur, P., and Yocke, M., *Assessment of the Feasibility of Modeling Wind Fields Relevant to the Spread of Brush Fires*, Report R74-15, Systems Applications, Inc., San Rafael, CA, 1974.
7. Pasquill, F., The Estimation of Dispersion of Windborne Material, *The Meteorological Magazine* **90**(1063), 33–49 (1961).
8. Turner, P. B., *Workbook of Atmospheric Dispersion Estimates*, Public Health Service Publication No. 999-AP-26, U.S. Department of Health, Education and Welfare, 1969.
9. U.S. Environmental Protection Agency, *Guideline on Air Quality Models*, OAQPS Guideline Series, Research Triangle Park, NC, 1980.
10. Briggs, G. A., Plume Rise Predictions, *Lectures on Air Pollution and Environmental Impact Analyses*, American Meteorological Society, Boston, MA, 1975.
11. Moses, H., and Kraimer, M. R., Plume Rise Determination—A New Technique Without Equations, *Journal of Air Pollution Control Association* **22**(8), 621–630 (1972).
12. Overcamp, T. J., A General Gaussian Diffusion-Deposition Model for Elevated Point Sources, *Journal of Applied Meteorology* **15**, 1167–1171 (1976).
13. Sehmel, G. A., Particle and Gas Dry Deposition: A Review, *Atmospheric Environment* **14**, 983–1011 (1980).
14. Horst, T. W., A Surface Depletion Model for Deposition from a Gaussian Plume, *Atmospheric Environment* **11**, 41–46 (1977).
15. Huber, A. H., and Snyder, W. H., Building Wake Effects on Short Stack Effluents, *Preprint Volume for the Third Symposium on Atmospheric Diffusion and Air Quality*, American Meteorological Society, Boston, MA, 1976.
16. Huber, A. H., Incorporating Building/Terrain Wake Effects on Stack Effluents, *Preprint Volume for the Joint Conference on Applications of Air Pollution Meteorology*, American Meteorological Society, Boston, MA, 1977.
17. Bowne, N. E., Londergan, R. J., Minott, D. H., and Murray, D. R., *Preliminary Results from the EPRI Plume Model Validation Project—Plains Site*, Report EPRI EA-1788, Electric Power Research Institute, Palo Alto, CA, 1981.
18. Londergan, R. J., Mangano, J. J., Bowne, N. E., Murray, D. R., and Borenstein, H., *An Evaluation of Short-Term Air Quality Models Using Tracer Study Data*, Vol. I, API Report No. 4333, American Petroleum Institute, Washington, DC, 1980.
19. MacCracken, M. C., Wuebbles, D. J., Walton, J. J., Duewer, W. H., and Grant, K. E., The Livermore Regional Air Quality Model, *Journal of Applied Meteorology* **17**, 254–272 (1978).

20. Reynolds, S. D., and Roth, P. M., Mathematical Modeling of Photochemical Air Pollution, *Atmospheric Environment* **7**, 1033–1061 (1973).

21. Stewart, D., and Yocke, M. A., *Users Guide to Reactive Plume Model*, RPM II, SAI Report No. EI-79-93R, Systems Applications, Inc., San Rafael, CA, 1979.

22. Lange, R., ADPIC—A Three-Dimensional Particle-in-Cell Model for the Dispersal of Atmospheric Pollutants and Its Comparison to Regional Tracers, *Journal of Applied Meteorology* **17**, 320–329 (1978).

23. Pasquill, F., *Atmospheric Diffusion*, Wiley, New York, 1974.

24. Londergan, R. J., Comparison of Dispersion Predicted by Gaussian Models with Observed Tracer Dispersion, *Paper 81-20.5*, 74th Annual Meeting of the Air Pollution Control Association, Philadelphia, PA, June 23, 1981.

25. Pasquill, F., *Atmospheric Dispersion Parameters in Gaussian Plume Modeling, Part II: Possible Requirements for Change in the Turner Workbook Values*, Report EPA-600/476-030b, U.S. Environmental Protection Agency, Research Triangle Park, NC, 1976.

26. Hewson, E. W., and Gill, G. C., *Meteorological Investigations in Columbia River Valley near Trail, B.C.*, U.S. Bureau of Mines Bulletin 453, Washington, D.C., 1944.

27. Lyons, W. A., and Olsson, L. E., Mesoscale Air Pollution Transport in the Chicago Lake Breeze, *Journal of Air Pollution Control Association* **22**(11) 876–881 (1972).

CHAPTER 35
ATMOSPHERIC CHEMISTRY

ALAN LLOYD

Environmental Research & Technology, Inc.
Westlake Village, California

1 ATMOSPHERIC CHEMISTRY

Our understanding of atmospheric chemistry has increased substantially in the last decade. At the same time, researchers have realized the importance of the coupling of the gas–aqueous phase chemistry under ambient conditions. This coupling and the chemistry occurring in the aqueous phase and ambient aerosols remain areas of significant uncertainty. Yet, greater understanding is required in order to understand the role of atmospheric chemistry in secondary pollutant and secondary aerosol formation involved in visibility reduction, acid precipitation, and health impacts to the general population.

This chapter attempts to summarize our current understanding of atmospheric chemistry. It deals exclusively with the lower atmosphere with emphasis on the polluted troposphere. Time and space dictate that only major points are presented and more extensive reviews need to be consulted for a fuller discussion of the various chemical reactions. Following a brief overview of the major gas phase processes occurring in the polluted troposphere, subsequent chapters address the chemistry of various classes of hydrocarbons in addition to the explicit reactions of the "inorganic" system involving O_2, O_3, NO, NO_2, SO_2, and radicals such as OH, HO_2, and so on. The possible role of biogenic hydrocarbon emissions on O_3 formation is discussed. The chemistry of formation of sulfates and nitrates is discussed, subsequently followed by a brief summary and an outline of areas of uncertainty.

1.2 Overall Aspects of Polluted Atmosphere Photochemistry

Our knowledge of the gas phase chemistry occurring in photochemical air pollution systems has advanced rapidly in the last 5 yr or so (Finlayson-Pitts and Pitts, 1977; Atkinson and Lloyd, 1981, 1983) and in most cases we now have a qualitative understanding of the gas phase chemistry involved. The atmospheric chemistry involving hydrocarbon oxidation in the presence of NO_x and SO_x is conveniently summarized in Figure 1. This shows the conversion of NO to NO_2 and the formation of O_3. The role of radical species in converting NO to NO_2 is shown, these radicals being formed by the attack of ozone or hydroxyl radicals on the various classes of hydrocarbons.

The organics present in ambient air can be categorized into three general classes of hydrocarbons, as shown in Table 1 (Arnts and Meeks, 1980): alkanes, such as n-butane, isobutane, and the pentanes; alkenes (ethene, propene, butenes); and aromatics (benzene, toluene, and xylenes). Oxygenates, such as the aldehydes RCHO shown in Figure 1, are another class. These oxygenates are mostly composed of aldehydes, but smaller amounts of ketones, alcohols, ethers, and esters are also present. Of these, only the aldehydes and ketones are photoactive, but the other species can also react to form radicals.

Figure 2 illustrates the relationship of the gas phase processes to those occurring in the aqueous phase for NO_x and SO_2 oxidation. The coupling of these processes is important and is receiving increasing attention (Chameides and Davis, 1982).

Our present knowledge of the gas phase chemistry of polluted atmospheres is best presented by consideration of each of these organic classes, together with the inorganic reaction mechanisms.

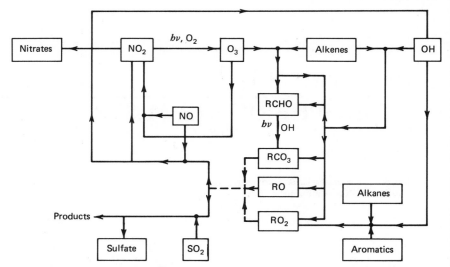

Fig. 1 Schematic representation of current knowledge of the overall feature of gas phase photochemical smog formation. Source: Roger Atkinson and Alan C. Lloyd

2   CHEMISTRY OF THE INORGANIC SYSTEM

2.1 O_3, Nitrogen Oxides, and Radicals

The inorganic reaction mechanisms applicable to the lower troposphere and air pollution chemistry now appear to be reasonably well understood (Hampson and Garvin, 1978; NASA, 1981; Atkinson and Lloyd, 1983). The basic reactions are

$$NO_2 + h\nu \rightarrow NO + O(^3P)$$
$$O(^3P) + O_2 + M \rightarrow O_3 + M$$
$$NO + O_3 \rightarrow NO_2 + O_2$$

followed by the important tropospheric OH radical production steps

$$O_3 + h\nu \ (\lambda \lesssim 310 \ nm) \rightarrow O('D) + O_2 \ ('\Delta g)$$
$$O('D) + M \rightarrow O(^3P) + M \ (M = air)$$
$$O('D) + H_2O \rightarrow 2OH$$

together with the reaction of O_3 with NO_2 leading to NO_3 and N_2O_5 chemistry.
    Reaction of hydroxyl radicals with CO:

$$OH + CO \xrightarrow{\quad O_2 \quad} HO_2 + CO_2$$

followed by

$$HO_2 + NO \rightarrow OH + NO_2$$

leads to regeneration of OH, and this chain oxidation of CO was first postulated around 1969 (Heicklen et al., 1969; Stedman et al., 1970). The inorganic chemistry "mechanism" is then rounded out by reactions of OH and HO_2 with O_3 and by the termination reactions of OH with NO and NO_2 and of HO_2 with NO_2:

$$OH + NO \xrightarrow{\quad M \quad} HONO \qquad (nitrous \ acid)$$
$$OH + NO_2 \xrightarrow{\quad M \quad} HNO_3 \qquad (nitric \ acid)$$
$$HO_2 + NO_2 \xrightarrow{\quad M \quad} HO_2NO_2 \qquad (pernitric \ acid)$$

Table 1 Hydrocarbon Composition of Tulsa Air (Arnts and Meeks, 1980)[a]

Hydrocarbons	Concentration (ppbc)	Hydrocarbons	Concentration (ppbc)	Hydrocarbons	Concentration (ppbc)
Alkanes		Alkenes		Aromatics	
Ethane	7.2	Ethene	7.3	Toluene	14.8
Propane	8.9	Propene	2.8	Ethylbenzene	2.5
Isobutane	12.9	Isobutene	4.3	p-Xylene	2.0
n-Butane	48.7	trans-2-Butene	4.7	m-Xylene	5.4
Isopentane	65.9	cis-2-Butene + butadiene	0	o-Xylene	2.5
n-Pentane	40.9	1-Pentene	2.2	Isopropylbenzene	0
Cyclopentane	4.6	2-Methyl-1-butene	3.1	n-Propylbenzene	0.9
2-Methylpentane	19.6	trans-2-Pentene	7.3	m + p-Ethyltoluene	0
3-Methylpentane	13.0	cis-2-Pentene	5.6	1,3,5-Trimethylbenzene	1.0
n-Hexane	14.0	2-Methyl-2-butene	6.0	o-Ethyltoluene	0
n-Heptane	5.8	Isoprene	0.3	1,2,4-Trimethylbenzene	1.3
Nonane	1.3	4-Methyl-2-pentene	0.8	m-Ethyltoluene	1.3
Decane	1.8	trans-2-Hexene	1.3		
Σ Alkanes	244.6	Σ Alkenes	45.8	Σ Aromatics	31.7

[a] 10:33–11:03 A.M., Tulsa Post Office, 7-27-1978.

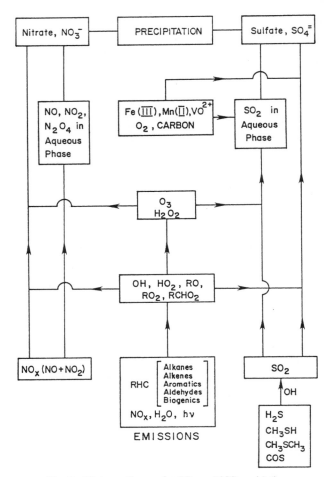

Fig. 2   Major pathways for SO_2 and NO_x oxidation.

Because of the efficient regeneration of OH + NO from HONO by photolysis and of HO_2 + NO_2 from HO_2NO_2 by thermal decomposition, under ambient lower tropospheric conditions, the OH + NO_2 termination reaction is the most important of these three.

The reactions occurring in the nitrogen oxides, ozone, and air system are discussed in detail by Atkinson and Lloyd (1983). Table 2 (Atkinson et al., 1982) summarizes the reaction pathways and rate constants for the major reactions occurring in the inorganic system in the polluted lower troposphere and these expressions are the best available in 1982 and are generally consistent with recommendations from the CODATA and NASA groups responsible for evaluating and recommending rate constant data applicable to the atmosphere.

2.2   Atmospheric Chemistry of Sulfur Dioxide (SO_2)

Calvert et al. (1978) have provided an excellent review of the gas phase photooxidation of SO_2 under ambient conditions. The recent review of gas phase oxidation processes for SO_2 by Atkinson and Lloyd (1980) concludes that only two species, the hydroxyl radical OH, and the Criegee intermediate $RCHO_2$, need be considered for the normal atmospheric conditions. The current status of these reactions is summarized below and is taken largely from Atkinson and Lloyd (1980).

$$OH + SO_2 + M \ (M = \text{air}) \rightarrow HSO_3 + M$$

Table 2   Reactions and Rate Constants in the Basic Organic Mechanism [a] (from Atkinson et al., 1982)

Reaction	Rate Constant (ppm min units)
$NO + O_3 \rightarrow NO_2 + O_2$	$k_2 = 1.0 \times 10^6 T^{-1} e^{-1450/T}$
$OH + NO \xrightarrow{M} HONO$	$k_4 = 8.7 \times 10^8 T^{-2}$
$OH + NO_2 \xrightarrow{M} HNO_3$	$k_5 = 1.5 \times 10^9 T^{-2}$
$HO_2 + NO_2 \xrightarrow{M} HO_2 + NO_2$	$k_7 = 3.7 \times 10^6 T^{-1}$
$HO_2 + NO_2 \xrightarrow{M} HO_2NO_2$	$k_8 = 1.5 \times 10^8 T^{-2}$
$HO_2NO_2 \xrightarrow{M} HO_2 + NO_2$	$k_9 = 7.8 \times 10^{15} e^{-10420/T}$
$HO_2 + HO_2 \rightarrow H_2O_2 + O_2$	$k_{10} = 3.4 \times 10^4 T^{-1} e^{1100/T} + 5.8$
	$\times 10^{-5} T^{-2} e^{5800/T} [H_2O]$
$OH + CO\ 2HO_2$	$k_{12} = 1.3 \times 10^5 T^{-1}$
$NO_2 + O_3 \rightarrow NO_3$	$k_{13} = 5.3 \times 10^4 T^{-1} e^{-2450/T}$
$NO + NO_3 \rightarrow 2NO_2$	$k_{14} = 8.4 \times 10^6 T^{-1}$
$NO_2 + NO_3 \xrightarrow{M} N_2O_5$	$k_{15} = 3.1 \times 10^7 T^{-1} e^{-1100/T}$
$N_2O_5 \xrightarrow{M} NO_2 + NO_3$	$k_{16} = 3.5 \times 10^{18} e^{-12280/T}$
$N_2O_5 + H_2O \rightarrow 2HNO_3$	$k_{17} \approx 1.3 \times 10^{-3} T^{-1}$
$OH + O_3 \rightarrow HO_2$	$k_{19} = 7.0 \times 10^5 T^{-1} e^{-940/T}$
$HO_2 + O_3 \rightarrow OH$	$k_{20} = 4.8 \times 10^3 T^{-1} e^{-580/T}$

[a] Photodissociation reactions are not included.

This reaction is the most important gas phase reaction for SO_2 oxidation. It is in the falloff region between second and third order kinetics at the pressures encountered under ambient conditions (Atkinson et al., 1976). From a review and evaluation of the literature data, Calvert et al. (1978) recommended a bimolecular rate constant at 1 atm of air and 298 K of

$$k_1 = (1.1 \pm 0.3) \times 10^{-12} \text{ cm}^3/\text{molecule-sec}$$

while Davis et al. (1979) have estimated, that at 1 atm of N_2 and at 298 K,

$$k_1 = 9 \times 10^{-13} \text{ cm}^3/\text{molecule-sec}$$

with an estimated uncertainty of a factor of -1.5 and -2.5.

The data of Davis et al. (1979) were included in the review and evaluation of Calvert et al. (1978). The most recent kinetic studies of this reaction have been carried out by Harris et al. (1980) and Wone and Ravishankara (1982), using flash photolysis–resonance fluorescence techniques.

Atkinson and Lloyd (1983) recommend that at atmospheric pressure

$$k_1 \ (M = \text{air}) = 9.0 \times 10^{-13} \text{ cm}^3/\text{molecule-sec}$$

at 298 K with an uncertainty of a factor ± 1.5, and

$$k_1 \ (M = \text{air}) = 9.0 \times 10^{-13} \ (T/298)^{-2.5} \text{ cm}^3/\text{molecule-sec}$$

The subsequent fate of the HSO_3 radicals has recently been discussed by Davis et al. (1979), and investigated, using long-path FTIR spectroscopy, by Niki et al. (1980).

Davis et al. (1979) suggest the following sequence of reactions of HSO_3:

$$HSO_3 + O_2 \xrightarrow{(M)} HSO_5$$
$$HSO_5 + H_2 \longrightarrow HSO_5 \cdot \quad (H_2O)$$
$$\downarrow H_2O$$
$$HSO_5 \cdot \quad (H_2O)_x$$

followed by possible reaction with NO, SO_2, and so on. These reactions are most likely gas–aerosol in nature and would contribute to visibility degradation. The FTIR spectroscopic study of Niki et al. (1980) is consistent with such a sequence of reactions. In irradiated Cl_2–H_2–NO–SO_2–air mixtures, where the dominant reactive species is the OH radical, SO_2 was observed to disappear, with the appearance of infrared absorption bands attributable to H_2SO_4 (dry and ~20% hydrated) in the aerosol phase. It was suggested (Niki et al., 1980) that formation of H_2SO_4 from the H_2SO_5 radical could occur via the reaction sequence

$$HSO_5 + NO \rightarrow HSO_4 + NO_2$$
$$HSO_4 + HO_2 \rightarrow H_2SO_4 + O_2$$

However, the study of Niki et al. (1980) was carried out at ppm (~10^{14} molecule/cm³) concentrations of H_2O vapor, and under ambient atmospheric conditions, where H_2O concentrations are $\geq 10^{17}$ molecule/cm³, the fate of the HSO_5 radical may be as suggested by Davis et al. (1979), in that hydration may predominate over reaction with NO. In summary, it appears that the rate determining step in the conversion of SO_2 to sulfate by reaction with OH is the initial step, that is, the addition of OH radicals to SO_2. Subsequent reactions of SO_2, NO, and NO_2 with hydrated HSO_5 species are uncertain and may involve gas–liquid or gas–aerosol chemistry.

For application to sulfate modeling studies, it is reasonable to assume that the HSO_3 radical rapidly reacts with O_2 and hydrates to (ultimately) form sulfate aerosol, that is,

$$OH + SO_2 \xrightarrow{(M)} HSO_3 \xrightarrow[fast]{O_2} SO_4^=$$

2.2.1 O_3–Alkene–SO_2 Mixtures

The oxidation of SO_2 by intermediates in O_3–alkene reaction mixtures may be a significant reaction in polluted urban atmospheres under certain conditions. It appears that SO_2 oxidation occurs via

$$R\dot{C}R'O\dot{O} + SO_2 \rightarrow RCHO + SO_2$$

(Niki et al., 1977; Su et al., 1980).

An earlier key investigation was that of Cox and Penkett (1972) who studied O_3–alkene–SO_2–H_2O–air systems at atmospheric pressure to elucidate the amount of thermalized biradicals found under these experimental conditions. They showed that an intermediate was capable of oxidizing SO_2 to (ultimately) H_2SO_4 aerosols, and it was noted that aldehyde yields were increased in the presence of SO_2, indicating that this intermediate reacts with SO_2 to yield the aldehyde and (presumably) SO_3. A substantial effect of relative humidity was noted, with the conversion of SO_2 to H_2SO_4 aerosol (relative to the O_3 + alkene reaction rate) *decreasing* with increasing relative humidity (at constant temperature). Calvert et al. (1978) have reviewed and reevaluated this data of Cox and Penkett (1972) and, by a slight extension of Cox and Penkett's (1972) suggested mechanism, postulated that [taking cis-2-butene as an example and simplifying the mechanism of Calvert et al. (1978)]:

$$O_3 + CH_3CH = CHCH_3 \rightarrow CH_3\underset{\overset{|}{\alpha\ CH_3\dot{C}HO\dot{O}\ +\ other\ products}}{CH}\overset{\overset{O}{\overset{O\diagup\ \diagdown O}{|}}}{\underset{}{}}CHCH_3$$

(where α is the fraction of thermalized $CH_3\dot{C}HO\dot{O}$ radicals formed), followed by reactions of thermalized $CH_3\dot{C}HO\dot{O}$ biradicals:

$$\begin{aligned} CH_3\dot{C}HO\dot{O} + SO_2 &\rightarrow CH_3CHO + SO_3 &\quad a\\ CH_3\dot{C}HO\dot{O} + H_2O &\rightarrow CH_3COOH + H_2O &\quad b\\ CH_3\dot{C}HO\dot{O} &\rightarrow other\ products &\quad c \end{aligned}$$

In the atmosphere, in addition to H_2O, other competing reactions are

$$CH_3CHOO + NO \rightarrow CH_3CHO + NO_2$$
$$CH_3CHOO + NO_2 \rightarrow CH_3CHO + NO_3$$

and it is not clear whether this SO_2 oxidation reaction is important under atmospheric conditions. Calvert et al. (1978), neglecting any reactions of this Criegee biradical with NO or NO_2, conclude that this route for SO_2 oxidation is minor. Inclusion of the $RR'C\dot{O}\dot{O} + NO_x$ reactions will further decrease the importance of this pathway.

In summary, the gas phase reactions involved in the atmospheric oxidation of SO_2 appear to be:

$$OH + SO_2 \xrightarrow{(M)} HOSO_2 \xrightarrow{H_2O} \xrightarrow{etc.} SO_4^=$$

and possibly

$$RR'CHO_2 + SO_2 \rightarrow products$$

3   CHEMISTRY OF ORGANICS

3.1   Alkanes

The alkanes make up a large fraction of the hydrocarbons present in the atmosphere. Methane is the largest component with a worldwide background of about 1.4 ppm. It is now well established from basic laboratory investigations, supported by smog chamber modeling studies (Hendry et al., 1978; Falls and Seinfeld, 1978; Carter et al., 1979; Whitten et al., 1979; Cox et al., 1980), that under ambient atmospheric conditions the sole chemical loss process of the alkanes is via reaction with the OH radical

$$OH + RCH_2R_1 \rightarrow H_2O + R\dot{C}HR_1$$

followed by, for the simple ($\lesssim C_3$) alkanes,

$$R\dot{C}HR_1 + O_2 \rightarrow RR_1CHOO \cdot$$
$$RR_1CHOO \cdot + NO \rightarrow RR_1CHO \cdot + NO_2$$
$$RR_1CHOO \cdot + NO_2 \leftrightarrows RR_1CHO_2NO_2$$
$$RR_1CHO \cdot + O_2 \rightarrow RCOR_1 + HO_2 \quad \text{(reaction with } O_2)$$
$$RR_1CHO \cdot \rightarrow \begin{Bmatrix} RCHO + R_1 \cdot \\ R_1CHO + R \cdot \end{Bmatrix} \quad \text{(decomposition)}$$
$$RR_1CHO \cdot + NO_2 \rightarrow RR_1CHONO_2$$

followed by the regeneration of the chain carrier, the OH radical:

$$HO_2 + NO \rightarrow OH + NO_2$$

While the above reaction scheme is applicable to the smaller alkanes, that is, methane, ethane, and propane, two other processes involving RO_2 and RO reactions can occur for the larger alkane systems.

1.   *Alkyl Nitrate Formation from RO_2 + NO.*   This reaction pathway

$$RO_2 + NO \rightarrow RONO_2$$

was postulated by Darnall et al. (1976) from smog chamber product studies of the NO_x photooxidations of *n*-butane, *n*-pentane, and *n*-hexane. It was proposed that the reaction of RO_2 radicals with NO proceed via

$$RO_2 + NO \begin{cases} \rightarrow RO + NO_2 \\ \rightarrow RO_2NO^* \end{cases}$$
$$RO_2NO^* \leftrightarrows RONO_2^*$$
$$RONO_2^* \rightarrow RO + NO_2 \quad \text{(decomposition)}$$
$$RONO_2^* \xrightarrow{M} RONO_2 \quad \text{(stabilization)}$$

with the ratios $k_a/(k_a + k_b)$ derived being given in Table 3, where k_a and k_b are the rate constants for the overall reactions (*a*) and (*b*)

$$RO_2 + NO \rightarrow RONO_2 \quad (a)$$
$$\rightarrow RO + NO_2 \quad (b)$$

As expected from unimolecular theory, the stabilization/decomposition ratio k_a/k_b increases with the number of degrees of freedom of the RO radical, and this ratio is less for branched than unbranched alkylperoxy radicals (Carter, 1979).

Table 3 Fractions of n-Alkanes Reacting to Form Alkyl Nitrates, That Is, the Rate Constant Ratio $k_a/(k_a + k_b)$, Under Atmospheric Conditions (from Atkinson et al., 1983)

n-Alkane	$k_a/(k_a + k_b)$
Ethane	<0.014[a]
Propane	0.036 ± 0.005[b]
n-Butane	0.077 ± 0.009[b]
n-Pentane	0.129 ± 0.019[b]
n-Hexane	0.223 ± 0.035[b]
n-Heptane	0.309 ± 0.050[b]
n-Octane	0.332 ± 0.034[b]

[a] From the Cl_2–NO–ethane–air irradiation.
[b] From CH_3ONO–NO–n-alkane–air irradiations.

2. *Alkoxy Radical Isomerization.* Isomerization of alkoxy radicals, well known from coll flames and hydrocarbon oxidations at elevated temperatures, was first proposed, on the basis of smog chamber product studies, as being of importance in room temperature NO_x–photooxidations by Carter et al. (1976), and subsequently by Baldwin et al. (1977).

Thermochemical calculations by both groups (Carter et al., 1976; Pitts et al., 1976; Baldwin et al., 1977) have shown that under ambient atmospheric conditions only the 1,4- and 1,5-H shifts of alkoxy radicals (via five- and six-membered ring transition states) are of importance (alkyl and alkyl peroxy radicals react with O_2 and NO, respectively, much faster than the corresponding isomerizations). Thus, the simplest alkoxy radical isomerization is that of the n-butoxy radical:

$$CH_3CH_2CH_2CH_2O^{\cdot} \rightarrow {}^{\cdot}CH_2CH_2CH_2CH_2OH$$

For $\geq C_4$ alkanes, this then means that the alkoxy radicals formed during their NO_x photooxidations can react via three routes (neglecting combination with NO_2 to form alkyl nitrates): reaction with O_2, unimolecular decomposition, or isomerization. The information available on the relative importance of these routes is summarized in Atkinson and Lloyd (1983).

For n-butane, the isomerization pathway only affects ~15% of the total reaction pathway with the OH radical, since the initial radicals are ~15% n-butyl and ~85% sec-butyl (Carter et al., 1979; Atkinson et al., 1979; Atkinson and Lloyd, 1983); while for the higher alkanes such as n-pentane, of the total number of alkoxy radicals formed, ~70% can undergo isomerization (Atkinson et al., 1979; Carter et al., 1976), and alkoxy radical isomerization probably dominates for the $\geq C_5$ straight chain alkanes (Carter et al., 1976). The reactions subsequent to the first isomerization were uncertain for a number of years, since in theory the hydroxy alkoxy radical formed subsequent to the first isomerization could isomerize further, as for example:

$$CH_3CH_2CH_2CH_2O \cdot \xrightarrow{\text{ISOM}} \cdot CH_2CH_2CH_2CH_2OH \xrightarrow[\substack{NO_2}]{\substack{NO \\ O_2}} \cdot OCH_2CH_2CH_2CH_2OH$$

$$\xrightarrow{\text{ISOM}} HOCH_2CH_2CH_2\overset{\cdot}{C}HOH$$

$$\xrightarrow[\substack{NO_2 \\ O \cdot}]{\substack{O_2 \\ NO}} HOCH_2CH_2CH_2\overset{\cdot}{C}HOH$$

$$\xrightarrow{\text{ISOM}} HO\overset{\cdot}{C}HCH_2CH_2CH(OH)_2$$

etc.

However, experimental data of Niki et al. (1978) and of Carter et al. (1979) have shown that the α-hydroxy radicals, $RR_1\overset{\cdot}{C}OH$, react with O_2 to form the corresponding carbonyl and HO_2:

$$RR_1\overset{\cdot}{C}OH + O_2 \rightarrow RCOR_1 + HO_2$$

and so terminate the isomerization sequence after a total of two isomerizations. This is illustrated below for the 2-pentoxy radical which forms a δ-hydroxy carbonyl (competing reactions, i.e., decomposition and reaction with O_2, of these alkoxy radicals are expected to be minor; reaction with NO_2 to form $RONO_2$ can also occur, of course):

$$CH_3\overset{\underset{\displaystyle |}{O \cdot}}{C}HCH_2CH_2CH_3 \xrightarrow{\text{ISOM}} CH_3CHOHCH_2CH_2CH_2 \cdot$$

$$\text{2-pentoxy} \qquad\qquad\qquad\qquad\qquad\qquad\qquad \Big\downarrow O_2$$

$$NO \xrightarrow{\hspace{3cm}} NO_2$$

$$\text{ISOM}$$

$$CH_3\dot{C}(OH)CH_2CH_2CH_2OH$$

$$\Big\downarrow O_2$$

$$CH_3COCH_2CH_2CH_2OH + HO_2$$

$$\text{5-hydroxy-2-pentanone}$$

Since these species retain the carbon skeleton, then, at least for the δ-hydroxyketones, further reaction with the OH radical will lead to progressively more polyfunctionality and may ultimately lead to aerosol formation.

The importance of the nitrate formation and isomerization of these long chain alkanes has increased with the growth of diesel-powered automobiles. Exhaust from the vehicles contains a higher proportion of long chain compounds than the gasoline equivalents (NRC, 1982).

3.2 Kinetic Data for Reactions of Alkanes with the OH Radical

Rate constant determinations for the reaction of alkanes with the OH radical are summarized in Atkinson et al. (1979). Grainer (1970) carried out an extensive study of many alkanes over a temperature range. He obtained mathematical expression to fit all his rate constant determinations. Darnall et al. (1978) and Atkinson et al. (1979) note constant data for $\geq C_3$ alkanes (apart from cyclopropane and cyclobutane) can be reasonably fit over the temperature range 300–500 K by the modified given formula:

$$^{k}OH + \text{alkane} = 1.0 \times 10^{-12} N_1 e^{-823/T} + 2.41 \times 10^{-12} N_2$$
$$e^{-428/T} + 2.10 \times 10^{-12} N_3 \text{ cm}^3/\text{molecule-sec}$$

where N_1, N_2, and N_3 are the number of primary, secondary, and tertiary C—H bonds, respectively.

3.3 Alkenes

3.3.1 *Reaction with the OH Radical*

Under ambient atmospheric conditions the alkenes react predominantly with ozone and OH radicals (Atkinson et al., 1979). The OH radical rate constants are now known reasonably well for most alkenes (Cox et al., 1980) and values for selected alkenes are shown in Table 4, taken from Atkinson and Lloyd (1980).

Table 4   Rate Constants for the Reaction of OH Radicals with Selected Alkenes (Atkinson and Lloyd, 1980)[a]

Alkene	cm³/molecule-sec	E/R	cm³/molecule-sec
C_2H_4	2.2×10^{-12}	-382	8.0×10^{-12}
C_3H_6	2.5×10^{-11}	-537	2.5×10^{-11}
1-C_4H_8	7.4×10^{-12}	-462	3.3×10^{-11}
t-2-C_4H_8	1.14×10^{-12}	-542	7.0×10^{-11}

[a] See Atkinson and Lloyd (1980) for citation of original references.

The initial reaction pathway for ethene and the methyl-substituted ethenes proceeds essentially totally via OH radical addition to the double bond (Atkinson et al., 1979; Herron et al., 1979). However, for alkenes containing longer ($\geq C_2$) side chains, a substantial amount [20 ± 5% for the case of 1-butene (Biermann et al., 1980)] of the overall OH radical reaction proceeds via H atom abstraction from the weak allylic C—H bonds (Atkinson et al., 1979; Atkinson et al., 1977; Biermann et al., 1980):

$$OH + CH_2{=}CHCH_2CH_3 \rightarrow HOCH_2\dot{C}HCH_2CH_3 + \dot{C}H_2CHOHCH_2CH_3 \quad (a)$$
$$\text{(OH radical addition)}$$
$$\rightarrow H_2O + CH_2{=}CH\dot{C}HCH_3 \quad\quad\quad\quad\quad (b)$$
$$\text{[H atom abstraction, with } k_b/$$
$$(k_a + k_b) = 0.20 \pm 0.05$$

at room temperature (Biermann et al., 1980)]

The fate of the β-hydroxyalkoxy radicals form from the OH–alkene adducts is uncertain at this time (Atkinson and Lloyd, 1981):

$$HOCH_2\dot{C}HCH_3 + O_2 \rightarrow HOCH_2\overset{\overset{\displaystyle OO\,\cdot}{|}}{C}HCH_3$$

$$NO \xrightarrow{\quad\quad} NO_2$$

$$HOCH_2\overset{\overset{\displaystyle O\,\cdot}{|}}{C}HCH_3$$

As with the simple alkoxy radicals, these β-hydroxyalkoxy radicals can react with O_2 or unimolecularly decompose:

$$CH_3\overset{\overset{\displaystyle O\,\cdot}{|}}{C}HCH_2OH + O_2 \rightarrow HO_2 + CH_3COCH_2OH$$

$$CH_3\overset{\overset{\displaystyle O\,\cdot}{|}}{C}HCH_2OH \rightarrow CH_3CHO + \dot{C}H_2OH$$

$$\downarrow O_2$$
$$HCHO + HO_2$$

Theoretical estimates of Baldwin et al. (1977), Golden (1979), and Batt (1979) indicate that reaction of O_2, at least for $\leq C_4$ β-hydroxyalkoxy radicals, should predominate, while the experimental data, from smog chamber studies (Carter et al., 1979) and from irradiated HONO–alkene–no-air mixtures (Niki et al., 1979), show conclusively that decomposition of these β-hydroxyalkoxy radicals predominates.

Another reaction of interest is that subsequent to H atom abstraction from alkenes. Thus, taking 1-butene as an example, ~20% of the OH radical reaction yields the allylic radical $CH_2{=}CHCHCH_3$ (Biermann et al., 1980), which can presumably add O_2 to form the peroxy radical

$$CH_2{=}CH\overset{\overset{\displaystyle OO\,\cdot}{|}}{C}HCH_3$$

This radical can, besides reacting with NO to yield the corresponding alkoxy radical (mainly), cyclize:

$$CH_2{=}CH\overset{\overset{\displaystyle OO\,\cdot}{|}}{C}HCH_3 \xrightarrow{\ NO\ } NO_2 + CH_2{=}CH\overset{\overset{\displaystyle O\,\cdot}{|}}{C}HCH_3$$

$$\downarrow \text{cyclization}$$

$$\begin{array}{c} CH_2{-}\dot{C}H \\ | \quad\quad \diagdown \\ | \quad\quad\quad CH{-}CH_3 \\ O{-\!-\!-\!-}O \end{array}$$

Experimental data on the NO_x photooxidation of isoprene (Arnts and Gay, 1979), where methylvinylketone and methacrolein are produced in the early stages, that is, before significant amounts of O_3 are formed, indicate that, at least for these particular radicals [$CH_2{=}CRC(OO\cdot)R_1CH_2OH$, where R, $R_1{=}H$ and CH_3, or CH_3 and H, formed after reaction of OH radicals with isoprene] the reaction with NO predominates (Atkinson et al., 1980).

3.4 Reaction with O₃

For the reactions of O_3 with the alkenes, despite well over a decade of investigation, there are still significant areas of uncertainty (Herron et al., 1979). Thus, while the rate constants for the initial reaction of O_3 with the 1-alkenes seem to be reasonably well known (Herron et al., 1979), larger uncertainties exist for the internal alkenes, for instance, the rate constants at 298 K for *trans*-2-butene from the two studies are 1.76×10^{-15} (Huie et al., 1975) and 2.60×10^{-15} (Japar et al., 1974) cm³/molecule-sec. This topic has been discussed by Herron et al. (1979), to which the reader is referred for detailed discussion.

Work by Niki and coworkers (1977), Calvert and coworkers (1980), and Herron and Huie (1977, 1978, 1982) has shown that the O_3–alkene reactions proceed mainly by the Criegee route:

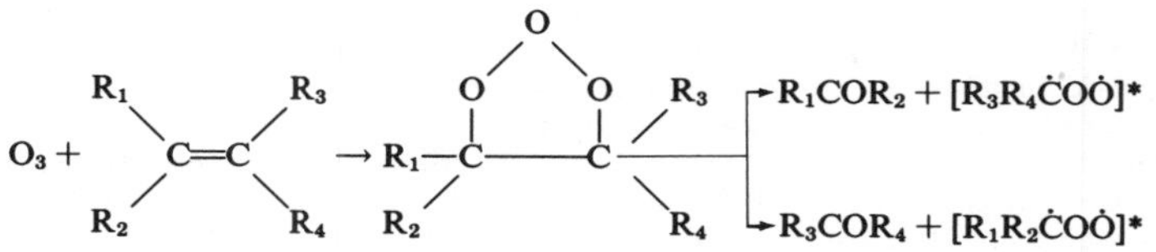

where the symbol [ ]* denotes an initially energy-rich species.

The observation of significant yields of secondary ozonide formation (Niki et al., 1977), including secondary ozonide formation from the ozone–ethene reaction (Su et al., 1980), shows that a significant fraction (~20–40%) of the initially energy-rich biradicals is thermalized under atmospheric conditions, with rearrangement and/or decomposition accounting for the other reaction pathways of the biradicals. Thus, for example, some of the recently proposed reactions (Dodge and Arnts, 1979) of the $CH_3\dot{C}HO\dot{O}$ biradical are shown below:

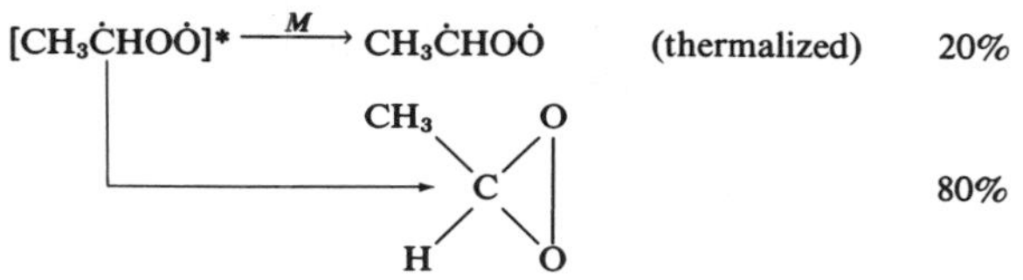

followed by

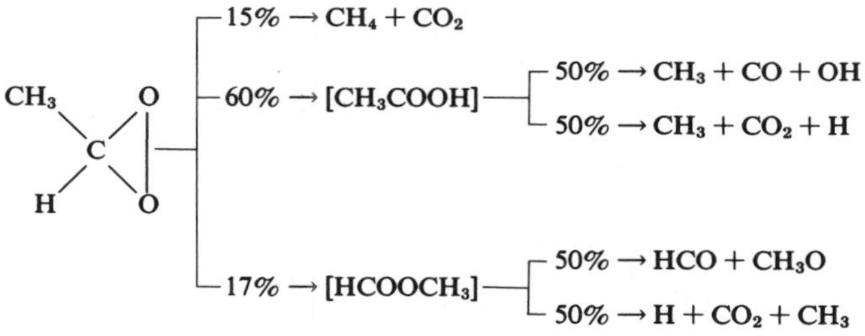

The reactions of the thermalized biradicals ($\dot{C}H_2O\dot{O}$, $CH_3\dot{C}HO\dot{O}$, etc.) need further investigation. They have been shown to react with aldehydes (Niki et al., 1977; Su et al., 1980) to form secondary ozonides; with SO_2 (Niki et al., 1977; Su et al., 1980; Calvert et al., 1978) presumably to form SO_3 and the corresponding aldehyde; and with H_2 (Calvert et al., 1978).

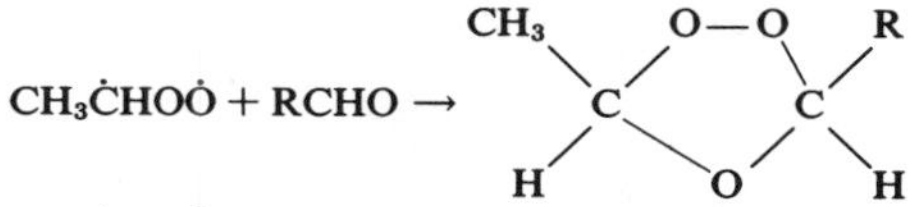

$$CH_3\dot{C}HO\dot{O} + SO_2 \rightarrow CH_3CHO(?) + SO_3$$
$$CH_3\dot{C}HO\dot{O} + H_2O \rightarrow \rightarrow CH_3COOH + H_2O$$

and further possible important reactions with NO and NO_2 need to be studied (Dodge and Arnts, 1979; Calvert et al., 1978; Atkinson and Lloyd, 1983).

$$CH_3\dot{C}HO\dot{O} + \begin{Bmatrix} NO \\ NO_2 \end{Bmatrix} \rightarrow CH_3CHO + \begin{Bmatrix} NO_2 \\ NO_3 \end{Bmatrix}$$

More work on these systems remains to be carried out. We have a reasonably good qualitative, and in some cases semiquantitative, understanding of the O_3–alkene systems in the absence of NO_x, but the finer points regarding the precise splits between different mechanistic pathways, and the relevant rate constant data for these pathways, need to be known (Atkinson and Lloyd, 1983).

3.5 Aromatics

The role of aromatic hydrocarbons in atmospheric chemistry has been the subject of substantial research efforts in the 1970s. Significant advances have been made during this time in our understanding of the atmospheric chemistry of the aromatics. It has been known since the early 1970s that the only important chemical loss process of the aromatic hydrocarbons under atmospheric conditions is via reaction with the OH radical, and data are now available concerning both the overall OH radical rate constants (Atkinson et al., 1979; Ravishankara et al., 1978; Perry et al., 1977), and the relative amounts of the two reaction pathways (Atkinson et al., 1979; Perry et al., 1977; Kenley et al., 1978; Hendry et al. in NBS Special Publication 557, 1977); OH radical addition to the ring and H atom abstraction, mainly from the substituent groups. Based on the mechanism of Atkinson et al. (1980), initial reactions for toluene are:

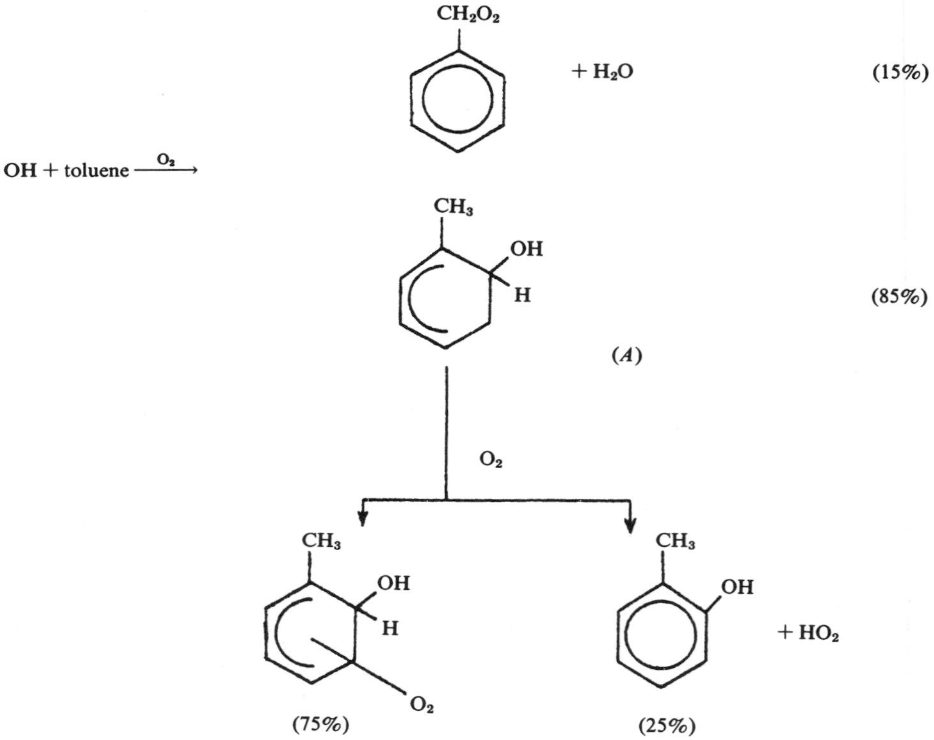

$$OH + toluene \xrightarrow{\;O_2\;}$$

(top pathway) CH_2O_2 benzene ring $+ H_2O$ (15%)

(lower pathway) CH_3 cyclohexadienyl ring with OH and H   (85%)

(A)

$\downarrow O_2$

(left product, 75%) CH_3 ring with OH, H, and O_2

(B)

(right product) CH_3 ring with OH $+ HO_2$ (25%)

The abstraction reaction for benzene is essentially zero (Atkinson et al., 1979), and is small for p- and m-xylene. For toluene as shown, benzaldehyde accounts for about 15% of the products. It is formed by the abstraction route. The major uncertainty lies in the fate of the adducts (A) and (B). Based on experimental data to date, it is believed that the route shown is reasonable for adduct (A) (Atkinson et al., 1980), although more data are required.

    The fate of the adduct (B) is not known with any certainty. However, based on current data from smog chamber studies (e.g., O'Brien et al., 1979; Pitts et al., 1979), one can postulate the following reactions (Atkinson and Lloyd, 1980). From the O_2 adduct (B), bicyclic radicals such as

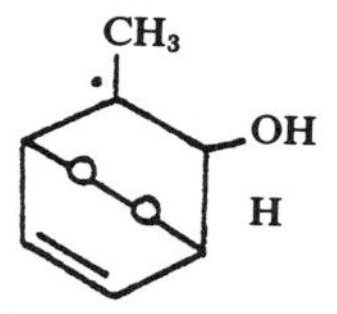

CH$_3$

OH
H

(C)

are formed. These bicyclic radicals can then add O$_2$ and react with NO similarly to alkyl radicals:

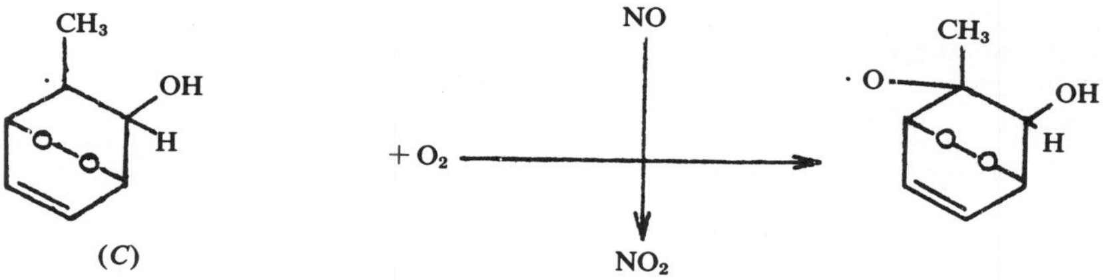

CH$_3$

OH
H

(C)

NO

$+ O_2$ →

NO$_2$

CH$_3$

· O

OH
H

followed by spontaneous decomposition of these bicyclic oxy radicals:

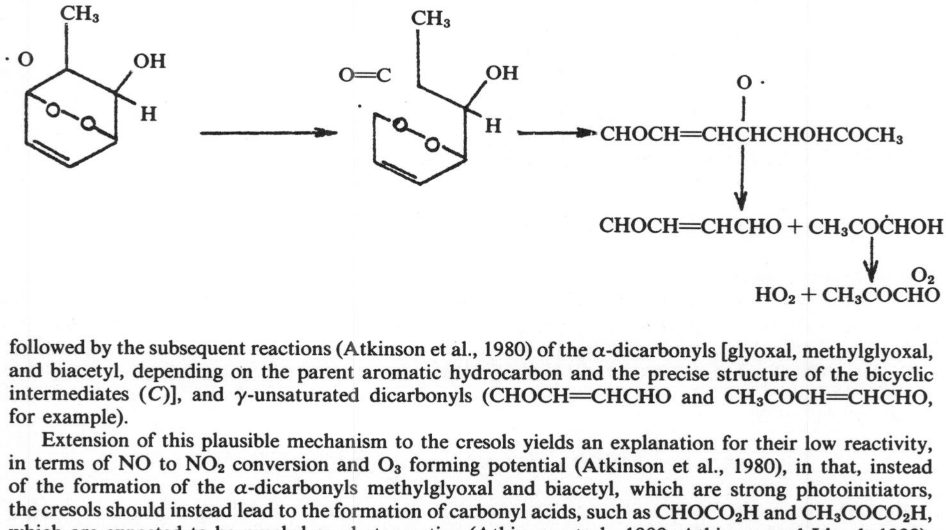

CH$_3$

· O

OH
H

→

CH$_3$

O=C

OH
H

→ CHOCH=CHCHCHOHCOCH$_3$

O ·

↓

CHOCH=CHCHO + CH$_3$COĊHOH

↓ O$_2$

HO$_2$ + CH$_3$COCHO

followed by the subsequent reactions (Atkinson et al., 1980) of the α-dicarbonyls [glyoxal, methylglyoxal, and biacetyl, depending on the parent aromatic hydrocarbon and the precise structure of the bicyclic intermediates (C)], and γ-unsaturated dicarbonyls (CHOCH=CHCHO and CH$_3$COCH=CHCHO, for example).

Extension of this plausible mechanism to the cresols yields an explanation for their low reactivity, in terms of NO to NO$_2$ conversion and O$_3$ forming potential (Atkinson et al., 1980), in that, instead of the formation of the α-dicarbonyls methylglyoxal and biacetyl, which are strong photoinitiators, the cresols should instead lead to the formation of carbonyl acids, such as CHOCO$_2$H and CH$_3$COCO$_2$H, which are expected to be much less photoreactive (Atkinson et al., 1980; Atkinson and Lloyd, 1980).

In this regard, since glyoxal apparently does not photodissociate rapidly, if at all, in the actinic region to form radicals [but rather photodissociates to yield HCHO + CO (Calvert and Pitts, 1966)] then benzene should also be less reactive, as regards NO to NO$_2$ conversion and O$_3$ forming potential, than may be expected solely on the basis of its OH radical rate constant. At this stage, more experimental data, under a variety of conditions, are required before any mechanism can be regarded as unambiguous.

The rate constants for reactions of the OH radical with various aromatic compounds are shown in Table 5. These are taken from the review of Atkinson et al. (1979). The activation energies and A factors are given for both the abstraction route and the addition route.

3.6 Oxygenated Hydrocarbons

The major oxygenated hydrocarbons of interest in atmospheric chemistry are the aldehydes and ketones. Attention here is focused on the aldehydes, although ketones such as methyl ethyl ketone (MEK) are important as radical precursors that undergo photodissociation under ambient conditions.

Aldehydes are major products in the oxidation of hydrocarbons. They can provide significant sources of free radicals: for example, HO$_2$ and RO$_2$ radicals are formed from the reaction of formaldehyde

Table 5   Arrhenius Parameters for the Reaction of OH Radicals with Aromatics[a] (from Atkinson et al., 1979)

Aromatic	$12E \log_{10} A_{Abs.}$, cm³/molecule-sec	$E_{Abs.}$, kcal/mole	$13E \log_{10} A_{Add.}$, cm³/molecule-sec	$E_{Add.}$, kcal/mole
Benzene	1.6 ± 1.6	4 ± 3	1.7 ± 0.7	$0.9 + 1.0$
Toluene	$0.7 + 0.5$	$0.9 + 1.0$	$0.5 + 0.7$	$-1.6 + 1.0$
o-Xylene	$0.7 + 0.8$	$0.3 + 1.5$	$1.6 + 1.4$	$-0.7 + 2.0$
m-Xylene	$1.7 + 0.8$	$2.3 + 1.5$	$2.3 + 1.4$	$-0.1 + 2.0$
p-Xylene	$1.8 + 0.8$	$2.4 + 1.5$	$1.8 + 0.7$	$-0.6 + 1.0$
1,2,3-Trimethylbenzene	$2.5 + 1.3$	$3.3 + 2.5$	$0.8 + 1.4$	$-2.3 + 2.0$
1,2,4-Trimethylbenzene	2.2 ± 0.8	2.8 ± 1.5	$1.3 + 1.4$	$-1.7 + 2.0$
1,3,5-Trimethylbenzene	$2.1 + 0.8$	$2.7 + 1.5$	$1.6 + 1.4$	$-1.7 + 2.0$
Methoxybenzene	$0.2 + 1.0$	-0.5 ± 2.0	$1.6 + 1.4$	$-0.8 + 2.0$
o-Cresol	1.7 ± 1.0	$1.8 + 2.0$	$1.2 + 1.4$	$-1.8 + 2.0$

[a] The indicated errors are the estimated overall error limits.

HCHO and the higher molecular weight aldehydes, RCHO. The chemistry of formaldehyde will be discussed along with that of acetaldehyde which will come as a representative of the higher molecular weight aldehydes.

Formaldehyde undergoes reaction in the atmosphere by photolysis and by reaction with the OH radical:

$$HCHO + h\nu \rightarrow H + HCO$$
$$HCHO + h\nu \rightarrow H_2 + CO$$
$$HCHO + OH \rightarrow HO_2 + CO + H_2O$$

The radical products from the first reaction react further to give HO_2 radicals:

$$H + O_2 + M \rightarrow HO_2 + M$$
$$HCO + O_2 \rightarrow HO_2 + CO$$

For years the photochemistry of formaldehyde under atmospheric conditions was uncertain (Lloyd, 1979), but the work of Moortgat et al. (1978) and Horowitz and Calvert (1978) have provided reliable data for the yields of photolysis. The absorption cross sections for HCHO have been measured by Bass et al. (1980) who obtained values significantly lower than previously used values. More detailed discussion is provided in Atkinson and Lloyd (1983) and in extensive review by Calvert et al. (1980).

Acetaldehyde, CH_3CHO, also reacts with the OH radical losing the aldehydic H alone to produce the acetyl radical CH_3CO

$$CH_3CO + O_3 + NO_2 \rightarrow CH_3CO_3NO_2$$

Within the last decade, PAN has been shown to decompose thermally according to the rate expression (Hendry and Kenley, 1977).

$$k = 1.95 \times 10^{16} e^{-13543/T}/sec$$
$$= 3.7 \times 10^{-4}/sec \text{ at } 298 \text{ K}$$

The photolysis of acetaldehyde has been the subject of much uncertainty until the recent work of Calvert and coworkers (Horowitz and Calvert, 1982; Horowitz, Kershner, and Calvert, 1982). These workers identified three primary processes:

$$CH_3CHO^* \rightarrow CH_3 + HCO$$
$$CH_3CHO^* \rightarrow CH_4 + CO$$
$$CH_3CHO^* \rightarrow H + CH_3CO$$

The first process, from a vibrationally rich, first-excited triplet state of acetaldehyde, was found to be much more important than either of the other two processes at 3000 Å. This study showed that radical generation from CH_3CHO is less than previously thought based on earlier work (Weaver et al., 1976/1977). Horowitz and Calvert (1982) state that CH_3CHO is far less efficient in generating free radicals in the lower troposphere than HCHO. The first reaction for CH_3CHO photolysis was estimated to be the only one of significance.

In addition to aldehydes, there are several other classes of oxygenated hydrocarbons known or suspected to be present in urban areas of the troposphere. These include ketones, alcohols, esters, and ethers. Their possible role in atmospheric chemistry will be discussed briefly.

The OH radical is the major species attacking aldehydes and is likely to be the most important intermediate for the remaining oxygenates. Of the four classes of oxygenated hydrocarbons mentioned above, only ketones undergo photolysis under ambient conditions.

The results of studies of the reactions of OH radicals with individual oxygenated hydrocarbons are summarized in Table 6 which is a modified version of that given by Atkinson et al. (1979). Several studies have been updated by Atkinson and Lloyd (1983). Most of these determinations are for one temperature (around 300 K) and were obtained using a relative rate technique. Absolute values were obtained as indicated in Table 6. The most reactive species are the ketones followed by the ethers with the acetates appearing least reactive.

Ketones play a similar role to aldehydes in that they photolyse to produce radicals that promote the oxidation of NO to NO_2 with the concomitant formation of photochemical smog.

Ketone photolysis has been described by Calvert and Pitts (1966) and the relevant studies on the photolysis and photooxidation of selected ketones have been summarized recently by Atkinson and Lloyd (1983). Due to uncertainty in the behavior of ketones under atmospheric conditions, radical production is often assumed to be 100% efficient. For example, for MEK, the reactions are

$$CH_3COC_2H_5 + h\nu \rightarrow CH_3CO + C_2H_5$$
$$\rightarrow CH_3 + COC_2H_5$$

Carter et al. (1979) suggest that triplet formation is the dominant initial process under ambient conditions and this species reacts with atmospheric O_2 to give ethyl radicals and acetyl peroxy radicals. Atkinson et al. (1982) include acetone photolysis in a chemical model suitable for application to the lower troposphere but acetone reaction with the OH radical was neglected because of the slow reaction rate (Cox et al., 1980).

However, major uncertainty remains concerning the photolysis of ketones under ambient conditions including the quantum efficiency of radical production in the presence of O_2.

3.7 Biogenic Hydrocarbons

The atmospheric chemistry in urban areas, particularly O_3 formation, is dominated by anthropogenic hydrocarbon emissions. Biogenic hydrocarbon emissions can represent a significant contribution to atmospheric hydrocarbon loadings in rural areas but our understanding of their chemistry remains incomplete. Our current understanding of the role of biogenic hydrocarbons in atmospheric chemistry is briefly discussed below.

The role of biogenic hydrocarbons [mainly isoprene and the monoterpenes (Graedel, 1979)] in influencing atmospheric ozone concentrations has been of interest for many years (Dimitriades and Altshuller, 1977; Coffey, 1977; Westberg, 1977; Graedel, 1979; Dimitriades, 1981). Recent field and laboratory studies (Rasmussen et al., 1976; Lonneman et al., 1978; Holdren et al., 1979; Zimmerman, 1979; Arnts and Gay, 1979; Arnts and Meeks, 1980; Kamens et al., 1981) have contributed significantly to the understanding of the emission rates, ambient concentrations, and the atmospheric chemistry of these naturally emitted hydrocarbons. Nevertheless, considerable controversy and uncertainty still exist as to the role of these compounds on ozone formation/destruction in rural and urban regions and the uncertainties have been discussed elsewhere (Dimitriades, 1981).

Since the first publication of natural hydrocarbon (HC) emission estimates (Went, 1960), researchers have debated whether natural HCs act as sources or sinks for ozone and as to the magnitude of such sources or sinks (Dimitriades and Altshuller, 1977; Coffey, 1977; Sandberg et al., 1978; Bufalini, 1980; Dimitriades, 1981). Their role has been difficult to assess because measurements of ambient concentrations and emissions (Graedel, 1979; Dimitriades, 1981).

Many of the natural hydrocarbons (HCs) have been shown to react rapidly with ozone (Grimsrud et al., 1975; Japar et al., 1974; Atkinson et al., 1982a) and with OH radicals (Atkinson et al., 1979, 1982b). Smog chamber experiments have shown that these natural hydrocarbons are efficient at forming ozone when photooxidized in the presence of NO_x (Arnts and Gay, 1979; Kamens et al., 1981).

Two naturally emitted hydrocarbons (isoprene and α-pinene) have received most study to date. Isoprene, a hemiterpene, is emitted in significant amounts from deciduous trees such as oaks and sycamore and the evergreen eucalyptus (Rasmussen, 1972; Sanadze and Dolidze, 1961; Sanadze and Kalandadze, 1966), while α-pinene is more commonly emitted from evergreen trees such as pines and firs (Rasmussen, 1972; Tingey et al., 1978; Westberg, 1980). Field measurements of emissions from trees have shown these two compounds frequently comprise large fractions of the measured HC emissions (Zimmerman, 1979). Laboratory studies have shown these two compounds are perhaps the most efficient ozone producers among the known biogenic HCs (Arnts and Gay, 1979; Kamens et al., 1981).

Table 6 Rate Constant Data and Arrhenius Parameters for the Reaction of OH Radicals with Selected Oxygen-Containing Organics, (from Atkinson et al., 1979; Atkinson and Lloyd, 1983)

Reactant	$10^{12} \times A$, cm³/molecule-sec[a]	E, cal/mole	$10^{12} \times k$, cm³/molecule-sec	At T K	Reference
Aldehydes					
HCHO	10.1	0	$10.1 \pm 0.8 + 25\%$	298	Atkinson and Lloyd (1983)
CH_3CHO	6.87	-510 ± 300	16.0 ± 1.6	298	Atkinson and Pitts (1978)
C_2H_5CHO			20.8 ± 0.8	298 ± 2	Niki et al. (1978) (relative to OH + C_2H_4 = 8.00×10^{-12})
C_6H_5CHO			12.8 ± 0.8	298 ± 2	Niki et al. (1978) (relative to OH + C_2H_4 = 8.00×10^{-12})
Ethers					
CH_3OCH_3	12.9	770 ± 300	3.50 ± 0.35	299	Perry, Atkinson, and Pitts (1977)
Diethyl ether			8.9 ± 1.8	305 ± 2	Lloyd et al. (1976) (relative to OH + isobutene = 4.80×10^{-11})
Di-n-propyl ether			16.3 ± 3.3	305 ± 2	Lloyd et al. (1976) (relative to OH + isobutene = 4.80×10^{-11})
Tetrahydrofuran			16 ± 2	298	Ravishankara et al. (1978)
$CH_2{=}CHOCH_3$	6.10	-1015 ± 300	33.5 ± 3.4	299	Perry, Atkinson, and Pitts (1977)

Alcohols					
CH$_3$OH			1.06 ± 0.10	296 ± 2	Overend and Paraskevopoulos (1978)
			1.0 ± 0.1	298	Ravishankara et al. (1978)
C$_2$H$_5$OH			3.74 ± 0.37	296 ± 2	Overend and Paraskevopoulos (1978)
n-Propanol			5.33 ± 0.53	296 ± 2	Overend and Paraskevopoulos (1978)
Isopropanol			5.48 ± 0.55	296 ± 2	Overend and Paraskevopoulos (1978)
Acetates					
Methyl acetate			0.18 ± 0.05	292	Campbell et al. (1976) (relative to OH + n-butane = 2.60 × 10^{-12})
Ethyl acetate			1.94 ± 0.22	292	Campbell et al. (1976) (relative to OH + n-butene = 2.60 × 10^{-12})
n-Propyl acetate			4.1 ± 0.8	305 ± 2	Winer et al. (1977) (relative to OH + isobutene = 4.80 × 10^{-11})
Methyl propionate			0.29 ± 0.10	292	Campbell and Parkinson (1978) (relative to OH + n-butane = 2.60 × 10^{-12})
Ethyl propionate			1.77 ± 0.25	292	Campbell and Parkinson (1978) (relative to OH + n-butane = 2.60 × 10^{-12})
Ketones					
Methyl ethyl ketone	3.85		0.88	298	Cox et al. (1982)
Methyl vinyl ketone		−906 ± 145	17.9 ± 2.8	298	Kleindienst et al. (1982)
Methacrolein	17.7	−347 ± 103	31.4 ± 4.9	300	Kleindienst et al. (1982)
Methylglyoxal			7.1 ± 1.6 } +0.08	297	Kleindienst et al. (1982)
Biacetyl			0.24 } −0.06	298 ± 2	Darnall, Atkinson, and Pitts (1978)

[a] Mean Arrhenius preexponential factor.

Lloyd et al. (1982) and Lurmann et al. (1982) have carried out a modeling study to develop a chemical mechanism for isoprene and α-pinene and to assess their potential importance in generating O_3 under various atmospheric conditions. Two of the major products of isoprene $[CH_2\!=\!CH\cdot C(CH_2)]$ photooxidation are methylvinylketone $(CH_2\!=\!CHCOCH_3)$ and methacrolein $[CH_2\!=\!C(CH_3)CHO]$ (Arnts and Gay, 1979; Cox et al., 1980; Kamens et al., 1981). The reactions illustrate that isoprene is effective in oxidizing NO_x to products through the formation of organic and inorganic radical species. Isoprene, and the two major initial products of methylvinylketone and methacrolein, react with both the OH radical and O_3. A PAN-type compound is predicted to be formed with a lifetime assumed to be similar to PAN. Methylglyoxal, CH_3COCHO, predicted to be formed in several reactions, photodissociates to produce HO_2 radicals and acetylperoxy radicals which produce PAN in the presence of NO_x. A mechanism has also been developed for α-pinene (Lloyd et al., 1982). However, greater uncertainties exist for its atmospheric photooxidation: these are discussed in Lloyd et al. (1982) and will not be repeated here.

In summary, certain biogenic hydrocarbons are quite reactive, and while their emissions are not large, they are significant and should probably be included in regional modeling studies.

4   AEROSOL FORMING REACTIONS

This section discusses briefly the reactions forming sulfates, nitrates, and organic aerosols and focuses on heterogeneous and aqueous phase processes for sulfates and nitrates. The gas phase reactions of SO_2 and NO_x have been discussed in the section on inorganic reactions. These reactions ultimately produce sulfates and nitrates from sulfuric and nitric acid, respectively.

4.1   Sulfate Formation

Two types of reactions are considered—oxidation of SO_2 on particles and in the aqueous phase. Several workers have studied the oxidation of SO_2 on different surfaces. In general, two types of reaction mechanisms occur: a capacity-limited reaction and a catalytic reaction. In both cases, the initial reaction on the surface produces a rapid loss of SO_2 from the gas phase followed by:

1. The reaction rate decreasing to zero for the capacity limited case.
2. The reaction rate levels off and then approaches zero for the catalytic route. This behavior is attributed to a pH decrease caused by sulfuric acid formation.

Urone et al. (1968) examined the effectiveness of a number of solid particles in removing SO_2, including ferric oxide, lead oxide, lead dioxide, calcium oxide, and aluminum oxide. Ferric oxide was found to be the most reactive species in removing SO_2 in the system studies by these authors.

Novakov et al. (1974) studied the oxidation of SO_2 over finely divided carbon (soot) particles and showed that increased oxidation of SO_2 occurred. This catalytic formation of sulfate on soot particles was identified before similar studies were carried out in the presence of water vapor. The results showed that graphite and soot particles oxidize SO_2 in air and that the process exhibits a significant saturation effect; that is, the efficiency of the catalytic conversion is considerably reduced with time. This phenomenon is attributed to the saturation of the particles by SO_2 vapor.

Judeikis and coworkers (1973, 1978) have also identified such saturation effects in their work. They studied the rate of removal of gaseous SO_2 over different solids whose composition was typical of those encountered in ambient aerosols. The relative humidity in these systems was found to be important in determining the total amount, but not the initial rate, of SO_2 uptake. The SO_2 uptake increases at higher humidity. The initial calculated SO_2 removal rates decreased in the order of MgO, Fe_2O_3, Al_2O_3, MnO_2, PbO, and NaCl.

Baldwin (1982) studied the rate of adsorption of SO_2 on a carbonaceous surface at less pressure in a flow reactor. Rapid saturation of the surface by SO_2 was found, suggesting that this mechanism is of limited importance. A much slower surface reaction, enhanced by the presence of NO_2 was found. Based on the experimental data, Baldwin calculates a maximum atmospheric loss of SO_2 by SO_2 adsorption of 1%/hr for a particle density of 100 $\mu g/m^3$. However, this loading is recognized as being very high and hence, the contribution of the reaction to the removal of SO_2 under atmospheric conditions is probably much lower than the calculation indicates.

Other studies including oxidation of SO_2 on particles have been carried out by Urone et al. (1968), Smith et al. (1969), Liberti et al. (1978), and Tartarelli et al. (1978).

In summary, current information on the heterogeneous surface catalyzed oxidation indicates that initial oxidation rates may be large but very quickly approach zero. For practical atmospheric applications on a regional scale, the only possible solid surface of importance appears to be carbon or soot.

In the aqueous phase, the rate of oxidation of SO_2 is a function of the chemical composition of the aqueous phase. The chemical composition, in turn, is dependent on the concentrations and types of chemical compounds found in the gas phase and in suspended particulate matter and their chemical and physical abilities to enter the aqueous phase. The equilibrium distribution of soluble gases between the gas and aqueous phases is defined by the Henry's law relationship. Focus in this chapter is on the kinetic processes.

While scavenging of preexisting sulfate aerosol was observed to account for nearly all cloud and precipitation water sulfate in one case study (Scott and Laulainen, 1979), it is usually necessary to invoke a pathway whereby sulfur dioxide gas is absorbed by cloud–water droplets and subsequently oxidized to sulfate in the aqueous phase in order to explain the observed sulfate concentrations in precipitation (e.g., Hales and Dana, 1979; Hegg and Hobbs, 1981).

The kinetics of aqueous phase oxidation of S(IV) pertinent to the atmosphere have been the subject of a number of reviews (Hegg and Hobbs, 1978; Beilke and Gravenhorst, 1978; Möller, 1980; Seinfeld, 1980; Middleton et al., 1980; Chang et al., 1981; Martin, 1982; and Schwartz, 1982). Of the various species which may be considered, the three potentially most important pathways are oxidation by O_3, H_2O_2, and NO_x. The effects of catalytic agents, for example, trace metals and elemental carbon, on these reactions and on the oxidation of S(IV) by O_2 are discussed subsequently.

Rates of formation of S(IV) in the atmosphere due to aqueous phase chemistry are given in the general form

$$R_i = \frac{dS(VI)}{dt} = -\frac{dS(IV)}{dt} = k^{(2)} K_{H_i} p_i [S(IV)(aq)]_T$$

where

R_i = the aqueous phase rate of formation of oxidized sulfur by species i in M/sec
$k^{(2)}$ = the second-order rate constant for reaction of species i with S(IV) in the aqueous phase, M/sec
K_H = the Henry's law constant in units of M/atm
p = partial pressure in units of atmospheres

The kinetics of the reaction of S(IV) with O_3 have been determined by several investigators (Erickson et al., 1977; Larson et al., 1978; Penkett et al., 1979; Maahs, 1982; and Martin, 1982). The second-order rate expression is

$$R_{O_3} = -\frac{d[S(IV)]}{dt} = k^{(2)}[O_3(aq)] [S(IV)(aq)]_T$$

The dependence of the second-order rate constant $k^{(2)}$ on solution pH is shown in Figure 3. There is considerable agreement among these data with exception of the data of Larson et al. (1978)

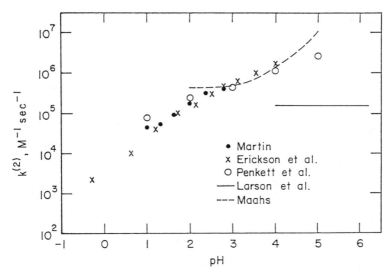

Fig. 3 Second-order rate constant, $k^{(2)}$, for oxidation of S(IV) by O_3 according to $-d[S(IV)]/dt = k^{(2)}[O_3(sq)][S(IV)aq]_T$ as a function of solution pH. Data of Erickson (1977), Larson et al. (1978), Penkett et al. (1979), Maahs (1982), and Martin (1982). Temperature = 25°C.

and that of Maahs (1982) and Penkett et al. (1979) at pH = 5. Young (1982) recommends that for low pH, 0 to 3, the rate constant of Martin (1982), $1.9 \times 10^4\ [H^+]^{-0.5}$, shall be used while at pH values of 3 to 7, the rate constant of Maahs (1982), $4.19 \times 10^5\ (1 + 2.39 \times 10^{-4}\ [H^+]^{-1})$, is appropriate.

Available data for the influence of metal ions as catalysts for this reaction are not definitive. Martin (1982) found no sensitivity of the ozone reaction to metal ion catalysis by Fe^{3+}, Mn^{2+}, VO^{2+}, or Cu^{2+} although Harrison et al. (1982) report an increase due to catalysis by manganese and iron.

The kinetics of the reaction of S(IV) *with H_2O_2*, have been determined by several investigators (Mader, 1958; Hoffman and Edwards, 1975; Penkett et al., 1979; and Martin and Damschen, 1981). A second-order rate expression can be written

$$R_{H_2O_2} = -\frac{dS(IV)}{dt} = k^{(2)}[H_2O_2(aq)]\ [SO_2(aq)]$$

The dependence of this second-order rate constant on solution pH is shown in Figure 4, corrected for temperature and the effect of buffer catalysis. These data agree well with the solid curve generated using the rate expression of Martin and Damschen (1981) with a value of $k^{(2)}$ of $(8 \pm 2 \times 10^4)\ (0.1 + [H^+])^{-1}$, except in the pH range 5 to 6 where the curve underpredicts the experimental data according to Mader (1958). The dashed portion of the curve includes a sulfite ion term.

Oxidation of S(IV) by nitrogen compounds has been studied less than the reactions of S(IV) with H_2O_2 and O_3. The species investigated include NO, NO_2, HNO_2, and HNO_3 in studies by Takeuchi et al. (1977), Nash (1979), Sato et al. (1979), Chang et al. (1981), Martin et al. (1981), Lee and Schwartz (1981), and Schwartz and White (1982). Martin et al. (1981) observed that NO(aq) and NO_3^- are essentially unreactive with S(IV) at representative atmospheric concentrations. They set an upper limit on the second-order rate constant of 0.01/M-sec for both reactions. Nitric acid was also found to be unreactive with S(IV) in the presence of Fe^{3+}, Mn^{2+}, Cu^{2+}, Co^{2+}, Pb^{2+}, or VO^{2+}.

Reaction of S(IV) with nitrous acid (HNO_2), studied by Oblath et al. (1981), Martin et al. (1981), and Chang et al. (1981) is considerably faster than reaction with NO or HNO_3. With reference to the rate expression

$$R_{HNO_2} = -\frac{dS(IV)}{dt} = k^{(2)}\ [N(III)(aq)]\ [S(IV)(aq)]_T$$

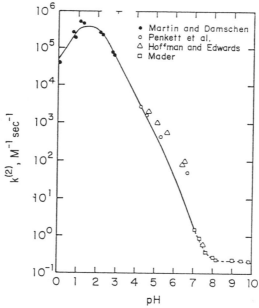

Fig. 4 Second-order rate constant, $k^{(2)}$, for oxidation of S(IV) by H_2O_2 according to $-d[S(IV)]/dt = k[H_2O_2(aq)]\ [S(IV)aq]_T$ as a function of solution pH. Data of Mader (1958), Hoffman and Edwards (1975), Penkett et al. (1979), and Martin and Damschen (1981). Temperature = 25°C. The effect of buffer catalysis has been removed.

Martin et al. determined a value for $k^{(2)}$ of $142[H^+]^{0.5}/M^{1.5}$-sec over the pH range 0 to 3. The reaction did not respond to catalysis by Fe^{3+}, Mn^{2+}, or VO^{2+}. Oblath et al., conducting experiments in the pH range 4.5 to 7.0, report $k^{(2)}$ to be $3.8 \times 10^3[H^+]$. These two values of $k^{(2)}$ are consistent and indicate a change in pH dependence near pH 3.

For atmospheric concentrations of 5 ppb SO_2 and 3 ppb HNO_2, the rate of S(IV) oxidation by HNO_2 is much slower than when ozone or peroxide are present (Martin et al., 1981), although it may be important at night if atmospheric droplet lifetimes are long.

Takeuchi et al. (1977), Nash (1979), Sato et al. (1979), Lee and Schwartz (1981), and Schwartz and White (1982) have studied reactions of S(IV) oxidation by NO_2. This reaction appears to be the most rapid of those with N species. Estimates of the second-order rate constant have been made by Schwartz and White based on the data of Takeuchi et al. They suggest values for $k^{(2)}$ of 3×10^5 and $1 \times 10^7/M$-sec for reaction of $NO_2(aq)$ with HSO_3^- and SO_3^{2-}, respectively.

Oxidation of S(IV) by molecular oxygen in the absence of catalysts is unimportant when compared to oxidation by O_3 or H_2O_2 at representative atmospheric concentrations (Penkett et al., 1979). Catalysis of this reaction by trace metals, for example, Fe^{3+}, Mn^{2+}, and Fe^{3+} plus Mn^{2+}, can substantially increase the rate of this reaction such that it becomes competitive with the stronger oxidants O_3 and H_2O_2. Martin (1982) provides a comprehensive review of the relevant work in this area and gives the relevant rate expressions.

The catalytic oxidation of S(IV) by O_2 on carbon particles in aqueous suspension has been investigated by Brodzinsky et al. (1980) and Chang et al. (1981). They found the reaction rate to be pH independent below pH 7.6, first-order in carbon and 0.69th-order in dissolved oxygen and a complex function of $[S(IV)(aq)]_T$.

Chang et al. (1981) evaluated oxidation rates of S(IV) by several mechanisms under representative atmospheric conditions and found catalytic oxidation by carbon particles to be an important process close to sources and in heavily polluted areas where soot concentrations are high and when the lifetime of fog or cloud droplets is long.

Schwartz (1982) has evaluated the oxidation rates of S(IV) by O_3 and H_2O_2 under representative atmospheric conditions of 30 ppb and 1 ppb, respectively. Results of the calculations are shown in Figure 5. The aqueous-phase oxidation rate, R, per ppb SO_2, is shown on the left-hand ordinate and

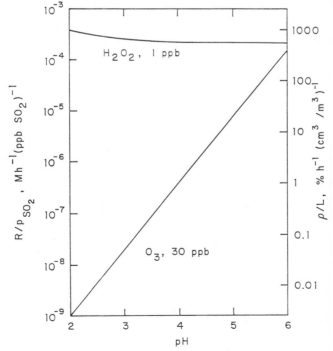

Fig. 5 Rate of aqueous-phase oxidation of S(IV) by O_3 (30 ppb) as a function of solution pH. Gas-aqueous equilibria are assumed for all reagents. R/P_{SO_2} represents aqueous reaction rate per ppb of gas phase SO_2. ρ/L represents rate of reaction referred to gas phase SO_2 partial pressure, per cm^3/m^3 liquid water volume fraction. (From S. Schwartz 1982, reproduced with permission.)

the gas-phase rate, R/L, per cm³/m³ liquid water volume fraction, is shown on the right-hand ordinate. These rates scale linearly with p_{O3}. A strong pH dependence is seen reflecting S(IV) solubility as well as the pH dependence of $k^{(2)}$. The transformation of SO₂ to sulfate is dominated by H₂O₂ oxidation over the pH range 1 to 5. The lack of pH-dependent behavior in the H₂O₂ system results from the balance between a decreasing $k^{(2)}$ as a function of pH and an increasing $[S(IV)(aq)]_T$ as a function of pH.

In addition to the H₂O₂ assumed to be formed by photochemical reactions in the gas phase and then transferred to the aqueous phase, Chameides and Davis (1982) have suggested the direct reaction of gas phase free radicals with aqueous aerosol. These workers carried out calculations to demonstrate that OH and HO₂ radicals in the gas phase can enter the aqueous phase and play an important role in oxidizing S(IV) to S(VI). The hypothesis is based on field measurements which suggest that aqueous phase chemical reactions within cloudwater can produce a significant fraction of the dissolved H₂O₂ (Zika et al., 1982; Zika and Saltzman, 1982; Heikes et al., 1982).

A recent study (Richards et al., 1983) has obtained results showing significant concentrations of H₂O₂ and S(IV) coexisting in cloud water in the Los Angeles, California air basin. The authors postulate that HCHO acts to retard the oxidation by the information of an adduct with sulfite to form hydroxy-methanesulfonic acid.

5 NITRATE FORMATION

Nitric acid is formed from NO₂ by the reactions discussed earlier

$$OH + NO_2 + M \rightarrow HNO_3 + M$$
$$NO_2 + O_3 \rightarrow NO_3 + O_2$$
$$ARH + NO_3 \rightarrow ARO_2 + HNO_3 \text{ (minor)}$$
$$RO_2 + NO_2 \rightarrow HNO_3 + RCO_3$$
$$NO_2 + NO_3 + M \rightarrow N_2O_5 + M$$
$$N_2O_5 + H_2O \rightarrow 2HNO_3$$

The last reaction may proceed by a heterogeneous mechanism and except for this reaction, heterogeneous routes for HNO₃ formation have received little study.

The HNO₃ produced in the above reactions is converted to particulate nitrate, NO₃⁻, if the nitric acid interacts with other atmospheric species to form nitrate, for example, reaction with ammonia, NH₃, will produce ammonium nitrate, NH₄NO₃.

The aqueous phase oxidation of NO and NO₂ is believed to be quite slow based on the work of Schwartz and coworkers (Schwartz and White, 1981, 1982; Lee and Schwartz, 1981, 1982). These workers report second-order rate constants for the following reactions.

$$2NO_2(aq) + H_2O \rightarrow 2H^+ + NO_3^- + NO_2^-$$
$$NO(aq) + NO_2(aq) \rightarrow 2H^+ + 2NO_2^-$$

of 7×10^7 and $3 \times 10^7/M$-sec, respectively, for atmospheric conditions.

Russell et al. (1982) have carried out a computer modeling study of the formation and transport of ammonium nitrate in the Los Angeles basin. Ammonium nitrate concentrations were found to be strongly influenced by ambient temperature and ammonia levels and computed concentrations agreed well with ambient observations.

6 SUMMARY AND CONCLUSIONS

The above brief description of atmospheric chemistry is not intended as a comprehensive review. Rather, it aims at addressing some of the more important topics and attempts to show the relationship between gas phase and aqueous phase chemistry. This facet of atmospheric chemistry has been brought into sharper focus as a result of applied problems in acid deposition, hazardous waste, and toxic chemicals in the environment.

Substantial advances in our knowledge of atmospheric chemistry have been made over the last decade. The continued emergence of new problems related to the environment puts additional stress on our knowledge in previously unexplored technical areas. Some examples where uncertainty remains in our knowledge of chemical processes applicable to the environment are the aqueous phase chemistry of sulfur, nitrogen and organic compounds, aromatic and polycyclic aromatic hydrocarbon photooxidation, and the atmospheric chemistry of biogenic hydrocarbon emissions. Frequent monitoring of the burgeoning literature relevant to atmospheric chemistry will be necessary for the scientist to keep abreast of progress in these and other areas.

REFERENCES

Arnts, R. R., and Gay, B. W., Jr. (1979), *Photochemistry of Some Naturally Emitted Hydrocarbons,* EPA-600/3-79-081, September.

Arnts, R. R., and Meeks, S. A. (1980), *Biogenic Hydrocarbon Contribution to the Ambient Air of Selected Areas,* EPA-600/3-80-023, January.

Atkinson, R., Darnall, K. R., Lloyd, A. C., Winer, A. M., and Pitts, J. N., Jr. (1979), *Adv. Photochem.* **11,** 375.

Atkinson, R., and Lloyd, A. C. (1980), *Evaluation of Kinetic and Mechanistic Data for Photochemical Smog Chamber Modeling,* EPA Contract No. 68-02-3280, ERT Document No. P-A040.

Atkinson, R., and Lloyd, A. C. (1981), Smog Chemistry and Urban Airshed Modeling, pp. 559–592, in *Oxygen and Oxy-Radicals in Chemistry and Biology,* Academic Press, New York.

Atkinson, R., Lloyd, A. C., and Dodge, M. C. (1983), Evaluation of Kinetic and Mechanistic Data for Photochemical Smog Chamber Modeling, *J. Phys. Chem. Ref. Data.*

Atkinson, R., Perry, R. A., and Pitts, J. N., Jr. (1976), *J. Chem. Phys.* **65,** 306.

Atkinson, R., Perry, R. A., and Pitts, J. N., Jr. (1977), *J. Chem. Phys.* **67,** 3170.

Atkinson, R., and Pitts, J. N., Jr. (1978), *J. Chem. Phys.* **68,** 3581.

Baldwin, A. C. (1982), Heterogeneous Reactions of Sulfur Dioxide with Carbonaceous Particles, *Int. J. of Chem. Kinet.* **14,** 269.

Baldwin, A. C., Barker, J. R., Golden, D. M., and Hendry, D. (1977), *J. Phys. Chem.* **81,** 2483.

Batt, L. (1979a), *Int. J. Chem. Kinet.* **11,** 977.

Batt, L. (1979b), Reaction of Alkoxy Radicals Relevant to Atmospheric Chemistry, *European Symposium on Physico-Chemical Behavior of Atmospheric Pollutants, Ispra.* Oct. 16–18.

Beilke, S., and Gravenhorst, G. (1978), Heterogeneous SO_2 Oxidation in the Droplet Phase, *Atmos. Environ.* **12,** 231–239.

Biermann, H. W., Harris, G. W., and Pitts., J. N., Jr. (1980), 14th Informal Conf. Photochem., Newport, CA, Mar. 30–Apr. 3.

Brodzinsky, R., Chang, S. G., Markowitz, S. S., and Novakov, T. (1980), Kinetics and Mechanisms for the Catalytic Oxidation of Sulfur Dioxide on Carbon in Aqueous Suspensions, *J. Phys. Chem.* **84,** 3354–3358.

Bufalini, J. J. (1980), *Impact of Natural Hydrocarbon on Air Quality,* EPA-600/2-80-86, U.S. Environmental Protection Agency, Office of Research and Development, Research Triangle Park, NC.

Calvert, J. G., and Pitts, J. N., Jr. (1966), *Photochemistry,* Wiley, New York.

Calvert, J. G., Su, F., Bottenheim, J. W., and Strausz, O. P. (1978), Mechanism of the Homogeneous Oxidation of Sulfur Dioxide in the Troposphere, *Atmos. Environ.* **12,** 197–226.

Campbell, I. M., McLaughlin, D. F., and Handy, B. J. (1976), *Chem. Phys. Lett.* **38,** 362.

Campbell, I. M., and Parkinson, P. E. (1978), *Chem. Phys. Lett.* **53,** 385.

Carter, W. P. L. (1979), Private communication.

Carter, W. P. L., Darnall, K. R., Graham, R. A., Winer, A. M., and Pitts, J. N., Jr. (1979), *J. Phys. Chem.* **83,** 2305.

Carter, W. P. L., Darnall, K. R., Lloyd, A. C., Winer, A. M., and Pitts, J. N., Jr. (1976), *Chem. Phys. Lett.* **42,** 22.

Carter, W. P. L., Lloyd, A. C., Sprung, J. L., and Pitts, J. N., Jr. (1979), *Int. J. Chem. Kinet.* **11,** 45.

Chamaides, W. L., and Davis, D. D. (1982), The Free Radical Chemistry of Cloud Droplets and Its Impact upon the Composition of Rain, *J. Geophys. Res.* **87,** 4863–4877.

Chang, S. G., Toosi, R., and Novakov, T. (1981), The Importance of Soot Particles and Nitrous Acid in Oxidizing SO_2 in Atmospheric Aqueous Droplets, *Atmos. Environ.* **15,** 1287–1292.

Cox, R. A., Derwent, R. G., and Williams, M. R. (1980), Atmospheric Photooxidation Reactions: Rates, Reactivity, and Mechanism for Reaction of Organic Compounds with Hydroxyl Radicals, *Environ. Sci. Technol.* **14,** 57–61.

Cox, R. A., Patrick, K. F., and Chart, S. A. (1981), *Environ. Sci. & Technol.* **15,** 587.

Cox, R. A., and Penkett, S. A. (1972), *J. Chem. Soc. Faraday Trans I* **68,** 1735.

Darnall, K. R., Atkinson, R., and Pitts, J. N., Jr. (1978), *Chem. Phys. Lett.*

Darnall, K. R., Carter, W. P. L., Winer, A. M., Lloyd, A. C., and Pitts, J. N., Jr. (1976), *J. Phys. Chem.* **80,** 1948.

Davis, D. D., Ravishankara, A. R., and Fischer, S. (1979), *Geophys. Res. Lett.* **6,** 113.

Dodge, M. C., and Arnts, R. R. (1979), *Int. J. Chem. Kinet.* **11**, 399.

Erickson, R. E., Yates, L. M., Clark, R. L., and McEwen, D. (1977), The Reaction of Sulfur Dioxide with Ozone in Water and Its Possible Atmospheric Significance, *Atmos. Environ.* **11**, 813–817.

Falls, A. H., and Seinfeld, J. H. (1978), *Environ. Sci. Technol.* **12**, 1398.

Finlayson-Pitts, B. J., and Pitts, J. N., Jr. (1977), *Adv. Environ. Sci. Technol.* **7**, 75.

Freiberg, J. E., and Schwartz, S. E. (1980), Oxidation of SO_2 in Aqueous Droplets: Mass-Transport Limitation in Laboratory Studies and the Ambient Atmosphere, *Atmos. Environ.* **15**, 1145–1154.

Golden, D. M. (1977), in *Chemical Kinetic Data Needs for Modeling the Lower Troposphere* (J. T. Herron, R. E. Huie, and J. A. Hodgeson, eds.), NBS Special Publication 557, August.

Hales, J. M., and Dana, M. T. (1979), Precipitation Scavenging of Urban Pollutants by Convective Storm Systems, *J. Appl. Meteor.* **18**, 294–316.

Hampson, R. F., Jr., and Garvin, D. (1978), Reaction Rate and Photochemical Data for Atmospheric Chemistry—1977, National Bureau of Standards Special Publication 513, May.

Harris, G. W., Atkinson, R., and Pitts, J. N., Jr. (1980), *Chem. Phys. Lett.* **69**, 378.

Harrison, H., Larson, T. V., and Monkton, C. S. (1982), Aqueous-Phase Oxidation of Sulfites by Ozone in the Presence of Iron and Manganese, *Atmos. Environ.* **16**, 1039–1041.

Hegg, D. A., and Hobbs, P. V. (1978), Oxidation of SO_2 Aqueous Systems with Particular Reference to the Atmosphere, *Atmos. Environ.* **12**, 241–253.

Heicklen, J., Westberg, K., and Cohen, N. (1969), Report No. 115-69, Center for Air Environment Studies, Pennsylvania State University, University Park, PA.

Heikes, B. G., Lazrus, A. L., Kok, G. L., Kunen, S. M., Gandrud, B. W., Gitlin, S. N., and Sperry, D. D. (1982), Evidence for Aqueous Phase Hydrogen Perioxide Synthesis in the Troposphere, *J. Geophys. Res.* **87**, 3045.

Hendry, D. G. (1977), in *Chemical Kinetic Data Needs for Modeling the Lower Troposphere* (J. T. Herron, R. E. Huie, and J. A. Hodgeson, eds.), NBS Special Publication 557, August.

Hendry, D. G., Baldwin, A. C., Barker, J. R., and Golden, D. (1978), *Computer Modeling of Simulated Photochemical Smog*, EPA-600/3-78-059, January.

Hendry, D. G., and Kenley, R. A. (1979), *Nitrogenous Air Pollutants* (D. Grosjean, ed.), Ann Arbor Press, Ann Arbor, MI.

Hoffman, M. R., and Edwards, J. O. (1975), Kinetics of the Oxidation of Sulfite in Hydrogen Peroxide in Acidic Solution, *J. Phys. Chem.* **79**, 2096–2098.

Horowitz, A., and Calvert, J. G. (1978), *Int. J. Chem. Kinet.* **10**, 805.

Horowitz, A., and Calvert, J. G. (1982), *J. Phys. Chem.* **86**, 3105–3114.

Horowitz, A., Kershner, C. J., and Calvert, J. G. (1982), *J. Phys. Chem.* **86**, 3094–3105.

Huie, R. E., and Herron, J. T. (1975), *Int. J. Chem. Kinet. Symp.* **1**, 165.

Japar, S. M., Wu, C. H., and Niki, H. (1974), *J. Phys. Chem.* **78**, 2318.

Judeikis, H. S., and Siegl, S. (1973), Particle-Catalyzed Oxidation of Atmospheric Pollutants, *Atmos. Environ.* **7**, 619–631.

Judeikis, H. S., Stewart, B. T., and Wren, A. G. (1978), *Atmos. Environ.* **12**, 1633.

Kenley, R. A., Davenport, J. E., and Hendry, D. G. (1978), *J. Phys. Chem.* **82**, 1095.

Kleindienst, T. E., Harris, G. W., and Pitts, J. N., Jr. (1982), *Environ. Sci. Technol.* **16**, 844–846.

Kok, G. L. (1980), Measurements of Hydrogen Perioxide in Rainwater, *Atmos. Environ.* **14**, 653–656.

Larson, T. V., Horike, N. R., and Harrison, H. (1978), Oxidation of SO_2 by O_2 and O_3 in Aqueous Solution: A Kinetic Study with Significance to Atmospheric Rate Processes, *Atmos. Environ.* **12**, 1597–1611.

Lee, Y.-N., and Schwartz, S. E. (1981), Evaluation of the Rate of Uptake of Nitrogen Dioxide by Atmospheric and Surface Liquid Water, *J. Geophys. Res.* **86**, 11971–11983.

Lee, Y.-N, and Schwartz, S. E. (1982), Kinetics of Aqueous Phase Reactions of NO_2 with NO at Low Partial Pressures.

Liberti, A., Biocco, D., and Possanzini, M. (1978), *Atmos. Environ.* **12**, 255.

Lloyd, A. C. (1979), in *Chemical Kinetic Data Needs for Modeling the Lower Troposphere* (J. T. Heron, R. E. Huie, and J. A. Hodgeson, eds.), NBS Special Publication 557, August, p. 34.

Lloyd, A. C., Darnall, K. R., Winer, A. M., and Pitts, J. N., Jr. (1976), *Chem. Phys. Lett.* **42**, 205.

Maahs, H. G. (1982), The Importance of Ozone in the Oxidation of Sulfur Dioxide in Nonurban Tropospheric Clouds, 2nd Symposium on the Composition of the Nonurban Troposphere, Amer. Meteorological Society, Williamsburg, VA, May.

Mader, P. M. (1958), Kinetics of the Hydrogen Perioxide-Sulfite Reaction in Alkaline Solution, *J. Am. Chem. Soc.* **80**, 2634–2639.

Martin, L. R. (1982), Kinetic Studies of Sulfite Oxidation in Aqueous Solution, to be published in *SO₂, NO and NO₂ Oxidation Mechanisms: Atmospheric Considerations* (J. G. Calvert, ed.), Ann Arbor Science, Woburn, MA.

Martin, L. R., and Damschen, D. E. (1981), Aqueous Oxidation of Sulfur Dioxide by Hydrogen Peroxide at Low pH, *Atmos. Environ.* **15**, 1615–1621.

Martin, L. R., Damschen, D. E., and Judeikis, H. S. (1981), The Reactions of Nitrogen Oxides with SO₂ in Aqueous Aerosols, *Atmos. Environ.* **15**, 191–195.

Middleton, P., Kiang, C. S., and Mohnen, V. A. (1980), Theoretical Estimates of the Relative Importance of Various Urban Sulfate Aerosol Production Mechanisms, *Atmos. Environ.* **14**, 463–472.

Möller, D. (1980), Kinetic Model of Atmospheric SO₂ Oxidation Based on Published Data, *Atmos. Environ.* **14**, 1067–1076.

Moortgat, G. K., Slemr, F., Seiler, W., and Warneck, P. (1978), *Chem. Phys. Lett.* **54**, 444.

NASA (1979), *Chemical Kinetic and Photochemical Data for Use in Stratospheric Modeling,* Evaluation Number 2, JPL Publication 79-27 and revisions thereof, Jet Propulsion Laboratory, Pasadena, April 15.

Nash, T. (1979), The Effect of Nitrogen Dioxide and of Some Transition Metals on the Oxidation of Dilute Bisulphite Solutions, *Atmos. Environ.* **13**, 1149–1164.

National Research Council (1982), Impacts of Diesel Powered Light Duty Vehicles, National Academy Press, Washington, D.C.

Niki, H., Maker, P. D., Savage, C. M., and Breitenbach, L. P. (1977), *Chem. Phys. Lett.* **46**, 327.

Niki, H., Maker, P. D., Savage, C. M., and Breitenbach, L. P. (1978), *J. Phys. Chem* **82**, 135.

Niki, H., Maker, P. D., Savage, C. M., and Breitenbach, L. P. (1980), *J. Phys. Chem* **84**, 14.

Novakov, T., Chang, S. G., and Harker, A. B. (1974), Sulfates as Pollution Particulates: Catalytic Formation on Carbon (Soot) Particles.

Oblath, S. B., Markowitz, S. S., Novakov, T., and Chang, S. G. (1981), Kinetics of the Formation of Hydroxylamine Disulfonate by Reaction of Nitrite with Sulfites, *J. Phys. Chem.* **85**, 1017–1021.

Overend, R., and Paraskevopoulos, G. (1978), *J. Phys. Chem.* **82**, 1329.

Penkett, S. A., Jones, B. M. R., Brice, K. A., and Eggleton, A. E. J. (1979), The Importance of Atmospheric Ozone and Hydrogen Peroxide in Oxidizing Sulfur Dioxide in Cloud and Rainwater, *Atmos. Environ.* **13**, 123–137.

Perry, R. A., Atkinson, R., and Pitts, J. N., Jr. (1977a), *J. Chem. Phys.* **67**, 611.

Perry, R. A., Atkinson, R., and Pitts, J. N., Jr. (1977b), *J. Phys. Chem.* **81**, 296.

Pitts J. N., Jr., Darnall, K., Carter, W. P. L., Winer, A. M., and Atkinson, R. (1979), *Mechanisms of photochemical Reactions in Urban Air,* Final Report, EPA-600/3-79-110, November.

Pitts, J. N., Jr., Grosjean, D. M., Winer, A. M., Lloyd, A. C., and Doyle, G. J. (1975/1976), Chemical Transformations in Photochemical Smog and Their Application to Air Pollution Control Strategies, Third Annual Progress Report, NSF-RANN Grant, ENV73-02904 A03, October 1, 1975–December 31, 1976.

Rasmussen, R. A., Chatfield, R. B., Holdren, M. W., Robinson, E. (1976), *Hydrocarbon Levels in Midwest Open-Forested Area,* Report submitted to the Coordinating Research Council, October.

Ravishankara, A. R., Wagner, S., Fischer, S., Smith, G., Schiff, R., Watson, R. T., Tesi, G., and Davis, D. D. (1978), *Int. J. Chem. Kinet.* **10**, 783.

Ravishankara, A. R., Wagner, S., Schiff, R., and Davis, D. D. (1978), Unpublished results.

Richards, L. W., Anderson, J. A., Blumenthal, D. L., McDonald, J. A., Kok, G. L., and Lazrus, A. L. (1983), *Atmos. Environ.*

Sanadze, G. A., and Dolidze, G. M. (1961), Mass Spectrographic Identification of Compounds of C₅H₈ (Isoprene) Type in Volatile Emissions from the Leaves of Plants, *Soobsch. Akad. Nauk. Gruz. SSR* **27**, 747–750.

Sanadze, G. A., and Kalandadze, A. N. (1966), Evolution of the Diene C₅H₈ by Poplar Leaves Under Various Conditions of Illumination, *Dokl. Bot. Sci.* **168**, 95–97.

Sandberg, J. S., Basso, M. J., and Okin, B. A. (1978), *Science* **200**, 1051.

Sato, T., Matani, S., and Okabe, T. (1979), The Oxidation of Sodium Sulfite with Nitrogen Dioxide, with Special Reference to Analytical Methods for Nitrogen-Sulfur Compounds Produced in the Reaction System, *Nippon Kaguku Kaishi* **1979**(7), 869–878. In Japanese. Also, Amer. Chem. Soc. 177th National Meeting, Honolulu, HI, April 1, 1979, paper INDE-210, 1979.

Schwartz, S. E. (1982), Gas-Aqueous Reactions of Sulfur and Nitrogen Oxides in Liquid-Water Clouds, in *SO₂, NO and NO₂ Oxidation Mechanisms: Atmospheric Conditions* (J. G. Calvert, ed.) Ann Arbor Science, Woburn, MA.

Schwartz, S. E., and White, W. H. (1981), Solubility Equilibria of the Nitrogen Oxides and Oxyacids in Dilute Aqueous Solution, in *Advan. Environ. Sci. Engng.*, Vol. 4 (J. R. Pfafflin and E. N. Ziegler, eds.), Gordon and Breach, New York.

Schwartz, S. E., and White, W. H. (1982), Kinetics of Reactive Dissolution of Nitrogen Oxides into Aqueous Solution, in *Advan. Environ. Sci. Technol.*, Vol. 12 (S. E. Schwartz, ed.), Wiley, New York.

Scott, B. C., and Laulainen, N. S. (1979), On the Concentration of Sulfate in Precipitation, *J. Appl. Meteor.* **18**, 138–147.

Seinfeld, J. H. (1980), Lectures in Atmospheric Chemistry, *Amer. Inst. of Chemical Engineers Monograph* **12**, Vol. 76.

Smith, B. M., Wagman, J., and Fish, B. R. (1969), Interaction of Airborne Particles with Gases, *Environ. Sci. Technol.* **3**, 558–562.

Stedman, D. H., Morris, E. D., Jr., Daby, E. E., Niki, H., and Weinstock, B. (1970), 160th National Meeting of the American Chemical Society, Chicago, IL, Sept. 14–18.

Su, F., Calvert, J. G., and Shaw, J. H. (1980), *J. Phys. Chem.* **84**, 239.

Takeuchi, H., Ando, M., and Kizawa, N. (1977), Absorption of Nitrogen Oxides in Aqueous Sodium Sulfite and Bisulfite Solutions, *Ind. Eng. Chem. Process Design. Dev.* **16**, 303–308.

Tartarelli, P. D., Morelli, F., and Corsi, P. (1978), *Atmos. Environ.* **12**, 288.

Tingey, D. T., et al. (1978), *Isoprene Emission Rates from Live Oak,* Final Report, EPA-904/9/78-004, U.S. Environmental Protection Agency, Research Triangle Park, NC, April.

Urone, P., Lutsep, H., Noyes, C. M., and Parcher, J. F. (1968), Static studies of Sulfur Dioxide Reactions in Air, *Environ. Sci. Technol.* **2**, 611–618.

Weaver, J., Meagher, J., and Heickler, J. (1976/1977), *J. Photochem.* **6**, 111.

Westberg, H. H. (1977), Part IV: The Issue of Natural Organic Emissions, in *International Conference on Oxidants—1976: Analysis of Evidence and Viewpoints.*

Whitten, G. Z., Hogo, H., Meldgin, M. J., Killus, J. P. and Bekowies, P. J. (1979), *Modeling of Simulated Photochemical Smog with Kinetic Mechanisms,* Vol. 1, Interim Report, EPA-600/3-79-001a, January.

Winer, A. M., Lloyd, A. C., Darnall, K. R., Atkinson, R., and Pitts, J. N., Jr. (1977), *Chem. Phys. Lett.* **51**, 221.

Zika, R., Saltzman, E., Chameides, W. L., and Davis, D. D. (1982), H₂O₂ Levels in Rainwater Collected in South Florida and the Bahama Islands, *J. of Geophys. Res.* **87**, 5015–5017.

Zika, R. G., and Saltzman, E. S. (1982), Interaction of Ozone and Hydrogen Peroxide in Water: Implications for Analysis of H₂O₂ in Air, *Geophys. Res. Lett.* **9**, 231–234.

CHAPTER 36
AIR POLLUTION STANDARDS AND REGULATIONS

DONALD J. HENZ

PEDCo Environmental, Inc.
Cincinnati, Ohio

1 INTRODUCTION

1.1 Purpose

Air pollution regulations are promulgated by the federal, state, and local governments. The federal regulations alone cover hundreds of pages, regulating not only owners of emission sources, but manufacturers of air pollution monitoring equipment, those who test and monitor pollution sources, and even those citizens who wish to file a lawsuit regarding some aspects of pollution control.

This chapter is designed to assist the reader in determining the extent to which the various regulations apply under a given set of circumstances. Since the main thrust of this work is to assist the reader in problem solving, emphasis is placed on the various regulations applicable to stationary sources of emissions. These are the types of sources about which engineers and managers must constantly be making decisions—decisions whether to start a new project, expand, modify, or even continue operation. Mobile sources, such as motor vehicles and aircraft, are, for the most part, subject to regulations that impact the engine manufacturer. While it is true that the owner must maintain the equipment and, in some cases, test for emissions, these regulations are not the type that call for decisions.

Although state and local regulations may apply to specific sources, they are too varied to treat in the same detail as the federal regulations. However, this chapter will provide the reader with a general knowledge of the basic types of state and local regulations one may encounter. The federal regulations, which greatly shape and impact the state and local regulations, will be covered in more detail.

1.2 The Regulatory Process

There was a time when human behavior was ordered strictly by the legislative branch of government. As society became more and more complex, however, the legislators found that they did not possess the expertise required to make their laws workable. As legislated matters grew further and further from the common pool of knowledge, the need for specialized regulatory bodies increased. This need, in the United States, grew out of the constitutionally protected requirement that limitations on human behavior be reasonably related to the amelioration of a public detriment that is deemed to exist. An abundant amount of case law supports this principle, but it is not within the purview of this chapter to study the common law aspects of regulation. Suffice to say that, in the United States, governmental limitations must have some nexus to the correction of a detrimental condition. Thus, although the legislature may prohibit harmful activities, it often lacks the knowledge to set reasonable limits.

Such is the case with environmental protection. With very few exceptions, the legislatures of this country state that it is their intent to maintain the integrity of the ambient air; and the power to determine the exact limitations required to achieve that goal is, through enabling legislation, invested in a regulatory agency. This legislation creates the regulatory agency and specifies its duties and its powers. The Clean Air Act (CAA) designates the U.S. Environmental Protection Agency (EPA) as the regulatory agency.

1.2.1 *General Regulatory Scheme Under the Clean Air Act*

An overall view of the regulatory scheme is helpful in sorting out the various regulations applicable to limiting air pollutant emissions.

First. Determine which pollutants endanger public health and welfare (the criteria pollutants). Then determine the maximum allowable ambient air levels for these pollutants [National Ambient Air Quality Standards (NAAQS)]. Next, each state adopts emission regulations (SIP regulations) designed to provide ambient air that meets the NAAQS for the various criteria pollutants. There are eight criteria pollutants:

1. Particulate matter (TSP)
2. Sulfur oxides (SO_x)
3. Carbon monoxide (CO)
4. Hydrocarbons (HC)
5. Oxidants
6. Nitrogen oxides (NO_x)
7. Lead (Pb)
8. Ozone

Second. There are some large stationary sources of pollution that significantly pollute. These require special attention since the expansion of these categories of sources, owing to industrial growth, will likely cause a violation of the NAAQS. Thus national emission standards are developed for these sources. These regulations, usually called New Source Performance Standards (NSPS), are generally more restrictive than the SIP regulations. Although the NSPS apply only to new and reconstructed or modified facilities, if the pollutant so regulated is not listed as a criteria pollutant or a hazardous pollutant, each state must adopt a limiting regulation applicable to similar, existing facilities.

Third. In addition to the more common pollutants that adversely affect health and welfare and those especially large and growing sources of criteria pollutants, there are hazardous pollutants. These result in an increase in mortality or an increase in serious irreversible, or incapacitating reversible, illness. These pollutants are not listed in criteria pollutants but, instead, become the subject of National Emission Standards for Hazardous Pollutants (NESHAP). The NESHAP regulations generally apply to new and existing emission sources alike.

Fourth. Control the emissions of aircraft and motor vehicles. Owing to the interstate nature of these pollutant sources, regulation is generally preempted by the EPA.

Figure 1 graphically depicts these four major areas of emission limitations under the Clean Air Act. In addition to the published emission limitations applicable to stationary sources, there are also those that may be considered as *negotiated* limitations. These are often negotiated when seeking a permit for new, modified, or reconstructed sources. They may be identical to the limitations published under the NSPS but more than likely they will be that control limit that is considered, for the particular installation under question, best available control technology (BACT) or lowest achievable emission reduction (LAER).

Note that emission limits applicable to a given facility may change from time-to-time with the development of new standards and/or major process changes or reconstruction. The progression of more and more restrictive emission limitations applicable to a given source is illustrated in Figure 2.

1.3 Types of Regulations

Regulations applicable to air pollution control are of two types: limiting and administrative. Limiting regulations prescribe a quantitative standard that one may not exceed. These types of regulations, described in detail in Section 3, may limit exhaust gas pollutant concentration, hourly pollutant discharge, smoke density, and the like. They are the regulations that determine the type of pollution control system one must use.

In some respects limiting regulations are only the tip of the iceberg. Section 301(a) of the Clean Air Act authorizes the administrator of EPA to prescribe such regulations as are necessary to carry out his or her function under the act. This authority empowers him or her to promulgate regulations relative to recordkeeping, testing, and permitting. Each of these areas has many facets and the costs associated with compliance may be substantial. Although limiting regulations apply only to sources of air contaminants, administrative regulations affect state and local governments, manufacturers of testing equipment, fuel suppliers, and even citizens who may wish to bring a lawsuit under the Clean Air Act.

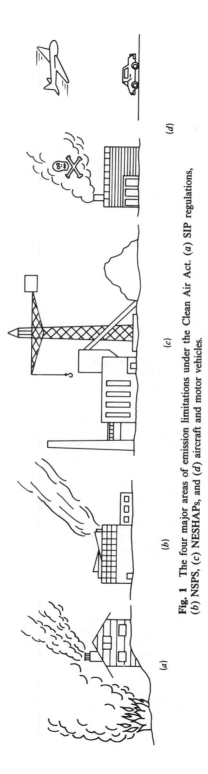

Fig. 1 The four major areas of emission limitations under the Clean Air Act. (*a*) SIP regulations, (*b*) NSPS, (*c*) NESHAPs, and (*d*) aircraft and motor vehicles.

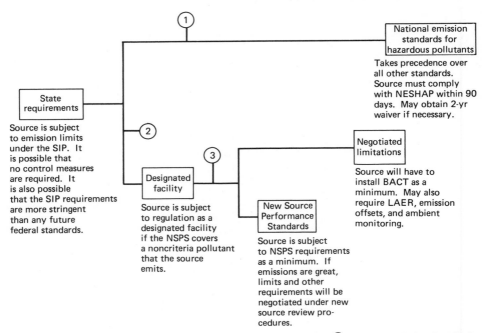

Fig. 2 Emission regulation progression for existing stationary sources. ① National Emission Standards for Hazardous Pollutants adopted, ② New Source Performance Standards adopted, and ③ source is modified or reconstructed.

1.4 Who Is Affected?

Table 1 lists the major federal regulations and indicates their reach and scope. It should be noted that the majority of the regulations presented in Table 1 are administrative-type regulations. This table may be used as a general guide to determine whether one is affected by a particular area of federal regulation.

In addition to federal regulations, one must account for state and local air pollution regulations. For new sources of emissions, these governmental entities are the source of administrative regulations—often in conjunction with the State Implementation Plan (SIP). The limiting regulations are generally similar to those promulgated by EPA. State and local air pollution regulations mainly affect facilities that, directly or indirectly, emit air pollutants. However, many states have additional requirements regarding testing and monitoring, and some jurisdictions regulate product design and product packaging. For example, in Los Angeles it is unlawful to sell, or use, house paint containing photochemically reactive solvent if it is packaged in containers of one-quart capacity or larger.

1.5 Sources of Reference

Federal air pollution control regulations are constantly being developed and modified. They literally cover hundreds of pages. In addition, there are state and local regulations—usually incorporated into the SIP—that apply. How does one determine exactly what regulations apply? In making this determination with respect to a specific emission source, discuss the matter with the local air pollution control agency. Some states grant extensive powers to local jurisdictions, cities, or counties. In such cases, these local jurisdictions may be more restrictive than the state. Locally applied limiting regulations are often incorporated in the SIP, but not necessarily so. Therefore, one should always determine what local regulations exist. In addition to emission regulations, the administrative requirements of both the state and local jurisdictions must also be determined. Failure to do so may be costly or, as a minimum, may result in a strained relationship with local authorities.

Next, be aware of state regulations. These are generally not published commercially and must be obtained from the control agency. It will not do to ask for a copy of the SIP, however, since an entire SIP is typically more than 6 in. thick and includes a great deal of plant-specific regulation. In those states where environmental regulations are published and updated by a commercial publisher, it is helpful to subscribe to that service.

Table 1 Major Areas of Federal Regulation of Air Pollution

Title 40 Code of Federal Regulations, Part Number	General Area of Regulation	Generally Affected Entities
50	Ambient air quality standards and prescribed monitoring methods	1. State agencies 2. All persons who monitor ambient air quality
51	Requirements for preparation of state implementation plans	1. State agencies
52	Contents of implementation plans	1. All persons regulated under the SIPs
53	Specifications and reference standards for ambient air quality monitoring equipment	1. Ambient air quality monitoring equipment manufacturers 2. All persons who monitor ambient air quality
54	Notice requirements prior to bringing suit under the Clean Air Act	1. Private citizen litigants
58	Prescribed methodology for ambient air quality monitoring	1. All persons who monitor ambient air quality
60	New source performance standards	1. State agencies 2. Owners and operators of affected facilities 3. All persons who test emission sources
61	National emission standards for hazardous air pollutants	1. States 2. Owners of source of hazardous pollutants 3. All persons who test and monitor sources of hazardous pollutants
62	Existing facilities similar to those affected by NSPS	1. Owners of the facilities
66	Noncompliance penalties and their calculation	1. Owners of emission sources
81	Air quality planning, defines AQCRs and attainment status	1. States 2. Owners of emission sources 3. Local government
85, 86	Motor vehicle emission limits	1. Manufacturers of motor vehicles
87	Aircraft emission limits	1. Manufacturers and owners of certain aircraft

Federal regulations can be monitored in the *Federal Register*. All federal regulations must be published in the *Federal Register*. This publication, printed daily by the U.S. Government Printing Office (USGPO), is the most expeditious way to obtain the text of proposed or promulgated regulations. After the regulations have been published therein, they are codified in the Code of Federal Regulations. This too may be obtained from the USGPO. However, this publication is not very timely and it is imperative that the subscriber keep it updated with clippings from the *Federal Register*.

All regulations issued by the federal government are officially cited by their Code of Federal Regulations (CFR) title and section number. For example, 40 CFR 52.21 is in the citation for the federal regulations regarding the prevention of significant deterioration. The nomenclature means:

40 — Title 40, "Protection of Environment"
52 — Part 52, Approval and Promulgation of Implementation Plans
.21 — Section 21, Prevention of Significant Deterioration of Air Quality

Very often regulations will be cited by their *Federal Register* publication. Thus, one may see a citation such as 45 FR 52676. This particular example cites the amendments to the Prevention of Significant Deterioration (PSD) regulations. The nomenclature means:

45 — Volume number
52676 — Page number

Note that when a federal regulation is amended, only the amended portions are published in the *Federal Register*. Because of this, one will find it helpful to subscribe to one of the commercial services that incorporates all amendments on a continuing basis. These services, however, do not replace the timeliness of the *Federal Register*. Where one's responsibilities require a knowledge of current regulations as well as regulatory trends, it is advisable to subscribe to both the *Federal Register* and a commercial service.

2 EVOLUTION OF AIR POLLUTION CONTROL REGULATIONS

Although air pollution control is now firmly entrenched as just another facet of doing business, it is a relatively new concept. In fact until recently smoke was considered a sign of progress. (In a lighter vein I might add that the state of Ohio's Professional Engineering Certificates, as late as 1969, incorporated a design depicting ships and factories with visible plumes.) Until the early 1970s, facilities could be built almost anywhere without any thought given to their effect on the ambient air.

2.1 Early Prohibitions

Prior to the end of World War II, relatively little thought was given to the regulation of air pollution. What little regulation existed was covered by local ordinances prohibiting nuisances. Although concern over air pollution control was growing throughout the country, the birth of modern-day regulatory programs resulted from attempts to rid Los Angeles of its smog. It was reasoned that if industrial sources of dust and smoke were required to control their emissions, the problem would be solved. This resulted in the development of the nation's first air pollution regulation based on control technology. First, air pollutant emissions were measured downstream from the control device. Then the results were plotted relative to the process' material input and a curve was then defined which indicated the degree to which a source of a given size (material throughput) could control its emissions. This regulation became known as Rule 54, or the Process Weight Regulation. A more detailed explanation of this regulation is given in Section 3.

2.2 The SIP and Beyond

Local initiatives were not deemed sufficient to curb the growing air pollution problem. Indeed, the Los Angeles effort did not solve the smog problem at all. There was growing pressure for federal entrance into the air pollution control field and in 1955 the Air Pollution Control Research and Technical Assistance Act was enacted. However, this legislation provided no regulatory authority. It was followed in 1963 by legislation that granted federal regulatory authority to hold abatement conferences in an attempt to force polluters in targeted areas to control emissions. The lack of success of this approach led to the Air Quality Act of 1967 which, again, failed to produce any appreciable results. By 1970 it became clear to Congress that if substantial gains were to be made in air pollution control, federal legislation providing ample regulatory authority and regulatory mandates was necessary. The result was the Clean Air Act of 1970.

The Clean Air Act of 1970 gave EPA the responsibility and authority to establish a nationwide program of air pollution abatement and air quality enhancement. In general, the legislation provided for a combination of state and federal stationary source emission standards designed to achieve and maintain federally prescribed ambient air quality. This prescribed air quality is defined for certain pollutants, called criteria pollutants, by standards called National Ambient Air Quality Standards (NAAQS).* Primary responsibility for implementing the program rests with the states, subject to federal supervision. The legislation further provided for federal enforcement should a state fail to adopt or implement a suitable air pollution control program. Thus was born the State Implementation Plan, or SIP.

Each state has its own SIP. It is a plan by which the state plans to achieve and maintain the NAAQS. Naturally, air quality varies from place to place, thus requiring different measures of control. Rather than consider an entire state as homogeneous, the nation was divided into Air Quality Control Regions (AQCRs), based on jurisdictional boundaries, urban–industrial concentrations, and other factors necessary to provide adequate implementation of the air quality standards. Some AQCRs are interstate while others are intrastate. There are approximately 240 AQCRs in the country. The boundaries of each of these areas are defined in 40 CFR 81. The attainment status of each AQCR, with respect to the NAAQS, is also defined in that section.

And so the control of air pollution initially centered on the NAAQS. The attainment of these standards is determined on a regional basis, each region being an AQCR. Each state has developed a plan, known as a SIP, whereby the NAAQS are to be achieved and maintained. Although the Clean Air Act Amendments of 1977 substantially broadened the federal program, this basic approach was

---

* See Section 3.1 infra.

left intact and the balance of the discussion in this chapter will refer to the latest legislation as the Clean Air Act (CAA).

3 LIMITING REGULATIONS

Limiting regulations, those that directly limit the discharge of emissions, may also incorporate administrative requirements. For example, the NSPS for vinyl chloride manufacturing includes a biannual compliance reporting and record-keeping requirement. However, since the primary thrust of these standards is to limit emissions, they are regarded here as limiting regulations. Table 2 lists the various types of limiting regulations and their application. All limiting regulations are based on one of, or a combination of, the generic types of regulations shown in Table 3. A detailed discussion of each type of generic regulation follows.

Table 2 Types of Limiting Regulations

Type of Regulation	Application	Primary Source Publication
State regulations	New and existing sources of air pollution not subject to other, specific regulation	SIP (40 CFR 52)
New Source Performance Standards (NSPS)	New or modified sources of air pollution specifically designated by regulation	40 CFR 60
Designated facility standards	Existing sources of air pollution that would be subject to NSPS if new or modified	SIP (40 CFR 52)
National Emission Standards for Hazardous Air Pollutants (NESHAP)	All specifically designated sources and all sources of any designated pollutant	40 CFR 61
Aircraft emissions	All aircraft; no variance between state standards and federal standards	40 CFR 87
Motor vehicle emissions	New motor vehicles; area completely preempted by federal government[a]	40 CFR 86
National Ambient Air Quality Standards (NAAQS)	All ambient air	40 CFR 50

[a] The state of California is excepted.

Table 3 Generic Groups of Limiting Regulations

Generic Group	Commonly Used Units
Equivalent opacity	Percent obscuration
Process weight rate	kg/hr
Mass rate	kg/hr
Exhaust gas concentration	g/m^3, ppm
Ambient air concentration	$\mu g/m^3$
Process design requirements	—
Process standards	—
Nuisance abatement	—

3.1 National Ambient Air Quality Standards

These standards are the cornerstone of all criteria pollutant limitations promulgated under the Clean Air Act, since all the emission regulations focus on achieving and maintaining these ambient concentrations. To most, the level of these standards is now academic but, as is discussed later, these limits affect the location of new sources of pollution and the modification of existing sources.

The National Ambient Air Quality Standards (NAAQS) are set at two levels, primary and secondary. The former is based on health criteria while the latter is based on welfare considerations such as the deleterious effect of a pollutant on architectural surfaces, vegetation, visibility, and the like. Except for sulfur oxides and particulate matter the secondary NAAQS for all criteria pollutants is identical to the primary standard. The NAAQS are shown in Table 4.

3.2   Generic Types of Regulations for Stationary Sources

3.2.1   *Equivalent Opacity*

Equivalent opacity regulations are an evolution of the Ringelmann regulation. This regulation is based on the methodology developed by Maxmillian Ringelmann, near the turn of the century, to determine the degree to which smoke is being emitted from fuel-burning facilities.

The Ringelmann chart consists of six sections numbered from 0 to 5, each section being approximately 5¾ × 8¾ in. in size. Section 0 is completely white while section 5 is completely black. These two sections are usually not included in the chart. Sections 1 to 4 consist of a white background with intersecting heavy black lines imprinted thereon, the lines growing progressively wider from section 1 to section 4. No. 1 Ringelmann is 20% black; No. 2 is 40%; No. 3 is 60%; and No. 4 is 80%. The chart was originally designed to be posted 50 ft from the observer. Estimates of the density of the plume were then made by comparing it with the intervening chart, selecting the section that most nearly resembles the smoke. However, this was awkward, if not impossible, in heavily trafficked areas and today trained observers estimate plume density without the aid of the chart.

Since the Ringelmann chart is based on the blackness of the plume, it is generally only applicable to inefficient fuel–combustion processes. As the science of fuel combustion and as industry developed, it became desirable to develop a convenient method for limiting nonblack plumes also. The result was the equivalent opacity regulation which is based on the Ringelmann principle and is usually defined in terms of Ringelmann. A typical ordinance reads as follows:

> *A person shall not discharge into the atmosphere from any single source of emission whatsoever any air contaminant for a period or periods aggregating more than 3 min in any 1 hr which is:*
>
> (a) *As dark or darker in shade as that designated as No. 2 on the Ringelmann chart, as published by the United States Bureau of Mines, or*
>
> (b) *Of such opacity as to obscure an observer's view to a degree equal to or greater than does smoke described in subsection (a) of this section.*

Thus the degree of obscuration of a nonblack plume is equated to the degree of obscuration of a black plume. Since opacity, rather than darkness, is measured, a chart cannot be used and trained observers must now estimate plume opacity by sight.

EPA has developed a specific methodology for taking opacity readings. The methodology, commonly referred to as Method 9, is found in Appendix A—Reference Methods, to 40 CFR 60. Appendix A specifies the method of taking readings, certification requirements, and performance characteristics of and calibration requirements for the equipment used to certify qualified "observers" or "smoke readers."

In general, the smoke reader must stand in clear view of the plume with the sun oriented in the 140° sector to his back. Additionally, his line of vision must be, as nearly as possible, perpendicular to the axis of the plume since any other line of sight will increase the apparent depth of the plume thereby increasing obscuration. See Figure 3.

To receive certification as a qualified observer, one must demonstrate the ability to assign opacity readings to 50 different plumes (of known opacity, from a smoke generator) with an error not to exceed 15% opacity on any one reading and a maximum average error of 7.5% opacity. Certification is valid for a 6-month period. The smoke test is administered by EPA and certified state and local air pollution control agencies at various times throughout the year.

Visible emission regulations offer a convenient means of enforcement to the air pollution control authorities. Almost every major jurisdiction in the country has such a regulation.

In recent years the trend has been to lower the allowable plume opacity. While most jurisdictions allow opacities up to 20%, quite a few have tightened their regulation to 10%, and some have even adopted regulations that allow *no* visible emissions. Zero visible emissions is the trend.

Most stationary sources of particulate emissions are subject to visible emission limitations in addition to the more specific limits based on process material input or outlet grain loading. However, the visible emission limitations are often more stringent than regulations based on process material input. For example, a maximum allowable visible emission equivalent to No. 1 Ringelmann would probably allow a maximum dust concentration of 0.1 to 0.05 gr/scf, or less, while the other applicable process

Table 4 National Primary and Secondary Ambient Air Quality Standards

Pollutant	Averaging Time	Primary Standards [a]	Secondary Standards [a]
Sulfur oxides	Annual arithmetic mean	80 μg/m³ (0.03 ppm)	
	24 hr	365 μg/m³ (0.14 ppm)	1300 μg/m³ (0.5 ppm)
	3 hr		
Particulate matter	Annual geometric mean	75 μg/m³	60 μg/m³
	24 hr	260 μg/m³	150 μg/m³
Carbon monoxide	8 hr	10 mg/m³	Same as primary standard
	1 hr	40 mg/m³ (35 ppm)	
Ozone (corrected for NO_2 and SO_2)	1 hr	240 μg/m³ (0.12 ppm)	Same as primary standard
Hydrocarbons (corrected for methane)	3 hr	160 μg/m³ (0.24 ppm)	Same as primary standard
Nitrogen oxides	Annual arithmetic mean	100 μg/m³ (0.05 ppm)	Same as primary standard
Lead	3 months	1.5 μg/m³	Same as primary standard
Ozone	1 hr	235 μg/m³ (0.12 ppm)	Same as primary standard

[a] Except for annual means, standards are not to be exceeded more than once a year.

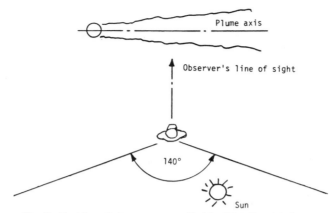

Fig. 3 Position of observer as specified by EPA Test Method 9.

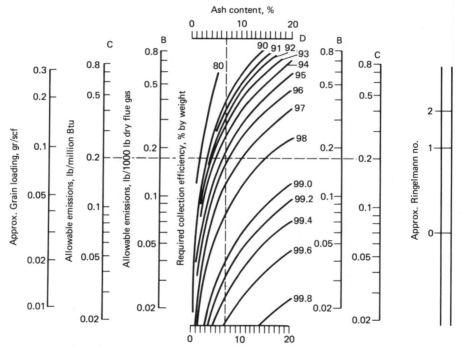

Fig. 4 Comparison of various particulate emissions limitations on required flyash collection efficiency for pulverized coal-fired steam generators.

regulation may allow a higher emission. But, since both regulations apply, the more stringent visible emission limitation dictates the degree of control required. Figure 4 illustrates this.

3.2.2 *Process Weight*

Although the Process Weight (PW) regulation was developed as a basis for control of metallurgical fumes,* its use has spread to encompass nearly every type of industrial process. Thus used, the PW

* See McCabe, L. C., et al., Dust and Fume Standards, *Industrial Engineering Chemistry* **41**, 2388 (Nov. 1949).

regulation is not even remotely based on demonstrated control technology. However, as a practical matter, the application of PW regulations to various nonmetallurgical processes has generally been successful. On its face the regulation is simple. A typical regulation may read as seen in Table 5.

This is a rather liberal PW regulation. For example, the South Coast Air Quality Management District's Rule 405 (California) allows a maximum emission rate of 30 lb/hr at a process weight rate of 500,000 kg/hr (551 tons/hr). These two regulations are graphically compared in Figure 5. The significance of the difference lies in the required control efficiencies. A plant having a 20 ton/hr through-put and uncontrolled emissions of 600 lb/hr requires a control efficiency of 95% to meet the more lenient regulation while an efficiency of nearly 97.7% is required to meet the more stringent.

Most PW regulations carefully define process weight. Note, for example, section (b) in the typical regulation in Table 5 excepts liquid and gaseous fuels and combustion air. The differences between

Table 5 Regulation of Particulates from Miscellaneous Industrial Processes

(a) No person shall cause, suffer, allow, or permit particulate matter caused by industrial processes for which no other emission control standards are applicable to be discharged from any stack, vent, or outlet into the atmosphere in excess of the hourly rate shown in the table in Subdivision (c) of this Regulation.

(b) Process weight per hour means the total weight of all materials introduced into any specific process that may cause any emission of particulate matter. Solid fuels charged are considered as part of the process weight, but liquid and gaseous fuels and combustion air are not. For a cyclical or batch operation, the process weight per hour is derived by dividing the total process weight by the number of hours in one complete operation from the beginning of any given process to the completion thereof, excluding any time during which the equipment is idle. For a continuous operation, the process weight per hour is derived by dividing the process weight for a typical period of time by the number of hours in that typical period of time.

(c) The allowable emission rate for particulate matter based on actual process weight rate is listed in the following table:

Process Weight Rate, lb/hr	Allowable Emission Rate for Particulate Matter		Process Weight Rate, lb/hr	Allowable Emission Rate for Particulate Matter	
	tons/hr	lb/hr		tons/hr	lb/hr
100	0.05	0.551	16,000	8	16.5
200	0.10	0.877	18,000	9	17.9
400	0.20	1.39	20,000	10	19.2
600	0.30	1.83	30,000	15	25.2
800	0.40	2.22	40,000	20	30.5
1,000	0.50	2.58	50,000	25	35.4
1,500	0.75	3.38	60,000	30	40.0
2,000	1.00	4.10	70,000	35	41.3
2,500	1.25	4.76	80,000	40	42.5
3,000	1.50	5.38	90,000	45	43.6
3,500	1.75	5.97	100,000	50	44.6
4,000	2.00	6.52	120,000	60	46.3
5,000	2.50	7.58	140,000	70	47.8
6,000	3.00	8.56	160,000	80	49.1
7,000	3.50	9.49	200,000	100	51.3
8,000	4.00	10.4	1,000,000	500	69.0
9,000	4.50	11.2	2,000,000	1,000	77.6
10,000	5.00	12.1	6,000,000	3,000	92.7
12,000	6.00	13.6			

(d) Calculation of the rate of emission for process weight rates up to 60,000 lb/hr shall be accomplished by use of the equation $E = 4.10$ times P to the power of 0.67 and for process weight rates in excess of 60,000 lb/hr shall be accomplished by use of the equation $E = 55.0$ times P to the power of 0.11 minus 40 where E = allowable emission rate for particulate matter in lb/hr and P = process weight rate in tons/hr.

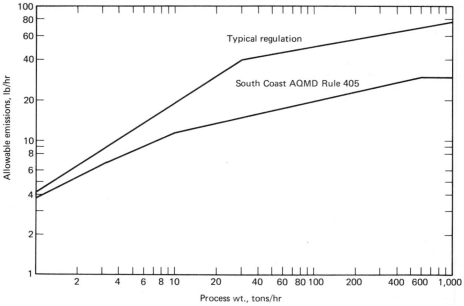

Fig. 5 Comparison of two process weight regulations applicable to miscellaneous industrial processes.

continuous and batch type processes are also delineated. However, how does one define process weight for a process that has several operations? The answer lies in the interpretation of the phrase, "specific process" used in section (b). The San Francisco Bay Area Air Pollution Control District uses the term, "source operation" and defines it as:

> the last operation preceding the emission of an air contaminant, which operation (a) results in the separation of the air contaminant from the process materials or in the conversion of the process materials into air contaminants, as in the case of combustion of fuel; and (b) is not an air pollution abatement operation.

Despite the early opposition to the PW regulation, it has become perhaps the most frequently used type of limiting regulation for particulate matter. It is very adaptable to specific processes and has been used as a basis for limiting emissions from such specific processes as ferrous jobbing foundries, chemical fertilizer manufacturing, feldspar processing, and hot mix asphalt plants. As an example, Figure 6 compares the PW regulations adopted by the state of North Carolina.

3.2.3 Mass Rate

Mass rate standards are somewhat similar to the PW regulation in that they generally allow a specified emission rate (lb/hr, lb/10⁶ Btu, lb/ton of product, etc.). The allowable unit rate of emissions usually decreases as the process throughput or potential emissions increase. There is a large variety of mass rate emission regulations applicable to gaseous as well as solid pollutants. Several of the more popular types of mass rate regulations will be discussed in this section.

The Potential Emission Rate regulation allows a mass rate of emissions based on potential emissions. Pennsylvania's regulation §123.13 is one such regulation, allowing emissions at the rate of $0.76E^{0.42}$ lb/hr, where E is the production or charging rate in units per hour multiplied by a specified process factor. This factor represents the theoretical uncontrolled emission rate. The regulation specifies 30 process factors. Table 6 shows calculated allowable emissions for several of these processes at typical production rates.

Note that the control efficiency required for compliance with this type of regulation increases with the production rate. Therefore it is necessary to determine the facility's production rate and operate at that rate when testing the slack. Another problem arises where two distinct, but similar, facilities exhaust to the same stack. Are the allowable emissions twice that allowed for one, or are they that which is permitted for a facility having a capacity of the two facilities combined? The difference

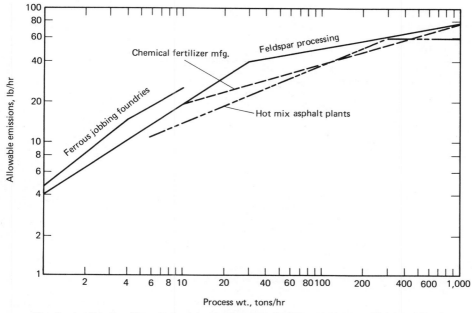

Fig. 6 Application of process weight regulations to specific industries in North Carolina.

is significant. For example, consider two iron ore sintering strands, each having a production rate of 60 tons/hr. The allowable emissions and the required control efficiencies are as follows:

Condition	Allowable Emissions	Required Control Efficiency, %
Both strands considered as one process	19.97	99.2
Each strand considered as a separate process	29.86	98.8

The 0.4% difference in control efficiency is significant in this range.

Table 6   Allowable Mass Rate of Emissions for Several Processes per Pennsylvania Regulation §123.13

Process	Process Factor	Production or Charging Rate, units/hr	Allowable Emissions, lb/hr	Required Control Efficiency, [a] %
Grain drying	200 lb/ton of product	5	13.83	98.6
		20	24.76	99.4
Byproduct coke mfg. pushing operation	1 lb/ton of coke pushed	20	2.67	86.6
		85	4.91	94.2
Secondary lead smelting	0.5 lb/ton of product	3	0.901	39.9
		7.5	1.32	64.8
Primary iron and/or steelmaking: Sintering-windbox	20 lb/ton of dry solids feed	125	20.32	99.2
		250	27.18	99.4

[a] Assumes the process factor accurately represents the uncontrolled emission rate.

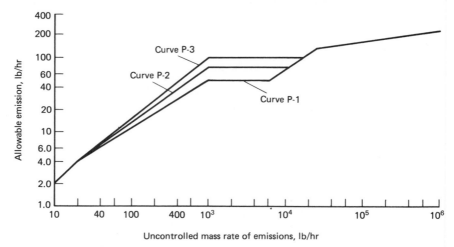

Fig. 7 State of Ohio's potential emission rate regulation.

Another type of Potential Emission Rate regulation is graphically shown in Figure 7. The basis for this regulation is similar to Pennsylvania's regulation §123.13 except (other than being less stringent) that it provides different allowable emission rates based on the geographic location of the source. For example, all industrial sources located in a "Priority I Region" must follow curve P-1. One of the obvious disadvantages of this type of regulation is that two tests—before and after the control device—are required to determine whether the source complies with the standard. On the other hand, some flexibility is given to account for the different ambient air qualities found throughout the state. The thrust of the standard is to only require that degree of emission removal necessary to protect air quality. Note however that Ohio requires that process sources also comply with a PW regulation and, where both are applicable, the more stringent applies. This is not an unusual situation and the reader is well advised to read thoroughly all regulations.

Mass rate standards have also been adopted for refuse incinerators and stationary indirect heat exchangers (boilers). New Jersey's regulation 7:27–4.2 bases allowable emissions from boilers on the unit's heat input rate. The regulation specifically states that the heat input rate is the sum of the input rates of all units discharging through a single stack. Although the regulation limits particulate emissions to an hourly rate, it is similar to the fuel-burning equipment regulation adopted by a number of jurisdictions throughout the country wherein allowable emissions are stated in terms of the heat input of the facility. This regulation is shown graphically in Figure 8. New Jersey's regulation is superimposed on the figure for purposes of comparison. Also shown are the approximate collection efficiencies. Not all jurisdictions define the heat input rate as does New Jersey. In such case the required collection efficiency would be lower.

3.2.4 Exhaust Gas Concentration

Regulations limiting the concentration of pollutants are commonly used either as a primary regulation or as a supplemental one. For example, a regulation may allow an emission rate of x units per hour or a gas concentration of y units per unit volume, whichever is greater. Or else the stated pollutant concentration in the exhaust gas may be permitted even though an accompanying regulation, such as PW or mass rate regulation, is more restrictive. The (San Francisco) Bay Area Air Pollution Control District generally limits visible emissions to 10% opacity but one is deemed to be in compliance with this limitation if it is shown that the concentration of particulate matter in the exhaust gas does not exceed $0.6/L$ gr/scf, where L is the significant dimension of the emission point, in feet (Regulation 2, Division 3). This allows the following particulate matter emissions:

Significant Dimension, ft	Allowable Grain Loading, gr/scf
2	0.300
4	0.150
6	0.100
10	0.060

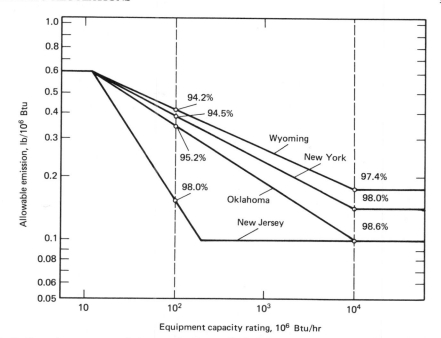

Equipment capacity rating, 10^6 Btu/hr

Fig. 8 Several mass rate regulations applicable to solid fuel-fired boilers. Percentages indicate approximate required control efficiency for a pulverized coal-fired, dry bottom boiler burning coal with an ash content of 10%.

For the most part, however, concentration standards are used as primary regulations. Approximately half the National Standards of Performance (NSP) limit pollutant concentrations. Particulate matter standards are usually expressed in terms of unit weight per unit volume (e.g., gr/scf), and gaseous pollutants in terms of a volumetric relationship [e.g., parts per million (ppm)]. Although this type of regulation is straightforward [(a) one need only measure the concentration after the control device and (b) larger processes are not required to have greater control efficiencies than are required for smaller processes], it can be circumvented by dilution of the gas stream. Therefore, virtually all regulations prohibit dilution. This may pose a problem for industrial processes whose exhaust gas flow rates are not linearly proportional to process size, however. For example, some boilers fire at 40% excess air while others may operate at 30% or less; and hooded work stations may require more open area between the emission source and the hood than do other hooded sources such as conveyor transfer points. Very often the design of fugitive emission capture systems is a matter of engineering judgment. One must therefore be able to demonstrate the engineering soundness of any design that is subject to variable exhaust gas rates.

That exhaust gas flow rate may vary from operation to operation is recognized by most regulations. Nearly all of them specify the conditions at which concentrations are to be determined. The most common condition applies to exhausts from refuse incinerators where the language generally reads:

No owner or operator shall cause to be discharged into the atmosphere any gases which contain particulate matter in excess of 0.18 g/dscm corrected to 12% CO_2.

Thus if tests show a particular concentration of only 0.15 g/dscm but the CO_2 concentration is only 8%, the actual concentration of particulate matter will be deemed to be

$$0.15 \text{ g/dscm} \times \frac{12\%}{8\%} = 0.225 \text{ g/dscm}$$

The specified CO_2 concentration is based on generally accepted operating practice for refuse incinerators. Although the example shown applies to refuse incinerators, the same principle can be applied to other processes as well, where owing to poor operating procedures, exhaust gas rates may exceed good engineering practice.

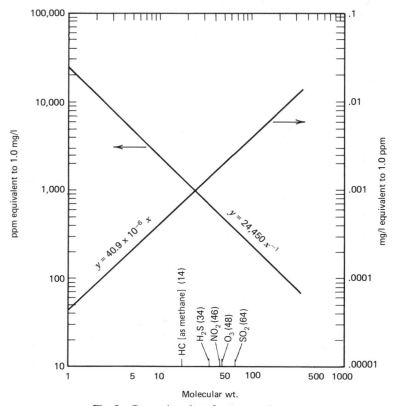

Fig. 9   Conversion chart for gases and vapors.

The conditions at which the concentration of particulate matter (or grain loading) is determined are important to note. "Standard" conditions are generally specified as 68°F and a pressure of 29.92 in. Hg. That is the NSP definition of "standard conditions." It also is important to note that gaseous concentrations are generally specified in terms of "dry" volumes. Source test results should always be given in terms of the units specified in the regulations to avoid confusion and misunderstandings. For gaseous pollutants, the relationship between ppm and mg/l is a function of the pollutants' molecular weight. This relationship is shown in Figure 9.

3.2.5 Design Requirements

Design requirements may be applied to either the pollution control system or the process components. One of the most common of these is seen in hydrocarbon emission regulation where storage tanks are required to have floating roofs. The basic design of the seal between the floating roof and the tank wall is often specified.

Incineration requirements often include design criteria with respect to fire-box temperature and retention time. Texas' limitation of carbon monoxide emissions from iron foundry cupolas states that the exhaust gases will be incinerated in an afterburner meeting certain design requirements as follows:

Control Requirements for Iron Cupolas.   *No person shall emit a vent gas stream from any iron cupola into the atmosphere unless the vent gas stream is properly burned at a temperature equal to or greater than 1300°F (704°C) in an afterburner having a retention time of at least ¼ of a second and having a steady flame that is not affected by the cupola charge and relights automatically if extinguished.*

Other examples of design standards are:

1.   The requirement that oil–water separators be covered.
2.   Submerged–fill requirements for gasoline loading.

3. Conveyor covers.
4. Minimum stack heights.

Very often, design requirements are used to ensure that minimum emission limits are met. However, these types of regulations have the disadvantages of stifling control technology development and exculpating the owner or operator if the design does not work. These disadvantages are overcome by allowing for alternate, equally effective means of control and limiting design standards to only tried-and-true requirements. Thus, if emissions from an oil–water separator could be controlled by a surfactant, as well as by a full cover, application could be made for approval of the use of the surfactant.

3.2.6 Process and Operating Standards

Sometimes, emissions can be reduced by a change in process materials, operating parameters, or even finished product specifications. One of the most common materials specification regulations is the restriction on the sulfur content of coal or fuel oil. It is also common to restrict the formulation of degreasing solvents and dry-cleaning solvents.

The NESHAP regulations specify that asbestos-containing materials may not be used for paving roadways. And certain operating standards apply to demolition of buildings containing asbestos insulation.

Some products, such as soybean oil and polyvinyl chloride, can emit pollutants owing to residual materials in the product. Soybean oil is extracted with hexane. To the extent the hexane is not driven off in the desolventizer–cooler, it remains in the meal to evaporate subsequent to the processing. Similarly, vinyl chloride can evaporate from the polyvinyl chloride product. The NESHAP for polyvinyl chloride plants addresses this problem by limiting residual vinyl chloride in the product to 2000 or 400 ppm, depending on the type of product.

3.2.7 Ambient Air Concentration

Some emission limits are a function of air quality. Since maximum ambient concentrations diminish with stack height, such regulations may ease the economic burden of control, where the incremental stack costs are less than the incremental pollution control hardware costs. (Note however, that tax benefits applicable to pollution control systems may not apply to the cost of constructing a higher stack.) An example of this type of regulation is seen in the state of Texas' regulation regarding sulfuric acid mist. That regulation, in part, reads:

> **131.04.03   Control of Sulfuric Acid .001.**   *No person may cause, suffer, allow or permit emissions of sulfuric acid from a source or sources operated on a property or multiple sources operated on contiguous properties to exceed:*
>
> *(a) A net ground level concentration of 15 μg/m³ of air averaged over any 24-hr period; or*
>
> *(b) A net ground level concentration of 50 μg/m³ of air averaged over a 1-hr period of time more than once during any consecutive 24-hr period; or*
>
> *(c) 100 μg/m³ of air maximum at any time.*

In many cases, compliance is determined by taking downwind ambient samples. Where background concentrations exist, samples must be taken both upwind and downwind. These regulations must consider how to determine compliance in the absence of reliable ambient data however, One method would be to make reliable ambient data a condition of compliance. Another would be to adopt a method of calculation based on an air dispersion model. In either case, a source operator will want to use a dispersion model when designing the control system to estimate allowable emissions. Texas bases its calculations on a standard equation. Allowable emissions are a function of stack height and gas volume, temperature, and velocity. Figure 10 compares allowable sulfuric acid mist emissions for stacks having varying heights and diameters, based on Texas' regulation §131.04.03. The advantage of a tall stack is readily seen. For example, a source having uncontrolled emissions of 240 lb/hr and a 2-ft diameter stack would require a 95% control efficiency if its stack were only 100 ft high, but less than 91% with a 200-ft high stack.

3.2.8 Nuisance Abatement

These are general regulations that very often are designed for those sources not readily subject to specific limiting regulations. Note, however, that these regulations may also apply to emission sources that *are* readily subject to specific regulations. Since these regulations, although subjective in nature, may actually be more restrictive than the applicable specific regulation, one should be aware of their

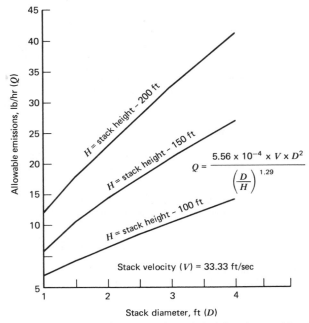

Fig. 10. Relative allowable sulfuric acid mist emissions based on ambient air impact.

The equation shown in the figure:

$$Q = \frac{5.56 \times 10^{-4} \times V \times D^2}{\left(\dfrac{D}{H}\right)^{1.29}}$$

existence. Nuisance abatement regulations generally require the use of good engineering practice or reasonable precautions. A nuisance abatement regulation may read as follows:

(A) No person shall cause or permit any materials to be handled, transported, or stored; or a building or its appurtenances or a road to be used, constructed, altered, repaired, or demolished without taking reasonable precautions to prevent particulate matter from becoming airborne. Such reasonable precautions shall include, but not be limited to, the following:

(1) Use, when possible, water or chemicals for control of dust in the demolition of existing buildings or structures, construction operations, the grading of roads, or the clearing of land;

(2) Application of asphalt, oil, water, or suitable chemicals on dirt roads, materials stockpiles, and other surfaces that can create airborne dusts;

(3) Installation and use of hoods, fans, and control equipment to enclose, contain, capture, and vent the handling of dusty materials. Adequate containment methods shall be employed during sandblasting or other similar operations;

(4) Cover, at all times when in motion, open bodied vehicles transporting materials likely to become airborne;

(5) Conduct agricultural practices such as tilling of land, application of fertilizers, etc., in such manner as to prevent dust from becoming airborne;

(6) The paving of roadways and their maintenance in a clean condition;

(7) The prompt removal of earth or other material from paved streets onto which earth or other material has been deposited by trucking or earth moving equipment or erosion by water or other means.

In the past, odors were always limited by nuisance standards, and very often still are. Now, however, odorous pollutants are generally controlled by hydrocarbon-specific emission limitations.

With respect to odorous emissions, one should also be aware that, in preventing a nuisance, one may run afoul of another regulation. If, for example, hydrogen sulfide is incinerated, large quantities of sulfur dioxide (a criteria pollutant) may be formed. The generalized language of these types of regulations renders them subjective. Thus owners or operators of existing emission sources must give thought to compliance not only with the specific limitations applicable to the process' emissions, but with the nuisance abatement regulations as well.

3.3 National Standards for Stationary Sources

3.3.1 National Standards of Performance

The Clean Air Act requires that EPA develop standards of performance for new stationary sources of significant air pollution. These standards, commonly known as New Source Performance Standards (NSPS), are based on the best system of continuous emission reduction that has been adequately demonstrated, taking into account such nonair quality impacts as economics and energy. Note that these regulations take the form of standards, not just emission limits. Thus an NSPS regulation may require monitoring, process modification, or even specific emission reduction methods.

New sources of emissions subject to NSPS are called "affected facilities." An affected facility may be an entire plant, as in the case of nitric acid manufacturing plants, or a specific entity such as a basic oxygen furnace in a steel-making shop. For each NSPS the affected facility or facilities are designated.

Modified sources of pollutants are also subject to NSPS regulations. Although "modification" is defined in the Clean Air Act, the definition is expanded and clarified by regulation to include both modifications and reconstructions. Very simply, a modification subjects a source to NSPS if it is a substantial change in the facility that results in an increase in emissions of those pollutants covered by the NSPS.

A reconstruction occurs if the change costs more than half the capital expenditure needed to construct an entirely new facility. Reconstructed sources may become affected facilities even though the regulated emissions are *not* increased. Whether a source is considered an affected facility owing to its reconstruction is determined on a case-by-case basis. Note, however, that if a reconstructed source results in an increase in the regulated emissions, it is also subject to the criteria relative to modified sources. Figure 11 outlines the process used to determine whether a modified or reconstructed source is an affected facility.

Through mid-1982, EPA has adopted NSPS for 30 major source categories and plans to develop standards for 44 others. These are shown in Table 7. A brief summary of the emission limits and the monitoring requirements for the 30 standards already promulgated is shown in Table 8. Do these NSPS affect existing facilities that have not been modified or constructed? The answer is "yes" if the NSPS limits a pollutant not designated as a criteria pollutant. Criteria pollutant-emitting sources, built prior to the promulgation of the applicable NSPS, are already subject to pollutant limitations under the SIP regulations. However, in the event the NSPS limits a noncriteria pollutant, each state is required to develop regulations (presumably less stringent than the NSPS) limiting emissions of that pollutant from all existing sources that, if new, would be subject to the NSPS. These regulations are to be developed within 9 months after EPA publishes the applicable guideline document. In general, existing sources affected by the regulations must comply within 12 months. Longer compliance schedules, when permitted, must include legally enforceable increments of progress. Application of the NSPS-related SIP requirements may be determined on a case-by-case basis [40 CRF 60.24 (F)] considering:

1. Cost (owing to plant age, location, or basic process design).
2. Physical impossibility.
3. Other factors that make application of the standard unreasonable.

3.3.2 National Emission Standards for Hazardous Air Pollutants

EPA has developed five national emission standards for hazardous air pollutants (NESHAP). These standards regulate emissions of asbestos, beryllium, mercury, and vinyl chloride. Two of these standards apply to beryllium. Standards for benzene, cadmium, arsenic, polycyclic organic matter (POM), and radionuclides are either being considered or are under development. These standards are a combination of limiting regulations and administrative regulations. The limitations include not only specified emission limits, but design and operation specifications as well. Unlike NSPS, NESHAP regulations apply to all facilities, existing and new. These standards are based on health considerations and do not require consideration of economic effects. Because of the severity of the problem of unchecked emissions, NESHAP reporting and monitoring requirements are extensive. Although the NESHAPs are pollutant specific, the regulations specify those sources to which the emission abatement, reporting, testing, and monitoring requirements apply. Table 9 summarizes the sources subject to NESHAP requirements.

Once a NESHAP is promulgated, the owner or operator of any existing source must comply with the standard within 90 days or request a waiver of compliance. Such waiver, if granted, may not exceed 2 yr from the effective date of the NESHAP. EPA has established specific requirements for

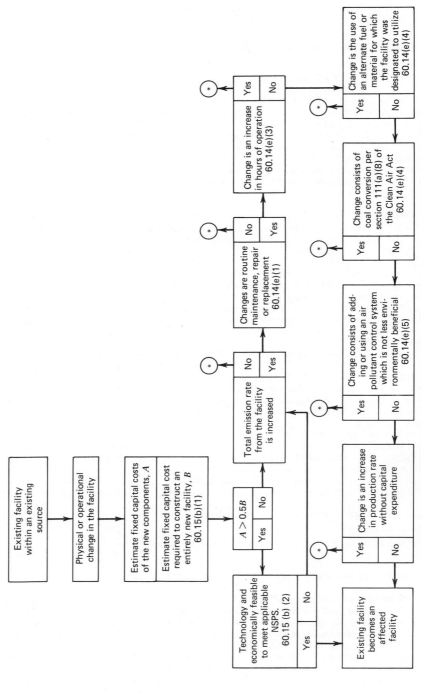

Fig. 11 Method of determining whether changes to existing facility constitute a modification or reconstruction under §§40 CFR 60.14 and 60.15. The emission rate refers only to pollutants regulated by the specific NSPS. Asterisks indicate facility that is not deemed to be modified and is not subject to NSPS.

Table 7 List of Major Source Categories for Which Standards of Performance Are to Be Promulgated [a]

1. Synthetic organic chemical manufacturing	20. By-product coke ovens
(a) Unit process	21. Synthetic fibers
(b) Storage and handling equipment	22. Plywood manufacture
(c) Fugitive emission sources	23. Industrial surface coating: large appliances
(d) Secondary sources	24. Crude oil and natural gas production
2. Industrial surface coating: cans	25. Secondary aluminum
3. Petroleum refineries: fugitive sources	26. Lightweight aggregate industry: clay, shale,
4. Industrial surface coating: paper	and slate
5. Dry cleaning	27. Gypsum
(a) Perchloroethylene	28. Sodium carbonate
(b) Petroleum solvent	29. Polymers and resins: phenolic
6. Graphic arts	30. Polymers and resins: urea-melamine
7. Polymers and resins: acrylic resins	31. Polymers and resins: polystyrene
8. Stationary internal combustion engines	32. Polymers and resins: ABS-SAN resins
9. Industrial surface coating: fabric	33. Fiberglass
10. Fossil fuel-fired steam generators: industrial	34. Polymers and resins: polypropylene
boilers	35. Textile processing
11. Nonmetallic mineral processing	36. Asphalt processing and asphalt roofing manu-
12. Metallic mineral processing	facture
13. Foundries: steel and gray iron	37. Brick and related clay products
14. Polymers and resins: polyethylene	38. Ammonium nitrate fertilizer
15. Charcoal production	39. Polymers and resins: polyester resins
16. Synthetic rubber	40. Ammonium sulfate
(a) Tire manufacture	41. Starch
(b) SBR production	42. Perlite
17. Vegetable oil	43. Uranium refining
18. Industrial surface coating: metal coil	44. Urea (for fertilizer and polymers)
19. Petroleum transportation and marketing	

[a] Listed in order of priority. At press time, standards have been promulgated for Sources 6, 18, 23, and 36. See 40 CFR 60, Subparts QQ, SS, TT, and UU, respectively.

the reporting and waiver-request. Within 90 days, the owner or operator of the existing source must supply the following information:

1. Name and address of the owner or operator
2. Source location
3. Type of hazardous pollutant
4. Detailed process description
5. Amount of hazardous materials processed
6. Description of hazardous pollutant control system

In addition the owner or operator must either state that he will comply with the standard within 90 days of its effective date or request a waiver of compliance. If a waiver of compliance is requested, the following additional information must be furnished:

1. Description of the proposed control system.
2. A compliance schedule showing the dates by which contracts will be let, equipment will be ordered, construction will commence and end, and final compliance will be achieved.
3. Description of interim control measures.

A specific format for reporting the above information and requesting a waiver is shown in Appendix A to 40 CFR 61. If the administrator grants a waiver, its continued application depends, in part, on the applicant's ability to meet the milestones set out in the compliance schedule.

New or modified sources are subject to the same requirements as existing sources except that there is no provision for a waiver of compliance. These sources must comply with the applicable NESHAP upon startup. Notice of planned construction or modification must be given prior to starting construction. This notice must generally contain the information cited above. The administrator must then approve or disapprove the construction within 60 days of receipt of sufficient information on

Table 8 Summary of New Source Performance Standards (40 CFR 60)

Source Category and Affected Facility	Emission Limits	Monitoring Requirements	Comments
Fossil fuel-fired steam generator: >250 million Btuhr input rate[a]	Particulate—0.10 lb/million Btu	Not required	(a) Size standard applies to each unit, not to aggregate; applies to all units firing any form of fossil fuel or wood residue
	Opacity—20% (27% for one 6-min period per hr)	Continuous, except not required for gas-fired units	
	SO₂—Oil—0.8 lb/million Btu	Continuous[b]	(b) See §60.45(b) for exceptions
	Coal—1.2 lb/million Btu	Continuous[b]	(c) See §60.44(a)(5)
	Oil and coal combined—see §60.43	Continuous[b]	(d) See also §60.44(c)
	NOₓ—Oil—0.30 lb/million Btu	Continuous[b]	
	Gas—0.20 lb/million Btu	Continuous[b]	
	Coal—0.70 lb/million Btu	Continuous[b]	
	Lignite—0.60 lb/million Btu	Continuous[b]	
	Certain lignite[c]—0.80 lb/million Btu	Continuous[b]	
	Solid fuel containing >25% coal refuse—no limit[d]	Not required	
	Cofiring—see §60.44(b)	Continuous[b]	
Electric utility steam generating units: >250 million Btuhr input rate.[a] (Boilers and combined cycle gas turbines that exhaust >250 million Btuhr to a steam generating unit.)	Particulate—0.03 lb/million Btu	Not required	(a) Size standard applies to each unit, not to aggregate; see §60.330 et seq. for stationary gas turbines
	Opacity—20% (27% for one 6-min period per hr)	Continuous, except not required for gas-fired units	
	SO₂[b]—Solid fuels[c]—1.20 lb/million Btu[d] and 90% reduction or 0.60 lb/million Btu and 70% reduction.	Continuous	(b) Compliance determined on 30-day rolling average basis except as otherwise noted
	Oil—0.80 lb/million Btu[d] and 90% reduction or 0.20 lb/million Btu and zero reduction	Continuous	(c) Except anthracite
			(d) No reduction required for noncontinental U.S. sources
	Gas—same as oil		
	Solvent Refined Coal (SRC-I)—1.20 lb/million Btu and 85% reduction[e]	Continuous	(e) 85% reduction determined on a 24-hr basis
	Anthracite—1.20 lb/million Btuhr	Continuous	(f) See §60.44a(a)
	Resource recovery—1.20 lb/million Btuhr	Continuous	
	Cofiring—see §60.43a(h)	Continuous	

Affected facility	Standard	Monitoring requirement	Remarks
	NO_x[b]—*Solid fuel containing >25% coal refuse—no limit*	Not required	(a) Size standard applies to each unit, not to aggregate
	Coal-derived solid fuel—0.50 lb/million Btu and 65% reduction	Continuous	
	Certain lignite[f]—0.80 lb/million Btu and 65% reduction	Continuous	
	Subbituminous coal—0.50 lb/million Btu and 65% reduction	Continuous	
	All other solid fuels—0.60 lb/million Btu and 65% reduction	Continuous	
	Coal-derived gas—0.50 lb/million Btu and 25% reduction	Continuous	
	All other gas—0.20 lb/million Btu and 25% reduction	Continuous	
	Coal-derived oil—0.50 lb/million Btu and 25% reduction	Continuous	
	Shale oil—0.50 lb/million Btu and 25% reduction	Continuous	
	All other oil—0.30 lb/million Btuhr and 25% reduction	Continuous	
Incinerators: >50 tons/day charging rate[a]	Particulate—0.08 gr/dscf corrected to 12% CO_2	Record daily charging rates and hours of operation	
Portland cement plants:			
Kiln	Particulate—0.30 lb/ton feed Opacity—20%	Record daily production rates and kiln feed rates	
Clinker cooler	Particulate—0.10 lb/ton feed Opacity—10%		
Fugitive emission points	Opacity—10%		
Nitric acid plants:			
Entire plant	Opacity—10% NO_x—3.0 lb/ton of acid produced	Not required Continuous	
Sulfuric acid plants:			
Entire plant	Opacity—10% Acid mist—0.15 lb/ton of 100% acid produced SO_2—4.0 lb/ton of 100% acid produced	Not required Not required Continuous	

Table 8 (Continued)

Source Category and Affected Facility	Emission Limits	Monitoring Requirements	Comments
Asphalt concrete plants:			
Entire plant	Particulate—0.04 gr/dscf	Not required	
	Opacity—20%	Not required	
Petroleum refineries:			
Fluid catalytic cracking unit catalyst regenerator	Particulate—1.0 lb/1000 lb coke burnoff	Not required	
	Opacity—30% (>30% during one 6-min period per hr)	Not required	
Fuel gas combustion devices	SO_2—0.10 gr/dscf	Not required	
	H_2S—230 mg/dscm	Not required	
Claus sulfur recovery plants >20 tons/day	250 ppm SO_2 at 0% O_2 on dry basis if controlled by incinerator	Not required	
	or		
	300 ppm TRS and 10 ppm H_2S at 0% O_2 on dry basis if not controlled by incinerator		
Storage vessels for petroleum liquids: >40,000 gal capacity;[a,b] for true vapor pressure of 1.5 to 11.1 psia	VOC—One of the following design standards:		(a) Does not apply to vessels having <420,000 gal capacity used to store liquids prior to custody transfer
	(1) External floating roof	Keep records of liquids stored	(b) See §60.110 et seq. for vessels constructed prior to May 18, 1980
	(2) Fixed roof with internal floating cover	Keep records of liquids stored	
	(3) Vapor recovery systems (95% reduction)	Not required	
	(4) Equivalent system (see §60.114a)	Keep records of liquids stored	
Secondary lead smelters:			
Pot furnaces >550 lb capacity	Opacity—10%	Not required	
Blast furnaces (cupolas) } Reverberatory furnaces }	Particulate—0.022 gr/dscf	Not required	
	Opacity—20%		
Secondary brass and bronze plants:			
Reverberatory furnaces ≧ 2205 lb production capacity	Particulate—0.022 gr/dscf	Not required	
	Opacity—20%	Not required	

Affected facility	Emission standard	Monitoring requirement	Notes
Electric furnaces ≥ 2205 lb production capacity	Opacity—10%	Not required	
Blast furnaces (cupolas) ≥ 550 lb/hr production capacity	Opacity—10%	Not required	
Iron and steel plants:			
Basic oxygen furnace (BOF)	Particulate—0.022 gr/dscf Opacity—10%; more than 10% but less than 20% allowed once per cycle	Record cycle time; monitor performance of venturi scrubber[a] Not required	(a) Monitoring of control system is required only for venturi scrubber
Sewage treatment plants:			
Incinerator[a]	Particulate—1.30 lb/ton dry sludge Opacity—<20%	Measure sludge feed rate Not required	(a) Each incinerator that combusts wastes containing more than 10% municipal sewage sludge, or each incinerator that charges more than 2205 lb municipal sewage sludge per day
Primary copper smelters: Dryer Roaster Smelting furnace Copper converter	Particulate—0.022 gr/dscf Opacity—20% SO₂—0.065% by volume[a] Opacity—20%[b]	Record weight and makeup of charge Continuous Continuous Not required	(a) Reverberatory smelting furnaces are exempt while processing feed containing a high level of volatile impurities (b) Applies only to affected facilities controlled by sulfuric acid plant
Primary zinc smelters: Roaster Sintering machine[b]	SO₂—0.065% by volume Opacity—20%[a] Particulate—0.022 gr/dscf Opacity—20%	Continuous Not required Not required Continuous	(a) Applies only to affected facilities controlled by sulfuric acid plant (b) May also be subject to SO₂ limitation; see §60.173
Primary lead smelters: Blast furnace Dross reverberatory furnace Sintering machine discharge end Sintering machine Electric smelting furnace Converter	Particulate—0.022 gr/dscf Opacity—20% SO₂—0.065% Opacity—20%[a]	Not required Continuous Continuous Not required	(a) Applies only to affected facilities controlled by sulfuric acid plant

Table 8 (*Continued*)

Source Category and Affected Facility	Emission Limits	Monitoring Requirements	Comments
Primary aluminum reduction plants:			
Potroom groups[a]	Total fluorides—2.0 lb/ton aluminum produced (Soderberg plants); 1.9 lb/ton aluminum produced (prebake plants)[b]	Record feed and production rates	(a) This is *not* necessarily one potroom
	Opacity—<10%		(b) See §60.192 for certain exceptions
Anode bake plants	Total fluorides—0.1 lb/ton aluminum equivalent		
	Opacity—<20%		
Phosphate fertilizer industry:	Total fluorides—	Measure and record material throughput	
Wet process phosphoric acid plants	0.020 lb/ton equivalent P_2O_5 feed	Measure and record pressure drop across scrubber	
Superphosphoric acid plants	0.010 lb/ton equivalent P_2O_5 feed		
Diammonium phosphate plants	0.060 lb/ton equivalent P_2O_5 feed		
Triple superphosphate plants	0.20 lb/ton equivalent P_2O_5 feed		
Granular triple superphosphate storage facilities	5.0×10^{-4} lb/hr-ton equivalent P_2O_5 stored		
Coal preparation plants:[a]			
Thermal dryer	Particulate—0.031 gr/dscf	Exhaust gas temperature; venturi scrubber performance	(a) Applies only to plants that process >200 tons/day
	Opacity—<20%		
Pneumatic coal cleaning equipment	Particulate—0.018 gr/dscf	Not required	
	Opacity—<10%	Not required	
Processing and conveying equipment	Opacity—<20%	Not required	
Storage systems		Not required	
Transfer and loading systems			
Ferroalloy Production Facilities:			
Electric submerged arc furnace Stack	Particulate—0.99 lb/MW-hr or 0.51 lb/MW-hr[a]	Process and power consumption; continuous volumetric flow rates	(a) Limit depends on type of alloy produced
	Opacity—<15%		(b) See §60.265(d)
	CO—<20% by volume	Not required	
Furnace fugitive	Opacity—invisible		
Tapping fugitive	Opacity[b]—no limit during not more than 40% of each tapping period; invisible balance of period		
Dust handling equipment	Opacity—<10%	Not required	

Source	Standard	Monitoring	Notes/Exceptions
Steel Plants:			
Electric arc furnace (EAF)	Particulate—0.0052 gr/dscf Opacity—<3% from control device; invisible from roof monitor[a]	Process cycle records Continuous Flow rates in hood Pressure in DSE system	(a) See §60.272(3) for exceptions
Dust handling equipment (for EAF)	Opacity—<10%	Not required	
Kraft pulp mills:[a]			(a) Includes kraft pulping combined with neutral sulfite semichemical pulping
Recovery furnace	Particulate—0.044 gr/dscf, corrected to 8% oxygen Opacity—<35%	Not required Continuous	(b) See §60.283 for exceptions
	Total reduced sulfur (TRS)—5 ppm, dry, corrected to 8% oxygen (straight furnace); 25 ppm, dry, corrected to 8% oxygen (cross furnace)	Continuous	
Smelt-dissolving tanks	Particulate—0.2 lb/ton black liquor solids (dry weight)	Venturi scrubber performance	
	TRS—0.0168 lb/ton liquor solids (dry weight)	Continuous	
Lime kiln	Particulate—0.067 gr/dscf corrected to 10% oxygen (gas); 0.13 gr/dscf corrected to 10% oxygen (oil)	Venturi scrubber performance (if used)	
	TRS—8 ppm, dry, corrected to 10% oxygen	Continuous	
	TRS—5 ppm, dry, corrected to 10% oxygen[b]	Continuous	
Digester systems			
Brown stock washer systems			
Multiple-effect evaporator systems			
Black liquor oxidation systems			
Condensate stripper systems			
Grain elevators:			
All loading and unloading stations, and handling operations	Particulate—0.01 gr/dscf	Not required	(a) See §60.302(a) for exceptions
Truck loading stations	Opacity—10%	Not required	
Truck unloading stations	Opacity—5%	Not required	
Rail loading stations	Opacity—5%	Not required	
Rail unloading stations	Opacity—5%	Not required	
Barge and ship loading stations	Opacity—20%	Not required	
Barge and ship unloading stations	Design standard [see §60.302(d)]	Not required	
Dryers	Opacity—invisible[a]	Not required	
Grain handling	Opacity—invisible	Not required	

945

Table 8 (*Continued*)

Source Category and Affected Facility	Emission Limits	Monitoring Requirements	Comments
Stationary gas turbines: Turbines > 10 mmBtuhr heat input	NO_x—Exhaust gas NO_x concentration limited by: Rated heat input, rated heat rate, and nitrogen content of fuel See §60.332 SO_2—0.015% volume adjusted to 15% oxygen, dry and 0.8% (weight) fuel sulfur content	Continuously monitor and record fuel consumption and water-to-fuel ratio[a] Record nitrogen content of fuel Record sulfur content of fuel	(a) Only required for turbines using water injection
Lime Manufacturing plants:[a] Rotary lime kiln	Particulate—0.30 lb/ton Opacity—10%	Continuous: pressure drop across scrubber and scrubbing liquid supply pressure[b] Continuous[c] Continuous: mass rate of limestone feed	(a) Not applicable to facilities used to manufacture lime at kraft pulp mills (b) Monitoring required only if kiln is controlled with a scrubber (c) Monitoring required only if kiln is controlled with a dry system Note: This standard was remanded to EPA for background data to support the limits applicable to the kiln. *National Lime Association v EPA*, 14 ERC 1509, May 19, 1980
Lime hydrator	Particulate—0.15 lb/ton	Continuous: scrubbing liquid flow rate, electric current usage, and mass rate of lime feed	
Glass manufacturing plants: Glass melting furnace[a]	Particulate—emission limits are in g/kg or glass produced as follows:[b] Furnace fired with gaseous fuel Furnace fired with liquid fuel	Not required	(a) Not applicable to hand glass melting furnaces, glass melting furnaces with less than 4550 kg/day capacity, and electric melters (b) See §60.292(a)(2) for limits on co-fired furnaces
Glass manufacturing plant industry segment			

946

Source category			Monitoring	Comments
Container glass	0.1	0.13		
Pressed and blown glass				
(a) Borosilicate recipes	0.5	0.65		
(b) Soda-lime and lead recipes	0.1	0.13		
(c) Other than borosilicate, soda-lime, and lead recipes (including opal, fluoride, and other recipes)	0.25	0.325		
Wood fiberglass	0.25	0.325		
Flat glass	0.225	0.225		
Lead acid battery mfg. plants:			Scrubber performance (if used) for affected facilities	(a) Applies to metal bodies only
Grid casting	Lead—0.000176 gr/dscf Opacity—0%			(b) Source emission test required on a monthly basis
Paste mixing	Lead—0.00044 gr/dscf Opacity—0%			(c) Quarterly reports required for incinerator operation
Three-process operation	Lead—0.00044 gr/dscf Opacity—0%			
Lead oxide mfg. process	Lead—0.010 lb/ton of lead feed Opacity—0%			
Lead reclamation	Lead—0.00198 gr/dscf Opacity—5%			
Any other lead-emitting operation	Lead—0.00044 gr/dscf Opacity—0%			
Automobile and light truck surface coating operations:			Incinerator performance (if used) for all affected facilities	
Prime coat operation	VOC—0.16 kg/l of applied coating solids			
Guide coat operation	VOC—1.40 kg/l of applied coating solids			
Top coat operation	VOC—1.47 kg/l of applied coating solids			
Phosphate rock plants:			Opacity—continuous except if scrubber is used	(a) Applies only to plants > 4 tons/hr capacity
Dryers	Particulate—0.06 lb/ton of phosphate rock feed Opacity—10%			

Table 8 (*Continued*)

Source Category and Affected Facility	Emission Limits	Monitoring Requirements	Comments
Calciners, unbeneficiated rock	Particulate—0.23 lb/ton of phosphate rock feed Opacity—10%	Scrubber—continuous	(b) Monitoring requirements do not apply to ground rock handling and storage
Calciners, beneficiated rock	Particulate—0.11 lb/ton of phosphate rock feed Opacity—10%		
Grinders	Particulate—0.012 lb/ton of phosphate rock feed Opacity—0%		
Ground rock handling and storage	Opacity—0%		
Ammonium sulfate manufacturing: Dryer	Particulate—0.30 lb/ton of product Opacity—15%	Feed material input—continuous Air pollution control device pressure drop—continuous	(a) Applies to the caprolactam by-product, synthetic, and coke oven by-product sectors of the industry only

Table 9 Summary of Sources Subject to NESHAP Requirements

Pollutant	Sources and Activities	Emission Limitations
Asbestos	Asbestos mills	0% opacity
	Roadway surfacing	0% opacity
	Manufacturing—textiles; cement products; fireproofing and insulating materials; friction products; paper, millboard, and felt; floor tile; paints, coatings, caulks, adhesives, and sealants; plastics and rubber materials; chlorine; shotgun shells; asphalt concrete	0% opacity
	Demolition and renovation	0% opacity
	Spraying	0% opacity
	Fabricating—cement building products; friction products; cement or silicate board	0% opacity
	Insulating	
	Waste disposal activity	0% opacity
	Waste disposal sites	0% opacity
Beryllium	Manufacturing—extraction plants; ceramic plants; foundries; incinerators; propellant plants; machine shops that process any metal containing more than 5% beryllium by weight	Not to exceed 10 g/24-hr period
	Rocket motor testing	Not to exceed 2 g/hr nor 10 g/day (also mandatory ambient air concentration limits)
Mercury	Mercury ore processing	Not to exceed 2300 g/24-hr period
	Chlorine gas and alkali metal hydroxide manufacture by use of mercury chlor-alkali cells	Not to exceed 2300 g/24-hr period
	Incinerating or drying wastewater treatment plant sludge	Not to exceed 3200 g/24-hr period
Vinyl chloride	Ethylene dichloride plant:	
	Purification process	Not to exceed 10 ppm
	Oxychlorination reactor	Not to exceed 0.2 g/kg of product
	Vinyl chloride plant	Not to exceed 10 ppm
	Polyvinyl chloride plant:	
	Reactor exhaust	Not to exceed 10 ppm
	Reactor opening	Not to exceed 0.02 g/kg of product
	Manual vent valve discharge	Emergency discharges only
	Stripper	Not to exceed 10 ppm
	Prestripping activity	Not to exceed 10 ppm
	Monomer recovery system exhaust	Not to exceed 10 ppm
	Poststripping activity	2 g/kg of product for dispersion resins (excluding latex resins) 0.4 g/kg of product for all other resins including latex resins
	Residual vinyl chloride in product	Not to exceed 2000 ppm (weighted average) for dispersion resins (excluding latex resins); not to exceed 400 ppm for all other resins including latex resins

which to base a judgment. Although construction or modification activities are initially approved, two further notices are required prior to startup as follows:

Notice	Due Date
Anticipated date of initial startup	Not > 60 nor < 30 days prior to anticipated date
Actual date of initial startup	Within 15 days of actual date

With respect to a modification, note that any change that results in any increase in emissions of hazardous pollutants, however small, is deemed a modification. Although de minimis values apply to these pollutants for purposes of PSD (see Section 4.1 infra), there is no minimal emission rate applied to NESHAP regulations. Thus all modifications, no matter what the quantity of hazardous pollutant emissions increase may be, must undergo NESHAP review.

The NESHAP testing and reporting requirements vary from standard to standard. Some standards require continuous testing, monitoring, or reporting. Demolition activity requires, in effect, a permit for each individual project. These requirements plus the various design and operation standards are too voluminous to discuss fully here. Owners or operators of sources subject to NESHAP regulations should read carefully the specific standard that applies to their process.

4 ADMINISTRATIVE REGULATIONS

Administrative regulations are the heart of the air pollution control implementation program. These regulations require information gathering, specify minimum requirements for tax relief, and direct the growth of major sources of pollution in the country. Although reporting and testing requirements apply to mobile sources, this section will focus on the application of the EPA's principal administrative regulations to stationary emission sources.

There are two types of stationary sources—direct and indirect. A direct source is one that, itself, generates the pollutants emitted. Indirect sources do not generate pollutants themselves, but cause an increase in emissions owing to the vehicular traffic that the source encourages. Examples of indirect sources are: shopping centers, amusement parks, and football stadiums.

There are nine major areas on which EPA's administrative regulations focus. These are shown in Table 10 as they apply to the various types of stationary sources. Not discussed here are the various administrative requirements (testing, monitoring, and reporting) that may apply to existing sources permitted under SIP regulations and those new, minor emission sources that are exempt from PSD or nonattainment review. These sources are, or will be, subject to a variety of regulations adopted by state or local authorities.

4.1 Prevention of Significant Deterioration and Nonattainment Regulations

These regulations are probably the most complex of the federal air pollution regulations. They address the preconstruction review of major new or modified emission sources, with an emphasis on ambient air quality in each AQCR. In a nutshell, with respect to a given pollutant, there are two kinds of

Table 10 Major Administrative Regulations Applicable to Stationary Sources

Regulatory Area	Existing Direct Source	Type of Stationary Source	
		New Direct Source	New Indirect Source
Prevention of Significant Deterioration (PSD)		✓	
Nonattainment (NA)		✓	
Bubble concept		✓	
Indirect sources			✓
Source Testing	✓	✓	
Ambient air monitoring		✓	✓
Visibility	✓	✓	
Noncompliance penalties	✓		
Tax considerations	✓	✓	

AQCRs: those having ambient air quality better than the applicable NAAQS (these are commonly called "clean areas") and those having ambient air concentrations in excess of the NAAQS (these are commonly called "nonattainment areas"). The thrust of the Prevention of Significant Deterioration (PSD) and Nonattainment (NA) regulations is to preserve the pristine character of the clean areas and ameliorate the poor air quality conditions that exist in the nonattainment areas, while allowing industrial growth. Thus a major new source of pollutants may locate in a clean area only upon showing that its emissions will not violate prescribed ambient air standards; it may locate in a nonattainment area only upon assuring that its emissions will be more than offset by emission reduction measures taken on other facilities.

PSD regulations have been promulgated for particulate matter and sulfur dioxide emissions only. EPA expects to promulgate regulations for the Set II pollutants (carbon monoxide, nitrogen dioxide, ozone, volatile organic compounds, and lead) in 1982. These regulations will be known as the "PSD Set II" regulations. The following discussion is limited to the Set I PSD regulations.

Much controversy has attended this regulatory scheme and the final PSD Set I regulations were molded by litigation, *Alabama Power Co.* v. *Costle,* 13 ERC 1225, 1979. Since there is a great deal of similarity between the PSD and NA regulations, both will be discussed in this subsection.

At the outset, it must be said that the applicable SIP or federal regulations [40 CFR 52.21 (PSD) and 40 CFR 52.24 (NA)] should be studied carefully when a specific facility is being considered. Although PSD and NA requirements are a part of each state's SIP, federal regulations dictate the minimum SIP requirements and most SIPs will follow the federal guidance. Therefore this discussion is based on the federal regulations. That the regulations are complex is seen in the fact that the preamble to the final amendments to 40 CFR Parts 51 and 52 covered more than 50 pages of the *Federal Register* (45 FR 52676–52729). The following discussion is outlined as follows:

1. General overview of the program
2. Sources subject to PSD and NA review
3. What constitutes a source
4. Major stationary source
5. Reconstruction
6. Major modification
7. PSD requirements
8. NA requirements

4.1.1   *General Overview of the Program*

Each Air Quality Control Region (AQCR)† is classified as Class I, II, or III with respect to particulate matter and sulfur dioxide. Congress designated all AQCRs as Class II [CAA §162 (b)] except for international parks, national wilderness areas, and memorial parks which exceed 5000 acres in size, national parks which exceed 6000 acres in size, and areas designated as Class I prior to the 1977 Clean Air Act Amendments. In general, all Class II areas may be redesignated as Class I or Class III by the states and Indian tribes. The CAA sets specific air quality deterioration increments for particulate matter and sulfur dioxide for each class. These increments are shown in Table 11. Major

Table 11   Allowable PSD Air Quality Increments

Pollutant/Time Period	Air Quality Increment over the Baseline,[a] $\mu g/m3$		
	Class I	Class II	Class III
Particulate Matter			
Annual geometric mean	5	19	37
24-hr maximum	10	37	75
Sulfur Dioxide			
Annual arithmetic mean	2	20	40
24-hr maximum	5	91	182
3-hr maximum	25	512	700

[a] Not to exceed the NAAQS.

† Some AQCRs are broken into subareas, each subarea having its own classification. See Section 2.2 supra for a discussion of air quality control regions.

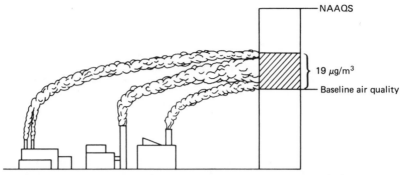

Fig. 12   Graphic portrayal of increment use in a Class II area—particulate matter.

new sources of air pollution may not impact beyond these limits and, in no case may a NAAQS be violated. All Class I, II, and III areas are clean areas. When the increment is used up, no more major new sources are permitted. Figure 12 graphically portrays this scheme for a Class II area. Note that if the baseline air quality is too high, not all the increment is available.

Many areas have not achieved the NAAQS. These are the nonattainment areas. Since there is no air quality increment available for growth, major sources wishing to locate in these areas must provide for emission reduction greater than one for one. This is the "emission offset." Through this means, air quality will eventually meet the NAAQS while, it is hoped, growth is permitted.

Not only new sources are subject to PSD and NA review.* Modifications to existing sources and, in the case of nonattainment areas, reconstruction of existing sources (notwithstanding the latter might not result in an emissions increase) are also subject to PSD or NA review. The overall regulatory scheme for PSD and NA review is shown in Figure 13.

4.1.2   Sources Subject to PSD and NA Review

Only new major stationary sources (MSS) or major modifications are subject to PSD or NA review. In addition, reconstruction of an existing major stationary source will cause that source to be treated as a new major stationary source for NA review purposes. It is easiest to think of three types of sources:

Source Type	Review Requirements
New major stationary source	PSD or NA
Major modification	PSD or NA
Reconstruction of existing major stationary source	NA only

In general, only sources that have potential to emit† 250 tons/yr of any pollutant governed by the Clean Air Act (or 100 tons/yr for certain designated sources) are subject to PSD review requirements and sources that have the potential to emit 100 tons/yr are subject to NA review requirements. In an effort to reduce the regulatory workload, existing major stationary sources that are modified are considered as major modifications only if the modification causes a substantial increase in emissions.

If the source is a major emitter, how does one determine whether the PSD or NA regulations apply? Remember, a source may emit more than one pollutant and an AQCR may be designated as attainment for some pollutants and nonattainment for others. At one point, some of the regional offices of EPA considered a source subject to PSD review even if the source did not substantially emit any pollutant for which the area was designated as an attainment area, on the basis that the AQCR was, after all, an attainment area with respect to *some* pollutant. EPA has clarified its stance in this regard. Now, a major stationary source must undergo PSD review with respect to the "attainment"

---

* NA review is also known as "new source review" and is often abbreviated NSR.
† Potential to emit is determined after the control device. If the source is one that is regulated under §111 (NSPS) or §112 (NESHAP), potential emissions include all fugitive emissions as well.

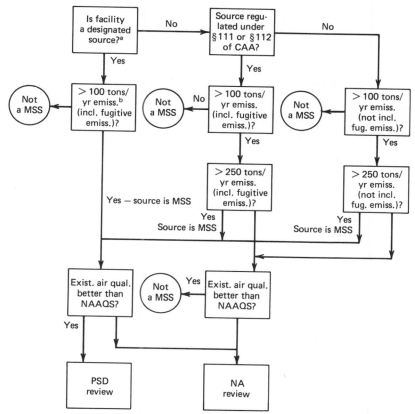

Fig. 13 PSD and NA review criteria. MSS = major stationary source. *a* See 52.21(b)(1)(i)(a) or 52.24(h) for a listing of designated sources. *b* Fugitive emissions not applicable to steam electric power plants.

pollutants it emits. If it also, or only, emits "nonattainment" pollutants, it undergoes NA review with respect to those pollutants.

4.1.3 What Constitutes a Source?

Source, as used here, means "stationary source." "Stationary source," for purposes of PSD, means all the pollutant-emitting activities that belong to the same industrial grouping, contiguously located, and operated by the same person. In short, groupings are by the first two digits of the Standard Industrial Classification (SIC) Code, [40 CFR 52.21(b)(6)]. Further, for purposes of NA review, stationary sources *also* include individual, identifiable pieces of process equipment. This is critical when considering reconstructions, which are discussed later in this subsection.

4.1.4 Major Stationary Sources

Whether a source is deemed to be a major stationary source subject to PSD review consideration depends on its potential to emit any pollutant subject to regulation under the Clean Air Act. Any source that has the potential to emit 250 tons/yr of any pollutant from its stack is a major stationary source. Sources subject to §111 (NSPS) or §112 (NESHAP) of the Clean Air Act are major stationary sources if potential emissions *including* fugitive emissions are 250 tons, or more, per year. Fossil fuel-fired steam electric plants (>250 mmBtuhr heat input) having potential emissions greater than 100 tons/yr are also major stationary sources. Also if a source is one of the 27 sources shown in Table 12, it is a major stationary source if its emission potential, *including* fugitive emissions, is 100 tons/yr. Once it has been determined that a source is "major" for PSD review consideration, determine if the area is "attainment" for the major pollutant.

Table 12 Designated Sources for PSD Review Consideration

Coal-cleaning plants (thermal dryers)
Kraft pulp mills
Portland cement plants
Primary zinc smelters
Iron and steel mill plants
Primary aluminum ore reduction plants
Primary copper smelters
Municipal incinerators capable of charging more than 250 tons refuse per day
Hydrofluoric acid plants
Nitric acid plants
Petroleum refineries
Lime plants
Phosphate rock processing plants
Coke-oven batteries
Sulfur recovery plants
Carbon-black plants (furnace process)
Primary lead smelters
Fuel conversion plants
Sintering plants
Secondary metal production facilities
Chemical process plants
Fossil fuel boilers of more than 250 million Btuhr heat input
Petroleum storage and transfer facilities w/capacity exceeding 300,000 bbl
Taconite ore processing facilities
Glass–fiber processing plants
Charcoal production facilities

If the source is located in a nonattainment area, it will be subject to NA review and *not* PSD review. What other sources are subject to NA review? Any source that emits from its stack 100 tons/yr of a regulated pollutant is a major stationary source subject to NA review. Also, sources subject to §111 or §112 are major stationary sources if potential emissions equal or exceed 100 tons/yr *including* its fugitive emissions. Note in calculating potential emissions, each pollutant is calculated separately. The reader is referred to Figure 13 for a graphic explanation of the above discussion.

4.1.5 Reconstruction

Reconstructed sources are *not* the same as modified sources. A reconstructed source is actually considered to be a new source. One might think of it as a worn out facility that has been replaced by the new (reconstructed) facility. Thus, for a source to be considered a reconstruction, it is not necessary that emissions be increased. Indeed a source may be deemed reconstructed even though emissions diminish. Can a reconstructed source be subject to PSD review? No. However, a reconstructed source may be deemed a major stationary source for NA review.

What determines whether a source is a reconstruction? Basically a reconstructed facility is considered as new. Thus a reconstruction is presumed to have taken place where the fixed capital cost of the new components exceeds 50% of the fixed capital cost of a comparable entirely new facility. Note that, for NA review purposes, a stationary source may mean not only an entire production process, but individual, identifiable pieces of process equipment as well. Thus substantial remodeling of a grain dryer at an existing terminal elevator may trigger an NA review, with its requirement for offsets.

Whether a source has been reconstructed is an accounting question in that "fixed capital cost" means the capital needed to provide all the depreciable components. Final decisions whether a reconstruction has occurred are made under 40 CFR 60.15(f)(1)–(3). This is discussed more fully under Section 3.3.1 on New Source Performance Standards.

4.1.6 Major Modification

Both the PSD and NA regulations apply to major modifications to existing major stationary sources. A major modification is any physical or operational change that results in a significant *net* emissions increase. Unlike the emission criterion for new sources, for purposes of major modifications significant increases are generally less than 100 tons/yr. The levels of emission deemed insignificant are referred to a *de minimis* in the regulations. The de minimis level of sulfur dioxide, for example, is 40 tons/yr. This means that if, in modifying a major stationary source, potential sulfur dioxide emissions will increase by less than 40 tons/yr, the emissions are "insignificant" and the modification is subject to neither PSD nor NA review. Again, bear in mind that "emissions" means potential emissions after the control device.

Note that the additional emissions that result from the modification are not added to the previous emissions to determine whether the source is a major stationary source. Two things are necessary for a major modification:

1. The facility is already a major stationary source.
2. The modification must result in a significant *net* increase in emissions.

Of course, the modification might constitute by itself a major stationary source. The process for determining whether a project is a major modification is outlined in Figure 14.

Finally, notice that the emphasis is on *net* increases in emissions. This definition employs a bubble concept, and any other increases or decreases in emissions are added or deducted to determine whether the change is a major modification. See Section 4.2 for a discussion of the bubble concept and emission reduction credits.

4.1.7 *PSD Requirements*

PSD permit applications require a large amount of information covering such things as existing ambient air quality and an analysis of the proposed facility's impact on ambient air quality, visibility, soil, and vegetation. In addition, postconstruction ambient air monitoring may be required and, if the source will have an impact on nonattainment areas or federal Class I areas, substantial other information may be required. There are certain exemptions available, however, and these are discussed below.

Each application must include sufficient information to analyze the applicant's air quality impact analysis. This includes design capacity, operating schedule, design drawings and specifications, construction schedule, and a detailed description of the proposed emission reduction system. Other data that must be included, unless the source is exempt from such requirements, are:

1. A showing that all existing emission sources at the location are in compliance with the applicable SIP, NSPS, or NESHAP regulations.
2. A showing that the proposed control system for the source/pollutant under PDS review represents BACT.
3. A showing that the increased emissions will not violate the applicable NAAQS nor the increment.
4. Continuous air quality monitoring data relative to the pollutants of concern. These data must cover the preceding 12 months, or lesser period (a minimum of four months) as the administrator may determine.
5. A plan for postconstruction air quality monitoring.
6. An analysis of the impairment to visibility, soils, and vegetation having significant commercial or recreational value.

Certain sources are exempt from some of the above items. The following sources are exempt from items (3), (4), (5), and (6):

1. Sources located in a Class II area.
2. Sources in existence prior to March 1, 1978.
3. Sources where the net increase in each pollutant, after application of BACT, is less than 50 tons/yr.

In addition, the administrator may exempt a new source or a modification of an existing source from items (4) and (5) above, if the new emissions will have ambient impacts less than those shown in Table 13, or if the existing concentrations of the pollutant in the affected area are less than those shown in Table 13, or if the pollutant is not listed in Table 13.

If the source will adversely affect the air quality of a nonattainment area, it is likely that four additional requirements must be met.

1. The source must meet an emission limitation equal to LAER.
2. The applicant must provide proof that all its existing major sources located within the state are in compliance with the applicable SIP, NSPS, or NESHAP.
3. An emissions offset must be obtained.
4. The applicant must show that the offset will ameliorate air quality in the nonattainment area.

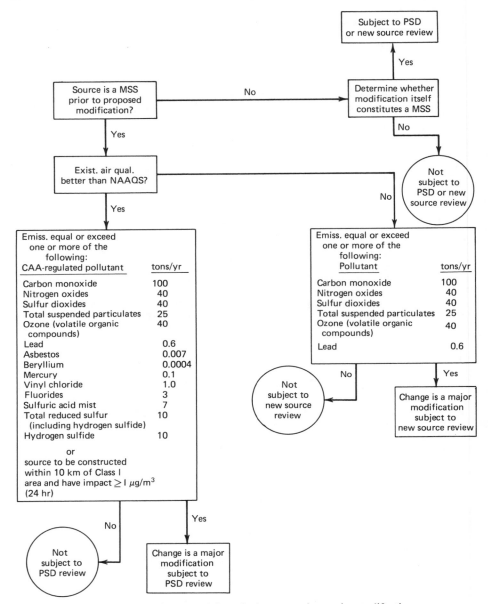

Fig. 14 Process for determining whether source is a major modification.

Obtaining a PSD permit is likely to take more than a year. First, as previously stated, ambient air data may be required covering a period up to 1 yr prior to the permit application. Once the application is made, the administrator must notify the applicant within 30 days of any deficiencies. Upon receipt of a complete permit application, the administrator has up to 1 yr in which to make a final decision. During that period the administrator must prepare a draft permit or a draft denial along with a statement of basis, or a fact sheet, that provides the information on which the preliminary decision was made. The administrator may, at his discretion, call for a public hearing. If the administrator does not call for a public hearing, a 30-day period is allowed for public comments, at which time any interested person may request a public hearing. Such request must state the nature of the issues proposed to be raised in the hearing. A 30-day notice of the public hearing is required. In the absence of a public hearing, the applicant has a 10-day period following the close of public comments to

Table 13   Maximum Ambient Pollutant Concentrations Applicable to Certain Exemptions to PSD Permit Requirements

Carbon monoxide	575 $\mu g/m^3$, 8-hr average
Nitrogen dioxide	14 $\mu g/m^3$, annual average
Total suspended particulate	10 $\mu g/m^3$, 24-hr average
Sulfur dioxide	13 $\mu g/m^3$, 24-hr average
Lead	0.1 $\mu g/m^3$, 24-hr average
Mercury	0.25 $\mu g/m^3$, 24-hr average
Beryllium	0.0005 $\mu g/m^3$, 24-hr average
Fluorides	0.25 $\mu g/m^3$, 24-hr average
Vinyl chloride	15 $\mu g/m^3$, 24-hr average
Total reduced sulfur	10 $\mu g/m^3$, 1-hr average
Hydrogen sulfide	0.04 $\mu g/m^3$, 1-hr average
Reduced sulfur compounds	10 $\mu g/m^3$, 1-hr average

address the issues raised. The public comment period may be lengthened for good reason and, if new, unforeseen issues arise after the close of the public comment period, it may be reopened. The intent of these regulations is to provide all the relevant background information available for the administrator's decision. A schematic diagram of the PSD application process is shown in Figure 15.

4.1.8   NA Requirements

New or modified major sources must comply with four conditions with respect to nonattainment pollutants:

1. The source must meet an emission limitation equal to LAER.
2. The applicant must provide proof that all its existing major sources located within the state are in compliance with the applicable SIP, NSPS, or NESHAP.

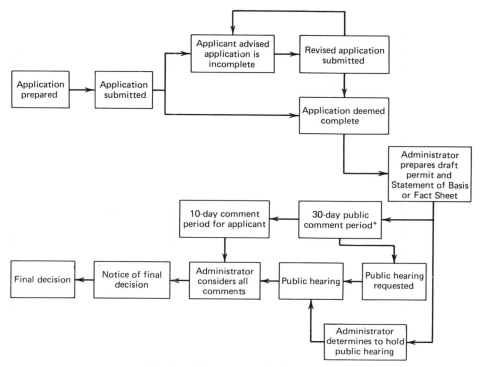

Fig. 15   PSD permit application process.

3. A greater than one-for-one emission offset must be obtained.

4. The applicant must show that the offset will ameliorate air quality.

Unlike PSD applications, no previous ambient air quality data are required, nor is postconstruction monitoring required. There are limited exemptions for conditions (3) and (4). These are available to resource recovery facilities burning municipal solid wastes and to certain sources that are required to switch fuels, provided:

1. Best efforts were made to obtain the required offsets and such efforts were unsuccessful.

2. All available offsets were secured.

3. The applicant will continue to seek the required offsets and will secure them as they become available.

4.2 Emissions Trading

4.2.1 The Bubble Concept

The bubble concept allows for the most cost-effective program of emission reduction. Under this concept, all the sources of a particular pollutant located in a defined area are enveloped by an imaginary bubble and the entire bubble is then treated as one source. Since process variables generally make it more expensive to control emissions from one process relative to another, the control cost curves assume different shapes. This is shown graphically in Figure 16. Assume the following with respect to these emission sources.

Source	Uncontrolled Emissions, tons/yr	Allowable Emissions, tons/yr	Control Efficiency, %
A	200	20	90
B	500	25	95
	700	45	

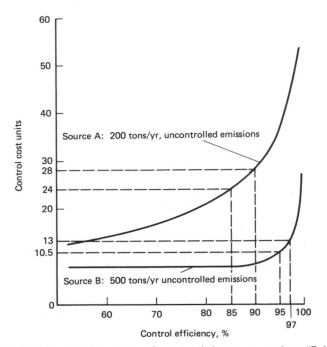

Fig. 16 Relative control cost curves for two emission sources under a "Bubble."

The relative control cost under these conditions is 38.5 units. Now suppose only 85% of source A's emissions are removed while the removal efficiency of source B is increased to 97%. Total emissions are still 45 tons/yr, but control costs drop to 37 units, a savings of 1.5 units. Bubbles can be placed over large geographical areas and can even include the facilities of different owners. All that is necessary is a demonstration that the surplus emission reduction is federally enforceable and that the impact on ambient air quality is at least equivalent to the nonbubbled scenario. As long as processes exist that exhibit relatively flat cost curves, the bubble concept can be advantageous.

Rather than adding pollution control equipment, one can also reduce emissions by ceasing operations. This, too, is a bona fide emission reduction method under a bubble concept. The primary thrust is to improve ambient air quality.

4.2.2 EPA's Emissions Trading Policy

Emissions trading is based on the bubble concept. This is a policy (not a regulation) and therefore will not be found in the Code of Federal Regulations. It is to be found only in the *Federal Register*, 47 FR 15076, April 7, 1982. Although the Emissions Trading Policy (ETP) is not a regulation, it is a very important and useful concept. It is extremely flexible, and, as a practical matter, is only limited by air quality: trading cannot result in substandard air quality at any location. This puts some restriction on the geographical size of the area in which the trading partners are located. Emissions trading consists of bubbles, netting, emission offsets, and emission reduction banking. In short, each of these allows that excess emission reduction for one source may be used to meet prescribed requirements for another source.

The bubble lets *existing* plants or groups of plants meet applicable SIP emission limits as if the bubbled emission sources were one. This gives plant owners the flexibility to apply the most efficient controls to the easiest or most cost-effective sources. Credit can also be obtained by terminating operation of a facility. Cost advantages of using the bubble are great and even reduction in fugitive emissions can be credited against stationary source emissions. There is no geographical restriction on the use of bubbles. They may be used in nonattainment areas as well as attainment areas.

Netting applies to existing plants that wish to expand or modernize. If a plant can show that plantwide emission increase due to the proposed activity are insignificant, the new source review requirements of the Prevention of Significant Deterioration regulations and the Nonattainment regulations can be avoided. As to what are deemed "insignificant" emission increases, see Section 4.1.6, supra. In no case may emissions be "netted" to avoid NSPS requirements for a new facility, however.

Emission offsets are required by regulation (40 CFR 51.18) for new or modified sources in nonattainment areas. They are, however, addressed in the ETP also.

Emission Reduction Banking allows firms to bank qualified emission reductions for future use in bubble, netting, or offset transactions. The "currency" is called Emission Reduction Credits (ERCs). These may actually be sold to other firms and used in accordance with the applicable bubble, netting, and offset requirements.

Only those emission reductions that are surplus, enforceable, permanent, and quantifiable may be converted to ERCs. Surplus emissions are those not currently required by law or regulation. This is established by each state in that they must determine the baseline against which "surplus" will be measured. There are several requirements stated in the Emissions Trading Policy. These address the baseline determination in both attainment and nonattainment areas. One of the simplest explanations as to how this baseline is used follows. Suppose plant A has reported uncontrolled emissions of 500 tons/yr, both the plant and the state agency being in agreement. Further, suppose the two parties agreed to an emission reduction of 96%. As such, the state used the balance, 20 tons/yr, in its uncontrolled emission inventory to demonstrate attainment of the applicable NAAQS by reducing total AQCR-wide controlled emissions to, say, 1470 tons/yr. Now at some future date if plant A wishes to close its facility, the ERC is limited to 20 tons/yr. Even if it is shown that uncontrolled emissions are actually 600 tons/yr *not* the 500 tons initially used in the state's emission inventory, plant A cannot obtain credit for 24 tons [$600 \times (1 - 0.96)$].

ERCs may be used by sources in bubble, netting, and offset transactions. Their use is limited by the attainment and maintenance of ambient air quality. The following uses and constraints apply to ERCs.

1. Emission trades must involve the same criteria pollutant.
2. All uses of ERCs must satisfy applicable ambient air tests.
3. Emission trades may not increase hazardous pollutants.
4. Emission trades may not be used to meet applicable technology-based requirements.
5. Bubbles may be used in nonattainment areas.
6. Bubbles may be used to meet SIP requirements.
7. Bubbles may involve open dust sources (such as roadways).

4.3 Indirect Sources

Indirect sources are facilities or operations that may attract mobile source activity that results in emissions of a pollutant for which there is a national standard. Included are such things as shopping centers, amusement parks, industrial parks, schools, highways, and airports. Whether construction or modification of an indirect source is subject to EPA approval depends on its size and location. Table 14 lists the conditions under which indirect sources are subject to preconstruction approval.

As a minimum, an application for approval of indirect sources having associated parking facilities must include the following information:

1. Name and address of the applicant.
2. A topographic map of the project area.
3. Description of use, activities, and operating schedule.
4. Site plan including traffic patterns and location and height of buildings.
5. Identification of principal roads, highways, and intersections, and their capacities, within ¼ mile of the indirect source.
6. Traffic volume estimates: daily average; maximum 1-hr; maximum 8-hr.
7. Availability of existing or proposed mass transit service.

Airports and highway projects require additional information specific to those types of activity.

It takes a minimum of 4 months to have an indirect source permit application processed. Once the application is submitted, the administrator must advise the applicant within 20 days whether additional information is needed. After the application is complete, a preliminary decision must be made to either approve, approve with conditions, or disapprove the project. Then a 30-day period is allowed for public comment followed by a 10-day period for written responses by the applicant. The final determination is made within 30 days thereafter. The indirect source permit application process is shown in Figure 17.

Whether the project is approved is determined by evaluating the anticipated concentration of carbon monoxide at exposure sites that will be affected by the mobile source activity expected to be attracted by the indirect source. For indirect sources other than highway projects and airports, approval will not be granted if the project will cause a violation of a SIP control strategy or cause or exacerbate a violation of the national ambient air standard for carbon monoxide at any receptor point. In making the determination, the administrator will rely on specified atmospheric diffusion modeling and/or any other reliable analytic method. Although it is not required, applicants are permitted to submit the

Table 14   Indirect Sources of Air Pollution Subject to EPA Approval (40 CFR 52.22)

Project	Geographic Location	
	Inside SMSA	Outside SMSA
Parking facilities[a]		
New	1000 or more parking spaces	2000 or more parking spaces
Modified	Increase of 500 or more parking spaces	Increase of 1000 or more parking spaces
Highway projects		
New	20,000 vehicles per day within 10 yr	—
Modified	10,000 vehicles per day within 10 yr	—
Airports		
New	—	Additional capacity of 50,000 operations[b]/yr or 1.6 million passengers/yr within 10 yr
Modified	—	

[a] Includes associated parking areas for all indirect sources.
[b] Operation means a takeoff or a landing.

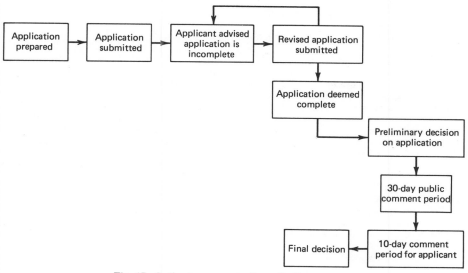

Fig. 17 Indirect source permit application process.

results of any modeling or other analytic methods and supporting verification. When this is done, the administrator must consider the applicant's favorable data before making a final determination.

4.4 Ambient Air Monitoring and Source Testing

Federal regulations have been promulgated for monitoring ambient air quality and testing sources of emissions. Those regulations related to ambient air monitoring are found in 40 CFR Parts 50, 53, and 58. The basic methods for sampling and analyzing the ambient air for air pollutants are specified in Appendixes to 40 CFR 50. These are:

Appendix	Pollutant
A	Sulfur dioxide
B	Particulate matter
C	Carbon monoxide
D	Ozone
E	Hydrocarbons
F	Nitrogen dioxide
G	Lead

These regulations delineate the sensitivity of each method and precision, accuracy, and stability. They also specify the reagents, equipment, analysis, and procedures to be used. In addition, calibration and calculation procedures are given and the literature references on which the method is based are given.

Only certified methods and equipment may be used for the collection and analysis of ambient air samples. The regulations found in 40 CFR 53 specify the manner in which one may apply for certification of a test method and the procedures for testing the method's performance characteristics. Applications must include:

1. A full description of the method consisting of the measurement principle, manufacturer's name, model number, listing of significant components, schematic diagrams, and a detailed description of the apparatus and measurement procedures.

2. An instruction manual providing operational and calibration procedures, safety information, maintenance and trouble-shooting procedures, and a parts list.

3. Test results including test data, records, and calculations as required by subparts B and C of 40 CFR 53.

4. A detailed quality control program for the test method.

If the test method is certified, the manufacturer: (1) must affix a label to each unit specifying the range or ranges for which the unit is certified; (2) must guarantee the unit will function properly for 1 yr; (3) must maintain records of all purchasers of the equipment; and (4) in case the designation is cancelled by EPA, must notify all purchasers of that fact.

Monitoring criteria including instrument siting, monitoring methods, operating schedules, and quality assurance procedures are specified in 40 CFR 58. These regulations apply to state and local control agencies that operate state or local air monitoring stations (SLAMS), sources or operators of sources that may operate an ambient monitoring system, and owners or operators of stations generating information for PSD purposes. In addition to providing complete specifications for the design and operation of monitoring systems, these regulations require systematic audits of the monitoring stations by EPA-certified individuals and equipment.

Test methods for stationary sources of emissions, other than hazardous pollutants, are specified in 40 CFR 60. These test methods, shown in Table 15, apply not only to sources subject to NSPS but all other sources of the designated pollutants as well. Similarly, test methods are specified in 40 CFR 61 for hazardous pollutants and in 40 CFR 87 for aircraft emissions. These regulations specify the methods of sampling, analysis, and reporting.

4.5 Visibility Regulations

These regulations are required by Section 169A of the Clean Air Act to protect and remedy visibility in Class I federal areas. Although visibility is addressed in the PSD regulations vis-a-vis Class I federal areas, its consideration under these regulations is much more extensive. This subject is so complex however, that technical and scientific limitations have prevented complete development of the regulations. Consequently EPA is promulgating the visibility regulations in phases. Phase I is discussed below.

Table 15   EPA Test Methods for Nonhazardous Pollutants from Stationary Sources

Method Number[a]	Subject Matter
1	Sample and velocity traversing
2	Determination of gas velocity and flow rate
3	Analysis for carbon monoxide, oxygen, excess air, and dry molecular weight
4	Determination of moisture content of gas
5	Determination of particulate emissions
6	Determination of sulfur dioxide emissions
7	Determination of nitrogen oxide emissions
8	Determination of sulfuric acid mist and sulfur dioxide emissions
9	Determination of opacity of emissions
10	Determination of carbon monoxide emissions
11	Determination of hydrogen sulfide content of fuel gas streams in refineries
13	Determination of total fluoride emissions
14	Determination of fluoride emissions from primary aluminum plants
15	Determination of hydrogen sulfide, carbonyl sulfide, and carbon disulfide emissions
16	Determination of sulfur emissions (semicontinuous)
17	Determination of particulate emissions (in-stack filtration)
19	Determination of sulfur dioxide removal efficiency and particulate, sulfur dioxide, and nitrogen oxide emissions (electric utilities)
20	Determination of nitrogen oxide, sulfur dioxide, and oxygen emissions (gas turbines)
24	Determination of volatile matter content and other properties of surface coatings

[a] There are no methods designated 12, 18, 21, 22, or 23.

Since these regulations are meant to remedy as well as protect visibility, they apply to both existing sources and new sources. These regulations are layered on all the other requirements of the various air pollution regulations and, indeed, can even impose additional pollution control burdens on existing sources that are properly operating under a previously obtained permit. Eventually visibility regulations may affect sources of emissions even smaller than those classified as "major" under the PSD regulations.

Phase I focuses on:

1. Remedying visibility impairment caused by existing major sources and small groups of sources.
2. Protecting visibility from impairment by new major sources.
3. Protecting and remedying visibility through state-implemented long-range plans that may affect small stationary sources, area sources, and even mobile sources.

4.5.1 *Existing Sources*

Major emission sources that impair visibility in Class I federal areas are identified. These sources are required to employ best available retrofit technology (BART). If BART is more efficient than the source's existing control system, the retrofit must be made. BART is defined on a case-by-case basis considering costs, energy usage, and other environmental effects of the more efficient control system. Under Phase I, very few existing sources are affected. EPA initially estimated that 12 sources would be required to employ BART. However, total costs of BART for these sources were expected to range from $25 to $125 million annually.

4.5.2 *New Sources*

PSD regulations require that the appropriate federal land manager be notified regarding the application of any major source that has an impact on a Class I federal area. The visibility regulations make it clear, however, that the notification is triggered not only by an adverse air quality impact. Since the visibility analysis includes consideration of cost, energy, and other environmental impacts, the applicant should include such an analysis with the application.

4.5.3 *Long-Range Plans*

EPA's visibility regulations require each of the 36 states that contain Class I federal areas to adopt long-range, 10–15 yr strategies for remedying existing, and preventing future, visibility impairment. These strategies must show promise of reasonable progress toward the visibility goal. None of the states have revised their SIPs as of this writing, but it can be expected that existing emission sources of all types, including mobile sources, will be affected.

4.6 Noncompliance Penalties

4.6.1 *General*

Congress has mandated that EPA assess penalties to noncomplying major sources of air emissions in an amount at least equal to the economic advantage of noncompliance. Regulations to effect this program are promulgated under §120 of the Clean Air Act. Affected are not only those sources that have never achieved compliance, but also those under consent decrees and those that have failed to continue complying. There are a limited number of exemptions available, falling into the following general categories:

1. Certain coal-burning facilities.
2. Facilities using innovative technology pursuant to a compliance order.
3. Impossibility of compliance through no fault of the source, and if the source is under certain compliance orders issued under §113 of the CAA.
4. Sources under an approved employment or energy emergency order.
5. Noncompliance is de minimis in nature and effect.

Within 45 days of receipt of the notice of noncompliance, the firm must:

1. Calculate the penalty and payment schedule and transmit the calculations and all supporting data to EPA, or
2. Submit a petition for reconsideration, alleging that the source is either in compliance or that an exemption applies.

If a petition for reconsideration is submitted, it may be amended anytime within the remainder of the initial 45-day period. After that, amendments are permitted only if based on unforeseeable conditions occurring after termination of the 45-day period or as otherwise permitted by the administrator of EPA.

Within 30 days after receiving a petition for reconsideration the administrator must notify the petitioner of one of the following:

1. Part or all of the requested relief is granted.
2. A hearing on the petition is denied.
3. A hearing is granted.
4. The information submitted is inadequate to determine that the petitioner is entitled to the relief requested.

If the information is inadequate, the administrator will specify the deficiencies and the petitioner has 30 days to rectify his petition.

In the event no petition for reconsideration is filed, and the source owner or operator submits a penalty calculation, the administrator will, within 30 days:

1. Accept the calculation subject to modification after compliance is achieved;
2. Reject the calculation and recalculate the penalty; or
3. Notify the owner or operator that there are deficiencies in the calculation.

In case of the latter, such information must be submitted within 30 days. If the penalty is recalculated by EPA, a petition for reconsideration must be submitted within 45 days.

4.6.2 *Illustration of Noncompliance Penalty Calculation*

Since the penalty is to reflect the present value of the net economic benefit derived from violation of an applicable air pollution control requirement, the impact of such things as control system operating costs, inflation, investment tax credits, depreciation, and the time value or opportunity cost of money is considered. For example, assume you have $100 in a savings account at a 5.5% annual interest rate. It will earn $5.50 over the year. However, if you are in the 39% tax bracket, the net benefit is only $3.36 and your opportunity cost is 3.36%. This is a simple example of the time value of money, taking taxes into account. Assume further that you are assessed $100 by your club or lodge at the beginning of the year, but intend to withhold payment until the end of the year. Although you will eventually comply with the assessment requirements, the net economic benefit of the late compliance is $3.36—you will be $3.36 better off than if you had paid the assessment at the outset—and the *present value* of that $3.36 is $3.25 (3.36 ÷ 1.0336).

In calculating the noncompliance penalties, one must consider the present value of the costs of compliance under two scenarios: compliance at time zero and compliance at such future time as it is estimated to occur. The difference is the economic benefit of noncompliance during the interim period. This is illustrated by the following scenario:

1. A source of pollution receives a notice of noncompliance. Its opportunity costs are 12%. This is the discount rate used in determining the present values of expenditures. The company will realize a 10% investment tax credit on its capital expenditures and its income tax rate is 46%. The inflation rate is 8.0%/yr and is expected to remain so for the next decade.
2. Compliance costs will be compared over a 10-yr period.* The total project cost to design, construct, and start up the control system is an estimated $3.6 million currently (because of inflation the cost will be an estimated $3.89 million 1 yr later). Operation and maintenance costs are currently $155,000/yr. This, too, increases with inflation.
3. It requires 2 yr to design and build the required control system. (Assume the system will be paid for in one lump sum at completion of the project.)
4. There are two types of costs associated with compliance: project costs and operation and maintenance costs.
5. There are two types of cash inflows associated with compliance: depreciation (which lowers the firm's income tax) and the equipment's salvage value, if any. It is assumed that the equipment has a useful life of 10 yr and that straight-line depreciation will be used.

Tables 16 and 17 show the various expenditures under the two scenarios. The present value (or cost) of delayed compliance is $333,000 less than that of timely compliance. This is the amount of the noncompliance penalty.

---

* EPA's noncompliance regulations actually require an analysis over a 30-yr period.

Table 16 Present Value of Compliance Costs Assuming Source Does Not Achieve Compliance until Year Two[a]

Budget Item	P.V.[b]	0	1	2	3	4	5	6	7	8	9	10
Project cost[c]	3013			3779								
Operation and maintenance[d]	0		0									
	0			0								
	69				98							
	67					105						
	65						114					
	62							123				
	60								133			
	58									143		
	56										155	
	54											167
Depreciation[e]	0		0									
	0			0								
	-124				-174							
	-110					-174						
	-99						-174					
	-88							-174				
	-79								-174			
	-70									-174		
	-63										-174	
	-56											-174
Salvage value[f]	-112											-348
Total costs	2703	0	0	3779	-76	-69	-60	-51	-41	-31	-19	-355

[a] Two years after notice of noncompliance is issued.
[b] Present value of funds (thousands of dollars).
[c] $3,600,00 less 10% investment tax credit (net project cost).
[d] Net O&M costs $\times$ $(1.08)^{n-1} \div (1.12)^n$, where net O&M = gross O&M $\times$ (1 − tax rate).
[e] 10% $\times$ net project cost $\times$ tax rate.
[f] Equipment still has useful life of 2 yr.

Table 17 Present Value of Compliance Costs Assuming Source Is in Compliance Beginning in Year Zero

Budget Item	P.V.[a]	\\ Year \\ 0	1	2	3	4	5	6	7	8	9	10
Project cost[b]	3240	3240										
Operation and maintenance[c]	75		84									
	72			90								
	69				98							
	67					105						
	65						114					
	62							123				
	60								133			
	58									143		
	56										155	
	54											167
Depreciation[d]	−133		−149									
	−119			−149								
	−106				−149							
	−95					−149						
	−84						−149					
	−76							−149				
	−67								−149			
	−60									−149		
	−54										−149	
	−48											−149
Salvage value[e]	0											−149
												0
Total costs	3036	3240	−65	−59	−51	−44	−35	−26	−16	−6	6	18

[a] Present value of funds (thousands of dollars).
[b] $3,600,000 less 10% investment tax credit (net project cost).
[c] Net O&M costs $\times (1.08)^{n-1} \div (1.12)^n$, where net O&M = gross O&M $\times$ (1 − tax rate).
[d] 10% $\times$ net project cost $\times$ tax rate.
[e] Equipment only has useful life of 10 yr.

4.7 Tax Considerations

Air pollution control facilities are subject to a variety of tax credits or exemptions. For example, many states exempt pollution control equipment from sales taxes. If the sales tax rate is 5%, after-tax savings for a $100,000 project might range $800–$1500. Substantial savings are also effected by property, or ad valorem, tax exemptions. Regulations for special tax treatment at the state and local levels are varied and they require special consideration.

On the federal level, some tax relief is found in the accelerated depreciation provisions applicable to air pollution control facilities. These facilities, if installed at a plant that was in operation before 1976, may be depreciated over a 60-month period. This can result in significant savings to a corporate taxpayer, with benefits running to six and seven figures. Table 18 compares the benefits of accelerated depreciation for two systems. Besides limiting this benefit to pre-1976 plants, there are two other considerations: only the first 15 yr of the system's useful life may be depreciated rapidly, and only "certified" systems are eligible.

Table 18 Tax Benefits of Accelerated Depreciation[a]

	Control System Cost	
	$100,000	$5,000,000
Annual income due to depreciation (net of taxes)		
5-yr depreciation	$ 9,200	$ 460,000
15-yr depreciation	3,070	153,000
Present value of annual income[b]		
5-yr depreciation	34,900	1,744,000
15-yr depreciation	23,400	1,164,000
Present value of benefit	11,500	580,000

[a] Assumes straight-line depreciation and 15-yr useful life.
[b] Assumes 10% discount rate.

The useful life limitation means that if the pollution control system has a 20-yr useful life, only 75% of its cost may be amortized over 5 yr. The remaining 25% is subject to normal depreciation over 20 yr. If the system has a useful life of less than 15 yr, but more than 5 yr, the accelerated depreciation period is still 5 yr.

Before a facility is eligible for accelerated depreciation it must be certified by the applicable state certifying authority and then the EPA regional administrator. Applications to EPA must include as a minimum, the following:

1. Applicant's name, address, and IRS identification number.
2. Complete description of the pollution control facility including drawings and a narrative.
3. Facility address.
4. General description of the process to be controlled by the facility.
5. Dates of construction and startup.
6. Facility cost and expected useful life.
7. An estimate of profits or other benefits that will flow to the applicant as a result of the facility.
8. Proof of state certification.

The facility will then be certified by the regional administrator if it is determined that it removes or prevents pollutants, the applicant is in compliance with all other federal regulations applicable to the facility, the facility is in compliance with local governmental emission requirements, and the facility does not, by more than 5%, increase plant capacity, extend its useful life, or reduce operating costs.

CHAPTER 37
AIR QUALITY MANAGEMENT

JOHN J. ROBERTS

Argonne National Laboratory
Argonne, Illinois

Selling the Right to Pollute
Maryland's Economic Development and Community Administration has asked the General Assembly to consider a proposal that would allow businesses in the state to buy, sell and trade the right to pollute the air.

The new plan would also give industry a powerful profit motive to develop new pollution control technologies that drastically reduce costs. There is an urgent need for such innovation.

WASHINGTON POST 8/18/80

1 INTRODUCTION AND OVERVIEW

Air quality increments, emission offsets, marketable permits, emission density limits: These recent entries into the lexicon of air pollution control terminology in the United States bear witness to the transition from the urban smoke control programs and ad hoc SO_2 and particulate control requirements of the 1960s to a comprehensive, systematic approach to the management of air quality as a limited and valuable resource.

Air quality management is not simply a euphemistic synonym for air pollution control; rather, it represents a well-defined process or rationale for establishing and enforcing regulations governing emissions of a wide array of pollutants from diverse urban and rural sources. Four phases of this process can be distinguished:

1. Establishment of federal policies
2. Development of nationwide standards and procedures
3. Development of regionally sensitive regulations and procedures
4. Implementation—the routine operations of a governmental regulatory agency

The essential elements of each of these phases are captured by the flow diagram in Figure 1. This chapter examines the first three of these principal steps in the process in terms of a theoretical foundation and one or more case studies.

The uppermost tier of Figure 1 outlines actions predominantly at the federal level. In practice the Congress has received input from governmental agencies, individuals, and corporate entities representing a wide spectrum of views on environmental matters. The legislative output of the attendant debate may include highly specific provisions (such as vehicle emission limits) or, more commonly, guidance to governmental agencies responsible for administering the law. In the latter, after acquisition of more extensive data, the cognizant governmental agency(ies) makes a second pass across the upper tier and emerges with specific standards and regulations, most important among which are the national ambient air quality standards.

Ambient air quality standards, which serve as targets to be attained by long-term strategies (e.g.,

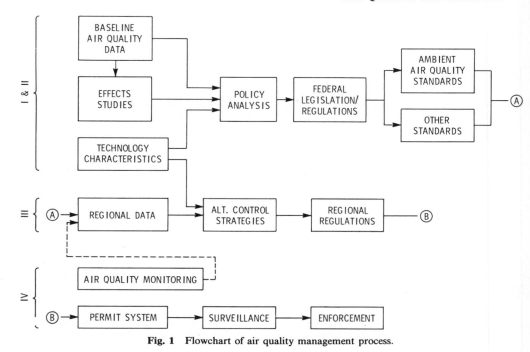

Fig. 1 Flowchart of air quality management process.

emission limits and associated control equipment; creation of the traffic-free zones with peripheral parking) and short-term tactics (e.g., curtailment of operations or fuel-switching during conditions of extremely high pollution levels), provide the foundation for the air quality management process. Such ambient air quality standards:

1. Invariably consider health effects.
2. Often consider physical damage to property as well as aesthetic impacts.
3. Are generally established for individual pollutants (but could in principle be keyed to a multipollutant index if the latter is clearly related to perceived health or welfare effects).
4. May be established after direct or implicit consideration of the costs of compliance and accrual of benefits associated with different pollutant levels and exposure times.

Emission limitations and other elements of a clean air attainment plan (typically for one or more major urban centers and environs—an air quality control region) are designed to achieve these ambient air quality standards.[1] In the model of Figure 1, substantial freedom is suggested in the design of such plans, for example to reflect relative costs of compliance and degree of culpability among contributing pollution sources.

Promulgation of specific regulations[2] signals the start of the implementation phase for which key elements include a permit system, field surveillance, monitoring of air quality, and enforcement, as needed.

In principle, the four major phases form a logical sequence; in practice, they have proceeded in an iterative fashion with substantial overlap in timing accompanied by substantial uncertainty at the compliance and enforcement end of the chain. Figure 2 shows this round robin nature quite clearly.

Congressional debate in the mid 1960s focused on the mounting evidence that air pollution in urban areas exceeded levels at which adverse health effects were known or suspected. The Air Quality Act of 1967 (CAA 1967) for the first time established the requirement for ambient air quality standards

---

[1] Throughout this chapter, except as otherwise noted, the terms "emission limit" or "emission control" serve as surrogates for a wide range of techniques for reducing air pollution, including add-on controls, process changes, revised operating practices, as well as tactics such as car pooling, public transportation, and other approaches to limiting vehicular pollution.

[2] More broadly, the State Implementation Plan (SIP), as required by §108 of the Clean Air Act Amendments of 1977, *amending* §110 of the Clean Air Act, 42 U.S.C. §7410 (Supp. II 1978).

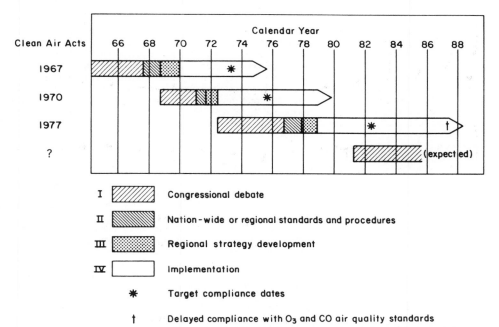

Fig. 2 Sequencing of the major phases of the air quality management process in the United States. The implementation phase invariably has proceeded in a climate of political uncertainty. a. The 1967 act did not set firm dates for compliance with state (or federally) promulgated ambient air quality standards as in subsequent acts. Schedule shown approximates "reasonable time" as called for in §108 (c) (1). b. Congressional debate preceded by hearings before the National Commission on Air Quality pursuant to §323 of the act as amended in 1977.

promulgated by the states, either statewide or for individual air quality control regions (AQCR) designated by the U.S. Secretary of Health, Education and Welfare and called on each state to adopt "a plan for the implementation, maintenance, and enforcement of such standards . . . within a reasonable time."[3] The states had extensive latitude in this procedure with federal intervention constrained by a tortuous process of conferences and hearings.

By 1970 Congress was persuaded that the mechanisms established in the 1967 act were ineffective, to wit: health-related ambient standards varying from region to region, a wide range of pollution control requirements among AQCRs, and ineffectual legal remedies.

Thus the 1970 act[4] reaffirmed the principles of protection of health and welfare, however to be achieved via a substantially stronger federal presence. The Environmental Protection Agency promulgated uniform national health- (primary) and welfare- (secondary) related ambient air quality standards (NAAQS) (Sec. 109);[5] enforced, with wide latitude for intervention, a state implementation planning (SIP) (Sec. 110)[6] and enforcement process (Sec. 113);[7] and established uniform national new source performance (emission) standards (NSPS) (Sec. 111) and national emission standards for hazardous air pollutants (NESHAP) (Sec. 112).[8]

[3] Pub. L. No. 90-148, §2, 81 Stat. 485,490-91 (1967) (amending 42 U.S.C. §1857c-2), *repealed by* Clean Air Amendments of 1970, Pub. L. No. 91-604, §4(a), 84 Stat. 1676, 1678 (1970).
[4] Clean Air Amendments of 1970, Pub. L. No. 91-604, 84 Stat. 1676 (1970) [current version at 42 U.S.C. §7410 et seq. (1976 & Supp. II 1978)] (hereinafter cited as CAA 1970).
[5] 42 U.S.C. §7409 (Supp. II 1978).
[6] U.S.C. §7410.
[7] U.S.C. §7413.
[8] U.S.C. §7412. Only in the setting of NSPS does the CAA 1970 explicitly call for consideration of compliance costs in the standard-setting process for stationary sources, see U.S.C. §7411(a)(1)(C), Clean Air Act §III. Congress also requested the EPA to provide periodic estimates of the cost of compliance with motor vehicle emission limitations established under the act, see U.S.C. §7521(b)(4), Clean Air Act §202(b)(4), and to consider such costs in its review of any applications for extension and for establishing interim limits in accord with U.S.C. §7521(b)(5), Clean Air Act §202(b)(5).

The conflict between ambitious levels and timing of compliance, and the inherent costs dominated the ongoing reappraisal during the mid-1970s even as the state and local governments struggled to enforce the newly developed SIPs. The Clean Air Act Amendments of 1977 (CAA 1977)[9] again underscored the high national priority for attainment of clean air goals. Though revised (delayed) compliance dates were authorized, the air quality management structure was retained and, in fact, embellished by concepts such as *prevention of significant deterioration.* Congressional concern for the cost of compliance appeared in two ways: implicitly through deferral of dates for attainment of NAAQS and explicitly (along with impacts on energy and other environmental values) in the establishment at the federal level of certain emission standards.[10] More generally, Section 317 of the 1977 act required the preparation of "economic impact assessment(s)" for a wide variety of federal regulatory actions[11] and Section 323 established a National Commission on Air Quality to report to the Congress on the overall effectiveness with substantial emphasis on economics of the air quality management process as currently being implemented, and alternatives thereto.[12]

Thus with the states under strong pressure to deliver on their pledges under the revised SIPs of 1979 (and in some cases challenging federal authority, especially regarding establishment of vehicle inspection/maintenance programs and other elements of transportation control plans), a report by the National Commission on Air Quality followed by a new round of congressional debate clouds the picture.

Nevertheless, this overlapping sequence of national clean air policies and regional implementation has consistently reaffirmed and strengthened its reliance on the air quality management model depicted in Figure 1. Although there are numerous variations on the basic theme of this process (e.g., taxation policies, new source performance standards, increments for prevention of significant deterioration of air quality) they invariably owe their political acceptability to the degree to which they enhance the fundamental goal of attaining and maintaining desirable levels of air quality.

2 AMBIENT AIR QUALITY STANDARDS

The criteria and procedures employed in selecting the specific levels and timetables for attainment of national ambient air quality standards reflect more than any other facet of the air quality management process the outcome of political debate. Over the past decade, three trends in the United States are apparent: (a) a shift from public-health-regardless-of-cost to a cost-benefit basis for establishing levels; (b) a stretchout of compliance schedules for existing sources,[1] largely in response to cost-related arguments, and (c) a tightening of control requirements for new stationary sources.[2]

The air quality management process depicted in Figure 1 revolves about the establishment of specific clean air goals—ambient air quality standards. For such standards to be meaningful they must be related to measured or inferred costs to society, costs that reflect increased mortality, morbidity, damage to materials, or deterioration of the aesthetic quality of the environment. In the absence of such undesirable effects and without quantitative relationships,[3] however crude, between cause (air pollution) and effect, there is no basis on which to establish such standards.[4]

Given the nature of clean air as a "public good"[5] and the attendant lack of incentive for the polluter to internalize the costs of control, air pollution policies understandably fall within the realm of politics for their establishment and government for their enforcement. Given the political origins of the air quality management process, it is reasonable to assume that the costs of pollution control have sooner or later *always* been a significant consideration in the setting of standards and timetables for implementation. At the heart of the air quality management process is therefore a balancing in

---

[9] Pub. L. No. 95-95, 91 Stat. 685 (1977) [current version at 42 U.S.C. §7401 et seq. (Supp. II 1978)] (hereinafter cited as CAA 1977).

[10] For example, New Source Performance Standards, see U.S.C. §7411(a)(1)(A)(ii), Clean Air Act §111(a)(1)(A)(ii), Best Available Control Technology, see U.S.C. §7479(3), Clean Air Act §169(3) and Best Available Retrofit Technology, see U.S.C. §7491, Clean Air Act §169A.

[11] 42 U.S.C. §7617 (Supp. II 1978).

[12] U.S.C. §7623, Clean Air Act §323.

[1] Including automotive sources and related transportation control plans.

[2] Increments for prevention of significant deterioration (PSD) contribute significantly to this trend.

[3] Not necessarily in monetary terms.

[4] As will be discussed further in Section 3, uniform national emission standards (e.g., new source performance standards) are not tied to specific, target ambient pollution levels but rather to the costs and effectiveness of differing degrees of severity.

[5] In contrast to private economic goods for which traditional, reasonably efficient markets exist, *public goods* are characterized by a lack of a well-defined market to attract individual buyers. [See, for example, Samuelson (1954).]

the *political* arena of the costs associated with the reduction of atmospheric pollutants and the benefits achieved thereby.

2.1 Generalized Cost-Benefit (C/B) Framework for Evaluation of Air Quality Standards

This section considers a generalized structure for establishing ambient air quality standards where such standards include the pollutants (or combination of pollutants), the allowable levels for given averaging times, and the schedules for attainment. Variations (generally simplifications) on this generalized statement will then be identified and related by specific case studies to current environmental policies in the United States.

In simplest (deceptively so) terms, one can quantitatively define the present value of future streams of benefits and costs:

$$B = \sum_{n=1}^{N} \frac{B_n}{(1+r)^n} \tag{1}$$

$$C = \sum_{n=0}^{N} \frac{C_n}{(1+r)^n} \tag{2}$$

where B_n is the monetized equivalent of the expected benefits in the current ($n = 1$) and future years over a time horizon of N years, C_n is the associated costs over this same period (with $C_0 =$ initial investment), and r is a suitable "discount rate."

The terms *benefit* and *cost* as employed in this chapter refer to the benefits enjoyed by the community when high air pollution levels are reduced and acceptably low levels maintained, and to the costs incurred principally by producers of pollution in controlling emissions. In the rhetoric surrounding the establishment of environmental policy, these terms could as well be reversed with costs signifying the impact on the public of the negative externalities associated with inadequate pollution control by the private sector, and benefits representing the monetary savings enjoyed by the private sector in the absence of adequate pollution control.

Probably the most straightforward aspect of evaluating the costs and benefits of alternative environmental policies and specific regulations is the enumeration of the potential sources of uncertainty. Not surprisingly, the estimation of benefits is considerably complicated by their diffuse nature (see Table 1); however, prediction of the costs of compliance have been consistently off by factors of two to four for reasons suggested in Table 2.[6] The errors inherent in each stage of the estimation process

Table 1   Typical Sources of Uncertainty in Benefit Estimates

1. Current exposure (baseline for future benefits)
   a. Accuracy of air quality data
   b. Relationship of air quality data to actual exposure
   c. Size and other characteristics of exposed population
2. Future exposure (additions to 1 above)
   a. Actual reduction in emissions (e.g., control equipment reliability)
   b. Prediction of atmospheric dispersion and aerochemical reactions
3. Damage functions—epidemiology
   a. Representativeness of air quality data
   b. Synergistic effects of multipollutant environment
   c. Controls for other correlated attributes (e.g., age, smoking, home heating system, workplace environment)
4. Damage functions—materials
   a. Extrapolation of controlled studies to multipollutant environment
   b. Synergistic effects of weather
5. Monetization
   a. Valuation of reduced mortality and morbidity
   b. Specification of the discount rate [r in Eq. (1)]

[6] Given the state of the art of assessment methodologies (including collection of data) and setting aside the "uncertainty" in monetizing changes in mortality, one should currently be able to estimate costs to within a factor of 2 (historically overestimated) and benefits to within an order of magnitude (historically underestimated) as a part of the normal legislative and rule-making steps in air quality management.

Table 2 Typical Sources of Uncertainty in Cost Estimates

1. Direct costs
   a. Selection of control equipment
   b. Potential for innovative, cost-saving approaches
   c. Inflationary impact of sudden increase in demand for control equipment and skilled labor
   d. Requirements for redundancy to assure adequate reliability
   e. Future operating and maintenance costs
   f. Potential for reducing more than one pollutant with same control scheme
   g. Specification of the discount rate [r in Eq. (2)]
2. Cost-related impacts[a]—Effects on:
   a. Prices
   b. Markets
   c. Employment
   d. Regional and national balance of trade

---

[a] These cost-related impacts reflect the distribution of the direct costs and thus will not, in general, be additive to the direct costs. Nevertheless, they represent important inputs to the political decision-making process.

---

can, for practical purposes, be considered to be independent (i.e., uncorrelated). Thus, symbolically, if the cost C is a function of several variables such as the equipment investment E, labor costs L, and annual fuel costs F, that is, $C = f(E, L, A)$, then ΔC, the range in error of C, can be approximated by

$$(\Delta C)^2 = \left(\Delta E \frac{\partial f}{\partial E}\right)^2 + \left(\Delta L \frac{\partial f}{\partial L}\right)^2 + \left(\Delta F \frac{\partial f}{\partial F}\right)^2 \tag{3}$$

One may also estimate the highly unlikely extremes in C or B by combining together first the most unfavorable extremes of each independent variable and then the most favorable extremal values. However, it is preferable to perform a sensitivity analysis, essentially a term-by-term examination of the generalized error equation stated above. Thus, the effect on the benefit tally of different valuations of a "statistical life" (variously judged as low as \$10,000 and as high as \$150,000,000) highlights for the decision maker the sensitivity of the mortality component to the rationale employed in monetization of this impact.

As shown symbolically in Figure 3 and from Eqs. (1) and (2), both costs and benefits will rise as the air quality standard and timetable for attainment are tightened. The principle of *economic efficiency* would select the air quality standard at point A where the *marginal* increase in benefit equals the *marginal* increase in cost, the point that maximizes $(B - C)$. In practice, the curves in Figure 3, especially that for benefits, will be characterized at best by broad bands reflecting substantial, inherent uncertainty. Thus, whether one sets the marginal value of $\Delta C/\Delta B = 1$, or chooses $C \simeq B$ (point A' in Figure 3), or simply uses gross estimates of C and B to assure that clean air policies are reasonable, is solely a matter for the political process.[7] Therefore, in the broadest sense, whether the entries in Eqs. (1) and (2) can be quantified accurately or crudely estimated to within an order of magnitude, the C/B calculus provides a useful background against which environmental policies can be developed, and further serves as a guide to needed improvements in data and methods.

The following subsections explore the degree to which one can transform Eqs. (1) and (2) into quantitative guideposts for decision makers.

---

[7] One may also choose to monetize some of the benefits (e.g., reduced materials damage and soiling) and leave others (particularly estimates of mortality or morbidity) in nonmonetary units. Aside from the methodological difficulty of monetizing health benefits, some (e.g., Kalman, 1980) argue strongly on moral grounds against such a utilitarian view of life. On the other hand, it can be argued that anytime decisionmakers make a judgment concerning expenditures of public and private funds to achieve certain health benefits, they implicitly assign a monetary value to a statistical life; thus, they should be cognizant of this aspect of the decision.

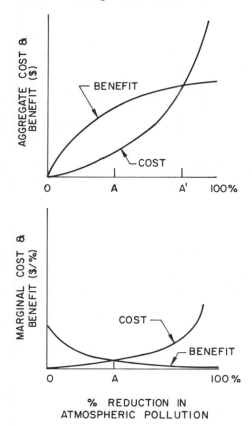

Fig. 3   Cost vs benefit.

2.2   Estimation of Benefits

The benefits associated with a given air quality standard flow to the public in terms of reduced costs associated with mortality, morbidity and damage to materials, and enhanced enjoyment of cleaner air (e.g., improved visibility).[8]

Equation (4) combines the elements necessary for such an evaluation:

$$B = \Delta\chi \ PEM \qquad (4)$$

where

B = projected benefit (dollars)

$\Delta\chi$ = change in concentration χ of a given air pollutant, or surrogate for a given pollutant (e.g., hydrocarbons for ozone) or group of pollutants (e.g., a multipollutant index)

P = "population" exposed (e.g., persons, acres of soybeans)

E = normalized effect function (e.g., number of days lost work/person/year/unit change in pollutant χ). E independent of the absolute level χ implies a linear model; most general information is $E = E(\chi)$

M = monetization coefficient (e.g., $/day of lost work)

---

[8] Standards and regulations governing air quality may yield benefits or disbenefits outside their primary purpose. For example, an automobile inspection/maintenance program should increase engine efficiency. Dollar savings would be counted in the C/B analysis but enhancement of a national goal of reduced oil imports is an additional benefit which should be identified. Conversely, stringent limits on emissions from coal-fired boilers could lead to increased costs as well as conversions to oil or gas.

Equation (4) can be adorned with subscripts for different pollutants (i), for different effects (j), and for populations representative of each geographical region (k). Clearly $B_{ij} = \sum_k B_{ijk}$. If the effects are disjoint or clearly separable, then $B_i = \sum_j \sum_k B_{ijk}$ (e.g., morbidity effects are additive to crop losses). However, morbidity effects or material damage will be subsumed in part if one also includes as a measure the effect of air pollutants on property values. Finally one should examine closely any calculation that evaluates the benefits B_k for reductions $\Delta\chi_{ik}$ of several pollutants in region k as $B_k = \sum_i B_{ik}$ since, at least to date, the ability of epidemiological and laboratory studies to characterize with reasonable certainty the simultaneous effects of differing levels of different pollutants is quite limited; further, the potential for synergistic effects (e.g., SO_2 and particulates) leave the sign of the relationship uncertain; that is, $B_k \overset{?}{\underset{\lessgtr}{}} \sum_i B_{ik}$.

The state of the art in benefit analysis is evolving rapidly. Expanded sources of data and refined calculational methods address the uncertainties illustrated in Table 1. For purposes of this chapter, it will be most useful to conclude by summarizing in terms of Eq. (4) a portion of the conclusions of one recent study of the mortality benefits associated with reduced levels of air pollution.[9]

Example: Sixty-City Mortality Study (Crocker, 1979, v.I)

Citywide 1970 mortality data for 60 U.S. cities was disaggregated by major disease categories (e.g., vascular disease, pneumonia and influenza, cancer) and statistically related to key population characteristics (including race, age, cigarettes per capita, doctors per capita, certain dietary variables, and air pollution). "Rather small but important associations are found between pneumonia and bronchitis and particulates in air and between early infant disease and sulfur dioxide air pollution" (Brookshire, 1979, p. 4).

Following the notation in Eq. (4), the mortality benefit associated with an assumed 60% reduction in ambient particulate and SO_2 levels becomes:

$\overline{\chi}_1 =$ mean urban particulate level $= 115 \ \mu g/m^3 \pm 34$

$\overline{\chi}_2 =$ mean urban SO_2 level $= 27 \ \mu g/m^3 \pm 22$

$\Delta\chi_1 = (0.6)(115) = 69 \ \mu g/m^3$

$\Delta\chi_2 = (0.6)(27) = 16 \ \mu g/m^3$

$P_1 = 1.5 \times 10^8$ persons

$P_2 = 3 \times 10^6$ births/yr

$E_1 = 1.4 \times 10^{-3}$ deaths/yr/thousand persons/unit increase in particulates

$E_2 = 4.4 \times 10^{-2}$ deaths/thousand births/unit increase in SO_2

$M = \$340,000–\$1,000,000/$death, a range for this parameter based on studies of consumer willingness
to pay for reduced risk [references: Thaler and Rosen (1975) and Smith (1974), respectively]

$B_1 = \$5.0 \times 10^9$ to $\$14 \times 10^9/yr$

$B_2 = \$0.7 \times 10^9$ to $\$2.2 \times 10^9/yr$

$B = \$5.7 \times 10^9$ to $\$16 \times 10^9/yr$

2.3 Estimation of Costs

In contrast to the foregoing, somewhat cursory look at the benefit side of the C/B equation, the following examination of costs will be far more detailed. The environmental engineer routinely faces the need to evaluate the costs of alternative approaches to pollution control but rarely if ever is responsible for developing *independent* estimates of the critical parameters E and M in the benefit calculation [Eq. (4)].

The specific purpose of any analysis of the costs of air pollution regulations will determine the overall methodology, the elements of cost to be considered, degree of sophistication of algorithms, level of detail of input data, and the evaluation, interpretation, and presentation of results. Typical purposes for a cost analysis include (in order of requirement for increased detail):

1. Cost-benefit assessment of alternative ambient air quality standards.
2. Design of emission limitations to achieve established ambient air quality standards.
3. Request by a regulated company or trade association for reduction in stringency (or delay in enforcement) of a specific emission limitation.

---

[9] In addition, see the cost-benefit example in Section 2.5, in particular Tables 9 and 10.

4. Determination of the monetary fine appropriate for a particular source not in compliance with applicable regulations.
5. Cost estimates associated with the engineering of a site-specific air pollution control project.
   a. Conceptual design
   b. Preliminary design
   c. Detailed (or final) design

Conceptually, it is useful to consider cost estimates for the broadest cost-benefit assessments (item 1) to be *scoping* in nature, estimates for items 2–5 (a) to entail some *study* of the overall characteristics of different classes of plants as well as nonindustry activity, and items 5 (b) and 5 (c) to require the most *definitive* analyses. Further, items 1 through 4 could include secondary cost considerations such as effects on international trade, national productivity, local and regional employment, and profitability of specific product lines.

Given that this chapter considers air quality management primarily from a strategic point of view, the methodologies presented for cost estimation will emphasize the *scoping* and *study* levels as distinguished from *definitive* analyses.[10]

2.3.1 Direct Costs of Pollution Control

The elements which taken together comprise the cost of an air pollution control system are most conveniently divided into (a) the initial capital investment as well as initial transactional costs, such as legal fees, and (b) ongoing, time-varying expenses for operation of the equipment. Table 3 lists the items typically considered in an estimation of the initial cost of the plant as they might appear in a first-cut study for control of a particular class of industrial sources. To the estimated costs of the main pieces of equipment (MPI) are added key ancillary elements, here estimated by standard multiplying factors which, taken together, imply that (total plant cost) $\simeq 5.1 \times$ (cost of delivered MPI).

Table 4 graphically displays the increase in detail required and the commensurate increase in accuracy for prediction of plant capital costs as one moves from crude conceptual studies toward more detailed design.

Tables 5 and 6 list typical annual operating and general administrative expenses along with representative factors for their estimation. These ongoing expenses should be adjusted to reflect (a) additional costs such as loss of production (e.g., lower profits due to decreased market for the product,[11] and (b) additional revenue such as from sale of recovered materials (e.g., sulfur) and end-of-life salvage value.

From the perspective of the investor, capital and operating costs are combined via Eq. (2) for the discounted present value with C_n = actual after-tax expenses incurred in year n, exclusive of annual

Table 3   Items That Comprise Total Plant Cost—A Typical List. Representative Factors for the Chilton (1973) Method Are Also Shown. (Source: Uhl, 1979a, p. 9)

Item No.	Item	Multiplying Factor	Operating on Item No.	Cost of Item
1.	Cost of delivered equipment (MPIs)	1.0	1	$
2.	Installed, erected equipment cost	1.60	1	
3.	Piping (includes insulation)	0.40	2	
4.	Instrumentation	0.15	2	
5.	Buildings and site development	0.20	2	
6.	Auxiliaries (electric, steam, etc.)	0.25	2	
7.	Other	0.12	2	—
8.	Total physical cost (items 2 through 7), or direct costs			$
9.	Engineering and construction	0.35	8	
10.	Contingency and contractor's fee	0.15	8	—
11.	Total plant cost (items 8 through 10)			$

[10] For a detailed presentation (with examples) of cost engineering of air pollution control systems, the reader should consult Chapter 14 of this handbook.
[11] To be considered under secondary costs in Section 2.3.2.

Table 4 Estimations Guide for Capital Investment (Source: Perry and Chilton, 1973, p. 25–15)

Required information		Order-of-magnitude estimate > ± 30% range	Study estimate ± 30% range	Budget authorization estimate ± 20% range	Project control estimate ± 10% range	Firm estimate ± 5% range
Site	Location		•	•	•	•
	General description		•	•	•	•
	Soil bearing			•	•	•
	Location & dimensions R.R., roads, impounds, fences			•	•	•
	Well-developed site plot plan & topographical map				•	•
	Well-developed site facilities					•
Process flow sheet	Rough sketches		•			
	Preliminary			•		
	Engineered				•	•
	Preliminary sizing & material specifications		•	•		
Equipment list	Engineered specifications					•
	Vessel sheets					•
	General arrangement				•	•
	(a) Preliminary			•		
	(b) Engineered				•	•
Building and structures	Approximate sizes & type of construction		•			
	Foundation sketches			•		
	Architectural & construction				•	•
	Preliminary structural design			•		
	General arrangements & elevations				•	
	Detailed drawings					•
Utility requirements	Rough quantities (steam, water, electricity, etc.)		•			
	Preliminary heat balance			•		
	Preliminary flow sheets			•		
	Engineered heat balance				•	•
	Engineered flow sheets				•	•
	Well-developed drawings					•
Piping	Preliminary flow sheet & specification		•	•		
	Engineered flow sheets				•	
	Piping layouts and schedules					•
Insulation	Rough specifications			•		
	Preliminary list of equipment & piping to be insulated				•	•
	Insulation specifications & schedules					•
	Well-developed drawings or specifications					•
Instrumentation	Preliminary instrument list			•		
	Engineered list & flow sheet				•	•
	Well-developed drawings					•
Electrical	Preliminary motor list – approximate sizes		•	•		
	Engineered list & sizes				•	
	Substations, number & sizes, specifications				•	
	Distribution specifications				•	
	Preliminary lighting specifications				•	
	Preliminary interlock, control, & instrument wiring specs.				•	
	Engineered single-line diagrams (power & light)					•
	Well-developed drawings					•
	Engineering & drafting		•	•	•	•
Man-hours	Labor by craft					•
	Supervision					•
Project scope–standard processes	Product, capacity, location & site requirements. Utility & service requirements. Building & auxiliary requirements. Raw materials & finished product requirements, & storage requirements.	•				

Most probable cost

Error range %

Table 5 Annual Operating Expense Items and Information for Their Estimation (Source: Uhl, 1979a, p. 13)

Raw materials	Consult current issue of *Chemical Marketing Reporter* (8) for rough estimate; secure quotes or consult commodity experts for lower contract prices
Operating (direct) labor	The number of full-time operating personnel times average earnings of \$12,000–\$15,000/yr[a] (includes shift differential and overtime)
Direct supervision	10–25% of earnings of operating labor. Annual earnings of supervisors, \$18,000–\$24,000/yr[a]
Maintenance	4–10% of total plant cost, I_F (supervision of maintenance is included in plant overhead)
Operating supplies	6% of earnings of operating labor or 15% of maintenance expenses
Labor additives (fringe benefits)	25–50% of operating labor earnings; may also include maintenance labor earnings
Utilities	Develop directly from energy balances plus an allowance for losses
Plant overhead	50–100% of direct operating and maintenance labor earnings, or 50% of direct operating and 25% of maintenance labor earnings, or 45–50% of operating labor earnings plus 1 to 5% of total plant cost, I_F
Control laboratory	\$40,000–\$50,000 per analyst,[a] or 10–20% of operating labor earnings
Technical and engineering	\$40,000–\$50,000/man;[a] may be included in plant overhead
Insurance and property taxes	1–2% of I_F
Depreciation	Varies but common rate is 10% of I_F or total depreciable investment for 10 yr

[a] 1977 rates.

Table 6 Annual General Expense Items and Factors for Their Estimation (Source: Uhl, 1979a, p. 14)

Administration	2–3% of sales or capital investment
Sales	Usually 2–6% of sales; but up to 30% for specialty items
Research	2–5% of sales or capital investment
Finance	Largely interest on bonds; often not considered in conceptual estimates

depreciation. However, as an input to public policy decisions, actual pretax expenditures are most meaningful.

For economic assessments of pollution control regulations, the year $n = 0$ can be any convenient reference year, preferably with costs expressed in constant dollars indexed to that year. On the other hand, a company evaluating the cost of pollution controls to commence operation in year M might choose to calculate the present value of costs for the year of startup:

$$C = \sum_{n=0}^{N+M} \frac{C_n}{(1+r)^{n-M}} \quad \text{or} \quad \sum_{n=-M}^{N} \frac{C_n}{(1+r)^n} \tag{5}$$

Results of a cost analysis can be presented in many different ways according to the purpose of the evaluation, as well as the type of organization to which the results apply. Table 7 lists several indices or figures of merit in common use.

Scoping studies in support of the development of air quality management strategies will require cost data aggregated to a substantial degree and presented via algebraic equations of relatively simple form. For example, an appropriate level of aggregation might describe typical hydrocarbon control costs (annualized or capital investment) for gasoline storage tanks in terms of the control option (i)

Table 7 Representation of Pollution Control Costs

Representation	Method of Calculation	Application/Interpretation
Present value (*PV*)	Direct substitution into Eq. (2) for individual or multiplicity of sources; use "representative" discount rate	Estimated cost of a control program or cost of a given emission regulation
Unit cost (e.g., $/kWe)	Divide *PV* by the total activity covered by the control program (e.g., kWe; tons of product)	Normalized cost of control for typical industrial processes. This, plus size-scaling laws, can be used to estimate control costs for other similar processes. Most suitable for tabulation
Annualized cost (*A*)	$A \sum\limits_{n=1}^{N} \dfrac{1}{(1+r)^n} = PV$ or unit cost Note: $\sum\limits_{1}^{N} \dfrac{1}{(1+r)^n} = \dfrac{1-(1+r)^{-N}}{r}$ $\triangleq \dfrac{1}{\text{capital recovery factor}}$	Levelized annual cost or unit cost; can be interpreted as a revenue requirement for a regulated public utility where *r* is the allowable return on equity and *PV* includes all allowable capital costs
Internal rate of return (IROR)	Value of *r* for which present value of costs equals present value of benefits	Comparison of two alternate investments where, for example, one entails larger first costs but returns a profit (net benefit). IROR then becomes the root *r* of the expression for incremental costs and benefits (i.e., alt. 2 − alt. 1)
Payback period (discounted)	For a prescribed *r* (or IROR), the value of *N* for which $PV = 0$; e.g., $\sum\limits_{n=1}^{N} \dfrac{B_n - C_n}{(1+r)^n} = C_0$	As with IROR, meaningful if comparing overall costs and benefits, or relative benefit of an alternative investment with higher first cost but increased profit (e.g., pollution control equipment vs shift to low polluting *and* potentially more profitable process)
Payback period (simple)	Value of *N* at which the accumulation of net benefits (*not* discounted to present value) equals the initial (depreciable) investment; $\sum\limits_{n=1}^{N} B_n - C_n = C_0$	Traditional but outmoded figure of merit for comparing alternative investment policies

and a multiplier which is a function of tank capacity:[12] $C_i = A_i F_i$ (capacity). Alternatively, the range of particulate control options for a cement plant associated with increasingly efficient removal devices might be characterized by a cost equation of the form:

$$[\text{annualized cost}] = [a][\text{removal efficiency}]^b [\text{gas flow rate}]^c$$

In these examples, the constants would be established by multilinear regression using actual cost data or the results of series of conceptual designs. Alternative functional forms for the generalized cost equations would be tested to improve the fit.

A prototypical example of such a scheme for representing costs of control equipment is the "emcost" equations (Babcock, 1973). Table 8 shows the results of a least-squares fit of published control-cost data for major particulate removal devices to the generalized equation:

$$\text{emcost} = aQ^b \left(\frac{R}{1-R}\right)^c \tag{6}$$

[12] For example, $C = a[\text{capacity}]^b$ as employed in *The Cost of Clean Air and Water* (EPA, 1979).

Table 8 Cost Equations—1965 Basis (Babcock, 1973)

		$R^{2\ a}$	Number of Data Points	Efficiency Range of Applicability, %
Wet collector	$\text{emcost} = 41.5 \times 10^{-6} Q^{0.91} \left(\dfrac{R}{1-R}\right)^{0.52}$	0.997	9	75–99
Low-voltage electrostatic precipitator	$\text{emcost} = 75.9 \times 10^{-6} Q^{0.90} \left(\dfrac{R}{1-R}\right)^{0.14}$	0.996	9	88–99
High-voltage electrostatic precipitator	$\text{emcost} = 520.5 \times 10^{-6} Q^{0.69} \left(\dfrac{R}{1-R}\right)^{0.18}$	0.982	9	90–99.5
Filter	$\text{emcost} = 119.5 \times 10^{-6} Q^{0.89}$	0.978	8	99.9
Dry centrifugal	$\text{emcost} = 18.7 \times 10^{-6} Q^{0.96} \left(\dfrac{R}{1-R}\right)^{0.12}$	0.999	9	50–95
Gravitational	$\text{emcost} = 3.2 \times 10^{-6} Q^{0.98} \left(\dfrac{R}{1-R}\right)^{1.31}$	0.987	9	42–72
Composite of above	$\text{emcost} = 15.5 \times 10^{-6} Q^{0.96} \left(\dfrac{R}{1-R}\right)^{0.30}$	0.814	53	0–99.5

^{a} (explained variation)/(total variation).

where

$$\text{emcost} = \text{annualized capital and operating cost (\$/hr)}$$
$$Q = \text{flue gas flow rate (acfm)}$$
$$R = \text{particulate removal efficiency (decimal)}$$
$$a,\ b,\ c = \text{empirical constants}$$

Note also the high statistical significance of the composite particulate control curve. Separate composite curves of superior reliability could readily be developed for the control options and associated efficiency ranges characteristic of each class of major emission sources.[13]

2.3.2 Impacts of Pollution Control Costs

Whereas the direct costs for installation and operation of pollution control equipment can be estimated with accuracy appropriate for decision making in the air quality management process, the impacts of such costs on the economy may be difficult to forecast, even qualitatively. For the individual plant or industry, such increases in the unit cost of production may be met by increased prices, reduced product quality, reduced output (possibly shutdown or relocation), deferral of plans for expansion or modernization, changes (+ or −) in employment, or combinations of the above. Customer response may be to accept increased costs or reduce consumption, the latter by lowering product quantity and/or quality or by direct substitution. For the region, cost impacts of regulations may take the form of changes in employment and/or inhibition of economic growth. For the nation, cost impacts may be seen in inflation, employment, and balance of trade.[14]

In most cases these effects are diffuse, complex, and, as with the benefit side of the C/B equation, very difficult to quantify; yet, an assessment, however tentative, of such cost impacts should be a part of the analysis of legislative initiative and major air pollution control regulations. In the following, techniques for analyzing several categories of cost impacts will be outlined.

2.3.2.1 Direct Effects on Plants or Major Industries.
Pollution control costs should be expressed as unit costs (e.g., \$/ton of product), annualized in accord with prevailing interest rates, and compared as a percent to the unit production cost or the unit sales price. As a minimum, such a calculation should distinguish low-impact (<1%) and high-impact (>10%) situations, the latter (as well as selected intermediate cases) warranting special study of the range of plausible responses by the affected industry and its customers. At this point questions to be addressed might include:

[13] As is shown for coking facilities in Section 2.3.6.

[14] For a discussion of how one state (Illinois) approaches the assessment of the economic impact of pollution control regulations, see Babcock et al. (1978).

1. What are the unit costs at competing (especially newer) plants, including foreign suppliers?
2. What are the costs of substitute products?
3. What is the nature of short-term (<2 yr) and long-term (>5 yr) price elasticities of demand,[15] the former reflecting the potential for immediate substitution (e.g., imported vs domestic fibers) and the latter, the nature of consumers' current fixed investments (e.g., home heating equipment)?
4. As a consequence of (1), (2), and (3), to what extent can the plant pass on or absorb increased costs? How might output be reduced?[16] Are any plant closings likely?[17]
5. What is the balance between reduced employment due to reduced output and increased employment associated with installation and maintenance of pollution controls?

2.3.2.2 State and Local Impacts. As a first cut at forecasting cost impacts at this level of geographical detail, one should sum the impacts on major industries, as discussed above.

It is widely recognized that changes in employment in one sector of a local economy will influence employment elsewhere. This effect is most pronounced when the primary industry is a member of the *export sector* satisfying demand outside the region.[18] Employment multipliers or more sophisticated functional relationships between primary and secondary employment will vary significantly according to the size of the affected community, relationship to neighboring communites, and mix of industries. Such information will normally be on hand at metropolitan, regional, and state planning agencies. The appropriate planning agency should be consulted in any evaluation of local or regional impacts.

2.3.2.3 National Impacts. As for local and state impacts, the first order of assessment of cost-impacts at the national level is the summation of industry-by-industry costs.[19] Secondly, detailed studies of industries with substantial international involvement (e.g., energy, steel, automobiles) should be conducted.

Finally, macroeconomic impacts of environmental legislation can be examined using input/output (I/O) analysis (Chiang, 1974; Leontief, 1966) coupled with long term forecasting of the economy in the absence of such legislation.[20]

For example, the direct costs attributable to federal air and water pollution control legislation have been translated into impacts on inflation, unemployment, and GNP for the U.S. (DRI, 1979).[21] Higher employment associated with a new pollution control activity (Figure 4) coupled with constant or decreasing useful output leads to a slight increase in inflation (Figure 5). The impacts on individual industrial sectors (e.g., automobile in Figure 6) readily fall out of the I/O analysis.

2.3.3 Cost Effectiveness (C/E)

In contrast to *cost benefit* analysis wherein a balance is sought between air quality goals and associated societal costs, *cost effectiveness* (C/E) methods seek optimal pathways to the attainment of predetermined goals[22] or, in the absence of quantitative goals, compare the costs associated with increasingly severe emission limitations.[23]

---

[15] Price elasticity of demand represents the percent change in demand for a product associated with a 1% increase in its price, other factors such as quality, prices of other goods, and purchasers' income remaining fixed.

[16] For example, Babcock et al. (1978) suggest establishing an upper bound of market reduction effects for plants with products having substantial transportation costs (e.g., coal) by equating the increased (pollution control) costs with a reduced radius of transportation.

[17] Discouraging the migration of plants to areas of the country with more lax emission limits is one of the primary effects of a stringent, uniform national new source performance standards (NSPS). 42 U.S.C. §7411 (Supp. II 1978), Clean Air Act §111.

[18] Clearly a steelworker and family will support secondary service industries such as supermarkets, but not the reverse.

[19] See for example the *Cost of Clean Air and Water* (EPA, 1979).

[20] For those states (e.g., Illinois) where I/O tables have been developed, similar analyses can be conducted. For example Leung *et al.* (1982) employed a 28-sector I/O model developed by the California Department of Water Resources to translate an estimated $94 million loss of agricultural output in the South Coast Air Basin due to ozone into a region-wide total economic impact on sales of $276 million and an employment loss of 9525 person-years.

[21] Note that beneficial effects were not folded into the analysis.

[22] The U.S. Congress implicitly endorsed a cost-effectiveness philosophy under the Clean Air Act of 1970 by mandating the establishment of primary air quality standards based exclusively on health considerations and allowing the states to choose pathways for attainment. On the other hand, Congress itself reentered the decision process once the costs and benefits of the 1970 act became clearer, choosing to extend the dates for compliance, thereby reducing both costs and benefits (thus the political process is seen employing, albeit post hoc, a cost-benefit philosophy).

[23] As in the establishment of New Source Performance Standards. 42 U.S.C. §7411 (Supp. II 1978), Clean Air Act §111.

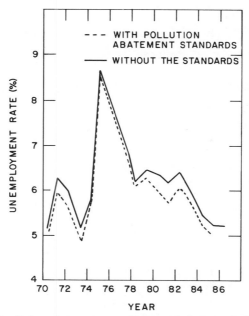

Fig. 4 The impact of pollution abatement expenditures: principal conclusions. Better employment prospects (source: DRI, 1979, p. 4).

Analytical approaches to C/E vary over the full spectrum of operations research methodologies, from direct enumeration of options to linear and integer programming. As a minimum, the analyst should develop several attainment strategies, evaluate and document their respective costs (including limitations on, or ranges of, accuracy) and any ancillary benefits or disbenefits, and display the results in tabular and/or graphical form for decision makers.

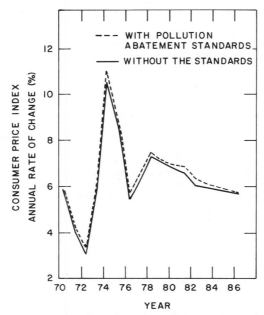

Fig. 5 The impact of pollution abatement expenditures: principal conclusions. Effect of higher inflation on the consumer price index (source: DRI, 1979, p. 4).

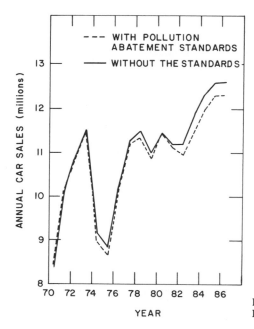

Fig. 6 Impact on consumer spending (source: DRI, 1979, p. 20).

Example: Cost-Effectiveness Model for Pollution Control at Coking Facilities (Kemner, 1979)

The system: Figure 7 characterizes each stage of a coking facility in terms of material flows (based on a 1-ton input of cleaned coal) and principal atmospheric emissions. This is a generic flow diagram; actual plants may not contain all elements and will employ a wide range of equipment.

Process and pollution control technology: Equipment typically employed throughout the coking industry is described in detail and potential emissions cataloged. Alternative pollution control technologies evaluated include both new and retrofit emission control devices as well as revised operating procedures. A total of 16 different major emission sources are considered, with up to eight control options per source.

Cost estimation for pollution control: Cost equations of the form [cost] = A[size]^B are developed from engineering conceptual designs. Costs are expressed in two ways: (1) initial capital investments and (2) annualized cost. The former include elements such as those listed in Table 3. The latter incorporate the initial capital costs via an annual capital recovery factor[24] to which is added typical annual costs such as enumerated in Tables 5 and 6. Retrofit costs are estimated at approximately 20% over costs for controls on new process equipment.

Emissions inventory: A total of 215 batteries associated with nine plants throughout the United States are characterized in sufficient detail to match alternate control technologies to each major emission point.

Optimization technique: The optimization is cast as an integer programming problem (Hillier and Lieberman, 1974). A matrix Y of integers (0, 1) act as "switches" to choose the particular control "device" for a given emission source at a given battery ($y = 1$), thereby setting $y = 0$ for all other possible "devices" for that source. Each matrix Y specifies a unique control strategy j for all source points i at all batteries under consideration. A cost matrix assigns a cost C_{ij} to each possible option j for source point i. Thus the cost optimization problem becomes

$$\min \sum_{i,j} c_{ij}\, y_{ij}$$

subject (1) to constraints on total allowable pollution from each battery or (2) to a prescribed percentage reduction in each pollutant for each battery. Note that the solution may include both uncontrolled and highly controlled sources, depending on the relative costs of available control options.

---

[24] See Table 7. The study employed a discount rate $r = 10\%$ and a project lifetime $N = 15$ yr.

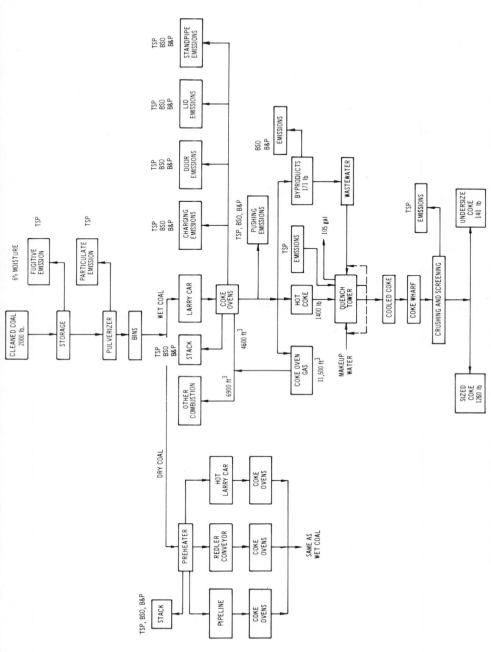

Fig. 7 Relationship of emission sources in a typical by-product coke plant. (source: Kemner, 1979, p. 45).

Modes of cost-effectiveness analysis

1. Determine cost for a given control strategy or set of strategies (i.e., prespecified control equipment and operating procedures) for one or more batteries.
2. Least cost—Select the control strategy that achieves a desired reduction of emissions at lowest cost.
3. Maximum control—Select the control strategy that achieves the maximum reduction in emissions for a given cost ceiling.

Sample results

1. Annualized costs of three alternative control systems for the coal-charging operation are shown in Figure 8 as a function of battery size: option 1 (uncontrolled, no cost); option 2—modified larry car with steam supply and smoke boot (80% efficiency for particulate control); option 3—new larry car with steam supply and smoke boot (99% efficiency); option 4—same as option 3 plus second collecting main for effluent (99.5% efficiency).
2. Annualized cost (nationwide) associated with stringent particulate control requirements is plotted in Figure 9.

2.4 The Social Discount Rate

The discount rate r in Eqs. (1) and (2) exerts tremendous leverage on the present value calculation of benefits and costs. Though understandably much of the debate over the application of C/B methods to the evaluation of public investment policies has centered on the identification and valuation of the elements that constitute the stream of benefits and costs, the importance of the discount rate has not been overlooked. Economists have published extensively on the subject in the last decade. Politicians have become keenly aware of the public policy implications embodied in the choice of the discount rates used to guide government investments in social welfare projects, thus the term *social discount rate*.

Tremendous leverage is exerted by the social discount rate. For example, the present value of a steady stream of annual benefits B/yr assumed to accrue at the end of each year yields

$$PV = \sum_{n=1}^{\infty} \frac{B}{(1+r)^n} = \frac{B}{r} \tag{7}$$

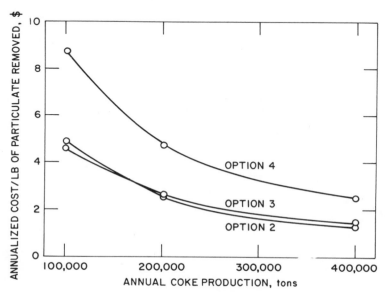

Fig. 8 Cost per pound of particulate removal for control options for wet coal charging (source: Kemner, 1979).

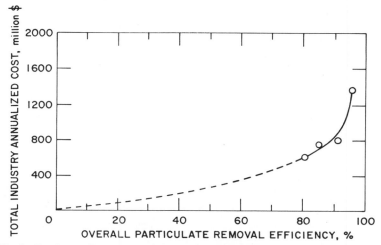

Fig. 9 Total annualized cost as a function of overall efficiency (source: Kemner, 1979).

Further the duration over which the C/B calculation is to be performed, that is, the year N at which the summation in Eqs. (1) and (2) is to be truncated, is important, especially when short-term, high-payoff investments are to be compared to ones of long duration and low payoff.[25]

Future costs and benefits are normally estimated under the assumption of stable prices (i.e., no inflation) with the social discount rate selected to reflect that assumption. To the extent that the discount rate can be viewed as a proper interest rate net of inflation, incorporation of inflation in a C/B calculation will simply scale numerator (costs or benefits) and denominator (by an appropriate inflation of r) with no net effect on the present value.[26] One should therefore be careful that the estimation of costs and benefits and the selection of the discount rate embody the same set of assumptions regarding inflation.

The following are examples of discount rates (their origin, values, and rationale) used (or recommended for use) in C/B analyses.[27]

Example 1: $r = 1$–2% (Arrow, 1966)

The discount rate here reflects the marginal rate of time preference (i.e., interest rate) net of inflation that the general public demands in compensation for foregoing current consumption by purchasing savings accounts or U.S. Treasury Certificates. Associated with this model is the assumption that the diversity of government projects produces a form of self-insurance or risklessness, in contrast with investments by private industry.

Example 2: $r = 6\frac{1}{8}\%$ (U.S. Water Resources Council, 1976)

The council, a federal interdepartmental group, in its "Principles and Standards for Planning Water and Related Land Resources" based its rate (updated annually) on the interest rate on federal securities with maturities of 15 yr or more.

---

[25] The conflicts inherent in such a choice have, in part, led some economists to combine Eqs. (1) and (2), setting benefits = costs, and calculate the value of r [thus termed an *internal rate of return* ($IROR$)]. Alternative projects are then compared in terms of their relative IRORs (e.g., see Seagraves, 1970).

[26] Capital budgeting in a corporate setting in the United States represents an important exception to this observation because future streams of tax-deductible depreciation writeoffs will be based on the initial capital outlay, whereas future streams of revenues and variable costs will respond to inflation. Thus, a realistic assessment of net present value, after tax, should employ expected rates of inflation along with a discount rate equal to the current cost of capital to the firm.

[27] Of greatest importance in these examples is the diversity of considerations upon which the recommended discount rates are based. Understandably, over time, the actual values of r consistent with each rationale will change, reflecting changes in local, national, and world economic conditions.

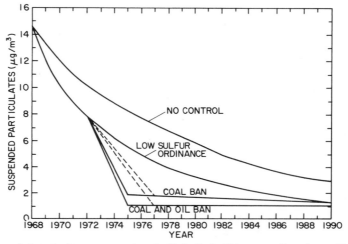

Fig. 10 Suspended particulate concentrations in Hyde Park, Chicago, attributed to residential sources (source: Cohen et al., 1974, p. 64).

Example 3: $r = 8\%$ (Townsend, 1978)

The social opportunity cost of capital is derived as a weighted average of the returns after inflation to corporate investors and private savers. Distortions induced by U.S. tax policies, relative risk, and the internal capital market are considered.

Example 4: $r = 10\%$ (U.S. Office of Management and Budget—OMB, 1972 Rev.)

OMB is responsible for overseeing the application of C/B analysis by most U.S. governmental agencies.

Example 5: $r > 30\%$ (Harberger, 1972)

In a country (or locale) where all workers to be employed by one or more competing new projects can be assumed to be drawn from a large pool of unemployed (e.g., in India), from a societal viewpoint the wages paid to labor should not be considered an economic cost. Thus the net annual benefit from a project will be higher. On this basis, the social opportunity cost of capital as derived from the marginal productivity of capital in the private sector is unusually high. The implication is that optimum social policies for such a country favor lower technology, labor intensive ones (e.g., hydro and solar power) versus capital intensive ones (e.g., nuclear power).

The discussion above has dealt with discrete discount rates which, as treated by Eqs. (1) and (2), assume that all costs and benefits associated with the nth year accrue on the last day of that year. Such costs and benefits are generally continuous in nature and can be handled via continuous (vs discrete) discount rates (e.g., see Uhl, 1979b) if one seeks a formulation wherein the discounting process is more sensitive to such timing. For example, the present worth of a uniform cash flow of $1/yr for 5 yr at a nominal annual interest rate of 10% is:[28]

1. $3.98 using a model of continuous discounting
2. $3.87 using Eq. (2) with $r = 0.1$

2.5 Cost-Benefit Analysis: Case Study (Cohen, Fishelson, and Gardner, 1974)

Analysis of high levels of particulates and SO_2 in residential areas of Chicago pointed to combustion of coal, oil, and garbage in one- to three-story buildings as the primary cause. Figure 10 shows the impact on particulate levels of several regulations under consideration. Coal use was expected to decline

[28] $PV_{cts} = \int_0^5 \dfrac{\$1}{(1+0.1)^n}\, dn$

$PV_{disc} = \displaystyle\sum_{n=1}^{5} \dfrac{\$1}{(1+0.1)^n}$

Table 9 Summary of Benefits (1972 Dollars) in Specific Damage Categories (Cohen et al., 1974, p. 75)

Benefit Category	Present Value per Household of a 1 $\mu g/m^3$ Reduction in		Principal Reference
	Particulates	Sulfur Dioxide[a]	
Life expectancy:			
Extension of earnings	5.40	0.0	Lave and Seskin (1970)
Defer expenses of final illness and burial	0.31	0.0	Schrimper (1973)
Reduced health care expenses	14.00	14.00	Lave and Seskin (1970)
Reduced frequency of materials maintenance	45.00	16.00	Barrett and Waddell (1973)
Expenditures diverted from other uses	17.00	0.0	
Total	82	30	

[a] The zero entries for SO_2 reflect an absence of adequate statistical data on the rate at which SO_2 reduction reduces death rates. A 10% discount rate was used throughout the analysis.

Table 10 Benefit Estimates (1972 Dollars) Using Two Alternative Methods (Cohen et al., 1974, p. 77)

Basis of Estimation	Reduction in Particulates[a]		Reduction in Sulfur Dioxide[a]	
	Per $1000 of Residential Property	Per Household	Per $1000 of Residential Property	Per Household
Property value method[b]				
Minimum estimate	1.75	36	0.0	0
Midrange estimate	2.50	51	1.75	36
Specific damages method (derived from Table 9)	4.00	82	1.45	30

[a] All figures are present values for an immediate 1 $\mu g/m^3$ improvement in air quality, continuing indefinitely into the future. A 10% discount rate was used throughout the analysis.
[b] Reference: Crocker (1970).

even in the absence of regulation due to the economic advantages (lower fuel and janitorial costs of conversion to natural gas). A low-sulfur ordinance (1% S by weight) would further accelerate this trend.

Benefits were evaluated via two separate approaches: (1) the aggregate value to each household of reduced mortality, morbidity, and soiling and corrosion of materials due to a given reduction in particulates and SO_2; and (2) the estimate of a family's willingness to pay for improved air quality as measured by differences in property values. Table 9 presents a breakdown of the specific damage factors; Table 10 compares these to coefficients reflecting the impact on property values. This comparison is presented in terms of the present value of an immediate and permanent reduction of 1 $\mu g/m^3$ in the level of each pollutant [see Eq. (4)].

In practice, the benefits of a control regulation will accrue gradually, as implied by Figure 7, and thus the present value of such benefits must be calculated via Eq. (1). Further, such a calculation must be performed for each separate geographical area and summed to estimate citywide or region-wide benefits in accord with Eq. (4) et seq. Table 11 summarizes the results of the benefit analysis based conservatively on the midrange property value coefficients in Table 10. The benefits to outlying counties of coal conversions occurring almost entirely within the city point to a broader, political issue: in many instances the segment of the population that bears the preponderance of cost does not enjoy a fully proportional share of the benefits.[29]

[29] Although such transfers generally do not affect the regional C/B accounts, they should be evaluated as part of a thorough regulatory analysis.

Table 11   Marginal Increase of Benefits (Present Value, Millions of 1972
Dollars) for Four Residential Pollution Control Strategies in Chicago[a] (Source:
Cohen et al., 1974, p. 79)

Policy Comparison	City of Chicago	Outlying Eight-County Area	Total
1% sulfur law vs no control	194	41	240
Coal ban in 1974 vs 1% sulfur law	51	13	64
Coal and oil ban in 1975 vs coal ban	27	8	35

[a] Basis: 10% discount rate; midrange property-value coefficients (Table 10);
benefits projected from 1973 to 1990, beyond which alternative policies lose
their distinction (Figure 7).

Costs associated with implementation of alternative residential fuel policies in Chicago included
the following elements: equipment conversion, increased rate of abandonment of marginal buildings,
operating and maintenance costs, fuel prices, and governmental administration and enforcement costs.
Fuel prices are influenced primarily by factors such as world oil prices having nothing to do with
the choice of the fuel-related air quality regulation for Chicago (exogenous price changes) and to a
much lesser extent by the increased demand for gas and oil associated with such regulations (endogenous
price changes).

Table 12 compares the marginal benefits and costs for four alternative residential fuel policies
(including "no control").

Table 12   Marginal Increases in Benefits and Costs (Present Value, Millions
of 1972 Dollars) for Four Residential Pollution Control Strategies in Chicago[a]
(Source: Cohen et al., 1974, p. 96)

Policy Comparison	Benefits	Costs	Net[b]
1% sulfur law vs no control	240	6	230
Coal ban 1975 vs 1% sulfur law	64	(7)[c]	71
Coal and oil ban 1975 vs coal ban 1975	35	38	(3)

[a] Basis: (see note a in Table 11).
[b] Net = Benefits − costs.
[c] The large saving on fuel costs with the estimated shift from low-sulfur coal
to natural gas accounts for this net savings. Thus one would expect a 1%
sulfur law to induce a substantial number of conversions.

The authors applied the same cost-benefit methodology to the complete set of regulations governing
emissions of particulates and SO_2 in the Chicago Metropolitan Air Quality Control Region (ANL,
1976). Table 13 compares these regulations from the 1972 Illinois State Implementation Plan with a
set of "best alternatives." Subject to the limitations of their analysis,[30] the costs of directly reducing
emissions from existing, large, isolated sources of SO_2 and particulates far outweigh the benefits.

[30] The principal limitation is the calculation of benefits based solely on reductions in long-term (annual)
levels, in contrast to short-term (1–24 hr), high concentrations. However, if damage coefficients are
constant so that impacts are directly proportional to the integral over time of dose to population
(i.e., a linear theory of damage), then the conclusion in Table 13 regarding electric utilities would be
substantially correct *as it pertains to regional policies* for control of ambient levels of SO_2.

Table 13   Comparison of Illinois SO_2 and Particulate Regulations with the "Best Alternative" for the Chicago Metropolitan Air Quality Control Region (ANL, 1976)

	Benefits and Costs (10^6/yr)[c]					
	Illinois Regulations			Alternative Regulations[a]		
Sector	Benefit	Cost	Net	Benefit	Cost	Net
Power plants	57.5	117.4	−59.9	57.6	6.3	51.3
Incinerators	6.7	0.4	6.3	7.5	0.4	7.1
Industrial process	269.3	2.7	266.6	285.6	3.2	282.4
Industrial fuel combustion	596.0	25.9	570.1	596.4	26.0	570.4
Residential/commercial	62.2	3.7	58.5	62.2	3.7	58.5
Total	991.7	150.1	841.6	1009.3	39.6	969.7

Difference[b] in max. SO_2 (annual arithmetic mean, $\mu g/m^3$) = 0.59
Difference in max. part (annual geometric mean, $\mu g/m^3$) = 0.36

[a] The following regulations were used: power plants—stack height increases except for Waukegan; incinerators—one-half the allowable emissions of Rule 203.e; industrial process—the allowable emissions of Rule 204.f and one-half the allowable emissions of Rule 203.a; industrial fuel combustion—the allowable emissions of Rule 204 and one-half the allowable emissions of Rule 203.g; residential/commercial—a coal ban for space heating, assumed benefits twice those of a residential coal ban and costs equal to the residential coal and oil ban.
[b] Difference = (max. air quality Illinois regulations) − (max. air quality alternative regulations).
[c] Annualized cost A is defined by $A \sum_{n=1}^{N} 1/(1 + r)^n$ = present value.

3   UNIFORM NATIONAL EMISSION LIMITATIONS

Technical and political realities in the United States have led to a set of uniform national emission limitations[1] established parallel to and independent of national ambient air quality standards and in certain instances with minimal consideration of specific health and welfare effects.[2]

There is no single rationale for such uniform national emission limits; rather they satisfy a wide range of purposes:

1. Protect health and welfare in the absence of specific ambient air quality standards,[3] especially where long-term, low-level, irreversible impacts are suspected but not suitably quantified.[4]

2. Protect against locally high concentrations of hazardous pollutants emitted from a small number of isolated sources.[5]

3. Establish and enforce emission limits at a limited number of points of manufacture of small, widely dispersed but in the aggregate, significant sources of air pollution.[6]

[1] Other countries as well, for example, Canada, Federal Republic of Germany.
[2] Most notably in the U.S.: New Source Performance Standards, 42 U.S.C. §7411 (Supp. II 1978), Clean Air Act §111, National Emission Standards for Hazardous Air Pollutants, U.S.C. §7412, Clean Air Act §112, and the Federal Motor Vehicle Control Program, U.S.C. §7521–§7551. In addition, the Environmental Protection Agency promulgates a wide range of guidelines which serve as quasienforceable standards against which State Implementation Plans for nonattainment areas can be judged.
[3] Prior to the CAA 1970, in the absence of national or regional ambient air quality goals, almost all state and local air pollution programs followed this principle and established emission limitations based on proved control technology.
[4] For example, a principal argument against the use of high stacks to enhance the dispersion of SO_2 from large combustion sources, independent of local and regional ground level concentrations, is the potential contribution of SO_2 to acid deposition.
[5] In principle, an ambient standard could be set for and enforced at the boundary of the plant.
[6] The U.S. Federal Motor Vehicle Control Program, 42 U.S.C. §7521–§7551 (Supp. II 1978) is the best example. The certification by many state and local agencies of incinerators according to manufacturer and model is somewhat analogous. If, as in the case of the FMVCP, states are in general prohibited from imposing more stringent control requirements, a conflict can arise where uniform national emission limits are set without full consideration of their impact on the economic and political feasibility of attainment of related national ambient air quality standards (e.g., carbon monoxide and ozone) on a prescribed schedule.

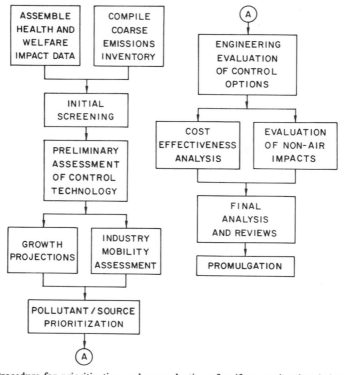

PHASE I - PRIORITIZATION   PHASE II - PROMULGATION
                                  (sequentially in order of
                                  pollutant/source priority)

Fig. 11 Procedure for prioritization and promulgation of uniform national emission standards.

4. Discourage the migration of industry to regions with lax emission restrictions or to areas with cleaner air.[7]

5. Provide a vehicle for imposing nonair policy considerations (e.g., energy and scarce fuel savings).[8]

The governmental agency faced with a myriad of pollutant-source combinations as candidates for uniform emission limitations should follow a systematic selection procedure, such as the two-phase process outlined in Figure 11.[9]

Growth projections[10] are critical for new source standards to minimize future, more costly and politically sensitive requirements for retrofit. An index of the relative importance of such standards among different source categories i for the same pollutant j can be formulated as follows (Monarch et al., 1978):

$$I_{i/j} = K(E_0 - E_n) \sum_{k=1}^{N} (N - k)(C_{rk} + C_{nk}) \tag{8}$$

[7] Closely related to and supportive of the concept prevention of significant deterioration, U.S.C. §7470–§7491.

[8] For example, CAA 1977, §109f, *amending* the Clean Air Act §111(a)(8), 42 U.S.C. §7411(a)(8) (Supp. II 1978), exempting from New Source Performance Standards certain existing oil- and gas-fired power plants converting to coal.

[9] For an extensive examination of this subject see *Priorities for New Source Performance Standards Under the Clean Air Act Amendments of 1977* (Monarch et al., 1978).

[10] Control agency activities such as emissions inventory and growth projections will be discussed in Section 6.

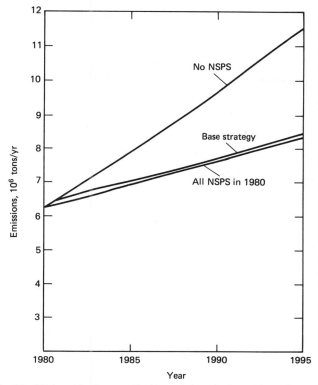

Fig. 12 Nationwide nitrogen dioxide emissions (source: Monarch, 1978).

where

C_{rk} = new industrial plants replacing obsolete current capacity in year k. A linear replacement rate for current capacity C_0 can often be assumed; for example, $C_{rk} = C_0/$[typical plant lifetime, yr].

C_{nk} = new industrial capacity expected in year k; for example, for compound growth at rate R, $C_{nk} = C_0 R (1 + R)^{k-1}$

E_0 = industry average emission rate under current standards

E_n = approximate level of emissions allowable if state-of-art control technology required for new sources

K = representative average load factor (operating/total capacity) for the industry

N = planning time horizon (yr)

Figures 12 and 13 show the results of such a nationwide analysis for nitrogen dioxide and particulates for three cases: (a) absence of federal New Source Performance Standards (NSPS); (b) all such standards promulgated immediately (an impossible, bounding case); and (c) the *base strategy*, a practical phase-in of NSPS in accord with priorities developed via Eq. (8).

One may translate the index $I_{i/j}$ into one (I_{ij}) for intercomparison of pollutants by introducing a weighting factor:[11]

$$I_{ij} = \frac{\chi_j}{S_j} I_{i/j}$$

where

[11] See, for example, Babcock et al. (1978) for development of indices for this and similar purposes.

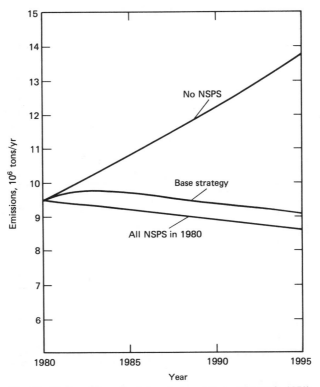

Fig. 13   Nationwide particulate emissions (source: Monarch, 1978).

S_j = applicable ambient standard or other measure of acceptable exposure (e.g., threshold limiting value for occupational health exposure)

χ_j = maximum ground level concentration per unit of plant capacity for a typical (model) plant[12]

Mobility in Figure 11 refers to the migration potential of existing sources. Factors such as age and production economics of existing plants, price competitiveness of alternate supplies (or supply routes) of critical raw materials, transportation costs to key markets, and labor availability should be examined to identify heavy industries prone for any reason to relocate in the near term.

The second phase of the process outlined in Figure 11 is self-explanatory. The parallel consideration of direct costs and cost impacts (see Section 2.3) as well as nonair impacts reflects the close coupling in many cases between the management of air quality and management of other resources (e.g., water, land, energy, skilled labor, capital).[13]

4   PREVENTION OF SIGNIFICANT DETERIORATION (PSD) OF AIR QUALITY

Ambient air quality standards are at the heart of the air quality management process. As described in Sections 1 and 2, such standards embody both the current knowledge of the health and welfare impacts of air pollution and to a limited extent the price that must be paid to enjoy the benefits of clean air. A set of levels, exposure times, and a schedule for attainment summarize the scientific and economic understanding of, and the political response to, the problem. As suggested in Figure 14, urban areas (curve A) with pollution levels above a primary, health-related standard should be under the greatest pressure to attain that goal. A more extended period may be appropriate for attainment of a secondary, welfare-related standard. Undeveloped areas of the country (curve B) would be allowed

---

[12] See Chapter 34 for discussion of techniques for modeling atmospheric dispersion.

[13] Indeed, establishment of NSPS in Section 109(c)(1)(a) of the CAA 1977 requires "taking into consideration the cost of achieving such emission reduction, and any nonair quality health and environmental impact and energy requirements." 42 U.S.C. §7411(a)(1)(A)(ii) (Supp. II 1978).

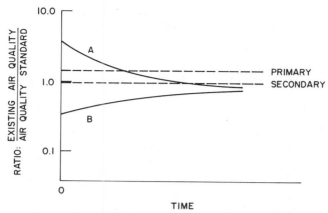

Fig. 14 Attainment and maintenance of primary and secondary ambient air quality standards (source: Stern, 1977).

to pursue industrial commercial growth, if they so choose, only to the extent that the secondary standard is maintained.

4.1 PSD—The Concept

For a number of reasons, the simplicity inherent in Figure 14, curve *B*, whereby undeveloped areas of the country can allow their superior air quality to deteriorate, if only to the level of the secondary standard, has been judged unacceptable. Thus a policy for the prevention of significant (air quality) deterioration (PSD) has emerged in the United States over the past decade.

Many different approaches have been proposed to achieve the aims of a national PSD policy, all of which employ in one form or another more stringent air quality standards and/or emission limits on new sources. According to the scheme embodied in Title I, Part C of the U.S. Clean Air Act as amended in 1977,[1] all areas of the country with air quality superior to the standard are classified initially either as Class I (deserving the greatest degree of protection and including "all (1) international parks, (2) national wilderness areas which exceed 5000 acres in size, (3) national memorial parks . . . , and (4) national parks. . . .")[2] or as Class II (allowing moderate increases in pollution levels). Redesignation of Class II areas to Class I or Class III (maximum allowable increases) can occur only in accord with proscribed procedures. For each area classification, maximum allowable pollution increments above baseline or reference levels[3] are defined in the law. A new source[4] must, in its application for a construction permit, provide air quality data and modeling estimates to assure that it, in combination with all other sources that commenced operation or received permits to operate since the reference date (January 6, 1975), will not violate any applicable increment. Further, the source must install the Best Available Control Technology, determined on a case-by-case basis with consideration given to cost, energy, and other environmental impacts to be at least as stringent as any applicable new source performance standard.[5] Table 14 lists the increments established for sulfur dioxide along with the corresponding secondary ambient air quality standard.[6]

---

[1] As with other issues in air quality management, it is not the purpose of this chapter to provide details concerning specific legislation and regulations governing PSD, but rather to explore in theory and by example the underlying logic of the concept and provide general guidance in the translation of a reasonable policy into practical clean air programs. The reader interested in a thorough history of PSD up to May 1977 should consult Stern (1977). An excellent review of legal issues is provided by Currie (1980).

[2] Section 127 of the CAA 1977, *amending* the Clean Air Act §162(a), 42 U.S.C. §7472 (Supp. II 1978).

[3] Taken by the U.S. EPA as levels known or assumed to exist as of January 6, 1975 [40 C.F.R. §52.21 (1982)].

[4] Generally with uncontrolled emissions in excess of 9.1 × kg/yr (100 tons/yr) or a like expansion (modification) of an existing source.

[5] See Section 3 of this chapter.

[6] The Clean Air Act offers little guidance as to the allocation of the allowable increments among competing sources. This matter is considered in Section 8 along with other aspects of air quality management related to land use.

Table 14   PSD Increments for Sulfur Dioxide [CAA 1977, §127, Amending the Clean
Air Act §163, 42 U.S.C. §7473 (Supp. II 1978)]

Averaging Time	Maximum Allowable Increment, $\mu g/m^3$			Secondary Air Quality[b] Standard, $\mu g/m^3$
	Class I[a]	Class II	Class III	
Annual average	2	20	40	60[c]
24-hr	5	91	182	260[c]
3-hr	25	512	700	1300

[a] The Clean Air Act §165, 42 U.S.C. §7475(d)(2)(D)(iii)(Supp. II 1978), provides certain exceptions to these Class I increments, most notably a relaxation allowing increases in the maximum 3-hr increment up to 130 $\mu g/m^3$ in low terrain areas and 221 $\mu g/m^3$ in high terrain areas to occur on no more than 18 days in any annual period.

[b] The secondary standard serves as a ceiling for all PSD areas. Thus, the allowable increment in a Class III area with a baseline 3-hr SO_2 level of 800 $\mu g/m^3$ is 500 $\mu g/m^3$.

[c] The annual secondary standard for SO_2 was revoked in 1973. The 24-hr value was always viewed as a guide and not enforceable [38 *Federal Register* 25,678 (Sept. 14, 1973)].

4.2   PSD—Rationale

The prevention of significant deterioration of air quality is a concept embodied in the opening stanzas of the Air Quality Act of 1967: "To protect and enhance the quality of the Nation's air resources so as to promote the public health and welfare and the productive capacity of its population" [Section 101(b) of the Air Quality Act of 1967 (CAA 1967)]. To examine more closely the several purposes behind PSD policy in the United States, one must turn to Section 127(a) of the CAA 1977:

> (*1*) *to protect public health and welfare from any actual or potential adverse effect which in the Administrator's judgment may reasonably be anticipated to occur from air pollution or from exposures to pollutants in other media, which pollutants originate as emissions to the ambient air), notwithstanding attainment and maintenance of all national ambient air quality standards;*

> (*2*) *to preserve, protect, and enhance the air quality in national parks, national wilderness areas, national monuments, national seashores, and other areas of special national or regional natural, recreational, scenic, or historic value;*

> (*3*) *to insure that economic growth will occur in a manner consistent with the preservation of existing clean air resources;*

To these officially stated purposes might be added one other:

> (*4*) *Abandonment of a policy of prevention of significant deterioration will encourage flight of industry—and jobs—from areas where pollution levels are approaching or exceed the minimum federal standards to cleaner areas requiring less controls on industry.*[7]

PSD provisions can thus be viewed as an addendum to the two fundamental elements of the air quality management process: health and welfare related ambient air quality standards (NAAQS) and uniform emission limits for new sources (NSPS). What additional benefits do these provisions provide that cannot otherwise be captured by NAAQS and NSPS? To what extent are PSD constraints a surrogate for larger economic and social policies? Each of the above-listed purposes will be very briefly examined (in numerical order) in the light of these questions to determine if there is a defensible rationale for, and a desirable form of, PSD requirements.

1.   As presented in Section 2, primary and secondary ambient air quality standards can be established with any degree of conservatism and associated ranges of benefits and costs satisfactory to the political process. The concept of PSD adds nothing that cannot be embodied in the NAAQS.

2.   Certain public lands, especially pristine wilderness areas, may deserve a greater degree of protection (i.e., a tighter secondary standard) where damage to wildlife or deterioration of scenic

[7] H.R. Rep. No. 1175, 94th Cong., 1st Sess. 111 (1976) [Committee on Interstate and Foreign Commerce, *Clean Air Amendments of 1976,* as quoted by deNevers (1980)].

beauty can be demonstrated or reasonably hypothesized.[8] Visibility is the principal concern in the latter, a quality of the atmosphere which can be related (however crudely at the present time) to pollutant emissions, and treated as a pseudo air quality standard.[9]

3. To the extent that economic growth is inconsistent with the goal of preservation of existing clean air resources, there are ample remedies elsewhere in the air quality management process (e.g. NAAQS, NSPS, permit requirements).

4. Air pollution control regulations designed to attain and maintain NAAQS no doubt provide some incentive for large emission sources to locate in areas with lower levels of air pollution. Such a dispersive tendency may run counter to regional land-use plans and to the desires of certain urban areas for geographically confined growth.[10] However, these are issues that can properly be addressed directly via traditional zoning and other land-use planning and approval procedures. The federal government is certainly stretching its authority when, in the name of clean air, it tells the city council of a rural town that they cannot choose to encourage industrial growth (in compliance with NSPS) to the degree that pollution levels rise to secondary ambient standards whereas elsewhere in the state developed urban areas have only to attain and maintain that same standard.[11]

This review of the various stated objectives of the PSD policy leaves only one valid purpose—the protection of select public lands—and even in this important case, the structural flexibility of a welfare-oriented secondary standard,[12] expanded to include provisions governing visibility, will satisfy this objective. Thus, in theory and in reasonable practice, PSD as a *separate* air quality management concept offers no unique advantages over the combination of primary and secondary ambient air quality standards, new source performance standards, and traditional zoning and land use controls.

5 TAX POLICIES AND MARKETABLE EMISSION RIGHTS

No discussion of alternative philosophies for air quality management would be complete without an examination of the potential of taxation as a means to attain clean air standards in a more cost-effective manner.[1] The concept in its modern form is considered to originate with the British economist A. C. Pigou, who recognized "a heavy uncharged loss on the community, in injury to buildings and vegetables, expenses for washing clothes and cleaning rooms, expenses for the provision of extra artificial light, and in many other ways" (Pigou, 1970). As such, this observation of the London scene provides overall justification for some form (not necessarily taxation) of government intervention in a market-place where the cost of air pollution is not internalized in the costs of production. Pigou went further, however, and advocated a system of emission taxes to resolve this disparity. More recently, economists (e.g., Baumol, 1972) and others in government and the private sector[2] have argued that an emission tax on pollutants such as SO_2 would be preferable to the myriad of federal, state, and local emission regulations. Such preference derives presumably from assumptions of greater economic efficiency in the expenditures for pollution control and reduced governmental costs for enforcement, both inherent in a market-oriented approach to resolve a market-related imbalance.[3]

Establishing the tax rate for an emitted pollutant is understandably a challenging proposition, since embodied in a single number (T \$/kg)[4] are several factors: public policy concerning setting of air quality standards; a model of the likely response of emitters; estimates of the cost and performance

[8] Equivalently a state might adopt a very stringent ambient standard applicable in areas where a pollutant-sensitive crop is dominant.

[9] As is implicitly done in §127 of the CAA 1977, amending §163 of the Clean Air Act, 42 U.S.C. §7473 (Supp. 1978).

[10] Most industrial states supported the PSD doctrine and many filed *amicus* briefs in support of the Sierra Club suit which gave this doctrine the force of law (deNevers, 1980).

[11] It may be that Class II increments established in the CAA 1977 will permit modest growth and that in many cases an area can be redesignated Class III. Such flexibility is of great value; yet it does not justify the establishment of the PSD increments in the first place.

[12] In effect Class I increments (and their special exceptions) in the Clean Air Act set an extremely low secondary standard for parks and wilderness areas.

[1] This discussion of taxation is derived primarily from two comprehensive studies of a national tax on sulfur emissions commissioned in 1973 and 1974 by the EPA: Bingham et al. (1973) and Bingham and Miedema (1974).

[2] Most vocally during the administration of President Nixon.

[3] Interestingly enough, private industry initially pressed the Congress for an SO_2 emission tax (vs emission limits) with public interest groups in opposition. Subsequently during congressional hearings the two sides switched positions on the issue.

[4] In principle, the tax rate T can be a function of the emission rate itself, and could reflect such factors as stack height and geographical location as is the case for emission-limiting regulations.

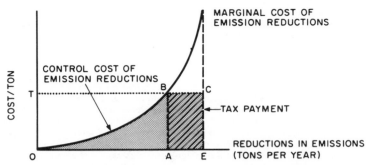

Fig. 15 Emission source behavior in response to a tax T ($/kg) on emissions (source: Bingham et al., 1973).

of current and future pollution control technologies; and the coupling between anticipated (reduced) emissions and ambient air quality.[5]

In the broadest sense, the tax represents the final output of a cost-benefit analysis and could in principle be derived from Figure 3. However, the inherently greater economic efficiency of an emission tax, at least in the ideal, would lower the marginal cost curve and thus lead to enhanced air quality at the intersection with the curve of marginal benefits. In a narrower sense, given an established air quality standard (e.g., as promulgated under the Clean Air Act of 1970), the tax could be viewed as a cost-effective approach to achieving the requisite reductions in emissions.

The most straightforward model of the response of an emissions source to a tax on emissions is that emissions will be reduced in a manner which minimizes the net (after tax) present value of total costs. As suggested by Figure 15, this point occurs when the marginal cost of emission reduction equals the tax T. Here the area OAB represents the total cost (present value) of the control strategy and area $ABCE$ the tax payment, where the initial (uncontrolled) emission rate is E and the final rate is $E - A$. Mathematically, the point A in Figure 15 can be established by selecting the pollution control/tax payment strategy j which minimizes the present value PV_j in an equation such as (Bingham and Miedema, 1974, p. 39):

$$PV_j = K_j - \frac{\Theta K_j}{L} \sum_{n=1}^{L} \left(\frac{1}{1+r}\right)^n \tag{9}$$

$$+ (1 - \Theta) \sum_{n=1}^{N} \left\{ \left(\frac{1}{1+r}\right)^n \left(V_{jn} + TE_{jn}\right) \right\}$$

where

K_j = initial investment cost of the jth strategy

N = planning horizon in years

Θ = marginal corporate income tax rate

L = number of years over which capital is depreciated on a straight line basis (an alternative form of depreciation can readily be substituted)

r = opportunity cost of funds to the industry

V_{jn} = annual variable cost associated with strategy j in year n, including costs of premium fuels, less value of recovered materials (e.g., sulfur) and other savings (e.g., enhanced efficiency of production through process changes)

T = emission tax ($/kg)

E_{jn} = emissions (subject to tax) in year n under strategy j

This formulation does not consider the effect of decreased demand (lost revenues) due to price increases associated with strategy j. Estimates of such changes in demand in response to a tax on sulfur emissions suggest that the effect will, in general, be of secondary importance (Bingham et al., 1973). Note also that PV_j is the net *private* cost of strategy j. Given that the tax payment TE_j is a transfer of funds

---

[5] Of course, it can be argued that most of these considerations are embodied in the setting of an ambient air quality standard. However, conceptually the standards approach is more satisfying because it provides a more readily understood statement of public policy.

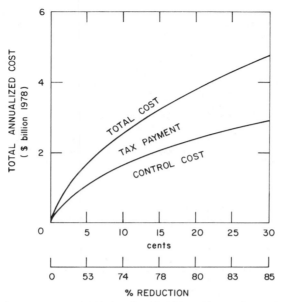

Fig. 16 Cost and effectiveness of a national tax on sulfur emissions (derived from Bingham et al., 1973). Tax (¢/lb of sulfur emissions), percent reduction in emission levels (see note 3).
Notes: 1. Totals for six industry classes together emitting over 90% of all SO₂. 2. Opportunity cost of capital $r = 0.12$ in Eq. (9). 3. New source performance standards for SO₂ are assumed in effect, regardless of sulfur tax. 4. Tax is stated in terms of lb. sulfur emissions (50% by weight of SO₂ emissions).

from one entity (private) to another (public), it can be argued that there is negligible social cost associated with such a payment. The social cost of government enforcement of the tax program is not considered by Eq. (9).

Figure 16 summarizes the cost and effectiveness of a tax on sulfur emissions in the United States from the five categories of sources which together comprise over 90% of such emissions: steam–electric power plants, residential/commercial fuel combustion, petroleum refineries, sulfuric acid plants, and primary nonferrous smelters. For each source category at least two (and in the case of steam–electric utilities about 1000) different strategies (j) were considered via Eq. (9).

Setting aside, for the moment, the matter of taxation and returning to the concept of explicit ambient air quality standards, the concept of *marketable emission rights* emerges as an economically efficient approach to attaining clean air goals.[6] More specifically, where the margin between current and maximum allowable levels of ambient air pollution is small compared to the demands of new pollution sources to consume the available increment, the "right to pollute" can become a marketable commodity. Such a condition can occur in the allocation of PSD increments, in urban areas implementing a rigid compliance plan to attain ambient air quality standards (so-called nonattainment areas), or in regions with pollution levels close to but below those same air quality standards.

A company wishing to initiate or expand existing operations in a given geographical area must develop an overall emission strategy which considers alternative processes and the associated emissions, pollution control options, and the costs of emission rights. The owner of an existing source of pollution in compliance with any applicable emission limits may choose to enter this market by comparing the costs of additional controls on the plant to the sales (or lease) value of emission rights generated thereby. In either case, Eq. (9) provides a framework for determining the breakeven point between investment in control equipment or process modifications and purchase (or sale) of emission rights. The taxation term TE_{jn} in Eq. (9) would then characterize the breakeven cost of emission rights.[7]

[6] The relationship between emission rights and land-use based concepts for allocation of emission rights (e.g., emission density zoning) will be discussed in Section 8.
[7] The single value T could be replaced by the time series T_n to describe a lease or lease–purchase arrangement. An outright purchase (or sale) would be characterized by the price T_0 and $T_n = 0$, $n \geq 1$.

There is thus a duality between the concepts of taxation of emissions and the marketing of emission rights. The former implicitly embodies the target air quality standard since presumably the tax would be set at a level sufficiently high to induce the desired reductions in emissions; the latter would be priced at a level reflecting the difference between current and maximum allowable emission levels.[8]

Implementation of a tax policy to achieve clean air goals will be difficult because one must predict the response of owners of pollution sources to a given tax T. On the other hand, establishment (a) of maximum allowable emission limits derived from desired ambient air quality levels (standards) and (b) a free-market mechanism via emission rights for the private sector to seek the most cost-effective solution is more direct and predictable in terms of protection of public health and welfare. Yet in principle a system of marketable emission rights will afford the same overall economic efficiency as a program of taxation. Both policies provide financial encouragement for the development of lower cost, higher efficiency processes for reducing emissions. Geographical sensitivity is necessary to ensure against excessively high, localized pollution levels. The practical difficulty of implementing a taxation program with such sensitivity is a major weakness. In an emission rights market, transactions would be tied to a permit system or emission reduction bank[9] so that one can in principle build in the desired geographical sensitivity.

6 DESIGN OF REGULATIONS FOR ATTAINMENT AND MAINTENANCE OF NATIONAL AMBIENT AIR QUALITY STANDARDS

The process for development of any national ambient air quality standard (NAAQS) should have included an assessment, however cursory, of the control programs likely to be required to attain alternative levels and schedules for that standard. Thus, the ensuing development of specific plans for attainment and maintenance of the NAAQS will be under way at the time the NAAQS are promulgated. As a minimum, preliminary data on air quality, meteorology, and emissions should have been assembled and models for translating emission reductions into air quality improvements developed and validated. In practice, the extent of available data and methods and the sophistication of the resulting analyses has increased significantly since the initial attempts at air quality management following the Air Quality Act of 1967 and the Clean Air Act of 1970.

This section describes the steps appropriate to the development of regulations for attainment and maintenance of NAAQS within a specific air quality control region (AQCR).[1] Examples emphasize stationary sources. Special attention is devoted to mobile source issues in Section 7. Relationships between pollution control policies and regional land use are examined in Section 8.

6.1 Intergovernmental Relations

Although the state agency designated by the governor as having prime responsibility for implementation of clean air programs will assume the lead in organizing ensuing planning activities, it is not necessarily the most appropriate agency for carrying out each of the tasks, especially where the range of measures for attainment and maintenance of the NAAQS overlaps regional land use and transportation policies formulated with objectives other than enhanced air quality.

As suggested by Figures 17 and 18, a very critical first stage of the planning process entails an inventory of affected governmental agencies, a review with key officials of that process, and *documented* agreements as to the role of each agency. Private sector (citizen, professional, industry) involvement must be coordinated as well during this initial planning stage.

6.2 Emission Inventory

With the evolution of more sophisticated and comprehensive approaches to air quality management, the emission inventory plays a role of ever increasing importance. As critical input to the design of regulations, establishment of emission offsets and issuance of PSD permits, official emission data files will be increasingly scrutinized for completeness and accuracy, and subject to court challenge. Thus all reasonable efforts should be made to assure that an inventory of high quality is developed and maintained.

Constant changes in emission source characteristics, inherent inaccuracies in emission factors and engineering analyses, human error, and perhaps above all practical limitations on available resources together imply that the management of an air pollution control agency make a careful assessment of the costs, risks, and benefits associated with the assembly and maintenance of inventories of varying

---

[8] More properly, ambient air pollution levels from which allowable emission levels can be derived via a suitable model of atmospheric chemistry and dispersion.

[9] See Section 8.4.3.

[1] This discussion of the design of regulations serves, as well, as a guide to acquisition and analysis of data suited for phase I of Figure 1, the development of NAAQS.

JURISDICTION RESPONSIBLE

POSSIBLE AQMP ELEMENT	FEDERAL	STATE	REGIONAL AGENCY	CITY	COUNTY	PRIVATE SECTOR
SOURCE CONTROL	O	● ■		O ■	O ■	□
LAND USE	O	O	O	● ■	● ■	□
TRANSPORTATION HIGHWAY TRANSIT	O O	O □ ■ O	● ● □ ■	O □ ■ O	O □ ■ O	

O   ADVISORY ROLE IN PLANNING FUNCTIONS
●   PRIMARY RESPONSIBILITY FOR PLANNING FUNCTIONS
□   RESPONSIBILITY FOR PHYSICAL DEVELOPMENT
■   PRIMARY RESPONSIBILITY FOR CONTROL OR ENFORCEMENT

Fig. 17 Summary of the distribution of responsibilities among agencies (two federal, four state, three regional interstate, and six local city and county) prior to the development in 1971 of the implementation plan for the St. Louis air quality control region (source: EPA, 1947a, p. 44).

scope, detail, and accuracy (level of confidence), such assessment to be performed *before* embarking on a new or substantially modified system. A framework for planning, assembling, and maintaining an emission inventory is outlined in Table 15; typical sources of error are enumerated in Table 16.

The emission inventory is first and foremost a descriptive catalogue of significant, individual sources of one or more air contaminants. These so-called *point sources* are typically industrial processes and fuel combustion activities exceeding a threshold based on: (a) actual emissions of a given pollutant (e.g., kg/hr), (b) potential emissions of a given pollutant in the absence of any pollution control equipment, (c) size or throughput of activity (e.g., all grain elevators with storage capacity greater than 1 million bushels; all coal-fired boilers with heat input greater than 100 million Btu/hr), or (d) type of source (e.g., asphalt batch plants). Understandably, any system thresholds should reflect as a minimum the toxicity of the pollutants and the size distribution of sources, thereby capturing at least 80% of regionwide emissions from stationary sources.

Smaller stationary sources as well as dispersed natural and anthropogenic sources are most conveniently grouped in classes (e.g., single family/residential/light-oil fuel combustion; fugitive dust) with emissions of like pollutants aggregated as *area sources* with areal sizes varying from 0.5 km² in a central business district to 100 km² for rural areas. Mobile source emissions are frequently merged into the same regional grid of area sources with major highway links accounted for separately as *line sources.*

Geographical coordinates for a point or area source should be expressed in the Universal Transverse Mercator (UTM) system (EPA, 1973) developed by the U.S. Army to provide for the projection of square grid zones with convenient measuring units (U.S. Army, 1958).

Given that state and regional emission inventory systems tend to progress over time from initial, crude surveys which capture in limited detail only major sources, homogenizing the rest in comparatively large areal blocks, to extensive documentation of the operating characteristics of tens of thousands of plants, each with one or more separately identified unit processes, there is no single way to design, compile, and maintain such systems. However, progression along this scale of complexity is governed primarily by the extent of resources devoted to the process rather than significantly different conceptual approaches. The following three subsections summarize the most common approaches to an inventory of industrial process activity, fuel combustion and incineration, and mobile sources.[2]

6.2.1 Industrial Processes

The inventory should focus on major point sources comprising in the aggregate at least 80% of the regionwide emission of a given pollutant. Certain low-level emitters may need to be included as well if implicated in locally high pollution levels. A phone or mail survey may be employed to develop rapidly a first-cut inventory, but followup site visits are essential. Background information of great value to the field engineer prior to a site visit includes a generalized flowchart, a description of the

[2] More extensive discussions of procedures can be found in EPA (1973).

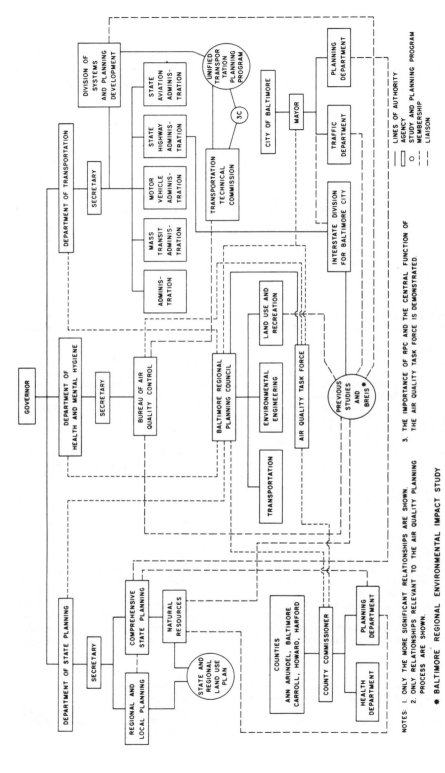

Fig. 18 Detailed map of relationships between governmental units with a significant interest in the development in 1971 of an implementation plan for the Baltimore air quality control region (EPA, 1974a, p. 96).

NOTES: 1. ONLY THE MORE SIGNIFICANT RELATIONSHIPS ARE SHOWN.
2. ONLY RELATIONSHIPS RELEVANT TO THE AIR QUALITY PLANNING PROCESS ARE SHOWN.
3. THE IMPORTANCE OF RPC AND THE CENTRAL FUNCTION OF THE AIR QUALITY TASK FORCE IS DEMONSTRATED.

* BALTIMORE REGIONAL ENVIRONMENTAL IMPACT STUDY

LINES OF AUTHORITY
☐ AGENCY
☐ STUDY AND PLANNING PROGRAM
○ MEMBERSHIP
--- LIAISON

Table 15   Phases in the Development of an Emission Inventory

---

I.   Strategic planning
    Establish goals: role(s) of inventory
    Establish objectives: overall specifications and schedule to satisfy goals
    Emissions survey: preliminary assessment of tasks
    Resource analysis: estimate staff and funding requirements and availability
    Organization: Assignment of responsibilities within the lead agency and among supporting orga-
        nizations.

II. Task planning
    Data acquisition: specification of sources; formats
    Analysis: establish and document procedures (including data processing requirements)
    Schedule: develop flowcharts
    Resource analysis: detailed assessment; staffing and funding plan finalized
    Organization: detailed assignment of responsibilities

III. Implementation—preparatory
    Supporting materials: preparation and testing
    Data processing: code development and testing
    Emission sources: prepare directories
    Staff: training and deployment

IV. Implementation—assembly
    Data acquisition: field surveys
    Analysis: (source-by-source)
    Validation: identify outriders; random checks
    Reporting: listings of sources and characteristics by type and location.

V.   Implementation—maintenance
    Periodic update
    Expansion: additional sources and/or detail
    Merge with permit and compliance files
    Validation
    Analysis: emission statistics; coupling with dispersion models
    Reporting: source listings; statistics

---

industrial process, and any data on file specific to the plant. The former may be found in Compilation of Air Pollutant Emission Factors (AP-42, EPA, 1978a) and in more extensive reports on the operating characteristics of, and pollution control options for, important source types. Site specific data should be available from regional source files and via computer printout from any prior emission inventory.

Table 17 shows a portion of the table of emission factors for iron and steel mills. Emission rates for uncontrolled processes are listed first, followed by those for representative types of control. All are expressed here in terms of weight of pollutant per unit weight of output. Of course, where available, site specific measurements or detailed engineering estimates of uncontrolled emissions and/or control equipment efficiency are preferable to the tabulated values. The emission factor ratings characterize the relative accuracy of the tables in accord with the following scheme: A—excellent, B—above average, C—average, D—below average, E—poor.

To incorporate the myriad of small industrial and commercial process emissions within a region-wide grid of equivalent areal sources requires artful interpretation of a wide array of data. Directories of manufacturers and trade associations are particularly useful. Correlations between levels of different types of commercial activity (e.g., dry cleaning establishments) and regional planning variables (e.g., population) are useful, particularly if one has an estimate of county- or regionwide activity and wishes to prorate among smaller area sources (e.g., see EPA, 1974b).

6.2.2   Fuel Combustion and Incineration

Major sources of air pollution from fuel combustion and incineration are treated similarly to industrial process point sources as described in Section 6.2.1. The most straightforward approach to residential, commercial, and light-industry space heating entails the development of an inventory of buildings grouped by size [number of dwelling units (DU) or gross square feet of floor area] and geographically distributed by census tract. Conversion to air emissions requires consecutive application of estimates

Table 16   Errors and Error Sources in the Emission Inventory Process (Source: Armentrout, 1979)

Error	Potential Sources
Missing facilities or sources	Permit and inventory systems out of phase; errors in estimating potential emissions; lost paperwork; problems with computer file updates
Duplicate facilities or sources	Name changes through corporate acquisitions; use of multiple data sources with different source numbering schemes
Missing operating or technical data	Ambiguous data request forms; intentional deletion by facility staff; inadequate followup procedures; inadequate project control, i.e., no tentative indication of inventory size
Erroneous technical data	Misinterpretation of data request instructions; assumed units, faulty conversions, etc.; intentional misrepresentation by the facility; poor handwriting
Improper facility location data	Recording coordinates of facility headquarters instead of the operating facility; inability of technicians to read maps; changes in UTM zones
Inconsistent area source categories or point source	Failure to designate inventory cutoffs
Inaccurate or outdated data	Mixed use of primary and secondary data without a standard policy
Errors in calculations	Transposition of digits; decimal errors; entering wrong numbers on a calculator; misinterpreting emission factor applications
Errors in emission estimates	Imprecise emission factors; applying the wrong emission factor; errors in throughput estimates; improper interpretation of combined sources; errors in unit conversions; faulty assumptions about control device efficiency; ranges of sulfur/ash contents in fuels
Reported emissions wrong by orders of magnitude	Recording the wrong source classification code for subsequent computer emission calculations; ignoring implied decimals on computer coding sheets; transposition errors; data coding field adjustment

of heating requirements on a per *DU* or per square foot basis for each size group, market shares of heating fuels for each size grouping, typical heating system efficiencies, and standard emission factors. As a check on such space heating estimates one may aggregate by county and compare with county-wide fuel-use data published elsewhere. The latter may also be employed to normalize the disaggregated estimates.

6.2.3   *Transportation*

Motor vehicles (light-duty gasoline, heavy-duty gasoline, and heavy-duty diesel), airplanes, railroad engines, and sea and river vessels fall under this general heading. In concept, the emission inventory procedure is straightforward: characterize the level of activity of each source category by location (e.g., area source) and apply suitable emission factors. Intense localized sources such as an interstate highway or rail yard are best treated as separate line, point, or compact area sources.

Emission factors for motor vehicles should reflect the distribution of vehicles by type and age, weighted by their relative use and corrected for variations in average speed. National averages are useful for a first cut but regionally sensitive values are ultimately required if a transportation control plan is necessary to attain one or more ambient air quality standards in an AQCR. For example, the stepwise procedure for developing a composite emission factor for light-duty gasoline vehicles consistent with emissions measured by the Federal Test Procedure is outlined in Table 18.[3] Calculations such as this should be computerized; standard codes are available (EPA, 1978b).

---

[3] Section 7.2 addresses the relationship to emission standards in the Clean Air Act (CAA 1977), deterioration of emissions over time, and the effectiveness of inspection/maintenance programs.

Table 17   Emission Factors for Iron and Steel Mill Emission Factor Ratings: A (Particulates and Carbon Monoxide), C (Fluorides)

	Total Particulates		Carbon Monoxide	
Type of Operation	lb/ton	kg/MT	lb/ton	kg/MT
Pig iron production				
Blast furnaces				
Ore charge, uncontrolled	110	55	1750 (1400–2100)	875 (700–1050)
Agglomerates charge, uncontrolled	40	20		
Total, uncontrolled	150 (130–200)	75 (65–100)	1750 (1400–2100)	875 (700–1050)
Settling chamber or dry cyclone	60	30		
Plus wet scrubber	15	7.5		
Plus venturi or electro-static precipitator	1.5	0.75		
Sintering				
Windbox, uncontrolled	20	10		
Dry cyclone	2.0	1.0		
Dry cyclone plus electrostatic precipitator	1.0	0.5		
Dry cyclone plus wet scrubber	0.04	0.02		

Table 18   Composite Emission Factor for Light-Duty Vehicles (Source: EPA, 1978b, p. 11)

Enpstwx = SUM(Cipn*Min*Ripstwx*Aip*Lp*Uipw*Hip)

Where all lower case letters are subscripts and:

SUM( ) = summation over model year (i), from the calendar year for which emission factors are being calculated ($i = n$) to the calendar year 19 yr previous ($i = n - 19$)

Enpstwx = Composite emission factor in g/mile for calendar year n, pollutant p, average speed s, ambient temperature t, fraction cold operation w, and fraction hot start operation x

Cipn = The FTP (1975 Federal Test Procedure) mean emission factor for the ith model year light-duty vehicles during a calendar year n, and for pollutant p

Min = The fraction of annual travel by the ith model year LDVs during calendar year n

Ripstwx = The temperature, speed, and hot/cold correction factor for the ith model year LDVs for pollutant p, average speed s, ambient temperature t, fraction cold operation w, and fraction hot start operation x

Aip = The air-conditioning correction factor for the ith model year LDVs, for pollutant p

Lp = The vehicle load correction factor for pollutant p

Uipw = The trailer towing correction factor for the ith model year LDVs, for pollutant p, and for fraction of cold operation x

Hip = The humidity correction factor for the ith model year LDVs, for pollutant p

6.3 Projections of Emissions

The emission inventory, especially when correlated with current air quality data, provides the status of significant point and area sources throughout the region. The combination over a 5- to 10-yr planning horizon of future pollution controls on existing sources and construction of new sources subject to federal, state, and possibly local new source performance standards will determine the degree of stringency of regulations necessary to attain and maintain the NAAQS or comply with PSD increments applicable within or proximate to the region.[4]

Procedures for projecting emissions are straightforward and, though time consuming in many aspects, follow the steps taken in assembling the initial inventory. It is recommended that projections of emissions from *existing* point and area sources reflect changes in emissions only to the extent that certain of these sources may, because of age or market factors, cease or significantly curtail operations or otherwise alter fuel use or process technology *in the absence of any current but as yet not implemented* control regulations. In subsequent analyses, alternative control regulations will then be imposed on this population of existing sources.

Projected *new* point sources (or expansions of existing sources) should be accounted for separately. Where a uniform national emission standard (Section 3) will govern such sources, the projected emissions should reflect all control requirements, or if as yet unspecified, a plausible value or range thereof.

Consultation with planning agencies[5] responsible for functions such as land use and transportation planning is a prerequisite for developing 5- and 10-yr projections of emissions. Ideally such agencies will take the lead in, or actively contribute to, these projections. They will have on hand most of the necessary macroeconomic data and, of great importance, will know how such data are interpreted and applied by the state and affected local governments in other related planning activities.

Concluding Section 6.3, each major category of emission source will be characterized by a preferred approach to emission projections and, if appropriate, a less demanding secondary one.

6.3.1 Industrial Processes

Preferred method: Focus on largest point sources responsible in the aggregate for at least 80% of all emissions of the pollutant(s) of interest. Establish specific plans for fuel or process changes, plant expansions, and other new construction via interviews with company officials. Important but more dispersed sources can be assessed via interviews with trade associations or by evaluation of surrogate variables (e.g., asphalt plants—projected highway construction and repair activities). Evaluate plans for establishment of, or expansions to, major industrial parks.

Secondary method:[6] Utilize national projections of employment and earnings disaggregated geographically (typically by Standard Metropolitan Statistical Area but also available by federally designated air quality control region) and by industrial group (typically over 35 groups at the two-digit SIC[7] level).[8] Again, for accuracy and consistency the preferred sources of such statistics are the cognizant regional and local planning agencies. Industry growth projections derived from related employment forecasts would be applied *in place*[9] to existing point and area sources to the extent that those source categories were not handled via the preferred method.

6.3.2 Fuel Combustion and Incineration

Preferred method: Projections of fuel combustion at industrial sources should follow the methods outlined in Section 6.3.1. Future emissions from central station electric utilities and large central heating plants should be developed on a plant-by-plant basis through interviews or capital expansion plans on file with the state or local public utility commission. Special attention must be paid to anticipated changes in fuel usage mandated or encouraged by national energy policies.

Residential and commercial fuel combustion for space heating, hot water, and miscellaneous small-

---

[4] For a detailed guide and workbook on the projection of emissions at county and subcounty levels see *EPA Guidelines*, Vol. 13, 1974 (EPA, 1974b) and *EPA Guidelines*, Vol. 7, 1975 (EPA, 1975a).

[5] Including the Metropolitan Planning Agency (MPA) as specified in §129(b) of the CAA 1977, *amending* the Clean Air Act §174, 42 U.S.C. §7504 (Supp. II 1978).

[6] To the extent that regional implementation plans to date have considered growth in the industrial sector, they have most often been based on this secondary method due to its simplicity.

[7] SIC—Standard Industrial Classification

[8] A standard data source is the OBERS projections developed originally by the Office of Business Economics (OBE—presently the Bureau of Economic Analysis) of the U.S. Department of Commerce and the Economic Research Service (ERS) of the U.S. Department of Agriculture (EPA, 1974).

[9] This assumption will generally lead to *overprediction* of ground-level air pollution. An important exception could result from the neglect of a new heavy industry park, especially if sited unfavorably with regard to terrain or regional meteorology.

process steam applications should be treated in a manner similar to the development of the initial emission inventory. Population, and thus residential/commercial density, are conveniently examined by census tract. Projections should be in terms of dwelling units or gross square feet of floor area, and thence to kilocalories (Btu's) where the conversion to thermal requirements reflects both building size and evolving energy conservation practices. Estimates of fuel use and thus emissions (via emission factors) entail examination and likely revision of current market shares of coal, oil types, natural gas, and electricity.[10] Specialized sources of data include local utility companies, fuel distributors, state and local public utility commissions, and energy departments at various levels of government.

Secondary method: As above except with a coarser degree of aggregation.

6.3.3 Transportation

As with the basic emission inventory, the level of detail required for projections of emissions from transportation-related activities depends on the problem under investigation and nature of algorithms characterizing aerochemistry and atmospheric dispersion in the air quality models to be employed. Thus, the projection of transportation-related emissions represents a direct extension of the initial inventory utilizing state and local transportation planning departments to forecast vehicle-kilometers traveled and average speed for existing and planned limited access highways (by designated line segments or links) and for subcounty areas. Emission factors for different classes[11] of vehicles must be revised to reflect the vehicle age distribution and associated emission control requirements.[12]

6.4 Evaluation of Alternative Control Regulations

The pieces of the air quality management puzzle come together as shown in Figure 19, a flowchart for the evaluation of control strategies (alternative sets of control regulations) designed to comply with applicable ambient air quality standards.

A reference (baseline) data base that captures emissions, ambient air pollutant levels, and pertinent meteorology over a time period representative of the range of atmospheric conditions likely to be experienced provides the input for validation of one or more air quality models (box E). As a minimum, such an exercise is necessary to give credence to the analytical procedures to be employed in translating proposed reductions in emissions into desired improvements in air quality. However, this step may be necessary to develop a statistical fit of model predictions to observed data through selection (typically on a least-squares basis) of one or more independent coefficients. This correlation procedure is shown in Figures 20 and 21 for the Climatological Dispersion Model (CDM)[13] of atmospheric dispersion as applied to the prediction of total suspended particulate (TSP) levels throughout the Chicago region.

Two reports derivable from the emission inventory will be of great value in designing regulations: (a) listings by pollutant of major sources in order of decreasing emissions; (b) listings by pollutant and by two (or more) digit SIC codes of industrial sources, with the categories ordered by decreasing aggregate emissions. Prominent emission source categories should then be studied to determine the state of the art of pollution controls, related costs, sensitivity of industry economics to pollution control costs, and other factors (e.g., average age of plants) which might bear on the regulatory decision-making process. Representatives of government, industry, and the general public should, time permitting, have substantial input in such studies, or as a minimum have the opportunity for careful review and comment *prior* to any publication of draft regulations. Taken together these studies constitute or provide the numerical input to an emission control data bank (box G).

Each control strategy (box H) will consist of a set of regulations governing emissions from most or all categories of sources of a given pollutant. Table 19 describes a number of approaches to defining allowable emissions or specific actions required to reduce emissions. Initially, the regulations should reflect the most cost-effective control requirements for each major source category as suggested by state-of-the-art surveys. Revisions, as shown symbolically by the two feedback loops in Figure 19, would be responsive to the likelihood of a given control strategy to attain and maintain applicable ambient standards and/or to specific information or inquiries which develop during intraagency review (box L) or at subsequent public hearings.

Each control strategy translates into a revised emission inventory, current and projected (box I). From a calculational standpoint, this step may be relatively straightforward and thus readily computerized (as for "process weight rate" or "% reduction" formulations) or may require extensive off-line analyses (as for modifications to a regional transportation network). In general, given the wide variety

---

[10] See Figure 10 for an example of the relative market shares of coal and natural gas changing in response to economic pressure independent of clean air constraints.
[11] Typically light-duty gasoline, heavy-duty gasoline, and heavy-duty diesel.
[12] Emission factors should reflect likely deterioration of automotive emission controls in the absence of an inspection/maintenance (I/M) program (Section 7.2) unless such a program is currently in force.
[13] See Busse and Zimmerman (1973) and Brubaker et al. (1977).

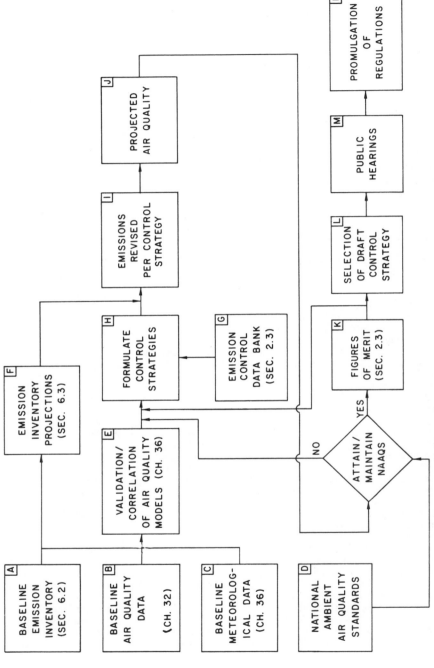

Fig. 19 Flowchart for development of regional strategy for attainment and maintenance of National Ambient Air Quality Standards.

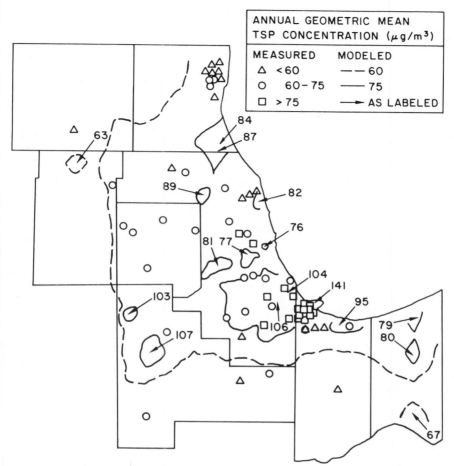

Fig. 20 Comparison of measured values of total suspended particulate (TSP) concentrations with estimates via the climatological dispersion model (source: Smith et al., 1980).

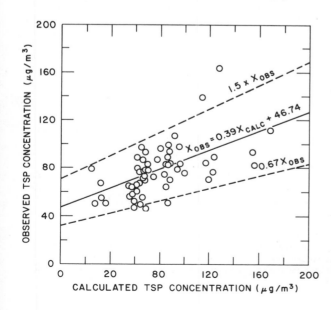

Fig. 21 Scatterplot of TSP concentrations at calibration receptors corresponding to Figure 20. Correlation $R^2 = 0.31$.

Table 19   Types of Emission Control Regulations

Type or Basis	Examples for Particulate Control[a]
Process weight rate	No person shall cause or allow the emission of particulate matter into the atmosphere from any portland cement clinker cooler to exceed 0.1 lb/ton of feed to the kiln
Heat rate	No person shall cause or allow the emission of particulate matter into the atmosphere in any 1-hr period from any new fuel combustion source using solid fuel exclusively, to exceed 0.1 lb of particulate matter per million Btu of actual heat input
Exhaust gas concentration	No person shall cause or allow the emission of particulate matter into the atmosphere from any incinerator burning more than 60,000 lb of refuse per hour to exceed 0.05 gr/scf of effluent gases corrected to 12% carbon dioxide
Visibility	No person shall cause or allow the emission of smoke or other particulate matter from any new portland cement process into the atmosphere having an opacity greater than 10%
"Picket fence" limits	Ambient concentrations of particulates on or outside the property line of any plant due solely to emissions from that plant shall not exceed 100 $\mu g/m^3$ for any 1-hr period. Determination of compliance shall be by measurement or calculational methods approved by the agency
Equipment efficiency (or % reduction)	Air contaminants collected from internal transfer operations for grain-handling facilities having a grain throughput exceeding 2 million bushels per year and located in a major population area shall be conveyed through air pollution control equipment which has a rated and actual particulate removal efficiency of not less than 98% by weight prior to release into the atmosphere
Equipment or design requirements	The largest effective circular diameter of transverse perforations in the external sheeting of a column grain dryer shall not exceed 0.094 in., and the grain inlet and outlet shall be enclosed
Emission density	No person shall cause or allow emissions of particulate matter into the atmosphere to exceed 100 tons/yr-acre of lot size in any area designated "heavy industry"
Outright prohibitions	No person shall cause or allow the use of beehive ovens in any coke manufacturing process
Work rules	All grain-handling and grain-drying operations must implement and use the following housekeeping practices: (a) floors shall be kept swept and cleaned from boot pit to cupola floor; (b) roof or bin decks and other exposed flat surfaces shall be kept clean of grain and dust that would tend to rot or become airborne . . . .

[a] Text adapted from Illinois Air Pollution Control Regulations, Illinois EPA, June 1976, except for "picket fence" and "land-use" examples.

of sources and options for emission reductions within a region, any computerized system for interconnecting boxes F, H, and I in Figure 19 will require a fair amount of off-line adjustments.

The projection of air quality (box J) will be greatly simplified if coupling coefficients can be developed for each source and receptor combination (e.g., $\mu g/m^3$ of SO_2 at receptor N per pound of SO_2 emitted by source M). Such a calculational scheme is generally applicable to inert pollutants (e.g., total suspended particulates or carbon monoxide) and to primary pollutants whose aerochemistry can be approximated by a linear relationship (i.e., independent of the concentration of the pollutant itself; e.g., sulfur dioxide). Important exceptions include volatile organic hydrocarbons and resultant ozone concentrations, and sulfates.

Similarly, the projection of control costs (box K) can be routinely performed by computer to the extent that simple algorithms such as described in Section 2.3 apply; however, the local and regional impacts of such costs will require a special assessment as discussed in Section 2.3.2. Figures of merit useful in ranking alternative regulations include: (a) $/kg of pollutant removed, (b) $/unit reduction of ambient air pollution, and (c) total cost.

Table 20   Summary of Organic Control Costs (Arledge and Pulaski, 1977, p. 86)

$/ton Removed	% of Category Emissions Removed	Cumulative % of Category Emissions Removed	Annualized Cost, $ × 10⁶	Cumulative Annualized Cost for Category, $ × 10⁶	Technique or Model Year Group
Petroleum production—no feasible controls					
Petroleum refining					
100	14	14	0.7	0.7	Vapor recycle
480	23	37	5.8	6.5	Vapor adsorption
1000	12	49	6.5	13.0	Secondary floating roof seals
Underground service station tanks					
100	55	55	1.24	1.2	Vapor recycle
480	90	90	9.76	9.8	Vapor adsorption or condensation
Automobile tank filling					
100	55	55	2.6	2.6	Vapor recycle
480	90	90	20.7	20.7	Vapor adsorption or condensation
Fuel combustion—no feasible controls					
Waste burning and other fires—no feasible controls					
Surface coating—heat treated					
518	90	90	2.0	2.0	Catalytic incineration

Example: Reactive Hydrocarbon Control Costs for Los Angeles (Arledge and Pulaski, 1977)

Analysis of air pollution in the Los Angeles basin led to the conclusion that approximately 95% control of reactive organic emissions would be required to attain the 1-hr national ambient air quality standard for ozone of 0.08 ppm. Table 20 summarizes a portion of the emission control data bank (box G in Figure 19). Costs for achieving various levels of control of emissions are shown in Figure 22 where marginal control costs vary from a low of $45/ton removed to a maximum of $2000/ton.

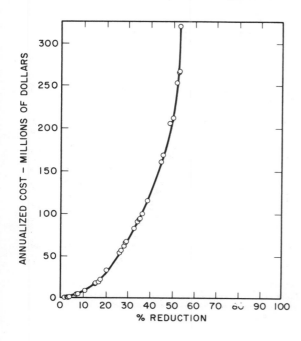

Fig. 22   Cost of achieving various levels of control of reactive organic emissions in the Los Angeles basin (Arledge and Pulaski, 1977, p. 91).

7   REDUCTION OF EMISSIONS FROM MOBILE SOURCES

No ambient air quality standard has generated as much controversy, challenging lawmakers, government agencies, scientists and engineers alike, as that for ozone. As originally established pursuant to the Clean Air Act of 1970, the maximum allowable 1-hr average of 0.08 ppm to be exceeded for at most 1 hr/yr placed virtually every major U.S. city in a nonattainment status as of 1976 (Figure 23). Even the recent relaxation of the standard to 0.12 ppm leaves over 60 cities in violation with some, such as Los Angeles, requiring over 50% reduction in nonmethane hydrocarbons as compared to projected nationwide reductions of approximately 26% when all federally mandated emission controls on automobiles and trucks and stringent emission limits for new stationary sources are coupled with anticipated growth (Table 21). Therefore, in addition to the need to impose increasingly severe controls on new and existing stationary sources, many of these metropolitan areas face the technically challenging task in design and the politically onerous requirement in implementation of county- or regionwide vehicle inspection/maintenance programs and transportation control strategies.

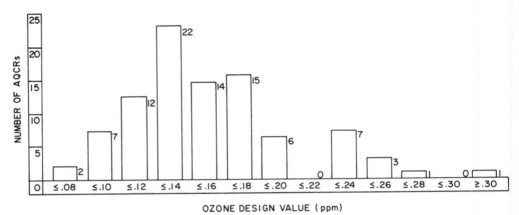

Fig. 23   Frequency distribution of design values for 90 nonattainment AQCRs, based on expected second high daily value (from air quality data for 1975–1977) (source: Lloyd, 1979, pp. 2–3).

Table 21   Projected Nonmethane Hydrocarbon (NMHC) Emissions in 1987 for 90 AQCRs (Millions of tons/yr) (Source: Lloyd, 1979, p. 3-3)

Source Category	1975 Emissions[a]	1987 Emissions with FMVCP and New Source Control[b]
Light-duty vehicles	5.36	1.82–2.09
Other highway vehicles	0.76	0.43–0.49
Nonhighway vehicles	1.22	1.38–1.64
Fuel combustion	0.13	0.15–0.17
Chemical manufacturing	0.71	0.67–0.73
Petroleum industries	0.64	0.41–0.42
Other industrial processes	0.61	0.87–0.98
Gasoline service stations	0.78	0.58–0.60
Petroleum storage and transport	0.82	0.61–0.63
Industrial solvent evaporation	1.47	1.24–1.34
Area solvent evaporation	5.46	4.36–4.65
Solid waste	0.34	0.27
Miscellaneous	0.49	0.49
Total	18.79	13.28–14.50

[a] This emission inventory represents the sum of emissions for each AQCR obtained from the National Emission Data System (NEDS) *Emission Summary Report* (NE204), October 1978.

[b] Range results from application of a range of assumed national growth rates.

Table 22 Transportation Control Measures (Source: EPA, 1974a, p. II-60)

Measures to Reduce Emission Rates [a]	Measures to Reduce Vehicle-Kilometers Traveled
Federal Motor Vehicle Control Program (FMVCP)	Traffic restrictions
	Street closings
Retrofit devices	Traffic-free zones
Vacuum spark advance disconnect with low idle	Partial traffic restriction
	Limited access zones
Air bleed to intake system	Idling restrictions
Oxidation catalysts	Gasoline rationing
Inspection/maintenance	Traffic restraints
Gaseous fuel conversion	Parking bans
Traffic flow improvements	Parking supply management
Better highway and interchange design	Parking surcharges
Signal progression	Road use of entry charges
One-way streets	Priority treatment for car pools
Reversible lanes	Increased gasoline taxes
Driver advisories	Increased vehicle registration fees
Loading regulations	Bikeways
Staggered work hours	Traffic avoidance
	Restricted road building
	Urban area bypasses
	Control of urban development; e.g., strategic planning and planned unit development
	Four-day work week
	Mass transit improvement
	Rapid rail
	Community rail
	Improved bus service
	Reduced mass transit fares
	Express bus lanes
	Employee mass transit incentives

[a] The technology for reduction of emissions from mobile sources is discussed in Chapter 18.

In contrast to the regionwide response required for attainment and maintenance of the ozone standard, excessive levels of carbon monoxide usually reflect highly localized emissions of this stable pollutant almost exclusively from automobiles in congested traffic situations. Transportation-related control strategies designed to attain ambient standards for ozone (and NO_2) will generally reduce CO levels; however, the latter may require special attention to local traffic patterns that have minimal impact on regional levels.

Table 22 separates transportation control measures into two major categories: those designed to reduce emission rates (emissions per vehicle-kilometer traveled) and those designed to reduce the aggregate numbers of vehicle-kilometers traveled (VKT).[1] This distinction is conceptually convenient yet clearly entails some overlap; in particular, many schemes to reduce VKT will enhance traffic flow and thus reduce the average emission rate. Further, there are feedback mechanisms inherent in many of these transportation control measures which may lead to highly counterproductive side effects. For example, establishing preferential bus and high-occupancy vehicle lanes on limited access highways can, if improperly designed, lead to excessive congestion elsewhere, with an overall increase in emissions and/or creation of local hot spots.

Finally it is evident that approaches to reduction of VKT and improvement of traffic flow coincide with interests of state and regional transportation planning agencies and thus invariably must be compati-

[1] Section 105 of the CAA 1977, *amending* the Clean Air Act §108, 42 U.S.C. §7408(f) (Supp. II 1978) sets forth an extensive list of transportation control actions which must be seriously considered as "reasonably available measures" by any state seeking an extension of up to 5 yr (up to December 31, 1987) for attainment of primary ozone or CO standards. Section 129(b) of the CAA 1977, *amending* §172 of the Clean Air Act, 42 U.S.C. §7502 (Supp. II 1978).

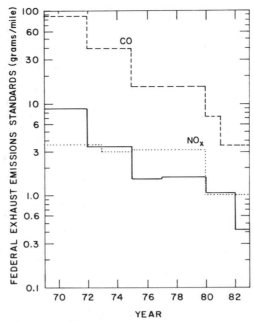

Fig. 24 Summary of the Federal Light Duty Vehicle (LDV) Standards since 1967 (source: English, 1980, p. 204).

ble with specific roadway, public transit, and other plans of such agencies[2] if they are to be realized without extreme political anguish.

7.1 Emission Limits

The progressive tightening of national emission standards for light-duty vehicles is captured in Figure 24 where, subject to the potential for further delay, the allowable emissions for 1981 and subsequent model years are 0.41 g hydrocarbons, 3.4 g CO, and 1.0 g NO_x per vehicle-mile (EPA, 1978b).[3] These emission limits are to be maintained throughout the useful life, which for light-duty vehicles is defined as 5 yr or 50,000 miles, whichever occurs first [CAA 1977, Sec. 202(d)]. Federal test procedures used to determine compliance examine a statistical sample of vehicles of all makes, models, styles, and engine types during an intensive 50,000-mile schedule within which periodic tuneups and other normal maintenance activities are performed.

However, the EPA has found that under conditions of driving and maintenance to which vehicles are subjected by the typical owner the rates of deterioration of emissions of hydrocarbons and CO will likely be substantially greater than indicated under the controlled tests for compliance with the Clean Air Act (Figure 25).

As shown in Figure 26, with the exception of NO_x,[4] the projected impact of the Federal Motor Vehicle Control Program (FMVCP) on emission from mobile sources is impressive notwithstanding the expected deterioration of vehicle emission rates: a 78% reduction in hydrocarbon and 60% reduction in carbon monoxide emissions by 1985 from their maxima in the late 1960s despite a 52% growth in total vehicles over the same span of time.

[2] See Sections 6.1 and 8.1.
[3] The figure does not tell the story of the successive deferral of deadlines and relaxation of emission limitations resulting from hearings before the federal EPA and the amendment of the act in 1977. For instance, the 1970 act established NO_x emissions at or below 0.4 g/vehicle-mile as of the 1976 model year, §6 of the CAA 1970, *amending* §202 of the Clean Air Act, 42 U.S.C. §7402 (Supp. II 1978).
[4] Section 201(a) of the CAA 1977, *amending* §202(b)(1)(A) of the Clean Air Act, also established a more stringent "research goal," 90% reduction of NO_x emissions below the 1971 baseline; i.e., a target of 0.4 g NO_x/vehicle-mile. 42 U.S.C. §7521 (Supp. II 1978).

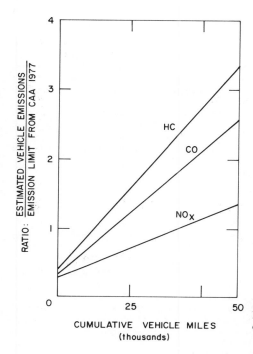

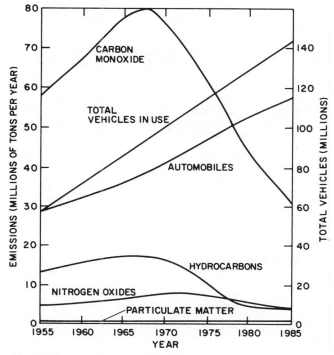

Fig. 25 Deterioration of post 1981 light-duty vehicle emission rates anticipated under normal conditions of driving and maintenance (source: EPA, 1978, Table I–1).

Fig. 26 Mobile source annual emission rates (source: English, 1980, p. 2–2).

7.2 Inspection/Maintenance (I/M) Programs[5]

As indicated in Figure 25, vehicle emission rates will deteriorate over time, especially in the absence of routine, prescribed maintenance of engine settings and control devices. Table 23 lists the primary causes of vehicles exceeding federal emission standards. Figure 27 demonstrates that substantial mobile source emission reductions will follow from implementation of a program of periodic (usually annual) inspection (testing) of all or most vehicles followed by maintenance as necessary to bring violators into compliance.

Such I/M programs are based on the concept that (a) regular annual tuneups preferably with special attention to manufacturers' specifications for low emission levels (e.g., carburetor settings) will counter the inherent deterioration of emission rates with time; (b) in the absence of an I/M program, many vehicle owners will permit emissions to deteriorate, either through lack of proper maintenance or deliberate tampering with pollution control systems.

Among the various structural approaches to this problem are:

1. Voluntary maintenance
2. Mandatory maintenance (without inspection)
3. Random inspections by government
4. Periodic inspections by government (or by a contractor to government)
5. Periodic inspections at licensed private garages

The impact of a voluntary program of vehicle maintenance is, of course, difficult to estimate. Even if coupled with the threat of random but highly infrequent inspections alongside highways, likely emission reductions will be substantially less than under mandatory programs. Similarly, elimination of the inspection (i.e., certification) step under a mandatory maintenance program will lead to substantial uncertainty as to the adequacy of servicing of vehicles. Thus, most states[6] have opted for, or are seriously considering, mandatory annual[7] inspections with a distinction between government-owned and licensed operations often reflecting compatibility with any ongoing vehicle safety inspection activities.

The Federal Environmental Protection Agency's Test Procedure (FTP) for certification of vehicles as to compliance with the Clean Air Act (EPA, 1978b, p. 9) entails a complex sequence of steps clearly beyond those appropriate for a regional I/M program processing hundreds of thousands to several millions of vehicles each year. Thus two *short tests* have been designed in which an emission analyzer measures CO and HC levels in the exhaust (EPA, 1977):

> Idle emissions test *or* idle test—*a test procedure for sampling exhaust emissions which requires operation of the engine in the idle mode only. At a minimum, the idle test should consist of the following procedure carried out on a fully warmed-up engine: a measurement of the exhaust emission concentrations for a period of time of at least 15 sec, shortly after the engine was run at 2000 to 2500 rpm with no load for approximately 60 sec.*

Table 23 Primary Causes of Vehicles Exceeding Federal Emission Standards (Source: U.S. Comptroller General, *Better Enforcement of Car Emission Standards—A Way to Improve Air Quality*, No. CED-78-180, January 23, 1979; as cited in IEPA, 1978b)

Cause	Occurrence, %
Maladjusted engine settings	47
Deterioration due to premature parts failure, the illegal use of leaded fuels, and improper car use	25
Tampering, or the removal or rendering inoperable of emission control systems	18
Lack of sufficient car maintenance	7
Manufacturer design and poor production practices	3

[5] See the review of I/M programs by Elston (1981) for a comprehensive treatment of this subject.
[6] For example, New Jersey (statewide), Arizona (Phoenix and Tucson), Oregon (Portland), Illinois (Chicago).
[7] A more modest version in California has required I/M upon change of vehicle ownership.

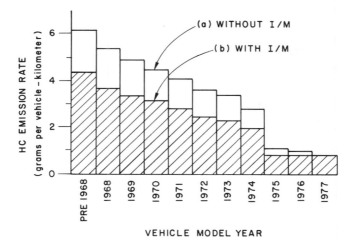

Fig. 27   Emissions of hydrocarbons from light-duty vehicles on the road in 1977 (a) without existence of an inspection maintenance program; (b) with an annual I/M in effect since 1973 ("stringency factor = 40%"). With model years weighted by relative population and VKT in 1977, a 27% reduction in HC emissions is achieved by the I/M program (source: EPA, 1977, p. 22182).

Load emissions test or loaded test—a test procedure for sampling exhaust emissions which exercises the engine under loading by use of a chassis dynamometer to simulate actual driving conditions. As a minimum requirement, the loaded test must include running the vehicle and measuring exhaust emissions at two speeds and loads other than idle (typically 50 and 30 mph).

Although somewhat more costly (approximately 25%) and time consuming (approximately 50%), the loaded emissions test is a better indicator of vehicle emissions under normal driving conditions; however, "available data indicate no overall difference in HC and CO emission reductions between the two tests," absent extensive training of mechanics in the proper use of loaded test diagnostic information (EPA, 1977, p. 22179). For these reasons, as well as the overall ease of training operators and duplicating equipment and results at many sites, states have favored the idle test mode coupled with an NO_x tamper check of the exhaust gas recirculation system.[8]

Estimation of emission reductions attributable to an I/M program proceed from an empirical determination of the *stringency factor*, "a measure of the rigor of a program, based on the estimated fraction of the vehicle population whose emissions would exceed cutpoints for either or both carbon monoxide and hydrocarbons were no improvements in maintenance habits or quality of maintenance to take place as a result of the program" (EPA, 1977, p. 22178). Tables 24 and 25 can then be entered to determine the percentage reduction in emissions in year L for vehicles of model year M under an annual I/M program which began in year N.[9] Second-order adjustments are then made to account for more frequent inspections and the implementation of a suitably rigorous mechanics' training program. The calculational procedure then follows that for determining a weighted emission factor for vehicles (see Section 6.2.3). In 1980 EPA revised its calculational procedures for estimation of the effectiveness of I/M programs. This so-called Mobile II model predicts greater reductions in emissions attributable to I/M programs than suggested by "Mobile I" (e.g., Tables 24 and 25), yet a lower total effectiveness if a training program for mechanics is instituted (Elston, 1981).

Costs for establishing, operating, and participating in an I/M program range between $8 and $15 (1982) per vehicle test plus any additional, unusual maintenance expenses borne by vehicle owners in anticipation of, or, if required, subsequent to, the inspection. Table 26 summarizes the cost projections for a contractor owned and operated I/M program providing 164 test lanes in a total of 46 stations located in the Chicago (primarily) and East St. Louis air quality control regions of Illinois. Costs for repair of failed vehicles can be expected to average about $35 (1982) based on the 1977 data shown in Table 27. Note that a state that exempts from compliance a vehicle requiring repairs in excess of

---

[8] Note that idle and loaded tests will in general fail a different, but overlapping, set of vehicles.
[9] Thus, for $L = 1977$, $M = 1974$, and $N = 1973$, the reduction in HC due to an I/M with a stringency factor of 40% is 30%, 10% from Table 24 plus 20% from Table 25 (as reflected in Figure 24).

Table 24 First Year of Program Credits (Source: EPA, 1977 p. 22178)

Stringency Factor (SF)	HC, % Model Years ≤1974	HC, % Model Years ≥1975	CO, % Model Years ≤1974	CO, % Model Years ≥1975
0.10	1	1	3	8
0.20	5	3	8	20
0.30	7	9	13	28
0.40	10	16	19	33
0.50	11	24	22	37

Table 25 Subsequent Years Program Credit (Source: EPA 1977, p. 22178)

Number of Inspections	Additive Credit HC, %	Additive Credit CO, %
2	7	8
3	14	15
4	20	19
5	25	23
6	30	27
7	33	30
8 or more............	36	35

Table 26 Summary of Final Cost Projections (Source: IEPA, 1980b, p. VI-6)

	Initial Outlay[a] ($ Million)	Program Average Total Annualized Costs[b] ($ Million)
Contractor		
Operating	$20.5	$23.8
G&A (15%)		3.6
Capital and start-up	60.2	11.1
Subtotal	80.7	38.5
Profit (10%)		3.8
Subtotal	80.7	42.3
State		
Operating	3.1	3.6
Capital and start-up	1.6	0.5
Subtotal	4.7	4.1
Total annualized cost		46.4
Average number of fee inspections per year		3,753,400
Average cost		$12.40

[a] 1982 for capital and start-up; 1983 for operating.
[b] Basis: Program life, 5 years; annual interest rate, 12%; inflation rates 10% land, buildings, equipment and 7% labor.

Table 27 I/M Program Repair Cost Data (for Failed Vehicles) (Source: IEPA, 1980a, p. 4-23)

I/M Program	Date of Implementation	Average Repair Cost	Median Repair Cost	Repair Cost Ceiling
New Jersey	02–01–74	$15.83	$ 7–10	None
Oregon	07–01–75	$20–25	$10	None
Arizona	01–01–76	$23.02	$11.25	$75
California	03–31–79	$21	N.R.	$50

N.R. = not reported.
Dollar values are 1977 dollars.

Table 28 Effects of Maximum Repair Cost in 1979[a] on Portland, Oregon I/M Program (Source: IEPA, 1980a, p. 4-25)

Cost Ceiling	% of Failed Vehicles Within Cost Limit	% of Total Vehicles Within Cost Limit	Effectiveness (% Emission Reduction[b]) HC	CO
No limit	100.0	100.0	32.5	42.5
$150	98.0	99.4	32.0	42.0
$100	93.5	98.1	25.0	40.5
$ 75	91.5	97.5	22.5	38.0
$ 50	85.5	95.7	20.0	34.0
$ 25	71.0	91.3	13.0	26.5

[a] Assumes a 30% failure rate, and a 50% catalyst and 50% noncatalyst vehicle distribution in 1979 according to national averages.
[b] Assumes no exemption for light-duty automobiles and trucks 14 years and older.

a prescribed cost ceiling[10] will lessen the effectiveness of its I/M program as indicated by a USEPA study in Portland, Oregon (Table 28).

In any estimation of I/M costs, credit should be taken for fuel savings due to improved maintenance, particularly regular tuneups. Typical annual fuel savings that may be achieved in those vehicles that (a) initially fail the I/M, (b) undergo repair, and (c) pass a second test may average 3% if repairs are conducted by reasonably skilled mechanics.[11] Though similar savings can in principle be attributed to vehicles which undergo pretest maintenance which, in the absence of I/M would not normally be performed, such an estimate would understandably be difficult to defend.

Is I/M a cost-effective control strategy?[12] The answer will be highly dependent on regional air quality, the relative culpability of stationary and mobile sources, and the availability of simple fixes to solve highly localized (typical CO) problems. Since I/M is a regionwide or countywide strategy, it will obviously be most cost effective in resolving environmental issues at the same geographical scale.[13] Calculations for the proposed Illinois I/M program suggest $1000/ton (1978) as a measure of cost effectiveness in reduction of hydrocarbon emissions, a value that falls within the range of Reasonable Available Control Technology (RACT) for stationary hydrocarbon sources (see, for example, Figure 22).

[10] A politically attractive but administratively cumbersome provision.
[11] Based on studies in California and Oregon. For example, gasoline savings are estimated at 27 gal/yr per failed vehicle for an I/M with a stringency factor of 30% in the Chicago and metro-East areas of Illinois (IEPA, 1980a).
[12] It certainly has been a politically unpopular one to the degree that it represents increased governmental spending and interference with the lives of private citizens.
[13] For example, a regional I/M program has been shown to have minimal effect on attainment of the 8-hr CO standard at street level within the Chicago central business district, compared to minor adjustments in traffic patterns and retrofit emission controls on older taxis and delivery vans.

7.3   Case Study: Implementation and Administration of Air Quality Transportation Controls: An Analysis of the Denver Colorado Area (Suhrbier et al., 1978)

To serve as a guide for cities facing requirements for transportation-related control strategies (i.e., beyond sole reliance on the FMVCP), this case study examined in depth the costs, effectiveness, and challenges to implementation of the six measures most frequently considered for reducing mobile source emissions:

1. Vehicle inspection/maintenance
2. Employer-based ride sharing
3. Preferential treatment of high occupancy vehicles
4. Parking management
5. Improved bicycle facilities
6. Improved transit

Although the specific conclusions are biased by special characteristics of the Denver area, the identification of key issues, methods of analysis, presentation of results, and in most instances the relative costs and benefits of alternative strategies offer a valuable framework for parallel studies elsewhere. Figure 28 outlines that framework. Developing an understanding of the wide range of potential impacts on transportation (e.g., travel volumes, modal splits), environment (e.g., air quality, energy

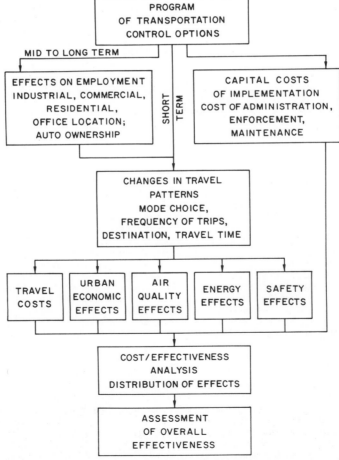

Fig. 28 Framework for conducting cost-effectiveness analysis of alternative transportation control strategies (source: Suhrbier et al., 1978, pp. 1–41).

consumption), regional economics (e.g., retail sales, land values), and governmental institutions (e.g., enforcement, revenues) requires cooperation of affected state and local agencies, sophisticated technical analysis, and a broadly based program of public information and public involvement.

The salient conclusion of this extensive analysis for the Denver area of the above-listed six transportation control measures is that, with the sole exception of I/M, they have minimal impact on air quality when considered in realistic, mutually supportive combinations. The FMVCP alone is expected to provide overall regionwide reductions of 41.7, 61.9, and 55.4% in HC, CO, and NO$_x$ emissions, respectively, despite substantial increases in work and nonwork travel associated with an anticipated population growth of 27% between 1975 and 1985. Further reductions of 3.9, 8.1, and 1.3% in HC, CO, and NO$_x$, respectively, would follow from an I/M program. The remaining vehicle-related measures were examined by grouping into three programs of increasing stringency (Table 29); results are summarized in Table 30. Not surprisingly, the most effective means of discouraging vehicle use are direct fiscal

Table 29 Alternative Transportation Control Program Packages for Denver AQCR (Source: Suhrbier et al., 1978, p. 11-11)

Measure	Program 1	Program 2	Program 3
Ride sharing	Employer promotion and matching in all firms with 50 or more employees	Employer promotion and matching in all firms with 50 or more employees; van pooling available in all firms with 250 or more employees; transit fare subsidy of 50% available to all workers; preferential car pooling	Same as no. 2
Transit	Improved frequency on CBD routes	Improved frequency on CBD routes; 20% reduction in in-vehicle travel time for CBD routes	25% areawide improvement frequency; 20% reduction in in-vehicle travel time for CBD routes
Parking	Increased commuter parking costs in CBD by $1.00 per day	Increase computer costs in CBD by $1.00 per day; reduce parking availability so that roundtrip walk times are increased by 10 min	Increase commuter costs areawide by $1.00 per day; reduce areawide parking availability so that roundtrip walk times are increased by 10 min
Preferential treatment	—	Improve areawide level of service for all vehicles by 5%	Same as no. 2
Pricing	—	—	Triple the price of fuel in terms of 1965 dollars

Table 30 Predicted Areawide Impacts of Alternative Transportation Control Program Packages for Denver (Source: Suhrbier et al., 1978, p. 11-13)

Program Package	Change, % VMT, miles/day Work	Nonwork	Total	Fuel Consumption, gal/day	Auto Emissions, kg/day HC	CO	NO$_x$
Base (av/day)	12.8	19.85	31.55				
1	−1.8	0.3	−0.5	−1.1	−0.8	−0.7	+0.6
2	−4.0	0.8	−1.0	−2.3	−1.6	−1.5	+1.1
3	−4.3	−22.0	−15.3	−18.0	−9.0	−9.0	−15.4

disincentives: parking fees, which are politically tough to implement, and gasoline prices, over which the state has some (via taxes), but generally limited, control.[14]

In conclusion, from the perspective of regional air quality, the Federal Motor Vehicle Control Program, escalating world oil prices, and I/M programs will achieve virtually all the gains that one can realistically expect during the next decade. High-efficiency engines, public transit, van pools, car pools, bicycle routes, and the like will provide a range of cost-competitive alternatives, but the impetus for their use will have little to do with air quality management.

8 LAND-USE ASPECTS OF AIR QUALITY MANAGEMENT

The first line of attack in air pollution has been the most straightforward one: clamp on control equipment or shift to cleaner fuels and processes. The difficulties emerge when the full panoply of economically reasonable technical fixes fail to reduce pollutant levels below the lid represented by the NAAQS or when that lid (or equivalently a PSD increment) is sufficiently tight to constrain or distort desirable geographical patterns of regional growth. Then an individual wishing to develop his land in accord with applicable zoning and other traditional property restrictions may find that privilege further encumbered because existing facilities have already consumed all or most of the available clean air resource.

8.1 A Case Study: The City of Renton, Washington (Felton and Rossano, 1976)

The city of Renton, Washington is situated toward the northern end of a lowland trough which stretches between Seattle on the north and Tacoma on the south. In the process of evaluating the application of a major oil company to construct a large petroleum products marketing and distribution plant, the city environmental and planning authorities became aware of ambitious development plans of a number of other area industries. It was immediately evident that the aggregate of expected nonmethane hydrocarbon (NMHC) emissions, when added to limited existing stationary and mobile sources, would lead to violations of ambient HC and ozone standards. The planning commission also recognized the danger of a "first-come-first-served" approach. Using a relatively unsophisticated model of atmospheric dispersion and aerochemistry, it was determined that on the average the regional airshed could tolerate emission rates of 1 ton NMHC per acre. Guidelines were then established for a limited number of parcels with high emission densities (9 tons/acre) with other areas to remain at or below 0.3 ton/acre so that ambient air quality standards were maintained. Enforcement was achieved by review of construction permits and periodic emission reports. It was recognized that such zoninglike controls (i.e., special-use permits) as implemented by the city would in the long run be ineffective unless incorporated into a comprehensive land use and air quality management plan for the Seattle–Tacoma AQCR.

8.2 Land-Use Plans and Air Quality Management

Most major urban areas and the communities therein have a general idea of how they wish to evolve over the next 10 to 20 yr. A broad articulation of goals, amplified by projections of gross measures such as population and employment, and detailed in a series of specific functional plans characterizing residential, commercial, and industrial development and related services such as transportation, public water supplies, and wastewater management together constitute a *comprehensive plan* for the region. *Land-use plans* refer to documents or documentation within the comprehensive plan dealing in *geographical* detail with projections of development and allocation of services.[1]

Two questions arise: (1) To what extent can regional land use and transportation plans serve as predictors of future levels of air pollution? (2) Given the regional nature of many air pollution problems, might not land use and transportation plans provide a framework over which to lay regional air quality management strategies?

The answer to the first of these questions lies with two matters acting in sequence: the variability in algorithms relating existing land-use classifications and airborne emissions; and the likelihood that current land-use plans constitute reliable representations of the future.

Correlations between land-use indices and existing industrial emissions have been, for example, developed for the Chicago AQCR (Kennedy et al., 1973) in the format: emissions per unit area as a function of zoning classification. An attempt to derive industrial land-use based *emission factors* from

---

[14] Confirmation of the minimal impacts on air quality of typical transportation-related strategies, other than the FMVCP and I/M, is shown in Figure 29 for the Baltimore AQCR.

[1] Two primary avenues of federal support for the development of comprehensive regional plans and the maintenance of an ongoing and increasingly sophisticated planning process concerned with housing, transportation, and general land use have been §701 of the Housing Act of 1954, Pub. L. No. 83-560, CH. 649, 68 Stat. 590, 640 [current version at 40 U.S.C. §461 (1976 & Supp. III 1979)] (HUD "701" grants) and the transportation-planning grants under the Federal-Aid Highway Act of 1962, Pub. L. No. 87-866, 76 Stat. 1145, 1148 (current version at 23 U.S.C. §134 (Supp. IV 1980) (the 3-C Process).

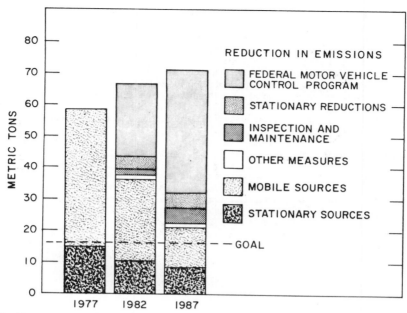

Fig. 29 Hydrocarbon emissions—morning peak period (6–9 A.M.) on a Friday in July (source: Hines, 1978, p. 2).

other traditional planning and census variables was also made (Baldwin and Kennedy, 1974). Cross checks of both of these approaches to estimating emissions with the official point and area source inventory, using alternative predictions of air quality levels for comparison, indicated limited confidence in such land-use based emission factors. Further, related studies such as the Hackensack Meadowlands Project (ERT, 1973) required extensive, locally specific data to tie together land use and existing or projected emissions. Thus, principally because of the variety of pollution sources and range of emission control devices that can fall within a given land-use classification, predictions of industrial emissions based on land use plans provide at most a crude guide to the future, but not one of sufficient precision upon which to base traditional air pollution regulations that specify emission limits for individual sources.[2]

The answer to the second question, whereby regional land-use and transportation plans provide a framework for air quality management, is more complex and depends on the degree to which such plans represent quasienforceable blueprints for regional development. As plans, they are first and foremost statements of *intent*, serving as guides rather than as enforceable instruments. Zoning, the traditional means by which land-use controls are established, is primarily a local affair with municipalities and counties within a region often acting quite independently of each other and of the comprehensive regional plan. Further, many suburban towns and unincorporated areas of counties deal with proposals for new or expanded residential, commercial, and industrial activity on an ad hoc basis as planned unit developments or via special-use permits. Not surprisingly, the record to date does not instill much confidence in regional land-use plans as enforceable instruments on which to rely for attainment of clean air objectives.[3]

[2] Substantial refinement of the concept of land-use based emission factors was a principal outcome of the GEMLUP (Growth Effects of Major Land Use Projects) studies (e.g., Benesh, 1976). Here planning and design variables such as "unit floor area" for commercial buildings and "dwelling unit heating-degree-days" for residences served as a basis for developing a highly disaggregated set of emission factors via multiple linear regression techniques.

[3] By no means atypical of the toothlessness of many metropolitan and regional planning agencies is the following: "Some of the most severely polluted towns in metropolitan Denver have shown the greatest growth in recent years. A 'carrying-capacity' based land use plan for Denver in the early 1970s called for minimal future growth of towns to the west and southwest of the city in large part because of the buildup of pollutants by diurnal winds. Yet these towns contain some of the most attractive residential sites available in the entire metropolitan area, and it has been impossible under existing law to prevent such growth (especially without the cooperation of the towns). Consequently, the number of people exposed to dangerous pollution concentrations has risen dramatically" (Suhrbier et al., 1978).

Increasingly, however, more comprehensive planning approaches are being followed in the development of major capital intensive elements of the regional infrastructure, particularly elements such as highways, public transit, and sewage systems funded in large by the federal government. A substantial body of federal and state law has evolved to provide encouragement for, and under limited circumstances, authority to enforce, such regional cohesiveness.[4]

The attainment and maintenance of clean air standards, that is, the management of a limited, regional clean air resource, is in some important aspects analogous to the development of the regional infrastructure;[5] thus, one might conclude that the thorny problem of long-term maintenance of the NAAQS *requires* a solution conceived and enforceable at the regional level. If so, it should be compatible with applicable regional land-use plans along with key functional plans, especially those associated with major public works projects.[6,7]

8.3 Land-Use Based Emission Limits

Management of the regional clean air resource in a manner compatible with other regionwide planning objectives entails a partitioning of allowable emissions across political jurisdictions, land-use classifications, and ultimately the explicit or implicit allocation of emission rights to individual parcels of land. How to design, no less carry out, such a grand scheme is quite a challenging undertaking, especially from a political point of view, since, despite the quantitative fuzziness of land-use based emission factors, it is clear that economic growth and emissions growth tend to go in consort, that is, the clean air resource is a *consumable*, a factor of production.

Brail (1975) has proposed a taxonomy of *emission quota strategies* to couple land use and air quality management:

1. Jurisdictional emission quotas
2. District emission quotas
3. Floating zone emission quotas
4. Emission density zoning ("unit area" emission quotas)

Figure 30 outlines the overall approach to designing such strategies. In this hierarchy, regionwide emissions consistent with maintenance of clean air standards[8] are first apportioned among the principal political jurisdictions (e.g., counties and/or major urban, suburban, and rural zones). Certain regionwide facilities such as major highways could well be treated as common resources; however, developments which induce significant vehicular traffic would be held accountable for the ensuing emissions (see Section 8.4). Such jurisdictional quotas offer to local government the maximum flexibility in tailoring

---

[4] For example, the National Environmental Policy Act of 1969 [42 U.S.C. §4321 et seq. (Supp. II 1978)]; the Federal Water Pollution Control Act [33 U.S.C. §1251 et seq. (1976 & Supp. II 1979)]; the Coastal Zone Management Act of 1972, 16 U.S.C. §1451 et seq. (1976 & Supp. III 1979); the California Coastal Act of 1976, Cal. Pub. Res. Code §3000 et seq. (West, 1977 and 1981, Supp.).

[5] Duality is perhaps a more appropriate term: the transportation network representing a regional investment to serve individual travel needs vs individuals investing in air pollution controls to achieve regional air quality objectives.

[6] Indeed, following a cosmetic deletion of the words *land use* from §110(a)(2)(B) of the Clean Air Act in §108 of the CAA 1977, 42 U.S.C. §7408(a)(2)(B) (Supp. II 1978), Congress proceeded to embed numerous provisions in the 1977 amendments which tie federal land-use and transportation related actions to clean air strategies. For example: (a) PSD Class I increments to protect federal lands of special scenic and historical value—§127(a) of the CAA 1977, *amending* §162 of the Clean Air Act, 42 U.S.C. §7472 (Supp. II 1978); (b) federal enforcement of regulations governing "federally assisted indirect sources and federally owned or operated indirect sources"—§108(e) of the CAA 1977, *amending* §110 of the Clean Air Act, 42 U.S.C. §7410(a)(5)(B) and (C) (Supp. II 1978); (c) conditioning of federal sewage treatment grants where a State Implementation Plan fails to adequately account for emissions of air pollutants which "may reasonably be anticipated to result directly or indirectly from the new sewage treatment capacity which would be created by such construction"—§306 of the CAA 1977, *amending* §306 of the CAA 1977, *amending* §316 of the Clean Air Act, 42 U.S.C. §7616(b)(2) (Supp. II 1978); (d) prohibition of federal highway construction grants to implement same in accord with a federally approved schedule—§129(b) of the CAA 1977, *adding* §176(b) and (c) of the Clean Air Act, 42 U.S.C. §7506 (Supp. II 1978); and (e) requirement that federally sanctioned metropolitan planning organizations deny "approval (normally via the A-95 Review Process) to any project, program, or plan which does not conform to a (State Implementation Plan) approved or promulgated under Section 110."

[7] For a general discussion of urban and regional land use and transportation planning and a summary of early studies relating these to air quality management, see EPA (1974c) and Roberts et al. (1976).

[8] As determined by appropriate models of dispersion and aerochemistry.

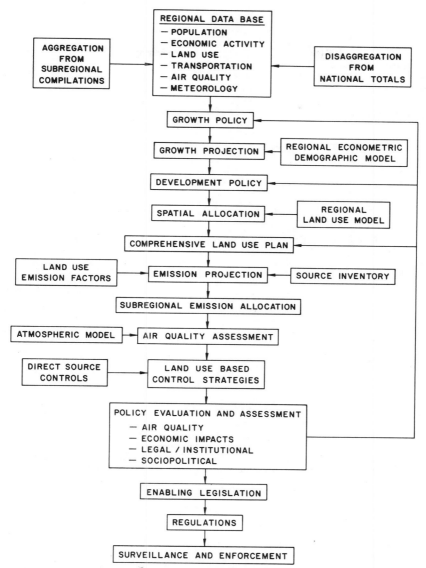

Fig. 30 The development and implementation of air pollution control strategies based on land use (source: APCA Critical Reviews, Roberts et al., 1976).

industrial, commercial, and residential growth to match local objectives. Enforcement would require a state or regional (e.g., council of governments) permitting authority for major new development.

A jurisdiction would be unlikely to spread its quota uniformly across all areas but rather would distribute allowable emissions somewhat in accord with current land use and transportation plans following an analysis of the compatibility of such plans with air quality constraints. Thus, as suggested by item 2 above, individual *districts* within the jurisdiction would be allocated portions of the overall quota. Floating zone emission quotas are applicable in the absence of governing land-use plans or zoning restrictions and are analogous to planned unit developments.

Lastly, emission density zoning (EDZ) entails a substantially greater degree of disaggregation by which, analogous to the floor-to-area ratios employed in traditional zoning ordinances, each parcel of land is assigned an allowable areal emission density. The most straightforward approach is to base such limits on existing zoning or land-use maps, as suggested in Figure 31. For example, an EDZ

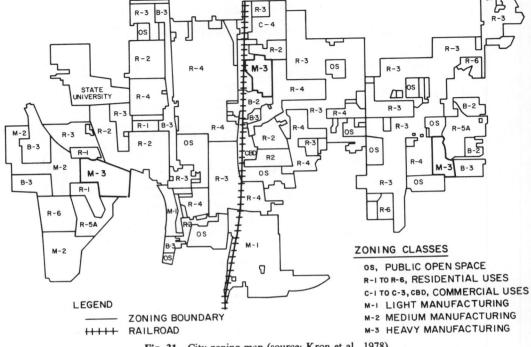

Fig. 31 City zoning map (source: Kron et al., 1978).

regulation might establish an upper limit of 1 kg/yr of reactive hydrocarbons (RHC) per square meter of land in M-3 zones designated for "heavy industry." Thus a firm with controlled[9] emissions of 1500 kg RHC/yr (1.6 tons/yr) would require an M-3 zoned site of at least 1500 m² (0.5 acre). If the plant could otherwise get a permit for location in an area zoned M-2, it most likely would require a larger minimum land parcel to accommodate the projected emissions. In the first instance, and possibly the second, a smaller site might be permissible if the owner were able to purchase additional *emission rights.*[10]

The set of emission density limits must satisfy the constraint Eq. (10) (Roberts et al., 1976):

$$\sum_i E_{ij} P_{ij}(x, y) + X_{p,j} \leq S_j(x, y) \tag{10}$$

where

E_{ij} = emission density limit for pollutant j in land-use class i (e.g., kg/km²-yr)

$P_{ij}(x, y)$ = "coupling coefficient" or "unit emission pattern" quantifying the contribution to levels of pollutant j at point (x, y) from all land of class i, assuming such land is fully developed; a uniform, unit emission rate (e.g., 1 kg/km²-yr) is assumed for all land of class i with zero emission assumed elsewhere

$X_{p,j}(x, y)$ = concentration at point (x, y) of pollutant j due to electric utility power plants on line or likely to be constructed within 10 yr (µg/m³)

$S_j(x, y)$ = applicable air quality standard for pollutant j at point (x, y), thereby reflecting the possibility of geographically dependent standards in accord with regulations governing prevention of significant deterioration (µg/m³)

---

[9] All sources, new and existing, would be required to comply with emission limits such as New Source Performance Standards.
[10] Refer to the discussion of "marketable emission rights" in Section 5 and to *offset* and *bubble* policies in Section 8.3.4.

Choice of specific values E_{ij} can be optimized by application of linear programming techniques to minimize an objective function of the form[11]

$$J = \sum_j \sum_i C_{ij} A_i (\hat{E}_{ij} - E_{ij}) \tag{11}$$

where

C_{ij} = a weighting factor such as regional value added per unit increase in pollutant j per acre in zoing class i

A_i = total land area in zoning class i

$\hat{E}_{ij}$ = maximum likely density of emissions of pollutant j estimated for zoning class i *under conditions of unconstrained growth* (i.e., subject only to existing point source emission limits without regard to ambient air quality)

8.4 Indirect Sources

When the north–south leg of the Chicago Crosstown Expressway was first proposed in the early 1970s, the accompanying land-use plan indicated a broad band of new light industry development along the 15-km corridor. However, the environmental impact statement (EIS) described the major highway link solely as an expediter of traffic flow, bypassing the central business district, substantially increasing average vehicle speed, reducing VKT and thus a boon to area clean air objectives. No mention was made in the EIS of the role of the highway in fostering new development likely to induce additional vehicular traffic that could, in less than a decade, turn the Crosstown at rush hour into a linear parking lot, so typical of urban freeway systems. To generalize: because land development (patterns and intensity) in U.S. metropolitan areas too often bears little resemblance to the land-use assumptions upon which transportation plans are based, potential gains in efficiency (e.g., reduced travel times, lower pollution) are lost as new transportation capacity induces suburban residential sprawl, decentralization of centers of employment, and associated traffic and air pollution.

Most State Implementation Plans (SIPs) developed pursuant to Section 110 of the Clean Air Act Amendments of 1970 failed to recognize that projections of future vehicle-related pollution based on regional transportation and land-use plans tended to be overly optimistic, and thus such SIPs were deficient to the extent that they lacked legally enforceable provisions to review and, if necessary, present the construction or modification of facilities which, while not significant direct sources of pollution, caused or induced substantial vehicular traffic.[12]

A so-called *indirect source* has been defined as "a facility, building, structure, installation, real property, road, or highway which attracts, or may attract, mobile sources of pollution" [CAA 1977, Par. 110(a)(5)]. In its proposed regulations (38 *Federal Register* 208, p. 29893, 30 October 1973) the Environmental Protection Agency required review of facilities such as new parking lots for 1000 cars or more, airports anticipating increases within 10 yr of 50,000 operations/yr or more, new roadways with anticipated average annual daily traffic (ADT) volumes within 10 yr of 20,000 or more vehicles per day, modified roadways with expected increases of 10,000 or more vehicles per day, and any other facility that induces 1000 or more vehicle trips in any 1-hr period or 5000 or more in any 8-hr period.

As with conventional point sources, evaluation of a proposed indirect source with respect to attainment and maintenance of the NAAQS can proceed on an ad hoc (i.e., first-come-first-served) basis or by requiring compatibility with an approved regional air quality management plan. The latter is preferable, though difficult to achieve ab initio because of the general reluctance of areawide planning agencies to assume the implicit permitting authority.

In either approach, the essence of a rational process for review and approval of proposed indirect sources is the *allocation of roadway capacity* (Roberts et al., 1975). Localized problems of traffic congestion (e.g., at points of ingress and egress) may call for additional expenses to be borne by the developer, as indicated in Table 31, but these are manageable.[13] However, as land along a transportation corridor develops, the carrying capacity of that corridor is consumed and air pollution increases, first linearly with increased vehicle activity and ultimately exponentially as additional vehicles create traffic congestion.

---

[11] The *Emission Density Zoning Guidebook* (Kron et al., 1978) provides a step-by-step procedure for designing a set emission density limit. It should be studied along with the companion document *Legal Issues of Density Zoning* (Jaffee et al., 1978).

[12] *Natural Resources Defense Council, Inc., et al.* vs *Environmental Protection Agency*, 475 F. 2d 968, 31 January 1973, and modified 12 March and 27 July 1973.

[13] USEPA Guidelines (EPA, 1975b) outline key design variables, basic traffic flow theories applicable to the analysis of vehicle movement, and resultant air quality in the immediate vicinity of an indirect source.

Table 31   Relative Marginal Cost of Compliance with Proposed Illinois Indirect Source Regulations: Summary of 17 Case Studies (Source: Cohen et al., 1974, p. 41)

Facility Type (No. Cases)	Compliance Costs A (10^3)	Total Facility Values B (10^3)	Percent (A/B) 100
Regional shopping centers (6)	5,160.0	243,000	2.1
Retail/commercial strip (2)	8.5	3,800	0.3
Office, government and misc. (4)	53.0	57,400	0.1
Manufacturing industry (1)	1.4	92,800	~0
Residential (2)	305.0	99,700	0.3
Recreation, sport, etc. (2)	3,970.0	29,300	13.5

The allocation problem can be visualized in Figure 32 where

S = the applicable national ambient air quality standard expressed as 100%

B = the background pollutant concentration attributable to sources other than those associated with the proposed highway and its indirect sources (= 10% of S, for this example)

E = existing air quality = background concentration (for this example)

E_1 = ambient pollution concentration in the vicinity of the highway absent any new indirect sources

D = the design capacity of a proposed highway, representing the maximum number of vehicles per hour for which an assumed design speed of 80 km/hr (50 mph) can be maintained

D' = the design figure for air quality purposes = the maximum number of vehicles per hour for which an indirect source permit can be issued = the number of vehicles per hour at a specified speed equivalent to a preset, maximum air quality increment above background $[0.8(S - E) = 0.72$ in this example$]$

CD' = ambient pollutant concentration at $D' = E + 0.8(S - E) = 0.82$

CD = ambient pollutant concentration at D

X = contribution that an indirect source, other than a public roadway, may make to ambient pollution levels, expressed as a percentage of S; $X \leq 0.3(S - E)$ is the allocation rule in this example. Three sequentially issued indirect source permits are shown in this example

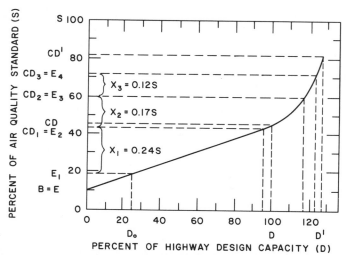

Fig. 32   Variation of air quality with highway traffic as applied to the issuance of permits for indirect sources.

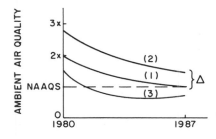

Fig. 33 Pathways toward attainment and maintenance of an ambient air quality standard.

Each new indirect source or collection of sources consumes portions X_1, X_2, and X_3 of the available air quality increment. In the absence of a regional plan to serve as a blueprint for granting permits, each indirect source might be allowed up to maximum fraction (e.g., 30%) of the remaining available increment, perhaps reserving the final 20% as a safety margin and/or to account for indirect sources too small to require separate review.

Though not explicitly stated above, the small-scale evaluations of the impacts of individual indirect sources on highway links and local roadways focus on CO, lead, and possibly short-term NO_x levels. However, the same fundamental principle is applicable to regionwide pollutant problems such as ozone and long-term NO_x levels.

One can face the problem up front and incorporate air quality objectives into land-use and transportation planning activities and embody the latter with adequate authority, or back into a solution via post hoc schemes such as emission offsets.

8.5 Emission Rights: Banking, Bubbles, and Offsets

Consider the three pathways toward attainment of the NAAQS shown schematically in Figure 33. For purposes here, the first may be considered to project the successful outcome of the retrofit in accord with published guidelines of reasonably available control technology to all existing sources with minimal allowance for new sources. The second characterizes the case where such generally prescribed controls are insufficient. The third pathway suggests a response sufficient to create a reserve for growth. An inflexible interpretation of the law under conditions (1) and (2) would lead to outright prohibition of significant new or expanded existing sources—a certain prescription for strong political backlash. Such is potentially the situation in many of the larger urban areas in the United States regarding attainment and maintenance of ambient standards for ozone, suspended particulates, and oxides of nitrogen.

Emission control regulations which provide the basis for projections such as those in Figure 33 capture the state of the art at some point in the recent past. Further, because of wide variations in the site- and process-specific features among sources faced with the same set of emission limits, costs can vary widely for the same degree of effectiveness; at the same time some sources may have options available for additional emission reductions at acceptable cost. A practical approach to smoothing out such wide variations in marginal control costs is to establish a market for trading in emission rights.

An *emission right* is the legally sanctioned privilege to emit under prescribed circumstances[14] a given quantity of a particular air pollutant. For example, as discussed in Section 8.3, if an emission density limit of 1 kg RHC/yr/m^2 is allowed on land zoned for heavy industry, a firm owning a 1500-m^2 (0.4-acre) site would simultaneously own emission rights of 1500 kg HC/yr. Emission density zoning, of course, requires an unusual degree of advanced planning not generally reflected in State Implementation Plans (SIPs). However, from another viewpoint, by virtue of the common first-come-first-served approach to allocating the regional clean air resource, every existing source "owns" emission rights of a magnitude equal to the level of emissions allowed in the SIP.

The lexicon of air pollution terminology has recently been augmented by three terms: *banking, bubbles,* and *offsets*. The first represents a necessary feature of every orderly market; the other two are creatures of federal regulation, variations on the concept of emission rights.

8.5.1 *Offsets*

The concept of an emission offset as first documented by the EPA (EPA, 1976) is straightforward: a proposed new source or expansion of an existing source which would represent a significant upward "blip" on pathways (1) or (2) of Figure 30 can receive a permit only to the extent that one or more

[14] Principally, compliance with all applicable emission limits.

existing sources reduce their emissions by an amount equal to [in case (1)], or greater than [e.g. 125% in case (2)] that required of them by applicable regulations, thereby providing an acceptable *offset*. This *interpretive ruling* of the EPA and subsequent amendments (in particular EPA, 1979a) define "significant," establish requirements for new or expanded existing sources to meet the "lowest achievable emission rate" (LAER), define the baseline against which existing sources can create offset credits, outline procedures for determining the net air quality benefit of the offset trade, and sketch out a criterion for "reasonable further progress" toward attainment of the NAAQS.

The following examples summarize several early, successful offset transactions.[15]

General Motors Corporation/Oklahoma City, Oklahoma. *This case centers around a proposed major new General Motors plant for the manufacture of Nova automobiles to be located in Oklahoma City. The approximate roof area of the new plant is to be 3.5 million ft², at an estimated cost of $400 million. Expected production is about 75 vehicles/hr. The area in which the plant is to be located is classified as nonattainment for photochemical oxidants and, in some sections, for particulates as well. Offsetting emissions for this plant were negotiated by the Oklahoma Chamber of Commerce, and voluntarily provided by various oil companies within an 85-mile radius and upwind of Oklahoma City. The oil companies installed floating roofs or vapor recovery systems on their existing fixed-roof petroleum storage tanks. The proposed plant in Oklahoma City had been under discussion for at least 6 yr; the necessary air permits were issued and approved in the fall of 1977; and the plant is currently under construction.*

Volkswagen Corporation of America/New Stanton, Pennsylvania. *This external offset case involves the construction and operating permits for an auto assembly plant in New Stanton, Pennsylvania. The plant is located in a structure built by Chrysler Corporation which Chrysler abandoned before it was put into operation. The new plant will result in the emission of increased hydrocarbons in an area classified by EPA as nonattainment for photochemical oxidants. Offsetting emissions in this case were provided through the use by the Pennsylvania Department of Transportation of water-based asphalt, rather than cutback asphalt which, when applied under certain conditions, results in the release of hydrocarbon emissions. The assembly plant started production in April 1978.*

Phillips Petroleum Company/Brazoria County, Texas. *This internal offset case involves major expansion of the Phillips Petroleum Company refinery which would double plant capacity. The plant is in an area that currently exceeds NAAQS photochemical oxidant standards. Offsetting emissions from the expansion were provided internally by Phillips, and were found through the installation of double seals on existing petroleum storage tanks, the abatement of a vacuum jet, and an inspection and maintenance plan. The proposed expansion is now under way.*

8.5.2 Bubbles

As with offsets, the *bubble* policy of the EPA is quite straightforward (EPA, 1979b). Applicable regulations (e.g., "reasonably available control technology") may require a plant to control a number of individual sources of a given pollutant, each to a specified degree. Let the plant owner redistribute the requirements for emission reductions among the various sources in what he deems is a most favorable (presumably most cost-effective) manner as long as the total plant emissions comply with the approved State Implementation Plan. In other words, the plant is viewed as operating within a bubble with a single hypothetical stack.[16]

Now consider expanding the bubble to include a neighboring plant as well or placing a zone of heavy industry within a single bubble where a formal, enforceable system is established to account for the various tradeoffs. The expanded bubble policy thereby converges toward a system of emission density quotas with transferable (marketable) emission rights.

8.5.3 Banking

If the concept of emission rights is to be extended beyond isolated offset deals and if a bubble policy is to be applicable beyond the boundary of a single plant, then a uniform system of accounts must be developed. Figure 34 enumerates the transactions to be conducted by an emission banking system established under governmental auspices. A sample ledger sheet is shown in Figure 35. Although stages 1 and 2 of the flowchart suggest that depositors are individuals in the private sector, the state or local governments in the region could in many instances control (i.e., create, deposit, and allocate)

---

[15] Summaries extracted verbatim from Foskett (1979, pp. 4, 5).

[16] For example, Armco Steelworks in Middletown, Ohio, has implemented an extensive $4 million program to control fugitive dust from coal piles, roadways, and other open areas on its 2600-acre site in lieu of $11.5 million for control of fugitive particulates from various steel-making processes.

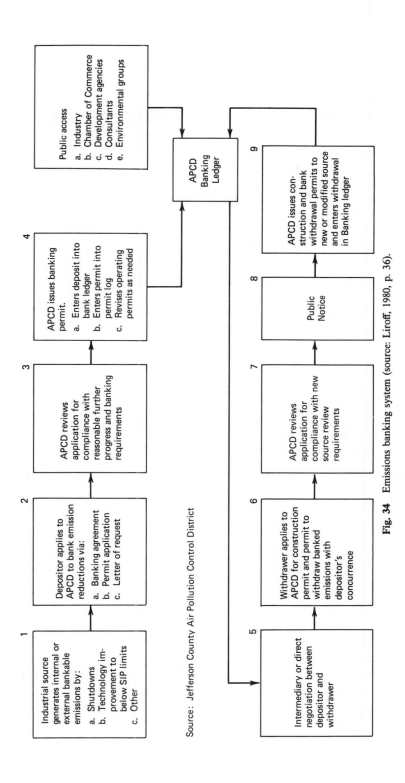

Source: Jefferson County Air Pollution Control District

Fig. 34 Emissions banking system (source: Liroff, 1980, p. 36).

JEFFERSON COUNTY AIR POLLUTION CONTROL DISTRICT

BANKED EMISSIONS LEDGER   □ ACTIVE
                             □ CLOSED

1.   Pollutant Description _____

2.   Deposit Code □ □ □ □ □ — □ □ □
3.a. Date of Deposit _____ b. Closing Date _____
4.a. Source of Banked Emissions _____
  b. Address _____
  c. Plant EIS _____
5.a. Description of process generating banked emissions and permit nos.
      _____
      _____
      _____
      _____
      _____

  b. What caused emissions to be available for banking?
      _____
      _____

6.   Banked emissions prior to discounting   _____ Tons/yr
7.   Initial discount   _____ Tons/yr
8.   Balance (subtract Line 7 from Line 6)   _____ Tons/yr
      For further explanation, see Note nos. _____

First Withdrawal
9.   Date _____ Buyer _____ Permit No. _____
10.   Emissions from source requiring offsets (but before applying
      offset ratio)   _____ Tons/yr
11.   Offset ratio _____ : 1
12.   Offset emissions (multiply Line 10 by Line 11)   _____ Tons/yr
13.   Balance (subtract Line 12 from Line 8)   _____ Tons/yr
      For further explanation, see Note nos. _____

Second Withdrawal
14.   Date _____ Buyer _____ Permit No. _____
15.   Emissions from source requiring offsets (but before applying
      offset ratio)   _____ Tons/yr
16.   Offset ratio _____ : 1
17.   Offset emissions (multiply Line 15 by Line 16)   _____ Tons/yr
18.   Balance (subtract Line 17 from Line 13)   _____ Tons/yr
      For further explanation, see Note nos. _____

Third Withdrawal
19.   Date _____ Buyer _____ Permit No. _____
20.   Emissions from source requiring offsets (but before applying
      offset ratio)   _____ Tons/yr
21.   Offset ratio _____ : 1
22.   Offset emissions (multiply Line 21 by Line 22)   _____ Tons/yr
23.   Balance (subtract Line 23 from Line 19)   _____ Tons/yr
      Enter here and on Line 24, page 2   _____ Tons/yr
      For further explanation, see Note nos. _____

(Author's note: This is the first page of a two page form. The second page provides room for notes on additional withdrawals.)

Fig. 35 Sample ledger sheet (source: Liroff, 1980, p. 35).

PRIVATE SECTOR

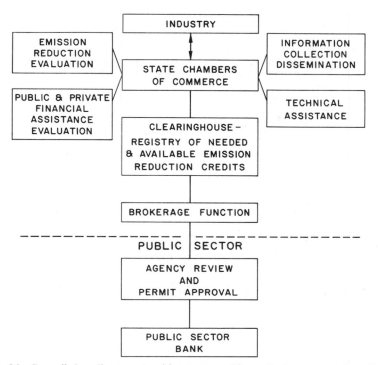

Fig. 36 Controlled trading system with separate public and private sector data files.

the disposition of certain emission rights, for example, those associated with the reserve in case (3) of Figure 33. Local government might also act as an intermediary, buying emission rights and assigning them to high-priority industrial development initiatives, much as property tax subsidies are used to attract new business.

The private sector may find it preferable to establish its own proprietary information system. For example, a "controlled trading" system is under development in Illinois with the State Chamber of Commerce promoting and managing a clearinghouse (Figure 36).

The detailed rules governing markets in emission rights, and thus banking transactions, will vary from region to region. Provisions governing claims for past emission reductions, limits to the time period an individual can bank emission rights, fees, and a myriad of other procedural aspects are subject to local discretion—as long as they do not subvert the Clean Air Act and related federal regulations.[17]

Probably the most critical issue with which any banking system must deal is the protection over time of the validity of emission rights or emission reduction credits deposited therein. Presumably such rights are based on a federally approved implementation plan designed to attain the NAAQS on schedule. However, if the plan falls short (as was the case with many such state plans prepared pursuant to the 1970 amendments) and additional control measures are required, those banked credits could be voided or substantially reduced in size.[18]

[17] See, for example, the EPA's *Emission Reduction Banking Manual* (EPA, 1980) for a highly readable overview of do's and don't's.

[18] Unless uncertainties in this regard are clearly resolved, the ambitious banking program being promoted by the EPA will falter. The agency is very much aware of the problem (EPA, 1980, p. 5) and has indicated it will act to protect the credibility of the banking program through provisions in forthcoming regulations (unpublished as of October 1980). However, it is not clear where in the 1977 amendments the EPA can derive such authority.

8.5.4 *Allocation of a Shortfall in Emission Reduction*

The somewhat obscure title to this section might be rephrased: What does one do in case (2) of
Figure 33 when the application of reasonably available control technology to existing sources falls
short of the NAAQS by the margin Δ? As suggested in the introductory remarks to Section 8.5 and
by subsequent examples, it is reasonable to assume that some slack exists in the system. Further, a
modest extension of the deadline for attainment will no doubt create additional slack as more advanced,
cost-effective control and process options evolve (especially under the impetus of a system of marketable
emission rights). Thus, the proposal by Foster (1979) appears quite reasonable: translate the shortfall
Δ into a uniform or possibly geographically sensitive requirement for additional control beyond that
mandated as *reasonably available* (e.g., approximately 25% in Figure 33). Then let a free market in
emission reductions redistribute the requirements for additional control in what presumably would
be the most cost-effecitve manner.[19]

8.5.5 *Case Study: A Statewide Bubble for SO_2 (Garvey et al., 1982)*

Quite likely the hottest scientific and political topic in the air quality management arena for the 1980s
will be the control of acid deposition via legislated, regionwide reductions of SO_2 emissions. Ninety-
two fossil fueled electric generating units on-line or anticipated to come on-line by the year 2000 in
Illinois are listed in Table 32 in terms of SO_2 emission limits associated with the approved State
Implementation Plan (SIP) and federal New Source Performance Standards (NSPS). The costs attributable to compliance with these regulations are estimated to be $639 million/yr (annualized basis: 1979
$) or approximately $760/ton of SO_2 removed by deliberate choice of low-sulfur coals or flue gas
desulfurization systems. A least-cost problem was solved by Garvey et al. in which no existing plant
would be permitted to exceed SIP limits but all others could raise emissions as high as 8 lb SO_2/
million Btu of actual heat input with the constraint that aggregate emissions be no greater than a
prescribed ceiling. For the "base case" corresponding to current regulations, total statewide utility
emissions were projected to be 1200 ktons SO_2/yr. The least-cost solution consistent with this ceiling

**Table 32   Illinois SO_2 Emission Limitations (Base Case)
(Garvey et al., 1982)**

SO$_2$ Limit		Units,	Capacity,
lb/10^6 Btu	Type^a	No.	MW
0.6^b	R	14	6,500
1.2	N	7	2,710
1.8	S	32	7,599
3.6	S	6	1,015
4.6–5.5	S	7	634
5.5–5.8	S	7	3,009
6.0	S	17	833
8.0	S	2	1,212
		92	23,512

^{a} Type R = revised NSPS of 1979
      N = 1971 NSPS
      S = SIP
^{b} The approximate emission rate from a source in compliance
with the 1979 NSPS in Illinois.
Note: The Illinois SIP requires the following emission limits:
      (1) 1.8 lb/10^6 Btu for sources in major metropolitan
      areas, and (2) 6.0 lb/10^6 Btu for sources in nonmajor
      metropolitan areas. Other values represent source-spe-
      cific limits, including certain variances if NAAQS are
      not exceeded.

[19] Raufer et al. (1978) examined alternative policies for reduction of volatile organic compounds in
the Twin Cities/St. Cloud, Minnesota area and concluded that cost savings on the order of 25% can
be achieved through a program of transferable emission reduction assessments as opposed to uniform
application of "reasonably available control technology" (RACT).

BUBBLE ANALYSIS - STATE OF ILLINOIS

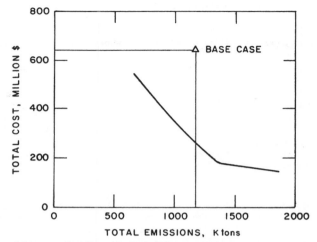

Fig. 37 Results of least-cost bubble analysis for SO_2 emissions in Illinois, year 2000 (costs in 1979 $s) (Garvey et al., 1982).

showed a significant reduction in cost to $335 million/yr ($400/ton) compared to the base case costs cited above. The savings result primarily from increases in emissions from new plants which would otherwise face the 0.6 lb/10^6 Btu limit and corresponding reductions in emissions from existing rural plants governed by the relatively lax standards set to insure maintenance of the National Ambient Air Quality Standards. Least-cost analyses demonstrate that substantial further reductions in state-wide SO_2 emissions can be achieved at costs well below the base case (Figure 37).

Two additional findings of the study are worthy of note. First, the least-cost case relaxes the requirement for scrubbing inherent in the NSPS under the 1979 Clean Air Act amendments, thus making the importation of low-sulfur coal from the Great Plains and Central Appalachia economically more attractive. Despite the potential savings for consumers, this shift would be politically unpopular in a coal-producing state such as Illinois. At the same time the potential magnitude of this shift is diminished significantly by the renewed attractiveness of certain midsulfur Illinois coals which under current regulations can be utilized only with scrubbers or coal blending.

Secondly, despite the shift in distribution of emissions among the utility plants under a least-cost approach, by constraining existing plants to compliance with the SIP and by virtue of the fact that new plants are almost all in rural, typically minemouth sites, local, 24-hr, and annual ambient standards are not violated.

Finally, it should be noted that there are no provisions in current Illinois utility regulatory law for the savings associated with the least-cost solutions to be distributed among the separate utility companies and their respective consumer populations.

REFERENCES

ANL (1976), Cohen, A. S., Tschanz, J. F., and Seyfried, R. K., *An Economic Evaluation of the Proposed Illinois Complex Source Regulation,* Argonne National Laboratory, ANL/ES-36, August 1974.

Armentrout, D. W. (1979), *Development of an Emission Inventory Quality Assurance Program,* PEDCo Environmental, Inc., prepared for the Environmental Protection Agency, EPA-450-4-79-006, June.

Arrow, K. J. (1966), Discounting and Public Investment Criteria, *Water Research,* Baltimore.

Babcock, L., Jr., et al. (1978), *A Handbook for the Assessment of Environmental Benefits and Pollution Control Costs,* Illinois Institute of Natural Resources.

Babcock, L. R., Jr. (1973), A Method for Estimation of Emission Control Costs, *Environmental Affairs* II(4), Spring.

Baldwin, T. E., and Kennedy, A. S. (1974), The Feasibility of Predicting Point Source Emissions Using Industrial Land Use Variables: A Path Analysis, presented at the 67th Annual Meeting of the Air Pollution Control Association, Denver, Colorado, June.

Barrett, L. B., and Waddell, T. E. (1973), *Cost of Air Pollution Damage: A Status Report,* U.S. EPA Publication No. AP-85.

Baumol, W. J. (1972), On Taxation and the Control of Externalities, *American Economic Review,* 307–322, June.

Benesh, F. (1976), *Growth Effects of Major Land Use Projects; Volume II—Compilation of Land Use Based Emission Factors,* EPA-450/3-76-012b, U.S. Environmental Protection Agency, Research Triangle Park, NC, September.

Bingham, T. H., et al. (1973), *A Projection of the Effectiveness and Costs of a National Tax on Sulfur Emissions—Final Report,* prepared by the Research Triangle Institute for the U.S. Environmental Protection Agency, November.

Bingham, T. H., and Miedema, A. K. (1974), *Final Report—Allocative and Distributive Effects of Alternative Air Quality Attainment Policies,* prepared by Research Triangle Institute for the U.S. Environmental Protection Agency, October.

Brail, R. K. (1975), Land Use Planning Strategies for Air Quality Maintenance, *Proceedings of the Long Term Maintenance of Clean Air Standards Specialty Conference,* Lake Michigan States Section, Air Pollution Control Association, February.

Brookshire, D. S., et al. (1979), *Methods Development for Assessing Air Pollution Control Benefits: Volume V, Executive Summary,* U.S. Environmental Protection Agency, EPA-600/5-79-001e.

Brubaker, K. L., Brown, P., and Cirillo, R. (1977), *Addendum to User's Guide for Climatological Dispersion Model,* U.S. EPA Report No. EPA-450/3-77-015, Research Triangle Park, NC, May.

Bussee, A. D., and Zimmerman, J. R. (1973), *User's Guide for the Climatological Dispersion Model,* U.S. EPA Report No. EPA-R4-73-024, Research Triangle Park, NC, December.

CAA (1967), Air Quality Act of 1967, P.L. 90-148, 81 Stat. 485.

CAA (1970), Clean Air Act Amendments of 1970, P.L. 91-604, 84 Stat. 1676.

CAA (1977), Clean Air Act Amendments of 1977, P.L. 95-95, 91 Stat. 685.

Chiang, A. C. (1974), *Fundamental Methods of Mathematical Economics,* 2nd edition, McGraw-Hill, New York.

Chilton, C. H. (1949), Cost Data Correlated, *Chemical Engineers' Handbook,* Vol. 56, No. 6, p. 97.

Cohen, A. S., Fishelson, G., and Gardner, J. L. (1974), *Residential Fuel Policy and the Environment,* Ballinger, Cambridge, Mass.

Crocker, T. D. (1970), *Urban Air Pollution Damage Functions: Theory and Measurement,* National Technical Information Service PB 197668, Springfield, VA, p. 23.

Crocker, T. D., et al. (1979), *Methods Development for Assessing Air Pollution Control Benefits,* Vol. 1, February.

deNevers, N. (1979), Some Alternative PSD Policies, *Journal of the Air Pollution Control Association* **29**(11), November.

DRI (1979), *The Macroeconomic Impact of Federal Pollution Control Programs: 1978 Assessment,* Data Resources, Inc., January.

Elston, J. C. (1981), Motor Vehicle Inspection Maintenance Programs, A Critical Review, *Journal of the Air Pollution Control Association* **31**(5), May.

English, T. D., Divita, E., and Lees, L. (1980), *Analysis of Air Quality Management with Emphasis on Transportation Sources,* Jet Propulsion Laboratory Publication 80-65, October.

EPA (1973), *Guide for Compiling a Comprehensive Emission Inventory,* APTD-1135, U.S. Environmental Protection Agency, March.

EPA (1974a), *Guidelines for Air Quality Maintenance Planning and Analysis, Vol. 5: Case Studies in Plan Development,* U.S. Environmental Protection Agency, EPA-450/4-74-006, December.

EPA (1974b), *Guidelines for Air Quality Maintenance Planning and Analysis, Vol. 13: Allocating Projected Emissions to Sub-County Areas,* U.S. Environmental Protection Agency, EPA-450/4-74-014, November.

EPA (1974c), *Guidelines for Air Quality Maintenance Planning and Analysis, Vol. 4: Land Use and Transportation Considerations,* U.S. Environmental Protection Agency, EPA-450/4-74-004, August.

EPA (1975a), *Guidelines for Air Quality Maintenance Planning and Analysis, Vol. 7: Projecting County Emissions, 2nd edition,* U.S. Environmental Protection Agency, EPA-450/4-74-008, January.

EPA (1975b), *Guidelines for Air Quality Maintenance Planning and Analysis, Vol. 9: Evaluating Indirect Sources,* U.S. Environmental Protection Agency, EPA-450/4-75-001, January.

EPA (1976), 41 *Federal Register* 5524, December 16.

EPA (1977), 42 *Federal Register* 22177, May 2.

EPA (1978a), *Compilation of Air Pollutant Emission Factors,* AP-42, U.S. Environmental Protection Agency.

EPA (1978b), *Mobile Source Emission Factors,* EPA-400/9-78-005, U.S. Environmental Protection Agency, Washington, D.C., March.

EPA (1979), *The Cost of Clean Air and Water, Report to Congress,* U.S. Environmental Protection Agency EPA-230/3-79-001, August.

EPA (1979a), 44 *Federal Register* 3274, January 15.

EPA (1979b), 44 *Federal Register* 71780, December 11.

EPA (1980), *Emission Reduction Banking Manual,* Pub. No. BG200, U.S. Environmental Protection Agency, Washington, D.C., September.

ERT (1973), *The Hackensack Meadowlands Air Pollution Study, Summary Report,* Environmental Research and Technology, October.

Felton, V. R., and Rossano, A. T. (1976), Air Quality Maintenance Through Land Use Planning— A Case Study, *Proceedings of the Air Pollution Control Association,* Portland, Oregon, June.

Foskett, W. H. (1979), *Emission Offset Policy at Work: A Summary Analysis of Eight Cases,* The Regulatory Center at the Performance Development Institute, Washington, D.C. April.

Foster, J. D. (1979), *Cleaning the Air for a Growing Economy,* U.S. Environmental Protection Agency.

Garvey, D. B., et al. (1982), Argonne National Laboratory, prepared for U.S. Dept. of Energy, Office of Environmental Assessments, May.

Harberger, A. C. (1972), On Discount Rates for Cost-Benefit Analysis, paper from *Project Evaluation* collected papers, pp. 70–93.

Hines, E., et al. (1978), *Baltimore TCP–1; Transportation Control Plan, Vol. 1, Summary,* September.

IEPA (1980a), *Evaluation of Motor Vehicle Emissions Inspection/Maintenance Programs for the State of Illinois, Task 3—Final Report,* Pacific Environmental Services, Illinois Environmental Protection Agency, January.

IEPA (1980b), *Design and Cost Study of an Emissions Inspection/Maintenance Program in Illinois,* Booz Allen & Hamilton, Illinois Environmental Protection Agency, September.

Jaffee, M. S., et al. (1978), *Legal Issues of Density Zoning,* U.S. Environmental Protection Agency, EPA-450/3-78-049, September.

Kalman, S. (1980), Cost-Benefit Analysis and Environmental, Safety, and Health Regulation Ethical and Philosophical Considerations, *The Analysis of Costs and Benefits in Environmental Regulation,* Conference co-sponsored by the Conservation Foundation and the Illinois Institute of Natural Resources, Chicago, Illinois, October 15–16.

Kemner, W. F. (1979), *Cost Effectiveness Model for Pollution Control at Coking Facilities,* U.S. Environmental Protection Agency EPA-600/2-79-185, August.

Kennedy, A. S., et al. (1973), *Methods for Predicting Air Pollution Concentrations from Land Use, Final Report,* U.S. Environmental Protection Agency, May.

Kron, F., Jr., et al. (1978), *Emission Density Zoning Guidebook, A Technical Guide to Maintaining Air Quality Standards Through Land-Use Based Emission Limits,* U.S. Environmental Protection Agency, EPA-450/3-78-048, September.

Lave, L. B., and Seskin, E. P. (1970), Air Pollution and Human Health, *Science* **139**(3947), 730.

Leontief, W. (1966), *Input-Output Economics,* Oxford University Press. London England.

Leung, S. K., Reed, W., and Geng, S. (1982), Estimations of Ozone Damage to Selected Crops Grown in Southern California, *Journal of the Air Pollution Control Association* **32**(2), 160–164.

Monarch, M. R., et al. (1978), *Priorities for New Source Performance Standards Under the Clean Air Act Amendments of 1977,* U.S. Environmental Protection Agency, EPA-450/3-78-019, April.

OMB (1972), Executive Office of the President, Office of Management and Budget, Circular A-94 Revised, March 27.

Perry, R. H., and Chilton, C. H. (1973), reference from *Chemical Engineers' Handbook,* 5th edition, McGraw-Hill, New York, p. 25-15.

Pigou, A. C. (1970), reference from text of *The Economics of Welfare,* Macmillan, London, pp. 160–161.

Raufer, R. K., et al. (1981), Emission Fees and Tera: An Evaluation of Policy Alternatives in the Twin Cities, *Journal of the Air Pollution Control Association* **31**(8), August.

Roberts, J. J., Croke, E. J., and Booras, S. (1975), A Critical Review of the Effect of Air Pollution Control Regulations on Land Use Planning, *Journal of the Air Pollution Control Association* **25**(5), May.

Roberts, J. J., Tamplin, S. A., and Melvin, G. L. (1976), Regulation of Indirect Sources, *Journal of the Air Pollution Control Association* **25**(3), March.

Samuelson, P. (1954), The Pure Theory of Public Expenditures, *Review of Economics and Statistics* **XXXVI**, 397.

Schrimper, R. A. (1973), *Investigation of Morbidity Effects,* University of Chicago, Chicago, IL pp. 13–14.

Seagraves, J. A. (1970), More on the Social Rate of Discount, *Quarterly Journal of Economics* **84,** 430–450.

Smith, A. E., et al. (1980), *Air Quality Impacts of Alternative Policies for the Review of Major New Stationary Sources: A Case Study,* Draft Final Report prepared for the Office of Planning and Evaluation, U.S. Environmental Protection Agency, Washington, D.C.

Smith, R. (1974), The Feasibility of an "Injury Tax" Approach to Occupational Safety, *Law and Contemporary Problems* (Summer–Autumn).

Stern, A. C. (1977), Prevention of Significant Deterioration, *Journal of the Air Pollution Control Association* **27**(5), May.

Suhrbier, J. H., et al. (1978), *Implementation and Administration of Air Quality Transportation Controls,* Cambridge Systematics, Inc., Mass.

Thaler, R., and Rose, S. (1975), The Value of Saving a Life: Evidence from the Labor Market, in *Household Production and Consumption* (N. E. Terlecky, Jr., ed.), Columbia University Press, New York, pp. 265–297.

Townsend, S. (1978), *The Social Opportunity Cost of Capital: Empirical Estimates,* Argonne National Laboratory, ANL/SPG-2, February.

Uhl, V. W. (1979a), *A Standard Procedure for Cost Analysis of Pollution Control Operations, Vol. I User Guide,* EPA-600/8-79-018a, June.

Uhl, V. W. (1979b), *A Standard Procedure for Cost Analysis of Pollution Control Operations, Vol. II, Appendices,* U.S. Environmental Protection Agency, EPA-600/8-79-018b, June.

U.S. Army (1958), *Universal Transverse Mercator Grid,* U.S. Department of the Army, Washington, D.C., Pub. No. TM5-241-8, July.

U.S. Water Resources Council (1976).

CHAPTER 38
INFORMATION RESOURCES

HAROLD M. ENGLUND

Air Pollution Control Association
Pittsburgh, Pennsylvania

To conclude a book of this magnitude with a chapter on information resources almost calls for a note of apology. Yet, the field of air pollution control technology is a fast developing one. Even as one reads these words, further research is taking place, regulations are changing, pilot studies are being reported, and thoughts are crystallizing to become part of tomorrow's literature on air pollution control. We must, therefore, identify some of those sources that provide a current awareness of developments in this fast moving field.

Air Pollution Control Association (APCA)
P.O. Box 2861
Pittsburgh, Pennsylvania 15230

The Air Pollution Control Association is a 77-year-old technical organization whose activities are directed to the collection and dissemination of authoritative information about air pollution and its control. Resources offered by APCA include the following:

Journal of the Air Pollution Control Association (JAPCA). This monthly journal, a benefit of membership in APCA, is in its 33rd year. Each issue includes feature articles, technical papers, note manuscripts, and newsletters. News departments cover business and the regulatory agencies, new products, professional development programs, and current literature, including abstracts of current periodicals. A recently inaugurated department, Control Technology News, serves a growing interest in applied air pollution control technology. The journal is indexed by subject and author annually. In addition, three cumulative indexes have been published: a 22-year index for volumes 1–22 (1951–1972), a 5-year index for volumes 23–27 (1973–1977), and a 5-year index for volumes 28–32 (1978–1982).

Product Guide. Annual directory of manufacturers of emission control equipment and air pollution instrumentation, listed by product and alphabetically.

Consultant Guide. Annual directory of air pollution control consultants, listed geographically within areas of competence.

Directory of Governmental Air Pollution Control Agencies. Annual directory of federal, state, regional, and local air pollution control agencies lists principals, contact information. Includes U.S. Environmental Protection Agency and Canadian federal and provincial agencies.

Continuing Education Programs. Continuing education is offered on a variety of technical subjects through one- to two-day short courses which are presented in conjunction with specialty meetings and the annual meeting of the association.

U.S. Environmental Protection Agency (EPA)
Office of Administration
Information Services Division
Research Triangle Park, North Carolina 27711

EPA Library Services constitute an important information resource and offer comprehensive literature searches and copies of EPA air pollution documents.

Copies of EPA Air Pollution Documents. Single copies of EPA documents are available on request from the Library Services Office as long as supplies last. When supplies are exhausted, ordering information for the Government Printing Office or the National Technical Information Service will be provided.

When a publication does not answer a question, the library serves as a switching station to suggest an appropriate information source.

Comprehensive Literature Searches. The following groups are eligible for free air pollution literature searches:

1. EPA employees
2. State and local government employees
3. Current contractors of EPA when approved by their project officer
4. Nonprofit citizens environmental groups
5. Foreign government employees

A computerized literature search draws on all the major relevant published information sources—books, journals, and technical documents. EPA can also locate research currently in progress.

EPA will search, as needed, the Air Pollution Abstracts (APTIC), Biological Abstracts, Engineering Index, Chemical Abstracts, Toxline, National Technical Information Service, and others. The result is a list of citations tailored to a specific interest. Many of these citations will include abstracts. For further information, contact the Library Services Office of the Environmental Protection Agency, Research Triangle Park, North Carolina 27711.

Items from EPA's Air Pollution Abstracts (APTIC) are available from EPA Library Services. Items cited in data bases are available from local universities, commercial information brokers, and the National Technical Information Service, Journal Article Service, Springfield, VA 22161.

If one is not eligible for free searches, many local universities and commercial "information" firms perform computerized literature searches for a fee. The search service vendors below can be contacted for information on searching services:

> Lockheed Information Systems
> Dept. 208, Building 201
> 3251 Hanover Street
> Palo Alto, California 94304
>
> System Development Corporation
> 2500 Colorado Avenue
> Santa Monica, California 90406
>
> Bibliographic Research Service
> Corporation Park, Building 702
> Scotia, New York 12302

National Technical Information Service (NTIS)
5285 Port Royal Road
Springfield, Virginia 22161

Two information services with relevance to air pollution control technology are offered by NTIS, operated by the U.S. Department of Commerce.

Environmental Pollution & Control. This abstract newsletter is published weekly and covers air, noise, solid wastes, environmental health and safety, and environmental impact statements. Papers abstracted can be ordered in paper copy, microforms, or magnetic tape. An annual subject index is distributed each January.

Information for Innovators. This interdisciplinary information service is published on alternate Tuesdays and covers engineering, product development, technical marketing requirements, U.S. and foreign technology, and management of R&D and advanced technology. To dig deeper, one can order the source document and names of people and organizations doing the work described.

Other Sources of Information

The Industrial Gas Cleaning Institute (IGCI) publishes annual new order booking totals for industrial air pollution control equipment in the United States and Canada. The total sales of member companies represent the value of bare apparatus, net F.O.B. shop, of the six basic types of equipment available: electrostatic precipitators, fabric filters, mechanical collectors, wet scrubbers, gaseous emission control devices, and FGD systems. The institute also publishes technical data on these products and provides information on performance capabilities of the industry. Contact: Industrial Gas Cleaning Institute, Inc., Suite 304, 700 North Fairfax Street, Alexandria, Virginia 22314.

The McIlvaine Company offers "knowledge networks" for air pollution control technology. Applications and new developments are covered in a series of specific and periodically updated manuals on wet scrubbers, flue gas scrubbing, fabric filters, electrostatic precipitators, and monitoring and sampling. These subjects are also covered in a series of specific monthly newsletters. Contact: The McIlvaine Company, 2970 Maria Avenue, Northbrook, Illinois 60062.

The American Chemical Society publishes *Environmental Science & Technology,* a monthly journal that reports on aspects of the total environment and its control by scientific, engineering, and political means. Contents include feature articles, critical reviews, current research papers, research notes, and correspondence. Contact: *Environmental Science & Technology,* 1155 16th Street, N.W., Washington, D.C. 20036.

Cambridge Scientific Abstracts publishes *Pollution Abstracts,* bimonthly abstracts journal providing access to literature relevant to environmental pollution research and related engineering studies. Publication is available on-line via Lockheed Information Systems and System Development Corporation and is also available on magnetic tape. Contact: Cambridge Scientific Abstracts, 5161 River Road, Washington, D.C. 20016.

Environment Information Center, Inc. (EIC) is one of the world's largest independent sources of energy and environmental information. EIC databanks, Energyline and Enviroline, provide access to the most significant facts, statistics, professional papers, journal articles, technical data on energy, and the environment. The same information is available in print abstract journals, Environment Abstracts, and Energy Information Abstracts. Other EIC products and services include legal handbooks, microfiche systems, directories, yearbooks, and magnetic tapes. Contact: EIC, 48 W. 38th Street, New York, New York 10018.

Abstracts on the Health Effects of Environmental Pollutants (HEEP) is a monthly publication of Biosciences Information Service. HEEP provides abstracts, content summaries, and a six-level indexing structure to thousands of research publications in air pollution and related fields. HEEP provides access to journal articles, conference papers, reviews, books, and reports which approach air pollution with emphasis on occupational health, industrial medicine, public health, and environmental science. Contact: Biosciences Information Service, 2100 Arch Street, Philadelphia, Pennsylvania 19103.

INDEX